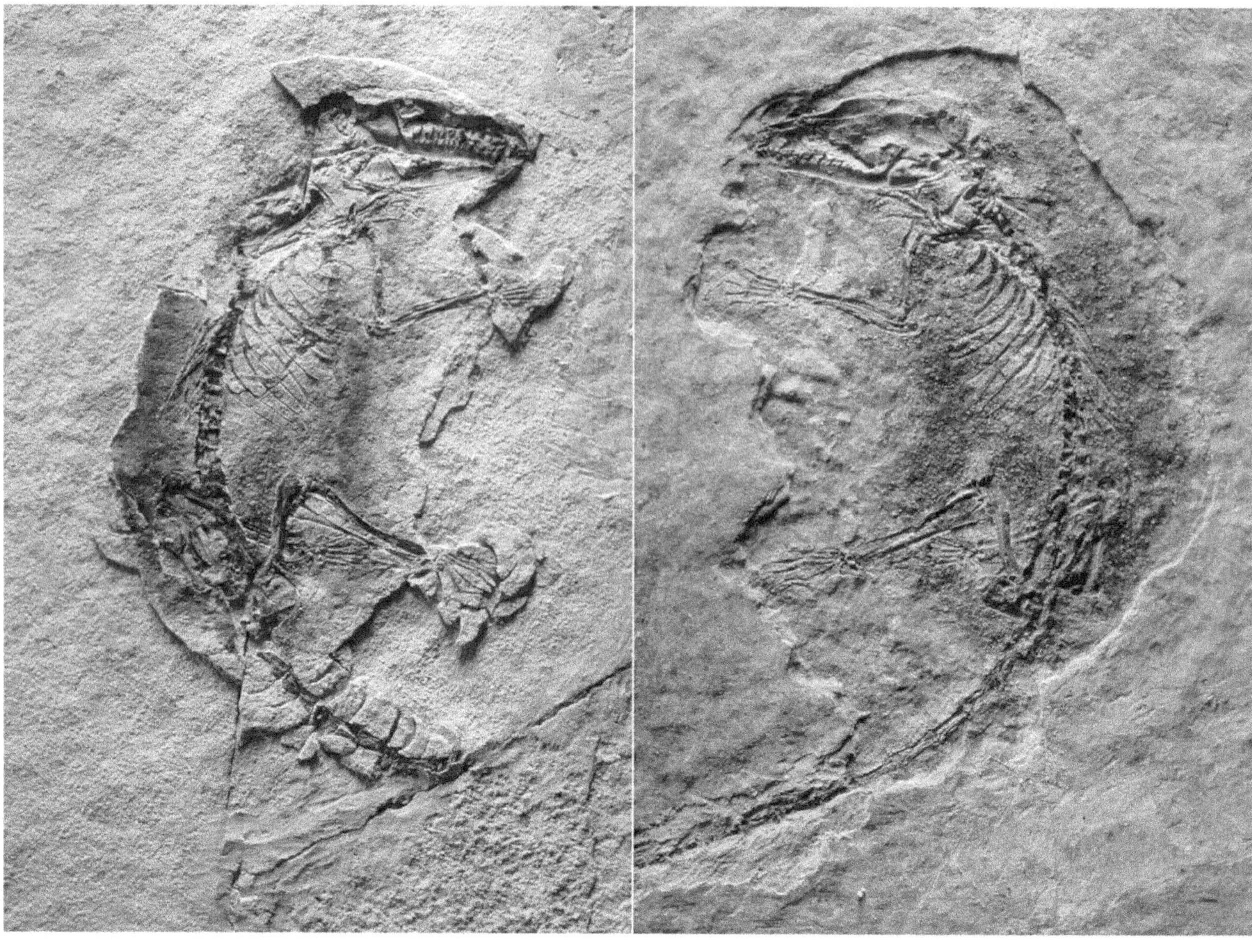

Eomaia scansoria (holotype specimen, part and counterpart: Chinese Academy of Geological Sciences [CAGS] 01-IG-1A, 1B). The fossil comes from the Yixian Formation (125 Ma), Dawangzhangzhi locality, Lingyuan County, Liaoning Province, China. Called the "dawn mother," *Eomaia scansoria,* the earliest-known mammal more closely related to modern placentals than to modern marsupials, was the size of a shrew (ca. 14 cm long, estimated weight 20–30 g) and had several features in its manus and pes for adaptation to scansorial life.

Mammals from the Age of Dinosaurs

ORIGINS, EVOLUTION, AND STRUCTURE

Zofia Kielan-Jaworowska,
Richard L. Cifelli, and
Zhe-Xi Luo

COLUMBIA UNIVERSITY PRESS
NEW YORK

Columbia University Press
Publishers since 1893
New York   Chichester, West Sussex

Library of Congress Cataloging-in-Publication Data

Kielan-Jaworowska, Zofia.
Mammals from the age of dinosaurs : origins, evolution, and structure / Zofia Kielan-Jaworowska, Richard L. Cifelli, and Zhe-Xi Luo.
p. cm.
Includes bibliographical references and index.
ISBN 0-231-11918-6 (cloth)
1. Mammals, Fossil. 2. Paleontology—Mesozoic. I. Cifelli, Richard. II. Luo, Zhe-Xi. III. Title.
QE881.K53 2004
569—dc22

2004052897

CONTENTS

FOREWORD

When most of us think about mammals from the Mesozoic Era we use our imaginary time-viewing machines to see furry little creatures with relatively large beady eyes and warm wiggly noses. We see them scampering through moonlit forests capturing insects, tearing apart fruits from early flowering plants, sometimes lapping nectar, and universally marking their territories with diverse body exudates. We think of them as denizens of a world much different from ours of today. The food chains of their generally warmer ecosystems are seen to be capped by dinosaurs, crocodiles, and birds, all of which were foraging within forests and shrub lands with wholly different floral compositions from ours, or across muddy and rocky landscapes devoid of grass. The Mesozoic continents themselves are observed to be markedly different in relative positions and shoreline configurations, and in the locations of plains and mountain belts. Usually we see much greater areas covered by shallow inland seas than exist today. We can picture broad diversification of reproductive modes among these mammals, with a range including egg-laying, marsupial-like patterns, and early permutations of prolonged internal development. We observe much more of a "sameness" among the mammals of the Mesozoic than today. For example, we miss great beasts having horns, tusks, fins, or bony armor. We also miss the almost ubiquitous modern diversity of rodents and bats. In our present view of all these general features of primary biological importance, little has been altered conceptually since the predecessor of this book was published in 1979.

Imaginary, time-viewing machines are only coarse-focus devices. When we look broadly at the diversity of studies done since 1979 using the fossil record itself, we are struck by the magnitude of paleobiological progress in understanding the roles of mammals in Mesozoic ecosystems. Particularly important has been extensive new documentation of evolutionary changes in macrofloras, insects, and other terrestrial invertebrates, which in turn have led to tests of hypotheses on their Mesozoic coevolution with mammals. Analyses of fossil plants also suggest greater degrees of climatic differentiation along latitudinal gradients during the age of reptiles than had been visualized before. Sophisticated analyses have documented changing compositions of total vertebrate faunas, especially for latest Cretaceous and earliest Cenozoic time in North America. Important, too, have been discoveries of a variety of mammals of moderate body size, members of a few species perhaps reaching masses of 20–30 kg; the earlier picture of almost universally minuscule mammals must be modified, at least a little, across the sweep of Mesozoic time.

And, oh-my, have there ever been expansions in the taxonomic detail and biological complexities applied to understanding evolution of these little animals! As one would expect following a quarter-century of research, the alpha-level diversity of known kinds (especially at the generic level) of Mesozoic mammals has skyrocketed. There have also been some big surprises, such as the discovery of Cretaceous mammals in South America having dentitions that look like they belonged in the Miocene, tribosphenic molars from Malagasy Jurassic strata, and haramiyid teeth attached to jaws from Greenland. The listing of names applied to cladistically based higher categories sounds progressively more exotic, including such grand terms as mammaliamorphans, gondwanatheres, and australodelphians. Probably much more importantly, some of the geographic and temporal chasms that characterized the sampling of Mesozoic mammals, although still yawning, have been narrowed significantly. Especially important faunal components have been discovered on the southern continents, in several parts of Asia, at multiple pinpoints in the realm of the Arctic, and even in new areas of eastern and western North America. The high degree of isolation of Southern versus Northern Hemisphere mammalian faunas and the biogeographically complex subdivisions of these regions are starting to emerge. Despite the wealth of new taxonomic detail, however, we remain in a position analogous to that of explorers of the fifteenth and sixteenth centuries: Anywhere we do new fieldwork outside the realms of the "known paleontological world" we invariably discover new and wondrous fossils that serve mainly to show us how little we really know about the evolution of mammals during the Mesozoic Era.

The study of Mesozoic mammals continues to be dominated by examination of dentitions. That bias is changing rapidly, however. More and more workers are honing their broader anatomical skills in the realization that greater progress in understanding adaptive radiations and relationships among early mammals will be gained using new details from cranial and postcranial skeletons. Despite our best efforts, however, discoveries of high-quality articulated skeletal remains of Mesozoic mammals continue to be extraordinarily rare. But once discovered, difficulties of their study related to anatomical complexity, small size, and hard rocks are being circumvented by spectacular advances in graphical technology. Key among these procedures is nondestructive study via high-resolution computerized tomographic imaging of fossils still encased in their original geologic vestments. To our personal delight, these forms of innovative technology are leading to a renaissance in anatomical expertise. Once again it is becoming essential for mammal-oriented paleontologists to really know skeletal anatomy—and its functional implications.

Since 1979, enormous strides in developmental biology have been made in understanding how morphological change and evolutionary innovation comes to be. A quarter-century ago knowledge among paleontologists of homeobox genes, distribution of their highly conserved subsets among disparate lineages of multicellular animals, and relationships of epigenetic properties to morphogenesis was nonexistent. Moreover, much greater appreciation for the ubiquity of homoplasy among mammals has developed since 1979, as recognized from more detailed study of the fossil record itself. All that is good for the discipline of paleontology from an intellectual point of view. Now, however, the realization that macroevolutionary jumps in morphology can have simple genetic causes and that homoplasia has a very high probability of occurring, in combination with awareness of the persistence of multitudinous gaps within the fossil record, is making the task of distinguishing phylogenetic relatedness from parallel evolution among early mammals all the more difficult. The quality of our cladograms can be only as good as our ability to recognize similarities derived from homoplasia. It is actually difficult for us to repress a real sense of pessimism at this point, despite the hugely enhanced appreciation for the diversity of Mesozoic mammals.

To add to the difficulty just noted, we see other analytical problems emerging, which might be thought of as kinds of "Lagerstätte effects." For example, since 1979 important phylogenetic analyses have been developed, but these were restricted to mammalian taxa known from more than dentitions. Most of these studies emphasized mammals known exclusively from brief intervals of Rhaeto-Liassic, Late Jurassic–Early Cretaceous, and very Late Cretaceous time. Exclusion from such studies of taxa known only by dentitions has been excused through special pleading related to problems resulting from missing data in cladistic analyses. But the effects upon interpretations of these analyses resulting from loss of data through exclusion of "dental" taxa have not been clearly acknowledged. Another variety of "Lagerstätte effect" might be seen in recently completed statistical analyses of diversity based upon available samples in museum collections. These studies have tended to conclude that we already know about most of the taxonomic diversity that existed among Mesozoic mammals. We predict that another quarter-century of fieldwork will show that conclusion is far from the truth.

The present book represents a greatly needed summary and updating of new knowledge about early mammalian history. Such a compendium can be especially useful in suggesting where the most glaring gaps remain within the story of Mesozoic mammals. For example, despite the wealth of newly discovered genera and their taxonomic interpretations, it is clear that the times and places of origin of almost all of the major groups remain unknown. Moreover, we are far from having achieved stable taxonomic schemes for mammals in general—or even for relationships within the major orders, Mesozoic and Cenozoic. The continuing uncertainty surrounding affinities of multituberculates is an especially telling example. Even the Mesozoic roots for almost any of the Cenozoic orders of mammals remain elements of uncertainty and lend continuing opportunity for productive discussion. Rather consistently, interpretations of the ages of known fossils suggest markedly younger origins of major groups than those estimated from molecular evidence drawn from living representatives. Why do these important discrepancies exist? Do they simply reflect artifacts caused by stratigraphic gaps and other weaknesses of the fossil record? Or could they perhaps derive from flaws in assumptions underlying molecular-clock methodology? To this point, recent advances in knowledge of Mesozoic mammalian diversification have allowed better recognition of discrepancies between the two methods—but not confidence in understanding their origins.

Where can paleontologists do the greatest good in strengthening the evidence upon which new analyses will depend? In two words, more fieldwork. There probably *is* more paleontological fieldwork being done per unit time now than in earlier days. But we also suggest that, relative to the comparatively huge number of active paleontologists, the proportion of effort being committed to hardcore field exploration for Mesozoic mammals is painfully small. As a first step, we need additional fossils, especially those including skeletons, from new places and from previously unsampled geologic intervals. We have to find more fossil localities kindred to the abundance of the Welsh fissures, Como Bluff, or Lance Creek, and at least some of them need to have the taphonomic excellence in preservation of skeletal remains of Guimarota, Liaoning, or many sites in Mongolia. Extraordinary localities must exist out there somewhere, just waiting for our enhanced efforts toward discovery.

Perhaps of equal importance to the discovery of new mammalian fossils is the study of stratigraphic and taphonomic contexts in which the fossils are preserved. As has become almost commonplace within selected Tertiary strata, applications of a diversity of geochemical analyses involving stable isotopes hold great promise within terrestrial Mesozoic sequences for recognition of progressive and short-term paleoclimatic changes that result in new opportunities for chemostratigraphic correlation. The sampling of Mesozoic mammal-bearing strata for associated fossils of ancient plants remains another activity deserving of greater focus.

But our main point is that future progress in genuinely understanding Mesozoic mammals depends upon studying the rocks and the fossils they entomb in new and innovative ways. That requires tedious, risky, energy-consuming field exploration followed by comparably difficult work in the laboratory—not the writing of additional reviews of evidence gathered from literature or existing contents of museum cases.

So what of value is left for future generations to learn about Mesozoic mammals? Again in two words, almost everything. This book provides both a milestone marking current progress in our understanding of these wonderful beasts and a starting line of challenges for future research.

Jason A. Lillegraven and
William A. Clemens

PREFACE

INTRODUCTION

The last (and only) general book on Mesozoic mammals, edited by Jason A. Lillegraven, Zofia Kielan-Jaworowska, and William A. Clemens, appeared in 1979. The need for an updated compendium on the subject has been clear for some time. Our basic source of information—a fossil record comprised of unbelievably tiny, delicate, and elusive fossils—has expanded by orders of magnitude over the past quarter-century. New specimens, some of them breathtakingly complete, have poured in from almost all parts of the globe. These discoveries, together with the application of new scientific approaches and techniques, have led to profound changes in our interpretation of early mammalian history—changes that are coming about at a greater pace today than ever before. Primary interest in the subject itself has grown considerably. Perhaps more significantly, however, Mesozoic mammals have come into their own as an important, rich source of information for evolutionary biology in general. Their record of episodic, successive radiations speaks to the pace and mode of evolution. Early mammals were small, but they provide key information on the morphological transformations that led to modern mammals, including our own lineage of Placentalia. Early mammals were significant and often rapidly evolving elements of the terrestrial biota for much of the Mesozoic—biota that are playing an increasingly important role in studies of paleoecology, faunal turnover, and historical biogeography. Finally, of course, the record of early mammals occupies center stage when it comes to testing molecular evolutionary hypotheses on the timing and sequence of mammalian radiations.

Given this proliferation of new information and multidisciplinary interest in Mesozoic mammals, it is not surprising that compilation of this volume was both an exciting and a humbling experience for us. We quickly found that we had given ourselves far too much credit when it came to knowledge of the subject matter. Along the way, we encountered some critical topics where additional first-hand experience was needed; in other cases, we ran into unresolved problems that would have to be faced if the book was to be completed. This led us to undertake several collateral studies, the most important of which were published as Kielan-Jaworowska et al. (1998), Kielan-Jaworowska and Hurum (2001), Luo, Cifelli, and Kielan-Jaworowska (2001), Kielan-Jaworowska, Cifelli, and Luo (2002), and Luo et al. (2002). Work on these projects inevitably delayed completion of the book, as did the pace of discovery and conceptual change itself. No sooner had a chapter been drafted, for example, when a new discovery or phylogenetic analysis would require its complete revision. But change is a measure of vitality, and we are convinced that the study of mammalian evolution is not only vitally alive, but that it is poised at the edge of paradigm shift. This is, perhaps, the most exhilarating time in history for students of early mammals, and there is no doubt that monumental changes will await a future volume that succeeds this one. In the meantime, we hope that our book will stimulate and foster those changes and that it might ultimately help sow the seeds of its own obsolescence.

BOOK CONTENTS AND FORMAT

We (see photo above) have endeavored to make this book as comprehensive as possible, within practical limitations.

Of the 15 chapters that follow, four (chapters 1–3 and 15) are of a general nature and apply to the full spectrum of early mammals, whereas the remaining 11 chapters deal with the anatomy, paleobiology, and systematics of the particular groups of mammals known from the Mesozoic, extinct and extant. In chapter 1, we provide some general background on mammals within the context of non-mammalian synapsids; on the successive radiations, biology, and ecological roles of mammals during their tenure of 155 million years (Ma) in the Mesozoic; on the history of study, including a summary of major advances in the past two decades; and on major conceptual issues, past and present.

Chapter 2, "Distribution," presents an account of Mesozoic mammals in space and time, summarizing their geological and geographical occurrences worldwide. Localities (or, in the case of local faunas, core areas) are plotted on generalized maps, and brief descriptions cover fossil yield, preservation, and diversity for each locality. Faunal lists are summarized in tables that include each site, and synoptic charts show the distribution of sites in time through the Mesozoic, continent-by-continent. The nature and amount of contextual information for published occurrences of Mesozoic mammals is highly variable; where possible, we have tried to give a brief account of depositional context and possible paleoenvironment. Finally, we comment briefly on any patterns of distribution, in both space and time, and their potential implications.

Chapter 3, "Origin of Mammals" provides an in-depth look at one of the fundamental aspects of mammalian evolution: their origin from among the nonmammalian synapsids. Using a phylogenetic and functional perspective, we treat each major anatomical system in the skull and postcranial skeleton, addressing the evolution of mammal-like characters among advanced cynodonts, appearance of mammalian characters at the cynodont-mammal transition, and transformation of character complexes among the earliest mammals. Finally, chapter 15, "Interrelationships of Mesozoic Mammals," gives a summary of the phylogenetic relationships among the individual groups of early mammals covered by the 11 systematic survey chapters. We discuss the often controversial relationships among the principal groups of mammals (illustrated by two alternative trees) and present a synthetic phylogeny that incorporates the full range of diversity among Mesozoic mammals, together with a sampling of the major living groups—monotremes, marsupials, and placentals. This chapter is based largely on the morphological data we have collected from all these Mesozoic mammal groups through our own observations and from the literature, as summarized in our recent study (Luo et al., 2002) and incorporates the results of a new parsimony analysis that includes additional taxa.

Mammalian phylogeny is the principle we have abided by for the book's organization. The sequence of the remaining 11 systematic survey chapters of major kinds of Mesozoic mammals follows the mammalian family tree presented in chapter 15. Each systematic survey chapter on a given group of Mesozoic mammals includes a brief characterization, an account of geologic and geographic distribution, comparative anatomy, paleobiology, and systematics. These chapters differ conspicuously in size, owing to the great variation in the number of described taxa and anatomical representation. The shortest of these, chapter 14, treats the poorly known, highly specialized Gondwanatheria, represented by only a few fossils and fewer described taxa. By far the longest chapter, "Allotherians" (chapter 8) includes the most diverse and incomparably best known of Mesozoic mammals—the multituberculates. Chapter length thus reflects, in part, limitation by chance representation in the fossil record, but probably, more importantly, it reflects real differences in the extent to which various groups flourished or floundered over the course of Mesozoic time.

A few comments on the systematics of chapters 4–14 are in order. As noted in chapter 1, the book's structure and the basic topology of the mammalian tree made it impractical to restrict the contents of chapters 4–14 to strictly monophyletic units; hence, some chapters deal with convenient sections of the tree, as delineated by the phylogeny of chapter 15. Chapter 4, "The Earliest-Known Stem Mammals," for example, covers Sinoconodontidae, Morganucodonta, and some genera of early mammals that clearly do not belong to either of these groups, but nonetheless are most appropriately treated with them. To emphasize this point and to make clear what is included in a given chapter, a small logo with a simplified mammalian tree appears on the first page of each of these chapters. Branches or clades covered in a chapter are indicated by boldfacing the appropriate part(s) of the mammalian tree. The names of paraphyletic groups cited in the book are given in quotation marks, for example, "symmetrodontans" or "eupantotherians."

The "Systematics" sections of chapters 4–15 include either stem- or node-based definitions, together with diagnoses for all relevant higher-rank taxa, with diagnoses extending down to the genus level. Type and referred species (those of Mesozoic age) are listed for all genera, and the literature cited (at the end of the book) gives the primary reference for each taxonomic name, down to the species level. We have included all taxa that were published as of March 31, 2002. Although we were unable to include all taxa published after that date, we have cited some papers published thereafter, in such cases as they were available to us while in press. New references published in the interval between April 1, 2002, and January 3, 2004, are listed as "Additional References" and follow the reference list.

Excepting only a few poorly known genera of doubtful status, for which no intelligible published illustration exists, all genera of Mesozoic mammals are illustrated here. Chapters 4–14 all include a table presenting a classification of the taxa covered. Relationships within groups of Mesozoic mammals are reasonably well understood in only a few cases. In order to avoid the implication of some preferred, within-group phylogenetic arrangement, we adopted a neutral system of arrangement whereby subordinate taxa are listed alphabetically within groups (except for type genus and type species, which are listed first under family- and genus-level headings, respectively). If, however, the described group has been recently revised and a phylogeny proposed, we follow the recently published phylogenetic scheme (e.g., some groups discussed in chapters 8 and 10). For the same reason, most of these chapters do not contain cladograms of the groups treated therein. An exception to this is chapter 8, for which we were able to include a hypothesis of relationships among higher taxa of multituberculates on the basis of recent work by Kielan-Jaworowska and Hurum (2001).

We call attention to the fact that some Mesozoic mammals are of uncertain or debatable affinities. Placing such forms among the chapters of this book called for arbitrary decisions, based on anatomical and historical considerations as well as varied opinions in the literature. "Obtuse-angled symmetrodonts," for example, could be variously placed: Kuehneotheriidae among stem mammals (chapter 4) or "symmetrodontans" (chapter 9), and Woutersiidae among stem mammals, "symmetrodontans," or docodontans (chapter 5). In each case, we have attempted to cross-reference among chapters and to point out alternative placements.

It will come as no surprise that preparation of this volume involved consultation of many references. This necessitated the adoption of some conventions, mainly aimed at streamlining or clarifying in-text literature citations. Most multiple-authored papers are cited in the text using the convention "et al." (e.g., Novacek et al., 1997). In the case of same-year, multiple-authored papers with the same first author, text citations list enough of the authors to distinguish the references (e.g., Cifelli, Gardner, et al., 1997 versus Cifelli, Kirkland, et al., 1997). There are a few cases of single- or first-authored works by different individuals bearing the same surname (e.g., Bonaparte, C. L., and Bonaparte, J. F.). In practically all cases, dates of publication fortunately do not overlap (as in the example given, where more than a century separates the respective published works) and initials of given names are omitted from in-text citations (although initials are used when the author's name is formally included as part of a taxonomic name). An exception to this occurs with E. F. Freeman and P. W. Freeman, both of whom are cited for works published in 1979; here we have included initials of given names in the in-text citations. We cite the published (Latinized) rather than the christened version of the name for the founder of modern systematic botany and zoology, Karl von Linné (e.g., Linnaeus, 1758, 1766).

CONVENTIONS

As discussed in chapters 1 and 3, we have adopted a traditional, broad-based definition of Mammalia that includes some wholly extinct groups that lie outside the crown clade formed by the three major living groups—monotremes, marsupials, and placentals. Under an alternative definition that restricts Mammalia to the mammalian crown group (see Rowe, 1988; McKenna and Bell, 1997), several groups included in this book, such as sinoconodontids, morganucodontans, docodontans, and *Hadrocodium,* would be considered "nonmammalian" mammaliaforms. The difference between these definitions is, of course, arbitrary. Irrespective of the varying opinions as to where the line between mammals and nonmammalian synapsids should be drawn, students of early mammals unanimously agree on two fundamental premises: sinoconodontids, morganucodontans, and docodontans are exceptionally well known, and they are critical to interpreting the history of early mammals. Regardless of what we call them, they are an indispensable part of our understanding of early mammalian evolution and must be included in any book on the subject.

With a few very minor exceptions, all data presented in this book are based on published references. We do not describe any new taxa, nor do we erect new higher systematic units. Along the same lines, we have adopted the traditional usage of Linnaean taxa, which we feel is more appropriate and accessible for present purposes (see discussions by Benton, 2000b; Dyke, 2002), in lieu of developing a new, Phylocode-based scheme. Linnaean classifications of the systematic chapters (4–14) list full taxon names (including author[s] and date) and taxa down to the species level. Chapters 1 (all mammals), 4, 9, and 10 each cover a nested hierarchy of several clades. For each of these chapters, we also provide a cladistic classification of the taxa covered in respective chapters, side-by-side with the traditional Linnaean classification table. The hierarchical scheme for the cladistic classifications given in chapters 1, 4, 9, and 10 is denoted by the established method of successive indentations; we refer readers to the description and example of McKenna and Bell (1997: 30) for an explanation of this procedure. For higher-rank taxa, we simply use the term "clade" with its attendant taxic definition, instead of the explicit ranks used by other authors, for example, Prothero (1981) and McKenna and Bell (1997). For lower taxonomic ranks, we use the traditionally accepted Linnaean units such as superorders, orders, families, and subfamilies. In a cladistic phylogeny it is also unavoidable that some taxa (often stem taxa) will cluster in polytomies in various types of consensus trees. These taxa are listed as *sedis mutabilis,* following the procedures recommended by Wiley et al. (1991).

Collective taxa are referred to in the plural, following Mayr and Ashlock (1991: 394–395), for example, "Eutheria are." In some cases the name for a higher taxonomic category, such as an order, bears the same root as a family and genus (e.g., Morganucodonta, Morganucodontidae, *Morganucodon*). In order to avoid confusion in vernacular usage (using the same example, "morganucodonts," which could be interpreted in significantly different ways), we refer to the higher category by adding the suffix "ans" ("morganucodontans").

We made an attempt to follow the anatomical nomenclature of Schaller's (1992) *Illustrated Veterinary Anatomical Nomenclature* (which is based on *Nomina Anatomica Veterinaria,* 1983) and *Miller's Anatomy of the Dog* (Evans, 1995). However, in some cases strict adherence to the terminology of the *Nomina Anatomica Veterinaria* proved to be impractical. Most problematic in this respect are the bones of the wrist (carpus) and ankle (tarsus). Pertinent literature, both paleontological and anatomical, almost universally employs the terminology of human anatomy with respect to these elements, and we have followed this custom.

Unfortunately, there is an important exception to this rule as well: the medial bone of the proximal row in the ankle. Here we have followed the long-standing and widespread practice of referring to this element as the astragalus (e.g., Jollie, 1962; Romer and Parsons, 1977) rather than the talus, as used in both the *Nomina Anatomica Veterinaria* and human anatomy. For structures that do not occur in domestic mammals, we follow the terminology of widely used paleontological and anatomical texts (Jollie, 1962; Romer, 1966; Romer and Parsons, 1986; Lillegraven et al., 1979; Kuhn-Schnyder and Rieber, 1986; Carroll, 1988), together with specialized monographs and articles, cited in particular chapters as appropriate. For the dentition, we use the standard abbreviations of I, C, P, and M for incisors, canines, premolars, and molars, respectively; upper and lower case letters designate the upper and lower dentitions, respectively. Where possible, dental formulae have been abbreviated by sequentially listing the number of incisors, canines, premolars, and molars in each quadrant, separated by periods; the formula for the upper dentition is given first and is separated from that of the lower dentition by a slash. Hence, the human dentition, which normally has two incisors, a canine, two premolars, and two to three molars in each upper and lower quadrant, would be abbreviated 2.1.2.2–3/2.1.2.2–3.

ACKNOWLEDGMENTS

We began work on this book on Mesozoic mammals at the beginning of 1999. With support from our colleagues Donald R. Prothero and Hans-Dieter Sues, we signed the contract with Columbia University Press in April 1999. During our five years of work on this book, we greatly benefited from the help and advice of numerous colleagues and friends, who generously offered us enormous assistance in various ways. Primary among them (cited in alphabetic order) are Alexander O. Averianov and William A. Clemens, who agreed to serve as official reviewers of the whole book, read all the chapters at various stages of our work, and offered most valuable comments and criticism. We acknowledge, with heartfelt thanks, the many colleagues who kindly offered invaluable help in reading and reviewing one or several chapters, and/or provided us with photographs, casts, or needed information: J. David Archibald, Michael Archer, K. Christopher Beard, Percy M. Butler, Zoltan Csiki, Brian M. Davis, Mary R. Dawson, Steven Diem, Paul C. Ensom, Tim F. Flannery, Emmanuel Gheerbrant, Pascal Godefroit, Cynthia L. Gordon, Gerhard Hahn, John Hunter, Jørn H. Hurum, Louis L. Jacobs, Harry J. Jerison, Wighart von Koenigswald, Wann Langston, Jason A. Lillegraven, Scott K. Madsen, Thomas Martin, Malcolm C. McKenna, Christian de Muizon, Anne Musser, Michael J. Novacek, the late Constantin Rădulescu, Thomas H. Rich, Kenneth D. Rose, Andrew Ross, Guillermo W. Rougier, Donald E. Russell, Bernard Sigé, Denise Sigogneau-Russell, Frederick S. Szalay, Tim Tokaryk, Michacl W. Webb, Anne Weil, John R. Wible, Dale A. Winkler, and André R. Wyss.

We are greatly indebted to the following colleagues, who graciously provided us with original drawings (based on specimens, casts, or photographs) that we publish herein: Alexander O. Averianov, Percy M. Butler, Cynthia L. Gordon, Nicholas J. Czaplewski, Yaoming Hu, Aleksandra Hołda, Michael J. Novacek, Constantin Rădulescu, and Karol Sabath. The artists Mark A. Klingler and Linda Lillegraven also kindly assisted with illustrative work, providing some of the reconstructions published in the book.

We obtained invaluable help from two colleagues at the Institute of Paleobiology in Warsaw. Andrzej Kaim cooperated patiently with us throughout the preparation of this book and processed digital images of all illustrations, unifying and greatly improving them. Aleksandra Szmielew shared with us her experience in making the first preliminary layout of the book.

We extend our sincere thanks to the following individuals for their assistance in final compilation and proofing of the manuscript: Cynthia L. Gordon, Brian M. Davis, Karol Sabath, and Wojciech Siciński. We thank our successive editors at Columbia University Press, Holly Hodder and Robin Smith, for their help and support throughout. It is also a pleasure to acknowledge, with thanks, Cyd Westmoreland of Princeton Editorial Associates, who spearheaded the effort to tame an unruly manuscript and who patiently fielded our sometimes naïve queries and niggling requests.

Our work was supported by the Institute of Paleobiology, Polish Academy of Sciences (PAN), in Warsaw (to ZK-J) and a collaborative program between the U.S. National Science Foundation (NSF) and PAN (ZK-J and RLC). Aspects of this work were directly supported or facilitated by grants from NSF (RLC and Z-XL), the Petroleum Research Fund of the American Chemical Society (RLC), the National Geographic Society (RLC and Z-XL), the National Science Foundation of China (NSFC), and the Carnegie Museum of Natural History (Z-XL).

CHAPTER 1

Introduction

The first 155 million years (Ma) of mammalian history occurred during the Mesozoic. During this vast time span, mammals diversified into many lineages and underwent enormous anatomical evolution. Humans are primates; primates are placentals; placentals are eutherians; and stem eutherians have a long history, which extends well back into the Early Cretaceous. Mesozoic mammals include the trunk and a bewildering bush of basal branches for the entire mammalian family tree. Their fossil records are indispensable for our understanding of the deep history that gave rise to extant mammals, including our own lineage of placentals.

Mammals have thrived in the world's biota for 65 Ma, since the beginning of the Cenozoic, and they are represented today in a spectacular diversity of more than 4,600 modern species (Nowak, 1991; Wilson and Reeder, 1993; MacDonald, 2002). These diverse Cenozoic and modern mammals are descendants of three major evolutionary lineages: eutherians (including placentals), metatherians (including marsupials), and australosphenidans (including monotremes). The surviving members of these lineages, however, represent only a tiny fraction of a dazzling array of some 25 distinctive mammalian clades (at the order or family level of the traditional Linnaean taxonomy) that existed in the Mesozoic. Modern mammals represent merely the tips of three twigs of a vast evolutionary bush, most of which was pruned by extinction in the Mesozoic (Clemens et al., 1979; McKenna and Bell, 1997; Cifelli, 2001; Luo et al., 2002).

PREMAMMALIAN SYNAPSIDS

The synapsid lineage that includes mammals split from other amniote vertebrates over 300 Ma ago, during the Late Carboniferous (Pennsylvanian) in the Paleozoic. Synapsids are characterized by the presence of a lower temporal fenestra in the skull, among other apomorphies. Their stem taxa, or pelycosaurs, thrived in the Late Pennsylvanian through the Early Permian (Kemp, 1982; Carroll, 1988; Hopson, 1994). Therapsids, a derived subgroup of synapsids, are characterized by enlargement of the temporal fenestra to accommodate enlarged jaw muscles, a distinctive notch in the angular bone of the lower jaw, several derived features in the limb girdles, and a more anteroventral (near parasagittal) posture of the hind limbs than in the pretherapsid synapsids. Therapsids first appeared at the beginning of the Late Permian and promptly split into several stem groups, which dominated the terrestrial biota in the Late Permian. Cynodonts, a derived subgroup of therapsids, appeared in the Late Permian. Cynodonts are distinguished from precynodont therapsids in having a better-developed secondary bony palate, to separate the nasal passage for breathing from the mouth cavity for feeding, and a larger coronoid process on the dentary, with a definitive masseteric fossa for strengthened jaw adductor muscles, among many other derived features (Kemp, 1982; Hopson, 1994). The advanced cynodonts, such as probainognathids, tritylodontids, and tritheledontids, achieved many mammal-like features in both the skull and postcranial skeleton (see chapter 3). Mammalia (see box, next page) are nested within the advanced cynodonts and are defined here as all descendants of the common ancestor of *Sinoconodon* and extant monotremes, marsupials, and placentals. Mammals so defined are diagnosed by the dentary condyle to the squamosal glenoid in the jaw hinge and the presence of the petrosal promontorium. The earliest-known and transitional taxa bearing these and other derived features appeared about 220 Ma ago, in the Norian age of the Triassic, possibly as early as the Carnian (Fraser et al., 1985; Lucas and Luo, 1993; Sigogneau-Russell and Hahn, 1994; Datta and Das, 1996; Jenkins et al., 1997; see also figure 1.1). The poorly known *Adelobasileus* Lucas and Hunt, 1990, is also tentatively included within the Mammalia.

A substantial body of evidence supports the hypothesis that therapsids achieved a higher level of overall activity and more elevated growth rates than pretherapsid synapsids. Brink (1956, 1980) speculated on the basis of skeletal evidence that some of the nonmammalian cynodonts (advanced "mammal-like reptiles") acquired hair and a diaphragm, which he suggested are indicative of an evolutionary trend toward endothermy. Bakker (1971, 1975) argued that predator-prey ratios of Late Permian–Early Triassic therapsid communities were closer to those of extant endothermic communities than to ectothermic communities. Based on a study of bone histology, Ricqlès (1974, 1976) concluded that therapsids had a high level of overall activity and high growth rates in comparison to pelycosaurs. He argued that nonmammalian therapsids probably had high metabolic rates and were likely to have achieved some kind of endothermy. In comparison to the pretherapsid pelycosaurs, advanced nonmammalian therapsids had a more upright posture, indicating a more active locomotory pattern (Kemp, 1982; Hopson, 1994). Several studies (e.g., Bennett and Ruben, 1986; Hillenius, 1992, 1994) have documented the presence of a maxillary ridge in the nasal cavity in nonmammalian therapsids. This provides indirect evidence for the development of maxillary turbinates, which in turn are related to thermoregulation in extant mammals (Moore, 1981). Hillenius (1994) proposed that thermoregulatory adaptation was developed among cynodonts and even in precynodont therapsids. Although there is broad consensus that therapsids had elevated levels of overall activity and growth rates, it remains unclear as to whether any nonmammalian therapsids (including cynodonts) developed the same homeothermic (constant body temperature) physiological functions as those seen in modern mammalian insectivores (Crompton et al., 1978).

> Mammalia are a clade defined by the shared common ancestor of *Sinoconodon*, morganucodontans, docodontans, Monotremata, Marsupialia, and Placentalia, plus any extinct taxa that are shown to be nested with this clade by parsimony analyses. This stem-based definition excludes from Mammalia such advanced cynodonts as chiniquodontids, probainognathids, tritylodontids, tritheledontids, or any taxa nested within a subset of the latter cynodont groups. The Mammalia concept here is equivalent to the Mammaliaformes of Rowe (1988: figure 4, but not that of Rowe,1993: figure 10.2).
>
> In their impressive book, *Classification of Mammals above the Species Level,* McKenna and Bell (1997) adopted a crown-based concept of Mammalia, originally proposed by Rowe (1988) and subsequently accepted by some other authors. Following the definition by Rowe (1988) and McKenna and Bell (1997), several groups regarded here to be mammals (e.g., Morganucodonta, Docodonta, Haramiyoidea) would be excluded from the Mammalia, instead being regarded as nonmammalian mammaliaforms. The status of several other groups—including the best known and most diverse of Mesozoic mammals, the Multituberculata—would be ambiguous under this definition, as their phylogenetic position relative to crown Mammalia has yet to be settled decisively (see chapter 15).
>
> There is a massive body of literature dealing with the merits and problems associated with crown-, stem-, or character-based definitions of Mammalia (see Lucas,1992; Rowe and Gauthier,1992; Bryant, 1994). The definition adopted by us is one of several alternatives. We prefer this inclusive definition of mammals because (1) it is consistent with widespread, traditional usage (e.g., Hopson, 1994); (2) it has the virtue of being relatively stable in its membership contents of both living and fossil taxa (Luo et al., 2002); and (3) it happens to be diagnosed by characters that we view as being biologically significant (see chapter 3). The first two of these criteria, at least, address a primary purpose of taxonomic nomenclature: to promote stability and commonality of usage.

SUCCESSIVE DIVERSIFICATIONS OF MESOZOIC MAMMALS

Most Mesozoic mammal species are represented only by teeth and incomplete jaws. Inadequate as they are for a comprehensive understanding of whole organisms, teeth and jaws nonetheless provide useful information about the taxonomic richness and diverse feeding adaptations of early mammals. Their records show several major episodes of diversification (figure 1.1).

The earliest diversification took place among stem mammals during the Late Triassic to Early Jurassic. From this initial burst of mammalian evolution arose several groups: haramiyidans with multirow, multicuspate, "multituberculate-like" molars; *Sinoconodon* and morganucodontans with "triconodont-like" molars showing three main cusps in alignment; and kuehneotheriids with triangulated, "symmetrodont-like" molars. All these mammals possessed the derived mammalian feature of the dentary

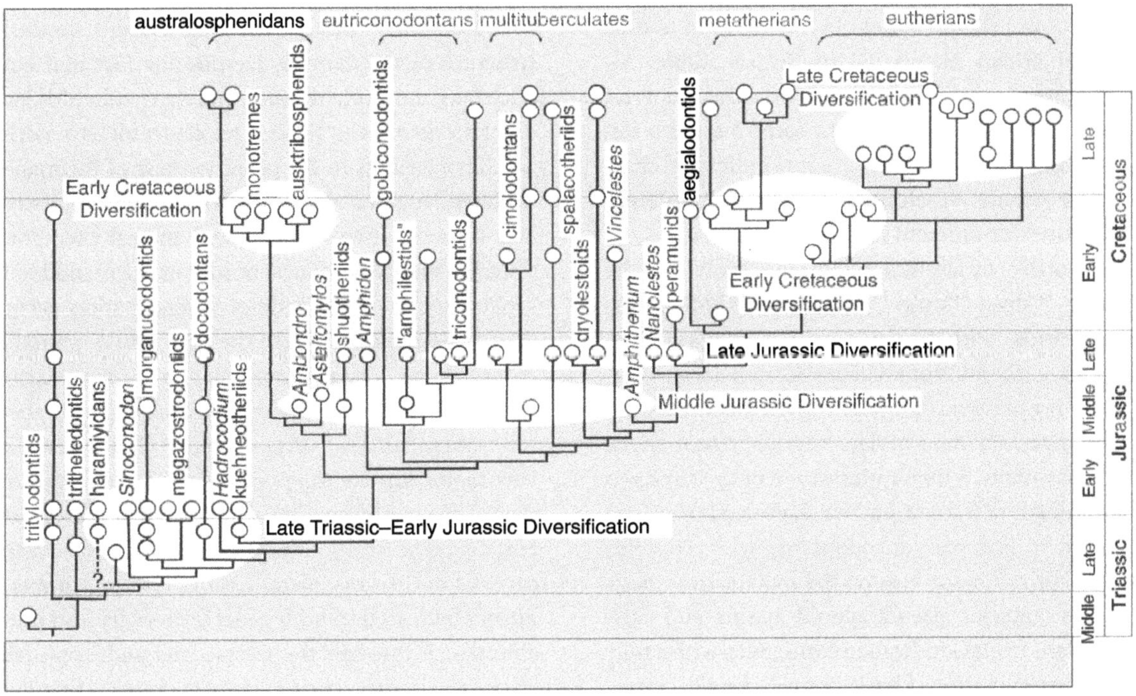

FIGURE 1.1. Overview of the temporal distribution and relationships among Mesozoic mammal clades. The great evolutionary bush of diverse Mesozoic mammalian clades is the dominating feature in their taxic evolutionary pattern. The more recent cladistic analyses of dental, cranial, and skeletal characters of all phylogenetically relevant clades have revealed a far greater array of taxa and ranks of clades than could be accommodated by the traditional views of four "long-branch" lineages to the Early Triassic (e.g., Simpson 1928a; figure 1.2A) or two prototherian versus therian divisions extending to the Rhaeto-Liassic (figure 1.2B). The almost fully resolved cladogram of all Mesozoic mammal groups, together with their improved records of temporal distribution of fossils, suggests that mammalian diversifications occurred episodically during the entire span of Mesozoic mammalian history. The five episodes of diversifications are as follows: (1) The earliest-known episode of diversification occurred in the Late Triassic–Early Jurassic on a global scale, when haramiyidans, morganucodontans, kuehneotheriids, and docodontans (if including *Woutersia*) appeared. (2) The next episode occurred globally in the Middle Jurassic, characterized by the appearance of shuotheriids, the earliest australosphenidans, eutriconodontans, putative multituberculates, and amphitheriid "eupantotherians." (3) The Late Jurassic diversification occurred primarily in Laurasia among eutriconodontans, spalacotheriids, dryolestoids, and peramurans. (4) The Early Cretaceous episode saw diversification within australosphenidans and toothed monotremes on the Gondwanan continents, and the basal splits of eutherians and metatherians and diversification of tricondontids on the Laurasian continents. (5) The Late Cretaceous episode witnessed diversification within metatherians, within eutherians, and within cimolodontan multituberculates on the northern continents and of gondwanatherians on the southern continents. The earliest diversifications of stem eutherians and stem metatherians, as documented here by the currently available fossil evidence, predate the likely time window estimated by molecular evolutionary studies, which indicates that some superordinal clades of placental mammals may extend back into the Cretaceous (e.g., split of the earliest placental superordinal clades around 108 ± 6 Ma ago; Murphy, Eizirik, Johnson, et al., 2001; Murphy, Eizirik, O'Brien, et al., 2001; and several other studies). Full test of the molecular estimates of the phylogenetic divergence would require a combined analysis of currently available molecular sequence data and morphological data of major extant clades, with the main characters for resolving Mesozoic mammalian relationships coded for modern clades. This has not been done and is beyond the scope of this book. Therefore, for the time being we believe that the issue of the timing of the origin of modern placental superordinal clades is still unresolved. *Source:* stratigraphic distribution from chapter 2; cladogram based on Luo et al. (2002) and Ji et al. (2002).

condyle to squamosal glenoid in jaw joint,[1] but also retained the plesiomorphous ("reptilian") features of an articular-quadrate jaw joint and the attachment of the postdentary bones (homologues of tympanic and middle ear bones of crown mammals) to the mandible. Although haramiyidans, *Sinoconodon,* morganucodontans, and kuehneotheriids have almost the same mandible design (notwithstanding variations in some individual characters), these groups developed astonishingly different molar structures for different feeding adaptations.

Six more order- or family-level lineages evolved in the next episode of diversification, during the Middle Jurassic. Amphilestids, with triconodont-like molars; eleutherodontans, with multituberculate-like molars; and amphitheriids and peramurids, with a triangulate trigonid plus a talonid heel on lower molars were all present on the Laurasian continents. A major innovation of this time was that of grinding molar function, which evolved in at least three emergent lineages: docodontans, with complex shearing and grinding surfaces on the molars; shuotheriids, with an anterior pseudotalonid basin; and australosphenidans from Gondwanan continents, with a fully developed posterior talonid basin for grinding function, in addition to the plesiomorphic shearing features on the molars. It is obvious that the adaptive molar structures related to grinding functions are homoplastic among these groups.

Consistent with this is the fact that, in juxtaposition with their derived, multifunctional molars, docodontans, shuotheriids, and at least some Middle Jurassic australosphenidans retained the primitive features of the postdentary trough on the mandible. Presence of these primitive mandibular features in the Jurassic southern mammals with multifunctional tribosphenic molars is strong evidence that they are not closely related to boreosphenidans, which evolved on the northern continents some 25 Ma later, in a separate lineage with far more derived mandibular features (Luo et al., 2002).

The third episode of diversification among early mammals occurred in the Late Jurassic, with five newly emerged clades: Multituberculata (*sensu stricto*), triconodontids, spalacotheriids, dryolestoids (dryolestids and paurodontids), and tinodontids. The most significant apomorphic feature of these emergent clades of the Late Jurassic is the absence of the postdentary trough on the mandible (Rowe, 1993), a primitive feature retained by almost all Early and Middle Jurassic lineages (except for amphilestids and *Hadrocodium*). Another striking pattern concerns the structure of the dentary. Despite the fact that eutriconodontans, multituberculates, and spalacotheriids have very different dentitions, these three clades all have fairly similar characteristics in the posterior part of the mandible: a rounded "angular" region grades into the dentary condyle, and there is often a prominent medial pterygoid crest ("shelf") along the ventral border of the mandible.

The last major episode of diversification occurred in the Early Cretaceous when the stem taxa of metatherians, eutherians, and basal boreosphenidans appeared in Laurasia, while toothed monotremes and some of their australosphenidan relatives appeared in Gondwana. The only new major lineage to appear in the Late Cretaceous was that of the gondwanatherians, a poorly known group occurring on several southern landmasses. The existing lineages of eutherians, metatherians, and multituberculates greatly increased in both generic diversity and numerical abundance through the Cretaceous and eventually survived (albeit with varied success) the mass extinction at the end of the period. However, spalacotheriids, eutriconodontans, and gondwanatherians had declined and went extinct. Dryolestoids (*Peligrotherium*) survived in South America into the Cenozoic (Gelfo and Pascual, 2001).

Our overall view is that Mesozoic mammal evolution is a great evolutionary bush. The main feature of their taxic macroevolutionary pattern is successive episodes of diversification. This differs from the traditional views of four "long-branch" lineages to Early Triassic (e.g., Simpson, 1928a) (figure 1.2A) or two prototherian versus therian divisions extending to the Rhaeto-Liassic (figure 1.2B).

EVOLUTION IN THE SHADOW OF DINOSAURS

The evolutionary fortune of mammals and the premammalian synapsids was quite different from those of other land vertebrates during the Mesozoic. Although phyletically diverse, Mesozoic mammals appear to have been relatively rare, even by comparison with other small, nondinosaurian vertebrates. During the Permian and most of the Triassic, premammalian synapsids thrived in terrestrial vertebrate faunas worldwide. By the Late Triassic, however, premammalian cynodonts were eclipsed by the rapidly rising dinosaurs. Throughout the rest of Mesozoic, nonavian dinosaurs completely dominated over mammals.

Mesozoic mammals were small to extremely small animals. The Late Triassic and Early Jurassic mammals are, on average, much smaller than the premammalian cyn-

[1] *Haramiyavia* is the best-represented member of haramiyidans (see chapter 8), and is known by the mandible and parts of the dentition. The posterior end of the mandible in the holotype of *Haramiyiavia clemmenseni* is not preserved, but Jenkins et al. (1997) inferred from the preserved posterior part of the mandible that this taxon had a dentary condyle. We follow their interpretation.

odonts. The cynodont-mammal transition involved a great decrease in body mass, which apparently took place in the Late Triassic. The smallest known Mesozoic mammal was *Hadrocodium,* which is estimated to have had a body mass of about 2 g (Luo, Crompton, and Sun, 2001). The largest of well-known Mesozoic mammals attained the size of a fox (e.g., *Repenomamus,* see Li et al., 2000) or a cat (e.g., the largest *Sinoconodon* specimens, Luo, Crompton, and Sun, 2001); isolated teeth suggest the presence of several kinds of mammals that were substantially larger (Clemens et al., 2003). The vast majority of early mammals, however, were shrew to mouse sized, a pattern that persisted through the entire 155-Ma history of mammals in the Mesozoic.

Dental structure and body size suggest that most Mesozoic mammals were insectivorous; some larger animals, such as *Sinoconodon* and gobiconodontid adults, were probably carnivorous, incorporating significant amounts of vertebrate prey (or carrion) in their diets; still others, such as multituberculates, were probably herbivorous to omnivorous.

By analogy with modern placental insectivores, Crompton et al. (1978) hypothesized that small, insectivorous early mammals achieved homeothermy (although not necessarily high resting metabolic rates). Homeothermy would have allowed the earliest mammals to be active and forage nocturnally, in places and at times that would have

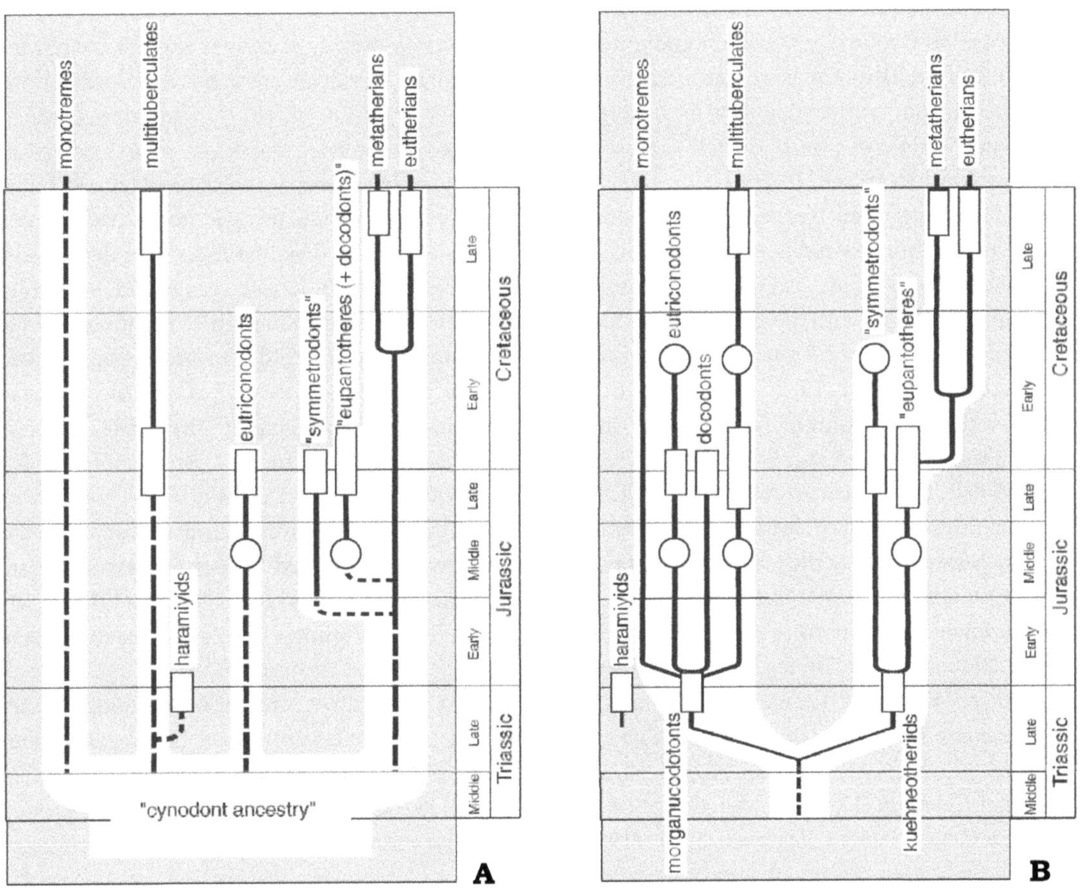

FIGURE 1.2. Changing interpretations of Mesozoic mammal phylogeny and the timing of their patterns of diversifications. A, An historical (now abandoned) view of the polyphyletic origin of the Class Mammalia via several independent lineages from premammalian cynodont ancestry, as advocated by Simpson (e.g., 1928a, 1959) and accepted by others in the 1940s–1960s (e.g., Olson 1959). In this view of mammalian evolution, there were four independent origins of mammals (in the white bands) of monotremes, multituberculates, triconodonts (= eutriconodontans), and extant therians. Stratigraphic ranges of mammal lineages here reflect the available fossil record up to the 1940s. B, Single origin of Mammalia but a fundamental dichotomy within mammals (e.g., Hopson and Crompton, 1969; Crompton and Jenkins, 1979). This was a prevalent view from the mid-1960s to the 1980s that was generally accepted by many fossil mammal workers. While the monophyletic origin of Mammalia is supported by current evidence, the early division of the "prototherians" versus therians (two white bands) within the Mammalia has been abandoned. Stratigraphic ranges of major lineages here reflect the available fossil record up to the 1980s. The current view of interrelationships of Mesozoic mammals is shown in figure 1.1. *Source:* modified from Cifelli (2001).

been largely impossible for poikilothermic vertebrates. Coupled with their more sensitive hearing and olfaction, larger brains for better sensory and motor coordination, and feeding adaptations for insectivory, homeothermy enabled early mammals to successfully invade the nocturnal niches of terrestrial ecosystems of the Mesozoic. Early mammals survived and diversified in the nocturnal niches that had not been previously exploited by other kinds of small, insectivorous vertebrates. The fact that more than half of all modern mammals, and the vast majority of all small mammals, live nocturnal lives (Ryszkiewicz, 1989) is suggestive of a mammalian ancestral niche from their long sojourn with Mesozoic nights.

The evolutionary causes by which mammals remained predominantly small throughout the rest of the Mesozoic are still not well understood. Several speculative scenarios have been developed to account for this phenomenon, including avoidance of predation by large nonmammalian vertebrates, as well as competitive ecological exclusion by small nondinosaur vertebrates and juvenile dinosaurs (critically reviewed by Lillegraven, 1979).

Crompton (1968) suggested that whereas cynodonts acquired hair as insulation, they did not develop the cooling mechanisms characteristic of most modern mammals (but not monotremes). Faced with the increasing temperatures of the Triassic, they evolved smaller sizes to avoid overheating. Bakker speculated (1971: 655): "Small body size and high surface-area–volume ratios would have eased the difficulties of overheating. Nocturnality would also reduce the difficulties—the mammals could be active in ambient temperatures usually at least a few degrees below the body temperature during the Mesozoic nights and could avoid exposure to direct solar radiation."

The earliest known eutherians were very small, ranging from 5 g for *Batodon tenuis,* to 7 to 8 g for *Prokennalestes* and *Montanalestes* (Wood and Clemens, 2001), and to about 20 g for *Eomaia scansoria* (Ji et al., 2002). The majority of stem metatherians of Cretaceous age were in a similar size range (chapter 12), with the exception of *Didelphodon* (Clemens, 1966). Lillegraven et al. (1987) noted that small body mass played an important role in the origin of the earliest eutherians and marsupials during the Early Cretaceous in relation to metabolic and reproductive constraints of these mammals. In fact, it is a general pattern that the earliest members of the mammalian clade tended to be smaller than their later and more derived phylogenetic relatives, as documented by a theoretical study of Cenozoic mammals (Alroy, 1998) and a case study of Late Cretaceous mammals of Wyoming (Lillegraven and Eberle, 1999).

It was only after the dinosaur extinction 65 Ma ago that mammals took off in a great evolutionary radiation, occupying diverse ecological niches, with many feeding adaptations and a great range of body sizes. Whereas all but two of the well-known Mesozoic mammals are estimated to be much less than 500 g, by the end of Paleocene the body mass of many mammals had reached hundreds of kilograms and by the Eocene the mass of some mammals could be measured in tons. In a great adaptive radiation that gave rise to most of the modern mammalian orders, Eocene mammals invaded all niches accessible for vertebrates, with bats capable of powered flight, dermopterans capable of gliding, whales and sirenians adapted to the aquatic environment, and so on.

MAJOR DISCOVERIES OF THE PAST TWO DECADES

The first Mesozoic mammal was discovered in 1764, although its significance was not appreciated until almost a century later (Owen, 1871). The first scientifically documented Mesozoic mammal, now known as *Phascolotherium bucklandi,* was discovered in 1812 in a masonry quarry in the Middle Jurassic Stonesfield slates near Headington, England. The fossil was later deposited at Oxford University, where it was recognized by Baron Georges Cuvier to be mammalian in 1818, announced as such by William Buckland (1824), and formally described by Broderip (1828). However, prior to the appearance of the definitive study by Richard Owen (1842), it was generally believed that no mammals existed before the Tertiary (see historical reviews by Desmond, 1984 and Rowe, 1999).

During the late nineteenth century, Mesozoic mammals were discovered in Europe, especially in England (Owen, 1871), and in North America (Marsh, 1879c, 1881, 1887, 1889a; Osborn, 1887a,b, 1888a,b). Knowledge of these fossils was summarized by George Gaylord Simpson in two of the most seminal paleontological monographs of the twentieth century: *A Catalogue of the Mesozoic Mammalia in the Geological Department of the British Museum* (1928a) and *American Mesozoic Mammalia* (1929a). In these monographs, Simpson collectively recognized 39 genera of Mesozoic mammals known from Europe and North America; of these, 12 "genera" are now regarded as junior synonyms or otherwise invalid. Fifty years later, Jason A. Lillegraven, Zofia Kielan-Jaworowska, and William A. Clemens (1979) edited the book *Mesozoic Mammals: The First Two-Thirds of Mammalian History,* in which they listed 116 valid genera of Mesozoic mammals worldwide, with all but a few from Europe, Asia, and North America.

The 23 years since publication of the Mesozoic mammal book by Lillegraven et al. (1979) have witnessed a truly phenomenal growth in knowledge of Mesozoic mammals. By our count, the number of known Mesozoic genera increased to 283 by the year 2000 (chapter 2). New Mesozoic mammals discovered in the past 20 years are one and one-half times more than the total of those known to science from the previous 200 years combined (figure 1.3).

Our knowledge of the earliest mammals has greatly improved in many ways apart from the total number of described taxa—notably anatomical information and geographic/temporal sampling. Many new Mesozoic mammals have been found in areas where the record was previously blank, or nearly so. Most intriguing of these discoveries are those from the Gondwanan continents. In 1979, the only evidence of Mesozoic mammals from the entire South American continent were the footprints designated *Ameghinichnus patagonicus* from the Callovian or Oxfordian of Santa Cruz Province, Argentina (Casamiquela, 1964), and some fragmentary fossils, which may be of Tertiary rather than Cretaceous age, from Laguna Umayo, Peru (chapter 2). Since that time, highly diverse Cretaceous and some Jurassic mammals have been discovered at several sites in Argentina, owing to the great enterprise of South American paleontologists, especially José F. Bonaparte, Rosendo Pascual, and their respective collaborators. Mesozoic mammals from South America now include purported Docodonta, australosphenidans, eutriconodontans, multituberculates, "symmetrodontans," dryolestoid "eupantotherians," prototribosphenidans, and enigmatic gondwanatherians (see Bonaparte and Rougier, 1987; Bonaparte, 1990, 1994; Kielan-Jaworowska and Bonaparte, 1996; Pascual et al., 2000; Pascual and Goin, 2001; Rauhut et al., 2002; and chapters 5–11 for reviews). Pascual and co-workers also discovered a toothed monotreme from the Paleocene of Argentina (Pascual et al., 1992a, 2002).

Other important discoveries have come from Australia, previously a totally blank region insofar as Mesozoic mammals were concerned. In 1985, Michael Archer and colleagues reported the discovery of the first toothed monotreme from the Cretaceous, *Steropodon;* two more Early Cretaceous monotremes (*Kollikodon* and *Teinolophos*) have since been described (Flannery et al., 1995; Rich, Vickers-Rich et al., 2001; see also chapter 6). Since 1997 the able team of Thomas Rich, Patricia Vickers-Rich, and their colleagues has been recovering an assemblage of tribosphenic mammals from the Lower Cretaceous of Australia. These authors have argued that these Australian tribosphenic mammals are placentals (Rich et al., 1997; see also Rich and Vickers-Rich, 1999 and references therein). Mammals with tribosphenic molars have also been reported from the Middle Jurassic of Madagascar by Flynn et al. (1999) and from the Middle or Late Jurassic of Argentina by Rauhut et al. (2002). Our team (Luo, Kielan-Jaworowska, and Cifelli, 2001; Luo et al., 2002; see also

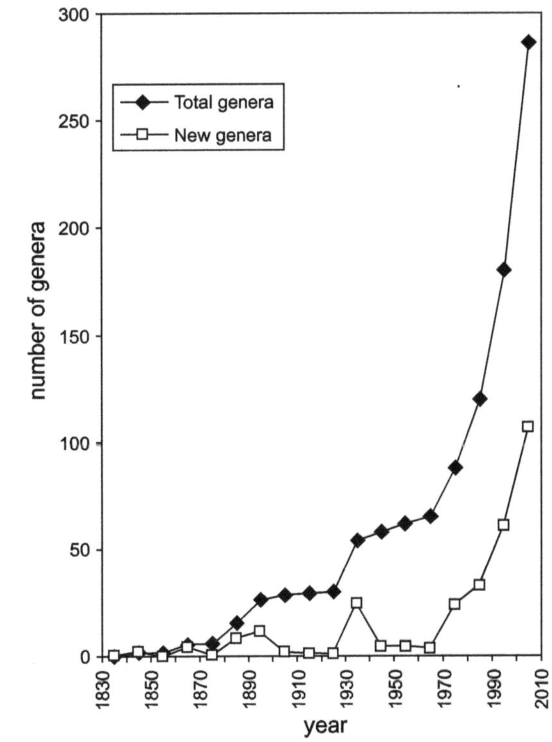

FIGURE 1.3. Number of Mesozoic mammal genera known through time. This graph shows the total number of described genera (line joined by diamonds) and the number described during each 10-year interval (line joined by squares) from 1830 to 2003. Three distinct episodes of intense scientific activity in this field can be recognized. The first, in the 1870s and 1880s, corresponds to Owen's major study (1871) of British mammals from the Purbeck and Stonesfield slates, together with publication, largely by Marsh, of the first known Jurassic and Cretaceous mammals from North America (e.g., Marsh, 1879b, 1887, 1889a). The second took place in the 1920s, when Simpson (1928a, 1929a) reviewed all Mesozoic mammal fossils then known and added a number of new genera, mainly based on previously unstudied or understudied fossils from Como Quarry 9. The third, which continues unabated, began in the 1960s and early 1970s (see References). *Source:* original.

Kielan-Jaworowska et al., 1998, 2002) challenged the placental affinities of southern Mesozoic mammals with tribosphenic dentitions. We argued that Gondwanan tribosphenidans are more closely related to toothed monotremes than to placentals and marsupials and that the functionally adaptive features on tribosphenic molars (a protocone on upper molars opposing a basined talonid on lowers) are homoplastic. According to this hypothesis, tribosphenic mammals have dual origins on the Laurasian and Gondwanan continents. The southern tribosphenic mammals belong to a clade in their own right, designated australosphenidans, as opposed to the northern Mesozoic mammals with tribosphenic molars, designated boreosphenidans, and we follow this interpretation (see chapters 6, 11, and 15).

Mesozoic mammals previously known from Africa include morganucodontans from the Liassic Stormberg Series of South Africa and a toothless dentary of *Brancatherulum* from the Upper Jurassic of Tendaguru in Tanzania. More recently Denise Sigogneau-Russell discovered and described an important collection of isolated teeth representing diverse Early Cretaceous mammals from Morocco (Sigogneau-Russell, 1989a,b, 1991c, 1992, 1995a,b, 1999; Sigogneau-Russell and Ensom, 1998 and references therein). Almost simultaneously, Louis Jacobs, Michel Brunet, and collaborators recovered mammalian fossils from the Early Cretaceous of Cameroon (Jacobs et al., 1989). In the 1990s and 2000s, David Krause and colleagues described several isolated teeth from the Late Cretaceous of Madagascar (Krause et al., 1997; Krause, 2001). Wolf-Dieter Heinrich (1998, 1999) recovered three specimens of mammals from the matrix associated with dinosaur bones (Upper Jurassic Middle Saurian Bed) obtained for the Museum für Naturkunde in Berlin by the 1909–1913 German Expeditions to Tendaguru, Tanzania. The newly found Tendaguru mammals belong to eutriconodontans (*Tendagurodon*), peramurans (*Tendagurutherium*), and haramiyidans (*Staffia*). Finally, Christopher E. Gow (1986a) published on new materials of morganucodontans from South Africa.

In the 1980s and 1990s, Mesozoic mammals were also found in the Late Triassic, Early Jurassic, and Late Cretaceous deposits of India, described by P. Yadagiri, D. P. Datta, G. V. R. Prasad, and their co-workers, adding significantly to the Mesozoic mammalian faunal lists of India (Yadagiri, 1985; Prasad and Sahni, 1988; Prasad et al., 1994; Datta and Das, 1996, 2001; Prasad and Manhas, 1997, 2002).

New discoveries over the past quarter century on the Laurasian continents have been numerous and, in some cases, spectacular. The vast territory of China has provided very well-preserved fossils. The classic vertebrate sites in the Early Jurassic Lower Lufeng Formation, Yunnan Province, have long been important for our understanding of basal mammals (Young, 1947; Patterson and Olson, 1961). The late British paleontologist Kenneth A. Kermack (1919–2000) and his co-workers published an exhaustive description of the Chinese Early Jurassic mammal *Morganucodon* (Kermack et al., 1973, 1981). New field explorations in the 1980s and 1990s yielded important new specimens (Sun et al., 1985; Luo and Wu, 1994). Recent studies make it clear that *Sinoconodon* retains many "reptilian" characteristics in the dentition (Crompton and Sun, 1985; Crompton and Luo, 1993; Zhang et al., 1998). Even more astonishing was the discovery in the Early Jurassic Lufeng Formation of a tiny skull dubbed *Hadrocodium*, described by Luo, Crompton, and Sun (2001). *Hadrocodium* has an enlarged brain cavity and no postdentary trough in the dentary, indicating separation of the middle ear bones from the mandible.

In 1982, the late Chinese paleontologist Minchen Chow (1918–1996), in cooperation with Thomas Rich from Australia, described an entirely new kind of mammal from the Middle Jurassic of China: the enigmatic *Shuotherium*, which has "pseudotribosphenic" lower molars. Their largely hypothetic model of occlusion received welcome corroboration by subsequent discoveries of upper teeth belonging to *Shuotherium* (Sigogneau-Russell, 1998; Wang, Clemens et al., 1998). We now interpret shuotheriids as the sister taxon of Australosphenida (Kielan-Jaworowska, Cifelli, and Luo, 2002; see chapter 6).

The Early Cretaceous Yixian Formation of Liaoning Province, China, has yielded a trove of new mammals. Most remarkable about the Yixian taxa is that most of the specimens are preserved as exceptionally complete skeletons, filling in many blank areas of the postcranial anatomy for some major clades of early mammals. Noteworthy discoveries include an almost complete skeleton of the "symmetrodont" *Zhangheotherium*, described by Chinese paleontologist Yao-Ming Hu and colleagues (Hu et al., 1997); a complete eutriconodontan skeleton designated *Jeholodens* and a nearly complete skeleton of an early eutherian known as *Eomaia*, described by Qiang Ji and colleagues (1999, 2002), and a new multituberculate named *Sinobaatar* (Hu and Wang, 2002a,b). The newly discovered gobiconodontid *Repenomamus* is represented by many skulls and skeletons (Li et al., 2000; Wang, Hu, Meng et al., 2001). In the 1980s and 1990s, the international teams of Zhi-Ming Dong, Philip Currie, and Pascual Godefroit collected well-preserved mammals from the Late Cretaceous Bayan Mandahu sites in Inner Mongolia (Dong, 1993; Smith et al., 2001).

The Gobi Desert of Mongolia is renowned for its unusually well-preserved Cretaceous mammals. Collections of Mongolian Mesozoic mammals assembled by the Polish-Mongolian Expeditions between 1963 and 1971 were partly reported by Lillegraven et al. (1979). Zofia Kielan-Jaworowska subsequently studied these materials, sometimes in cooperation with Boris A. Trofimov, Demberlyin Dashzeveg, Percy M. Butler, Alfred W. Crompton, Cecile Poplin, Robert Presley, P. P. Gambaryan, and Jørn H. Hurum.

Russian-Mongolian Expeditions (1969–1996) also assembled an important collection of Cretaceous mammals, described mostly by B. A. Trofimov, often in cooperation with Z. Kielan-Jaworowska. A major contribution from the Russian-Mongolian Expeditions was a rich collection of Early Cretaceous (Aptian–Albian) mammals (see, e.g., Kielan-Jaworowska, Dashzeveg, and Trofimov, 1987). These expeditions also discovered the first marsupials from the

Late Cretaceous of Asia, described by B. A. Trofimov in cooperation with American paleontologist Frederick S. Szalay (Trofimov and Szalay, 1994; Szalay and Trofimov, 1996). The Mongolian paleontologist Demberlyin Dashzeveg, working by himself, assembled a notable collection of Early and Late Cretaceous mammals (e.g., Dashzeveg and Kielan-Jaworowska, 1984; Kielan-Jaworowska and Dashzeveg, 1989; Sigogneau-Russell et al., 1992).

In 1990, the American Museum of Natural History and the Institute of Geology of the Mongolian Academy of Sciences began to carry out expeditions in Mongolia, amassing spectacular collections (several hundred skulls and several dozen skeletons). Some of these have been prepared and studied by Demberlyin Dashzeveg, Inez Horovitz, Malcolm C. McKenna, Michael J. Novacek, Guillermo W. Rougier, John R. Wible, and others (see review by Kielan-Jaworowska et al., 2000). The preliminary studies published thus far include the discovery of the marsupial (epipubic) bone in Late Cretaceous eutherian mammals (Novacek et al., 1997), a new eutherian (*Ukhaatherium*), and new materials of *Deltatheridium* that support the hypothesis of deltatheroidan-marsupial affinities (Rougier et al., 1998). Several papers on the cranial anatomy of multituberculates and the ankle of eutherian mammals have also been published (Horovitz, 2000; Wible and Rougier, 2000).

Exercising remarkable enterprise and drive, the late Russian paleontologist Lev A. Nessov (1947–1995) organized a series of expeditions to Uzbekistan, Kazakhstan, Tadzhikistan, Kirgizstan, and adjacent areas. Overcoming very harsh conditions, these expeditions amassed a significant collection of fossils, including a number of mammals from a hitherto almost unknown time interval, the early Late Cretaceous (see posthumously published review by Nessov, 1997; and Averianov, 2000). After Lev Nessov's death, Alexander Averianov continued exploration in these areas, often in collaboration with J. David Archibald and other colleagues from both Russia and elsewhere. The mammalian faunas are dominated by ungulate-like "zhelestids," and hence are quite different from the multituberculate-dominated assemblages of the Gobi Desert. Russian paleontologists have also discovered promising new Cretaceous mammal localities in Siberia (Maschenko and Lopatin, 1998; Averianov and Skutschas, 1999b, 2001; Averianov, 2000; Gambaryan and Averianov, 2001).

The last quarter-century saw both new discoveries and publication of important, previously collected early mammals from Western Europe. Significant among the latter is a huge, Late Jurassic assemblage from the famous Guimarota coal mines in Portugal (see Martin and Krebs, 2000), discovered in 1961 by the legendary German fossil hunter Walter G. Kühne (1911–1991). Gerhard Hahn (often accompanied by his wife Renate Hahn) completed a long series of monographs and descriptive papers on the multituberculates from Guimarota, as well as other Jurassic and Early Cretaceous localities on the Iberian Peninsula. The late Bernard Krebs (1934–2001) worked on dryolestoids from Guimarota and elsewhere, a project that is now being continued by Thomas Martin and his students. The late Georg Krusat (1938–1998), accompanied by Jason A. Lillegraven, described the docodontans from Guimarota.

William A. Clemens (1980a) published a review of existing collections of Late Triassic mammals from the Keuper of Switzerland and Germany. Most significant among newly collected fossils from the Keuper are those from Saint-Nicolas-de-Port, France, published in a series of contributions by Denise Sigogneau-Russell (e.g., 1983b, 1989b). Late Cretaceous mammals (multituberculates) were found in Romania (e.g., Rădulescu and Samson, 1996; Csiki and Grigorescu, 2000), France, and Spain (Gheerbrant et al., 1997; Gheerbrant and Astibia, 1999).

In 1986, Paul Ensom began assembling an impressively large, diverse collection of mammals from the Purbeck Limestone Group (now regarded as of Early Cretaceous age) near Dorset, England. Ensom studied the multituberculates with Z. Kielan-Jaworowska and worked on other mammals with D. Sigogneau-Russell. Among British paleontologists, Kenneth A. Kermack and coauthors continued their work on various Mesozoic mammals; the last paper by this team was published in 1998 (Kermack et al., 1998). An experienced researcher of early mammals, Percy M. Butler, continued to publish on the function of teeth in Mesozoic mammals and their relationships (e.g., Butler, 2000).

Among the most important and exotic discoveries of early mammals in recent years was that made by Farish A. Jenkins, Jr., and colleagues in the Late Triassic of northwestern Greenland. Among others (Jenkins et al., 1994), they recovered associated dentary, cranial, and postcranial remains of a haramiyidan, a group represented until then only by isolated teeth (Jenkins et al., 1997).

A virtual flood of new fossil mammals has poured in from the Mesozoic of North America, and space constraints permit us to mention only a small sample here (see chapter 2 for a fuller account). Some of these discoveries have helped to fill in knowledge for periods of geologic time that were hitherto not represented by the fossil record of this continent. Geologically oldest of these are *Dinnetherium* (a morganucodontan) and several other taxa recovered from the Early Jurassic of Arizona by Jenkins et al. (1983). Somewhat younger mammals, mainly

?eutriconodontans, were discovered in the ?Middle Jurassic of Mexico by James M. Clark (see summary by Montellano et al., 1998). Significant inroads have been made into the record for the Early Cretaceous and early Late Cretaceous Aptian–Albian mammals, including a reasonably well-represented eutherian (Cifelli, 1999b), have been recovered from the Cloverly Formation by Richard L. Cifelli, W. Desmond Maxwell, and colleagues. Additional Early Cretaceous mammals have been collected in the Trinity Group, Texas and Oklahoma (Jacobs et al., 1989; Winkler et al., 1990; Cifelli, 1997a); and Thomas R. Lipka made the seemingly improbable discovery of relatively well-preserved Aptian mammals along the East Coast, in Maryland (Cifelli, Lipka, et al. 1999; Rose et al., 2001). Mammalian faunas, in some cases diverse and well-represented, have been recovered in southern and central Utah by Cifelli and Jeffrey G. Eaton in rocks spanning much of the first half of the Late Cretaceous (see Cifelli, Kirkland, et al., 1997; Eaton, Cifelli, et al., 1999). Finally, work in northwestern Colorado by J. David Archibald and his students has shed new light on the poorly understood "Edmontonian" (Diem, 1999).

New discoveries have also significantly expanded geographic sampling of early mammals from North America. Several sites have yielded Late Cretaceous mammals from the Eastern Seaboard; of these, the Ellisdale locality, New Jersey, under investigation by Barbara Grandstaff and her colleagues, has shown most promise to date (e.g., Grandstaff et al., 1992). In the South and Southwest, Timothy B. Rowe, Anne Weil, Cifelli, and collaborators have recovered a diverse assemblage of Judithian age from the Aguja Formation, Texas (see Rowe et al., 1992); and Cifelli, Jeffrey G. Eaton and others have amassed significant Judithian- and Aquilan-aged faunas from southwestern Utah (Eaton, Cifelli et al., 1999). Without doubt, the greatest geographic range extension for North American Cretaceous mammals was made by William A. Clemens and his students, who collected a small faunule from the North Slope of Alaska (Clemens, 1995).

Finally, the North American record of Mesozoic mammals has benefited from sampling (involving large numbers of specimens in some cases) new local faunas of similar age to those already known. Such additions to knowledge are important because they can be placed within the context of an enormous body of existing data. George Engelmann, Scott Madsen, and co-workers have collected significant new mammalian fossils from the Morrison Formation of Dinosaur National Monument, Utah (Engelmann and Callison, 1998). Important new local faunas from the Cretaceous include: Judith River and Hell Creek formations, Montana, collected by William A. Clemens and students (e.g., Archibald, 1982; Montellano, 1992); "Mesaverde," Lance, and Ferris formations, Wyoming, by Jason A. Lillegraven and students (e.g., Lillegraven and McKenna, 1986; Eberle and Lillegraven, 1998a,b; Webb, 2001); and Frenchman and/or Ravenscrag formations, Saskatchewan, by Richard C. Fox, John E. Storer, and associates (e.g., Johnston and Fox, 1984; Fox, 1989; Storer, 1991).

NEW MORPHOLOGICAL STUDIES

The past two decades also witnessed the development of important new venues for interpreting mammalian evolution. Among these are significant advances in understanding of mammalian ontogenesis and morphogenesis, owing to a resurgence of interest in the 1980s and 1990s. The integration of descriptive embryology with interpretation of soft tissue anatomy of the skull and skeleton has produced many successful case studies that are instructive for understanding mechanisms underlying character transformation (MacPhee, 1981; Kuhn and Zeller, 1987a; Zeller, 1989a,b, 1993; Smith, 1996, 1997; Sánchez-Villagra et al., 2002; Sánchez-Villagra and Maier, 2002). Detailed studies on the ossification of the lateral wall from the embryonic precursor of the sphenoobturator membrane provided essential background for interpreting the homologies of the braincase wall in Mesozoic mammals (Presley, 1981; Kuhn and Zeller, 1987b; Maier, 1987; Hopson and Rougier, 1993). Observation on the relationships of soft tissues (such as vasculature and innervation) to skeletal features in extant mammals served as the basis for vascular reconstruction in stem taxa (Wible, 1984, 1987; Kielan-Jaworowska et al., 1986; Rougier et al., 1992; Wible and Hopson, 1993, 1995; Wible and Rougier, 2000). Studies on the development of Meckel's cartilage and the middle ear bones have brought new perspective on the timing and possible epigenetic mechanism for development of the middle ear (Maier, 1990, 1993; Zeller, 1993; Rowe, 1996a,b; Sánchez-Villagra et al., 2002). Finally, studies of the pectoral girdles of monotremes and marsupials have contributed to the much-needed understanding of the ontogeny of some important mammalian features, such as the supraspinous fossa of the scapula (Klima, 1973, 1987; McKenna, 1996; Sánchez-Villagra and Maier, 2002). Kielan-Jaworowska and Peter P. Gambaryan provided muscular reconstruction for Cretaceous multituberculates and their masticatory and locomotory functions (Kielan-Jaworowska and Gambaryan, 1994; Gambaryan and Kielan-Jaworowska, 1995, 1997).

Techniques of serial sectioning and computer tomography have made it possible to study the internal structures of the skull of early mammals, which would otherwise be difficult or impractical to access (e.g., Kielan-Jaworowska et al.,

1986; Graybeal et al., 1989; Luo and Ketten, 1991; Hurum, 1994, 1998a; Rowe et al., 1994, 1997; Luo et al., 1995; Cifelli et al., 1996; Rowe, 1996a,b; Luo, 2001). Studies of the enamel microstructure of Mesozoic mammals started in the late 1980s (e.g., Grine and Vrba, 1980; Carlson and Krause, 1985; Fosse et al., 1985) and have been continued by numerous researchers (see Koenigswald and Sander, 1997a; Wood et al., 1999, for reviews).

IMPROVEMENT IN THE QUALITY IN MORPHOLOGICAL DATA

Before the 1970s, most studies of Mesozoic mammals concentrated on dental morphology, partly because teeth offer a surprisingly rich source of morphological information, but more importantly because little else was available for study. For nearly a century, it was generally held that the major features of early mammalian evolution were those documented by changes in molar pattern (Osborn, 1907; Patterson, 1956; Kermack, 1967b)—a perspective that was inevitable, given that the fossil record had yielded mainly teeth and tooth-bearing jaws. Recovery of Mesozoic mammals by screen washing, which blossomed in the 1950s and 1960s, greatly expanded our knowledge of taxonomic diversity and geographic distribution, but morphological data other than teeth and jaws remained stagnant. The late Alfred Sherwood Romer (1968) observed this problem with a sense of humor: "So great has been this concentration on dentitions that I often accuse my 'mammalian' colleagues, not without some degree of justice, of conceiving of mammals as consisting solely of molar teeth and of considering that mammalian evolution consisted of parent molar teeth giving birth to filial molar teeth and so on down through the ages." Before the 1980s, there were few examples in which cranial data were used to infer phylogeny of Mesozoic mammals, except for the well-known case of the lateral wall of the braincase, which was used to support the grouping of "prototherians" (Kermack, 1963; Hopson, 1964).

Remedy for this conspicuous paucity of cranial characteristics for phylogeny began in the 1980s, when skull anatomy became known for several key taxa of early mammals. By the early 1990s, enough basicranial features were known for major clades of Mesozoic mammals that it became possible for some workers to attempt phylogenetic analyses on the basis of this character complex alone (e.g., Wible and Hopson, 1993; Wible et al., 1995; Rougier, Wible, and Hopson, 1996). Embryological studies of cranial features have also provided important insight into how to interpret the orbitotemporal and rostral structures of stem mammals of the Mesozoic (e.g., Presley, 1981; Kuhn and Zeller, 1987b; Maier, 1987, 1989, 1999; Wible, 1987; Zeller, 1989a, 1993; Wible et al., 1990; Hopson and Rougier, 1993; Hillenius, 2000).

Although the skeletal anatomy of cynodonts and some stem mammals was well described in the 1970s (e.g., Jenkins, 1971b; Jenkins and Parrington, 1976), incorporation of skeletal characteristics into phylogenetic inference only began with two later studies by Kemp (1983) and Rowe (1988). These were followed by several rounds of reevaluation and further expansion of data based on skeletal characteristics, as more complete skeletons belonging to additional clades of Mesozoic mammals and relevant cynodonts were published (Sues, 1985b; Jenkins and Schaff, 1988; Kielan-Jaworowska and Gambaryan, 1994; Sereno and McKenna, 1995; Hu et al., 1997; Ji et al., 1999, 2002; Luo et al., 2002).

By the late 1990s, comparative anatomical datasets for parsimony-based phylogenetic analysis of Mesozoic mammals had been expanded to include many informative features of the dentition, mandible, basicranium and other cranial structures, and postcranial skeleton. Inevitably, various discrepancies and inconsistencies have arisen among existing data matrices. We expect that these issues will be resolved, as explicitly coded characters can be reexamined and clarified, reinterpreted, or corrected as needed, and that the currently used character matrices (e.g., Rowe, 1988; Wible and Hopson, 1993; Rougier, Wible, and Hopson, 1996; Rougier et al., 1998; Hu et al., 1997; Ji et al. 1999, 2002; Luo, Kielan-Jaworowska, and Cifelli, 2001; Luo, Crompton, and Sun, 2001; Luo et al., 2002) will continue to be revised, improved, and replenished. Given the availability of such a wealth of data, it is no longer excusable to propose hypotheses of relationships among Mesozoic mammals on the basis of some single-character complexes: such issues must be framed and tested in the context of all available morphological evidence from all relevant taxa.

INTEGRATION OF MOLECULAR AND MORPHOLOGICAL DATA FOR EXTANT CLADES

Knowledge of Mesozoic mammals has increased enormously since the publication of Lillegraven, Kielan-Jaworowska, and Clemens's book (1979), resulting in a wider range of taxonomic diversity, new records in previously blank geographic areas, and better documentation of anatomical characteristics. The morphological data for Mesozoic taxa are far more extensive than they were two decades ago, and recent studies have resulted in explicit, comprehensive hypotheses of relationships among most major groups of Mesozoic mammals. But new problems have arisen, especially in the area of molecular versus morphological estimates for the timing and sequence of origination for extant mammalian clades having their roots in the Mesozoic.

Rapid recent advances in molecular evolutionary studies have made it possible to estimate the sequence and timing of the splits of major extant mammalian lineages, based on the assumption of a molecular clock. This has presented a welcome opportunity for molecular evolutionists to challenge the widely accepted outline for Mesozoic mammal evolution as established by paleontologists and morphologists (e.g., Kumar and Hedges, 1998; Murphy, Eizirik, Johnson et al., 2001; Murphy, Eizirik, O'Brien, et al., 2001). Conversely, improvements in the fossil record have allowed paleontologists to test molecular hypotheses on early mammalian evolution by reference to the fossil record of the Mesozoic (see, e.g., Benton, 1999; Foote et al., 1999; Novacek, 1999; and compare with Kumar and Hedges, 1998; Esteal, 1999; Gee, 1999). Morphologists have countered that assumption of constancy of rate in molecular evolution, on which the earlier time estimates for divergence were established, are not necessarily reliable. It is also possible that calculated rates are biased, owing to calibration on the basis of spurious evidence from the fossil record (e.g., Benton, 1999; Foote et al., 1999).

There are great differences between the molecular trees of modern mammal orders and those based on morphological data. The latest molecular studies of placental mammals achieved a strong consensus on the phylogenies of superordinal clades of placentals. Of the four superordinal clades, Afrotheria (proboscideans, hyracoids, macroscelideans, and some others) represent the earliest-diverging clade, followed by Xenarthra, Laurasiatheria (artiodactyls, perissodactyls, carnivorans, lipotyphlans, and others), and Euarchonta (primates, chiropterans, dermopterans). The grouping of afrotheres and their placement as a basal clade among crown Placentalia by molecular evidence is contradicted by morphological phylogenies for these mammals (e.g., Novacek, 1992a; Fischer and Tassy, 1993; Prothero, 1993; Shoshani and McKenna, 1998; Allard et al. 1999; Asher, 1999). The case study of the ungulate-whale relationship represents another widely discussed example of incongruences in the morphological and molecular evidence. The paraphyly of artiodactyls suggested by molecular evidence has been criticized by morphologists (O'Leary and Geisler, 1999; Gatesy and O'Leary, 2001), although most recently, the finding from Pakistan in 2001 by two teams of paleontologists provided further evidence for the nesting of whales among artiodactyls (Gingerich et al., 2001; Thewissen et al., 2001), as suggested by molecular evolutionists.

Moreover, recent molecular studies have postulated a much earlier diversification of the superordinal placental clades than predicted from fossil records. The current molecular dating suggests that ordinal diversification of living placental groups ranged from 104 to 60 Ma (e.g., Waddell et al., 1999; Eizirik et al., 2001; Madsen et al., 2001; Murphy, Eizirik, Johnson, et al., 2001; Murphy, Eizirik, O'Brien, et al., 2001; but see Kumar and Hedges, 1998), whereas recognizable members of most living orders do not appear until the late Paleocene or early Eocene. A thorough assessment as to whether any of the earliest fossils of eutherians can be placed in one of the superordinal placental groups on the basis of paleontological data is the key to resolving the conflict between molecules and fossils for the time estimate for the earliest placental diversification. There is no doubt that paleontologists will have to continue their difficult task of searching for ever earlier fossils of modern lineages. At the same time the molecular evolutionists might rethink the assumptions and their datasets for their time estimates, especially about the constancy and calibration of the molecular clock.

Relatively speaking, there is less discordance between molecular-based time estimates of marsupial diversification and the appearance of relevant clades of crown Marsupialia in the fossil record. Springer (1997) estimated a divergence of "Ameridelphia" from various groups of Australidelphia between 78 and 70 Ma ago and, within "Ameridelphia," a split of Paucituberculata from "Didelphimorphia" between 65 and 61 Ma ago. The earliest known fossil taxa (Stagodontidae) that can be placed within the "Ameridelphia" crown group are about 80 Ma old, and the earliest putative member of Paucituberculata (*Glasbius*) is about 70 Ma old (chapter 12). According to Christian de Muizon, the earliest known South American didelphimorphs are from the early Paleocene, about 63 Ma old (Marshall and Muizon, 1995; Muizon et al., 1997; Muizon, 1998; Muizon and Cifelli, 2001), although Rougier et al. (1998) placed these Late Cretaceous and Paleocene taxa outside the Marsupialia crown group.

There is better concordance between molecules and fossils in the placement of monotremes within mammals and in the timing of the divergence of monotremes from crown therians. Molecular phylogenies of the relationship of monotremes to other mammals differ depending on the types of molecules and genes examined. Although the debate is ongoing, the balance of evidence clearly favors the traditional hypothesis of a monophyletic therian group to the exclusion of monotremes.

Studies of DNA-DNA hybridization and mitochondrial genomes suggest that monotremes may be the sister taxon to marsupials, to the exclusion of placental mammals—a relationship advocated by Janke, Kirsch, and their colleagues. This led some molecular evolutionists to argue for the resurrection of Gregory's (1947) classic "Marsupionta" hypothesis (Janke et al., 1996, 1997, 2002; Penny and Hasegawa, 1997; Kirsch et al., 1997). From a morphological point of view, this hypothesis was too poorly supported to be taken seriously (see discussions by Zeller, 1999a; Szalay and Sargis, 2001; and Luo et al., 2002; among others).

All studies on nuclear genes and most studies on protein sequences have clearly rejected the "Marsupionta" hypothesis, instead uniting placentals and marsupials to the exclusion of Monotremata. The therian hypothesis is supported by an earlier study on protamine DNA, corroborated by studies of neurotrophin genes, β-globin gene sequences, and the latest studies on imprint genes and retroposons (reviewed by Killian et al. 2000, 2001). A recent comparative study on the nuclear versus mitochondrial genomes by Springer et al. (2001) shows that the nuclear exon genes are more informative than the protein-coding mitochondrial genes for resolving deep mammal phylogeny, weakening the relative strength of the mitochondrial evidence for recovering higher-level phylogenetic history in mammals, at least for the case of deep phyletic splits of monotremes, marsupials, and placentals.

Although an earlier study on the combined protein sequences yielded an unresolved trichotomy for the three groups of living mammals (Czelusniak et al., 1990), the more recent evidence from the protamine protein sequence, the α-lactalbumin protein sequence, and the IgM proteins excluded monotremes from a monophyletic Theria. In summary, although some types of molecular data have provided limited support for the "Marsupionta" hypothesis, the prevailing evidence from nuclear genes, retroposons, and protein sequences favors the traditional hypothesis of a monophyletic crown Theria (Mammalia of Linnaeus).

The study of molecular evolutionary rates by α-lactalbumin protein and IgM protein sequences suggests that modern monotremes split from crown therians about 170 to 168 Ma ago (Messer et al., 1998; Belov et al., 2002). This molecular time estimate is in concordance with the latest morphological hypothesis incorporating relevant fossils. Recent analyses place the Middle Jurassic *Ambondro* and the Middle-Late Jurassic *Asfaltomylos* as stem taxa related to monotremes, with shuotheriids (also of Middle Jurassic age) as a sister group to australosphenidans (Luo, Kielan-Jaworowska, and Cifelli, 2001; Luo et al., 2002; Rauhut et al., 2002; Kielan-Jaworowska, Cifelli, and Luo, 2002). As such, the split sequence and time estimate by molecules are in good agreement with the morphological hypothesis based on fossils.

The mammalian evolutionary tree is fundamental to comparative mammalian genomics, as differences and similarities in the gene and protein sequences of diverse modern mammalian species are accumulated through their evolutionary history. The cross illumination of molecular evolutionary biology and mammalian paleontology can only enhance our understanding of the deep evolutionary history of mammals.

PARADIGM SHIFTS IN MESOZOIC MAMMAL EVOLUTIONARY STUDIES

A significant conceptual advance from recent studies of early mammals is the understanding that the greatest taxonomic diversification and morphological divergence tends to appear in the earlier periods of each major clade (Rowe, 1993; Cifelli, 2001; Luo et al., 2002). The most significant diversification of Mesozoic mammals occurred long before the splits of modern lineages of monotremes, marsupials, and placentals. It is near the root of the mammal tree that many basal branches split off, so as to generate a very bushy branching pattern of the stem clades in the Jurassic (Rowe, 1999). Based on the almost fully resolved tree of all phylogenetically relevant taxa, Mesozoic mammals have some 26 or more clades, which appeared in the fossil record by several successive bursts of diversification of stem lineages. Most of these stem lineages are dead-end evolutionary experiments that did not survive the Cretaceous-Tertiary extinction. Of all 26 or more clades (figure 1.1), only four major lineages—the monotremes, cimolodontan multituberculates, marsupials, and placentals—have a significant presence after the Cretaceous-Tertiary extinction.[2]

Most of the orders and families of the Late Triassic and Early Jurassic, such as sinoconodontids, morganucodontans, and kuehneotheriids, represent a hierarchical series of stems at the very base of the mammalian tree. They are distantly related to modern mammals and did not directly give rise to the major orders and families of the mammalian crown group of the Middle and Late Jurassic. (It remains unclear whether haramiyidans, which are among the earliest mammals, are related to multituberculates, which appeared and diversified considerably later; see Rowe, 1993; Butler, 2000; and chapters 8 and 15.) Similarly, the major mammalian groups of the Late Jurassic—dryolestoids, docodontans, Multituberculata (*sensu stricto*), and triconodontids—represent evolutionary dead ends in

[2] Monotremes, cimolodontans, marsupials, and placentals are the four main groups that survived the Cretaceous-Tertiary extinction. It has been reported that one dryolestoid, otherwise known only from Jurassic-Cretaceous, is found in the Paleocene of South America (Gelfo and Pascual, 2001). There is also another report of a therapsid taxon from the Paleocene of North America (Fox et al., 1992), but this has been debated (Sues 1992).

TABLE 1.1. A Cladistic Classification of Mesozoic Mammals at the Suborder to Family Levels

- Class Mammalia Clade of (*Sinoconodon* + Crown Mammalia)
  - Family Sinoconodontidae
  - Family *incertae sedis Adelobasileus* (*sedis mutabilis*)
    - Clade (*Morganucodon* + Crown Mammalia)
      - Order Morganucodonta
      - Order *incertae sedis,* family Kuehneotheriidae (*sedis mutabilis*)
      - Clade (Docodonta + Crown Mammalia)
        - Order Docodonta
          - Order and family *incertae sedis Reigitherium* (*sedis mutabilis*), *Woutersia* (*sedis mutabilis*)
          - Clade (*Hadrocodium* + Crown Mammalia)
            - Order and family *incertae sedis, Hadrocodium*
          - Clade of mammalian crown group
            - Clade Yinotheria (*Shuotherium* + Crown Monotremata)
              - Order Shuotheridia
              - Subclass Clade Australosphenida
                - Order and family *incertae sedis, Asfaltomylos, Ambondro* (*sedis mutabilis*)
                  - Order Ausktribosphenida
                    - Stem ausktribosphenidans *Ausktribosphenos, Bishops* (*sedis mutabilis*)
                      - Order Monotremata
                        - Family Kollikodontidae (*sedis mutabilis*)
                      - Clade (*Steropodon* + Crown Monotremata)
                        - Family Ornithorhynchidae
                        - Family Steropodontidae
                        - Family Tachyglossidae
            - Clade (Eutriconodonta + Allotheria + Trechnotheria)
              - Order *incertae sedis,* family Tinodontidae (*sedis mutabilis*)
              - Order Eutriconodonta
              - Clade (Allotheria + Trechnotheria)[1]
                - Subclass Allotheria
                  - Order Haramiyida (*sensu* Butler, 2000)
                  - Order Multituberculata
                    - Suborder "Plagiaulacidans"
                    - Suborder Cimolodonta
                - Clade Trechnotheria (Spalacotheriidae + Crown Theria)
                        - Family Spalacotheriidae
                      - Clade Cladotheria (Dryolestoidea + Crown Theria)
                    - Superorder Dryolestoidea
                    - Order Dryolestida
                      - Family Dryolestidae
                      - Family Paurodontidae
                    - Order *incertae sedis*
                      - Family Vincelestidae
                    - Order Amphitheriida
                      - Family Amphitheriidae
                    - Clade Zatheria (Peramuridae + Crown Mammalia)
                          - Order and Family *incertae sedis, Arguimus Afriquiamus, Magnimus, Minimus, Nanolestes, Palaeoxonodon* (*sedis mutabilis*)
                            - Family Peramuridae
                            - Subclass Boreosphenida
                              - Order Aegialodontia
                                - Family Aegialodontidae

TABLE 1.1. Continued

- Infraclass Metatheria
  - Cohort Deltatheroida
  - Cohort Marsupialia
    - Superorder Asiadelphia
    - Superorder "Ameridelphia"
      - Order "Didelphimorphia"
      - Order Paucituberculata
    - Superorder Australidelphia
- Infraclass Eutheria
  - Order and family *incertae sedis, Eomaia, Montanalestes, Murtoilestes, Prokennalestes (sedis mutabilis)*
    - Clade (Asioryctitheria + Placentalia)
      - Superorder Asioryctitheria
      - Clade Placentalia
        - Superorder Anagalida
        - Superorder Archonta
        - Superorder Insectivora
          - Order Leptictida
          - Order Lipotyphla
        - Superorder Ferae
          - Order Cimolesta
        - Superorder Ungulatomorpha
          - Order "Condylarthra"
          - ?Order Notoungulata

Subclass *incertae sedis*
- Order Gondwanatheria

[1] We tentatively accept in chapter 8 the grouping of Allotheria, including the poorly known Haramiyida and Multituberculata. The relationship between the two allotherian groups is, however, uncertain (see chapter 15 and figures 15.1 and 15.2), and the possibility that haramiyidans and multituberculates are not directly related cannot be excluded. As long as fossils of haramiyidans are (with the exception of *Haramiyavia*) limited to isolated teeth, this issue cannot be unequivocally resolved. On the other hand, significant differences in mandibular features between *Haramiyavia* and multituberculates do not support the haramiyidan-multituberculate relationship (see Jenkins et al., 1997). An alternative cladistic classification strictly consistent with the cladogram would be to place the Haramiyida as subclass *incertae sedis* under the class Mammalia.

their own right and are not related to monotremes, nor to crown therians. The greatest morphological divergence in the mammalian dentition occurred in the Middle to Late Jurassic, including the first of two convergent appearances of the tribosphenic molar pattern (among some southern mammals).

A series of studies in the 1980s and 1990s also demonstrated that many northern tribosphenic mammals from the late Early Cretaceous (Aptian–Albian) could not be reliably placed within either the metatherian or eutherian clades, as previously believed (compare the earlier views of Slaughter, 1968a,b, to those of Kielan-Jaworowska, Eaton, and Bown, 1979; Clemens and Lillegraven, 1986; and Cifelli, 1993b). Rather, these stem boreosphenidans appear to represent a burst of diversification of what became blind evolutionary lineages that diverged from crown Theria near the time of the marsupial-placental dichotomy (Cifelli, 1993b).

These scenarios of Mesozoic mammal evolution are not based solely on molar features, as was the case for most phylogenetic hypotheses before the 1980s; rather, they are supported by analyses of comprehensive datasets that include craniomandibular and postcranial characters, as well as those based on molars (Rowe, 1988, 1993; Lillegraven and Krusat, 1991; Wible, 1991; Rougier, Wible, and Hopson, 1996; Luo, Kielan-Jaworowska, and Cifelli, 2001; Luo et al., 2002).

Our current understanding of Mesozoic mammal evolution (figure 1.1) is significantly different from what was envisioned by Simpson, Romer, or Olson in the first half of the twentieth century, wherein the mammalian grade was achieved independently by various evolving lineages of cynodonts (e.g., Simpson, 1959; Olson 1959) (figure 1.2A). The first significant paradigm shift occurred in the 1960s, when the traditional Simpsonian view of mammalian polyphyly was replaced by the concept of monophyletic origin for mammals (e.g., Hopson and Crompton, 1969). With the abandonment of a polyphyletic origin for Mammalia came general recognition of a fundamental early dichotomy of the class into "prototherian" and "therian" groups (figure 1.2B).

Interpretation of early mammalian evolution underwent two important conceptual changes in the 1970s and 1980s. The traditionally recognized groupings of "Symmetrodonta" and "Eupantotheria" (then regarded as basal therians or, later, "holotherians") were challenged by McKenna (1975), whose cladistic interpretation suggested instead that they represent paraphyletic grades. McKenna's phylogenetic hierarchies for "therians" were largely corroborated by further studies. In the 1980s, a framework for cladistic relationships of all synapsids was established. A significant conclusion of studies by Kemp and by Rowe was that the "prototherian" grouping widely accepted in the 1970s is paraphyletic. Some aspects of these hypotheses were controversial and relevant morphological evidence has been reevaluated (Sues, 1985b; Hopson, 1991, 1994; Wible, 1991; Luo, 1994). Nonetheless, the contributions by McKenna (1975), Kemp (1983), and Rowe (1988) helped to frame the further inquiry of early mammalian evolution in cladistic terms.

The most recent conceptual shift in understanding of mammalian evolution during the Mesozoic was prompted by discoveries of new, highly derived fossil mammals from the previously blank geographic areas of Gondwanan landmasses in the late 1990s. One piece of the puzzle, the Australian Early Cretaceous *Steropodon*, was discovered as early as 1985, but its full significance was not generally appreciated until much later. Cladistic and other comparative studies make it clear that the new fossils represent some previously unknown mammalian radiation(s) (e.g., Bonaparte, 1990; Krause et al., 1997; Rich et al., 1997; Flynn et al., 1999; Rich, Vickers-Rich et al. 2001; Sigogneau-Russell et al., 2001; Rauhut et al., 2002). It is equally clear that these recently discovered taxa cannot be accommodated into the prevalent view—based on evidence limited to Laurasian landmasses—that tribosphenic mammals are monophyletic and originated in the Northern Hemisphere. Beyond this, no consensus has yet emerged. In our view, Southern tribosphenic mammals of the Middle Jurassic–Early Cretaceous are not closely related to marsupials and placentals, but rather represent a radiation endemic to Gondwanan landmasses, with a probable relationship to monotremes. Hence, divergence of modern monotremes from modern therians may well be a vicariant event (Luo, Kielan-Jaworowska, and Cifelli, 2001; Sigogneau-Russell et al., 2001; Luo et al., 2002; Rauhut et al., 2002).

SYSTEMATIC ISSUES

We employ either stem-based or node-based definitions, together with diagnostic characters, for higher-rank monophyletic groups comprised solely of extinct genera. We also used stem-based definitions for the three major clades that include living mammals—Monotremata, Metatheria, and Eutheria. For example, eutherians are defined as all mammals that are more closely related to the common ancestor of crown Placentalia, such as humans, than to crown Marsupialia, such as kangaroos. In this book, we restrict usage of formal taxonomic names to those higher-rank groups for which monophyly can be demonstrated on the basis of evidence now available (even in cases where such evidence is weak). If several names are in use for a monophyletic taxon, we have given preference to the name that, in our judgment, is most widely understood and meaningful.

A fully resolved phylogeny with a large number of terminal taxa can greatly proliferate the resultant hierarchical categories or cladistic ranks (figure 1.1). Each well-supported node represents a cladistic hierarchy; ideally, each node would be defined formally and supported by an attendant diagnosis based on synapomorphies. However, this becomes impractical and undesirable in the case of a phylogeny comprised of numerous terminal taxa (some of which are obscure, poorly represented, or both) resolved into many dichotomous nodes. Our goal herein is to summarize current knowledge of Mesozoic mammals within the context of a general systematic review and framework. For the purposes of this synopsis, it would be cumbersome if not outright impractical to formally name and diagnose each node of the mammalian phylogenetic tree. Indeed, given the instability of a number of nodes and the diversity of tree topologies in extant phylogenetic hypotheses, we are skeptical that such an exercise would be a useful endeavor.

In compiling this book, we quickly learned that it is both unrealistic and linguistically inconvenient to entirely avoid all uses of names for some traditional groupings that are now recognized as paraphyletic; in many cases, no simple, explicit *nomen* based on monophyly exists or can be defensibly proposed. Moreover, mention of some paraphyletic groupings is unavoidable because they have been historically associated with certain familiar, yet symplesiomorphic features (such as "triconodonts," named after a molar pattern shared by what are now considered to be separate clades). In such cases, we follow the convention of placing these historical and widely understood (but nonmonophyletic) names in quotes (such as "symmetrodontans"). Finally, for the practical

TABLE 1.2. Simplified Traditional Linnaean Classification of the Class Mammalia Linnaeus, 1758, above the Suborder Level[1]

- Subclass and order *incertae sedis*
  - Family Sinoconodontidae Mills, 1971 (chapter 4)
- Subclass *incertae sedis*
  - Order Morganucodonta Kermack et al., 1973 (chapter 4)
  - Order Docodonta Kretzoi, 1946 (chapter 5)
- Subclass Australosphenida Luo, Kielan-Jaworowska, and Cifelli, 2001, new rank (chapter 6)
  - Order Ausktribosphenida Rich et al., 1997
  - Order Monotremata C.L. Bonaparte, 1837
- Subclass *incertae sedis*
  - Order Shuotheridia Chow and Rich, 1982 (chapter 6)
- Subclass *incertae sedis*
  - Order Eutriconodonta Kermack et al., 1973 (chapter 7)
- Subclass Allotheria Marsh, 1880 (chapter 8)
  - Order Haramiyida Hahn et al., 1989
  - Order Multituberculata Cope, 1884
- Subclass and order *incertae sedis* ("Symmetrodontans") (chapter 9)
  - Family Kuehneotheriidae Kermack et al., 1968 (Comment: In addition to Kuehneotheriidae, which might not be related to any other "symmetrodontans" and which probably are not closely related to at least Tinodontidae and Spalacotheriidae [see chapter 15], there are seven poorly known "symmetrodontan" families, of which only one [Spalacotheriidae[2]] can be supported as monophyletic.)
- Subclass and order *incertae sedis* ("eupantotherians") (chapter 10)
  - Superorder Dryolestoidea Butler, 1939, emended[3]
    - Order Dryolestida Prothero, 1981
    - Order Amphitheriida Prothero, 1981
  - Superorder Zatheria McKenna, 1975
    - Stem-lineage Zatheria Martin, 2002
    - "Peramurans" (former order Peramura McKenna, 1975)
- Subclass Boreosphenida Luo, Kielan-Jaworowska, and Cifelli, 2001, new rank ("Tribotherians," chapter 11)
  - Order Aegialodontia Butler, 1978
  - Order *incertae sedis*
- Infraclass Metatheria Huxley, 1880 (chapter 12)
  - Cohort Deltatheroida Kielan-Jaworowska, 1982
  - Cohort Marsupialia Illiger, 1811
    - Superorder "Ameridelphia" Szalay, 1982, new rank
      - Order Asiadelphia Trofimov and Szalay, 1994
      - Order "Didelphimorphia" Gill, 1872
      - Order Paucituberculata Ameghino, 1894
- Infraclass Eutheria Gill, 1872 (chapter 13)
  - Magnorder Epitheria McKenna, 1975
    - Superorder and order *incertae sedis*
    - Superorder Asioryctitheria Novacek et al., 1997
    - Superorder Anagalida Szalay and McKenna, 1971
    - Superorder Archonta Gregory, 1910, new rank
    - Superorder Insectivora Bowdich, 1821
      - Order Leptictida McKenna, 1975
      - Order Lipotyphla Haeckel, 1866
    - Superorder Ferae Linnaeus, 1758
      - Order Cimolesta McKenna, 1975
    - Superorder Ungulatomorpha Archibald, 1996b
      - Order "Condylarthra" Cope, 1881
      - ?Order Notoungulata Roth, 1903
- Subclass *incertae sedis*
  - Suborder Gondwanatheria Mones, 1987 (chapter 14)

[1] The suborders and orders *incertae sedis* are omitted. For full lists of all the genera and *incertae sedis* high-rank taxa, see tables 4.1–14.1.

[2] McKenna and Bell (1997) assigned a part of "symmetrodonts," including Spalacotheriidae (but not Kuehneotheriidae), to the superlegion Trechnotheria McKenna, 1975.

[3] Superorder Dryolestoidea has been assigned by McKenna and Bell (1997) to the legion Cladotheria McKenna, 1975.

purpose of organizing this book into a manageable number of chapters, some chapters must cover multiple hierarchical levels in the mammalian tree—that is, portions of the tree that we now know to be made up of paraphyletic groups, or grades. These include chapter 4 (earliest stem mammals), chapter 9 ("symmetrodontans" or stem trechnotherians), chapter 10 ("eupantotherians" or stem cladotherians), and chapter 11 ("tribotherians" or stem boreosphenidans).

Taxonomic classification is essential to a compendium such as this, and here we faced the dilemma of which scheme to adopt. There is a large body of literature on the pros and cons of traditional Linnaean taxonomy versus a strictly cladistic hierarchy that closely mirrors genealogy (see the recent review of the history of classification by McKenna and Bell, 1997). It is beyond the scope of this book to resolve these long-standing problems in biological classification. Rather, we decided to offer two alternative schemes of Mesozoic mammal classification (tables 1.1 and 1.2).

Table 1.1 presents a cladistically based classification of mammals that uses indentation of successively lower ranks to reflect the intended scheme of phyletic sequencing (Wiley, 1979, 1981; Nelson and Platnick, 1981; McKenna and Bell, 1997). Because the main purpose of the table is to provide a broad outline, it includes mainly higher-rank taxa; however, some stem genera are included for their importance in representing the successive hierarchies of phylogeny. The sequence of the classification in table 1.1 is illustrated in figure 1.1 (see also Luo et al., 2002 and figure 15.1 herein).

Table 1.2 presents a simplified Linnaean taxonomy of Mesozoic mammals. The taxonomic arrangement mainly follows the phylogenetic sequence resulting from our parsimony analyses, presented in chapter 15 and figures 15.1 and 15.2. Reference to the chapters in which descriptions of corresponding taxa appear are given in parentheses following each taxon name. The complete systematic lists, including genera and species, are presented as individual tables in systematic chapters 4–14.

CONCLUDING REMARKS

The study of Mesozoic mammals has made great leaps forward in the past two decades. A vast number of new fossils have been discovered and are now available for study. Some traditional ideas have been abandoned, while many new ideas have emerged—some of which are gaining increasingly wide acceptance, whereas others are hotly debated. This virtual explosion of new information, and interpretation based on these data, poses a daunting challenge for anyone attempting to comprehend all of it. Since the publication of the previous comprehensive book on Mesozoic mammals by Lillegraven, Kielan-Jaworowska, and Clemens in 1979, nearly a thousand papers on Mesozoic mammals have appeared in many different languages, scattered in numerous journals, symposium proceedings, and books published around the globe. We have endeavored to gather this scattered information in one book, so that readers may use this compendium as a gateway to a vast body of literature from a rapidly developing field of evolutionary biology. We hope that this volume will be a useful tool not only for students of Mesozoic mammals, but also for those who work on Tertiary and extant mammals and on terrestrial ecosystems of the Mesozoic.

CHAPTER 2

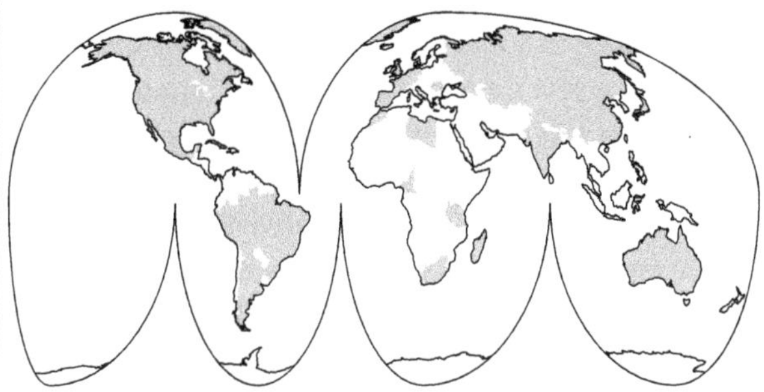

Distribution: Mesozoic Mammals in Space and Time

INTRODUCTION

Following our traditional concept of Mammalia, and thus including a number of forms known only by fossils of the Mesozoic, the earliest mammals are from the Late Triassic, a bit over 220 Ma old (figure 2.1). The Mesozoic record thus spans about 155 Ma of mammalian history—more than twice the duration of the entire Cenozoic. Yet the Mesozoic record is frustratingly sparse.[1] Reasonably complete specimens—skulls and skeletons—are rare and are known for only a few taxa, while for many groups the discovery of a partial jaw is a notable occurrence. The record is also discontinuous, being punctuated by large gaps in time for which little or nothing is known. A third deficiency of the existing fossil data is that diversity is often sampled poorly or unevenly. As a result, our understanding of Mesozoic mammals in their faunal context and of the roles they played in their faunas is sketchy. Finally, the geographic representation of Mesozoic mammals is rather checkered. For some significant landmasses, particularly the southern continents, the record is blank, or nearly so, during much or all of the Mesozoic. Obviously, these deficiencies in the fossil record place limitations on what we can reasonably infer about the paleobiology and relationships of Mesozoic mammals. But it must also be appreciated that the deficiencies serve as factors that influence interpretation.

We return to further discussion of these factors in the summary at the end of this chapter. However, we must point out here that there is another side to this coin. Many major advances in knowledge of Mesozoic mammals have been made over the past several decades. In our judgment, most of these advances have come about because of new fossil discoveries. Indeed, the limited nature of the existing fossil record provides an outstanding and, to us, inspirational opportunity for contributing to knowledge. In few other fields of investigation can a single new specimen provide such significant insight into age-old controversies or reveal such unimagined diversity.

Our purpose here is to review the record of Mesozoic mammals—their fossil representation and their geographic, geologic, and paleoenvironmental occurrences. The nature and amount of information for each occurrence varies considerably, depending on what is available in published accounts and on the significance of the occurrence. Although we have relied on colleagues and our own personal knowledge in making faunal lists as up-to-date as possible, we have otherwise relied strictly on published literature and theses in assembling this compendium. In many instances, a mammalian fauna or a local fauna (an assemblage reasonably interpreted as being made up of species that lived at more or less the same time and in more or less the same place) is known by specimens from one site only. In such cases, each species' occurrence is identical with its presence at that site. In other instances, however, species have been collected from a number of sites, closely spaced stratigraphically and geographically,

[1] The map in the logo above shows symbolically the distribution of Mesozoic mammals among continents, marked by shading of those countries with a varying record of Mesozoic mammal fossils. The shading is not an accurate representation of the actual Mesozoic mammal sites, which are always sparse, far smaller, and more localized in geographic area than the countries where they are known.

PERIOD	EPOCH		AGE	Ma
				65
CRETACEOUS	LATE		MAASTRICHTIAN	71.3
			CAMPANIAN	83.5
			SANTONIAN	85.8
			CONIACIAN	89.0
			TURONIAN	93.5
			CENOMANIAN	99.0
	EARLY		ALBIAN	112
			APTIAN	121
		NEOCOMIAN	BARREMIAN	127
			HAUTERIVIAN	132
			VALANGINIAN	137
			BERRIASIAN	144
JURASSIC	LATE		TITHONIAN	151
			KIMMERIDGIAN	154
			OXFORDIAN	159
	MIDDLE		CALLOVIAN	164
			BATHONIAN	169
			BAJOCIAN	176
			AALENIAN	180
	EARLY		TOARCIAN	190
			PLIENSBACHIAN	195
			SINEMURIAN	202
			HETTANGIAN	206
TRIASSIC	LATE		RHAETIAN	210
			NORIAN	221
			CARNIAN	227
	MIDDLE		LADINIAN	234
			ANISIAN	242
	EARLY		OLENEKIAN	245
			INDUAN	248

FIGURE 2.1. Geological time scale for the Mesozoic. *Source:* modified after Palmer and Geissman (1999).

for example, those of the type Lance Formation, eastern Wyoming (Clemens, 1963b). We find it neither practical nor desirable to list all such occurrences independently, and instead use composite local faunas (admittedly arbitrary in some cases) for such assemblages. Faunal lists are grouped together by age and region; individual sites or local faunas are keyed to corresponding maps by numbers.

Where possible, we use European marine ages (figure 2.1), or subdivisions thereof; absolute ages follow the geologic time scale of Palmer and Geissman (1999). However, the chronostratigraphic framework is poor for many occurrences; in some cases the epoch is not well established and in a few even the period is questionable. In North America, a sequence of land-mammal "ages" has achieved wide usage for faunas of Campanian and Maastrichtian age, despite some problems with definition (Russell, 1975; see especially reviews by Lillegraven and McKenna, 1986; and Cifelli et al., 2004). Land-vertebrate or land-mammal "ages" have been proposed for earlier assemblages in North America (L. S. Russell, 1964, 1975; Lucas, 1993) and for Asia (Jerzykiewicz and Russell, 1991; Lucas, 1996; Lucas and Estep, 1998) and South America (Bonaparte et al., 1987). These "ages" are not widely used, and as they are of limited utility in the current context, we do not employ them, although we refer to them as appropriate. Transliteration of stratigraphic and place names for Russia, Middle Asia, and Mongolia follows Benton (2000a).

LATE TRIASSIC–EARLY JURASSIC

Mammals are known from a number of Upper Triassic and Lower Jurassic sites, mainly in Western Europe, but also in Greenland, China, Africa, India, and North America (possible mammals of this age are represented in South America only by footprints). Collectively, these sites span the Carnian through the Liassic or possibly later. Although the faunas differ in composition, they share broad points of taxonomic similarity. Owing to this similarity and to the fact that they are collectively separated from well-known Middle Jurassic (Bathonian) assemblages by a substantial hiatus, it is convenient to treat them together. During the Late Triassic–Early Jurassic, the continental landmasses were largely united into the supercontinent Pangaea. Faunal continuity among regions that are now geographically disparate is indicated by the fact that many of Earth's terrestrial vertebrates—therapsids, mammals, dinosaurs, and others—are remarkably similar, often at the level of genus.

CONTINENTAL EUROPE

Late Triassic through Early Jurassic mammal sites of Western Europe (table 2.1, figure 2.2) fall into two distinct geographic and depositional settings: those in Britain occur in fissure fillings, whereas those of continental Europe generally occur in more conventional settings, such as stratified fluvial, nearshore marine, or deltaic deposits. Stratigraphically, the continental sites are placed in the Keuper Formation (see reviews by Clemens, 1980a;

TABLE 2.1. Late Triassic and Early Jurassic Mammals of Western Europe and Eastern Greenland (see figures 2.2, 2.3, 2.8; note that numbers do not correspond between figures and tables). Localities: 1, Halberstadt, Germany; 2, Württemberg, Germany; 3, Hallau, Switzerland; 4, Saint-Nicolas-de-Port, France; 5, Medernach, Luxembourg; 6, Syren, Luxembourg; 7, Attert, Belgium; 8, Hachy (Sagnette), Belgium; 9, Habay-la-Vieille (Gaume), Belgium; 10, Varangéville, France; 11, Holwell, Britain; 12, Emborough, Britain; 13, St. Bride's Island (Ewenny, Pontalum, Duchy, and Pant quarries); 14, Jameson Land, Greenland

- Mammalia *incertae sedis* (7)
    - *Hallautherium schalchi* (3)
  - Sinoconodontidae
    - aff. *Sinoconodon* sp. (10)
  - Family *incertae sedis*
    - Gen. et sp. indet. (14)
- Morganucodonta
  - Megazostrodontidae
    - *Brachyzostrodon coupatezi* (4)
    - *Brachyzostrodon maior* (4)
    - *Brachyzostrodon* sp. 1 (4)
    - *Brachyzostrodon* sp. 2 (4)
    - cf. *Brachyzostrodon* sp. (14)
  - Morganucodontidae
    - *Eozostrodon parvus* (11, 13)
    - *Helvetiodon schutzi* (3)
    - *Morganucodon watsoni* (13)
    - *Morganucodon peyeri* (3)
    - *Morganucodon* sp. (4, 10)
    - aff. *Morganucodon* sp. (10)
    - Gen. et sp. indet. (4, 5, 6, 10)
- ?Docodonta, fam. *incertae sedis*
    - *Delsatia rhupotopi* (4)
- Haramiyida
  - Haramiyaviidae
    - *Haramiyavia clemmenseni* (14)
  - Haramiyidae
    - *Thomasia antiqua* (2, 4, 6, 9, 10, 11)
    - *Thomasia* cf. *antiqua* (10)
    - *Thomasia hahni* (1)
    - *Thomasia moorei* (3, 4, 11)
    - *Thomasia woutersi* (7, 9)
    - *Thomasia* sp. (2, 11, 13)
  - Theroteinidae
    - *Theroteinus nikolai* (4)
    - *Theroteinus* sp. (4)
- ?Multituberculata *incertae sedis*
    - *Mojo usuratus* (9)
- Archaic "symmetrodontans"
  - Kuehneotheriidae
    - *Kuehneon duchyense* (13)
    - *Kuehneotherium praecursoris* (13)
    - *Kuehneotherium* sp. (4, 11, 13, 14)
    - *Kuehneotherium* sp. nov. (6)
  - Woutersiidae
    - *Woutersia butleri* (4)
    - *Woutersia mirabilis* (4, 10)

Sigogneau-Russell and Hahn, 1994). This unit spans the Upper Triassic, but mammalian fossils are restricted to its upper part. Thus far, specimens consist only of isolated teeth or parts thereof. Given the great antiquity of these remains, the affinities of the taxa they represent are sometimes debatable. *Tricuspes*, for example, occurs at several sites in the Keuper. It may be a mammal (Clemens, 1980a, 1986), but is more generally considered a cynodont (Godefroit and Sigogneau-Russell, 1995), and so we have omitted it from our listing here, but for the sake of documentation we discuss it in chapter 4. The geologically oldest site yielding an uncontested mammal from the Keuper is of Norian age and is near **Halberstadt**, Germany. The haramiyid *Thomasia hahni* is found here. Haramiyids are enigmatic mammals that may be related to multituberculates, although there is not universal agreement on this point (chapter 8). They are common members of Late Triassic mammalian faunas and, if related to multituberculates, were separated from that group by a considerable hiatus.

A cluster of sites in the upper Keuper occurs in **Württemberg** (Baden-Württemberg), southern Germany. These

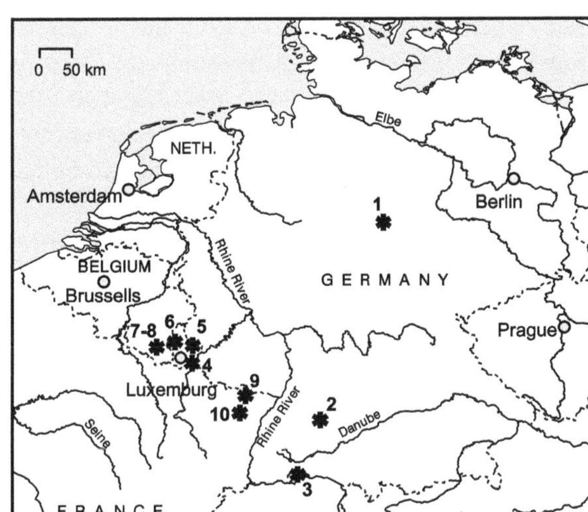

FIGURE 2.2. Late Triassic, continental Europe (all in the upper Keuper; Late Triassic). Localities or local faunas: **1, Halberstadt,** Germany; **2, Württemberg,** Germany; **3, Hallau,** Switzerland; **4, Syren,** Luxembourg; **5, Medernach,** Luxembourg; **6, Attert,** Belgium; **7, Hachy** (Sagnette), Belgium; **8, Habay-la-Vieille** (Gaume), Belgium; **9, Varangéville,** France; **10, Saint Nicolas-de-Port,** France.

are younger, possibly later Rhaetian in age. The fossils, which include a mixture of terrestrial and marine taxa, are commonly abraded and show signs of transportation (Sigogneau-Russell and Hahn, 1994). A mammal tooth—again belonging to the haramiyid *Thomasia*—was discovered near Stuttgart (in Degerloch, then a neighboring village) as early as 1847. Additional sites in the area, Olgahain and Gaisbrunnen, were worked by Friedrich von Huene and Otto Schindewolf in the first half of the twentieth century. To date, the haramiyid *Thomasia* (perhaps represented by two species) is the only generally accepted mammal from the Württemberg sites.

A larger, better represented assemblage of mammals is known from **Hallau,** Switzerland, just south of the German border. Like the Württemberg sites, it may be later Rhaetian in age (Clemens, 1986). Though Hallau was discovered in the 1870s, it was not worked extensively until Bernhard Peyer began screenwashing sediments from the site in the 1940s, work he was to continue into the 1960s. In 1956, he published a monograph on the materials collected to that date (Peyer, 1956). His collection, housed in Zürich, was restudied by William A. Clemens, who also reviewed the German collections from the Keuper (Clemens, 1980a). In addition to the haramiyid *Thomasia,* several "triconodonts" are known from Hallau. The morganucodontid *Morganucodon* is represented by an endemic species, *M. peyeri. Morganucodon* is the best known of Late Triassic–Early Jurassic mammals, being represented elsewhere by relatively complete specimens and large series of jaws and isolated teeth (chapter 4). It is also broadly distributed, being known from the Keuper, the fissure fills of Britain, China, Greenland, and (perhaps) North America. Clemens (1980a) described two other taxa from Hallau, *Helvetiodon schutzi* and *Hallautherium schalchi,* that are also considered to be morganucodontans (Hahn et al., 1991).

In the past two decades, numerous microvertebrate sites have been worked in the Keuper of Belgium and Luxembourg, largely owing to the efforts of Georges Wouters, Jean-Claude Lepage, and Dominique Delsate. To date five of these sites have yielded mammals. An unidentified morganucodontid has been reported from **Medernach,** Luxembourg (Cuny et al., 1995; Godefroit and Sigogneau-Russell, 1995). The site is probably of Norian age. The locality of **Syren,** Luxembourg, is assigned a Rhaetian age based on palynomorphs (Godefroit et al., 1998). The vertebrate fauna includes three mammals: a haramiyid (*Thomasia antiqua*), an unidentified morganucodontid, and an apparently new species of *Kuehneotherium* (see Godefroit et al., 1998). From the Lorraine region of Belgium, two other sites include single records of Rhaetian mammals: **Attert,** from which the haramiyid *Thomasia woutersi* is known, and **Hachy** (Sagnette), which has yielded a single, unidentified mammal tooth (Sigogneau-Russell and Hahn, 1994). Two species of *Thomasia* have also been reported from the nearby site of **Habay-la-Vieille** (Gaume), Belgium. The most extraordinary occurrence at this site is that of *Mojo usuratus.* Represented by a worn, incomplete tooth, this species has been referred to the multituberculate family Paulchoffatiidae (Hahn, 1987a; Hahn et al., 1989). This occurrence, if verified, would represent a colossal range extension for the family, which otherwise appears in the Late Jurassic. If multituberculates, haramiyids, and theroteinids form a monophyletic group (chapter 8), this occurrence implies a remarkably early divergence of these lineages.

Other fossils from the Keuper sites in Belgium and Luxembourg further illustrate the difficulties of identifying early mammals based on tooth structure, as mentioned previously in connection with *Tricuspes.* Chiniquodontoid cynodonts have a "triconodont" cusp pattern and, in some instances, divided tooth roots. Small (mouse-sized) chiniquodontoids from the Keuper include *Pseudotriconodon, Gaumia, Lepagia,* and *Meurthodon* (see discussion in Sigogneau-Russell and Hahn, 1994). *Pseudotriconodon* is known from Medernach and from the Late Triassic of New Mexico (Lucas and Oakes, 1988), and may be similar to *Microconodon* and *Dromatherium* from North America (Hahn et al., 1984; Godefroit and Sigogneau-Russell, 1995; Sues, 2001). Mammalian affinities for the last two genera (both known from the Triassic Cunmock Formation of North Carolina, with *Microconodon* possibly present in the Dockum Group of Texas) have been variously debated (Osborn, 1887b; Simpson, 1926d; Clemens et al., 1979). Based on their probable therapsid affinities, we omit them from further discussion.

The final sites in the Keuper yielding mammalian fossils are in the Lorraine region of France. **Saint-Nicolas-de-Port** is the best represented of the two. The age of the site is uncertain, but it is probably at least as old as early Rhaetian (Sigogneau-Russell, 1983). An earlier, Norian age was advocated by Buffetaut and Wouters (1986) on the basis of similarity of the herpetofauna to that of Halberstadt (uppermost middle Keuper, discussed earlier), but the biostratigraphic constraints are imprecise (Hahn et al., 1989). Fossils occur in a conglomeratic layer at the base of the local section, interpreted to have been deposited in a nearshore tidal environment. The locality has been known since 1851. Large-scale fossil recovery, using underwater screenwashing techniques, was begun in 1976 by Denise Sigogneau-Russell and Donald E. Russell after the discovery of a mammal-like tooth the previous year by G. Wouters. Shortly thereafter, Clemens et al. (1979: 11) presciently observed, "The locality could easily become the most prolific source of Rhaetian mammalian fossils in continental Europe." Saint-Nicolas-de-Port has fully lived up to this expectation: thus far, it has yielded more than 1,000 mammal teeth, more than three times the rest of the Keuper sites combined. The haramiyid teeth (Sigogneau-Russell, 1989b, 1990), representing two species, were later reviewed by Butler and MacIntyre (1994) in connection with a study of British haramiyids. The sample is sufficiently large (more than 200 specimens) to permit recognition of two types—those previously referred to *Thomasia* and *Haramiya*—as lower and upper teeth of the same animal (Sigogneau-Russell, 1989b), thereby solving a long-standing enigma and permitting revision of the group. Variations among teeth in the sample have tentatively been ascribed to positional differences; wear patterns also suggest a multituberculate mode of chewing (see Butler, 2000, and chapter 8).

Another notable occurrence at Saint-Nicolas-de-Port is represented by the Theroteinidae (Sigogneau-Russell et al., 1986), known by *Theroteinus nikolai* and *Theroteinus* sp. Teeth of these mammals bear some similarity to certain multituberculates (Paulchoffatiidae) and to haramiyids, but are distinct enough that Hahn et al. (1989) created a separate order for them, Theroteinida, placed with these other groups in the Allotheria (see also Butler and MacIntyre, 1994; Butler, 2000, and discussion in chapter 8).

The most abundant morganucodontan at Saint-Nicolas-de-Port is the widely distributed *Morganucodon* (Hahn et al., 1991). An undescribed taxon (Sigogneau-Russell and Hahn, 1994) and *Brachyzostrodon*, represented by two named species and perhaps as many as two others (Sigogneau-Russell, 1983b; Hahn et al., 1991), are also present. *Brachyzostrodon* is rather large, by morganucodontan standards and is generally similar to South African *Megazostrodon*, though the two differ in occlusal pattern and presumed diet (Hahn et al., 1991).

Several "symmetrodontans" are known from Saint-Nicolas-de-Port. *Kuehneotherium*, first described from Late Triassic–Early Jurassic fissure fillings in Wales (see later), is represented by one or more species, based on a large and highly variable sample of isolated molars (see Sigogneau-Russell and Hahn, 1994; Godefroit and Sigogneau-Russell, 1999). Also present is the highly distinctive *Woutersia* with two species, *W. mirabilis* and *W. butleri*. *Woutersia* is broadly similar to *Kuehneotherium*, but is characterized by broad molars suggestive of some crushing function similar to that of the later docodontans (Sigogneau-Russell, 1983a; Sigogneau-Russell and Hahn, 1995). The final member of this diverse mammalian fauna is *Delsatia rhupotopi*, a possible docodontan (Sigogneau-Russell and Godefroit, 1997, see also chapter 5). *Delsatia* is based on molars that superficially resemble those of *Woutersia*. Docodontans, which are characterized by precociously specialized molars (Jenkins, 1969b), otherwise make their first appearance in the Middle Jurassic (Waldman and Savage, 1972; Kermack et al., 1987). Affinities of docodontans are problematic but, given that they retain many primitive features, the Docodonta may have diverged early in mammalian history (Lillegraven and Krusat, 1991; Sigogneau-Russell and Hahn, 1995). Thus, presence of some putative relatives of theirs in the Triassic may not be surprising.

The site of **Varangéville** is about 10 km southeast of Nancy and 4 km north of Saint-Nicolas-de-Port. Like the latter, fossils occur in a conglomerate, and the age of the horizon is not well understood. Lithostratigraphically, it corresponds to the base of the Rhaetian, which appears to be time transgressive (Godefroit, 1997). The mammalian fauna of Varangéville includes the haramiyid *Thomasia antiqua*, as many as three morganucodontids (including *Morganucodon* itself), and the "symmetrodontan" *Woutersia mirabilis*, all of which are known from Saint-Nicolas-de-Port. The most notable occurrence is that of the basal mammal group Sinoconodontidae (represented by a form similar to *Sinoconodon*), otherwise known from the Early Jurassic of China (Godefroit, 1997).

BRITAIN

The other major European occurrence of Late Triassic–Early Jurassic mammals is in southern England and Wales, from sites on either side of the Bristol Channel (figure 2.3). We have drawn from the excellent review of Evans and Kermack (1994) in summarizing these occurrences. During the Late Triassic and Early Jurassic, shallow seas covered parts of the British Isles, transgressing through time, so that major islands transformed into an archipelago. Karst topography developed on the subaerially exposed Carboniferous limestone, and vertebrate remains were carried into the systems of fissures and caves together with clastic sediments. The age of these fissure deposits was long uncertain, with the result that they were simply termed Rhaeto-Liassic in age (e.g., Kermack et al., 1973). Palynological and other evidence, including reference to rocks marking the transgression that defines the beginning of the Rhaetian (Fraser et al., 1985), has led to general agreement on the ages of most of the mammal-yielding localities. One grouping of sites occurs in Glamorgan, near Brigend, Wales. The principal mammal-yielding sites (Ewenny, Pontalun, Duchy, and Pant Quarries) are part of an assemblage of localities called St. Bride's Island. They are closely grouped geographically, probably representing the same small paleoisland (Robinson, 1957), and are considered to be of Early Jurassic age. The other

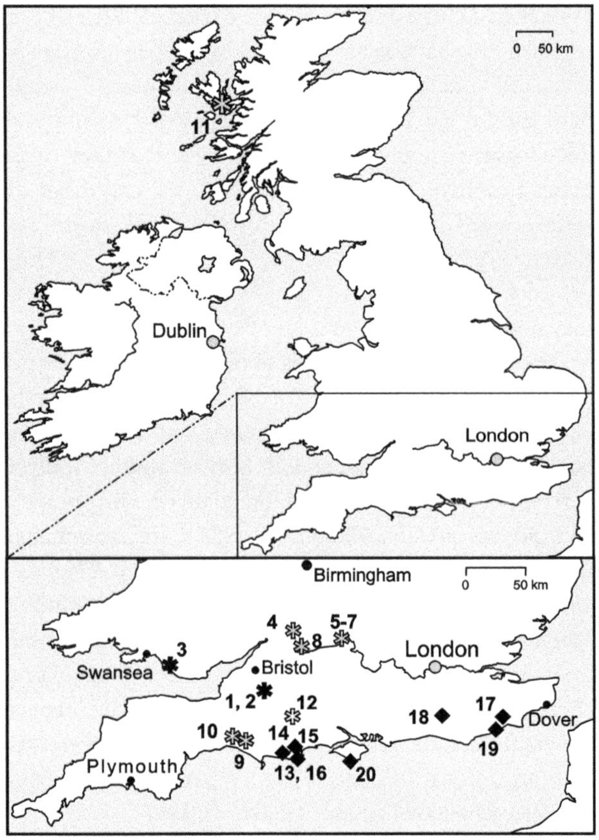

FIGURE 2.3. Mesozoic mammals of Britain. Asterisks, Late Triassic–Early Jurassic (black; localities 1–3), Middle Jurassic (gray; localities 4–11), and Late Jurassic (white; locality 12). Diamonds, Early Cretaceous (?Berriasian, localities 13, 14; Valanginian and ?Valanginian, localities 15–20). Localities or local faunas: **1, Holwell** (?late Norian or Rhaetian, fissure filling); **2, Emborough** (Norian, fissure filling); **3, St. Bride's Island** (Ewenny, Pontalun, Duchy, and Pant quarries; fissure fillings, Liassic); **4, Hornsleasow** (Chipping North Formation; early Bathonian); **5, Kirtlington** (Forest Marble; late Bathonian); **6, Woodeaton** (Hampen Marly Formation; middle Bathonian); **7, Stonesfield** (Stonesfield Member, Sharps Hill Formation; early middle Bathonian); **8, Tarlton Clay Pit; 9, Swyre; 10, Watton Cliff** (Forest Marble; late Bathonian); **11, Loch Scavaig** and **Elgol,** Isle of Skye (Ostracod Limestone; middle Bathonian); **12, Chicksgrove** (Portland Stone Formation, Tithonian); **13, Durlston Bay; 14, Sunnydown Farm** (Lulworth Formation, Purbeck Limestone Group; Berriasian); **15, Belle Vue; 16, Acton** (Durlston Formation, Purbeck Limestone Group; Berriasian); **17, Tighe Farm** (Wadhurst Formation, Wealden Group; early Valanginian); **18, Paddockhurst Park** (Grinstead Clay, Wealden Group; middle Valanginian); **19, Cliff End** (Wadhurst Formation, Wealden Group; early Valanginian); **20, Isle of Wight** (Wealden Group, ?Valanginian).

major grouping of sites is to the east of the Bristol Channel, in the Bristol-Mendip region of southern England. Two of these sites, Emborough and Holwell, have yielded mammals; both are considered to be of Late Triassic age; and Emborough, at least, is placed in the Norian (Fraser et al., 1985).

The first mammals to be recovered from the Late Triassic–Early Jurassic fissures of Britain were isolated haramiyid teeth found by Charles Moore in 1858 at the Holwell site (Simpson, 1928a). Commercial mining of the Carboniferous limestone resulted in development of quarries and exposure of the contained fissures (Evans and Kermack, 1994). These quarries were exploited with great success by paleontologists beginning in 1939, when Walter Kühne visited Holwell. Kühne washed the clays (as had Moore before him) and, without optical aid, recovered two morganucodontid teeth and twelve belonging to haramiyids (Kermack, 1988). He later discovered mammalian specimens at one of the quarries (Duchy) on St. Bride's Island. From the early 1950s onward, work in the fissures on both sides of the Bristol Channel was continued by researchers from University College, London, including Kenneth A. Kermack, Pamela L. Robinson, Frances Mussett, Susan Evans, and their students and successors. This long-term research program led to the discovery of many additional sites and to the recovery of an extraordinary wealth of specimens (mainly from St. Bride's Island) that have provided key insights into the early history of mammals.

The haramiyids from **Holwell** have been studied in detail (Butler and MacIntyre, 1994). Perhaps as many as three species may be present—*Thomasia antiqua, T. moorei,* and *T.* sp. Two morganucodontid teeth from this quarry were originally described as representing distinct species, *Eozostrodon parvus* and *E. problematicus* (see Parrington, 1941), but these were later synonymized (Parrington, 1971). In the meantime, Kühne (1949) had described the rather similar *Morganucodon watsoni* from one of the St. Bride's quarries (Duchy). Confusion arose in the 1970s because of an extended controversy as to whether *E. parvus* and *M. watsoni* are synonymous (e.g., Parrington, 1971, 1973), a complicating factor being that the type of the former is not generally considered to be diagnostic (e.g., Kermack et al., 1973). Here we follow Clemens's (1979b) suggestion that the name *Eozostrodon* be restricted to the specimens from Holwell and that the name *Morganucodon* be applied to material from Wales (chapter 4).

Only one mammal has been reported from **Emborough,** also in the Bristol-Mendip region. The occurrence is an important one, however: the fissure can be dated as Norian in age because of overlying basal Rhaetian transgression deposits and the mammal is *Kuehneotherium* (Fraser et al., 1985). *Kuehneotherium* has a "symmetrodontan" molar pattern and was long considered to be the oldest therian *sensu lato* (see, e.g., Crompton and Jenkins, 1967; Kermack et al., 1968); however, the affinities of "symmetrodontans" are problematic (chapter 9). Regardless, *Kuehneotherium* is also known from the Liassic of St. Bride's Island and hence had a wide temporal range.

The fissures of **St. Bride's Island,** Wales, have yielded truly enormous collections of Early Jurassic mammals, including not just the usual isolated teeth, but jaws and skeletal elements as well. The bones are disassociated but are well preserved and show no evidence of reworking. Some fissure-fill matrix, which varies from clay to marl, is so fossiliferous that it might well be described as a bone coquina. For the most part (see later), the quarries on St. Bride's Island have yielded a consistent assemblage, termed the *Hirmeriella* association for the abundant conifer of that name. Curiously, this plant is not known from the other fissure deposits of the region. The vertebrate fauna of this assemblage includes only three principal taxa: the lepidosaur *Gephyrosaurus bridensis* and the mammals *Morganucodon watsoni* and *Kuehneotherium* sp. *Morganucodon* is the larger and by far the more abundant of the two mammals. Almost every element of the skeleton is represented in existing collections, and as a result *Morganucodon* is incomparably the best known of Late Triassic–Early Jurassic mammals (e.g., Kermack et al., 1973, 1981; Jenkins and Parrington, 1976). Except for one pocket in Pontalun Quarry (Kermack et al., 1968), *Kuehneotherium* is rather rare, and is known mainly by isolated teeth and jaws. *K. praecursoris* is the described species of the genus; there is evidently another kuehneotheriid represented (Kermack et al., 1968; Mills, 1984) in the fauna, and we list this simply as *Kuehneotherium* sp. Whether either of these is the same as *Kuhneon duchyense,* from Duchy Quarry, cannot be determined because the type and only specimen of this species, an isolated molar, has been lost.

Because it typically includes only three vertebrates, the fauna of the *Hirmeriella* association has been characterized as depauperate (Fraser, 1989). One exception to this is found in a single fissure at Pant Quarry, which has yielded a diverse assemblage of reptiles and mammals, the latter including the haramiyid *Thomasia, Morganucodon watsoni, Kuehneotherium,* and an undescribed large morganucodontid (Evans and Kermack, 1994). Both the incredible volume of vertebrate remains and the fact that only three taxa are generally present in the assemblages of St. Bride's Island could be explained by accumulation through action of a selective predator—perhaps the large morganucodontid from Pant (Evans and Kermack, 1994).

ASIA

An extraordinary wealth of fossil vertebrates has been collected from the **Lufeng Basin,** Yunnan Province, southern China (figure 2.4). The systematics, assemblages, and stratigraphic and geologic occurrences of the small tetrapods have recently been reviewed by Luo and Wu (1994). The most productive localities are northeast of Lufeng, in the Dachong-Dawa area, in the lower Lufeng Formation, which is about 400 m thick in this region. The lower Lufeng Formation is divided into two distinct lithologic bodies, each characterized by a distinct fauna (Sun et al., 1985). The Dull Purplish Beds, below, contain mostly saurischian dinosaurs and the therapsid *Bienotherium.* The overlying Dark Red Beds are of greater interest in the current context: they have yielded a fauna comprised mainly of ornithischians, a diversified assemblage of therapsids (but not *Bienotherium*), and early mammals (table 2.2). The vertebrate fauna, most importantly some of the mammals and tritylodontid therapsids, is similar to those of the Welsh fissure fills and the Kayenta Formation, United States, indicating an Early Jurassic age (Luo and Wu, 1994; see also Sigogneau-Russell and Sun, 1981; Sun et al., 1985), with biochronologic constraints suggesting that the mammals are no older than Sinemurian and possibly extend into the Pliensbachian (Luo and Wu, 1995). The Dark Red Beds include mainly fluvial and overbank deposits made up of mudstones and siltstones; most of the small vertebrate fossils, including the mammals, are found in calcareous nodules that probably represent paleosols. The fossils are well preserved, usually consisting of partial or complete skulls, sometimes with associated skeletal material. Major localities in the Dark Red Beds of the Lufeng Basin are **Dahuangtian, Dadi, Dawa, Heiguopeng,** and **Zhangjiawa** (figure 2.4).

The first mammal to be described from the lower Lufeng Formation is *Sinoconodon rigneyi* (see Patterson and Olson, 1961). *Sinoconodon* is similar to *Morganucodon* in many ways, but differs from the latter and other

TABLE 2.2. Early Jurassic Mammals of the Lower Lufeng Formation, Yunnan, China (see figure 2.4)

Mammalia *incertae sedis*
Hadrocodium wui
Kunminia minima
Sinoconodontidae
Sinoconodon rigneyi
Morganucodonta
Morganucodontidae
Morganucodon oehleri
Morganucodon heikuopengensis

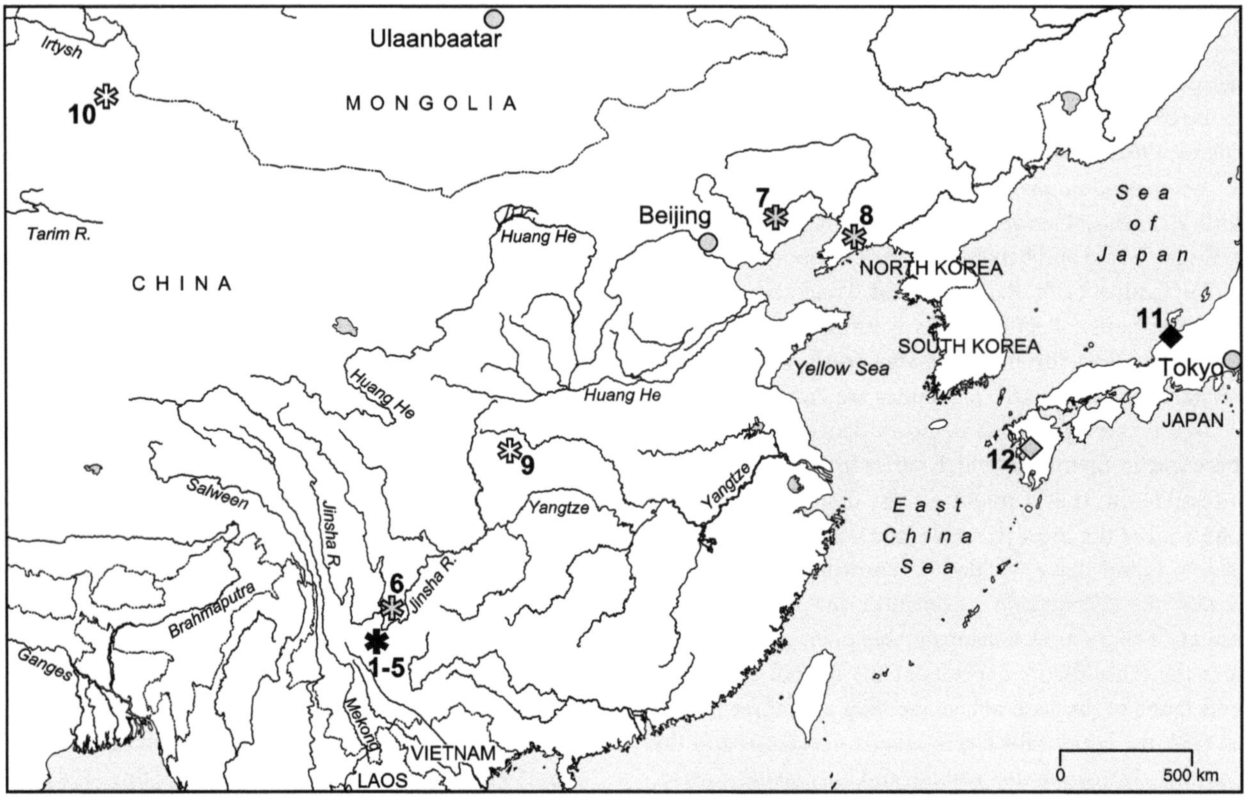

FIGURE 2.4. Jurassic (China) and Cretaceous (Japan) mammal localities. Asterisks, Early (black; localities 1–5), Middle–Late (gray; localities 6–9), and Late (white; locality 10) Jurassic; diamonds, Early (black; locality 11) and Late (gray; locality 12) Cretaceous. Localities or local faunas: **1, Dahuangtian; 2, Dawa; 3, Heiguopeng; 4, Zhangjiawa; 5, Dadi** (Dark Red Beds, lower Lufeng Formation; Early Jurassic, Yunnan Province); **6, Luchang** (lower Yimen Formation; Early Jurassic, Sichuan Province); **7, Fangshen** (Haifanggou Formation; ?Middle or Late Jurassic, Liaoning Province); **8, Zha-zy-ao Mine** (Wa-fang-dian Formation; ?Middle Jurassic–Early Cretaceous, Liaoning Province); **9, Shilongzhai** (Shaximiao Formation; ?Middle–Late Jurassic, Sichuan Province); **10, Jianshan Wash** (Shishuguo Formation; Late Jurassic, Xinjiang Province); **11, Kaseki-kabe** (Kuwajima Formation; late Hauterivian, Ishikawa Prefecture, Japan); **12, Amagimi Dam** ("Upper Formation" of the Mifune Group; late Cenomanian–early Turonian, Kumamoto Prefecture, Japan).

Late Triassic–Early Jurassic mammals in its dentition: upper and lower molars do not have a one-to-one correspondence, the postcanines do not develop occlusal wear facets, and some of the teeth, at least, were replaced continuously (Crompton and Luo, 1993, see chapter 4). *Sinoconodon* is now known by a number of skulls. Several other taxa subsequently described from the lower Lufeng Formation, "*S. parringtoni*," "*S. yangi*," and "*Lufengoconodon changchiawaensis*" (see Young, 1982a; Zhang and Cui, 1983), are thought to be synonyms of *Sinoconodon rigneyi*, which evidently has a broad range of ontogenetic variation (Crompton and Luo, 1993; Luo and Wu, 1994). The lower Lufeng Formation has also yielded several skulls of *Morganucodon*, with two species, *M. ohleri* and *M. heikuopengensis*, currently recognized (see Rigney, 1963; Young, 1978). These fine specimens have provided the basis for a detailed understanding of the cranial anatomy of this important early mammal (Kermack et al., 1981; Crompton and Luo, 1993). *Kunminia minima*, earlier thought to be a therapsid (Young, 1947), is probably a mammal, though its affinities are uncertain. *Hadrocodium wui* is the most recently described mammal from the lower Lufeng Formation (Luo, Crompton, and Sun, 2001). *Hadrocodium* is known by a nearly complete skull and articulated mandible, and appears to be structurally intermediate between stem mammals (e.g., morganucodontids, docodontans) and monotremes (see chapter 4).

INDIA

A possible mammal, *Gondwanadon tapani*, has been described from near the village of **Tiki,** Madhya Pradesh (figure 2.5, table 2.3). *Gondwanadon* is based on an incomplete, two-rooted, molariform tooth, said to be similar to

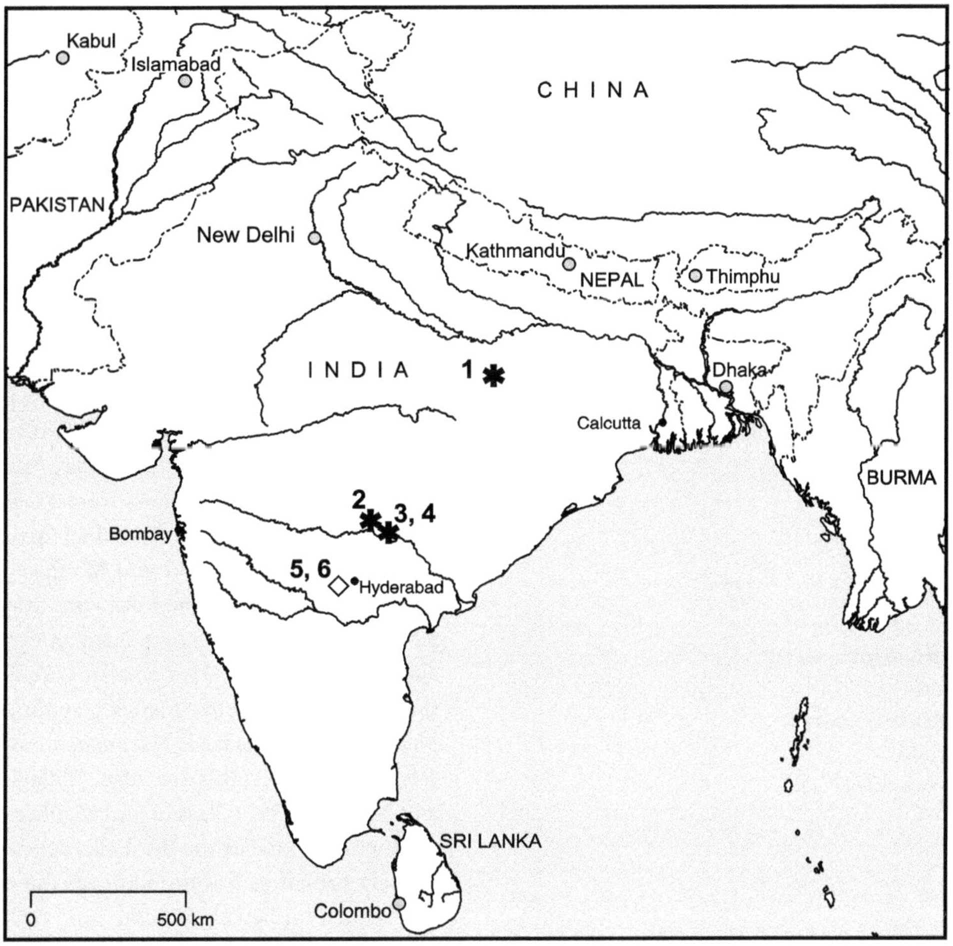

FIGURE 2.5. Mesozoic mammals of India. Asterisks, Late Triassic–Early Jurassic (localities 1–4); diamond, Cretaceous (localities 5, 6). Localities or local faunas: **1, Tiki** (Tiki Formation; Carnian, Madhya Pradesh); **2, Paikasigudem; 3, Manganpalli; 4, Yamanapalli** (Kota Formation; ?Early Jurassic, Andhra Pradesh); **5, Naskal; 6, Rangapur** (Intertrappean Beds; Maastrichtian, Andhra Pradesh).

those of *Morganucodon* except in relative cusp development (Datta and Das, 1996). The occurrence is in the lower part of the Tiki Formation, Gondwana Supergroup, and is believed to be Carnian in age, based on a comparison of the flora and fauna (Maleri fauna, see Jain and Roy-Chowdhury, 1987) to the Keuper. Correlations suggest that this occurrence may be slightly older than that of *Adelobasileus* from the Dockum of Texas (Lucas and Luo, 1993). If the correlations and taxonomic identity are upheld, *Gondwanadon* may prove to be the oldest identified mammal (Datta and Das, 1996).[2]

The Kota Formation, India, has yielded an important assemblage of fossil vertebrates, including several early mammals. The unit may have a wide stratigraphic range. The age of the vertebrate fauna is not well understood; fishes and palynomorphs suggest Early Jurassic, whereas ostracods are more similar to those of the Middle Jurassic (see references in Prasad and Manhas, 1997). We follow the prevailing view (e.g., Jain and Roy-Chowdhury, 1987) and tentatively accept an Early Jurassic (Liassic) age for the mammals of the Kota Formation.[3]

[2] Footprints named *Cynodontipus,* from the Early to Middle Triassic of France, show traces of hair. They have been alternately attributed to a cynodont (Ellenberger, 1976) or a mammal (Sarjeant, 2000). Given the equivocal nature of the evidence, the affinities of this ichnotaxon remain unresolved.

[3] However, a recent abstract (Prasad and Manhas, 1999) suggests that the mammal-bearing part of the unit may actually be as young as Early Cretaceous (Barremian–Aptian). Clearly, interpretation of the biogeographic and biostratigraphic significance of the Kota mammals must be approached with extreme caution and should not be based solely on comparison of isolated mammalian teeth.

TABLE 2.3. Late Triassic–Early Jurassic Mammals of India (see figure 2.5; locality numbers do not correspond). Localities: 1, Tiki (Tiki Formation; Carnian, Late Triassic); 2, Manganpalli; 3, Paikasigudem; 4, Yamanapalli (all Kota Formation, ?Early Jurassic)

Mammalia *incertae sedis* (3)
Morganucodonta
?Megazostrodontidae
Indozostrodon simpsoni (4)
?Morganucodontidae
Gondwanadon tapani (1)
Indotherium pranhitai (3)
Eutriconodonta
Family *incertae sedis*
Dyskritodon indicus (3)
?"Amphilestidae"
Paikasigudodon yadagirii (3)
Gen. et sp. indet. (3)
Archaic "symmetrodontans"
Family *incertae sedis*
Trishulotherium kotaensis (3)
?Amphidontidae
Nakunodon paikasiensis (3)
?Kuehneotheriidae
Kotatherium haldanei (2)

The Kota Formation includes limestones and sandstones, with intercalated claystone horizons, but mammals and most of the other terrestrial vertebrates have been collected from the claystones (Prasad and Manhas, 1997). A review of the vertebrate assemblage is given by Jain (1980).

Mammalian remains from the Kota are reported to include postcranial and skull elements (Datta et al., 1978), but thus far only isolated teeth have been described. Three localities have yielded mammals, all in the Pranhita-Godavari Valley, Adilabad District, Andhra Pradesh (figure 2.5). The first is near the village of **Manganpalli,** which has yielded upper molars of *Kotatherium haldanei* (see Datta, 1981). These appear to belong to some sort of primitive "symmetrodontan" and are distinct from the kuehneotheriids of the European Late Triassic–Early Jurassic. Whether or not *K. haldanei* is related to Tinodontidae (Datta, 1981; Prasad and Manhas, 1997, see chapter 9), otherwise known from the Late Jurassic–Early Cretaceous, is uncertain. A second, nearby site in the vicinity of **Yamanapalli** has yielded a single tooth referred to the megazostrodontid morganucodontan, *Indozostrodon simpsoni.* The remaining mammals of the Kota Formation are from a locality near the village of **Paikasigudem.** Most striking of these is the eutriconodont *Dyskritodon indicus,* known by two incomplete lower molars. The species shares an unusual cusp pattern (in which cusp b is strongly reduced) with *D. amaghizi* from the Berriasian of Morocco (Prasad and Manhas, 2002). *Paikasigudodon yadagirii,* based on an upper molar with a slightly triangulated cusp pattern, may represent an "amphilestid" eutriconodontan (chapter 7). *Indotherium pranhitai,* known by upper molar fragments, was initially considered to be a "symmetrodontan" (Yadagiri, 1984). Restudy by Prasad and Manhas (1997, 2002) showed that it compares more favorably with morganucodontid "triconodonts," such as *Morganucodon watsoni* and *Eozostrodon parvus.*

Trishulotherium kotaensis is known by a single, worn lower molar (Yadagiri, 1984). It has a "symmetrodontan" cusp arrangement and differs from *Kuehneotherium* in a number of ways, but little more than this can be said of the species at present (Prasad and Manhas, 1997). The final described mammal of the Kota Formation is *Nakunodon paikasiensis,* which is known from an upper molar that is almost monocuspid. This pattern is analogous to that of the Late Jurassic North American "symmetrodontan" *Amphidon,* and on this basis *Nakunodon* has been tentatively referred to the Amphidontidae (Yadagiri, 1985; Prasad and Manhas, 1997). Prasad and Manhas (1999) reported another mammal from the Paikasigudem locality. It is vaguely similar to docodontans on the one hand, to the "pseudotribosphenic" *Shuotherium* (chapter 6) on the other, and possibly represents a hitherto unknown mammalian lineage.

AFRICA

Only two Early Jurassic mammals, both morganucodontans (chapter 4), are known from Africa, but they are represented by extraordinary specimens: each is based on a nearly complete skull and skeleton from the southern part of the continent (figure 2.6). Two occurrences in Lesotho are in the Red Beds of the Stormberg Group, which is the source for a remarkable fauna of terrestrial vertebrates, including a remarkable diversity of therapsids. The Red Beds and the remainder of the Stormberg Group above the Molteno Beds are regarded as being of Early Jurassic (Liassic) age (see Clemens et al., 1979; Clemens, 1986, and references therein). The third occurrence is from the Elliot Formation, also in the Lower Jurassic part of the Stormberg Group (Olsen and Galton, 1984).

Megazostrodon rudnerae is a large, *Morganucodon*-like mammal, placed by some with Docodonta (e.g., McKenna and Bell, 1997). It differs from *Morganucodon* in its pattern of molar occlusion (Crompton and Jenkins, 1968; Crompton, 1974). *Megazostrodon* was first described from the **Pokane** locality in Lesotho. A second specimen, con-

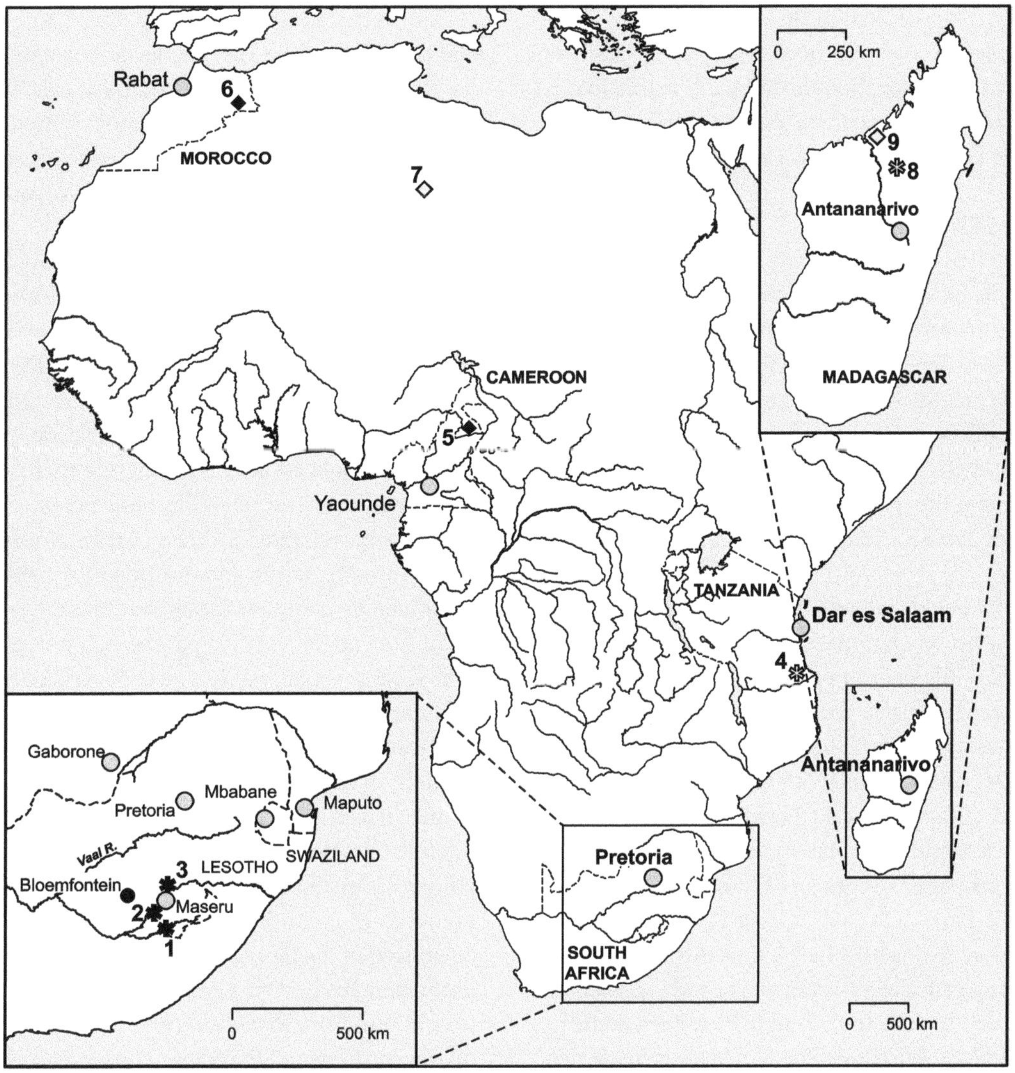

FIGURE 2.6. Mesozoic mammal sites of Africa and Madagascar. Asterisks, Early (black; localities 1–3), Middle (gray; locality 8), and Late (white; locality 4) Jurassic; diamonds, Early (black; localities 5, 6) and Late (gray; localities 7, 9) Cretaceous. Localities or local faunas: **1, Pokane; 2, Mafeteng** (Red Beds of the Stormberg Group; Early Jurassic, Lesotho); **3, Gertruida Farm** (Elliott Formation; Early Jurassic, Orange Free State, South Africa); **4, Tendaguru** (Tendaguru Beds; Kimmeridgian–Tithonian, Tanzania); **5, Koum Basin** (Grès de Gaba Member, Koum Formation; Barremian, Cameroon); **6, Synclinal d'Anoual** (Séquence B des Couches Rouges; Berriasian, Morocco); **7, Draa Ubari** (Mizdah Formation; Santonian–Campanian, Libya); **8, Ambondromamay** (Isalo level IIIb, Isalo "Group"; Bathonian, Madagascar); **9, Berivotra** (Maeverano Formation; Maastrichtian, Madagascar).

sisting of a distorted but nearly complete skull with associated vertebrae, was described by Gow (1986a), from the **Gertruida Farm** near Clocolan, Orange Free State. The second specimen is of interest in that it documents replacement of molariform teeth in Megazostrodontidae and in that it has a so-called pseudangular process on the jaw, a feature also seen in North American *Dinnetherium* (see Jenkins et al., 1983). *Erythrotherium parringtoni*, based on a specimen of a young individual (Crompton, 1964b), is from the **Mafeteng** locality. The dental and skeletal anatomy of both *Megazostrodon* and *Erythrotherium* have been described and analyzed in great detail (Crompton and Jenkins, 1968; Crompton, 1974; Jenkins and Parrington, 1976).

Trackways possibly belonging to mammals are also known from the Stormberg Group in Lesotho (see Clemens et al., 1979, and references therein). A variety of footprint types, some with possible traces of hair, were described by Ellenberger (e.g., 1972, 1974, 1975) and referred to various groups of Mesozoic mammals. As noted by Sarjeant (2000), the affinities of these ichnotaxa remain debatable.

Archaeodon reuningi, based on a supposed jaw fragment with part of a tooth from the Late Triassic of Namibia, was previously thought to be a possible mammal, but restudy of the specimen showed it to be of nonorganic origin (Hopson and Reif, 1981).

SOUTH AMERICA

No body fossils of uncontested mammals from the Late Triassic–Early Jurassic of South America have yet been found. The South American record for this period consists of tracks and trackways, some of uncertain age and some of debatable affinities. We have drawn from the remarkably complete, meticulous monograph of Leonardi (1994) in summarizing the possible mammalian ichnofossils from the Late Triassic–Early Jurassic of South America (figure 2.7).

Stipanichnus bonetti is based on tracks said to belong to either a therapsid or mammal from an unnamed Upper Triassic (Norian–Rhaetian) unit at **Los Menucos,** Río Negro Province, Argentina. *Ameghinichnus patagonicus,* from the "Lias de **Piedra Pintada,**" is from Neuquén Province, Argentina. Leonardi (1994) referred the tracks to the Jurassic; Rainforth and Lockley (1996) ascribed them to the La Matilde Formation, which they indicated to be Middle Jurassic. Assignment of these tracks to Mammalia has not been contested. Casamiquela (1964) ascribed the tracks to a "pantothere"; on the basis of an inferred galloping gait, Kielan-Jaworowska and Gambaryan (1994) suggested that the tracks might have been made by a multituberculate. Whatever the case, they provide evidence of alternating gaits and hopping in an early mammal (Rainforth and Lockley, 1996). Much larger tracks from the Newark Supergroup, eastern North America, have been ascribed to *Ameghinichnus* (see Olsen, 1980). To our knowledge, these have not yet been fully described and compared with the tracks from Argentina.

GREENLAND

A hard-won vertebrate fauna, including mammals, has been collected by Farish A. Jenkins, Jr., and associates from Upper Triassic rocks of **Jameson Land,** northeastern Greenland, at more than 71° N latitude (figure 2.8). Thus far, the assemblage includes dinosaurs and other archosaurs, turtles, labyrinthodonts, and fishes, in addition to four mammals. The fauna is very similar to that of the Norian part of the Keuper in Western Europe, sharing such taxa as *Gerrothorax, Cyclotosaurus, Proganochelys, Aetosaurus,* and *Plateosaurus;* current evidence suggests that it is at least as old as mid-Norian (Jenkins et al., 1994). The mammals are thus some of the very earliest known.

The occurrences are in the Malmros Klint and Ørsted Dal members of the Fleming Fjord Formation. This unit includes playa-mudflat systems, loess beds, sand flats, flat pebble conglomerates, and paleosols. Climate is inferred to have varied from humid to dry (with seasonal rainfall) to arid and to have been controlled by Milankovich cycles (Jenkins et al., 1994).

Haramiyavia clemmenseni (originally placed in the Haramiyidae but now referred to its own, monotypic family) is based on a partial skeleton with associated jaws and cranial elements (Jenkins et al., 1997; see also Butler, 2000). Given the fact that haramiyids and dentally similar mammals (Allotheria) of the Late Triassic–Early Jurassic are known only by isolated teeth, the find is of great significance—whatever the affinities of *Haramiyavia* might prove to be (chapter 8). The specimen unambiguously shows the orientation and position of the cheek teeth and is sufficiently complete to suggest that jaw movement was mainly orthal, rather than palinal, as it is in multituberculates (Jenkins et al., 1997, but see Butler, 2000).

The other mammals (table 2.1) known from the Fleming Fjord Formation are represented only by isolated teeth (Jenkins et al., 1994). Two incomplete molariforms testify to the presence of a rather large, unidentified "triconodont." *Brachyzostrodon,* a morganucodontan otherwise known from the Late Triassic of Saint-Nicolas-de-Port, France (Sigogneau-Russell, 1983b; Hahn et al., 1991), may be present in the fauna, though the fragmentary nature of the specimen (half of a molar) leaves positive identification in question. *Kuehneotherium,* a rare but widespread "symmetrodontan" otherwise known from both Triassic and Jurassic deposits of Western Europe (see earlier), is probably represented in the Fleming Fjord Formation as well.

NORTH AMERICA

Several previously mentioned occurrences of possible mammals in the Late Triassic of North America—specifically North Carolina and Texas (Clemens et al., 1979)—are now thought more likely to represent therapsids, as noted earlier (Sues, 2001). In the meantime, however, more secure records of mammals from both the Late Triassic and Early Jurassic have been reported from the continent. *Adelobasileus cromptoni* (see Lucas and Hunt, 1990) is known by the posterior part of a skull from **Home Creek,** Crosby County, Texas (figure 2.8, table 2.4). The occurrence is in the Tecovas Member of the Dockum Formation. Various lines of evidence, summarized by Lucas and Luo (1993), suggest a late Carnian age; hence, *Adelobasileus* may be the oldest known mammal (but see earlier comments on *Gondwanadon* from the Late Triassic of

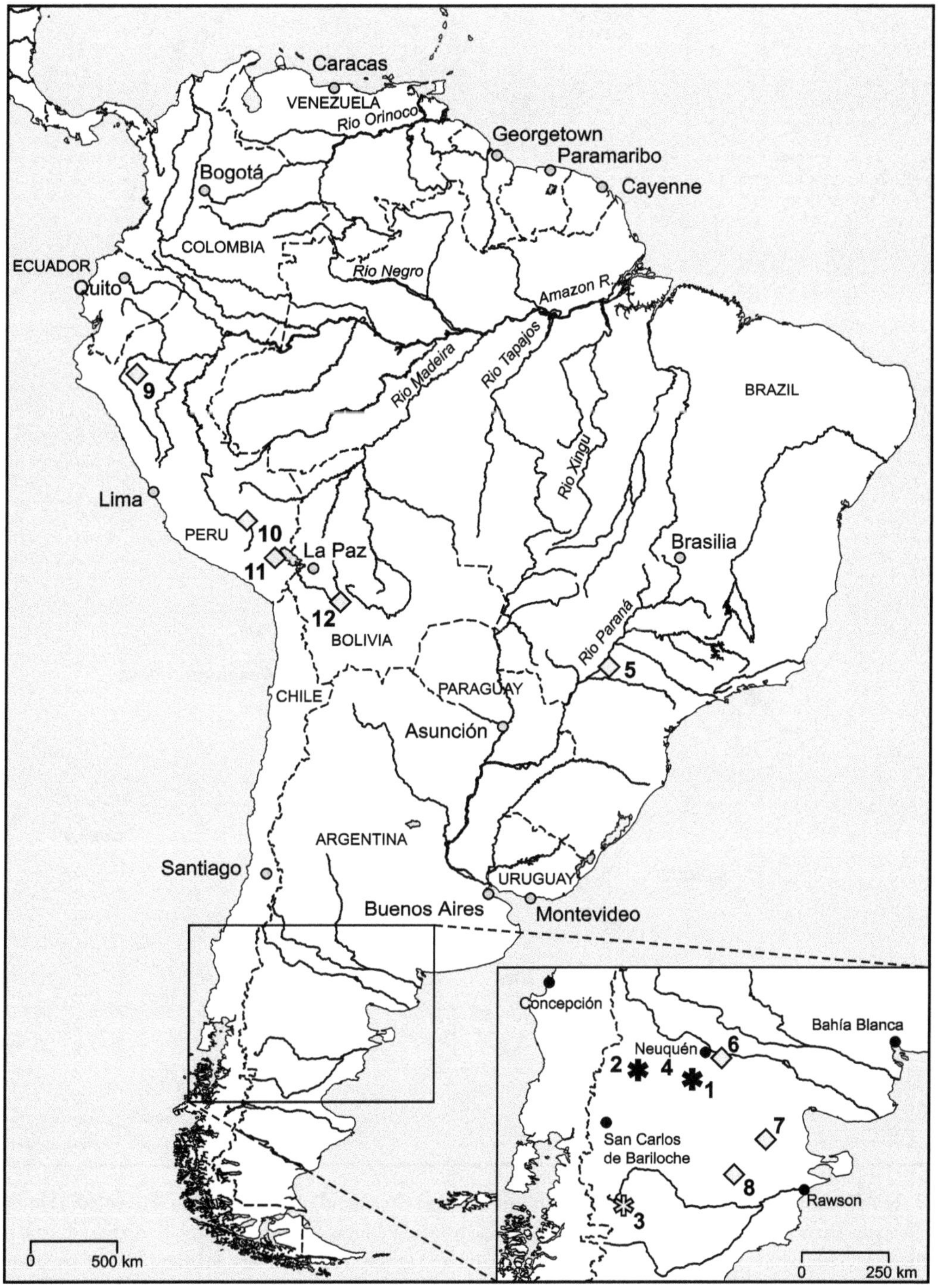

FIGURE 2.7. Mesozoic mammal localities of South America. Asterisks, Late Triassic or Jurassic (localities 1–3); diamonds, Early (white; locality 4) and Late (gray; localities 5–12) Cretaceous. Localities: **1, Los Menucos** (unnamed unit, Norian–Rhaetian, Río Negro Province, Argentina); **2, Piedra Pintada** ("Lias de Piedra Pintada," Early or Middle Jurassic, Neuquén Province, Argentina); **3, Queso Rallado** (Cañadón Asfalto Formation, Callovian–Oxfordian, Chubut Province, Argentina); **4, Zapala** (La Amarga Formation, Hauterivian–Barremian, Neuquén Province, Argentina); **5, Paraná** (Adamantina Formation, Late Cretaceous; São Paulo, Brazil); **6, Paso Córdoba** (Río Colorado Formation, Campanian or Maastrichtian, Río Negro Province, Argentina); **7, Los Alamitos** (Los Alamitos Formation, Campanian or Maastrichtian, Río Negro Province, Argentina); **8, Meseta de Somuncura** (La Colonia Formation, Campanian–Maastrichtian, Chubut Province, Argentina); **9, Fundo el Triunfo** (Fundo el Triunfo Formation, late Campanian–early Maastrichtian, Peru); **10, Paruro** (unspecified unit, Late Cretaceous; Cuzco Department, Peru); **11, Laguna Umayo** (Vilquechico Group, ?late Maastrichtian, Puno Department, Peru); **12, Pajcha Pata** (El Molino Formation, middle Maastrichtian, Bolivia).

FIGURE 2.8. Late Triassic to Early–Middle Jurassic (North America and Greenland) and Late Cretaceous (Alaska) mammal localities. Asterisks, Late Triassic–Early Jurassic (black; localities 1–3) and Early–Middle Jurassic (gray; locality 4); diamond, Late Cretaceous (locality 5). Localities or local faunas: **1, Jameson Land** (Malmros Klint and Ørsted Dal members, Fleming Fjord Formation; ?middle Norian, Greenland); **2, Home Creek** (Tecovas Member, Dockum Formation; late Carnian, Texas); **3, Gold Springs** (Kayenta Formation; Sinemurian or Pliensbachian, Arizona); **4, Huizachal Canyon** (La Boca Formation; Early or Middle Jurassic, Tamaulipas, Mexico); **5, Colville River** (Prince Creek Formation; ?late Maastrichtian: Lancian, Alaska).

India). The skull, found by bulk screenwashing of rock matrix, unfortunately lacks a dentition. *Adelobasileus* shares a number of cranial features in common with Liassic mammals; in other respects, it appears to be structurally intermediate between cynodont therapsids and mammals (Lucas and Luo, 1993, see chapter 4).

Early Jurassic mammals have been recovered from the Kayenta Formation at **Gold Springs** (figure 2.8), northern Arizona (Jenkins et al., 1983). Evidence constraining the age of the unit is sparse; it is currently considered to be of Liassic (Sinemurian to Pliensbachian) age, based on the presence of the dinosaur *Scelidosaurus* (see Padian, 1989).

TABLE 2.4. Late Triassic and Early Jurassic Mammals of North America (see figure 2.8; locality numbers do not correspond between table and figure). Localities or local faunas: 1, Home Creek, Texas (Dockum Formation, Carnian); 2, Gold Springs, Arizona (Kayenta Formation, Sinemurian–Pliensbachian); 3, Huizichal Canyon, Mexico (La Boca Formation, ?Early or Middle Jurassic)

Mammalia *incertae sedis*
Adelobasileus cromptoni (1)
Morganucodonta
Megazostrodontidae
Dinnetherium nezorum (2)
cf. *Dinnetherium* sp. (3)
Morganucodontidae
Morganucodon sp. (2)
Eutriconodonta
?"Amphilestidae"
Gen. et sp. indet. (3)
?Triconodontidae
Gen. et sp. indet. (3)

Several elements of the fauna are known at the generic level from other parts of the world. The theropod *Syntarsus*, for example, is also known from the Stormberg Group of southern Africa (Rowe, 1989); and the tritylodontid therapsid *Oligokyphus* is known from the Liassic of Britain (Sues, 1985a).

The best-represented mammal from the Kayenta Formation is the "triconodont" *Dinnetherium nezorum*, known by associated upper and lower jaws, which is similar in many respects to *Morganucodon* (Crompton and Luo, 1993); the latter is also present in the fauna. Available material, consisting of a crushed skull, several isolated teeth, and some postcranial elements, shows that the species is probably distinct from those represented in the Late Triassic–Early Jurassic of Western Europe and from the Liassic of China (Jenkins et al., 1983). The final mammal known from the Kayenta Formation is a possible haramiyid, represented by a single molariform tooth.

Several mammals are known from **Huizachal Canyon,** Tamaulipas, Mexico (figure 2.8). The occurrence is in the La Boca Formation; the fossiliferous part of the unit in Huizichal Canyon is not well constrained biochronologically and may be of Early or Middle Jurassic age. Preliminary interpretation of radiometric dates from an underlying level suggests that the vertebrate fauna is at least as old as 186 Ma, or late Early Jurassic (Fastovsky et al., 1998). A relatively recent summary of the fossils and their occurrence was given by Clark et al. (1994). The rock consists of well-indurated siltstones and sandstones, which may represent debris flows (Fastovsky et al., 1987). Fossil preservation is evidently quite variable, but the available sample includes associated skeletal elements and, in a few instances, nearly complete skulls for some of the vertebrate fauna. The assemblage includes a lepidosaur, rhynchocephalians, several archosaurs (including a well-preserved pterosaur skeleton), and rather abundant remains of the highly derived tritylodontid *Bocatherium* (see Clark and Hopson, 1985). Interestingly, all the specimens represent rather small animals.

The mammals have not yet been fully described or named. Six specimens (dentulous jaw fragments: five dentaries and one maxilla) are known to date, all with molars of a "triconodont" pattern (Montellano et al., 1998). At least three taxa are represented. One, known from a dentary fragment with several teeth, is broadly similar to *Dinnetherium* from the Early Jurassic of Arizona (Jenkins et al., 1983). It clearly has a "triconodont" molar pattern of rather primitive design, and the presence of a postdentary trough (a primitive feature, see chapter 4) suggests that postdentary elements may have been attached, as in morganucodontids. A second "triconodont" is known by two specimens, a partial dentary and a partial skull. Dental features (Fastovsky et al., 1987) suggest reference to the Triconodontidae, a family not otherwise known prior to the Late Jurassic. Moreover, unlike *Dinnetherium*, the back of the jaw lacks a postdentary trough, and several other derived features of the jaw and ear region are present. Indeed, the advanced morphology of this mammal suggests that the fauna of Huizichal Canyon may be younger than Early Jurassic (Clark et al., 1994). The final mammal is a diminutive form, known by both maxillary and mandibular fragments. It is similar, insofar as is known, to "Amphilestidae."

MIDDLE JURASSIC

The Middle Jurassic is the most poorly represented epoch in mammalian history. Most of the known record comes from Britain, where several faunules and localities have been reported. Several isolated occurrences of probable or possible Jurassic age are known from China; and the La Boca Formation, Mexico (discussed earlier), has yielded three mammals that may be Early or Middle Jurassic in age. A single specimen recently described from Madagascar is Bathonian in age. Middle Jurassic mammalian assemblages include both primitive (Morganucodontidae) and advanced ("Amphilestidae") "triconodonts," aberrant "symmetrodontans," and, notably, the first appearance of "eupantotherians"—a group that possibly gave rise to northern tribosphenic mammals. Docodonta, possibly present as early as the Late Triassic, are surely represented in Middle Jurassic assemblages of Britain. Of Allotheria, the Late Triassic–Early Jurassic haramiyids and theroteinids are no longer present (though haramiyids survived in Africa [see chapter 8] until at least the Late Jurassic); instead, the group is represented by the enigmatic eleutherodontids (known only from the Middle Jurassic) and, possibly, Multituberculata, a group that achieved prominence by the Late Jurassic and remained as an important element of terrestrial faunas through the Mesozoic, well into the Tertiary.

EUROPE

An excellent review of the Middle Jurassic microvertebrates of Britain and their geologic context was given by Evans and Milner (1994), and we have drawn from their account in summarizing mammal occurrences (table 2.5) there. During the Middle Jurassic, most of Britain was covered by an epicontinental sea. A brief regression during the Bathonian provided estuarine, lagoonal, and deltaic depositional environments suitable for the preservation of terrestrial vertebrate fossils, including mammals. For the most part, the age of these occurrences can be determined with some precision through reference to nearby marine units and their contained molluscs. The first specimens to be discovered, those from the Stonesfield Slate, were recovered *in situ* during the course of quarrying. These provide a wealth of information, but unfortunately they represent only a few species. In recent years, bulk processing and concentration techniques (Freeman, 1982; Ward, 1984; Kermack et al., 1987) have been employed on unconsolidated clays and marls. Though virtually all of the mammalian fossils recovered are isolated teeth, the result of this effort has been a far larger, more diverse sample of mammals and other microvertebrates. We review the Middle Jurassic sites of Britain (figure 2.3) in approximate stratigraphic order.

The geologically oldest occurrence of mammals in the Middle Jurassic of Britain is in the **Hornsleasow Quarry**, which is in the Chipping Norton Formation and is of early Bathonian age (figure 2.3). Microvertebrates, including unspecified mammals, have been collected from a claystone within the Chipping Norton limestone.

The first mammals to be found and described from any rocks of Meosozic age are those of the Stonesfield Slate, from the **Stonesfield** quarries (figure 2.3). Rock from these quarries is actually a calcareous sandstone, not a slate, and has long been valued for construction purposes because it splits well along bedding planes. A mammalian specimen was evidently found by a stonemason working at Stonesfield as early as 1764, but the occurrence did not receive scientific attention until decades later when Buckland (1824), encouraged by the great Georges Cuvier, mentioned them. Hitherto, mammals had not been known from the Mesozoic and the announcement caused considerable controversy (see Simpson, 1928a; Desmond, 1984). The quarries lie in the Stonesfield Member of the Sharps Hill Formation, which overlies the Chipping Norton Formation, and are of early middle Bathonian age. The fauna includes marine elements, and the depositional setting is

TABLE 2.5. Middle Jurassic Mammals of Britain (see figure 2.3; locality numbers do not correspond between figure and table). Localities: 1, Hornsleasow Quarry (Chipping Norton Limestone, early Bathonian); 2, Loch Scavaig and Elgol, Isle of Skye (Ostracod Limestone, middle Bathonian); 3, Stonesfield (Stonesfield Member of the Sharps Hill Formation, middle Bathonian); 4, Woodeaton (Hampen Marly Formation, middle Bathonian); 5, Tarlton Clay Pit; 6, Swyre; 7, Watton Cliff; 8, Kirtlington (all Forest Marble Formation, late Bathonian)

Mammalia *incertae sedis* (1, 4, 5, 6, 7)
Morganucodonta
Family *incertae sedis* (7)
Morganucodontidae
Wareolestes rex (8)
?*Wareolestes* sp. (5)
Docodonta
Family *incertae sedis*
Gen. et sp. indet. (5, 7)
Cyrtlatherium canei (8)
Docodontidae
Borealestes serendipitus (2)
Simpsonodon oxfordensis (8)
Simpsonodon sp. (7)
Shuotheridia
Shuotheriidae
Shuotherium dongi (8)
Shuotherium kermacki (8)
Shuotherium sp. (8)
Eutriconodonta
"Amphilestidae"
Amphilestes broderipii (3, 7)
Phascolotherium bucklandi (3)
Haramiyida
Eleutherodontidae
Eleutherodon oxfordensis (7, 8)
Stem Cladotheria ("eupantotherians")
Family *incertae sedis* (2, 7, 8)
?Amphitheriidae
?*Amphitherium* sp. (8)
Amphitheriidae
Amphitherium prevostii (3, 7)
Amphitherium rixoni (3)
Dryolestidae indet. (8)
Peramuridae
Palaeoxonodon ooliticus (8)
Palaeoxonodon sp. (7)

thought to have been nearshore marine or estuarine. Four mammals are known from Stonesfield, each represented by several comparatively good specimens consisting of dentulous jaws generally preserved in slabs. Two, *Amphilestes broderipii* and *Phascolotherium bucklandi*, are "amphilestid triconodonts"; the others, *Amphitherium prevostii* and *A. rixoni* (see Butler and Clemens, 2001), are "eupantotherians" that may be stem members of the lineage that ultimately gave rise to boreosphenidan mammals (chapter 11). A footprint from Stonesfield that possibly belongs to a mammal, *Pooleyichnus burfordensis*, was described by Sarjeant (1975).

Continuing stratigraphically upward, mammals have been reported from **Woodeaton,** in the "Monster Bed" of the Hampen Marly Formation (middle Bathonian). Details on the fauna are contradictory. Clemens et al. (1979) list seven taxa, including an allotherian, a morganucodontid, a docodontan, a "symmetrodontan" similar to *Kuehneotherium,* and at least three "eupantotherians." A more recent compendium (Evans and Milner, 1994) mentions only a possible mammal incisor.

Two mammal localities, **Loch Scavaig** and **Elgol,** are known from the Isle of Skye, western Scotland (figure 2.3). They are in the Ostracod Limestone, also considered to be middle Bathonian. Two taxa, the docodontan *Borealestes serendipitus* and an unidentified "eupantotherian," have been reported from the Isle of Skye (Waldman and Savage, 1972; Savage, 1984).

By far the greatest diversity of Middle Jurassic mammals comes from the Forest Marble, which is upper Bathonian and thus represents the youngest in Britain's sequence of Middle Jurassic assemblages. To date, four localities in the Forest Marble have yielded mammals: **Swyre, Tarlton Clay Pit, Watton Cliff,** and **Kirtlington** (figure 2.3). The paleoenvironments of Kirtlington and Tarlton Clay Pit are believed to be similar and to represent a swampy inshore brackish or freshwater lagoon, perhaps similar to the Florida Everglades (Evans and Milner, 1994). E. F. Freeman (1979) suggested that the bone accumulation at Kirtlington may have been the result of predator activity, through "coprocoenosis"—concentration in feces or regurgitata (see also Mellett, 1974). Watton Cliff and Swyre are believed to have had a higher-energy depositional environment. Though the fossils are often abraded, the fauna is similar in known respects to that of the other mammal-bearing sites in the Forest Marble. Swyre has yielded one unidentified mammal; three taxa are known from Tarlton Clay Pit, including a morganucodontid and a docodontan. The assemblages of Watton Cliff and, especially, Kirtlington are much more diverse. Two taxa, the aforementioned eutriconodontan *Amphilestes* and the "eupantotherian" *Amphitherium* (both also known from the Stonesfield Slate), are present at Watton Cliff but not Kirtlington (but see E. F. Freeman, 1979). Otherwise, the fauna of Kirtlington is considerably better sampled and thus deserves special mention.

Microvertebrates were discovered in the Old Cement Works Quarry at Kirtlington by Eric Freeman in the 1970s (E. F. Freeman, 1976a,b, 1979). The productive horizon, termed the Mammal Bed, is a thin marly clay that is astonishingly fossiliferous. The quarry was later worked by Kermack and associates; the total collection includes some 700 mammal teeth, together with skeletal elements (Kermack et al., 1998). Some of the information has been published, but much of the mammalian fauna remains undescribed at this writing. The collection is currently under study by Denise Sigogneau-Russell.

Whereas the eutriconodontans of the Stonesfield Slate (*Amphilestes* and *Phascolotherium*) are advanced, particularly with respect to their jaw structure, the presence of *Wareolestes rex* at Kirtlington documents the survival of morganucodontids at least into the Bathonian (E. F. Freeman, 1979). "Amphilestids," curiously, have not yet been reported from Kirtlington. Early reports (e.g., E. F. Freeman, 1979) suggested the presence of true multituberculates in the Forest Marble, based on fragmentary teeth. The occurrence is notable because undoubted multituberculates do not otherwise appear in the fossil record until the Late Jurassic. Kermack et al. (1998) referred all allotherian teeth (13 specimens) from Kirtlington and Watton Cliff to *Eleutherodon oxfordensis.* The authors referred *Eleutherodon* to a new higher category of Allotheria, Eleutherodontida (see chapter 8). However, Butler (2000) reportedly has identified a few teeth from Kirtlington that he considers referable to the Multituberculata. On this basis, we tentatively retain multituberculates in the faunal list for the Forest Marble. The docodontans *Cyrtlatherium canei* (originally described as a kuehneotheriid "symmetrodontan," see E. F. Freeman, 1979; Sigogneau-Russell, 2001) and *Simpsonodon oxfordensis* (see Kermack et al., 1987) have also been described from Kirtlington; *Simpsonodon* is also known from Watton Cliff. Perhaps the most remarkable mammalian occurrence at Kirtlington is that of *Shuotherium,* represented by as many as three species (Sigogneau-Russell, 1998). *Shuotherium* is characterized by lower molars that have a "talonid" developed anterior to the trigonid and upper molars with a cusp pattern similar to (but evidently evolved separately from) that of tribosphenic mammals (Sigogneau-Russell, 1998; Wang, Clemens, et al., 1998; Kielan-Jaworowska et al., 2002, see chapter 6). The genus is otherwise known only from the Late Jurassic of China (Chow and Rich, 1982); one of the taxa from Kirtlington is so similar to the Chinese form that it is placed in the same species, *S. dongi* (see

Sigogneau-Russell, 1998). "Eupantotherians" of Kirtlington are noteworthy for their diversity: five or more taxa, not all yet described, may be represented (E. F. Freeman, 1979; Evans and Milner, 1994), as compared to the single species, *Amphitherium prevostii*, known from Stonesfield. *Palaeoxonodon oolithicus* is referred to the "eupantotherian" family Peramuridae, a group that is widely believed to be closely related to northern tribosphenic mammals (Kraus, 1979). Its occurrence at Kirtlington is the oldest for the family. Another oldest record for the site is that of Dryolestidae, represented by an unnamed taxon. Dryolestids later became abundant, diverse elements of the mammalian faunas of the Late Jurassic and earliest Cretaceous (Simpson, 1928a, 1929a).

A final occurrence of a Middle Jurassic (late Bathonian) mammal from this region is that of an isolated femur from a fissure-fill deposit at the **Peski Quarry,** Kolomensk District, Russia, about 100 km southeast of Moscow (figure 2.11). The specimen is well preserved and was referred to Morganucodontidae by Gambaryan and Averianov (2001). Given the poor state of knowledge of the skeleton in most groups of stem mammals, we follow the referral with question.

ASIA

Several sites in Asia (figure 2.4, table 2.6) have yielded single specimens of mammals that may be Jurassic, perhaps Middle Jurassic, in age. One of these is **Luchang** (Luzhang), in Sichuan Province, China, where a fragmentary tooth with two roots has been reported (Chow and Rich, 1984b). The occurrence is in the lower Yimen Formation and is of Early Jurassic age. A second occurrence is that of *Manchurodon simplicidens,* from the **Zha-zy-ao Mine,** Liaoning Province. The species is based on a well-preserved dentary and a scapula that may be associated (Yabe and Shikama, 1938); unfortunately, the specimen appears to be lost. There is also confusion as to the age of this occurrence. Yabe and Shikama (1938) described the fossil as having come from the Husin Series, which they believed to be Jurassic in age on the basis of plant fossils. As discussed later, the Husin (Fuxin) is generally regarded as Lower Cretaceous and, indeed, a number of noteworthy mammals of this age occur in western Liaoning (Wang et al., 1995, Wang, Hu, Li, et al., 2001). However, the occurrence of *Manchurodon* is now generally believed to be in the Wa-fang-dian Formation, which is reported to be Jurassic, perhaps Middle or Upper Jurassic (Zhang, 1984; Zhou et al., 1991). In light of the new dates for the Lower Cretaceous of Liaoning (Swisher et al., 1999) and the possibility that the age for much of Mesozoic sequence may be revised upward, this estimate may be too old.

TABLE 2.6. Middle Jurassic (?) Mammals of Asia and European part of Russia (see figures 2.3, 2.11; locality numbers do not correspond between table and figures). Localities or local faunas: 1, Luzhang, China (lower Yimen Formation, Early Jurassic); 2, Zha-zy-ao Mine, China (Wa-fang-dian Formation, ?Middle or Late Jurassic); 3, Fangshen, China (Haifanggou Formation, ?Middle Jurassic); 4, Kalmakerchin; 5, Tashkumyr (both in the Balabansay Formation; Callovian, Kirghizia); 6, Peski Quarry, Russia (fissure fillings, Bathonian, Middle Jurassic)

:Mammalia *incertae sedis*
Family *incertae sedis*
Gen. et sp. indet. (1)
?Morganucodonta
?Morganucodontidae
Gen. et sp. indet. (6)
?Docodonta
Family *incertae sedis*
Gen. et sp. indet. (4)
Docodonta
Docodontidae
Gen. et sp. indet. (5)
Archaic "symmetrodontans"
?Amphidontidae
Manchurodon simplicidens (2)
Eutriconodonta
"Amphilestidae"
Liaotherium gracile (3)

Liaotherium gracile represents another Jurassic, perhaps Middle Jurassic, mammal from China. From the **Fangshen** locality, Haifanggou Formation (= Jiulongshan Formation), Lingyuan County, Liaoning Province, the specimen consists of a dentary with part of the last molar; *Liaotherium* is tentatively referred to the eutriconodontan family "Amphilestidae" (see Zhou et al., 1991, and chapter 7).

The final mammal occurrences in the Middle Jurassic of Asia are from the Balabansay Formation, Fergana Depression, Kirghizia. The site of **Kalmakerchin** has yielded a possible docodontan, based on a badly broken upper molar and an ulna (Nessov, Kielan-Jaworowska, et al., 1994). Also known from the site are salamanders of the family Karauiidae, known only from the Jurassic; an overlying unit has yielded leaves reported to be of Middle Jurassic age (Averianov, 2000). A second, nearby site, **Tashkumyr,** has yielded a lower molar referred to Docodonta, as a possible relative of *Tegotherium* Tatarinov, 1994, from the Late Jurassic of Mongolia (see chapter 5 and Martin and Averianov, 2001).

MADAGASCAR

A single mammal, *Ambondro mahabo*, has recently been described (Flynn et al., 1999) from the **Ambondromamy** Region of the Mahajanga Basin, northwestern Madagascar (figure 2.6). Stratigraphically, the occurrence is near the top of Isalo level IIIb in the Isalo "Group," and is placed in the Bathonian on the basis of a rich invertebrate fauna from this unit. *A. mahabo* is known by a jaw fragment with three teeth, of which the molars have a remarkably advanced, basined heel or talonid, indicating that, morphologically at least, they are of tribosphenic grade. Other features of this extraordinary animal are unlike what is encountered among "typical" early tribosphenic mammals, and we recognize it as a member of an endemic Gondwanan clade (Australosphenida) that independently acquired a complex heel on lower molars (Luo, Cifelli, and Kielan-Jaworowska, 2001, see chapter 6).

SOUTH AMERICA

One Jurassic mammal has been described from the South American continent: *Asfaltomylos patagonicus*, represented by a dentary with much of the postcanine dentition (Rauhut et al., 2002). *Asfaltomylos* has a tribosphenic molar pattern and peculiar specializations of the premolars and molars; it has been referred to Australosphenida, an endemic Gondwanan group that includes monotremes (chapter 6). The specimen was collected from the Cañadón Asfalto Formation at **Queso Rallado,** Chubut, Argentina. The occurrence is regarded as being of late Middle to early Late Jurassic (Callovian–Oxfordian) age.

LATE JURASSIC

Pangaea had begun to break up by the Late Jurassic, but Gondwana remained essentially intact. Some communication with northern faunas remained, however, as similar genera of dinosaurs are found in Late Jurassic rocks of North America and Africa. Asia was separated from Western Europe and shows a certain degree of endemism in its terrestrial fauna (Russell, 1993). Late Jurassic mammals are known from Western Europe, Asia, Africa, and North America. As previously mentioned, multituberculates are diverse in Late Jurassic faunas. Most characteristic are the diverse suites of "plagiaulacidan" multituberculates and "eupantotherians"; docodontans, "symmetrodontans," and "triconodonts" are also present, but are generally less abundant and not so diverse.

WESTERN EUROPE

The mammals of the Purbeck Limestone, England, have traditionally been considered to be of Late Jurassic age and, in fact, they are rather similar to those of the Upper Jurassic Morrison Formation, North America (Simpson, 1928a, 1929a). Recent work has shown that the Purbeck and its contained mammalian fauna is probably Berriasian, or Lower Cretaceous (Allen and Wimbledon, 1991), and it will be so regarded here (see the excellent discussion in Clemens et al., 1979). In the meantime, however, another site in Britain, **Chicksgrove,** Wiltshire (figure 2.3), discovered by Bill Wimbledon, has yielded mammals of Late Jurassic age. Chicksgrove lies in the Portland Stone Formation, which is of Portlandian (Tithonian) age. The only published information on the locality indicates that it includes ?"triconodont," multituberculate, and "eupantotherian" remains (Kermack, 1988).

The major Late Jurassic mammal site in Western Europe is the **Guimarota Coal Mine** in Leiria, Portugal (figure 2.9). The unit is unnamed and is referred to simply as the Guimarota Beds (T. Martin, pers. comm.). The mammalian fauna is probably somewhat older than that of the Morrison Formation, United States (Martin, 1995). Mohr (1989) reported an Oxfordian age for the site, based on palynomorphs. Schudack (1993) indicated that charophytes from Guimarota could be of either Kimmeridgian or Oxfordian age and favored the former correlation because it is suggested by ostracods, which are considered more reliable. Finally, Schudack (2000) provided evidence for a Kimmeridgian age of the Guimarota Beds. We follow this assessment here. There is some marine influence; the coals of Guimarota are thought to have been deposited in a swampy upper coastal plain environment (Erve and Mohr, 1988; Mohr, 1989). A recent summary of Guimarota's fossils and their geological context, together with a comprehensive bibliography for the site, is given in a volume edited by Martin and Krebs (2000).

Paleontological investigations at Guimarota were begun by Walter Kühne in the 1960s (Kühne, 1961a,b, 1968) and continued by Siegfreid Henkel, Bernard Krebs, and others. Most of the operations were conducted at considerable expense and effort because commercial mining had ceased and they had to pump the mine out and partly rebuild it (Krebs, 1980). Enormous volumes of coal were quarried out, resulting in the recovery of more than 700 mammal jaws, partial multituberculate skulls, a docodontan skull, and three skeletons, one of which (belonging to the "eupantotherian" *Henkelotherium*) has been described. These specimens are badly flattened, but they have provided a wealth of morphological information.

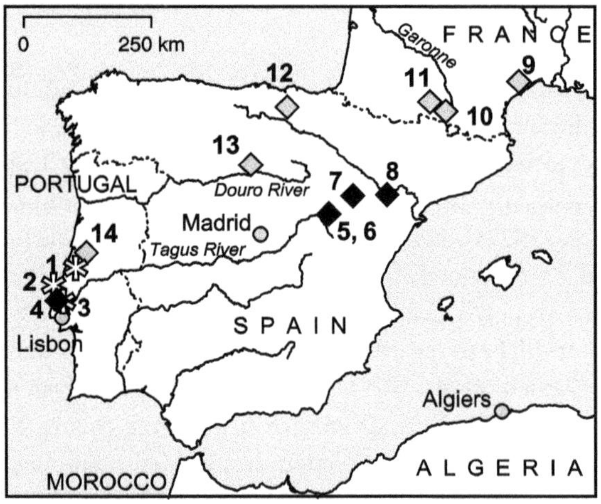

FIGURE 2.9. Mesozoic mammal localities of southern France and the Iberian Peninsula. Asterisks, Jurassic (localities 1–3); diamonds, Early (black; localities 4–8) and Late (gray; localities 9–14) Cretaceous. Localities or local faunas: **1, Guimarota** (Guimarota Beds; Kimmeridgian, Portugal); **2, Pai Mogo; 3, Porto das Barcas** (Lourinhã Formation; late Kimmeridgian or Tithonian, Portugal); **4, Porto Pinheiro** (Lourinhã Formation; ?Berriasian, Portugal); **5, Uña; 6, Pié Pajarón** (Camarillas Formation; early Barremian, Cuena Province, Spain); **7, Galve** (Camarillas Formation; early Barremian, Teruel Province, Spain); **8, Vallipón** (Artoles Formation; late Barremian, Teruel Province, Spain); **9, Champ-Garimond** (Campanian, Gard, France); **10, Barrage Filleit** (Grès de Labarre Formation; late Campanian or early Maastrichtian, Ariège, France); **11, Peyrecave** (Marnes d'Auzas Formation; late Maastrichtian, Haute Garonne, France); **12, Laño** (unspecified unit; early Maastrichtian, Basque Country, Spain); **13, Quintanilla del Coco** (Calizas de Lynchus Formation; Maastrichtian, Burgos Province, Spain); **14, Taveiro** (unspecified unit; ?Maastrichtian, Portugal).

Follow-up investigations using screenwashing techniques yielded only one taxon not known by jaws, suggesting that the fauna has been rather completely sampled (Martin, 1999b).

The mammalian fauna of Guimarota (table 2.7) is partly an endemic one. Notably, it lacks triconodontids and "symmetrodontans" (Spalacotheriidae and Tinodontidae), which are reasonably common in both the Purbeck (lowest Cretaceous of Europe) and Morrison (Upper Jurassic of North America). One of the most notable aspects of the fauna from Guimarota is its remarkably diverse, well-represented assemblage of multituberculates, described in a meticulously detailed series of publications by Gerhard and Renate Hahn (e.g., Hahn, 1969, 1971, 1977a,b, 1978a,b, 1981, 1985, 1987b, 1988, 2001; Hahn and Hahn, 1994, 1998a,b,c,d). They described 21 taxa and left another 12 in open nomenclature. The list is surely somewhat inflated because some unidentified taxa probably belong to named species and also because, in some cases, upper and lower dentitions described under separate names may belong together (see chapter 8). Nonetheless, it is clearly a diverse multituberculate assemblage; all are placed in the "Plagiaulacida," a primitive, probably paraphyletic assemblage known mainly from the Late Jurassic through earliest Cretaceous.

One species, *Proalbionbaatar plagiocyrtus,* is referred to the Albionbaataridae, which are otherwise known only from the Purbeck (Kielan-Jaworowska and Ensom, 1994) and, possibly, the Early Cretaceous locality of Badaohao in Liaoning, China (Wang et al., 1995). The remaining multituberculates of Guimarota are placed in the Paulchoffatiidae, a group known from the Berriasian of England (Kielan-Jaworowska and Ensom, 1992); the Early Cretaceous of Galve and Uña, Spain (e.g., Hahn and Hahn, 1992); and, with grave doubt, the Late Triassic of Belgium (Hahn, 1987a; Hahn et al., 1989). With the exception of this last record, and for the possible presence of Albionbaataridae in China, the multituberculates show affinities only with those of the earliest Cretaceous of England, suggesting that the Iberian Peninsula had at least an ephemeral connection with this region in the Late Jurassic or Early Cretaceous.

The docodontan *Haldanodon exspectatus,* originally based on dentulous jaw fragments (Kühne and Krusat, 1972), is now known by a complete skull (Krusat 1973, 1980). Detailed study of this specimen shows that docodontans are remarkably primitive in a number of cranial features, suggesting that they occupy a remote phylogenetic position among mammals (Lillegraven and Krusat, 1991, see chapter 5).

Most of the "eupantotherians" of Guimarota were poorly known until recently. In a preliminary paper, Kühne (1968) mentioned and in some cases figured taxa (see Clemens et al., 1979), but did not formally describe them, creating something of a nomenclatorial time bomb. "*Butlerigale* sp." was later referred to under the name *Dryolestes* sp. nov. (Martin, 1995) and formally designated *D. leiriensis* (see Martin, 1999a). An ontogenetic series representing this species allowed Martin (1997) to reconstruct the tooth replacement pattern, the first ever documented for a "eupantotherian" (see also Butler and Krebs, 1973). "*Simpsonodon splendens*" of Kühne (1968) was later named as *Henkelotherium guimarotae* by Krebs (1991); in the interim, the generic name *Simpsonodon* was used for a docodontan from the Middle Jurassic of Britain (Kermack et al., 1987). *Henkelotherium* is known by a rather complete skeleton that is said to show arboreal adaptations (Krebs, 1987, 1991). Finally, Kühne's (1968) "*Guimarota frey*" is based on upper molars probably referable (T. Martin, pers.

TABLE 2.7. Jurassic Mammals of Portugal (see figure 2.9). Localities: 1, Guimarota Coal Mine (Guimarota Beds, Kimmeridgian); 2, Pai Mogo; 3, Porto Das Barcas (both Lourinhã Formation; late Kimmeridgian or Tithonian)

- Mammalia *incertae sedis* (3)
- Docodonta
  - Docodontidae
    - *Haldanodon exspectatus* (1)
- Multituberculata
  - Albionbaataridae
    - *Proalbionbaatar plagiocyrtus* (1)
  - Paulchoffatiidae
    - *Bathmochoffatia hapax* (1)
    - *Guimarotodon leiriensis* (1)
    - *Henkelodon naias* (1)
    - *Kielanodon hopsoni* (1)
    - *Kuehneodon barcasensis* (3)
    - *Kuehneodon dietrichi* (1)
    - *Kuehneodon dryas* (1)
    - *Kuehneodon guimarotoensis* (1)
    - *Kuehneodon hahni* (2)
    - *Kuehneodon simpsoni* (1)
    - *Kuehneodon uniradiculatus* (1)
    - *Kuehneodon* sp. (1)
    - *Meketibolodon robustus* (1)
    - *Meketichoffatia krausei* (1)
    - *Meketichoffatia* sp. (1)
    - *Paulchoffatia delgadoi* (1)
    - *Paulchoffatia* sp. (1)
    - *Plesiochoffatia peperethos* (1)
    - *Plesiochoffatia staphylos* (1)
    - *Plesiochoffatia thoas* (1)
    - *Pseudobolodon krebsi* (1)
    - *Pseudobolodon oreas* (1)
    - *Pseudobolodon* sp. nov. (1)
    - ?*Pseudobolodon* sp. indet. (1)
    - *Renatodon amalthea* (1)
    - *Xenachoffatia oinopion* (1)
    - Gen. A, sp. indet. (1)
    - Gen. B, sp. indet. (1)
    - Gen. C, sp. indet. (1)
    - Gen. D, sp. 1 (1)
    - Gen. D, sp. 2 (1)
    - Gen. D, sp. 3 (1)
    - Gen. D, sp. 4 (1)
- Stem Cladotheria ("eupantotherians")
  - Dryolestidae
    - *Dryolestes leiriensis* (1)
    - *Guimarotodus inflatus* (1)
    - *Krebsotherium lusitanicum* (1)
  - Paurodontidae
    - *Drescheratherium acutum* (1)
    - *Henkelotherium guimarotae* (1)
  - Stem-lineages of Zatheria
    - *Nanolestes drescherae* (1)

comm.) to a species later named by Martin (1999a) as *Krebsotherium lusitanicum.* The "eupantotherian" tentatively designated as cf. *Peramus* by Kühne (1968; Martin, 1999b) has been named *Nanolestes drescherae;* the genus is also known from Porto Pinheiro (Martin, 2002). Other "eupantotherians" of Guimarota are *Guimarotodus inflatus* (see Martin, 1999a) and *Drescheratherium acutum* (see Krebs, 1998). *Dryolestes* and *Krebsotherium* are Dryolestidae; *Dryolestes* itself is known from the Late Jurassic of North America. *Henkelotherium* and *Drescheratherium* were referred to an endemic family Henkelotheriidae by Krebs (1991, 1998). Other workers (e.g., McKenna and Bell, 1997) referred them to the Paurodontidae, a "eupantotherian" group considered to be closely related to boreosphenidan mammals (Prothero, 1981, see chapter 10). Paurodontids are also known from the Purbeck and Morrison.

Unidentified mammals were reported some time ago (see Clemens et al., 1979, and references cited therein) from the Lourinhã Formation at **Porto das Barcas,** close to Porto Pinheiro (mentioned later). The multituberculate *Kuehneodon barcasensis* has recently been described on the basis of two teeth (one tentatively referred) from Porto das Barcas. The species is similar to *K. uniradiculatus* from Guiomarota, suggesting that the two sites may be of equivalent age (Hahn and Hahn, 2001a).

The final Jurassic mammal locality worth noting in Portugal is **Pai Mogo,** which is on the coast about 6 km north-northwest of Lourinhã and also in the Lourinhã Formation. The site is notable in that it has yielded eggs and small bones attributed to theropod dinosaurs; a single mammal jaw with two teeth is referred to its own species, *Kuehneodon hahni,* by Antunes (1998). The occurrence is near the top of the Porto Novo Member of the Lourinhã Formation, dated by ostracods as late Kimmeridgian or Tithonian. *Kuehneodon,* just mentioned, is otherwise known from the geologically older sites of Guimarota and Porto das Barcas.

ASIA

Omitting some mammals described earlier in the section on Middle Jurassic, there are three possible occurrences of Late Jurassic mammals in Asia. One of these is at **Shar Teeg,** Mongolia (figure 2.12). An account of this locality, its setting, and fauna is given by Gubin and Sinitza (1996).

The fauna and flora are quite diverse, including a large assemblage of insects. Vertebrates include fishes, labyrinthodonts, turtles, crocodilians, and dinosaurs, as well as one mammal. Overall, the biota most resembles that of the Late Jurassic, with some suggestion of an affinity to Chinese faunas of that age. The strata are predominantly of lacustrine origin. The mammal from Shar Teeg is *Tegotherium gubini,* thought by its describer to represent a new higher category of therian mammals (Tatarinov, 1994), but considered by others to be a docodontan (e.g., Hopson, 1995; Kielan-Jaworowska et al. 2000; see also chapter 5).

The other two Late Jurassic mammal sites of Asia are both in China. **Shilongzhai** is in Sichuan Province, in the Shaximiao Formation. The age of the unit is not settled. An early report (Chow and Rich, 1982) indicated that it is Callovian, or Middle Jurassic; in rough agreement, Lucas (1996) placed the Shaximiao Formation within his "Tuojiangian" vertebrate faunachron, which he considers to be of late Middle Jurassic age (Bathonian–Callovian). However, another recent report considers the unit to be Upper Jurassic, probably Oxfordian or Kimmeridgian (Wang, Clemens, et al., 1998).

The first mammal to be described from Shilongzhai is *Shuotherium dongi,* which documents a previously unknown group of mammals (see chapter 6). The primary three lower molar cusps are arranged in the typical triangular pattern characteristic of therians, but the heel is developed on the front of the tooth, rather than the back as in tribosphenic mammals (Chow and Rich, 1982). Later discoveries of shuotheres include another species from Shilongzhai, *S. shilongi,* as many as three species of *Shuotherium* from the Jurassic of England, and upper molars referable to that genus (Sigogneau-Russell, 1998; Wang, Clemens, et al., 1998). The British occurrence is well established as late Bathonian and, given the apparent presence there of one of the same species (*Shuotherium dongi*) as at Shilongzhai, it may provide support for a Middle Jurassic correlation for the Shaximiao Formation. A mammal reported from the Kota Formation, India, by Prasad and Manhas (1999), as "shuothere-like" may belong to Docodonta (G. V. R. Prasad, pers. comm.).

The other possible Late Jurassic mammal site in China is **Jianshan Wash,** in Xinjiang Province (figure 2.4). The site is in the Shishuguo Formation, which was first (Chow and Rich, 1984a) reported to be Middle or Upper Jurassic and later (Dong, 1993) referred to the Upper Jurassic. The mammal described from Jianshan Wash is *Klamelia zhaopengi,* an "amphilestid triconodont" said to be rather similar to *Gobiconodon* of the Early Cretaceous (see Kielan-Jaworowska and Dashzeveg, 1998, and comments under Early Cretaceous, later).

AFRICA

Only one site in Africa, **Tendaguru,** in eastern Tanzania (figure 2.6) has thus far yielded Late Jurassic mammals. Tendaguru was worked for its dinosaurs, most notably by a German team between 1909 and 1913, and yielded *Brachiosaurus* and *Barosaurus* (both also found in the Morrison Formation of the western United States), among others. In addition to individual skeletal elements, the expedition brought back rock matrix containing dinosaur bones—a fortunate circumstance as the rock proved to contain fossils of mammals and other small vertebrates in addition to the specimens then considered of primary interest. The Tendaguru beds include three units that have yielded dinosaurs, the lower, middle, and upper saurian beds (Janensch, 1914), which are considered to represent brackish to limnic deposits and to be Kimmeridgian to Tithonian, Late Jurassic (Russell et al., 1980; Heinrich, 1998). A review of the geology and paleontology of Tendaguru is given by Heinrich et al. (2001).

The first mammal to be reported from Tendaguru is *Brancatherulum tendagurense,* from the upper saurian bed (Heinrich, 1998), described by Dietrich (1927) and later discussed by Simpson (1928e) and Heinrich (1991). *Brancatherulum* has been referred to the "eupantotherian" family Peramuridae (see discussion by Kraus, 1979), which are thought to be close relatives of northern tribosphenic mammals (chapter 10). Given that the only known specimen is an edentulous dentary, the affinities of *Brancatherulum* are uncertain.

Additional mammals from Tendaguru have recently been described by Heinrich (1998), based on specimens from the middle saurian bed. *Tendagurodon janenschi,* known from a lower molariform tooth, is a "triconodont" tentatively considered here as a member of "Amphilestidae" (chapter 7). *Tendagurutherium dietrichi* is represented by the posterior part of a dentary with most of the last tooth intact; it is tentatively referred to the Peramuridae (see Heinrich, 1998), which are otherwise known from the Early Cretaceous of Africa and Britain, the Middle Jurassic of Britain, and the Late Jurassic of Portugal. *Staffia aenigmatica* is known by three teeth—a lower premolar, a lower molar, and an upper molar (Heinrich, 2001). This taxon is significant in that it is referred to the Haramiyidae (see chapter 8), which are otherwise known only from the Late Triassic–Early Jurassic of northern landmasses (Heinrich, 1999). Presence of this group in the Jurassic of Africa provides another possible source for enigmatic possibly allotherian groups of the southern continents, such as gondwanatherians.

NORTH AMERICA

North America's Late Jurassic mammal assemblage (table 2.8) is the most diverse of this age in the world. All the specimens come from one unit, the Morrison Formation, and the vast majority of published specimens come from one site at Como Bluff, Wyoming (Simpson, 1926c). Notwithstanding some reports to the contrary (e.g., Bilbey, 1998), an overwhelming body of radiometric, paleomagnetic, palynological, and other data indicate the Morrison Formation to be an exclusively Upper Jurassic unit, ranging from the uppermost Oxfordian or lowest Kimmeridgian to the lowest Tithonian (e.g., Kowallis et al., 1998; Litwin et al., 1998; Steiner, 1998). The Morrison Formation is geographically widespread, extending from Arizona and New Mexico northward to Montana, and reaches its easternmost extent in the Oklahoma panhandle. The unit includes a variety of mainly terrestrial deposits, predominantly of fluvial and lacustrine origin, but also contains evaporites, paleosols, and eolian sandstones. The paleoclimate is interpreted to have been semiarid, with seasonal rainfall and subtropical westerly winds (Demko and Parrish, 1998).

Without doubt, the most important mammal locality in the Morrison Formation is **Como Quarry 9,** Como Bluff, southern Wyoming (figure 2.10). Work at Como Bluff began in 1877 by field parties under the direction of Othniel C. Marsh, following discovery of dinosaurs by William H. Reed. The recovery of a single mammal specimen (later described as the "eupantotherian" *Dryolestes*

TABLE 2.8. Mammalian Fauna of the Morrison Formation (Late Jurassic, United States; figure 2.10). Localities: 1, Como Quarry 9; 2, Chuck's Prospect; 3, Delta T; 4, Breakfast Bench; 5, Garden Park; 6, Sundance; 7, Fruita; 8, Dry Mesa; 9, Dinosaur National Monument; 10, unspecified Como site; 11, Small Quarry; 12, Ninemile Hill

- Mammalia *incertae sedis* (2, 8)
- Docodonta
  - Docodontidae
    - *Docodon victor* (1, 6, 12)
    - *Docodon* sp. (5, 11)
- Eutriconodonta
  - "Amphilestidae"
    - *Aploconodon comoensis* (1)
    - *Comodon gidleyi* (1)
    - *Triconolestes curvicuspis* (9)
  - Triconodontidae
    - *Priacodon ferox* (1)
    - *Priacodon grandaevus* (1)
    - *Priacodon lulli* (1)
    - *Priacodon robustus* (1)
    - *Priacodon fruitaensis* (7)
    - *Trioracodon bisulcus* (1)
    - Gen. et sp. indet. (9)
- Multituberculata
  - Allodontidae
    - *Ctenacodon laticeps* (1)
    - *Ctenacodon scindens* (1)
    - *Ctenacodon serratus* (1)
    - *Ctenacodon* sp. (9, 12)
    - *Psalodon fortis* (12)
    - *Psalodon marshi* (1)
    - ?*Psalodon marshi* (6)
    - *Psalodon potens* (1)
    - ?*Psalodon* sp. (9)
    - Gen. et sp. indet. (2, 3)
  - ?Plagiaulacidae
    - "*Ctenacodon*" *brentbaatar* (4)
  - Zofiabaataridae
    - *Zofiabaatar pulcher* (4)
  - Family *incertae sedis*
    - *Glirodon grandis* (7, 9)
- Archaic "symmetrodontans"
  - Amphidontidae
    - *Amphidon superstes* (1)
  - Tinodontidae
    - *Tinodon bellus* (1)
    - *Tinodon* sp. (9)
- Stem Cladotheria ("eupantotherians")
  - Dryolestidae
    - *Amblotherium gracile* (1, 9)
    - *Amblotherium minimum* (1)
    - *Amblotherium* sp. (= *Kepolestes coloradensis*) (5)
    - *Amblotherium* sp. (2, 9)
    - *Dryolestes priscus* (1, 2, 10)
    - *Dryolestes* cf. *priscus* (12)
    - *Laolestes eminens* (1)
    - *Laolestes* sp. (12)
    - *Laolestes goodrichi* (1)
    - *Laolestes oweni* (1)
    - Gen. and sp. indet. (2, 9, 11)
  - Paurodontidae
    - *Araeodon intermissus* (1, 9)
    - *Archaeotrigon brevimaxillus* (1)
    - *Archaeotrigon distagmus* (1)
    - *Comotherium richi* (1)
    - *Euthlastus cordiformis* (1)
    - *Foxraptor atrox* (4)
    - *Paurodon valens* (1)
    - *Tathiodon agilis* (1)

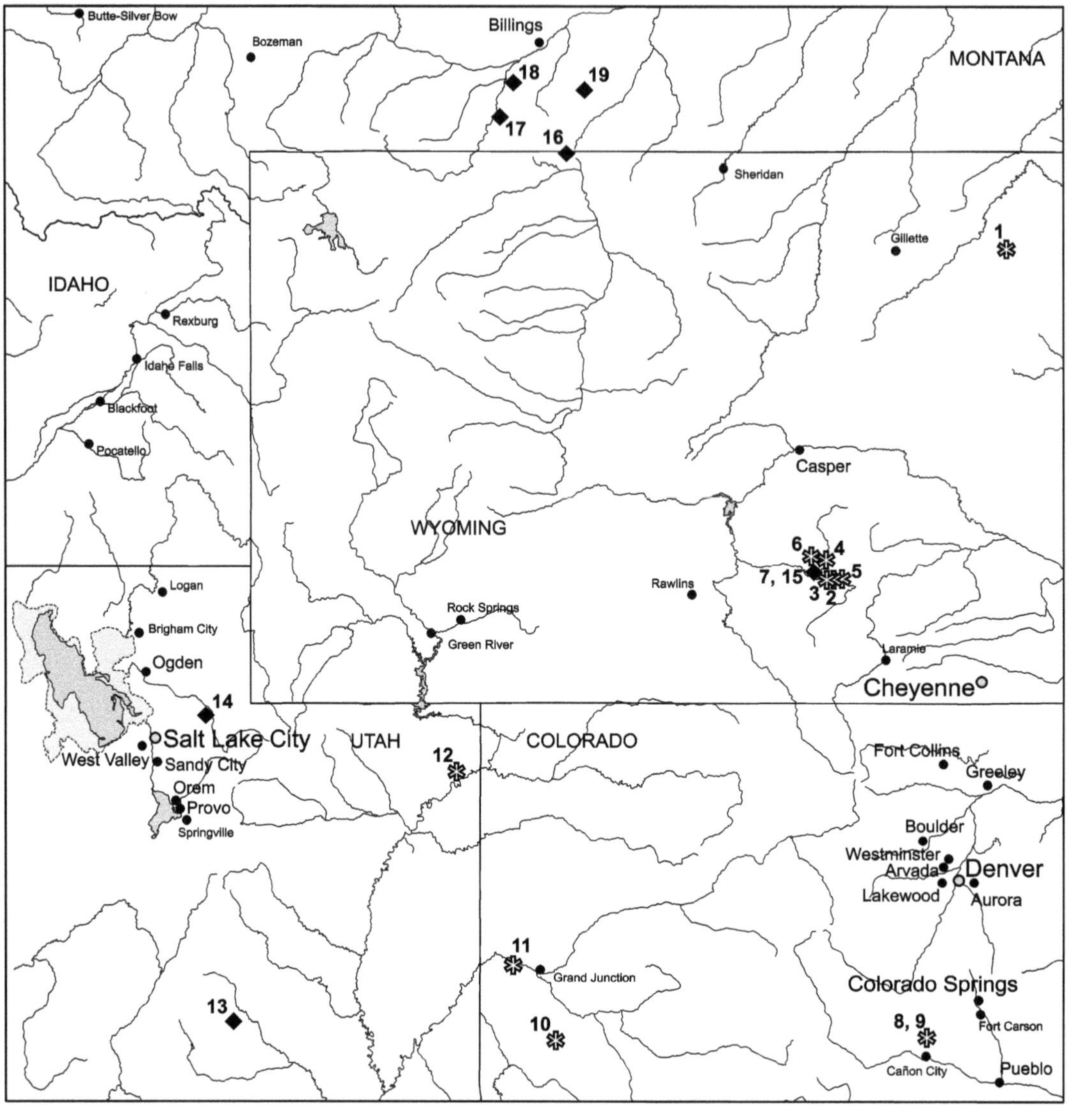

FIGURE 2.10. Late Jurassic and Early Cretaceous mammal localities of western North America. Asterisks, Late Jurassic (all Morrison Formation; Kimmeridgian–Tithonian; localities 1–12); diamonds, Early Cretaceous (localities 13–19 Localities or local faunas: **1, Sundance; 2, Como Quarry 9; 3, Delta T; 4, Chuck's Prospect; 5, Breakfast Bench; 6, Freezout Hills** (all Wyoming); **7, Ninemile Hill** (horizon in the Morrison Formation), **8, Garden Park, 9, Small Quarry; 10, Dry Mesa; 11, Fruita** (all Colorado); **12, Dinosaur National Monument,** Utah; **13, Mussentuchit** (Mussentuchit Member, Cedar Mountain Formation, Albian-Cenomanian, Utah); **14, Coalville** (Kelvin Formation, Aptian–Albian, Utah); **15, Ninemile Hill** (horizon in Cloverly Formation, Wyoming); **16, Crooked Creek; 17, Bridger; 18, Cottonwood Creek; 19, Cashen Ranch** (Cloverly Formation, Aptian–Albian, Wyoming and Montana).

priscus) in 1878 prompted a wider search, and Quarry 9 was discovered in 1879 (Simpson, 1929a). The quarry was worked continuously by Marsh's parties until 1889, resulting in the recovery of more than 300 mammal specimens; a single specimen was found at nearby Quarry 11. Reopening of the quarry in 1897 by a field party from the American Museum of Natural History resulted in recovery of a single jaw, later described by Simpson (1937b). A joint venture by the American Museum of Natural History and Yale University, led by Charles R. Schaff and Thomas H. Rich, worked the Como Bluff area from 1968 to 1970 (see Prothero, 1981). More specimens were collected from Quarry 9, and several additional sites, mentioned later, were discovered.

Despite the appearance of recent summaries of mammals from the Morrison Formation, including some of those from Quarry 9 (e.g., Prothero, 1981; Chure et al., 1998; Engelmann and Callison, 1998), much of the current taxonomy dates back to Simpson (1929a) and is in need of revision. For example, some taxa based, respectively, on upper and lower dentitions are almost surely synonymous (Simpson, 1927c; Clemens et al., 1979; see Martin, 1999a); further, other genera are probably oversplit (e.g., Jenkins, 1969b). Fortunately, the largest group, the "eupantotherian" Dryolestidae, has recently been revised on the basis of comparison with associated materials from Guimarota (Martin, 1999a). Regardless, there is no doubt that the mammalian fauna is an exceptionally diverse one, particularly by comparison with those from older units: we list 30 taxa for this one site alone (table 2.8). Eutriconodontans include two genera (*Aploconodon, Comodon*) of "amphilestids" and two (*Priacodon, Trioracodon*) of the advanced family Triconodontidae; the latter make their first unequivocal appearance in the Upper Jurassic Morrison Formation (possible triconodontids are known from the ?Early Jurassic of Mexico and the Late Jurassic of Africa). The Docodonta are represented by a number of specimens referable to *Docodon* itself. "Plagiaulacidan" multituberculates (five species representing two genera, *Ctenacodon* and *Psalodon*) were traditionally placed in the Plagiaulacidae and compared favorably with plagiaulacids of Britain (e.g., Simpson, 1929a). More recent work (Kielan-Jaworowska and Hurum, 2001, see chapter 8) places most of the Morrison multituberculates in the Allodontidae, an endemic family restricted to the North American Late Jurassic. Uncontested Plagiaulacidae are known only from the Early Cretaceous of Western Europe.

"Symmetrodontans" include the Tinodontidae (one or possibly two genera), which have a more advanced jaw structure than the Kuehneotheriidae of earlier faunas, and the peculiar *Amphidon*, which may have a relative in the ?Early Jurassic of India (see earlier). Much of the mammalian fauna of Como Quarry 9 is comprised of "eupantotherians" (perhaps 12 genera and 17 species), which evidently had undergone major radiation since the Middle Jurassic. Two families, Dryolestidae and Paurodontidae (the former is possibly known from the Bathonian of Britain; both are known from the Jurassic–Early Cretaceous of Western Europe), are represented. *Dryolestes* itself is also known from the Late Jurassic of Portugal (Martin, 1999a).

The joint American Museum–Yale field parties discovered several other sites at Como Bluff, or in the vicinity, that yielded mammals from the Morrison Formation. The most diverse assemblage is from **Chuck's Prospect** (figure 2.10), discovered by Schaff in Bone Cabin Draw, northwest of Como Bluff. As many as three dryolestids and an unidentified plagiaulacidan multituberculate are known from Chuck's Prospect (Prothero, 1981); an unidentified mammalian petrosal has also been described from the site (Prothero, 1983). **Delta T,** 2 km west of Quarry 9, is apparently higher in the Morrison Formation. One unidentified multituberculate is known from this site. Two other sites mentioned by Prothero (1981), each of which yielded a single mammal specimen, are close to Quarry 9 and do not deserve special commen t. **Ninemile Hill,** also in the immediate area, has yielded a small mammalian fauna from the Morrison Formation (Trujillo, 1999).

Breakfast Bench (figure 2.10), at Pine Tree Ridge east of Como Bluff, was discovered by Robert T. Bakker in 1977. Though it has yielded only three mammals, each is represented by a jaw (Bakker and Carpenter, 1990; Bakker, 1998). The mammals, including the multituberculates *Zofiabaatar* and "*Ctenacodon*"[4] and the paurodontid "eupantotherian" *Foxraptor,* are significant because they may be the stratigraphically highest known from the Morrison Formation, and they are not represented at other sites (Carpenter, 1998).

One unidentified mammal tooth was found in the Morrison Formation in the **Freezout Hills,** northwest of Como Bluff, by Kenneth Carpenter (Clemens et al., 1979).

Three mammalian jaws were recovered from one of Marsh's dinosaur quarries in the Morrison Formation near **Garden Park,** Colorado (figure 2.10). Two are referred to an unidentified species of *Docodon;* the third was the basis for "*Kepolestes coloradensis,*" a dryolestid "eupantotherian" known from this single specimen (Simpson, 1929a). The specimen is now considered indeterminate and we refer to it as *Amblotherium* sp. Engelmann and Callison (1998) indicate that Garden Park is in the lower part of the Brushy Basin Member of the Morrison Formation, whereas all other mammal-yielding sites are in the upper part of that member. A nearby site, **Small Quarry,** has yielded microvertebrate fossils, including *Docodon* and an unidentified dryolestid (Small and Carpenter, 1997).

An important mammal locality in the Morrison Formation was discovered in 1975 by James A. Clark, a member of a field party under the direction of George Callison, near **Fruita,** Colorado (figure 2.10). Fossils occur at a variety of levels in fluvial overbank deposits. The vertebrate fauna includes a variety of lizards, sphenodonts, small dinosaurs, and other terrestrial taxa in addition to mammals; amphibians are scarce (Callison, 1987). A sizable, diverse assemblage of mammal specimens has been collected, including skulls and associated skeletal elements.

[4] As discussed in chapter 8, "*Ctenacodon*" *brentbaatar* may not be referable to this genus and may, in fact, be a plagiaulacid, whereas *Ctenacodon* is placed in the Allodontidae.

Unfortunately, most of the material remains unpublished at this writing. The multituberculate *Glirodon grandis* and the triconodontid *Priacodon fruitaensis*, each known from a partial skull, are the only two mammals described from Fruita thus far (Rasmussen and Callison, 1981; Rougier, Wible, and Hopson, 1996; Engelmann and Callison, 1999).

A mammalian faunule has been collected at **Dinosaur National Monument**, northeastern Utah (figure 2.10), by George Engelmann, Scott Madsen, and colleagues. Two main sites have apparently yielded mammals (Engelmann and Callison, 1998), but no further information is available. Most of the specimens consist of isolated teeth obtained through processing of bulk rock samples, but one partial multituberculate skull and one "triconodont" jaw were found by quarrying. At least 12 taxa are known, including most of the more common mammals known from the Morrison (table 2.8). One new taxon, the "amphilestid triconodont" *Triconolestes curvicuspis*, was described from this assemblage (Engelmann and Callison, 1998), but others may be represented in the existing collection (S. Madsen, pers. comm.). The multituberculate *Glirodon grandis* is shared with the assemblage from Fruita (Engelmann and Callison, 1999) but no other Morrison faunules.

The **Dry Mesa Quarry**, western Colorado (figure 2.10), is another fossil locality in the Morrison Formation that has been worked primarily for dinosaurs; indeed, it has yielded the most diverse dinosaur fauna from the Late Jurassic of North America (see, e.g., Britt, 1991). The only purported mammalian specimen known from this site is the distal half of a humerus, said to be of "prototherian" design (Prothero and Jensen, 1983), which might belong to a lizard.

The final occurrence of mammals in the Morrison Formation is near **Sundance**, Wyoming (figure 2.10). Fossils have been recovered from two closely spaced sites, the Mammal and Little Houston quarries. Two taxa have been reported (Martin and Foster, 1998), ?*Psalodon* and *Docodon*, both of which are reasonably common in the Morrison Formation.

EARLY CRETACEOUS

The Early Cretaceous is an important interval for several reasons. It is the longest epoch of the Mesozoic, spanning more than 45 Ma. Terrestrial life, both plants and animals, changed radically over the course of the Early Cretaceous, both taxonomically and morphologically. In part, this has been attributed to reciprocal biotic interactions involving the appearance and radiation of flowering plants, or angiosperms (see Wing and Tiffney, 1987; and papers in Friis et al., 1987), but other factors such as climatic change, sea level, and intercontinental dispersal (e.g., Cifelli, Kirkland, et al., 1997) are probably involved as well. Many "modern" groups—cimolodontan multituberculates, marsupials, and placentals among mammals, for example—first appear or are suspected to have originated during the Early Cretaceous (Lillegraven et al., 1987; Kielan-Jaworowska and Dashzeveg, 1989; Cifelli, 1993b; Kielan-Jaworowska and Hurum, 2001). On northern continents, archaic groups, such as "triconodonts," "symmetrodontans," and "eupantotherians," dwindled in importance or became extinct. Significant separation of Laurasia and Gondwana occurred, with the result that their respective faunas became more distinct than they had been earlier in the Mesozoic.

When Simpson summarized the known record of Early Cretaceous mammals in 1928, only three specimens, all isolated teeth from the Wealden of England, were known from this 45-Ma interval (Simpson, 1928a). Given the near absence of Early Cretaceous fossils, the taxonomic and morphologic differences between mammals of the Late Jurassic and Late Cretaceous were magnified—particularly because the early Late Cretaceous was poorly known as well. In this context, discoveries in Lower Cretaceous rocks during the second half of the twentieth century collectively rank as one of the major advances in understanding the early history of mammals. Early Cretaceous mammals are now known from Western Europe, Russia (Siberia), various parts of Asia including Japan, Africa, South America, Australia, and North America—every major landmass except Antarctica. To be sure, many of these are incompletely known, but some are represented by magnificently complete skulls and skeletons.

WESTERN EUROPE

Thus far, Early Cretaceous mammals of Western Europe are known from England (figure 2.3), Spain, and Portugal (figure 2.9). Most or all of these occurrences are older than those from the eastern part of Eurasia, being of Berriasian to Barremian age.

Britain

Early Cretaceous mammals are known from several horizons contained within two principal units in Britain, the Purbeck Limestone Group and the Wealden Supergroup. The stratotypes lie within separate depositional subbasins and they partially overlap in age. The base and top of the Purbeck are placed at the Portlandian-Berriasian (generally but not universally considered to be the Jurassic-Cretaceous boundary) and Berriasian-Valanginian boundaries, respectively. The Wealden ranges from upper Berriasian to lower Aptian (Allen and Wimbledon, 1991; Rasnitsyn et al., 1998). A useful review of these strata and their mammal faunas is given by Clemens et al. (1979), from which we have drawn in preparing this account.

The Purbeck Limestone Group is often divided into two units: the Durlston Formation above (the lowest unit of which is the Cinder Member) and the Lulworth Formation below (e.g., Casey, 1963; Allen and Wimbledon, 1991). Purbeck mammals come from several horizons within the uppermost two members of the Lulworth Formation, the Marly and (overlying) Cherty Freshwater members. The depositional environment is thought to have consisted largely of mudflats along a lake with little or no salinity (e.g., Kielan-Jaworowska and Ensom, 1994). Based on current evidence, we consider all mammals of the Purbeck to be of early Berriasian age.

The first mammalian fossils from the Purbeck were collected in 1854 from the cliffs of **Durlston** (also called Durdlestone) **Bay,** Dorset (Owen, 1854). A large collection, including a number of important mammal specimens, was subsequently made by Samuel Beckles; many of these were also described by Owen (1871; see Simpson, 1928a). Several specimens were collected in later years (e.g., Willett and Willett, 1881), but the main quarry appears to be exhausted and it is in any case no longer accessible (Kermack, 1988). That quarry was in the so-called "Mammal Bed," contained within the Marly Freshwater Member of the Purbeck (Clemens et al., 1979). The exact horizon is uncertain (Kermack, 1988), but it has been linked to bed DB 83 of Clements (1993).

The mammals of Durlston Bay are of more than historical interest in that they include morphologically informative jaws and at least one partial skull. The fauna (mostly described by Simpson, 1928a) is generally similar to that of the Morrison Formation (Upper Jurassic) of North America, with one or more shared genera. There are at least 17 mammals known from Durlston Bay (table 2.9). As with the Morrison, several triconodontids and multituberculates are present, with docodontans and "symmetrodontans" being less diverse and the majority of taxa being dryolestoid "eupantotherians." Notably, the common "symmetrodontan," *Spalacotherium,* is more derived than its counterpart in the Morrison, *Tinodon* (see Patterson, 1956; Cifelli and Madsen, 1999). Another notable element of the Durlston fauna is the "eupantotherian" *Peramus,* which appears to be closely related to boreosphenidan mammals (chapter 11).

In 1986, Paul Ensom began recovering vertebrate microfossils from the Purbeck through underwater screenwashing and related procedures. The mammals are almost entirely represented by isolated teeth, but the collective sample is large, including some 800 specimens. Vertebrate microfossils have been found from several localities, but currently the most important by far is **Sunnydown Farm** (figure 2.3) on the Isle of Purbeck, about 5 km east of Durlston Bay, near Langton Matravers, Dorset (Ensom et al., 1994). Two horizons were sampled at this site, both in the Cherty Freshwater Member of the Lulworth Formation, not far below the Cinder Bed (e.g., Sigogneau-Russell and Ensom, 1994). The mammals of Sunnydown Farm thus are slightly higher, stratigraphically, than those from Durlston Bay (see Clements, 1993).

At least 16 mammal taxa have been described or mentioned from Sunnydown Farm (see Kielan-Jaworowska and Ensom, 1992, 1994; Sigogneau-Russell and Ensom, 1994, 1998; Ensom and Sigogneau-Russell, 1998, 2000; Sigogneau-Russell, 1999; Sigogneau-Russell and Kielan-Jaworowska, 2002). This approximates the diversity known from Durlston Bay, but only two genera, the "symmetrodontan" *Spalacotherium* and the docodontan *Peraiocynodon,* are shared between the two localities. On the other hand, the mammals of Sunnydown Farm strengthen links of the Purbeck fauna with those known from elsewhere. The "symmetrodontan" *Tinodon* and the docodontan *Docodon* are known from the Morrison Formation of North America. Multituberculates (*Albionbaatar, Sunnyodon, Gerhardodon*) share similarity, at the family level at least, with the Late Jurassic (Albionbaataridae, Paulchoffatiidae) and Early Cretaceous (Paulchoffatiidae, Pinheirodontidae) of the Iberian Peninsula (Hahn and Hahn, 1992, 1998a,b,d, 1999a). The aforementioned "symmetrodontan" *Spalacotherium* is also known from the Early Cretaceous of the Iberian Peninsula (Krebs, 1993). The bizarre "symmetrodontan" *Thereuodon* and early boreosphenidan mammals are shared with the Berrisian of Morocco (Sigogneau-Russell, 1994a; Sigogneau-Russell and Ensom, 1998). The "eupantotherians" *Chunnelodon* and *Magnimus* from Sunnydown Farm do not have obvious close relatives elsewhere (Ensom and Sigogneau-Russell, 1998; Sigogneau-Russell, 1999).

Most recently, mammals have been reported from two sites in the Intermarine Member of the Durlston Formation near Swanage, Dorset. These represent the stratigraphically highest mammals found to date in the Purbeck Limestone Group. One of the sites, in a quarry near **Belle Vue,** appears to represent a horizon laterally equivalent to beds DB 116–117 of Clements (1993). Finds reported thus far include a dentary fragment with teeth resembling deciduous premolars of a spalacotheriid and an upper molar of the dryolestid cf. *Kurtodon* (Ensom, 2000). The horizon of the second site, a quarry near **Acton,** is correlated with DB 115–116 of Clements (1993). The single specimen reported to date from Acton is a battered lower molar of the tribosphenic mammal *Tribactonodon bonfieldi* (see Sigogneau-Russell et al., 2001).

TABLE 2.9. Early Cretaceous Mammals of Britain (see figure 2.3; locality numbers do not correspond between map and table). Localities: 1, Durlston Bay; 2, Sunnydown Farm; 3, Belle Vue; 4, Acton (all Purbeck Limestone Group, Berriasian); 5, Cliff End; 6, Tighe Farm (Wadhurst Formation, early Valanginian); 7, Paddockhurst Park (Grinstead Clay, middle Valanginian); 8, Isle of Wight (Wealden Group, ?Valanginian)

- Docodonta
  - Docodontidae
    - cf. *Docodon* sp. (2)
    - *Peraiocynodon inexpectatus* (1, 2)
- Multituberculata
  - Albionbaataridae
    - *Albionbaatar denisae* (2)
  - Eobaataridae
    - *Loxaulax valdensis* (5)
    - *Loxaulax* sp. (8)
  - Paulchoffatiidae
    - *Sunnyodon notleyi* (2)
  - Pinheirodontidae
    - *Gerhardodon purbeckensis* (2)
  - Plagiaulacidae
    - *Bolodon crassus* (1)
    - *"Bolodon" elongatus* (1)
    - *Bolodon falconeri* (1)
    - *Bolodon minor* (1)
    - *Bolodon osborni* (1)
    - *Plagiaulax becklesii* (1)
- Eutriconodonta
  - Triconodontidae
    - *Triconodon mordax* (1)
    - *Trioracodon ferox* (1)
    - *Trioracodon major* (1)
    - *Trioracodon oweni* (1)
  - Family *incertae sedis*
    - Gen. et sp. indet. A (2)
    - Gen. et sp. indet. B (2)
- Archaic "symmetrodontans"
  - Thereuodontidae
    - *Thereuodon taraktes* (2)
  - Tinodontidae
    - *Tinodon micron* (2)
  - Family *incertae sedis*
    - Gen. et sp. indet. (2)
- Stem Trechnotheria
  - Spalacotheriidae
    - *Spalacotherium tricuspidens* (1, 2, 6)
    - *Spalacotherium taylori* (7)
    - *Spalacotherium evansae* (2)
    - Gen. et sp. indet. (3)
- Stem Cladotheria ("eupantotherians")
  - Dryolestidae
    - *Amblotherium pusillum* (1)
    - *Kurtodon pusillus* (1)
    - cf. *Kurtodon* sp. (3)
    - *Laolestes hodsoni* (5, 6)
    - *Peraspalax talpoides* (1)
    - *Phascolestes mustelulus* (1)
  - Paurodontidae
    - *Dorsetodon haysomi* (2)
  - Peramuridae
    - *Peramus tenuirostris* (1)
  - Family *incertae sedis*
    - *Chunnelodon alopekodes* (2)
    - *Magnimus ensomi* (2)
- Stem Boreosphenida ("tribotherians")
  - Aegialodontidae
    - *Aegialodon dawsoni* (5)
  - Family *incertae sedis*
    - *Tribactonodon bonfieldi* (4)
    - Gen. et sp. indet. (2)

Initial discoveries of mammals in the Wealden of Britain date to the latter part of the nineteenth and early twentieth centuries and include only three specimens (see Simpson, 1928a). In the 1960s, investigations by Kermack, Clemens, and colleagues resulted in a substantial improvement to the meagre record from the Wealden. Like the Purbeck Group, the Wealden was deposited in a number of subbasins across Western and Central Europe, but thus far mammals are known only from exposures in southern England. Correlation of the Wealden is discussed by Allen and Wimbledon (1991), and mammal occurrences and their stratigraphic context are summarized by Clemens et al. (1979, and references cited therein). In the Wealden subbasin, which includes most of the mammal localities, the basal Wealden overlaps with the uppermost Purbeck. The Wealden has two major divisions, the Hastings Beds (below) and the Weald Clay (above). All mammal sites lie within the Hastings Beds, which span the upper Berriasian through the Valanginian–Hauterivian boundary. In turn, the Hastings Beds include (from lowest to highest) the Ashdown Formation, the Wadhurst Formation, the Lower Tunbridge Wells Formation, the Grinstead Clay (locally), and the Upper Tunbridge Wells Formation.

The most diverse assemblage of Wealden mammals is from **Cliff End** (figure 2.3). The site is on the coast near Hastings (its name in older literature) and is the source for the first mammals known from the unit (e.g., Woodward,

1911). Cliff End is in the Wadhurst Formation and is of early Valanginian age. Four mammals are known from Cliff End (table 2.9). Of these, the "symmetrodontan" *Spalacotherium tricuspidens* is also known from the Purbeck, and the "eupantotherian" *Melanodon* is recorded from the Morrison Formation of the western United States. The most celebrated mammal from Cliff End is *Aegialodon dawsoni*, represented by a lower molar with a basined heel—that is, the tribosphenic design seen in marsupials and placentals (Kermack et al., 1965). For many years *Aegialodon* was the oldest known tribosphenic mammal, until the discovery of boreosphenidans from the Berriasian of Morocco and Britain (Sigogneau-Russell, 1991a, 1992; Sigogneau-Russell and Ensom, 1994).

The remaining three mammal sites in the Wealden have each yielded only one or two taxa. **Tighe Farm,** northeast of Cliff End, is also in the Wadhurst Formation (lower Valanginian) and shares the "eupantotherian" *Melanodon hodsoni* with that site. **Paddockhurst Park,** Sussex, is in the Grinstead Clay, which is stratigraphically higher than Cliff End and Tighe Farm. Judged by the correlations presented in Allen and Wimbledon (1991), Paddockhurst Park is probably of middle Valanginian age, somewhat older than suggested by Clemens et al. (1979). The "symmetrodontan" *Spalacotherium* is represented here by *S. taylori* (known only from the Wealden), rather than *S. tricuspidens*, which is found in both the Purbeck and the lower Wealden. Lastly, the multituberculate *Loxaulax*, also known from Cliff End, has been reported from a site in the Wealden on the **Isle of Wight** (Butler and Ford, 1977). The mammal specimens come from one of three lignitic claystones; unfortunately, there is no published information as to the stratigraphic placement of the site relative to others in the Wealden.

Spain and Portugal

Three sites on the Iberian Peninsula have yielded substantial remains of mammals from the latest Jurassic or Early Cretaceous; two others have yielded isolated teeth or single specimens (table 2.10). Most were developed by personnel of the Freie Universität Berlin, initially under the leadership of Walter Kühne and continued by Siegfreid Henkel, Bernard Krebs, and colleagues (see brief account in Krebs, 1980). The geologically oldest of these is **Porto Pinheiro** (sometimes referred to as Porto Dinheiro), on the coast near Lourinhã, Portugal (figure 2.9). The age of the site, which is in the Lourinhã Formation, is not well understood; it may be Late Jurassic or Early Cretaceous. Palynomorphs, which indicate a well-differentiated flora and a fluvial setting, suggest a Tithonian–Berriasian age

TABLE 2.10. Early Cretaceous Mammals of the Iberian Peninsula (Spain and Portugal; see figure 2.9; locality numbers do not correspond between map and table). Localities: 1, Porto Pinheiro (?Berriasian, Portugal); 2, Pié Pajarón; 3, Galve; 4, Uña (all early Barremian, Spain); 5, Vallipón (late Barremian, Spain)

- Eutriconodonta
  - Gobiconodontidae
    - *Gobiconodon* sp. nov. (5)
  - Triconodontidae
    - *Priacodon* sp. (1)
- Multituberculata
  - Family *incertae sedis*
    - Gen. et sp. indet. (5)
  - Eobaataridae
    - *Eobaatar hispanicus* (3, 4)
    - ?*Eobaatar pajaronensis* (2)
    - *Parendotherium herreroi* (3)
    - Gen. et sp. indet. (5)
  - Paulchoffatiidae
    - *Galveodon nannothus* (3, 4)
    - Gen. et sp. indet. (5)
  - Pinheirodontidae
    - *Bernardodon atlanticus* (1)
    - *Ecprepaulax anomala* (1)
    - *Iberodon quadrituberculatus* (1)
    - *Lavocatia alframbrensis* (3)
    - *Pinheirodon pygmaeus* (1)
    - *Pinheirodon vastus* (1)
    - *Pinheirodon* sp. (1)
- Archaic "symmetrodontans"
  - Family *incertae sedis*
    - Gen. et sp. indet. (5)
  - Tinodontidae
    - ?*Tinodon* sp. (1)
- Stem Trechnotheria
  - Spalacotheriidae
    - *Spalacotherium henkeli* (3)
- Stem Cladotheria ("eupantotherians")
  - Dryolestidae
    - *Crusafontia cuencana* (3, 4)
    - *Laolestes andresi* (1)
    - *Portopinheirodon asymmetricus* (1)
  - Peramuridae
    - *Nanolestes krusati* (1)
  - Family *incertae sedis*
    - Gen. et sp. indet. (5)

(Mohr, 1989, see also Krusat, 1989). Hahn and Hahn (1992) consider Porto Pinheiro to be of Berriasian age, an assessment we tentatively follow here.

Mammalian fossils from Porto Pinheiro were recovered through screenwashing (Kühne, 1971) and include more than 800 isolated teeth. Some two-thirds of the specimens are reported to belong to "eupantotherians"; of these, only the so-called "Porto Pinheiro molar" has been described (Krusat, 1969; Kraus, 1979). This enigmatic tooth is peramuran-like in some respects (chapter 10), and Martin (2002) placed it in the species *Nanolestes krusati*. The triconodontid *Priacodon* and, possibly, the "symmetrodontan" *Tinodon*, both of which are known from the Upper Jurassic Morrison Formation of the United States, are reported to be present at Porto Pinheiro (Krusat, 1989). Best known from this site are the multituberculates, which include four genera and six species of plagiaulacidans. These are referred to their own family, Pinheirodontidae (see Hahn and Hahn, 1999a), which are also recorded from the Barremian of Spain (*Lavocatia*) and the Berriasian of Britain (*Gerhardodon*).

The remaining Early Cretaceous sites on the Iberian Peninsula are in Spain (figure 2.9). The **Galve** locality, Teruel Province, is in the Camarillas Formation, considered to be a Wealden facies (Canudo and Cuenca-Bescós, 1996).[5] Fossils, including mammals, have been collected from a number of sites and stratigraphic levels within the Camarillas Formation (see reviews by Canudo et al., 1996a,b, 1997; Cuenca-Bescós et al., 1996). As for the geologically older site of Guimarota, Portugal, the palynomorphs indicated a well-differentiated flora and a fluvial regime; Galve is estimated to be of early Barremian age (Mohr, 1989). Mammalian fossils representing six taxa (Crusafont-Pairó and Adrover, 1966; Henkel and Krebs, 1969; Crusafont and Gibert, 1976; Krebs, 1985, 1993; Hahn and Hahn, 1992)[6] were collected through screenwashing. Of these, four represent last occurrences: Paulchoffatiidae (*Galveodon*) are otherwise known from the Late Jurassic and earliest Cretaceous (Guimarota, Purbeck) and Pinheirodontidae (*Lavocatia*), European Dryolestidae (*Crusafontia*), and *Spalacotherium* from earlier in the Cretaceous. These taxa may be relictual, their persistence on the Iberian Peninsula reflecting geographic isolation (Krebs, 1985). On the other hand, it should be noted that the remaining mammalian genera known from Galve, *Eobaatar* and *Parendotherium*, belong to a family known from both the Wealden and the Early Cretaceous of Asia. Indeed, *Eobaatar* itself was first described from Höövör (referred to previously as Khoboor or Khovboor), Mongolia (Kielan-Jaworowska, Dashzeveg, and Trofimov, 1987), indicating that some interchange was possible.

The site of **Uña**, Cuena Province, Spain, is also in beds sometimes termed "Weald," but here the vertebrate fauna derives from a coal. Palynomorphs and charophytes provide somewhat conflicting age assessments ranging from Hauterivian to late Barremian. The similarity of the vertebrate assemblage to that of Galve suggests that they are of the same age; that is, early Barremian (Krebs, 1995). The flora is a specialized, coastal swamp assemblage (Mohr, 1989), and the paleoenvironment is interpreted as having been a swampy alluvial plain near the margin of a lake (Krebs, 1995). The fossils, obtained by a combination of screenwashing and quarrying the coal, include several mammal jaws and a number of isolated teeth. Lower vertebrates are considerably more abundant than mammals. Only three mammal species are known from Uña, all of which are also known from Galve. Significantly, a well-preserved dentary of the dryolestid *Crusafontia* was recovered from Uña.

Pié Pajarón is near Uña and is probably also of Barremian age. Two isolated anterior upper premolars have been described as ?*Eobaatar pajaronensis*. Referral to the genus is tentative but noteworthy, if confirmed: as just noted, *Eobaatar* is also known from the Early Cretaceous of Mongolia as well as Western Europe.

The site of **Vallipón**, Teruel Province, lies in the Artoles Formation, in part of the section interpreted to represent a beach facies, transitional between continental and marine depositional settings (Canudo et al., 1996b; Cuenca-Bescós and Canudo, 1999). The fauna, of late Barremian age, includes abundant chondrichthyans, osteichthyans, archosaurs, lepidosaurs, and several mammals. Of these, the only one identified to genus level is *Gobiconodon*, known by an upper molar said to represent a new species (Cuenca-Bescós and Canudo, 1999). As noted elsewhere, *Gobiconodon* was evidently widespread, both geographically and temporally, and has been reported from beds spanning much of the Early Cretaceous, including units in North America, Europe, and Asia.

ASIA

With the exception of Mongolia, Asia has been a virtual *tabula rasa* for mammals of Early Cretaceous age until re-

[5] Contributions on the fossil mammals, cited below, give the impression of a single locality at Galve. A chart by Canudo et al. (1996a: figure 3) suggests that the fossils were collected from multiple levels, including at least three horizons in the Camarillas Formation, as well as one near the top of the underlying Castellar Formation.

[6] Passing mention should be made of "*Pocamus pepelui*," described from Galve by Canudo and Cuenca-Bescó's (1996) and later shown to be based on the fourth upper premolar of *Crusafontia cuencana* (see Martin, 1998).

cently. New discoveries, many of which remain to be described, show that fossils from central and eastern Asia will play a key role in interpreting mammalian history during this pivotal interval. We have drawn from several of the useful summaries of these discoveries (Nessov, Sigogneau-Russell, and Russell, 1994; Wang et al., 1995; Nessov, 1997; Novikov et al., 1998; Averianov, 2000), together with original sources (the most important of which are cited in those works) and other information in providing the following account.

Uzbekistan and Russia

Most of the relatively recent advances in knowledge of early mammals from Middle Asia are occurrences of Late Cretaceous age (summarized below), with one exception: *Bobolestes zhenge,* from the **Khodzhakul** locality in the Khodzhakul Formation, Kizylkum Desert, Uzbekistan (figure 2.11). The age is believed to be late Albian, based on the shark fauna. *Bobolestes* is a rather primitive eutherian and was first referred to the Pappotheriidae (Nessov, 1985a) before being placed in its own family (see Averianov, 2000).

The site of **Shestakovo** (Ilek Formation), in the Kemerovo region of Siberian Russia (figure 2.12), has thus far yielded two mammals: the eutriconodontan *Gobiconodon borissiaki* (see Maschenko and Lopatin, 1998; Alifanov et al., 1999) and an as yet unidentified "symmetrodontan." The site is notable in that it preserves articulated dinosaur remains; the fauna reportedly includes two species of *Psittacosaurus,* the protosuchian crocodilian *Tagarosuchus,* and a tritylodont; on the basis of the last two Novikov et al. (1998) suggested an earliest Cretaceous (Berriasian) age for the site. However, other workers (Averianov and Skutschas, 2000a) indicated that Shestakovo is of Albian age. Given the occurrence of the same species of *Gobiconodon* at Höövör, Mongolia, and the projected age of that site (see later), we believe that this latter estimate may be nearer to the mark. Tritylodontids were long thought to have become extinct in the Jurassic; in addition to the occurrence at Shestakovo (Tatarinov and Maschenko, 1999), an Early Cretaceous tritylodontid is now known from Japan (Setoguchi, Matsuoka, and Matsuda, 1999).

A second site in Russia, the **Mogoito** locality in the Murtoi Formation, Transbaikalia (figure 2.12), has thus

FIGURE 2.11. Jurassic (asterisks, localities 1, 2, 12) and Cretaceous (black diamond, locality 3, Early Cretaceous; gray diamonds, localities 4–11, Late Cretaceous) mammal sites of middle Asia and Russia (part). Localities or local faunas: **1, Kalmakerchin; 2, Tashkumyr** (both Balabansay Formation; Callovian, Middle Jurassic, Kirghizia); **3, Khodzhakul** (Khodzhakul Formation; late Albian, Uzbekistan); **4, Khodzhakulsay; 5, Sheikhdzheili; 6, Chelpyk** (all Khodzhakul Formation; early Cenomanian, Karakal-pakistan, Uzbekistan); **7, Ashikol** (unnamed unit; ?early Turonian, Chimkent District, Kazakhstan); **8, Dzharakuduk** (late Turonian and Coniacian localities in the Bissekty Formation and Santonian, Aitym Formation; Uzbekistan); **9, Kansay** (early Santonian, Khodzant District, Tajikistan); **10, Zhalmauz Well** (lower Bostobe Formation; Santonian, Kyzyl-Orda Province, Kazakhstan); **11, Grey Mesa** (= Alymtau; Darbasa Formation, early Campanian); **12, Peski Quarry** (fissure filling; late Bathonian, Kolomensk District, Russia). (Not shown: **Yantardakh;** Khets Formation, Santonian, Siberia.)

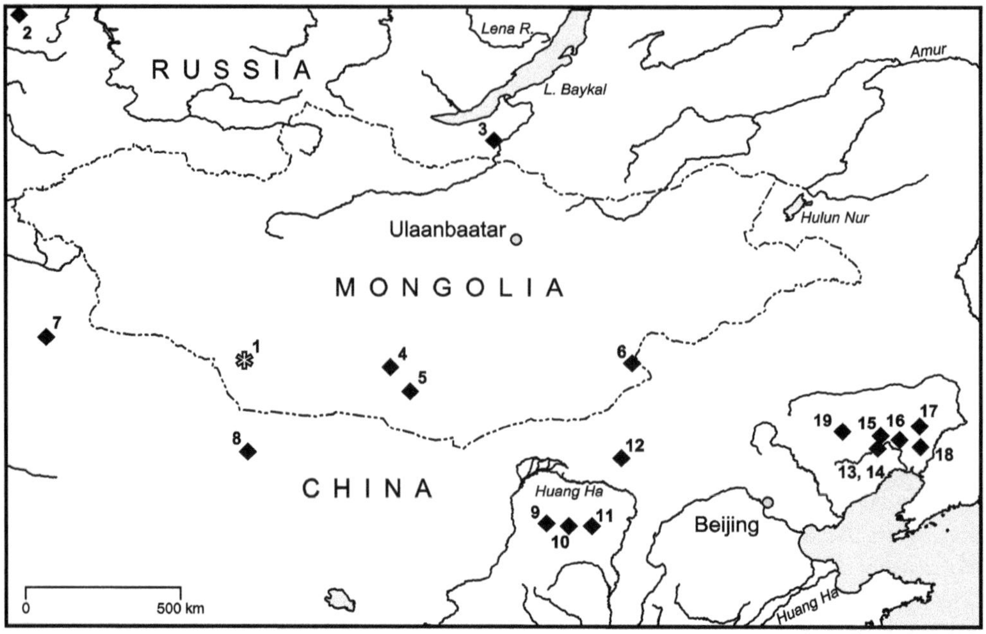

FIGURE 2.12. Late Jurassic (Mongolia) and Early Cretaceous (Russia, Mongolia, and China) mammal sites. White asterisk (locality 1), Late Jurassic; diamonds (localities 2–19), Early Cretaceous. Localities or local faunas: **1, Shar Teeg** (Late Jurassic, Mongolia); **2, Shestakovo** (Ilek Formation; Albian, Kemerovo District, Russia); **3, Mogoito** (Murtoi Formation; ?Barremian–Aptian or Aptian–Albian, Transbaikalia, Russia); **4, Höövör** ("Höövör Beds"; ?Aptian–Albian, Mongolia); **5, Oshih** (Oshih Formation; ?Valanginian, Mongolia); **6, Khamaryn Us** (Khukhtyk or Dzunbain Formation; ?Aptian–Albian, Mongolia); **7, Huang-Ni-Tan** (Shengjinkou Formation; Early Cretaceous, Xinjiang, China); **8, Mazongshan** (lower Xinmingbao Group; ?Barremian–Albian, Gansu, China); **9, Laolonghuoze; 10, Hangjin-Qi; 11, Yanhaizi;** (all Eijinhoro Formation; ?Barremian, Inner Mongolia, China); **12, Elesitai** (Bayan Gobi Formation; ?Aptian–Albian, Inner Mongolia, China); **13, Jianshangou; 14, Sihetun; 15, Lu-Jia-Tun** (all Yixian Formation; middle Barremian, Liaoning, China); **16, Xindi; 17, Xinqiu** (both Fuxin Formation; ?Aptian or Albian–Cenomanian, Liaoning, China); **18, Badaohao** (Shahai Formation; ?Aptian, Liaoning, China); **19, Dawangzhangzi** (Yixian Formation; middle Barremian, Liaoning, China).

far yielded an upper and a lower molar, first placed in the species *Prokennalestes abramovi* and later transferred to the new genus, *Murtoilestes* (Averianov and Skutschas, 2000a, 2001). The occurrence is a notable one because it is believed to be Barremian–Aptian and, hence, perhaps older than Höövör, the source of *Prokennalestes.*

Mongolia

A useful, relatively recent summary on Mongolian vertebrate faunas and their age, relationships, and stratigraphic context is provided by Jerzykiewicz and Russell (1991). They pointed out that the fossil assemblages, as characterized by molluscs and vertebrates (Martinson, 1982), are analogous to the land-mammal ages of North America and can be used as the basis for Mongolian land-vertebrate ages. The only Early Cretaceous "age" relevant in the current context is the Khukhtekian, which has been characterized by the presence of harpymimid theropods and *Iguanodon,* as well as abundant *Psittacosaurus* (which is also common to older assemblages). The best represented Early Cretaceous mammal assemblage of Mongolia (figure 2.12), that of Höövör (Trofimov 1978, 1980; Dashzeveg, 1979, 1994; Dashzeveg and Kielan-Jaworowska, 1984; Kielan-Jaworowska, Dashzeveg, and Trofimov, 1987; Kielan-Jaworowska and Dashzeveg, 1989; Sigogneau-Russell et al., 1992), is referred to the Khukhtekian. Jerzykiewicz and Russell (1991) considered the Khukhtekian to be Aptian–Albian in age, based on plant megafossils and elasmobranchs, as well as on the absence of trionychid turtles, and this was corroborated by Shuvalov (2000). The section at Höövör is more than 100 m thick, and the lacustrine and lacustrine-fluvial (deltaic) facies predominate. Shuvalov (2000) pointed out that the Aptian–Albian deposits in the region are more than 500 m thick. The beds that crop out at Höövör have not been formally designated, and we refer to them for the purpose of this book informally as "Höövör Beds."

The Aptian–Albian age of "Höövör Beds" accords with a similar estimate, based on the shared presence of *Gobiconodon* in both Höövör and in the Cloverly Formation of

Montana (Jenkins and Schaff, 1988; Kielan-Jaworowska and Dashzeveg, 1998). It should be pointed out, however, that the age of the Cloverly itself is only loosely constrained and is based mainly on similarity of its vertebrate fauna to that of the Trinity Group, Texas, which has interbedded marine horizons (Jacobs et al., 1991; Jacobs and Winkler, 1998). Moreover, some elements of the Höövör mammal assemblage, such as the eutriconodontan *Gobiconodon,* are now being reported from units said to be Barremian (e.g., Cuenca-Bescós and Canudo, 1999) or even older (Maschenko and Lopatin, 1998; Novikov et al., 1998; Rougier et al., 2001). On the other hand, recent palynological evidence suggests that at least one Khukhetian vertebrate fauna is no older than than middle–late Albian (Hicks et al., 1999). Given the equivocal nature of the available data, an Aptian–Albian age for the Höövör assemblage must be considered tentative.

The **Höövör** fauna, discovered by Soviet-Mongolian Expeditions, includes some 17 mammalian taxa (table 2.11). It is of great interest because it contains some apparently primitive species, as well as early members of advanced groups and bizarrely specialized forms. Of the six or so multituberculates, most are comparable to "plagiaulacidans" more characteristic of Late Jurassic and earliest Cretaceous faunas, while the uniquely derived *Arginbaatar* seems to fall between "plagiaulacidans" and the

TABLE 2.11. Early Cretaceous Mammals of Asia and Asiatic Part of Russia (see figure 2.12; locality numbers do not correspond between map and table). Localities: 1, Khodzhakul, Uzbekistan (Khodzhakul Formation, late Albian); 2, Shestakovo, Russia (Ilek Formation, ?Albian); 3, Mogoito, Russia (Murtoi Formation, ?Barremian–Aptian); 4, Höövör, Mongolia ("Höövör Beds," referred to also as Guchin Us Beds, ?Aptian–Albian); 5, Khamaryn Us, Mongolia (Khukhtyk or Dzunbain Formation, ?Aptian–Albian); 6, Elesitai, Inner Mongolia, China (Bayan Gobi Formation, ?Aptian–Albian); 7, Laolonghuoze; 8, Yanhaizi; 9, Hangjin-Qi, Inner Mongolia, China (all Eijinhoro Formation, ?Barremian); 10, Mazongshan, Gansu, China (Xinmingbao Group, Barremian–Albian); 11, Huang-Ni-Tan, Xinjiang, China (Shengjinkou Formation, Early Cretaceous); 12, Xinqiu; 13, Xindi, Liaoning, China (both Fuxin Formation, ?Aptian–Albian or Cenomanian); 14, Badaohao, Liaoning, China (Shahai Formation, ?Aptian); 15, Jianshangou; 16, Sihetun; 17, Lu-Jia-Tun; 18, Dawangzhangzi, Liaoning, China (all Yixian Formation, middle Barremian); 20, Oshih, Mongolia (Oshih Formation, ?Valanginian); 21, Kaseki-kabe, Ishikawa Prefecture, Japan (Kuwajima Formation, Hauterivian)

- Mammalia *incertae sedis* (6, 7, 8, 11)
- Eutriconodonta
  - "Amphilestidae"
    - Gen. et sp. indet. (5, 21)
  - Gobiconodontidae
    - *Gobiconodon borrissiaki* (2, 4)
    - *Gobiconodon hoburensis* (4)
    - *Gobiconodon hopsoni* (20)
    - *Gobiconodon* sp. nov. A (17)
    - *Gobiconodon* sp. nov. B (20)
    - *Hangjinia chowi* (9)
    - *Repenomamus robustus* (17)
    - Gen. et sp. nov. (10)
  - Family *incertae sedis*
    - *Jeholodens jenkinsi* (16)
- Multituberculata
  - ?Albionbaataridae
    - cf. *Albionbaatar* sp. (14)
  - Arginbaataridae
    - *Arginbaatar dmitrievae* (4)
  - Eobaataridae
    - *Eobaatar magnus* (4)
    - *Eobaatar minor* (4)
    - *Eobaatar* sp. A (4)
    - *Eobaatar* sp. B (4)
    - *Monobaatar mimicus* (4)
    - *Sinobaatar lingyuanensis* (18)
    - Gen. et sp. indet. (21)
  - ?Plagiaulacidae
    - Gen. et sp. indet. (14)
  - Family *incertae sedis*
    - Gen. et sp. indet. (13)
- Archaic "symmetrodontans"
  - Family *incertae sedis*
    - *Gobiotheriodon infinitus* (4)
    - Gen. et sp. indet. (14)
- Stem Trechnotheria
  - Spalacotheriidae
    - *Zhangheotherium quinquecuspidens* (15)
- Stem Cladotheria ("eupantotherians")
  - Arguitheriidae
    - *Arguitherium cromptoni* (4)
  - Arguimuridae
    - *Arguimus khosbajari* (4)
- Stem Boreosphenida ("tribotherians")
  - Aegialodontidae
    - *Kielantherium gobiensis* (4)
    - Gen. et sp. indet. (14)
- Eutheria
  - Family *incertae sedis*
    - *Endotherium niomii* (12)
    - *Eomaia scansoria* (18)
    - Gen. et sp. indet. (4, 15)
  - Bobolestidae
    - *Bobolestes zenge* (1)
  - Kennalestidae
    - *Murtoilestes abramovi* (3)
    - *Prokennalestes minor* (4)
    - *Prokennalestes trofimovi* (4)

Cimolodonta (Kielan-Jaworowska, Dashzeveg, and Trofimov, 1987, see chapter 8). *Gobiconodon,* a eutriconodontan genus shared with the Cloverly Formation of North America (among others elsewhere) and thus implying some degree of interchange among continents, is represented by two species (Trofimov, 1978; Kielan-Jaworowska and Dashzeveg, 1998). Other noteworthy elements of the fauna include a "symmetrodontan" (Trofimov, 1980, 1997), one or two peramurid-like "eupantotherians" (Dashzeveg, 1979, 1994), and *Kielantherium,* a remarkably primitive boreosphenidan that may be related to Deltatheroida (Dashzeveg and Kielan-Jaworowska, 1984). Last but certainly not least, Höövör has yielded one of the earliest generally accepted eutherians, represented by both upper and lower dentitions, *Prokennalestes* (see Belyaeva et al., 1974; Kielan-Jaworowska and Dashzeveg, 1989; Sigogneau-Russell et al., 1992); a petrosal has also been referred to *Prokennalestes* (Wible et al., 2001).

Another locality in Mongolia, **Khamaryn Us** (figure 2.12), is in the Khukhtyk or Dzunbain Formation. It apparently was discovered in 1977 and is thought to be of the same age as Höövör, that is, Aptian–Albian (Tumanova, 1987). The fauna includes *Psittacosaurus,* a variety of other dinosaurs, and varanoid lizards. An "amphilestid" mandible with m2–3 has apparently been collected at Khamaryn Us (Reshetov and Trofimov, 1980; Averianov, 2000).

The locality of **Oshih** (Aishle), long known for its assemblage of Early Cretaceous dinosaurs (e.g., Osborn, 1923), has recently yielded fossil mammals (see Rougier et al., 2001, chapter 7). Two species of *Gobiconodon* are represented at Oshih: *G. hopsoni* (known by fragments of the maxilla and dentary), which is as large or larger than North American *G. ostromi,* and an apparently new species (known by two dentary fragments, possibly belonging to the same individual), similar in size to *G. borissiaki,* known from Höövör and elsewhere.

China

Early Cretaceous mammals, as yet mostly unidentified, have been discovered at four sites in Inner Mongolia, China (figure 2.12): Elesitai, Laolonghuozi, Hangjin Qi, and Yanhaizi. **Elesitai** is in the Bayan Gobi Formation, locally made up of claystones, mudstones, and sandstones. The modest fauna, said to include psittacosaurs, is thought to be of Aptian–Albian age. The mammalian specimen, as yet underscribed, is a well-preserved lower jaw with three teeth (Dong, 1993; Lucas and Estep, 1998; Averianov and Skutschas, 2000a).

The remaining three sites (Averianov and Skutschas, 2000a) are all near Hangjin-Qi and are in the Eijinhoro Formation (synonymous with the Luohandong Formation). **Laolonghuoze** is 30 km west of Hangjin-Qi. The herpetofauna includes an atoposaurid crocodilian (Wu et al., 1996); a mammalian humerus has been collected from the site (Dong, 1993). **Yanhaizi** has yielded a maxilla of an unidentified mammal (Averianov and Skutschas, 2000a). **Hangjin-Qi** is the source of a mandible described as belonging to an "amphilestid" mammal, *Hangjinia chowi* (see Godefroit and Guo, 1999). Though the tooth crowns are broken off, X-rays show that molariforms were replaced, as they were in *Gobiconodon* (see chapter 7).

Early Cretaceous mammals have recently been reported, though not yet described, from the **Mazongshan** area, Xinmingbao Group (= Chi-Jin-Bao Group), Gansu Province (figure 2.12). Mammalian fossils have been collected from at least two sites in the lower part of the unit (lower Red Beds and middle Gray Beds). Combined sedimentologic and paleontologic data suggest a fluviolacustrine depositional setting and paleoclimatic conditions that were semiarid and subtropical (Tang et al., 2001). The lower part of the Xinmingbao Group is estimated to be of Barremian–Albian age, based on invertebrates, vertebrates, plant macrofossils, and palynomorphs (Tang et al., 2001; see also Chow and Rich, 1984b; Averianov and Skutschas, 2000a). All mammalian fossils collected to date reportedly represent a new taxon of Gobiconodontidae (Tang et al., 2001).

Xinjiang Autonomous Region is home to at least one occurrence of an Early Cretaceous mammal (Chow and Rich, 1984b). The locality, **Huang-Ni-Tan** (figure 2.12), is in the Ke-la-mei-li area of the Junggur Basin. The fossil, consisting of a dentary with five broken teeth belonging to an unidentified mammal, is from the Lower Cretaceous Shengjinkou Formation, Tugulu Group. Another site in Xinjiang Region, Jianshan Wash, was noted earlier under Late Jurassic. The described taxon from this site, *Klamelia,* is an "amphilestid triconodont" that is rather similar to Early Cretaceous *Gobiconodon,* suggesting the possibility that Jianshan Wash may eventually prove to be of Early Cretaceous age as well. The upper stratigraphic constraint of the site is reported to consist of horizons referable to the Upper Cretaceous (Chow and Rich, 1984a).

The majority (as well as the most spectacular) of Early Cretaceous mammal occurrences in China are from Liaoning Province. The mammal-bearing sequence lies in the Jehol Group. The basal unit of the Jehol Group is the Yixian Formation, which is mainly lacustrine, with interbedded volcanic tuffs and lavas. Beds 6 and 8 of the Yixian Formation (see Wang, Wang, et al., 1998) have yielded mammals. A radiometric date of 124.6 Ma (Swisher et al., 1999) from an underlying horizon (bed 4) is believed to closely approximate the age of the Yixian mammals, placing them in the middle Barremian. Overlying the Yixian

Formation is the Shahai Formation.[7] The Shahai Formation includes a mixture of coal, fluvial, and lacustrine deposits. The age of the Shahai Formation, from which mammals are also known, is not well understood. Given the radiometric dates from the underlying Yixian Formation, the mammals of the Shahai Formation must be younger than middle Barremian—perhaps much younger. We tentatively regard it as Aptian. The Fuxin Formation, in turn, overlies the Shahai Formation. The Fuxin Formation consists of coals, sandstone, and sandy conglomerates (Wang et al., 1995). Like the rest of the Jehol Group, this unit was at one time considered to be of Jurassic age (e.g., Yabe and Shikama, 1938), but is now referred to the Lower Cretaceous. Age estimates for the Fuxin Formation range from Barremian–Aptian (Nessov, Sigogneau-Russell, and Russell, 1994), to an upper limit of Aptian (Wang et al., 1995), to much younger. We tentatively regard it as Aptian–Albian and suspect that it may eventually prove to be younger, perhaps early Late Cretaceous. As noted earlier, the "symmetrodontan" *Manchurodon*, from the Zhadyzhao Mine, was incorrectly ascribed to the Fuxin Formation and may well be of Jurassic age.

The most remarkable newly discovered Early Cretaceous mammals are, without doubt, those from the Yixian Formation (see the review of stratigraphic sequence and fossil horizons by Wang, Wang, et al., 1998). To date, mammalian fossils have been reported from four sites in this unit. Three are especially noteworthy: the "symmetrodontan" *Zhangheotherium quinquecuspidens*, from the **Jianshangou** Valley (bed 8); the "triconodont" *Jeholodens jenkinsi*, from the nearby site of **Sihetun** (bed 6) (Hu et al., 1997, 1998; Ji et al., 1999); and the eutherian *Eomaia scansoria*, from the **Dawangzhangzi** locality, farther to the west (Ji et al., 2002). All three are represented by specimens that, though flattened, are breathtaking in their completeness, including the skull and skeleton (the holotype of *Eomaia scansoria* also includes remnants of some soft tissues, such as costal cartilages and fur). The Dawangzhangzi locality has also yielded the partial skull and skeleton of the eobaatarid multituberculate, *Sinobaatar lingyuanensis* (see Hu and Wang, 2002a,b).

The basal member of the Yixian Formation has yielded mammalian fossils from near the village of **Lu-Jia-Tun.** Skulls with articulated mandibles, together with some postcranial remains, are known for two taxa: *Gobiconodon* (reportedly represented by a new species) and its apparent close relative *Repenomamus* (Li et al., 2000, 2001; Wang et al., 2001b).

[7] Some workers (e.g., Wang et al., 1995) separate the Shahai Formation into the Shahai, *sensu stricto*, and the underlying Jiu-Fuo-Tang Formation.

The principal site in the Shahai Formation is **Badaohao,** Heishan (figure 2.12), from which at least five mammalian taxa are known (Wang et al., 1995). Two "plagiaulacidan" multituberculates are present. One of these is similar to *Albionbaatar,* otherwise known from the Berriasian of England; the other is a plagiaulacid, represented by a dentary. Other taxa from Badaohao include a "symmetrodontan," an extremely primitive, aegialodont-like boreosphenidan, a possible eutherian (Wang et al., 1995; see also Averianov and Skutschas, 2000a), and an unspecified type of "triconodont."

Two sites in the Fuxin Formation (sometimes referred to in older literature as the "Husin Series" [see Clemens et al., 1979]) have yielded mammals (figure 2.12). *Endotherium,* a probable eutherian represented by jaw and associated postcranials (Shikama, 1947), is from the **Xinqiu** (Hsinchiu) Mine in the Fuxin Formation. The specimen was unfortunately lost during World War II, but illustrations suggest a rather advanced morphology. A single tooth of an unidentified multituberculate is known from a similar horizon at another locality, near **Xindi** (Wang et al., 1995).

Japan

Discoveries of two Early Cretaceous mammals in Japan have been reported, though neither has yet been described. The fossil occurrences, at **Kaseki-kabe** in Ishikawa Prefecture (figure 2.4), are in the Kuwajima Formation and are believed to be of late Hauterivian age. One specimen is a dentary bearing molars of "amphilestid" design (Rougier et al., 1999) and the other is a dentary (bearing one tooth, p4; another tooth of uncertain identity may be associated) of an eobaatarid-like multituberculate (Takada et al., 2001).

AUSTRALIA

An island continent with a highly endemic biota sharing certain elements with other southern continents, Australia has long fascinated students of evolution and biogeography. Until recently, though, no mammalian fossils older than Miocene were known from this important landmass down under. Discoveries in the Early Cretaceous of Australia (table 2.12) are now beginning to shed light on the antiquity of mammals there and are raising fundamental new questions on the origin and relationships of some major mammal groups.

The first Early Cretaceous mammal to be reported from Australia is from **Lightning Ridge,** New South Wales (figure 2.13). The occurrence is in the Wallangulla Sandstone Member of the Griman Creek Formation in a whitish claystone interbedded with clayball conglomer-

TABLE 2.12. Early Cretaceous Mammals of Australia (see figure 2.13). Localities: 1, Lightning Ridge, New South Wales (Griman Creek Formation, Albian); 2, Dinosaur Cove (Eumeralla Formation, Aptian–Albian); 3, Flat Rocks, Victoria (Wonthaggi Formation, Aptian)

?Mammalia *incertae sedis* (1)
Mammalia *incertae sedis* (1)
Australosphenida
Ausktribosphenidae
Ausktribosphenos nyktos (3)
Bishops whitmorei (3)
Gen. et sp. indet. (3)
Kollikodontidae
Kollikodon ritchei (1)
Steropodontidae
Steropodon galmani (1)
?Steropodontidae
Teinolophos trusleri (3)
Family *incertae sedis*
Gen. et sp. indet. (2)

ates, representing a shallow estuarine environment (Archer et al., 1985; Flannery et al., 1995). Palynomorphs and other evidence indicate a middle Albian age for the unit. A diverse vertebrate fauna, including dinosaurs (see, e.g., Weishampel, 1990), is known from Lightning Ridge. Many of the fossils have been transported. Notably, the fossils are pseudomorphs: the bone has been replaced by opal, which is mined from the unit. The fossils are highly prized by collectors. *Steropodon galmani,* known by a dentary with the last three molars, was first described as a monotreme possibly belonging to the platypus family (Archer et al., 1985). It was later placed in its own family, Steropodontidae, and considered to represent a sister taxon to the two living families of monotremes, Ornithorhynchidae and Tachyglossidae (Flannery et al., 1995). Monotremes are generally considered to represent a clade that diverged extremely early in mammalian evolution. However, the teeth of *Steropodon* are remarkably advanced, suggesting a closer relationship to therians (*sensu lato*) than previously thought (Archer et al., 1985; Kielan-Jaworowska, Crompton, and Jenkins, 1987; Luo, Cifelli, and Kielan-Jaworowska, 2001). A second mammal from Lightning Ridge, *Kollikodon ritchei,* was originally based on a dentary with three molars. A well-preserved maxilla with teeth was subsequently found and is figured herein as figure 6.5D_2 (courtesy of Anne Musser). Flannery et al. (1995) argued that the morphology of the dentary indicates that *Kollikodon* might be a monotreme, but we assign it to Monotremata only tentatively (see chapter 6). Both *Steropodon* and *Kollikodon* are rather large by the standard of Mesozoic mammals, possibly reflecting cold conditions in the Early Cretaceous of Australia (Rich et al., 1989). Monotremes clearly achieved some degree of diversity in the Australian Mesozoic. Ornithorhynchidae are now known from the Tertiary of South America, suggesting possible dispersal to that continent no later than early Paleocene (Pascual et al., 1992a).

Two other specimens from Lightning Ridge deserve passing mention. One, an edentulous maxilla, belongs to a mammal but is not further identifiable. The other is a complex tooth of debatable affinities. If mammalian, the specimen most resembles the pattern seen in upper molars of Dryolestidae (Clemens et al., 2003). Dryolestids survived through much of the Late Cretaceous in South America (see later); hence their presence in the Early Cretaceous of Australia might be expected. If mammalian (dryolestid or not), this specimen is notable for its large size.

Early Cretaceous mammals have been recovered from two other sites in Australia, both on the coast of Victoria (figure 2.13). **Dinosaur Cove,** about 200 km west of Melbourne, is in the Eumeralla Formation. Two undescribed specimens, a monotreme humerus and a fragment of a mammal tooth, have been reported from this site (Rich et al., 1999).

By far the most significant and eye-opening mammalian fossils from the Cretaceous of Australia hail from the **Flat Rocks** site, about 120 km southeast of Melbourne. The site is under active investigation by Thomas H. Rich, Patricia Vickers-Rich, and associates (see Rich and Vickers-Rich, 2000). Flat Rocks, which lies in the Wonthaggi Formation, consists of fluvial volcaniclastic sandstones and mudstones. It is placed in the Aptian, based on combined palynological and radiometric data. Evidence of cryoturbation stratigraphically below the fossiliferous horizon attests to cold temperatures and seasonally frozen ground (Rich et al., 1999). First to be described from Flat Rocks is *Ausktribosphenos nyktos,* based on several dentaries. The cheek teeth have an advanced morphology that belies their age, suggesting the possibility of eutherian affinities (Rich et al., 1997, 1999, 2001a). Other evidence, including jaw structure (Kielan-Jaworowska et al., 1998), suggests that *Ausktribosphenos* may represent an early, divergent clade that also includes monotremes (Luo, Cifelli, and Kielan-Jaworowska, 2001; Luo et al., 2002; Rauhut et al., 2002, see chapter 6). Monotremes were clearly differentiated by the Aptian, as shown by another element of the Flat Rocks fauna: *Teinolophus trusleri,* a possible relative of *Steropodon,* mentioned earlier (see Rich et al., 1999; Rich, Vickers-Rich, et al., 2001). The most recent addition to the roster of named mammals from Flat Rocks is *Bish-*

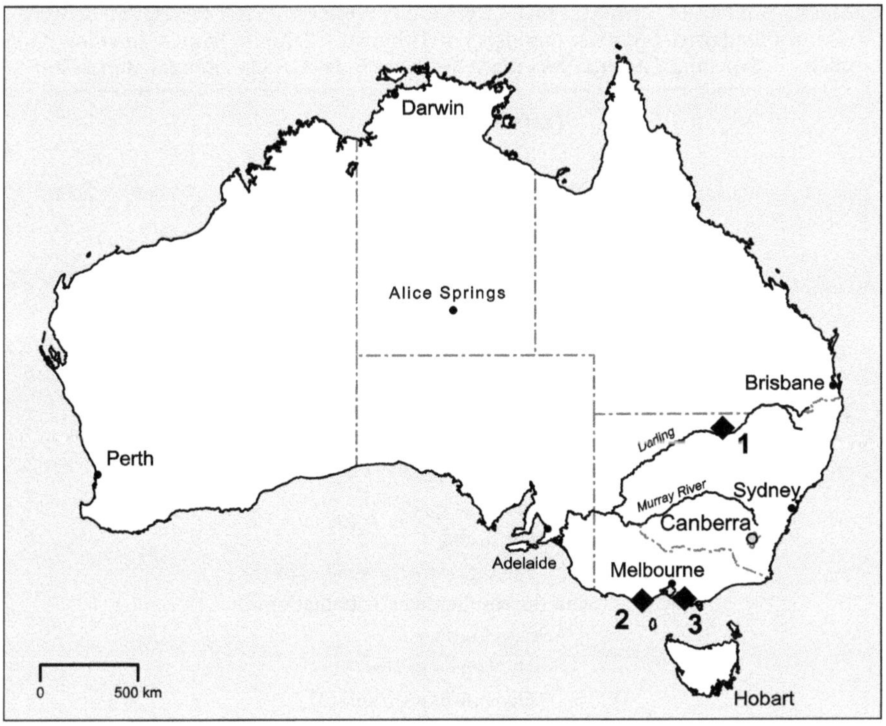

FIGURE 2.13. Mesozoic mammals localities of Australia (all Early Cretaceous; black diamonds). Localities: **1, Lightning Ridge** (Wallangulla Sandstone Member, Griman Creek Formation; Albian, New South Wales); **2, Dinosaur Cove** (Eumeralla Formation; ?Aptian–Albian, Victoria); **3, Flat Rocks** (Wonthaggi Formation; Aptian, Victoria).

ops whitmorei. Known by several dentaries and the entire postcanine dentition, *Bishops* is an ausktribosphenid and, like *Ausktribosphenos,* its affinities are contended.

AFRICA

The first Early Cretaceous mammals to be reported from Africa are from the **Koum Basin,** Cameroon (figure 2.6). The fossils occur in the Grès de Gaba Member of the Koum Formation; palynologic data suggest a hot, dry climate and a Barremian age (Brunet et al., 1990). The vertebrate fauna is moderately diverse, including fishes as well as numerous amphibians and reptiles. Mammalian specimens from the site include an edentulous jaw and several isolated teeth (Brunet et al., 1988, 1990; Jacobs et al., 1988). At least three taxa, all apparently nontribosphenic therians, are present. Of these, only one, *Abelodon abeli,* has been named. *Abelodon* is known from isolated upper and lower molars and is referred to the Peramuridae, a group of "eupantotherians" thought to be close to the ancestry of boreosphenidan mammals (chapter 10).

A diverse and extremely important Early Cretaceous fauna has been recovered from the **Synclinal d'Anoual,** about 100 km east of Talsinnt, Morocco (figure 2.6). The site was worked by Sigogneau-Russell and Russell in the late 1980s, and publication of the information is still continuing. The fossils occur in the Séquence B des Couches Rouges in an incredibly rich lens (e.g., Sigogneau-Russell, 1991b). Adjacent marine beds indicate a deltaic depositional environment; a Berriasian age is suggested on the basis of calcareous nannofossils (Sigogneau-Russell et al., 1998).

At least 15 kinds of mammals (table 2.13) have been described or explicitly mentioned from the Synclinal d'Anoual and three or more additional taxa are known from the site (Sigogneau-Russell et al., 1998), making it one of the most diverse known mammal assemblages from the Mesozoic. One or more "amphilestid triconodonts" appear to be present (Sigogneau-Russell et al., 1990). Two rather bizarre eutriconodontans from the Synclinal d'Anoual are of uncertain familial affinities; *Dyskritodon amazighi* and *Ichthyoconodon jaworowskorum* (see Sigogneau-Russell, 1995). One multituberculate, *Hahnodon taqueti* (referred to its own monotypic family), documents the presence of this group in the Early Cretaceous of Africa (Sigogneau-Russell, 1991c). Three "symmetrodontans," five "eupantotherians," and as many as three tribosphenic mammals are also known from the fauna (Sigogneau-Russell, 1989a, 1990, 1991a,b,d, 1992, 1994a, 1998). The tribosphenic mammals, *Hypomylos* sp. and *Tribotherium,* are of special interest in that they are some of the earliest known. Some elements of the fauna are known from the Late Jurassic or earliest Cretaceous of Western Europe. The "symmetrodontan" *Thereuodon,*

TABLE 2.13. Late Jurassic and Early Cretaceous Mammals of Africa (see figure 2.6; locality numbers do not correspond between map and table). Localities: 1, Tendaguru, Tanzania (Tendaguru beds; Kimmeridgian or Tithonian); 2, Koum Basin, Cameroon (Grès de Gaba Member, Koum Formation; Barremian); 3, Synclinal d'Anoual, Morocco (Séquence B des Couches Rouges, Berriasian)

- Mammalia
  - Family *incertae sedis*
    - Gen. et sp. indet. A (2)
    - Gen. et sp. indet. B (2)
- Eutriconodonta
  - ?"Amphilestidae"
    - Gen. et sp. indet. (3)
    - *Tendagurodon janenschi* (1)
  - Family *incertae sedis*
    - *Dyskritodon amazighi* (3)
    - *Ichthyoconodon jaworowskorum* (3)
- Haramiyida
  - Haramiyidae
    - *Staffia aenigmatica* (1)
- Multituberculata
  - Hahnodontidae
    - *Hahnodon taqueti* (2)
- Archaic "symmetrodontans"
  - Family *incertae sedis*
    - *Atlasodon monbaroni* (3)
    - *Microderson laaroussii* (3)
  - Thereuodontidae
    - *Thereuodon dahmanii* (3)
- Stem Cladotheria ("eupantotherians")
  - Family *incertae sedis*
    - Gen. et sp. indet. (3)
    - *Afriquiamus nessovi* (3)
    - *Minimus richardfoxi* (3)
    - *Brancatherulum tendagurense* (1)
  - Donodontidae
    - *Donodon perscriptoris* (3)
  - Peramuridae
    - *Abelodon abeli* (2)
    - *Peramus* sp. (3)
  - ?Peramuridae
    - *Tendagurutherium dietrichi* (1)
- Stem Boreosphenida ("tribotherians")
  - Aegialodontidae
    - *Hypomylos micros* (3)
    - *Hypomylos phelizoni* (3)
  - Family *incertae sedis*
    - *Tribotherium africanum* (3)

the "eupantotherian" *Peramus*, and an early tribosphenic mammal are present in the Purbeck (Sigogneau-Russell and Ensom, 1994). Though Morocco was presumably separated from Western Europe by the Tethys Sea during the Early Cretaceous and the fauna of the Synclinal d'Anoual has a high degree of endemism, these ties suggest some kind of recent immigration between the two areas (Sigogneau-Russell et al., 1998).

SOUTH AMERICA

The South American continent is, thus far, home to only one site for Early Cretaceous mammals. The site, near **Zapala** in Neuquén Province, Argentina (figure 2.7), is in the La Amarga Formation, of Hauterivian or Hauterivian–Barremian age (Bonaparte, 1986a; Rougier et al., 1992). Only a single taxon, *Vincelestes neuquenianus*, is known from the site, but it is a highly significant occurrence: *Vincelestes* is represented by unusually complete materials, including 17 dentaries, six nearly complete skulls, and various postcrania (see, e.g., Bonaparte and Rougier, 1987; Rougier et al., 1992; Rougier, 1993). *Vincelestes* is of special interest because it is so completely known and because it appears to be a proximal relative of boreosphenidan mammals (see chapter 10). One other occurrence in the Early Cretaceous of South America should be mentioned here for the sake of completeness. Several teeth from the Lower Cretaceous (Albian) Itapecuru Formation near Itapecuru-Mirim, Maranhno, Brazil, were described as a "triconodont," *Candidodon itapecuruense* (see Carvalho and Campos, 1988). Closer study revealed that the teeth belong to a notosuchian crocodile with heterodont dentition (Carvalho, 1994).

Footprints attributed to a mammal have been reported from rocks of unspecified age at Nanhuel-Huapi, Argentina. The site, not shown here, is included on a map with sites of Cretaceous age by Leonardi (1994: 11).

NORTH AMERICA

Bona fide Early Cretaceous mammals are known from four regions and stratigraphic units in North America: the Trinity Group, Texas and Oklahoma (figure 2.14); the Cloverly Formation, Wyoming and Montana; the Kelvin Formation, Utah (figure 2.10); and the Arundel Clay, Maryland (figure 2.15). A fifth, the Cedar Mountain Formation, Utah (figure 2.10), straddles Albian–Cenomanian time (Early–Late Cretaceous) and is arbitrarily discussed under this heading as well.

The Trinity Group in north-central Texas includes (in ascending order) the Twin Mountains, Glen Rose, and Paluxy formations. Invertebrates from the Glen Rose and

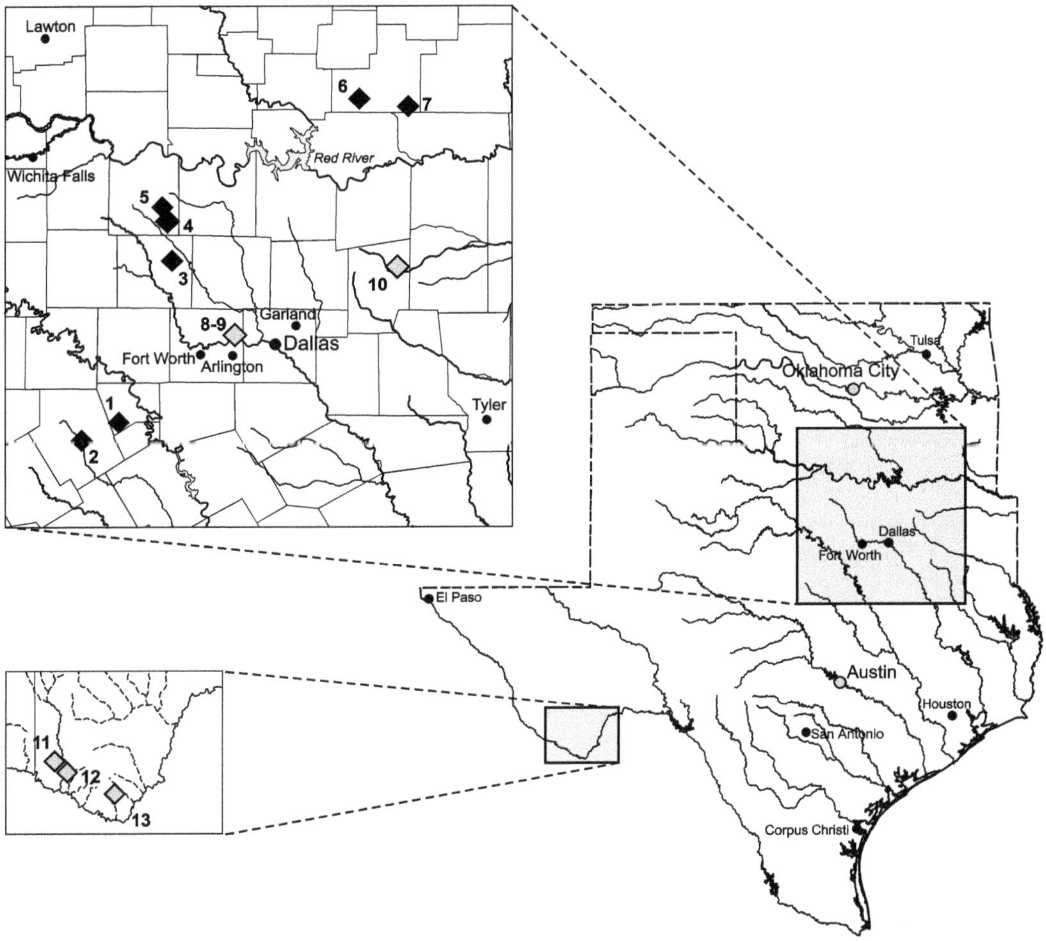

FIGURE 2.14. Cretaceous mammal localities of Texas and Oklahoma. Black diamonds, Early Cretaceous (localities 1–7); gray diamonds (localities 8–14), Late Cretaceous. Localities or local faunas: **1, Paluxy Church** (Twin Mountains Formation; late Aptian); **2, Pecan Valley Estates** (Paluxy Formation; early Albian); **3, Butler Farm; 4, Greenwood Canyon; 5, Willawalla; 6, Umsted Farm; 7, McLeod Honor Farm** (all Antlers Formation; Aptian–Albian); **8, Carter Field; 9, Bear Creek** (Woodbine Formation; Cenomanian); **10, Commerce** (Kemp Clay; Maastrichtian); **11, Terlingua; 12, Running Lizard; 13, Talley Mountain** (upper shale member, Aguja Formation; late Campanian: Judithian).

other marine units constrain the age of the Twin Mountains Formation as Aptian and that of the Paluxy Formation as early to middle Albian (Jacobs et al., 1991; Jacobs and Winkler, 1998). The Twin Mountains and Paluxy formations are terrigenous and are comprised of sandstones, claystones, and siltstones of fluviodeltaic origin, deposited on a coastal plain. Northward and eastward, in far northern Texas and southern Oklahoma, the Glen Rose pinches out. Here the terrigenous rocks of the Trinity Group are referred to the Antlers Formation and, lacking the marine invertertebrates needed for more precise correlation, are simply regarded as being of Aptian–Albian age. An excellent summary of the geology and faunas from localities in Texas is given by Winkler et al. (1990).

Thus far, only three mammals—an unidentified multituberculate, an unidentified therian, and the triconodontid *Astroconodon denisoni*—are known from beds unequivocally belonging to the Aptian Twin Mountains Formation. All of these are from a single site, **Paluxy Church,** Hood County, Texas (figure 2.14, table 2.14). The only mammal definitely known from a locality in the Paluxy Formation is *Comanchea hilli,* a somewhat aberrant tribosphenic mammal represented by an upper molar (Jacobs et al., 1989) from **Pecan Valley Estates** in Erath County, Texas (see chapter 11).

All the remaining mammals from the Trinity Group come from the Antlers Formation. The initial discovery was made in 1949, when personnel from the Field Museum recovered two jaws of the triconodontid *Astroconodon denisoni* from the surface at **Greenwood Canyon** (see Patterson, 1951) (figure 2.14). Follow-up investigations by Bryan Patterson and colleagues led to the recov-

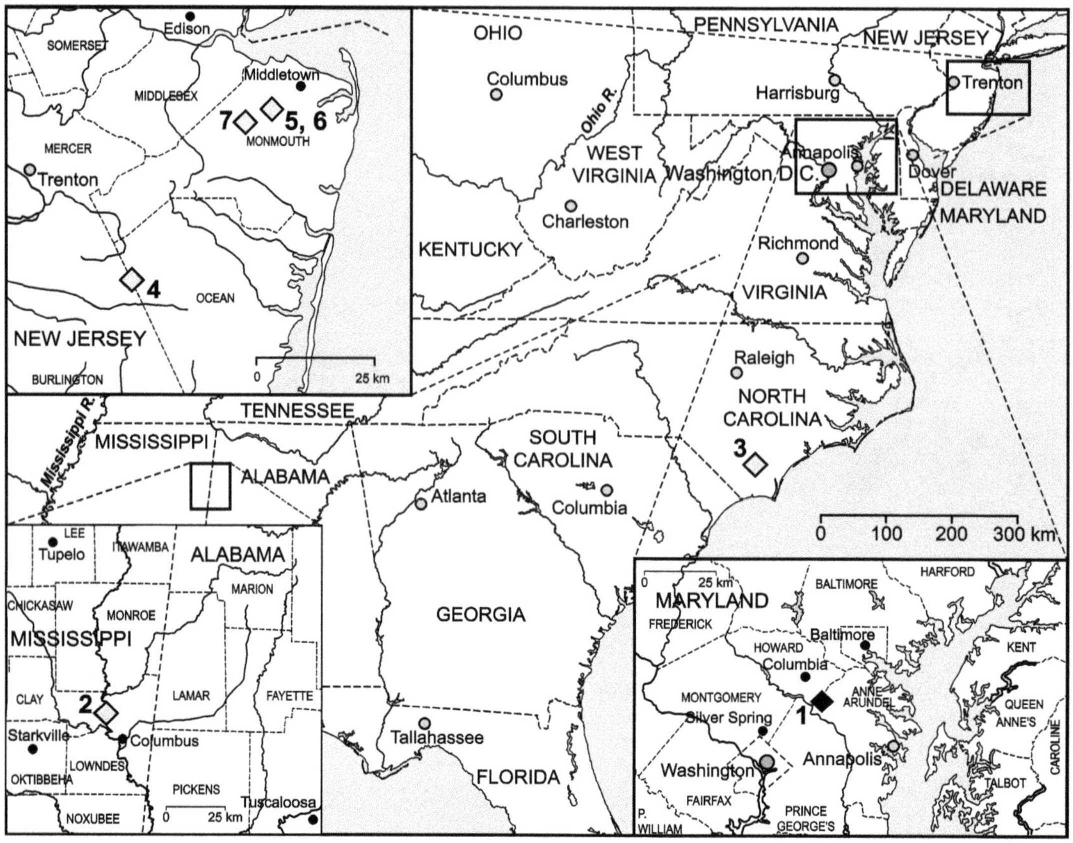

FIGURE 2.15. Cretaceous mammal localities of the eastern United States. Black diamond, Early Cretaceous (locality 1); gray diamonds, Late Cretaceous (localities 2–7). Localities or local faunas: **1, Muirkirk** (Arundel Clay facies, Potomac Group, ?middle Aptian, Maryland); **2, Vinton Bluff** (Tombigee Sand, Eutaw Formation, late Santonian, Mississippi); **3, Elizabethtown Dump** (Black Creek Group; ?Campanian, North Carolina); **4, Ellisdale** (Marshalltown Formation, late Campanian, New Jersey); **5, Hop Brook; 6, Ramanssein Brook** (both Mount Laurel Formation, Maastrichtian, New Jersey); **7, Monmouth Brook** (?Campanian, New Jersey).

ery of fossils belonging to additional taxa, including at least one multituberculate (herein tentatively assigned to the primitive cimolodontan ?*Paracimexomys crossi*), the "symmetrodontan" *Spalacotheroides bridwelli,* and tribosphenic mammals, two of which (*Holoclemensia texana* and *Kermackia texana*) were later given formal names (see Butler, 1978a).[8] The so-called "Trinity therians" occupy a prominent position among early mammals because their study led to the current interpretation of molar evolution during the Mesozoic (Patterson, 1956; see also Butler, 1939). Patterson employed underwater screenwashing and associated techniques to recover the mammalian fossils, as have all subsequent paleontologists working in the Trinity Group. As a result, virtually all the specimens are isolated teeth or parts thereof. Notwithstanding an earlier statement (Patterson, 1956) to the contrary, there is no definitive evidence for "plagiaulacidan" multituberculates in the Trinity Group of Texas or, for that matter, from other Aptian–Albian faunas of North America. All specimens recovered thus far appear to represent primitive cimolodontans referable to the *Paracimexomys* group (Krause et al., 1990; Cifelli, 1997a). A nearby site to the north, **Willawalla,** was worked by personnel from the Field Museum (figure 2.14). Patterson (1956) mentioned unspecified mammals, but to our knowledge, the only published occurrence from this site (also known as King's Creek) is *Astroconodon,* reported by Winkler et al. (1990).

During the 1960s, Bob Slaughter, of Southern Methodist University, made a significant collection of mammalian fossils from several sites at **Butler Farm,** Wise County, Texas (figure 2.14). At least seven taxa are represented, four of which are tribosphenic therians. Slaughter (e.g., 1971) believed that both Marsupialia (*Holoclemensia*) and

[8] We omit *Adinodon pattersoni,* based on an edentulous dentary (Hershkovitz, 1995), which we consider indeterminate (see Cifelli and Muizon, 1997).

TABLE 2.14. Early Cretaceous Mammals of North America (see figures 2.10, 2.14, 2.15; locality numbers do not correspond between table and maps). Localities: 1, Paluxy Church (Twin Mountains Formation, Aptian); 2, Pecan Valley Estates (Paluxy Formation, Albian); 3, Greenwood Canyon; 4, Butler Farm; 5, Willawalla; 6, Umsted Farm; 7, McLeod Honor Farm (all Antlers Formation, Aptian–Albian); 8, Bridger (two sites, undifferentiated here); 9, Cashen Ranch; 10, Cottonwood Creek; 11, Crooked Creek; 12, Ninemile Hill (all Cloverly Formation, Aptian–Albian); 13, Coalville (Kelvin Formation, Aptian–Albian); 14, Muirkirk (Arundel Clay, Aptian)

Mammalia *incertae sedis* (11, 12, 13)	Stem Boreosphenida ("tribotherians")
Eutriconodonta	Family *incertae sedis*
Gobiconodontidae	*Comanchea hilli* (2)
Gobiconodon ostromi (8, 9)	*Slaughteria eruptens* (4)
Triconodontidae	Gen. et sp. indet. (1, 7, 9, 10, 11)
Gen. et sp. indet. (10, 11)	Kermackiidae
Arundelconodon hottoni (14)	*Kermackia texana* (3)
Astroconodon denisoni (1, 3, 4, 7)	*Trinititherium slaughteri* (4)
Astroconodon cf. *denisoni* (5, 6)	Pappotheriidae
cf. *Astroconodon* sp. (8, 9)	Gen. et sp. indet. (7)
Corviconodon montanensis (9)	*Holoclemensia texana* (3, 4)
Multituberculata	*Pappotherium pattersoni* (4)
Family *incertae sedis*	?Deltatheroida
Gen. et sp. indet. (1, 7, 9)	Family *incertae sedis*
?*Paracimexomys crossi* (7)	*Atokatheridium boreni*
cf. *Paracimexomys* sp. (3, 4, 11)	cf. *Atokatheridium boreni*
Stem Trechnotheria ("symmetrodontans")	Eutheria
Spalacotheriidae	Family *incertae sedis*
Gen. et sp. indet. (10)	*Montanalestes keeblerorum* (10)
Spalacotheroides bridwelli (3)	
cf. *Spalacotheroides* sp. (7, 10, 11)	

Eutheria (*Pappotherium*) were present in the fauna, but evidence for this is tenuous (Butler, 1978a; Jacobs et al., 1989; Cifelli, 1999b), and they are treated later (chapter 11) as "tribotherians." Both Greenwood Canyon and Butler Farm are reportedly no longer accessible (Winkler et al., 1989), but ongoing investigations at other localities of the Trinity Group of Texas are being carried out by Louis L. Jacobs, Dale A. Winkler, and associates, also at Southern Methodist University (e.g., Jacobs et al., 1989; Jacobs and Winkler, 1998; Winkler et al., 1990).

The Antlers Formation also crops out in a narrow band along southern Oklahoma. Field parties led by Richard L. Cifelli have recovered mammals from two sites, **Umsted Farm** and the **McLeod Honor Farm,** both in Atoka County (figure 2.14). Mammals described thus far include the multituberculate ?*Paracimexomys crossi;* an unidentified (probably new) tribosphenic mammal (Cifelli, 1997a); and a possible deltatheroidan, *Atokatheridium boreni* (see Kielan-Jaworowska and Cifelli, 2001, and chapter 12). Several other mammals and a host of other microvertebrates are known from the sites (Cifelli, Gardner, et al., 1997), and work is continuing.

The Cloverly Formation, exposed along the margins of the Bighorn Basin and other areas of Montana and Wyoming, is well known for its dinosaur fauna (Ostrom, 1970). Age constraints on the unit are imprecise; it is older than the overlying Thermopolis Shale, which is estimated on the basis of marine invertebrates to be early late Albian (Jacobs et al., 1991; Jacobs and Winkler, 1998). The Cloverly is generally assigned an Aptian–Albian age based on the similarity of its vertebrate assemblage with that of the Trinity Group (see, e.g., Brinkman et al., 1998). The Cloverly consists of fluvially deposited claystones, siltstones, and occasional sandstones, with color banding suggestive of paleosol development. Three units of the Cloverly Formation, V–VII, are recognized in Ostrom's (1970) stratigraphic sequence, and each of these has yielded mammals. Though fewer taxa are known than from the Trinity Group, the Cloverly Formation is significant because it has yielded some rather complete specimens through surface prospecting and quarrying procedures, as well as isolated teeth obtained through bulk rock processing.

Field parties led by Farish A. Jenkins, of Harvard University, made the initial discovery of mammals in the Cloverly Formation in the 1970s, and work by Cifelli and associates is in progress. Lewis Pocket, southeast of **Bridger** (figure 2.10), yielded a remarkable series of a triconodontid similar to *Astroconodon*, represented by nearly complete skulls and postcranial elements (Jenkins and Crompton, 1979; Crompton and Sun, 1985). The site is near the famous Yale *Deinonychus* Quarry (Ostrom, 1970); fossils occur in limonitic nodules in unit V. About 2 km to the east, a second site in the Bridger area yielded jaws and postcrania of the large eutriconodontan *Gobiconodon ostromi* (see Jenkins and Schaff, 1988). The site is in the base of unit VII.

Fossil mammals have also been recovered from the vicinity of **Cashen Ranch** (figure 2.10), southern Montana, in the base of unit VII of the Cloverly Formation. In addition to *Gobiconodon* (see Jenkins and Schaff, 1988), the triconodontid *Corviconodon* (see Cifelli et al., 1998) and several undescribed taxa are present (RLC, unpubl. data).

A third area for mammals in the Cloverly Formation is in the **Cottonwood Creek** drainage, east of Edgar, Montana (figure 2.10). The productive zone, again in the base of unit VII, has thus far yielded jaws of several undescribed taxa, as well as the tribosphenic mammal *Montanalestes keeblerorum* (see Cifelli, 1999b). *Montanalestes* appears to be a eutherian, which are known elsewhere in the Early Cretaceous only from Asia (Averianov and Skutschas, 2000a; Ji et al., 2002). Eutheria are otherwise unknown from North America until the early Campanian (Fox, 1984b; Cifelli, 1990e), which may not be surprising considering the sparseness of the fossil record from the early Late Cretaceous. A second mammal from Cottonwood Creek is a spalacotheriid symmetrodontan similar to *Spalacotheroides* from the Antlers Formation of Texas. The fossil from Cottonwood Creek is a remarkably complete dentary, including the canine and all postcanine teeth (Cifelli et al., 2000).

A small microvertebrate assemblage, obtained through screenwashing, is known from unit V of the Cloverly Formation at **Crooked Creek,** northern Wyoming (figure 2.10). Two mammals are known from this site; one is a triconodontid otherwise unknown from the Cloverly Formation, so it documents a considerable diversity of these primitive mammals from the unit—at least four eutriconodontans are known from the Cloverly (Cifelli et al., 1998). A single mammalian tooth, a lower premolar, has been recovered from beds presumed to represent the Cloverly Formation at **Ninemile Hill,** Como Bluff, Wyoming (Trujillo, 1999).

A single mammal specimen, an unidentified petrosal, has been recovered from a site near **Coalville,** Utah (figure 2.10). The specimen is from the Kelvin Formation, reported to be of Aptian–Albian age (Prothero, 1983). The site has also yielded dinosaur eggshell material (Jensen, 1970).

The Arundel Clay, exposed in Maryland (figure 2.15) and adjacent parts of the U.S. Eastern Seaboard, is another Lower Cretaceous unit known primarily for its dinosaurs. Most such fossils were collected in the nineteenth century, while the clay was being mined for the sedimentary iron ore contained within it (Kranz, 1996), and consist of isolated, largely indeterminate elements (see, e.g., Gilmore, 1921). The Arundel Clay is considered to be a facies within the Potomac Group and appears to represent swamp deposits in a system of fluvial oxbows (Kranz, 1998). The Potomac Group is well known for its fossil flora; palynomorphs suggest an Aptian, probably middle Aptian, age for the Arundel Clay (Doyle, 1992). The only mammal known from the Arundel Clay is the triconodontid *Arundelconodon hottoni,* based on a dentary found by Thomas R. Lipka at a site near **Muirkirk,** Maryland (a second, edentulous dentary from the same site is tentatively referred to the same species, see Rose et al., 2001). This species clearly belongs to an endemic North American clade of the group, but is more primitive than other taxa known from the Early Cretaceous of this continent (Cifelli, Lipka, et al., 1999).

The Cedar Mountain Formation, Utah (figure 2.10), is of interest because it appears to include rocks spanning the Barremian through the uppermost Albian. Several superposed dinosaur faunas are known (Kirkland et al., 1997) but, thus far, mammals are only known from the uppermost part of the Cedar Mountain Formation. Field studies by Jeffrey G. Eaton and Michael Nelson in 1983 led to recovery of the first mammals known from the unit (Nelson and Crooks, 1987; Eaton and Nelson, 1991). Later investigations by Cifelli and colleagues, still continuing, have resulted in a large, well-sampled collection, including some 80 taxa of vertebrates (Cifelli, Kirkland, et al., 1997). The fossil assemblage is from a restricted stratigraphic level in the Mussentuchit Member of the Cedar Mountain Formation and is termed the **Mussentuchit** local fauna (Cifelli, Nydam, Weil, et al., 1999). Mammals, known from eight sites, were collected mainly through bulk matrix processing, though some well-preserved specimens were found through quarrying. Most of the sites are in fluvial overbank deposits. A radiometric date of 98.39 ± 0.07 Ma places the Mussentuchit local fauna at the Albian-Cenomanian (Early–Late Cretaceous) boundary (Cifelli, Kirkland, et al., 1997).

The fauna of some 28 mammal taxa (table 2.15) includes at least three triconodontids (two genera, *Astroconodon* and *Corviconodon,* are shared with the Aptian–Albian faunas of North America) and four "symmetrodontans" (one genus, *Spalacotheridium,* is shared with younger North American faunas), indicating a surprising diversity of these archaic groups. Both the triconodontids and the "symmetrodontans" belong to endemic North American clades and are distantly related to those known from else-

TABLE 2.15. Albian-Cenomanian Mammals of the Mussentuchit Local Fauna, Utah (see figure 2.10)

- Eutriconodonta
  - Triconodontidae
    - *Astroconodon delicatus*
    - *Corviconodon utahensis*
    - *Jugulator amplissimus*
- Multituberculata
  - ?"Plagiaulacida," family *incertae sedis*
    - *Ameribaatar zofiae*
    - *Janumys erebos*
    - cf. *Janumys erebos*
    - Gen. et sp. indet.
  - Cimolodonta, family *incertae sedis*
    - *Bryceomys intermedius*
    - *Bryceomys* cf. *intermedius*
    - *Cedaromys bestia*
    - *Cedaromys* cf. *bestia*
    - *Cedaromys parvus*
    - *Cedaromys* cf. *parvus*
    - ?*Paracimexomys perplexus*
    - ?*Paracimexomys* cf. *perplexus*
    - ?*Paracimexomys robisoni*
    - ?*Paracimexomys* cf. *robisoni*
  - Neoplagiaulacidae
    - ?*Mesodma* sp.
- Stem Trechnotheria ("symmetrodontans")
  - Spalacotheriidae
    - *Spalacolestes cretulablatta*
    - *Spalacolestes inconcinnus*
    - *Spalacotheridium noblei*
  - Family *incertae sedis*
    - Gen. et sp. indet.
- Stem Boreosphenida ("tribotherians")
  - Picopsidae
    - Gen. et sp. indet.
  - Family *incertae sedis*
    - Gen. nov., sp. A
    - Gen. nov., sp. B
- Marsupialia
  - ?Stagodontidae
    - *Pariadens mckennai*
  - Family *incertae sedis*
    - *Adelodelphys muizoni*
    - *Kokopellia juddi*
    - *Sinbadelphys schmidti*

where in the world (Cifelli and Madsen, 1998, 1999). As many as 15 taxa of multituberculates, collectively represented by more than 1,000 specimens, are known from the Mussentuchit local fauna (Eaton and Cifelli, 2001). Several may represent surviving "plagiaulacidans." Most of the multituberculates of the Mussentuchit local fauna belong to the *Paracimexomys* group of basal cimolodontans (see chapter 8); *Bryceomys*, otherwise known from the Turonian of Utah, makes its first appearance in this fauna. Also notable is the first appearance of Neoplagiaulacidae, in the form of ?*Mesodma*, a genus otherwise known from Aquilan and younger assemblages in the Late Cretaceous of North America. The Mussentuchit local fauna includes six boreosphenidan mammals, four of which are marsupials: *Adelodelphys, Kokopellia, Pariadens,* and *Sinbadelphys* (Cifelli, 1993a; Cifelli and Muizon, 1997; Cifelli, 2004). *Pariadens*, a possible stagodontid, is also known from the Cenomanian of Utah (Cifelli and Eaton, 1987; Eaton, 1993b). Despite the fact that they dominate the therian component of the Mussentuchit local fauna, marsupials of the Cedar Mountain Formation are rare and low in diversity (both taxonomic and morphologic) compared to Campanian–Maastrichtian faunas of North America. Currently, these four taxa are the oldest known marsupials. Their primitiveness, low diversity, and modest morphologic differentiation suggest that they may lie near the base of North America's Cretaceous marsupial radiation (Cifelli, 2004). There is as yet no definitive evidence for Eutheria in the Mussentuchit local fauna, despite their earlier (Aptian–Albian) presence in North America. Curiously, while the mammals seem to represent North American clades with no obvious close ties to groups known from elsewhere, other taxa in the Mussentuchit local fauna, particularly dinosaurs, suggest a proximal origin from Asia (Cifelli, Kirkland, et al., 1997).

Two footprints attributed to a mammal, *Duquettichnus kooli*, have been described from the Early Cretaceous of the Peace River Canyon, British Columbia. They are reported to show syndactyly of pedal digits II and III, so they have been referred to the Marsupialia (Sarjeant and Thulborn, 1986). Additional tracks from the Gates Formation, Alberta (Albian), are believed to represent several types of mammals (Sarjeant, 2000).

LATE CRETACEOUS

As with the Early Cretaceous, the past two decades have witnessed a remarkable improvement in the record of Late Cretaceous mammals. Specimens or faunas are now known from most major landmasses, Australia and Africa being the most notable and most unfortunate exceptions. Advances in knowledge have been particularly welcome from the pre-Campanian, or early Late Cretaceous, part of the record. Twenty years ago only two known occurrences —both consisting of practically indeterminate specimens from North America—fell within this time interval (Clemens et al., 1979). Significant early Late Cretaceous specimens and faunas are now known from both Asia and North America.

Eutherians (and, in North America, marsupials) and multituberculates underwent diversification on the northern continents. Late Cretaceous records now available from Western and Central Europe suggest some degree of endemicity during this epoch (Rădulescu and Samson, 1997; Gheerbrant and Astibia, 1999). The existing pre-Maastrichtian record for South America lacks boreosphenidan mammals, which apparently did not arrive there until the end of the Cretaceous. Instead, endemic groups or derived members of taxa more typical of the Jurassic are present (Bonaparte, 1990).

WESTERN EUROPE

Late Cretaceous mammals of Western Europe are now known from France, Spain, and Portugal (table 2.16). Most sites or occurrences are listed as being of Campanian–Maastrichtian or Maastrichtian age, with the exception of Champ-Garimond, France, which is believed to be Campanian. As summarized by Le Loeff (1991), Europe during the Late Cretaceous was an archipelago. Of the four major noninundated areas, Late Cretaceous mammals are known from the Ibero-Armorican and Transylvanian landmasses. As a whole, the European fauna shows a complex mixed pattern involving both Laurasian and Gondwanan elements, some of which may have been inherited from earlier faunal continuity of the landmasses.

TABLE 2.16. Late Cretaceous mammals of Western Europe (see figure 2.9; locality numbers do not correspond between map and table). Localities: 1, Champ-Garimond (Campanian, France); 2, Barrage Filleit (Grès de Labarre Formation, Campanian–Maastrichtian, France); 3, Peyrecave (Marnes d'Auzas Formation, Maastrichtian, France); 4, Taveiro (Late Cretaceous, Portugal); 5, Quintanilla del Coco (Calizas de Lynchus Formation, Maastrichtian, Spain); 6, Laño (Maastrichtian, Spain)

?Marsupialia[1]
- ?Alphadontidae
  - cf. *Alphadon* sp. (4)
- ?"Pediomyidae"
  - cf. "*Pediomys*" sp. (4)

Eutheria
- ?Nyctitheriidae
  - cf. *Leptacodon* sp. (4)[1]
- ?Palaeoryctidae
  - Gen. et sp. indet. A (5)[1]
  - Gen. et sp. indet. B (5)[1]
- Family *incertae sedis*
  - *Labes garimondi* (1)
  - *Labes quintanillensis* (5)
  - ?*Labes* sp. (6)
  - *Lainodon orueetxebarriai* (6)
  - *Lainodon* sp. nov. (6)
  - Gen. and sp. indet. (2, 3)

[1] Occurrences of extraordinary interest if confirmed; presently considered highly dubious.

Champ-Garimond, Gard, France (figure 2.9), yielded a lower molar first described by Ledoux et al. (1966). Long known as the "Champ-Garimond molar," the specimen was formally referred to its own species, *Labes garimondi,* by Sigé (in Pol et al., 1992). *Labes,* also known from the Late Cretaceous (Maastrichtian) of Spain, is believed to represent a eutherian group that is endemic to Europe and is related to "zhelestids" of middle Asia (Gheerbrant and Astibia, 1999). "Zhelestids" are of importance because they are believed to be basal members of the ungulate radiation (Nessov, Archibald, and Kielan-Jaworowska, 1998, see chapter 13). More recently collected material from Champ-Garimond, including another lower molar, is mentioned by Pol et al. (1992) and Sigé et al. (1997).

Work by Emmanuel Gheerbrant and colleagues has resulted in the discovery of two additional Late Cretaceous mammal sites in France, both in the Petites Pyrénées (figure 2.9). **Barrage Filleit** (figure 2.9) is in the Grès de Labarre Formation, which is placed in the upper Campanian to lower Maastrichtian. The Barrage Filleit site includes both macrovertebrates and microvertebrates, deposited in a fluvial setting. The only mammal specimen is the distal metatarsal of a primitive eutherian or metatherian (Gheerbrant et al., 1997). The site of **Peyrecave** (figure 2.9) is in the Marnes d'Auzas Formation, which is upper Maastrichtian. It includes paleoenvironments ranging from marginal marine to lagoonal to continental, with paleosols and fluvial deposits. Microvertebrates were recovered by screenwashing and acid-bathing matrix. The Peyrecave site has yielded fragmentary mammal teeth from two levels, the "marnes rousses" and the "lumachelle à huîtres" (Gheerbrant et al., 1997).

A site at the clay quarry of the Cerâmica de Montego, **Taveiro,** Portugal (figure 2.9), has yielded five mammal teeth or tooth fragments of Late Cretaceous age (Antunes et al., 1986). An insectivore said to be similar to *Leptacodon* (otherwise known from the early Tertiary of North America and Europe) is present; at least two possible marsupial taxa, one said to be similar to *Alphadon* and the other to "*Pediomys,*" have also been reported (Antunes et al., 1986). Both of these are otherwise known only from the Late Cretaceous of North America and, if the records from Taveiro are verified, they will be of great biogeographic interest. Gheerbrant and Astibia (1999) indicate that one of the tribosphenic mammals from Taveiro may represent the endemic eutherian clade that includes *Lainodon* and *Labes.* A report of a possible "symmetrodontan" from the site is based on a lower tooth that is now lost. Its identity is not determinable; from the available figure (Antunes et al., 1986), we tentatively suggest that the tooth represents a lower premolar of a eutherian.

Two other sites on the Iberian Peninsula, both in Spain (figure 2.9), have yielded Late Cretaceous mammals. **Quintanilla del Coco,** Burgos Province, is in the Calizas de Lynchus Formation and is reported to be of Maastrichtian age (Pol et al., 1992). Three mammals are known from the site; two are reported to be possible "palaeoryctid" eutherians and the third was named as *Labes quintanillensis.* As noted earlier, *Labes* is also known from France and is thought to be related to Asiatic "zhelestids" and, ultimately, ungulates (Gheerbrant and Astibia, 1994, 1999). The site of **Laño,** in the Basque region, has also yielded at least three mammal taxa, all represented by isolated teeth, through screenwashing. The stratigraphic unit for the site is not given, but it lies above a Campanian marine unit and is thought to be late Campanian or early Maastrichtian in age (Astibia et al., 1991; Gheerbrant and Astibia, 1999). *Laindon orueetxebarriai* is the best-represented member of the fauna, which includes another possible species of this genus and one possibly referable to *Labes* (Gheerbrant and Astibia, 1994, 1999). Interestingly, all of the mammal specimens from this site—and, for that matter, other Late Cretaceous sites of Western Europe—represent therians. This is quite unlike the situation encountered in North America, where there are abundant multituberculates, or in Romania, where the limited record thus far includes only multituberculates.

EASTERN EUROPE

Only one region in Eastern Europe has thus far yielded mammals of Late Cretaceous age, the Haţeg Basin, Romania (figure 2.16, table 2.17). A useful account of the geologic setting and microvertebrate fauna in general is given by Grigorescu et al. (1999). As noted, Europe assumed the form of an archipelago during parts of the Late Cretaceous. Theropods known from Transylvania are generally small, and this seems to be true of other dinosaurs as well, suggesting that the region was, indeed, insular (Weishampel et al., 1991; Csiki and Grigorescu, 1998). Mammals are known from five sites in the Haţeg Basin. Two are near Sînpetru (also called the Sibişel Valley) and are in the Sînpetru Formation. The locality of **După Râu** (figure 2.16) has yielded an unidentified cimolodontan (Grigorescu, 1984; Grigorescu and Hahn, 1987). A partial skull is known from the nearby site of **Tamaşel.** *Kogaionon ungureanui,* based on this specimen, is referred to its own family, Kogiaononidae, known only from Europe (Kielan-Jaworowska and Hurum, 2001, see chapter 8). Also in the Sînpetru Formation, the locality of **Pui** (figure 2.16), in the eastern part of the Haţeg Basin, has yielded three multituberculate taxa (Grigorescu et al., 1985; Rădulescu and Samson, 1986, 1997). The only named species is *Barbatodon transylvanicum,* which is a basal cimolodontan referred to the so-called *Paracimexomys* group (see chapter 8).

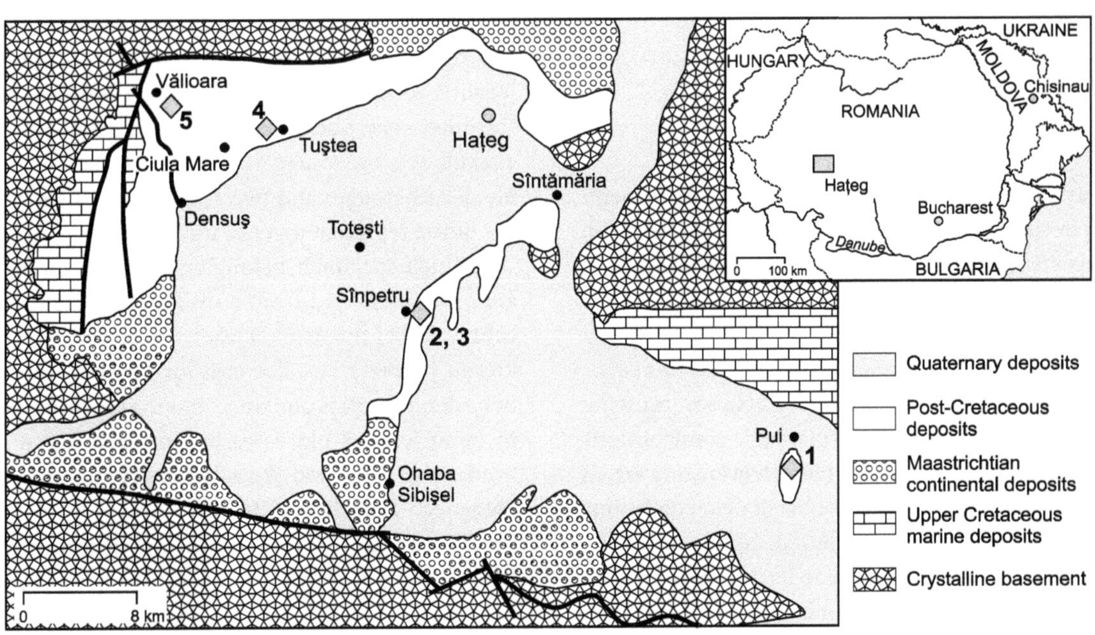

FIGURE 2.16. Late Cretaceous (Maastrichtian) mammal localities in the Haţeg Basin, Romania. Localities: **1, Pui; 2, După Râu; 3, Tamaşel** (Sînpetru Formation); **4, Tuştea; 5, Fântânele** (Densuş-Ciula Formation). *Source:* redrawn from an illustration provided by Z. Csiki.

TABLE 2.17. Late Cretaceous Mammals of Romania. Localities (all in the Haţeg Basin, Romania; see figure 2.16; locality numbers do not correspond between map and table): 1, După Râu; 2, Tamaşel; 3, Pui (Sînpetru Formation, Maastrichtian); 4, Tuştea; 5, Fântânele (Densuş-Ciula Formation, Maastrichtian)

Multituberculata
Family *incertae sedis*
Barbatodon transylvanicum (3)
Gen. et sp. indet. ("ptilodontoid") (3)
Gen. et sp. indet. ("taeniolabidoid") (3)[1]
Gen. et sp. indet. (1, 4, 5)
Kogaionidae
Hainina sp. A (5)
Hainina sp. B (5)
Kogaionon ungureanui (2)

[1] Occurrence of great significance if confirmed; presently considered highly dubious.

The remaining two sites are in the northwest part of the Haţeg Basin (see Csiki and Grigorescu, 2000). They are both in the Densuş-Ciula Formation, which is believed to be a correlative of the Sînpetru Formation. A single incisor of an unidentified cimolodontan is known from the site of **Tuştea.** The **Fântânele** locality, west of Tuştea and near Vălioara, has yielded three multituberculate taxa, two of which are referable to *Hainina.* This genus appears to be related to *Kogaionon;* its occurrence in the Densuš-Ciula Formation is notable because *Hainina* is otherwise known from the Paleocene of Western Europe (Vianey-Liaud, 1979, 1986).

ASIA

Middle Asia

The vast area of Russia and adjacent countries formerly included in the Soviet Union was long a complete blank for Cretaceous mammals. A single specimen of a Late Cretaceous eutherian from Kazakhstan was described by Bazhanov (1972), but it remained for Lev A. Nessov to more fully exploit middle Asia for fossil mammals in a series of expeditions beginning in 1978. Nessov must be given the greatest credit for his remarkable contributions to knowledge of early mammals. His achievements are all the more impressive in light of the harsh field conditions and virtual lack of support he endured; among other difficulties, his team occasionally had to hitchhike to the field area and dig 3-m-deep holes to find drinking water (see, e.g., Averianov, 2000). Nessov described more than 25 species of Cretaceous mammals, and part of the collection has not yet been published; research is being continued by Alexander O. Averianov, J. David Archibald, and colleagues. The mammal assemblages of middle Asia (table 2.18) are noteworthy for several reasons. First, the fossils are reasonably complete, including many jaws and, in one case, a skull. Second, they sample an important time interval—the early part of the Late Cretaceous, which is otherwise poorly represented by mammalian fossils. Finally, the assemblages are important zoogeographically (see Nessov, 1992). The mammals are quite different from those to the east, in the Late Cretaceous of Mongolia, which are presumed to be arid-adapted and are in this sense specialized. Among others, the mammal assemblages of middle Asia appear to include primitive relatives of ungulates (Archibald, 1996b)—eutherians that would not flourish or radiate widely until the early Tertiary. Most of the Late Cretaceous mammals of middle Asia—Kazakhstan, Uzbekistan, Tadjikistan, and Kirghizia—are included in several recent summary papers (Nessov, Sigogneau-Russell, and Russell, 1994; Nessov, 1997; Nessov, Archibald, and Kielan-Jaworowska, 1998; Averianov, 2000). We review the occurrences in stratigraphic order.

Three mammal localities are known from the Khodzhakul Formation, Karakalpakistan, Uzbekistan (figure 2.11). The mammal localities are in the upper part of the unit, which is believed to be lower Cenomanian based on correlation with a marine regressive phase and on some of the lower vertebrate fauna (Averianov, 2000). Undescribed specimens, consisting largely of edentulous jaws of unidentified therians, are known from the site of **Khodzhakulsay** (figure 2.11), which is interpreted to represent a channel connecting two brackish reservoirs (Nessov, Sigogneau-Russell, and Russell, 1994). Unidentified mammal remains have also been reported from the **Chelpyk** locality, which may be slightly higher in the section than the other sites. **Sheikhdzheili,** to the northeast of Khodzhakulsay, is the source for three mammals; a large possible deltatheroidan and two eutherians, which are among the oldest representatives of this group.

A single specimen, belonging to the eutherian *Sorlestes kara,* was recovered from a drill core at the **Ashikol** "locality," east of the Ashikol lakes, Chimkent District, Kazakhstan (figure 2.11). The unit for the fossil horizon (500 m underground) is unnamed, but the specimen is believed to be at least as old as early Turonian, based on palynomorphs recovered from an overlying unit (Nessov, Sigogneau-Russell, and Russell, 1994).

The vast majority of the mammals known from the Late Cretaceous of middle Asia are from the Bissekty Formation, Kizylkum Desert, Uzbekistan (figure 2.11). Fossil mammals have been recovered from several horizons in this unit at **Dzharakuduk** (figure 2.11). The oldest mammal-yielding level may be late Turonian, based on sharks (Nessov, Sigogneau-Russell, and Russell, 1994). Six or

TABLE 2.18. Late Cretaceous Mammals of Russia and the Commonwealth of Independent States (see figure 2.11); locality numbers do not correspond between map and table). Localities: 1, Sheikhdzheili; 2, Khodzhakulsay; 3, Chelpyk (Karakalpakistan, Uzbekistan; Khodzhakul Formation, early Cenomanian); 4, Ashikol (Chimkent District, Kazakhstan; early Turonian); 5, Dzharakuduk (Kizylkum Desert, Uzbekistan; Bissekty Formation, Turonian); 6, Dzharakuduk (Bissekty Formation, Coniacian); 7, Dzharakuduk (Aitym Formation, Santonian); 8, Zhalmauz Well (Kyzyl-Orda District, Kazakhstan; Bostobe Formation, Santonian); 9, Kansay (Khodzant District, Tadzhikistan; Yalovach Formation, Santonian); 10, Yantardakh (Yakutia, Russia; Khets Formation, Santonian); 11, Grey Mesa (Chimkent District, Kazakhstan; Darbasa Formation, early Campanian)

CENOMANIAN
Mammalia
  Family *incertae sedis*
    Gen. et sp. indet. (2, 3)
?Deltatheroida
  Family *incertae sedis*
    Oxlestes grandis (1)
Eutheria
  Otlestidae
    Otlestes meiman (1)
  "Zhelestidae"
    Eozhelestes mangit (1)
TURONIAN
Stem Trechnotheria ("symmetrodontans")
  Spalacotheriidae
    ?*Shalbaatar bakht* (5)
Eutheria
  Zalambdalestidae
    Kulbeckia rara (5)
  "Zhelestidae"
    Aspanlestes aptap (5)
    Sorlestes kara (4)
  Family *incertae sedis*
    Daulestes kulbeckensis (5)
    Daulestes inobservabilis (5)
CONIACIAN
Multituberculata
  Family *incertae sedis*
    Uzbekbaatar kizylkumensis (6)
    Gen. et sp. indet. (6)
Deltatheroida
  Deltatheridiidae
    Deltatherus kizylkumensis (6)
    Sulestes karakshi (6)
    Sulestes sp. (6)
    Gen. et sp. indet. (6)
Marsupialia
  Asiatheriidae
    Marsasia aenigma (6)
    Marsasia sp. (6)
Eutheria
  Kennalestidae
    ?*Kennalestes uzbekistanensis* (6)
    Sailestes quadrans (6)
  "Nyctitheriidae"
    Paranyctoides aralensis (6)
  Zalambdalestidae
    Kulbeckia kulbecke (6)
  "Zhelestidae"
    Eoungulatum kudukensis (6)
    cf. *Eoungulatum kudukensis* (6)
    Kumsuperus avus (6)
    Ortalestes tostak (6)
    Parazhelestes minor (6)
    Parazhelestes robustus (6)
    Sorlestes budan (6)
    Zhelestes temirkazyk (6)
    Gen. et sp. indet. (6)
  Family *incertae sedis*
    Bulaklestes kezbe (6)
    Daulestes nessovi (6)
    Daulestes kulbeckensis (6)
    "*Sazlestes tis*" (6)
SANTONIAN
Mammalia
  Family *incertae sedis*
    Gen. et sp. indet. (10)
Multituberculata
  Family *incertae sedis*
    Uzbekbaatar wardi (7)
Eutheria
  Zalambdalestidae
    Kulbeckia kansaica (9)
  ?Zalambdalestidae
    Beleutinus orlovi (8)
  "Zhelestidae"
    Gen. et sp. indet. (7)
  Family *incertae sedis*
    Gen. et sp. indet. (8, 9)
EARLY CAMPANIAN
Multituberculata
  Family *incertae sedis*
    Bulganbaatar sp. (11)
Deltatheroida
  Deltatheridiidae
    Deltatheridium nessovi (11)
Eutheria
  Kennalestidae
    Gen. et sp. indet. (11)
  Zalambdalestidae
    Alymlestes kielanae (11)
    ?*Alymlestes* sp. (11)
  "Zhelestidae"
    ?*Aspanlestes* sp. (11)

more mammals are known from this level. One of these, *Shalbaatar bakht* (named but not figured by Nessov, 1997), appears to be a "symmetrodontan"; the remainder are eutherians, including the "palaeoryctid"-like *Daulestes* and one of the oldest of the "Zhelestidae" (*Aspanlestes*), which are thought to be related to ungulates.

A dozen or more localities of Coniacian age have yielded mammals from the Bissekty Formation at **Dzharakuduk** (figure 2.11, table 2.18). For convenience, we treat them collectively here, though several separate levels are recognized (see Nessov, Sigogneau-Russell, and Russell, 1994; Averianov, 2000). Some 25 or more kinds of mammals have been reported from these sites. The multituberculate *Uzbekbaatar* (see Kielan-Jaworowska and Nessov, 1992) is assigned to Cimolodonta *incertae sedis* (see chapter 8). Deltatheroida, a predominantly Old World group that may be related to Marsupialia (chapter 12), are represented by at least three taxa (Kielan-Jaworowska and Nessov, 1990; Nessov, 1997). Of special interest among the fauna is *Marsasia,* which appears to be a true marsupial and thus is the earliest representative of that group in Asia (Averianov and Kielan-Jaworowska, 1999). Marsupialia had long been thought to be of North American origin, not reaching the Old World until the Tertiary. *Asiatherium,* mentioned later, and *Marsasia* challenge this interpretation and suggest a much earlier presence in Asia than previously envisaged (see chapter 12). Some of the dentally primitive eutherians from the Coniacian of Dzharakuduk also have interesting biogeographic implications. One is possibly referable to *Kennalestes,* known from the Campanian of Mongolia; another, *Paranyctoides aralensis,* belongs to an insectivoran genus otherwise restricted to the Campanian–Maastrichtian of North America (Fox, 1979d, 1984b; Cifelli, 1990e; see detailed treatment by Archibald and Averianov, 2001b). Most of the eutherians from the Coniacian of Dzharakuduk are referred to the "Zhelestidae." These are especially noteworthy because they are believed to be related to ungulates (Archibald, 1996b; Nessov, Archibald, and Kielan-Jaworowska, 1998). As such, they represent the most compelling evidence for early Late Cretaceous radiation of a modern placental group (Archibald, 1999). Additionally, most eutherian groups from the Cretaceous of Asia lack obvious ties with more recent taxa, particularly those from North America (Novacek et al., 1997; Cifelli, 1999a). These ungulate-like "zhelestids" are the best known exception to this rule and thus provide a viable source for the later radiation of ungulates in North America. A final noteworthy occurrence in the Coniacian of Dzharakuduk is that of *Daulestes nessovi* (the genus is also known from the Turonian part of the Bissekty Formation), which is known by the oldest reasonably complete skull of a eutherian mammal (McKenna et al., 2000).

Fossil mammals have recently been reported from the Aitym Formation at Dzharakuduk (Averianov, 1999; Archibald and Averianov, 2001b). This unit, which overlies the Bissekty Formation, is of probable Santonian age and includes a fauna dominated by marine taxa, including chondrichthyans, plesiosaurs, turtles, and molluscs. Three mammals are known thus far from the Aitym Formation: a large eutherian represented by an edentulous jaw fragment, an insectivoran *Paranyctoides* sp., and the multituberculate *Uzbekbaatar wardi* (known from a p4), which may be a descendant of *U. kizylkumensis* from the Bissekty Formation (Averianov, 1999).

Two additional sites of Santonian age are known from middle Asia. **Zhalmauz Well,** Kyzyl-Orda Province, Kazakhstan (figure 2.11), is in the lower Bostobe Formation, with a Santonian correlation based on turtles and chondrichthyans (Averianov, 2000). *Beleutinus orlovi,* described from this site by Bazhanov (1972), may be related to a peculiar eutherian group known from Mongolia, the Zalambdalestidae (Nessov, Sigogneau-Russell, and Russell, 1994). The only other mammal specimen known from this site is a large, unidentifiable cervical vertebra (Nessov and Khisarova, 1988). The **Kansay** site, Khodzant District, Tadzhikistan (figure 2.11), has also yielded two mammal specimens. The occurrences are in the Yalovach Formation, believed to be early Santonian on the basis of fishes and turtles (Nessov, Sigogneau-Russell, and Russell, 1994). The only described mammal from Kansay is *Kulbeckia kansaica* (see Nessov, 1993); this genus is also represented in older horizons (Turonian and Coniacian, Bissekty Formation) of middle Asia.

Another Santonian mammal occurrence, well out of the region under discussion but mentioned here for convenience, is in the Khets Formation at **Yantardakh,** Siberia (not figured). The specimen consists of mammalian hair embedded in amber, and is notable for its high (70° N) paleolatitudinal occurrence (Nessov, Sigogneau-Russell, and Russell, 1994).

The youngest Late Cretaceous mammals of middle Asia are from **Grey Mesa,** near the Alymtau Range, Kazakhstan. The mammals, including some six taxa, come from the lower part of the Darbasa Formation, which is believed to be early Campanian (Nessov, 1993; Averianov and Nessov, 1995; Averianov, 2000). The fauna is thus important because it is close to the age of the oldest Late Cretaceous mammals known from Mongolia, which occur in the Djadokhta Formation and presumed equivalents. Two genera, the multituberculate *Bulganbaatar* and the deltatheroidan *Deltatheridium,* are also known from the Djadokhta Formation, and the placental families Kennalestidae and Zalambdalestidae are also represented in both the Grey Mesa and Mongolian faunas (see Averianov and Nessov, 1995; Averianov, 1997, 2000; Archibald et al., 2001).

Mongolia

Without doubt, the best known of all Mesozoic mammals come from the Late Cretaceous of the Gobi Desert, Mongolia (table 2.19). Skulls and, commonly, articulated skeletons are the rule, and isolated fragments the exception. This is in stark contrast with the record from the Late Cretaceous of North America (and virtually everywhere else), where fossils generally consist of jaw fragments or isolated teeth, usually recovered through screenwashing. Indeed, though North America boasts many more Late Cretaceous mammal sites and specimens, not a single reasonably complete skull of this age has yet been published from anywhere on the continent. The extraordinary preservation of the Mongolian specimens is due to an unusual depositional setting characterized by arid conditions and extensive dune systems (see later).

Initial discoveries in Late Cretaceous strata of the Gobi Desert were made in the 1920s by the Central Asiatic Expeditions of the American Museum of Natural History, led by Roy Chapman Andrews (see informative and readable account by Andrews, 1932). The major fossil area for Late Cretaceous mammals (e.g., Gregory and Simpson, 1926; Simpson, 1928d) and dinosaurs discovered by these expeditions is near the so-called Flaming Cliffs, at a place then called Shabarakh Usu and now termed Bayan Zag (spelled Bayn Dzak in most recent literature). The fossil-

TABLE 2.19. Late Cretaceous Mammals of Mongolia, China, and Japan (see figures 2.4, 2.17; locality numbers do not correspond between maps and table). Localities or local faunas: 1, Bayan Zag (Djadokhta Formation, Campanian); 2, Tögrög (Tögrög Beds, equivalent of the Djadokhta Formation, Campanian); 3, Ukhaa Tolgod ("Ukhaa Tolgod Beds," not formally named , Campanian); 4, Khulsan (Baruungoyot Formation, Campanian); 5, Nemegt (Baruungoyot Formation, Campanian); 6, Hermiin Tsav (Red Beds of Hermiin Tsav, Campanian); 7, Udan Sayr (unnamed unit, Campanian); 8, Khaichin Uul (?Nemegt Formation, ?Maastrichtian); 9, Guriliin Tsav (?Nemegt Formation, ?Maastrichtian), Mongolia; 10, Tsondolein-Khuduk, Gansu, China (unit uncertain; ?Cenomanian); 11, Bayan Mandahu, Inner Mongolia, China (Bayan Mandahu Formation, Campanian); 12, Amagimi Dam, Kiyushu, Japan ("Upper Formation" of the Mifune Group, late Cenomanian or early Turonian)

- Multituberculata
  - Family *incertae sedis*
    - *Bulganbaatar nemegtbaataroides* (1)
    - *Bulganbaatar* cf. *nemegtbaataroides* (3)
    - *Chulsanbaatar vulgaris* (4, 5, 6)
    - *Chulsanbaatar* cf. *vulgaris* (3)
    - *Nemegtbaatar gobiensis* (3, 4, 5, 6)
    - Gen. nov., sp. A (3)
    - Gen. nov., sp. B (3)
  - ?Cimolomyidae
    - *Buginbaatar transaltaiensis* (8)
  - Djadochtatheriidae
    - *Catopsbaatar catopsaloides* (4, 6)
    - *Djadochtatherium matthewi* (1, 2, 11)
    - ?*Djadochtatherium matthewi* (3)
    - *Kryptobaatar dashzevegi* (1, 2, 3, 6)
    - *Kryptobaatar mandahuensis* (11)
    - *Tombaatar sabuli* (3, ?11)
  - Sloanbaataridae
    - *Kamptobaatar kuczynskii* (1)
    - *Kamptobaatar* cf. *kuczynskii* (3)
    - *Nessovbaatar multicostatus* (6)
    - *Sloanbaatar mirabilis* (1)
    - *Sloanbaatar* cf. *mirabilis* (3)
- Deltatheroida
  - Deltatheridiidae
    - *Deltatheridium pretrituberculare* (1, 3, 6)
    - *Deltatheroides cretacicus* (1)
    - *Khuduklestes bohlini* (10)
    - Gen. et sp. nov. (9)
- Marsupialia
  - Asiatheriidae
    - *Asiatherium reshetovi* (7)
- Eutheria
  - Family *incertae sedis*
    - *Hyotheridium dobsoni* (1)
    - cf. *Hyotheridium* sp. (3)
  - Asioryctidae
    - *Asioryctes nemegetensis* (4, 5, 6)
    - cf. *Asioryctes* sp. (3)
    - *Ukhaatherium nessovi* (3)
  - Kennalestidae
    - *Kennalestes gobiensis* (1)
    - *Kennalestes* cf. *gobiensis* (3)
    - *Kennalestes* sp. (11)
  - Zalambdalestidae
    - *Barunlestes butleri* (5, 6)
    - *Zalambdalestes lechei* (1, 2, 3)
    - *Zalambdalestes* sp. (11)
  - "Zhelestidae"
    - *Sorlestes mifunensis* (12)

yielding unit is the Djadokhta Formation. This unit and many other geographic and stratigraphic terms of the region are variously spelled and referenced (see, e.g., Gradziński et al., 1968; Gradziński and Jerzykiewicz, 1974; Novacek et al., 1994). Polish-Mongolian Expeditions (1963–1971), led by Zofia Kielan-Jaworowska, worked these exposures and discovered many additional fossils at Bayan Zag and at some nearby newer sites. They also discovered a number of other exposures, most notably those to the west of the Nemegt Basin. These are significant in that they include mammals from younger horizons—the Baruungoyot Formation and its presumed equivalents (see, e.g., Jerzykiewicz and Russell, 1991). Fossil mammals collected by the Polish-Mongolian Expeditions have been described in a long list of publications, most of which are cited later. The region was also worked by Soviet-Mongolian Expeditions, mainly in the late 1960s and 1970s (see review by Kurochkin and Barsbold, 2000). New sites for Late Cretaceous mammals were discovered but, to our knowledge, specimens from only three of these have been published (see Kielan-Jaworowska and Sochava, 1969; Kielan-Jaworowska and Nessov, 1990; Szalay and Trofimov, 1996). Several other institutions have sent field parties to the Gobi Desert in search of Late Cretaceous mammals, the most notable of these being the current American Museum–Mongolian Academy Expeditions, which began in 1990. The most significant discovery of these expeditions to date is a new locality, Ukhaa Tolgod, in the Nemegt Basin, where literally hundreds of skulls and, in many cases, associated skeletons were discovered (Dashzeveg et al., 1995). The assemblage, which includes a few previously unknown taxa (Novacek et al., 1997; Rougier et al., 1997, see also Wible and Rougier, 2000), appears to be similar to that of the Djadokhta Formation, although it also contains the elements previously known only from the Baruungoyot Formation and its equivalent the Red Beds of Hermiin Tsav (see later).

The Gobi Basin includes three major subbasins. For the Upper Cretaceous rocks, syndepositional block faulting, discontinuous sedimentation, lack of laterally extensive marker horizons, and discontinuity of exposures, among other features, make lithostratigraphic correlation among geographically widespread fossil localities problematic. Three units are central to a discussion of the Late Cretaceous fossil vertebrate sequence in the Gobi Desert (lower to higher): the Djadokhta, Baruungoyot, and Nemegt formations. The type area of each unit includes one or more localities with significant assemblages of fossil vertebrates. Correlation of fossiliferous strata outside the type areas is based on similarity in lithology, inferred depositional facies, fauna, and other considerations. We refer readers to the excellent summaries of Gradziński et al. (1977), Jerzykiewicz and Russell (1991), and Jerzykiewicz (1995).

Correlation of the Upper Cretaceous strata of the Gobi Desert with vertebrate sequences from elsewhere is also problematic. Marine strata are lacking, so that it is not possible to establish correlation on the basis of standard marine stages. Similarly, there are currently no radiometric, paleomagnetic, or other data to establish a chronostratigraphic framework. Historically, age estimates have varied widely—with the Djadokhta mammals, for example, being considered as old as Cenomanian or as young as Tertiary (see the excellent summary and references cited by Lillegraven and McKenna, 1986). Comparisons based on the vertebrates themselves are fraught with difficulty because the Mongolian taxa are arid adapted and highly endemic (see, e.g., Osmólska, 1980). Based on the stage of evolution (e.g., Fox, 1978) and inferred climatological changes in the sequence (Jerzykiewicz and Russell, 1991), we follow Lillegraven and McKenna (1986) in considering the faunas of the Djadokhta and Baruungoyot formations to correlate with the Campanian. We also tentatively accept the early Campanian age for the Djadokhta and late Campanian for the Baruungoyot Formation. Similarity in the mammalian assemblages of the Djadokhta and Darbasa (Kazakhstan, discussed earlier) formations led Averianov (1997) to propose a lower Campanian correlation for the Djadokhta Formation. This correlation would imply that the Baruungoyot Formation is also of early Campanian or, more probably, late Campanian age. The Nemegt Formation is generally considered to be Maastrichtian, perhaps middle Maastrichtian (Jerzykiewicz and Russell, 1991), in age.

The stratotype of the Djadokhta Formation is at Bayan Zag (see Gradziński et al., 1977) in the Ulan Nur Basin. The unit consists mainly of fine poorly cemented and mostly aeolian sandstones with interbedded conglomerates and caliche horizons. The depositional environment is interpreted to have included extensive dune systems with sporadic and ephemeral lakes and the occasional development of arid paleosols (see, e.g., Loope et al., 1998). The Baruungoyot Formation was defined on the basis of exposures at Khulsan in the Nemegt Basin, nearly 200 km west of Bayan Zag (Gradziński and Jerzykiewicz, 1974). Its superpositional relationship with the Djadokhta Formation cannot be determined directly. The lower boundaries of both units are covered in their type areas; the Djadokhta is overlain unconformably by Tertiary rocks (Khashaat, or "Gashato" Formation), whereas the Cretaceous Nemegt Formation overlies the Baruungoyot. However, the Djadokhta Formation is generally believed to be lower, at least in part, than the Baruungoyot Formation (*contra* Dashzeveg et al.,

1995). Furthermore, the units are lithologically distinct, and Djadokhta-facies rocks are widespread in the Gobi Basin (Jerzykiewicz and Russell, 1991), including the Nemegt Basin itself (Dashzeveg et al., 1995). Like the Djadokhta, the Baruungoyot Formation consists largely of poorly cemented sandstones, but it includes more erosional channels and interbedded claystones, and caliches are less common. While the depositional setting is mainly aeolian, the lack of paleosols and the greater abundance of lacustrine facies suggest a wetter, more humid environment than for the Djadokhta Formation.

The stratotype for the Nemegt Formation, also in the Nemegt Basin, is at the Nemegt locality, just west of Khulsan (Gradziński and Jerzykiewicz, 1974; Gradziński et al., 1977). The lithology is predominantly sandstone, like the underlying Baruungoyot Formation, but the Nemegt is interpreted as consisting predominantly of alluvial floodplain and channel deposits, thereby indicating development of a perennial fluvial system and, hence, a wetter environment. Exposures of the Nemegt Formation and presumed equivalents are widespread.

The vertebrate faunas of the Djadokhta and Baruungoyot formations, both of which include mammals, are rather similar. Both are included in the "Barungoytian land-vertebrate age" (Jerzykiewicz and Russell, 1991), although older "Djadokhta" and younger "Khulsan" subdivisions are sometimes recognized (e.g., Gradziński et al., 1977; Osmólska, 1980). Predictably, the differences seem to reflect environmental change, with aquatic taxa being rare or absent in the Djadokhta Formation. Mammals are less diverse in the Baruungoyot Formation, but the composition is rather similar; djadochtatheroidean multituberculates dominate, with eutherians and deltatheroidans being represented by only a few species. Four or more species may be shared between the assemblages, depending on the correlation of the Ukhaa Tolgod locality and the identity of its fauna (see later). The fauna of the Nemegt Formation differs considerably from those of the Djadokhta or Baruungoyot formations and is assigned its own land-vertebrate "age," known as the Nemegtian (Jerzykiewicz and Russell, 1991). Aquatic vertebrates such as fish are present, reflecting the wetter, fluvially dominated depositional setting (data for charophytes and ostracods are given by Szczechura and Błaszyk, 1970; Karczewska and Ziembińska-Tworzydło, 1970, 1981; Szczechura, 1978). Large dinosaurs, such as the tyrannosaurid *Tarbosaurus*, are abundant, but microvertebrates are notoriously scarce. Mammals have not yet been reported from the Nemegt Formation in its type area, but two occurrences are in strata referred to the unit. In compiling mammal distributions from localities of the Gobi Desert, we have referred to our own unpublished observations, to existing compendia (Gradziński et al., 1977; Jerzykiewicz and Russell, 1991), and to the original literature (Kielan-Jaworowska and Sochava, 1969; Kielan-Jaworowska, 1970a, 1974a, 1975b,c, 1984b; Kielan-Jaworowska and Dashzeveg, 1978; Kielan-Jaworowska and Trofimov, 1980, 1981; Trofimov and Szalay, 1994; Dashzeveg et al., 1995; Rougier, Wible, and Novacek, 1996; Szalay and Trofimov, 1996; Kielan-Jaworowska and Hurum, 1997; Novacek et al., 1997; Rougier et al., 1997, 1998; Kielan-Jaworowska et al., 2003).

The classic locality for mammals from the Djadokhta Formation is **Bayan Zag** (figure 2.17), where 10 species are known. One of the multituberculates, *Djadochtatherium matthewi*, is similar to North American members of the Taeniolabididae and for a time was included in a North American genus (Kielan-Jaworowska and Sloan, 1979). Currently, all of the multituberculates from Bayan Zag and virtually all from the Late Cretaceous of Mongolia are referred to Djadochtatherioidea, which represent a largely Asiatic group (Kielan-Jaworowska and Hurum, 1997, 2001; Rougier et al., 1997, see chapter 8). At least two multituberculate families are represented at Bayan Zag, Djadochtatheriidae (*Djadochtatherium*, *Kryptobaatar*) and Sloanbaataridae (*Sloanbaatar*, *Kamptobaatar*); the affinities of the fifth genus, *Bulganbaatar*, are uncertain. Possibly two deltatheroidans, *Deltatheridium* and *Deltatheroides*, and *Hyotheridium* (of doubtful status) are present in the fauna. Interestingly, *Deltatheridium*, or something strikingly similar, is also known from the Late Cretaceous of North America (Fox, 1974a), suggesting Campanian dispersal between the two continents (Cifelli and Gordon, 1999). The affinities of the eutherians, *Zalambdalestes* and *Kennalestes*, are not settled; one or both may represent endemic Asiatic clades that left no descendants, or they may be related to groups that later achieved success either in Asia or elsewhere (e.g., Novacek et al., 1997; Rougier et al., 1998; Archibald et al., 2001; Fostowicz-Frelik and Kielan-Jaworowska, 2002, see chapter 13).

Two other mammal localities may be equivalent in age to Bayan Zag, though neither is formally placed in the Djadokhta Formation. The **Tögrög** (referred to also as Toogreeg, Toogreek, Tugrig, Tugrugeen Shireh) locality, some 40 km northwest of Bayan Zag (figure 2.17), is in the Tögrög Beds, which are similar to strata of the Djadokhta Formation and are thought to be equivalent (Gradziński et al., 1977). The occurrence of monospecific dinosaur accumulations at Tögrög and the fact that dinosaur skeletons commonly appear to be in a struggling death pose suggest that death occurred during episodic sandstorm events (Jerzykiewicz et al., 1993). Three mammals are

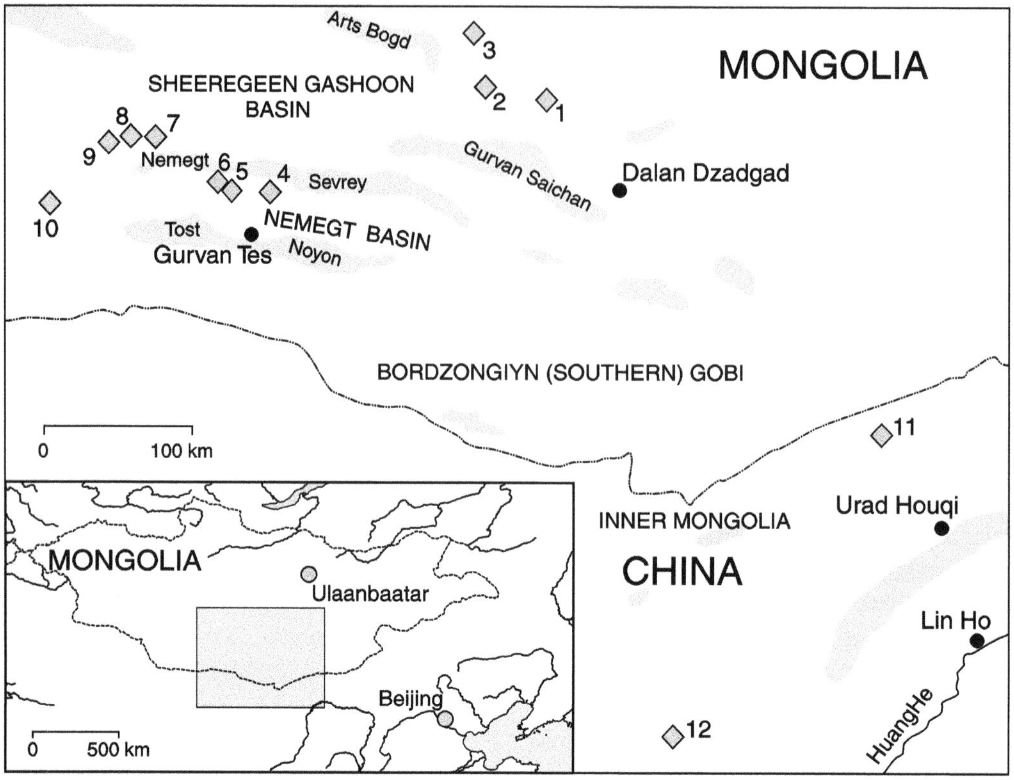

FIGURE 2.17. Late Cretaceous mammal localities of Mongolia and northern China. **1, Bayan Zag** (Djadokhta Formation, ?early Campanian); **2, Tögrög** (?equivalent to Djadokhta Formation, ?early Campanian); **3, Udan Sayr** (?equivalent to Baruungoyot Formation, ?late Campanian); **4, Ukhaa Tolgod** (?equivalent to Djadokhta Formation, ?early Campanian); **5, Khulsan** (Baruungoyot Formation, ?late Campanian); **6, Nemegt** (Baruungoyot Formation, ?late Campanian); **7, Guriliin Tsav** (?Nemegt Formation, ?Maastrichtian); **8, Bugiin Tsav** (?Nemegt Formation, ?late Maastrichtian or early Tertiary); **9, Khaichin Uul** (?Nemegt Formation, ?late Maastrichtian); **10, Hermiin Tsav** (Hermiin Tsav Red Beds, ?equivalent to Baruungoyot Formation, ?late Campanian); **11, Bayan Mandahu** (equivalent to Djadokhta Formation, ?early Campanian; Inner Mongolia, China); **12, Tsondolein-Khuduk** (unknown unit, ?Cenomanian; Gansu, China).

known from this site and all three are present at Bayan Zag. (*Kryptobaatar saichanensis,* originally described as a separate species by Kielan-Jaworowska and Dashzeveg, 1978, is now regarded as a junior synonym of *K. dashzevegi,* see Wible and Rougier, 2000, and chapter 8.)

A much more diverse assemblage is known from **Ukhaa Tolgod,** which (as noted) has yielded a virtual treasure trove of mammal skulls and skeletons. Lithologically, the unnamed beds at Ukhaa Tolgod are quite similar to those of the Djadokhta Formation (e.g., Rougier et al., 1997) and are generally considered to be equivalent. However, in view of faunistic differences (see Kielan-Jaworowska et al., 2003) and the lack of geological studies demonstrating the continuity of beds that crop out at Ukhaa Tolgod with those of the Djadokhta Formation in other places (e.g., Bayan Zag), for the purpose of this book we refer to these beds provisionally as "Ukhaa Tolgod Beds." Ukhaa Tolgod is in the Nemegt Basin and a few tens of kilometers east of two mammal sites in the Baruungoyot Formation, Khulsan and Nemegt (figure 2.17). As at Tögrög, death and burial of the fauna at Ukhaa Tolgod may have occurred during sandstorms, with the mammals perhaps being buried in their burrows (Dashzeveg et al., 1995). Somewhat surprisingly, fossorial adaptations have been suggested for only one of the mammalian taxa from the Djadokhta Formation, a multituberculate (Kielan-Jaworowska, 1989). To date, hundreds of mammal skulls, many with associated skeletons, have been collected at Ukhaa Tolgod, but the faunal list for the site (Dashzeveg et al., 1995) remains preliminary; thus far, only four of the mammal species from this locality have been described (Rougier, Wible, and Novacek, 1996; Rougier et al., 1997, 1998; Novacek et al., 1997; Kielan-Jaworowska, 1998; Wible and Rougier, 2000). Dashzeveg et al. (1995) make the interesting observation that multituberculate specimens far outnumber those of therians at Ukhaa Tolgod—

nearly 90% of the skulls belong to multituberculates. Notable among Ukhaa Tolgod mammals is the eutherian *Ukhaatherium nessovi*, represented by a skeleton (Novacek et al., 1997). Multituberculates of Ukhaa Tolgod include Sloanbaataridae (*Sloanbaatar, Kamptobaatar*), Djadochtatheriidae (*Djadochtatherium, Kryptobaatar,* and *Tombaatar*) (see Rougier et al., 1997; Kielan-Jaworowska et al., 2003), and *Chulsanbaatar,* currently assigned to family *incertae sedis.*

Mammalian specimens have been published from five areas in the Baruungoyot Formation or strata thought to be equivalent. All but two (Udan Sayr and Hermiin Tsav) are in the Nemegt Basin. **Khulsan,** the stratotype area for the Baruungoyot Formation itself, has yielded three mammals, two multituberculates and one therian, the latter unknown from Ukhaa Tolgod. Each of the species at Khulsan is also known from the nearby site of **Nemegt,** which also includes the eutherian *Barunlestes butleri.* **Hermiin Tsav** (referred to also as Khermeen Tsav) is farther to the west and the fossiliferous strata cannot be directly correlated with named units. However, the Hermiin Tsav Red Beds are thought to be time equivalent with the Baruungoyot Formation (Gradziński et al., 1977); Jerzykiewicz and Russell (1991) simply refer the fossil strata at Hermiin Tsav to this unit. There are two mammal sites in the area, Hermiin Tsav I and II; because they are close geographically and stratigraphically, our list of seven mammals is a composite. The multituberculate *Nessovbaatar multicostatus* is known only from Hermiin Tsav (Kielan-Jaworowska and Hurum, 1997). **Udan Sayr** (Ivakhnenko and Kurzanov, 1988) lies about 60 km west-southwest of Bayan Zag. It occurs in rocks that are lithologically similar to those of the Baruungoyot Formation (Jerzykiewicz and Russell, 1991), and we provisionally consider the fauna to be time equivalent. One mammal, *Asiatherium reshetovi,* has been described from Udan Sayr (Trofimov and Szalay, 1994; Szalay and Trofimov, 1996). *Asiatherium,* represented by a partial skull and skeleton, is of particular interest because it is thought to be a marsupial (but see Fox, 1997a), suggesting a much earlier presence of Marsupialia in Asia than previously envisaged (Trofimov and Szalay, 1994; Szalay and Trofimov, 1996). Though the dentition is rather specialized, it is marsupial-like in certain respects (Cifelli and Muizon, 1997, see also chapter 12).

Two mammalian occurrences are known from either the Nemegt Formation or possibly equivalent strata (figure 2.17). The locality of **Khaichin Uul** (Khaicheen Ula) lies about 100 km west-northwest of the type Nemegt area and 20–25 km southwest of Bügiin Tsav. Initially, the age of the fossiliferous stratum (believed to be equivalent to the highest unit exposed in the badlands of Bügiin Tsav) was cautiously estimated as either latest Cretaceous or early Paleocene (Kielan-Jaworowska and Sochava, 1969). Later work suggested correlation with (Gradziński et al., 1977) or placement in (Jerzykiewicz and Russell, 1991) the Nemegt Formation. The only mammal specimen collected at Khaichin Uul consists of the rostrum and dentaries of a multituberculate, *Buginbaatar transaltaiensis* (see Kielan-Jaworowska and Sochava, 1969; Trofimov, 1975). *Buginbaatar* has been referred to various families and continues to elude definitive taxonomic placement. Herein it is tentatively referred to the otherwise North American family Cimolomyidae (superfamily *incertae sedis,* see chapter 8), thus implying some Late Cretaceous interchange between North America and Asia. Ptilodontoid multituberculates, which flourished during the Late Cretaceous of North America are not known from the Late Cretaceous of Mongolia (Kielan-Jaworowska, 1979, see also Kielan-Jaworowska and Hurum, 2001).

The second mammal site that may be in the Nemegt Formation is **Guriliin Tsav,** about 75 km northwest of the type area for the formation and near Bügiin Tsav. Like Khaichin Uul, the fossil horizon at Guriliin Tsav has been considered to be either correlative to the Nemegt Formation (Kielan-Jaworowska and Nessov, 1990) or within that unit (Jerzykiewicz and Russell, 1991). The skull of a large, dentally advanced metatherian is known from Guriliin Tsav. The specimen, which has not yet been described or named, is referred to simply as the "Guirliin Tsav skull."[9] The specimen is of considerable importance in that it is the only known skull of a Cretaceous metatherian that preserves substantial morphology of the basicranial region (see Szalay and Trofimov, 1996). One feature of the Guriliin Tsav skull, an auditory bulla with a contribution from the alisphenoid bone, is marsupial-like (Kielan-Jaworowska and Nessov, 1990), but the significance of this character is uncertain (Muizon, 1994).

China

The record of Late Cretaceous mammals in China (table 2.19) is rather limited. The oldest documented occurrence, both in terms of geological age and time of discovery, is at **Tsondolein-Khuduk,** Gansu Province (figure 2.17). An isolated mammalian axis vertebra from this site was reported by Bohlin (1953) and later designated the type of *Khuduklestes bohlini* by Nessov, Sigogneau-Russell, and Russell (1994), who follow Nessov et al. (1989) in considering Tsondolein-Khuduk to be of Cenomanian age, based on occurrence there of primitive protoceratopsian

[9] The specimen was referred to "*Eodeltatheridium kurzanovi*" and "*Neodeltatheridium*" (no species designation) by Gambaryan (1989); each is a *nomen nudum.*

dinosaurs and other lower vertebrates. However, the rock unit and geographic position of the site are not known with certainty. The map location given by Bohlin (1953) appears to be close to at least three Early Cretaceous localities found by personnel of the Institute of Vertebrate Paleontology and Paleoanthropology, Beijing (see Dong, 1993: figure 1). Hence there is a possibility that the specimen of *Khuduklestes*—which in any case is not very informative—may be of Early Cretaceous age. We tentatively refer *Khuduklestes* to Deltatheroida, based mainly on its large size.

Unidentified mammalian fossils have been reported from the Quantou Formation, Jilin Province (Wood et al., 2001). Age for the occurrence is not well constrained, with estimates ranging from Aptian to Cenomanian. Comment must be deferred until the materials have been described, though the fossils promise to be of great interest.

More extensive finds of Late Cretaceous mammals have been made in the Gobi Desert, Inner Mongolia. The site of **Bayan Mandahu** (figure 2.17) has yielded about 50 mammal skulls, some with associated skeletons. The site has been worked by Chinese-Belgian Expeditions and much of the material is under study by Pascal Godefroit and Thierry Smith, with Chinese colleagues. The specimens come mainly from caliche nodules in a unit that is similar to the Djadokhta Formation (Dong, 1993); indeed, Bayan Mandahu was originally placed within this unit (Jerzykiewicz et al., 1989), but it occurs within a different depositional basin and is now simply considered a lateral equivalent (Jerzykiewicz et al., 1993). As at Bayan Zag, the type locality for the Djadokhta Formation, the most common fossils are protoceratopsian dinosaurs. Poses of the dinosaur skeletons (some of which occur in monospecific bone beds) suggest that they died and were buried in situ—some struggling to free themselves—during sandstorm events. As noted, this is a common finding for Djadokhta or Djadokhta-equivalent sites, and the same appears to be the case for Tögrög and Ukhaa Tolgod in Mongolia. The mammals from Bayan Mandahu remain largely unstudied; Wang et al. (2001a) list three taxa, the eutherians *Kennalestes* and *Zalambdalestes,* and the multituberculate *Kryptobaatar* (subsequently described as *K. mandahuensis* by Smith et al., 2001). Kielan-Jaworowska et al. (2003, based on a personal communication from T. Smith and Y.-M. Hu) listed also *Djadochtatherium matthewi* and ?*Tombaatar sabuli.* One would guess that several more taxa are represented, and our compilation (table 2.19) thus almost certainly underrepresents mammalian diversity at Bayan Mandahu.

Japan

A Late Cretaceous mammal was recently reported from Japan (Setoguchi, Tsubamoto, et al., 1999). The locality is **Amagimi Dam** near Mifune, Kumamoto Prefecture, Kiyushu (figure 2.4); the fossil horizon is in the upper part of the "Upper Formation" of the Mifune Group. Marine beds in the sequence permit age bracketing for the occurrence based on molluscs. The underlying "Middle Formation" is middle Cenomanian and the overlying Gankaizan Formation is lower Santonian; the Amagimi Dam site is thus believed to be of late Cenomanian to early Turonian age. The mammal specimen is a dentary fragment with one molar, referred to the new species *Sorlestes mifunensis.* This genus is otherwise known by two species from middle Asia, where it ranges from the early Turonian to the Coniacian (see Nessov, 1985a, 1993; Nessov, Sigogneau-Russell, and Russell, 1994). *S. mifunensis* is the oldest known member of "Zhelestidae," thought to be related to ungulates, and shows that the group has a greater antiquity and distribution than previously thought (Setoguchi, Tsubamoto, et al., 1999).

India

A diverse fauna and flora has been recovered from sedimentary rocks associated with the Deccan Basalts, Andhra Pradesh, India. The major fossiliferous horizon lies within the basalt sequence, in the so-called intertrappean beds. Long thought to be of Tertiary age, a diverse body of evidence indicates a Maastrichtian age, with radiometric dates indicating a 1- to 3-Ma eruptive phase for the basalts (Prasad et al., 1994). Thus far, the site of **Naskal** (figure 2.5, table 2.20) has yielded the bulk of mammalian fossils. Sedimentologic and taphonomic study (Khajuria and Prasad, 1998) indicates that Naskal represents a drought-induced mass death assemblage situated at a distal floodplain lake. The bones and teeth are corroded, with some hydrologic sorting, but there is no evidence of digestion or scavenging; instead, the damage has been attributed to subaerial exposure. As many as four mammals are known from Naskal. Three are assigned to *Deccanolestes* (see Prasad and Sahni, 1988; Prasad et al., 1994). The affinities of this genus, which have significant bearing on eutherian biogeography and faunal relations of India, are enigmatic (see chapter 13). *Deccanolestes* was described as a "palaeoryctid." However, Godinot and Prasad (1994) suggest that it may be related to Archonta—a major clade of eutherians including primates, bats, and suspected allies. Interestingly, study of foot bones referred to *Deccanolestes* led these authors to conclude that it had some degree of arboreal specialization.

The fourth mammal from Naskal is as yet unnamed, but is of great significance because it appears to belong to the highly specialized Gondwanatheria, a group of bizarre mammals previously assigned to Multituberculata but now classified as Mammalia *incertae sedis* (see chapters 8

TABLE 2.20. Late Cretaceous Mammals of India (see figure 2.5; locality numbers do not correspond between map and table). Localities: 1, Naskal; 2, Rangapur (both Intertrappean Beds; Maastrichtian ?and Paleocene, Andhra Pradesh)

Basal Boreosphenida ("tribotherians")	
Family *incertae sedis*	
Gen. et sp. indet. (2)	
Eutheria	
Family *incertae sedis*	
Deccanolestes hislopi (1)	
Deccanolestes cf. *hislopi* (1)	
Deccanolestes robustus (1)	
Deccanolestes sp. (2)	
Gondwanatheria	
Sudamericidae	
Gen. et sp. indet. (1)	

and 14). This unnamed mammal appears to be closely related to taxa known from the Late Cretaceous of South America and Madagascar. Thus, it provides evidence of pangondwanan distribution among some southern taxa, at least, which may have dispersed between South America and Indo-Madagascar via Antarctica (Krause, Prasad, et al., 1997). A second mammal site in the intertrappean beds of Andhra Pradesh has been reported from the vicinity of **Rangapur** (Wilson and Rana, 2001). Fossils collected to date (some 20 mammalian teeth) evidently represent two taxa, *Deccanolestes* sp. and a primitive boreosphenidan. The age of the site is uncertain; it may be Late Cretaceous or Paleocene.

AFRICA

The most significant Late Cretaceous mammals recovered from the African region are from Madagascar (figure 2.6). We include Madagascar in the section on the modern African region, but point out that it actually maintained some limited contact with India during the early part of the Late Cretaceous epoch. A diverse vertebrate fauna has been recovered from the Anembalema Member of the Maeverano Formation, near the village of **Berivotra** in the Mahajanga Basin, northwestern Madagascar (see summary by Krause, Hartman, and Wells, 1997). The Anembalema Member is estimated to be of Maastrichtian age. Four mammals, only one of which is named, are known from the unit. One is unidentifiable but significant in its large size; represented by a tooth that is more than 7.5 mm in length, it is one of the largest of the Mesozoic mammals (Krause et al., 1994). Each of the remaining taxa may have profound biogeographic implications if current identifications can be confirmed by more complete specimens. One is a possible multituberculate (Krause et al., 1999), otherwise known mainly from Laurasia (chapter 8). A second, known by part of a lower molar, is similar in known respects to early marsupials (Krause, 2001, but see Averianov et al., 2003). The third, and the only named mammal species from Berivotra, is *Lavanify miolaka*, known by two incomplete cheek teeth. These are curved, have flat occlusal wear facets, and are extraordinarily high crowned, making them reminiscent of Tertiary rodent teeth. *Lavanify* is referred to the Gondwanatheria, also known from the Cretaceous of India and the Cretaceous–Early Paleocene of South America (Pascual et al., 1999, see chapter 14).

A mammal represented by a caudal vertebra is known from **Draa Ubari**, western Libya (figure 2.6). The occurrence is in an unnamed sandstone lying between the Thala and Mazuzah members of the Mizhad Formation and is believed to be Santonian–Campanian in age (Nessov, Zhegallo, and Averianov, 1998). The associated fauna is comprised mainly of marine taxa (including a sea snake), but freshwater fishes are also present.

SOUTH AMERICA

Mammals of Late Cretaceous age are known from scattered sites in Brazil, Argentina, Bolivia, and Peru (figure 2.7, table 2.21). Collectively, these occurrences rank the Late Cretaceous mammal record from the South American continent as far superior to that of all other Gondwanan landmasses combined. Yet we still lack sufficient evidence for a reasoned interpretation of mammalian biogeography and faunal dynamics, both within South America and among the Gondwanan continents. It is especially important to recall that, aside from a few isolated occurrences, the pre-Maastrichtian record is limited to a single site in Patagonia. Then, as now, the continent spanned well over 50° of latitude south of the equator and presumably was characterized by great diversity in climate and habitat. It is conceivable, even likely, that major groups of mammals inhabited South America during the Late Cretaceous and remain unsampled in the existing record. Nonetheless, some significant new facts have emerged from this record—the cosmopolitan distribution of one group (Gondwanatheria, mentioned earlier) among several southern landmasses being a good example. Undoubtedly the most important insight that has emerged in recent years is that, prior to the Maastrichtian, at least, South America's mammal assemblage differed fundamentally from those of northern continents. By the Campanian, Laurasian assemblages were dominated by multituberculates, eutherians, and (mainly in North America) marsupials. The pre-Maastrichtian record of South America, by contrast, reveals an extraordinary diversity of

TABLE 2.21. Late Cretaceous Mammals of South America (see figure 2.7; locality numbers do not correspond between map and table). Localities or local faunas: 1, Paraná Basin, Brazil (Adamantina Formation, Late Cretaceous); 2, Paso Córdoba, Río Negro Province, Argentina (Río Colorado Formation, Campanian–Maastrichtian); 3, Meseta de Somuncura, Chubut Province, Argentina (La Colonia Formation, Campanian–Maastrichtian); 4, Los Alamitos, Chubut Province, Argentina (Los Alamitos Formation, Campanian–Maastrichtian); 5, Fundo el Triunfo, Peru (Fundo el Triunfo Formation, Campanian–Maastrichtian); 6, Laguna Umayo, Peru (Vilquechico Group, Maastrichtian); 7, Paruro, Peru (Couches Rouges Formation, late Maastrichtian); 8, Pajcha Pata, Bolivia (El Molino Formation, middle Maastrichtian)

Mammalia *incertae sedis* (1, 2, 5, 6, 7, 8)
?Docodonta
  Reigitheriidae
    Reigitherium bunodontum (3, 4)
Eutriconodonta
  Austrotriconodontidae
    Austrotriconodon mckennai (4)
    Austrotriconodon sepulvedai (4)
?Multituberculata
  Family *incertae sedis*
    Gen. et sp. indet. (4)
Archaic "symmetrodontans"
  Bondesiidae
    Bondesius ferox (4)
Stem Cladotheria ("eupantotherians")
  Dryolestidae
    Groebertherium novasi (4)
    Groebertherium stipanicici (4)
    Leonardus cuspidatus (4)
    Gen. et sp. indet. (8)
  Mesungulatidae
    Mesungulatum houssayi[1] (4)
  Brandoniidae
    Brandonia intermedia (4)
    ?*Casamiquelia rionegrina* (4)
Marsupialia
  ?Peradectidae
    Gen. et sp. indet. (6)
    ?*Peradectes austrinum* (6)
  ?"Pediomyidae"
    Gen. et sp. indet. (6)
Eutheria
  Family *incertae sedis*
    Perutherium altiplanense (6)
    Gen. et sp. indet. (8)
Gondwanatheria
  Sudamericidae
    Gondwanatherium patagonicum (4)
  Ferugliotheriidae
    Ferugliotherium windhauseni (4)

[1]We consider *Quirogatherium major* to be a probable synonym of *Mesungulatum houssayi* (see chapter 10).

mammals (in some cases, highly derived) that appear to be members of archaic groups more common in the Jurassic of Laurasia—"triconodonts," "symmetrodontans," dryolestoid "eupantherianss," and (perhaps) "plagiaulacidan" multituberculates and docodontans. Tribosphenic mammals, which clearly had an early (Berriasian) presence in northern Africa (Sigogneau-Russell, 1991a, 1992), radiated widely on northern continents through the entire Late Cretaceous. The limited evidence at hand suggests that marsupials and placentals, if they were ever present in South America prior to the Maastrichtian, were not very successful and were subordinate to the flourishing lineages of archaic therian groups.

There is clear evidence to suggest interchange of terrestrial vertebrates between the Americas in the latest Cretaceous and Early Tertiary. In North America, for example, the sauropod family Titanosauridae made a brief appearance in southwestern faunas of the Maastrichtian, presumably as immigrants from the south (Lucas and Hunt, 1989). Marsupials appeared in South America in the Late Cretaceous, and by the Early Tertiary they had achieved some diversity on the continent (Marshall, 1987; Muizon, 1991). Eutherians are also recorded in the Late Cretaceous of South America (Grambast et al., 1967; Gayet et al., 2001), though close resemblances between early Paleocene faunas of the two continents (Van Valen, 1988) suggest that some of the interchange, at least, may have occurred during the Tertiary. Late Cretaceous mammalian interchange among southern continents has only recently been documented (Krause, Prasad, et al., 1997). The Ornithorhynchidae (platypuses), a group long thought to be restricted to Australia (chapter 6), have been reported from the early Paleocene of Argentina (Pascual et al., 1992a), and it is possble that they were present in South America even earlier—perhaps much earlier.

Brazil

A mammal specimen was found by Christian de Muizon in the **Paraná Basin** (figure 2.7), São Paulo. The occurrence is in the Adamantina Formation, the age of which is not well constrained, other than the fact that it is Late Cretaceous. The specimen is a dentary fragment with several alveoli and a premolar (Bertini et al., 1993); it appears to belong to a tribosphenic mammal, but further identification is problematic.

Argentina

Three sites in Argentina, all in the southern part of the country (figure 2.7), have yielded Late Cretaceous mammals. A site at **Paso Córdoba,** Río Negro Province, lies in the Río Colorado Formation, which may be either Campanian or Maastrichtian in age. The single specimen from this site is a dentary fragment with alveoli for the last two molars, described as possibly belonging to a marsupial (Goin et al., 1986). The limited information presented by this specimen leaves its identity in doubt (Marshall and Muizon, 1988). Another Argentine occurrence with a single described Late Cretaceous mammal is at the locality of **Meseta de Somuncura,** Chubut Province. The site is in the middle part of the La Colonia Formation. The unit is not well constrained in terms of age: it may span Campanian (or pre-Campanian) to the Lower Tertiary; the site is believed to lie in the Campanian–Maastrichtian part of the sequence, which has also yielded remains of the bizarre theropod *Carnotaurus.* Pascual et al. (2000) report a jaw of *Reigitherium bunodontum* from Meseta de Somuncura. *Reigitherium* was initially described as a dryolestoid (Bonaparte, 1990). Pascual et al. (2000) consider it to be a member of Docodonta—a group that went extinct elsewhere by the Early Cretaceous. However, the relationships of *Reigitherium* remain uncertain (see chapter 5). An isolated mammalian petrosal described as "placental-like" has also been reported from Meseta de Somuncura (Pascual et al., 2000), but it has not yet been published.

The most diverse assemblage of Mesozoic mammals known from the South American continent is from **Los Alamitos,** Río Negro Province, a site discovered by José F. Bonaparte and colleagues. The site is in the Los Alamitos Formation, placed in the Campanian–Maastrichtian (Bonaparte, 1990; Pascual et al., 2000). The depositional environment is interpreted as predominantly lacustrine, with some brackish water influence (Andreis, 1987).

Some 14 mammals[10] are known from Los Alamitos (Bonaparte and Soria, 1985; Bonaparte, 1986a,b, 1990, 1992), which is the basis for the "Alamitan" South American land-mammal age (Bonaparte et al., 1987). There are no tribosphenic mammals; rather, the assemblage includes "triconodonts," a ?multituberculate, "symmetrodontans," the aforementioned *Reigitherium,* dryolestoid "eupantotherians," and a group of uncertain affinities (Gondwanatheria). Many of the included taxa are highly derived members of their respective groups, suggesting that the fauna is an endemic one, separated from Laurasian assemblages since perhaps the Late Jurassic. The eutriconodontans (*Austrotriconodon* spp.) were initially placed in the Triconodontidae, but later referred to their own family, of enigmatic affinities (Bonaparte, 1992). Multituberculata may be represented at Los Alamitos by a dentary with p4 referred to originally as *Ferugliotherium* by Kielan-Jaworowska and Bonaparte (1996), and a few upper premolars, but now placed by us (see chapter 8) in Multituberculata *incertae sedis.* Other teeth, especially upper molars, referred to *Ferugliotherium* (e.g., Krause, Kielan-Jaworowska, and Bonaparte, 1992; Krause and Bonaparte, 1993) are now assigned to Gondwanatheria (see chapter 14). Upper molars of *Ferugliotherium* show striking similarities to those of the Late Cretaceous *Gondwanatherium* and Paleocene *Sudamerica,* assigned to Gondwanatheria. The discovery by Pascual et al. (1999) of a dentary of *Sudamerica* with four molar loci precludes its assignment to Multituberculata (see also Pascual and Goin, 2001, and references therein, and chapter 14). At present, we can tentatively conclude that at least *Gondwanatherium* is related to taxa from the Late Cretaceous of Madagascar and India (Krause, Prasad, et al., 1997). Most of the remaining taxa from Los Alamitos are "symmetrodontans" and "eupantotherians" (six species). Three species are included in the Dryolestidae, which are rather common elements of Late Jurassic–earliest Cretaceous faunas of Laurasia, but which (with one exception) had become extinct elsewhere long before the onset of the Late Cretaceous. *Mesungulatum houssayi,* referred to its own monotypic family, is a relatively large species with rather bunodont molars that, interestingly, appear to be functionally convergent on those of "condylarth" eutherians (Bonaparte, 1990). Three additional genera and species of dryolestoids ("*Rougiertherium tricuspes,*" "*Alamitherium bishopi,*" and "*Paraungulatum rectangularis*") from Los Alamitos were noted by Bonaparte (1999). Although they have not yet been formally described or illustrated and are therefore *nomina nuda,* they suggest a greater diversity of "eupantotherian" taxa from this site than currently recognized.

Peru

French teams, including Bernard Sigé and colleagues, have recovered fossil mammals from several Late Cretaceous and/or Early Tertiary sites in the Peruvian Andes (figure 2.7). A tooth fragment belonging to an unidentified therian is known from **Fundo el Triunfo** south of Bagua in the Bagua syncline. The occurrence is in the basal Red Beds of the Fundo el Triunfo Formation, first estimated to be late Santonian to Campanian (Mourier et al., 1986), and later considered to be late Campanian to early Maastrichtian, based on charophytes and other evidence (Mourier et al., 1988). The associated fauna includes amphibians, turtles, and titanosaurid dinosaurs. Dinosaur eggshells have also been reported from this formation (Vianey-Liaud et al., 1995).

The most renowned mammal site in the Peruvian Andes is **Laguna Umayo,** Department of Puno. The site is in

[10] *Barberenia araujoae* appears to be based on milk teeth of *Brandonia intermedia* (see Martin, 1999a), and there may be other such cases.

the Vilquechico Group (formerly called the Vilquechico Formation, see Jaillard et al., 1993). The Late Cretaceous age originally reported for Laguna Umayo (e.g., Grambast et al., 1967) has been disputed. Van Valen (1988), in particular, found the evidence presented by charophytes to be equivocal. Studies of dinosaur eggshell from Laguna Umayo (Kerourio and Sigé, 1984; Vianey-Liaud et al., 1995) support a Late Cretaceous assignment. Jaillard et al. (1993) place the site in the upper part of the Vilquechico Group and, based on sedimentologic, lithologic, palynologic, and other paleontologic data, indicate it to be of probable Maastrichtian age.

At least five mammalian taxa, all represented by fragmentary remains, are known from the site. All are boreosphenidan mammals; one is a eutherian and at least three marsupials are represented. Initially, at least, the marsupials were compared favorably with those from the Late Cretaceous of North America (see chapter 12). A possible "pediomyid" was reported from Laguna Umayo (Sigé, 1972). Another species was initially described as belonging to the North American genus *Alphadon,* as *A. austrinum* (see Sigé, 1971), but later transferred to *Peradectes* (Crochet, 1980). Species referred to *Peradectes* are widespread, occurring in the Early Tertiary of North America, Europe (Crochet, 1980), and Africa. Given the limited nature of the evidence, the significance of this astonishingly broad distribution is difficult to interpret. The eutherian from Laguna Umayo is *Perutherium altiplanense,* known by a dentary fragment with parts of two molars (see chapter 13). *Perutherium* has sometimes been considered to be a "condylarth" (Grambast et al., 1967; Sigé, 1972), even a member of the otherwise North American family Periptychidae (Van Valen, 1978; see summary by Cifelli, 1983). Alternatively, *Perutherium* may be an early member of one of South America's indigenous groups, the most likely being the Notoungulata (Marshall et al., 1983).

This variety of opinion serves to underscore the fact that there is presently insufficient evidence for definitive placement. Accepting the Maastrichtian age assessment for Laguna Umayo, we consider this faunule significant because it helps "bracket" the Cretaceous turnover in South American mammal faunas. Whatever the specific affinities of its included taxa, the faunule from Laguna Umayo is clearly of "modern" aspect; like later faunas of South America, it includes marsupials and, probably, an ungulate. In these respects, it is completely different from the somewhat older (Campanian–early Maastrichtian) fauna of Los Alamitos, Argentina, which is comprised exclusively of nontribosphenic mammals. We close our discussion of Laguna Umayo with brief mention of another site with which it has frequently been compared—Tiupampa in Bolivia.

A diverse, stunningly well-represented assemblage of mammals has been recovered from Tiupampa, including complete skulls and skeletons of marsupials (Muizon, 1994; Marshall and Sigogneau-Russell, 1995; Muizon et al., 1997). Tiupampa was initially placed in the El Molino Formation and considered to be Late Cretaceous in age (see, e.g., Marshall and Muizon, 1988). Later studies refer the site to the Santa Lucía Formation (see review by Gayet et al., 1991) and show it to be of early Paleocene age (Van Valen, 1988; Muizon, 1998).

Another Peruvian site of uncertain age is **Chulpas,** which lies in a microconglomerate in the upper sandstone levels of the Umayo Formation, some 200 m higher in the section than Laguna Umayo. There is no associated flora and as yet no constraint on the age of Chulpas, though the upper sandstones of the Umayo Formation are conformably overlain by volcanic rocks that might be datable (Crochet and Sigé, 1993). Eleven varieties of mammals are known from Chulpas (Crochet and Sigé, 1993, 1996). At least three of these are eutherians; two or more are notoungulates, and a proteutherian (*sensu* Romer, 1966), said to be similar to a *Cimolestes*-like taxon from Tiupampa (see Marshall and Muizon, 1988), are present. The remainder are marsupials. In addition to several unidentified didelphids, these include a representative of a new, monotypic family of polydolopoids (Sillustaniidae), a caenolestoid, and two caroloameghiniids. Interestingly, one of the marsupials of Chulpas (*Chulpasia*) may be closely related to *Thylacotinga* from the early Eocene of Australia, suggesting Early Tertiary dispersal between the continents (Sigé et al., 1995). Aware of the dangers of circular reasoning, we point out that the marsupial groups of Chulpas, like notoungulates, are generally considered to be characteristic of South America's Tertiary faunas. We herein consider Chulpas to be of Early Tertiary age—somewhat younger than the Cretaceous-Tertiary transition, where it is currently placed (Crochet and Sigé, 1993, 1996).

A final possible occurrence of a Late Cretaceous mammal in Peru is based on footprints from the Couches Rouges Formation at **Paruro,** Cuzco. The tracks are described as "mammaloid" and rather large, the size of a dog or a cat (Leonardi, 1994). A variety of dinosaur tracks has been reported from this and other sites in the Cuzco-Sicuani Basin. The exact age of the tracks is unknown; the unit spans the Santonian-Paleocene, and the tracks are placed in the Cretaceous part of the section (Noblet et al., 1995).

Bolivia

A moderately diverse vertebrate assemblage, including mammals, has been reported from **Pajcha Pata,** near Cochabamba, Bolivia (Gayet et al., 2001). The fossils were recovered from the lower member of the El Molino Formation; combined geochronologic and fauna evidence points to a middle Maastrichtian age for the assemblage. Pajcha Pata is thus of great interest, as it is almost surely younger than Los Alamitos, Argentina (which contains an endemic mammalian fauna, comprised of highly derived members of archaic groups), and older than earliest Tertiary assemblages of South America (which are comprised of marsupials and placentals—groups that are presumed to have immigrated from North America sometime in the latest Cretaceous). Mammalian fossils known from Pajcha Pata include only two teeth and one tooth fragment. Nonetheless, they suggest a fauna comprised of both autochthonous and allochthonous elements: at least one dryolestid and one eutherian are represented.

NORTH AMERICA

The record of Late Cretaceous mammals in North America has long consisted almost exclusively of faunas from the Campanian and Maastrichtian of the northern and central Rocky Mountain region (see summary by Clemens et al., 1979). In recent decades, sampling—both temporal and geographic—has vastly improved. Reasonably informative faunas are now known from the Albian-Cenomanian boundary, the Cenomanian, and the Turonian; a few occurrences are known for the Coniacian–Santonian, and the possibilities for improved knowledge from this interval are excellent. Geographically, the greatest improvement in the Late Cretaceous record of North America has been increased sampling from more southerly and eastern areas: southern Utah, New Mexico, the Big Bend region of Texas and, remarkably, New Jersey.

Most notable in both geographic and temporal respects is a sequence of faunas from southern and central Utah, which collectively sample much of the Late Cretaceous (Cifelli, Nydam, Eaton, et al., 1999; Cifelli, Nydam, Weil, et al., 1999; Eaton, Cifelli, et al., 1999). These faunas are treated individually in the sections that follow, but we note some general comments here. The Cretaceous system in the Kaiparowits region of southern Utah includes, in ascending order, the Dakota, Tropic, Straight Cliffs, Wahweap, Kaiparowits, and Canaan Peak formations. The Tropic and parts of the Dakota and Straight Cliffs formations are of marine origin, providing reasonable age constraints; the upper part of the Straight Cliffs and overlying units are terrigenous, and their correlations are based largely on palynomorphs and vertebrate fossils (Eaton, 1991). Sampling for mammals and other microvertebrates is poor in parts of the section, but reasonably good faunas are known from the Dakota Formation (upper Cenomanian), Smoky Hollow Member of the Straight Cliffs Formation (Turonian), Wahweap Formation (lower Campanian), and Kaiparowits Formation (upper Campanian). To the west, on the Paunsaugunt Plateau, the Markagunt Plateau, and extreme southwestern Utah, several additional Late Cretaceous mammal specimens and faunules are known. Correlation with the sequence on the Kaiparowits Plateau is not straightforward in some cases, but it is likely that localities in these regions include occurrences of Turonian, Coniacian–Santonian, and Campanian age (Eaton, 1999a,b; Eaton, Diem, et al., 1999). Other geographically nearby faunas are known from the Albian-Cenomanian Cedar Mountain Formation (discussed earlier under Early Cretaceous) and the Maastrichtian North Horn Formation, both in central Utah. The assemblages from Utah are therefore significant in that they collectively span the entire Late Cretaceous.

Most of the fossils of North American Late Cretaceous mammals resulted from screenwashing and other bulk processing techniques, which tend to yield fossils in vast quantities but are rather destructive. A continuing and regrettable feature of the record of Late Cretaceous mammals from North America is poor morphological representation: many taxa are known only by isolated teeth, few are represented by reasonably complete jaws, and exceedingly few are known by anything more than isolated jaws. This is in marked contrast to the record from Mongolia, where, as noted, Late Cretaceous mammals are commonly represented by skulls and, in some cases, essentially complete skeletons.

North American Late Cretaceous land-mammal ages have long been employed for faunas dating from approximately the beginning of the Campanian onward. Three such ages are widely recognized, the Aquilan (early Campanian), Judithian (late Campanian), and Lancian (late Maastrichtian). A fourth land-mammal age, the "Edmontonian," may represent the time intervening (early Maastrichtian) between Judithian and Lancian, but its definition is problematic (see discussions by Lillegraven and McKenna, 1986; Cifelli et al., 2004). Our treatment of local faunas (and, where practical, individual sites) follows this sequence of land-mammal ages. Geologically older assemblages or individual occurrences are referred to by approximate corresponding marine stage, as we have done elsewhere in this chapter.

North American Late Cretaceous mammal faunas typically include a large proportion of multituberculates. Morphological diversity is low among multituberculates of pre-Campanian faunas (Cifelli, Kirkland, et al., 1997b), with most forms being assigned to the *Paracimexomys* group of primitive cimolodontans (chapter 8). Faunas throughout the North American Late Cretaceous also include a significant component of Marsupialia. Early differentiation of major clades is apparent (Fox, 1981; Cifelli and Eaton, 1987), though problems in phylogenetic interpretation persist, and certain dental morphologies may represent iterative patterns in marsupial evolution (Fox, 1987b).

Eutherians appear to have been in North America during the Aptian–Albian (Cifelli, 1999b), but from that time until the Campanian, evidence for their presence is mysteriously lacking. The fossil record remains poor, and it is possible that they maintained a restricted distribution on the continent during the Cenomanian–Santonian. Nonetheless, it is fair to conclude that eutherians were not major players in North American faunas prior to the late Campanian. If they were present, the fact that marsupials flourished during this time offers a challenge to the notion that eutherians are competitively superior to marsupials and that they will dominate when the two groups co-occur. In recent decades, it has also been shown that surviving members of archaic groups also played roles, albeit limited, in North America's Late Cretaceous faunas. Triconodontidae survived until the Aquilan (Fox, 1969) and spalacotheriid "symmetrodontans" until the Aquilan (Fox, 1976a; Cifelli and Madsen, 1986) or Judithian (Eaton, Cifelli, et al., 1999); a "eupantotherian" has been recorded in the Judithian (Lillegraven and McKenna, 1986). Sampling is extremely limited, but there appear to be patterns in the distributions of these taxa, perhaps reflecting biogeographic provinciality in the Late Cretaceous of North America (Cifelli and Gordon, 1999; see also Lehman, 1987, 1997).

Cenomanian–Santonian

Cenomanian mammals are known from two sites near the Dallas–Fort Worth Airport, Tarrant County, Texas (figure 2.14, table 2.22). Both are in the Woodbine Formation, the age of which is well established because of the marine invertebrates contained in it and overlying units (Jacobs and Winkler, 1998). The **Carter Field** locality has yielded two mammals: an unidentified tribosphenidan and an unidentified multituberculate (McNulty and Slaughter, 1968; Krause and Baird, 1979). The **Bear Creek** locality (figure 2.14) has yielded a single mammalian molar belonging to a marsupial. The occurrence is significant because it lies east of the North American epicontinental seaway, which had been established by the late Albian, and shows that marsupials were present in both eastern and western North America by the Cenomanian (Jacobs and Winkler, 1998).

A diverse mammalian assemblage has been collected from the Dakota Formation, southern Utah, by Jeffrey G. Eaton. The mammals come from four principal sites in adjacent areas of the Paunsaugunt and Kaiparowits plateaus (figure 2.18), in the upper 10–20 m of the middle member of the unit. For convenience, they are here treated collectively as the **Dakota** fauna. The Dakota Formation includes marine facies; molluscs indicate that the entire unit is upper Cenomanian in the area (Eaton, 1995). The mammal-bearing sites are mainly in floodplain beds of fluvial origin (Eaton, 1993b). Some 21 mammals are known from the Dakota fauna (Cifelli and Eaton, 1987; Eaton, 1993b, 1995; Eaton, Cifelli, et al., 1999). Of the nine multituberculates, most (*Dakotamys*, *Paracimexomys*, and several unnamed forms) belong to the *Paracimexomys* group, *Cimolodon* being a notable exception. This genus is typical of Campanian and later faunas of the North American Cretaceous and apparently has a remarkably long stratigraphic distribution. Indeed, the species from the Dakota fauna is closely similar to or identical with *C. similis*, described from the Aquilan of Canada and also possibly represented in the entire sequence (through the Judithian) in southern Utah. An unidentified "symmetrodontan" and perhaps two "Theria of metatherian-eutherian grade" are known from the assemblage. Curiously missing are Triconodontidae, which are diverse in the slightly older and nearby Mussentuchit local fauna (Cifelli and Madsen, 1998) but are not known from any later assemblage except that of the upper part of the Milk River Formation, Alberta (Fox, 1969). Given the fact that even fragments of triconodontid teeth are readily identified as such, local extirpation in Utah appears to be the most likely explanation. Perhaps six marsupials are known from the Dakota fauna, including a possible early record of Stagodontidae, a family of mainly large species characterized by teeth with carnivorous and durophagous specializations (Fox and Naylor, 1995). Most notable among the marsupials are two or more species assigned to the genus *Alphadon* (see Eaton, 1993b). This genus is characteristic of Judithian and Lancian faunas (see review by Johanson, 1996b), but is not otherwise known in earlier faunas. The apparent disjunct distribution of *Alphadon* in the Late Cretaceous of North America could reflect changing zoogeographic patterns or, perhaps, morphological convergence of unrelated lineages.

Several mammalian postcranial elements were reported from the Wall Creek Member of the Frontier Formation near Como Bluff, Wyoming (Clemens et al., 1979, see figure 2.10). The horizon is Cenomanian–Turonian in age. Nothing further has been collected from the site, which is omitted from our faunal list and figure.

Returning to southern Utah, a reasonably diverse mammalian assemblage has been collected from the Smoky Hollow Member of the Straight Cliffs Formation by Cifelli and Eaton. The fossils come from several localities on adjacent parts of the Kaiparowits and Paunsaugunt

TABLE 2.22. Cenomanian–Santonian Mammals of North America (see figures 2.14, 2.15, 2.18; locality numbers do not correspond between maps and table). Localities or local faunas: 1, Carter Field; 2, Bear Creek (both Woodbine Formation; Cenomanian, Texas); 3, Dakota (Dakota Formation, Cenomanian, Utah); 4, Smoky Hollow (Smoky Hollow Member of the Straight Cliffs Formation, Turonian, Utah); 5, John Henry (John Henry Member of the Straight Cliffs Formation, Coniacian–Santonian, Utah); 6, Cedar Canyon (?Straight Cliffs Formation, Coniacian, Utah); 7, Cedar Canyon (Straight Cliffs Formation, Santonian, Utah); 8, Pine Valley Mountains (Iron Springs Formation, Turonian, Utah); 9, Pine Valley Mountains (Iron Springs Formation, Coniacian–Santonian, Utah); 10, Parowan Canyon (Iron Springs Formation, Coniacian–Santonian, Utah); 11, Vinton Bluff (Eutaw Formation, Santonian, Mississippi)

- Multituberculata
  - Family *incertae sedis*
    - Gen. et sp. indet. (1, 6, 10)
    - Gen. et sp. indet. A (3)
    - Gen. et sp. indet. B (3)
    - Gen. et sp. indet. C (4)
    - Gen. et sp. indet. D (4)
    - *Bryceomys fumosus* (4)
    - *Bryceomys* cf. *fumosus* (4)
    - *Bryceomys hadrosus* (4)
    - *Bryceomys* sp. (8)
    - *Bryceomys* sp. nov. (7)
    - *Dakotamys malcolmi* (3)
    - ?*Dakotamys* sp. (3)
    - *Paracimexomys* sp. nov. A (5)
    - *Paracimexomys* sp. nov. B (7)
    - *Paracimexomys* cf. *priscus* (9)
    - *Paracimexomys* cf. *robisoni* (3, 4)
    - *Paracimexomys* sp. (3)
    - cf. *Paracimexomys* sp. (3)
  - Cimolodontidae
    - Gen. et sp. indet. (3)
    - *Cimolodon nitidus* (7)
    - *Cimolodon similis* (7)
    - *Cimolodon* cf. *similis* (3, 5)
    - *Cimolodon* sp. (7)
  - Cimolomyidae
    - ?*Cimolomys* sp. nov. (7)
    - *Meniscoessus intermedius* (7)
- Stem Trechnotheria
  - Spalacotheriidae
    - Gen. et sp. indet. (3)
    - *Spalacotheridium mckennai* (4)
    - *Symmetrodontoides oligodontos* (4)
    - *Symmetrodontoides* cf. *oligodontos* (5)
    - *Symmetrodontoides* sp. (7)
- Stem Boreosphenida ("tribotherians")
  - Picopsidae
    - Gen. et sp. indet. (4)
    - cf. *Picopsis* sp. (7)
  - Family *incertae sedis*
    - Gen. et sp. indet. (1, 11)
    - Gen. et sp. indet. A (3)
    - Gen. et sp. indet. B (3)
    - Gen. et sp. indet. C (3)
    - Gen. et sp. indet. D (4)
    - Gen. et sp. indet. E (4)
    - *Dakotadens morrowi* (3)
    - *Dakotadens* sp. (3)
- Deltatheroida
  - Deltatheridiidae
    - Gen. et sp. indet. (4)
- Marsupialia
  - Family *incertae sedis*
    - ?*Anchistodelphys delicatus* (4)
    - Gen. et sp. indet A (2)
    - Gen. et sp. indet. B (4)
    - Gen. et sp. indet. C (4)
    - Gen. et sp. indet. D (10)
  - "Alphadontidae"
    - *Alphadon clemensi* (3)
    - *Alphadon lillegraveni* (3)
    - *Alphadon* sp. (3, 4, 6, 9)
    - *Alphadon* sp. nov. A (7)
    - *Alphadon* sp. nov. B (7)
    - *Protalphadon* sp. (4)
    - *Protalphadon* sp. nov. (3)
    - Gen. et sp. indet. (3, 5)
  - "Pediomyidae"
    - Gen. et sp. nov. (7)
  - ?Stagodontidae
    - *Pariadens kirklandi* (3)
    - Gen. et sp. indet. (4, 5)
- ?Eutheria
  - Family *incertae sedis*
    - Gen. et sp. indet. (4)

plateaus (figure 2.18) and, for convenience, are collectively termed the **Smoky Hollow** fauna. The lower part of the Smoky Hollow Member includes some nearshore marine and brackish water deposits as well as coal; the mammal localities are in the upper part of the unit, where strata were deposited in floodplain and lacustrine environments. The Smoky Hollow Member is of Turonian age (Eaton, 1987, 1991). Multituberculates of the Smoky Hollow fauna (Eaton, 1995) include some six taxa, all rather primitive cimolodontans, assigned to the *Paracimexomys*

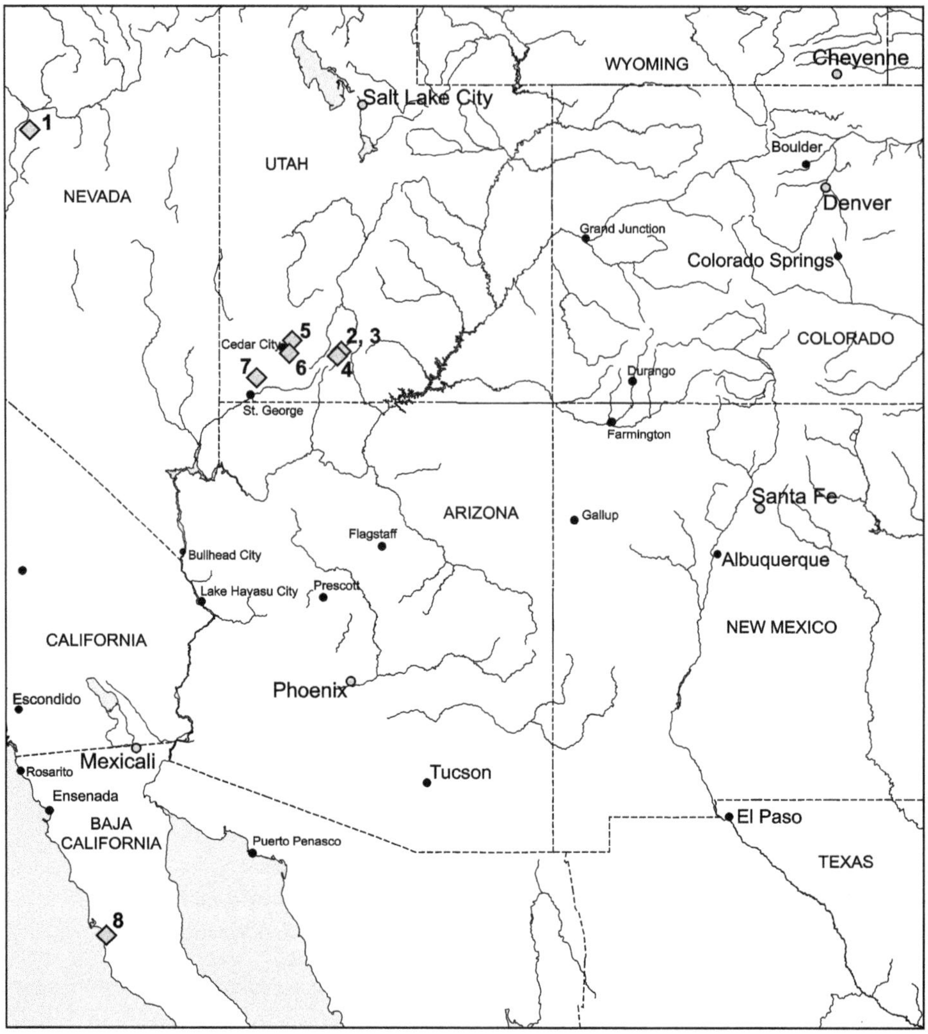

FIGURE 2.18. Cenomanian–Santonian and ?Campanian (Utah), Campanian (Baja California), and unspecified Late Cretaceous (Nevada) mammal localities, southwestern United States and Mexico. Localities or local faunas: **1, Eureka County** (Newark Canyon Formation, Late Cretaceous, Nevada); **2, Smoky Hollow** (Smoky Hollow Member, Straight Cliffs Formation, Turonian, Utah); **3, John Henry** (John Henry Member, Straight Cliffs Formation, Coniacian–Santonian, Utah); **4, Dakota** (Dakota Formation, late Cenomanian, Utah); **5, Parowan Canyon** (Iron Springs Formation, Coniacian–Santonian, Utah); **6, Cedar Canyon** (four horizons; lowest equivalent to ?Coniacian; middle, Santonian; upper, ?Campanian, possibly Judithian; Utah); **7, Pine Valley** Mountains (Iron Springs Formation; two horizons, lower ?Turonian, upper Coniacian–Santonian; Utah); **8, El Gallo** (El Gallo Formation, Campanian, Baja California, Mexico).

group. The multituberculate assemblage much resembles that of the Dakota Formation but is less diverse. Therians, on the other hand, differ more substantially from those of the Dakota Formation, which is only slightly older (Cifelli, 1990c; Eaton, 1993b). "Symmetrodontans" are reasonably abundant, with two species present (Cifelli and Gordon, 1999); one represents the last occurrence of *Spalacotheridium*, also known from the Albian-Cenomanian, while the other represents the earliest occurrence of *Symmetrodontoides*, which ranges into the Aquilan. The marsupials are, surprisingly, more generally primitive (e.g., ?*Anchistodelphys delicatus*) than those of the Dakota Formation. A possible deltatheroidan is represented by a fragmentary tooth.

The John Henry Member of the Straight Cliffs Formation overlies the Smoky Hollow Member. Only a few mammalian fossils, from several sites on the Kaiparowits Plateau (figure 2.18) worked by Cifelli and Eaton, are known for the **John Henry** assemblage. Of Coniacian–Santonian age, the John Henry Member contains marine rocks, particularly on the eastern part of the plateau. Fossil localities tend to yield many euryhaline-tolerant species with a scarcity of terrestrial vertebrates, suggesting a brackish water environment (Eaton, Cifelli, et al., 1999). Only five mammal taxa are known from the John Henry fauna; most are unremarkable, except for a multituberculate similar to *Cimolodon similis,* which appears to be wide ranging.

A preliminary report by Eaton, Diem, et al. (1999) establishes the presence of a sequence of mammalian faunas from ?Coniacian, Santonian, and Campanian horizons in **Cedar Canyon,** the Markagunt Plateau, southwestern Utah (figure 2.18). The lower part of the sequence in Cedar Canyon can be correlated with the lower part of the Straight Cliffs Formation to the east. Mammal-yielding sites are known from four horizons within the upper 600 m of the local section, which consists of relatively homogeneous mudstones and sandstones. The two lowest sites, grouped here for convenience, may be of Coniacian age. Collectively, they include only an unidentified multituberculate and the wide-ranging marsupial *Alphadon.* A somewhat better-sampled fauna higher in the section is thought to be of Santonian age, based on palynomorphs. Twelve mammal taxa are known from this horizon, seven of which are multituberculates. Conflicting ages are suggested by the included taxa: *Cimolodon nitidus,* for example, is characteristic of Lancian faunas, whereas a new species of *Paracimexomys* is most similar to a Turonian form. Most notable among the therians is the marsupial "*Pediomys,*" which is otherwise rare or absent in faunas of the Southwest. Five mammals, one multituberculate and the rest marsupials, are known from the highest site in the section at Cedar Canyon. The age of the horizon is uncertain, though the mammals suggest that it may be Campanian, perhaps even Judithian.

Several horizons in the Iron Springs Formation of extreme southwestern Utah (figure 2.18) have yielded a few mammalian specimens (Eaton, 1999a). The unit is approximately 1 km thick and the span of time it represents is not well understood. Two mammal sites are known from the **Pine Valley Mountains.** One is low in the section and may be of Turonian age because it directly overlies a sandstone-mudstone sequence deposited in brackish water and thus is interpreted as representing maximum transgression of the Cenomanian-Turonian Greenhorn Sea (Eaton et al., 1997). At least one marsupial and one multituberculate are known from this horizon; the only identifiable taxon is *Bryceomys,* a primitive cimolodontan from the Turonian of the Straight Cliffs Formation (Eaton, 1995). The second site in the Pine Valley Mountains is in a horizon near the top of the Iron Springs Formation, which may be Coniacian–Santonian in age. The site has yielded fragmentary teeth possibly referable to the marsupial *Alphadon,* as well as a molar similar to that of the primitive multituberculate *Paracimexomys priscus,* which is otherwise known only from the Judithian and Lancian. The final mammal locality in the Iron Springs Formation is in **Parowan Canyon.** Stratigraphically, it is also at the top of the unit and hence is presumably of Coniacian–Santonian age. Thus far, only an unidentified multituberculate and a marsupial are known from the site, but the excellent preservation suggests that, with additional effort, informative specimens might be recovered here.

A single mammalian fossil has been reported from **Vinton Bluff,** Clay County, Mississippi (figure 2.15). The occurrence is in the Tombigbee Sand, the uppermost member of the Eutaw Formation, which is of late Santonian age (Emry et al., 1981). The specimen consists of a partial lower molar that, though fragmentary, is of interest in its age and in that it represents one of only a few occurrences east of the Late Cretaceous interior seaway in North America. It is generally thought to belong to a eutherian with low-crowned teeth, perhaps similar to *Protungulatum* (Emry et al., 1981) or to "Zhelestidae," which are considered to be basal ungulates (Nessov, Archibald, and Kielan-Jaworowska, 1998). We find it unidentifiable, other than noting the fact that it belongs to some kind of tribosphenic mammal.

Campanian- and ?Campanian-Aged Assemblages Not Assigned to Land-Mammal Age

Several Late Cretaceous mammal occurrences or assemblages (table 2.23) cannot be assigned to land-mammal age, owing to poor age constraints, geographic location, or other factors. We begin with the **El Gallo** fauna from Baja California, Mexico (figure 2.18). The fossils come from three localities at various levels in the middle part of the "El Gallo Formation." All of these lie above a tuff dated at 73.59–74.87 Ma (Lillegraven, 1972; Renne et al., 1991). Ammonites from within and above the "El Gallo Formation" suggest a Campanian age (Clemens et al., 1979); Morris (1981) cites marine invertebrates from the unit as being Campanian-Maastrichtian. Taking the date at face value, we find that the El Gallo mammals appear to be younger than at least part of the Judithian land-mammal age (Goodwin and Deino, 1989) and may prove to be "Edmontonian" in age.

Five mammals have been reported from the "El Gallo Formation" (Lillegraven, 1972, 1976); three multituberculates, a marsupial, and a eutherian mammal of uncertain affinities (*Gallolestes*). One of the multituberculates is similar to *Mesodma formosa,* an abundant and widespread neoplagiaulacid characteristic of Lancian assemblages. Another is of interest in that it is referred to *Stygimys* (see Weil and Clemens, 1998), which, with the removal of Bug Creek assemblages to the Paleocene (Lofgren, 1995), is

TABLE 2.23. Campanian, ?Campanian, and Mammals of Uncertain Age, North America (see figures 2.15, 2.18; locality numbers do not correspond between maps and table). Localities or local faunas: 1, El Gallo, Baja California ("El Gallo Formation," Campanian, Mexico); 2, Cedar Canyon, Utah (uncertain unit, ?Campanian, Utah); 3, Ellisdale local fauna, New Jersey (Marshalltown Formation, late Campanian); 4, Monmouth Brook, New Jersey (unspecified unit, ?late Campanian); 5, Elizabethtown Dump, North Carolina (Black Creek Group, ?Campanian); 6, Eureka County, Nevada (Newark Canyon Formation, Upper Cretaceous)

- Multituberculata
  - Family *incertae sedis*
    - Gen. et sp. indet. (2, 6)
  - Cimolodontidae
    - Gen. et sp. indet. (3)
  - Cimolomyidae
    - *Cimolomys* sp. (3, 5)
    - cf. *Cimolomys clarki* (3)
  - Eucosmodontidae
    - ?*Stygimys* sp. (1)
    - Gen. et sp. indet. (1, 4)
  - Neoplagiaulacidae
    - *Mesodma* cf. *formosa* (1)
- Marsupialia
  - "Alphadontidae"
    - *Alphadon* cf. *attaragos* (2)
    - *Alphadon* cf. *sahnii* (2)
    - *Protalphadon lulli* (3)
  - "Pediomyidae"
    - cf. *Iqualadelphis* sp. (2)
    - "*Pediomys*" sp. (1)
    - cf. "*Pediomys*" sp. nov. (2)
  - Stagodontidae
    - ?*Didelphodon* sp. (3)
- Eutheria
  - Family *incertae sedis*
    - *Gallolestes pachymandibularis* (1)

otherwise rare or absent in the Cretaceous. These multituberculates hint at a younger age for El Gallo than has previously been envisaged, or a diachronous appearance for these taxa. The therian *Gallolestes* has been alternatively interpreted as a therian of metatherian-eutherian grade "tribotherian" in this book (Clemens, 1980b) or a eutherian (Lillegraven, 1976; Cifelli, 1994); confounding the problem is the fact that one of the teeth in the type of *G. pachymandibularis* may be deciduous (Butler, 1990a). Nessov, Archibald, and Kielan-Jaworowska (1998) believed that *Gallolestes* may be a basal ungulate, perhaps related to "Zhelestidae"; recent analysis suggests a possible relationship to *Paranyctoides* (see discussion by Archibald and Averianov, 2001b).

Five mammalian taxa have been reported from near the top of a 600-m section of terrigenous Upper Cretaceous rocks in **Cedar Canyon** (figure 2.18) on the Markagunt Plateau, southwestern Utah (Eaton, Diem, et al., 1999). As with a similar site lower in the section, a "pediomyid" marsupial is present, which is unusual for the southerly faunas. Two of the tentatively identified species of the marsupial *Alphadon* (*A.* cf. *sahnii* and *A.* cf. *attaragos*) are otherwise known from the Judithian (Lillegraven and McKenna, 1986), whereas the marsupial *Iqualadelphis* is restricted to the Aquilan (Fox, 1987a).

The **Ellisdale** local fauna, New Jersey (figure 2.15), includes a diverse assemblage of marine and terrestrial vertebrates, including four mammals. The site lies in the Marshalltown Formation, which, based on foraminiferans, is upper Campanian (Grandstaff et al., 1992). A date of 71.6 Ma, from glauconite, would suggest placement at or near the Campanian-Maastrichtian boundary (Gradstein et al., 1995). As is the case for El Gallo, the radiometric date is younger than determinations for the Judithian from the Judith River Formation itself (Goodwin and Deino, 1989). The Marshalltown Formation represents a nearshore marine environment; fossils occur in a coarse, sandy horizon with clayballs, interpreted as a storm deposit (Grandstaff et al., 1992). Four mammals are known from Ellisdale (Grandstaff et al., 1992, 2000): an unidentified cimolodontid multituberculate, a cimolomyid (known by an incisor and an upper premolar) said to be similar to *Cimolomys clarki*, and the marsupials *Protalphadon lulli* (represented by an upper molar) and ?*Didelphodon* sp. (several unspecified teeth). Meager as these materials would appear, they are significant in being the most informative Late Cretaceous mammal specimens from the eastern United States. The close similarity to species known from the Western Interior suggests that the epicontinental seaway may have been less of an impediment to terrestrial animals than has been supposed (Grandstaff et al., 1992).

Two additional sites in eastern North America (figure 2.15) have yielded mammals that may be of Campanian age (Grandstaff et al., 2000). An isolated multituberculate incisor, possibly belonging to a eucosmodontid, has been reported from **Monmouth Brook**, New Jersey (rock unit not specified). The **Elizabethtown Dump** site, Bladen County, North Carolina, has yielded a P4 assigned to the multituberculate *Cimolomys*. The site apparently lies in the Black Creek Group.

A final site discussed under this heading for convenience is in **Eureka County**, Nevada (figure 2.18). Two unidentified multituberculate teeth were collected by screenwashing matrix from an anthill. The occurrence is in the Newark Canyon Formation, which is constrained no more precisely than Upper Cretaceous (Clemens et al., 1979).

Aquilan

The Aquilan is the oldest of the North American Late Cretaceous land-mammal ages and is defined on the basis of a mammalian fauna from the upper parts of the Milk River Formation, Alberta, Canada. The fauna is highly distinctive, sharing few species with the younger Judithian land-mammal age (see list and discussion in Lillegraven and McKenna, 1986). One problem in defining the Aquilan is the lack of well-represented assemblages that are only slightly older. In addition to the fauna from the upper parts of the Milk River Formation, several other assemblages (all from Utah) have been referred to the Aquilan in recent years (table 2.24). These greatly increase geographic representation of this land-mammal age. Perhaps most notable among first appearances (keeping in mind the earlier caveat) is that of the nyctitheriid insectivoran, *Paranyctoides* (see Fox, 1984b; Cifelli, 1990e). Except for

TABLE 2.24. Aquilan Mammals of North America (see figures 2.19, 2.20; locality numbers do not correspond between maps and table). Sites or local faunas: 1, Verdigris Coulee, Alberta (Milk River Formation); 2, Wahweap, Utah (Wahweap Formation); 3, Campbell Canyon, Utah (Wahweap Formation); 4, Masuk, Utah (Masuk Formation)

- Eutriconodonta
  - Triconodontidae
    - *Alticonodon lindoei* (1)
    - Gen. et sp. indet. (1)
- Multituberculata
  - Family *incertae sedis*
    - *Cimexomys antiquus* (1)
    - *Cimexomys* cf. *antiquus* (2)
    - *Paracimexomys magister* (1)
    - *Paracimexomys* sp. nov. (2, 4)
    - *Paracimexomys* sp. (2)
    - *Viridomys orbatus* (1)
  - Cimolodontidae
    - *Cimolodon electus* (1, 2)
    - *Cimolodon similis* (1, 2)
    - *Cimolodon* sp. (2, 3)
    - ?*Cimolodon* sp. (2)
    - ?*Cimolodon* sp. A (1)
    - ?*Cimolodon* sp. B (1)
  - Cimolomyidae
    - *Cimolomys* cf. *clarki* (2)
    - *Cimolomys* sp. (2)
    - ?*Cimolomys* sp. A (1)
    - ?*Cimolomys* sp. B (1)
    - *Meniscoessus ferox* (1)
    - ?*Meniscoessus* sp. (2)
  - ?Cimolomyidae
    - Gen. et sp. indet. (2)
  - Neoplagiaulacidae
    - *Mesodma senecta* (1)
    - *Mesodma* cf. *formosa* (2)
    - *Mesodma* sp. (1)
    - ?*Mesodma* sp. (2, 3, 4)
- Stem Trechnotheria ("symmetrodontans")
  - Spalacotheriidae
    - *Symmetrodontoides canadensis* (1)
    - *Symmetrodontoides foxi* (2)
- Basal Boreosphenida ("tribotherians")
  - Family *incertae sedis*
    - Gen. et sp. indet. (1)
    - *Potamotelses aquilensis* (1)
    - *Zygiocuspis goldingi* (2)
  - Picopsidae
    - *Picopsis pattersoni* (1)
    - *Picopsis* sp. (1)
- Marsupialia
  - Family *incertae sedis*
    - *Anchistodelphys archibaldi* (2)
    - *Anchistodelphys* sp. (2)
    - *Iugomortiferum thoringtoni* (2)
    - cf. *Iugomortiferum* sp. (2)
    - Gen. et sp. indet. (2)
  - "Alphadontidae"
    - *Albertatherium primum* (1)
    - *Albertatherium secundum* (1)
    - *Alphadon* sp. (1)
    - ?*Alphadon* sp. (3)
    - *Varalphadon creber* (1)
    - *Varalphadon crebreforme* (2)
    - *Varalphadon wahweapensis* (1)
  - "Pediomyidae"
    - *Aquiladelphis incus* (1)
    - *Aquiladelphis minor* (1)
    - *Iqualadelphis lactea* (1)
    - "*Pediomys*" *exiguus* (1)
    - cf. "*Pediomys*" *exiguus* (4)
    - Gen. et sp. indet. (3)
  - Stagodontidae
    - *Eodelphis* sp. (1)
- ?Eutheria, *incertae sedis* (1)
- Eutheria
  - Family *incertae sedis*
    - Gen. et sp. indet. (1)
    - Gen. et sp. indet. A (1)
  - "Nyctitheriidae"
    - *Paranyctoides maleficus* (1)
    - *Paranyctoides* sp. A (2)
    - *Paranyctoides* sp. B (2)

Aptian–Albian *Montanalestes* (see earlier), this represents the oldest occurrence of an uncontested eutherian in North America and follows a significant (ca. 30 Ma) "eutherian hiatus" on the continent. Intriguingly, a species referred to this genus has been described from the Coniacian of Uzbekistan (Nessov, 1993), suggesting the possibility of a pre-Campanian dispersal event between Asia and North America (Cifelli, 2000c; Archibald and Averianov, 2001b). Another important difference between Aquilan and older faunas in North America is that multituberculates appear to be significantly more diverse, both morphologically and taxonomically.

The type fauna for the Aquilan is from **Verdigris Coulee** in the upper parts of the Milk River Formation, just north of the U.S. border (figure 2.19). The fauna is highly significant: it is diverse, certain key taxa are represented by comparatively complete specimens, and it was the first well-represented pre-Judithian assemblage known from the North American Late Cretaceous. As such, it serves as an important benchmark. Credit for the recovery of this fauna goes to Richard C. Fox, who began work at Verdigris Coulee in the 1960s and described the mammals and some other vertebrates in a long series of publications (e.g., Fox, 1968, 1969, 1970, 1971a,b, 1972a,b, 1974b, 1976a, 1980b, 1982, 1984a,b, 1985, 1987a). The upper part of the Milk River Formation has conventionally been assigned to the lower Campanian (see discussion of Santonian-Campanian boundary by Lillegraven, 1991). However, a different interpretation has been presented by Leahy and Lerbekmo (1995) based on magnetostratigraphy and ammonite zonation. If their correlation is correct, as the balance of evidence suggests (see summary by Braman, 2001), then most or all of the Milk River Formation would be of Santonian age.

Thirty-one mammalian taxa have been reported from Verdigris Coulee. A significant aspect of the fauna is that it demonstrates the survival of archaic clades into the Late Cretaceous—triconodontids, "symmetrodontans," and primitive tribosphenic mammals (Fox, 1972b, 1976a, 1980b, 1982, 1984a). The fauna is also notable for first appearances and for significant differences in composition from earlier assemblages. Of the nine multituberculates (Fox, 1971a), only one, *Paracimexomys magister*, is referable to the *Paracimexomys* group of primitive cimolo-

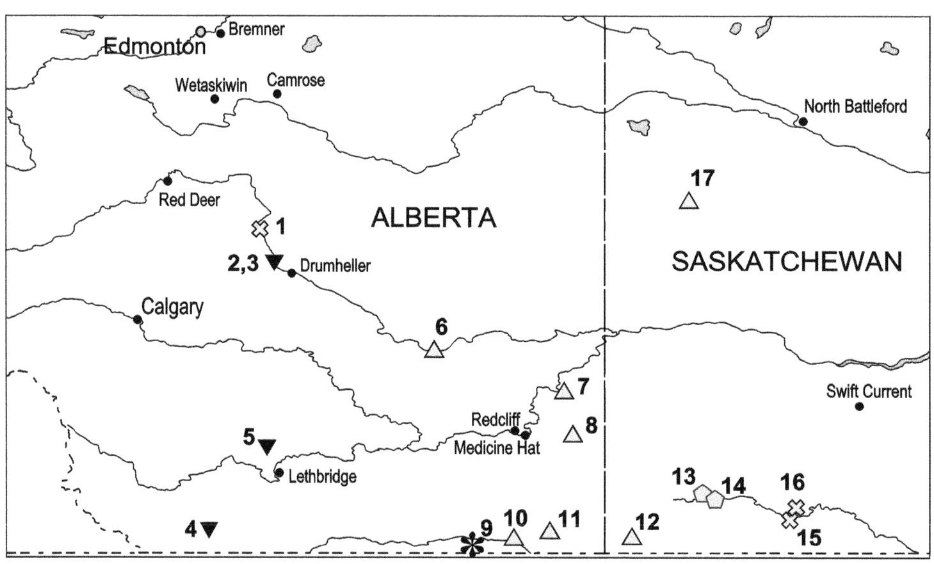

FIGURE 2.19. Cretaceous mammal localities of Canada. Asterisk, Aquilan (locality 9); gray triangles, Judithian (localities 6–8, 10–12, 17); inverted black triangles; "Edmontonian" (localities 2–5); crosses, Lancian (localities 1, 15–16); pentagons, ?Late Cretaceous, Puercan land-mammal age (localities 13–14). Localities, assemblages of localities, or local faunas: **1, Trochu** (lower Scollard Formation, Lancian; several localities, Alberta); **2, Michichi Creek; 3, Newcastle** (both Horseshoe Canyon Formation; "Edmontonian," Alberta); **4, Lundbreck; 5, Scabby Butte** (both St. Mary's River Formation; "Edmontonian," Alberta); **6, Dinosaur Provincial Park** (various localities in the Dinosaur Park and Oldman formations, Judithian, Alberta); **7, South Saskatchewan River** (Oldman Formation, Judithian, Alberta); **8, Irvine** (Dinosaur Park Formation, Judithian, Alberta); **9, Verdigris Coulee** (upper parts of Milk River Formation; Aquilan, Alberta); **10, Milk River Valley** (Foremost and Oldman formations, Judithian, Alberta); **11, Manyberries–Onefour** (Oldman Formation, Judithian, Alberta); **12, Woodpile Creek** (Dinosaur Park Formation, Judithian, Saskatchewan); **13, Long Fall** (lower Ravenscrag Formation, Puercan, Saskatchewan); **14, Fr-1** (Puercan, Saskatchewan); **15, Gryde; 16, Wounded Knee** (Lancian, upper Frenchman Formation, Saskatchewan); **17, Unity** (Dinosaur Park Formation, Judithian, Saskatchewan).

dontans; the remainder are cimolodontids and neoplagiaulacids, groups that would also dominate in later assemblages of the North American Late Cretaceous. Notable in this context is the first undoubted appearance of the neoplagiaulacid *Mesodma,* an abundant and characteristic genus of the Judithian and Lancian. Among marsupials, important first occurrences are those of "pediomyids," or "pediomyid"-like taxa (Fox, 1971b, 1987b), characterized by a reduced anterior stylar shelf on the upper molars; and the stagodontid *Eodelphis,* which is also known from the Judithian (Fox, 1981). As noted earlier, the Aquilan fauna from Verdigris Coulee also includes a eutherian, *Paranyctoides maleficus* (see Fox, 1984b).

The most diverse of the probably Aquilan mammalian assemblages from southern Utah is that from the Wahweap Formation, Garfield and Kane counties (figure 2.20), under investigation by Cifelli and Eaton since 1983. There are several known productive localities on the Kaipariwits Plateau (e.g., Eaton and Cifelli, 1988), distributed through the lower and upper part of the unit; the mammalian fossils are collectively grouped herein as the **Wahweap** assemblage. The Wahweap Formation overlies the Drip Tank Member of the Straight Cliffs Formation and is entirely terrigenous, including mainly fluvially deposited sediments. Most of the more productive localities occur in well-indurated, sandy clayball conglomerates, indicating relatively high-energy deposition. Given the modest size of existing collections from the Wahweap Formation, the large number (13 taxa) of multituberculates is somewhat surprising; this may be partly due to the fact that existing information is preliminary (see Eaton, 1987). As with Verdigris Coulee, the multituberculates of the Wahweap Formation differ from older assemblages in that *Paracimexomys*-like taxa are rare, with more modern cimolodontans (e.g., Cimolodontidae, Neoplagiaulacidae) dominating. The "symmetrodontan" *Symmetrodontoides* (see Cifelli and Madsen, 1986; Cifelli and Gordon, 1999) and the eutherian *Paranyctoides* (see Cifelli, 1990e) are elements common to the Verdigris Coulee and Wahweap faunas, though they are represented by different species. The Wahweap assemblage includes some boreosphenidan mammals of uncertain affinities, as well as primitive marsupials (Cifelli, 1990b,d). Interestingly, Stagodontidae and "Pediomyidae," which are known from the Aquilan of

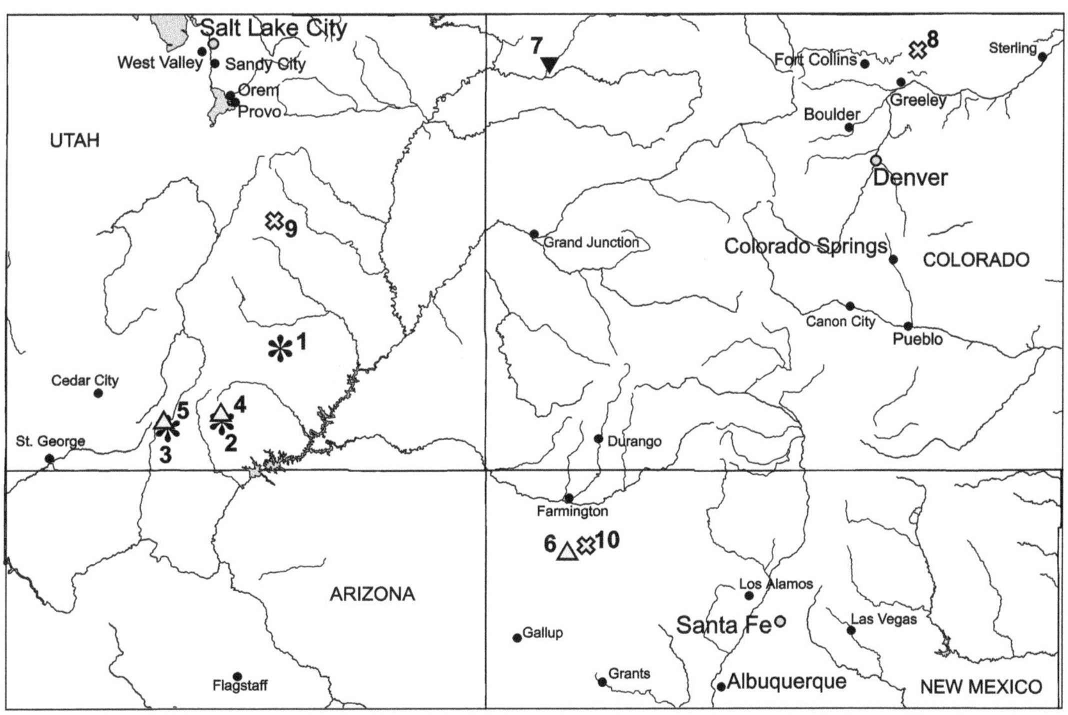

FIGURE 2.20. Aquilan–Lancian mammal localities, southwestern United States. Asterisks, Aquilan (localities 1–3); gray triangles, Judithian (localities 4–6); inverted black triangle, "Edmontonian" (locality 7), crosses, Lancian (localities 8–10). Sites, local faunas, or assemblages: **1, Masuk** (Masuk Formation; ?Aquilan, Utah); **2, Wahweap** (Wahweap Formation; Aquilan, Utah); **3, Campbell Canyon** (?Wahweap Formation; ?Aquilan, Utah); **4, Kaiparowits** (Kaiparowits Formation; Judithian, Utah); **5, Paunsaugunt** (?Kaiparowits Formation; ?Judithian, Utah); **6, Lower Hunter Wash** (Fruitland-Kirtland formations; Judithian, New Mexico); **7, Williams Fork** (Williams Fork Formation; "Edmontonian," Colorado); **8, Weld County** (Laramie Formation; Lancian, Colorado); **9, North Horn** (North Horn Formation; Lancian, Utah); **10, Alamo Wash** (Naashoibito Member, Kirtland Formation; Lancian, New Mexico).

Canada, have not yet been reported from the Wahweap Formation, suggesting the possibility of geographic restriction of these taxa.

Two other faunules probably or possibly of Aquilan age are known from Utah.[11] The first is in the uppermost Cretaceous rocks of the Paunsaugunt Plateau, Bryce Canyon National Park (figure 2.20). Correlation with the units on the Kaiparowits Plateau, to the east, is problematic (Eaton, 1993a). The site, **Campbell Canyon,** is believed to lie in the Wahweap Formation or its lateral equivalent (Eaton et al., 1998). Four taxa are known from Campbell Canyon, two multituberculates (*Cimolodon* sp., ?*Mesodma* sp.) and two marsupials (*Alphadon* sp. and a possible but unidentified "pediomyid"). The second faunule is from the Henry Basin, southeastern Utah (figure 2.20). Fragmentary mammalian fossils were recovered by Eaton from the lower and middle members of the Masuk Formation, which is believed to represent coastal floodplain deposits broadly equivalent in age to the Wahweap and the upper part of the Milk River formations (Eaton, 1990). Five localities have yielded mammals; the assemblage is collectively termed the **Masuk** faunule. Two multituberculates (*Paracimexomys* sp. nov., ?*Mesodma* sp.) and a "pediomyid" marsupial (cf. "*Pediomys*" *exiguus,* a species known from the Aquilan of Verdigris Coulee) are the only mammals known thus far from the Masuk Formation.

Judithian

The Judithian land-mammal age was defined by Lillegraven and McKenna (1986) on the basis of a mammalian fauna from the Judith River Formation in north-central Montana, north of the Missouri River in Blaine and Choteau counties (see Sahni, 1972). The Judithian is better represented geographically than any other Late Cretaceous land-mammal age in North America: in addition to the type, correlative faunas or individual specimens are known from other areas in Montana (Montellano, 1988, 1992; Fiorillo, 1989), Alberta (see references cited in Lillegraven and McKenna, 1986), Wyoming (Lillegraven and McKenna, 1986), Utah (Eaton, 1993a, 1999a), New Mexico (Clemens, 1973b; Flynn, 1986; Rigby and Wolberg, 1987), and Texas (Standhardt, 1986; Rowe et al., 1992; Weil, 1992; Cifelli, 1994; Sankey, 1998). Hence, reasonably well-represented Judithian faunas range from southern Canada to nearly the Mexican border, spanning more than 20° of latitude.

Despite tie-in with marine units (generally to the east, but sometimes in the same stratigraphic sections), correlation of the Judithian land-mammal age to European marine stages is problematic because of differing interpretations based on ammonites and foraminiferans (see discussions and references in Lillegraven and McKenna, 1986; Lillegraven and Ostresh, 1990). Lillegraven and McKenna (1986) placed the Judithian in the late Campanian to early Maastrichtian, somewhat younger than conventional estimates. Radiometric determinations from the Judith River Formation, Hill County, Montana, give a date of ca. 78 Ma (Goodwin and Deino, 1989), considerably older than the Campanian-Maastrichtian boundary, placed at 71.3 Ma by Gradstein et al. (1995). Hence, the Judithian is almost certainly older than Campanian-Maastrichtian boundary and probably spans several million years. We follow Weil (1999), who recognized the Judithian as spanning the interval from approximately 79 to 74 Ma (see also Cifelli et al., in press).

As noted by Lillegraven and McKenna (1986), the Judithian is a distinctive fauna (table 2.25), sharing few species-level taxa with the Lancian. Certain mammals are not diagnostic morphologically, or are restricted to one or a few Judithian local faunas and are of little biostratigraphic utility. Among multituberculates, probably the most useful first appearances are those of *Meniscoessus major* and *M. intermedius* (both of which appear to range into the "Edmontonian"), *Mesodma primaeva,* and *Cimolomys clarki* (which is geographically widespread in the Judithian; it also may be present in the Aquilan). Most distinctive first appearances among boreosphenidans mammals are those of marsupials of the genus *Turgidodon,* which (as the name implies) are a clade characterized by inflated molar cusps, and the leptictoid eutherian *Gypsonictops.*

Given the great latitudinal spread of Judithian assemblages, it is not surprising that there is a significant degree of heterogeneity in composition among local faunas (Lillegraven and Ostresh, 1990). Lehman (1997) recognized two discrete faunal provinces among dinosaur assemblages of Judithian age: a northern province, which was humid, temperate, and had more closed-canopy evergreen forest (the *Aquilapollenites* palynoflora); and a southern province, characterized by a warm, nonseasonal climate and open-canopy forest with widely spaced trees (the *Normapolles* palynoflora). Detailed comparison of Judithian mammal assemblages has not yet been undertaken, but it is clear that there are distinct southern and northern assemblages among mammals as well. In partic-

[11] Besides these, two other faunules from Utah may prove to be referable to the Aquilan when they become better known. Both are in Cedar Canyon on the Markagunt Plateau. The lower faunule is thought to be Santonian based on palynomorphs and is described in that section, whereas the upper faunule is probably Campanian and is noted under undifferentiated faunas of that age.

TABLE 2.25. Judithian Mammals of North America (see figures 2.14, 2.19, 2.21; locality numbers do not correspond between maps and table). Local faunas or localities: 1, Type Judithian, Montana; 2, Hill County, Montana; 3, Musselshell River, Montana (all Judith River Formation); 4, Egg Mountain, Montana (Two Medicine Formation); 5, Foremost-Oldman, Alberta (Foremost and Oldman formations); 6, Dinosaur Park, Alberta (Dinosaur Park Formation); 7, Wind River Basin, Wyoming; 8, Bighorn Basin, Wyoming ("Mesaverde" Formation); 9, Kaiparowits Plateau, Utah (Kaiparowits Formation); 10, Paunsaugunt Plateau, Utah (?Kaiparowits Formation); 11, Lower Hunter Wash, New Mexico (Fruitland-Kirtland formations); 12, Terlingua, Texas; 13, Talley Mountain, Texas; 14, Running Lizard, Texas (all Aguja Formation)

- Multituberculata
  - Family *incertae sedis*
    - Gen. et sp. indet. (2, 13, 14)
    - Gen. et sp. indet. A (14)
    - Gen. et sp. nov. A (7)
    - Gen. et sp. nov. B (12)
    - ?*Bryceomys* sp. nov. (9)
    - ?*Cimexomys gregoryi* (10)
    - *Cimexomys* cf. *antiquus* (5)
    - *Cimexomys judithae* (1, 2, 4, 11)
    - *Cimexomys* cf. *judithae* (9)
    - cf. *Cimexomys* sp. (5, 13)
    - *Paracimexomys magnus* (1)
    - *Paracimexomys priscus* (2, 7)
    - *Paracimexomys* sp. (10–13)
    - *Paracimexomys* sp. A (9)
    - *Paracimexomys* sp. B (9)
  - Cimolodontidae
    - *Cimolodon electus* (11)
    - *Cimolodon* cf. *electus* (12)
    - *Cimolodon* cf. *nitidus* (9, 10)
    - *Cimolodon* cf. *similis* (9)
    - *Cimolodon* sp. nov. A (9)
    - *Cimolodon* sp. nov. B (9)
    - *Cimolodon* sp. nov. C (11)
    - *Cimolodon* sp. nov. D (12)
    - *Cimolodon* sp. (1)
    - ?*Cimolodon* sp. (5, 10)
  - Cimolomyidae
    - *Cimolomys clarki* (1, 2, 5, 7, 8, 12)
    - *Cimolomys* cf. *clarki* (2)
    - *Cimolomys milliensis* (10)
    - *Cimolomys* sp. (13)
    - *Cimolomys* sp. nov. A (9)
    - *Cimolomys* sp. nov. B (9)
    - ?*Cimolomys* sp. nov. (9)
    - ?*Essonodon* sp. nov. (11)
    - *Meniscoessus intermedius* (5, 7, 11)
    - *Meniscoessus major* (1, 5, 6)
    - *Meniscoessus* sp. (9, 13)
    - *Meniscoessus* sp. nov. (12)
    - Gen. et sp. indet. (14)
  - ?Cimolomyidae
    - Gen. et sp. indet. (10)
  - Eucosmodontidae
    - Gen. et sp. indet. (14)
  - Neoplagiaulacidae
    - *Mesodma primaeva* (1, 2, 6, 7)
    - *Mesodma* cf. *primaeva* (5)
    - *Mesodma* cf. *formosa* (9, 10)
    - *Mesodma* cf. *hensleighi* (10)
    - *Mesodma* cf. *senecta* (9, 11)
    - *Mesodma* sp. (3, 7, 8, 10)
    - *Mesodma* sp. A (2)
    - *Mesodma* sp. B (2)
    - *Mesodma* sp. C (9)
    - *Mesodma* sp. D (9)
    - *Mesodma* sp. E (9)
    - cf. *Mesodma* sp. (11)
  - Ptilodontidae
    - *Kimbetohia campi* (11)
- Stem Cladotheria ("eupantotherians")
  - Dryolestidae
    - Gen. et sp. indet. (8)
- Stem Boreosphenida ("tribotheridans")
  - Family *incertae sedis*
    - *Bistius bondi* (11)
    - *Falepetrus barwini* (2, 7)
    - *Palaeomolops langstoni* (12)
    - Gen. et sp. indet. (12)
    - Gen. et sp. indet. A (2)
    - Gen. et sp. indet. B (2)
- Deltatheroida
  - Deltatheridiidae
    - cf. *Deltatheridium* sp. (6)
- Marsupialia
  - "Alphadontidae"
    - *Aenigmadelphys archeri* (9)
    - *Alphadon attaragos* (8, 9)
    - *Alphadon* cf. *attaragos* (2, 10)
    - *Alphadon halleyi* (1, 2, 4, 6, 7, 8, 9, 11)
    - *Alphadon* cf. *halleyi* (12)
    - *Alphadon* cf. *marshi* (11)
    - *Alphadon perexiguus* (12)
    - *Alphadon sahnii* (1, 6, 7, 8, 9)
    - *Alphadon* cf. *sahnii* (9, 12)
    - *Alphadon* cf. *wilsoni* (10, 11)
    - *Alphadon* sp. (3, 5, 13, 14)
    - *Alphadon* sp. A (13)
    - *Alphadon* sp. nov. (11)
    - *Protalphadon lulli* (8)
    - *Turgidodon lillegraveni* (9)
    - *Turgidodon* cf. *lillegraveni* (9, 12)
    - *Turgidodon madseni* (9)
    - *Turgidodon parapraesagus* (11)
    - *Turgidodon praesagus* (1, 2, 5, 6)

(continued)

TABLE 2.25. Continued

- *Turgidodon russelli* (1, 2, 5, 6, 7, 8)
- *Turgidodon* cf. *russelli* (10)
- *Turgidodon* sp. (9, 10)
- ?*Turgidodon* sp. (10)
- *Varalphadon wahweapensis* (9)
- Gen. et sp. indet. (12)

"Pediomyidae"
- *Aquiladelphis paraminor* (11)
- "*Pediomys*" *clemensi* (1, 2, 6)
- "*Pediomys*" cf. *cooki* (11)
- "*Pediomys*" *prokrejcii* (1, 2, 5, 6)
- *Pediomys* cf. *elegans* (6)
- "*Pediomys*" *fassetti* (11)
- "*Pediomys*" cf. *hatcheri* (6)
- "*Pediomys*" sp. (5, 6, 8)

Stagodontidae
- *Eodelphis browni* (6)
- *Eodelphis cutleri* (1, 6)
- *Eodelphis* sp. (1, 2, 3, 5, 6)
- ?*Eodelphis* sp. (11)

Eutheria

Family *incertae sedis*
- *Avitotherium utahensis* (9)
- *Gallolestes agujaensis* (12)
- Gen. et sp. indet (11)

Cimolestidae
- *Cimolestes* sp. (5, 9)
- Gen. et sp. indet. (6)

Gypsonictopsidae
- *Gypsonictops clemensi* (11)
- *Gypsonictops* sp. (1, 2, 6, 7)
- *Gypsonictops* cf. *lewisi* (5, 11)
- *Gypsonictops* sp. A (9)
- *Gypsonictops* sp. B (9)
- *Gypsonictops* sp. C (9)

"Nyctitheriidae"
- *Paranyctoides maleficus* (2)
- *Paranyctoides megakeros* (8)
- *Paranyctoides sternbergi* (5, 6)
- *Paranyctoides* cf. *sternbergi* (11)
- *Paranyctoides* sp. (2)
- *Paranyctoides* sp. A (9)
- *Paranyctoides* sp. B (9)

ular, the faunas of the Kaiparowits Formation, Utah, and the Aguja Formation, Texas, share species not found in the local faunas from Alberta, Montana, and Wyoming, and vice versa (Rowe et al., 1992; Cifelli, 1994; Sankey, 1998; Weil, 1999).

The area of the **type Judithian** in north-central Montana (figure 2.21) is of historical significance in that F. V. Hayden collected the first North American dinosaur fossils from there in the 1850s (Leidy, 1856, 1860), and subsequently (Hayden, 1871) defined the Judith River Formation in this area (see historical reviews by Sahni, 1972; Montellano, 1992). Most workers (e.g., Lillegraven and Ostresh, 1990) now follow McLean (1971) in considering the Oldman Formation of the Belly River Group, Alberta, as a junior synonym of the Judith River Formation. We follow Montellano (1992) in continuing to use "Oldman" in reference to the composite local fauna from Canada. Depositionally, the Judith River Formation is thought to represent a deltaic environment, with nearshore marine to alluvial and lacustrine facies (McLean, 1971).

The type Judithian mammal fauna was collected from three principal localities in the early 1960s: Ankylosaur Point, Clayball Hill, and Clambank Hollow. This last-named site, by far the most productive of the three, occurs about 12 m from the top of the Judith River Formation, in a deposit presumed to represent a river meander or oxbow lake (Sahni, 1972). By comparison to more recently collected Judithian faunas, the composite assemblage of 13 species (six multituberculates, six marsupials, and one eutherian) from the type Judithian is only moderately diverse. One multituberculate, *Paracimexomys magnus*, is known only from the type Judith River. The other taxa seem to fall into either of two categories of distribution, though doubtful identification in many cases may either provide the exception or fit the pattern, depending on interpretation. Some of the species appear to be rather cosmopolitan in distribution, being tentatively recorded in both southern and northern faunas: *Cimolomys clarki*, *Alphadon halleyi*, and *A. sahnii* (and perhaps also *Cimexomys judithae* and *Gypsonictops lewisi*). Others appear to be more restricted to northern assemblages of Judithian age: *Mesodma primaeva*, *Turgidodon praesagus*, *T. russelli*, "*Pediomys*" *clemensi*, and *Eodelphis cutleri*. "Pediomyidae" and Stagodontidae (represented, respectively, by the last two taxa) are generally scarce or absent in southern faunas of the Late Cretaceous. Some taxa, including *Meniscoessus major* and *Turgidodon russelli*, apparently range into the "Edmontonian." The distinctive eutherian *Gypsonictops* makes its first appearance in the Judithian, being represented in the type area by the Judithian species *G. lewisi*.

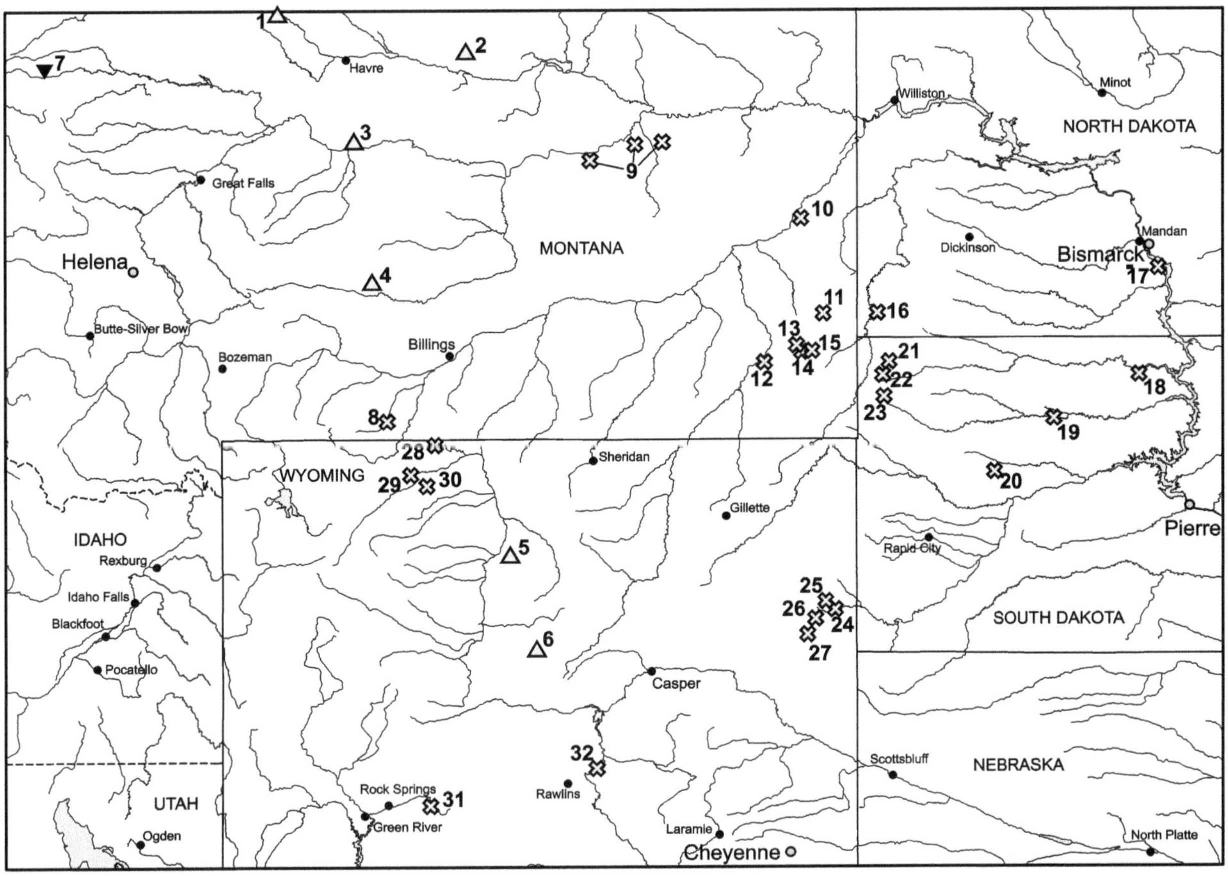

FIGURE 2.21. Judithian–Lancian mammal localities, northern United States. Gray triangles, Judithian (localities 1–6); inverted black triangle, "Edmontonian" (7); crosses, Lancian (localities 8–32). Sites, local faunas, or assemblages: **1, Hill County** (Judith River Formation, Judithian, Montana); **2, Egg Mountain** (upper Two Medicine Formation, Judithian, Montana); **3, Type Judithian** (upper Judith River Formation, Judithian, Montana); **4, Mussellshell** (Judith River Formation, Judithian, Montana); **5, Bighorn Basin** (middle–upper "Mesaverde" Formation, Judithian, Wyoming); **6, Wind River Basin** (middle "Mesaverde" Formation; Judithian, Wyoming); **7, Shell Hell** (lower St. Mary River Formation; "Edmontonian," Montana); **8, Red Lodge** (?Lance Formation or equivalent, Lancian, Montana); **9, Hell Creek; 10, Muddy Tork; 11, Fallon; 12, Powderville; 13, Claw Butte; 14, Blacktail; 15, Ekalaka** (all upper Hell Creek Formation, Lancian, Montana); **16, Little Missouri Badlands; 17, Stumpf** (both upper Hell Creek Formation, Lancian, North Dakota); **18, Black Horse** (upper Hell Creek Formation, Lancian, South Dakota); **19, Iron Lightning; 20 Red Owl** (both Fox Hills Formation, Lancian, South Dakota); **21, Joe Painter Quarry, 22, Harding County; 23, Eureka Quarry** (all upper Hell Creek Formation, Lancian, South Dakota); **24, Cheyenne River; 25, Mule Creek Junction; 26, Greasewood Creek; 27, Type Lance** (all Lance Formation, Lancian, Wyoming); **28, Dumbbell Hill; 29, Hewitt's Foresight One** (both Lance Formation, Lancian, Wyoming); **30, southwestern Bighorn Basin** (?Lance Formation or equivalent, Lancian, Wyoming); **31, Black Butte Station** (Lance Formation, Lancian, Wyoming); **32, Hanna Basin** (Ferris Formation, Lancian, Wyoming).

A larger mammalian assemblage was collected from the Judith River Formation, **Hill County,** Montana (figure 2.21), by parties led by William A. Clemens, in the 1970s and 1980s. Two separate areas, Kennedy Coulee and the Havre badlands, yielded mammalian fossils from a total of three sites, though only one (in Kennedy Coulee) was extensively sampled (Montellano, 1992). We group the assemblages collectively as the Hill County fauna. Twenty-three species are recorded, slightly fewer than half (nine) of which are multituberculates.[12] In addition to the distributions mentioned earlier, we note that *Alphadon attaragos* appears to be a fairly cosmopolitan mammal for Judithian faunas, whereas *"Pediomys" prokrejcii* (like other "pediomyids") may have been restricted to more northerly realms. Other noteworthy occurrences include

[12] See comments on taxonomy of the multituberculates by Weil (1999).

at least one "tribotherian," *Falepetrus barwini*, and perhaps two species of the nyctitheriid eutherian, *Paranyctoides*. Representation of the latter in the Hill County fauna by *P. maleficus*, an otherwise Aquilan species (Fox, 1984b), potentially represents a significant range extension.

Two other Judithian faunules are known from Montana (figure 2.21). The first of these is at the **Egg Mountain** locality, northern Montana. The occurrence is in the upper part of the Two Medicine Formation, a lateral equivalent of the Judith River Formation. Widely renowned for its remarkable accumulations of dinosaur eggs and juveniles (Horner and Makela, 1979; Horner, 1982), Egg Mountain has yielded three mammal specimens through quarrying procedures. Two of these are jaws (one well preserved) of the common Judithian marsupial *Alphadon halleyi* (see Montellano, 1988). The third specimen, referable to the multituberculate *Cimexomys judithae*, is exceptional: it consists of the entire lower jaw and rostrum, together with several associated lumbar vertebrae (Montellano et al., 2000). A final occurrence worth passing mention is that of possible mammal burrows in the Two Medicine Formation (Martin, 2001).

The other mammalian assemblage is from scattered sites in the Judith River Formation near the **Musselshell River** (figure 2.11), Wheatland and Golden counties, in the south-central part of Montana (Fiorillo, 1989; Fiorillo and Currie, 1994). Three mammals, none identified to species or particularly diagnostic at the genus level (*Mesodma* sp., *Alphadon* sp., *Eodelphis* sp.) are known from this fauna.

Mammalian fossils were first reported from what later became known as the Oldman Formation by Lambe (1902). Additional specimens trickled in through ensuing decades (e.g., Russell, 1952), usually as a result of dinosaur-oriented field investigations. In 1966, Richard C. Fox began systematic and intensive field studies aimed at the recovery of fossil mammals from the Oldman Formation. A number of localities were worked, mainly along the Red Deer and South Saskatchewan rivers in the vicinity of Steveville, Medicine Hat, and Manyberries, southeastern Alberta. The collection is of great significance because of its large size (including several thousand specimens) and because many taxa are represented by rather complete jaws—in some cases, series of jaws. Later work was done in the vicinity of Dinosaur Provincial Park and the Milk River Valley, Alberta, by Donald B. Brinkman and associates.

The sequence including Judithian mammals in Alberta and Saskatchewan is now recognized as the Judith River Group, including (from lowest to highest) the Foremost, Oldman, and Dinosaur Park formations (Eberth and Hamblin, 1993). Mammals have been found in each of these units. The Oldman and Dinosaur Park formations correspond, respectively, to the "lower Oldman" and "upper Oldman" of Fox (1976b, 1979b,c,d, 1980a, 1981). In the following account, we treat only the mammals that have been included in theses or primary literature. Faunal lists for these units are also given by Braman et al. (1995).

Mammals from the lower part of the Canadian Judith River Group (which we collectively call the Foremost-Oldman fauna) are not well known. A preliminary report lists 14 taxa from the Foremost and Oldman formations in the eastern part of the **Milk River Valley** (figure 2.19) (Peng, 1997; Peng et al., 2001). Mammals have been collected from several sites in the Oldman Formation. One, discovered in 1974, is about 50 km north of Medicine Hat, along the **South Saskatchewan River** (figure 2.19). The site is just above the contact with the underlying Foremost Formation and more than 75 m below the productive zone in the upper part of the Oldman Formation. Only two mammals have been reported, the multituberculate *Meniscoessus intermedius* and an unidentified species of the marsupial *Alphadon* (Fox, 1976b). *M. intermedius* is also known from the Judithian of Wyoming and, perhaps, the Judithian of New Mexico (Flynn, 1986) and the "Edmontonian" of Colorado (Diem, 1999). Fox (1976b) listed a specimen of *M. intermedius* from a site some 50 km west of Medicine Hat. The unit for this site is not given, but the species distribution that is listed implies that it may be in the Foremost Formation. Other sites in the Oldman Formation are known from far southeastern Alberta, in the **Manyberries-Onefour** area and the eastern part of the Milk River Valley (Peng, 1997; Peng et al., 2001); and in **Dinosaur Provincial Park** (figure 2.19), along the Red Deer River southeast of Wardlow (Brinkman, 1990).

The Dinosaur Park Formation has yielded the bulk of mammalian fossils published to date from the Judith River Group of Canada. We refer to them collectively as the Dinosaur Park fauna. Some 45 sites were discovered by field parties under Fox's direction (see Fox, 1997b); all are presumably located in the same fossiliferous zone high in the Oldman Formation. The principal sites are in **Dinosaur Provincial Park** (the most important of which is at the Steveville Railway Grade) and the **Irvine** locality, near the Alberta-Saskatchewan border southeast of Medicine Hat. Eighteen mammals are recorded from the upper Oldman Formation (Fox, 1974a, 1979b,c,d, 1980a, 1981; Lillegraven and McKenna, 1986). Published accounts list only two multituberculates for the Dinosaur Park fauna; others, as well as tribotheres from this horizon, remain to be described (Fox, 1997c). The two reported multituberculates, *Meniscoessus major* and *Mesodma primaeva*, have been commented on earlier. A noteworthy occurrence in the Dinosaur Park fauna is that of a deltatheroidan that resembles *Deltatheridium* and *Deltatheroides*; a similar occurrence is known for the Lance Formation of Wyoming (see Fox, 1974a). The close similarity of the Canadian fossil to these otherwise Asiatic genera (Kielan-Jaworowska, 1975c) suggests a proximal dispersal between Asia and North America, rather than derivation from a geologically older taxon in North America (Cifelli and Gordon, 1999). The Dinosaur Park fauna is also remarkable for the great abundance and diversity of marsupials belonging to the genus "*Pediomys*": as many as five species may be present (Fox, 1979c). Of these, two are similar to taxa known from the Lancian, *P. elegans* and "*P.*" *hatcheri*. Similarly, the Stagodontidae are comparatively abundant in the Dinosaur Park fauna (Fox, 1981). Finally, a single tooth documents the presence of a *Cimolestes*-like cimolestid, representing the first appearance of this eutherian group that became moderately diverse in the Lancian.

Sparse mammalian fossils have also been reported from the upper part of the Judith River Group (presumably in the Dinosaur Park Formation) in Saskatchewan (Storer, 1993). The **Unity** site (see Eberth et al., 1990) has yielded a single specimen of the multituberculate *Mesodma*. Fragmentary remains possibly belonging to the marsupial *Alphadon halleyi* are known from **Woodpile Creek,** in far southwestern Saskatchewan (figure 2.19).

Proceeding southward, mammalian assemblages of Judithian age have been reported from two areas in Wyoming. In both cases, the fossils derive from an unnamed rock unit erroneously referred to as the "Mesaverde Formation," and this name is therefore used in quotes. Correlation based on marine molluscs suggests that these mammalian assemblages are essentially contemporaneous with those of the Judith River and Oldman formations (Lillegraven and McKenna, 1986). Fossils were recovered from various localities; we follow Lillegraven and McKenna (1986) in grouping them into two assemblages, Wind River Basin and Bighorn Basin.

Mammalian fossils of the **Wind River Basin** fauna were collected from a single horizon at a site called Barwin Quarry and its lateral extension, Fales Rock, in the Wind River Basin east of Ervay, Natrona County, Wyoming (figure 2.21). Stratigraphically, these lie in the lower part of the middle member of the "Mesaverde" Formation, which includes beds interpreted as being fluvial, lagoonal, and coastal swamp deposits. Eleven varieties of mammals, six of which are multituberculates, are known for the Wind River Basin fauna. Most included taxa are typical of the Judithian; some are shared only with other northern assemblages (e.g., *Mesodma primaeva, Turgidodon russelli*), whereas others are fairly cosmopolitan (e.g., *Cimolomys clarki, Alphadon halleyi*). The primitive cimolodontan *Paracimexomys priscus* is a widespread Lancian species, otherwise known only from the Hill County fauna among Judithian assemblages.

The **Bighorn Basin** fauna was collected from a number of sites northeast of Worland, Washakie County, and a single site northwest of Thermopolis, Hot Springs County (figure 2.21). Most of the fossils come from friable channel sandstones in the middle to upper part of the "Mesaverde" Formation. The fauna is comparable in diversity to that of the Wind River Basin and includes 10 taxa. The marsupial "*Pediomys*," lacking in the Wind River Basin fauna, is known here by only a half of a lower molar. In this respect, the fauna of the "Mesaverde" Formation is strikingly different from northern assemblages, where "*Pediomys*" is common, and similar to those to the south, where it is rare or absent (Clemens, 1973b; Lillegraven and McKenna, 1986; Cifelli, 1994). We also note that stagodontids are rare or absent, another similarity to Judithian assemblages of the southwest.

Yet the Bighorn Basin fauna, like that of the Wind River Basin, also has certain species (e.g., *Turgidodon russelli*) that are more characteristic of northern assemblages, suggesting that the fossils of the "Mesaverde" Formation may be faunally as well as geographically intermediate. The Bighorn Basin fauna includes a species otherwise restricted to the Lancian, the marsupial *Protalphadon lulli*. Another potential link to younger faunas is *Paranyctoides megakeros*, which may be referable to Lancian *Alostera* (see Fox, 1989). The most remarkable occurrence in the Bighorn Basin assemblage, however, is that of a dryolestid "eupantotherian," representing a group not otherwise known in North America after the Late Jurassic. Dryolestids, like "symmetrodontans," "triconodonts," and certain archaic groups of boreosphenidan mammals, evidently survived through most of the Cretaceous in limited numbers and, perhaps, restricted areas in North America. This occurrence and the even more astonishing recent discovery of a nonmammalian synapsid in the North American Paleocene (Fox et al., 1992) provide ample warning to interpret the known record of Mesozoic mammals cautiously.

Mammalian assemblages of Judithian age were discovered in south-central Utah by Cifelli and Eaton in 1983, with work continuing. Most of the localities and specimens are from the Kaiparowits Formation on the Kaiparowits Plateau. A few additional mammalian fossils have been recovered from uppermost Cretaceous rocks on the nearby Paunsaugunt Plateau, to the west. Because of problems in lithostratigraphic correlation, we list the Paunsaugunt assemblage separately.

Mammalian fossils have been recovered from approximately the middle half of the Kaiparowits Formation on the Kaiparowits Plateau (figure 2.20), Garfield and Kane counties (Cifelli, 1990a). The unit, which is comprised of rather homogeneous fluvially deposited sandstones, siltstones, and mudstones, is about 850 m thick in the northwest part of the Kaiparowits region, and the mammal-yielding sites thus span about 470 m of section. Though some minor compositional differences have been noted among the sites (Cifelli, 1990a,e), they are quite similar, and the mammalian fauna, derived primarily from five sites, is treated as a unit, the **Kaiparowits** assemblage.

The Kaiparowits fauna includes some 36 mammal taxa, placing it among the most diverse of all mammalian assemblages known from the Mesozoic. Part of this may be artificial, stemming from the fact that identification of multituberculates remains preliminary and includes a high proportion of poorly identified taxa; additionally, it may eventually prove useful to subdivide localities stratigraphically, as was previously attempted (Eaton and Cifelli, 1988). Still, it is clear that the mammalian assemblage is diverse by the standards of the North American Late Cretaceous. It is also highly distinctive, reflecting perhaps the geographic separation of the Kaiparowits region from most well-known Judithian assemblages. Three described species (*Aenigmadelphys archeri, Turgidodon madseni,* and *Avitotherium utahensis*) are known only from the Kaiparowits fauna, perhaps nine additional multituberculates are new, and certain species of the eutherians *Gypsonictops* and *Paranyctoides* may also represent new taxa (Eaton, 1987; Cifelli, 1990a,e; Cifelli and Johanson, 1994; Eaton, Cifelli, et al., 1999). Several taxa are shared with older faunas: the primitive cimolodontan (*Paracimexomys* group) *Bryceomys* is otherwise known from the Turonian (Eaton, Cifelli, et al., 1999), and the marsupial *Varalphadon wahweapensis* (see Johanson, 1996b) and the neoplagiaulacid multituberculate *Mesodma senecta* (tentatively recorded in the Kaiparowits fauna) are otherwise known from the Aquilan. *Symmetrodontoides,* a "symmetrodontan" otherwise ranging from the Turonian through Aquilan, was reported from the fauna as well (Eaton, Cifelli, et al., 1999), but the specimen in question appears not to be identifiable. Geographically, two cosmopolitan species (*Alphadon sahnii* and *A. halleyi*) appear to be shared with both northern and southern assemblages of the Judithian, and one each with otherwise northern (*Alphadon attaragos*) and southern (*Turgidodon lillegraveni*) assemblages. Species of the eutherians *Gypsonictops* and *Paranyctoides* are unidentified, but nonetheless suggest faunal change through the section (Cifelli, 1990e), as does at least one of the marsupials (Cifelli, 1990a). A final noteworthy occurrence in the Kaiparowits fauna is that of the eutherian *Avitotherium,* which appears to be a basal ungulatomorph and, as such, is the most ancient member of this group in North America (Cifelli, 1990e; Nessov, Archibald, and Kielan-Jaworowska, 1998).

The **Paunsaugunt** assemblage was collected by Eaton from a single locality in Bryce Canyon National Park on the Paunsaugunt Plateau, to the west of the Kaiparowits Plateau (figure 2.20). Differences in thickness and rock unit representation hamper correlation with the formations established for the Kaiparowits Plateau (see Gregory and Moore, 1931; Gregory, 1951), but the locality in question is believed to be in the Kaiparowits Formation (Eaton, 1993a, 1999b). The fauna, not yet well known and still under study, provisionally includes 15 taxa, of which a relatively large number (nine) are multituberculates. Two species, both multituberculates, are endemic to the Paunsaugunt fauna, *Cimolomys milliensis* and ?*Cimexomys gregoryi* (see Eaton, 1993a), the latter possibly referable to the Turonian genus *Bryceomys* (Eaton et al., 1998). The fauna appears unusual in other respects. Of the seven remaining taxa provisionally identified to species, four are rather common; two of these (*Cimolodon* cf. *nitidus, Alphadon* cf. *wilsoni*) are known from both "Edmontonian" and Lancian sites. *Mesodma* cf. *formosa* is known only from the Lancian, whereas *M.* cf. *hensleighi* is known from both the Lancian and Puercan (Eberle and Lillegraven, 1998a). These elements suggest that the Paunsaugunt fauna may be slightly younger than other Judithian assemblages. To complicate matters, two other species, *Turgidodon* cf. *russelli* and *Alphadon* cf. *attaragos,* are typical of the Judithian, whereas the "symmetrodontan" *Symmetrodontoides* is more characteristic of Aquilan and older faunas.

The San Juan Basin, northwestern New Mexico, has long been known for its terrestrial vertebrate faunas (Simpson, 1981). Late Cretaceous mammals from the area were first reported by Clemens (1973b), and a number of other reports have been issued in the years that have followed. The relevant part of the terrestrial sequence includes the Fruitland Formation and the overlying Kirtland Formation; between the two lies a transitional zone, so that fossils from this part of the section are sometimes described as coming from Fruitland-Kirtland strata. Fossil mammals have been recovered from two general levels:[13] the uppermost part of the Fruitland Formation and the transitional zone; and the upper part, or Naashoibito Member, of the Kirtland Formation. We follow Flynn (1986) in grouping the mammal faunas into two composite assemblages, the Lower Hunter Wash (below) and Alamo Wash local faunas. The Alamo Wash local fauna,

[13] Sullivan (1999) and Williamson and Sullivan (1998) recognized a separate (third) vertebrate assemblage from the San Juan Basin, based entirely on dinosaurs—the Willow Wash local fauna, from the upper shale (De-na-zin) Member of the Kirtland Formation, which underlies the Naashoibito Member. Unfortunately, the Willow Wash local fauna is not yet known to contain mammals.

including only a few mammals, is widely accepted as being of Lancian age (see, e.g., Lehman, 1984). The age of the Lower Hunter Wash assemblage, originally considered to be "Edmontonian" (Clemens, 1973a), has proven more contentious, depending on alternative interpretations of magnetostratigraphy and correlations based on mammals and on invertebrates contained by adjacent marine units. Butler et al. (1977) and Butler and Lindsay (1985) correlated the part of the section containing this fauna with magnetic anomalies 30 and 31, placing it in the late Maastrichtian, hence implying that the mammals are of Lancian age (see discussion by Flynn, 1986). On the other hand, Rigby and Wolberg (1987) reinterpreted the paleomagnetic data based on mammalian biostratigraphy, preferring correlation with chron 33, and a Campanian (Judithian) age for what is now known as the Lower Hunter Wash fauna. Consideration of marine invertebrates and the migrational evolution of the shoreline of the Western Interior seaway led Lillegraven (1987) and Lillegraven and Ostresh (1990) to return to a more traditional interpretation; that is, an intermediate ("Edmontonian") age for the Lower Hunter Wash local fauna. Revised interpretation of biostratigraphy together with radiometric dates (Fassett and Obradovich, 1996; Fassett and Steiner, 1997) constraining the age of at least some sites now suggest that the Lower Hunter Wash local fauna is probably Judithian in age.

The composite assemblage recognized at the **Lower Hunter Wash** local fauna comes from a variety of localities, most of which are east of the former Bisti Trading Post and on the south side of Hunter Wash, northwestern New Mexico (figure 2.20). The fauna (see Clemens, 1973b; Armstrong-Ziegler, 1978; Flynn, 1986; Rigby and Wolberg, 1987), including about 25 taxa, may be somewhat inflated because it is not clear in all cases which taxa correspond between earlier and later studies (Clemens, 1973b; Rigby and Wolberg, 1987) and because some taxa are of doubtful status.[14] The Lower Hunter Wash local fauna includes elements known from Judithian (and older) faunas on the one hand and Lancian assemblages on the other. There are several endemic taxa, of which the large, enigmatic therian *Bistius bondi* is perhaps the most distinctive (see Clemens and Lillegraven, 1986). Another notable occurrence is that of the multituberculate *Kimbetohia campi*. *Kimbetohia* is otherwise restricted to the Tertiary.

Farther to the east and south, mammals of probable Judithian age have been reported from localities in the Aguja Formation, in and near the Big Bend National Park, Brewster County, Texas (figure 2.14). The most significant of these is the **Terlingua** local fauna, northeast of Study Butte, which was worked by field parties led by Cifelli and Timothy B. Rowe in the late 1980s and early 1990s. Fossils were collected from a single locality near the base of the upper shale member of the Aguja Formation. The site lies at an interface of coastal swamp and coastal floodplain facies, with deposition in an estuarine environment (Weil, 1992). The diverse vertebrate fauna includes a mixture of terrestrial and marine taxa; fossil wood from the site commonly bears *Teredo* borings, indicating brackish to salt water. Marine invertebrates from the lower part of the Aguja Formation and the underlying Pen Formation indicate that the Terlingua assemblage is late Campanian in age, and the fauna itself suggests referral to the Judithian (Rowe et al., 1992). Multituberculates have received only preliminary treatment (Weil, 1992); therians were reviewed by Cifelli (1994).

The Terlingua local fauna includes 14 mammalian taxa; not surprisingly, a high number of these (*Meniscoessus* sp. nov., Multituberculata gen. et sp. nov., *Palaeomolops langstoni, Alphadon perexiguus, Gallolestes agujaensis*) appear to be unique, at least at the species level. Several taxa (*Cimolomys clarki* and, tentatively identified, *Alphadon halleyi* and *A. sahnii*) are shared with numerous other Judithian assemblages and, as noted, may be useful as "index" fossils for the age. Others (tentatively identified *Turgidodon lillegraveni* and *Cimolodon* sp. nov.) are otherwise known only from southern assemblages, hence suggesting some degree of provincialism. At higher taxonomic levels, the eutherian *Gallolestes* is also known from Baja California, and *Palaeomolops* may be related to *Iugomortiferum* from southern Utah (Cifelli, 1994). Recognizing that absence data must be interpreted with care, it is nonetheless worthwhile to point out that the multituberculate *Mesodma*, "pediomyid" and stagodontid marsupials, and the eutherian *Gypsonictops*, which are common elements of other Judithian assemblages, are lacking at Terlingua.

A second mammalian assemblage from the Aguja Formation is from the area of **Talley Mountain,** in Big Bend National Park (figure 2.14). The fossils come from several localities in channel deposits spanning 20 m of section in the lower part of the upper shale member of the unit. Six taxa (four multituberculates and two marsupials) are known from Talley Mountain (Sankey, 1998). Unfortunately, the fossils are extremely fragmentary, and identification at the species level is problematic. All genera from Talley Mountain (*Cimolomys, Meniscoessus, Paracimexomys, Alphadon*) are typical constituents of Judithian and Lancian faunas of North America.

[14] These include *Ectocentrocristus foxi*, apparently based on dP3 of a large species of *Turgidodon;* and *"Cimolestes" lucasi*, based on a lower molar probably belonging to *Alphadon* (see Cifelli, 1990e).

Several mammalian fossils have been collected from the **Running Lizard** locality (figure 2.14), near Dawson Creek, on the west side of Big Bend National Park (Standhardt, 1986). The site is in the upper shale member of the Aguja Formation and hence is in the same general part of the stratigraphic section as the Terlingua and Talley Mountain localities. Four kinds of mammals are known from Running Lizard, three unidentified multituberculates and the marsupial *Alphadon.*[15]

"Edmontonian"

As noted by Lillegraven and McKenna (1986), there is general agreement that a considerable gap in time separates typical Judithian and Lancian assemblages, but that recognition of an intervening land-mammal age ("Edmontonian," see L. S. Russell, 1964, 1975) is hampered by two problems. First, the time interval in question (late Campanian, early Maastrichtian, or both, depending on definition) is one of marine transgression, so that terrigenous rocks of this age are rare in the North American Western Interior. Lillegraven (1987) pointed out, for example, that this interval in Montana corresponds to the time of deposition of the Bearpaw Shale, which separates the largely terrestrial Judith River and Hell Creek formations. Second, most of the few known mammals (table 2.26) coming from rocks that apparently represent the "Edmontonian" are either conspecific with those known from the Lancian or are inadequately known and thus not temporally diagnostic. In recent years, discovery of new species apparently unique to the "Edmontonian" (Fox and Naylor, 1986; Lillegraven, 1987) suggests that it may eventually prove to be definable in the same fashion as the Aquilan, Judithian, and Lancian land-mammal ages. In the meantime, we follow Lillegraven and McKenna (1986) in citing the "Edmontonian" in quotes. The first mammalian assemblage to be referred to the "Edmontonian" is the Scabby Butte local fauna of the St. Mary River Formation, Alberta (Sloan and Russell, 1974; Russell, 1975). A fauna from the upper Fruitland and lower Kirtland formations, San Juan Basin, New Mexico, has alternatively been considered to be of "Edmontonian" or Judithian age; as noted above, we tentatively place it in the Judithian. A single species known from the Horseshoe Canyon Formation, Alberta, appears to be referable to the "Edmontonian" (Fox and Naylor, 1986), as is a small fauna from the Williams Fork Formation, Colorado (Diem, 1999; see also Archibald, 1987; Lillegraven, 1987). We place the "Edmontonian" in the early Maastrichtian (see Lerbekmo and Coulter, 1985; Lillegraven and McKenna, 1986; Lillegraven, 1987).

The **Scabby Butte** local fauna (see Sloan and Russell, 1974; Russell, 1975; Fox, 1997c) includes 12 mammal taxa,

TABLE 2.26. "Edmontonian" Mammals of North America (see figures 2.19, 2.21; locality numbers do not correspond between maps and table). Localities and local faunas: 1, Scabby Butte, Alberta; 2, Lundbreck, Alberta (both St. Mary River Formation); 3, Michichi Creek, Alberta; 4, Newcastle, Alberta (both Horseshoe Canyon Formation); 5, Williams Fork, Colorado (Williams Fork Formation); 6, Shell Hell, Montana (St. Mary River Formation)

Multituberculata
Family *incertae sedis*
?*Paracimexomys* sp. (5)
Cimolodontidae
Cimolodon nitidus (1, 2, 5)
Cimolodon sp. nov. (5)
Cimolomyidae
Cimolomys gracilis (1)
?*Cimolomys* sp. (5)
Meniscoessus collomensis (5)
Meniscoessus aff. *intermedius* (5)
Meniscoessus major (1, 5)
Meniscoessus robustus (1)
Neoplagiaulacidae
Mesodma thompsoni (5)
Mesodma cf. *thompsoni* (1)
Marsupialia
"Alphadontidae"
Aenigmadelphys sp. nov. (5)
Alphadon marshi (5)
Alphadon wilsoni (5)
Turgidodon ?*parapraesagus* (6)
Turgidodon rhaister (5)
Turgidodon cf. *rhaister* (1)
Turgidodon russelli (5)
"Pediomyidae"
Aquiladelphis incus (5)
"Pediomys" cooki (5)
"Pediomys" cf. *cooki* (1)
"Pediomys" cf. *krejcii* (1)
Stagodontidae
Didelphodon coyi (3, 4)
Didelphodon sp. (1)
Eodelphis sp. (5)
?*Eodelphis* sp. (1)
Eutheria
Cimolestidae
Cimolestes sp. (1)
Gypsonictopsidae
Gypsonictops sp. (1)

[15] One of the multituberculates, which we list as Cimolomyidae gen. and sp. indet., was called *"Jas tejana"* by Standhardt (1986). To our knowledge, this name remains unpublished and is thus a *nomen nudum.*

collected from the lower part of the St. Mary River Formation at an exposure northeast of Nobleford, south-central Alberta (figure 2.19). As has been previously observed (Flynn, 1986; Lillegraven and McKenna, 1986), most of the taxa identified to species are typical of Lancian assemblages: the multituberculates *Cimolodon nitidus, Meniscoessus robustus,* and *Mesodma thompsoni* are found at many localities; the multituberculate *Cimolomys gracilis* and the marsupials *Turgidodon rhaister, "Pediomys" cooki,* and *"Pediomys" krejcii* are nearly as broadly distributed. The presence of the multituberculate *Meniscoessus major,* otherwise known only from the Judithian, is suggestive of an older age. A more compelling reason to believe that the fauna of Scabby Butte differs from typical Lancian assemblages is provided by the large durophagous marsupials of the family Stagodontidae: the primitive *Eodelphis* is otherwise known only from the Aquilan through the Judithian; and a distinctive new (unnamed) species of *Didelphodon* is more primitive than the species known from the Lancian (Fox and Naylor, 1986).

A second locality in the St. Mary River Formation but in the upper part of the unit is in far southwestern Alberta (figure 2.19), on the bank of the Oldman River north of **Lundbreck** (Clemens et al., 1979). A single mammal tooth, referable to the common Lancian multituberculate *Cimolodon nitidus,* has been reported from this site (Russell, 1975). Six mammalian taxa have been reported from the lower part of the St. Mary River Formation at the **Shell Hell** locality near Browning, Montana (Heinrich et al., 1998). Most notable here is the presence of two *Paracimexomys*-group taxa, one similar to *?P. robisoni* (otherwise known from the Albian-Cenomanian) and the other to *P. priscus* (Judithian–Lancian). One of the marsupials is similar to *Turgidodon parapraesagus* from the Judithian of New Mexico and the other is a typical Lancian species, *"Pediomys" krejcii.*

Two sites in the Horseshoe Canyon Formation exposed in the valley of the Red Deer River near Drumheller, Alberta (figure 2.19), have each yielded a fossil mammal referred to the "Edmontonian" (Fox and Naylor, 1986). Both are low in the formation. The **Michichi Creek** locality, which lies 107 m above the contact with the underlying Bearpaw Formation, is the source of the type jaw of *Didelphodon coyi,* a rather distinctive and specialized stagodontid marsupial. Referred specimens of this species were collected at a site near **Newcastle,** some 76 m above the contact with the Bearpaw Formation.

In northwestern Colorado, the Williams Fork Formation (figure 2.20) has yielded a fauna suspected to be of "Edmontonian" age. A multituberculate jaw with two molars was found at the **Jubb Creek** locality, 6.4 km west of Axial, Moffat County, by Kenneth Carpenter in 1983 (Lillegraven, 1987). Follow-up investigations and further collecting were done by J. D. Archibald and colleagues (Archibald, 1987; Diem and Archibald, 2000). Several additional mammal localities were discovered; we refer to the composite assemblage (see Diem, 1999) from these sites as the **Williams Fork** local fauna (figure 2.20). The age of the assemblage can be established with some degree of precision through reference to molluscs contained in adjacent marine units. Correlation to the *Baculites compressus* cephalopod zone establishes the age as intermediate between typical Lancian and Judithian faunas (Lillegraven, 1987). The paleoenvironmental setting was a well-vegetated coastal plain. Sixteen mammals are known from the Williams Fork fauna, half of which are multituberculates. The first mammal reported from the fauna is the multituberculate *Meniscoessus collomensis,* which Lillegraven (1987) considered to be structurally intermediate between Judithian and Lancian members of the genus and, hence, a potentially useful species for characterizing the "Edmontonian." Diem (1999) subsequently recognized *M. collomensis* from the Lancian Laramie Formation, Colorado, based on material described by Carpenter (1979). Whether or not this record is verified, the first appearance of the species may nonetheless prove useful in characterizing the "Edmontonian."

To an even greater extent than Scabby Butte, the mammalian fauna of the Williams Fork Formation appears to include a mixture of Lancian and Judithian taxa. Typical Lancian taxa include the multituberculates *Cimolodon nitidus* and *Mesodma thompsoni* and the marsupials *Alphadon marshi, A. wilsoni, Turgidodon rhaister,* and *"Pediomys" cooki.* Several of these are also recorded at Scabby Butte and in the Hunter Wash local fauna (Judithian) of New Mexico, a pattern suggesting that these are stratigraphically long-ranging species rather than randomly occurring Lancian taxa in the "Edmontonian." Taxa apparently shared with Judithian or older faunas include the multituberculates *Meniscoessus* aff. *intermedius, M. major,* and an unnamed species of the multituberculate *Cimolodon* and the marsupials *Turgidodon russelli, Aquiladelphis incus, Eodelphis* sp., and *Iqualadelphis* sp.

At present, the only named species unique to the "Edmontonian" is the marsupial *Didelphodon coyi,* though *Didelphodon* sp. (intermediate between Judithian and Lancian taxa, Fox and Naylor, 1986) and *Meniscoessus collomensis* (unless present in the Laramie Formation) may be similarly restricted. Possible first occurrences include *Cimolodon nitidus, Mesodma thompsoni, Meniscoessus collomensis, M. robustus, Alphadon marshi, A. wilsoni, Turgidodon rhaister, "Pediomys" cooki,* and *"P." krejcii.* Possible last appearances are *Meniscoessus major, Aenigmadelphys, Turgidodon russelli,* and *Aquiladelphis incus.*

Lancian

The latest Cretaceous Lancian holds the dual distinction of including the world's earliest discovered Cretaceous mammals (Van Valen, 1967a) and of being the most heavily sampled geographic/temporal interval known for Mesozoic mammals: more than one-fifth of all known occurrences are in the Lancian, which is generally regarded as being of late Maastrichtian age (Clemens, 1973a). Some of the first specimens were found by Edward D. Cope and/or Jacob L. Wortman in the Hell Creek Formation of the Dakotas (see review by Wilson, 1965), but the first significant collection was made by J. B. Hatcher in beds now referred to the Lance Formation in its type area in what is now Niobrara County, east-central Wyoming (see reviews by Simpson, 1929a; Clemens, 1963b). Hatcher and his associates worked the region between 1889 and 1892 and accumulated a large collection for Othniel C. Marsh of Yale University. This and subsequent collections made through the early 1950s were accomplished mainly through surface collecting on scattered badlands and "blowouts" (exposures maintained by wind erosion), often focusing on anthill accumulations and occasionally using a flour sifter to concentrate the fossils (Hatcher, 1896). Because of the great lithologic similarity of the containing strata and the fact that mammals were almost invariably found in the same beds that contained horned dinosaurs, the rock unit was generally referred to as the "Laramie Formation" (a term now restricted to a unit in the Denver Basin, see Carpenter, 1979) or the "*Ceratops* Beds." Strata in eastern Montana and adjacent parts of North Dakota are now generally referred to the Hell Creek Formation (named for a tributary of what is now the Fort Peck Reservoir in Montana) and those in Wyoming to the Lance Formation (for Lance Creek, in the type area, see Clemens, 1963b). As reviewed later, fossil mammals are known from presumably equivalent units in Wyoming, South Dakota, and elsewhere.

A milestone in the collection and study of North American Late Cretaceous mammals came in 1956, when Malcolm C. McKenna and others visited the type Lance area, discovering what would prove to be one of the most productive sites there. This and other sites were worked by Clemens and colleagues in subsequent years and, through the then-innovative method of underwater screenwashing of quarried rock matrix, an astonishingly large, well-represented mammalian fauna was collected (Clemens, 1963b, 1966, 1973a). Soon thereafter similar investigations were initiated by Clemens and Jason A. Lillegraven in what is now known as the Scollard Formation, Red Deer River Valley, Alberta, with similar results (Lillegraven, 1969). The unprecedented wealth of fossils resulting from these programs represented a radical departure from all previous field activities in Upper Cretaceous strata of North America. It may be fairly said that these two projects ushered in a new era in the study of Mesozoic mammals, characterized by an explosive expansion in basic data—the fossil record itself.

With a few exceptions, local faunas representing the Lancian are confined to the northern and central parts of the Rocky Mountain region and hence (despite the density of sampling) are not as broadly distributed as those of Judithian age. In addition to the type Lance and Scollard faunas just mentioned, local faunas have been described from Saskatchewan (Johnston and Fox, 1984; Fox, 1989, 1997a,b; Storer, 1991), Wyoming (Clemens et al., 1979; Breithaupt, 1982; Whitmore, 1985; Whitmore and Martin, 1986; Eberle and Lillegraven, 1998a,b; Lillegraven and Eberle, 1999), Montana (Clemens et al., 1979; Archibald, 1982; Lofgren, 1995; Hunter et al., 1997), North Dakota (Hoganson et al., 1994; Murphy et al., 1995; Hunter and Pearson, 1996), South Dakota (Wilson, 1965; Lillegraven, 1969; Sloan and Russell, 1974; Clemens et al., 1979; Wilson, 1983, 1987), and Colorado (Carpenter, 1979). Farther from this central area, small faunas are known from Alaska (Clemens, 1995), central Utah (Cifelli, Nydam, Eaton, et al., 1999), and northwestern New Mexico (Flynn, 1986). A few specimens of late Maastrichtian age, possibly equivalent to the Lancian, are known from as far away as Texas and New Jersey (Krause and Baird, 1979; Tokaryk, 1987).

The fauna of the Bug Creek anthills (BCA), Montana, has been problematic since it was first described (Sloan and Van Valen, 1965; Van Valen and Sloan, 1965; see review by Clemens et al., 1979). The assemblage includes a mixture of mammalian taxa, some of which are typical of Lancian faunas and others that are more of Paleocene aspect—particularly certain eutherians such as ungulates and (possible) primates. Stratigraphic placement is hampered by the fact that Bug Creek lies a considerable distance from the nearest complete exposed section, and as William A. Clemens (pers. comm.) observed, "all that lies between is grass and air, neither of which is stratigraphically useful." Archibald (1982) regarded the BCA assemblage as a faunal facies within the Hell Creek Formation, correlative with a "typical" Lancian fauna. Other studies have suggested that the composition of BCA reflects faunal succession—that it is temporal and not ecologic in nature[16]—and have recognized BCA as representing a dis-

[16] It should be noted that ecological versus temporal differences between Late Cretaceous and Paleocene mammal faunas were debated long before discovery of BCA; see, for example, Matthew (1937).

crete time interval (Sloan, 1987; Archibald and Lofgren, 1990). Recent work has shown that the Bug Creek is probably a temporally mixed assemblage: the sediments containing it appear to represent channels of the Paleocene Tullock Formation, incised into the underlying Cretaceous Hell Creek Formation. As a result, the Bug Creek fauna is now interpreted as an assemblage including syndepositional Paleocene fossils together with those reworked from Cretaceous deposits (Lofgren et al., 1990; Lofgren, 1995). Accordingly, we have omitted Bug Creek taxa believed to be of Paleocene age, following the criteria and listing given by Lofgren (1995). Mammals of "Paleocene" aspect, mainly ungulates, have also been reported from the Frenchman and Ravenscrag formations, Saskatchewan (Johnston, 1980; Johnston and Fox, 1984). The age of these occurrences has been disputed; we tentatively follow Fox (1997b) in regarding them to be Late Cretaceous. However, the primitive ungulate *Protungulatum*, a first appearance datum for the Puercan land-mamal age (Archibald and Lofgren, 1990), is present at the two main sites. This introduces the paradox of recognizing Puercan mammal assemblages in the Late Cretaceous of Canada but not in the well-sampled, intensively studied Late Cretaceous of Montana (see summary by Clemens, 2002). Hopefully, this discrepancy will be resolved by future field investigations (see discussion by Hunter et al., 1997). We have also omitted mammals reported from the Javelina Formation in the Big Bend region of Texas. It was first thought that the mammal-bearing horizon was of Cretaceous age (Lawson, 1972; Clemens et al., 1979), but subsequent work has shown it to be Paleocene (Standhardt, 1986).

As noted by Clemens et al. (1979), mammalian faunas of Lancian age (table 2.27) are generally similar to those of the Judithian, differing principally in having a greater diversity of eutherians. Most notable in this respect is a moderate diversity of cimolestids (*Cimolestes, Batodon,* and similar taxa), which are almost surely present but exceedingly rare in the Judithian and "Edmontonian" (see discussion by Clemens, 2002). The Lancian also continues a trend, established by the Judithian, of increased body size within some groups, the opossum-sized marsupials *Turgidodon rhaister* and *Didelphodon vorax,* and the fisher-sized eutherian *Cimolestes magnus* being good examples. Lists of first, last, and unique occurrences for the Lancian land-mammal age are given by Lillegraven and McKenna (1986) and Cifelli et al. (2004). Our coverage of the Lancian is organized around the best-known faunas (those of the Lance, Hell Creek, Scollard, and Frenchman formations), together with smaller and presumably equivalent faunules, followed by "outlier" specimens and assemblages. Although two problematic "transitional" assemblages from Canada are formally excluded from the Lancian, for convenience they are treated in this section as well.

The **type Lance** fauna is the most thoroughly documented assemblage of mammals known from the Mesozoic of North America (Clemens et al., 1979). It comes from exposures of the Lance Formation in the Lance Creek drainage, about 40 km northwest of Lusk, Niobrara County, eastern Wyoming (figure 2.21). Some 40 or more localities are known, virtually all from the upper part of the unit, but the vast majority of specimens was collected from three sites (Clemens, 1963b). Fossils most commonly occur in fine, poorly consolidated sandstones of fluvial origin. The fauna of the type Lance includes 29 kinds of mammals: 9 multituberculates, 1 deltatheroidan, 13 marsupials, and 6 eutherians. Some taxa (*Cimolodon nitidus, Protalphadon lulli, Pediomys elegans, "P." hatcheri*) possibly or probably extend back to the Judithian, whereas others (*Cimolomys gracilis, Mesodma thompsoni, Alphadon marshi, A. wilsoni, Turgidodon rhaister, "Pediomys" cooki*) probably are common with the "Edmontonian." Though not as diverse as the therians, multituberculate fossils tend to be more commonly encountered in this and other Lancian faunas, which explains the fact that the majority of the more cosmopolitan taxa (*Cimolodon nitidus, Meniscoessus robustus, Mesodma formosa, M. thompsoni, Alphadon marshi, Pediomys elegans*) are multituberculates. Eutherians, on the other hand, are generally rare as fossils in the Lancian (*Gypsonictops* being an exception), with the result that the included species are common to only a few other faunas. *Clemensodon megaloba, Glasbius intricatus,* and *Telacodon laevis* are known only from the type Lance.

Two mammalian assemblages have been reported from sites to the north of Lance Creek, in northern Niobrara County. The **Greasewood Creek** locality has yielded a single mammal, a marsupial possibly referable to the common "*Pediomys*" *krejcii* (see Whitmore and Martin, 1986). Two additional sites, located near the Cheyenne River northwest of **Mule Creek Junction,** have collectively yielded five kinds of mammals, all typical of the Lancian (Whitmore, 1985).

A second reasonably well-represented local fauna from the Lance Formation was collected in the vicinity of the **Black Butte Station** (figure 2.21), Sweetwater County, southern Wyoming (Breithaupt, 1982). Fossils were collected from four sites, with the overwhelming majority coming from a locality discovered earlier by Douglas Lawson (Clemens et al., 1979). All sites are in the upper 30 m of the Lance Formation, with most occurring in a mudstone—a difference from the type area, where fossils are most commonly encountered in friable, fine sandstone. Therians are poorly represented in this fauna of 11 mammalian taxa, though this may be an artifact of small

TABLE 2.27. Lancian, Presumed Lancian Equivalent, and ?Late Cretaceous Puercan Mammals of North America (see figures 2.14, 2.15, 2.19–2.21; locality numbers do not correspond between maps and table). Localities and faunas: 1, Type Lance; 2, Greasewood Creek; 3, Mule Creek Junction; 4, Black Butte Station; 5, Hewitt's Foresight; (all Lance Formation, Wyoming); 6, Cheyenne River, Montana; 7, Dumbbell Hill, Wyoming; 8, Red Lodge, Montana (all ?Lance Formation); 9, Hell Creek, Montana; 10, Little Missouri Badlands, North Dakota; 11, Ekalaka, Montana; 12, Claw Butte, Montana; 13, Blacktail Creek, Montana; 14, Fallon County, Montana; 15, Powderville, Montana; 16, Muddy Tork, Montana; 17, Stumpf, North Dakota; 18, Black Horse, South Dakota; 19, Harding County, South Dakota; 20, Eureka Quarry, South Dakota; 21, Joe Painter Quarry, South Dakota (all Hell Creek Formation); 22, Red Owl; 23, Iron Lightning (Fox Hills Formation, South Dakota); 24, Trochu (Scollard Formation, Alberta); 25, Wounded Knee; 26, Gryde; 27, Fr-1 (both Frenchman Formation, Saskatchewan); 28, Long Fall (Ravenscrag Formation, Saskatchewan); 29, Colville River (Prince Creek Formation, Alaska); 30, Hanna Basin (Ferris Formation, Wyoming); 31, Weld County (Laramie Formation, Colorado); 32, North Horn (North Horn Formation, Utah); 33, Alamo Wash (Kirtland Formation, New Mexico); 34, Commerce (Kemp Clay Formation, Texas); 35, Hop Brook; 36, Ramanssein Brook (Mount Laurel Formation, New Jersey)

- Mammalia
  - Family *incertae sedis*
    - Gen. et sp. indet. (8, 10, 12)
- Multituberculata
  - Family *incertae sedis*
    - *Bubodens magnus* (22)
    - *Cimexomys minor* (1, 4, 5, 9, 28)
    - *Cimexomys* sp. (22)
    - *Cimexomys* cf. *gratus* (28)
    - cf. *Cimexomys* sp. (7)
    - *Paracimexomys priscus* (9, 24, 25, 28)
    - *Paracimexomys* sp. (32)
    - Gen. et sp. indet. (10, 23, 31, 35, 36)
  - Cimolodontidae
    - *Cimolodon nitidus* (1, 3, 5, 7, 9, 10, 12, 13, 22, 24–26, 28–30)
    - *Cimolodon* cf. *nitidus* (20, 28, 30)
    - *Cimolodon* sp. (10)
    - ?*Cimolodon* sp. (25)
  - Cimolomyidae
    - *Cimolomys gracilis* (1, 5, 7, 9, 13, 24, 26)
    - *Cimolomys* cf. *gracilis* (20, 28)
    - *Cimolomys trochuus* (24)
    - *Cimolomys* sp. (31, 32)
    - *Essonodon browni* (1, 9, 12, 33)
    - *Essonodon* sp. (4, 25)
    - ?*Essonodon browni* (12)
    - *Meniscoessus collomensis* (31)
    - *Meniscoessus conquistus* (19)
    - *Meniscoessus robustus* (1, 3–5, 9, 10, 12–17, 21–23, 25)
    - *Meniscoessus* cf. *robustus* (20, 27, 28)
    - ?*Meniscoessus robustus* (26)
    - *Meniscoessus seminoensis* (30)
    - *Meniscoessus* sp. (10, 33)
    - ?*Meniscoessus* sp. (10)
  - Eucosmodontidae
    - *Clemensodon megaloba* (1)
    - *Stygimys cupressus* (28)
    - Gen. et sp. indet. (28)
  - Neoplagiaulacidae
    - *Mesodma formosa* (1, 4, 5, 7, 12, 13, 20–22, 24–26, 33)
    - *Mesodma* cf. *formosa* (9, 12)
    - *Mesodma hensleighi* (1, 4, 5, 9, 12, 13, 21, 22, 24, 26)
    - *Mesodma* cf. *hensleighi* (20, 32)
    - *Mesodma thompsoni* (1, 4, 5, 9, 10, 12, 13, 15, 20, 21, 24, 28)
    - *Mesodma* cf. *thompsoni* (16, 22, 25, 26)
    - *Mesodma* sp. (9, 12, 16, 28)
    - ?*Mesodma* sp. (12)
    - ?*Neoplagiaulax burgessi* (5, 9)
    - *Parectypodus foxi* (26)
    - Gen. et sp. indet. (9)
  - Ptilodontidae
    - *Kimbetohia campi* (1)
    - Gen. et sp. indet. (24)
  - Taeniolabididae
    - *"Catopsalis" johnstoni* (28)
    - *"Catopsalis"* cf. *joyneri* (27, 28)
- Deltatheroida
  - Deltatheridiidae
    - cf. *Deltatheridium* sp. (1, 24)
- Marsupialia
  - Family *incertae sedis*
    - Gen. et sp. indet. (10, 23)
  - "Alphadontidae"
    - *Alphadon eatoni* (32)
    - *Alphadon jasoni* (1, 5, 9, 24–27)
    - *Alphadon marshi* (1, 4, 5, 9, 10, 12, 13, 22, 24, 28, 33)
    - *Alphadon* cf. *marshi* (7, 12, 21)
    - *Alphadon wilsoni* (1, 9, 13, 24)
    - *Alphadon* cf. *wilsoni* (12)
    - *Alphadon* sp. (2, 17, 20, 28, 32)
    - *Protalphadon foxi* (9)
    - *Protalphadon lulli* (1, 5, 12, 13, 30)
    - *Protalphadon* ?*lulli* (22, 30)
    - *Turgidodon petaminis* (26)
    - *Turgidodon rhaister* (1, 5, 9, 12, 13, 16, 24)
    - *Turgidodon* cf. *rhaister* (1)
    - Gen. et sp. indet. (30)
  - Glasbiidae
    - *Glasbius intricatus* (1, 5)
    - *Glasbius twitchelli* (9, 12, 25)
    - *Glasbius* cf. *twitchelli* (12)
  - "Pediomyidae"
    - *"Pediomys" cooki* (1, 5, 9, 12, 13, 22)
    - *Pediomys elegans* (1, 4, 5, 9, 12, 13, 20, 21, 24–26, 28, 29)

TABLE 2.27. Continued

"*Pediomys*" *florencae* (1, 5, 9, 10, 12, 13, 16, 20)	*Cimolestes propalaeoryctes* (5, 9, 13, 24)
"*Pediomys*" cf. *florencae* (9, 10, 15)	*Cimolestes* cf. *propalaeoryctes* (25)
"*Pediomys*" *hatcheri* (1, 3, 9, 12, 13, 16, 20, 21, 24, 32)	*Cimolestes stirtoni* (1, 9, 13, 25)
"*Pediomys*" cf. *hatcheri* (12, 13, 15, 26)	*Cimolestes* cf. *stirtoni* (28)
"*Pediomys*" *krejcii* (1, 9, 12, 13, 21, 24, 26)	cf. *Cimolestes* sp. (3, 30)
"*Pediomys*" cf. *krejcii* (2, 13, 16, 22)	*Procerberus* cf. *formicarum* (28)
"*Pediomys*" sp. (9, 12, 13, 15, 16, 25, 32)	*Telacodon laevis* (1)
cf. "*Pediomys*" sp. (23)	Gen. et sp. indet. (6, 13, 32)
Stagodontidae	Gypsonictopsidae
Didelphodon padanicus (18)	*Gypsonictops hypoconus* (1, 4, 5, 9, 12, 20–22, 24)
Didelphodon vorax (1, 3, 5, 9, 10, 11, 15 16, 24–26)	*Gypsonictops illuminatus* (5, 9, 10, 24, 26, 28)
Didelphodon cf. *vorax* (20, 22)	*Gypsonictops* cf. *illuminatus* (15, 25)
Didelphodon sp. (10)	*Gypsonictops* sp. (12, 13, 29, 30, 31)
cf. *Didelphodon* sp. (17, 21)	?Gypsonictopsidae
Eutheria	Gen. et sp. indet. (6, 7)
Family *incertae sedis*	"Nyctitheriidae"
Alostera saskatchewanensis (9, 24–26)	cf. *Paranyctoides* sp. (13)
Gen. et sp. nov. A (25)	"Arctocyonidae"
Gen. et sp. nov. B (25)	*Baioconodon* sp. (28)
Gen. et sp. indet. (32, 34)	*Oxyprimus* cf. *erikseni* (28)
Cimolestidae	*Protungulatum* cf. *donnae* (27, 28)
Batodon tenuis (1, 9, 12, 13, 24, 26)	?Hyopsodontidae
Cimolestes cerberoides (24)	Gen. et sp. indet. (28)
Cimolestes cf. *cerberoides* (9, 28)	?Periptychidae
Cimolestes incisus (1, 5, 9, 13, 26)	*Mimatuta* sp. (28)
Cimolestes cf. *incisus* (25)	Gen. et sp. indet. (27)
Cimolestes magnus (1, 5, 9–11, 24, 26)	

sample size (see earlier comments on type Lance). The assemblage is unremarkable, except for the possible occurrence of the primitive ungulatomorph *Protungulatum*, represented by a fragmentary molar.

Five other sites in Wyoming and southern Montana (figure 2.21) have yielded fossil mammals from what may prove to be the Lance Formation (see Clemens et al., 1979). One, probably north of the **Cheyenne River,** northeastern Wyoming, was worked by Wortman and O. A. Peterson in 1892. Only three types of mammals have been reported (Van Valen, 1967a), but given that all are eutherians, it appears probable that marsupials and multituberculates are also represented in the available collection. An assemblage including perhaps six taxa is known from **Dumbbell Hill** near Powell, Park County, northwestern Wyoming (figure 2.21). All are common Lancian taxa; notable is the fact that several multituberculate jaws were found in a coprolite. A third assemblage possibly from the Lance Formation has been collected from sites along Cottonwood and Kirby creeks in the **southwestern Bighorn Basin,** Wyoming. No specimens have yet been identified. **Hewitt's Foresight One** and nearby sites, also in the southwestern part of the Bighorn Basin, are more securely placed in the Lance Formation. Fieldwork led by Lillegraven, University of Wyoming, has been conducted at this site since 1981, resulting in the collection of an impressive 1,600 mammal specimens. The specimens reportedly come from fine-grained channel sandstones and include representation of extremely small mammals (Webb, 1998, 2001). Finally, an unidentified mammal tooth has been reported from a site near **Red Lodge,** Carbon County, Montana (figure 2.21)(Clemens et al., 1979).

The Hell Creek Formation is recognized as distinct from the Lance only on the basis of discontinuity in exposure between the two (Clemens, 1963b; Clemens et al., 1979). Exposures of the Hell Creek are distributed mainly along adjacent areas of eastern Montana, western North and South Dakota, and northeastern Wyoming. The main assemblage from the unit, the **Hell Creek** fauna, was collected by Clemens and colleagues from the vicinity of Fort Peck Reservoir, Garfield and McCone counties, Montana (figure 2.21). Many localities are known, but excluding the Bug Creek Anthills and other mixed Cretaceous-Paleocene assemblages (see earlier), a substantial sample is known from only one, which is included with four other sites in the Flat Creek local fauna (Archibald, 1982). Most of the mammal-yielding sites occur in the upper parts of the Hell Creek Formation (Clemens, 2002). Thirty-five kinds of mammals have been reported for the Hell Creek fauna: 12 multituberculates, 14 marsupials, and 9 eutherians. The area is geographically intermediate between the type Lance to the south and the Trochu and other assemblages to the north. As noted by Archibald (1982), this geographic intermediacy is reflected in the distribution of mammals in the three areas (but see Hunter and Archibald, 2002). No species are shared by Trochu and the type Lance to the exclusion of Hell Creek; however, the Hell Creek fauna shares certain taxa only with the Lance and other southern faunas and others only with the Trochu and other northern assemblages. "Southern" Lancian mammals appear to include, at least, the multituberculates *Essonodon browni* (which has also been reported from the San Juan Basin, New Mexico) and *Cimexomys minor;* and the marsupials *Alphadon lulli* and "*Pediomys*" *florencae.* The most convincing of "northern" Lancian mammals found in the Hell Creek fauna are all eutherians: *Gypsonictops illuminatus, Cimolestes cerberoides, C. propalaeoryctes,* and *Alostera saskatchewanensis.* The Hell Creek fauna includes at least one more Lancian species apparently shared with Judithian faunas, the multituberculate *Paracimexomys priscus.*

Mammals have been reported from 11 additional areas or individual sites (figure 2.21) in the Hell Creek Formation, of which three are diverse faunas and four are small assemblages, with the remaining four representing single occurrences. A cluster of sites occurs in southeastern Montana and adjacent parts of the Dakotas. Sixteen mammalian taxa typical of the Hell Creek Formation have been collected from sites in the **Little Missouri badlands** (Hunter and Archibald, 2002). The most prolific site, which is in Bowman County, southwestern North Dakota (Hunter and Pearson, 1996), occurs in sandstone lenses about 35 m below the contact with the overlying Ludlow Formation. Several areas yielding mammals from the Hell Creek Formation in the nearby Powder River Basin, southeastern Montana, are known (Clemens et al., 1979). One specimen, a maxilla of the eutherian *Cimolestes magnus,* has been described from a site about 15 km south of **Ekalaka** (see Clemens, 1973a), in Carter County. Four additional closely spaced sites in Carter County occur about 65 m below the top of the Hell Creek Formation and have collectively yielded the **Claw Butte** local fauna. This is a typical but very rich Lancian assemblage, including nearly 30 mammalian taxa (Hunter and Archibald, 2002). A final, noteworthy site in Carter County has yielded the **Blacktail Creek** local fauna, from about 61 m below the top of the Hell Creek Formation. The assemblage from Blacktail Creek includes more than 25 mammals (Hunter and Archibald, 2002). Clemens et al. (1979) reported mammals from the Hell Creek Formation, from an area 40–48 km north-northeast of Ekalaka, in **Fallon County** (figure 2.21). To our knowledge, only one specimen has been described from this area: a jaw of the multituberculate *Meniscoessus robustus* (see Archibald, 1982). Farther to the west, a small assemblage of mammals has been collected from the upper part of the Hell Creek Formation, 8 km south of **Powderville.** Seven taxa are known from this area (Clemens et al., 1979). Well to the north of these sites around Ekalaka, the **Muddy Tork** local fauna was collected from several sites in the Hell Creek Formation near Glendive, Dawson County, eastern Montana (Hunter et al., 1997). The most productive of these, Muddy Tork, lies in a fluvial sandstone that is placed about 11 m below the local Cretaceous-Tertiary boundary. Nine taxa are known from the fauna, which is somewhat atypical in including only three multituberculates, with the remaining six being marsupials.

A Lancian faunule of three taxa is known from the **Stumpf** site (figure 2.21) in the Hell Creek Formation of western North Dakota (Hoganson et al., 1994; Murphy et al., 1995). Two of Cope's types were apparently collected from the Hell Creek Formation of South Dakota, but their exact positions are uncertain. One, **Black Horse,** is thought to be along the Grand River, about 6 km southeast of the town of Black Horse (Wilson, 1965), and is the site where the type (probably including two individuals) of the marsupial *Didelphodon padanicus* was collected (see Clemens, 1966; Archibald, 1982; Fox and Naylor, 1986). The other site, the source of the type of the multituberculate *Meniscoessus conquistus,* is probably in **Harding County,** northwestern South Dakota (see Wilson, 1965; Sloan and Russell, 1974; Archibald, 1982; Lillegraven, 1987). Several additional sites in South Dakota have yielded mammals from the Hell Creek Formation. The **Eureka Quarry,** 20 km southwest of Buffalo and hence probably near the Harding County site just noted, was worked by Robert W. Wilson and associates from 1962 to 1965. The mammalian assemblage of 12 taxa resembles that of the larger type Lance fauna, except that *Cimolomys* is more common and *Alphadon* less so (Wilson, 1983). **Joe Painter Quarry,** northwest of Buffalo, Harding County (figure 2.21), was collected by Wilson and associates in 1966 and 1970. Ten taxa are known from this site (Wilson, 1983).

Two small mammalian faunas are known from the Fox Hills Formation, South Dakota (figure 2.21). The better known of these is near **Red Owl,** Meade County. Fourteen taxa (including one endemic multituberculate[17]) are known from this site. According to Wilson (1987), the productive horizon at Red Owl is probably equivalent to the Fox Hills Formation in the Lance Creek area, where it underlies the Lance Formation. Hence, the mammalian assemblage from Red Owl may be older than typical faunas from the Lance and Hell Creek formations. This view is supported by correlations based on marine invertebrates and paleomagnetic stratigraphy (Cifelli et al., 2004). A site in the Fox Hills Formation near **Iron Lightning,** Ziebach County, has yielded four taxa, of which only one, the multituberculate *Meniscoessus robustus,* has been identified at the species level (see Waagé, 1968; Clemens et al., 1979).

To the north in Canada, a major Lancian fauna is known from what is now regarded as the lower part of the Scollard Formation (see Clemens et al., 1979, for a review of literature and confusing terminology) in the Red Deer River valley, Alberta (figure 2.19). A number of localities were worked, but the vast majority of mammal specimens were recovered from a single site, KUA-1. The rock here is a well-indurated mudstone that was worked through a combination of quarrying (through which a number of important jaws were discovered) and underwater screen-washing (Lillegraven, 1969). The assemblage, which remains one of the best-represented of Mesozoic mammal faunas, has come to be known as the **Trochu** fauna, named for a town near the southwest extent of the exposures. Twenty-two kinds of mammals are known for the Trochu fauna. Of these, a relatively high proportion—nearly one-third—are eutherians, a departure from more southerly Lancian faunas (see Hunter and Pearson, 1996, and references cited therein). As noted earlier in connection with the Hell Creek fauna, there is some indication that mammals of Lancian age may, to some extent, have been distributed along a latitudinal gradient. Notably lacking at Trochu are the multituberculates *Meniscoessus* and *Essonodon,* as well as the marsupials *Glasbius, "Pediomys" cooki,* and "*P.*" *florencae.* One species, the multituberculate *Cimolomys trochuus,* is endemic to the Trochu fauna; unnamed taxa represented by isolated teeth also appear to be unique (Lillegraven, 1969).

Important Late Cretaceous mammalian assemblages have been reported from the Cypress Hills region, southwestern Saskatchewan (see summary papers by Fox, 1997a,b). The sites are distributed through the upper Frenchman and lower Ravenscrag formations in the valley of the Frenchman River, south and southwest of Shaunavon (figure 2.19), and are notable in that the faunas include those typical of the Lancian, as well as superposed "transitional" Lancian-Puercan and Puercan assemblages.

The **Wounded Knee** locality (figure 2.19), found by Paul Johnston and Richard C. Fox in 1979, lies in a poorly consolidated sandstone in the upper Frenchman Formation, 25 m below the Ferris Coal, which represents the local Cretaceous-Tertiary boundary (see Fox, 1989). The fauna includes 20 kinds of mammals, of which a relatively high proportion (35%) are eutherians (see earlier comments on Trochu). Closest resemblance is to the Flat Creek local fauna of the Hell Creek Formation, Montana, which shares 15 of the species known from Wounded Knee. Like the Flat Creek fauna, the assemblage from Wounded Knee appears to show a blend of northern and southern influences (Fox, 1989). For example, *Meniscoessus, Essonodon,* and *Glasbius,* lacking at Trochu, are present at Wounded Knee, whereas *Paracimexomys priscus, Gypsonictops illuminatus,* and *Cimolestes propalaeoryctes* are shared with Trochu but not the type Lance. Several unnamed species from Wounded Knee are unknown elsewhere, which suggests a relatively high level of endemism.

The **Gryde** locality, source of the Gryde local fauna (figure 2.19), was discovered by John E. Storer and Tim Tokaryk in 1984, and lies on the south side of the Frenchman River, 3.4 km from the Wounded Knee site. The Gryde locality is also in the upper Frenchman Formation, but it is probably lower in the section than Wounded Knee, from which it further differs in occurring in a fine siltstone, representing an overbank deposit, rather than in a sandstone. The Gryde local fauna includes 19 mammalian taxa, represented by more than 1,000 teeth and dentulous jaws (Storer, 1991). Like Wounded Knee, the fauna includes both northern and southern taxa, but the Gryde fauna differs in having a lower proportion (20%) of eutherians. Two taxa are known only from Gryde: the marsupial *Turgidodon petaminis* and the multituberculate *Parectypodus foxi.* The latter is of interest in that the genus is otherwise Paleocene in distribution.

Whereas the faunas of Wounded Knee and Gryde are typical of the Lancian, the remaining two Late Cretaceous mammal sites in southwestern Saskatchewan have yielded assemblages that include a mixture of Lancian mammals with those of "Paleocene aspect."[18] *Protungulatum* is

[17] *Bubodens magnus.* A second species described by Wilson (1987), *Meniscoessus greeni,* was synonymized with *M. robustus* by Eberle and Lillegraven (1998a).

[18] Additional Lancian sites of Saskatchewan are listed by Storer (1993). Details are provided for only one, the Glen McPherson local fauna, which has yielded a single mammal, the multituberculate *Mesodma* cf. *garfieldensis.* In view of the difficulties in identifying species of *Mesodma,* and of the probability that *M. garfieldensis* is probably a strictly Puercan species in any case (Lofgren, 1995), we omit these sites.

found at both sites, and for this reason they are formally referred to the Puercan land-mammal age, despite their apparent Cretaceous age. The **Fr-1** locality is on the north side of the Frenchman River and west of Wounded Knee and Gryde, just north of Ravenscrag (figure 2.19). The site occurs in thinly bedded, loosely consolidated sandstones, siltstones, and mudstones in the upper part of the Frenchman Formation (Fox, 1989). The Ferris coal is not present in this area, but a Cretaceous age is indicated by palynomorphs (Fox, 1997b). The site is said to differ from the BCA, which are now interpreted as a mixture of Paleocene and reworked Cretaceous fossils (Lofgren, 1995), in not occurring in a channel and in unambiguously showing "Paleocene aspect" mammals to co-occur with dinosaurs. An articulated dinosaur is known from the same horizon as Fr-1, several hundred meters distant (Fox, 1989). Seven mammals are known from Fr-1; of these, four are "typical" Lancian taxa, whereas (except for the Long Fall site) the multituberculate *Catopsalis* and two ungulates (*Protungulatum* and a possible periptychid) are more generally encountered in the Paleocene.

The **Long Fall** site (figure 2.19), just northwest of Fr-1, was discovered by Johnston in 1979. The stratigraphic position and age of Long Fall have been disputed, with some workers placing it in the Paleocene (Lerbekmo, 1985; Sloan, 1987). It is currently placed in the lower part of the Ravenscrag Formation and, like Fr-1, is considered to be younger than Wounded Knee and Gryde, but still Cretaceous in age (see discussion in Fox, 1989, 1997b). Unfortunately, the regional marker bed for the Cretaceous boundary, the Ferris coal seam (Lerbekmo and Coulter, 1985), is not found in the vicinity of Long Fall, and palynomorphs were not recovered (Fox, 1997b). An early Paleocene (Puercan) fauna occurs at the Rav W-1 site, a few meters above Long Fall (Johnston and Fox, 1984). The Long Fall fauna includes 21 mammalian taxa, of which only three are marsupials and nearly half (nine) are eutherians—a striking difference from other assemblages of the North American Cretaceous and a similarity to those of the Paleocene. Mammals of "Paleocene aspect" at Long Fall include the multituberculates *Catopsalis* and *Stygimys*, and (especially) primitive ungulates, of which there appear to be at least four kinds.

Farther north, a small assemblage of Lancian age is known from a site by the **Colville River,** north of the Brooks Range, Alaska (figure 2.8). The occurrence is in the Prince Creek Formation and is of special interest because of its extremely high paleolatitude, which was closer to the paleopole of rotation than its present position near 70° N latitude (Clemens, 1995). Three kinds of mammals are known from the Colville River site: the Lancian multituberculate *Cimolodon nitidus,* the marsupial *Pediomys elegans,* and the Judithian–Lancian eutherian *Gypsonictops.* It appears likely that the site lies stratigraphically below volcanic rocks dated at 69.1 ± 0.03 Ma, and hence it is probable that the assemblage from Colville River is older than most Lancian faunas (Cifelli et al., 2004).

Field parties led by Lillegraven, University of Wyoming, have collected a superposed sequence of Cretaceous–Paleocene faunas from the **Hanna Basin,** near the North Platte River, about 30 km northeast of Sinclair, south-central Wyoming (figure 2.21). All of the assemblages are in the Ferris Formation; six localities yielded mammals of Lancian age, collectively including eight taxa (Eberle and Lillegraven, 1998a,b; Lillegraven and Eberle, 1999). Notable among these are *Cimolodon nitidus* and *Protalphadon lulli,* perhaps the southernmost occurrences of these species, and the endemic *Meniscoessus seminoensis,* which apparently does not co-occur geographically with its more common Lancian sister species, *M. robustus* (see Eberle and Lillegraven, 1998a).

Farther to the south, a small Lancian fauna has been reported from **Weld County,** northeastern Colorado (Carpenter, 1979). The assemblage comes from a single site in the upper part of the Laramie Formation. This formation, once conceived to include many Upper Cretaceous rock units in the Western Interior, is now restricted to the Denver Basin and is a lateral equivalent of the Hell Creek, Lance, and several other formations. Unlike many fossil localities in the Hell Creek and Lance formations, which occur in channel sandstones, the Weld County site is in a silty claystone, interpreted as representing an overbank deposit (Carpenter, 1979). Four mammals are known from the Weld County site. Most noteworthy among these is *Meniscoessus collomensis,* otherwise known only from the "Edmontonian" of Moffat County, Colorado, and previously considered as a possibly diagnostic taxon for that land-mammal age (see Lillegraven, 1987; Diem, 1999).

The North Horn Formation, central Utah (figure 2.20), includes both Cretaceous and Paleocene horizons (Gazin, 1941b; Gilmore, 1946). It is of interest in that it lies geographically between assemblages known to the north (in Wyoming, Montana, and Canada) and to the south (in New Mexico) (Cifelli, Nydam, Eaton, et al., 1999). Most notable among the dinosaurs from the unit is the sauropod *Alamosaurus,* which apparently represents a dispersal of Titanosauridae from South America into North America in the Late Cretaceous. *Alamosaurus* is known only from Late Cretaceous faunas of the southwestern United States (Lucas and Hunt, 1989). Mammals, including an unidentified multituberculate and the Lancian marsupial "*Pediomys*" *hatcheri,* were first reported from the Cretaceous part of the North Horn Formation by Clemens (1961; Clemens et al., 1979). These specimens were collected at the famous "lizard locality" in South Dragon Canyon, Emery County (Gilmore, 1942). Fossil mammals were later collected by quarrying at this site and by screenwashing rock from sites at approximately the same stratigraphic horizon on the flanks of North Horn Mountain, about 6 km to the north. Unfortunately, the rock is difficult to disaggregate and the fossils tend to become badly broken in the process, despite the fact that relatively complete specimens of small vertebrates are known from the unit. The composite **North Horn** fauna (figure 2.20) includes nine mammals (Cifelli, Nydam, Eaton, et al., 1999). Of these, the endemic marsupial *Alphadon eatoni* is noteworthy in being represented by a well-preserved jaw of a juvenile individual—a rare occurrence in the Cretaceous of North America (Cifelli et al., 1996; Cifelli and Muizon, 1998a).

As noted earlier, mammals widely regarded as being of Lancian age are known from the upper part (Naashoibito Member) of the Kirtland Formation in the San Juan Basin, northwestern New Mexico (figure 2.20). The assemblage from this horizon is referred to as the **Alamo Wash** local fauna (e.g., Lehman, 1981; Flynn, 1986). Despite its geographic importance (see Lehman, 1987), the Alamo Wash fauna is poorly represented by mammals; thus far, only four taxa (three multituberculates and a marsupial) are known. Ongoing field investigations by Anne Weil and Thomas Williamson promise to greatly improve knowledge of this assemblage (Weil and Williamson, 2000).

Finally, several occurrences of late Maastrichtian mammals, probable or possible age equivalent to the Lancian, are known from more easterly parts of North America. A single lower premolar of a eutherian has been recovered from the Kemp Clay Formation near **Commerce,** Texas (Tokaryk, 1987, see figure 2.14). The Kemp Clay Formation, deposited in an estuarine environment, is primarily known for its prolific yield of marine fossils, mainly isolated elements of chondrichthyans and mosasaurs. The occurrence of a mammal in this context is somewhat surprising but not unprecedented. A complicating factor is that fossils from the Kemp Clay are commonly found in reworked deposits together with fossils of much younger age. The other occurrences are in the Mount Laurel Formation, Monmouth County, New Jersey (figure 2.15). The **Hop Brook** site has yielded a predominantly marine fauna; the single mammal specimen is a fragment of a femur believed to belong to a multituberculate, which may have found its way to the sea within the stomach of a crocodilian (Krause and Baird, 1979). A similar nearby site, **Ramanssein Brook,** has yielded a femur reported to belong to a multituberculate (Mehling, 2001).

CONCLUDING REMARKS

Despite significant discoveries over the past several decades, major features of the record of Mesozoic mammals continue to be its general sparseness and its uneven coverage through time and among major landmasses (figures 2.22–2.25). There are few areas where sampling of a given time interval is sufficient to suggest that negative evidence (i.e., absence of a given group of mammals) is meaningful (Clemens et al., 1979). Even where coverage appears to be reasonably good, new surprises emerge from the fossil record—as illustrated by the discoveries of a Late Cretaceous North American "eupantotherian" (Lillegraven and McKenna, 1986) and Late Cretaceous Asiatic marsupials (Trofimov and Szalay, 1994; Averianov and Kielan-Jaworowska, 1999).

In terms of the geological time scale, the most recent epochs of the Mezosoic—the Late Cretaceous, Early Cretaceous, and Late Jurassic, in that order—are incomparably the best known in terms of mammalian occurrences. The Early Cretaceous, which spans some 45 Ma, is of great interest because modern groups—cimolodontan multituberculates, boreosphenidans, and possibly monotremes—arose or underwent their earliest radiations. Except for important local faunas from Western Europe (figure 2.22) and a few specimens from China (figure 2.23) and South America (figure 2.24), the Early Cretaceous record comes largely from near the end of the epoch. We expect major advances in interpretation to come from new fossils that predate the Aptian–Albian in the Early Cretaceous. Worldwide, the most poorly represented epoch in the history of Mesozoic mammals is the Middle Jurassic: only a few local faunas from Britain (figure 2.22) plus several occurrences in Eastern Europe, Asia (figure 2.23), and Madagascar (and possibly North America; figures 2.24–2.25) are known for this interval. Given the controversies over origins of multituberculates and tribosphenic mammals, to name only two examples, it is clear that major advances are to be expected from the Middle Jurassic as well. Similarly, many enigmas remain for Late Triassic–Early Jurassic mammals, although here, at least, several taxa are known by rather complete skulls and, in some cases, skeletons.

Geographically, the greatest persisting deficiency in the record of Mesozoic mammals is the scarcity of fossils from the southern continents (figure 2.24). A few faunules or individual specimens are known from scattered intervals for several of the landmasses, but reasonably well-sampled local faunas are known only for the Early Cretaceous of Morocco and the Late Cretaceous of Patagonia.

The converse is that the record of Mesozoic mammals is heavily biased toward the northern continents,

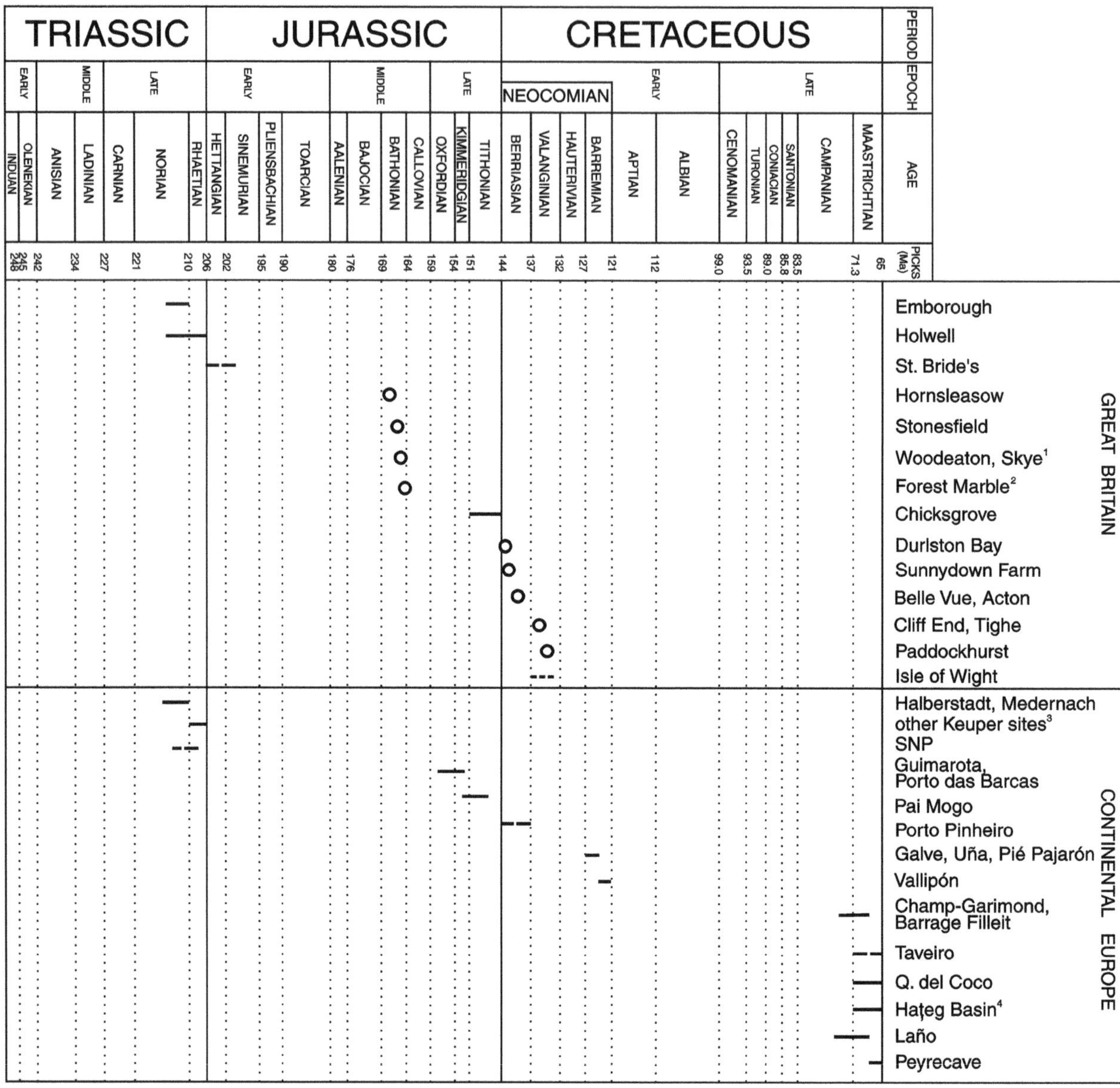

FIGURE 2.22. Temporal distribution of major Mesozoic mammal localities and local faunas of Britain and continental Europe (time scale after Palmer and Geissman, 1999). Circles indicate reasonably well-established chronostratigraphic control; solid lines indicate probable range; dashed lines indicate estimated range. 1, Loch Scavaig, Elgol; 2, Kirlington, Swyre, Watton Cliff, Tarlton Clay pit; 3, Württemburg, Hallau, Hachy, Habay-la-Vieille, Attert, Varangéville, Syren; 4, După Râu, Tamaşel, Pui, Tuştea, Fântânele.

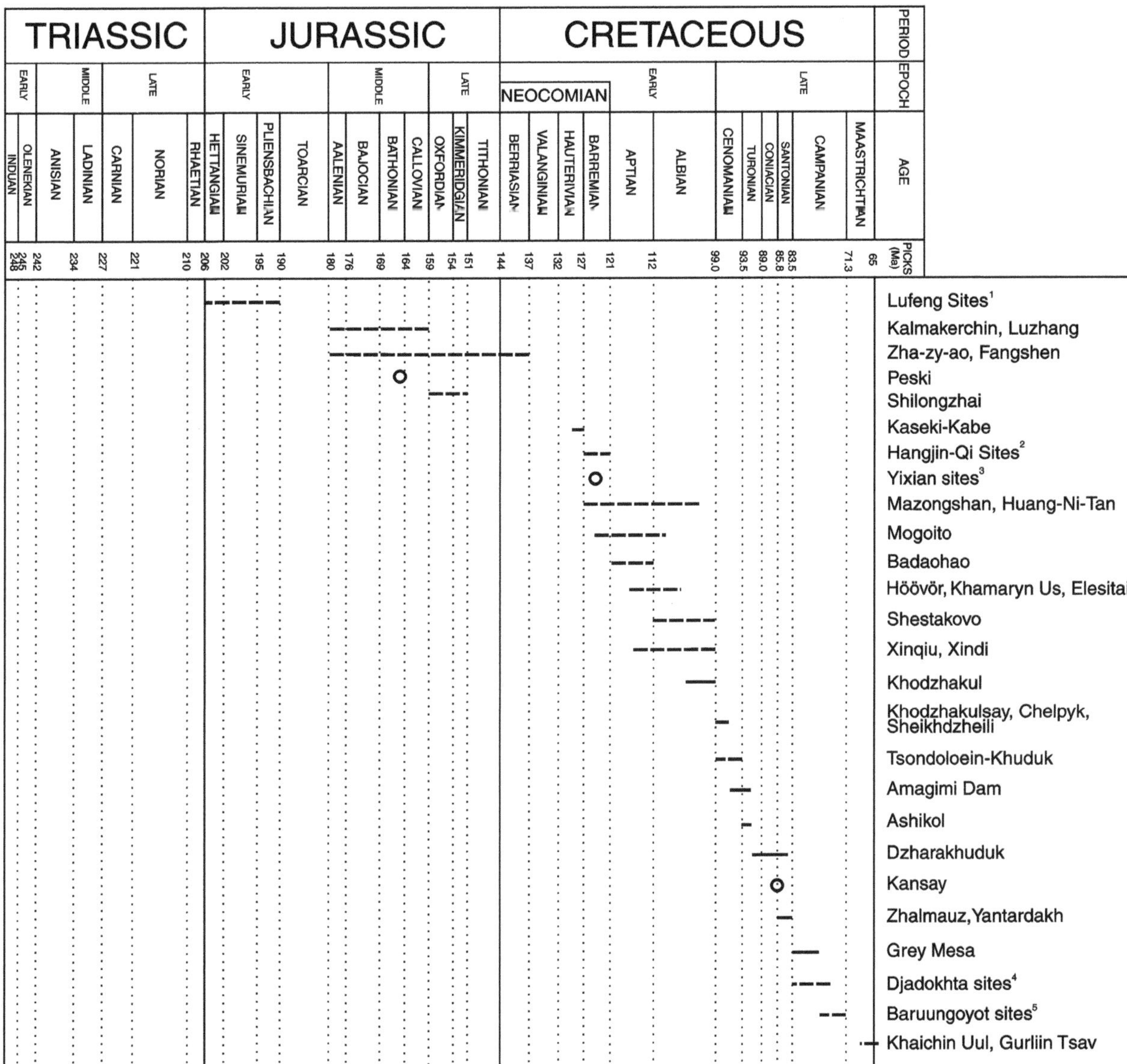

FIGURE 2.23. Temporal distribution of major Mesozoic mammal localities and local faunas of Russia and middle-east Asia (time scale after Palmer and Geissman, 1999). Circles indicate reasonably well-established chronostratigraphic control; solid lines indicate probable range; dashed lines indicate estimated range. 1, Dahuangtian, Dawa, Heiguopeng, Zhangjiawa, Dadi; 2, Hangjin Qi, Laolonghuoze, Yanhaizi; 3, Jianshangou, Sihetun, lu-Jia-Tun, Dawangzhangzi; 4, Djadochta and presumed equivalents: Byan Zag, Tögrög, Ukhaa Tolgod, Bayan Mandahu; 5, Baruungoyot and presumed equivalents: Khulsan, Nemegt, Hermiin Tsav, Udan Sayr.

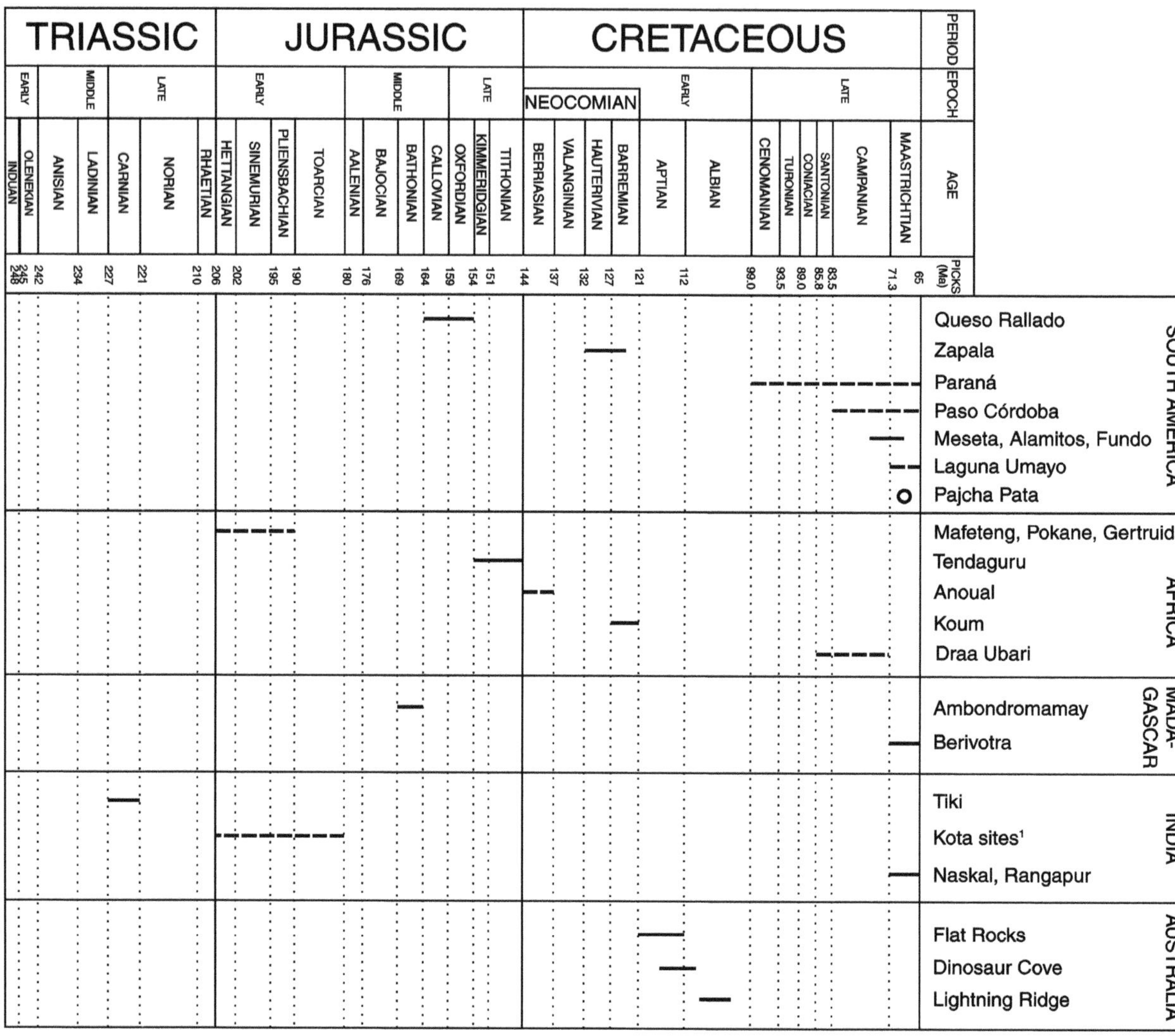

FIGURE 2.24. Temporal distribution of major Mesozoic mammal localities and local faunas of southern landmasses (time scale after Palmer and Geissman, 1999). Circle indicates reasonably well-established chronostratigraphic control; solid lines indicate probable range; dashed lines indicate estimated range. 1, Manganpalli, Yamanapalli, Paikasigudem.

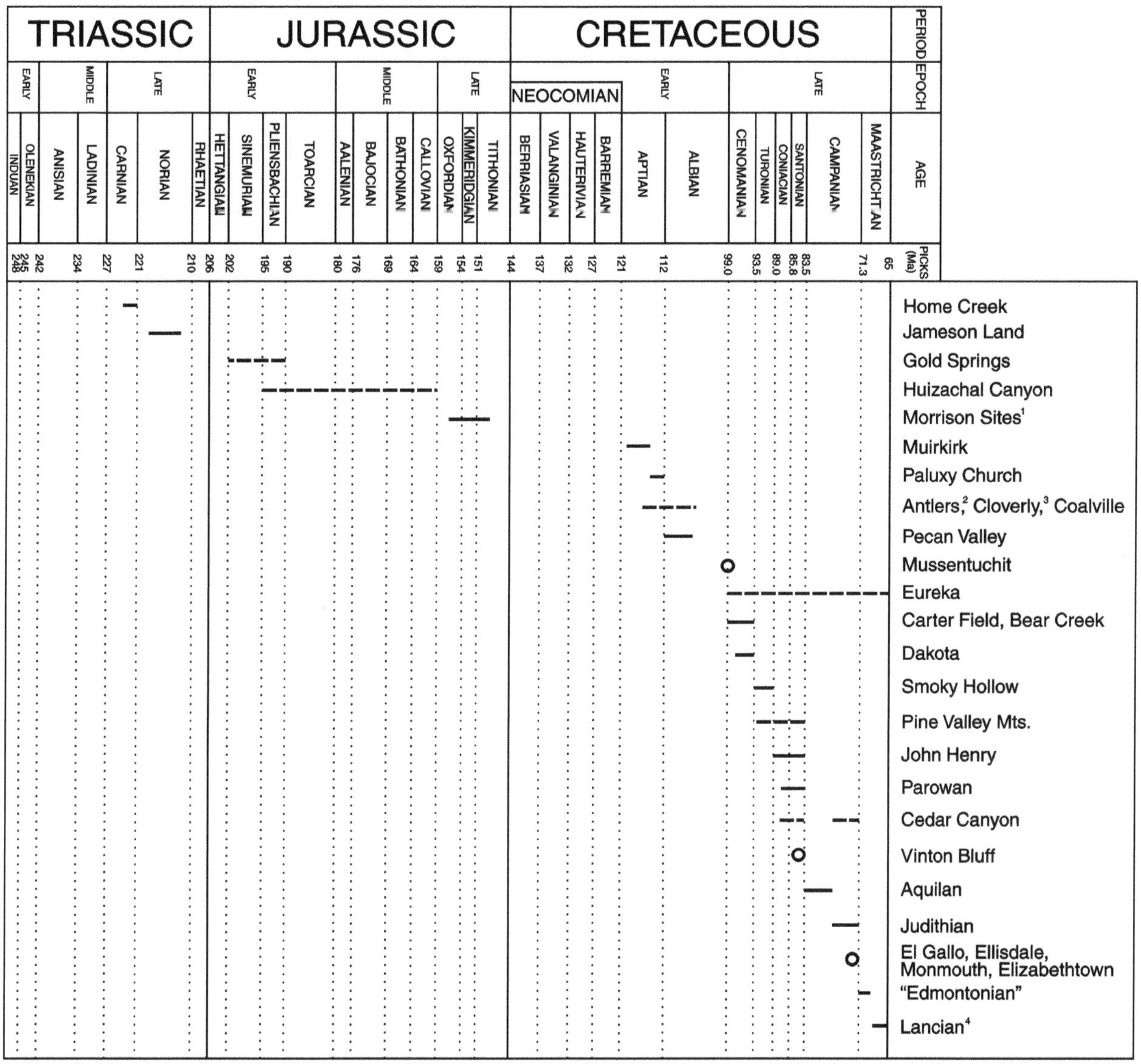

FIGURE 2.25. Temporal distribution of major Mesozoic mammal localities and local faunas of Greenland and North America (time scale after Geissman and Palmer, 1999). Circles indicate reasonably well-established chronostratigraphic control; solid lines indicate probable range; dashed lines indicate estimated range. 1, Como Quarry 9, Chuck's Prospect, Delta T, Breakfast Bench, Freezout Hills, Garden Park, Fruita, Dinosaur National Monument, Dry Mesa, Sundance, Small Quarry, Ninemile Hill; 2, Greenwood Canyon, Willawalla, Butler Farm, Umsted Farm, McLeod Honor Farm; 3, Bridger, Cashen Ranch, Cottonwood Creek, Crooked Creek, Ninemile Hill; 4, Also Hop Brook, Ramanssein Brook, Coalville.

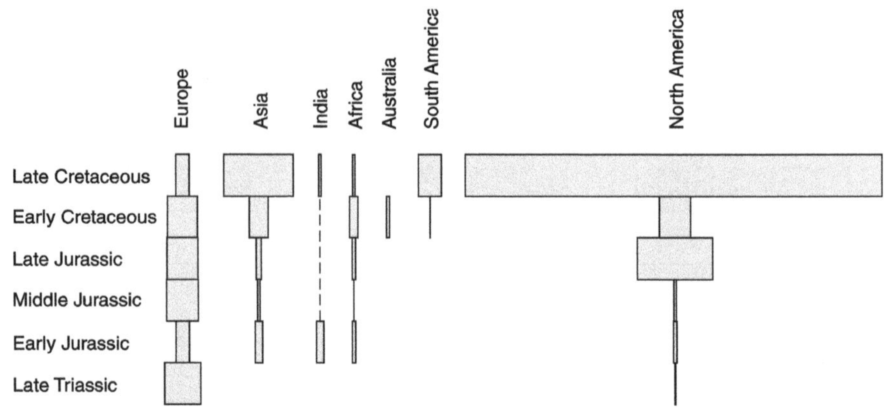

FIGURE 2.26. Temporal distribution of Mesozoic mammal occurrences by major landmass (Greenland included with North America; Madagascar with Africa). "Occurrence" refers to a record at lowest identified taxonomic level within a locality or local fauna. For most pre-Late Cretaceous occurrences and virtually all occurrences outside of North America, the totals reflect all taxa from all known sites and in many cases reflect most known individual specimens. Lumping occurrences by local faunas results in underrepresentation for the North American Late Cretaceous, which nonetheless includes a disproportionately large number of occurrences.

particularly North America (figure 2.26). This fact and the concomitant negative evidence from the southern continents at least partly explain the so-called "Sherwin-Williams effect," whereby biogeographers commonly tend to envision northern origin for many mammal groups, with later dispersal southward (Hershkovitz, 1972). Western Europe (figures 2.22, 2.26) probably has the most continuous overall record of Mesozoic mammals, with major faunas known for all epochs, although very little is known for the Late Cretaceous of the region. Asia, by contrast, is the source of some small assemblages and individual specimens of Jurassic age, but is primarily known for its Cretaceous fossils, particularly those from the Late Cretaceous. Reasonably well-sampled assemblages of Late Jurassic and Early Cretaceous age are known from North America, but the remarkable feature here is the relatively enormous record for the Late Cretaceous: more than 50% of all known Mesozoic mammal occurrences are in Upper Cretaceous strata of North America.

In terms of composition as well as temporal and geographic variation, more is known for North America's Late Cretaceous than for any other part of the Mesozoic mammal record. Interestingly, though, this bias is not universal, because North American Late Cretaceous mammals are known primarily by teeth and jaws, often obtained by the somewhat destructive process of screenwashing. Phylogenetic and paleobiological interpretation, however, is generally dependent upon the amount of morphological information available, and it is here that the North American record is wanting. Multituberculates, which for present purposes can be considered to be confined to the Late Jurassic–Late Cretaceous of the northern continents, serve as a useful example. In a study of multituberculate relationships, Simmons (1993) derived 28% of her information on Mesozoic taxa from those of Late Jurassic age (11% North America, 17% Europe), 11% from Early Cretaceous taxa (3% Europe, 8% Asia), and 60% from Late Cretaceous fossils. Of those from the Late Cretaceous, however, only 17% of the total came from North American multituberculates, as opposed to a staggering 43% from those of Asia. These figures are readily explained by the fact that Asiatic taxa are represented by skulls and skeletons and are incomparably better known than those of North America.

Similarly, only one Early Cretaceous mammal is known from all of South America. But that mammal, *Vincelestes*, is exceptionally well represented and, owing to its apparent proximity to boreosphenidan mammals, it has played a far greater role in recent phylogenetic interpretation than has any mammal from the Cretaceous of North America. Clearly major improvements have to be made in the fossil record from the Late Cretaceous of North America, despite the fact that it has been heavily sampled and studied.

CHAPTER 3

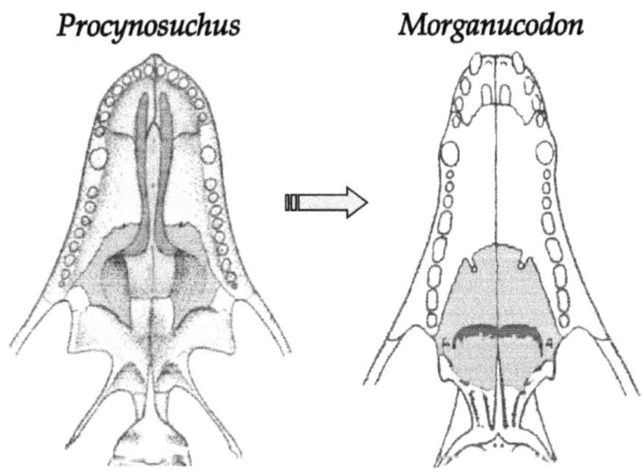

Origin of Mammals

INTRODUCTION

Mammals are distinctive from nonmammalian vertebrates in many derived features that, in turn, are correlated with important biological adaptations. They are unique among vertebrates, for example, in having fur, which provides insulation for maintaining an elevated (homeothermic) metabolism, and mammary glands, which provide milk for neonates. The important reproductive feature of nursing the neonate by mother's milk is correlated with mammalian growth patterns and numerous apomorphies of the dentition and skull, as discussed by Pond (1977), Tyndale-Biscoe and Renfree (1987), Jenkins (1990), and Zeller (1999a).

The origin of mammals was accompanied by the evolution of a number of derived osteological and dental apomorphies (table 3.1): (1) Modifications of the braincase in the frontal and parietal regions, in correlation with enlargement of the brain. (2) A temporomandibular joint formed by the dentary condyle and the squamosal glenoid, important for mammalian mastication. (3) Formation of the entire bony housing of the inner ear by a single bone, the petrosal, and the presence of the petrosal promontorium to accommodate an elongated cochlear canal for better hearing of high-frequency sounds. (4) Separation of the angular (ectotympanic) and articular (malleus) from the mandible and incorporation of the entire middle ear into the basicranium for more sensitive hearing. (5) Diphyodont dental replacement, in correlation with determinate skull growth and maintenance of continuity in molar function. (6) Development of precise molar occlusion, which is, in turn, correlated with the advancement of the dentary-squamosal temporomandibular joint (TMJ), as pointed out by Brink (1956), Kermack and Kermack (1984), Gow (1985), and Crompton and Hylander (1986).

Mammals differ from their phylogenetic relatives, the nonmammalian cynodonts, in these derived features. Enormous morphological evolution must have occurred through the origin of mammals from their cynodont ancestry. This chapter reviews the patterns of acquisition of these mammalian characteristics through the cynodont-mammal transition.

The diagnostic osteological features that have traditionally been associated with modern mammals did not evolve together in the fossil record (Luo, 1994; Sidor and Hopson, 1998), although Gow (1985) argued for a sudden emergence of mammalian characters. Some mammalian apomorphies first appeared in the fossil record with the earliest known stem mammals in the Late Triassic. Other derived features appeared among Jurassic mammals. But it now turns out that all of the so-called "mammalian features" have some kind of precursor condition among premammalian cynodonts, as shown by extensive studies of many well-preserved fossils belonging to relevant taxa in the past two decades. The derived characters of modern mammals were accumulated in stepwise fashion over a prolonged interval of synapsid-mammal evolution (Kemp, 1982; Hopson and Barghusen, 1986; Rowe, 1993; Luo, 1994; Sidor and Hopson, 1998; Sidor, 2001).

The first step toward an understanding of anatomical evolution at the cynodont-mammal transition involves a survey of the distribution of mammalian characters among stem mammals and their plesiomorphous precursor conditions in premammalian cynodonts (table 3.1). From these patterns, we may infer the historical evolution

TABLE 3.1. Major Osteological and Dental Apomorphies of Mammalia and Their Plesiomorphic Precursor Conditions in Premammalian Cynodonts

Character	Mammalian crown condition[1]	Mammalian condition	"Mammaliamorph" precursor condition	Eucynodont precursor condition
Craniomandibular joint (CMJ)	CMJ formed ***exclusively*** by the squamosal glenoid and the dentary condyle	Articulation of the squamosal glenoid and the dentary condyle	Incipient contact of squamosal with dentary (in tritheledontids) ("the secondary joint")	CMJ formed primarily by quadrate-articular ("the primary joint"), although there may be a surangular-squamosal contact in some taxa
Incudomalleus joint (middle ear)	Immobilized by retention of embryonic condition	Plesiomorphous as in eucynodonts →	Plesiomorphous as in eucynodonts →	Mobile joint functioning as jaw hinge
Ectotympanic and malleus attachment	Ectotympanic and malleus separated from dentary[2]	Plesiomorphous as in eucynodonts →	Plesiomorphous as in eucynodonts →	Attached to dentary as angular and articular-prearticular complex
Pars cochlearis of petrosal	Promontorium enlarged and bulbous	Pars cochlearis emergent from basicranium as promontorium	Plesiomorphous as in eucynodonts →	Pars cochlearis covered by the basisphenoid wing and basioccipital
Cochlear structure	Cochlear canal elongate, curved, or coiled	Plesiomorphous as in mammaliamorphs →	Presence of cochlear canal	Presence of a cochlear "recess"
Cavum epiptericum of trigeminal ganglion	The cavum incorporated into braincase by the absence of pila antotica[3]	Complete bony floor to cavum epiptericum	Plesiomorphous as in eucynodonts →	No bony floor to cavum epiptericum, which is separated from braincase by pila antotica in adult
Sidewall of cavum epiptericum	Plesiomorphous as in mammalian condition[4] →	Formed by enlargement of anterior lamina of petrosal	Plesiomorphous as in eucynodonts →	No enlargement of anterior lamina of petrosal
Orbital wall in orbitosphenoid region	Plesiomorphous as in tritylodontids →	Plesiomorphous as in tritylodontids →	Ossification of orbitosphenoid to form the medial wall of orbit (only in tritylodontids)	Absence of the medial wall of orbit ("orbital vacuity")
Floor of frontal region of braincase	Plesiomorphous as in mammalian condition →	Expanded orbitosphenoid floor in frontal region of braincase	Orbitosphenoid forming the floor in frontal region of braincase	Absence of a floor in the frontal region of the braincase
Sphenoidal structures	Sphenoidal sinus present but variable[5]	Presence of subcerebral space (= sphenoidal extension of nasal cavity)	Posterior extension of nasal cavity within the sphenoid (only in tritylodontids)	Absence of full ossification of the orbito- and presphenoid structures
Ethmoid structures	Presence of cribriform plate and ethmoid turbinates in addition to transverse lamina	Presence of transverse lamina within nasal cavity[6]	Presence of ethmoid internasal septum (only in tritylodontids)	Lack of ossification of ethmoid structures

Maxillary ridge for nasal turbinate	Full development	Present in variable condition	Incipient	Incipient
External nares	Confluent[7]	Plesiomorphous as in eucynodonts →	Plesiomorphous as in eucynodonts →	Separated by internarial process of premaxilla
Septomaxilla	Lost, or reduced without horizontal shelf	Plesiomorphous as in eucynodonts →	Plesiomorphous as in eucynodonts →	Present with horizontal shelf
Posterior wall of the nasal cavity	Plesiomorphous as in tritylodontids →	Plesiomorphous as in tritylodontids →	Ascending ("orbital") process of palatine enclosing nasal cavity (in tritylodontids)	Orbital process of palatine is small, and does not enclose the posterior nasal cavity wall
Secondary bony palate	Plesiomorphous as in mammaliamorphs →	Plesiomorphous as in mammaliamorphs →	Secondary bony palate extending to posterior end of tooth row	Secondary bony palate ending anterior to the end of tooth row
Maxilla in subtemporal margin	Plesiomorphous as in mammaliamorphs →	Plesiomorphous as in mammaliamorphs →	Maxilla participating in the subtemporal margin	Maxilla excluded by jugal from the subtemporal margin
Palatine in subtemporal margin	Plesiomorphous as in tritylodontids →	Plesiomorphous as in tritylodontids →	Palatine forming subtemporal margin of bony roof of pharynx (only in tritylodontids)	Palatine excluded lateral borders of the pharyngeal roof
Pterygoid hamulus or transverse process	Hamulus reduced, without contact to the mandible	Plesiomorphous as in mammaliamorphs →	Presence of pterygoid muscle fossa on lateral side of hamulus	Transverse process of pterygoid contacting mandible, without lateral muscular fossa
Molariform versus premolariform postcanines	Plesiomorphous as in mammalian condition	Differentiation of molariforms versus molariforms	Plesiomorphous as in eucynodonts →	Lack of differentiation of molariforms from premolariforms
Postcanine replacement	No replacement of molariforms	No replacement of molariforms (except for *Sinoconodon*)	Plesiomorphous as in eucynodonts (except for tritylodontids)	Alternating and multiple replacements (except for gomphodonts)
Replacement of antemolars	Diphyodont of all antemolars	Diphyodont of all antemolars (except for *Sinoconodon*)	Plesiomorphous (except for tritylodontids)	Alternating and multiple replacements
Cusp-supported wear facet on molariforms	Plesiomorphous as in the mammalian condition →	Present (except for *Sinoconodon*)	Absent (except for tritylodontids)	Absent (except for gomphodonts)
Jaw rotation during occlusion	Plesiomorphous as in the mammalian condition →	Present (in triangular trajectory)	Absent	Absent
Atlas	Fully fused atlas ring	Plesiomorphous as in mammaliamorphs →	Loss of altas postzygapophyses but unfused as atlas ring	Presence of atlas poszygopophyses
Axis	Plesiomorphous as in mammaliamorphs →	Plesiomorphous as in mammaliamorphs	Development of dens on axis (unknown in tritheledontids)	Absence of dens in axis
Cervical vertebral rib and transverse foramen	Rib fused to cervical vertebral body in most groups	Plesiomorphous as in eucynodonts →	Plesiomorphous as in eucynodonts →	Rib not fused to cervical body

(continued)

TABLE 3.1. Continued

Character	Mammalian crown condition[1]	Mammalian condition	"Mammaliamorph" precursor condition	Eucynodont precursor condition
Lumbar vertebrae	Without ribs[8]	Plesiomorphous as in eucynodonts →	Plesiomorphous as in eucynodonts →	With ribs
Sacral vertebrae	Two or three[9]	Two or three	Four or five	Four or five
Coracoid	Reduced to hooklike coracoid process (except for monotremes)	Plesiomorphous as in eucynodonts →	Plesiomorphous as in eucynodonts →	Large, forming half of the glenoid
Procoracoid	Absent in most crown mammals (present in monotremes but without foramen)	Plesiomorphous as in eucynodonts →	Plesiomorphous as in eucynodonts →	Procoracoid present, with a coracoid foramen
Scapular glenoid	Ball-in-socket (except for monotremes)	Plesiomorphous as in eucynodonts →	Plesiomorphous as in eucynodonts →	Saddle shaped
Supraspinous fossa	Presence of laterally facing supraspinous fossa (except for monotremes)	Plesiomorphous as in eucynodonts →	Supraspinous fossa on the cranial aspect of scapula	Absence of supraspinous fossa
Ilium	Plesiomorphous as in mammaliamorphs →	Plesiomorphous as in mammaliamorphs →	Triangular in cross section, anterodorsally oriented without posterior process	Platelike with relative large posterior process
Pubis	Plesiomorphous as in mammalian condition →	Reduced, length shorter than the acetabular diameter	Plesiomorphous as in eucynodonts →	Large, length longer than acetabular diameter

[1] By some cladistic phylogenies, eutriconodontans (e.g., *Triconodon*), and multituberculates are nested within the mammalian crown group. Given the present evidence, there is still a less probable and alternative placement in which multituberculates and eutriconodontans are excluded from the mammalian crown group.

[2] Fossilized Meckel's cartilage is present in the gobiconodontid *Repenomamus*, indicating the ectotympanic and malleus was still linked anterior to the mandible, although the two bones are separated mediolaterally from the mandible (Wang, Hu, Meng, et al., 2001).

[3] The pila antotica is present in multituberculates, which would constitute an exception if multituberculates are nested within the mammalian crown group. The pila antotica is absent in *Hadrocodium*, which is placed outside the mammalian crown group in our current phylogeny.

[4] The primitive condition of the mammalian crown group is an enlarged anterior lamina of petrosal. The absence of this structure is a secondarily derived condition within therians (Hopson and Rougier, 1993).

[5] The sphenoidal sinus shows considerable variation among extant mammals (Paulli, 1900).

[6] The presence of the ethmoid transverse lamina is only confirmed for *Sinoconodon* and *Morganucodon*.

[7] Confluence of external nares in all crown mammals, with the possible exception (if multituberculates lie within the crown) of the multituberculate *Lambdopsalis* (see Miao, 1988).

[8] Lumbar ribs are absent in all living mammals, though they are present in *Gobiconodon* (see Jenkins and Schaff, 1988).

[9] Sacral vertebrae are limited to two or three in most extant mammals except for placental xenarthrans.

of mammalian characters. The origin of mammals can be best elucidated by mapping the distribution of the primitive "cynodont-like" characters and the derived "mammal-like" characters onto a robust evolutionary tree (see chapter 15), as provided by many phylogenetic studies (Kemp, 1982; Hopson and Barghusen, 1986; Rowe, 1988; Luo, 1994; Rougier, Wible, and Hopson, 1996; Luo, Cifelli and Kielan-Jaworowska, 2001). The major hierarchies of this phylogenetic framework are explained below.

Relationships of Mammalian Groups. As mentioned in chapter 1, we define Mammalia as the clade including the common ancestor of *Sinoconodon*, extant mammals, and all of the fossil taxa nested within this group. During the first half of the twentieth century there was considerable debate as to whether mammals were polyphyletic (Simpson, 1928a, 1929a, 1945; Olson, 1959) or monophyletic (Gregory, 1910; Weber, 1927, 1928). But this issue has long since been settled by overwhelming evidence in favor of a monophyletic Mammalia (see also chapter 1). Here we consider Mammalia to be monophyletic, following the universal consensus of more recent workers on mammalian evolution (e.g., Hopson and Crompton, 1969; McKenna, 1975; Kemp, 1982). The group includes Rhaeto-Liassic taxa such as *Sinoconodon*, morganucodontans, haramiyidans, kuehneotheriids, and probably also the incompletely known Carnian taxa *Adelobasileus* and *Gondwanadon*, the Early Jurassic *Indozostrodon*, as well as the stem taxa nested between these Rhaeto-Liassic taxa and extant mammals. This definition corresponds to Mammaliaformes Rowe, 1988 (see also Rowe and Gauthier, 1992; McKenna and Bell, 1997). The mammalian crown group is defined by the common ancestor of monotremes (with their closely related stem australosphenidans and shuotheriids, Kielan-Jaworowska, Cifelli, and Luo, 2002; Luo et al., 2002), marsupials and placentals, plus all descendants of this common ancestor (Mammalia of Rowe, 1988; McKenna and Bell, 1997). The inclusion of eutriconodontans and multituberculates in the mammalian crown group is favored by strict parsimony of the currently available anatomical characters (see the phylogeny in figure 15.1), but the support for such inclusion is not overwhelmingly strong by comparison with their alternative placements outside the mammalian crown group (e.g., see figure 15.2). The mammalian crown group, with a minimum age of Middle Jurassic, is a subgroup of mammals.

Within the mammalian crown group, the therian crown group (marsupials + placentals) is nested within boreosphenidans (or northern tribosphenic mammals, Luo, Cifelli, and Kielan-Jaworowska, 2001; Luo et al., 2002). The earliest record of the therian crown group is dated to the Barremian age of the Early Cretaceous (Ji et al., 2002), and the history of boreosphenidans can be traced to the Berriasian age of the earliest Cretaceous (Sigogneau-Russell et al., 2001). The boreosphenidan clade belongs to Cladotheria, which have a geological history extending to the Middle Jurassic. The cladotherian clade is a part of Trechnotheria (McKenna, 1975; Prothero, 1981; Sigogneau-Russell and Ensom, 1998), whose earliest members also date back to the Middle Jurassic.

Relations of Mammals to Major Nonmammalian Cynodont Groups. An understanding of the earliest anatomical evolution of mammals depends on a larger phylogenetic framework that must include nonmammalian cynodonts, in which mammals are nested. The relationship of mammals to the successively more distant cynodont groups is crucial for assessing the evolutionary patterns of the earliest mammals.

For this review of anatomical evolution, we adopt the following phylogenetic hierarchies (see figures 15.1, 15.2). First, Mammalia are nested within Mammaliamorpha, a clade defined by the common ancestor of tritylodontids, tritheledontids, and mammals. The members of this clade share many derived characters (apomorphies) in the skull and skeleton (Kemp, 1983; Rowe, 1988; Luo et al., 2002; but see different opinions by Crompton, 1972a; Sues, 1985b; Hopson and Barghusen, 1986; Hopson and Kitching, 2001). We consider the tritheledontids and tritylodontids to be more closely related to mammals than other advanced cynodonts (Rowe, 1988; Wible, 1991; Luo, 1994). The earliest representative of this clade is of Late Triassic age (Bonaparte, 1980). Within mammaliamorphs, we follow the strong consensus that tritheledontids are the sister taxon to mammals, a hypothesis that was first expounded by Hopson and Barghusen (1986) and has since been supported by a long series of more recent studies (e.g., McKenna, 1987; Shubin et al., 1991; Crompton and Luo, 1993; Luo, 1994; Hopson and Kitching, 2001).

In a successively more distant order, mammaliamorphs are nested within Eucynodontia, an advanced cynodont clade defined by the common ancestor of *Cynognathus* and mammals (see Hopson and Barghusen, 1986; Hopson and Kitching, 2001). The eucynodont clade appeared in the fossil record by the Middle Triassic (Kemp, 1982; Rowe, 1993). Eucynodonts, *Thrinaxodon* and galesaurids belong to the epicynodont clade (*sensu* Hopson and Kitching, 2001), whose history extends to the Early Triassic (Kemp, 1982; Rowe, 1993). Epicynodonts, *Procynosuchus*, and *Dvinia* form the clade Cynodontia (Kemp, 1982; Hopson and Barghusen, 1986). The origin of the cynodont clade has been dated to the Late Permian (Kemp, 1982).

When dromatheriids and tritylodontids were still represented by very incomplete fossils at the beginning of the twentieth century, these groups were designated by some as mammals (Owen, 1884; Osborn, 1886a,b, 1887b; Broom, 1910, 1914). As recently reviewed by Luo et al. (2002), tritylodontids may be related to mammals, but are not mammals themselves. By the currently much better evidence, dromatheriids are only remotely related to mammals at best, as reemphasized recently by Sues (2001) and Bonaparte and Barberena (2001). Two other groups were once proposed as being closely related to mammalian ancestry: Thrinaxodontidae (Hopson, 1969; Hopson and Crompton, 1969; Barghusen and Hopson, 1970; Fourie, 1974) and Probainognathidae (Romer, 1970a; Crompton and Jenkins, 1979; Hopson, 1994). The proposal for a close relationship of *Thrinaxodon* to mammals is not supported by any of the more recent studies (Kemp, 1983; Crompton and Sun, 1985; Hopson and Barghusen, 1986; Rowe, 1988), and the proposal for a close relationship between probainognathids and mammals, to the exclusion of tritylodontids and tritheledontids, also presents some difficulties, as pointed out by Kemp (1983). Most recently, Hopson and Kitching (2001) proposed that the Probainognathidae are a sister taxon to the clade of tritheledontids and mammals and that probainognathids, tritheledontids, and mammals belong to a clade of probainognathians.

A persistent disagreement concerning the relationships of advanced cynodonts centers on the placement of the highly specialized and herbivorous tritylodontids, either from the more traditional cladistic phylogenies (Kemp, 1983, 1988; Sues, 1985b; Hopson and Barghusen, 1986; Hopson, 1991) or from matrix-based parsimony analyses (Rowe, 1988, 1993; Hopson and Kitching, 2001). Nonetheless, there is a consensus on many clades in cynodont phylogeny, as outlined above.

EVOLUTION OF THE SNOUT

PALATAL STRUCTURES
(figure 3.1)

Basal Cynodont Condition. The bony secondary palate, formed primarily by the maxilla and palatine in mammals, separates the nasal cavity, used in olfaction and respiration, from the oral cavity, which is concerned primarily with feeding (figures 3.1, 3.2). The posterior edge of the bony secondary palate extends to the midlevel of the postcanine row in basal cynodonts, forming an arch-shaped margin (Kemp, 1982; Hopson and Barghusen, 1986). By comparison, the secondary palate is more fully developed, with a straight margin, and is extended to the posterior end of the postcanine rows in such stem mammals as *Sinoconodon*, *Morganucodon*, and *Haldanodon*. It has been hypothesized by various authors that the posterior extension of the bony secondary palate may be correlated either with a better respiratory function while ingesting food (Lessertisseur and Sigogneau, 1965; Romer, 1970b; Starck, 1982; Hillenius, 1994) or with a better capacity for suckling and swallowing food (Maier, 1999). It is possible that the posterior extension of the secondary hard palate was related to both functions.

It is generally accepted that a soft secondary palate continued posteriorly from the bony secondary palate in such cynodonts as *Thrinaxodon* (figure 3.1B). Barghusen (1986) and Crompton (1995) suggested that the pterygoid transverse process (or the hamulus) was reduced and slightly separated from the skull in the tritheledontid *Pachygenelus* and the mammal *Morganucodon* (figure 3.1E,F), so that the muscles in the lateral fossa of the pterygoid transverse process could invade the soft palate to muscularize the velum in these derived taxa. Some authors have speculated that these muscles would make it possible for some mammal-like functions of swallowing and suckling (e.g., Maier, 1999), although the additional pharyngeal arch structures necessary for modern mammalian functions are still unknown from the fossil record (Smith, 1992). Maier (1999) went further to propose that these derived mammalian functions of the soft palate and related velum could have occurred in basal cynodonts or even in advanced precynodont therapsids.

Maier (1999) hypothesized that the closure of the maxillary secondary palate may have evolved separately in precynodont therocephalians and in cynodonts (including mammals). In primitive therocephalians from the Permian, the maxillary secondary bony palate is not fully developed (Hopson and Barghusen, 1986; Hillenius, 1994; Maier et al., 1996). In the derived Triassic taxa of therocephalians with a maxillary secondary palate, there is no incisive fissure between the premaxilla and maxilla. Based on this evidence, Maier suggested that the complete maxillary secondary palate would have started in the anterior part of the palate and then proceeded in an anteroposterior direction within the therocephalian clade. By contrast, in the cynodont-mammal clade, the maxillary secondary palate may have developed first in the middle portion of the palate and then proceeded in the posteroanterior direction (figure 3.1). Maier's hypothesis implies that the incisive foramen in the secondary palate of the cynodont-mammal clade is a vestigial structure for retention of the nasopalatine (incisive) duct, which is a conduit between the oral cavity and the vomeronasal organ inside the nasal cavity, and possibly serves an olfactory function (Hillenius, 2000). Presence of the incisive foramen and the greater palatine foramen on the bony palate therefore may be apomorphies for cynodonts (Hopson and Barghusen, 1986).

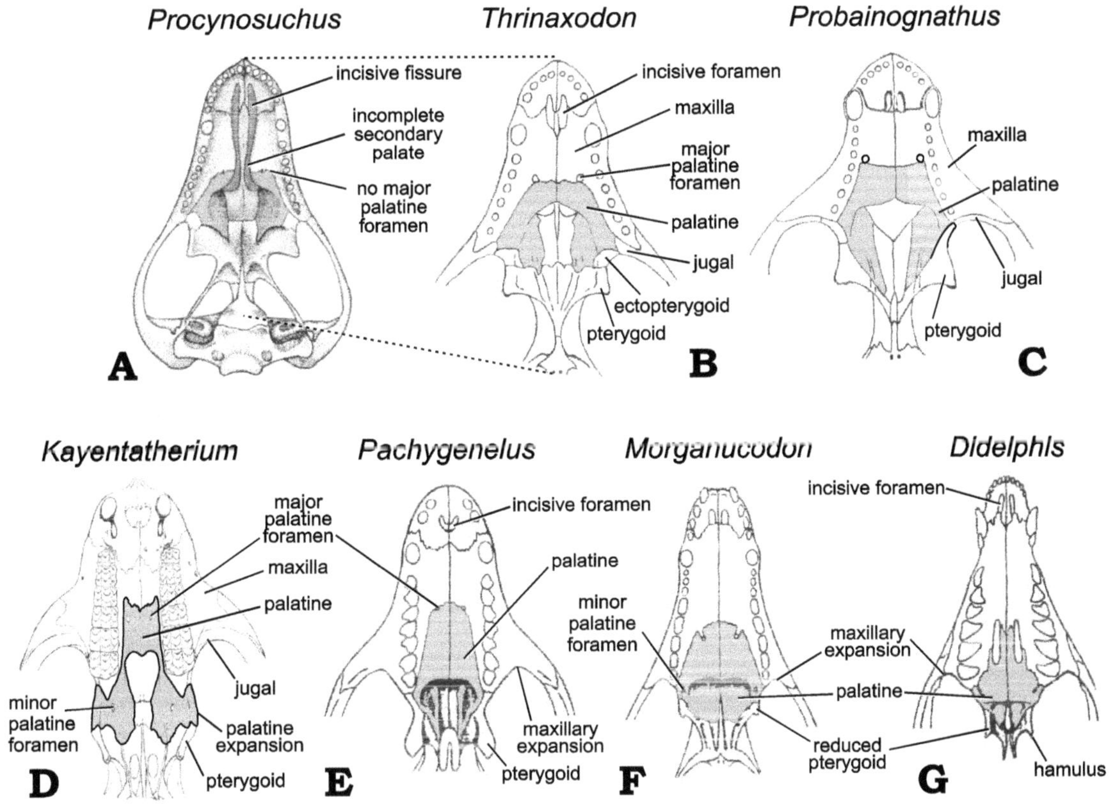

FIGURE 3.1. Pattern of the secondary bony palate in nonmammalian cynodonts and mammals. A, *Procynosuchus.* B, *Thrinaxodon.* C, *Probainognathus.* D, *Kayentatherium.* E, *Pachygenelus.* F, *Morganucodon.* G, *Didelphis.* Acquisition of apomorphies: fully developed maxillary bony palate, formation of the incisive foramen, formation of a short palatine bony palate, and presence of the greater palatine foramen: the clade of nonmammalian epicynodonts (*sensu* Hopson and Barghusen, 1986; Rowe, 1988: *Thrinaxodon* through *Didelphis*). Palatine bony palate extending near the posterior end of the tooth row: Mammaliamorpha (*sensu* Rowe, 1988: tritylodontids, tritheledontids, and mammals). Participation of the maxilla in the subtemporal border (at the expense of the anterior process of the jugal): the clade of tritheledontids + mammals (Hopson and Barghusen, 1986; Luo, 1994). Participation of the palatine bone in the subtemporal border; the lesser palatine foramen enclosed by the palatine: mammals (although also present in tritylodontids). Not to scale. *Source:* A, modified from Maier (1999); B, C, E, F, from Allin and Hopson (1992); D, from Sues (1986); G, original.

The palatine forms the posterior part of the secondary bony palate in cynodonts and mammals (figure 3.1). The palatine part of the secondary bony palate has several character states in nonmammalian cynodonts and mammals (Kemp, 1983; Sues, 1985a; Hopson and Barghusen, 1986; Maier et al., 1996). In such basal cynodonts as *Thrinaxodon,* the palatines meet in the midline of the palate, forming the posterior edge of the internal choana; but the palatine's contribution to the secondary bony palate is small relative to that of the maxilla. In the derived condition of tritylodontids and tritheledontids, the palatine makes up a far greater proportion of the palate, up to 50% of the length of the postcanine row; the posterior (choanal) margin of the palatine extends near the posterior end of the tooth row.

In *Probainognathus* and tritheledontids, the palatine is excluded by the neighboring bones from the subtemporal margin in the orbital wall (Luo, 1994). The palatine is small in the lateral view of the skull (figures 3.2A, 3.3A). The palatine is so much larger in tritylodontids that it expands onto the subtemporal margin of the orbit and displaces the transverse flange ("hamulus") of the pterygoid posteriorly (figure 3.3B) (Kemp, 1983; Sues, 1986; Luo, 1994). Both these features are derived characters shared by *Sinoconodon, Morganucodon, Haldanodon,* and the mammalian crown group (Kemp, 1983; Rowe, 1988; Lillegraven and Krusat, 1991; Luo, 1994).

NASAL STRUCTURES
(figure 3.2)

The posterior wall of the nasal cavity in mammals is formed by the maxilla, the orbital (ascending) process of the palatine, and the orbitosphenoid (Jollie, 1962; No-

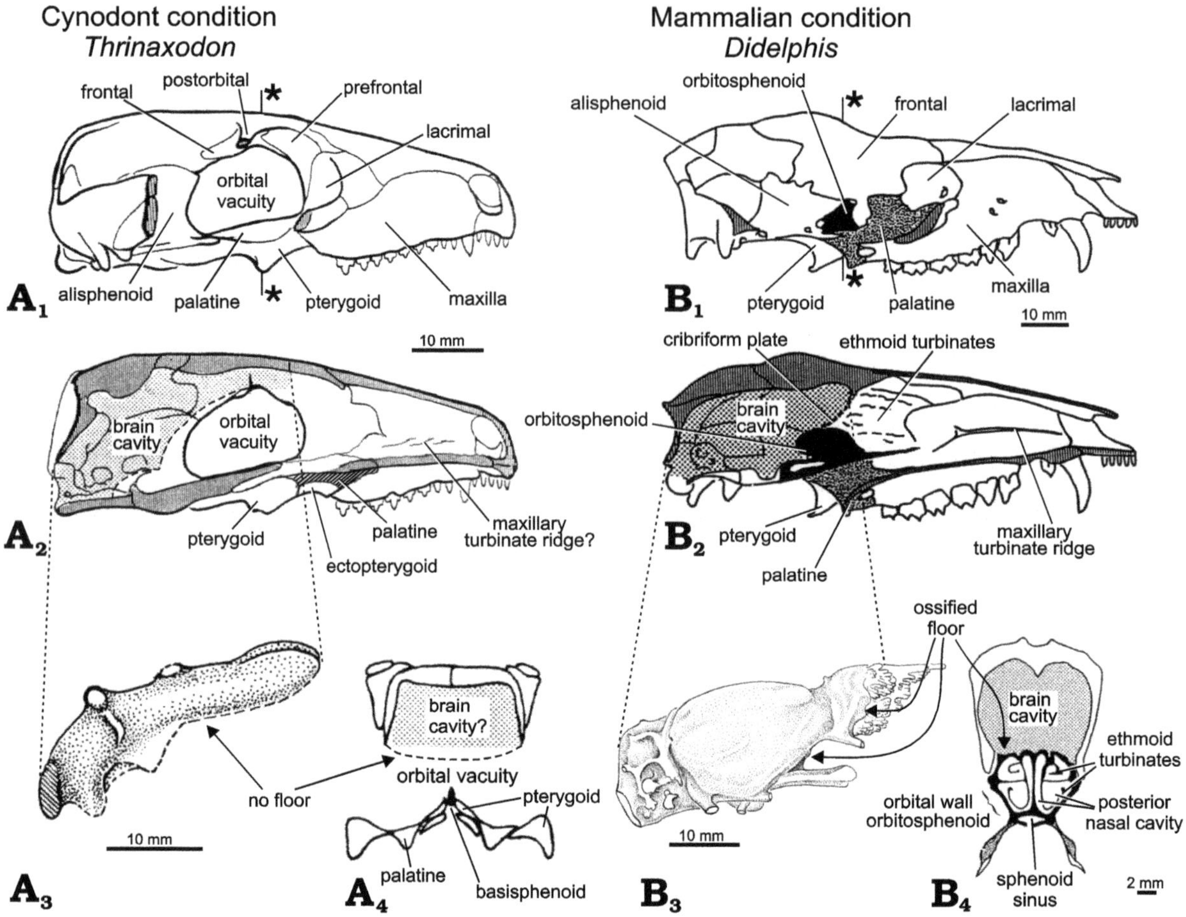

FIGURE 3.2. Comparison of the sphenoidal complex and the anterior braincase of nonmammalian cynodonts and mammals. A, The nonmammalian cynodont *Thrinaxodon:* lateral view (A_1); sagittal section of the skull with shading indicating the extent and position of the brain (A_2); reconstruction of brain endocast (A_3); transverse section at the level marked by asterisk in A_1 (A_4). B, Marsupial *Didelphis:* lateral view (B_1); sagittal section of the skull, with shading indicating the extent and position of the brain (B_2); brain in lateral view (B_3); transverse section at the level marked by asterisk in B_1 (B_4). Primitive cynodont features (exemplified by *Thrinaxodon*): presence of the orbital vacuity and absence of an ossified medial orbital wall; no ossified orbitosphenoid floor to the anterior braincase (dashed line); absence of the transverse lamina in the nasal cavity. Derived mammalian features (exemplified by *Didelphis*): ossified medial orbital wall and complete orbitosphenoid floor for the anterior braincase; presence of the transverse lamina in the nasal cavity and ossification of ethmoidal structures, such as the cribriform plate and ethmoturbinates. The ethmoidal structures are known only in some stem lineages of mammals. *Source:* A_1, A_2, A_4, from Fourie (1974); A_3, original, based on a specimen of the Institute of Paleobiology, Warsaw; B_1, B_4, original; B_2, modified from Hillenius (1994); B_3, modified from Rowe (1996a).

vacek, 1986b; Hurum, 1994; Evans, 1995). By comparison, in the primitive condition of such cynodonts as *Thrinaxodon* (figure 3.2A) and *Probainognathus* (figure 3.3A), the orbital plate of the palatine is small. It contributes to the posterior wall of the nasal cavity but is not large enough to fully enclose the nasal cavity. Only in tritylodontids is this part of the posterior nasal wall fully formed by the orbital process of the palatine plus the orbital process of the frontal, as well as a fully ossified orbitosphenoid (figure 3.3B), as described by many authors (Kühne, 1956; Sun, 1984; Sues, 1986). The tritylodontid condition (figure 3.3B) is very similar to the derived mammalian condition (Kermack et al., 1981; Lillegraven and Krusat, 1991; Hurum, 1994; Luo, 1994).

Within the nasal cavity, the vomer and the internasal septum are the only internal structures consistently preserved in the cynodonts and stem mammals that have been studied (Hillenius, 1994; Hurum, 1994). The other major feature is the turbinate ridge on the inside of the facial portion of the maxilla, which supports the maxillary turbinates. The maxillary turbinate ridge is best developed in therian mammals and is also present in the monotreme *Tachyglossus* (Kuhn, 1971). Their related maxillary turbinates are important for thermoregulation (Paulli, 1900;

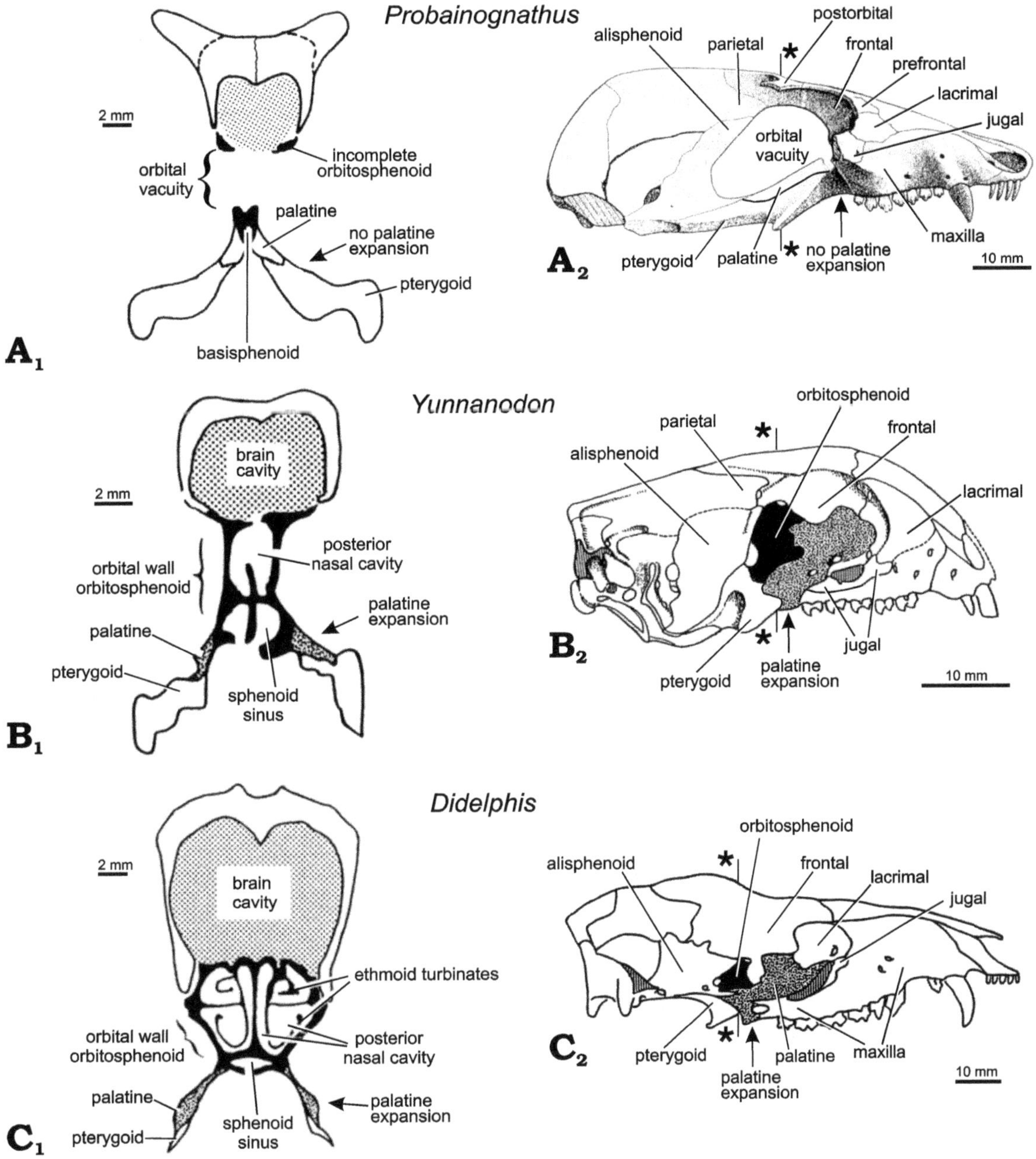

FIGURE 3.3. Structures of the medial orbital wall, posterior nasal cavity, and anterior braincase of nonmammalian cynodonts and mammals. A, Cynodont *Probainognathus*. B, Tritylodontid *Yunnanodon*. C, Marsupial *Didelphis*. A_1, B_1, and C_1 are sections across the brain cavity at the level of the pterygoid bone; A_2, B_2, and C_2 show lateral view of the skulls. *Yunnanodon* and mammals are characterized by the derived conditions of full ossification of the sphenoid floor for the anterior braincase, expansion of the palatine on the subtemporal margin, reduction of the pterygoid, and posterior displacement of the pterygoid hamulus by the palatine. Not to scale. *Source:* A_1, original (section from a broken skull in Museum of Comparative Zoology); A_2, modified from Romer (1970a); B_1, B_2, based on original section of a skull, courtesy of Ai-Lin Sun; C_1, C_2, original.

Moore, 1981; Bennett and Ruben, 1986; Hillenius, 1992). Hillenius (1994) suggested that the maxillary turbinate ridge first developed in precynodont therapsids. This primitive structure is retained in various conditions in several nonmammalian cynodonts, *Morganucodon* (see Kermack et al., 1981), and some multituberculates (Miao, 1988; Hurum, 1994). The presence of the maxillary ridge has been regarded as indirect evidence for the development of maxillary turbinates. Based on this evidence, Hillenius (1994) inferred that thermoregulatory adaptation was developed in stem mammals, cynodonts, and even precynodont therapsids.

Of other nasal structures in cynodonts, Hillenius (2000) hypothesized that the nasolacrimal duct entered the nasal cavity from the lacrimal foramen in the orbit, in close association with the maxillary turbinate ridge, then passed through the septomaxillary foramen to connect with the vomeronasal organ in the nasal cavity. The nasolacrimal fluid passing through the duct would supply the vomeronasal organ. In stem mammals with a septomaxilla but without a septomaxillary canal and in extant mammals without a septomaxilla, the nasolacrimal duct would become disconnected from the vomeronasal organ, and nasolacrimal fluids would be rerouted in the nasal cavity. It is likely that some remnants of the paraseptal structures associated with the vomeronasal organ, as seen in extant mammals (Sánchez-Villagra et al., 2002), are also present in some stem mammals, albeit in very incomplete condition (Z-XL, pers. obs.).

ORBITOSPHENOIDAL AND ETHMOIDAL STRUCTURES

Condition in Mammalian Crown Group (figure 3.3). In living mammals, the sphenoidal complex is developed from several bones: the basisphenoid, the orbitosphenoid, the presphenoid, and the alisphenoid (Paulli, 1900; Jollie, 1962; Kuhn, 1971; Zeller, 1989b; Maier, 1993). Sutures between the orbitosphenoid, presphenoid, and ethmoidal elements are often fused in adult skulls, making the sphenoidal complex an integral structure of the cranium.

The orbitosphenoid is an important element for mammalian cranial structure because it forms the floor of the frontal region of the braincase, the medial wall of the orbit, and part of the posterior wall for the ethmoid recess of the nasal cavity (Maier, 1993), which is a blind space or extension in the posterior part of the nasal cavity (figure 3.2B). The ethmoidal recess is bounded posteriorly by the cribriform plate (the lamina cribrosa) and ventrally by the posterior transverse lamina, which separates the recess above from the nasopharyngeal passage below. The ethmoidal recess, filled with ethmoturbinates, provides the function of retaining air for olfaction (Negus, 1958; Kuhn, 1971; Moore, 1981; Maier, 1993; Hillenius, 1994). These derived ethmoidal structures contribute to the strong development of olfactory function in mammals (Zeller, 1999b).

A varying degree of pneumatization, or secondary development of hollow spaces within bones, may excavate posteriorly from the ethmoidal recess into the base of the chondrocranium to form the sphenoid paranasal sinus within the presphenoid portion of the sphenoidal complex in extant mammals (Paulli, 1900; Cave and Haines, 1940; Moore, 1981). The development of a sphenoidal paranasal sinus may be variable in different orders and families of modern mammals.

Condition in Primitive Cynodonts. The orbitosphenoid is not fully developed and does not form a complete floor for the anterior braincase in most nonmammalian cynodonts, including tritheledontids (figures 3.2, 3.3). The orbitosphenoid is absent in *Thrinaxodon* (see Fourie, 1974), although it is partially ossified in *Exaeretodon* (see Bonaparte, 1966), *Massetognathus*, and *Probelesodon*. In those cynodonts in which the orbitosphenoid is partially ossified, it does not form a complete medial wall of the orbit, leaving a large orbital vacuity in the lateral view of the skull (Luo, 1994).

Condition in Tritylodontids and Early Fossil Mammals. The orbitosphenoid in tritylodontids is significantly different from the primitive cynodont condition (Hopson, 1964; Sues, 1986; Luo, 1994). This has phylogenetic implications, as tritylodontids are an advanced cynodont group that, according to some authors, may be close to mammals (e.g., Kemp, 1983; Rowe, 1988; Wible, 1991). Studies of serial sections and CT scans of the sphenoidal region and rostrum of tritylodontids (figure 3.3B), *Sinoconodon, Morganucodon, Hadrocodium,* and multituberculates revealed some orbital and nasal structures that are intermediate between those of primitive cynodonts and extant mammals (Kielan-Jaworowska et al., 1986; Hurum, 1994, Z-XL, pers. obs.).

The orbital wall in tritylodontids is more similar to those of mammals than other nonmammalian cynodonts (figure 3.3B, also see Hopson, 1964; Sues, 1986; Witschel, 1999). Formed by the orbitosphenoid and the orbital process of the palatine, the medial orbital wall encloses the orbital vacuity seen in other cynodonts (figure 3.3). The fused orbitosphenoid and presphenoid form the ventral floor to the frontal region of the braincase and the posterior wall to the ethmoidal recess of the nasal cavity. These are derived characters otherwise known only for modern therians (Novacek, 1986b; Maier, 1993; Evans, 1995) and the monotreme *Tachyglossus* (Kuhn, 1971). However, the posterior transverse lamina of the ethmoid, a typical feature of living mammals, cannot be confirmed with certainty in either *Yunnanodon* or *Tritylodon.* Tritylodontids have an incomplete ethmoidal internasal septum, a feature that is not developed at all in *Probainognathus, Probelesodon,* and *Massetognathus.*

Morganucodon has developed the posterior transverse lamina of the ethmoid that is the ventral floor of the ethmoidal recess (Z-XL, pers. obs.). The lamina is a derived feature that is not present in any nonmammalian cynodonts, including tritylodontids. Other aspects of the orbital and nasal structures are essentially the same in *Sinoconodon* and *Morganucodon* as they are in tritylodontids.

Transformation of Orbital and Nasal Structures. Several steps of transformation can be recognized by mapping of the previously mentioned orbital and nasal characters among nonmammalian cynodonts and mammals (figures 3.3, 3.4). Tritylodontids and stem mammals have a fully ossified orbitosphenoid wall and a large palatine. The posterior wall of the nasal cavity is completely formed. None of these apomorphies is present in most other cynodonts. *Morganucodon* and multituberculates are more derived than tritylodontids in their development of the posterior transverse lamina that separates the ethmoidal recess from the rest of the nasal cavity and in the more expanded and wider orbitosphenoid floor for the frontal braincase. The internasal septum may be present but is normally incomplete in *Yunnanodon, Sinoconodon,* and *Morganucodon* (Z-XL, pers. obs.); however, it is better developed in multituberculates (Hurum, 1994). A paranasal sinus can be developed in the sphenoid structure at least in *Yunnanodon* (figures 3.3B, 3.4B), some stem mammals, and multituberculates, although the distribution of the sphenoid sinus structures among these taxa is homoplastic. In serial sections or CT scans of *Sinoconodon, Morganucodon,* and multituberculates, some remnants of the ethmoidal structures are present in the posterior part of the nasal cavity. However, these taxa did not develop a fully ossified cribriform plate, or any fully ossified ethmoidal turbinate (Kielan-Jaworowska et al., 1986; Hurum, 1994, Z-XL, pers. obs.).

The cribriform plate, the fully ossified ethmoidal internasal septum, and ethmoturbinates are apomorphies of the mammalian crown group (figure 3.4), as they have been documented in *Tachyglossus* and numerous living therians (Paulli, 1900; Cave and Haines, 1940; Kuhn, 1971; Novacek, 1986b). Thus the complete separation of the nasal cavity and the frontal braincase by the cribriform plate is the last major apomorphy to evolve among many bony features that are crucial for modern mammalian olfactory and respiratory functions (Hillenius, 1992, 1994, 2000; Maier, 1993; Maier et al., 1996). The only significant exception among living mammals is *Ornithorhynchus,*

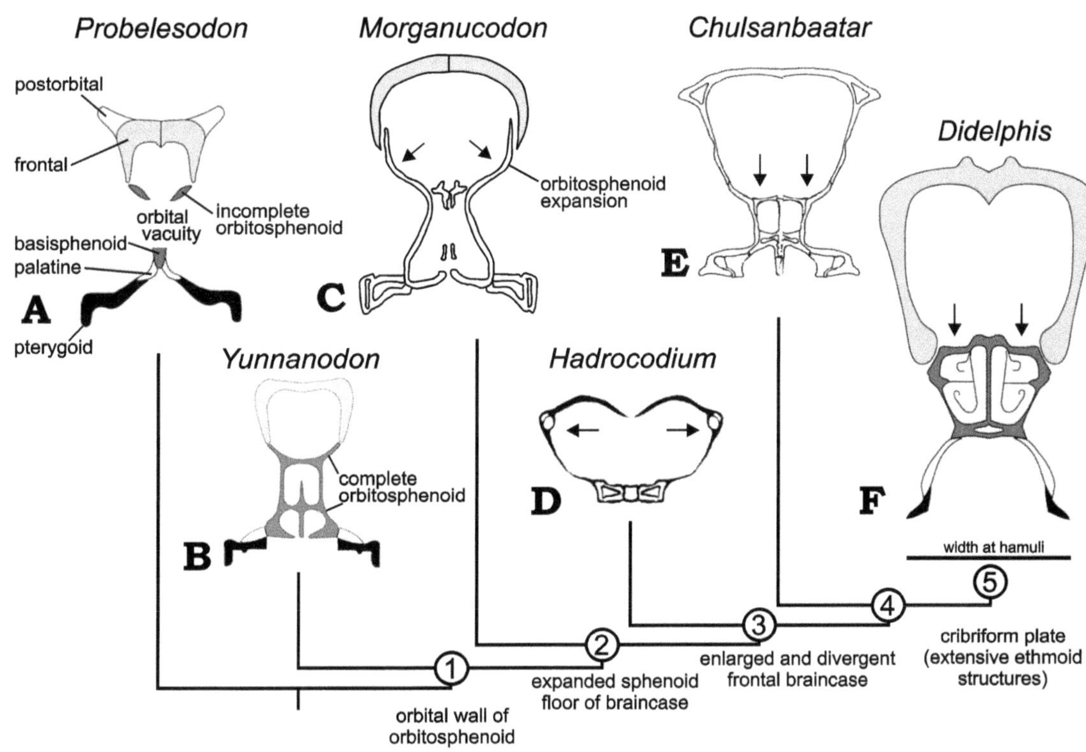

FIGURE 3.4. Comparison of the sphenoethmoidal floor and the expansion of anterior braincase in cynodonts and mammals (illustrated by transverse section through or near the pterygoid hamulus). A, *Probelesodon.* B, *Yunnanodon.* C, *Morganucodon.* D, *Hadrocodium.* E, *Chulsanbaatar.* F, *Didelphis.* Transverse sections of different skulls standardized to the pharyngeal width at the pterygoid hamuli. Arrows within brain cavity indicate the direction of expansion of the anterior braincase. Acquisition of apomorphies—Node 1: ossification of most sphenoid structures. Node 2: expansion of the floor of the anterior braincase. Node 3: enlargement and divergence of the frontal lobe of cerebrum. Node 4: expansion of the olfactory tubercle region and more extensive orbitosphenoid floor of the braincase. Node 5: full ossification of the cribriform plate as the floor to the olfactory lobe, better ossification of all ethmoid structures. Not to scale. *Source:* A, B, C, D, and F, original, based on skulls, sections, and/or CT scans; E, from Hurum (1994).

which has secondarily reduced the cribriform plate, the posterior transverse lamina, and the ethmoturbinates, probably owing to aquatic adaptation. The loss of these olfactory structures may be compensated by the specialized sensory functions in the duckbill via the trigeminal nerves (Zeller, 1988).

CHANGES IN THE LATERAL WALL OF THE BRAINCASE

ALISPHENOID-PETROSAL REGION

Condition in Nonmammalian Cynodonts (figure 3.5). The lateral wall of the braincase, including the cavum epipterícum for the trigeminal ganglion, shows three distinctive patterns among nonmammalian cynodonts, monotremes, various Mesozoic mammal groups, and the therian crown group (Watson, 1916; Kermack and Mussett, 1958a; Kermack, 1963; Hopson, 1964; Kermack and Kielan-Jaworowska, 1971; Crompton and Jenkins, 1979; Kielan-Jaworowska et al., 1986; Hopson and Rougier, 1993; Wible and Hopson, 1993, 1995). In nonmammalian cynodonts (figure 3.5), the ascending process of the alisphenoid (or epipterygoid) forms most of the side wall of the braincase between the orbital vacuity, where the ophthalmic branch of the trigeminal nerve (V_1) exited the braincase, and the large trigeminal foramen for the maxillary (V_2) and mandibular (V_3) branches of the trigeminal nerve. The trigeminal foramen ($V_{2,3}$) is situated at the suture of the alisphenoid and the anterior lamina of the prootic (or "petrosal"). The cranial portion of the squamosal is anteroposteriorly narrow and is excluded from the primary braincase wall.

Condition in Mammalian Crown Group. In the therian crown group, the ascending process of the alisphenoid is large and encircles the foramen for the mandibular branch of the trigeminal nerve (Kuhn and Zeller, 1987a; Maier, 1987; Zeller, 1987). The ascending process of the alisphenoid borders posteriorly on the cranial portion of the squamosal, which is a part of the sidewall of the braincase. The petrosal lacks an equivalent to the anterior lamina of nonmammalian cynodonts and other Mesozoic mammals. The petrosal does not border on the orbitosphenoid anteriorly, as in monotremes (see later).

In monotremes, the anterior lamina of the petrosal is formed by the fusion of the embryonic lamina obturans to the otic capsule, the embryonic precursor of the adult petrosal (Watson, 1916; Kuhn, 1971; Zeller, 1989b; Hopson and Rougier, 1993). This large anterior lamina bounds laterally the foramen of the mandibular branch of the trigeminal nerve and extends anteriorly to border on the orbitosphenoid. The cranial portion of the squamosal is superficial to the chondrocranial bones, such as the petrosal and occipitals, and is not part of the primary wall of the braincase (Kuhn, 1971; Kuhn and Zeller, 1987b).

The latest consensus is that the ascending process of the alisphenoid and the anterior lamina of the petrosal are distinctive structures (e.g., Kuhn and Zeller, 1987a; Maier, 1987; Hopson and Rougier, 1993). However, historically, there have been different interpretations of these structures in extant mammals. Watson (1916) suggested that monotremes are distinctive in that the lateral wall of the braincase is formed by the intramembranous ossification of the anterior lamina of the petrosal. By contrast, he characterized the therian group, consisting of marsupials, placentals, and their close allies, as having the lateral wall of the braincase formed by the alisphenoid bone (instead of the petrosal). The anterior lamina of the petrosal has been used to support grouping of several clades of Mesozoic mammals in the "Prototheria" (e.g., Kermack, 1963; Hopson, 1964; Kermack and Kielan-Jaworowska, 1971; but see MacIntyre, 1967), as distinguished from therians. Watson's characterization of the lateral wall of the braincase in extant mammals was challenged by a later observation by Presley and Steel (1976; Presley, 1980, 1981). Presley pointed out that it is formed within the sphenoobturator membrane in both monotremes and marsupials. He noted that the ossification of the lateral wall within the sphenoobturator membrane began with several ossification centers. These ossifications may be fused either with the otic capsule, as in the anterior lamina of the petrosal of monotremes, or alternatively with the ala temporalis, as in the ascending process of the alisphenoid. The early embryonic development appears to be the same for both living therians and monotremes. Hence the distinction of the braincase structure between adult therians and monotremes appears to be far less significant than previously thought (Watson, 1916). Further and more detailed embryological studies (Kuhn and Zeller, 1987b; Maier, 1987, 1989; Zeller, 1989b) show that there is a clear difference between therians and monotremes in the timing and sequence of fusion of these embryonic structures to those that surround them (see review by Hopson and Rougier, 1993).

Most stem mammals, such as *Adelobasileus, Sinoconodon, Morganucodon,* and *Hadrocodium,* have an enlarged anterior lamina of the petrosal (Kermack, 1963; Kermack et al., 1981; Crompton and Luo, 1993; Hopson and Rougier, 1993; Lucas and Luo, 1993; Wible and Hopson, 1993; Luo, Crompton, and Sun, 2001). At least one of the foramina for the trigeminal nerve is encircled by the anterior lamina. Among the fossil lineages of the mammalian crown group, triconodontids have retained the morphology of the anterior lamina encircling the trigeminal foramen

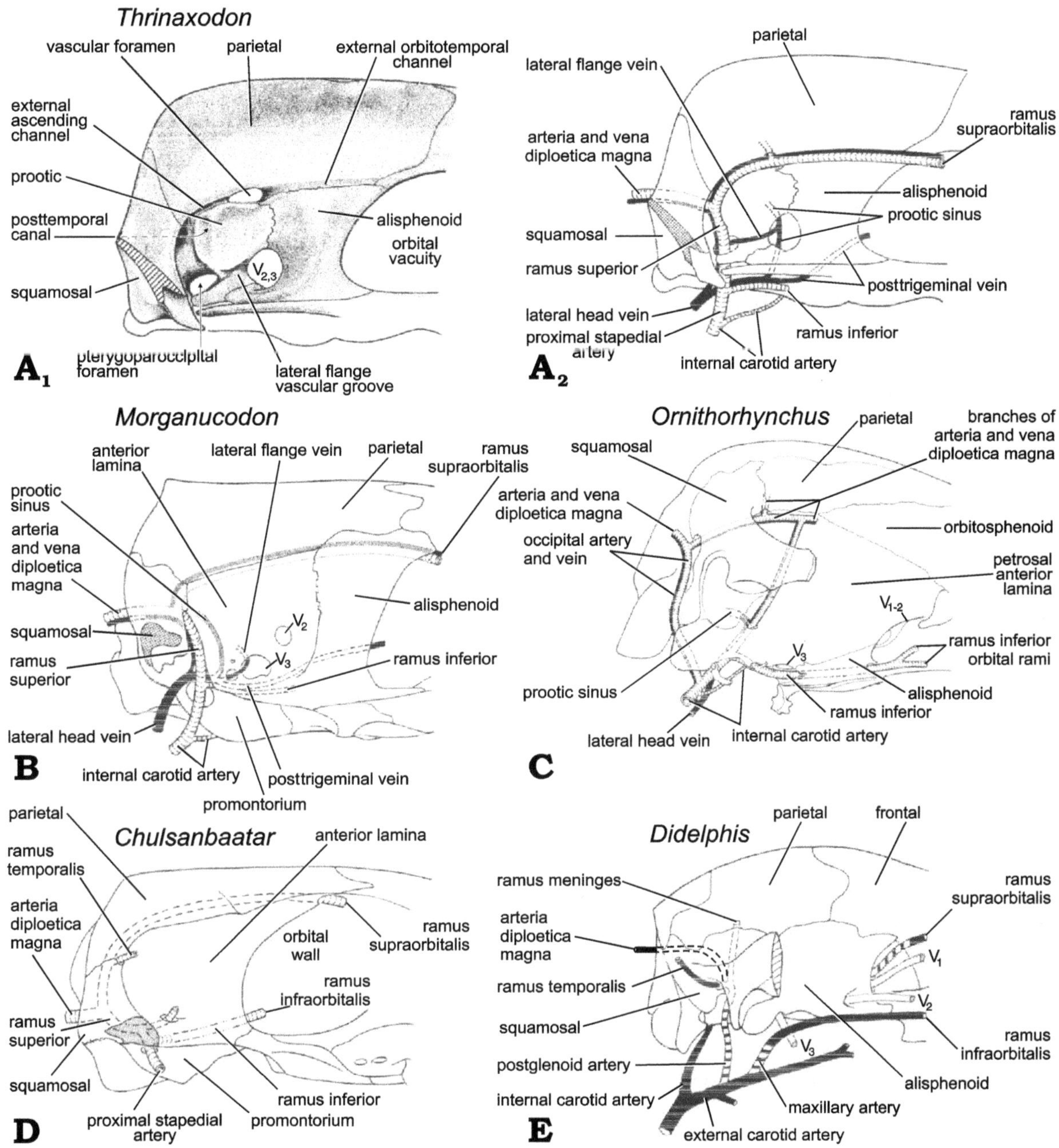

FIGURE 3.5. Structure of the lateral wall of the braincase in the region of the alisphenoid and petrosal and vascular reconstruction of the temporal region in cynodonts and mammals (vessel homology following Wible, 1987, and Wible and Hopson, 1995). A, *Thrinaxodon;* sidewall of braincase in petrosal and alisphenoid regions (A_1); vascular reconstruction (A_2). B, *Morganucodon.* C, *Ornithorhynchus.* D, *Chulsanbaatar.* E, *Didelphis.* See also figure 6.5 for structure of the lateral wall of the braincase in *Morganucodon, Ornithorhynchus,* and *Chulsanbaatar;* figure 10.14A_2 for the lateral wall of *Vincelestes;* and figure 10.15B for cranial vasculature in *Vincelestes.* The arteria and vena diploetica magna in the posttemporal canal and their tributary branches supplying the orbitotemporal region are primitive features of nonmammalian cynodonts. The open external channel for the ascending artery (a branch of arteria diploetica) and its associated vein is a primitive cynodont feature. The enclosure of the ascending artery and vein is a derived feature of most mammals (except for *Ornithorhynchus*). Abbreviations: V_{1-2}, foramen for ophthalmic and maxillary branches of the trigeminal nerve; V_2, V_3, foramen for the maxillary and mandibular branches of the trigeminal nerve. Not to scale. *Source:* A–D, adapted from Rougier et al. (1992) and Wible and Hopson (1995); E, modified from Wible (1987).

(Kermack, 1963, also see chapter 7), as does the stem therian *Vincelestes* (see Rougier et al., 1992). Multituberculates have multiple trigeminal foramina within the anterior lamina (Simpson, 1937a; Kielan-Jaworowska, 1971; Sloan, 1979; Miao, 1988; Luo, 1989; Wible and Hopson, 1995; Hurum, 1998a; Wible and Rougier, 2000). A large anterior lamina encircling one or more trigeminal foramina is a derived mammalian feature by comparison with nonmammalian cynodonts. However, it is a primitive character within Mammalia (Hopson and Rougier, 1993).

The anterior lamina of the petrosal is present but reduced in the Early Cretaceous eutherian *Prokennalestes* (Wible et al., 2001, see also figure 13.6B). In the Late Cretaceous eutherians for which complete skulls are known (Kielan-Jaworowska and Trofimov, 1980; Kielan-Jaworowska, 1981; McKenna et al., 2000, see figures 13.2, 13.5, 13.6A), the anterior lamina is absent, as it is in other eutherians (MacIntyre, 1972; MacPhee, 1981; Cifelli, 1982; Novacek, 1986b; Zeller, 1987; Geisler and Luo, 1998). In all metatherians for which the petrosal is known, the anterior lamina is absent (Clemens, 1966; Wible, 1990; Marshall and Muizon, 1995; Muizon, 1998, see also chapter 12 and figures 12.1–12.3). The trigeminal foramen is formed by the posterior extension of the alisphenoid, either in part or in entirety. These are derived characters known so far only for the therian crown group (Wible, 1990; Hopson and Rougier, 1993; Marshall and Muizon, 1995; Gaudin et al., 1996).

Homology of Ascending Process of Alisphenoid. There has also been an evolution in opinions about the homology of the ascending process of the therian alisphenoid. The majority of workers accept that the ascending process of the epipterygoid of nonmammalian amniotes is homologous with the ascending process of the alisphenoid (Goodrich, 1930; Maier, 1987, 1989; Hopson and Rougier, 1993). However, an alternative view was espoused by Gaupp (1902), Watson (1916), and de Beer (1926), who recognized that the lamina ascendens of the alisphenoid in extant therian mammals differs from the ascending process of nonmammalian amniotes in its topographical relations to the maxillary branch of the trigeminal nerve. They suggested that the lamina ascendens of the mammalian alisphenoid may not be homologous to the reptilian epipterygoid. Gaupp's view was supported by Presley and Steel (1976), Presley (1981), Kemp (1982, 1983), and Kuhn and Zeller (1987b).

There is a strong resemblance between the ascending process of the epipterygoid of cynodonts and the alisphenoid of most Mesozoic mammals, including the prototribosphenidan *Vincelestes* (Hopson and Rougier, 1993, see figure 3.5). Most stem mammals have two separate foramina for the exits of trigeminal nerves from the cavum epiptericum, which houses the trigeminal ganglion. It is widely accepted that the mandibular branch of the trigeminal (V_3) was enclosed by the anterior lamina of the petrosal. The maxillary nerve (V_2) was either enclosed by the anterior lamina of the petrosal or lay at the suture of the anterior lamina and the ascending process of the alisphenoid (Kermack, 1963; Gow, 1986b; Crompton and Luo, 1993; Wible and Hopson, 1993; Hurum, 1998a; Wible and Rougier, 2000). If it is assumed that one of the two separate foramina exiting the cavum epiptericum must have housed the maxillary nerve (V_2), the ascending process of the alisphenoid would have to be anterior to the maxillary branch of the trigeminal nerve in stem mammals. If so, the maxillary nerve and the ascending process would have topographic relations similar to those in nonmammalian amniotes. This supports the homology proposed by Goodrich (1930) and reaffirmed by Maier (1987). It is more likely that the posterior pathway of the maxillary nerve (V_2) to the ascending process of alisphenoid is a primitive condition for mammals as a whole. The pathway of the maxillary nerve anterior to the ascending process in many extant therians is a secondarily derived and homoplastic condition (Hopson and Rougier, 1993). Given these assumptions, a reduced ascending process of the alisphenoid would be a derived feature for monotremes and multituberculates (but see the exception of *Lambdopsalis*, Miao, 1988; and different interpretation by Kuhn and Zeller, 1987b).

CHANGES IN CRANIAL VASCULATURE

Cranial vasculature of the mammalian crown group has been thoroughly studied by, among others, Bugge (1974, 1979), Presley (1979), MacPhee (1981), Wible (1984, 1986, 1987), Wible and Hopson (1995), Asher (1999, 2001), and Wible et al. (2001). Comparative studies include reconstructed cranial blood vessels for nonmammalian cynodonts (e.g., Watson, 1920; Fourie, 1974; Sun, 1984; Gow, 1986b; Rougier et al., 1992; Wible and Hopson, 1995), the stem mammal *Morganucodon* (Kermack et al., 1981; Wible and Hopson, 1995), multituberculates (Kielan-Jaworowska et al., 1986; Miao, 1988; Luo, 1989; Wible and Hopson, 1995; Wible and Rougier, 2000), the prototribosphenidan *Vincelestes* (Rougier et al., 1992), stem metatherians (Wible, 1990), and stem eutherians (MacIntyre, 1972; Wible, 1983, 1986; Wible et al., 2001). Osteological correlates of the vasculature have also been described in a variety of fossil placentals, such as primates (Szalay, 1975; MacPhee and Cartmill, 1986), archontans (MacPhee, 1981; Novacek, 1986b), tenrecids (Asher, 1999,

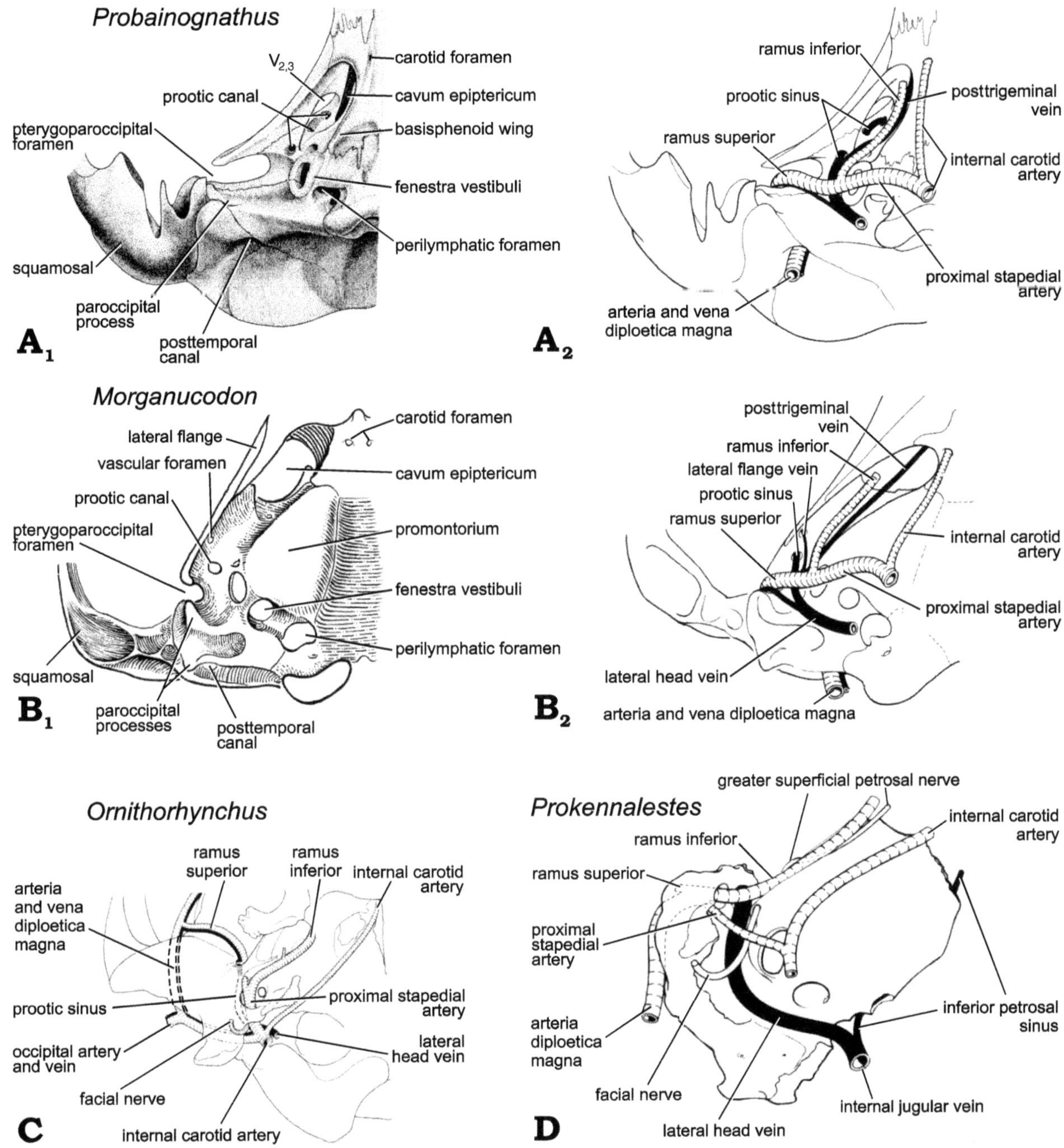

FIGURE 3.6. Vascular patterns of the petrosal ("periotic") region in cynodonts and mammals. A, *Probainognathus:* right basicranium in ventral view (A_1); vascular reconstruction of the same (A_2). B, *Morganucodon:* right basicranium in ventral view (B_1); vascular reconstruction of the same (B_2). C, *Ornithorhynchus.* D, Vascular reconstruction of *Prokennalestes.* See also partial cranial vascular reconstructions of multituberculates (figure 8.18), *Vincelestes* (figure 10.15), the petrosal of an unidentified marsupial from the Late Cretaceous (figure 12.3), and the petrosal structure of *Prokennalestes* (figure 13.6B_1, B_2). Abbreviations: V_{2-3}, maxillary and mandibular branches of the trigeminal nerve. Not to scale. *Source:* A, B_2, C, modified from Wible and Hopson (1995); B_1, modified from Luo, Crompton, and Sun (2001) and Kermack et al. (1981); D, from Wible et al. (2001).

2001), various carnivores, ungulates, and whales (Matthew, 1909; Cifelli, 1982; Geisler and Luo, 1998; Luo and Gingerich, 1999).

It has been hypothesized that the cranial vasculature of some nonmammalian cynodonts and stem mammals was primarily supplied by the following arteries (figures 3.5, 3.6): (1) Arteria diploetica magna, a tributary of the occipital artery (Kielan-Jaworowska et al., 1986; Wible, 1987; Wible and Hopson, 1995). This vessel entered the skull at the posttemporal canal. (2) The stapedial artery and its branches. The stapedial artery entered the skull in the auditory region and traversed the stapedial foramen of the stapes and the pterygoparoccipital foramen between the lateral flange and the paroccipital process of the prootic (homologous to part of the mammalian petrosal). (3) The internal carotid artery. This artery entered the cranial cavity through the internal carotid foramen in the basisphenoid in stem cynodonts, probainognathids, tritheledontids, and mammals. But the internal carotid foramen is absent in *Cynognathus*, Diademodontidae, Traversodontidae, and Tritylodontidae. In these latter groups, considered to be a clade by Hopson (Hopson and Barghusen, 1986; Hopson and Kitching, 2001), the internal carotid artery presumably entered the cranial cavity through the ventral opening of the cavum epiptericum, which is universally present in all cynodonts and most stem mammals.

Because these arteries are correlated to osteological landmarks, such as the foramina, grooves, and canals that are preservable in fossils, their distribution can be inferred for nonmammalian cynodonts and stem mammals (recently summarized by Wible and Hopson, 1995; Wible et al., 2001). The vertebral artery and the external carotid artery and its tributary branches also supplied the skull, but these vessels lack specific correlates in cranial bones that can be studied in fossil taxa.

Vasculature in the Temporal Region (figure 3.5). The posttemporal canal is a prominent channel in the skulls of cynodonts and stem mammals that connects the occiput to the temporal region. In most stem mammals its posterior opening is located between the squamosal and the mastoid region of the petrosal, whereas in nonmammalian cynodonts it is most often encircled by the tabular bone, but is occasionally found between the opisthotic and the tabular. Based on the comparative evidence from extant mammals and nonmammalian amniotes, Wible (1987) proposed that the arteria diploetica magna is the primary occupant of the posttemporal canal and would have entered the canal from the occiput. Anteriorly, the vessel anastomosed with the superior ramus of the stapedial artery, which entered the temporal region through the pterygoparoccipital foramen from the tympanic region. An ascending vessel arose from the confluence of the arteria diploetica magna and the ramus superior of the stapedial artery. It followed an ascending channel and extended anteriorly into the orbit to become the ramus supraorbitalis. This ascending vessel gives rise to the ramus meninges and the ramus temporalis in the monotreme *Ornithorhynchus*. Based on comparison with the vascular pattern of monotremes, it is hypothesized that similar vascular patterns were also present in the orbitotemporal region of nonmammalian cynodonts such as *Probainognathus* and stem mammals such as *Morganucodon* (Rougier et al., 1992; Wible and Hopson, 1995). In *Ornithorhynchus*, the ramus temporalis supplies the temporalis muscle. Wible (1987) inferred that this vessel was not only developed in stem mammals but was also present in some various conditions in nonmammalian cynodonts, where it presumably supplied the large temporalis muscles.

The lack of enclosure of the ascending channel is a primitive feature for both cynodonts and mammals. In the nonmammalian cynodonts *Thrinaxodon* (figure 3.5A), *Probainognathus*, *Pachygenelus*, and tritylodontids, the ascending vessel and the ramus superior of the stapedial artery occupied open grooves on the external wall of the braincase. A similar condition is also present in *Ornithorhynchus*. In the traversodontid *Massetognathus*, the course for this vessel is partially enclosed inside the lateral wall of the braincase, comparable to the pattern of *Tachyglossus* (Hopson, 1964; Sun, 1984; Rougier et al., 1992; Wible and Hopson, 1995).

A major apomorphy of these vascular channels in mammals is their enclosure within the wall of the braincase. In stem mammals such as *Adelobasileus*, *Sinoconodon*, and *Morganucodon* the temporal portion of the ascending channel is enclosed within the braincase wall by the parietal, the alisphenoid, and the anterior lamina of the petrosal (figure 3.5). A portion of the ascending channel is endocranial in therians and probably also in multituberculates (Kielan-Jaworowska et al., 1986; Miao, 1988; Lucas and Luo, 1993; Wible and Hopson, 1995). The shift of the topographic relations of the temporal portion of the ascending vessel relative to the lateral wall of the braincase is probably correlated with the enlargement of the braincase in the more derived mammals, in comparison to the narrow braincase in nonmammalian cynodonts (figure 3.4).

The channel for the superior ramus of the stapedial artery may show in one of two conditions. In the primitive condition seen in tritylodontids, *Pachygenelus*, *Sinoconodon*, and *Morganucodon*, the ramus superior was accommodated in an open notch at the pterygoparoccipital foramen (Kermack et al., 1981; Crompton and Luo, 1993; Wible and Hopson, 1993). In the more derived *Hadrocodium*, multituberculates, and *Vincelestes*, the pterygoparoccipital foramen and its associated channel for the ramus superior is enclosed by bone (Wible and Hopson, 1995; Luo, Crompton, and Sun, 2001), although *Adelobasileus* is an exception. The enclosure of the channel for the ramus superior of the stapedial artery is known for many placentals in which this vessel is present.

Wible (1987) proposed that the enclosures of the ramus temporalis and the ramus superior of the stapedial artery were caused by the anterior expansion of the cranial moiety of the squamosal medial to the temporalis muscle. A part of the temporalis muscle originated from the external surface of the petrosal in nonmammalian cynodonts and stem mammals (Crompton and Parker, 1978; Barghusen, 1986; Crompton and Hylander, 1986). This muscle was supplied by the temporalis ramus of the ramus superior of the stapedial artery, as in monotremes (Kuhn, 1971; Zeller, 1989a). In therian mammals, the cranial moiety of the squamosal is interposed between the temporalis muscle on the one hand and the petrosal on the other. This helped to enclose the ramus superior and/or its derivative vessels, resulting in the ramus temporalis piercing the squamosal to supply the temporalis muscle. One example of this primitive pattern can be seen in the prototribosphenidan mammal *Vincelestes* (see Rougier et al., 1992).

Vasculature in the Petrosal (figure 3.6). The internal carotid artery and branches of the stapedial may have osteological correlates on the petrosal in mammals (MacIntyre, 1972; Cifelli, 1982; Wible, 1983, 1986, 1990; Geisler and Luo, 1998; Wible et al., 2001). Among extant mammals, the monotreme *Ornithorhynchus* has a small internal carotid artery that travels medial to the petrosal promontorium and enters the cranial cavity through the internal carotid foramen. The proximal stapedial artery diverges from the internal carotid outside the tympanic cavity. Within the tympanic cavity, the proximal stapedial artery branches into the ramus superior and the ramus inferior. The former enters a foramen that is homologous to the pterygoparoccipital foramen in cynodonts and stem mammals. The latter traverses the lateral trough of the petrosal, enters the alisphenoid (Vidian) canal at the junction of the pterygoid and the alisphenoid, and continues anteriorly as the ramus infraorbitalis (figure 3.5).

In the placental insectivore *Erinaceus* (MacPhee, 1981; Wible, 1984), the internal carotid artery has a transpromontorial course before it enters the internal carotid foramen at the junction of the basisphenoid and the petrosal. The proximal stapedial artery splits from the internal carotid artery medial to the fenestra vestibuli and then traverses the stapedial foramen of the stapes before it bifurcates into the ramus superior and ramus inferior. The ramus superior enters the braincase through a foramen in the roof of the tympanic cavity. The ramus inferior runs ventral to the tympanic roof anteriorly before it enters the piriform fenestra. This condition is interpreted by Wible et al. (2001) to be applicable to *Prokennalestes*, one of the most primitive eutherians of the Early Cretaceous (see chapter 13). The transpromontorial course of the internal carotid artery is a primitive character for stem eutherians (MacIntyre, 1972; Wible, 1986; Wible et al., 2001), although the Late Cretaceous Mongolian eutherians *Asioryctes, Kennalestes,* and *Zalambdalestes* represent some exceptions (Kielan-Jaworowska, 1981, 1984b; Wible, 1986). In Crown Theria, the internal carotid artery may show a variety of character states among different orders of placentals (MacPhee, 1981; Wible, 1986).

In marsupials (figure 3.6, see also Wible, 1990; Wible and Hopson, 1995) the stapedial artery and its branches are lost in the adult. The internal carotid artery is medial to the promontorium and outside the tympanic bulla, thus leaving no osteological correlates on the promontorium in most marsupials. However, *Vombatus* has a small groove for the internal carotid artery on the promontorium (Sánchez-Villagra and Wible, 2000). *Pucadelphys* and *Mayulestes,* two marsupials from the Paleocene of South America, have a similar short groove (or indentation) near the anterior pole of the promontorium, possibly for the internal carotid artery (Marshall and Muizon, 1995; Muizon, 1998). The condition in the wombat is not typical of marsupials as a whole, nor is it present in the putative stem metatherian *Deltatheridium* (Rougier et al., 1998). The absence of the ramus superior system in the adults of most marsupials is a derived condition within the crown group of mammals (Wible, 1987, 1990).

EVOLUTION OF THE BRAIN

(tables 3.2 and 3.3)

CRANIAL ENDOCASTS AND BRAINCASE

Evolution of a larger brain from the nonmammalian cynodonts to mammals has been documented in many studies (Jerison, 1973; Hopson, 1979; Quiroga, 1980, 1984; Kielan-Jaworowska, 1986; Rowe, 1996a,b). The larger brain capacity of mammals compared to that of nonmammalian vertebrates (figure 3.7) is primarily attributable to the growth of the cerebral neocortex, although growth in the striatum region of the brain also may have contributed to the expansion of the forebrain (Ulinski, 1986).

The advanced features of the mammalian brain are significant because they are functionally correlated with

TABLE 3.2. Comparison of Encephalization Quotients of Nonmammalian Cynodonts and Primitive Mammals

Taxa	EQ[1]	EQ[2]	Source
Cynodonts[3]			
Thrinaxodon[4]	0.1		Calculated from Jerison (1973)
Diademodon[4]	0.21	0.14	EQ(1) calculated from Jerison (1973), EQ(2) from Quiroga (1980)
Massetognathus[5]	0.22	0.21	Quiroga (1980)
Probelesodon[5]	0.18	0.16	Quiroga (1980)
Exaeretodon[5]	0.15	0.125	Quiroga (1980)
Probainognathus[5]	0.17	0.16	Quiroga (1980)
Mammals[4]			
Triconodon		0.49	Kielan-Jaworowska (1983)
Ptilodus		0.49	Kielan-Jaworowska (1983)
Chulsanbaatar		0.55	Kielan-Jaworowska (1983)
Kryptobaatar		0.71	Kielan-Jaworowska and Lancaster (2004)
Kennalestes		0.36	Kielan-Jaworowska (1984c)
Asioryctes		0.56	Kielan-Jaworowska (1984c)
Zalambdalestes		0.70	Kielan-Jaworowska (1984c)
Tenricid placentals		0.4–1.03	Bauchot and Stephan (1966)
Didelphid marsupials		0.5–1.09	Eisenberg and Wilson (1981)

[1] Calculated by excluding the olfactory bulb (Jerison, 1973).
[2] Calculated by including the olfactory bulb (Quiroga, 1980).
[3] Quiroga (1980) used a varying degree of percentage (75~ 90%) to discount the meninges external to the brain and to account for the incomplete ossification of the bony braincase (see footnote 2).
[4] Based on: EQ = (Brain mass)/0.055(Body mass)$^{3/4}$.
[5] Based on: EQ = (Brain mass)/0.12(Body mass)$^{2/3}$.

more elaborate neural control of the skeletomuscular system for mastication and locomotion, as well as for perfected perception—especially for auditory, olfactory, and tactile information, but also visual perception for nocturnal life (Jerison, 1973, 1991; Ulinski, 1986; Butler, 1994; Zeller, 1999b). The enlargement of the brain through the cynodont-mammal transition may be correlated with a number of newly evolved biological adaptations of mammals, as discussed by a number of scholars.

Jerison (1973) believed that the invasion of the earliest mammals into nocturnal niches was accompanied by a number of changes in sensory functions. Mammals in a nocturnal environment would have had to rely more heavily on hearing, olfaction, tactile sense, and night vision (based on rods in the retina) than the daylight vision (based on cones in the retina) in a diurnal environment. The evolution of more sensitive ear structures (see later) and more olfactory structure in the nasal cavity would have allowed hearing and olfaction to compensate for the loss of spatial perception that would otherwise be gained by daylight vision. Jerison suggested that enlargement of the brain in mammals was correlated with (or even necessitated by) the improved capacity for processing information from auditory and olfactory senses that became the primary perception over daylight vision in a dark environment. He further argued that enlargement of the brain made it possible for mammals to invade the nocturnal and arboreal niches (Jerison, 1973).

Enlargement of the brain may also be correlated to mammalian energetics. As Hopson (1977: 443) suggested, the overall size of the brain could be related to activity level, as "the neural requirements of a vertebrate are, to a significant degree, a function of the total level of activity and therefore its total energy budget." At the tissue and organ levels, Else and Hulbert (1985) showed that the mammalian brain, visceral organs, and skeletal muscles are relatively larger than those of extant reptiles. The tissues of these organs contain a proportionately higher volume of mitochondria. Larger tissue mass of the brain and these organs is possible only with the higher metabolism in mammals as compared with extant reptiles. Beyond these functional correlates of a larger brain, it has been speculated that evolution of the metabolically expensive neural tissues of neocortex in the mammalian brain may be related to the development of endothermy (Quiroga, 1980; Allman, 1990; Aiello and Wheeler, 1995).

Enlargement of the Braincase and Brain Cavity. The evolutionary trend of brain enlargement from nonmammalian cynodonts to the mammalian crown group is clearly shown in a comparison of the encephalization quotients of nonmammalian cynodonts and mammals. Encephalization is the evolutionary increase in the rela-

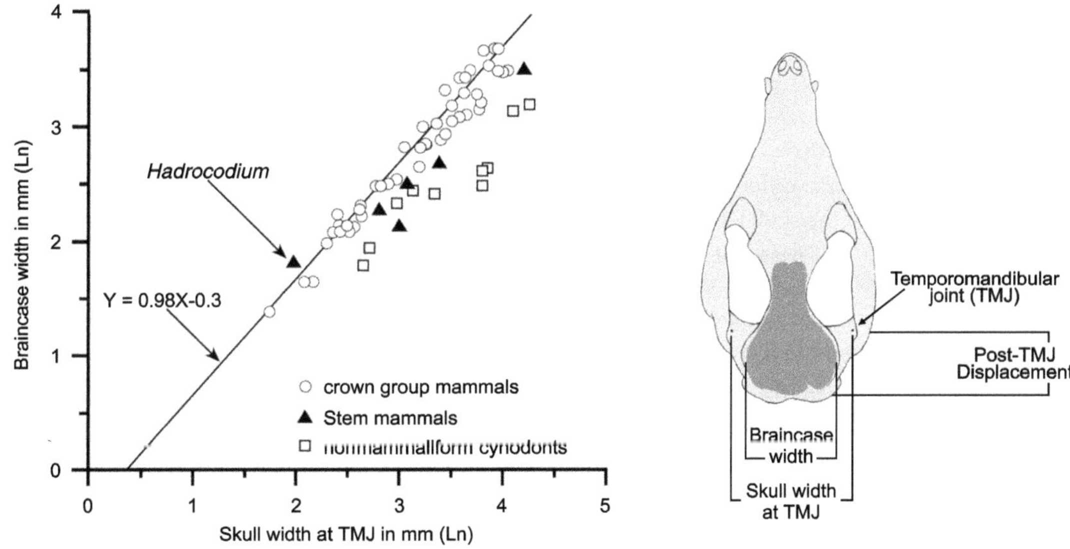

FIGURE 3.7. Scaling of the width of braincase (Y-axis), measured at its widest point (at the anterior suture of cranial moiety of the squamosal) versus skull width (X-axis), measured as the distance between the left and right TMJs. Measurements in mm; units of both axes in natural log scales, to illustrate braincase width relative to overall width of skull. The allometric equation [$y = 0.98x - 0.31$ ($r^2 = 0.715$)] is based on 37 living and 8 fossil species of the mammalian crown group. Data from cynodonts and stem mammals are added secondarily for comparison with the regression based on extant and fossil species of the mammalian crown group. After correction for skull size (approximated by skull width at TMJ), the stem mammals *Sinoconodon, Morganucodon, Haldanodon,* and a Cretaceous gobiconodontid (solid triangles) are intermediate between the extant mammals (open circles) plus *Hadrocodium* and nonmammalian cynodonts (open squares). This reflects the trend of increasing braincase size relative to skull size through the cynodont-mammal evolution. In mammals with a large brain, a significant part of the braincase extends posterior to the TMJ. This post-TMJ displacement is illustrated by the skull of *Didelphis. Source:* modified from Luo, Crompton, and Sun (2001).

tive size of the brain (Jerison, 1977). Encephalization quotient, or EQ (Jerison, 1973), is the ratio of actual brain size to the average predicted brain size for a living mammal with similar body mass. Hence, EQ is a measure of increase in brain size beyond what is expected from the allometric relations of the brain to the body. Brains of Mesozoic mammals are twofold larger than those of the advanced nonmammalian cynodonts (table 3.2), as measured by the volumetric EQ on brain endocasts (Jerison, 1973; Quiroga, 1980; Kielan-Jaworowska, 1983, 1984c; Kielan-Jaworowska and Lancaster, 2004). Jerison (1973, 1991) considered brain size to be the most important determinant of variation in morphological traits of the brain, as it can account for most of the morphological variations in its components and even for development of some external features such as sulci and gyri. The usefulness of relative brain size as a key feature for describing morphological evolution is endorsed by some authors (Hopson, 1979) but has been questioned by others (e.g., Radinsky, 1982). Early mammals had a brain cavity (as inferred from endocasts) twice as large as those of nonmammalian cynodonts, which is an important evolutionary apomorphy for mammals (table 3.3).

The difference in brain size between cynodonts and mammals is also reflected in the width of the braincase relative to the total width of the skull (figure 3.7). The brain size of nonmammalian cynodonts is intermediate between that of mammals and diapsid reptiles (Quiroga, 1980) after allometric correction. Among mammals, the braincase in the stem taxa *Sinoconodon, Morganucodon,* and *Haldanodon* is narrower than that of mammalian crown taxa, but wider than that of nonmammalian cynodonts (figure 3.7). The only exception among stem taxa is *Hadrocodium,* which is similar to living mammals and more derived than other stem mammals (Luo, Crompton, and Sun, 2001).

The earliest known mammals are much smaller than most nonmammalian cynodonts. But the body size difference by itself is not enough to account for the difference in brain size between mammals and nonmammalian cynodonts (table 3.2), as was shown by several scaling analyses (Jerison, 1973; Quiroga, 1980). The larger relative brain size in early mammals can only be attributed to the evolution of the cerebral neocortex, as discussed by Jerison and others (Jerison, 1973; Quiroga, 1980; Ulinski, 1986; Rowe, 1996a,b). Several morphological changes in the braincase are correlated with the increase in relative size of the brain through the cynodont-mammal transition.

Among marsupials and placentals, growth of the brain shows some heterochrony relative to the skeletal and muscular systems of the skull (Smith, 1996, 1997). The growth of the brain is also correlated with differential and allometric growth of other parts of the skull (Rowe, 1996a,b; Luo, Crompton, and Sun, 2001). These features include: (1) Posterior displacement of the braincase relative to the TMJ (figure 3.7). (2) Increase in the depth of the frontal region of the braincase (figure 3.4) related to enlargement of the olfactory bulb of the forebrain (figure 3.2B) (Kielan-Jaworowska, 1983; Hurum, 1994). (3) Widening of the parietal region, in correlation with growth of the neocortex in the cerebral hemispheres (figure 3.8) (Quiroga, 1980; Kielan-Jaworowska, 1986; Ulinski, 1986). (4) Widening of the occipital region of the braincase, related to the differentiation of the cerebellum. (5) Convergent overlap on the midbrain among major mammalian groups (Kielan-Jaworowska, 1984c, 1986, 1997). Because the transformation of the cranial endocast is intimately correlated with the surrounding structure of the braincase, we discuss them together in what follows.

Frontal Braincase and Forebrain Endocast Features (figures 3.2, 3.4, 3.9). In *Thrinaxodon* the frontal region of the braincase was narrow and tubular in shape (figure 3.2A) (Fourie, 1974; Hopson, 1979; Rowe, 1996a). The floor of the frontal part of the braincase and its related orbital structure are not ossified. The floor for the anterior part of the brain, which otherwise would be formed from the orbitosphenoid and the ethmoid, was presumably formed by the cartilaginous structure of the chondrocranium in life (Fourie, 1974; Rowe et al., 1994). Therefore, the ventral boundary of the forebrain endocast cannot be established in nonmammalian cynodonts without a fully ossified orbitosphenoid bone.

The forebrain endocast of cynodonts is long and narrow. The casts of olfactory bulbs are discernible owing to a slight constriction between the olfactory bulbs and the forebrain. The olfactory regions are an anterior extension of the cerebrum seen on the endocast, although the transition between the olfactory and the cerebral regions is gradual and the boundary between them is not well marked on the endocasts (Jerison, 1973; Hopson, 1979). An earlier reconstruction of distinctive olfactory bulbs and cerebral hemispheres by Simpson (1927a) for the

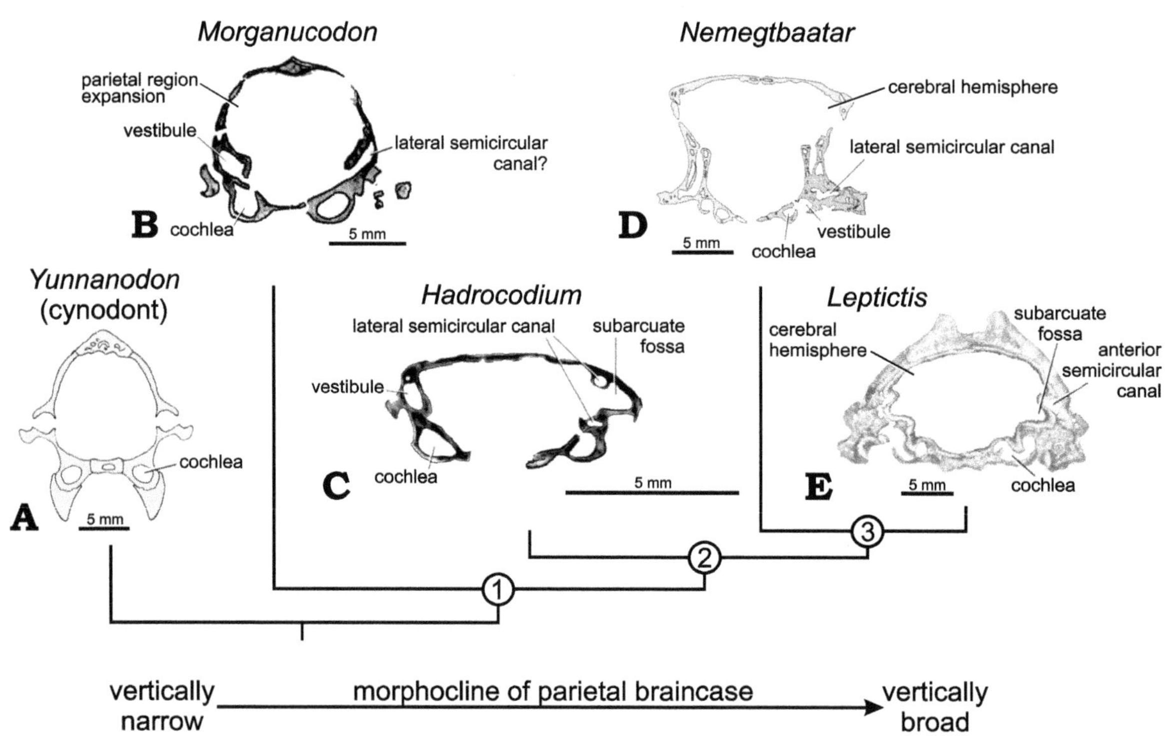

FIGURE 3.8. Morphocline of the posterior (parietal) braincase in derived cynodonts and mammals. A, Cynodont (tritylodontid) *Yunnanodon.* B, Stem mammal *Morganucodon.* C, Stem mammal *Hadrocodium.* D, Multituberculate *Nemegtbaatar.* E, Eutherian *Leptictis.* The braincase is broad in the parietal region in *Hadrocodium, Nemegtbaatar,* and *Leptictis,* but narrow in the more primitive *Morganucodon* and the cynodont *Yunnanodon.* Scale bars 5 mm. *Source:* A, B, C, original, based on sections and CT scans; D, from Hurum (1998a); E, from Novacek (1986b).

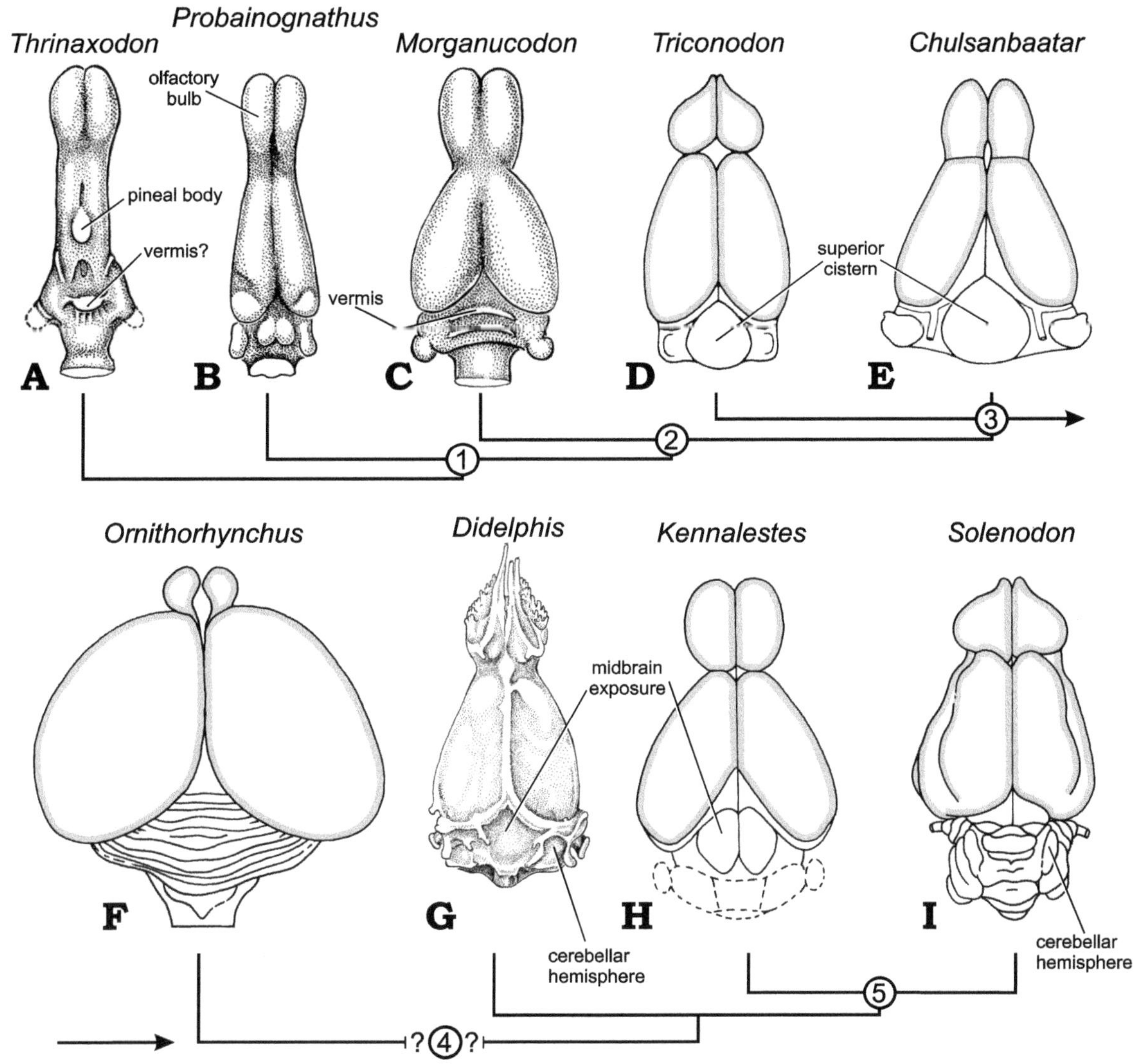

FIGURE 3.9. Brain endocasts in cynodonts and mammals. A, Cynodont *Thrinaxodon.* B, Cynodont *Probainognathus.* C, Stem mammal *Morganucodon.* D, *Triconodon.* E, Multituberculate *Chulsanbaatar.* F, Monotreme *Ornithorhynchus.* G, Marsupial *Didelphis.* H, Eutherian *Kennalestes.* I, Placental *Solenodon.* Endocasts standardized to approximately the same anteroposterior length; actual scales vary among taxa. Features of the cranial endocasts are mapped onto a simplified cladogram that leaves the relationships of eutriconodontans, multituberculates, and monotremes unresolved. Note the new interpretation of eutriconodontan (D) and multituberculate (E) endocasts, based on Harry J. Jerison's suggestion and Kielan-Jaworowska and Lancaster (2004). The correlated evolution in the wall structures of the braincase is presented in table 3.3. Acquisition of apomorphies—Node 1 (eucynodonts): loss of parietal eye and pineal body. Node 2 (mammals): incipient expansion of cerebrum. Node 3 (mammalian crown group and close relatives): spherical or near spherical olfactory bulbs, enlarged cerebral hemispheres with distinctive median division by longitudinal fissure. Node 4: lateral expansion of cerebral hemispheres. Node 5 (therian crown group): larger cerebellar hemispheres distinctive from vermis and paraflocculi, the dorsally exposed midbrain of basal therians, reversed in many groups of placental mammals. *Source:* A, B, C, G, original drawings and interpretation of Z-X Luo; A, based on an endocast specimen from the Institute of Paleobiology, Warsaw; B, based on figures of Quiroga (1984) and on a cast; modified from Rowe (1996a) on the basis of horizontal CT scans (C); D, E, H, from Kielan-Jaworowska (1986, interpretation modified); F, I, from Starck (1964).

Thrinaxodon-like *Nythosaurus* has been modified by more recent descriptions of Hopson (1979: figure 29c) and Kielan-Jaworowska (1986). *Nythosaurus* and *Thrinaxodon* did not have distinctive olfactory bulbs and cerebral hemispheres, as in more derived cynodonts.

Among epicynodonts, the chiniquodontid *Probelesodon* and the "gomphodonts" *Tirachodon* and *Massetognathus* developed the orbitosphenoid floor for the frontal region of the braincase. With a partial orbitosphenoid floor it is possible to establish the ventral extent of the anterior cranial endocast for these taxa and their relationship to the surrounding structures of the skull (figures 3.2, 3.3). In *Tirachodon* (see Hopson, 1979), *Massetognathus*, and *Probelesodon* (see Quiroga, 1979b), the floor for the narrow and tubular forebrain is separated from the base of the skull by a large orbital vacuity (figures 3.3, 3.4). The olfactory bulbs were more developed in *Probelesodon* and *Massetognathus* and better separated from each other than in *Thrinaxodon* (figure 3.9A) and *Nythosaurus* (Hopson, 1979). The olfactory bulbs graded into the cerebrum and the longitudinal fissure between the two cerebral hemispheres was absent. Both of these are primitive features of basal cynodonts.

Probainognathus (figure 3.9B) had several derived cerebral features in comparison to other cynodonts, including better-developed cerebral hemispheres separated by a weakly developed longitudinal fissure (Quiroga, 1980). The posterior poles of the cerebral hemispheres are divergent, and each hemisphere widened posteriorly, with a distinctive border from the cerebellar region. None of these are present in *Thrinaxodon* (figure 3.9A), diademodontids (Hopson, 1979), or chiniquodontids (Quiroga, 1979b). Quiroga (1980) interpreted a sulcus on the posterior and dorsal aspect of the cerebral hemisphere as the posterior extent of a neocortex, although it is possible that this only represents a vessel of the external surface of the endocast (Kielan-Jaworowska, 1986). It is also possible that this represents structures other than the neocortex, as suggested by MacLean (1986). Jerison (1990) urged the use of the rhinal fissure as a more reliable endocast landmark for inferring the extent of the neocortex in the brain of fossil mammals. Interpretation of other surface features of the endocast of *Probainognathus* is less than certain, as reviewed by Kielan-Jaworowska (1986).

Morganucodon is the only taxon among stem mammals for which some brain endocast features are known, as revealed by CT scans. The mammalian apomorphies recognized are therefore limited to comparison of *Morganucodon* with nonmammalian cynodonts. *Morganucodon* (figure 3.9C) has two derived conditions in the frontal region of the cranial endocast: (1) The olfactory bulb casts are enlarged and nearly oval in shape. At its broadest point, the bulb is as wide as the width of the posterior nasal cavity. By contrast, in nonmammalian cynodonts, the olfactory bulbs, if differentiated at all, were significantly smaller and were narrower than the posterior nasal cavity. (2) The division between the olfactory bulbs and the cerebral hemispheres is well defined. In nonmammalian cynodonts the olfactory bulbs are separated far from the cerebral hemispheres, so that the olfactory tracts form the "peduncle" between the two structures, as in the brain of many living reptiles (Hopson, 1979). Further enlargement of the frontal region of the braincase is a shared derived feature of *Hadrocodium* and modern mammals. *Hadrocodium* and the mammalian crown group are characterized by a conspicuous divergence of the enlarged olfactory bulbs as seen on endocasts, and the anterior portion of their cerebral hemispheres are expanded more laterally than in *Morganucodon*.

Parietal Braincase and Endocast Features. The posterior braincase of most nonmammalian cynodonts (except for small tritylodontids) is narrow on the parietal roof but slightly broader at the basioccipital floor. Three endocast features are prominent in this region: first, the large paraflocculus[1]; second, a small, vermislike structure of the cerebellum that may be homologous to the vermis on the well-preserved endocasts of many early mammals (see later); and third, the absence of the pineal body in eucynodonts. It should be noted, however, that the triangular vermislike bulge in multituberculate and eutriconodontan endocasts, originally referred to as the vermis (Kielan-Jaworowska, 1986, 1997), has recently been reconsidered to be the cast of a large vessel, the superior cistern (Jerison, pers. comm.; Kielan-Jaworowska and Lancaster, 2004;

[1] In older paleoneurological literature the cerebellar part housed in the fossa subarcuata was often referred to as the "flocculus," and the terms "flocculus" and "paraflocculus" have sometimes been used as synonyms, which may cause confusion. According to *Nomina Anatomica Veterinaria* (Schaller, 1992: 428), the flocculus corresponds to the nodulus vermis, whereas the paraflocculus consists of two parts: the paraflocculus dorsalis, which corresponds to a part of the pyramis vermis, and the paraflocculus ventralis, which corresponds to the uvula vermis. Most of the paraflocculi dorsalis and ventralis are located in the fossa subarcuata, although they may not be visible on fossil endocranial casts. The flocculus (smaller than the paraflocculus) is located on the ventral side of paraflocculus ventralis. Separation of the flocculus from the paraflocculus is not visible on endocranial casts. For these reasons, we refer to the structure housed in the fossa subarcuata as the paraflocculus (although it may include a part of the flocculus), both in mammals and cynodonts, and we do not use the term flocculus in endocranial casts.

see figures 3.9, 7.3, 8.19 and discussion in chapter 8). The cerebellar endocasts show well-developed paraflocculi in both *Thrinaxodon* (figure 3.9A; Rowe, 1996a,b) and the galesaurid *Nythosaurus* (Hopson, 1979). This cerebellar feature corresponds to the subarcuate fossa encircled by the bony anterior semicircular canal and is formed by the periotics and supraoccipital in *Thrinaxodon* (Fourie, 1974). The parafloculus is present in the endocast of several Middle Triassic cynodonts (Hopson, 1979; Quiroga, 1979b, 1984). For example, in *Massetognathus* (see Quiroga, 1979a), the cerebellar parafloccular lobes are well developed and protrude into the anterior semicircular canal. A well-developed paraflocculus is derived for cynodonts, by contrast to the poorly developed parafloccular lobes in precynodont therapsids, such as dicynodonts (Olson, 1944; Cox, 1962; Cluver, 1971) and gorgonopsians (Olson, 1944; Sigogneau, 1974). In a natural cranial and inner ear endocast of *Dicynodon* (see Cox, 1962), the parafloccular lobes are absent from the cerebellar endocast surface. In dicynodonts the endocast of the paraflocculus and its corresponding subarcuate fossa in the braincase may vary with skull size (Hopson, 1979). We cannot exclude the posssibility that the flocculi and paraflocculi were present on the cerebellum as soft tissue in some therapsids, but were not reflected on the endocast owing to the poor ossification of the subarcuate fossa floor in fossils, as is often the case in many precynodont therapsids (Olson, 1944; Jerison, 1973; Sigogneau, 1974). In gorgonopsians, the wall of the anterior semicircular canal and the vestibule are not fully ossified, so that the subarcuate fossa is only partially formed (Sigogneau, 1974). With the caveat that the absence of the parafloccular lobe on an endocast could be due to true absence of the paraflocculus on the brain or to lack of ossification in the bony subarcuate fossa area in the braincase wall, we provisionally recognize the presence of a well-defined parafloccular lobe on the cerebellar endocast with a corresponding well-ossified bony subarcuate fossa as an apomorphy of cynodonts.

The vermis is represented by a protuberance on the middorsal cerebellum in *Thrinaxodon* (figure 3.9A) and *Nythosaurus*. Previous workers observed this structure in the cerebellum of therapsids, but most followed Olson (1944) in identifying this feature as an artifact of the filling of a cartilaginous space in the supraoccipital (e.g., Quiroga, 1979b). We consider this protuberance to be a natural anatomical feature of the cerebellar endocast for several reasons. (1) This protuberance has continuous surfaces with surrounding endocast features in a well-preserved natural endocast of *Thrinaxodon* (figure 3.9A). (2) In the braincase of *Thrinaxodon*, this endocast structure is enclosed mostly by the parietal and interparietal, both being membranous bones (Fourie, 1974: figures 27, 29; Crompton and Jenkins, 1979). The endocranial surfaces of the parietal and interparietal are smooth (Fourie, 1974: figures 6, 29). (3) This feature is consistently preserved in the interparietal part of the natural endocasts in diverse cynodonts (*Thrinaxodon, Nythosaurus, Probelesodon, Trirachodon,* and *Massetognathus*). It is unlikely that this consistent distribution in many unrelated taxa would be an irregular artifact resulting from the unossified supraoccipital, as previously believed. (4) This cerebellar protuberance corresponds to the vermis on the cerebellar endocasts of several Cretaceous eutherians (Kielan-Jaworowska, 1986), but not of multituberculates and *Triconodon,* where, as noted above, it is obscured by the cast of the superior cistern, an endocranial vascular feature (Kielan-Jaworowska and Lancaster, 2004, see also chapters 7 and 8). The vermis is also present in *Morganucodon,* as revealed by CT scans.

We propose that the middorsal protuberance of the cerebellum of cynodonts is homologous to the vermis of mammals. This is based on the similarity of the vermis in morphology and identical topographical relationship to other endocast features and on the preponderance of such a consistent systematic distribution in a wide range of cynodonts and mammals. We suggest that the presence of a cerebellar vermis is a primitive feature of cynodonts, including mammals, although the small vermis in cynodonts is different from the larger vermis of most mammals.

The pineal body is absent on endocasts of mammals and some derived cynodonts, a derived condition by comparison to basal cynodonts. In interpreting the endocasts of cynodonts, it is important to note that the parietal eye may be present without a foramen, as in some lizards, or the pineal body may be retained without the parietal eye and foramen, as in turtles (Roth et al., 1986). Thus, absence of the pineal foramen in the parietal portion of the skull roof does not mean absence of the pineal body. One of the functions of the pineal body and its associated parietal eye is to monitor the light and temperature periodicity in those living lizards that retain these features (Quay, 1979; Roth et al., 1986).

In galesaurids, *Thrinaxodon,* and diademodontids, a parietal (pineal) foramen is present in the roof of the braincase between the parietals. Corresponding to this foramen is the parietal (pineal) protuberance on the endocast (Fourie, 1974; Hopson, 1979; Quiroga, 1979b; Roth et al., 1986). The parietal foramen indicates the presence of the parietal eye (Edinger, 1955; Roth et al., 1986). The presence of pineal foramen in the skull roof is a primitive condition of cynodonts, shared by most (but not all) precynodont therapsids and by all pelycosaurs (Roth et al., 1986).

The absence of the parietal foramen is a derived feature, at least of the skull roof, for derived cynodonts

(Kühne, 1956; Bonaparte, 1962, 1966; Romer, 1967, 1969a, 1970a; Sues, 1986; Rowe, 1988; Wible, 1991). The absence of the parietal foramen and its associated pineal structure on the endocast may be interpreted as the loss of the parietal "eye" (although not necessarily the loss of a pineal body). From the currently available endocast fossils, a pineal feature is absent on the endocasts of chiniquodontids and traversodontids (Quiroga, 1979b), tritylodontids, and mammals. We also regard the endocast pineal structure as absent in *Probainognathus.* Although a pineal area was previously identified on the endocast of *Probainognathus* (see Quiroga, 1980), this has been reinterpreted to represent either the midbrain or the overlap of the midbrain by the cerebellum (Kielan-Jaworowska, 1986). In the light of the new interpretation of the endocasts of multituberculates and *Triconodon* (Kielan-Jaworowska and Lancaster, 2004), it remains possible that, like these early mammals, some parts of the brain might also have been obscured on the endocasts of *Probainognathus* and related forms, by the impressions of the meninges and vessels that surround the brain. The absence of an endocast pineal structure is certainly different from (and more derived than) the endocast pineal structure that is present in *Thrinaxodon* and diademodontids.

The mammalian condition of the cerebrum, such as can be seen in *Morganucodon,* is characterized by a greater degree of posterior divergence and a greater degree of lateral expansion in each hemisphere, in correlation with a more expanded external braincase than that in nonmammalian cynodonts.

The cerebral hemispheres are more expanded in the parietal region of *Morganucodon* (figure 3.8). The cerebellar region has developed transverse gyres (figure 3.9C). In *Morganucodon,* the parietal roof is more convex externally, allowing a larger endocranial space than in most tritylodontids, tritheledontids, and *Sinoconodon.* In transverse section at the level of the cavum epiptericum, the parietal aspect of the braincase is more enlarged and rounded than in tritylodontids (figure 3.8).

However, *Morganucodon* retains several primitive conditions. In transverse section through the cerebellar region of *Morganucodon,* the braincase is broadest at the level of the subarcuate fossa (housing the paraflocculus) but much narrower near the sagittal line of the parietal (e.g., tritylodontid *Yunnanodon,* figure 3.8A). This is primitive for many cynodonts, such as *Thrinaxodon* (Fourie, 1974), *Probelesodon, Massetognathus,* and *Probainognathus* (Quiroga, 1979b, 1980, and pers. obs. of sections).

The parietal roof of the posterior braincase is wider in *Hadrocodium* and the crown mammalian group (figure 3.8C) than in *Sinoconodon, Morganucodon* (figure 3.8B), and *Haldanodon* (Luo, Crompton, and Sun, 2001) in both relative size to the skull (figure 3.7) and transverse section profile of the frontal region (figure 3.4). However, *Hadrocodium* does not have the cerebellar expansion above the paraflocculus (that occupies the subarcuate fossa) that is characteristic of the crown mammalian group in the parietal region of braincase (figure 3.8).

Endocast Features of Crown Mammalian Group. Extant mammals are distinctive from other nonmammalian vertebrates in the following external features of the brain: (1) Enlargement of the cerebral hemispheres in the more derived Tertiary mammalian groups (Jerison, 1973). (2) Transverse widening of the cerebellum by the extensive development of cerebellar parts between the vermis and paraflocculi in therians, as pointed out by Ariëns Kappers et al. (1960), and to a smaller degree in monotremes (Holst, 1986). (3) Dorsal exposure of the midbrain occurs in many basal therians (figure 3.9), but the midbrain is overlapped by the cerebral hemispheres convergently in different lineages of extant mammals, both monotremes and therians.

Brain endocasts of Late Cretaceous eutherians preserve the vascular channels of the transverse sinus that demarcates the posterior pole of the cerebral hemispheres from the midbrain on the endocast (Kielan-Jaworowska, 1984c, 1986). Hence the midbrain dorsal exposure on the endocast is not a postmortem difference of the brain from the endocast made of the brain plus its enveloping meninges.[2] Similar cases have been documented in the extant placental insectivores *Tenrec* and *Rhynchocyon* (Bauchot and Stephan, 1966, 1967). This type of cranial endocast is widespread in Cretaceous and Paleocene eutherians, in many extant insectivores, and in didelphid marsupials. It likely represents the basal condition of Late Cretaceous–Early Tertiary eutherian mammals (Edinger, 1964; Russell and Sigogneau, 1965; Radinsky, 1976, 1977a,b, 1981).

[2] Cranial endocasts of fossil mammals preserve the external morphology of the meninges, the connective tissues, and the meningeal vasculature that enveloped the brain. External features of the brain tend to be underrepresented in the external features of endocasts, even to the extent of being obscured completely by the meninges, as demonstrated by several studies (Bauchot and Stephan, 1966, 1967; Jerison, 1990). Therefore, absence of a feature on an endocast may not be counted as the absence of the corresponding feature in the brain itself; but the inference of the presence of a brain feature from the endocast tends to be less problematic. Moreover, the volume of the endocast is larger than the volume of the brain, especially in nonmammalian cynodonts, where ossification of the braincase may not be as complete as in mammals (Hopson, 1979; Quiroga, 1980). Several workers have developed procedures to take into account such differences in measuring fossil endocasts for comparative studies of the brain (e.g., Quiroga, 1980).

The transversely wide cerebellum in crown therians may be correlated to the widening of the space in the brain cavity above the level of the paraflocculus (figure 3.8) on the brain endocasts of early eutherians (Kielan-Jaworowska, 1984c, 1986) and many Tertiary placentals (Jerison, 1973; Radinsky, 1976, 1981). This can be seen in the serial sections of many fossil (Novacek, 1986b) and extant placentals (e.g., Zeller, 1987). As pointed out earlier (Ariëns Kappers et al., 1960; Kielan-Jaworowska, 1986, 1997), the greater development of the cerebellar structures, especially the widening of the cerebellum, is a derived condition of all extant mammals.

The transversely wide cerebellum in crown therians can be contrasted with the primitive condition of stem mammals (e.g., *Morganucodon,* figure 3.9), where the space for the midportion of the cerebellum in the lambdoidal region of the braincase is less developed. The lack of lateral expansion in the lambdoidal region of the braincase that accommodated the cerebellar parts of the brain is a primitive character of stem mammals and nonmammalian cynodonts (figure 3.8).

The well-developed cerebellum is separated from the cerebrum by the tentorium cerebelli, a meningeal septum in many extant mammals. Along this meningeal septum in the cranial cavity, an ossified septum, known as the tentorium osseum, may form in some extant therian taxa. The tentorium osseum is a derived condition among the therians. If preserved in a fossil, it can be used to infer a separation of the cerebral and cerebellar cavities. However, this feature is variably present in some lineages. For example, nonmammalian cynodonts certainly lack this feature, based on extensive study of the skull roofs of disarticulated skulls, as well as serial sections and CT scans (Fourie, 1974; Rowe et al., 1994; Luo, unpublished data). It is not clear if *Morganucodon* has a tentorium osseum, as claimed by Kermack et al. (1981: figure 100; see also Kielan-Jaworowska, 1997, for discussion).

Multituberculates and *Triconodon* (figure 3.9D,E). In multituberculates and *Triconodon,* the brain endocasts do not have any large parts of cerebellar hemispheres between the superior cistern and paraflocculi. The superior cistern is located between the laterally divergent cerebral hemispheres. Multituberculates and *Triconodon* are obviously different from extant mammals in these brain endocast characters.

Until recently multituberculate and *Triconodon* endocasts have been interpreted as having a greatly enlarged vermis that overlapped the midbrain posteriorly, and in lacking exposure of the cerebellar hemispheres between the vermis and very large paraflocculi (Simpson 1937; Hahn 1969; Jerison 1973; Kielan-Jaworowska 1983, 1986, 1997; Kielan-Jaworowska et al. 1986; Krause and Kielan-Jaworowska 1993). Kielan-Jaworowska and Lancaster (2004) offered a different interpretation of multituberculate and *Triconodon* endocasts. They regarded the large vermislike triangular bulge as the cast of a large vessel,the superior cistern, the cast of which obscures the vermis and apparently the midbrain, originally exposed on the dorsal side of the brain. It is therefore possible that multituberculate and eutriconodontan brains did not differ so dramatically from the brains of early therian mammals, as previously believed.

SUMMARY ON BRAIN EVOLUTION AND BRAINCASE STRUCTURES
(table 3.3)

(1) There is a tendency toward increasing brain size in the evolution of cynodonts and mammals, accompanied by a posterior shift of the braincase relative to other skull structures, such as the TMJ. (2) The cerebral hemispheres are much better developed in the successively more derived groups of stem mammals. (3) The cerebellum is better developed among the clades of the mammalian crown group, with differentiation of the vermis and enlargement of the cerebellar hemispheres. The transverse enlargement of the cerebellar components between the vermis and paraflocculi is an apomorphy for metatherians, eutherians, and monotremes. (4) Convergent overlaps of the midbrain by cerebral hemispheres occurred in monotremes, and secondarily in many groups of marsupials and placentals (Starck, 1964; Kielan-Jaworowska, 1986)

EVOLUTION OF THE LOWER JAW SUSPENSORIUM
(table 3.4)

The incus, malleus, and ectotympanic of mammals are homologous to the quadrate, articular, and angular, respectively, of nonmammalian tetrapods (figure 3.10). This homology was established by Reichert (1837) and Gaupp (1913), who recognized that the mammalian malleus and incus and the reptilian quadrate and articular were all derived from the embryonic Meckel's cartilage. The craniomandibular joint, or the jaw hinge, of nonmammalian vertebrates is formed by the articular and the quadrate (figure 3.10). By contrast, the TMJ of mammals is formed by the squamosal glenoid and the dentary condyle. This "secondary" jaw hinge replaces the "primary" jaw hinge between the incus (quadrate) and the malleus (articular) as the functioning hinge for closing the jaws. In nonmammalian tetrapods, the angular and the articular are parts

TABLE 3.3. Correlation of Apomorphies Preserved on Cranial Endocasts and in Braincase Wall Structures through the Cynodont-Mammal Transition

Node[1]	Phylogenetic hierarchy	Apomorphies of cranial endocast[2]	Apomorphies in braincase structure[3]
	Cynodontia	Presence of cerebellar paraflocculus, a small vermis in cerebellum, and incipient olfactory bulbs	Subarcuate fossa in periotic
1	Post-*Thrinaxodon* cynodonts	Incipient separation of olfactory bulbs from cerebral part of the endocast	Partial orbitosphenoid floor for anterior braincase
	Probainognathus + more derived taxa	Incipient cerebral hemispheres; differentiation of olfactory tract; loss of pineal eye	Enclosure of parietal foramen
	Mammaliamorpha		Full ossification of orbitosphenoid floor for anterior braincase
2	Mammalia	Olfactory bulbs enlarged and readily distinguished from cerebral hemispheres; expansion and divergence of cerebral hemispheres	Enlargement of frontal region of the braincase
	Post-*Morganucodon* clade	Hypertrophied olfactory bulbs; enlargement of pyriform region of cerebral hemispheres	Initiation of the post-TMJ displacement of braincase
3	Multituberculates +*Triconodon* + crown mammals	Transverse enlargement of cerebellum and lambdoidal regions	Expansion of braincase in both parietal
	Crown mammals	Presence of rhinal sulcus of cerebellar hemispheres; trigeminal ganglion incorporated in braincase	Loss of pila antotica; formation of cribriform plate
	Monotremes	Hypertrophy cerebral hemispheres	
5	Therians	Gyrencephalic cerebral hemispheres	Fully developed post-TMJ displacement of braincase

[1] Nodes here correspond to the cladogram nodes in figure 3.9.
[2] Some of these apomorphies are illustrated in figure 3.9.
[3] Some of these apomorphies are illustrated in figures 3.4, 3.7, and 3.8.

of a complex of postdentary bones on the mandible (figure 3.10). In mammals, the ectotympanic (angular), malleus (articular), and goniale part of malleus (prearticular) are attached to the mandible only in embryonic stages; they are separated from the mandible and relocated in the tympanic cavity in adults.

Through the transition from cynodonts to mammals, the quadrate (incus) and the articular (malleus) transformed from being parts of the jaw hinge (for feeding) to the mechanism of the middle ear (for hearing). As these bones are so delicate and their function is so vital, the totality of the evolutionary changes in their structure and function in transforming the craniomandibular joint and the middle ear bones of mammals has been a great case study of vertebrate evolutionary morphology (table 3.4).

The jaw hinge in many stem mammals, such as *Sinoconodon*, morganucodontans (*Morganucodon*, see figure 3.10C, and *Megazostrodon*), and the docodontan *Haldanodon* (figure 5.3B_4), has an intermediate condition, with a so-called compound jaw joint. In these stem mammals the modern mammalian jaw joint between the dentary and the squamosal has been established. However, the articular (malleus) still maintained a mobile articulation with the quadrate (incus), a primitive condition of nonmammalian amniotes. The postdentary trough still accommodated the homologues of the middle ear bones, as in nonmammalian cynodonts. The primitive "reptilian" jaw joint between the articular and the quadrate is present and is positioned medial to the new mammalian jaw joint (figure 3.10). Together, these articulations formed a double or compound jaw joint, and both were mobile during jaw movement.

Here we use the term craniomandibular joint (CMJ) for the cynodont jaw hinge of the articular and the quadrate (the so-called "primary or reptilian" jaw hinge). For the derived jaw hinge of the dentary condyle and squamosal glenoid in the mammalian crown group, we use the term temporomandibular joint (TMJ). In the mammalian crown group, the squamous portion of the

TABLE 3.4. Structural Differences of the Mammalian Middle Ear Bones from Their Cynodont Homologues

Plesiomorphies of nonmammalian cynodonts	Apomorphies of Mammalia	Apomorphies of mammalian crown group
Mobile joint between articular and quadrate	←Plesiomorphous as in cynodonts	Immobile joint between malleus and incus
Quadrate with simple dorsal plate	Incus with twisted dorsal plate (crus longus) separated by neck	Gracile crus longus of incus
Quadrate suspended by multiple bones	Incus suspended by petrosal only (at crista parotica)	←As in stem mammals
No stapedial process of the incus	Stapedial process of the incus	←As in stem mammals
Angular and surangular attached the dentary	←Plesiomorphous as in cynodonts	Ectotympanic and malleus separated mediolaterally from dentary
Short retroarticular process of the articular	←Plesiomorphous as in cynodonts	Long manubrium of the malleus
Large postdentary bones	Reduction of postdentary bones	Small middle ear ossicles

"temporal" bone contributes to the joint. The primary or "reptilian" joint ceased to function as the malleus and incus joint became immobilized, owing to retention of the embryonic condition, and the malleus becomes separated from the mandible in the adult. However, for stem mammals, the quadrate-articular joint was still mobile although no longer the main joint for jaw function. Therefore, we would use the CMJ and the TMJ interchangeably for those stem mammals that already have the mammalian apomorphy, while still retaining the primitive hinge, at least in part.

Evolution of the mandibular middle ear to the cranial middle ear involved: (1) transformation of the quadrate-incus; (2) modification of the articulation of the quadrate-incus to the cranium; (3) immobilization of the incus-malleus joint; (4) detachment of the prearticular-articular from the mandible (in correlation with the loss of the postdentary trough); and (5) detachment of the angular from the mandible (in correlation with the loss of the medial concavity on the angular region of the mandible).

Transformation of Quadrate-Incus (figure 3.11). Although the mobile joint of the articular and the quadrate is essentially "reptilian" in nature among stem mammals, their articulation of the quadrate (incus) to the cranium was already modified and significantly more derived than the condition in nonmammalian cynodonts. Moreover, the mammalian characteristics of the quadrate were assembled in a stepwise pattern through cynodont-mammal transition.

In the primitive, nonmammalian cynodont *Thrinaxodon* (figure 3.11A), the quadrate consists of a cylindrical trochlea and a dorsal plate, both in the same plane. The quadrate in the mammal *Morganucodon* (figure 3.11F) differs from that of *Thrinaxodon* in several features. The dorsal plate is twisted about 90° in relation to the trochlea; it is separated from the trochlea by a neck from which arises a neomorphic stapedial process for articulation with the stapes. A series of the nonmammalian cynodont clades successively more closely related to mammals show several intermediate character states between *Thrinaxodon* and *Morganucodon* (figure 3.11). The concave surface and the rotation of the quadrate are present in the clade of *Probainognathus*, tritylodontids, tritheledontids, and mammals. The neck between the dorsal plate and the trochlea is a shared derived character of tritylodontids, tritheledontids, and mammals. The stapedial process of the incus is either a shared derived character of *Morganucodon* and tritylodontids (if the latter group is considered to be the sister taxon to mammals, Rowe, 1993) or independently derived characters of *Morganucodon* and tritylodontids (Sues, 1985a; Luo and Crompton, 1994).

Modification of Articulation of the Quadrate/Incus to Cranium (figure 3.12). In correlation with the transformation of the quadrate, the cranial articulation of the quadrate-incus is also much more simplified in mammals than in nonmammalian cynodonts (figure 3.12). In *Thrinaxodon* (figure 3.12A), a more rigid articulation of the quadrate to the cranium is accomplished by a total of five bones: the lateral flange of the prootic, the quadrate ramus of the pterygoid, the squamosal, the quadratojugal, and the opisthotic (= petrosal). The quadrate shows no twisting of the dorsal plate. By comparison, *Morganucodon* has a more mobile articulation for the quadrate, with participation by only one bone (petrosal) (figure 3.13). The quadrate dorsal plate is twisted to fit the crista parotica, a ridge on the paroccipital process of the petrosal. Relative

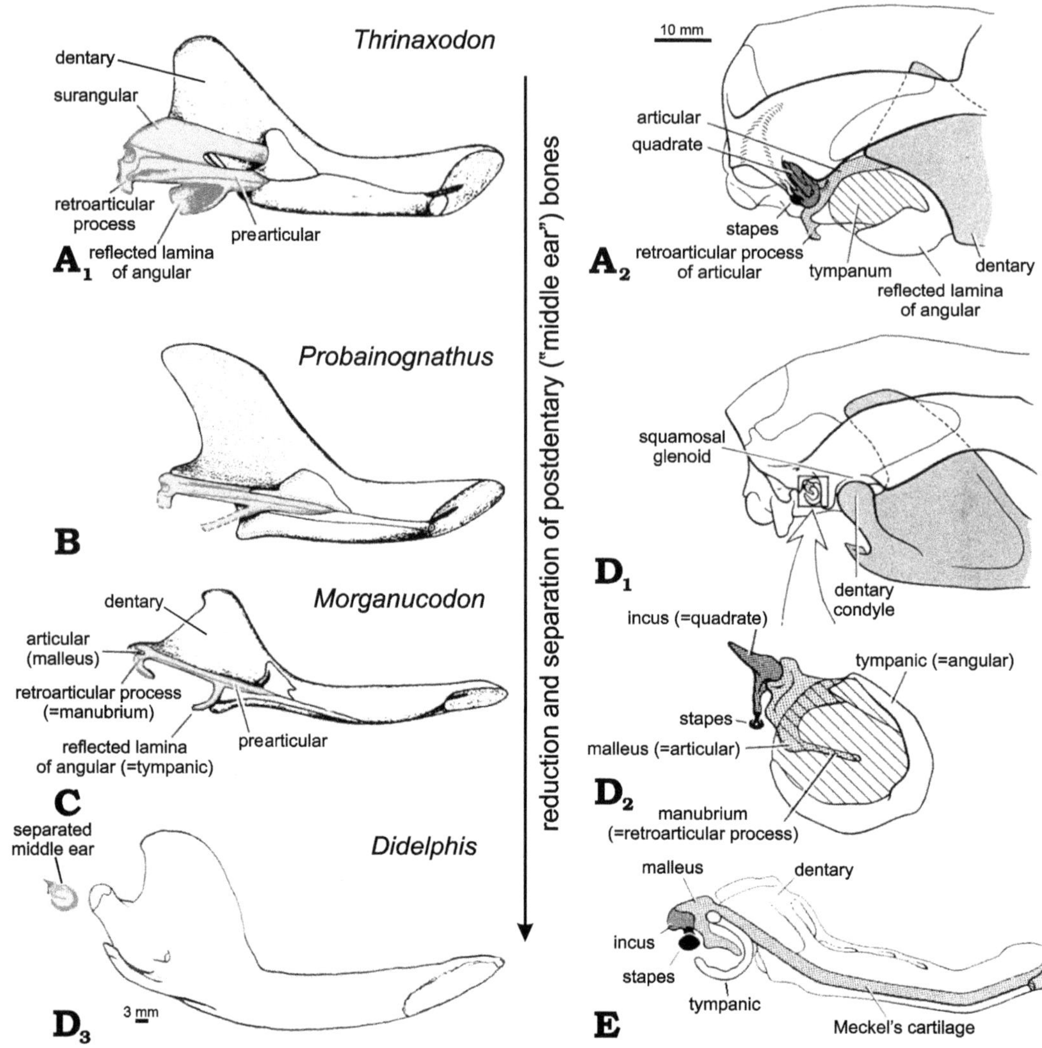

FIGURE 3.10. Comparative morphology of the CMJ of cynodonts and TMJ of mammals and the reduction of the postdentary elements in the cynodont-mammal evolution. A, Cynodont *Thrinaxodon:* medial view of mandible (A_1); lateral view of CMJ and the postdentary elements attached to the dentary (A_2). B, Cynodont *Probainognathus* (medial view of mandible, showing the reduction of postdentary or "middle ear" bones). C, Stem mammal *Morganucodon* (medial view of mandible, showing further reduction of postdentary or "middle ear" bones). D, Marsupial *Didelphis:* TMJ in lateral view (D_1); mammalian middle ear in lateral view, showing homology of middle ear bones (D_2); mandible in medial view, separation of middle ear bones from the dentary (D_3). E, Mandible in medial view, early embryonic stage of a placental (showing the attachment of Meckel's cartilage to the dentary). *Source:* A_1, B, C, from Allin and Hopson (1992); A_2, D_1, D_2, from Crompton and Jenkins (1979); D_3, original; E, modified from Sues (1986, originally from Gaupp, 1913).

to the crista parotica, the quadrate could rotate and rock with far greater mobility than the quadrates of cynodonts. These are all derived characters of the mammal crown group (table 3.4).

In *Thrinaxodon,* the quadrate-cranium articulation was reinforced by the quadratojugal, a bone that is also retained in more derived cynodonts (figure 3.13A), including tritylodontids (Kühne, 1956; Sues, 1986). In several derived cynodonts, such as *Probainognathus* and *Massetognathus* (see Crompton, 1972b), the surangular extends posteriorly so that it may come into contact with the squamosal; the posterior peduncle of the dentary is also enlarged. The enlarged posterior end of the dentary and the surangular encroach on the quadratojugal. In tritheledontids, which are considered by some to be the sister taxon to mammals (Hopson and Barghusen, 1986; Shubin et al., 1991; Luo, 1994), the newly established dentary-squamosal contact displaced or diminished the quadratojugal that was associated with the primitive articulation of the quadrate in basal cynodonts (figure 3.13A). Loss of the quadratojugal further contributed to greater mobility of the quadrate for increased efficiency in hearing function

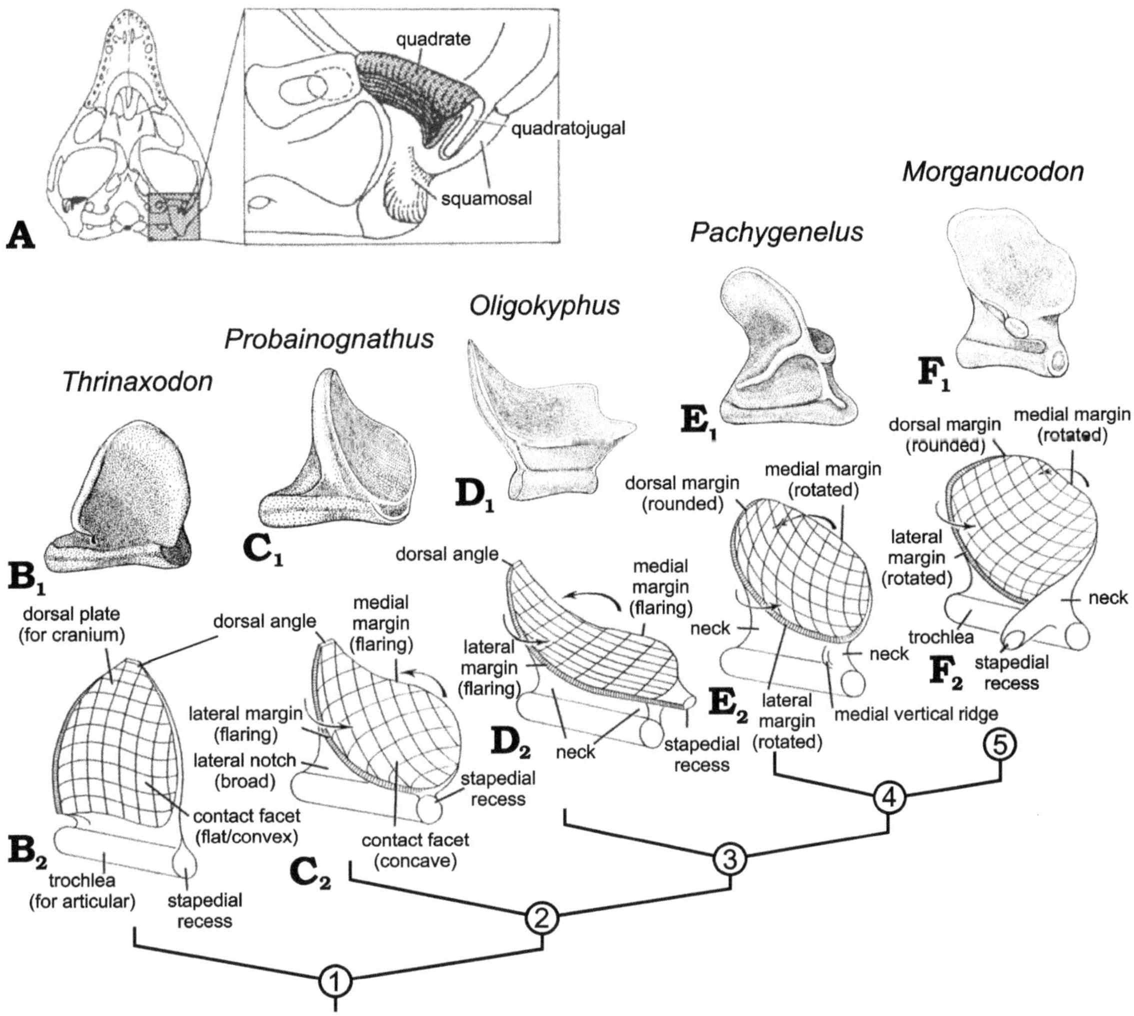

FIGURE 3.11. Comparative morphology of quadrates in nonmammalian cynodonts and stem mammal *Morganucodon*. A, Cynodont *Thrinaxodon:* location of the quadrate in the skull, in ventral view. B, *Thrinaxodon*. C, *Probainognathus*. D, *Oligokyphus*. E, *Pachygenelus*. F, *Morganucodon*. B_1–F_1, quadrates in posterior view. B_2–F_2, generalized models of quadrates in posteromedial view. Primitive cynodont condition, exemplified by *Thrinaxodon* (Node 1), is represented by a trochlea in the same plane as the posteriorly convex dorsal plate. Acquisition of apomorphies—Node 2: Concave posterior surface of dorsal plate and twisting of dorsal plate relative to the trochlea. Node 3: Presence of a neck between the twisted dorsal plate and trochlea. Node 4: Loss of dorsal angle. Node 5: Presence of stapedial recess (also present in tritylodontids). *Source:* B_1, C_1, D_1, E_1, F_1, from Luo (1994); A, B_2, C_2, D_2, E_2, F_2, from Luo and Crompton (1994).

(Luo and Crompton, 1994), as part of a morphological transformation of the middle ear in therapsid evolution (Allin, 1975; Allin and Hopson, 1992). In mammals such as *Sinoconodon* and *Morganucodon*, the dentary condyle established its contact with the squamosal glenoid joint (see previous section on the "compound" jaw joint in stem mammals). Concurrently, the incus, smaller than its homologues in cynodonts, was pushed medially onto the petrosal. Therefore, the simplification of the incus-cranial articulation and the migration of the incus are secondary consequences of the evolution of a CMJ necessitated by increasingly mammal-like feeding functions.

Detachment of Angular and Prearticular/Articular (Ectotympanic and Goniale) from the Mandible. There has been an ongoing debate as to whether the final separation of the middle ear from the dentary occurred only once (Kemp, 1983; Rowe, 1988, 1996a,b; Luo, Crompton, and Sun, 2001; Wang, Hu, Meng, et al., 2001), or separately in different phylogenetic lineages (Allin, 1975; Crompton and Parker, 1978; Crompton and Jenkins, 1979; Allin and Hopson, 1992).

Many evolutionary modifications of the middle ear (postdentary) bones toward the modern mammalian condition had already occurred in cynodonts and stem mam-

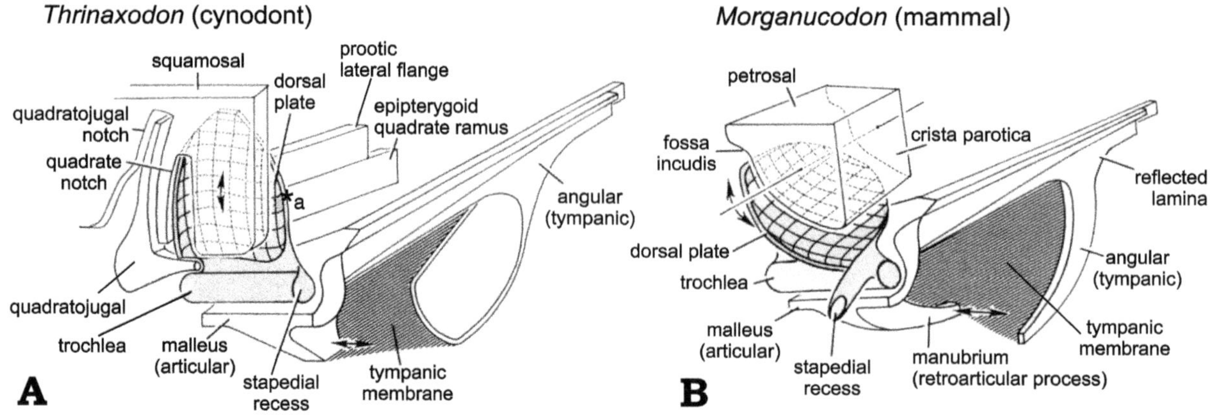

FIGURE 3.12. Comparison of the quadrate-cranium articulation in cynodonts and mammals (posteromedial views). A, Non-mammalian cynodont *Thrinaxodon*. B, Stem mammal *Morganucodon*. Cranial articulation of the quadrate in cynodonts, exemplified by *Thrinaxodon*, is complex and consists of five different bones or bony elements: quadratojugal, squamosal, lateral flange of prootic, quadrate ramus of pterygoid, paroccipital process of prootic (not illustrated, represented by *a). This complex articulation contributes to a more rigid suspension of the quadrate in cynodonts than in mammals. Cranial articulation of the quadrate in mammals is entirely by the crista parotica on the petrosal. It is greatly simplified by comparison to those of cynodonts. The mammalian cranioincus joint is more mobile than in cynodonts owing to the conforming geometry of the quadrate dorsal plate and its corresponding petrosal crest, the crista parotica, and the simplification of this joint. *Source:* from Luo and Crompton (1994).

mals. The angular bone and the prearticular-articular complex probably already functioned to transmit sound as part of the "mandibular middle ear" in cynodonts and precynodont therapsids, as hypothesized by Allin (1975, 1986). The ear of stem mammals, such as *Morganucodon*, could even have achieved some better sensitivity to high-frequency hearing (Rosowski and Graybeal, 1991; Rosowski, 1992). However, the evolution of a completely "mammalian" middle ear required further modification, involving separation of the malleus (including goniale) and ectotympanic from the mandible and their incorporation into the basicranium in a condition known as the "cranial middle ear" (Maier, 1987; Zeller, 1993; Rowe, 1996a,b; Sánchez-Villagra et al., 2002).

The angular (ectotympanic) and the prearticular-articular complex (partially homologous to the goniale) are tightly nested within the postdentary trough in cynodonts and the stem mammals *Sinoconodon*, *Morganucodon*, and *Haldanodon* (Kermack et al., 1973; Kermack and Mussett, 1983; Lillegraven and Krusat, 1991; Crompton and Luo, 1993). It is generally assumed that in the more derived mammalian taxa without the postdentary trough, the middle ear elements would have achieved some degree of separation from the dentary (Allin and Hopson, 1992; Rowe, 1996a,b; Luo, Crompton, and Sun, 2001), if not complete severance from the latter (Wang, Hu, Meng, and Li, 2001). The earliest known mammal without a postdentary trough and Meckel's groove is the Early Jurassic *Hadrocodium*. By the currently available evidence, the middle ear elements had already separated from the dentary in *Hadrocodium* (figure 3.15). Rowe (1996a,b) and Luo, Crompton, and Sun (2001) espoused the view that the separation of the mammalian middle ear from the dentary occurred only once.

However, the pattern of phylogenetic evolution for the complete detachment of the middle ear elements from the dentary may be more complex among Mesozoic mammals (Allin and Hopson, 1992; Wang, Hu, Meng, et al., 2001, see also chapter 7). In the gobiconodontid *Repenomamus* (Li et al., 2001; Wang, Hu, Meng, et al., 2001), Meckel's cartilage is fossilized and still attached to the Meckel's sulcus of the dentary. Although the ectotympanic and the malleus themselves are not preserved, it is highly likely that, in life, these were still connected to the dentary via the ossified Meckel's cartilage in *Repenomamus* (Wang, Hu, Meng, et al., 2001). Meckel's cartilage is well separated from the coronoid region of the dentary. This resembles the embryonic pattern of monotremes (Maier, 1993) and placentals (MacPhee, 1981), in which Meckel's cartilage and its derivative elements (such as the incus and malleus) are positioned on the basicranium and well separated from the dentary very early in ontogeny. It differs from marsupials, in which Meckel's cartilage and its derivative elements are closely attached to the dentary until the final separation from the latter at a rather late postnatal stage (Maier, 1993; Rowe, 1996a,b; Sánchez-Villagra et al., 2002).

The attachment of Meckel's cartilage to the dentary in *Repenomamus* is also consistent with an earlier model by

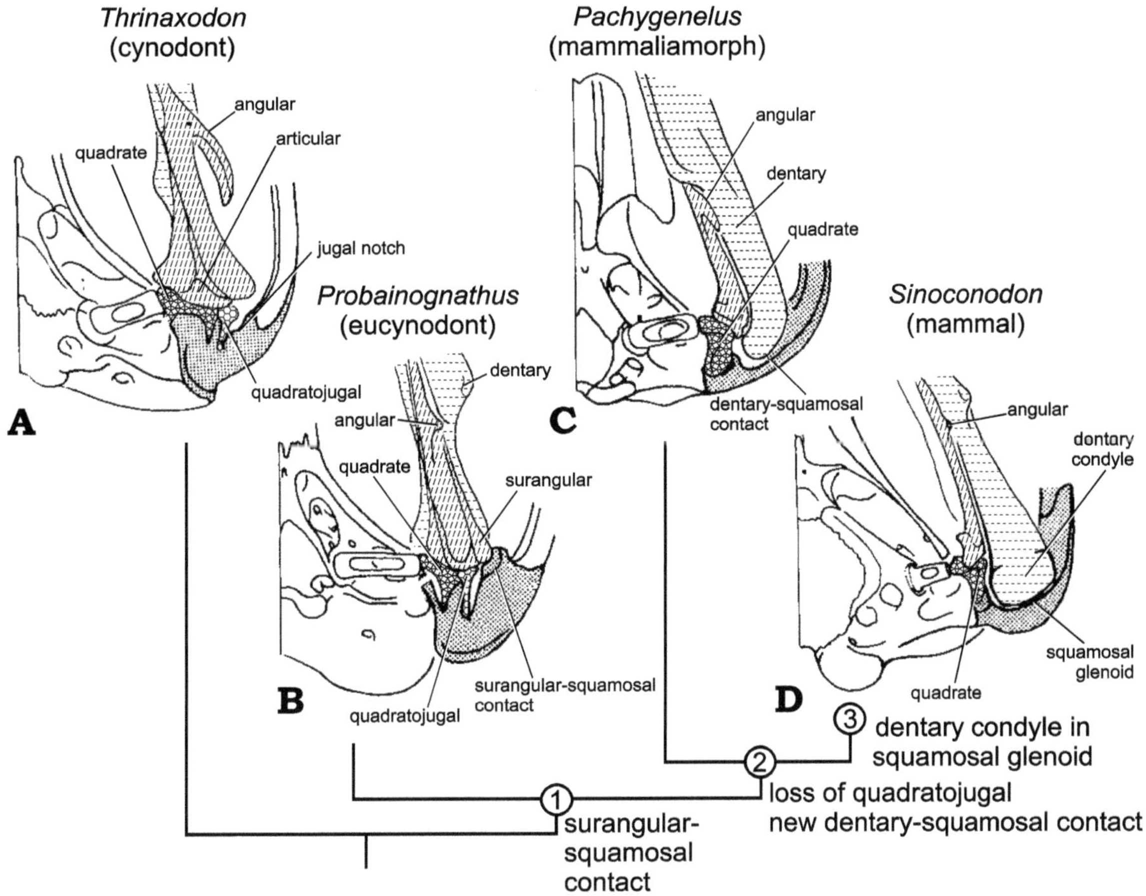

FIGURE 3.13. Comparison of CMJs in cynodonts and mammals (ventral view). A, *Thrinaxodon.* B, *Probainognathus.* C, *Pachygenelus.* D, *Sinoconodon.* Acquisition of apomorphies—Node 1 (eucynodonts, such as *Probainognathus*): surangular-squamosal contact (only in *Probainognathus* is there a squamosal facet; other eucynodonts may have a surangular contact to the squamosal without a squamosal facet). Node 2 (*Pachygenelus* + mammals): dentary-squamosal contact (although neither a true dentary condyle nor a true squamosal glenoid is present in *Pachygenelus*); posterior end of dentary extends into the ventral notch between cranial and zygomatic parts of squamosal, possibly displacing the qudratojugal. Node 3 (mammals): mammalian TMJ with the dentary condyle and the squamosal glenoid. *Source:* A, B, modified from Crompton (1972a) and Crompton and Hylander (1986); C, modified from Allin and Hopson (1992); D, modified from Crompton and Luo (1993) and Luo, Crompton, and Sun (2001).

Edgar Allin, who proposed that the ectotympanic would have been attached to Meckel's groove on the dentary (Allin, 1975; but see alternative interpretation by Rougier, Wible, and Hopson, 1996) in one of two possible arrangements. First, the ectotympanic could have been connected to the dentary through a vestigial Meckel's cartilage, as hypothesized for the "eupantothere" *Amphitherium* by Allin and Hopson (1992). The fossilized Meckel's cartilage in *Repenomamus,* as shown by Wang, Hu, Meng, and Li (2001), is consistent with this model for *Amphitherium.* In a second possible arrangement, a hypothetical ligament could have linked the ectotympanic to the dentary, as illustrated by Allin (1975) for a dryolestoid "eupantotherian." In either arrangement, the ectympanic would have retained its anterior connection, via a cartilage or a ligament, to Meckel's groove, but it became separated mediolaterally from the posterior part of mandible owing to the loss of the postdentary trough. The tympanic membrane supported by the ectotympanic bone in such an intermediate position could have received sound through the external auditory meatus, accommodated partially by the pterygoid fossa in the mandible.

The attachment of Meckel's cartilage to the dentary in the adults of *Repenomamus* helps to show that the absence of the postdentary trough in the dentary is not sufficient evidence to demonstrate complete disconnection of the angular and the articular from the dentary (Wang, Hu, Meng, et al., 2001). The ectotympanic and malleus can be mediolaterally separated from the mandible, while the ectotympanic (angular) could still be connected via a vestigial cartilage to Meckel's groove. Morphological evolution of the mammalian middle ear likely underwent two transformational steps, the first being an intermediate condition involving partial and sideways separation of the angular and articular bones from the angular and condylar regions of the dentary, with the final step involving anterior disconnection of the angular and articular bones, owing to the distrophy of Meckel's cartilage.

Gobiconodontids are part of Eutriconodonta (see chapter 7, Kermack et al., 1973; Jenkins and Crompton, 1979; Ji et al., 1999; Wang, Hu, Meng, et al., 2001; Luo et al., 2002). Eutriconodontans have alternately been included within (Rowe, 1988; Rougier, Wible, and Novacek, 1996a) or placed outside (Hopson, 1994; Hu et al., 1997; Ji et al., 1999) the mammalian crown group, or at an unresolved polytomy with the rest of the mammalian crown group (Luo, Crompton, and Sun, 2001; Wang, Hu, Meng, et al., 2001). According to the currently available morphological dataset of Mesozoic mammal clades, these alternative placements of eutriconodontans do not differ significantly (Luo et al., 2002, see also chapter 15).

However, these alternative placements do have different ramifications for mapping the separation of middle ear ossicles from the dentary. Detachment of the middle ear ossicles probably occurred multiple times in early mammal evolution, as argued by Crompton and Jenkins (1979) and Allin and Hopson (1992), if eutriconodontans are placed within the mammalian crown group or are closer to the mammalian crown group than *Hadrocodium* (Luo, Crompton, and Sun, 2001). But convergent separations would be less likely if eutriconodontans turn out to be an early divergent stem clade placed outside the mammalian crown group (Hu et al., 1998; Ji et al., 1999; Wang, Hu, Meng, et al., 2001).

Several hypotheses have been proposed to explain the mechanism for the separation of the middle ear (postdentary) elements from the dentary—the last crucial step in the transformation from the "mandibular ear" of nonmammalian cynodonts to the "cranial ear" of living mammals. Maier (1987, 1990) proposed that the initial movement of the dentary in later development stages of neonates could result in the disruption of Meckel's cartilage, leading to the separation of the proximal Meckel's cartilage and its associated malleus and incus from the dentary. In marsupial neonates, Meckel's cartilage forms a movable contact with the cranium (Maier, 1987; Sánchez-Villagra et al., 2002), although not a synovial joint (Filan, 1991, see also figure 3.10D_1). As such, the proximal portion of Meckel's cartilage is likely to provide mechanical support for minor movement and stress incurred during suckling (Kuhn, 1971; Maier, 1987; Sánchez-Villagra et al., 2002). After the dentary-squamosal joint is established at a later growth stage, the movement of the dentary could cause the disruption of Meckel's cartilage.

Herring (1993a,b) proposed an alternative scenario. She showed that teratological attachment of masticatory muscle to Meckel's cartilage could delay the resorption of Meckel's cartilage in abnormal human development, as the mandibular arch and masticatory muscles are interdependent during development (Kay, 1986). By inference, the resorption of Meckel's cartilage may be associated with the lack of mechanical loading from the adductor muscle, as the muscles shifted attachment to the dentary during normal development of the mammalian skull. The lack of mechanical stress by the masticatory muscles may be a contributing factor in the resorption of Meckel's cartilage during normal development.

Rowe (1996a,b) suggested that negative allometrical growth of the middle ear relative to the cranium and the brain is related to the detachment of the middle ear from the mandible. During the ontogeny of didelphid marsupials, the ectotympanic ring and other middle ear bones ossify and terminate their growth earlier than the surrounding basicranial structures (Maier, 1987; Sánchez-Villagra et al., 2002). The ectotympanic reaches its adult size after its ossification in the third neonate week. The malleus undergoes ossification during the fourth and fifth weeks. In subsequent development, these middle ear structures, with their size fixed upon ossification, show negative allometry relative to the skull and brain. The stapes is permanently attached to the fenestra vestibuli (Rowe, 1996a,b) and the incus articulates with the crista parotica (Maier, 1993; Luo and Crompton, 1994). After the sizes of these middle ear bones are fixed, the ectotympanic ring moves away from the jaw, as the jaw joint becomes further separated from the basicranial attachment points for the middle ear bones (the fossa incudis and the fenestra vestibuli) during prolonged peramorphic cranial growth. Therefore, the detachment of the middle ear bones from the mandible is ultimately correlated with the negative allometry of the middle ear elements and the peramorphic growth of the rest of the skull, driven by the growth of the brain (Rowe, 1996a,b, see figure 3.14).

Thus far, this development has only been documented in marsupials, but it is consistent with the observation of the negative ontogenetic allometry of the middle ear relative to the skull for monotremes and some placentals (Zeller, 1987, 1989a, 1993). The negative allometry of the middle ear bones relative to the skull in adults is also known for diverse placentals of a wide range of body sizes (Nummela, 1995, 1997). Rowe (1996a,b) further attributed the peramorphic growth of the cranium to brain enlargement during growth.

Hadrocodium is the earliest-known mammal without a postdentary trough or Meckel's groove, indicating that its

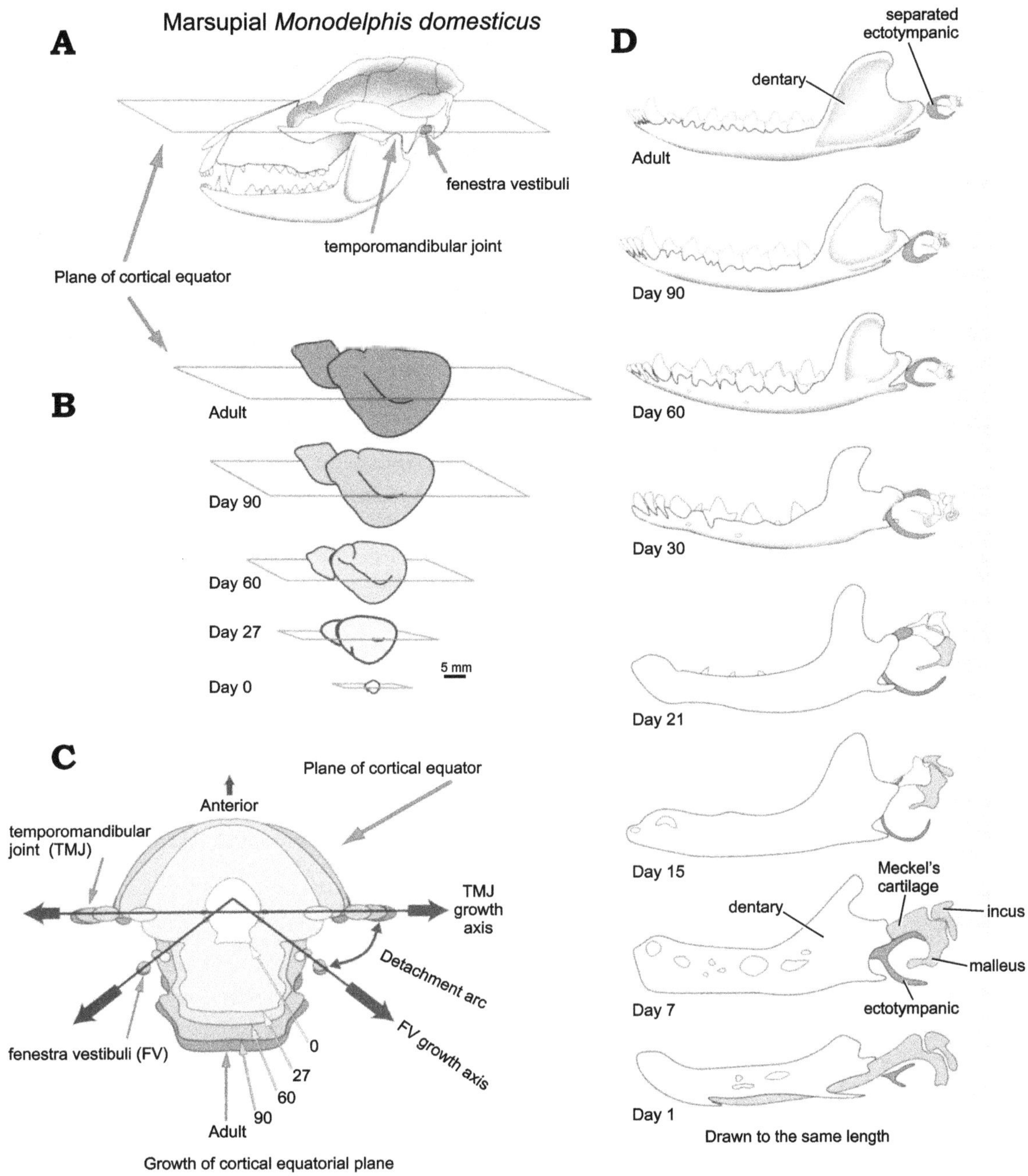

FIGURE 3.14. Ontogeny of the brain, mandible, and middle ear in the marsupial *Monodelphis:* growth in brain size, negative allometric growth of middle ear bones, and correlation of brain growth with the detachment of the postdentary elements from the jaw. A, Position of the brain relative to the TMJ and the fenestra vestibuli in the adult skull. B, Growth of the brain from neonate stage to adult. C, Distance between the fenestra vestibuli and the temporomandibular joint ("detachment arc") increases as the brain grows in volume during ontogeny, as an example of the relative displacement of basicranial structures during ontogeny. D, Ontogeny of the dentary and the middle ear ("postdentary") elements. The ectotympanic reaches its adult size at the end of the third week, and ossification of the malleus begins in the fourth and fifth weeks. The middle ear bones are fixed in size long before cessation of proportional changes among the basicranial structures, such as the fenestra vestibuli and TMJ, owing to the brain growth. The middle ear bones show negative allometry relative to the mandible, basicranium, and brain. Relocation of the middle ear bones may be correlated to displacement by the growth of the brain and the basicranium. *Source:* from Rowe (1996a,b).

middle ear had already been separated from the mandible. *Hadrocodium* also has a proportionally larger braincase than its contemporary *Sinoconodon, Morganucodon,* and *Megazostrodon,* and than the Late Jurassic *Haldanodon.* The concurrence of the expanded brain and the separation of middle ear from the mandible in *Hadrocodium* would suggest a correlation of the detachment of the middle ear from the mandible and the peramorphic growth of the brain (Luo, Crompton, and Sun, 2001). Many Triassic and Jurassic mammals with a mandibular middle ear lack the postglenoid region behind the TMJ. By contrast, *Hadrocodium* has a well-developed postglenoid region behind the TMJ (figure 3.15), similar to those of most mammals in the mammalian crown group. In marsupials (Rowe, 1996a,b), the growth of the basicranium and braincase behind the TMJ is correlated with the detachment of the middle ear from the mandible. The absence of the mandibular attachment of the middle ear in *Hadrocodium* may be correlated with the posterior displacement of the basicranium relative to the TMJ (figure 3.15).

In summary, the separation of the angular and articular from the dentary and smaller ear bones was necessitated by the negative allometry of the middle ear after the early fixation of their size in relation to the rest of the basicranium, which grows with the size of the brain. Detachment of the middle ear bones from the dentary could occur by a ventral displacement of middle ear bones relative to the TMJ (Rowe, 1996a,b) (figure 3.14), by a posterior displacement of the incus-cranial articulation relative to the (TMJ) as a postglenoid region developed in the skull of more derived mammalian taxa (Luo, Crompton, and Sun, 2001) (figure 3.15), or by the medial displacement of the middle ear bones from the mandible as shown in embryonic stage of monotremes (Maier, 1993) and adult gobiconodontids (Wang, Hu, Meng, et al., 2001). It is likely that the initial separation of middle ear bones from the dentary correlated with a combination of all these factors, which are not necessarily mutually exclusive. These observations on differential growth rates of the middle ear (Rowe, 1996a,b; Luo, Crompton, and Sun, 2001) are not incompatible with the biomechanical interpretations of Maier (1993) and Herring (1993a,b).

EVOLUTION OF THE INNER EAR STRUCTURE AND ITS BONY HOUSING

Cynodont Condition (figures 3.16A, 3.17A). The bony structure surrounding the inner ear differs between mammals and nonmammalian cynodonts. The bony housing for the inner ear in cynodonts is formed by multiple bones, as has been documented in great detail for many nonmammalian cynodonts by Olson (1944), Estes (1961), Fourie (1974), Quiroga (1979a), Allin (1986), and Rowe et al. (1994). The osseous cochlear cavity is formed by the prootic, which is overlapped superficially by the basisphenoid wing. The vestibular part of the inner ear of cynodonts is enclosed by the prootic and the opisthotic (known collectively as the periotic bones, or "periotics"). The semicircular canals are enclosed by the opisthotic, the exoccipital, and the supraoccipital. None of these bones are fused. The prootic and the opisthotic bones in *Thrinaxodon* correspond to their homologues in extant amniotes, which are developed from the ossifications of the embryonic auditory capsule. The bony housing of the inner ear in adult *Thrinaxodon* is formed by a mosaic of several bones, which can be either endochondral or intramembranous in embryonic origin among extant amniotes.

Many features of the inner ear bony housing of *Thrinaxodon* (figure 3.16A) are primitive because they are also present, albeit in a slightly different condition, in precynodont therapsids (Olson, 1944; Cox, 1962; Sigogneau, 1974). The basisphenoid wing (parasphenoid ala), a large component of the parabasisphenoid complex, overlaps the prootic and indirectly contributes to the cochlear housing. The basisphenoid wing borders on the fenestra vestibuli, and the fenestra has a thickened rim of bone. The basioccipital overlaps extensively on the prootic (figures 3.16, 3.17). Part of the posterior semicircular canal is enclosed by the exoccipital and even, possibly, by the supraoccipital (Olson, 1944).

The cochlear cavity of the osseous inner ear is distinctive from the vestibule in primitive cynodonts such as *Thrinaxodon* (Estes, 1961; Fourie, 1974). This cochlear cavity is small and globular; it does not extend anterior to the fenestra vestibuli in the reconstructed inner ear endocast (figure 3.16A_1). This bony cochlea is too short and too small to be termed the cochlear canal in *Thrinaxodon* in which the cochlea is more developed than in such precynodont therapsids as dicynodonts (Cox, 1962) and gorgonopsians (Sigogneau, 1974). In these therapsids, the sacculocochlear cavity (or recess) is not differentiated from the vestibular cavity. The fenestra vestibuli is located at the anterior end of this sacculocochlear cavity.

The Mammaliamorph Condition (figures 3.17–3.19). The pars cochlearis, or the cochlear part of the petrosal, is completely ossified in the tritylodontid *Yunnanodon,* as first noted by Sun and Cui (1987). The pars cochlearis is superficially covered by the basioccipital on the medial side and by a hypertrophied basisphenoid wing anteriorly (Luo, 2001). Because of coverage by the basioccipital and basisphenoid, there is no external (tympanic) exposure of the pars cochlearis in the ventral view of the basicranium.

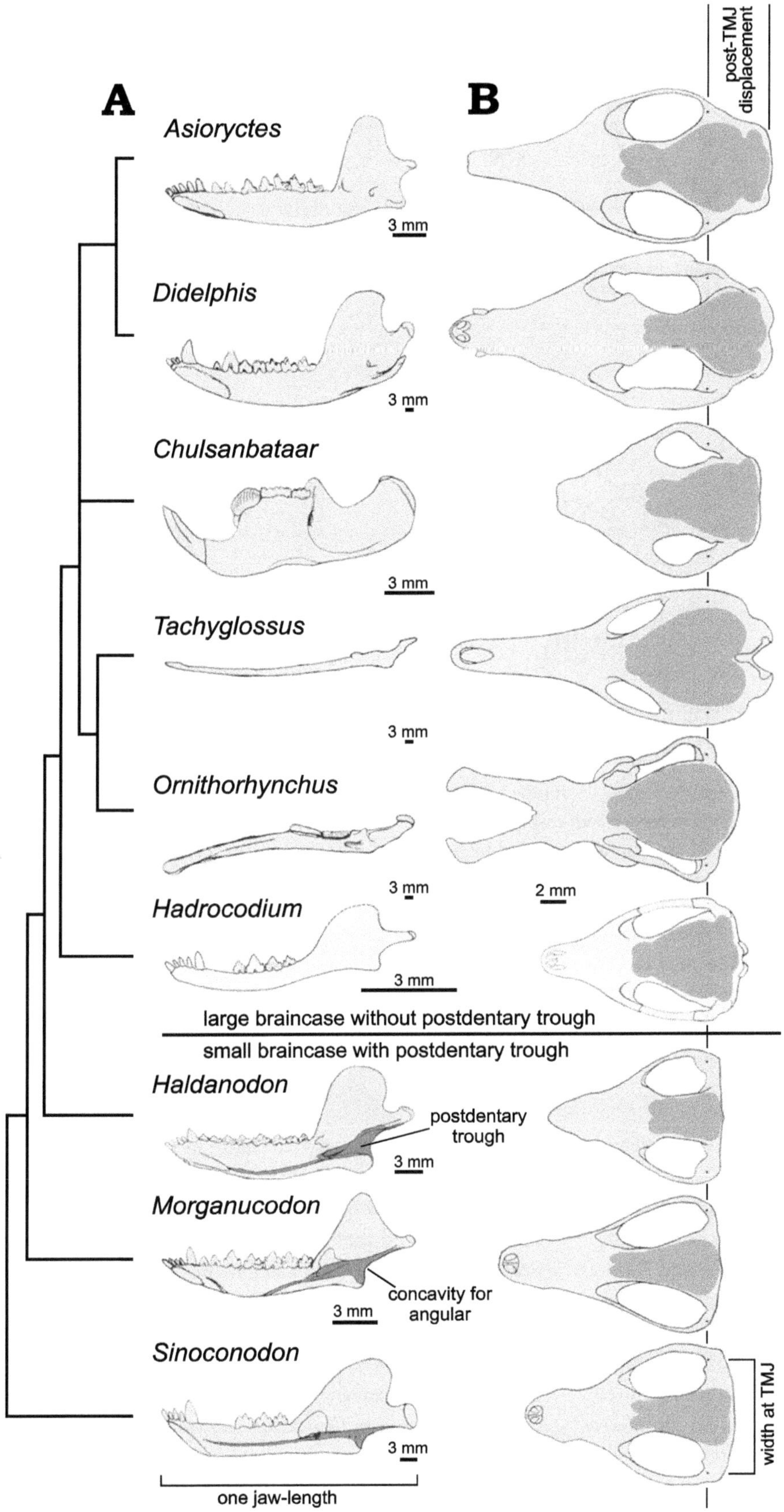

FIGURE 3.15. Evolution of braincase and mandibular structure in mammals: correlation of braincase expansion and loss of the postdentary trough and medial concavity of mandibular angle for the angular (ectotympanic) bone. A, Mandibles in medial view (scales vary among mandibles of different taxa): mandibles are of standardized length; the postdentary trough, the medial concavity, and the Meckel's groove on the mandibular angle are darkened. B, Cranium in dorsal view: crania of different sizes are standardized to the same width between the left and right TMJs; scales vary among the crania of different taxa. The darkened areas represent the approximate extent of the brain endocasts including olfactory lobes. Stem mammals in the basal part of the tree, such as *Sinoconodon*, *Morganucodon*, and *Haldanodon*, have the postdentary trough and medial concavity of the mandibular angle (for postdentary "ear" elements) as well as a small braincase. *Hadrocodium* and more derived taxa, with larger braincases, show the separation of the middle ear bones from the mandible and also more extension of the braincase posterior to the TMJ. *Source:* modified from Luo, Crompton, and Sun (2001).

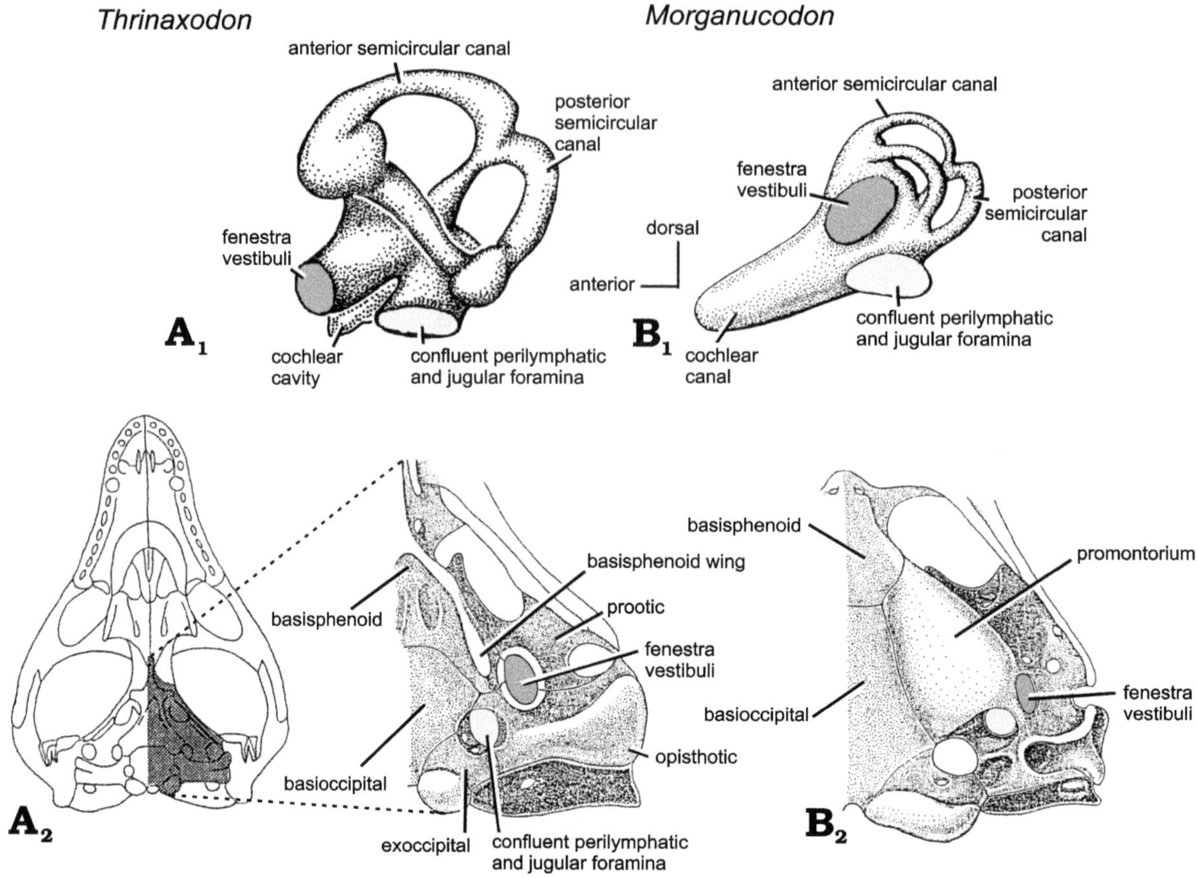

FIGURE 3.16. Comparison of inner ears and their bony housing among nonmammalian cynodonts and mammals. A, Cynodont *Thrinaxodon:* reconstructed inner ear endocast in ventrolateral view (A_1); inner ear bony housing in the basicranial region, in ventral view (A_2). B, Stem mammal *Morganucodon:* reconstructed inner ear endocast in ventrolateral view (B_1); inner ear bony housing in the basicranial region, in ventral view (B_2). The inner ear of cynodonts is enclosed by five bones or bony elements: prootic, opisthotic, basisphenoid wing, basioccipital, and exoccipital. The mammalian inner ear is enclosed by a single bone, the petrosal, which is homologous to the fused prootic and opisthotic. The petrosal is much larger and has displaced and excluded other bones from the immediate housing of the inner ear. The mammalian pars cochlearis, or the bony housing for the cochlear duct, forms a ventral eminence, the promontorium, to accommodate the elongate cochlear canal of the inner ear. The promontorium and the cochlear canal are derived features by comparison to the nonmammalian cynodont *Thrinaxodon,* where there is no promontorium and the cochlear recess is poorly developed. *Source:* A_1, modified from Fourie (1974) with sketches courtesy of E. F. Allin; A_2, B_2, modified from Luo et al. (1995); B_1, from Luo (2001).

Hence, tritylodontids lack a "promontorium" (Gow, 1986a; Luo, 2001).

The cochlear canal in *Yunnanodon* (figure 3.17A, 3.18B_1, 3.19E) is significantly longer and more tubular than in *Thrinaxodon* (figure 3.19B), *Probainognathus,* and *Massetognathus.* Unlike the condition in *Thrinaxodon,* the cochlear canal of *Yunnanodon* extends anterior to the fenestra vestibuli. The presence of a tubular cochlear canal has also been reported in other tritylodontids (Kühne, 1956; Crompton, 1964a; Hopson, 1965). Unlike *Thrinaxodon,* the semicircular canals of *Yunnanodon* are entirely enclosed by the petrosal.

The bony housing of the inner ear and the relationships among the pars cochlearis, the basisphenoid, and the basioccipital are essentially the same in the tritheledontid *Pachygenelus* (see Crompton, 1995) and tritylodontids (figures 3.17A, 3.18B). Tritylodontids and tritheledontids are more derived than *Probainognathus, Massetognathus, Probelesodon,* and *Thrinaxodon* in having the fused prootic and opisthotic and a basisphenoid wing that is smaller, withdrawn anteriorly, and no longer bordering on the periphery of the fenestra vestibuli, in contrast to the pattern of *Thrinaxodon* (see figure 3.16A).

The Mammalian Condition (figure 3.16B). The bony housing of the inner ear in stem taxa of mammals is formed exclusively by the petrosal, which is the single bone homologous to the fused prootic and opisthotic elements in nonmammalian cynodonts (Kermack et al., 1981; Rowe,

1988; Luo et al., 1995). The pars cochlearis is much larger than the prootic of cynodonts and forms a ventrolateral eminence known as the promontorium, which is a very conspicuous external feature of the mammalian basicranium (Kermack et al., 1981; Gow, 1985; Hopson and Barghusen, 1986; Rowe, 1988; Lillegraven and Krusat, 1991; Wible and Hopson, 1993; Luo et al., 1995). In derived therian mammals, the promontorium forms a bulbous structure on the tympanic surface of the pars cochlearis (Williams et al., 1989; Schaller, 1992), with some vascular features (Rougier et al., 1992; Wible et al., 2001).

In mammals, an enlarged petrosal excludes the parabasisphenoid complex, the basioccipital, and the exoccipital from the bony housing for the inner ear. This has been documented in a wide range of stem mammals, including *Sinoconodon* (see Luo et al., 1995), morganucodontans (Kermack et al., 1981; Gow, 1985; Graybeal et al., 1989; Luo and Ketten, 1991), docodontans (Lillegraven and Krusat, 1991), *Hadrocodium* (Luo, Crompton, and Sun, 2001), and triconodontids (Kermack, 1963; Crompton and Luo, 1993; Rougier, Wible, and Hopson, 1996). The bony structure housing the entire inner ear is essentially the same in stem mammals (Luo et al., 1995), monotremes (Kuhn, 1971; Zeller, 1989a; Luo and Ketten, 1991; Fox and Meng, 1997), multituberculates (Miao, 1988; Luo and Ketten, 1991; Lillegraven and Hahn, 1993; Meng and Wyss, 1995; Fox and Meng, 1997; Hurum, 1998b), and in some stem therians (Wible et al., 1995; Hu et al., 1997).

The cochlear canals in the Jurassic stem mammals (figure 3.19F,G) are more derived than in *Thrinaxodon* (figures 3.16A, 3.19B) in being more elongate and distinctively differentiated from the saccular cavity. An elongate cochlear canal is a derived feature shared by diverse Mesozoic mammal groups, as documented in multituberculates (Miao, 1988; Luo and Ketten, 1991; Lillegraven and Hahn, 1993; Meng and Wyss, 1995; Fox and Meng, 1997; Hurum, 1998b), docodontans (Lillegraven and Krusat, 1991), and possibly in stem therians (Wible et al., 1995; Hu et al., 1997). The elongate bony cochlear canal suggests a better-developed cochlear duct. This may indicate more sensitivity to high-frequency sound, which is very important in the hearing function of all extant mammals and was probably important for at least some of the earliest mammals (Rosowski and Graybeal, 1991; Rosowski, 1992; Hurum, 1998b).

***Adelobasileus* and *Sinoconodon* (figure 3.17B,C).** *Adelobasileus* has an incipient promontorium on the petrosal and retains the primitive condition in which the basisphenoid wing partially covers the anterior part of the pars cochlearis (Lucas and Luo, 1993). Although the suture between the pars cochlearis and the basioccipital is not clear, it appears that the basioccipital is broad (figure 3.17B), suggestive of an overlap of the basioccipital on part of the cochlear housing, as in *Sinoconodon* and *Yunnanodon.*

In *Sinoconodon* and other more derived mammals, the basisphenoid wing is completely absent (figures 3.17, 3.18). The pars cochlearis is enlarged and fully exposed on the ventral side of the basicranium as the promontorium. However, the lateral part of the basioccipital overlaps on the medial side of the pars cochlearis (figure 3.17C). Therefore, part of the cochlear bony housing has a composite structure. All semicircular canals of the inner ear are enclosed by the petrosal, to the exclusion of the exoccipital and the supraoccipital (Luo et al., 1995).

The cochlear canal inside the pars cochlearis is slightly curved toward its apex in *Sinoconodon.* The canal is oriented more anteromedially, rather than in the ventromedial direction as in nonmammalian cynodonts (Fourie, 1974; Allin, 1986), and extends anteriorly for less than half the length of the promontorium. The presence of a short cochlea within the fully developed promontorium in such early mammals as *Sinoconodon* suggests that the formation of an enlarged pars cochlearis with a promontorium may be a precursor condition for the further elongation of the cochlea in the later transformation of mammalian ear structure.

***Morganucodon* and *Hadrocodium*.** The pars cochlearis forms a larger promontorium in *Morganucodon, Hadrocodium,* and multituberculates than in *Sinoconodon* and *Adelobasileus* (figure 3.18). The enlarged pars cochlearis has displaced the basioccipital, and the lappet of basioccipital on the pars cochlearis is no longer present (figures 3.17D,E, 3.18). The promontorium is also more inflated and expanded in *Morganucodon* (Kermack et al., 1981; Luo et al., 1995), *Hadrocodium* (Luo, Crompton, and Sun, 2001), and multituberculates (Hahn, 1988; Hurum, 1998b) than in *Sinoconodon* and *Adelobasileus* (Luo et al., 1995).

Elongation of the cochlear canal in these stem mammals is correlated with the expansion of the pars cochlearis (Luo et al., 1995). In *Morganucodon* and *Haldanodon,* the cochlear canal reaches anteriorly about three-quarters the length of the promontorium (Graybeal et al., 1989; Lillegraven and Krusat, 1991; Luo et al., 1995). Relative to the length of the skull, the promontorium is not significantly longer in *Morganucodon* and *Haldanodon* than in *Sinoconodon,* but their cochlear canals are significantly longer than that of *Sinoconodon* if the cochlear length is standardized to skull length (Luo et al., 1995). Multituberculates (Meng and Wyss, 1995; Hurum, 1998b) and eutriconodontans are similar to *Morganucodon* rather than to *Sinoconodon* in the proportions of the promontorium and cochlear canals.

Crown Mammals. A major feature of inner ear evolution is the convergent coiling of the cochlear canal in dif-

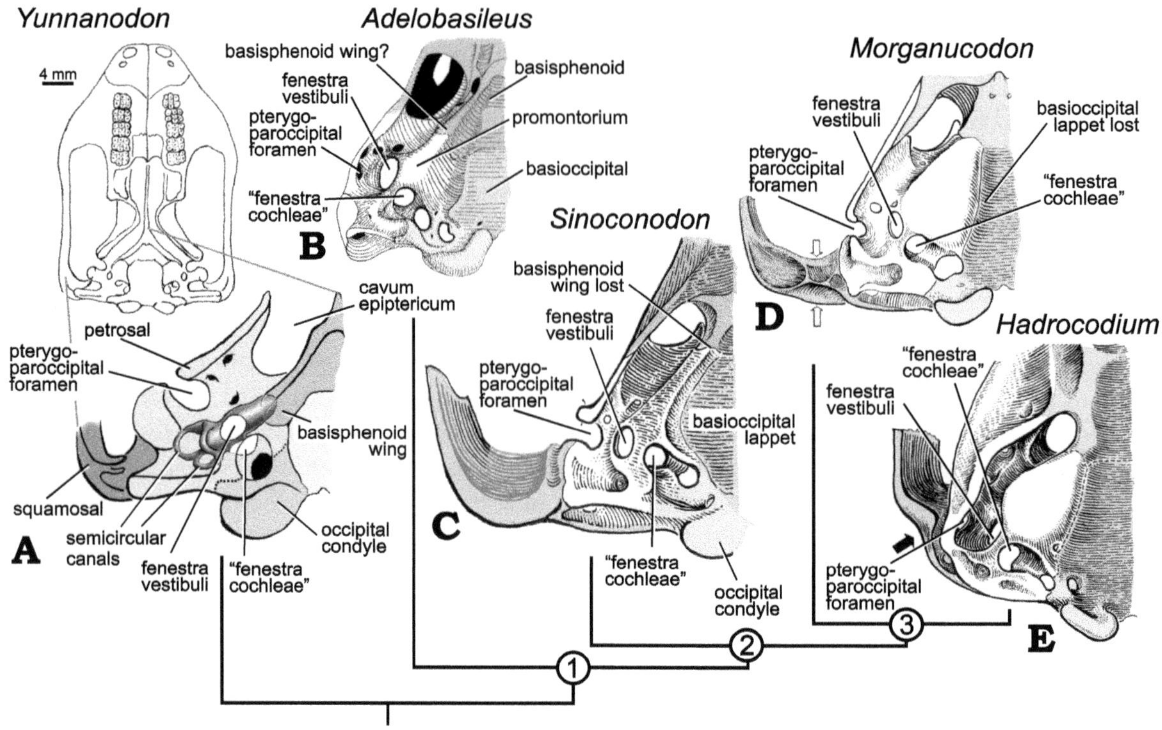

FIGURE 3.17. Transformation of the bony housing of the inner ear in mammaliamorph-mammal evolution (ventral view): the enlarged petrosal, especially its pars cochlearis, excludes other bones from the inner ear bony housing. A, Tritylodontid *Yunnanodon.* B, *Adelobasileus.* C, *Sinoconodon.* D, *Morganucodon.* E, *Hadrocodium.* Acquisition of apomorphies—Node 1: emergence of pars cochlearis as an incipient promontorium. Node 2: promontorium displacing the basisphenoid wing from the inner ear housing. Node 3: promontorium displacing the lateral lappet of the basioccipital. *Source:* modified from: A, Luo (2001); B, Lucas and Luo (1993); C, E, Luo, Crompton, and Sun (2001); D, Kermack et al. (1981).

ferent lineages of the mammalian crown group. The coiling of the bony labyrinth and membranous labyrinth in crown mammals shows a considerable degree of homoplasy (figure 3.20). The living monotremes have a coiled cochlear duct (membranous labyrinth), but they lack the corresponding coil of the bony cochlear canal (bony labyrinth) (Zeller, 1989b; Luo and Ketten, 1991; Fox and Meng, 1997). Multituberculates have a slightly curved bony cochlear canal without a coil (Luo and Ketten, 1991; Meng and Wyss, 1995; Hurum, 1998b), as in the stem therian *Zhangheotherium* (Hu et al., 1997). In the pretribosphenic mammal *Vincelestes,* the cochlear canal is coiled for about 270° (Rougier, 1993). In the early eutherian *Prokennalestes,* it is coiled for about 360° (Wible et al., 2001). Only metatherians and eutherians have a fully coiled membranous labyrinth (cochlear duct) that is intricately associated with the coiled bony labyrinth (cochlear canal) (Zeller, 1989a; Luo and Ketten, 1991; Meng and Fox, 1995b; Fox and Meng, 1997).

Multituberculates are considered by many to be a part of the mammalian crown group. *Zhangheotherium* is more closely related to crown therians than to either multituberculates or monotremes. Given these relationships, the pattern of cochlear coiling presents a conspicuous case of homoplasy (figure 3.20). Either the coiling of the membranous cochlear duct in living monotremes has to be considered as convergent to those of living therians; or, less likely, the uncoiled cochlear canals of eutriconodonts, multituberculates, and *Zhangheotherium* have to be regarded as an atavistic reversal to the condition of stem mammals (Hu et al., 1997).

SUMMARY OF INNER EAR EVOLUTION (figures 3.16–3.20)

Mammals are most derived among living vertebrates in their complex ear structures and related hearing adaptation. The pars cochlearis containing a cochlear canal is one of the most complex character systems of the mammalian skull and is crucial for sensitive hearing function, especially for high-frequency sound. The assembly of such a complex character system with significant functional adaptation occurred in several incremental steps during the morphological evolution of nonmammalian cyn-

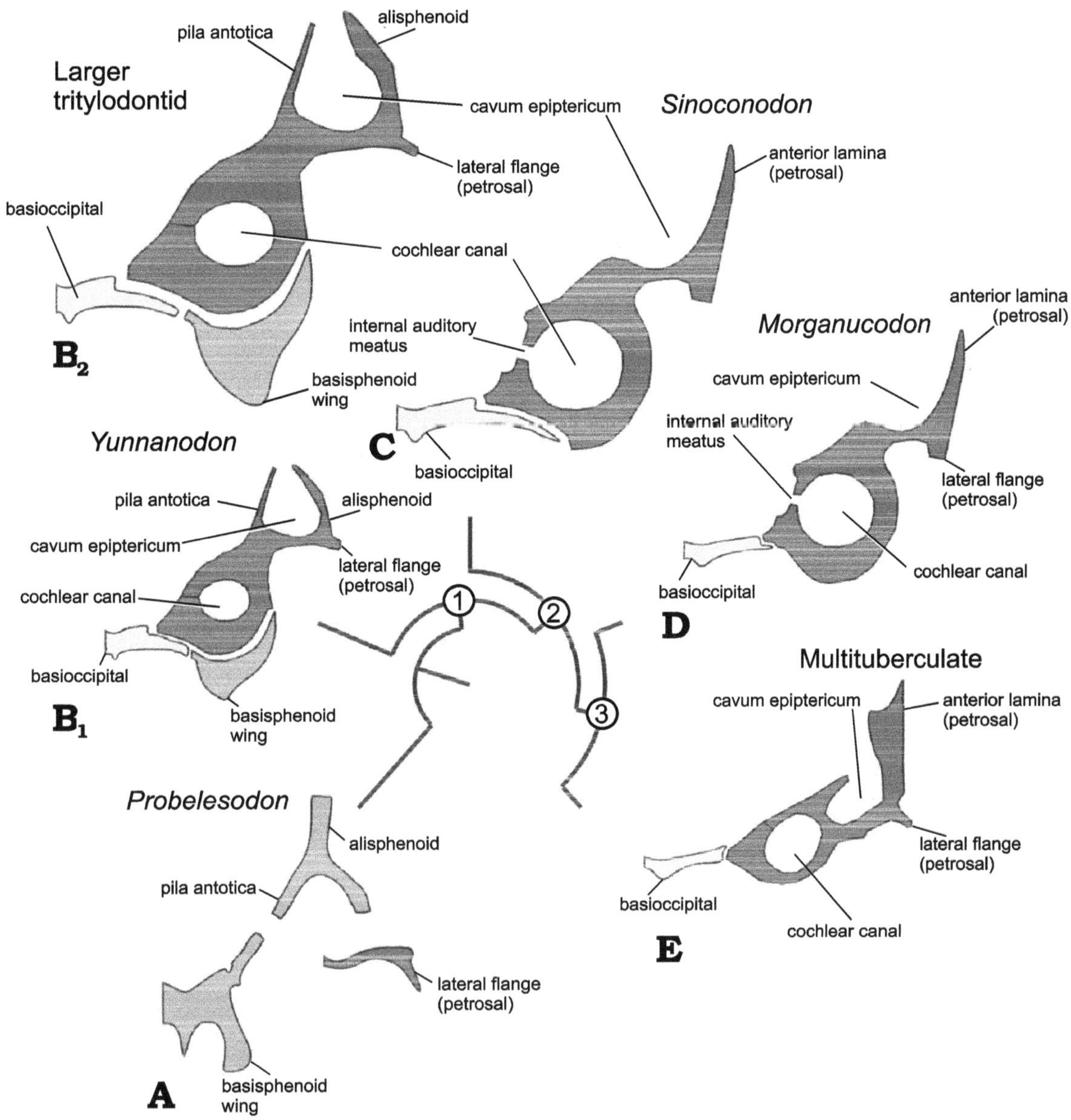

FIGURE 3.18. Transformation of the cochlear bony housing ("pars cochlearis") through the cynodont-mammalian transition (transverse section across the ear region near the cavum epipterricum). A, Cynodont *Probelesodon.* B, Tritylodontids: *Yunnanodon,* a small tritylodontid (B_1) and a large tritylodontid (B_2), to illustrate the allometric size difference of the basisphenoid wing relative to skull size. C, *Sinoconodon.* D, *Morganucodon.* E, Multituberculate. Acquisition of apomorphies—Node 1: formation of a pars cochlearis for the cochlear canal. Node 2: enlargement of pars cochlearis of the petrosal, and its emergence as the promontorium, at the expense of the basisphenoid wing. Node 3: promontorium displacing the lateral lappet of basioccipital. *Source:* A–D, from Luo (2001); E, from Hurum (1998b).

odonts and early mammals (figure 3.20). The differences between mammals and cynodonts in the inner ear and its bony housing are the result of their correlated structural transformation. Emergence of the pars cochlearis as the promontorium and enlargement of the promontorium are correlated with the elongation of the cochlear canal. The enlarged pars cochlearis displaced the neighboring sphenoid complex and basioccipital bone. Inflation of the bulbous promontorium in the mammalian crown group is associated with coiling of the cochlear canal.

EVOLUTION OF THE DENTAL REPLACEMENT AND SKULL GROWTH PATTERN

The diphyodont replacement in modern eutherian mammals is characterized by a single replacement of a deciduous (milk) dentition by permanent incisors, canines, and premolars. The molars are part of the permanent dentition and are never replaced. The number of successional teeth per tooth locus in mammals is far smaller than in the

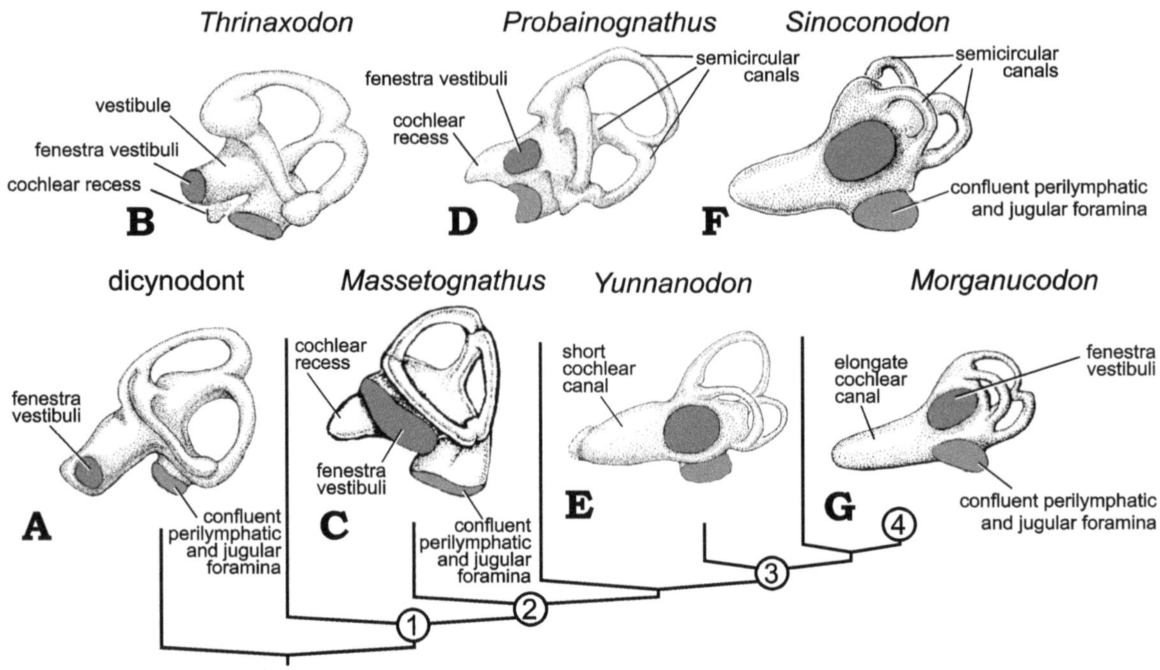

FIGURE 3.19. Comparison of inner ear endocasts of cynodonts and primitive mammals. A, Generalized inner ear structure of a dicynodont. B, *Thrinaxodon.* C, *Massetognathus.* D, *Probainognathus.* E, *Yunnanodon.* F, *Sinoconodon.* G, *Morganucodon.* Acquisition of apomorphies—Node 1 (epicynodonts): presence of cochlear recess. Node 2 (eucynodonts): anterior position of the cochlear recess. Node 3 (mammaliamorphs): presence of a well-defined cochlear canal. Node 4 (*Morganucodon* and more derived mammals): elongation of the cochlear canal. Not to scale. *Source:* A, modified from Olson (1944) and Cox (1962); B, modified from Fourie (1974) and from sketches courtesy of E. F. Allin; C, modified from Quiroga (1979a); D, modified from Allin (1986); E, F, from Luo (2001); G, modified from Graybeal et al. (1989) and Luo and Ketten (1991).

polyphyodont replacement of cynodonts (figure 3.21) because there is a delay in eruption of the deciduous teeth in mammalian neonates, resulting from lactation after birth, and there is an early termination of dental replacement, owing to determinate skull growth. Therefore, diphyodonty is a major mammalian apomorphy, as has been discussed by many mammalian biologists and paleontologists (Brink, 1956; Hopson and Crompton, 1969; Romer, 1970b; Ziegler, 1971; Pond, 1977; Kermack and Kermack, 1984; Crompton and Sun, 1985; Gow, 1985; Luckett, 1993; Luo, 1994; Luo et al. in press).

A delay in dental eruption in the mammalian diphyodont pattern is possible because lactation allows toothless mammalian neonates to achieve a considerable amount of cranial growth at a rapid rate before the teeth erupt (Brink, 1956; Hopson, 1973; Pond, 1977; Luckett, 1993). Mammalian diphyodont dental replacement is also correlated with the determinate growth pattern of the skull. Prior to weaning, the rate of skull growth exceeds that of the postcranial skeleton, but it slows down after weaning. The end of skull growth (determinate pattern) usually coincides with the eruption of the last molar (Pond, 1977; Smith, 2000). Therefore, dental replacement in early mammals can be correlated with skull growth pattern and may even allow an indirect inference of lactation in extinct mammalian lineages.

Replacement Pattern in Extant Mammals. Placental mammals have a diphyodont dentition in which all deciduous antemolars (including the premolars or milk molars) are replaced by successional and permanent teeth. In marsupials (figure 3.21), only the ultimate premolar (P3/p3) has both an erupted deciduous tooth and a replacing successional tooth (Cifelli, 1993b; Luckett, 1993; Cifelli et al., 1996). In marsupials other antemolars have only one generation of functional teeth. The first and second premolars of marsupials retain the erupted deciduous teeth in the adult, without replacement. The incisors and canines have rudimentary and nonerupting deciduous precursors that are lost or resorbed, followed by the accelerated development of successional permanent teeth (Luckett, 1993; Luckett and Woolly, 1996; Luckett and Hong, 2000). It has been well established by Luckett (1977, 1993) that the highly transformed diphyodont replacement in marsupial neonates is correlated with their perinatal specialization for prolonged lactation by fixation on the maternal nipples (Maier, 1993; Zeller, 1999a).

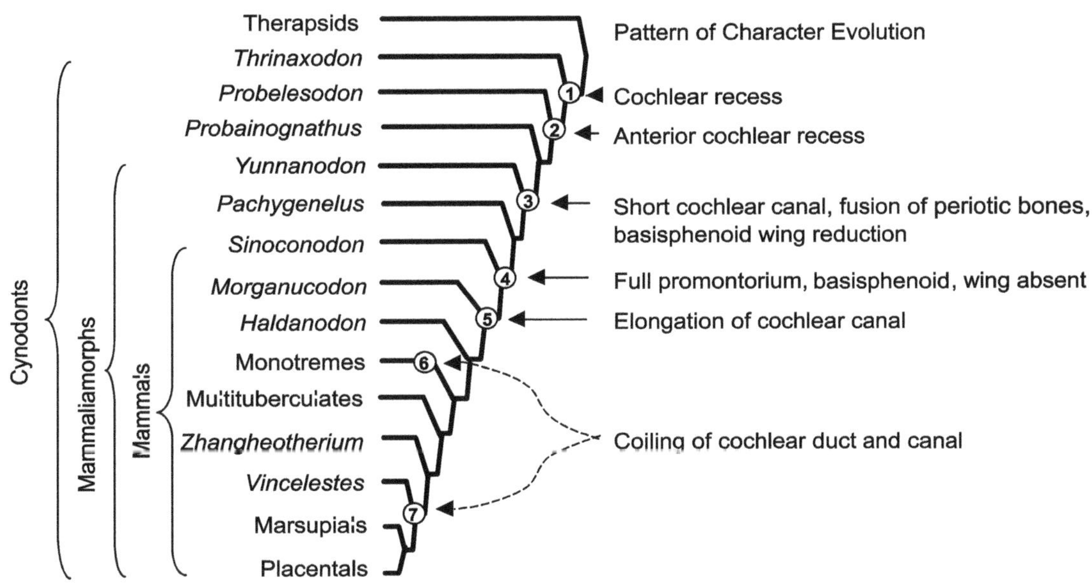

FIGURE 3.20. Stepwise transformation of the bony housing of the inner ear and cochlear structures from nonmammalian cynodonts to mammals. The primitive condition in precynodont therapsids is the lack of differentiation of the cochlea from the saccule of the vestibule. Acquisition of apomorphies—Node 1 (epicynodonts): presence of the cochlear recess. Node 2 (eucynodonts): cochlear recess positioned anteromedial to the fenestra vestibuli. Node 3 (mammaliamorphs): short but distinctive cochlear canal. Node 4 (mammals): full promontorium, loss of the basisphenoid wing. Node 5 (the clade of *Morganucodon* + derived mammals): elongation of the cochlear canal. Nodes 6 (monotremes) and 7 ("eupantotherians" and more derived therians): convergent coiling of the bony cochlear canal (dashed lines). Although the membranous labyrinth of the cochlea (cochlear duct) is coiled in all living mammals, the bony labyrinth of the cochlea (cochlear canal) lacks the corresponding coil in monotremes. The derived prototribosphenidan *Vincelestes*, some stem metatherians, and some stem eutherians have a coil of bony labyrinth (cochlear canal) of 270° in correlation with the coiled membranous labyrinth (cochlear duct). Only in crown marsupials and crown placentals is the cochlea coiled to or beyond a full turn. At least two groups nested within the crown mammalian group, multituberculates and the "symmetrodont" *Zhangheotherium*, have a straight or a slightly curved cochlear canal without a coil. Either the coiled membranous cochlear duct in living monotremes has to be considered as convergent to those of living therians, or the uncoiled cochlear canals of multituberculates and *Zhangheotherium* have to be regarded as an atavistic reversal to the condition seen in stem mammals. The coiled cochlear structures within the pars cochlearis are homoplastic among the main lineages of the mammalian crown group. *Source:* modified after Luo (2001).

Thus documentation of dental replacement in Late Cretaceous marsupials (Cifelli et al., 1996) sheds light on the timing of evolution of their reproductive and developmental specializations.

The monotreme *Ornithorhynchus* has deciduous teeth in early growth stages but these are replaced by horny dental pads in adults (Simpson, 1929b; Green, 1937; Woodburne and Tedford, 1975). Of the two deciduous premolars, dp1 has no replacing successor, whereas dp2 has a successor dental lamina (Luckett and Zeller, 1989). In this regard, premolar replacement in *Ornithorhynchus* retains a vestige of the diphyodont premolar replacement presumed to be primitive for living therian mammals (Green, 1937; Luckett and Zeller, 1989; Archer et al., 1993).

Dental Replacement and Skull Growth in Diapsid Amniotes. Most of the toothed nonmammalian amniotes are characterized by multiple generations of dental replacement (polyphyodonty). In living diapsids, such as lizards, hatchlings must be capable of independent feeding with a full set of functional teeth. As demonstrated by Edmund (1960), Osborn (1971, 1973), Westergaard and Ferguson (1986, 1987), and Berkowitz (2000), the teeth erupt early in growth and are subsequently replaced in a continuous, alternating pattern throughout life. Indeterminate growth of the jaws in extant diapsids is correlated with this continued replacement, because continued lengthening of the jaws is required for the larger successor teeth to replace the smaller predecessor teeth, as the size and morphology of teeth are fixed at eruption. Continuous dental replacement in adults is correlated with the lengthening of the jaws, as part of the indeterminate growth in most living diapsids (e.g., Osborn, 1971; Westergaard and Ferguson, 1986, 1987). Modern reptiles with indeterminate growth lack the early rapid growth of juvenile mammals nourished by maternal milk. These dental replacement characteristics of extant diapsids are also

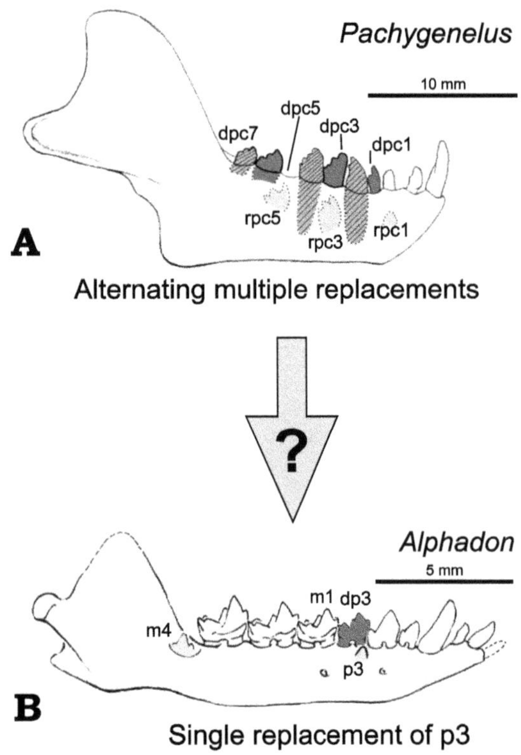

FIGURE 3.21. Comparison of dental replacement patterns in nonmammalian cynodonts and mammals. A, Cynodont *Pachygenelus.* B, Metatherian *Alphadon.* Most (but not all) nonmammalian cynodonts, as exemplified by *Pachygenelus,* are characterized by alternating replacement of all teeth for multiple generations, in correlation with indeterminate growth of the jaw. The deciduous postcanines are replaced at every other or every third tooth position in the same replacement wave. As the successor tooth tends to be larger than the precursor tooth in each tooth position, the jaw had to grow in length to accommodate the larger replacement teeth in cynodonts, as in extant diapsids. Mammals, as exemplified by *Alphadon,* have limited replacement of teeth in correlation with lactation and determinate growth of the jaw. Abbreviations: dp, deciduous premolars; dpc, deciduous postcanines; rpc, replacing postcanines. *Source:* A, modified after Crompton and Luo (1993); B, based on Clemens (1966), Luckett (1993), and Cifelli et al. (1996).

present in dinosaurs (Erickson, 1996; Erickson and Tumanova, 2000).

Condition of Cynodonts (figure 3.21). Most cynodonts are similar to diapsids in having an alternate dental replacement pattern and indeterminate jaw growth (*Thrinaxodon,* Crompton, 1963; Osborn and Crompton, 1971; *Pachygenelus,* Gow, 1980; Crompton and Luo, 1993); notable exceptions are the diademodontids (Hopson, 1971) and tritylodontids (Cui and Sun, 1987). Postcanines in *Thrinaxodon* are replaced up to three times per locus for the posterior tooth loci. Although lower than in living diapsids, the replacement rate is much higher than for any mammal. There is little doubt that most cynodonts had a reptilian pattern of indeterminate growth because their jaws continued to lengthen as smaller precursor teeth were replaced by larger successor teeth at the posterior loci in the large individuals. Because dental replacement begins in the smallest known cynodont individuals (Crompton, 1963; Hopson, 1971; Osborn and Crompton, 1971; Osborn, 1973; Crompton and Luo, 1993), it is unlikely that nonmammalian cynodonts developed lactation.

***Sinoconodon* (figure 3.22).** *Sinoconodon* has the differentiation of premolars from molars and a single replacement of premolars (Zhang et al., 1998; *contra* Crompton and Luo, 1993). Both are derived features of modern mammals. Replacement of the postcanines in *Sinoconodon* is sequential in the anteroposterior direction. This is a derived character for *Sinoconodon* and other stem mammals because it differs from the alternating replacement in most cynodonts (except for diademodontids and tritylodontids).

However, other dental replacement characteristics are very primitive for mammals. Incisors in *Sinoconodon* show an alternate replacement pattern, and the canines were replaced at least three times (Crompton and Luo, 1993; Zhang et al., 1998), as seen in many nonmammalian cynodonts.

The youngest known individuals of *Sinoconodon* (figure 3.22) have two premolars, which were replaced once before being permanently lost in the larger (thus older) specimens. The anterior molars (M1–M2) of the smaller individuals were probably lost without replacement. The loss of anterior postcanines (premolars and molars) resulted in a postcanine diastema that becomes increasingly larger in older individuals. The loss of anterior postcanines, coupled with successive addition of the newly erupted molariforms at the posterior end of the tooth row, results in a posterior shift of the functional tooth rows in the jaws (figure 3.22). These are primitive features known for several cynodonts, including *Thrinaxodon, Probainognathus,* diademodontids, and tritylodontids.

This pattern of loss of the anterior postcanines occurs, although to a lesser extent, in *Morganucodon, Hadrocodium,* and possibly also in *Kuehneotherium* (Mills, 1971, 1984; Parrington, 1971; Luo, Crompton, and Sun, 2001). Few other mammals have this pattern of tooth loss during growth, except where it was presumably acquired secondarily, as in proboscideans of the Tertiary.

The posterior molars (M3–M5) have one replacement in the larger (presumably older) individuals of *Sinoconodon.* The replacement of a smaller ultimate molar by a larger successor, followed by eruption of another ultimate molar in an older individual (figure 3.22) is observed only

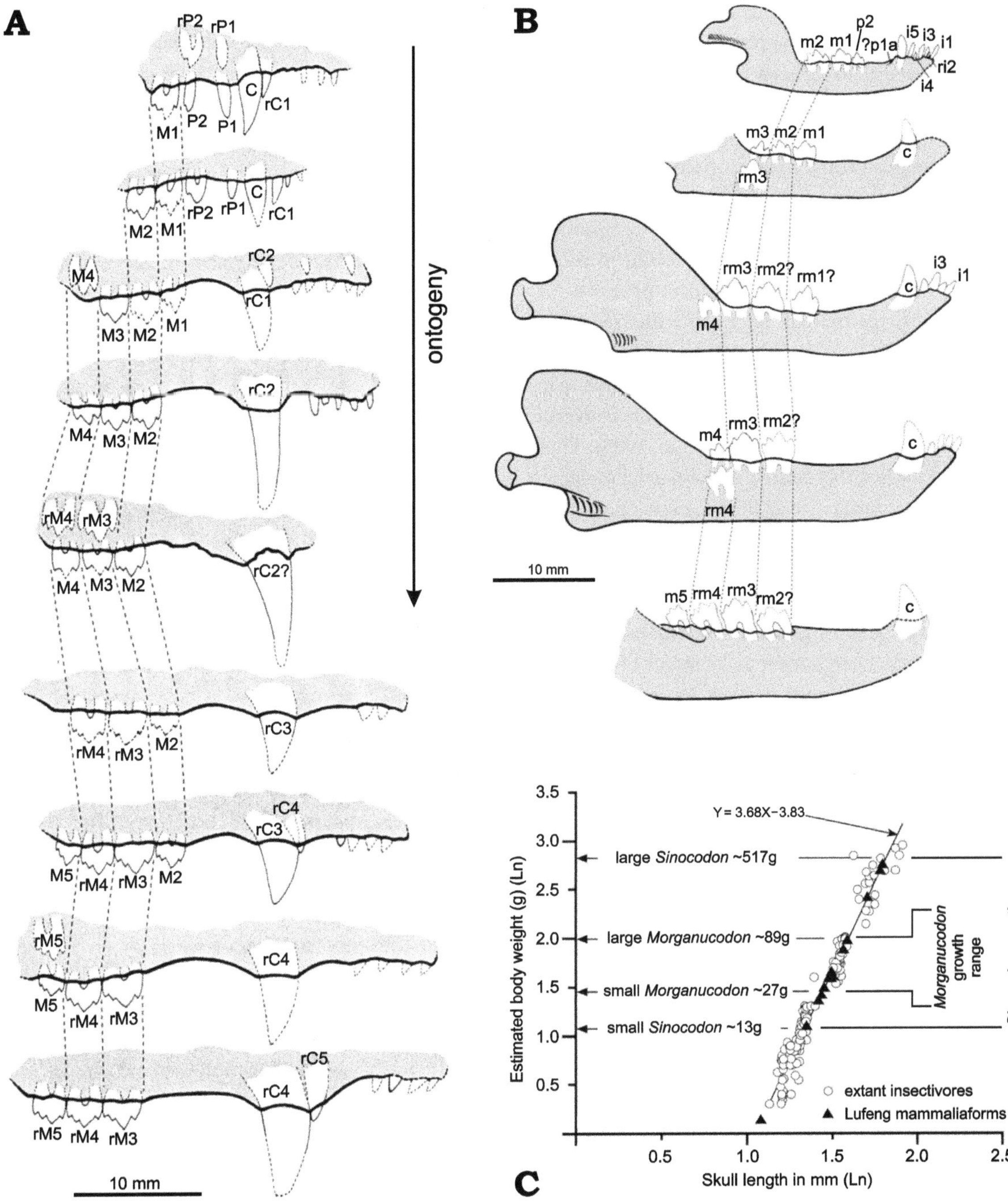

FIGURE 3.22. Continuous dental replacement during skull growth in *Sinoconodon*. A, Replacement of upper dentition and growth of maxilla and premaxilla. B, Replacement of the lower dentition and growth of mandible. C, Ranges of skull size of currently known specimens in *Sinoconodon* and *Morganucodon* and estimates of their body masses based on scaling data for extant insectivores. At least four generations of replacement canines occurred throughout the entire observed size range of *Sinoconodon* skulls collected from the Early Jurassic local fauna of the lower Lufeng Formation, China. In correlation with the continuous replacement of smaller precursor teeth by larger successors, the jaws and skulls increased in size (the skull length ranging from 22 to 62 mm; estimated body masses ranged from about 13 to about 517 g) in *Sinoconodon*. By comparison, *Morganucodon* skulls from the same Lufeng fauna ranged from 27 to 38 mm (with estimated body masses ranging from 27 to 89 g). This suggests that *Sinoconodon* had continuous skull growth accompanied by continuous tooth replacement well into late stages of adult life, a characteristic of the indeterminate growth pattern seen in extant nonmammalian amniotes and nonmammalian cynodonts: solid triangles, *Morganucodon* and *Sinoconodon* skulls; open circles, extant insectivore skull and body mass data ($y = 3.68x - 3.83$). Abbreviations: i, c, p, m: incisors, canine, premolars, and molars; r, replacing (successional) teeth at each tooth locus. *Source:* A, B, modified from Zhang et al. (1998); C, from Luo, Crompton, and Sun (2001), with scaling data courtesy of P. D. Gingerich.

in *Sinoconodon* and diademodontids (Crompton, 1963; Hopson, 1971). *Sinoconodon* also resembles diademodontids in that the smaller deciduous postcanines with simpler crowns are replaced by erupting postcanines that are larger and more molariform. By contrast, in the cynodont *Thrinaxodon,* the postcanine precursor tends to be more molariform and complex, and the successive replacing teeth are progressively simpler (Osborn and Crompton, 1971). A convergent pattern can also be found in most mammals in which a deciduous molariform precursor is usually replaced by a permanent premolar with a less molariform crown.

Sinoconodon lacked precise dental occlusion because it did not have a consistent pattern of opposition between upper and lower molars (Crompton and Sun, 1985). This primitive feature is possibly related to the partial replacement of the posterior molars and the successive posterior shift of the functional molariform row as part of the indeterminate growth pattern of the skull.

The currently available sample of *Sinoconodon* specimens shows a large range of growth, from the smallest individual, with an estimated body mass of about 13 g, to the largest, with an estimated body mass of more than 500 g (figure 3.22C). During this growth, the posterior molariforms were being replaced while the upper and lower jaws continued to lengthen in the successively older individuals.

From these characteristics, we infer that *Sinoconodon* experienced indeterminate growth at least in its skull, accompanied by continuous tooth replacement as in nonmammalian cynodonts (Crompton, 1963; Hopson, 1971; Osborn and Crompton, 1971; Osborn, 1974b) and modern diapsid reptiles (Osborn, 1971, 1974a; Westergaard and Ferguson, 1986, 1987). Moreover, it is possible (although difficult to prove) that *Sinoconodon* did not develop lactation (Zhang et al., 1998) because the polyphyodont replacement of its anterior teeth had already begun with the smallest known individual (figure 3.22A,B).

Sinoconodon is considered to be the sister taxon to all other mammals (Crompton and Sun, 1985; Crompton and Luo, 1993; Rowe, 1993; Hopson, 1994; Luo, 1994; Wible et al., 1995; Rougier, Wible, and Hopson, 1996; Luo, Cifelli, and Kielan-Jaworowska, 2001; Luo et al., 2002). Given its position in the cynodont-mammal transition, it is parsimonious to regard the dental replacement of *Sinoconodon* as representing an intermediate stage in the character evolution from the primitive pattern of polyphyodont replacement seen in most cynodonts to the derived diphyodont replacement of mammals. The combination of the mammalian dentary-squamosal joint and the cynodont-like multiple replacement of incisors and canines in *Sinoconodon* indicates that the mammalian TMJ evolved before the typical mammalian dental replacement pattern (Luo, 1994). The replacement features of *Sinoconodon* also suggest that the reduction in postcanine replacement preceded the reduction of the replacement of incisors and canines. In other words, the suppression of dental replacement occurred first in the postcanines and subsequently in the anterior dentition.

Morganucodon, Haldanodon,* and *Gobiconodon. *Morganucodon* had a single replacement of the posterior premolars (Mills, 1971; Clemens and Lillegraven, 1986). Parrington (1971, 1973, 1978) suggested that *Morganucodon* ("*Eozostrodon*") had a typically mammalian replacement of the incisors and canines, and subsequent observations are consistent with this suggestion (Kermack et al., 1973, 1981; Gow, 1985; Crompton and Luo, 1993). The mode of replacement of premolars is sequential in the anteroposterior direction, similar to that in *Sinoconodon.*

It is more difficult to rule out the nonreplacement of any molariform in morganucodontans. Gow (1986a) suggested that *Megazostrodon,* which is closely related to *Morganucodon,* might have replaced its m2 because a specimen shows that the m2 was more worn than the m1. But the sample size of *Megazostrodon* is too small to be certain about this. Parrington (1971, 1973) described the groove for the replacing dental lamina in some dentaries of *Morganucodon.* This groove is usually present in cynodonts with ongoing postcanine replacement. But dissection of the posterior dentary did not reveal any replacement teeth. The ultimate molars in *Morganucodon* are variable in size, which was considered by Parrington (1973) to be due to dimorphic or polymorphic variation.

The consensus of those who have studied the dentition of morganucodontans is that the molars are not replaced (Mills, 1971; Parrington, 1973; Crompton, 1974; Crompton and Parker, 1978; Clemens and Lillegraven, 1986; Luo, 1994). In the series of complete mandibles of *Morganucodon oehleri* (Young, 1982b; Crompton and Luo, 1993) and *Dinnetherium* (Jenkins et al., 1983; Crompton and Luo, 1993), there is little evidence for replacement of ultimate molars. While it is safe to suppose that *Morganucodon* achieved the typical mammalian diphyodont dental replacement pattern in both antemolars and anterior molars, the hypothesis of nonreplacement of the ultimate molars in morganucodontans remains to be tested by new data from a more complete sampling of skull growth series. If the variation in size and morphology of posterior molars of *Morganucodon* is shown to be correlated with growth stages, then the mode of skull growth of *Morganucodon* must be reconsidered.

We accept that *Morganucodon* had developed a typical diphyodont replacement pattern in all of its antemolars and that its anterior molars were not replaced. We tentatively accept the hypothesis that the ultimate molar was not replaced, with the caveats mentioned earlier. This working hypothesis (Parrington, 1971; Kermack and Kermack, 1984; Crompton and Luo, 1993; Luo, 1994) is consistent with the skull growth pattern of *Morganucodon* as seen in the currently available (although still small) sample of specimens (figure 3.22C).

Morganucodon shows far smaller size range in the skull than the contemporary *Sinoconodon* (figure 3.22C). The eight relatively complete skulls of *Morganucodon* discovered so far range from 27 to 38 mm in length (Rigney, 1963; Young, 1982b; Crompton and Luo, 1993; Luo et al., 1995; Zhang et al., 1998). This corresponds to a body size range from about 27 to 89 g (Luo, Crompton, and Sun, 2001), in strong contrast to a much wider range of variation in estimated body size of *Sinoconodon:* from 13 to over 500 g. This indicates that the adult skulls grew far less in *Morganucodon* than in *Sinoconodon.* Most maxillae and mandibles of *Morganucodon* from the Rhaeto-Liassic fissure deposits of Britain represent adults, and few are juveniles (Parrington, 1971, 1978; Kermack et al., 1973, 1981). Based on this Gow (1985) suggested that *Morganucodon* probably had a very short juvenile stage, a view supported by Luo (1994). The skull growth series and dental replacement indicate that *Morganucodon* grew to adult size during a short period of rapid growth, at least by comparison to *Sinoconodon* (figure 3.22) and cynodonts. *Morganucodon*'s growth pattern is closer to the determinate growth pattern typical of extant mammals than those of *Sinoconodon* and nonmammalian cynodonts. It is possible that *Morganucodon* had achieved determinate growth in its skull, as seen in living mammals.

Dental replacement of the docodont *Haldanodon* is now documented by a large sample of juvenile jaws (Krusat, 1980; Martin and Nowotny, 2000; Nowotny et al., 2001). The molars were not replaced. The incisors, canines, and premolars show diphyodont replacement that proceeded sequentially from front to back (Martin and Nowotny, 2000). This is a mammalian apomorphy. *Haldanodon* is more derived than *Sinoconodon, Morganucodon,* and *Kuehneotherium* in lacking the increasingly large postcanine diastema in older individuals, owing to the ontogenetic loss of the anterior premolars.

Gobiconodontids (figure 3.23B) show replacement of anterior molariform postcanines (Jenkins and Schaff, 1988; Kielan-Jaworowska and Dashzeveg, 1998; Wang, Hu, Meng, et al., 2001). *Gobiconodon* is unique in that the successor permanent tooth is similar to its deciduous precursor in the complexity of molariform morphology. By contrast, in most mammals the replacing successional tooth is simpler in crown morphology than its deciduous precursor at the same locus (Butler, 1952, 1995). By definition, postcanine loci with replacement should be considered to be premolars, whether the permanent teeth have a molariform or premolariform crown. The wave of replacement of the anterior postcanines was sequential and continuous from front to back, as shown by Jenkins and Schaff (1988) and corroborated by Wang, Hu, Meng, and Li (2001).

Multituberculates (figure 3.23B). The incisors and at least some premolars are diphyodont in their replacement (Szalay, 1965; Hahn, 1978b; Clemens and Kielan-Jaworowska, 1979; Greenwald, 1988; Hahn and Hahn, 1998a). On the basis of *Taeniolabis* and data from other multituberculates of the North American Tertiary, Greenwald (1988) proposed that multituberculates had a diphyodont replacement pattern similar to that seen in most placental mammals and that tooth eruption and replacement apparently occurred in an anteroposterior sequence (figure 3.23B). Hahn and Hahn (1998a) suggested that the Late Jurassic paulchoffatiid *Kielanodon* is an exception to this typical pattern: replacement of premolars could have occurred in an alternating mode, in two waves, and in the posteroanterior direction, as in *Thrinaxodon.* There are three possible interpretations. First, the posteroanterior and alternate replacement in paulchoffatiids could be the basal condition of all multituberculates, as proposed by Hahn and Hahn (1998a). In this case, multituberculates ancestrally would have had a *Thrinaxodon*-like replacement pattern. The anteroposterior sequential replacement would thus be interpreted as being secondarily derived for the North American Tertiary multituberculates. Another possible interpretation is that the anteroposterior sequential replacement is primitive for multituberculates as a group, as proposed by Greenwald (1988) and that the condition of paulchoffatiids is secondarily convergent to the distant cynodonts. A third possibility is that both paulchoffatiids and Tertiary multituberculates have independently derived conditions, neither of which was derived from the other.

Trechnotheria (*sensu* McKenna, 1975). The few stem trechnotherians in which tooth replacement is known (or can be reasonably inferred) show an alternating, anteroposterior pattern of diphyodont premolar replacement (figure 3.24). The alternating replacement in trechnotherians is limited to only one generation of successor. By contrast, the alternating replacement in many cynodonts is in the posteroanterior direction and includes multiple generations of successors.

Juvenile and subadult specimens of the spalacotheriid "symmetrodontan" *Zhangheotherium* show alternate replacement in the anteroposterior direction at least in the lower jaws (figure 3.24A). It appears that p1 erupted first, followed by the eruption of p3, and lastly by p2. The eruption of p3 occurred around the time of the eruption of m5.

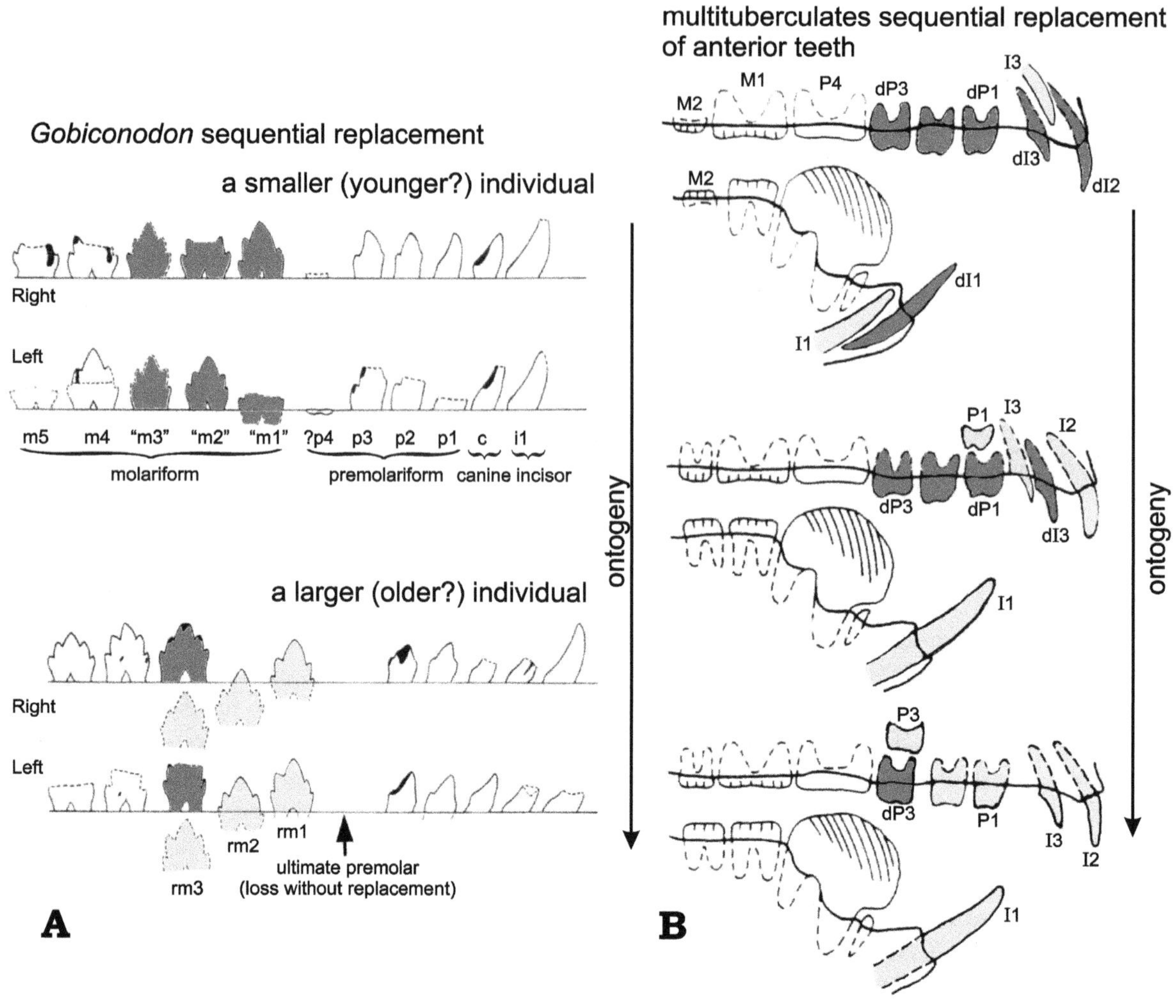

FIGURE 3.23. Sequential replacement of the anterior postcanines in multituberculates and gobiconodontids. A, Growth series of *Gobiconodon:* at least two anterior molariform postcanines were replaced, each by a molariform successor; both multituberculates and gobiconodontids retain the primitive anteroposterior sequence for replacing the antemolars. B, Multituberculate replacing sequence: I1 → I2 → I3 → P1 → P2 → P3 → P4(?). *Source:* A, from Jenkins and Schaff (1988); B, modified from Greenwald (1988).

The replacement of p2 occurred around the time of the eruption of m6. Thus the sequence of replacement was: p1 → p3 → p2. We suggest that the alternate pattern of premolar replacement in *Zhangheotherium* could be applicable to other spalacotheriids for which replacement of premolars is known (see Cifelli, 1999a). There is no evidence that molars were replaced in any spalacotheriids.

Dental replacement in the "eupantotherian" *Dryolestes* (see chapter 10) is documented in detail by extensive data. Deciduous teeth have long been recognized in dryolestids (Butler, 1939; Butler and Krebs, 1973). Martin (1997, 1999a) further demonstrated that all antemolar teeth were replaced in *Dryolestes.* The replacement, at least in the lower jaw, occurred in two anteroposterior series or waves. The first replacing series consisting of i2, i4, p1, p3 was followed by the second series of i1, i3, c, p2, p4. The p4 was the last premolar to erupt, and it did so just prior to the eruption of the sixth molar (m6). The premolar eruption is characterized as p1 → p3 → p2 → p4.

The stem boreosphenidan *Slaughteria* (assigned to an informal group, "tribotheres," see chapter 11) apparently had a similar replacement sequence of premolars: p3 → p2 → p4 (Kobayashi et al., 2002). Therefore, the alternating sequence of premolar replacement (p3 → p2 → p4) is present from spalacotheriids to dryolestids to stem boreosphenidans, representing three different hierarchies of therian mammal phylogeny, which makes a compelling case for this being a diagnostic condition for the trechnotherian lineage (including the basal eutherians, see later). This is different from the primitive pattern of antero-

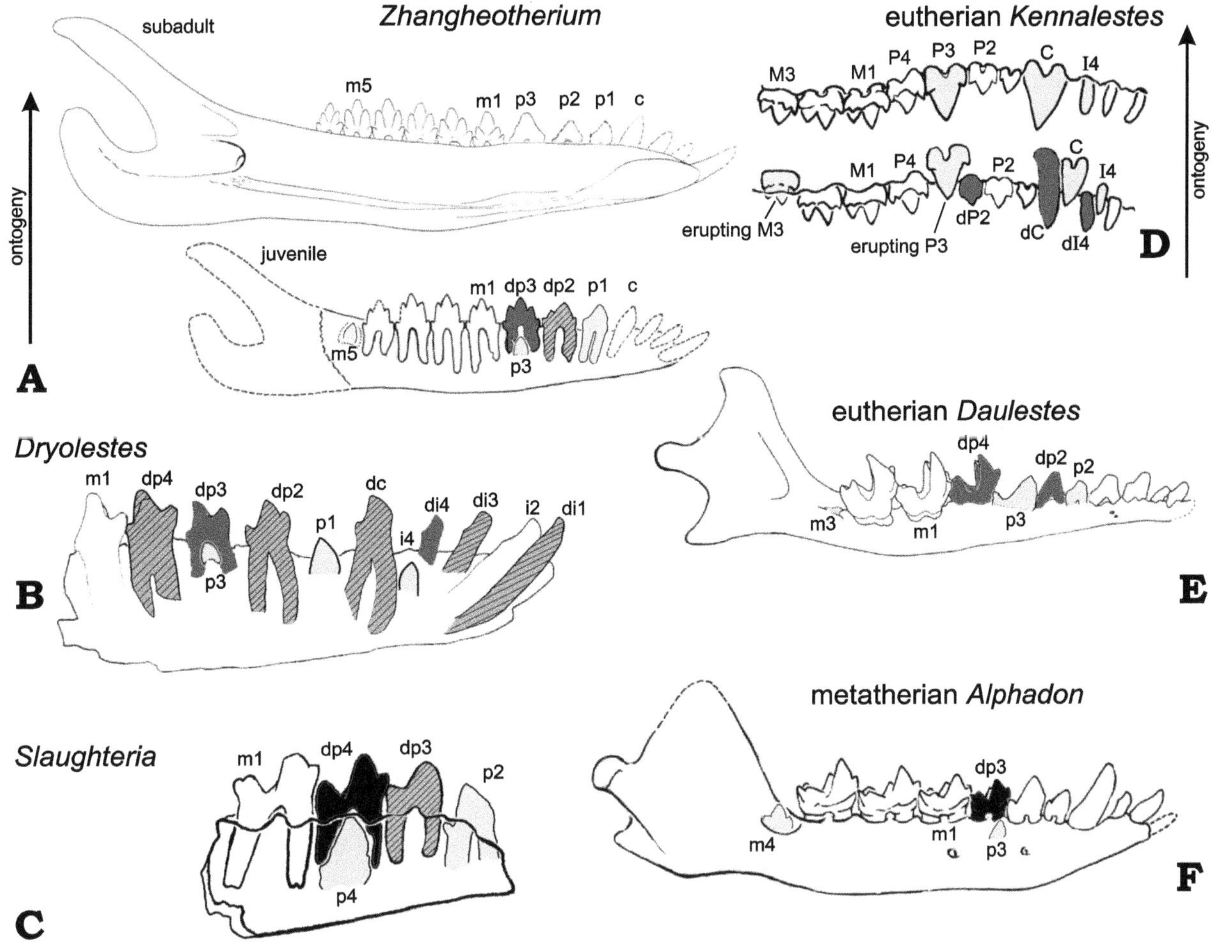

FIGURE 3.24. Alternating premolar replacement in therian mammals, including "symmetrodontans" (stem trechnotherians), "eupantotherians" (stem cladotherians), and "tribotherians" (stem boreosphenidans). A, *Zhangheotherium.* B, *Dryolestes.* C, *Slaughteria.* D, eutherian *Kennalestes.* E, Eutherian *Daulestes.* F, Metatherian *Alphadon.* By comparison to the anteroposterior sequential replacement of premolars in gobiconodontids, multituberculates, and *Sinoconodon,* the premolar replacement in Trechnotheria (clade of *Zhangheotherium* and extant therians) occurred alternately, characterized by P1 → P3 → P2 → P4. This pattern appears to have been retained in basal eutherians. It is only in the placental crown group that premolar replacement occurs sequentially, which is hypothesized to be a secondarily derived condition. In marsupials, postcanine replacement only occurs in the ultimate premolar, P3/p3. *Source* from: A, Luo, Ji, and Ji (2001); and Luo et al. (in press); B, Martin (1999a); C, Kobayashi et al. (2002). D, Kielan-Jaworowska (1981), E, McKenna et al. (2000); F, based on Clemens (1966), Luckett (1993), and Cifelli et al. (1996).

posterior sequential replacement of postcanines in all mammal groups outside the trechnotherian clade, such as most multituberculates, *Gobiconodon, Haldanodon, Morganucodon,* and *Sinoconodon.* We regard the alternating replacement of premolars (at least for p3 → p2 → p4) to be a synapomorphy of the trechnotherian clade.

Eutherians (figure 3.24). Replacement of the anterior dentition is known for three Late Cretaceous eutherians: *Daulestes* (see McKenna et al., 2000), *Kennalestes* (Kielan-Jaworowska, 1975b, 1981), and *Gypsonictops* (Lillegraven, 1969; Clemens, 1973a; Novacek, 1986b) (see chapter 13). Among these, specimens of *Daulestes* and *Kennalestes* preserve evidence for dental replacement; both can be interpreted as having a similar alternate replacement of premolars in the anteroposterior direction, as seen in the stem "therians" *Zhangheotherium, Dryolestes,* and *Slaughteria.*

The juvenile specimen of *Daulestes* has four premolar loci. Its permanent p2 had erupted before permanent p3, which in turn erupted before p4 (figure 3.24). The sequence of eruption of the permanent teeth in *Daulestes* is consistent with an anteroposterior sequential replacement. However, if we consider that permanent p2 and deciduous dp2 coexisted with each other because permanent p2 failed to dislodge dp2 (also see discussion by Luckett, 1993, on

Kennalestes), then replacement would be characterized differently. The shedding of dp3 occurred before the shedding of dp2, forming the sequence dp3 → dp2 → dp4.

A juvenile specimen of the Cretaceous eutherian *Kennalestes* also had dP3 → dP2 → dP4 as the shedding sequence of the deciduous teeth (figure 3.24), as revealed by comparison of the juvenile skull to the adult skull (Kielan-Jaworowska, 1975b, 1981; Luckett, 1993). If the premolar before the erupting successional p3 is indeed dp2, as interpreted by Luckett (1993) and accepted by McKenna et al. (2000) and Cifelli (2000a), then the replacement dp3 → dp2 → dp4 of *Daulestes* and *Kennalestes* would be identical to the alternating, anteroposterior replacement in *Slaughteria* (see Kobayashi et al., 2002), *Dryolestes* (see Martin, 1997), and *Zhangheotherium*. However, regardless of whether the replacement is anteroposterior sequential (for eruption of permanent teeth) or anteroposterior alternating (for shedding of deciduous teeth), the replacement in the Cretaceous *Daulestes* and *Kennalestes* was different from the pattern in living eutherians.

Extant Placentals. Most extant placentals have typical diphyodont replacement, except that replacement at the first premolar locus has been documented in only a few instances (Luckett, 1993). However, in the extreme case of small insectivores, none of the premolars is replaced and the entire postcanine series is monophyodont (Osborn, 1971; Bloch et al., 1998). The same monophyodont condition also appears to be present in some geolabidid insectivores of the early Tertiary (Lillegraven et al., 1981), although not in others (Bloch et al., 1998). But this monophyodont pattern seems to be an exception to the generalized diphyodont replacement of placentals as a whole.

The sequence of replacement of premolars is also variable among various extant placental groups. In some talpids, chrysochlorids, and macroscelidids, the eruption of deciduous premolars and their replacement occur earlier at the ultimate premolar locus than at the more anterior premolar loci (Kindahl, 1963, 1967; Ziegler, 1971; Luckett, 1993). This posteroanterior sequence of premolar replacement was previously considered to be a typical pattern for placental insectivores (Osborn, 1973; Luckett, 1993). Primates as a whole have posteroanterior sequential premolar replacement (Smith, 2000). Among primates there are exceptions among stem anthropoids and australopithecines in which the upper and lower premolars are replaced in an alternating fashion (Kay and Simons, 1986; Smith, 1994).

However, the sequence of premolar replacements in placental ungulates, carnivores, and other placental fossil taxa is in the opposite anteroposterior direction (Schmid, 1972; Smith, 2000). The erupting sequence of p3 and p4 (in an anteroposterior direction) for the premolars is variable among the Early Tertiary plesiadapiform and microsyopid placentals (Bloch et al., 2002). Given the distribution of these characteristics, it is clear that primitively the placental crown group as a whole had sequential replacement of premolars, but it is not clear whether the ancestral state involved anteroposterior or posteroanterior replacement. Among extant placentals, only derived anthropoids show alternating replacement of premolars; but this is clearly an atavistic reversal to the condition of Cretaceous eutherians.

Metatherians (figures 3.21B, 3.24F). The perinatal adaptation in marsupial fetuses by fixation of the mouth to the nipple for prolonged lactation (Tyndale-Biscoe and Renfree, 1987; Zeller, 1999a; Zeller and Freyer, 2001) is directly correlated with the accelerated growth of the skeletomuscular system of the skull (Maier, 1993) relative to the brain (Smith, 1996, 1997). It is also correlated with the highly transformed diphyodont replacement of marsupials, which is limited to a single postnatal replacement of the ultimate premolar (Luckett, 1977, 1993). The dental replacement pattern of marsupials can be traced back at least to the Late Cretaceous (Cifelli et al., 1996).

Evolution of Dental Replacement in Cynodonts and Mammals. The complex phenomenon of dental replacement can be reduced to several basic morphological elements. (1) Frequency of replacement: number of successional teeth per tooth locus and whether replacement occurred in the posterior molariforms; these are directly correlated with the patterns of skull growth. (2) Mode of replacement: alternating versus sequential. (3) Direction or sequence of replacement: anteroposterior versus posteroanterior. We hypothesize that the following transformations occurred in the evolution of mammalian dental replacement (figure 3.25).

The replacement frequency through the cynodont-mammal transition shows a two-step reduction, first in the premolars of *Sinoconodon* (figure 3.25: node 1) and then in the incisors, canines, and posterior molariforms (= molars) of *Morganucodon* (figure 3.25: node 2). Once the apomorphic diphyodont replacement pattern and determinate skull growth occurred in the stem mammals of the Early Jurassic, there was not a single case of a derived mammalian group that subsequently reversed to polyphyodont replacement (of the entire dentition) and indeterminate skull growth. This is among the most fundamental transformations in the biological adaptation of mammals, as it represents a significant shift in growth pattern, probably correlated with the origin of lactation (Brink, 1956; Hopson, 1973; Pond, 1977; Gow, 1985; Luo, 1994; Zhang et al., 1998).

The mode and direction of the replacement waves appear to show some degree of homoplasy in the cynodont-

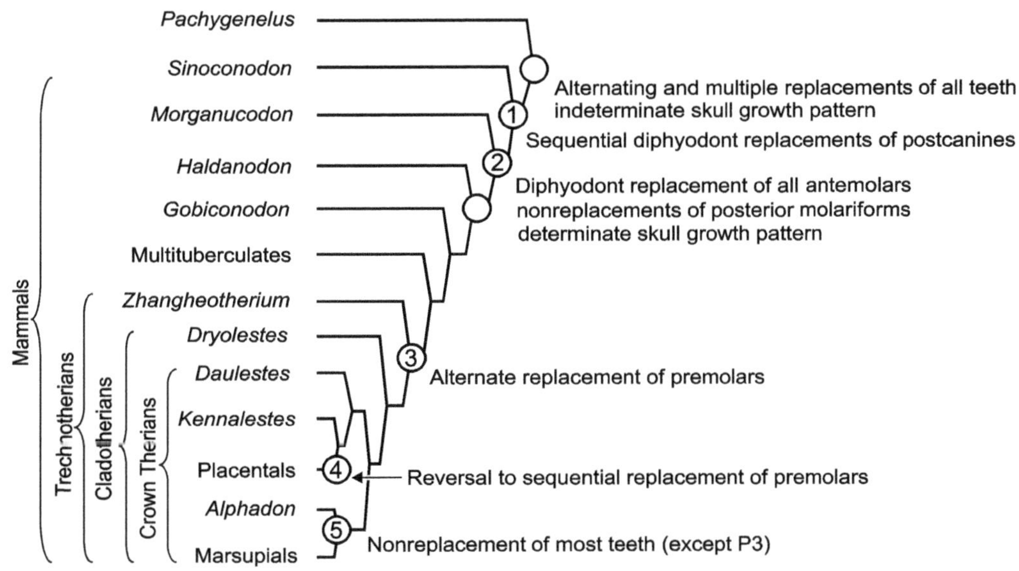

FIGURE 3.25. Evolution of dental replacement and skull growth pattern in the early history of mammals. The primitive nonmammalian cynodonts have multiple and alternate replacements of teeth, whereas extant placentals generally have diphyodont, sequential replacement of antemolars. Acquisition of apomorphies—Node 1 (mammals): differentiation of premolars and molars, single replacement of premolars, single replacement of some posterior molariforms in anteroposterior sequence. Node 2 (clade of *Morganucodon* and extant mammals): single replacement of incisors and canines (possibly associated with lactation), anteroposterior sequence for premolar replacement, molars without replacement; determinate skull growth pattern. Node 3 (Trechnotheria): alternating, single replacement of premolars. Node 4 (extant placentals): reversal to the anteroposterior sequential replacement of premolars. Node 5 (Metatheria): nonreplacement of most teeth (except P3/p3). *Source:* modified from Luo et al. (in press).

mammal evolution. The alternating mode of replacement is primitive for many cynodonts, including tritheledontids (Parrington, 1936; Crompton, 1963; Osborn and Crompton, 1971; Gow, 1980; Crompton and Luo, 1993). Two cases of homoplasy to this pattern involve diademodontids and traversodontids with the "gomphodont" postcanines, where postcanine replacement was sequential in an anteroposterior direction (Hopson, 1971; Osborn, 1974b). In tritylodontids, there was no replacement of postcanines; they erupted by sequential addition at the posterior end of the tooth row, and the worn postcanines were shed anteriorly (Kühne, 1956; Hopson, 1965; Cui and Sun, 1987). If tritheledontids are the sister taxon to mammals, as is preferred by the majority of workers, then the origin of mammals can be characterized by a shift from the primitive multiple alternating replacement of all postcanines in most cynodonts to a derived pattern of sequential single replacement of postcanines in mammals (figure 3.25: node 1). In the more derived clade comprised of *Morganucodon, Haldanodon,* and living mammals, the anteroposterior sequential replacement only occurred among premolars, not in posterior molariforms (molars) (figure 3.25: node 2). The incisors and canines were also limited to one replacement, as in most extant therians, instead of multiple replacements, as in *Sinoconodon* (Crompton and Luo, 1993; Zhang et al., 1998) and cynodonts. Gobiconodontids replaced the premolars and anterior molariforms ("molars") (Jenkins and Schaff, 1988; Wang, Hu, Meng, et al., 2001), but not the posterior molars, and are therefore consistent with the derived mammalian dental replacement pattern.

EVOLUTION OF THE POSTCRANIAL SKELETON

Most diagnostic characters for mammals (Mammaliaformes of Rowe, 1988), as frequently cited in leading textbooks and monographs, are from the skull and dentition (see, e.g., Romer, 1966; Crompton and Jenkins, 1979; Kemp, 1982; Kermack and Kermack, 1984; Kuhn-Schnyder and Rieber, 1986; Kuhn and Zeller, 1987a; Benton, 1988, 1990a; Carroll, 1988, and Sigogneau-Russell, 1994b). There are, however, relatively few derived postcranial skeletal characters diagnostic for mammals. Our understanding of skeletal anatomy is improving as better preserved fossils sampling the cynodont-mammal transition have been discovered; it appears that many functionally important postcranial features for modern mammals originated among mammaliamorphs and are primitive for mammals.

The postcranial osteological features are well documented by Jenkins and Parrington (1976) for morganucodontans, and some limited information on postcranial anatomy is also available for the stem mammals *Sinoconodon* (see Young, 1982b) and *Haldanodon* (Krusat, 1991). The postcranial features of these stem mammals are mostly primitive characters that are also present, albeit in modified conditions, in tritylodontids (Kühne, 1956; but see Sues, 1983) and tritheledontids (Rowe, 1993; Gow, 2001; Hopson and Kitching, 2001). A few other postcranial features of stem mammals are present in a more inclusive group of eucynodonts, such as *Exaeretodon* (see Bonaparte, 1963), *Probelesodon* (see Romer, 1969a), *Massetognathus* (see Jenkins, 1970a), and *Probainognathus* (see Romer and Lewis, 1973).

Vertebral Column. Extant mammals differ from nonavian sauropsid reptiles and primitive precynodont synapsids in having a greater degree of regional differentiation of the vertebral column. The anatomical modification in the cervical vertebrae, especially the first two, is correlated with a greater degree of cranial mobility, as pointed out by Jenkins (1969a, 1970b). The atlas (first cervical) and axis (second cervical) vertebrae of extant mammals are characterized by several functionally important features. The first of these is the bicondylar atlantooccipital joint. The divided occipital condyles and the atlas form a double ball-and-socket joint that facilitates the flexion and extension of the cranium but does not permit any rotation or lateral flexion, thereby protecting the spinal cord and medulla oblongata from tensional stress. The second feature is the absence of the postzygapophysis on the atlas ring, allowing the atlas to rotate. The third feature is the dens (odontoid process) of the axis, a neomorphic feature developed from part of the atlas centrum anlage, but synostosed (or completely fused) to the axis in extant mammals. Together with the transverse ligament of the atlas, which constrains the dens of the axis to the atlantal ring, the dens substitutes for the lost atlantoaxial zygapophysis in maintaining the atlas-axis articulation (Jenkins, 1969a, 1970b).

All three features have a more ancient history among premammalian cynodonts, and none is a synapomorphy for mammals. Some degree of division of the two occipital condyles is a primitive feature of Cynodontia (Jenkins, 1969a; Kemp, 1982; Hopson and Barghusen, 1986). Fusion of the dens to the axis is a derived character of mammals and tritylodontids (Kühne, 1956; Jenkins, 1969a; Sues, 1983). It certainly has a more ancient origin because an incipient odontoid process is already present in some cynodonts (Jenkins, 1969a, 1970b) (epicynodonts of Hopson and Kitching, 2001). The atlas postzygapophysis is absent both in mammals and in tritylodontids (Sues, 1983; Rowe, 1993). However, it should be noted that the atlas and axis of tritheledontids are not adequately known (Gow, 2001), so it is uncertain whether these derived features are also plesiomorphies shared by tritheledontids or separately derived features of tritylodontids and mammals. Mammals (Jenkins and Parrington, 1976), tritylodontids (Kemp, 1982; Sues, 1983; Rowe, 1988), and tritheledontids (Gow, 2001; Hopson and Kitching, 2001) have the platycoelous centra of the cervical vertebrae and the anteroposteriorly shortened centra and neural arches for increased mobility of the cervical vertebral column (Rowe, 1993). All these functionally important characters of cervical vertebrae in extant mammals have a premammalian origin in cynodonts.

The mammalian crown group has some derived features that are absent in cynodonts and stem mammals. These include the fusion of the cervical ribs to enclose the transverse foramen on cervical vertebrae and the fusion of the atlas neural arches and body to form a complete atlas ring. Both features are likely to have evolved within the mammalian crown group, as the stem mammal *Morganucodon* still has the unfused atlas ring and cervical ribs.

Extant mammals are also distinguished from nonavian sauropsids and primitive precynodont synapsids in several characteristics of the lumbar and sacral vertebrae. These are related to a greater capacity of flexion and extension of the vertebral column for locomotion. A distinctive lumbar region is present, marked by the reduction (or total loss) of the lumbar ribs. The lumbar ribs were first reduced in stem cynodonts (Jenkins, 1971b; Kemp, 1982) and are lost completely in crown therians. The mammalian sacrum typically has two or three sacral vertebrae (although there are some exceptions among the living placental and marsupial orders, such as xenarthrans). By comparison, in many cynodonts there are four or five sacral vertebrae (Jenkins, 1971b; Romer and Lewis, 1973; Kemp, 1980). Reduction of the sacral vertebrae to the "mammalian" sacral vertebral count occurred in tritylodontids and tritheledontids (Rowe, 1993) and is primitive for mammals.

Shoulder Girdle. Basal cynodonts such as *Procynosuchus* and *Thrinaxodon* had a sprawling forelimb posture, although incipient rotation of the limbs below the trunk had occurred (Brink, 1956; Jenkins, 1971b; Kemp, 1980). These primitive features are also retained in tritylodontids (Sues, 1983; Sun and Li, 1985) and, to a lesser extent, also in tritheledontids (Gow, 2001). The only derived feature of the forelimb for tritylodontids and tritheledontids is a larger hemispherical head of the humerus, as in mammals.

Pectoral girdle features of morganucodontans (Jenkins and Parrington, 1976), *Sinoconodon,* and monotremes (Klima, 1973) are "cynodont-like" and primitive. Large interclavicles are present in *Sinoconodon,* as in monotremes, although the preservation of this bone in *Morganucodon* is uncertain (Evans, 1981). The clavicle and the interclavicle form an immobile articulation. A large coracoid forms part of the glenoid of the shoulder joint. The procoracoid, with coracoid foramen, is present in *Morganucodon* and *Megazostrodon.* The scapula has a laterally turned acromion. The supraspinous fossa is positioned on the cranial (dorsal) margin of the scapula. All these features are also present, albeit in slightly modified conditions, in tritylodontids (Kühne, 1956; Sues, 1983; Sun and Li, 1985) and tritheledontids (Rowe, 1993; Gow, 2001; Hopson and Kitching, 2001). The basic design of the shoulder girdle is largely the same for all mammaliamorphs, and is plesiomorphic for mammals.

The interclavicle, a primitive cynodont feature, is now known to be present in multituberculates, in the "symmetrodontan" *Zhangheotherium* (Meng and Miao, 1992; Sereno and McKenna, 1995; Hu et al., 1997), and in the eutriconodontans *Repenomamus* and *Jeholodens* (Ji et al., 1999). But unlike those of cynodonts, the interclavicles of these mammals are reduced in size, and the clavicle-interclavicle joint is mobile, to facilitate a pivotal movement of the clavicle relative to the sternal apparatus, as in extant therians (Jenkins, 1974b; Jenkins and Weijs, 1979). The pivotal clavicle-interclavicle joint is present in eutriconodontans, multituberculates, and in the "symmetrodontans" *Zhangheotherium* (Sereno and McKenna, 1995; Hu et al., 1997, 1998; Ji et al., 1999). If eutriconodontans and multituberculates are nested within the mammalian crown group (a possibility that cannot be ruled out), then the pivotal clavicle joint can be added to a list of derived features shared by eutriconodontans, multituberculates, and trechnotherians. The interclavicle is lost in the derived cladotherian groups, as exemplified by *Henkelotherium* (see Krebs, 1991), *Vincelestes* (Rougier, 1993), marsupials, and placentals (Klima, 1987; Ji et al., 2002; note, however, that a well-formed separate interclavicle ossification is present in the Paleocene marsupial *Pucadelphys;* see Marshall and Sigogneau-Russell, 1995).

The laterally facing supraspinous fossa and its related median scapular spine along the length of the scapula are likely to be homoplastic among different lineages of mammals. They are fully developed in the eutriconodontans *Gobiconodon* and *Jeholodens* (Jenkins and Schaff, 1988; Ji et al., 1999) and trechnotherians, but are absent in stem mammals (Jenkins and Parrington, 1976) and monotremes.

Multituberculates have an incipient supraspinous fossa, as described by Kielan-Jaworowska and Gambaryan (1994) and Sereno and McKenna (1995, see also chapter 8). The supraspinous fossa is more similar in size and orientation to those of morganucodontans and monotremes than to that of therians. If multituberculates are a part of the mammalian crown group (again, a possibility that cannot be ruled out with currently available evidence, see chapter 15), then either the fully developed and "therian-like" supraspinous fossa and scapular spine are convergences in eutriconodontans and therians, or the absence of a fully developed, laterally facing supraspinous fossa in multituberculates and monotremes would have to be secondary atavistic reversals (Ji et al., 1999).

Sánchez-Villagra and Maier (2001) suggested that only the anterior (acromional) portion of the scapular spine of modern therians is homologous to the so-called scapular spine in monotremes and stem mammals. The posterior (or distal) portion of the scapular spine in extant therians is formed by a neomorphic ossification of the intermuscular septum between the anlages of the supraspinous and the infraspinous muscles, not by the rotation of the endochondral spine, as in monotremes (Cheng, 1955). This suggests that at least the dorsal part of the supraspinous fossa of extant therians is not formed by the rotation of the scapular spine and is not homologous to that of monotremes. The scapular spine and the supraspinous fossa certainly have different developmental histories in monotremes and in extant therians (Sánchez-Villagra and Maier, 2001, 2002), underlining a more ancient and complex evolutionary history of scapular structures than had been accepted previously.

Pelvic Girdle and Hindlimb. The cumulative acquisition of "mammalian" pelvic girdle and hindlimb characters is correlated with evolution from a more sprawling to a more upright limb posture among nonmammalian synapsids (Lessertisseur and Sigogneau, 1965; Jenkins, 1970b, 1971b; Kemp, 1982; Rowe, 1993).

The ilium in stem cynodonts is a broad, spatulate plate in contact with four or five sacral vertebrae. It has a large posterior iliac process posterior to the acetabulum (the hip joint) and a dorsally convex profile (Brink, 1956; Jenkins, 1971b; Kemp, 1980; Hopson and Kitching, 2001). By comparison, the mammalian pelvis has a long, anterodorsally directed ilium with a flat or concave dorsal profile and is positioned entirely anterior to the acetabulum. The ilium has a triangular outline in transverse section, with an external ridge dividing the iliac external surfaces (Jenkins and Parrington, 1976; Kemp, 1982; Rowe, 1993). The elongate ilium has narrower contact with the sacrum, in correlation with fewer (two or three) sacral vertebrae. The pubis is more reduced in mammals.

A varying degree of reduction of the posterior iliac process is known for a number of eucynodonts (Bonaparte, 1962; Romer, 1969b; Jenkins, 1970b; Romer and Lewis, 1973; Rowe, 1993; Hopson and Kitching, 2001), by comparison to the basal cynodont *Thrinaxodon.* The anterodorsal orientation of an elongate ilium with triangular cross section, a reduced pubis, and a posteriorly positioned acetabulum are derived features shared by tritylodontids (Kühne, 1956; Kemp, 1983; Sues, 1983), tritheledontids (Rowe, 1993; Hopson and Kitching, 2001), and mammals (Jenkins and Parrington, 1976; Jenkins and Schaff, 1988; Kielan-Jaworowska and Gambaryan, 1994; Ji et al., 1999). Related to the modifications in the pelvis, the femur has developed a relatively spherical head, a better differentiated greater trochanter, and a lesser trochanter situated on the medial side of the femoral shaft (Kühne, 1956; Kemp, 1983; Hopson and Kitching, 2001). Therefore, all pelvic features of mammals are primitive and are possibly retained from the common ancestor of tritylodontids, tritheledontids, and mammals.

The mammalian ankle is characterized by the presence of the tuber calcanei of the calcaneal bone and a slight overlap of the astragalus over the calcaneus (Jenkins and Parrington, 1976; Krause and Jenkins, 1983; Szalay, 1993b; Kielan-Jaworowska and Gambaryan, 1994; Ji et al., 1999). But these features were already present in tritylodontids, albeit in slightly different condition (Kühne, 1956; Sues, 1983). The complete superposition of the astragalus over the calcaneus, regarded as a "mammal-like" characteristic, was apparently developed in parallel in marsupials (Szalay, 1994) and placentals (Horovitz, 2000; Ji et al., 2002, see, e.g., chapters 4, 8, 12, and 13).

In summary, most of the so-called "mammalian" characters in the sacral vertebrae, pelvis, and hindlimb were acquired by premammalian cynodonts, and did not originate with either mammals or the mammalian crown group. However, for the ankle, modifications of the cynodont-like and primitive features occurred quite late in mammalian evolution, possibly convergently after the split of main lineages of extant therians.

MACROEVOLUTIONARY PATTERN OF CYNODONT-MAMMAL EVOLUTION

If mapped on our current phylogeny (chapter 15), most of the complex mammalian characters can be shown to have intermediate character states in advanced nonmammalian cynodonts (Crompton, 1972b; Gow, 1985; Allin, 1986; Allin and Hopson, 1992; Rowe, 1993; Luo, 1994; Luo, Crompton, and Sun, 2001). These characteristics are therefore far less distinctive than previously thought (Brink, 1956; Gow, 1985). The "mammalian" features in the braincase, palatal and orbital structures, inner and middle ears, CMJ, and dental replacement and mode of skull growth, were all assembled incrementally, at several hierarchical levels of cynodont-mammal phylogeny, and were accumulated by stepwise evolutionary transformation (Hopson and Barghusen, 1986; Benton, 1990b; Luo, 1994; Sidor and Hopson, 1998; Luo, Crompton, and Sun, 2001; Sidor, 2001). However, this does not, by any means, diminish the robustness of the mammalian clade (defined as the common ancestor of *Sinoconodon* + Monotremata + Theria), which is the best-supported monophyletic group in the synapsid-mammal phylogeny. We conclude that all evolutionary innovations of modern mammals are assemblages of accumulative morphological changes through the cynodont-mammal evolution.

The predominant patterns of macroevolution for the diagnostic features of mammals (table 3.1), as can be optimized on the current phylogeny, are stepwise, incremental, and even-paced. There has been no single event or episode of co-evolution of a large number of apomorphies during the evolutionary history of cynodonts and the earliest known mammals (Sidor and Hopson, 1998; Luo, Crompton, and Sun, 2001; Sidor, 2001).

CHAPTER 4

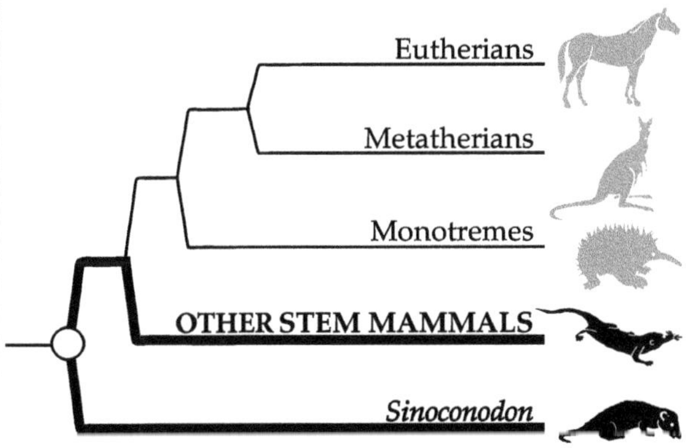

The Earliest-Known Stem Mammals

INTRODUCTION

Mammals are defined by the common ancestry of *Sinoconodon* and extant mammals[1] (see chapter 3). Mammals so defined include those synapsids that are more closely related to monotremes, marsupials, and placentals than to tritheledontids ("ictidosaurs"), tritylodontids, probainognathids, dromatheriids, or any combination of these cynodont groups. The Mammalia under this stem-based definition can be diagnosed by the derived characters of a craniomandibular joint formed by the dentary condyle and the squamosal glenoid, a promontorium in the petrosal, a well-developed separation of the jugular foramen from the hypoglossal foramen (except for *Ornithorhynchus*), and a host of other characters. A number of stem taxa from the Late Triassic to Early Jurassic are morphologically more primitive than the crown group of mammals and are placed outside the mammalian crown group by cladistic relationships. Nonetheless, these earliest-known taxa are qualified for this definition of Mammalia, as they are diagnosed by a combination of these derived "mammalian" characters, which are absent in the most advanced nonmammalian cynodont groups.

In order to organize this book into a manageable number of chapters, we decided to summarize in chapter 4 all anatomical and taxonomic information of the earliest-known stem mammals from the Late Triassic to Early Jurassic (except for haramiyidans, described in chapter 8, and kuehneotheriids, described in chapter 9). In the best available phylogenetic trees (see figures 1.1, 15.1, and 15.2) these stem taxa are outside the Crown Mammalia. Sinoconodontidae, Morganucodonta, and *Hadrocodium* discussed in this chapter are now resolved to form a series of clades that are successively closer to the Crown Mammalia (table 4.1). In a fully resolved cladistic phylogeny and a complete cladistic classification (table 1.1), these taxa (Sinoconodontidae, Morganucodonta, and *Hadrocodium*) would occupy different positions in the classification scheme, separated by intervening ranks of clades defined by kuehneotheriids and docodontans (described in chapter 5), which are also placed outside the Crown Mammalia in a comprehensive phylogenetic tree (figure 15.1). We tentatively include the very incomplete Late Triassic *Adelobasileus* in this chapter, which has diagnostic mammalian features and is from Upper Triassic rocks of North America (Carnian, ca. 225 Ma, Lucas and Hunt, 1990; Lucas and Luo, 1993). The currently available evidence suggests that it occupies a more basal position than Sinoconodontidae and Morganucodonta in the cynodont-mammal phylogeny. We are cognizant of the fact that the position of *Adelobasileus* cannot be fully resolved within the stem-defined Mammalia.

The earliest-known member of the Morganucodonta was discovered in Upper Triassic rocks of India (Datta and Das, 1996, 2001). However, these stem taxa are represented by very incomplete fossils and are placed within Mammalia tentatively, pending confirmation of the more complete fossils to be discovered in the future. The next oldest records of stem mammals are presented by several morganucodontans, kuehneotheriids, and haramiyidans. These

[1] As discussed in chapter 1, in this book we accept the traditional concept of the class Mammalia and we assign the earliest known stem mammals described in this chapter to Mammalia, rather than to Mammaliaformes, as proposed by Rowe (1988, 1993) and accepted by McKenna and Bell (1997).

appeared in the Norian sediments of Britain and Greenland (Fraser et al., 1985; Jenkins et al., 1994; Jenkins et al., 1997) and in the Rhaetian Beds of Europe (Clemens, 1980a; Sigogneau-Russell, 1989b; Hahn et al., 1991; Sigogneau-Russell and Hahn, 1994). In the Early Jurassic, morganucodontans achieved a global distribution and are known from the Liassic sediments of Britain (Evans and Kermack, 1994), the Hettangian–Sinemurian of North America (Jenkins et al., 1983; age estimate by Sues, 1986), Asia (Patterson and Olson, 1961; Kermack et al., 1973; Young, 1982a; Zhang and Cui, 1983; Sun and Cui, 1986; Luo and Wu, 1994, 1995), southern Africa (Crompton, 1964b, 1974; Crompton and Jenkins, 1968), and the Early Jurassic of India (Prasad and Manhas, 1997, 2002; Datta and Das, 2001).

These earliest-known stem mammals do not belong to a monophyletic group. *Adelobasileus, Sinoconodon,* morganucodontans, docodontans, and *Hadrocodium* have been resolved into a series of clades that are related to the mammalian crown group in a successively closer order by analyses of a large body of comparative morphological evidence (e.g., Rowe, 1993; Rougier, Wible, and Hopson, 1996; Luo, Cifelli, and Kielan-Jaworowska, 2001; Wang, Hu, Meng, et al., 2001; Luo et al. 2002; see also figures 15.1, 15.2). Because the heterogeneous groups described in this chapter are not characterized by synapomorphies (except for the apomorphies common to them and to all other mammals), we do not provide a "Brief Characterization" for these groups as a whole, but rather describe each of these stem groups separately.

In this chapter we also treat the Liassic *Hadrocodium* from China, which is more advanced than other Rhaeto-Liassic mammals and is placed closer to the mammalian crown group than its contemporary stem mammal groups mentioned earlier and closer as well than *Haldanodon* (chapter 5, Docodonta), as proposed by Luo, Crompton, and Sun (2001).

Two other stem mammalian groups from the Rhaeto-Liassic are the Haramiyida and Kuehneotheriidae. We tentatively place Haramiyida (*sensu* Butler, 2000) in the subclass Allotheria and describe haramiyidans in chapter 8. Kuehneotheriids (*Kuehneotherium* and *Woutersia*) are described in chapter 9, which is devoted to "symmetrodontans." The Docodonta, a stem mammal group from the Middle Jurassic with highly specialized dentition, described in chapter 5, have been placed outside the mammalian crown group in recent phylogenetic studies (Lillegraven and Krusat, 1991; Luo, 1994; Rougier, Wible, and Hopson, 1996; Luo, Cifelli, and Kielan-Jaworowska, 2001; Luo et al., 2002). Several taxa represented by isolated teeth, such as *Hallautherium* Clemens, 1980, or by incomplete skulls, such as *Adelobasileus* Lucas and Hunt, 1990, have been assigned to Mammalia, order and family *incertae sedis.* These putative mammalian taxa are also included here for the sake of convenience.

Several cynodont taxa from the Late Triassic are characterized by molariform teeth with a laterally compressed crown, a single main cusp flanked by smaller mesial and distal cusps in alignment. Some of these taxa have molariforms with roots that are divided to varying degrees (Bonaparte and Barberena, 1975; Russell et al., 1976; Lucas et al., 2001; and Shapiro and Jenkins, 2001). At one time or another, these taxa were considered to be closely related to mammals because of these "mammal-like" dental features (Bonaparte and Barberena, 1975; Russell et al., 1976; Hahn et al., 1984, 1987, 1994; Lucas and Oakes, 1988; Sigogneau-Russell and Hahn, 1994). Some of them used to be assigned to Mammalia: *Dromatherium* Emmons, 1857, *Microconodon* Osborn, 1886, *Tricuspes* Huene, 1933 (e.g., Owen, 1871; Osborn, 1886b, 1887b; Peyer, 1956). But the current consensus regarding these taxa is that they are nonmammalian cynodonts (Simpson, 1926a; Clemens, 1980a), with possible affinities to chiniquodontids or dromatheriids (Hahn et al., 1994; Sigogneau-Russell and Hahn, 1994; Bonaparte and Barberena, 2001; Shapiro and Jenkins, 2001; Sues, 2001). The latest reviews suggest that none of these groups is close to being the sister taxon to mammals (Bonaparte and Barberena, 2001; Hopson and Kitching, 2001; Sues, 2001; Luo et al., 2002), so we do not consider them any further here.

TERMINOLOGY

In descriptions of teeth we adopt the terminology proposed by Crompton and Jenkins (1968) and used by Crompton (1971, 1974) in subsequent studies of these groups. Upper-case letters denote the cusps on upper cheek teeth and lower-case letters on lower cheek teeth. The four main cusps on upper molars were designated A, B, C, D, and E, with A being the largest central cusp, D the distal cusp, E the mesial cusp (if present). Similarly, a, b, c denote the main (noncingulid) cusps on lower molars (figure 4.1). Additional lingual cusps on lower molars are designated d (if present on the cingulid at the distal end of tooth), e (if present on the cingulid at the mesiolingual corner of the tooth), f (if present on the cingulid at the mesiolabial corner of the tooth), g (if present on the distal part of the lingual cingulid).

Class Mammalia Linnaeus, 1758
Subclass and Order *incertae sedis*
Family Sinoconodontidae Mills, 1971
(see tables 4.1 and 4.2)

Comments. In their original description of *Sinoconodon,* Patterson and Olson (1961) placed it in the Tricon-

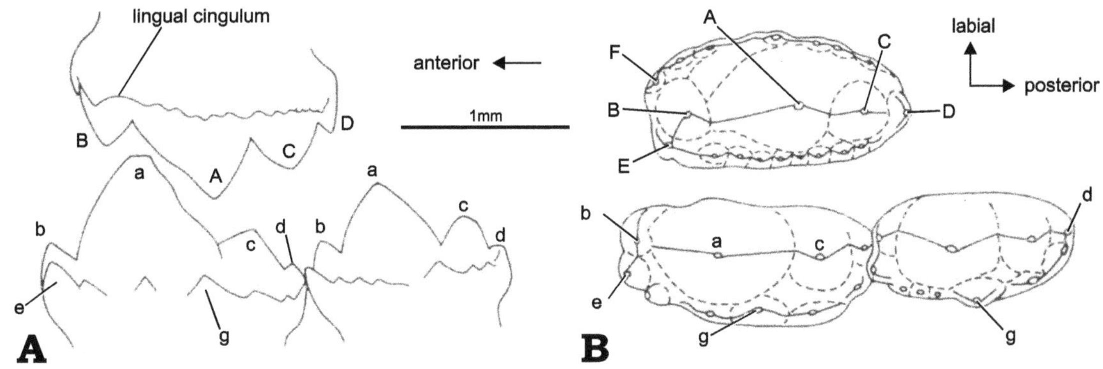

FIGURE 4.1. Nomenclature of stem mammal postcanine teeth (upper teeth at top and lower at bottom), as exemplified by *Morganucodon watsoni*. A, Lingual view (right side). B, Crown view (right side, with upper molar inverted). *Source:* modified from Crompton (1974).

odontidae because of its "triconodontid-like" molar crown and some features of the skull. Their rationale for this assignment was consistent with the prevailing taxonomic approach of the 1950s. However, in current phylogenetic systematics, the primitive characters for linking *Sinoconodon* with the Triconodontidae are not phylogenetically informative because they are shared not only by triconodontids, "amphilestids," and morganucodontans, but also by some derived nonmammalian cynodonts. Hence *Sinoconodon* was subsequently placed in the Morganucodontidae ("Eozostrodontidae") together with *Morganucodon, Megazostrodon, Erythrotherium* (Hopson and Crompton, 1969), largely based on the shared primitive features of the mandible and dentition.

Mills (1971) erected the family Sinoconodontidae to include *Sinoconodon* and *Megazostrodon* based on his interpretation that *Sinoconodon* had the embrasure occlusion between the upper and the lower molars, as established for *Megazostrodon* by Crompton and Jenkins (1968). Crompton (1964b, 1974) was the first to note that *Sinoconodon* has several mandibular features that are similar to those of the nonmammalian cynodonts and that it lacks the cingulid and the cingulid cusps typical of *Megazostrodon* and has no molar interlock mechanisms. Based on these considerations Crompton (1974) excluded *Megazostrodon* from Sinoconodontidae, an opinion endorsed by Zhang and Cui (1983).

Crompton and Sun (1985) demonstrated that the upper and lower postcanines of *Sinoconodon* (figure 4.2) may differ in size and that their rows have a variable number of teeth, at least in some individuals. The upper and lower postcanines do not correspond in a one-to-one fashion; neither could they have occluded precisely with each other as in other Rhaeto-Liassic mammals (figure 4.3). As a consequence, there are no consistent wear facets on the cusps.

The lack of occlusion is a primitive feature, which led Crompton and Sun (1985) to suggest that *Sinoconodon* is more primitive than many of its contemporary stem mammals: *Morganucodon, Megazostrodon, Kuehneotherium,* docodontans, and haramiyidans. Based on additional characters in the dental replacement, the craniomandibular joint, and the ear region, Crompton and Luo (1993) proposed that *Sinoconodon* is the sister taxon to a clade of all other Rhaeto-Liassic and more derived mammalian groups. Sinoconodontidae represent one of the earliest-diverging lineages from the remaining Late Triassic and Early Jurassic mammals (Crompton and Sun, 1985; Crompton and Luo, 1993; Luo, 1994). This most basal position of Sinoconodontidae in mammalian phylogeny has been corroborated by many large-scale phylogenies of Mesozoic mammals (see Luo et al., 2002, for a relatively recent review).

BRIEF CHARACTERIZATION

Sinoconodontids (figures 4.2, 4.3) have primitive "triconodont-like" cheek teeth that were adapted for insectivory or carnivory. *Sinoconodon* has a wide range of ontogenetic variation of skull sizes within the currently available specimens from the Lower Lufeng Formation mammalian fauna: the smallest-known skull is about 20 mm long, the largest is more than 60 mm long (Zhang et al., 1998; Luo, Crompton, and Sun, 2001; figures 3.22, 4.2, 4.3). The limited postcranial fossils of *Sinoconodon* show that it had a robust forelimb and shoulder girdle, in a sprawling posture not unlike those of nonmammalian cynodonts (Jenkins, 1971b; Kemp, 1982; Sues, 1983) and extant monotremes.

Sinoconodon is unambiguously placed in Mammalia because it shares many derived characters with other mammals (including the mammalian crown group), such as the craniomandibular joint formed by the dentary

TABLE 4.1. Cladistic Classification above the Species Level of the Earliest-Known Stem Mammals from the Late Triassic to the Early Jurassic

- Class Mammalia Linnaeus, 1758
  - Clade (*Sinoconodon* + Crown Mammalia)
    - Order and family *incertae sedis*
      - *Adelobasileus* Lucas and Hunt, 1990 (*sedis mutabilis*)
    - Order *incertae sedis*
      - Family Sinoconodontidae Mills, 1971
        - *Sinoconodon* Patterson and Olson, 1961
  - Clade (*Morganucodon* + Crown Mammalia)
    - Order Morganucodonta Kermack et al., 1973
      - Family Morganucodontidae Kühne, 1958
        - *Morganucodon* Kühne, 1949
        - *Eozostrodon* Parrington, 1941
        - *Erythrotherium* Crompton, 1964
        - *Helvetiodon* Clemens, 1980
      - Family ?Morganucodontidae (*sedis mutabilis*)
        - *Gondwanadon* Datta and Das, 1996
        - *Indotherium* Yadagiri, 1984
      - Family Megazostrodontidae Gow, 1986
        - *Megazostrodon* Crompton and Jenkins, 1968
        - *Dinnetherium* Jenkins et al., 1983
        - *Brachyzostrodon* Sigogneau-Russell, 1983a
        - *Wareolestes* E. Freeman, 1979
      - ?Family Megazostrodontidae
        - *Indozostrodon* Datta and Das, 2001
  - Intervening clade of the family Kuehneotheriidae (see chapter 9)
  - Clade (Docodonta + Crown Mammalia)
  - Intervening clade of the order Docodonta (see chapter 5)
  - Clade (*Hadrocodium* + Crown Mammalia)
    - Order and family *incertae sedis*
      - *Hadrocodium* Luo, Crompton, and Sun, 2001
  - Clade of Crown Mammalia

condyle and the squamosal glenoid, the petrosal promontorium, an enlarged anterior lamina of petrosal, and the bony floor of the cavum epiptericum for the trigeminal ganglion. However, the dentition of *Sinoconodon* is primitive. Sinoconodontidae have retained the primitive alternating and multiple replacements of the incisors and canines seen in many nonmammalian cynodonts. Its postcanine replacement is characterized by the ultimate molariform tooth of a smaller (younger) individual being replaced by a larger successional tooth in a larger (older) individual, to be followed by the addition of another ultimate molariform to the posterior end of the tooth row. In the meantime, the anteriormost postcanines are successively lost, resulting in an increasingly large postcanine diastema (Crompton and Luo, 1993; Zhang et al., 1998, see also figure 3.22). The postcanine replacement features are identical to those of gomphodont cynodonts (Hopson, 1971).

The primitive dental replacement in the postcanines of *Sinoconodon* is correlated with their lack of precise upper-to-lower occlusion and with the indeterminate mode of skull growth as reflected in the wide range of body sizes within the same species. The mandibular and middle ear structures are essentially the same as in other stem mammals, such as *Morganucodon* and *Megazostrodon*, with a primitive postdentary trough (albeit more reduced than in *Morganucodon*) for the accommodation of the middle ear elements.

Diagnosis and Distribution. The family Sinoconodontidae is monotypic. Its diagnosis and distribution are the same as of the type genus *Sinoconodon*.

ANATOMY

As Sinoconodontidae include only a single genus *Sinoconodon* (see comments), the dental, cranial, and postcranial features of this family are based entirely on *Sinoconodon*. The description is summarized from Patterson and Olson (1961), Kermack et al. (1981), Young (1982a), Zhang and Cui (1983), Crompton and Sun (1985), Crompton and Luo (1993), and Zhang et al. (1998).

Skull and Mandible (figures 4.2, 4.3, see also Kermack et al., 1981; Crompton and Sun, 1985; Crompton and Luo, 1993). The premaxilla has an internasal process that wedges into the median notch between the two nasals. Its palatal process has an incisive fissure and a large pit for receiving the tip of the lower canine. The septomaxilla is large and forms the lateral border of the external nares. It has a transverse shelf overlapping the nasal side of the palatal process of the premaxilla, with a septomaxillary foramen at the maxilla-septomaxilla suture. The maxilla has a large canine eminence and two infraorbital foramina, with a third infraorbital foramen at the maxilla-jugal suture. The lacrimal is large with extensive facial exposure. The medial orbital wall is fully ossified with contributions from the orbitosphenoid and the orbital process of the palatine. The ascending process of the alisphenoid is a broad and vertical plate. The anterior lamina of the petrosal is broad, with two foramina for the mandibular branches (V_3) of the trigeminal nerve and a foramen at the alisphenoid-petrosal suture for the maxillary branch (V_2). The bulbous anterior paroccipital process is very similar to the homologous structure of tritylodontids; it is laterally covered by the squamosal and lacks the well-defined crista parotica.

In both these features *Sinoconodon* differs from morganucodontans. The anterior and posterior paroccipital processes are separated by a broad groove extending from the stapedial muscle fossa; the latter lacks a ventral projection, as seen in *Morganucodon* and *Dinnetherium*

TABLE 4.2. Linnaean Classification of the Earliest-Known Stem Mammals[1,2]

- Class Mammalia Linnaeus, 1758
  - Subclass and order *incertae sedis*
    - Family Sinoconodontidae Mills, 1971
      - *Sinoconodon* Patterson and Olson, 1961
        - *S. rigneyi*[3] Patterson and Olson, 1961
  - Subclass *incertae sedis*
    - Order Morganucodonta Kermack et al., 1973
      - Family Morganucodontidae Kühne, 1958
        - *Morganucodon* Kühne, 1949
          - *M. watsoni* Kühne, 1949, type species
          - *M. oehleri* Rigney, 1963
          - *M. heikoupengensis* (Young, 1978)
          - *M. peyeri* Clemens, 1980
        - *Eozostrodon* Parrington, 1941
          - *E. parvus* Parrington, 1941
        - *Erythrotherium* Crompton, 1964
          - *E. parringtoni* Crompton, 1964
        - *Helvetiodon* Clemens, 1980
          - *H. schutzi* Clemens, 1980
      - Family ?Morganucodontidae
        - *Gondwanadon* Datta and Das, 1996
          - *G. tapani* Datta and Das, 1996
        - *Indotherium* Yadagiri, 1984
          - *I. pranhitai* Yadagiri, 1984
      - Family Megazostrodontidae Gow, 1986
        - *Megazostrodon* Crompton and Jenkins, 1968
          - *M. rudnerae* Crompton and Jenkins, 1968
        - *Dinnetherium* Jenkins et al., 1983
          - *D. nezorum* Jenkins et al., 1983
        - *Brachyzostrodon* Sigogneau-Russell, 1983a
          - *B. coupatezi* Sigogneau-Russell, 1983a, type species
          - *B. maior* Hahn et al., 1991
        - *Wareolestes* E. Freeman, 1979
          - *W. rex* E. Freeman, 1979
      - Family ?Megazostrodontidae
        - *Indozostrodon* Datta and Das, 2001
          - *I. simpsoni*[4] Datta and Das, 2001
    - Order Morganucodonta
      - Family *incertae sedis*
        - *Hallautherium* Clemens, 1980
          - *H. schalchi* Clemens, 1980
  - Subclass, order and family *incertae sedis*
    - *Hadrocodium* Luo, Crompton, and Sun, 2001
      - *H. wui* Luo, Crompton, and Sun, 2001
- Class ?Mammalia Linnaeus, 1758
  - Subclass, order, and family *incertae sedis*
    - *Adelobasileus* Lucas and Hunt, 1990
      - *A. cromptoni* Lucas and Hunt, 1990
  - Subclass, order, and family *incertae sedis*
    - *Tricuspes* Huene, 1933
      - *T. tubingensis* Huene, 1933, type species
      - *T. sigogneauae* Hahn et al., 1994
      - *T. tapeinodon* Godefroit and Battail, 1997

[1] Mammals here are defined as those taxa more closely related to living mammals than tritheledontids, tritylodontids, dromatheriids, probainognathids, or a combination thereof. This definition is similar to the traditional Mammalia (similar to Mammaliaformes Rowe, 1988; McKenna and Bell, 1997). Stem mammals are the taxa that are placed outside the Crown Mammalia.
[2] Among other stem mammals, docodontans are described in chapter 5; *Kuehneotherium* and related taxa are described in chapter 9.
[3] Including *Lufengoconodon changchiawaensis* Young 1982a ("*Sinoconodon changchiawaensis*" Crompton and Sun, 1985); *Sinoconodon parringtoni* Young 1982a; *Sinoconodon yangi* Zhang and Cui, 1983.
[4] Prasad and Manhas (2002) consider *Indozostrodon simpsoni* to be a junior synonym of *Indotherium pranhitai* Yadagiri, 1984.

(Crompton and Luo, 1993). The lateral flange of the petrosal is overlapped by the quadrate ramus of the epipterygoid. Its posterior and vertical portion is perforated by two vascular foramina. The lateral trough of the petrosal has the opening of the prootic canal, the facial foramen, and possibly the fissure for the greater petrosal nerve. The trough forms the posterior periphery of a large ventral opening of the cavum epiptericum. The petrosal promontorium is smaller than those of morganucodontans. It has a flat medial surface, which in an intact skull is overlapped by the lateral process of the broad basioccipital. The basisphenoid has two small internal carotid foramina, but the basisphenoid wing is absent. The palatine has a large orbital process to form the posterior wall of the nasal cavity and a palatal process that contributes to the posterior part of the secondary bony palate, bearing the greater and the lesser palatine foramina. The palatine is significantly larger than those of many nonmammalian cynodonts and forms the margins of the pharyngeal passage, displacing the pterygoid hamulus posteriorly. The bony roof of the pharynx has a median crest and two pterygopalatine crests. The pterygoid is small by comparison to nonmammalian cynodonts. It has the transverse process (or hamulus) bearing a crescent muscular fossa on its lateral surface.

Dentition (figure 4.3, also see figure 3.22). The dental formula is: 5.1.2.5/5.1.2.5. All tooth types are variable in number among the skull specimens owing to the cynodont-like continuous dental replacements (see later) in correlation with the indeterminate skull growth. Sinoconodontidae have retained the alternating replacements of incisors and canines seen in most nonmammalian cynodonts. The canines were replaced at least four times, starting from the smallest-known skull (Crompton and Luo, 1993;

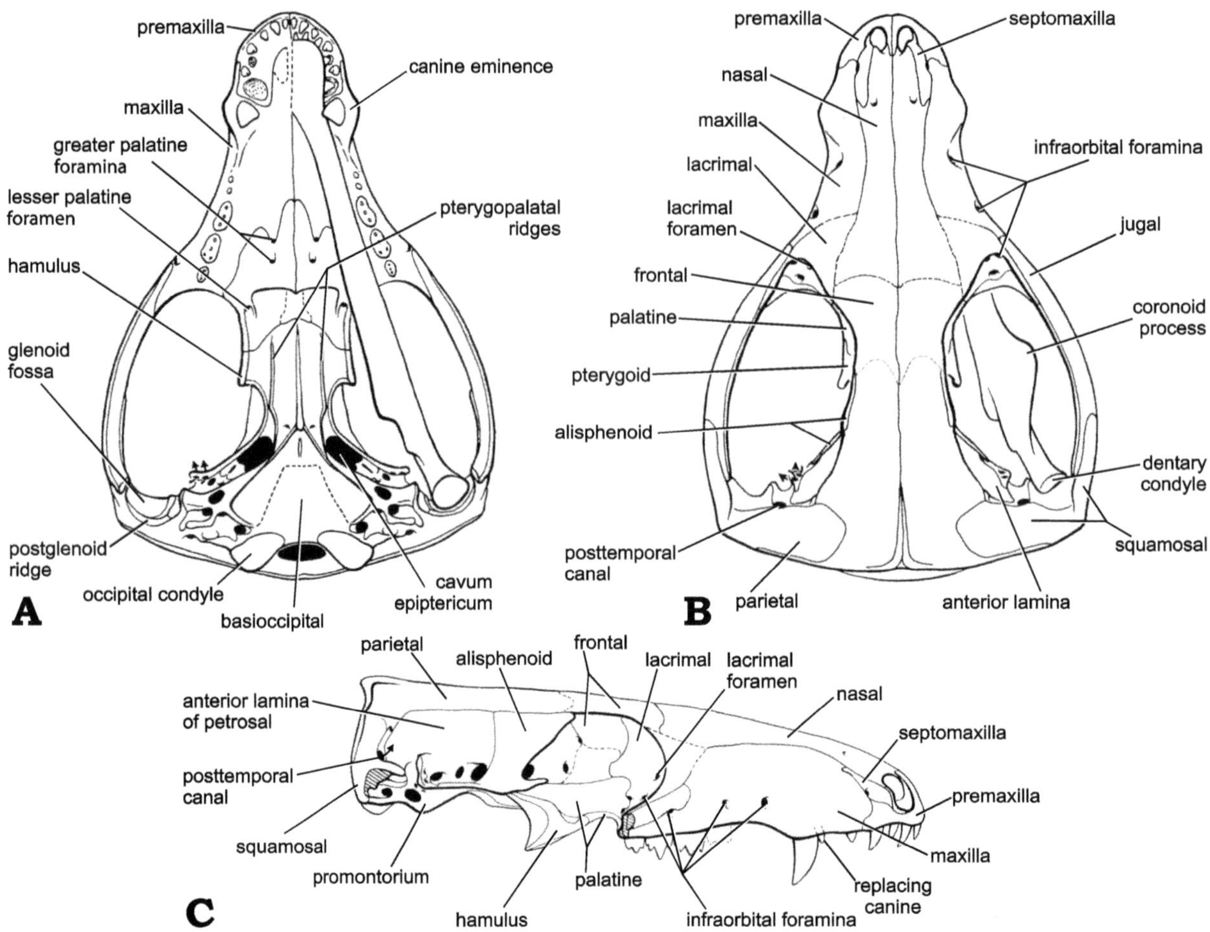

FIGURE 4.2. Skull reconstruction of *Sinoconodon*. A, Ventral view. B, Dorsal view. C, Lateral view (with zygoma and mandible removed to show the orbit wall). The skulls range from about 23 mm to more than 60 mm in length, so the scale can be variable. *Source:* modified from Crompton and Luo (1993).

Zhang et al., 1998). As a result, the size of the functional canine is highly variable among different individuals (Crompton and Luo, 1993). In its fullest erupted form the canine is daggerlike. The upper canine is so large that it reaches beyond the lower border of the mandible (Young, 1982a). But in the early stage of eruption, the erupted portion of the canine crown can be short and conical.

The incisors are peglike and there are five possible incisor loci. However, the actual number of functioning incisors is variable. Some individuals can have as few as three functional incisors, as a result of ongoing incisor replacements. The incisors have alternately large and small sizes in several skulls, starting from the smallest-known individuals (Crompton and Luo, 1993; Luo, 1994; Zhang et al., 1998). The alternating arrangement of small and large teeth is typical of the nonmammalian cynodonts, which had continuous and alternating replacement of teeth. This indicates that its incisors had an alternating replacement typical of cynodonts (Crompton and Luo, 1993; Luo, 1994). Overall, replacement of anterior dentition of *Sinoconodon* is consistent with a cynodont-like pattern of multiple and alternating replacement.

There are two premolar loci and the premolariforms are only present in juvenile individuals (Zhang et al., 1998). At the first premolar locus, the tooth is peglike and appears to be lost without replacement in older individuals. The second premolar locus has a peglike deciduous premolar in younger individuals, which was then replaced by a bicuspate successor tooth that, in turn, was permanently lost in the larger (thus older) specimens (figure 3.22).

The lower molariform postcanines are generally "triconodont-like" but show a wide range of morphological variation, possibly related to their replacements (figure 4.3A,B). In the ultimate lower molariform postcanine of smaller (presumably younger) skulls (figure 4.3B), cusp a is considerably higher than cusps b and c, with a distal cusp d but without mesial cusp e. By comparison, cusps b

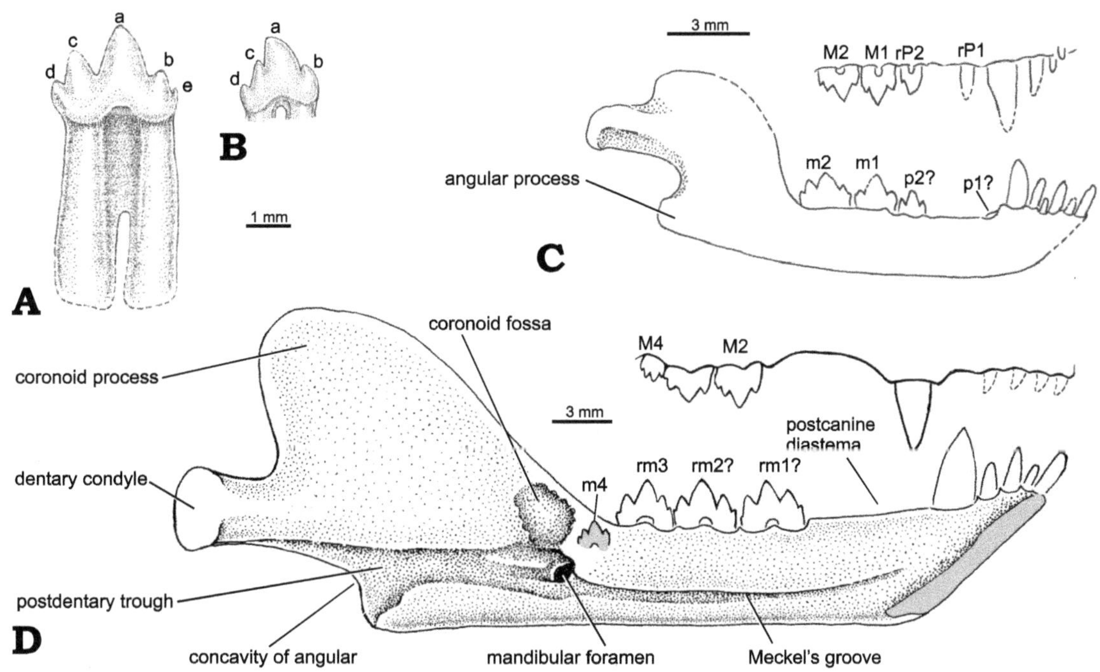

FIGURE 4.3. Molariform postcanines and jaws of *Sinoconodon*. A, Lingual view of a penultimate lower molariform of a large individual (late growth stage). B, Lingual view of an erupting ultimate molariform of a small individual (early growth stage). C, Composite upper and lower jaws of early growth stage of the skull. D, Composite upper and lower jaws of a late growth stage of the skull. *Source:* based on A, Luo (1994); B, Patterson and Olson (1961); C, Zhang et al. (1998); D, Crompton and Luo (1993), and Luo (1994).

and c are much larger in the penultimate molariforms of larger skulls. The penultimate molariform of some of the larger skulls has the mesial cingulid cusp e that is absent in the molariform tooth of the smaller skulls (figure 4.3A). Cusp e is a highly variable feature and may be absent in the anterior postcanines. There is no overlap of the distal cusp d of a preceding molariform with the mesial cusp e of the succeeding molariform (*contra* Patterson and Olson, 1961). The upper molariform of the smaller individuals has a "triconodont-like" tricuspate crown with cusps A, B, and C. By contrast, the upper molariform of larger individuals tends to develop an additional distal cusp D (Young, 1982a; Crompton and Luo, 1993). As mentioned earlier, there is no one-to-one occlusion between the upper and lower postcanines in *Sinoconodon* owing to these variations in tooth size and morphology.

There are likely five molariform loci in the upper and lower jaws in the entire growth series of skulls of *Sinoconodon* (Crompton and Luo, 1993). However, there are usually no more than three functional molariforms at a given replacement stage because the anterior two molariforms are lost without further replacement in older individuals as new teeth are added to the posterior end of the postcanine row (Crompton and Luo, 1993; Zhang et al., 1998). The size and morphological difference of the ultimate molariform of the smaller skulls from the corresponding penultimate molariform of the larger skulls suggest that the former may be replaced, followed by eruption of a new tooth in the succeeding position (Crompton and Luo, 1993; Zhang et al., 1998). The loss of the premolars and first molariform resulted in a prominent postcanine diastema in larger individuals. This is also present, although to a lesser extent, in morganucodontans and possibly also in *Kuehneotherium* and *Hadrocodium* (Parrington, 1973, 1978; Luo, Crompton, and Sun, 2001). The overall postcanine replacement sequence of *Sinoconodon* is probably convergent to that of cynodont *Diademodon* (Crompton, 1963; Hopson, 1971).

Postcranial Skeleton. Only very limited information concerning postcranial anatomy is available from *Sinoconodon* (Patterson and Olson, 1961; Young, 1982a). An incomplete pectoral girdle is known for a *Sinoconodon* specimen (Cui, pers. comm.). Most of the preserved features of *Sinoconodon* are similar to those of monotremes (Klima, 1973) and almost identical to the pectoral features of *Bienotheroides*, as described by Sun and Li (1985). A large interclavicle is present, with an immobile articulation of clavicle and interclavicle, as in monotremes. A large coracoid forms part of the glenoid of the shoulder joint. The scapula has a laterally turned acromion and a supraspinous muscle fossa positioned on the cranial (dorsal) margin of the scapula. The humerus of *Sinoconodon* has a spherical head and a greater tuberosity, which is continuous distally with the deltopectoral crest for about half the length of the humerus. Only the femur is known of all the pelvic and hindlimb elements. It is characterized by a "mammal-like" femoral head, separated from the greater trochanter by a deep notch. An incomplete lesser trochanter is preserved on the medial side of the femoral shaft. The patellar groove is absent from the distal end of the femur (Patterson and Olson, 1961). All these features are also present, albeit in slightly modified conditions, in tritylodontids (Kühne, 1956; Sues, 1983; Sun and Li, 1985) and tritheledontids, and are therefore plesiomorphic for mammals (Rowe, 1993; Gow, 2001; Hopson and Kitching, 2001).

Genus *Sinoconodon* Patterson and Olson, 1961
(figures 3.13, 3.15, 3.17C, 3.18C, 3.19F, 3.22, 4.2, 4.3)

Synonym: *Lufengoconodon* Young, 1982a.

Emended Diagnosis. Other than the apomorphies shared by the remaining stem mammals, *Sinoconodon* lacks autapomorphies of its own. Nonetheless it can be distinguished by a combination of characters, many of which are primitive among mammals. In the petrosal, *Sinoconodon* differs from all nonmammalian cynodonts in the presence of a promontorium, an enlarged anterior lamina, and the floor of the trigeminal ganglion. In the mandible, it differs from all nonmammalian cynodonts in having a well-developed dentary condyle. The mandible has a more robust dentary condyle than in morganucodontans, kuehneotheriids, and *Hadrocodium.* It also differs from morganucodontans, kuehneotheriids, and docodontans in that Meckel's groove is parallel to the ventral border of the mandible, instead of converging on the ventral border (Luo, 1994). *Sinoconodon* is derived in that the medial ridge overhanging the postdentary trough is greatly reduced. On dental features, it differs from all morganucodontans in the absence of cingula and cingulids on most of the postcanines (except for the posterior postcanines of some larger individuals—plesiomorphy). It differs from all other mammals in lacking the one-to-one correspondence of the upper and lower postcanines and from morganucodontans, *Hadrocodium,* kuehneotheriids, docodontans, and all other mammals in lacking wear facets on the molars (plesiomorphy). It differs from morganucodontans, kuehneotheriids, and docodontans (but not *Hadrocodium*) in the presence of an exceedingly large postcanine diastema. It differs from kuehneotheriids in retaining the primitive linear alignment of three main cusps of teeth and differs from allotherians in lacking the multiple rows of cusps on the cheek teeth.

Comments. Based on the work by Crompton and Luo (1993) on the ontogenetic series of *Sinoconodon* skulls, Luo and Wu (1994) pointed out that the different fossil specimens assigned to various "sinoconodontid species" actually represent different growth stages of a single taxon. This taxonomic assessment has been now accepted by all workers who have examined these fossils (Zhang et al., 1998).

"*Lufengoconodon changchiawaensis*" Young, 1982a ("*Sinoconodon changchiawaensis*" according to Crompton and Sun, 1985) was separated from *Sinoconodon rigneyi* because Young (1982a) believed that they had different dental formulae. He also erected "*Sinoconodon parringtoni*" and separated it from *Sinoconodon rigneyi* on the basis of three characters: a short and robust rostrum, a larger canine, and a vertically deep skull. In the same light, Zhang and Cui (1983) separated "*Sinoconodon yangi*" from *Sinoconodon rigneyi* by the difference in the number of incisors. When a more complete sample of *Sinoconodon* skulls became available (Crompton and Luo, 1993; Luo and Wu, 1994; Zhang et al., 1998), it was clear that the differences between these taxa, as cited by various authors, are due to ontogenetic differences among individuals at different dental replacement stages. As described earlier, the number of each type of tooth may vary accordingly. This is further supported by the observation that all specimens of *Sinoconodon* are from localities in close geographic proximity and at the same stratigraphic level (Stratum 6, Upper Red Beds, Lower Lufeng Formation, Luo and Wu, 1994, 1995). Here we follow Luo and Wu (1994) in considering "*Lufengoconodon changchiawaensis,*" "*Sinoconodon parringtoni,*" and "*Sinoconodon yangi*" as junior synonyms of *Sinoconodon rigneyi* (see Zhang et al., 1998).

Species. *Sinoconodon rigneyi* Patterson and Olson, 1961, type species by monotypy.

Distribution. Early Jurassic (Sinemurian): China, Yunnan (several localities within Upper Red Beds, Lower Lufeng Formation).

Subclass *incertae sedis*
Order Morganucodonta Kermack, Mussett, and Rigney, 1973

INTRODUCTION

Morganucodonta include some of the earliest-known taxa in the Late Triassic to Middle Jurassic mammalian faunas worldwide (Clemens, 1980a; Sigogneau-Russell, 1983a,c; Datta and Das, 1996). They are most diverse taxonomically and have a global distribution on Laurasian continents and in Africa and India among the Gondwanan landmasses. The wide geographic distribution of the species of *Morganucodon* during the Early Jurassic makes it a useful taxon for stratigraphic correlation (Shubin et al., 1991; Luo and Wu, 1995). Thanks to numerous and exhaustive studies on *Morganucodon*'s dentition, skull, and postcranium, it has long been a crucial taxon for comparative studies of the anatomical evolution and phylogenetic relationships of stem mammals (Kermack, 1963; Crompton, 1971, 1974; Mills, 1971; Kermack et al., 1973, 1981; Parrington, 1973, 1978; Jenkins and Parrington, 1976; Crompton and Luo, 1993; Luo, 1994).

In 1949 Walter Georg Kühne, a legendary explorer of Mesozoic mammals, erected the genus *Morganucodon*, with a single species *M. watsoni*, based on a well-preserved lower molar, from the fissure fillings in Duchy Quarry at Glamorgan in Wales. In 1958 he described numerous isolated teeth of *M. watsoni*, reconstructed its tooth series, and erected the family Morganucodontidae. In the 1950s Kenneth A. Kermack and co-workers began to explore the fissure fillings in various quarries at Glamorgan (Kermack et al., 1956; Kermack and Mussett, 1958a). The number of jaws with teeth, isolated teeth, and bones amounted to thousands of specimens (Parrington, 1971). The age of fissure fillings at Glamorgan has been established as Liassic, possibly Sinemurian (Kermack et al., 1981).

Kühne was also the first modern worker to collect at the fissure fillings of Holwell Quarry in England. This collection was studied by F. R. Parrington, who, in 1941, erected the genus *Eozostrodon*, including two species: *Eozostrodon parvus*, on the basis of an upper premolar, and *Eozostrodon problematicus*, on an incomplete m1. Parrington (1973) later considered *E. problematicus* as a junior synonym of *E. parvus* and made the former type specimen of *E. problematicus* the first referred specimen of *E. parvus*. This was endorsed by subsequent workers (Clemens, 1979b).

For many years, there was disagreement as to the taxonomic status of these species from the Late Triassic–Early Jurassic of Britain. Parrington (1967, 1971) argued that *Morganucodon* was a junior synonym of *Eozostrodon*, while Kermack and co-workers suggested that *E. parvus* was indeterminate as a taxon because it was based on a premolar that lacks distinguishable features from either *M. watsoni* (Kermack et al., 1973) or *Kuehneotherium praecursoris* (Kermack et al., 1968). Clemens (1979b) concluded that there are differences in size and in some morphological features of the molariform teeth of these taxa (also see Mills, 1971). He suggested that *E. parvus* be restricted to the specimens from Holwell Quarry in England and that *M. watsoni* be retained as the name for a distinct taxon of morganucodontid from Wales. This is consistent with the known taxonomic heterogeneity of the vertebrate faunas that yielded these species. At the same time Clemens (1979b) regarded the generic difference between the two species as controversial.

Rhaeto-Liassic mammals were also discovered from the Rhaetian sediments of Hallau, Switzerland (see Peyer, 1956, for earlier description of isolated teeth and Clemens, 1980a, for detailed description of this mammalian fauna). In the early 1980s, additional morganucodontan taxa were also described from the Rhaetic of the St. Nicholas-de-Port of France (Sigogneau-Russell, 1983a). Other morganucodontans from Britain and Europe include *Wareolestes* E. F. Freeman, 1979, and *Helvetiodon* Clemens, 1980. At the generic level, morganucodontans are far more diverse in the Rhaeto-Liassic fauna of Europe than anywhere else in the world (Clemens, 1986).

The next discovery of morganuconodontids came from China. In 1948, E. T. Oehler of the Catholic University in Beijing found a complete skull of *Morganucodon* from the richly fossiliferous sites in the Lower Lufeng Formation of Yunnan, discovered earlier by the team of Bien (1941) and Young (1940, 1947). This skull was briefly described by H. W. Rigney (1963) as *M. oehleri*. Kermack et al. (1973, 1981) published two exhaustive monographs on *Morganucodon*, the first one on the lower jaw and the second on the skull, based on materials of *M. watsoni* from fissure fillings in Wales and the skull of *M. oehleri* from the Lower Lufeng Formation of China. More skulls were discovered through the field explorations of the Institute of Vertebrate Paleontology and Paleoanthropology of Beijing in the 1970s and 1980s (Young, 1978, 1982a; Zhang, 1984), which provided additional information on skull anatomy and dental character variations (Crompton and Luo, 1993; Luo, 1994).

Jenkins et al. (1983) were the first to discover morganucodontans from North America. *Dinnetherium* Jenkins et al., 1983, from the Early Jurassic of Arizona, is represented by most of the dentition and some skull materials (Jenkins, 1984; Crompton and Luo, 1993). Also discovered in Arizona are isolated teeth of *Morganucodon* sp. (Jenkins et al., 1983). Incomplete morganucodontan material, possibly belonging to *Brachyzosotrodon*, was also found in Greenland (Jenkins et al., 1994). A possibly morganucodontan femur was described from the Middle Jurassic of central Russia (Gambaryan and Averianov, 2001).

For the Gondwanan continents, two morganucodontan genera from the Early Jurassic of southern Africa (originally referred to as the Late Triassic) were established in the 1960s: *Erythrotherium* Crompton, 1964, and *Megazostrodon* Crompton and Jenkins, 1968. Crompton (1974) described the dentition of *Megazostrodon* in detail and compared it with that of *Morganucodon* (referred to as *Eozostrodon*). Crompton (1974) also described the mandible and dentition of the juvenile specimen of *Erythrotherium*.

Jenkins and Parrington (1976) studied the postcranial skeletal elements of *Megazostrodon* and *Erythrotherium*, as well as the isolated bones of *Morganucodon* (referred to as *Eozostrodon*). Gow (1986a) described a new skull of *Megazostrodon* and erected the family Megazostrodontidae. In recent years, isolated teeth referred to the morganucodontans *Gondwanadon* Datta and Das, 1996, and *Indozostrodon* Datta and Das, 2001, have been discovered from the Late Triassic and Early Jurassic of India.

When *Morganucodon* was first discovered, it was widely accepted that it was ancestral to docodontans because the lingual cingulid structure of the lower molar in *Morganucodon* bears some resemblance to that of docodontans (Patterson, 1956; Kermack and Mussett, 1958a). A series of studies on the molar wear facets in the 1960s and 1970s revealed that the wear facets of *Morganucodon* are more similar to eutriconodontans (Crompton and Jenkins, 1968; Crompton, 1971, 1974; Mills, 1971; Kermack et al., 1973). Originally based on similarities in dental characters, the suborder Morganucodonta (or family Morganucodontidae) were placed within the order Triconodonta in a series of papers in the 1970s and 1980s (Mills, 1971; Kermack et al., 1973; Parrington, 1973; Jenkins and Crompton, 1979; Clemens, 1980a; Jenkins et al., 1983; Sigogneau-Russell, 1983a).

Subsequent cladistic analyses (e.g., Rowe, 1988; Wible and Hopson, 1993; Rougier, Wible, and Hopson, 1996; Luo, Cifelli, and Kielan-Jaworowska, 2001) placed morganucodontans far away from eutriconodontans (see also chapter 7). This was reflected in the elevation of the suborder Morganucodonta to the rank of an order by Stucky and McKenna (1993). Also our recent and more comprehensive analysis (Luo et al., 2002) of major groups of Mesozoic mammals and some extant forms clearly demonstrated the paraphyly of "Triconodonta" *sensu lato* (see figures 15.1, 15.2). We follow these results and regard Morganucodonta as a stem mammalian order of mostly Late Triassic and Early Jurassic forms placed outside the crown group of Mammalia and not closely related to Eutriconodonta.

BRIEF CHARACTERIZATION

Morganucodontans have primitive "triconodont-like" cheek teeth. The current evidence indicates that most morganucodontans may have had a typical diphyodont dental replacement, although there is still some uncertainty in this feature for *Megazostrodon* (Parrington, 1973, 1978; Gow, 1985). The best available data from the skull specimens show that morganucodontans have a narrower range of ontogenetic size variation than Sinoconodontidae and nonmammalian cynodonts (figure 3.22C; Luo, Crompton, and Sun, 2001). Individuals of smaller species, such as *Megazostrodon rudnerae*, are estimated to weigh about 20 to 30 g (Jenkins and Parrington, 1976). Individuals of larger species, such as *Morganucodon oehleri* from Lufeng, may have a range of body mass from about 27 to 80 g, as estimated from skull size (Luo, Crompton, and Sun, 2001). The limited postcranial fossils suggest that morganucodontans were small gracile mammals capable of actively climbing on uneven substrate (Jenkins and Parrington, 1976).

Morganucodontans are more closely related to the crown mammalian group than is *Sinoconodon*, in that the upper and lower molars have one-to-one occlusal relationship and the opposing molars develop wear facets from this precise dental occlusion. Morganucodontans are also more similar to the crown mammalian group than *Sinoconodon* in a number of features of the craniomandibular joint (Crompton and Luo, 1993; Luo, 1994), a longer cochlear canal (Luo et al., 1995), and a slightly larger cranial capacity.

The mandibular and middle ear structures are essentially the same as in other stem mammals, such as *Sinoconodon* and docodontans. On the medial side of the dentary there is a primitive postdentary trough for the accommodation of the middle ear elements. Both the dentary-squamosal joint and the quadrate-articular joint functioned as parts of the jaw hinge. Primarily based on these primitive cranial features and other postcranial features (to be detailed later), morganucodontans have been excluded from Crown Mammalia in most of the recent large-scale phylogenetic studies (Rowe, 1988; Wible and Hopson, 1993; Rougier, Wible, and Hopson, 1996; Luo et al., 2002).

ANATOMY

Skull (figure 4.4). Morganucodontans are small (shrew to rat size) mammals with a skull range from 27 to 38 mm in length (Luo, Crompton, and Sun, 2001). Many skull features of morganucodontans are generalized mammalian characters and are primitive for morganucodontans themselves. The dentary condyle is in full articulation with the squamosal glenoid, the foremost diagnostic mammalian feature. Medial to this mammalian joint, the reptilian jaw joint between the articular and quadrate is retained (plesiomorphy), the two joints together forming a double (compound) jaw joint. The orbitosphenoid is fully ossified as in all mammals and tritylodontids. The lateral lappet of the basioccipital is displaced by a fully exposed promontorium. The anterior paroccipital process has a distinctive crista parotica for the articulation of the incus, and the incus has a stapedial process. The posterior paroccipital process has a ventral projection. The straight cochlea is more elongate than in *Sinoconodon*, tritheledontids, and tritylodontids. Beyond these, most other cranial characters of morganucodontans are primitive for all mammals.

Two of the most important diagnostic features of mammals from nonmammalian cynodonts are the presence of an enlarged anterior lamina of the petrosal for the lateral wall of the braincase, pierced by two foramina (foramen ovale and foramen rotundum) for the trigeminal nerves (figure 3.5B), and the presence of the bony floor of the cavum epiptericum for the trigeminal ganglion. The significance of these petrosal features was first noted by Kermack (1963) and Hopson (1964) in their studies of *Morganucodon, Triconodon,* and nonmammalian tritylodontids. A prevailing hypothesis in the 1960s and early 1970s was that several Mesozoic mammalian lineages and modern monotremes could be grouped into "prototherians" because these groups shared the derived condition of an enlarged anterior lamina in the lateral wall of the braincase (e.g., Watson, 1916; Kermack, 1963; Hopson, 1964; Kermack and Kielan-Jaworowska 1971; Crompton and Jenkins, 1979), by contrast to therians in which the alisphenoid is a large element for the lateral wall of the braincase and the anterior lamina of the petrosal is absent.

In modern monotremes, the anterior lamina of the petrosal is derived from the embryonic lamina obturans, one of several ossification centers within the spheno-obturator membrane that is the embryonic precursor to the adult braincase wall in all extant mammals (Presley, 1980). There is a clear difference between therians and monotremes in the timing and sequence of these embryonic structures' fusion to the surrounding structures; in monotremes, the lamina obturans is fused first with the otic capsule to become the anterior lamina of the petrosal (Kuhn and Zeller, 1987b; Zeller, 1989b). Based on embryonic development in monotremes, most experts on Mesozoic mammals now agree that the anterior lamina of the petrosal is a neomorphic feature of mammals (figure 3.5), but not a distinguishing feature of stem mammals and monotremes from the braincase wall of therians because the anterior lamina of the petrosal is also present in stem therians (Rougier et al., 1992; Hopson and Rougier, 1993).

All stem mammals (and some more advanced groups) for which the petrosal is complete have an enlarged anterior lamina encircling the foramina for the trigeminal nerve branches, including *Adelobasileus* (Lucas and Luo, 1993), *Sinoconodon* (Kermack et al., 1981; Crompton and Luo, 1993), *Morganucodon* (Kermack, 1963; Wible and Hopson, 1993), *Hadrocodium* (Luo, Crompton, and Sun, 2001), triconodontids (Kermack, 1963; Wible and Hopson, 1993; Rougier, Wible, and Hopson, 1996), and multituberculates (Hurum, 1998b; Wible and Rougier, 2000, and literature cited therein). However, an enlarged anterior lamina is also present in the stem therian *Vincelestes* (Rougier et al., 1992; Hopson and Rougier, 1993; Wible and Hopson, 1993, see also chapter 10). This indicates that the pretribosphenic eupantotherians also have this primitive condition. Recently, it was reported that the anterior lamina of the petrosal is present but reduced in the Early Cretaceous eutherian *Prokennalestes* (Wible et al., 2001, see also figure 13.6B). This shows that the enlarged anterior lamina of the petrosal is primitive for mammals as a whole.

By contrast, the ascending process of the alisphenoid is the primary element in the lateral wall of the braincase in metatherians (Rougier et al., 1998) and all Late Cretaceous and Tertiary eutherians (MacIntyre, 1972; MacPhee, 1981; Novacek, 1986b; Kuhn and Zeller, 1987b). The absence of the anterior lamina of the petrosal is a derived condition of the crown therian group (figure 3.5E, see also chapters 12 and 13).

Morganucodontans also have a great number of other primitive characters. The cranial moiety of the squamosal is a narrow strap of bone. It is positioned superficial to the petrosal and parietal and does not contribute to the primary wall of the parietal braincase. The septomaxilla, a primitive feature, is present in the rostrum.

Mandible. Morganucodontans have a well-defined dentary condyle in articulation with the squamosal glenoid, a mammalian apomorphy. The mandible retained a postdentary trough for the postdentary ("middle ear") bones, with a functional, mobile joint of the quadrate and articular in the so-called compound or double jaw joint. Presence of the postdentary trough and the attachment of the postdentary bones to the mandible are primitive characters of nonmammalian cynodonts and a wide range of other stem mammals.

The postdentary trough is a prominent feature on the medial side of the mandible (Kermack et al., 1973). It starts anteriorly from below the posterior end of the tooth row and extends posteriorly to near the dentary condyle. The dorsal border along much of the postdentary trough is formed by the medial ridge that overhangs the trough. The broadest part of the postdentary trough is below the coronoid process, where the floor of the trough has separate scars that correspond to the angular bone and the prearticular bones (Kermack et al., 1973). The opening of the mandibular canal is positioned in the anterior part of the trough. The trough forms a concavity on the medial side of the mandible that accommodates the reflected lamina of the angular bone (Crompton and Luo, 1993). Meckel's groove extends anteriorly from the postdentary trough. Much thinner and shallower than the trough, it begins near the posterior opening of the mandibular canal and curves to intersect with the ventral edge of the dentary near the middle of the tooth row (figure 4.4C_1). Morganucodontans differ from the nonmammalian cynodonts and *Sinoconodon* in this feature (Luo, 1994).

The angular region of the dentary is variable in the currently known morganucodontans. In morganucodontids, the dentary angle forms a pointed process (figure 4.4C) and its medial side is a concavity for the angular bone. In this feature morganucodontids are similar to *Sinoconodon*

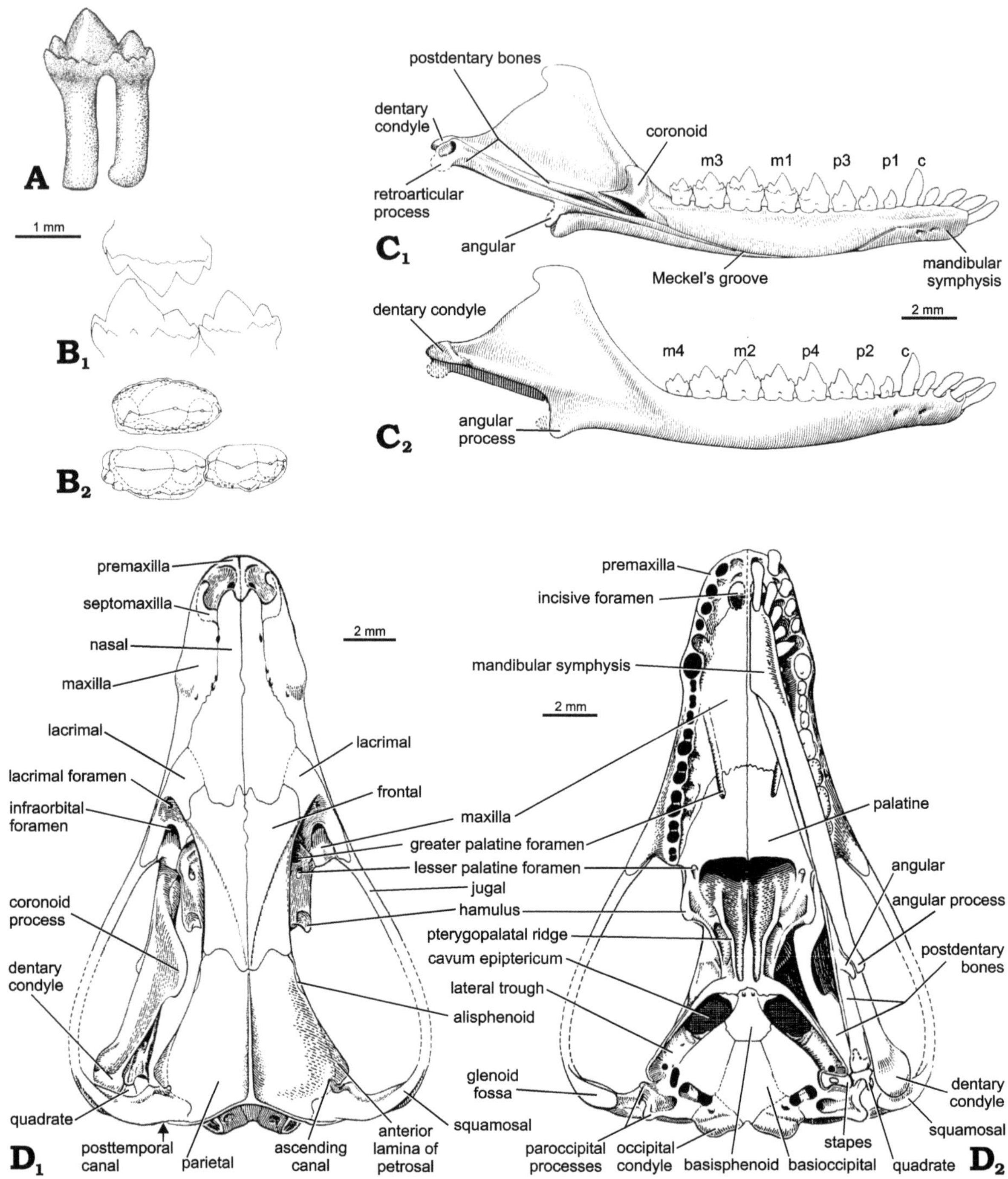

FIGURE 4.4. Anatomy of *Morganucodon watsoni*. A, Generalized anterior lower molar. B, Molar occlusion in lingual (B_1) and crown (B_2) views. C, Mandible in medial (C_1), and lateral (C_2) views. D, Reconstruction of the skull in dorsal (D_1), and ventral (D_2) views. *Source:* A, from Luo (1994); B, redrawn from Crompton (1974); C, D, modified from Kermack et al. (1973 and 1981, respectively); in D modifications are based on original observations of the Chinese specimens.

and docodontans. However, the angular region is much reduced, with a nearly rounded outline, lacking a pointed angle and its medial concavity in *Megazostrodon* and *Dinnetherium;* both are included in Megazostrodontidae herein (figure 4.8). In this feature *Megazostrodon* and *Dinnetherium* are more similar to *Kuehneotherium.*

All morganucodontans have a very shallow symphyseal region in the anterior part of the mandible. The symphysis is mobile. The coronoid bone is present and attaches to the anterior base of the coronoid process of the dentary. Overall the postdentary bones (angular, surangular, prearticular, and articular) are more gracile than in nonmammalian cynodonts (figure 3.10).

Dentition. The most important apomorphies of morganucodontans are in their dentition. In numerous dental features morganucodontans are significantly more derived than *Sinoconodon* and nonmammalian cynodonts. Morganucodontans do not have the primitive, multiple replacement of anterior teeth, as in *Sinoconodon* and cynodonts. Their upper and lower molars have one-to-one correspondence and have developed wear facets for precise occlusion of the upper and lower molar cusps, which are absent in *Sinoconodon.* The precise occlusion of molars and the rotation of the lower jaw in a triangular trajectory during occlusion, made possible by the mobile mandibular symphysis, are the most important evolutionary innovations for mammalian feeding (Crompton and Jenkins, 1968; Crompton, 1971, 1974, 1995; Mills, 1971; Kermack et al., 1973; Crompton and Hylander, 1986).

The cheek teeth are divided into premolars and molars, unlike *Sinoconodon,* whose older individuals have only molariform postcanines. Morganucodontans differ from *Kuehneotherium* and "symmetrodontans" in retaining the primitive linear alignment of the principal molar cusps, rather than forming an obtuse-angle triangle. Lower molars have three main cusps a, b, and c, plus a distal cingulid cusp d. There is a distinctive lingual cingulid that may develop a series of minute cuspules. Among the lingual cingulid cusps, the mesial cusp (e) and the cusp slightly behind the midline (g, referred to by Parrington, 1967, as kühnecone, often misspelled as kuhneocone) are the largest. Upper molars are similarly dominated by four main cusps, A, B, C, D, in straight alignment; the prominent cingula supporting cusps are present along the labial and lingual margins of the crown (figures 4.1, 4.4). In the Chinese species of *Morganucodon,* distal cusp D and mesial cingular cusp E are highly variable. A cingular cusp E may be present on the mesiolingual of the upper tooth. A small postcanine diastema is present in the larger (older) individuals in some morganucodontans as the anterior premolariform teeth are lost (Parrington, 1973).

Morganucodontans have two types of molar cusp occlusions (Crompton, 1974; Jenkins and Crompton, 1979). In *Morganucodon* the principal cusp a of the lower molar occludes between cusps A and B of the upper molar (see figure 4.1). This feature is also present in some eutriconodontans (Crompton, 1974). In *Megazostrodon,* the principal cusp a of the lower molar occludes between cusp B of the opposing upper molar and cusp C of the preceding upper molar (Crompton, 1974; Gow, 1986a). The latter type of occlusion is also present in amphilestids and in the obtuse-angle symmetrodonts, such as in *Kuehneotherium.* While the development of these wear facets are derived characters of mammals as a whole, the occlusal relationship of the upper and lower cusps seems to have some convergence among different groups of early mammals. Morganucodontans and kuehneotheriids tend to develop the precisely matching upper and lower wear facets by beveling after a certain amount of enamel is worn off the crowns by the initial contact. By comparison, in more derived mammals the upper and lower wear facets can match upon eruption (Crompton and Jenkins, 1968; Mills, 1971, 1984; Hahn et al., 1991).

Postcranial Skeleton. The postcranial skeletons of morganucodontans were thoroughly described by Jenkins and Parrington (1976), with supplementary observations from Gow (1985). Most postcranial characters of morganucodontans are primitive features seen in tritheledontids and tritylodontids, and some are even present in other eucynodonts outside Mammaliamorpha. In the axial skeleton, the atlas elements are not fully fused together and the suture between the dens and axis is present. The cervical ribs are not fused to the centra (Jenkins and Parrington, 1976); the lumbar ribs are synostosed to the centra to form the transverse processes; and there is a clear distinction between the thoracic and the lumbar regions of vertebral column.

In the shoulder girdle, the procoracoid, coracoid, and scapula are very similar to those of monotremes, tritylodontids (Kühne, 1956; Sues, 1983; Sun and Li, 1985), and tritheledontids (Gow, 2001). The humerus has a modern mammal-like spherical head and greater and lesser tuberosities, but retains a cynodont-like spiral ulnar condyle (Jenkins and Parrington, 1976). The overall morphology of the shoulder girdle and forelimb is similar to that of cynodonts.

The pelvis and hindlimb are more derived, although these structures also share some features with tritylodontids (Kühne, 1956; Kemp, 1983; Rowe, 1993). The ilium is elongate and directed anterodorsally, with a triangular cross section. The ilium, ischium, and pubis are not fused at the acetabulum (Jenkins and Parrington, 1976). The astragalus lacks the distinctive head of the crown therians and is in juxtaposition to the calcaneus, which has a short tuber calcanei. Jenkins and Parrington (1976) reconstructed *Megazostrodon* to be a gracile animal, capable of climbing on uneven substrate.

SYSTEMATICS

Order Morganucodonta Kermack, Mussett, and Rigney, 1973

Originally erected as a suborder, but McKenna (in Stucky and McKenna, 1993) elevated Morganucodonta to ordinal rank, which we follow.

Diagnosis. More derived than *Sinoconodon* and most cynodonts in having the derived feature of cusp occlusion between the upper and lower molars and a typical diphyodont replacement of the antemolars; but more primitive than most of the crown mammalian taxa in that the wear facets are not matched at tooth eruption, but are developed by beveling of enamel after a significant amount of enamel is worn off by the opposing teeth. Morganucodontans differ from kuehneotheriids and other obtuse-angle "symmetrodontans" in retaining the primitive character of linear alignment of the three main cusps of molars. Distinguishable from the more derived triconodontids, "amphilestids," and spalacotheriids in retaining the primitive characters of a postdentary trough and an anteriorly placed angular process (pseudangular process of some authors) and in lacking a pterygoid fossa on the medial side of the mandible and the associated structures that are typical of eutriconodontans and "symmetrodontans." For other cranial and postcranial features see "Brief Characterization" above.

Distribution. Late Triassic–Middle Jurassic: Europe, Britain, Belgium, Luxembourg, France, Germany, and Switzerland; Late Triassic–Early Jurassic: India; Late Triassic: Greenland; Early Jurassic: Asia, China; Africa, South Africa; North America, Arizona.

Family Morganucodontidae Kühne, 1958

Diagnosis. The diagnosis of the family is the same as of *Morganucodon*, which is the most completely known genus of the family.

Genera. *Morganucodon* Kühne, 1949, type genus; *Eozostrodon* Parrington, 1941; *Erythrotherium* Crompton, 1964; *Gondwanadon* Datta and Das, 1996; *Helvetiodon* Clemens, 1980; *Wareolestes* E. F. Freeman, 1979.

Distribution. Late Triassic (?Carnian, Norian, and Rhaetian) to Middle Jurassic: Europe (including Britain), Asia, North America, India, and Africa.

Genus *Morganucodon* Kühne, 1949

(figures 3.1F, 3.4C, 3.5B, 3.6B, 3.8B, 3.9C, 3.10C, 3.11F, 3.12B, 3.15, 3.16B, 3.17D, 3.18D, 3.19G, 4.4, 4.5)

Morganucodon is the type and the best known genus of Morganucodontidae, represented by thousands of skull fragments with teeth, isolated teeth, and postcranial fragments of *M. watsoni* and by many complete skulls and some incomplete postcranial elements of *Morganucodon oehleri.* The remaining species are represented by fragmentary jaws and isolated teeth. As other morganucodontid genera are known mostly from teeth and (at best) the dentaries, we restrict the diagnosis of *Morganucodon* to dental and mandibular characters.

Emended Diagnosis. Dental formulae: 3.1.4?.4/4.1.4.5 (*M. watsoni*); 4.1.4.3/4.1.5.3–4 (*M. oehleri*); 5.1.4–5.3/5.1.4–5.3 (*M. heikuopengensis*). *Morganucodon* is readily distinguished from *Sinoconodon* and nonmammalian cynodonts by derived characters, but is only distinguishable from other morganucodontids by a combination of primitive characters also present in some other stem mammals. All *Morganucodon* species are characterized by primitive features of linear alignment of three main cusps, presence of the lingual cingulid and related cusps of the lower, and presence of the lingual and labial cingula of the upper. By these primitive characters *Morganucodon* can be separated from all eutriconodontans. The adjacent lower molars are interlocked by the placement of distal cusp d in the notch between mesial cusps e and b of the succeeding tooth, thereby distinguishing *Morganucodon* from *Dinnetherium, Kuehneotherium,* and eutriconodontans. All species of *Morganucodon* have a characteristic molar occlusal pattern: the central cusp a of lower molar occludes on the lingual side of the embrasure between cusps B and A of the upper. The central cusp A of the upper occludes on the labial side of the embrasure between cusps a and c (Crompton and Jenkins, 1968; Mills, 1971; Butler, 1972b; Crompton, 1974). *Morganucodon* is distinguishable from *Megazostrodon, Brachyzostrodon,* and *Indozostrodon* by these occlusal features. *Morganucodon* has many typical mandibular features of stem mammals: presence of a mobile symphysis (derived for mammals); presence of the postdentary trough (primitive), the fossa for the coronoid, and the concavity for the angular (primitive); the external masseteric fossa is weakly developed (primitive); the presence of a robust dentary condyle (in *M. oehleri* and *M. heikuopengensis*) or a gracile condyle that is dorsoventrally compressed (in *M. watsoni*). The conditions of the dentary condyle are derived for *Morganucodon* by comparison to other stem mammals (Crompton and Luo, 1993; Luo, 1994).

Species. *M. watsoni* Kühne, 1949, type species (referred to by some authors as *Eozostrodon watsoni*); *M. heiku-*

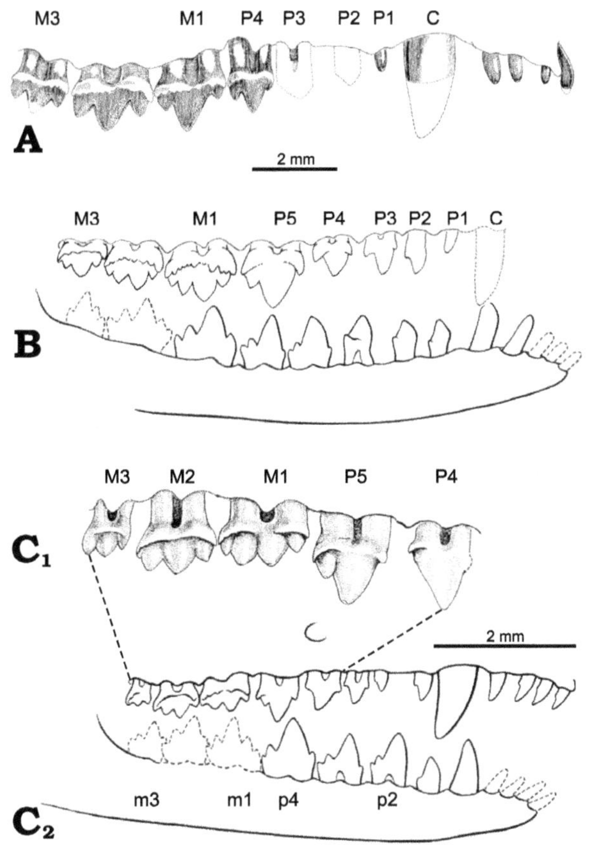

FIGURE 4.5. Variation in dental morphology of the Chinese *Morganucodon* species. A, *Morganucodon oehleri*, reconstruction of the upper dentition of a large (older?) specimen, in lateral view. B, *Morganucodon oehleri*, nearly full upper and lower dentition of a smaller (ontogenetically younger?) specimen, in lateral view. C, *Morganucodon heikuopengensis*; details of upper postcanines (C_1), lateral view, and full dentition (C_2). The upper scale bar is for A and B. *Source:* A, modified from Luo et al. (1995); B, from Crompton and Luo (1993); C_1, original; C_2, from Crompton and Luo (1993).

opengensis Young, 1978 (originally described as *Eozostrodon heikuopengensis*); *M. oehleri* Rigney, 1963; *M. peyeri* Clemens, 1980; unnamed species from Arizona (Jenkins et al., 1983).

Distribution. Late Triassic (Lower Rhaetian): France, Lorraine (Saint-Nicolas-de-Port, Varangéville); Late Triassic (Rhaetian): Switzerland, Kanton Schaffhausen (Hallau bonebed); Early Jurassic ("Liassic," ?Sinemurian): Britain, Wales (fissure fillings in Glamorgan), China, Yunnan (several localities within Lower Lufeng Formation), and United States, Arizona (Gold Springs, Kayenta Formation).

Taxonomic Comments. The two Chinese species of *Morganucodon* are clearly the more derived taxa within the *Morganucodon* clade (Luo and Wu, 1994, 1995). By comparison to *Morganucodon watsoni* (Kermack et al., 1973), they have more reduced posterior molars, more robust dentary condyles, and a greatly reduced medial ridge over the postdentary trough (Crompton and Luo, 1993).

As has been noted by several workers, *Morganucodon* species show a great range of variability in many dental and mandibular features. The dental specimens assigned to various *Morganucodon* species have a distribution over widely separated regions of Pangea in the Triassic-Jurassic transition (Britain, continental Europe, Arizona, and China). *Erythrotherium* from southern Africa also closely resembles *Morganucodon* (Crompton, 1964b, 1974; Mills, 1971; Kermack et al., 1973), and is even considered by some to be a synonym of *Morganucodon* (e.g., Kermack et al., 1973, although we consider *Erythrotherium* a valid genus). Also adding to the possible geographic variations of *Morganucodon* species are the populational variations in the large samples from the Jurassic of Britain and China. We briefly summarize below the observations of many workers (e.g., Mills, 1971; Parrington, 1973, 1978; Crompton and Luo, 1993; Luo and Wu, 1994) on the ontogenetic variation in dental formulae and variability of molar morphology.

1. Ontogenetic variation in dental formulae. Although dental formulae have been used as part of the diagnosis in previous taxonomic studies of morganucodontan taxa, the premolar and molar counts are not reliable characters if the sample size is small and the growth series unknown. The dental formulae, especially the premolar counts, can vary according to the stage of replacement (Crompton and Luo, 1993; Luo and Wu, 1994). In the type specimen of *M. oehleri*, only four premolars are present (Kermack et al., 1973), whereas there are five premolars in other specimens that presumably represent younger individuals (Crompton and Luo, 1993; Luo and Wu, 1994). The difference in the number of premolars is correlated with the loss of the anterior premolars in larger (older) individuals, a dental replacement pattern well documented in *M. watsoni* (Mills, 1971; Parrington, 1973; Crompton, 1974), in *Sinoconodon* (Crompton and Luo, 1993), and in nonmammalian cynodonts (e.g., Hopson, 1971). The dental formulae differ between *M. watsoni* of Britain and the *Morganucodon* species of China. It is possible that some of these differences may be attributable to the ontogenetic variations within samples of each of these two geographic species, but such variations in dental formula by themselves are not sufficient to account for the differences between *M. watsoni* and *M. oehleri*. The implications of this cynodont-like variability in dental formulae in *Morganucodon* must be considered in the treatment of generic taxonomy in future systematic studies.

2. Variability of molar morphology. Parrington (1973, 1978) noted that the ultimate functional molar in *Mor-*

ganucodon ("*Eozostrodon*") *watsoni* is variable in size and morphology. There are also qualitative observations that the occlusal features are variable within the dental samples of *M. watsoni* (Crompton and Jenkins, 1968; Mills, 1971). Among the two Chinese taxa, the labial cingulum and the embrasure between cusp A and cusp B in the upper molars are much better developed in *M. oehleri* than in *M. heikuopengensis* and *M. watsoni.* The difference in the height between cusp A and cusp B and the development of the cingular cusps are much more pronounced in *M. oehleri* than in *M. heikuopengensis* (Luo and Wu, 1994; Luo et al., 1995). In addition, the largest postcanine tooth is M2 in *M. oehleri,* M1 in *M. heikuopengensis,* but the ultimate premolar (P4) in *M. watsoni.* M3 is much smaller in *M. heikuopengensis* than in *M. oehleri* (figure 4.5). Other variable features include the size of upper cusp D and its topographic relation to cusp C and to distal cingular cuspule. Cusp D can be present on M2 of some specimens but not on others. However, the sample of the posterior molars is far smaller for the Chinese *Morganucodon* species than for *M. watsoni* from Britain. At present it is not possible to establish whether the Chinese *Morganucodon* species have a pattern of polymorphic variation similar to that of *Morganucodon watsoni.* Given these variations, we cannot exclude the possibility that some *Morganucodon* species can be separated into different genera. In this chapter, we provisionally maintain the species that are currently assigned to *Morganucodon* and postpone the resolution of the generic and specific taxonomy to a future, comprehensive study.

Genus *Eozostrodon* Parrington, 1941
(figure 4.6A,B)

Comment. *Eozostrodon* is herein considered to be a valid genus, based on two teeth of *Eozostrodon parvus* (including "*Eozostrodon problematicus*"). The holotype is an upper premolar with a large cusp A, a small cusp C, and an even more reduced cusp B on the anterior edge. The labial cingulum is present along the posterior half of the tooth. By comparison to the more complete dental series of *Morganucodon watsoni* and *M. oehleri,* the upper premolars tend to have a better developed cingulum on the labial side than on the lingual side. The first referred specimen is an incomplete anterior lower molar that has a more gracile and posteriorly reclined cusp c and a more reduced kuhnecone, distinguishable from a more robust and rounded cusp c and a much larger kuhnecone in the m1 of *M. watsoni* (Parrington, 1971; Clemens, 1979b). The posterior root of lower molar of *E. parvus* is curved and tapering toward the distal end, unlike the posterior root of lower molars of *M. watsoni,* which tends to develop a "club foot" (Parrington, 1971).

Diagnosis. *Eozostrodon* differs from *Morganucodon watsoni* in having a better-developed labial cingulum on the upper premolar, a more gracile and posteriorly reclined cusp C, and a more tapering distal end of the posterior root (Parrington, 1971; Clemens, 1979b).

Species. *Eozostrodon parvus* Parrington, 1941 (including *Eozostrodon problematicus* Parrington, 1941), type species by monotypy. Parrington (1971) synonymized *E. parvus* with *E. problematicus* and designated the latter as a junior synonym. Two species that were formerly referred by some to *Eozostrodon, E. watsoni,* and *E. heikuopengensis,* are now assigned to *Morganucodon* (see Clemens, 1979b, for earlier discussion on the taxonomic issues and solution on *Morganucodon* and *Eozostrodon*).

Distribution. Late Triassic (?Norian): Britain, Oxfordshire (fissure fillings in Holwell Quarry).

Genus *Erythrotherium* Crompton, 1964
(figure 4.6C)

Comment. *Erythrotherium* is represented by the holotype of the type species (*E. parringtoni*), which includes a crushed skull associated with the mandible and a large part of the dentition, which includes both deciduous and permanent teeth. The holotype represents a juvenile individual (Crompton, 1964b, 1974; Crompton and Jenkins, 1968). Mills (1971) and Kermack et al. (1973) suggested that the main diagnostic characters of *Erythrotherium* might be within the range of variation in the large sample of teeth of *Morganucodon watsoni.* However, Crompton (1974) demonstrated that several features of the permanent teeth (p4 and m1) of *Erythrotherium* separate this southern African taxon from *M. watsoni.* We consider *Erythrotherium* to be a valid genus (*contra* Kermack et al., 1973), although very closely related to *Morganucodon* in that m2 and m3 are identical in the occlusal relationship of the major cusps to those of *Morganucodon* (Mills, 1971; Crompton, 1974).

Diagnosis. The dental formula of *E. parringtoni* is: 4.1.4.4(?)/3.1.4.4. The dentition of *Erythrotherium* very closely resembles that of *Morganucodon;* the difference is that in *Erythrotherium* the ultimate upper premolar is smaller than the first upper molar, whereas in *Morganucodon* the first upper molar is of the same size or even slightly smaller than the ultimate premolar. Another difference is that in *Erythrotherium* cusp g is absent on the first molar, whereas it is present in *Morganucodon.*

Species. *Erythrotherium parringtoni* Crompton, 1964, type species by monotypy.

Distribution. Early Jurassic ("Liassic"): southern Africa, Lesotho (Mafeteng, Cave Sandstone, Clarens Formation, Stormberg Group).

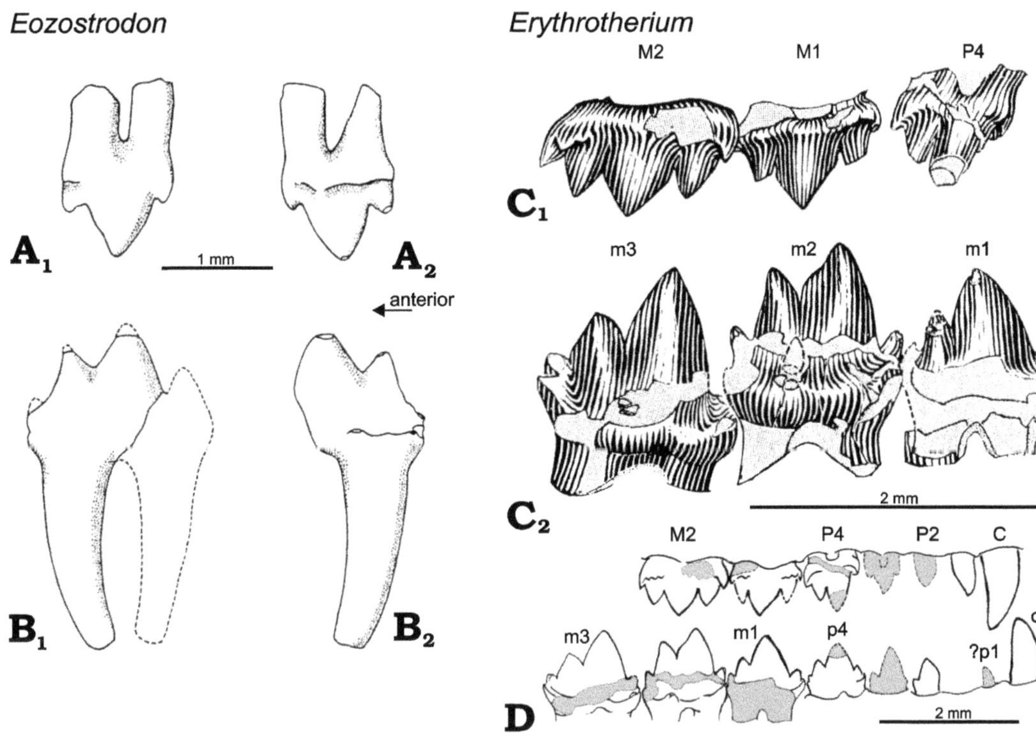

FIGURE 4.6. Teeth of *Eozostrodon* (A, B) and *Erythrotherium* (C, D). A, *Eozostrodon parvus*, right posterior upper premolar (holotype) in labial (A_1) and lingual (A_2) views. B, Right anterior lower molar in labial (B_1) and lingual (B_2) views. C, *Erythrotherium parringtoni*, right P4–M2 in labial view (C_1); left m1–m3 in lingual view (C_2). D, Composite reconstruction of the right upper and left lower dentition. *Source:* A, B, from Clemens (1979b); C, D, modified from Crompton (1974).

Genus *Helvetiodon* Clemens, 1980
(figure 4.7A)

Diagnosis (modified from Clemens, 1980a). *Helvetiodon* is based on isolated teeth thought to be upper molariforms and on an additional lower premolariform tentatively referred to this taxon. Molariforms are one and a half times the size of those of *Morganucodon, Megazostrodon,* and *Erythrotherium,* and in the size range of *Brachyzostrodon* and *Wareolestes.* They differ from *Tricuspes* and nonmammalian cynodonts in the more extensive development of labial and lingual cingula and cusps. Presumed upper molariforms of *Helvetiodon* differ from those of *Megazostrodon* and *Brachyzostrodon* in having a very reduced anterior cusp B and in having more massive and bulbous morphology of cingular cusps on the lingual cingulum. The main cusp A differs from those of *Morganucodon* in having wrinkled enamel. In this feature *Helvetiodon* is similar to *Brachyzostrodon* and *Wareolestes.*

Species. *Helvetiodon schutzi* Clemens, 1980, type species by monotypy.

Distribution. Late Triassic (Rhaetian): Switzerland, Kanton Schaffhausen (Hallau Bonebed).

Family ?Morganucodontidae Kühne, 1958
Genus *Gondwanadon* Datta and Das, 1996
(figure 4.7B)

Diagnosis. *Gondwanadon* is based on a single right lower molar, the holotype of the type species (*G. tapani*). The genus differs from other morganucodontid genera in having a very robust cusp b of almost equal size to cusp a, larger and more robust than cusp c. Cusp b is positioned in the most mesial part of the tooth. Differs from most morganuconodontids and megazostrodontids, but not *Dinnetherium* or *Kuehneotherium,* in that mesial cusps e and f are present (this part of the tooth was broken during preparation of photograph) (P. M. Datta, pers. comm.). Differs also from *Morganucodon* in having much smaller kuhnecone (cusp g).

Comment. Two characters of *G. tapani* are reminiscent of morganucodontans: presence of lingual cingulid cusps g and h in the posterior part of the tooth and the presence of wear facets between cusps a and c and between cusps a and b (a primitive feature shared by triconodontids). Because the holotype specimen is incomplete, we tentatively accept the assignment of this tooth to Morganucodontidae until corroboration by better fossil materials.

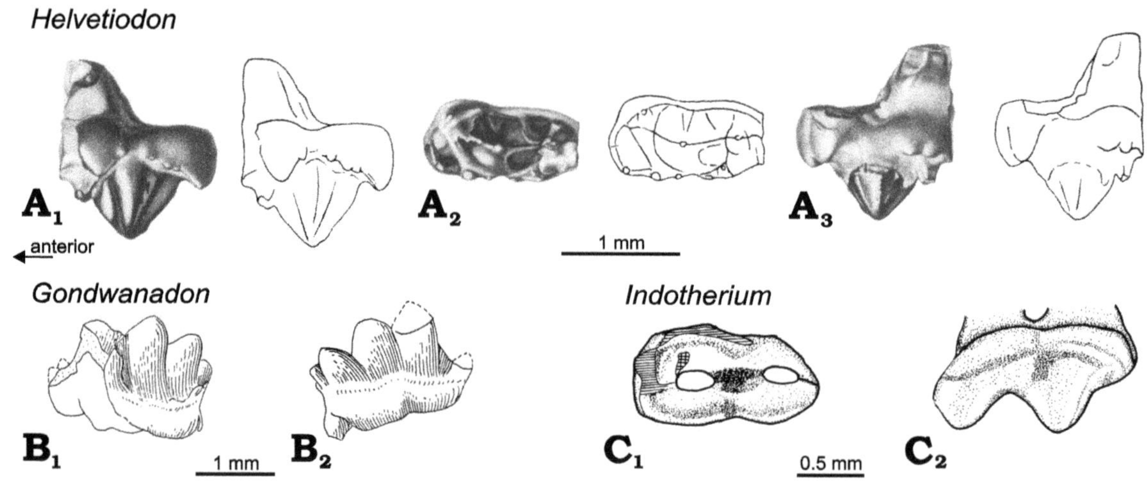

FIGURE 4.7. A, Right upper molariform of *Helvetiodon* in lingual (A_1), crown (A_2), and labial (A_3) views. B, Lower right molariform of *Gondwanadon* in lingual (B_1) and labial (B_2) views. C, Upper right molar of *Indotherium:* in occlusal (C_1) and labial (C_2) views. *Source:* A, shaded drawings from Clemens (1980a); outline drawings from Sigogneau-Russell (1983a); B, modified from Datta and Das (1996); C, modified from Prasad and Manhas (2002).

Species. *Gondwanadon tapani* Datta and Das, 1996, type species by monotypy.

Distribution. Late Triassic (Carnian): India, Madhya Pradesh (Tiki Formation). *Gondwanadon* is the oldest known mammal if its affinities to morganuconodontids can be confirmed by additional materials.

Genus *Indotherium* Yadagiri, 1984
(figure 4.7C)

Emended Diagnosis (modified from Prasad and Manhas, 2002). *Indotherium* is based on two isolated upper molars. The primary diagnostic character is the linguolabially compression of upper cusps A and C. The genus is distinguishable from all other morganucodontans in having a much smaller anterior cusp B and a much deeper notch between cusps A and C. The ridges in the A–C embrasure have a pattern of transverse wear, similar to *Indozostrodon* but different from most morganucodontans.

Comment. Both the type and the referred specimens of *Indotherium* (Yadagiri, 1984; Prasad and Manhas, 1997) are isolated teeth, previously misidentified as lower molars. Prasad and Manhas (2002) suggested that these are upper molars, based on the presence of both labial and lingual cingula. The specimens of *Indotherium* are likely to be ultimate upper molars because that tooth tends to have a more asymmetrical outline in the labial view in other more complete specimens of *Morganucodon oehleri.* The primary character by which *Indotherium* is regarded to be more closely related to *Morganucodon* than to *Megazostrodon* is the inferred cusp occlusion, in which the lower cusps occluded into the valleys between cusps A and C and anterior to cusp A of the upper molars. The wear of the A–C embrasure, however, indicates a degree of transverse wear, differing from any known morganucodontans except for *Indozostrodon.* The outline of the upper molars in lateral view is strikingly similar in *Indotherium* and *Indozostrodon* from the same formation. Based on these characters it is likely that these two genera are closely related. Prasad and Manhas (2002) suggested that the two taxa are synonymous. Here we tentatively accept the assignment of *Indotherium* to Morganucodontidae (Prasad and Manhas, 2002), pending further study when more materials are discovered.

Species. *Indotherium pranhitai* Yadagiri, 1984; type species by monotypy.

Distribution. Early (to early Middle?) Jurassic: India, Andhra Pradesh, Pranhita-Godavari Valley (Kota Formation).

Family Megazostrodontidae Gow, 1986

Comments. When *Megazostrodon* was first described, it was considered to be a "triconodont" with close affinities to *Morganucodon* (Crompton and Jenkins, 1968). Although Mills (1971) considered *Megazostrodon* to be closely related to *Sinoconodon,* other workers accepted that *Megazostrodon* belongs among morganucodontans (Crompton, 1974; Clemens, 1979b; Crompton and Jenkins, 1979; Gow, 1986a), and that *Sinoconodon* is not closely related to either *Megazostrodon* or *Morganucodon* (see the earlier discussion). McKenna and Bell (1997) referred *Megazostrodon* to Docodonta. Gow (1986a) recognized that *Megazostrodon* shares some unique mandibular features with *Dinnetherium* (Jenkins, 1984) and proposed the family Megazostrodontidae for both genera. We follow Gow's taxonomic scheme here.

Diagnosis. Megazostrodontidae are distinctive from morganucodontids and sinoconodontids (although not from kuehneotheriids or haramiyidans) in having a reduced angular process (or pseudangular process of other authors), and in having a strongly flaring ridge of the dentary peduncle (which was designated as a neomorphic angular process by some authors, Gow, 1986a). It differs from gobiconodontids, triconodontids, "amphilestids," and spalacotheriid symmetrodontans in lacking a fully rounded angular process with a pterygoid shelf. It differs from most morganucodontids in that most upper molars have well-developed labial cingular cusps and that the upper labial cingulum is divided into the distinctive anterior and posterior lobes (Crompton and Jenkins, 1968).

Genera. *Megazostrodon* Crompton and Jenkins, 1968, type genus; *Brachyzostrodon* Sigogneau-Russell, 1983; *Wareolestes* E. F. Freeman, 1979; *Dinnetherium* Jenkins et al., 1983; and possibly *Indozostrodon* Datta and Das, 2001.

Distribution. Late Triassic (Rhaetian) to Middle Jurassic (Bathonian): Europe, France and Britain; Late Triassic: Greenland; Early Jurassic: North America, United States, southern Africa.

Possible Distribution. Early (?Middle) Jurassic: India.

Genus *Megazostrodon* Crompton and Jenkins, 1968 (figure 4.8A,B)

Emended Diagnosis. Dental formula 4.1.5.5./4.1.5.5 (Gow, 1986a). *Megazostrodon* is characterized by an extremely well-developed labial cingulum with more robust cusps than morganucodontids. The main cusp A of upper molars occludes in between cusp c of the respective lower molar and cusp b of the succeeding lower molar (the "embrasure occlusion"). The occlusion of its upper main cusps is more posterior relative to main cusps of the lower molars than in morganucodontids. It differs from *Dinnetherium* in having a hypertrophied kuhnecone (Jenkins et al., 1983) and from *Brachyzostrodon* and *Wareolestes* in lacking a wrinkled enamel surface and in being significantly smaller.

Species. *Megazostrodon rudnerae* Crompton and Jenkins, 1968; type species by monotypy.

Distribution. Early Jurassic ("Liassic"): Lesotho (Red Beds, Stormberg Group) and South Africa, Orange Free State (Upper Elliot Formation).

Genus *Brachyzostrodon* Sigogneau-Russell, 1983a (figure 4.9A,B)

Diagnosis. *Brachyzostrodon* is known only from isolated teeth of two named and three unnamed species (Sigogneau-Russell, 1983a; Hahn et al., 1991; Jenkins et al., 1994). The genus is unique among its contemporary mammals in showing heavy abrasion at the apices of the main cusps (Sigogneau-Russell, 1983a; Hahn et al., 1991), a wear pattern that is analogous to the unrelated Cretaceous *Gobiconodon* (Jenkins and Schaff, 1988). Molars of *Brachyzostrodon* have a wrinkled enamel surface, a rare condition among morganucodontans, except for *Wareolestes* and possibly also for *Helvetiodon* (Sigogneau-Russell, 1983a; Hahn et al., 1991; Sigogneau-Russell and Hahn, 1994), but typical for docodontans. *Brachyzostrodon* is similar to *Megazostrodon,* but distinguished from all species of morganucodontids in having a hypertrophied lingual cusp g on lower molars, and in having separated mesial and distal lobes of the upper labial cingulum (Hahn et al., 1991). It differs from *Megazostrodon* and *Morganucodon* in having no continuous lingual cingulid of the lower molar, nor any additional cingulid cusps between lingual cusp g and the mesiolingual cusp e. Differs from *Morganucodon* in having a relatively smaller cusp a and better development of cusp b. The latter may be either small, or even absent in some specimens of *Morganucodon.*

All species of *Brachyzostrodon* are distinctly larger than any species of *Morganucodon, Erythrotherium,* and *Megazostrodon:* the lower molar of the holotype of *Brachyzostrodon coupatezi* is 1.9 mm in length, 1.2 mm in width. Lower molar cusps b and c are well separated from large cusp a. To the rear of cusp c there is a small cusp d, which forms the labialmost cusp of the distal cingulid (consisting of three cusplike crenulations), which wraps around the tooth distally. No intact molar series is known in *Brachyzostrodon,* but it can be safely inferred that molars are interlocked by distal cusp d of a preceding molar fitting between cusps b and e of a succeeding molar. But the notch between cusps b and e is shallow, so the interlock is loose.

Comment. Hahn et al. (1991) pointed out some similarities of *Brachyzostrodon* and *Megazostrodon.* Given the balance of evidence, we accept *Brachyzostrodon* to be more closely related to *Megazostrodon* than *Morganucodon.* This would imply that the similarity in the cusp occlusion between *Brachyzostrodon* and *Morganucodon* is either primitive or homoplastic, as the same occlusal pattern is present in more derived triconodontids. Hahn et al. (1991) further pointed out that the Middle Jurassic *Wareolestes* shares the wrinkled enamel and a large, broad cusp a on the molar. We also tentatively place *Wareolestes* in Megazostrodontidae.

Species. *Brachyzostrodon coupatezi* Sigogneau-Russell, 1983a, type species. *Brachyzostrodon maior* Hahn et al., 1991; also two unnamed species mentioned by Hahn et al. (1991).

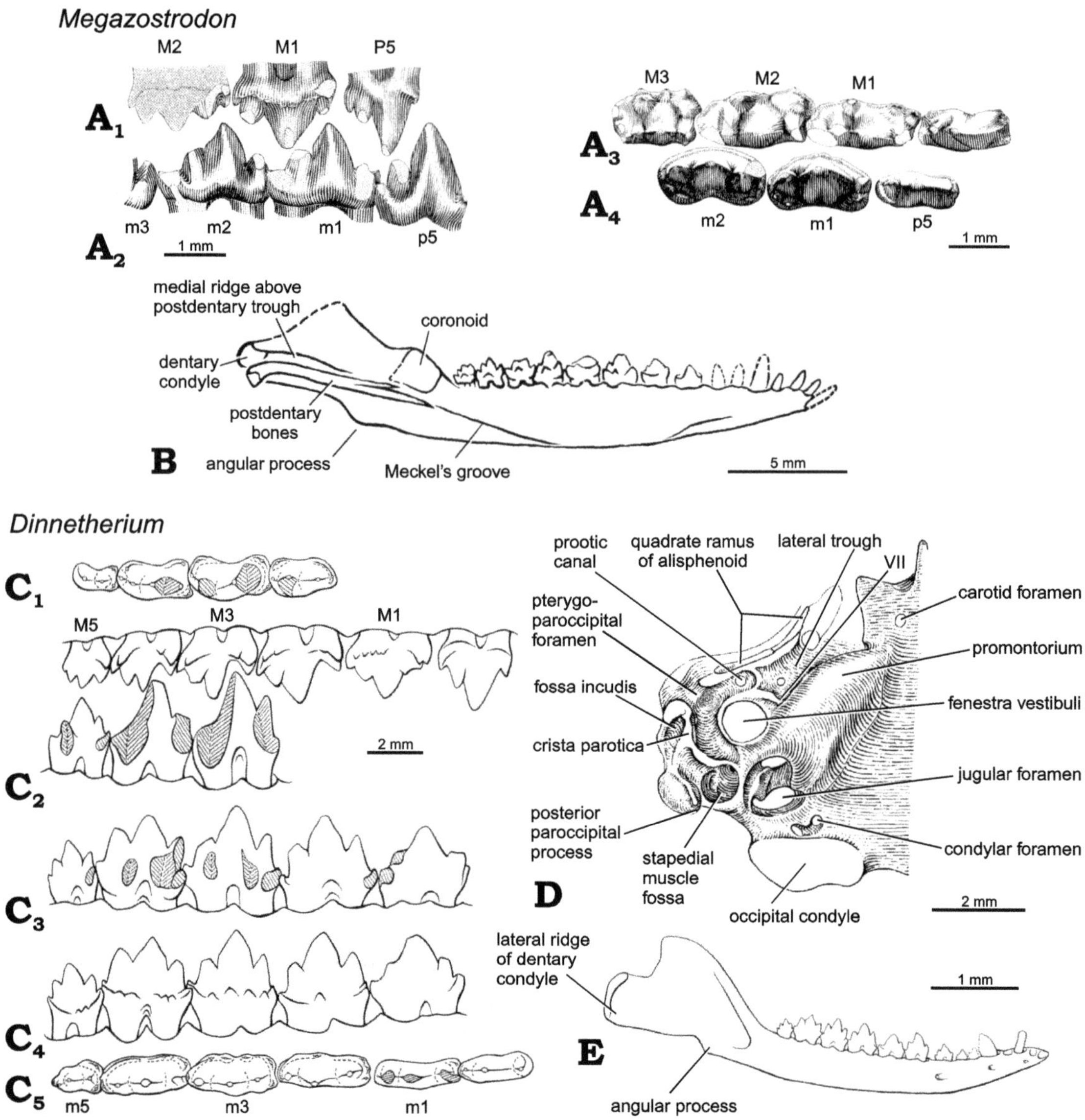

FIGURE 4.8. *Megazostrodon* and *Dinnetherium*. A, B, Dentition and mandible of *Megazostrodon rudnerae:* lingual view of P5–M2 (A_1); labial view of p5–m2 (A_2); crown view of P5-M3 (A_3); crown view of p5–m2 (A_4); composite reconstruction of the mandible (B). C–E, Structure of *Dinnetherium nezorum.* M2-M5 in crown view (C_1); mode of occlusion (C_2); lower molars in labial (C_3), lingual (C_4); and crown (C_5) views (in C_5 p5 is also shown); right petrosal in ventral view (D); mandible in lateral view (E). *Source:* A, from Crompton (1974); B, composite drawing based on photos and illustration of Gow (1986a) and Rowe (1986); C_{1-2} and E, from Jenkins et al. (1983) and Crompton and Luo (1993); C_{3-5}, modified from Jenkins (1984); D, modified from Crompton and Luo (1993).

Distribution. Late Triassic (early Rhaetian): France, Lorraine (Saint-Nicolas-de-Port).

Possible Distribution. Late Triassic (Norian): Greenland, Jameson Land (Fleming Fjord Formation).

Genus *Dinnetherium* Jenkins, Crompton, and Downs, 1983
(figure 4.8C–E)

Diagnosis. Dental formula: ?.?.2+?.5/4.1.4.5. *Dinnetherium* is known from dentaries with complete dentition and upper premolars and molars, plus disarticulated basicranial elements. The molar occlusion is similar to that of *Morganucodon,* but the central cusps A and a are taller. The difference is that in *Dinnetherium* the lower molar cusps b and c are of equal height in unworn crowns and are positioned symmetrically on either side of a, whereas in *Morganucodon* cusp b is lower than c and closer to a than c. The wear facets on the lower molars of *Dinnetherium* are close to vertical, whereas those on the upper are close to horizontal. Despite the different orienta-

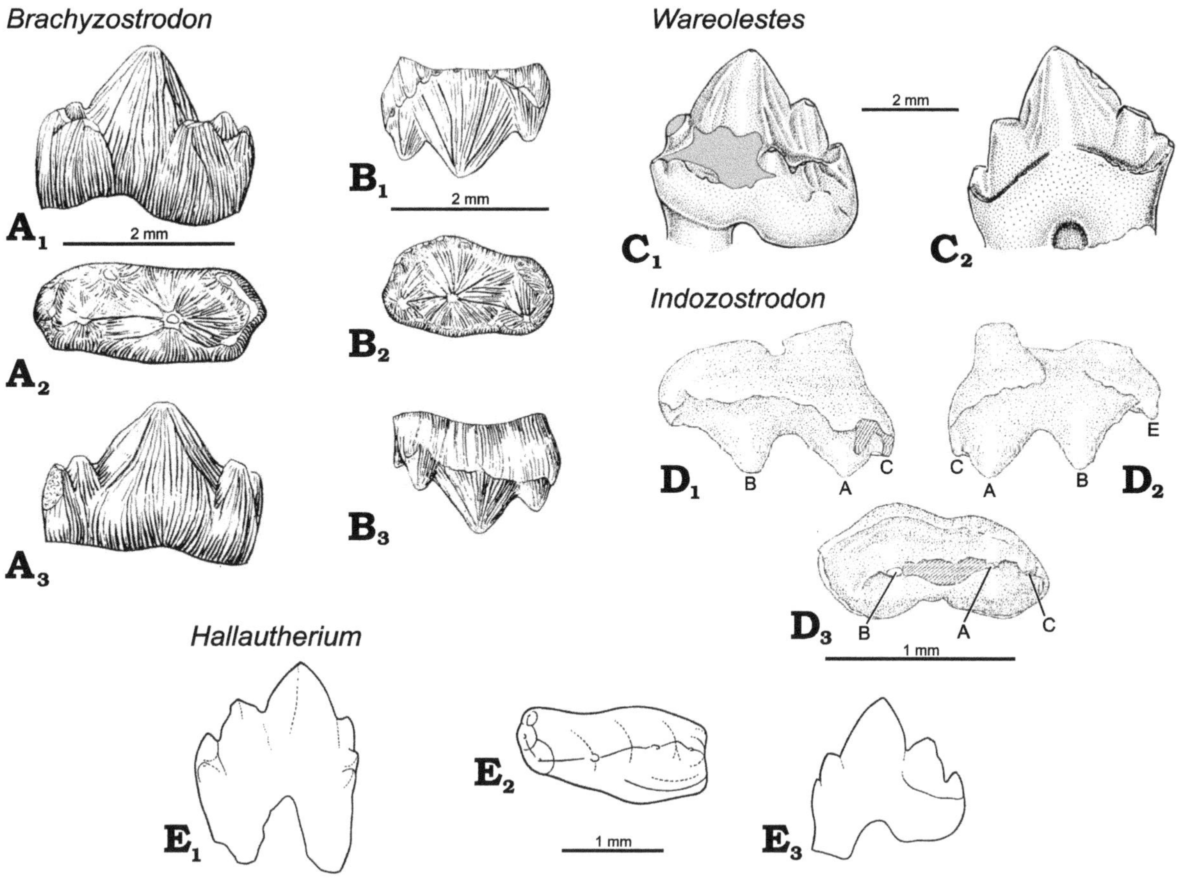

FIGURE 4.9. Teeth of *Brachyzostrodon, Wareolestes, Indozostrodon,* and *Hallautherium.* A, *Brachyzostrodon maior,* right lower molar, holotype, in lingual (A_1), occlusal (A_2), and labial (A_3) views. B, *Brachyzostrodon* sp., right upper molar in lingual (B_1), occlusal (B_2), and labial (B_3) views. C, *Wareolestes rex,* lower molar, holotype, in lingual (C_1) and labial (C_2) views. D, *Indozostrodon simpsoni,* upper molar, holotype in lingual (D_1), labial (D_2), and occlusal (D_3) views. E, *Hallautherium schalchi* (holotype), right lower molariform in lingual (E_1), occlusal (E_2), and labial (E_3) views. *Source:* A, B, from Hahn et al. (1991); C, drawings based on SEM micrographs of Freeman (1979); D, from Datta and Das (2001); E, from Clemens (1980a).

tion of the upper and lower wear facets, they could still contact by rotation of the lower jaw during occlusion. *Dinnetherium* differs from other morganucodontans and *Megazostrodon* but resembles kuehneotheriids, "amphilestids," and gobiconodontids in having the distal cusp d interlocking between the mesial cusps e and f of the succeeding molar (Jenkins et al., 1983). By comparison, in *Morganucodon* and *Megazostrodon,* the interlock between adjacent molars is by cusp d of the preceding molar and cusps e and b of the succeeding molar.

The dentary of *Dinnetherium* differs from that of *Morganucodon* in having a less prominent angular process (referred to originally by Jenkins et al., 1983, as the pseudangular process) and more extensive development of the lateral ridge below the condylar process of the dentary (referred to as the true angular process, Jenkins, 1984). In this respect it is similar to the second, adult specimen of *Megazostrodon* (Gow, 1986a). The postdentary trough is smaller than in *Morganucodon.*

Comment. Jenkins et al. (1983) pointed out that the mandibular features are more derived in *Dinnetherium* than in morganuconodontids. Gow (1986a) noted the strong resemblance of the mandible of *Dinnetherium* to that of *Megazostrodon* and assigned *Dinnetherium* to Megazostrodontidae on the basis of the derived mandibular characters. Jenkins and Schaff (1988) further observed some resemblance of *Dinnetherium* to *Gobiconodon.* Hopson (1994) went further to propose that *Dinnetherium* may represent an intermediate condition in the derivation of the rounded angular process of eutriconodontans from the distinctive angular process (pseudangular process of some authors) of a morganucodontid. In our cladistical phylogeny (figure 15.1), *Dinnetherium* appears to be more closely related to *Megazostrodon* than to *Morganucodon.*

The similar occlusion of upper and lower molar cusps of *Dinnetherium* and morganuconodontids is also present in Jurassic and Early Cretaceous tricodontids (Jenkins and Crompton, 1979; Jenkins et al., 1983; Crompton and Luo, 1993). Given the character distribution, the type of molar occlusion may be primitive or homoplasic in *Dinnetherium* and *Morganucodon*. For these reasons, we follow Gow (1986a) and place *Dinnetherium* tentatively as a member of Megazostrodontidae.

Species. *Dinnetherium nezorum* Jenkins et al. 1983; type species by monotypy.

Distribution. Early Jurassic (?Sinemurian): United States, Arizona (Gold Springs, Kayenta Formation).

Genus *Indozostrodon* Datta and Das, 2001
(figure 4.9D)

Diagnosis. The diagnosis is limited to the holotype, which is a single isolated tooth with three cusps in approximately linear alignment, a pattern that is primitive and seen in many stem mammals such as morganucodontans. Differs from all morganucodontids in having subequal heights of cusps A and B and a small cusp C positioned on the posterior cingulum and twinned to cusp A. Differs from all morganucodontans except *Dinnetherium* and from kuehneotheriids in having a hypertrophied valley between cusps A and B.

Comment. Datta and Das (2001) assigned *Indozostrodon* to Megazostrodontidae on two characters. Some wear is developed anterolingual to both cusps A and C. From this it was inferred that the main cusp (a) of the lower molar occluded in the embrasure between cusp C of the preceding molar and cusp B of the succeeding molar. The well-developed labial cingulum is divided into the anterior and posterior lobes. Both features are typical of megazostrodontids (Crompton, 1974) and kuehneotheriids among contemporary mammals, but distinctive from morganucodontids as exemplified by *Morganucodon*. We tentatively accept this assignment.

The strongly developed wear in the A–B valley of the upper molar in *Indozostrodon* is a rare feature that is otherwise seen only on a referred specimen of *Indotherium* among all Late Triassic and Early Jurassic mammals. *Indozostrodon* and *Indotherium* also have a strikingly similar outline in the lateral view of their upper teeth. *Indotherium* has been referred by Prasad and Manhas (2002) to Morganucodontidae. *Indozostrodon* and *Indotherium* are distinctive from each other in the cingulid cusps. These genera may be closely related, and one cannot rule out the possibility that the two are synonymous (Prasad and Manhas, 2002) as they are from the same formation in the same geographic area. Whether *Indozostrodon* and *Indotherium* are synonymous or should be placed together in Megazostrodontidae versus Morganucodontidae will have to be resolved by more complete materials than the currently available isolated teeth.

Species. *Indozostrodon simpsoni* Datta and Das, 2001; type species by monotypy.

Distribution. Early (to early Middle?) Jurassic: India, Andhra Pradesh, Pranhita-Godavari Valley, Yamanapalli (Kota Formation).

Genus *Wareolestes* E. F. Freeman, 1979
(figure 4.9C)

Diagnosis. The genus is represented by a single right lower molar, the holotype of the type species (*Wareolestes rex*). However, Hahn et al. (1991) speculated that the holotype may be an upper tooth because it has both the lingual and labial cingula, whereas lower molars of this pattern tend to develop only the lingual cingulid. *Wareolestes* is similar to *Morganucodon* in having a very wide and tall cusp a, much higher than cusps b and c. It differs in a more anterior position of cusp g, which is situated opposite cusp a, rather than more posteriorly to the level of the valley between cusps a and c as in other morganucodontid genera. The lingual cingulid curves around the distal end of the tooth, where cusp d arises from it, being situated directly distal to cusp c. Differs from *Morganucodon* in having a poorly defined, noncuspidate labial cingulid. The type specimen of *Wareolestes* is 2.31 mm in length and 1.24 mm in width, about one-and-a-half times most morganucodont teeth. Differs from most morganucodontans, but resembles *Brachyzostrodon* (and also *Helvetiodon* to a lesser extent) in having wrinkled enamel surface of the main cusp.

Species. *Wareolestes rex* E. F. Freeman, 1979; type species by monotypy.

Distribution. Middle Jurassic (late Bathonian): Britain, England, Oxfordshire, Kirtlington (Forest Marble Formation). The youngest megazostrodontid and morganucodontan genus.

Family *incertae sedis*
Genus *Hallautherium* Clemens, 1980
(figure 4.9E)

Diagnosis. Poorly known genus represented by two lower molariform teeth (Clemens, 1980a: plate 12, figure 5). Clemens (1980a: 85) provided the following diagnosis of *Hallautherium:* "Only known from lower molariforms with a main row of four cusps and one or more anterior and posterior lingual cusps. These molariforms lack a basal lingual cingulum [cingulid] or kuhnecone [cusp g]. Unlike any known morganucodontids, a large, posterior buccal [labial] basin is present on the type (a small depression on the referred specimen). Teeth are in size range of *Morganucodon watsoni* and *M. peyeri*, but are smaller than those of *Tricuspes* and *Helvetiodon*."

Species. *Hallautherium schalchi* Clemens, 1980, type species by monotypy.

Distribution. Late Triassic (Rhaetian): Switzerland, Kanton Schaffhausen (Hallau Bonebed).

Class Mammalia Linnaeus, 1758
Order and Family *incertae sedis*
Genus *Hadrocodium* Luo, Crompton, and Sun, 2001a (figures 3.4D, 3.8C, 3.15, 4.10)

Comments. *Hadrocodium* is a very small mammal that represents a distinct, hitherto unrecognized stem lineage from the Early Jurassic (Luo, Cifelli, and Kielan-Jaworowska, 2001). It is more closely related to the mammalian crown group (inclusive of eutriconodontans) than to docodontans, morganucodontans, or *Sinoconodon*. Although it retains the typical triconodont-like molar pattern, it has very derived cranial and mandibular characters by comparison to its contemporary stem mammals. *Hadrocodium* helps to extend the record of several derived mammalian features of the brain cavity and separation of middle ear from the mandible to the Early Jurassic.

Diagnosis. The holotype represents a subadult individual with a skull length of about 12 mm (after correction for slight distortion) and an estimated body weight of about 2 g (Luo, Crompton, and Sun, 2001). Dental formula: 5.1.2.2/5.1.2.2. Each molar has three main cusps and two accessory cusps in alignment on the laterally compressed crown. Primary cusp a of the lower occludes in the embrasure between the opposite upper molars (figure 4.10D,E). This occlusal pattern differs from those of morganucodontids, docodontans, and triconodontids, in which the cusp a of the lower molar occludes between cusps A and B of the upper. Differs from *Megazostrodon* in lacking prominent labial cingular cusps of the upper molars and from kuehneotheriids in lacking the triangulation of molar cusps. Differs from morganucodontids, eutriconodontans, and kuehneotheriids (but not *Sinoconodon*) in the presence of a much larger postcanine diastema. Differs from *Sinoconodon* and most cynodonts in the one-to-one precise occlusion of the upper and lower molars. It lacks the multicuspate rows on the teeth of Haramiyida and multituberculates and the multiple-ridged teeth of docodontans. *Hadrocodium* has the following skull characters that distinguish it from its contemporary stem mammals and docodontans: an enlarged cranial cavity and absence of the postdentary trough on the mandible. It differs from all other stem mammals and most crown mammal lineages in having a slightly inflected dentary angle, which is an autapomorphy. The squamosal glenoid for the jaw hinge is situated on the medial side of the squamosal zygoma, similar to *Sinoconodon* but different from most morganucodontans. It is more derived than docodontans, morganucodontans, and *Sinoconodon* in having a postglenoid region posterior to the temporomandibular joint.

ANATOMY

Hadrocodium has a smooth periosteal surface on the medial side of the mandible and lacks the postdentary trough and its associated medial ridge, Meckel's groove, and the medial concavity of the angular process. The postdentary trough and the medial concavity on the angular process accommodated, respectively, the prearticular/surangular and the reflected lamina of the angular in other stem mammals (Kermack et al., 1973; Lillegraven and Krusat, 1991) that are the homologues to the mammalian middle ear bones. The absence of these structures indicates that the postdentary bones ("middle ear ossicles") was separated from the mandible (see the caveat in chapter 3).

Hadrocodium has an enlarged braincase that is wider and more expanded in the frontal and the parietal regions than those of all other stem mammals and all other Jurassic mammals (see chapter 3). The braincase in the parietal region in *Hadrocodium* is comparable to those of the mammalian crown group. By allometric scaling of a large sample of living and fossil mammals, the brain case of *Hadrocodium* is larger than expected for mammals with a comparable skull width and far wider than in any other Triassic–Jurassic stem mammals. This indicates that the small size of *Hadrocodium*, in and by itself, is not sufficient to explain its large brain. Related to the expansion of the brain cavity, its cerebellar portion is expanded more posteriorly than the level of temporomandibular joint. The occipital (posterior) wall of the brain cavity is convex posteriorly beyond the lambdoidal crest (figure 4.10B), and the anterior part of the brain cavity clearly shows the divergence of the frontal lobes.

The temporomandibular joint (TMJ) in *Hadrocodium* is positioned anterior to the level of the fenestra vestibuli and to the occipital condyles, with the zygoma of squamosal swinging anteriorly from the cranial moiety of the squamosal. A postglenoid depression is present on the lateral aspect of the squamosal between the zygoma and the cranial moiety. The postglenoid region posterior to the TMJ (figure 4.10A) may be correlated with the posterior displacement of the basicranium and braincase behind the TMJ.

The petrosal of *Hadrocodium* has a prominent promontorium, more inflated than in other stem mammals, eutriconodontans, and most multituberculates and non-tribosphenic therians. The large promontorium may be

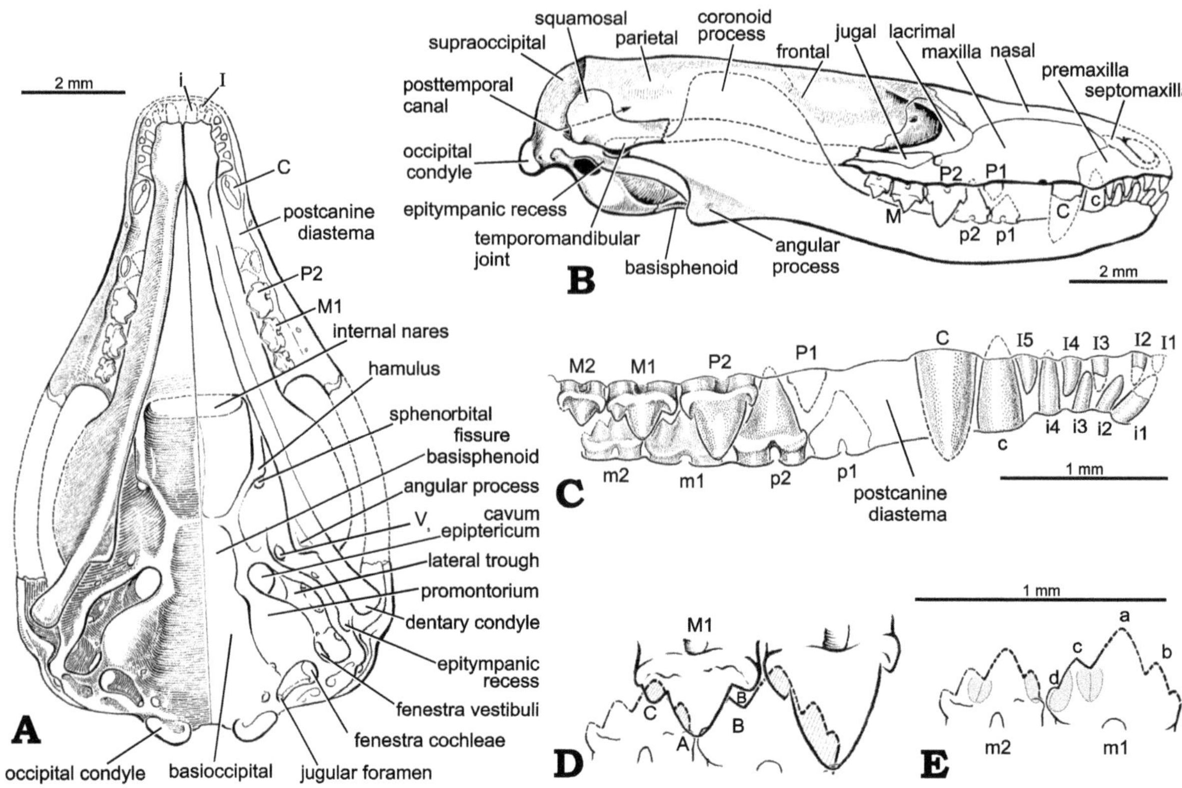

FIGURE 4.10. Skull and dentition of *Hadrocodium*. Ventral (A) and lateral (B) views of reconstructed skull and mandible. C, Full dentition (lateral view). D, Occlusion of molars. E, Development of wear facets on the lower molars. *Source:* modified from Luo, Crompton, and Sun (2001).

inversely correlated to the small size of the skull. The epitympanic recess for attachment of the incus is on the lateral side of the crista parotica, similar to the conditions of the monotreme *Tachyglossus* and multituberculates, but it lacks the distinctive fossa incudis of the monotreme *Ornithorhynchus* and triconodontids (Zeller, 1989b; Luo and Crompton, 1994; Rougier, Wible, and Hopson, 1996). The fossa is more posteriorly positioned relative to the TMJ than in these taxa. The pterygoparoccipital foramen is completely enclosed by the petrosal, different from morganucodontans and *Sinoconodon*, but similar to *Adelobasileus*, monotremes, and multituberculates. The paroccipital process lacks the ventrally projecting posterior paroccipital process seen in morganucodontids, triconodontids, multituberculates, and *Ornithorhynchus*.

The bony roof of the oropharyngeal passage is broad, flat, and almost featureless. There are no pterygopalatine ridges and no median ridge of the basisphenoid (figure 4.10A), all of which are primitive characters of nonmammalian cynodonts. The small hamulus of the pterygoid is similar to the condition in *Haldanodon*, *Ornithorhynchus*, and multituberculates, but more reduced than the transverse flange of most nonmammalian cynodonts, *Sinoconodon*, and *Morganucodon*. It is inferred that the secondary bony palate extends posterior to the tooth row.

Species. *Hadrocodium wui* Luo, Crompton, and Sun, 2001a; type species by monotypy.

Distribution. Early Jurassic (Sinemurian): China, Yunnan (Upper Red Beds within Lower Lufeng Formation).

Class ?Mammalia Linnaeus, 1758
Subclass, Order, and Family *incertae sedis*
Genus *Adelobasileus* Lucas and Hunt, 1990

INTRODUCTION

Adelobasileus Lucas and Hunt, 1990 (figure 4.11) is characterized by numerous mammalian cranial apomorphies. Its early age (Carnian) helps to extend the geological range of some crucial mammalian characteristics. As such the taxon is highly relevant to the understanding of the earliest mammalian cranial evolution (Lucas and Luo, 1993; Luo, 1994). Several recent phylogenetic analyses have

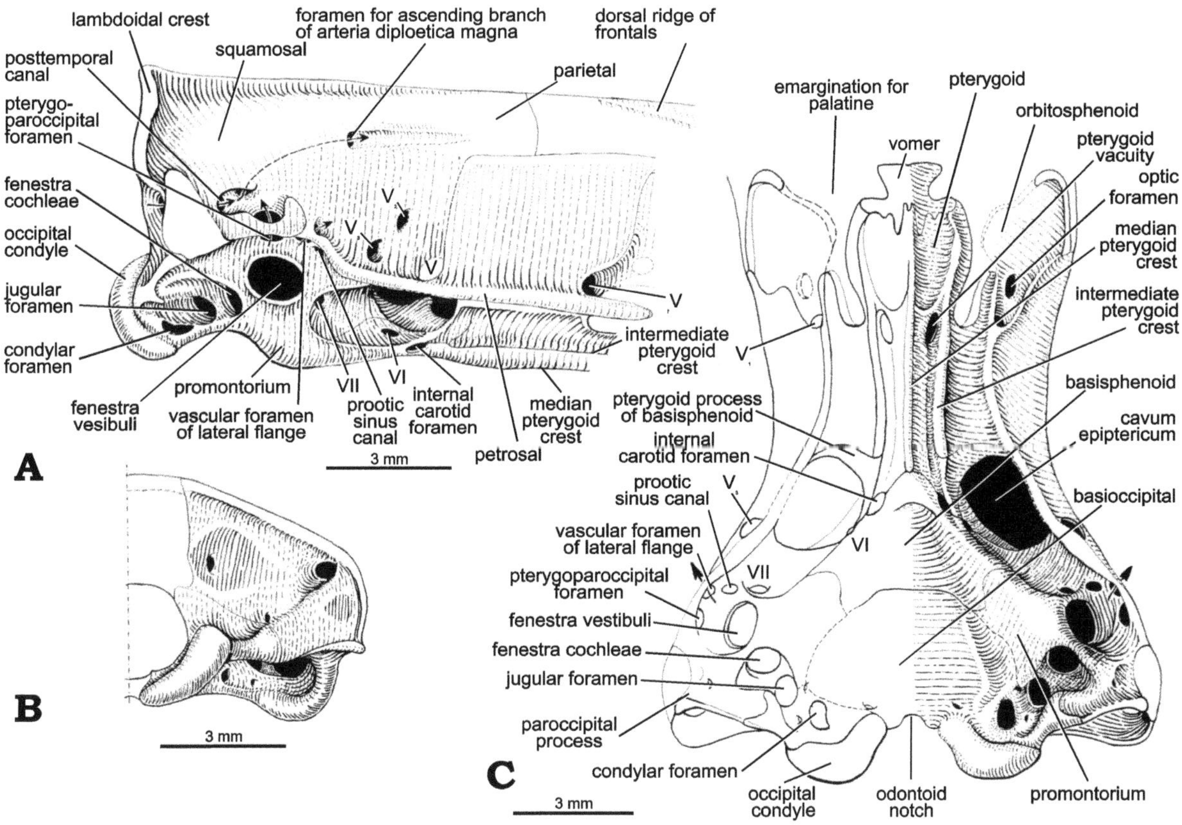

FIGURE 4.11. Reconstruction of the posterior part of the cranium of *Adelobasileus cromptoni,* in lateral (A), occipital (B), and ventral (C) views. *Source:* original based on the holotype (made available by Spencer G. Lucas).

placed *Adelobasileus* in an intermediate position between *Sinoconodon* and other mammaliamorphs (Rougier, Wible, and Hopson, 1996; Luo, Cifelli, and Kielan-Jaworowska, 2001; Luo et al., 2002). In this chapter, we tentatively list *Adelobasileus* within Mammalia. However, this placement is based mostly on the published basicranial characteristics: the holotype includes only the posterior half of the cranium, without dentition and mandible. It is entirely likely that *Adelobasileus* will turn out to be a taxon outside the clade of *Sinoconodon* and extant mammals, although almost certainly closely related to the clade of (*Sinoconodon* + crown mammals).

BRIEF CHARACTERIZATION

Adelobasileus shares with other stem mammals the following characters: an incipient promontorium (bony housing for the inner ear); an enlarged anterior lamina of the petrosal for the lateral wall of braincase encircling the foramina of the maxillary and mandibular branches of trigeminal nerve; an ossified floor to the cavum epiptericum for the trigeminal ganglion; full enclosure of the pterygoparoccipital foramen (for the stapedial artery) by bone and partial enclosure of the ascending channel for orbitotemporal branch of the arteria diploetica magna; separation of the cochlear foramen (fenestra cochleae) from the jugular foramen and separation of the hypoglossal (condylar) foramen from the jugular foramen. These derived features help to distinguish *Adelobasileus* from all known nonmammalian cynodonts and place it closer to mammals than to cynodonts (Lucas and Luo, 1993; Luo, Crompton, and Sun, 2001).

Other derived features of *Adelobasileus* may be equivocal because they are present in some (but certainly not all) nonmammalian cynodonts. These features are: the presence of an ossified orbitosphenoid in the medial wall of the orbit (in tritylodontids); the absence of the tabular bone (which is also absent in tritylodontids); the position of the posttemporal canal near the lateral margin of the occiput (between the squamosal and the mastoid region of the petrosal); a reduced basisphenoid wing; and the absence of the thickened rim of the fenestra vestibuli.

Comment. *Adelobasileus* is easily distinguishable from any nonmammalian cynodont. However, none of its mammalian features is a derived apomorphic character within mammals that can indicate a cladistic relationship between *Adelobasileus* and other stem mammals. For practical purposes of taxonomy, *Adelobasileus* is distinguishable from other stem mammals by a combination of primitive characters: lack of the stapedial muscle fossa, absence of the separation of the anterior and the posterior paroccipital processes, and presence of a vestigial basisphenoid wing. Lucas and Luo (1993) suggested that *Adelobasileus* could be an isolated braincase fossil of a "dromatheriid," such as *Microconodon*, a taxon known only by teeth and disarticulated jaws of similar geological age. *Adelobasileus* overlaps the lower end of the size range of the currently known mandibular fossils of *Microconodon*, as documented by Sues (2001). However, Sues (2001) showed that *Microconodon* is clearly a nonmammalian cynodont by dental and mandibular features. If an unambiguous association can be established between the derived braincase fossil of *Adelobasileus* and the primitive mandibular and dental remains of "dromatheriids" from the same fauna (Lucas and Oakes, 1988), then it is almost certain that *Adelobasileus* and *Microconodon* will be excluded from the clade of *Sinoconodon* and extant mammals.

Species. *Adelobasileus cromptoni* Lucas and Hunt, 1990; type species by monotypy.

Distribution. Late Triassic (late Carnian): United States, Texas, Home Creek (Tecovas Formation, Dockum Group).

Class ?Mammalia Linnaeus, 1758
Order and Family *incertae sedis*
Genus *Tricuspes* Huene, 1933

Comment. *Tricuspes* is represented by isolated teeth assigned to three species: *Tricuspes tubingensis* Huene, 1933, type species, *T. sigogneauae*, Hahn et al., 1994, and *T. tapeinodon* Godefroit and Battail, 1997, all from the Late Triassic (Rhaetian) of continental Europe. The molariform postcanines are tricuspate or tetracuspate with incipiently divided roots (Clemens, 1980a; Hahn et al., 1994) and may have a large range of size variation (Godefroit and Battail, 1997). Most dental characters of these isolated teeth are indistinguishable from the features of *Microconodon* (Sues, 2001) and other dromatheriids (Hahn et al., 1994). Clemens (1980a: 71) stated that the "Tentative allocation of *Tricuspes* to Mammalia is . . . based on inconclusive evidence." Hahn et al. (1994) suggested that dromatheriids (including *Tricuspes*) could be close to the ancestry of mammals but are not mammals themselves. Sues (2001) pointed out that all features of the isolated teeth of *Tricuspes* are primitive. He suggested that the taxa formerly assigned to dromatheriids may represent a heterogeneous group that lacked proper diagnostic characters. Given these assessments of taxa that may be related to *Tricuspes*, we considered the phylogenetic status of *Tricuspes* as indeterminate. We listed it here merely for the sake of complete documentation.

CHAPTER 5

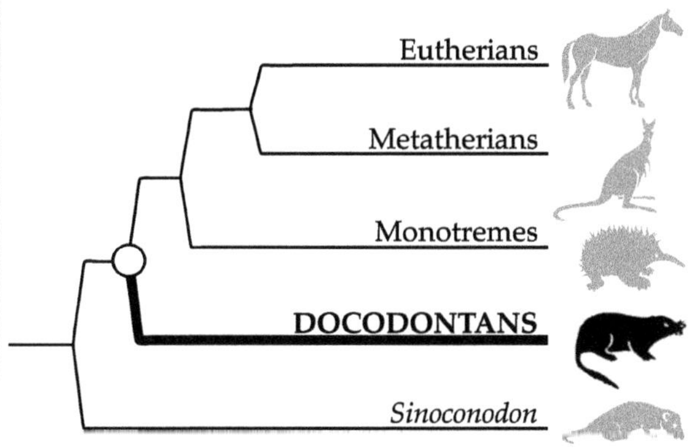

Docodontans

INTRODUCTION

Docodontans are a monophyletic group defined by the common ancestor of *Docodon, Simpsonodon,* and all other Mesozoic mammal taxa more closely related to *Docodon* and *Simpsonodon* than to morganucodontans, *Shuotherium,* and cladotherians (stem "eupantotherians"). This group consists of mole-sized or smaller mammals. Uncontested docodontans are known from the Middle Jurassic to Early Cretaceous of the Laurasian continents. One putative taxon may extend their range to the Late Triassic (Sigogneau-Russell and Godefroit, 1997, an assignment challenged by Butler, 1997). On the other hand, Pascual et al. (2000) reported a docodontan dentary with three molars in the Late Cretaceous of Argentina. We assign this specimen to Mammalia, subclass and order *incertae sedis* and describe it in this chapter at the end of the section on "Systematics."

The first docodont was discovered from the latest Jurassic Morrison Formation (see chapter 2) of North America and described by Marsh, who erected the genus *Docodon* for it in 1881, based on a fairly complete dentary with teeth. For nearly five decades *Docodon* was the only docodontan genus known, until Simpson (1928a) established *Peraiocynodon* from the Purbeck of England, regarded subsequently by Butler (1939, 1956), Simpson (1961), Kermack et al. (1987), and others as a juvenile *Docodon.* We follow these opinions and regard *Peraiocynodon* as a junior synonym of *Docodon* (but see Krusat, 1980).

Marsh assigned *Docodon* to his family Diplocynodontidae Marsh, 1887, and Osborn (1888b) placed it in the Dicrocynodontidae; but Simpson (1929a) regarded both family names as invalid and replaced them with the Docodontidae (see Kron, 1979, for review). However, A. O. Averianov (pers. comm.) argued that the replacement done by Simpson (1929a) was not valid according to rules of the International Commission on Zoological Nomenclature (ICZN). The oldest family name for this group is Diplocynodontidae Marsh, 1887, which is not valid according to ICZN article 39 because it was based on the generic name, which is a junior homonym. According to the same article, this family name should be replaced by the oldest available family group name, in this case Dicrocynodontidae Osborn, 1888, which should be considered the valid name for the family Docodontidae Simpson, 1929a. As the family name Docodontidae has a long tradition, Averianov (pers. comm.) has applied to the ICZN to retain the name Docodontidae Simpson, 1929a. Pending a decision by the Commission, we tentatively use Docodontidae Simpson, 1929a as a valid name.

For many years the Docodontidae were assigned to Pantotheria Marsh, 1880 (e.g., Marsh, 1887, and references therein; and Simpson, 1928a, 1929a, 1945). Butler (1939) followed Simpson in assigning docodontans to Pantotheria, but he pointed out important differences between the molars of the Docodontidae and Dryolestidae. Kretzoi (1946) elevated the family Docodontidae to the order Docodonta. The concept of Kretzoi's Docodonta was different from that in current use, as he included three genera: *Peramus* (now assigned to the monophyletic Zatheria, or "eupantotherians," see chapter 10) and two docodontan genera, *Peraiocynodon* and *Docodon.* Patterson (1956) accepted the order Docodonta, but, in addition to uncontested Docodontidae, assigned to it *Morganucodon,* which was subsequently removed from Docodonta

(Kermack and Mussett, 1958a; Simpson, 1961; Hopson, 1970; and others). Finally, McKenna and Bell (1997) assigned the separate orders Morganucodonta and Docodonta (limited to Docodontidae) to Mammaliaformes (see discussion in chapters 4 and 15).

Several docodontid genera were established subsequently, including *Borealestes, Haldanodon, Cyrtlatherium* (originally assigned to Kuehneotheriidae, see Sigogneau-Russell, 2001), *Simpsonodon,* and *Tegotherium* (originally assigned to Symmetrodonta by Tatarinov, 1994; but see Hopson, 1995, and Kielan-Jaworowska et al., 2000). In addition, we tentatively assign to Docodonta the genus *Delsatia* Sigogneau-Russell and Godefroit, 1997. *Reigitherium* Bonaparte, 1990, was assigned to Dryolestoidea by its author, but to Docodonta by Pascual et al. (2000). We do not regard *Reigitherium* as a member of the Docodonta, but we describe it in this chapter for lack of a better place. Most of the docodontans are represented only by isolated teeth and/or dentaries. Of all the docodontans currently known, *Haldanodon* Kühne and Krusat, 1972, is best preserved, represented by skulls, large samples of dental and jaw specimens, and a partial postcranial skeleton, which has received only preliminary description (Krusat, 1980, 1991; Lillegraven and Krusat, 1991; Martin and Nowotny, 2000).

Haldanodon caused controversy in the interpretation of the position of docodontans among mammals. The docodontan dentition is specialized among Jurassic and Early Cretaceous mammals. On the basis of the dentition, Crompton and Jenkins (1968) argued that docodonts originated from morganucodonts, while Butler (1997) suggested that docodonts might be a sister group of the Woutersiidae (see chapter 9). On the other hand, Lillegraven and Krusat (1991) demonstrated that *Haldanodon*'s skull possesses many plesiomorphic characters; on this basis they suggested that the skull structure of docodontans is more primitive than that of morganucodontans and regarded them as the plesiomorphic sister group of all other mammals. Wible and Hopson (1993) and Luo (1994) found basicranial evidence to be inconsistent with that conclusion. Instead they argued that the docodontid *Haldanodon* is more closely related to triconodontids and other mammals than are morganucodontids and *Sinoconodon.* In subsequent parsimony analyses with more extensive data (Wible et al., 1995; Rougier, Wible, and Hopson, 1996), *Haldanodon* was placed at a slightly higher position on the mammalian tree, closer to the mammalian crown group than morganucodontids and *Sinoconodon.* This position has been confirmed by most recent analyses (Luo, Cifelli, and Kielan-Jaworowska, 2001; Luo et al., 2002), which place Docodonta (see figures 15.1, 15.2) immediately above the Morganucodonta.

BRIEF CHARACTERIZATION

Docodontans are very small mammals with a modest generic diversity, whose undisputed taxa are known only from the Jurassic and earliest Cretaceous. They have a mixture of advanced characters in the structure of the molars and fairly primitive postcranial and cranial features. Docodontan molars were capable of shearing and possibly also of grinding. The lower molars are elongated longitudinally, with two rows of cusps, whereas the uppers superficially resemble those of advanced therians. Another characteristic feature of docodontan molars is the development of strong, transverse crests connecting the cusps. They are convergent to Shuotheriidae in some molar features and stem therians in others. Upper molars in occlusal view are transversely developed, roughly rectangular or figure-eight shaped, with four main cusps; the lower molars are longitudinally elongated, with two rows of two cusps each, of which the posterolabial cusp is the highest. The lingual part of the upper molar occluded between two adjacent lower molars. The mesial part of the lower molar has a crushing or grinding function that is otherwise known among Jurassic mammals only in *Shuotherium* Chow and Rich, 1982, *Ambondro* Flynn et al., 1999, and *Asfaltomylos* Rauhut et al., 2002 (see chapter 6).

The skull, as represented by *Haldanodon,* has expanded zygomatic arches and a roughly triangular outline in dorsal and ventral views; the hard palate extends far posteriorly and the glenoid fossa is extensive. The mandible has a large coronoid process, a fossa for the coronoid bone, and a well-developed dentary condyle, placed on a peduncle and situated far behind and above the angular process. The postdentary trough in the mandible is well developed, accommodating the articular and angular; there is also a fossa for the splenial. These bones thus appear to have remained associated with the mandible. Meckel's groove is present, extending far anteriorly. The articular and quadrate were presumably parts of the primary (reptilian) jaw joint, although the quadrate is not preserved. Only the stapes (incompletely preserved in *Haldanodon*) is present in the middle ear.

The incomplete postcranial skeleton (of *Haldanodon*) shows that the limb bones (especially humerus and femur) are very stout; the expanded proximal and distal ends of the humerus indicate either a fossorial or, more probably, a semiaquatic mode of life.

DISTRIBUTION

Middle Jurassic (middle Bathonian): Scotland, Loch Scavaig; Middle Jurassic (latest Bathonian): England, Oxford shire, Kirtlington, Watton Cliff; Late Jurassic (Kimmeridgian): Portugal, Leiria, Guimarota Coal Mine (Guimarota Beds); Late Jurassic (late Kimmeridgian–early Tithonian): United States, Wyoming and Colorado (Morrison Formation); Late Jurassic: Mongolia, Trans-Altaian Gobi, Shar Teeg; Early Cretaceous (Berriasian): England, Durlston Bay (Purbeck Limestone Group); Early Cretaceous (Aptian or Albian): Mongolia, Gobi Desert ("Hövöör Beds"); Early Cretaceous (Aptian or Albian): Russia, western Siberia (Ilek Formation). The two last cited Early Cretaceous occurrences concern undetermined docodontan teeth and are not cited under "Systematics"; they are based on Agadjanian (1999) and Leshchinskiy et al. (2001).

ANATOMY

SKULL AND MANDIBLE
(figures 5.1–5.3)

The docodont skull, which has been preserved only in *Haldanodon*, was first described by Krusat (1980) and subsequently in more detail by Lillegraven and Krusat (1991). The description that follows is based on these two papers.

Skull as a Whole (figures 5.1, 5.2). The skull of *Haldanodon exspectatus* as reconstructed by Lillegraven and Krusat (1991) is about 34 mm long. In dorsal view it has a triangular snout that gradually widens posteriorly; the skull reaches the greatest width across the posterior part of the strongly expanded zygomatic arches, in front of the glenoid fossa. The zygomatic arches are deep (in lateral view) and the occiput is oriented obliquely, sloping forward and up from the prominent occipital condyles. At the boundary between the cranial roof and the occipital plate are prominent lambdoidal crests. In lateral view the snout is relatively low and the skull strongly increases in height posteriorly, reaching the highest level at the lambdoidal crests. As the skull increases in depth posteriorly, an extensive part of the cranial roof is also exposed in an anterior view (figure 5.2A). The bones of the cranial roof are very thick and compact.

Rostrum and Zygoma. The snout is long and (measured to the anterior margin of the zygomatic arches) extends for half the skull length. In dorsal view (figure 5.1A) it forms almost an isosceles triangle, slightly bulging by the swollen canine eminences of the maxilla. The nasals are narrow anteriorly and very broad posteriorly, so they are roughly T-shaped in anterior view (figure 5.2A); they contact the frontal by a roughly transversal suture. Each nasal bears two nasal foramina, the anterior one small and rounded and the posterior one larger and elongated.

The premaxilla is small, barely exposed in dorsal and lateral views, and obscured by the anterolateral process of the maxilla. Between the premaxilla and the nasal, as exposed in lateral view, there is a space for a relatively large septomaxilla as reconstructed by Lillegraven and Krusat (1991), but only a small part of this bone is preserved in *Haldanodon* (figure 5.1C,D). The septomaxilla is also visible in anterior view (figure 5.2A). The maxilla is very extensive in comparison with the small premaxilla. There are three infraorbital foramina. Two anterior foramina are situated within the maxilla, the posterior one is located at the boundary between the maxilla and lacrimal. The anteriormost foramen is the largest. In addition, in front of the anterior infraorbital foramen, there are two small foramina near the canine eminence of the maxilla. The lacrimal is extensive in lateral and dorsal views and has two openings for the lacrimal canal.

The zygomatic arch is slender and laterally compressed in dorsal and ventral views, but deeper in lateral view. The anterior end of the zygomatic arch is formed by the lacrimal in dorsal view and by the maxilla in ventral view. The anterior one-third of the zygomatic arch is formed mostly by the jugal. The posterior two-thirds is formed by the anterior process of the squamosal along the dorsal margin of the zygoma and by the jugal along its ventral margin. The squamosal has a large, obliquely oriented glenoid fossa that, in ventral view, is roughly oval in outline. The glenoid has a convex anterior margin and a short postglenoid crest posteromedially, which borders on the bony external auditory meatus (the squamosal sulcus of Allin, 1986). The glenoid fossa in *Haldanodon* is more extensive and more derived in many features than in other early mammals with a double jaw joint, such as *Morganucodon* (Kermack et al., 1981) and *Sinoconodon* (Crompton and Luo, 1993, see also chapter 4 and figures 4.2, 4.4).

Palate. The palatal part of the premaxilla in *Haldanodon* is more extensive than the facial part. Each premaxilla sends posteriorly two narrow, pointed processes: the shorter ventrolateral process forms the lower margin of the upper jaw and the longer palatal process is exposed on the palate. There are alveoli for four incisors in the premaxilla, and the two distal incisors are situated within the maxilla, surrounded on both sides by the two premaxillary processes. There is a large incisive foramen which contacts the suture with the maxilla; a small part of the vomer is exposed between two of its counterparts. The palatal processes of the maxilla and palatine are very extensive and the hard palate extends far posteriorly behind the last molar. The transverse palatine suture is situated opposite the M1-M2 embrasure and is pierced by a large major palatine foramen. The palatine-maxillary suture turns laterally in its posterior course, reaching the alveolus of the last molar; the minor palatine foramen is situated at the posterolateral corner of the transverse part of the palatine

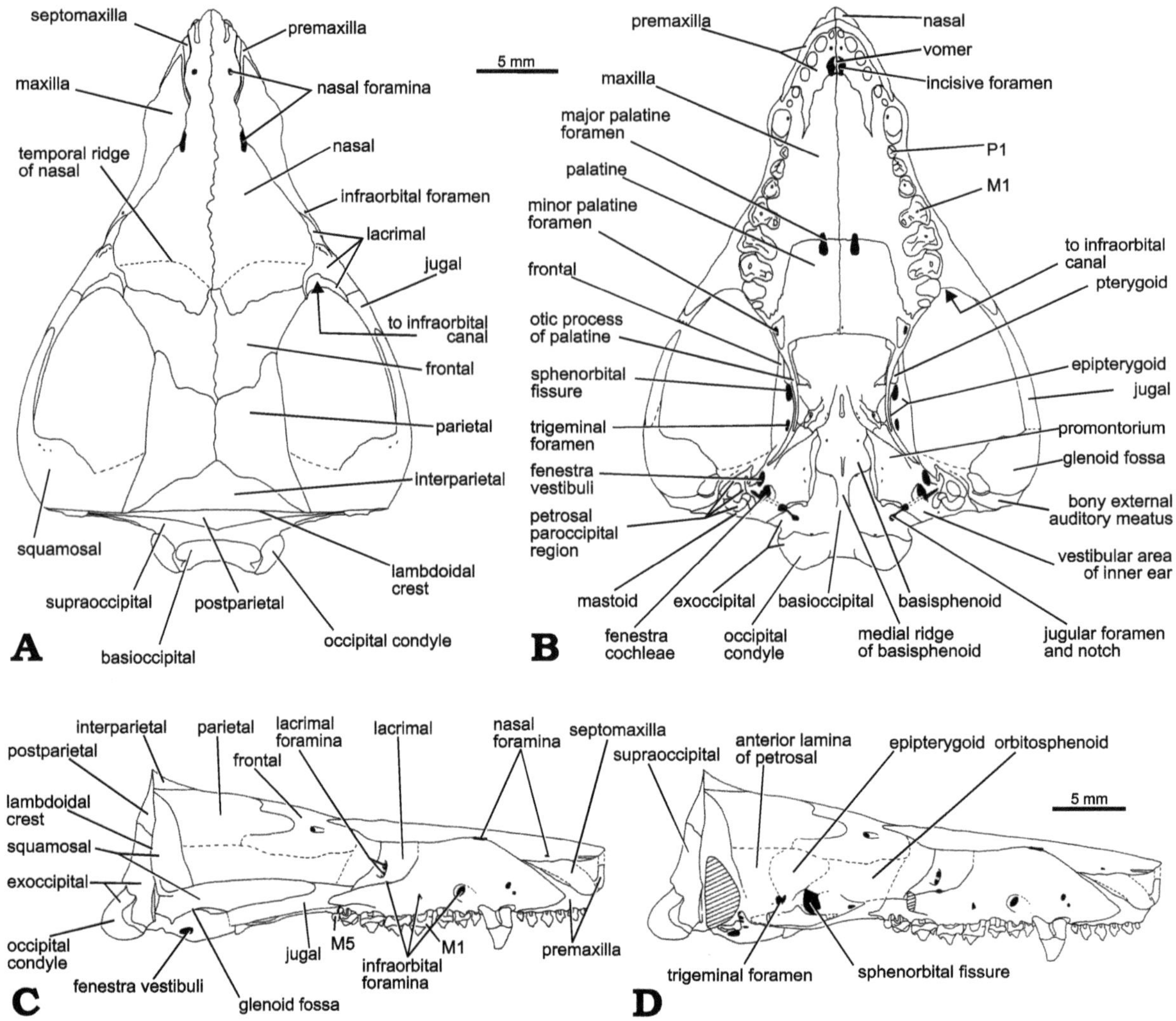

FIGURE 5.1. Reconstruction of a docodontan skull, exemplified by *Haldanodon exspectatus*, in dorsal (A), ventral (B), and lateral views (C, D). In C the zygomatic arch is retained; in D the zygomatic arch is removed to show the orbitotemporal region. *Source:* modified and simplified from Lillegraven and Krusat (1991).

bone. The choanal region has not been preserved and could not be reconstructed.

Cranial Roof. The nasal overhangs the premaxilla so that a small part of the nasal is visible in ventral view in front of the premaxilla (figure 5.1B). The nasofrontal suture is roughly transverse. The frontals insert between the nasals along the midline of the skull roof. The frontoparietal suture on either side of the skull roof is convex anteriorly, which is a primitive condition seen in *Sinoconodon* (figure 4.2A), *Morganucodon* (figure 4.5A), and some djadochtatherioid multituberculates (see figure 8.4B). The posterior part of the cranial roof has an unpaired interparietal with a triangular outline wedged between the two parietals in the dorsal view.

Occiput. The occiput forms a broad, roughly half-circular plate, sloping posteroventrally, with the postparietal constituting its upper part, separated from the extensive supraoccipital region by a transverse suture (figure 5.2B). The suture between the supraoccipital and exoccipital has not been recognized. The foramen magnum is surrounded laterally and ventrally by very prominent occipital condyles. Lateral to the condyle and separated from it by part of exoccipital, the mastoid portion of the petrosal is exposed in posterior view. The lambdoid crest is formed by the squamosal (its suture with the supraoccipital is not clear), which continues laterally as the squamosal zygoma. The posterior side of the squamosal zygoma has a faint groove for the bony external auditory meatus.

Orbit and Temporal Region. The orbit is widely open posteriorly, confluent with the temporal fossa. In the temporal region, a small part of the lateral wall of the braincase is exposed. In the anterior orbitotemporal region, the

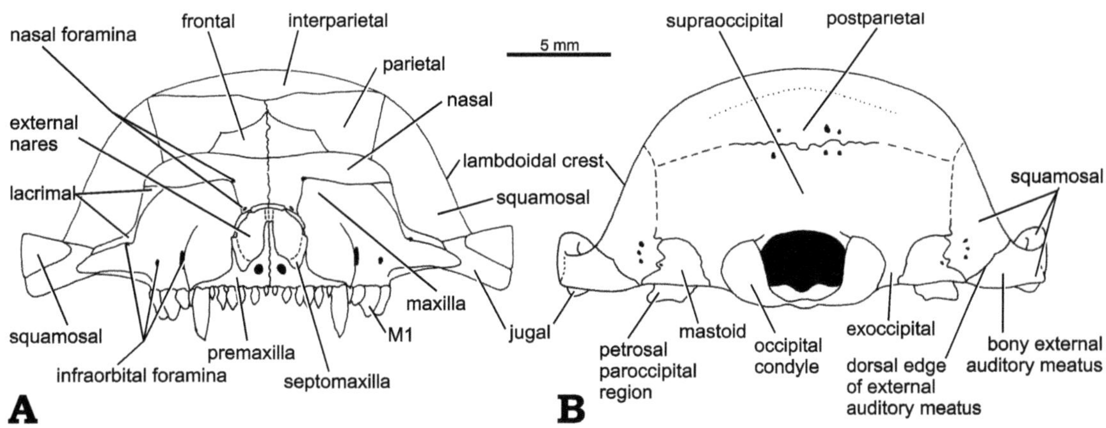

FIGURE 5.2. Reconstructions of a docodontan skull, exemplified by *Haldanodon exspectatus*, in anterior (A) and occipital (B) views. *Source:* modified and simplified from Lillegraven and Krusat (1991).

sutures of bones have not been identified. There is a relatively large sphenorbital fissure (figure 5.1D), which lies at the boundary of epipterygoid and orbitosphenoid; ventrally it contacts the pterygoid bone. The posterior part of the orbitotemporal region has a prominent anterior lamina of the petrosal. The undivided trigeminal foramen is placed at the boundary between the anterior lamina of the petrosal and the epipterygoid (alisphenoid).

Basicranium and Ear. The basicranium (figure 5.1B) is wide, with its central part occupied by the swollen basisphenoid; the choanal area in front of it has been poorly preserved and could not be interpreted. A fragment of the pterygoid bone and the posterior process of the palatine, covered dorsally by the restored quadrate ramus of the pterygoid bone, delimit the choanal region laterally. Posterior to the basispheniod is a large basioccipital with two prominent occipital condyles that protrude posteriorly. Lateral to the basisphenoid is a triangular promontorium of the petrosal, pointed anteriorly, with the fenestra vestibuli at its posterolateral part; the fenestra cochleae is close and posteromedial to the fenestra vestibuli. At the boundary of the promontorium and basioccipital there is a relatively small jugular notch with the jugular foramen. On the ventral surface of the paroccipital process is the crista parotica, which would presumably have supported the quadrate (not preserved, see Lillegraven and Krusat, 1991). Between the paroccipital region of the petrosal and the large glenoid fossa of the squamosal there is a slight constriction of the squamosal, a derived feature by comparison with *Sinoconodon* and *Morganucodon* (Luo, 1994). Only one ear ossicle, the stapes, has been found in *Haldanodon.* It has a large, oval footplate, which fits the fenestra vestibuli; laterally the bases of the anterior and posterior crura have been preserved.

Mandible and Articular Complex (figure 5.3). The structure of the docodont mandible (see later), as preserved in *Docodon* and *Haldanodon,* shows clearly that docodontans retained a double jaw joint, despite the fact that the quadrate is not preserved. Lillegraven and Krusat (1991) argued on the basis of a detailed anatomical reconstruction that a large quadrate was present in *Haldanodon,* whereas the presence of the quadratojugal could not be demonstrated with any certainty.

Simpson (1928a) described the fragmentary mandible in "*Peraiocynodon*" (now considered to be a juvenile of *Docodon*) and reconstructed (Simpson,1929a) the complete mandible of *Docodon* from the inner side (figure 5.3A). According to this reconstruction the dentary is long and slender, with a long symphysis. The condyle is situated at the level of the occlusal surface of the molars and faces posteriorly. The coronoid process is large, with a large scar for the coronoid bone. The angular process is situated anteriorly, and there is an extensive postdentary trough, which contains the posterior opening of the mandibular canal. Anteriorly, Meckel's groove extends up to the symphysis. Simpson (1928b) did not recognize the presence of postdentary bones in docodontans, but incorrectly regarded the trough as housing nerves and blood vessels. The presence of attached postdentary elements in docodontans was first recognized by Kermack and Mussett (1958). Waldman and Savage (1972) described the partial mandible of *Borealestes,* of which the posterior part is missing.

Krusat (1973) provided the first detailed description of *Haldanodon* from the Kimmeridgian Guimarota Coal Mine of Portugal. In a subsequent monograph Krusat (1980) reconstructed the mandible of *Haldanodon* (figure $5.3B_{1-3}$), comparing it with that of *Morganucodon,* de-

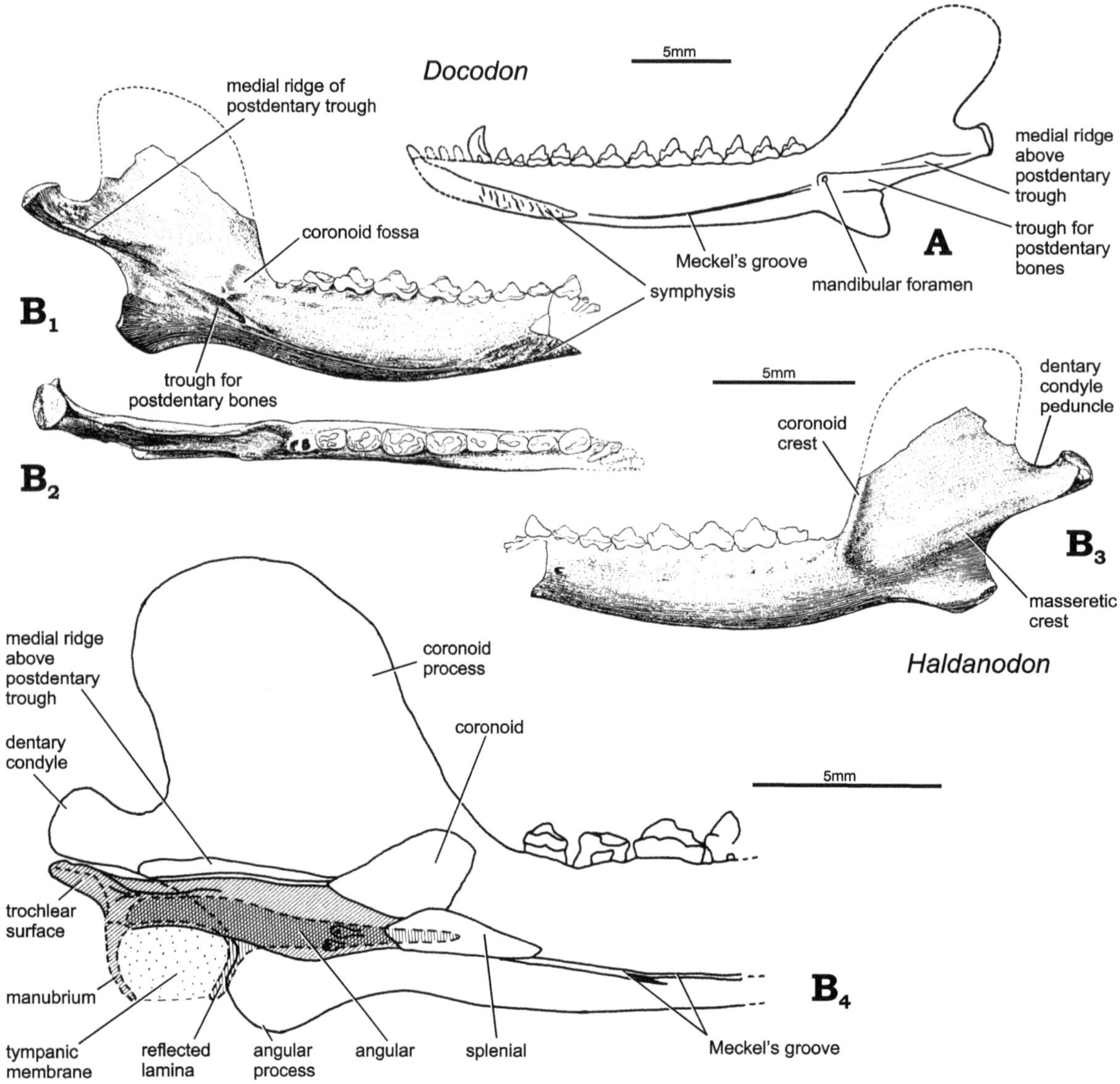

FIGURE 5.3. Structure of the mandible in Docodonta. A, Reconstruction of the right mandible of *Docodon* sp. in lingual view. B, Reconstruction of the left mandible of *Haldanodon exspectatus* in lingual (B_1), occlusal (B_2), and labial (B_3) views; the posterior part of the same in lingual view with reconstruction of the postdentary bones (B_4). Vertical hatching pattern in B_4 represents the approximate extent of the angular, and diagonal hatching pattern represents the articular complex. *Source:* modified from: A, Simpson (1929a); B_{1-3}, Krusat (1980); B_4, Lillegraven and Krusat (1991).

scribed in great detail by Kermack et al. (1973, see also chapter 4). Finally, Lillegraven and Krusat (1991) offered a more detailed, albeit putative, reconstruction of the articular complex in *Haldanodon,* based on facets on the dentary and isolated fragments of the postdentary bones (figure 5.3B_4).

The mandible of *Haldanodon* is missing the anterior part of the dentary and is less complete than that of *Docodon* (figure 5.3A). As reconstructed by Krusat (figure 5.3B_{1-3}), it differs from that of *Docodon* in having the condyle placed higher than the occlusal level of the molars, on a longer peduncle of the articular process and facing upward. The symphysis is very long (as in *Docodon*), extending posteriorly below the penultimate premolar; Meckel's groove is present; the coronoid process larger and arises more steeply than in *Docodon.* The labial side shows a large masseteric fossa, situated posterior to the last molar and delimited anterodorsally by a sharp coronoid crest; a weak lateral ridge extends from the dentary peduncle into the masseteric fossa (figure 5.3B_3).

In the paradentary complex Lillegraven and Krusat (1991) reconstructed (figure 5.3B_4) a large coronoid and a splenial by inference from the facets on the dentary. In the relatively large postdentary trough with a prominent overhanging medial edge, they reconstructed the approximate extension of the angular and its reflected lamina. The angular occupied the ventral part of the trough and the medial aspect of the mandibular angle. The articular complex has the manubrium, which together with the reflected lamina of the angular embraced the tympanic membrane. The articular complex extends posteriorly below the dentary condyle. The articular complex differs from that of *Morganucodon* (Kermack et al., 1973) in having a tripartite structure, with three facets (for contact with the quadrate) facing posteriorly and dorsomedially, instead of being bipartite, with two facets facing directly posteriorly as in *Morganucodon*. Another difference from other mammals with a double jaw joint is that the articular complex in *Haldanodon* is situated more ventrally relative to the dentary peduncle and condyle, whereas in *Morganucodon* and *Sinoconodon* it is medial to the dentary condyle (Kermack et al., 1973; Crompton and Sun, 1985; Crompton and Luo, 1993).

DENTITION

The upper and lower docodont dental formulae have been reconstructed for two genera, *Docodon* and *Haldanodon*, but associated upper and lower dentitions of *Docodon* have yet to be found. Simpson (1929a) estimated the dental formula of *Docodon* as ?.1.3.5+?/3.1.3–4.7–8 (figure 5.3A). Jenkins (1969b) reconstructed the upper postcanine dentition in *Docodon superus* as including a canine, three premolars and five molars; and the lower, on the basis of *Docodon victor*, as a canine, four premolars, and six molars. Jenkins (1969b) also stated that it is possible that *Docodon* molars varied in number within a species (see later section "Systematics" for comments on *Docodon* species). The estimate of the number of particular teeth in *Haldanodon* also changed with the progress of investigations. Recently, on the basis of studies of large collections of *Haldanodon* teeth, Martin and Nowotny (2000) concluded that the dental formula in *Haldanodon* was 6.1.3.5/4.1.3.5–6.

Authors have different opinions about the cusp homology of molars in docodontans, and this has resulted in different notations for the same cusps (e.g., Crompton and Jenkins, 1968; Jenkins, 1969b; Gingerich, 1973; Krusat, 1980; Kermack et al., 1987; Butler, 1988a, 1997; Sigogneau-Russell and Godefroit, 1997; Pascual et al., 2000; Sigogneau-Russell, 2001). In this situation when referring to particular molar cusps, one should always state, for example, "cusp g (*sensu* Krusat, 1980)," or "cusp g (*sensu* Butler, 1997)." This is cumbersome, but necessary, as these authors gave the notation "cusp g" to different cusps. As docodont relationships and cusp homology with other mammals cannot be established with any certainty, we follow the purely descriptive terminology of molar cusps introduced by Kermack et al. (1987) and completed by Sigogneau-Russell (2001). The method of cusp numbering is shown in figure 5.4. The drawback of this method is that the same terms are used for cusps in the upper and the lower molars. To avoid misunderstanding, Sigogneau-Russell used upper-case letters for upper molar cusps names and lower-case letters for cusps of lower molars. We follow this method.

The docodont upper molars (figure 5.4) are roughly rectangular (or figure-eight shaped), widened transversely, with two large cusps in the labial row and two in the lingual row. The mesiolabial cusp, situated near the middle of the labial row, is the highest cusp on the tooth; the distolabial cusp is posterior to it, and there is a small stylar cusp at the mesiolabial corner of the crown. The mesiolingual cusp is directly lingual to the mesiolabial cusp and connected to it by a strong transverse ridge; the distolingual cusp is smaller. The upper molars are three-rooted and the lowers two-rooted.

The lower molar is elongated longitudinally with two rows of cusps; the highest is the main (central) cusp in the labial row. In front of the main cusp is a lower mesiolabial cusp; there is also a small distolabial cusp. Of the two cusps in the lingual row, the posterior one (distolingual cusp) is usually placed more posteriorly than the main cusp, to which it is connected by an obliquely transverse ridge. The mesiolingual cusp is smaller, except in *Simpsonodon* and *Tegotherium*, where it is equal to the distolingual cusp and also connected by a ridge to the main cusp; this ridge is weak or absent in other genera. In all docodontans the mesiolingual and mesiolabial cusps are connected by a ridge; a second ridge runs from the distolingual cusp along the posterior edge of the crown to the small distolabial cusp. In some genera (especially in *Simpsonodon*) there is a small basin between the ridge that connects the mesiolingual and mesiolabial cusps and the anterior margin of the tooth, referred to by Kermack et al. (1987) as the pseudotalonid.

The enamel is ornamented by minor ridges or flutings, which differ in character in the different genera. In *Docodon* the ornamentation takes the form of sharp ridges running down the slopes of the cusps; in *Simpsonodon* the ridges are more irregular and, especially on the pseudotalonid, they anastomose to provide a roughened, pitted surface; in *Haldanodon* the ornamentation is weaker, though present.

The occlusion and wear facets of docodont molars have been investigated by Jenkins (1969b), Gingerich (1973), Kermack et al. (1987), and Butler (1988a). Jenkins (1969b)

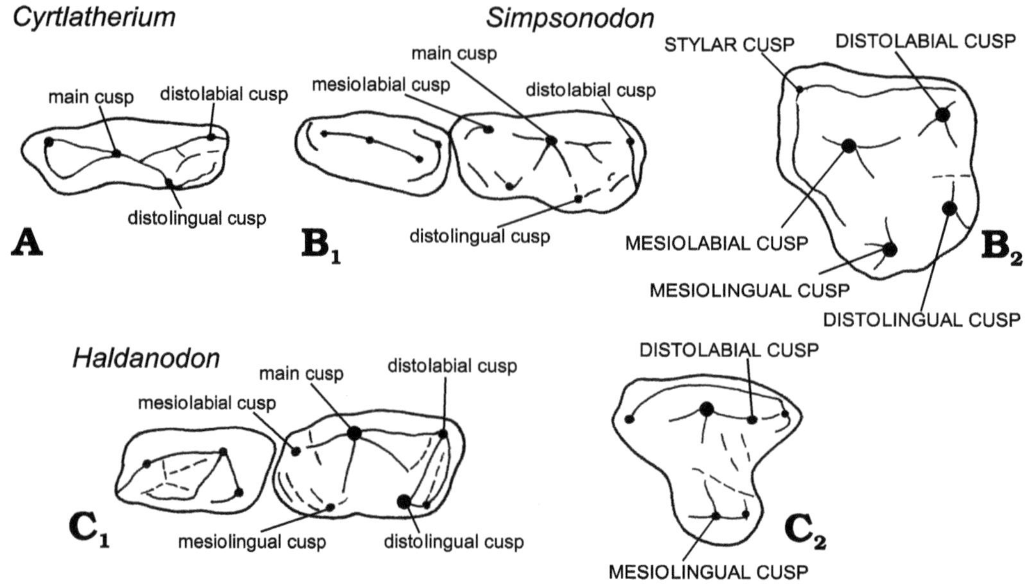

FIGURE 5.4. Generalized molars of docodontans in occlusal view to illustrate their molar cusp nomenclature. Anterior is to the left. A, *Cyrtlatherium canei*, right lower molar (holotype). B, *Simpsonodon oxfordensis*, right last lower premolar and first molar (B_1), right upper molar, reversed (B_2). C, *Haldanodon exspectatus*, last right lower premolar ?d3 and first lower molar m1 (C_1); left upper molar (C_2). Upper-case letters are used to denote cusps on the upper molars and lower-case letters on the lower molars. *Source:* modified from Sigogneau-Russell (2001).

demonstrated that the labial halves of the upper molars in *Docodon* passed labial to the lower molar row, whereas the lingual halves occluded in the lower intermolar basins formed by the adjacent halves of the lower molars. He concluded that molar function was primarily shearing, rather than crushing.

Gingerich (1973) repeated investigation on *Docodon* wear facets and concluded that there were two stages in mandibular movement. In the first stage the lower jaw moved upward and backward in a puncturing action, followed by an upward and forward shearing stroke. He did not find evidence of a grinding movement, but did not preclude a grinding function.

Kermack et al. (1987) studied the morphology and occlusion in *Simpsonodon.* They compared the pseudotalonid in *Simpsonodon* with that of *Shuotherium* Chow and Rich, 1982, and regarded the latter as being related to docodontans, an idea that has been rejected by subsequent authors (see chapter 6). They also compared the occlusal pattern of *Simpsonodon* with that of a quadritubercular extant insectivore (*Erinaceus*) and found striking similarities in their mode of occlusion, the only difference being the presence of an anterior pseudotalonid in *Simpsonodon,* in contrast to a posterior talonid in *Erinaceus.* As noted by Kermack et al. (1987), there is no doubt that a similar type of occlusion developed independently in each of the two groups.

Finally, Butler (1988a) compared the occlusion in *Docodon* with that of the extinct eutherian *Leptictis,* which possesses tribosphenic molars with an incipient hypocone. He also found a great similarity in the mode of occlusion of docodontans and primitive eutherians, which he again regarded as a striking example of convergence. He suggested that the lingual part of the upper molar could occlude with an expanded lingual cingulid of the lower molar in a crushing action. He considered that *Woutersia* (which we regard as a symmetrodont, see chapter 9) has a cusp on the lingual cingulum that could be ancestral to the hypertrophied lingual structure in the upper molar of docodontans (Butler, 1997).

POSTCRANIAL SKELETON

The docodont postcranial skeleton is known only from *Haldanodon* (Krusat, 1991) and still awaits detailed description. An incomplete, disarticulated skeleton of *Haldanodon* is associated with an incomplete skull and dentaries (figure 5.5). The axial skeleton has not been preserved, except for the three double-headed ribs, but the limb bones are more complete. Disarticulated bones of other individuals have been also found.

Krusat (1991: 37) stated: "The limb bones are unusually stout and strong for such a small animal. The longest femur available has an overall length of 20.0 mm and a

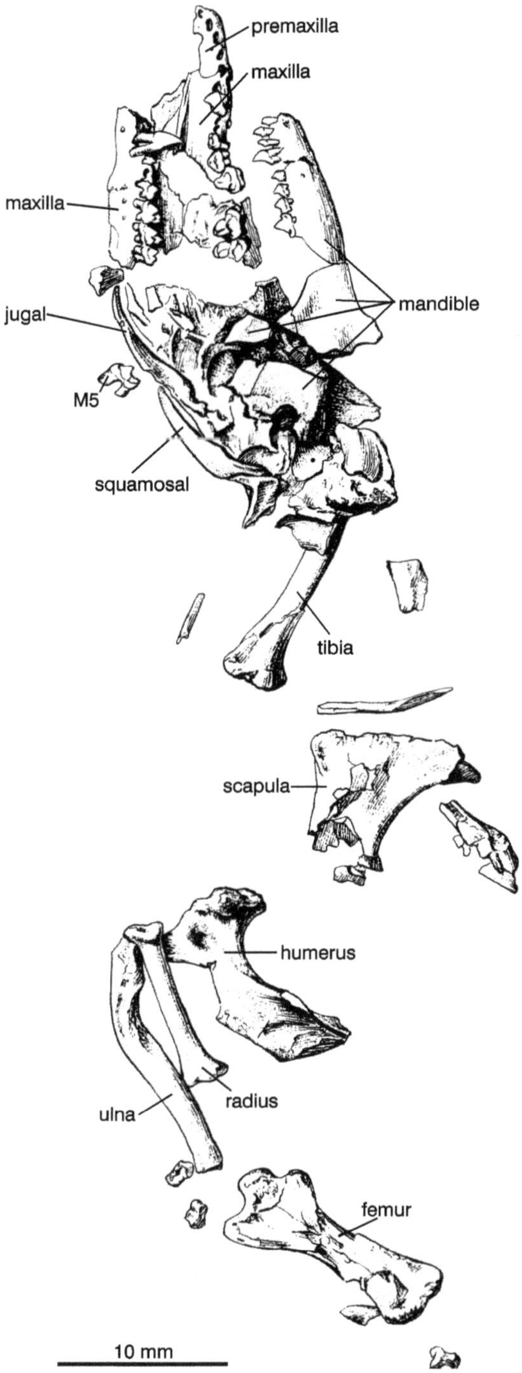

FIGURE 5.5. Skeleton of *Haldanodon exspectatus* as preserved. *Source:* modified from Krusat (1991).

width between trochanters of 9.7 mm. Remarkable are two large triangular trochanters, which prove by their presence and form that there was a well developed hip and gluteal musculature, giving the leg strength to push the body firmly forward. The very large bulbous femoral head (caput femoris) is situated at an angle to the elongation of the shaft and is inflected slightly to the lesser trochanter, thus showing that in life the thigh was positioned more or less horizontally away from the body allowing only a 'reptilian gait'."

The shoulder girdle has been preserved only in the form of incomplete scapulocoracoids. The suture between the scapula and metacoracoid is situated within a large glenoid fossa. There is also a large acromion, which indicates the presence of a clavicle.

The three main bones of the forelimb have been completely preserved. Krusat (1991: 38) described them as follows: "The humerus is a short, stout and powerful bone, even stronger and thicker than the femur of the same individual. The humerus bears a very large deltopectoral crest. The shoulder and elbow joints are elongated and twisted relative to each other at an angle of about 60°, giving the bone a waisted hourglass shape. There is a canalis entepicondyloideum formed at the greatly broadened distal joint. A fossa olecrani is not preserved. The ulna is the longest bone of the forelimb equipped with a very long olecranon process, comprising about 27% of the bone's length. So the ulna, having had strong muscles, inserted at the olecranon process was a very effective lever, capable of transmitting great force into the movement of the arm and hand. The much shorter radius is equal in thickness to the ulna and has widened proximal and distal joint ends."

PALEOBIOLOGY

Krusat (1991) compared the morphology of the hindlimb bones of *Haldanodon* to those of *Tachyglossus,* pointing to a great similarity. He also concluded that the structure of the appendicular skeleton, in particular of the forelimbs, suggests a fossorial mode of life for *Haldanodon.* This interpretation is consistent with such skull features as the compact and strongly ossified bones of the skull roof, as described by Lillegraven and Krusat (1991). The mode of life of *Haldanodon* has also been discussed by Martin and Nowotny (2000, see also our figure 5.6). When reconstructing the lifestyle of docodontans, one should take into account that postcranial specializations of small mammals to fossorial and semiaquatic lifestyles can be very similar. Martin and Nowotny (2000) noted some striking similarities between the skeleton of *Haldanodon* and that of the modern semiaquatic mole *Desmana.* Given that *Haldanodon* comes from the coal swamps of Guimarota, a semiaquatic, desmanlike lifestyle might possibly be more likely than a fossorial, talpidlike lifestyle for this docodont.

SYSTEMATICS

(Table 5.1)

The order Docodonta is small, including six uncontested genera. All of them are assigned to the family Docodontidae.

FIGURE 5.6. Reconstruction of *Haldanodon exspectatus. Source:* from Martin and Nowotny (2000).

In addition to the Jurassic–Early Cretaceous genera, two other genera have been assigned to Docodonta. The first is European Late Triassic *Delsatia* Sigogneau-Russell and Godefroit, 1997; we assign *Delsatia* to Docodonta only tentatively. The second is *Reigitherium* Bonaparte, 1990, from the Late Cretaceous of Argentina, which was considered by Pascual et al. (2000) to be a docodontan. We disagree and assign *Reigitherium* to Mammalia, subclass and order *incertae sedis.*

Class Mammalia Linnaeus, 1758
Order Docodonta Kretzoi, 1946
Family Docodontidae Simpson, 1929a

Synonym: Tegotheriidae Tatarinov, 1994.

Diagnosis. Only family of the order Docodonta, characterized by transversely elongated upper molars, figure-eight shaped or roughly rectangular in occlusal view, with four main cusps, connected with transverse ridges. Lower molars longitudinally elongated, with four main cusps in two rows, the largest being the main cusp of the labial row (figure 5.4). The two crests link two main rows of cusps, the anterior crest being preceded by a basin (pseudotalonid) developed to varying degrees in different genera. Postdentary elements remained associated with the dentary (plesiomorphy); articular complex differs from that of other stem mammals retaining this configuration in that the articular complex has three (rather than two) facets for the quadrate, lying in a different plane and lower with respect to the craniomandibular joint.

Genera. *Docodon* Marsh, 1881, type genus; *Borealestes* Waldman and Savage, 1972; *Cyrtlatherium* E. F. Freeman, 1979; *Haldanodon* Kühne and Krusat, 1972; *Simpsonodon* Kermack et al., 1987; *Tegotherium* Tatarinov, 1994.

Distribution. Middle Jurassic (middle Bathonian): Scotland, Loch Scavaig; Middle Jurassic (late Bathonian): England, Oxfordshire, Kirtlington, Watton Cliff (Forest Marble); Late Jurassic (Kimmeridgian): Portugal, Leiria, Guimarota Coal Mine (Guimarota Beds); Late Jurassic (late Kimmeridgian to early Tithonian): United States, Wyoming and Colorado (Morrison Formation); Late Jurassic: Mongolia, Trans-Altai Gobi, Shar Teeg; Early Cretaceous (Berriasian): southern England, Dorset, Durlston Bay, Sunnydown Farm (Purbeck Limestone Group).

Genus *Docodon* Marsh, 1881 (figure 5.7A)

Synonyms: *Diacynodon* Schlosser, 1890; *Dicrocynodon* Osborn, 1888b; *Diplocynodon* Marsh, 1880; *Ennacodon* Marsh, 1890; *Enneodon* Marsh, 1887; ?*Peraiocynodon* Simpson, 1928a.

Comment. Simpson (1928a) identified the four teeth preserved in the only known specimen of *Peraiocynodon* (*P. inexpectatus* Simpson, 1928a) as m1–m4, but in Butler's (1939) reinterpretation they represent deciduous premolars dp1–dp4, which appears to be more probable (but see Krusat, 1980, for an opposing opinion).

Diagnosis. *Docodon* differs from *Simpsonodon* in having upper molars strongly constricted in the middle of the width, giving the tooth a figure-eight-shaped appearance (shared to some extent by *Haldanodon*), rather than being roughly rectangular; it differs from *Haldanodon* in having the lingual length of the molars longer with respect to the labial length. It shares with *Simpsonodon* ornamentation of grooves on the posterior face of the lower molars, but

TABLE 5.1. Linnaean Classification of Docodontan Mammals

Class Mammalia Linnaeus, 1758
- Order Docodonta Kretzoi, 1946
  - Family Docodontidae Simpson, 1929a
    - *Docodon* Marsh, 1881 (including *Peraiocynodon* Simpson, 1928a)
      - *D. victor* (Marsh, 1880), type species
      - *D. affinis* (Marsh, 1887)
      - *D. crassus* (Marsh, 1887)
      - *D. striatus* Marsh, 1881
      - *D. superus* Simpson, 1929a
    - *Borealestes* Waldman and Savage, 1972
      - *B. serendipitus* Waldman and Savage, 1972
    - *Cyrtlatherium* E. F. Freeman, 1979
      - *C. canei* E. F. Freeman, 1979
    - *Haldanodon* Kühne and Krusat, 1972
      - *H. exspectatus* Kühne and Krusat, 1972
    - *Simpsonodon* Kermack et al., 1987
      - *S. oxfordensis* Kermack et al., 1987
    - *Tegotherium* Tatarinov, 1994
      - *T. gubini* Tatarinov, 1994
- Order ?Docodonta Kretzoi, 1946
  - Family *incertae sedis*
    - *Delsatia* Sigogneau-Russell and Godefroit, 1997
      - *D. rhupotopi* Sigogneau-Russell and Godefroit, 1997

Class Mammalia Linnaeus, 1758
- Subclass and order *incertae sedis*
  - Family Reigitheriidae Bonaparte, 1990
    - *Reigitherium* Bonaparte, 1990
      - *R. bunodontum*[1] Bonaparte, 1990

[1] Emended to reflect the gender (neuter) of the genus.

differs from that genus in the lack of pits. There is a very small pseudotalonid, much less distinct than in *Simpsonodon.* The lower molars are more rectangular than in other genera and bear more posterior cingulid cusps.

Species. Engelmann and Callison (1998) followed Simpson (1929a) in recognizing five species of *Docodon.* Four of these species are represented only by the lower dentition: *D. victor* (Marsh, 1880), type species, *D. affinis* (Marsh, 1887); *D. crassus* (Marsh, 1887); and *D. striatus* Marsh, 1881. *D. superus* Simpson, 1929a includes all known upper dentition of *Docodon* from the Morrison Formation. Jenkins (1969b) believed that there may be fewer species of *Docodon* than previously established by Simpson (1929a), but recognition of synonymy requires additional material and study. On the other hand, Chure et al. (1998) recognized in the Morrison Formation only two *Docodon* species: *D. victor* and *Docodon* sp. We leave the question of the division of *Docodon* species as an open one. As noted earlier, *Peraiocynodon inexpectatus* Simpson, 1928a, is almost certainly represented by a juvenile individual with deciduous teeth. We tentatively follow the consensus that *Peraiocynodon* is probably a synonym of *Docodon,* recognizing that the teeth of the two cannot be compared directly.

Distribution. Late Jurassic (late Kimmeridgian to early Tithonian): United States, Wyoming and Colorado (Morrison Formation); ?Early Cretaceous (Berriasian): Southern England, Dorset, Durlston Bay, Sunnydown Farm (Purbeck Limestone Group).

Genus *Borealestes* Waldman and Savage, 1972 (figure 5.7B)

Diagnosis. Number of premolars in *Borealestes* is unknown, possibly three; there are at least six molars. The genus differs from *Simpsonodon* in having the mesio lingual cusp of lower molars much lower than the distolingual and a much smaller pseudotalonid. The ornamentation of grooves and/or pits characteristic of *Docodon, Simpsonodon,* and *Haldanodon* is not seen in published figures of *Borealestes.*

Species. *Borealestes serendipitus* Waldman and Savage, 1972, type species by monotypy, based on a dentary with teeth.

Distribution. Middle Jurassic (middle Bathonian): Scotland, Isle of Skye, Loch Scavaig.

Genus *Cyrtlatherium* E. F. Freeman, 1979 (figures 5.4A, 5.8B)

Comment. This genus, represented by a single lower cheek tooth, was originally assigned by E. F. Freeman (1979) to the Kuehneotheriidae, but Sigogneau-Russell (2001) demonstrated that it is a docodont. The morphology of the single specimen strongly suggests that it is a deciduous premolar. It is highly likely that *Cyrtlatherium canei* and *Simpsonodon oxfordensis* (from the same locality) represent the same species; we tentatively retain them as distinct pending clarification based on additional fossils.

Diagnosis (based on Sigogneau-Russell's, 2001, description). The lower cheek tooth of *Cyrtlatherium* differs from those of other docodontid genera in being narrower transversely, having two crests (the lingual one forming an arch) leading from the dominant main cusp to a small mesiolabial cusp, and having a third crest to a distolingual cusp (as usual with docodontans); the mesiolingual cusp characteristic of other genera is absent. It shares with *Simpsonodon* the presence of a large pseudotalonid basin and posterior ornamentation.

Species. *Cyrtlatherium canei* E. F. Freeman, 1979, type species by monotypy.

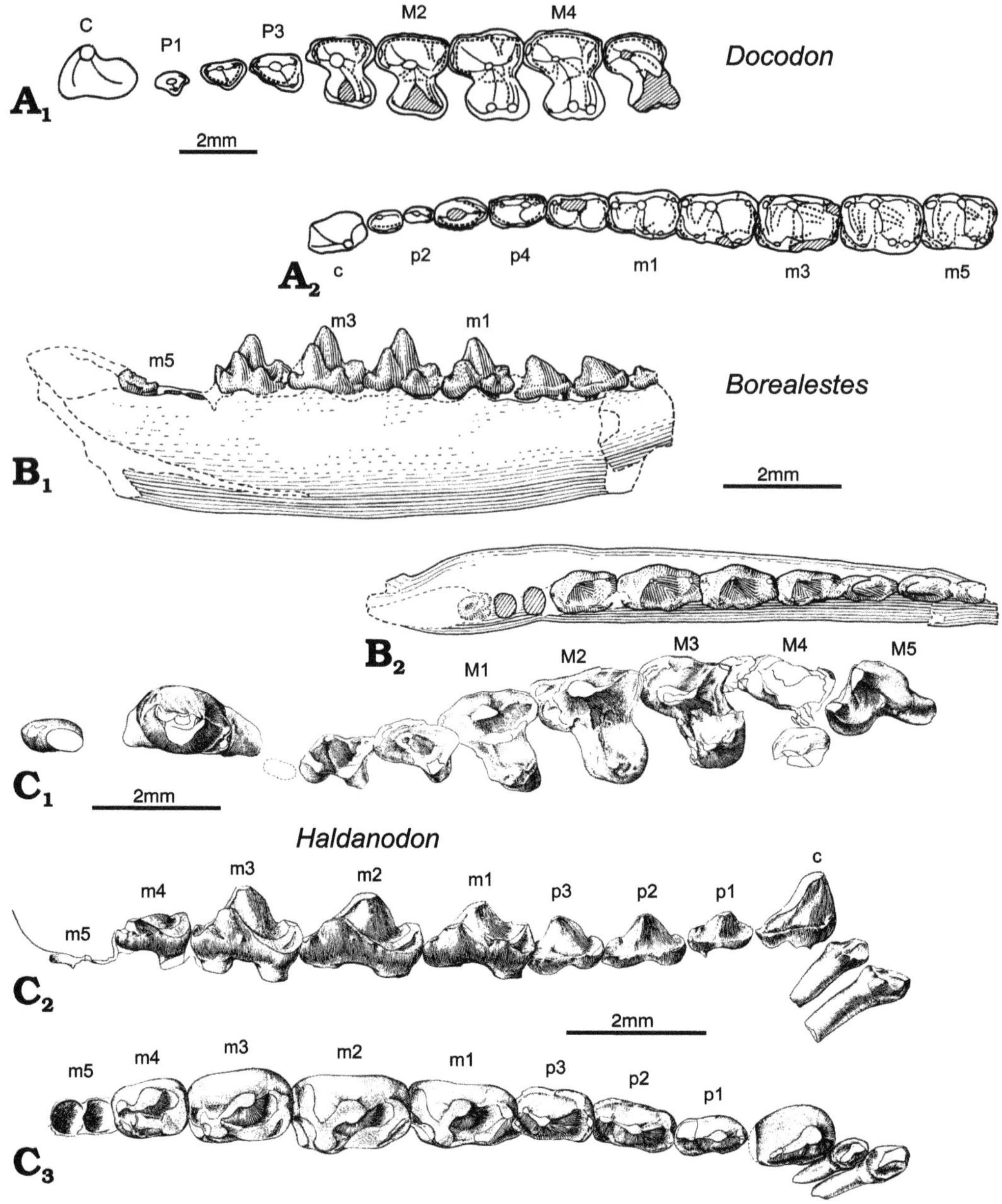

FIGURE 5.7. Dentition of selected docodontans. A, Reconstruction of the right upper (A_1) and lower (A_2) postcanine dentition of *Docodon* in occlusal view; upper dentition based on *D. superus*, lower on *D. victor.* B, Dentary (holotype) of *Borealestes serendipitus*, in lingual (B_1) and occlusal (B_2) views. C, Dentition of *Haldanodon exspectatus;* a partial left upper dentition (with a single upper incisor preserved): I4, C, P1 (alveolus), P2–3, M1–5, all in occlusal view (C_1); a partial left lower dentition: i3–4, c, p1–3, m1–4, m5 (alveolus), in lingual (C_2) and occlusal (C_3) views. For designation of cusps in *Haldanodon* see figure 5.4. *Source:* modified from: A, Jenkins (1969b); B, Waldman and Savage (1972); C, Krusat (1980).

Distribution. Middle Jurassic (late Bathonian): England, Oxfordshire, Kirtlington (Forest Marble).

Genus *Haldanodon* Kühne and Krusat, 1972
(figures 5.1–5.3B, 5.4C, 5.5, 5.6, 5.7C)

Diagnosis. *Haldanodon* differs from *Docodon* and *Simpsonodon* in having the labial part of the upper molars much longer than the lingual part and shares with *Docodon* a roughly figure-eight-shaped outline of the upper molars (albeit less rectangular than in *Docodon*). Differs from *Simpsonodon* and *Tegotherium* in having the distolingual cusp of the lower molars higher than the mesiolingual ones and in this respect resembles *Borealestes* and *Docodon.* Differs from *Simpsonodon* in having a shorter pseudotalonid. Shares with *Docodon* ornamentation of grooves on the lower molars, albeit much less conspicuous than in

Docodon, and differs in this respect from *Simpsonodon,* which has ornamentation consisting of grooves and pits, and from *Tegotherium* and *Borealestes,* in which ornamentation seems to be absent or very weak.

Species. *Haldanodon exspectatus* Kühne and Krusat, 1972, type species by monotypy, represented by incomplete skulls and skull elements, dentaries, and fragmentary postcranial skeleton (see earlier descriptions under "Anatomy").

Distribution. Late Jurassic (Kimmeridgian): Portugal, Leiria, Guimarota Coal Mine (Guimarota Beds).

Genus *Simpsonodon* K. A. Kermack, Lee, Lees, and Mussett, 1987
(figures 5.4B and 5.8B)

Synonym: *Borealestes* Waldman and Savage, 1972 (*partim,* only upper molar described by E. F. Freeman 1979: 160, plate 21, figures 3–5 as ?*Borealestes* sp.).

Diagnosis. *Simpsonodon* differs from *Docodon* and *Haldanodon* in having an ornamentation of pits and grooves on the lower molars, rather than only grooves. The three main cusps are arranged in a triangle, in front of which there is a distinct pseudotalonid basin that is much better developed than in other docodontid genera. Differs from *Tegotherium* in having a well-developed mesiolabial cusp (rather than two cuspules in *Tegotherium*) and from *Haldanodon* and *Borealestes* in having a larger mesiolingual cusp. The upper molars are less constricted in the middle than those in *Docodon,* but more rectangular than those of *Haldanodon.*

Species. *Simpsonodon oxfordensis* K. A. Kermack, Lee, Lees, and Mussett, 1987, type species by monotypy, represented by a dentary with a premolar and two molars and several isolated upper and lower molars, all of the type species.

Distribution. Middle Jurassic (late Bathonian): England, Oxfordshire, Kirtlington, Watton Cliff (Forest Marble).

Genus *Tegotherium* Tatarinov, 1994
(figure 5.8A)

Diagnosis. *Tegotherium* is generally similar to *Simpsonodon,* with which it shares the presence of a pseudotalonid that is narrower and less distinct than in *Simpsonodon.* There are three main cusps, rather than four as in other docodontans. The mesiolabial cusp is replaced by two

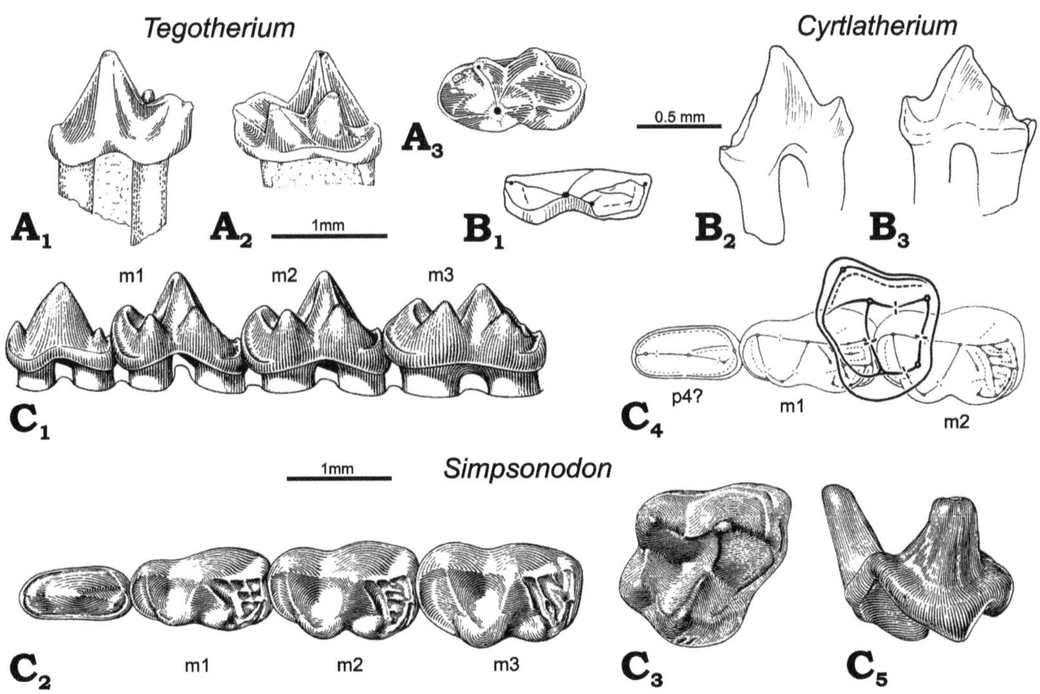

FIGURE 5.8. Dentition of selected docodontans. A, *Tegotherium gubini,* right lower molar, holotype in labial (A_1), lingual (A_2), and occlusal (A_3) views. B, *Cyrtlatherium canei,* right lower tooth (holotype), probably deciduous premolar; occlusal (B_1), labial (B_2), and lingual (B_3) views. C, *Simpsonodon oxfordensis,* last lower premolar and three molars in lingual (C_1) and occlusal (C_2) views; reconstruction showing lower penultimate premolar and two first molars, the latter in occlusion with upper molar, in occlusal (C_4) view; upper molar shown as transparent (C_4); right upper molar in occlusal (C_3), and lingual (C_5) views. For designation of cusps in *Cyrtlatherium* and *Simpsonodon* see figure 5.4. *Source:* A, original, courtesy of L. P. Tatarinov; B, and $C_{1-3,\,5}$, modified from Kermack et al. (1987); C_4, from Sigogneau-Russell (2001).

cuspules at the mesiolabial margin. Differs from *Simpsonodon* in having relatively higher cusps, especially the main cusp (see Tatarinov, 1994; Hopson, 1995; and Kielan-Jaworowska et al., 2000, for description and comments).

Species. *Tegotherium gubini* Tatarinov, 1994, type species by monotypy, represented by a single lower molar.

Distribution. Late Jurassic: Mongolia, Trans-Altai Gobi, Shar Teeg.

Order ?Docodonta Kretzoi, 1946
Family *incertae sedis*
Genus *Delsatia* Sigogneau-Russell and Godefroit, 1997
(figure 5.9A)

Comments. Sigogneau-Russell and Godefroit (1997) described *Delsatia* as the oldest representative of Docodonta and compared the arrangement of its cusps with those in the docodontan *Haldanodon* and of the symmetrodontan *Woutersia,* which, according to Butler (1997), is related to docodontans. Butler (1997) argued that there is no reason to regard *Delsatia* as a docodont, as it is more similar to symmetrodonts (e.g., *Tinodon*) in dental pattern. In view of the ambiguity we only tentatively place *Delsatia* in Docodonta.

Diagnosis. *Delsatia* differs from most docodontan genera (but not from *Simpsonodon* and *Tegotherium*) by having mesiolingual and distolingual cusps of the same size, linked to the main cusp by angulated crests, and by the presence of a well-developed mesiolabial cuspule. The main cusp is well developed, but there is neither a posterior basin nor a distolabial cusp.

Species. *Delsatia rhupotopi* Sigogneau-Russell and Godefroit, 1997, type species by monotypy, based on isolated lower molars and premolars.

Distribution. Late Triassic (early Rhaetian): France, Lorraine, Saint-Nicolas-de-Port.

Class Mammalia Linnaeus, 1758
Subclass and order *incertae sedis*
Family Reigitheriidae Bonaparte, 1990

Diagnosis. As for the genus.

Genus *Reigitherium* Bonaparte, 1990
(figure 5.9B)

Diagnosis. Differs from other Mesozoic mammals, including docodontans, in that the lower molars decrease in length and notably increase in width posteriorly, so that the tooth identified as m2 is rectangular, transversely elon-

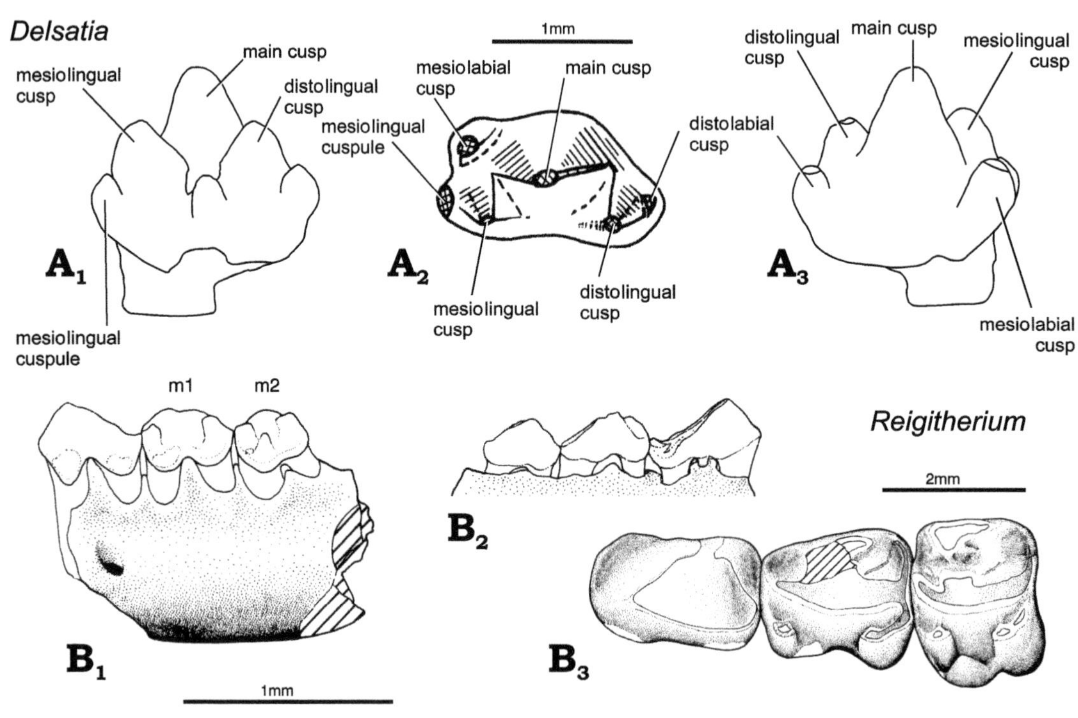

FIGURE 5.9. *Delsatia,* tentatively assigned to Docodonta, and *Reigitherium* (Mammalia, subclass and order *incertae sedis*). A, *Delsatia rhupotopi,* right lower molar in lingual (A_1), occlusal (A_2), and labial (A_3) views, showing wear facets. B, *Reigitherium bunodontum,* fragment of the left dentary, with p4–m2 in labial view (B_1); dentition of the same in the same scale in lingual view (B_2); enlarged p4–m2 of the same in occlusal view (B_3). *Source:* A, modified from Sigogneau-Russell and Godefroit (1997); B, modified from Pascual et al. (2000).

gated. The transverse widening of m2 is due to the presence of an extensive lingual cingulid with three cusps in this tooth, incorporated into the occlusal surface. The p4 and m1 are longer than wide, m1 wider than p4, almost quadrangular. There are two cylindrical cusps in the labial row in m1, and m2 has three cylindrical cusps, the middle one being the smallest, protruding labially, and the distal and mesial ones more lingually placed and higher. In the middle of the tooth there is a central cusp (identified by Pascual et al., 2000, as the protoconid), flanked mesially and distally by poorly identified cusps. Nine cusps are tentatively recognized on m2, but their homology with those in other mammals remains doubtful.

Species. *Reigitherium bunodontum* Bonaparte, 1990, type species by monotypy, represented by dentary fragment with three teeth and an isolated lower molar.

Distribution. Late Cretaceous (Campanian–Maastrichtian): Argentina, Río Negro Province, Los Alamitos (Los Alamitos Formation); Late Cretaceous (?Campanian–Maastrichtian): Argentina, Chubut Province, Meseta de Somuncura (La Colonia Formation).

Comments. *Reigitherium bunodontum* from the Late Cretaceous Los Alamitos Formation of Argentina was initially described by Bonaparte (1990) as an upper molar of dryolestid affinities. Pascual et al. (2000) described a second specimen of *Reigitherium* from the Late Cretaceous La Colonia Formation of Argentina, represented by a dentary fragment with three worn teeth, the most posterior of which (m2) is identical to the holotype of *R. bunodontum.* The authors suggested docodontan affinities for *Reigitherium* based on three features: (1) the presence of crests on the last premolar, extending from the main cusp posterolingually and posterolabially (but absent on the molars); (2) the presence of V-shaped intermolar embrasures between the labial parts of the lower molars (which in docodontans occur on the lingual side); and (3) the presence of a single "crenulation" on the lingual side of the molars.

In our opinion, the enamel crenulation is different from the dense ornamentation of minor ridges or fluting often associated with the pits characteristic of most docodontan genera. We believe that the unusual shape of the lower molars and the arrangement of cusps differentiate *Reigitherium* from known docodont taxa at the ordinal level. As the affinities of *Reigitherium* to other orders of mammals remain obscure, we assign it to Mammalia, subclass and order *incertae sedis,* pending discovery of more materials that may shed light on its affinities.

CHAPTER 6

Australosphenidans and *Shuotherium*

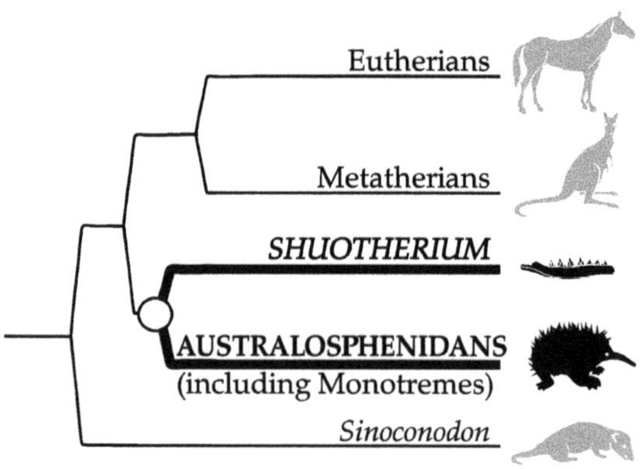

INTRODUCTION

Some of the most important new paleontological discoveries of the past two decades have been made on the southern continents. The first of these was the 1985 discovery of the first Mesozoic mammal from Australia—the Early Cretaceous toothed monotreme *Steropodon galmani,* represented by a partial dentary with three molar teeth (Archer et al., 1985). The authors stated that the dental terminology used by them for *S. galmani* (p. 363): "assumes that the teeth are tribosphenic and the cusps are homologues of topographically analogous cusps in therian mammals" (see chapter 11 for a discussion on the origin of tribosphenic molars). Kielan-Jaworowska, Crompton, and Jenkins (1987) agreed with Archer et al. (1985) that *Steropodon* is a therian *sensu lato,* but held that its teeth were not fully tribosphenic. They argued that *Steropodon* appears to have been derived from the therian stock before the tribosphenic molars made their appearance, possibly during the Jurassic, from forms close to *Peramus*-like ancestors (see chapter 10). Even in this case, however, the dentition of *Steropodon* was considerably more advanced than would have been predicted according to the prototherian-therian model of early mammalian evolution (see chapter 1).

The late 1990s brought still more exciting findings on the southern continents. Flannery et al. (1995) reported the discovery of another probable Cretaceous monotreme, *Kollikodon.* Rich et al. (1997) reported finding a dentary with four teeth of *Ausktribosphenos nyktos,* also from the Early Cretaceous of Australia (figure 6.1). The authors recognized three last molariform teeth as molars and the preceding tooth as a premolar, and concluded that *A. nyktos* is a placental. Rougier and Novacek (1998) briefly criticized this idea in a popular review, while Archer (in Musser and Archer, 1998) suggested that *Ausktribosphenos* may share a relationship with peramurids or possibly with monotremes, but did not elaborate on this thesis. Finally Kielan-Jaworowska et al. (1998) argued that *Ausktribosphenos* is a primitive stem mammal, possibly with its roots among early "symmetrodontans." It is not related to placentals because its mandible is far more primitive than those of all eutherians (including placentals) and metatherians, retaining several mandibular features otherwise only known from Early Jurassic mammals. However, Rich et al. (1998) continued to hold with their original idea. In a later report on another specimen of *Ausktribosphenos nyktos,* in which the angular process is preserved, Rich et al. (1999) argued that *Ausktribosphenos* is similar to the Erinaceidae in some dental features, such as the ultimate premolar having a fully developed three-cusped trigonid.

Rich et al. (1999) also reported the discovery of a new genus *Teinolophos,* which they tentatively considered a "eupantothere." After further preparation of the fossil, this taxon turned out to be a toothed monotreme (Rich, Vickers-Rich, et al., 2001). The preserved mandible of *Teinolophos* provides new information on the mandibular structure of toothed monotremes. Rich, Flannery, Trusler, Constantine et al. (2001) described a second ausktribosphenid genus, *Bishops,* based on a nearly complete dentary with six premolars, three molars, and three other dentary fragments (figure 6.2).

The next major discovery came from Madagascar. Flynn et al. (1999) reported a new mammal, *Ambondro mahabo* (figure 6.1A), from Middle Jurassic (Bathonian)

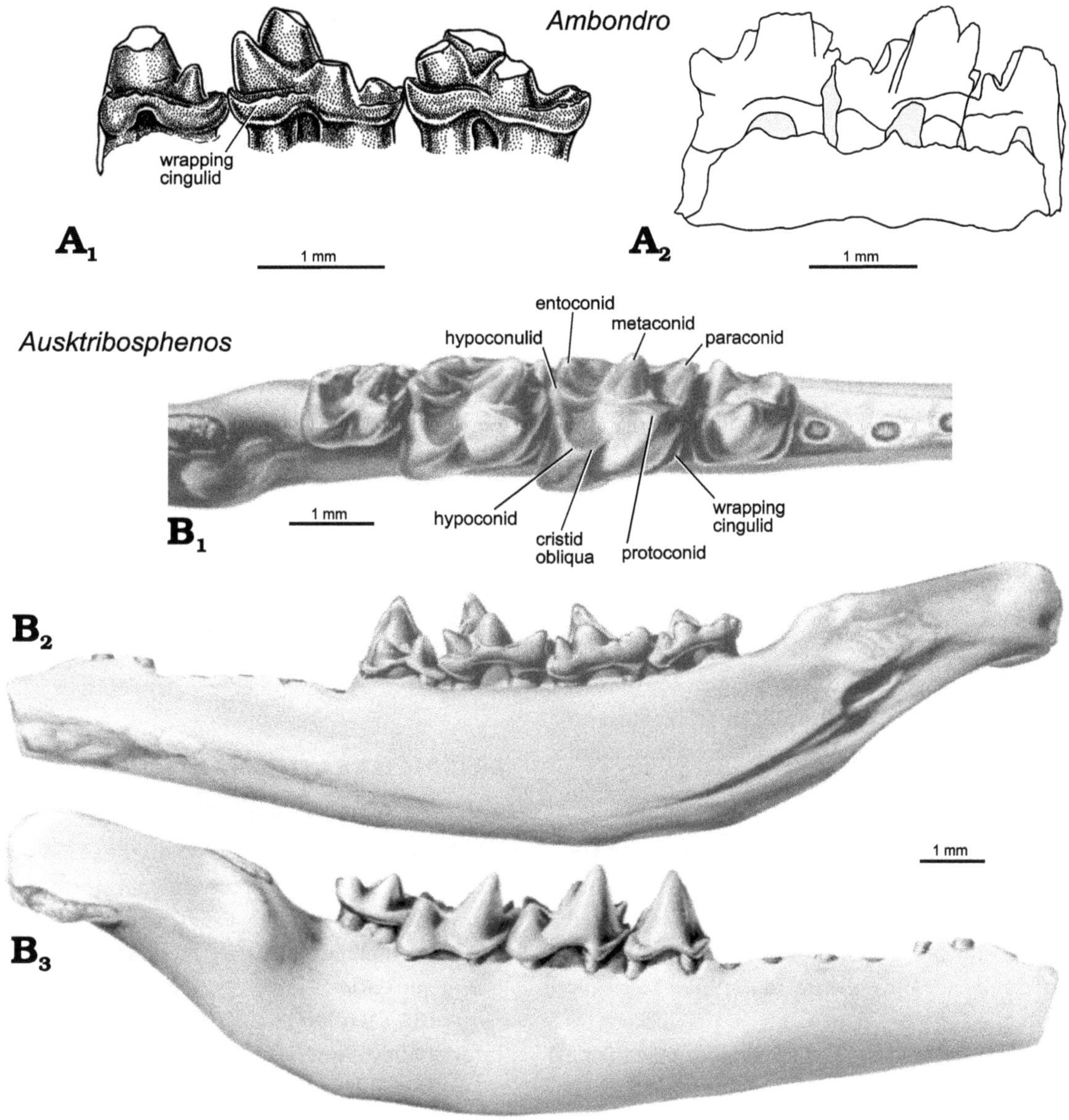

FIGURE 6.1. A, *Ambondro mahabo*, dentary fragment (holotype) with ultimate premolar and first two molars in lingual (A_1), and labial (A_2) views. B, *Ausktribosphenos nyktos*, incomplete dentary with ultimate premolar and three molars (holotype) in lingual (B_2) and labial (B_3) views; enlarged dentition of the same in occlusal view, anterior is to the right (B_1). *Source:* A, original, based on photographs in Flynn et al. (1999). B, modified from Rich et al. (1999).

beds. This taxon is represented by a dentary fragment with three teeth, the ultimate premolar and two anterior molars with a tribosphenic pattern. This discovery extended the existence of mammals with tribosphenic molars ("tribosphenidans") back by some 25 Ma. More significantly, the discoveries from Australia and Madagascar collectively challenged the long-held view that mammals with tribosphenic molars and their descendants, including all living placentals and marsupials, arose on northern continents (see Rich and Vickers-Rich, 1999). The initial debate about the affinities of these southern tribosphenic mammals centered on traditional and functionally important characteristics of the molar talonid. This begged the need to evaluate the characteristics of the premolar and mandible, as well as the nonocclusal features of molars that are also known for these southern tribosphenic mammals. Moreover, it was important to place these new fossils in the larger phylogenetic context of all mammals, including representatives of the three living groups and the major extinct groups of the Mesozoic.

Luo, Cifelli, and Kielan-Jaworowska (2001) made an attempt to estimate the phylogenetic relationships of these new Gondwanan fossils by all currently known morphological evidence and by comparing them to all known clades of Mesozoic mammals. Luo et al. (2002) expanded the scope of this parsimony analysis by sampling 46 taxa of all major groups of Mesozoic mammals and by including 275 osteological and dental characters. According to these authors *Ausktribosphenos* and *Ambondro* are unrelated to northern higher mammals; rather, they represent a previously unknown group derived from an archaic (Jurassic) ancestor and are more likely to be related to monotremes than to extant therians. An implication for morphological evolution is that the hallmark specialization of higher mammals—the derived talonid features in the multifunctional tribosphenic molars—arose not once but twice. The phylogenetic conclusions of Luo, Cifelli, and Kielan-Jaworowska (2001) resulted in the erection of two new infraclasses (now subclasses) within the subclass Holotheria Wible et al., 1995[1]: the Boreosphenida, which include mammals with tribosphenic dentition which arose on the northern continents and include "tribotherians," marsupials, and placentals (see chapters 11, 12, 13); and the Australosphenida, which arose and diversified on Gondwanan landmasses and are survived by the extant monotremes. The biogeographic scenario of this newly proposed phylogeny is consistent with the traditional idea for an origin of placentals, marsupials, and their close relatives on northern continents (e.g., Lillegraven, 1974). But it also offers a new hypothesis that the recently discovered southern tribosphenic mammals represent a previously unsuspected and endemic radiation of southern Mesozoic mammals.

Rauhut et al. (2002) described the first Jurassic mammal from Argentina, *Asfaltomylos patagonicus,* which is represented by a well-preserved dentary with three premolars and three molars (figure 6.3). They showed that *Asfaltomylos* is an australosphenidan and that the clade Australosphenida is strengthened by adding the new taxon to the phylogenetic dataset of Luo, Cifelli, and Kielan-Jaworowska (2001). The authors argued that the discovery of *Asfaltomylos* shows that australosphenidans were diversified and spread on Gondwana before the end of the Jurassic and that the mammalian fauna of the Southern Hemisphere by the Middle–Late Jurassic was noticeably different from that of the Northern Hemisphere.

This chapter also includes an aberrant taxon *Shuotherium* from the Jurassic of China (Chow and Rich, 1982; Wang, Clemens, et al. 1998) and Britain (Sigogneau-Russell, 1998), characterized by the presence of a pseudotalonid anterior to the trigonid. We regard *Shuotherium* to be the sister taxon of Australosphenida (Luo, Cifelli, and Kielan-Jaworowska, 2001; Kielan-Jaworowska et al., 2002; Luo et al., 2002; but see criticism of Averianov, 2002). This clade (Yinotheria of Chow and Rich, 1982) is defined by the common ancestor of shuotheriids and australosphenidans, to the exclusion of morganucodontans, eutriconodontans, multituberculates, and therians. Yinotheria may be diagnosed by the following combination of characters: differ from Eutriconodonta, Multituberculata, and Trechnotheria in the presence of postdentary trough (plesiomorphy) and from all other Mesozoic mammals in the presence of a sharp morphological break between penultimate and ultimate premolars. Last lower premolar, where known, differs from that of other Mesozoic mammals in having strongly molarized trigonid, with fully formed, obtusely triangulated protoconid, paraconid, and metaconid; little or no talonid, and strongly developed mesial cingulid. Lower molars differ from those of other Mesozoic mammals in strong development of mesial cingulid, tending to wrap around to the lingual side of the tooth.

BRIEF CHARACTERIZATION

Australosphenida (Latin *australis*—southern; Greek *sphen* —wedge, allusion to their origin on the Gondwanan continents and to tribosphenic molars) are a subclass of mammals erected by Luo, Cifelli, and Kielan-Jaworowska (2001) to include *Ambondro* Flynn et al., 1999; Ausktribosphenida Rich et al., 1997; and Monotremata Bonaparte, 1837. Rauhut et al. (2002) added the Middle–Late Jurassic South American *Asfaltomylos* to this group.

Australosphenida can be characterized by several derived characteristics (Luo, Cifelli, and Kielan-Jaworowska, 2001; Rauhut et al., 2002). The ultimate lower premolar has a fully triangulated trigonid but no talonid, a derived condition of autralosphenidans (weakly developed on the premolar of *Obdurodon* but not preserved in other toothed monotremes) that is not seen in boreosphenidans of the Early Cretaceous. Australosphenida (including toothed Monotremata) differ from boreosphenidans in having lower cheek teeth with a continuous, shelflike mesial cingulid rather than individualized cingulid cuspules. Owing to the lack of individualized mesial cingulid cuspules, australosphenidans lack the interlocking mechanism between molars that is present in various conditions in eutriconodontans, "eupantotherians," "tribotherians," and stem boreosphenidans. The shelflike mesial cingulid wraps around the mesiolingual corner of the trigonid to extend to the lingual side of the lower molar in *Ambondro, Ausktribosphenos,* and *Bishops,* although this is less obvious in the Jurassic *Asfaltomylos,* where the cin-

[1] For reasons discussed in chapters 9 and 15 we do not use the taxon Holotheria in this book.

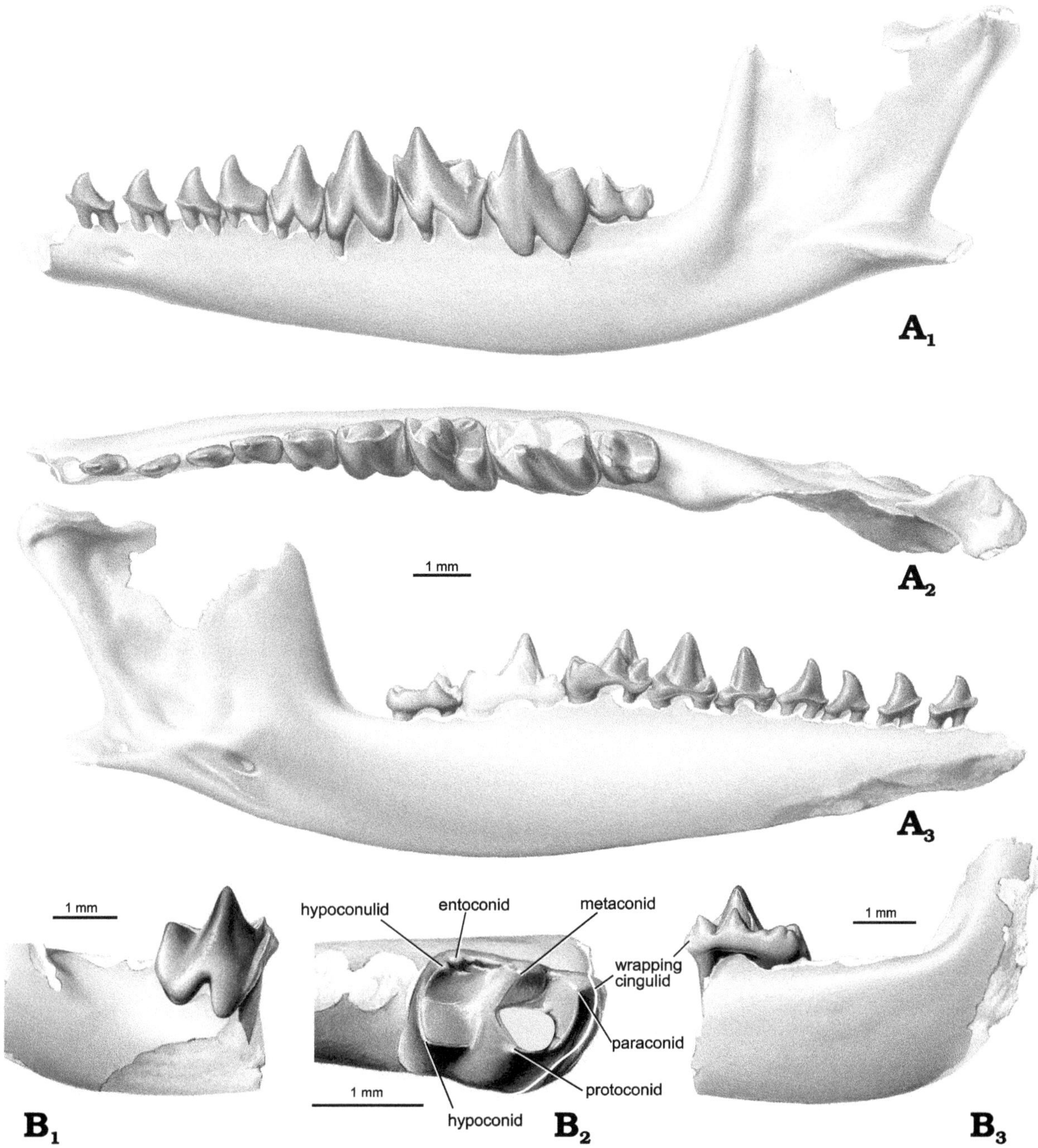

FIGURE 6.2. *Bishops whitmorei.* A, Dentary of the holotype with nine teeth in labial (A_1), occlusal (A_2), and lingual (A_3) views. B, Dentary fragment with well-preserved m2 of the referred specimen in labial (B_1) and lingual (B_3) views; enlarged m2 of the same in occlusal view, anterior is to the right (B_2). *Source:* modified from Rich, Flannery, Trusler, Constantine et al. (2001).

gulid is limited to the lingual side. In toothed monotremes the mesial cingulid is very prominent, but does not extend on the lingual side. The group differs further from Early Cretaceous boreosphenidans in having the hypoconulid that is procumbent anteriorly rather than slightly reclined posteriorly.

Australosphenida are also easily distinguishable from northern tribosphenic mammals in lacking many significant mandibular features of the latter, despite their superficial similarities in having a talonid basin capable of grinding. The group (at least in Ausktribosphenidae and *Asfaltomylos*) not only retains the primitive feature of postdentary trough on the dentary, but also has a derived feature of an elevated angular process, rather than one that is downturned as in early boreosphenidans (for these two features see Luo et al., 2002: figures 3 and 5, respectively).

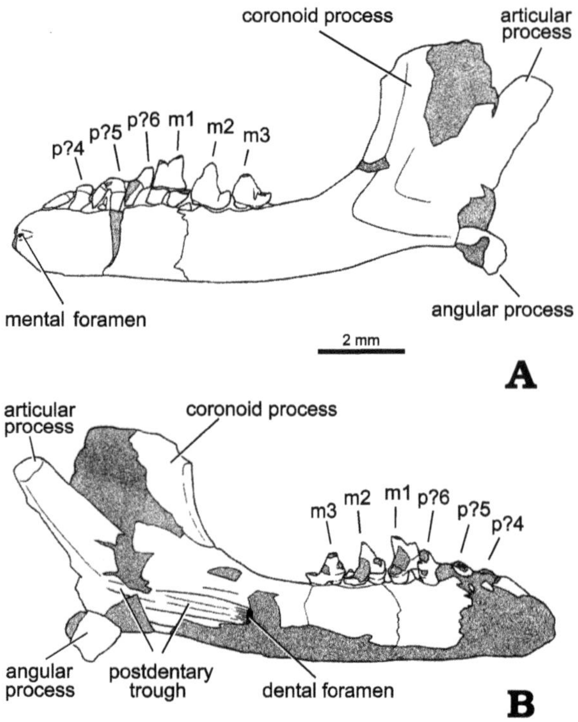

FIGURE 6.3. *Asfaltomylos patagonicus,* dentary of the holotype in labial (A) and lingual (B) views. The broken angular process probably originally pointed posteriorly, rather than posteroventrally as shown in the figure. *Source:* modified from Rauhut et al. (2002).

DISTRIBUTION

Middle Jurassic: Madagascar; Middle or Late Jurassic and Paleocene: South America; Early Cretaceous to Recent: Australian Region. The oldest australosphenidan is *Ambondro* from the Middle Jurassic of Madagascar.

SYSTEMATICS

(table 6.1)

Clade Yinotheria Chow and Rich, 1982
Subclass Australosphenida Luo, Cifelli, and Kielan-Jaworowska, 2001a, new rank
Order and Family *incertae sedis*
Genus *Ambondro* Flynn, Parrish, Rakotosamimanana, Simpson, and Wyss, 1999
(figure 6.1A)

Diagnosis. A tiny australosphenidan mammal, known only from the holotype specimen (dentary fragment with three teeth) of the type species, with strong lingual metacristid on the last premolar and on two preserved molars, and "open" trigonids on m1–2, with the three cusps arranged at an angle of approximately 90° (not acute). Shares with *Ausktribosphenos* and *Steropodon* the presence of a continuous, shelflike mesial cingulid that extends to the lingual side of the lower molars, and differs in this respect from tribosphenic molars of boreosphenidans. Shares with *Ausktribosphenos* an ultimate lower premolar with fully triangulated trigonid (unknown in *Steropodon*), and differs in this respect from peramurans and all Mesozoic boreosphenidans.

Species. *Ambondro mahabo* Flynn, Parrish, Rakotosamimanana, Simpson, and Wyss, 1999, type species by monotypy.

Distribution. Middle Jurassic (Bathonian): northwest Madagascar, Mahajanga Basin, Ambondromamy Region (Isalo "Group").

TABLE 6.1. Linnaean Classification of Mesozoic Australosphenidan Mammals and *Shuotherium*

Clade Yinotheria Chow and Rich, 1982
Subclass Australosphenida Luo, Cifelli, and Kielan-Jaworowska, 2001
Order and family *incertae sedis*
Ambondro Flynn et al., 1999
A. mahabo Flynn et al., 1999
Asfalotomylos Rauhut et al., 2002
A. patagonicus Rauhut et al., 2002
Order Ausktribosphenida Rich et al., 1997
Family Ausktribosphenidae Rich et al., 1997
Ausktribosphenos Rich et al., 1997, type genus
A. nyktos Rich et al., 1997
Bishops Rich, Flannery, Trusler, Constantine et al., 2001
B. whitmorei Rich, Flannery, Trusler, Constantine et al., 2001
Order Monotremata C. L. Bonaparte, 1837
Family Steropodontidae Flannery et al., 1995
Steropodon Archer et al., 1985, type genus
S. galmani Archer et al., 1985
Teinolophos Rich et al., 1999
T. trusleri Rich et al., 1999
?Order Monotremata C. L. Bonaparte, 1837
Family Kollikodontidae Flannery et al., 1995
Kollikodon Flannery et al., 1995, type genus
K. ritchei Flannery et al., 1995
(intervening ranks of clades in table 1.1)
Subclass *incertae sedis*
Order Shuotheridia Chow and Rich, 1982
Family Shuotheriidae Chow and Rich, 1982
Shuotherium Chow and Rich, 1982
S. dongi Chow and Rich, 1982, type species
S. kermacki Sigogneau-Russell, 1998
S. shilongi Wang et al., 1998

Order and Family *incertae sedis*
Genus *Asfalotomylos* Rauhut, Martin, Jaureguizar, and Puerta, 2002
(figure 6.3)

Diagnosis (modified from Rauhut et al., 2002). Australosphenidan mammal represented by a fairly complete, minute left dentary (holotype of the type species) with three molars and three posterior premolars (apparently p4–p6). Molar talonids fully basined with hypoconid and hypoconulid. Trigonids open lingually. The m1–m3 with faint lingual cingulid at the base of the paraconid. Dentary with large coronoid process, articular process positioned high, angular process present, postdentary trough shallow but extensive, subdivided by ridges, indicating the presence of postdentary bones. Differs from *Ambondro* and members of Ausktribosphenidae in having a weaker lingual cingulid that does not wrap around the mesial side. Also differs from *Ausktribosphenos* and *Bishops* in the lack of additional cristids on the talonid.

Species. *Asfaltomylos patagonicus* Rauhut, Martin, Jaureguizar, and Puerta, 2002, type species by monotypy.

Distribution. Middle or Late Jurassic (Callovian-Oxfordian): south Argentina, Chubut province (Cañadón Asfalto Formation).

Order Ausktribosphenida Rich, Vickers-Rich, Constantine, Flannery, Kool, and van Klaveren, 1997
Family Ausktribosphenidae Rich et al., 1997 (as previously)

Ordinal and Family Diagnosis. Small australosphenidan mammals with slender dentary, strong coronoid process, deep masseteric fossa, angular process and postdentary trough. Postcanine dental formula (as preserved in *Bishops*) is possibly p6, m3. Additional longitudinal ridges on the talonid of lower molars are an autapomorphy of the family. Share with *Ambondro* and *Steropodon* the presence of a continuous, shelflike mesial cingulid that extends to the lingual side of the lower molars, which differ in this respect from the tribosphenic molars of boreosphenidans. Differ from *Steropodon* in having a greater difference in the height between the trigonid and the talonid. Share with *Ambondro* structure of the ultimate lower premolar, with fully three-cusped trigonid (unknown in *Steropodon*) and differ in this respect from peramurans and all Mesozoic boreosphenidans.

Genera. *Ausktribosphenos* Rich et al., 1997, type genus; *Bishops* Rich, Flannery, Trusler, Constantine et al., 2001.

Distribution. Early Cretaceous (Aptian): Australia, Victoria (Wonthaggi Formation).

Genus *Ausktribosphenos* Rich, Vickers-Rich, Constantine, Flannery, Kool, and van Klaveren, 1997
(figure 6.1B)

Diagnosis. Monotypic ausktribosphenid genus, represented by two specimens of the type species, the holotype with three teeth, and the referred specimen with four teeth preserved. Differs from *Bishops* in having a larger postdentary trough with facets for postdentary bones and molars decreasing in size posteriorly, m1 being the largest (taller, wider, and longer than the remaining ones), whereas in *Bishops* the m2 is by far the largest.

Species. *Ausktribosphenos nyktos* Rich, Vickers-Rich, Constantine, Flannery, Kool, and van Klaveren, 1997, type species by monotypy.

Distribution. As for the family.

Genus *Bishops* Rich, Flannery, Trusler, Constantine, Kool, van Klaveren, and Vickers-Rich, 2001a
(figure 6.2)

Diagnosis. Monotypic ausktribosphenid genus known from the type species, represented by the holotype (nearly complete dentary with three molars and six premolars, partial coronoid, complete condylar and angular processes) and three less complete referred specimens. Differs from *Ausktribosphenos* in having shorter and shallower postdentary trough, without facets for postdentary bones, and mandibular foramen close to the ventral margin of the mandible. Further differences concern the different proportions among the lower molars, of which m2 is distinctly larger than m1 and m3.

Species. *Bishops whitmorei* Rich, Flannery, Trusler, Constantine, Kool, van Klaveren, and Vickers-Rich, 2001a, type species by monotypy.

Distribution. As for the family.

Order Monotremata C. L. Bonaparte, 1837

INTRODUCTION

Several ordinal names have been used for monotremes, of which Ornithodelphia de Blainville, 1834, and Monotremata Bonaparte, 1837, are most common (see McKenna and Bell, 1997: 35, for the full list of synonyms). We choose the Monotremata for this order (Gr. *monos*—single, *trema*—opening; an allusion to the presence of a cloaca), following Simpson (1945: 168), who stated: "The ordinal name Monotremata is preferred to the prior Ornithodelphia because it is in much more common use, both in this form and in the vernacular monotremes." In light of the phylogenetic schemes proposed by Luo, Cifelli, and Kielan-Jaworowska (2001); and Luo et al. (2002, see also chapter 15), we abandon the subclass Prototheria.

Monotremes are represented today by three genera, assigned to two families: Ornithorhynchidae (*Ornithorhynchus* of Australia and Tasmania) and Tachyglossidae (*Tachyglossus* of Australia, Tasmania, and New Guinea, and *Zaglossus* of New Guinea). Remnants of all three genera are known from the Pleistocene of Australia, Tasmania, and New Guinea (*Zaglossus*, Griffiths, 1968, 1978; Mahoney and Ride, 1975; Archer and Bartholomai, 1978; Murray, 1978a; Archer, 1981; McKenna and Bell, 1997; Archer et al., 1999). Miocene (originally referred to as Pliocene) monotremes were first reported from Australia in the nineteenth century (Dun, 1895; see also Rich, 1991). Woodburne and Tedford (1975) described *Obdurodon insignis*, represented by two six-rooted teeth from two disparate localities of late Oligocene age in South Australia. The teeth, originally described as upper molars (see also Clemens, 1979c) were subsequently recognized as the lowers (Archer et al., 1992). Allocation of *Obdurodon* to Ornithorhynchidae was based on a comparison with ephemeral, deciduous teeth of juvenile *Ornithorhynchus* (Simpson, 1929b; Green, 1937). Archer et al. (1978) described an edentulous dentary and a fragment of ilium from the locality yielding the holotype of *O. insignis* and attributed it to the same species. The next fossil assigned to Ornithorhynchidae was the skull with a dentary of *Obdurodon dicksoni* from the middle Miocene of Riversleigh, Queensland (Archer et al., 1992, 1993; Musser and Archer, 1998). In addition to the skull, several isolated molars and edentulous dentaries, some from early Miocene deposits, are also known (see Rich, 1991, for discussion of age of the mammal-bearing beds of Australia). The skull resembles that of Recent *Ornithorhynchus*, differing in being larger, retaining a large septomaxilla, nonvestigial dentition, and a prominent angular process on the dentary, and having a more anteriorly placed glenoid fossa. The oldest ornithorhynchid is represented by several isolated teeth of *Monotrematum sudamericanum*, from the early Paleocene of Patagonia (Argentina) (Pascual et al., 1992a,b; Pascual et al., 2002), which is the only monotreme outside the Australian region.

Tertiary tachyglossids are known from the Pliocene and are represented by skulls of *Megalibgwilia ramsayi* (Murray, 1978b; Griffiths et al., 1991). There are skulls from Pleistocene beds of Naracoorte in South Australia and from Pleistocene cave deposits in Australia and Tasmania. Like other tachyglossids, *Megalibgwilia* is edentulous. It differs from *Zaglossus* and *Tachyglossus* in the proportions of the skull elements, especially in the shape and size of the rostrum.

Monotremata also include two stem families: Steropodontidae and Kollikodontidae. Flannery et al. (1995) described *Kollikodon ritchei*, a large mammal from the same locality (Lightning Ridge, of Albian age) as *Steropodon*. They allocated *Kollikodon* to its own family, Kollikodontidae, and erected another family, Steropodontidae, to contain *Steropodon*. *Kollikodon* was based on a dentary fragment with three molars, but now a maxilla with teeth is also known. Both *Steropodon* and *Kollikodon* are among the largest Mesozoic mammals. A final Mesozoic record of Monotremata is that of a humerus from the Albian Dinosaur Cove site, Victoria, Australia (Rich and Vickers-Rich, 2000).

BRIEF CHARACTERIZATION

Monotremes share with other extant mammals (marsupials and placentals) such exclusively mammalian characters as hair, milk glands, three ear ossicles, a single bone (dentary) in the lower jaw, and a cribriform plate in the nasal region (or a transient embryonic lamina cribrosa as in *Ornithorhynchus*). They differ from other Recent mammals by retaining several plesiomorphic characters, such as oviparity and laying cleidoic eggs, which are incubated for some time before hatching. The relative positions of the ureters and the genital ducts differentiate monotremes from therians (Renfree, 1993). In monotremes the ureter opens into the urogenital sinus opposite the urethral openings of the bladder, whereas in therians the ureter migrates from a dorsal position adjacent to the Wolffian duct to a direct connection to the bladder. Other primitive features of the soft anatomy are the lack of a scrotum in the males and the presence of a cloaca (characteristic of reptiles and birds and among living mammals shared only with the Tenrecidae). The cloaca is a common chamber that is the receptacle for the discharges of the digestive system and urogenital tract.

The skull retains several plesiomorphic characters (figure 6.4). Monotremes are the only living mammals with an "egg-tooth" (the prenasal processes of the premaxillae, which meet at the midline between the external nares, bear an egg-tooth for cutting the shell at the time of hatching; tooth and processes then disappear, figure 6.4A_2). The ectopterygoid and septomaxilla (fused with the premaxilla in the adult, Wible et al., 1990) are reptilian bones present only in monotremes among living mammals. Archer et al. (1993) noted that an enormous septomaxilla, not fused with the premaxilla, is present in an adult Miocene ornithorhynchid *Obdurodon dicksoni*. According to these authors the septomaxilla is hypertrophied, apparently as an adaptation supporting elaborate sensory organs and the bill in ornithorhynchid monotremes. There is no lacrimal and the jugal is reduced or absent. There is no auditory bulla and the middle ear is partially surrounded by a horizontal tympanic ring; the cochlea is coiled for 270°.

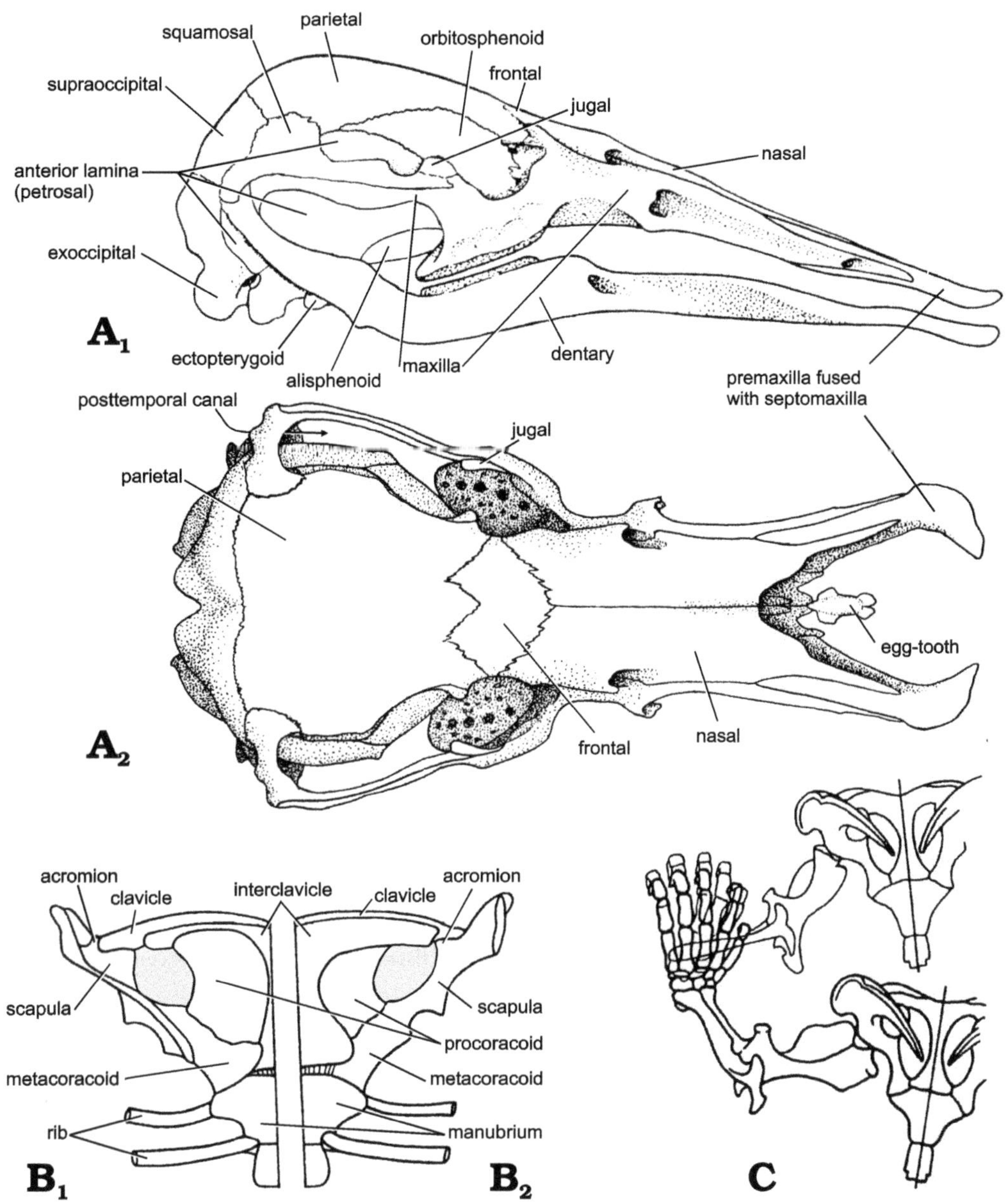

FIGURE 6.4. Selected skeletal elements and limb posture of Recent *Ornithorhynchus anatinus.* A, Skull and dentary in lateral (A_1) and dorsal (A_2) views. B, Pectoral girdle in dorsal (B_1) and ventral (B_2) views. C, Dorsal view of the forelimb during the propulsive phase; the thick line denotes the beginning of the phase, the thin line the end; note the sprawling limb posture. *Source:* A, modified from Jollie (1962); B, C, modified from Gambaryan (1998), based partly on Pridmore (1985).

The so-called anterior lamina of the petrosal (figure 3.5C) forms most of the lateral wall of the braincase (Watson, 1916). Monotremes share this character with multituberculates and *Vincelestes* (Kermack and Kielan-Jaworowska, 1971; Kielan-Jaworowska, 1971; Hopson and Rougier, 1993).

Zeller (1989a,b) pointed out on embryological evidence that the lateral wall of the braincase in monotremes is formed by the lamina obturans, a membrane ossification in the secondary lateral wall but not an anterior growth from the embryonic otic capsule that forms the adult petrosal (see also Kuhn, 1971). Hopson and Rougier (1993) regarded the terms anterior lamina and lamina obturans as synonymous. Whichever term is used, the structure of the lateral wall of monotreme braincase is different from that in Theria *sensu stricto,* in which the wall is built of the squamosal and alisphenoid (figure 3.5). The zygomatic arch is built mostly of the squamosal dorsally and maxilla ventrally, the jugal being reduced to a small element placed at the anterodorsal part of the arch. In adults, when fusion obliterates the sutures, the jugal is reduced to a small postorbital process of the zygomatic arch (figure 6.4A).

The monotreme skeleton shows many primitive, cynodont-like features. Monotremes differ from extant mammals in having a sprawling limb posture (figure 6.4C), with the proximal segments of the limbs, the humerus and femur, extending roughly transversely from the body, which limits their running ability (Jenkins, 1970c). They are plantigrade and have claws. The unfused cervical ribs are present and the thoracic ribs lack tubercles. In the pectoral girdle, the supraspinous fossa of the scapula is unexpanded; the interclavicle, clavicle, procoracoid, and coracoid are prominent (figure 6.4B), and both clavicle and scapula are much less mobile than they are in therians. The epipubic (marsupial) bones are present in the pelvic girdle. This latter character is shared with eutriconodontans, multituberculates, "symmetrodontans," marsupials, and even some Cretaceous eutherian mammals (Kielan-Jaworowska, 1969a, 1975c; Hu et al., 1997; Novacek et al., 1997; Ji et al., 1999, 2002). A character, unknown in other extant mammals, is the possession of a horny spur on the hind foot, found predominantly in males. This element is present in at least three groups of Mesozoic mammals, including eutriconodontans, multituberculates, and "symmetrodontans" (see chapters 7, 8, 9).

Monotremes display several specialized features. Recent taxa are toothless (except for juvenile *Ornithorhynchus*). The fossil taxa have some autapomorphies of m1, including a two-cusped trigonid without a paraconid. Skulls of ornithorhynchids have a ducklike bill, with an elongated rostrum built of the premaxilla fused with septomaxilla, and nasals covered by fleshy, highly innervated soft tissue and skin.[2] The echidnas are specialized for feeding on ants and termites and are efficient diggers. The platypus inhabits streams and bank burrows and is well adapted to semiaquatic life, having an elongated secondary palate, fusiform body, reduced external ears, and webbed feet, but the digits retain claws and are used for burrowing (Griffiths, 1968, 1978; Jenkins, 1970c).

Distribution. Early Cretaceous to Recent: Australian Region; Paleocene: South America.

[2] *Ornithorhynchus anatinus* first came to the attention of Western scientists in the late eighteenth and early nineteenth centuries, via dried specimens shipped from Australia. In mummified form, the fleshy tissue of the rostrum on the platypus resembles the keratinous covering of a bird's beak, and a controversy ensued as to whether the platypus was a naturally occurring chimera or an elaborate hoax.

SYSTEMATICS OF MONOTREMATA

The order Monotremata includes four families, two of which, Ornithorhynchidae and Tachyglossidae, are not known from the Mesozoic and are not discussed here. In what follows we describe Steropodontidae and tentatively assigned Kollikodontidae.

Family Steropodontidae Flannery, Archer, Rich, and Jones, 1995

Diagnosis (based on Rich, Vickers-Rich et al., 2001, emended). Differ from the Kollikodontidae by the presence of a distinct trigonid and talonid on lower molars rather than having a bunodont dentition. Differ from the Ornithorhynchidae by lower molars with only two elongated roots and from the Tachyglossidae by the presence of teeth. Steropodontidae (as represented by *Teinolophos*) differ from Tachyglossidae and Ornithorhynchidae by having a deep mandible with a condyle well above the dorsal margin of the horizontal ramus; a prominent ascending ramus with both medial and lateral flanges on its anterior border; and a more dorsally placed masseteric fossa. This family lacks the separation of the coronoid process from the internal coronoid process and the pocketlike pterygoid fossa of ornithorhynchids.

Genera. *Steropodon* Archer et al., 1985, type genus; *Teinolophos* Rich et al., 1999.

Distribution. Early Cretaceous (Aptian through Albian): Australia.

Genus *Steropodon* Archer, Flannery, Ritchie, and Molnar, 1985 (figure 6.5A,C_1)

Diagnosis. Relatively large steropodontid, represented by a single specimen, a holotype of the type species, which is a large fragment of a dentary with three molars and alveolus for the ultimate premolar. Differs from *Teinolophos* in having lower molars more elongated mesiodistally, the trigonids less compressed mesiodistally, a smaller difference between the heights of the trigonid and talonid, and larger mesial and distal cingulids.

Species. *Steropodon galmani* Archer, Flannery, Ritchie, and Molnar, 1985, type species by monotypy.

Distribution. Early Cretaceous (middle Albian): Australia, New South Wales (Griman Creek Formation).

Genus *Teinolophos* Rich, Vickers-Rich, Constantine, Flannery, Kool, and van Klaveren, 1999 (figure 6.5B,C_2)

Diagnosis. Differs from *Steropodon* by notably smaller dimensions, by having the lower molar (the single one known) less elongated mesiodistally, with strongly com-

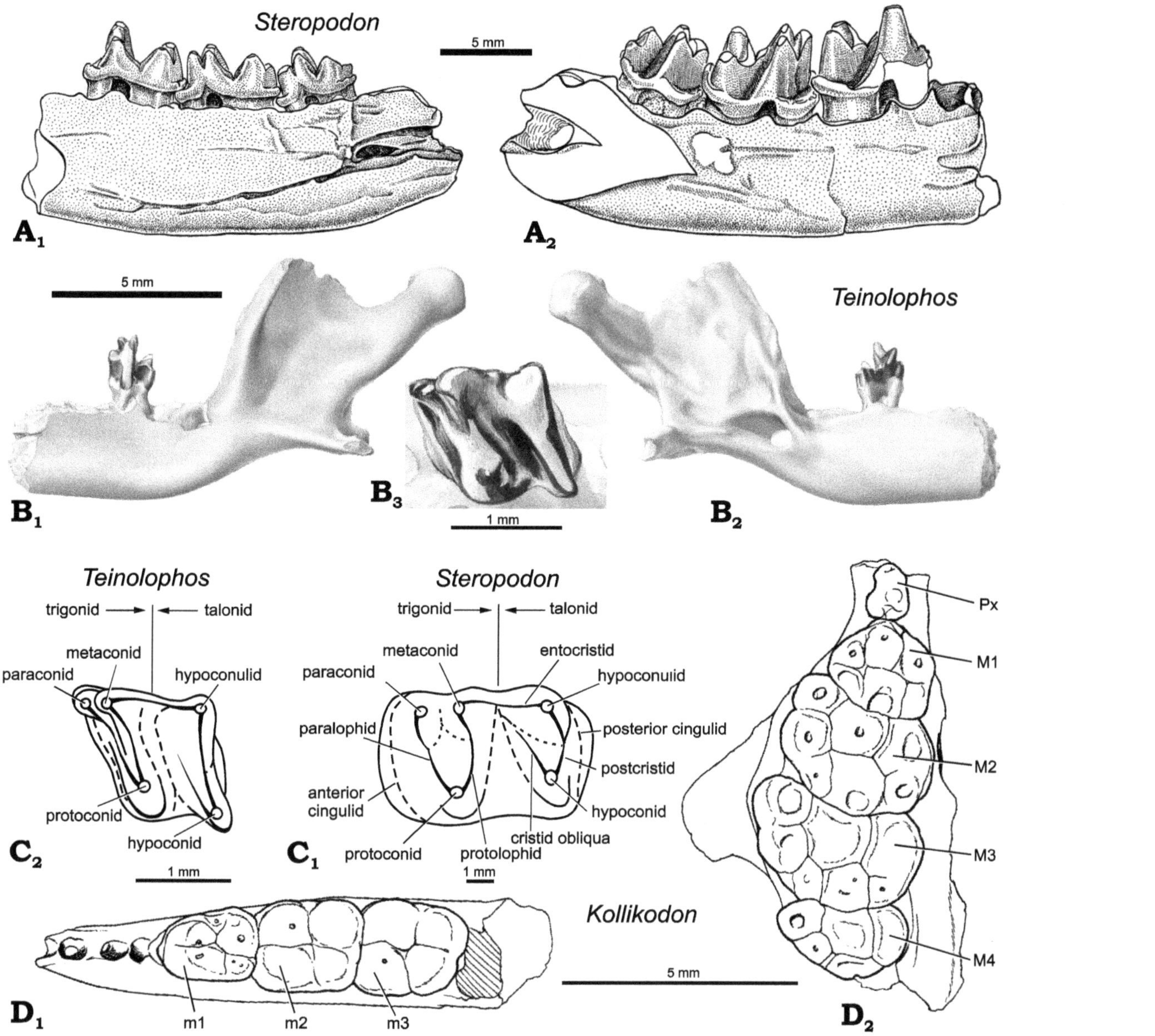

FIGURE 6.5. Cretaceous monotremes. A, *Steropodon galmani,* right dentary fragment of the holotype in lingual (A_1) and labial (A_2) views. B, *Teinolophos trusleri,* dentary of the holotype in labial (B_1) and lingual (B_2) views; enlarged m2 of the same in occlusal view (B_3). C, Schematic comparison of m2 of *Steropodon* (C_1) and *Teinolophos* (C_2) in occlusal views. D, *Kollikodon ritchei,* right dentary fragment (holotype), with m1–m3 and alveoli for p1–p2 and m4, in occlusal view (D_1); right maxilla of the same with penultimate premolar and m1–m4 in occlusal view (D_2). *Source:* A, based on cast and photographs of Archer et al. (1985); B, modified from Rich, Flannery, Trusler, Constantine et al. (2001); C, modified from Rich, Vickers-Rich et al., 2001; D, drawings, courtesy of Anne Musser.

pressed trigonid, and much narrower mesial and distal cingulids.

Species. *Teinolophos trusleri* Rich, Vickers-Rich, Constantine, Flannery, Kool, and van Klaveren, 1999, type species by monotypy.

Distribution. Early Cretaceous (Aptian): Australia, Victoria (Wonthaggi Formation).

Order ?Monotremata C. L. Bonaparte, 1837
Family Kollikodontidae Flannery, Archer, Rich, and Jones, 1995

Type genus. *Kollikodon* Flannery et al., 1995, the only genus known.

Diagnosis and Distribution. As for the type genus.

Genus *Kollikodon* Flannery, Archer, Rich, and Jones, 1995 (figure 6.5D)

Diagnosis (based on Flannery et al., 1995; A. Musser, pers. comm.). Differs notably from the uncontested monotremes in having at least four molars and bunodont molar morphology. The lower molars are almost regularly quadrangular in occlusal view, with four cusps and prominent labial cingulid. There are no lophs or vertical crests. The two first cusps ("trigonid") on m1, identified by Flannery et al. (1995) as the "protoconid" (smaller) and the "metaconid," are concave; in front there is also a large mesial cingulid. Four cusps on posterior molars are of subequal height. Penultimate upper premolar is notably smaller than the molars, narrow with ?two cusps. Upper molars are very large, especially M2, M3, with five (M4), six (M1), or seven (M2, M3) cusps, of structure unknown in any other mammal.

Species. *Kollikodon ritchei* Flannery, Archer, Rich, and Jones, 1995, type species by monotypy, represented by the holotype (a right dentary fragment with m1–m3, alveoli for p1–p2 and m4) and referred partial right maxilla with ultimate premolar and four molars, from the same locality.

Distribution. Early Cretaceous (middle Albian): Australia, New South Wales (Griman Creek Formation).

Comment. Flannery et al. (1995) assigned Kollikodontidae to Monotremata on the basis of a great size difference between the premolars and molars, shared by ornithorhynchids (e.g., *Obdurodon dicksoni,* see Archer et al., 1993: figure 7.4), an anteroposteriorly compressed trigonid on m1, lacking the paraconid (autapomorphic for Monotremata among the Mesozoic mammals), and very large dental (mandibular) canal, characteristic of all the monotremes and related to the development of the bill. McKenna and Bell (1997) assigned *Kollikodon* to Mammaliaformes Rowe, 1988 (unnamed rank). Because of great differences in the structure of the dentition between *Kollikodon* and other monotremes in which the teeth are known, for the lack of contrary evidence, we only tentatively assign Kollikodontidae to Monotremata.

RELATIONSHIPS WITHIN MONOTREMATA

The molecular data (e.g., Westerman and Edwards, 1992) suggest that the ornithorhynchids and tachyglossids diverged in the Early Tertiary or even Late Cretaceous. Zeller (1989a, 1993) argued that ornithorhynchids are more plesiomorphic than tachyglossids in details of the structure of the ear region and contents of the cavum epipterícum (see also Kuhn and Zeller, 1987b). Archer et al. (1985) originally assigned *Steropodon* to the ?Ornithorhynchidae, while Archer et al. (1992) omitted the question mark in front of the Ornithorhynchidae. Flannery et al. (1995), however, argued that in view of molecular data (which indicate the Early Tertiary divergence of the monotremes) placing *Steropodon* in the Ornithorhynchidae would make this family paraphyletic. Therefore, they proposed a separate family Steropodontidae, including at that time only *Steropodon.* Rich, Vickers-Rich et al. (2001) assigned the Early Cretaceous (Aptian) Australian mammal *Teinolophos,* originally described as "eupantotherian" to the Steropodontidae. Kollikodontidae Flannery et al., 1995, are the fourth ?monotreme family. They differ from uncontested monotremes in lack of vertical cutting surfaces and in having bunodont molars. They are specialized, but they retain four molars, regarded by Flannery et al. (1995) as a primitive feature. These authors suggested that kollikodontids might be the sister group of a steropodontid-tachyglossid-ornithorhynchid clade (Flannery et al., 1995: figure 3).

The Mesozoic and Early Tertiary records of monotremes are poor, making it difficult to infer much about relationships within the group. A possibility exists that *Kollikodon* (tentatively attributed by us to Monotremata), when better known, may even prove not to be a monotreme. Pending further discoveries we provisionally accept the cladogram of interrelationships within monotremes proposed by Flannery et al. (1995: figure 3).

Subclass *incertae sedis*
Order Shuotheridia Chow and Rich, 1982

INTRODUCTION

Chow and Rich (1982) described *Shuotherium dongi* from the Late Jurassic of Sichuan in China, based on a dentary with an extensive postdentary trough and five complete teeth, two very fragmentary, and one alveolus. They interpreted the preserved and reconstructed teeth as three premolars and four molars (figures 4 and 5 in their paper, see our figure 6.6A_{1-2}). The structure of the molars is intriguing. They consist of a well-developed high three-cusped trigonid and a low basined "talonid" situated not to the rear of the trigonid, as characteristic of tribosphenic molars (see chapter 11), but in front of the trigonid. The authors designated this "anterior talonid" as the pseudotalonid. Chow and Rich (1982: figure 2, see our figure 6.6A_3) also recognized a "talonid" posterior to the trigonid, although the talonid is smaller than the pseudotalonid, more cingulid-like, and appears to have had no occlusal contact with the upper molar. They offered a hypothetical reconstruction of an upper molar in *Shuotherium,* including reconstruction of corresponding

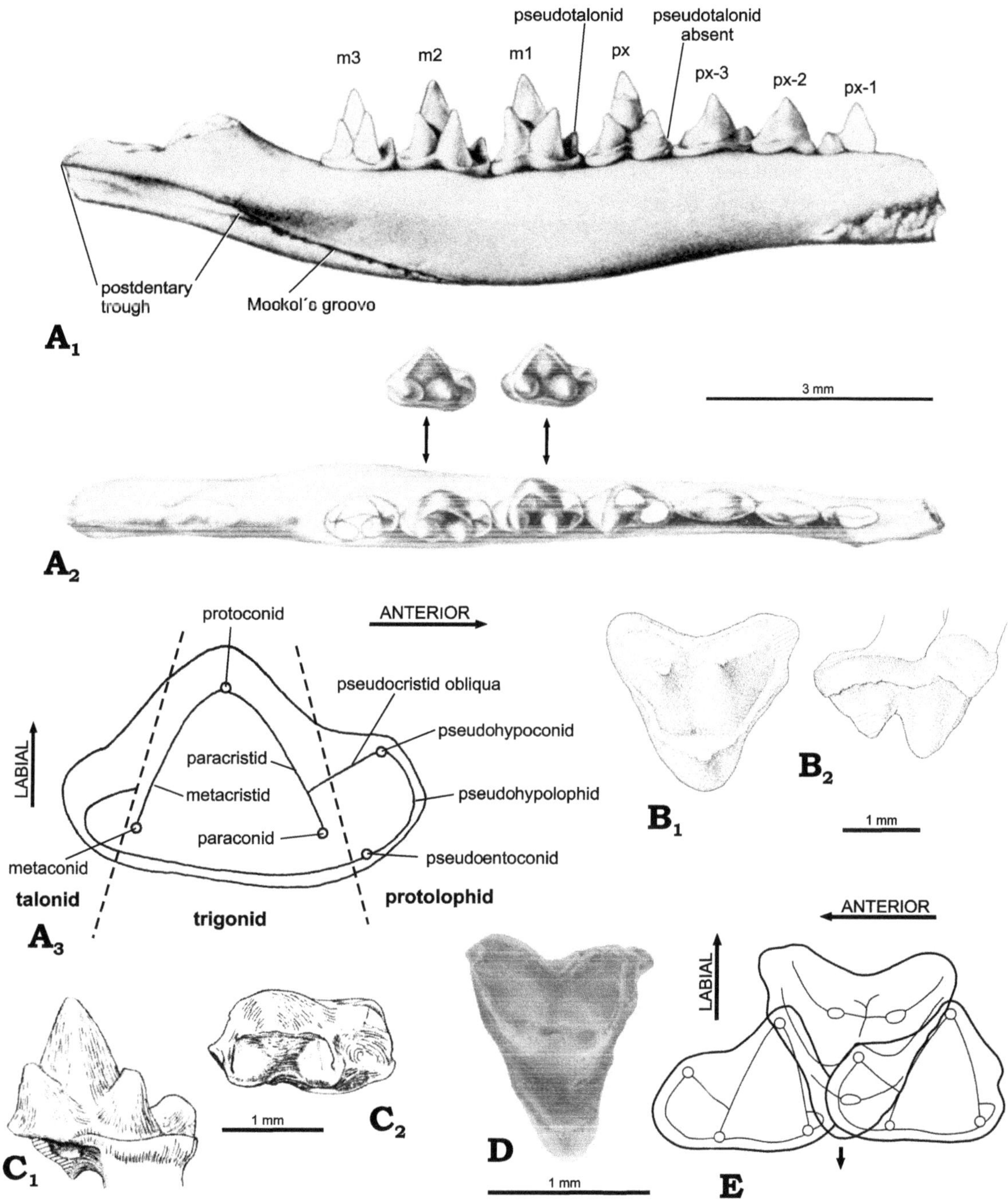

FIGURE 6.6. Representatives of the Shuotheriidae. A, *Shuotherium dongi,* left dentary (holotype), partly reconstructed in lingual view (A_1), the same in occlusal view (A_2), with designation of the teeth accepted in this book; schematic drawing of a molar in occlusal view (A_3), showing terminology used by Chow and Rich (1982). B, *Shuotherium shilongi,* right upper molar (holotype) in occlusal (B_1) and labial (B_2) views. C, *Shuotherium kermacki,* left lower molar (holotype) in lingual (C_1) and occlusal (C_2) views. D, Right upper molar of ?*Shuotherium* from the Bathonian of England. E, Reconstruction showing occlusion in *Shuotherium,* based upon upper molar of *S. shilongi* and lower molars of *S. dongi.* Drawings of lower molars have been enlarged to match the size of the uppers. *Source:* modified from: A, Chow and Rich (1982); B, E, Wang, Clemens et al. (1998); C, D, Sigogneau-Russell (1998).

shearing surfaces on the lower and upper molars, based on the wear facet system devised by Crompton (1971).

Chow and Rich (1982) erected a legion Yinotheria, order Shuotheridia, and family Shuotheriidae for *Shuotherium*. Although they recognized the postcanine dental formula as three premolars and four molars, with respect to m1 (Chow and Rich, 1982: 132) they stated: "It is possible that the tooth here regarded as M_1 should be more probably designated P_4. It does differ from the undoubted molars behind it in lacking a pseudo-talonid and having a more antero-posteriorly expanded trigonid. However, the sharpest change in the form of adjacent postcanine teeth occurs between the simple, somewhat blade-like P_3 and the highly molariform M_1."

Sixteen years after Chow and Rich's discovery, Sigogneau-Russell (1998) described several isolated lower and upper molars of *Shuotherium* from the Bathonian of England, assigning one of the lower teeth to *Shuotherium dongi* and others to *Shuotherium kermacki* or to *Shuotherium* sp. She also described and figured two upper molars, tentatively assigned to ?*Shuotherium*. Sigogneau-Russell (1998) followed McKenna and Bell (1997), who placed Shuotheriidae in Symmetrodonta. In the same year Wang, Clemens et al. (1998) described, from the locality yielding *S. dongi*, an isolated upper molar of *Shuotherium shilongi* (too large to be assigned to the type species of *Shuotherium*). This recently discovered tooth helped to confirm the reconstruction of *Shuotherium* upper molar and shearing surfaces, as proposed by Chow and Rich. They regarded Yinotheria as a sister taxon of Cladotheria McKenna, 1975. Kermack et al. (1987) suggested that *Shuotherium* might be related to docodonts, based on the presence of an anterior pseudotalonid, which also occurs in some docodonts. Subsequent authors have not accepted this idea.

The analysis of Luo et al. (2002: figures 1, 2, see our figures 15.1 and 15.2) placed *Shuotherium* as the sister taxon of the Australosphenida. These authors stated (2002: footnote to p. 23): "The 'm1' of *Shuotherium dongi* lacks the pseudo-talonid, characteristic of the succeeding molars. It cannot be ruled out that the tooth is, instead, an ultimate premolar; if this proves to be true, then the two australosphenidan apomorphies [mesial cingulid and ultimate lower premolar with trigonid, lacking the talonid] may be shifted to be the apomorphies of (*Shuotherium* + Australosphenida)."

Kielan-Jaworowska, Cifelli, and Luo (2002) further developed the latter idea by pointing out more mandibular and dental similarities between *Shuotherium* and nonmonotreme australosphenidans (*Ausktribosphenos, Bishops, Ambondro,* and *Asfaltomylos*). These authors revised the lower dentition of *Shuotherium* as having four premolars and three molars (figure 6.6A). They suggested that the ultimate premolar (previously identified as the first molar by Chow and Rich, 1982) of *Shuotherium* is similar to the ultimate premolar in australosphenidans in having a fully triangulated trigonid and in the presence of the precingulid and postcingulid. There is a sharp morphological boundary between these two teeth (designated as m1–2 by Chow and Rich, 1982; ultimate premolar and m1 herein) of *Shuotherium dongi* (figure 6.6A). A similar morphological break between the ultimate premolar and the first molar is also present in some stem australosphenidans (figure 6.2). Based on this evidence, Kielan-Jaworowska, Cifelli, and Luo (2002) proposed that *Shuotherium* might have descended from common ancestors with Australosphenida. We accept this view, but see Averianov (2002) for criticism.

Family Shuotheriidae Chow and Rich, 1982
Genus *Shuotherium* Chow and Rich, 1982
(figure 6.6)

Comment. The order Shuotheridia and family Shuotheriidae are monotypic and include the single genus *Shuotherium*.

Ordinal, Familial, and Generic Diagnosis. The holotype of *Shoutherium dongi* is represented by most of a dentary with seven teeth (two of which are broken). Additional species of *Shuotherium* are represented by isolated lower and upper molars. The dental formula may be tentatively reconstructed for lower postcanines as: < p4, m3. Differ from all other mammals by having pseudotribosphenic molars, in which the mesial cingulid is strongly expanded to form a pseudotalonid, while the distal talonid is underdeveloped, cingulid-like. Share with nonmonotreme members of Australosphenida a slender dentary with postdentary trough, but differ from Ausktribosphenidae in having a more extensive postdentary trough, similar to that in *Asfaltomylos*. Differ from nonmonotreme Australosphenida in the structure of the lower molars, but share with them the unique structure of the lower premolars. The ultimate premolar has a completely developed trigonid, with a dominating protoconid and small paraconid and metaconid spread from one to another, with only a tiny posterior cingulid-like talonid and no pseudotalonid. There is a sharp morphological break between the ultimate and penultimate premolars, the latter (as all other anterior premolars) are simple two-rooted teeth, without traces of molarization. Upper molars (at least in *S. shilongi*) of primitive "pseudotribosphenic" structure, longer than wide, with three dominating cusps and a small stylar area. The parastylar area lacks the parastylar groove typical of true tribosphenic mammals and "eupantotherians." The cusp designated pseudoprotocone occluded into the pseudotalonid of the succeeding lower molar.

Species. *Shuotherium dongi* Chow and Rich, 1982, type species; *S. kermacki* Sigogneau-Russell, 1998; *S. shilongi* Wang, Clemens, Hu, and Li, 1998; *Shuotherium* sp. and ?*Shuotherium* upper molars (Sigogneau-Russell, 1998).

Distribution. Middle Jurassic (late Bathonian): England, Oxfordshire (Forest Marble); Late or Middle Jurassic: China, Sichuan Province, Shilongzhai (Shaximiao Formation).

CHAPTER 7

Eutriconodontans

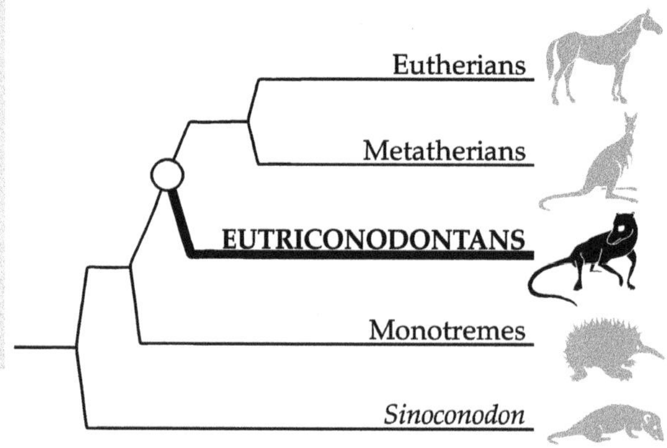

INTRODUCTION

Eutriconodontans comprise a diverse, well-known group of Mesozoic mammals. Molar structure and other craniodental features indicate a strictly animalivorous diet; body size spanned a considerably broader range than most other groups and included some of the largest of all mammals known from the entire Mesozoic (figure 7.1). As a taxonomic group, Eutriconodonta are named after the most conspicuous, albeit primitive, characteristic of their dentition: a molar pattern consisting of three main cusps placed in anteroposterior alignment on a crown that is somewhat compressed transversely (figure 7.2). The distinctiveness of these characteristics was formally recognized by early workers (e.g., Marsh, 1887; Osborn, 1888a,b), resulting in the erection of the order Triconodonta. Simpson's studies of the 1920s (1925b,c, 1928a, 1929a) provided detailed understanding of the morphology and systematics of then-known taxa (six genera, ranging from Middle Jurassic through earliest Cretaceous, all placed by Simpson in a single family). The concept of the group was later broadened to accommodate newly discovered, geologically older (Late Triassic–Early Jurassic) taxa, such as morganucodontids (Parrington, 1941, 1947; Kühne, 1949, 1958) and *Sinoconodon* (Patterson and Olson, 1961).

Despite similarities in molar pattern, morphological heterogeneity of an expanded Triconodonta soon became apparent, and Kermack et al. (1973) consequently divided the group into suborders, erecting Eutriconodonta to contain genera treated by Simpson (1928a, 1929a) and Morganucodonta for Late Triassic–Early Jurassic taxa.[1] Detailed study of the skull (e.g., Kermack et al., 1981; Crompton and Luo, 1993; Wible and Hopson, 1993; Luo, 1994; Rougier, Wible, and Hopson, 1996) and skeleton (e.g., Jenkins and Parrington, 1976; Jenkins and Crompton, 1979; Jenkins and Schaff, 1988; Ji et al., 1999) reveals fundamental differences between morganucodontans and eutriconodontans. Recent analyses consistently result in separation of the two groups, with morganucodontans occupying a basal position among mammals and eutriconodontans placed among, or proximal to, the living groups (e.g., Rowe, 1988; Rougier, Wible, and Novacek, 1996; Luo, Cifelli, and Kielan-Jaworowska, 2001; Luo, Crompton, and Sun, 2001; Luo et al. 2002). "Triconodonta" *sensu lato,* based largely on a plesiomorphic molar pattern, are therefore paraphyletic (see discussion in chapter 4). To avoid confusion, we use this term informally, in quotes, to refer to a general structural grade.

Eutriconodonta as a clade are defined as the common ancestor of the Triconodontidae (*sensu stricto*) plus any taxa more closely related to the Triconodontidae than to morganucodontans, spalacotheriids, tinodontids, and multituberculates. The robustness of monophyly of the

[1] Docodonta Kretzoi, 1946, were included as a third suborder based on retention of attached postdentary elements in *Docodon* (Kermack and Mussett, 1958a), and the plausible derivation of docodont molar structure from a *Morganucodon*-like ancestor (Patterson, 1956; Patterson and Olson, 1961, see chapter 5).

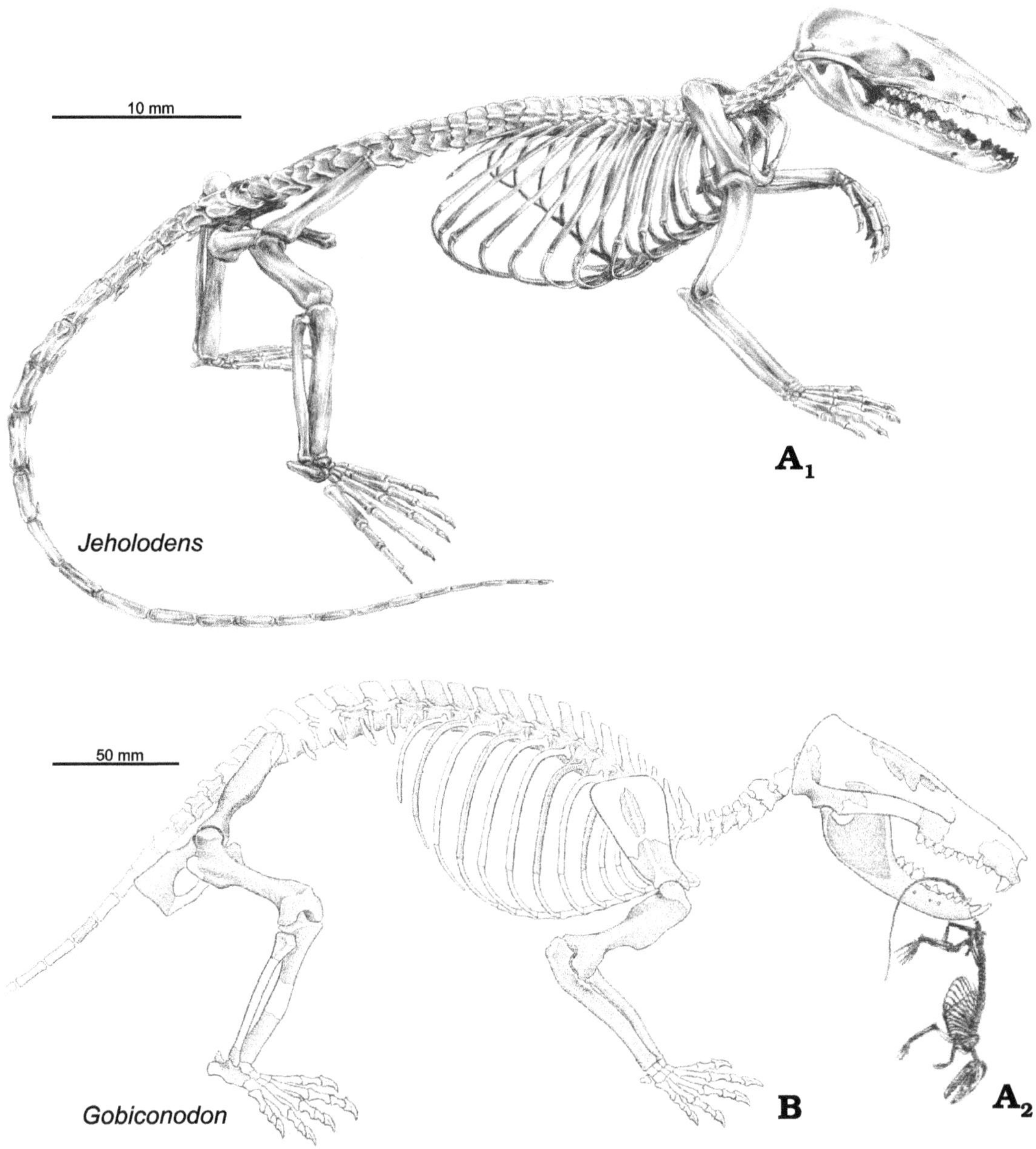

FIGURE 7.1. Restored skeletons of Eutriconodonta. A, *Jeholodens jenkinsi* (family *incertae sedis*). B, *Gobiconodon ostromi* (Gobiconodontidae). These are the only two eutriconodontans for which reasonably complete skeletons have been published. *Jeholodens jenkinsi* (restored skull length about 22 mm) and *Gobiconodon ostromi* (restored skull length about 100 mm) lie near the lower and upper parts, respectively, of the known size range for eutriconodontans. This range can be appreciated when the restored skeletons are reproduced at the same scale (B, A_2). Eutriconodontans are interpreted to have been faunivorous; at larger body sizes, diet presumably included vertebrate prey (B). *Source:* A, from Ji et al. (1999); B, from Jenkins and Schaff (1988), emended by adding A_2.

Eutriconodonta (conceptually similar to Triconodonta as circumscribed by Simpson 1928a,b with the addition of many genera described since that time) is not overwhelmingly strong. Nonetheless, we provisionally accept the group as monophyletic (see results and discussion of analyses by Luo et al., 2002). As for so many groups of Mesozoic mammals, the foundation for systematics within Eutriconodonta was laid by Simpson's landmark studies (1928a, 1929a), wherein a dichotomous grouping of Amphilestinae and Triconodontinae (both within Triconodontidae) was proposed. That basic structure remains part of our classification (table 7.1), though we follow most other recent treatments in restoring family rank for the two groups, as originally proposed by Osborn

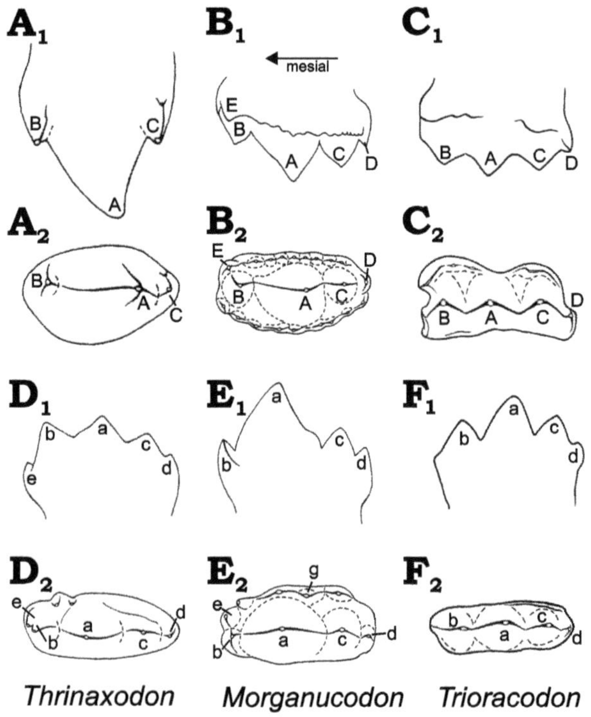

FIGURE 7.2. The triconodont molar pattern. This configuration of three principal, linearly arranged molar cusps (designated A–C on upper molars and a–c on lowers) is primitive for Mammalia and therefore uninformative as to relationships. The triconodont pattern, already present in advanced cynodonts (A, D), is seen in many of the earliest mammals such as morganucodontans (B, E), as well as in eutriconodontans (C, F). A–C, upper molars in lingual (A_1–C_1) and occlusal (A_2–C_2) views. D–F, lower molars in labial (D_1–F_1) and occlusal (D_2–F_2) views. A,D, *Thrinaxodon liorhinus.* B, E, *Morganucodon watsoni.* C, F, *Trioracodon bisulcus.* Not to scale. *Source:* A, B, D, E, modified from Jenkins (1984).

(1888b) and Marsh (1887), respectively. This change in rank is intended, in part, to reflect significant morphological differences among known genera, many of which were described on the basis of fossils unavailable to Simpson in the 1920s. Further recent work has provided new data supporting monophyly of one clade, Triconodontidae (e.g., Jenkins and Crompton, 1979; Cifelli et al., 1998), but cast doubt on integrity of the other. Accordingly, we treat "Amphilestidae" as a structural grade of primitive eutriconodontans (see Ji et al., 1999; Rougier et al., 1999; Luo et al., 2002).

Recent decades have witnessed a great improvement in the fossil record of Eutriconodonta, as for most groups of Mesozoic mammals. As a result, 24 genera, collectively including some 37 named species (table 7.1), are now known, placing eutriconodontans among the most diverse of Mesozoic mammal groups. Two genera known only by isolated teeth are treated herein as family *incertae sedis* and are referred to Eutriconodonta with question, as are Austrotriconodontidae (one genus). We restrict "Amphilestidae" to nine genera, recognizing *Gobiconodon* and suspected relatives (*Hangjinia, Repenomamus*) as a monophyletic group, Gobiconodontidae (see Chow and Rich, 1984a; Jenkins and Schaff, 1988; Kielan-Jaworowska and Dashzeveg, 1998). *Jeholodens,* from the Early Cretaceous of China, appears to be related to Triconodontidae (see Ji et al., 1999), but is not formally included in the family. Within Triconodontidae (eight genera), we recognize two monophyletic clades: Triconodontinae (Late Jurassic–earliest Cretaceous, North America and Western Europe) and Alticonodontinae (Cretaceous, North America).

BRIEF CHARACTERIZATION

Eutriconodontans are generally large mammals relative to other members of their respective faunas, characterized by molars with three or four trenchant, serially arranged cusps (plesiomorphy, figure 7.2) that are more laterally compressed, with lesser development of accessory cuspules, than in Morganucodontidae (apomorphies). The most noteworthy apomorphies distinguishing eutriconodontans from other "triconodonts" (e.g., morganucodontids) are in the dentary, which has a well-developed pterygoid fossa and lacks (1) an angular process together with the posteroventral emargination of the dentary, and (2) a postdentary trough with overhanging ridge, associated with support of postdentary elements (articular, prearticular, angular, surangular) by the dentary (Allin and Hopson, 1992; Rowe, 1996a, see also chapter 3). Presence of the postdentary trough and associated features are retained primitive features, shared by morganucodontans with the more distantly related nonmammalian cynodonts (Hopson and Barghusen, 1986; Rowe, 1988; Wible, 1991; Allin and Hopson, 1992; Luo, 1994); their absence in eutriconodontans is shared with crown mammals. The dentary of eutriconodontans is more robustly built than in most contemporary mammals, further differing in the presence of a medial pterygoid ridge, sometimes called a pterygoid shelf or pteryoid crest (Simpson, 1925b; Ji et al., 1999), also present in multituberculates (referred to as pterygoideus shelf) and some spalacotheriid "symmetrodontans" (Miao 1988; Gambaryan and Kielan-Jaworowska, 1995; Hu et al., 1998), lateral development of the inferior margin of the masseteric fossa (Simpson, 1928a; also seen in some Spalacotheriidae, Cifelli and Madsen, 1999), and strong development of the coronoid process.

TABLE 7.1. Linnaean Classification of Eutriconodontan Mammals

- ?Order Eutriconodonta Kermack et al., 1973
  - Family *incertae sedis*
    - *Dyskritodon* Sigogneau-Russell, 1995
      - *D. amazighi* Sigogneau-Russell, 1995, type species
      - *D. indicus* Prasad and Manhas, 2002
    - *Ichthyoconodon* Sigogneau-Russell, 1995
      - *I. jaworowskorum* Sigogneau-Russell, 1995
  - Family Austrotriconodontidae Bonaparte, 1992
    - *Austrotriconodon* Bonaparte, 1986
      - *A. mckennai* Bonaparte, 1986, type species
      - *A. sepulvedai* Bonaparte, 1992
- Order Eutriconodonta Kermack et al., 1973
  - Family "Amphilestidae" Osborn, 1888
    - *Amphilestes* Owen, 1859, type genus
      - *A. broderipii* (Owen, 1845)
    - *Aploconodon* Simpson, 1925
      - *A. comoensis* Simpson, 1925
    - *Comodon* Kretzoi and Kretzoi, 2000[1]
      - *C. gidleyi* (Simpson, 1925)
    - *Klamelia* Chow and Rich, 1984
      - *K. zhaopengi* Chow and Rich, 1984
    - *Liaotherium* Zhou et al., 1991
      - *L. gracile* Zhou et al., 1991
    - ?*Paikasigudodon* Prasad and Manhas, 2002
      - *P. yadagirii* (Prasad and Manhas, 1997)
    - *Phascolotherium* Owen, 1838
      - *P. bucklandi* (Broderip, 1828)
    - *Tendagurodon* Heinrich, 1998
      - *T. janenschi* Heinrich, 1998
    - *Triconolestes* Engelmann and Callison, 1998
      - *T. curvicuspis* Engelmann and Callison, 1998
  - Family Gobiconodontidae Chow and Rich, 1984[2]
    - *Gobiconodon* Trofimov, 1978,[3] type genus
      - *G. borissiaki* Trofimov, 1978, type species
      - *G. hoburensis* (Trofimov, 1978)
      - *G. hopsoni* Rougier et al., 2001
      - *G. ostromi* Jenkins and Schaff, 1988
    - *Hangjinia* Godefroit and Guo, 1999
      - *H. chowi* Godefroit and Guo, 1999
    - *Repenomamus* Li et al., 2000
      - *R. robustus* Li et al., 2000
  - Family *incertae sedis*
    - *Jeholodens* Ji et al., 1999
      - *J. jenkinsi* Ji et al., 1999
  - Family Triconodontidae Marsh, 1887
    - Subfamily Tricondontinae Marsh, 1887
      - *Triconodon* Owen, 1859, type genus
        - *T. mordax* Owen, 1859
      - *Priacodon* Marsh, 1887
        - *P. ferox* (Marsh, 1880), type species
        - *P. fruitaensis* Rasmussen and Callison, 1981
        - *P. grandaevus* Simpson, 1925
        - *P. lulli* Simpson, 1925
        - *P. robustus* (Marsh, 1879)
      - *Trioracodon* Simpson, 1928
        - *T. ferox* (Owen, 1871), type species
        - *T. bisulcus* (Marsh, 1880)
        - *T. major* (Owen, 1871)
        - *T. oweni* Simpson, 1928
    - Subfamily Alticonodontinae Fox, 1976
      - *Alticonodon* Fox, 1969, type genus
        - *A. lindoei* Fox, 1969
      - *Arundelconodon* Cifelli, Lipka et al., 1999
        - *A. hottoni* Cifelli, Lipka et al., 1999
      - *Astroconodon* Patterson, 1951
        - *A. denisoni* Patterson, 1951, type species
        - *A. delicatus* Cifelli and Madsen, 1998
      - *Corviconodon* Cifelli et al., 1998
        - *C. montanensis* Cifelli et al., 1998, type species
        - *C. utahensis* Cifelli and Madsen, 1998
      - *Jugulator* Cifelli and Madsen, 1998
        - *J. amplissimus* Cifelli and Madsen, 1998

[1] For *Phascolodon* Simpson, 1925b, preoccupied by *Phascolodon* Stein, 1859 (a ciliophoran). *Phascolotheridium* Cifelli and Dykes, 2001, is a junior synonym of *Comodon* Kretzoi and Kretzoi, 2000.
[2] Gobiconodontinae Chow and Rich, 1984a, *non* Gobiconodontidae Jenkins and Schaff, 1988; Repenomamidae Li et al., 2000, are a junior subjective synonym of Gobiconodontidae.
[3] Includes *Guchinodon* Trofimov, 1978 (see Kielan-Jaworowska and Dashzeveg, 1998).

DISTRIBUTION

Eutriconodontans hold the distinction of including both the first known specimen (a dentary now referred to the "amphilestid" *Amphilestes broderipii*, see Simpson, 1928a) and first described species (the "amphilestid" *Phascolotherium bucklandi*, credited to Broderip, 1828) of Mesozoic mammals. Both are from the Stonesfield Slate, Britain, of Bathonian (Middle Jurassic) age. Both taxa are known by multiple, fairly complete dentaries and hence are the geologically oldest eutriconodontans that are reasonably well represented. Other occurrences of eutriconodontans in beds possibly or probably of Middle Jurassic age include a dentary of the "amphilestid" *Liaotherium gracile* in the Haifanggou Formation, Liaoning Province, China (Zhou et al., 1991); and two undescribed taxa from the La Boca Formation, Tamaulipas, Mexico: an "amphilestid" and a possible triconodontid, each known by both maxillary and dentary fragments (Clark et al., 1994; Montellano et al., 1998). *Paikasigudodon*, from the Kota Formation

(?Early Jurassic) of India (Prasad and Manhas, 2002), is probably somewhat older, but is known only by a single tooth, and its referral to "Amphilestidae" is very tentative.

Eutriconodontans were most abundant and diverse in the Late Jurassic–earliest Cretaceous, where they are known from North America, Western Europe, Asia, and Africa. Their robust construction, the large size of the remains, and taxonomic diversity all contributed to their relatively good fossil representation (e.g., Osborn, 1907: figure 4). Triconodontidae make their first unambiguous appearance in the Late Jurassic (note possible occurrence in the ?Middle Jurassic La Boca Formation, mentioned previously). Two structurally similar genera of triconodontids, *Priacodon* and *Trioracodon*, are known from the Late Jurassic of the United States. Both are well represented by fossils; cranial and postcranial remains are known for *Priacodon* (Simpson, 1929a; Rasmussen and Callison, 1981; Rougier, Wible, and Hopson, 1996). *Trioracodon* is shared with the Purbeck (basal Cretaceous) of Britain (Simpson, 1928a), and *Priacodon* has been tentatively reported from rocks of that age in Portugal (Krusat, 1989). The Purbeck genera, *Trioracodon* and *Triconodon*, are each represented by large series of specimens; *Triconodon* is known by cranial remains (Kermack, 1963) and an ontogenetic series documenting tooth eruption and replacement (Simpson, 1928a). In light of the well-documented faunal similarities between the Morrison and Purbeck, it is worth pointing out that three genera of "amphilestids" (*Aploconodon, Comodon, Triconolestes*) are known from the Morrison, but none are known from the Purbeck.

Three named eutriconodontans, each based on isolated teeth and each of highly uncertain affinities, are known from Africa: *Tendagurodon janenschi*, from Tendaguru (Late Jurassic), Tanzania (Heinrich, 1998); and *Dyskritodon amazighi* and *Ichthyoconodon jaworowskorum*, from the Berriasian of Talsinnt, Morocco (Sigogneau-Russell, 1995b). We tentatively place *Tendagurodon* among eutriconodontans of "amphilestid" grade; the other two are morphologically divergent, though molars of more typical "amphilestid" pattern have also been reported from Talsinnt (Sigogneau-Russell et al., 1988; Sigogneau-Russell et al., 1990). Also from a Gondwanan landmass and of highly enigmatic affinities is *Austrotriconodon*, known by two species (each represented by isolated teeth) from the Campanian–Maastrichtian Los Alamitos Formation, Argentina (Bonaparte, 1986a, 1992).

The record from the Late Jurassic–Early Cretaceous of Asia includes several occurrences of Eutriconodonta. Most noteworthy among these is *Jeholodens jenkinsi*, a triconodontid or proximal relative of Triconodontidae, known by a spectacularly complete skeleton from the Barremian of Liaoning Province, China (Ji et al., 1999). Several "amphilestids," represented by dentaries and incomplete lower dentitions, are also known (e.g., Chow and Rich, 1984a; Zhou et al., 1991; Rougier et al., 1999), as are three genera of Gobiconodontidae. Of these, *Hangjinia* is represented by a single dentary (Godefroit and Guo, 1999), but skulls and skeletal material from the Early Cretaceous of China are known for *Gobiconodon* and *Repenomamus* (Li et al., 2001; Wang, Hu, Meng et al., 2001). *Gobiconodon* itself is of particular interest because of its broad geographic and temporal distribution; originally described from the ?Aptian–Albian of Mongolia (Trofimov, 1978; Kielan-Jaworowska and Dashzeveg, 1998), occurrences now include: Valanginian, Mongolia (Rougier et al., 2001); Barremian, China (Wang, Hu, Meng et al., 2001) and Spain (Cuenca-Bescós and Canudo, 1999); Albian, Russia (Maschenko and Lopatin, 1998), Albian, Gansu Province of China (Tang et al. 2001), and Aptian–Albian, Montana (Jenkins and Schaff, 1988). Parts of the skull and postcranial skeleton have been described for North American *Gobiconodon ostromi*.

On Laurasian landmasses, at least (the record from southern continents remains too poor to comment on) *Gobiconodon* appears to represent a specialized, remnant lineage of eutriconodontans that lasted to the late Early Cretaceous, when they briefly flourished in Asia. A growing body of evidence suggests that Triconodontidae enjoyed a modest radiation in North America well after they had become extinct in Europe and Asia. Data in hand suggest that these triconodontids form a monophyletic clade, Alticonodontinae, that was endemic to the Cretaceous of North America. One genus (*Arundelconodon*) has been reported from the Aptian of Maryland (Cifelli, Lipka et al., 1999; Rose et al., 2001); another (*Astroconodon*) is known from Aptian–Albian faunas of Texas and Oklahoma (Patterson, 1951; Slaughter, 1969; Turnbull and Cifelli, 1999). The contemporaneous Cloverly Formation, Montana, has yielded at least three taxa (*Astroconodon, Corviconodon*, and an unidentified triconodontid, Cifelli et al., 1998), as has the slightly younger (Albian–Cenomanian) Cedar Mountain Formation, Utah (*Astroconodon, Corviconodon*, and *Jugulator*, Cifelli and Madsen, 1998). Triconodontids are absent in geologically younger faunas of the southern part of the Western Interior (Cifelli and Gordon, 1999), making their last North American appearance in the early Campanian of Alberta (*Alticonodon*, Fox, 1969, 1976a).

ANATOMY

SKULL

The cranium of eutriconodontans is poorly known. The single specimen of *Jeholodens jenkinsi* includes a rather complete but badly flattened skull; more fragmentary materials are known for *Triconodon, Trioracodon, Priacodon,* and *Gobiconodon* (Simpson, 1928a; Kermack, 1963; Jenkins and Schaff, 1988; Rougier, Wible, and Hopson, 1996; Engelmann and Callison, 1998). A few preliminary observations have been published on an undescribed taxon, currently under study, from the Cloverly Formation (e.g., Crompton and Jenkins, 1979; Jenkins and Crompton, 1979). Limited by the available fossils, published information on the eutriconodontan skull has been largely confined to the petrosal and surrounding parts of the basicranium (Kermack, 1963; Crompton and Jenkins, 1979; Crompton and Sun, 1985; Crompton and Luo, 1993; Wible and Hopson, 1993; Rougier, Wible, and Hopson, 1996). Most recently, Wang, Hu, Meng et al. (2001) have reported the exciting discovery of remarkably complete, undistorted skulls of the gobiconodontids *Gobiconodon* and *Repenomamus.* The specimens remain largely undescribed at this writing, but some details may be gleaned from their illustrations and data matrix. The following account is based on the foregoing references.

The skull of known eutriconodontans is notable in its massive construction and large size relative to the skeleton (figures 7.1, 7.3A,B). The septomaxilla, an ossification forming the inferior and/or posterior border of the external naris, is present in gobiconodontids. This element, present in nonmammalian synapsids, various Mesozoic mammals (e.g., *Sinoconodon,* docodontans, morganucodontids, *Vincelestes,* see Wible et al., 1990; Lillegraven and Krusat, 1991), and monotremes is lacking in living therians. The secondary palate extends well posterior to the end of the maxillary tooth row (figure 7.3B). As in all

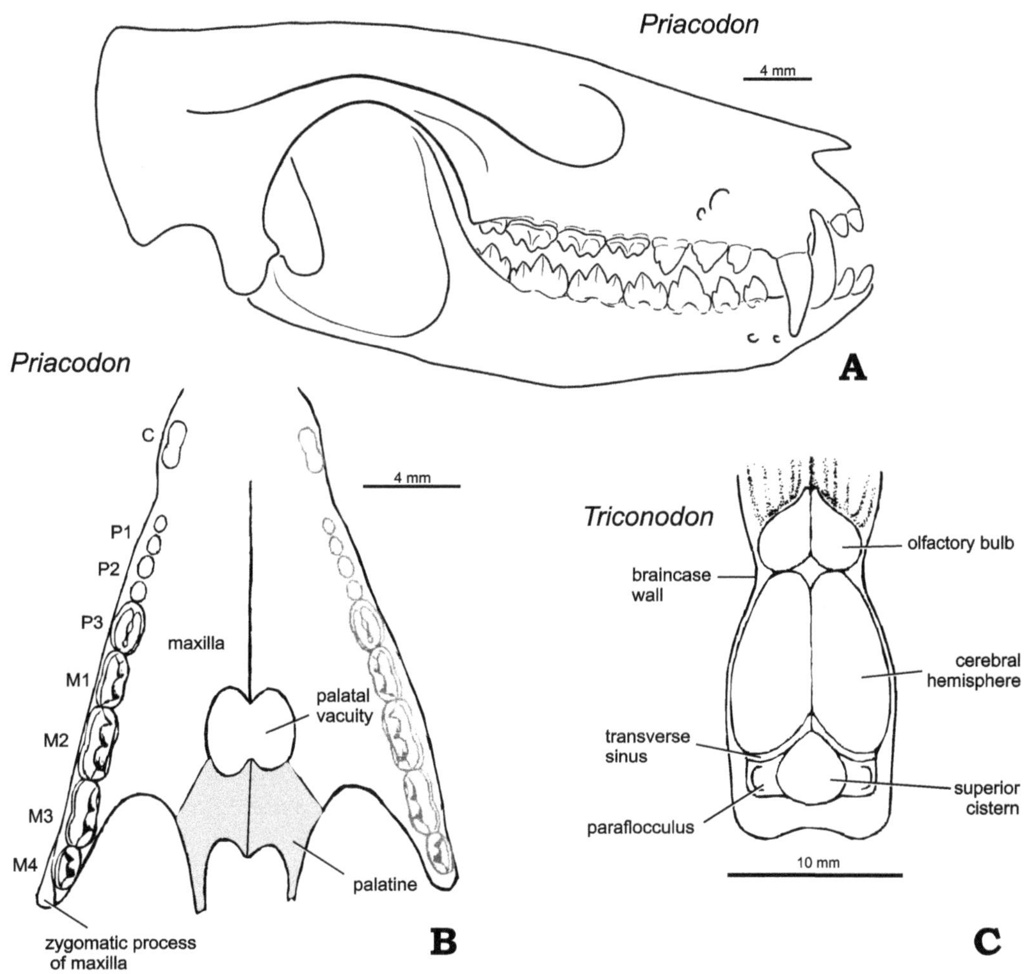

FIGURE 7.3. Skull and endocast of eutriconodontans. A, Skull of *Priacodon* (composite restoration based on *P. ferox* and *P. fruitaensis*). B, Palate of *Priacodon* (restoration based mainly on *P. ferox*). This restoration, though incorrect in some details, clearly shows the zygomatic process of the maxilla (which supports about half of the molar series), a diagnostic feature of Triconodontidae. C, Endocast of *Triconodon mordax* in dorsal view; note the large superior cistern, so that the mesencephalon and the vermis are not exposed in dorsal view. *Source:* B, modified from Simpson (1929a); C, modified from Kielan-Jaworowska (1986).

mammals, the orbital sidewall is completely ossified and the maxilla forms a well-defined edge along the subtemporal margin; a large sphenopalatine foramen as seen in tritylodontids is lacking. The braincase is rather narrow in *Repenomamus*, in this respect resembling stem mammals such as morganucodontids and differing from the more expanded braincase seen in Triconodontidae, multituberculates, and extant Mammalia. Posterolaterally, the squamosal does not extend into the cranial wall, being excluded by the petrosal. This is a primitive condition shared with nonmammalian cynodonts and most early mammals. By contrast, the cranial moiety of the squamosal contributes to the cranial wall in extant Theria and the "eupantotherian" *Vincelestes.*

Considerable, perhaps excessive, attention has been paid to the construction of the sidewall of the braincase in early mammals (see discussion in chapter 3). In eutriconodontans, a large part of this is made up of an extension, or anterior lamina of the petrosal (Kermack, 1963; Crompton and Jenkins, 1979; see figure 7.4A), as is also the case in stem mammals, multituberculates, and monotremes (see Watson, 1916; Kielan-Jaworowska, 1970b). In living therians, by contrast, the anterior lamina of the petrosal is vestigial or absent (figure 7.4B), and the sidewall of the braincase is formed by an enlarged dorsal part, or ascending process, of the alisphenoid (which surrounds important orienting structures, the foramina for the trigeminal nerve). This difference and molar pattern were among the primary bases for recognition of a Late Triassic "prototherian-therian" dichotomy in mammalian evolution, with placement of triconodontids among "prototherians" or "atherians" (e.g., Kermack, 1967a; Kermack and Kielan-Jaworowska, 1971). In light of embryological observations demonstrating a common development of the two structures (anterior lamina of the petrosal and ascending process of the alisphenoid), the significance of this distinction remains open to question (Presley and Steel, 1976; Presley, 1980, 1981). Furthermore, it now appears that reduction of the anterior lamina of the petrosal and concomitant development of an enlarged ascending process of the alisphenoid took place later in mammalian evolution than previously appreciated; a large anterior lamina of the petrosal is present in the prototribosphenidan *Vincelestes* (see chapter 10).

Early observations (Kermack, 1963) also suggested that "triconodonts" resemble advanced cynodonts in the persistence of a ventral (or anteroventral) extracranial opening for the cavum epiptericum (a space housing the trigeminal ganglion). The cavum epiptericum retains a limited anteroventral opening in triconodontids, but lacks a large opening for the trigeminal ganglion and is partially enclosed by the petrosal. In these respects it resembles the conditions in multituberculates and the prototribosphenidan *Vincelestes* and is distinct from what is seen in cynodonts, morganucodontids, and monotremes (e.g., Crompton and Sun, 1985; Crompton and Luo, 1993; Wible and Hopson, 1993; Rougier, Wible, and Hopson, 1996). This flooring for the cavum epiptericum is reflected in the development of a prominent lateral trough in the eutriconodontan petrosal. The lateral trough is virtually ubiquitous among early mammals, though lacking in boreosphenidans (a presumed apomorphy of the latter group). The lateral trough of eutriconodontans is perforated by an opening (hiatus Fallopii) for the facial nerve (figure 7.4C), as in morganucodontids and monotremes; in multituberculates and living therians, the hiatus Fallopii generally lies at the anterior margin or roof of the petrosal. A related feature is the pila antotica, an ossification of the braincase that separates the cavum epiptericum from the cranial cavity in cynodonts, morganucodontids, docodontans, and multituberculates, which is absent in triconodontids, as it is in living mammals (Wible and Hopson, 1993).

The petrosal of eutriconodontans (figure 7.4C) has a prominent lateral flange, developed as a ventrally directed crest, which is a plesiomorphic feature found in all mammals except Crown Theria. As in stem mammals (and in contrast to multituberculates and living mammal groups), the alisphenoid of eutriconodontans has a prominent quadrate ramus that overlaps the lateral flange of the petrosal. The paroccipital process is bifurcate and the posterior paroccipital process projects ventrally, both of which are primitive conditions shared with most nontrechnotherian mammals. A distinct crista parotica, seen in all mammals except *Sinoconodon,* is present. The lateral flange and crista parotica are broadly separated in eutriconodontans, a synapomorphy according to Rougier, Wible, and Hopson (1996). The crista interfenestralis, which is tall and horizontal, extends to the base of the paroccipital process, a primitive condition seen in nonmammalian synapsids and in all mammals except for cimolodontan multituberculates and Trechnotheria (see chapter 9). A posttympanic recess and caudal tympanic process, both apomorphies of Trechnotheria (Luo et al., 2002), are lacking.

The glenoid fossa of the temporomandibular joint is well developed, present either on the ventral side of the zygoma or on a platform of the zygoma. A postglenoid depression for the external auditory meatus, lacking in morganucodontids, monotremes, and multituberculates, is variable among eutriconodontans, being present in *Repenomamus* but absent in *Jeholodens.* The temporomandibular joint retains a primitive position, lateral to the fenestra vestibuli (also seen in morganucodontans, doco-

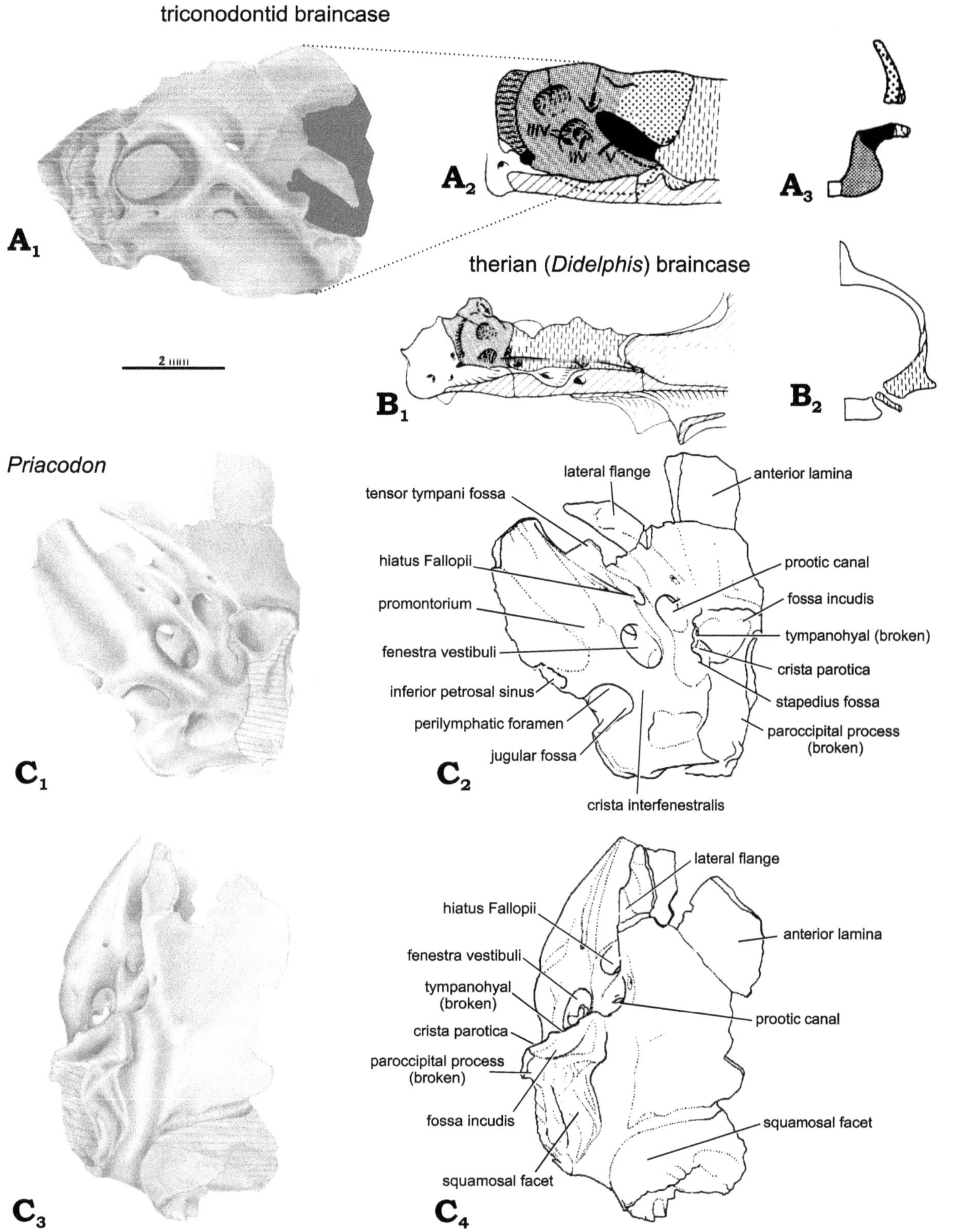

FIGURE 7.4. Side wall of braincase and petrosal. A, Triconodontidae (composite). B, *Didelphis* sp. Construction of the sidewall of the braincase (right side) and relationship to cranial nerve landmarks (A_{1-2}, B_1, internal view, anterior is to the right; A_3, B_2, cross section). In Triconodontidae (A), note the presence of an anterior extension (anterior lamina: stippled pattern) of the petrosal, forming part of the side wall of the braincase and incorporating an exit foramen for at least one of the branches (V_3) of the trigeminal nerve, and the presence of a floor for the cavum epiptericum (black). Both of these features are plesiomorphies lacking in *Didelphis,* where the side wall of the braincase is formed by the alisphenoid (vertical dashed lines). C, Petrosal of *Priacodon fruitaensis* in ventral (C_{1-2}), and lateral (C_{3-4}) views. *Source:* A_{2-3}, B, modified from Crompton and Jenkins (1979); A_1, C, modified from Rougier, Wible, and Hopson (1996).

dontans, monotremes, and some multituberculates), rather than being located anterior to it (as is the case for *Hadrocodium*, some multituberculates, and trechnotherians). This condition appears to be related to relative development of the brain in early mammals (Rowe, 1996a,b; Luo, Crompton, and Sun, 2001). A related feature is that the fossa incudis lies immediately medial to the temporomandibular joint (Wang, Hu, Meng et al., 2001). The presence of the fossa incudis, together with evidence from the lower jaw, indicates that the postdentary elements of eutriconodontans were still connected anterior to the horizontal ramus of the dentary via Meckel's cartilage but had separated mediolaterally from the angular region of the dentary. This is a precursor condition to the typically mammalian condition of the middle ear. The position of the fossa incudis, at least, appears to be structurally intermediate between the condition seen in stem mammals (such as *Morganucodon*), which retain the incus (quadrate) as part of a dual joint between the cranium and the mandible, and that of more advanced taxa (such as Crown Mammalia). A notable correlated plesiomorphy, seen in *Repenomamus*, is the fact that the incus (quadrate) retained at least some contact with the squamosal, as in nonmammalian synapsids, *Morganucodon*, and *Sinoconodon* and unlike all other mammals for which the condition is known.

Like other mammals and in contrast to nonmammalian synapsids, the pars cochlearis of the eutriconodont petrosal (figure 7.4C) is developed ventrolaterally into a distinct, well-exposed promontorium. The promontorium is elongate and fingerlike, with a steep lateral wall, as it is in almost all noncladotherian mammals (Luo et al., 2002), contrasting with the more bulbous-shaped promontorium of living therians and their proximal relatives. Presence and development of the promontorium are related to the size and configuration of the cochlea and, in turn, reflect hearing function. As inferred from the presence of a promontorium, eutriconodontans and stem mammals may have had some capacity for high-frequency hearing (Graybeal et al., 1989; Rosowski and Graybeal, 1991; Rosowski, 1992), but the cochlea presumably was straight or only slightly curved, not coiled as it is in living mammals. The promontorium of eutriconodontans, like that of all mammals save eutherians and *Vincelestes*, lacks a sulcus (or canal) for the stapedial artery and a transpromontorial sulcus for the internal carotid artery.

The brain endocast is known for *Triconodon* (Simpson, 1927a, 1928a, see our figure 7.3C). The olfactory lobe is large, with a teardrop-shaped outline. The cerebral hemisphere is long, oval, and flat, lacking the inflated appearance that is typical of monotremes, multituberculates, and therians (Kielan-Jaworowska, 1986). The cerebrum is neither expanded anteriorly to overlap the posterior part of the olfactory lobe, nor is it hemispherical. The endocast of *Triconodon* resembles that of multituberculates (Kielan-Jaworowska, 1997) in the presence of a large, roughly triangular, vermislike bulge, now recognized as a cast of a large vessel, the superior cistern (Kielan-Jaworowska and Lancaster, 2004, see also chapters 3 and 8). The superior cistern obscures the vermis and midbrain on the endocasts. The midbrain apparently was exposed on the dorsal side of the brain as in all primitive mammals (Bauchot and Stephan 1967; Jerison, 1973; Kermack et al., 1981; Kielan-Jaworowska, 1984c, 1986) and cynodonts (Quiroga, 1979a,b, 1980).

MANDIBLE

Most or all of the dentary is known for the eutriconodontan *Jeholodens*, the "amphilestids" *Amphilestes* and *Phascolotherium* (Owen, 1871; Osborn, 1888b; Simpson, 1928a), the gobiconodontids *Gobiconodon, Hangjinia,* and *Repenomamus* (Trofimov, 1978; Jenkins and Schaff, 1988; Kielan-Jaworowska and Dashzeveg, 1998; Godefroit and Guo, 1999; Wang, Hu, Meng et al., 2001), and the triconodontids *Corviconodon, Priacodon, Triconodon,* and *Trioracodon* (Owen, 1871; Osborn, 1888b; Simpson, 1925b,c, 1928a, 1929a; Cifelli et al., 1998; Ji et al., 1999). Dentaries of two taxa (*Jeholodens* and *Hangjinia*) probably belong to juveniles, so that some of the following generalizations cannot be thoroughly evaluated for them.

By comparison to stem mammals (figure 7.5A), "symmetrodontans," and "eupantotherians," the dentary of eutriconodontans (figure 7.5B–D) is of robust construction and the tooth-bearing part generally occupies less than 60% of the total length of the element (a notable exception is *Amphilestes*, a small taxon with a comparatively gracile dentary bearing a proportionately longer tooth row that is approximately 67% of the total dentary length). In Gobiconodontidae and some Triconodontidae, the dentary increases in depth dorsoventrally as it approaches the anterior margin of the coronoid process, which is posteriorly recumbent and, except for *Amphilestes*, is notable in its relatively great height and anteroposterior length. The masseteric fossa is deep and well marked, especially ventrally, where the inferior margin of the dentary is reflected laterally (e.g., Simpson, 1928a). Medially, a large, well-defined pterygoid fossa is present, with its inferior border developed into a shelflike crest that extends posteriorly to the condyle. Meckel's groove is generally well defined, though it is lacking in most Alticonodontinae in which the condition can be judged (Cifelli, Lipka et al., 1999). There

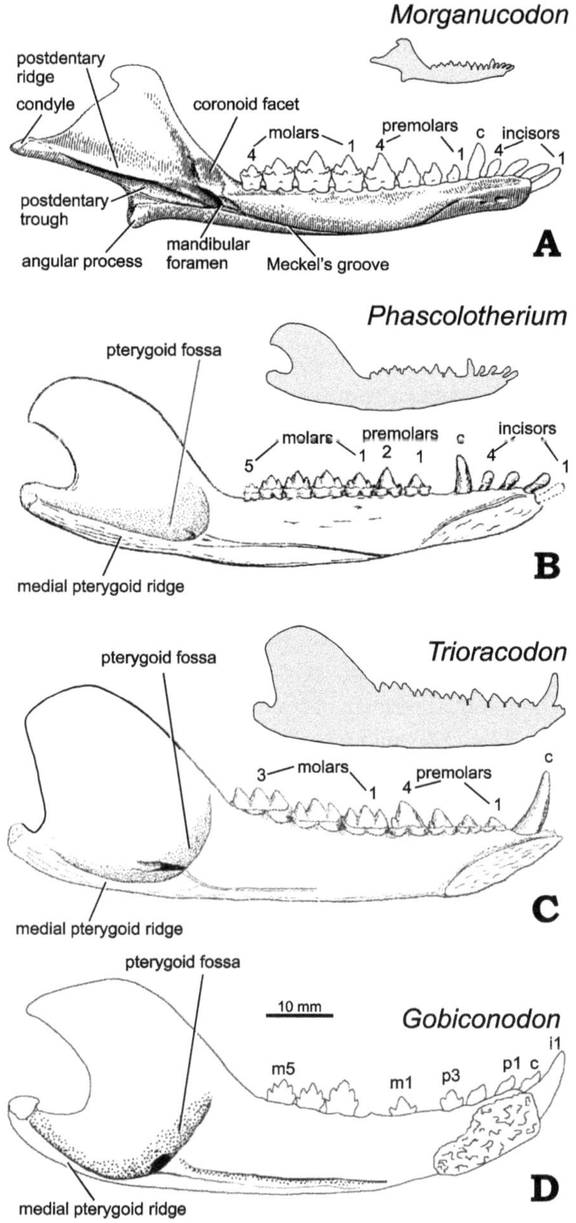

FIGURE 7.5. Tooth series and dentary in lingual view. A, *Morganucodon watsoni* (Morganucodonta). B, *Phascolotherium bucklandi* (Eutriconodonta, "Amphilestidae"). C, *Trioracodon ferox* (Eutriconodonta: Triconodontidae). D, *Gobiconodon ostromi* (Eutriconodonta, Gobiconodontidae). *Gobiconodon* (D) is depicted at actual size; silhouettes to the upper right of drawings A–C also show actual size. The loss of the postdentary trough, ridge, and associated features sharply distinguish Eutriconodonta (B–D) from stem mammals such as morganucodontans (A). Many eutriconodontans retain Meckel's groove (especially broad in *Gobiconodon*, wherein Meckel's cartilage was apparently ossified, see Wang, Hu, Meng et al., 2001); the presence of a pterygoid fossa and pterygoid shelf are apomorphies with respect to stem mammals such as Morganucodonta. *Source:* A, modified from Kermack et al. (1973); B, C, modified from Simpson (1928a); D, modified from Jenkins and Schaff (1988).

is no development of the angular process in any position, with the possible exception of *Amphilestes.*[2]

Meckel's groove, which was apparently lost within Eutriconodonta, as it was in many other groups of early mammals (e.g., Cifelli and Madsen, 1999), is particularly broad in Gobiconodontidae, wherein it housed a substantial element, interpreted as an ossified Meckel's cartilage (Wang, Hu, Meng et al., 2001).[3] Loss of the coronoid and the splenial (paradentary elements), as evidenced by absence of the attachment scars on the medial surface of the dentary, is another iterative theme in the evolution of Mesozoic mammals. The coronoid appears to have been present in *Gobiconodon* and *Hangjinia* (Jenkins and Schaff, 1988; Kielan-Jaworowska and Dashzeveg, 1998; Godefroit and Guo, 1999), but not in *Repenomamus* (Wang, Hu, Meng et al., 2001) or in any triconodontid. Loss of this element, at least, might be taken as an equivocal synapomporphy of the latter family. More definitively, and of greater phylogenetic significance, is the fact that there is no evidence for persisting attachment of the postdentary elements (articular, prearticular, angular, surangular) to the dentary. This is a substantial departure from the condition in morganucodontids and other "triconodont-like" among stem mammal lineages, wherein the integration of postdentary elements with the dentary is indicated by attachment scars (or preservation of the elements themselves), the postdentary trough and overhanging ridge, and emargination of the posteroventral border of the dentary to accommodate the reflected lamina of the angular (e.g., Allin and Hopson, 1992; Crompton and Luo, 1993).

[2] Early illustrations (e.g., Owen, 1871: pl. I, figure 25; Osborn, 1888b: pl. VIII, figure 1; Osborn, 1907: figure 5) of the holotype of *A. broderipii* show a posteriorly placed angular process, as also suggested by Owen's description: "In the present jaw the condyloid and coronoid processes are wanting, but have left their impressions on the matrix; there is the same wide and shallow groove near the lower margin of the hind part of the ramus . . . as in *Amphitherium Prevostii*" (Owen, 1871: 16). The issue has not been revisited since the authoritative view of Simpson (1928a: 72) on the matter: "The angular region is not well shown, but certainly was, as in other triconodonts, without any true angle but with a low, sharp pterygoid crest running back to the condyle."

[3] Homology of the groove on the medial side of the dentary in early mammals, now widely accepted as having housed Meckel's cartilage and therefore termed Meckel's groove, was long uncertain. Simpson (1928b) advocated usage of the neutral term "internal mandibular groove" for this structure, though he favored the interpretation that it housed an artery and associated nerve.

DENTITION

Incisors, Canines, and Premolars. Incisors are poorly known for Eutriconodonta, as they are for the vast majority of Mesozoic mammals. *Phascolotherium* (figure 7.5B) had four stout, columnar incisors with subspatulate tips (Simpson, 1928a). Only the distalmost of an unknown number of incisors is known for *Amphilestes;* the last incisor is similar to that of *Phascolotherium.* Similarly, only the distalmost lower incisor, also a subspatulate tooth, is known for *Triconodon;* Simpson (1928a) speculated that three or four may have been present. The maximum number of known incisors among eutriconodontans occurs in *Jeholodens* (figure 7.13E), in which there are four each in the upper and lower dentition; the incisors are uniquely derived in having spoon-shaped crowns with a median ridge. *Priacodon* (figure 7.3A) had two lower incisors (Rasmussen and Callison, 1981). In alticonodontines, the single lower incisor is anteriorly procumbent (Turnbull and Cifelli, 1999); in *Jugulator* (figure 7.16F), at least, there is a single, rather large, complex, and presumably procumbent lower incisor with a sharp, mittenlike crown (Cifelli and Madsen, 1998). The canine of most eutriconodontans is large to very large and recurved, with sharp edges; the number of roots is variable: the lower canine is single-rooted in *Amphilestes, Priacodon,* and *Triconodon* (though the dc is single-rooted in *Triconodon*).

The anterior dentition of Gobiconodontidae (figure 7.5D) is uniquely specialized, and tooth homologies are problematic; herein we follow the interpretation of Jenkins and Schaff (1988). In *Gobiconodon,* the single lower incisor is greatly enlarged, procumbent, and caniniform, with its root extending well back into the dentary, to a point under p2. A small diastema separates i1 from the canine, which is much smaller and more erect. The i1 fit into a deep palatal fossa in the premaxilla. The anterior upper dentition includes a small I1 and an enlarged and caniniform I2. As with the lower dentition, the upper canine is separated by a diastema from the incisor(s) and is short and semierect. *Repenomamus* and *Hangjinia* each had three lower incisors, with the first somewhat enlarged; the canine is reduced (Godefroit and Guo, 1999; Wang, Hu, Meng et al., 2001).

Amphilestes has four lower premolars, increasing in size posteriorly through the series. In lateral view (figure 7.6A), they are rather simple and symmetrical in appearance; the crown is formed by a tall, sharp cusp a, with small basal and subequal cusps b and c present on each (see figures 4.1 and 7.2 for an explanation of the cusp-numbering system). Cingulids are lacking. Only two premolars are present in *Phascolestes;* here, lingual cingula are present and the crowns are rather similar to those of *Amphilestes,* though cusp c is somewhat larger than cusp b. *Jeholodens* is similar to *Phascolestes* in number and morphology of premolars, except that the lingual cingulum is lacking. Premolars in Triconodontidae vary in number from three (*Priacodon*) to four or more (*Triconodon, Trioracodon,* and Alticonodontinae, where known). All are four-cusped, two-rooted, and generally have partial or complete labial cingula/cingulids. Cusps B/b and D/d tend to be small basal cusps associated with the cingulum/cingulid, but cusp C/c is placed higher on the crown, particularly on the penultimate and ultimate premolars (where cusps B/b tend to assume an even more basal position than cusps D/d), so that the teeth have a pronounced asymmetry in lateral view (figure 7.6B). The teeth increase in height through the series, with the penultimate and ultimate premolars having a tall, piercing cusp A/a; in alticonodontines, tooth length remains much the same through the premolar series. *Gobiconodon* has three or four premolars, which are simple, single-rooted (except for the last premolar of *G. hoburensis*), and slightly procumbent anteriorly, becoming more erect, with better-defined basal cuspules in the posterior part of the series (see Jenkins and Schaff, 1988; Kielan-Jaworowska and Dashzeveg, 1998).

Molars. The basic eutriconodont molar pattern, as seen in "amphilestids" (figure 7.6A) consists of three serially arranged cusps, A–C/a–c, placed on a laterally compressed crown that is supported by robust long roots. A cingulum is usually present labially on upper molars and a cingulid lingually on lowers; basal cusp D/d is present posteriorly, and cuspules e and f are primitively present (though lost in one or more groups) at the mesiolingual and mesiolabial corners, respectively, of the lower molars. Other cingular cusps, commonly seen in morganucodontids, are lacking. Cusp A/a is much the largest cusp on "amphilestid" molars; cusps B/b and C/c are much lower and are subequal in development. The pattern is similar in Gobiconodontidae, except that cusp C/c tends to be larger and placed higher on the crown than cusp B/b, recalling somewhat the appearance of triconodontid premolars. Variants include Austrotriconodontidae, wherein cusp B is the tallest of upper molar cusps (Bonaparte, 1992); *Dyskritodon,* with lower molars dominated by cusps a and c (cusp b is greatly reduced); and *Ichthyoconodon,* in which cusp c is tallest on lower molars (Sigogneau-Russell, 1995). Gobiconodontid upper molars (known for *Gobiconodon* and *Repenomanus*) are unique among eutriconodontans in being noticeably more labiolingually transverse than the lowers, with a strong cingulum present both lingually and labially, and in having a slight angulation of primary cusps A–C, increasing from M3 to M5. In triconodontids (figure 7.6B), cusps A–C and a–c are subequal in

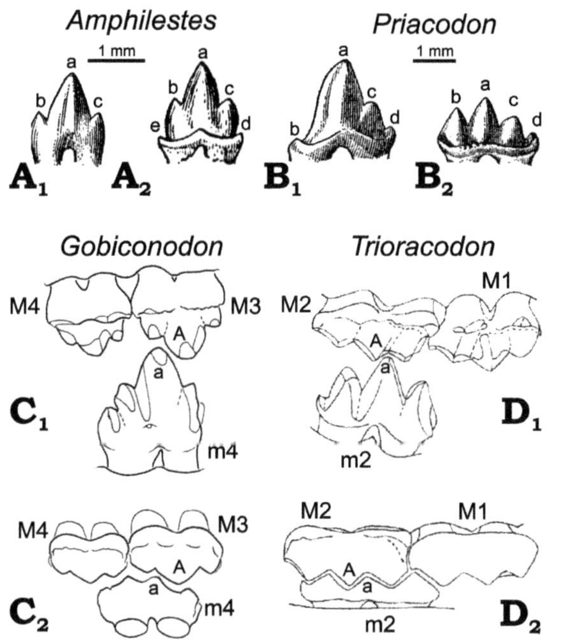

FIGURE 7.6. Cheek teeth and occlusion in Eutriconodonta. A, B, Comparative morphology of posterior premolars and molars in "Amphilestidae." A, *Amphilestes broderipii.* B, Triconodontidae (B_1, *Priacodon robustus;* B_2, *Priacodon* sp.), lingual view. Note the relative symmetry of cusps in the last premolar of *Amphilestes* (A_1) compared to *Priacodon* (B_1), where cusp b is much lower on the crown than cusp c. The molar pattern of "amphilestids" (A_2) is primitive with respect to that of triconodontids (B_2) in that cusps b and c are significantly lower than principal cusp a, anterolingual cusp e is retained, and cusp d is small and basally placed. C,D, Molar occlusion in *Gobiconodon ostromi* (C) and *Trioracodon bisulcus* (D); anterior is to the right. C_1, D_1, Oblique view showing relationships of occlusal surfaces (labial side of lower molars, lingual side of uppers, depicted as left and right teeth, respectively). C_2, D_2, Labial view parallel to shearing surfaces, showing molar series in occlusion. *Trioracodon* illustrates the one-to-one pattern, wherein lower molar cusp a occludes immediately mesial to upper molar cusp A. Each principal cusp bears two shearing facets (D_1) that occlude with counterparts in the opposing molar series, forming a zigzag pattern of shearing surfaces (D_2). Two-to-one (embrasure) occlusion is seen in *Gobiconodon,* where cusps A, a occlude more or less between molars of the opposing series. Figures C and D are not to scale. *Source:* A, B, from Simpson (1929a); D, from Crompton (1974).

size; *Jeholodens* is intermediate between gobiconodontids and triconodontids in this respect (Ji et al., 1999). Among advanced triconodontids, subfamily Alticonodontinae, cusp d of the lower molars becomes enlarged and the crown is effectively comprised of four cusps that are subequal in height.

Interlocking of the lower molar series is almost invariably present in Eutriconodonta, but is achieved in somewhat different ways among the eutriconodontan taxa. In *Dyskritodon* and some "amphilestids," cusp d fits into a small embayment between cusps e (the mesiolingual cingular cusp) and b of the succeeding tooth, as it does in *Morganucodon* and *Megazostrodon.* In other "amphilestids" (such as *Amphilestes,* see Crompton, 1974; E. F. Freeman, 1979) and Gobiconodontidae, cusp d fits between cusps e and f (mesiolingual and mesiolabial cingular cusps, respectively), of the succeeding molar (as is also the case in *Dinnetherium*). In *Klamelia,* the molars imbricate, with cusp d being placed labial to cusp b of the succeeding tooth. The most distinctive pattern of molar interlocking is found among *Jeholodens* and Triconodontidae, where the mesiolingual cuspules are lacking and cusp d fits into an underhanging embayment in the front of cusp b of the succeeding molar. This feature is moderately expressed in Triconodontinae, but in Alticonodontinae, the embayment of cusp b is formed into a sharp, V-shaped notch. The mesial part of the underlying root has a prominent groove throughout its length (figure 7.16E), and the posterior edge of the distal molar root bears a similarly well-developed, longitudinal ridge. The roots are therefore secured into their respective alveoli through tongue-in-groove joints. These features are highly distinctive, allowing alticonodontines to be recognized by the shape of their molar alveoli alone. The extensive interlocking system appears to be related to a derived pattern of occlusion seen in these taxa (see later).

Molar occlusion in eutriconodontans is rather simple on account of the simplicity of the molar pattern itself. Two major patterns can be recognized, the major differences in the two being related to the mesiodistal offset of corresponding upper and lower molars. The lingual sides of upper molar cusps occlude with labial sides of the lower molars. Lower molars are always positioned mesial to their counterparts in the upper jaw, but the extent of offset varies. In triconodontids (figure 7.6D), upper molar cusp A occludes between cusps a and c of the corresponding lower molar; lower molar cusp a occludes with the concavity formed between upper molar cusps A and B. Each cusp thus develops two wear facets, corresponding to the faces of the two adjacent cusps forming the embayment of the opposing tooth. The result is a series of shearing surfaces arranged in a zigzag pattern (figure 7.6D_2), analogous to the cutting edges of pinking shears. This pattern (often referred to as one-to-one opposition) is similar to what is seen in the stem mammal *Morganucodon* (see chapter 4 and figure 4.1). In advanced triconodontids (within Alticonodontinae), lower molar cusp d becomes enlarged and is fully incorporated into the functional part of the tooth crown. The proximolingual face of upper molar cusp C occludes with the labial face of this enlarged cusp d, and the distolingual face of cusp C occludes with the labial face of cusp b on the succeeding lower molar, so that there is full continuity of shearing surfaces in the molar series and upper molars bridge adjacent lower molars.

A different pattern of occlusion occurred in "amphilestids" and Gobiconodontidae (figure 7.6C). The offset between corresponding upper and lower molars is greater, so that the center (cusp A) of an upper molar occludes more or less between the corresponding lower molar and the molar that succeeds it. In "amphilestids," cusp A passes distal to lower molar cusp c and occludes with that cusp and the mesiolabial face of cusp d or with cuspule f of the succeeding lower molar. In gobiconodontids (Jenkins and Crompton, 1979; Jenkins and Schaff, 1988; Kielan-Jaworowska and Dashzeveg, 1998), the distolingual face of cusp A occludes with the mesiolabial face of cusp b and with the labial surface of cusp f on the succeeding lower molar. This pattern of occlusion is commonly termed two-to-one molar opposition, or embrasure occlusion, as it is shared by *Kuehneotherium* and other mammals with a reversed-triangle configuration of principal molar cusps; it is also known in the "triconodont-like" stem mammal *Megazostrodon* (Mills, 1971; Crompton, 1974; Jenkins and Crompton, 1979).

Tooth Replacement. Published data on tooth replacement in eutriconodontans are extremely scant. A small series of the triconodontine *Triconodon mordax* enabled Simpson (1928a) to document the replacement of dp4 and dc by successors, with p4 fully emerged and c just coming into use when three (of four) molars were fully erupted (figure 7.7). Replacement of premolars in *Arundelconodon,* and perhaps other Alticonodontinae, appears to have taken place at an earlier ontogenetic age (judged by eruption of the molar sequence) than in *Triconodon* (Rose et al., 2001). The holotype of *Jeholodens jenkinsi* (presumed to be a sister taxon to Triconodontidae) preserves i2 in the process of replacing di2, as m4 is erupting (Ji et al., 1999).

Tooth replacement in Gobiconodontidae is best documented for *Gobiconodon.* Jenkins and Schaff (1988) showed that sequential (front to back) replacement of molariforms occurred in *G. ostromi,* and the same appears to have been true of other species of *Gobiconodon* (Kielan-Jaworowska and Dashzeveg, 1998; Rougier et al., 2001). A similar pattern of molariform replacement has been proposed by Godefroit and Guo (1999) for *Hangjinia chowi,* known from a single rather incomplete specimen from the Early Cretaceous of China. Tooth homologies for this specimen are not clear (see account of *Hangjinia* in "Systematics," later), but the last tooth position represented is probably that of a molariform. Replacement at this locus, at least, is indicated by the presence of a fully formed tooth crown in its crypt, underlying two clearly defined alveoli for the shed tooth that was in the process of being replaced. Finally, though description and illustrations remain to be published, the data matrix of Wang, Hu, Meng, et al. (2001: supplementary information) indicates that the gobiconodontid *Repenomamus* had alternate multiple replacement of the incisors and canines, as in *Sinoconodon* and nonmammalian synapsids (Crompton and Luo, 1993; Zhang et al., 1998). In sum, the limited information available suggests remarkable diversity in patterns of tooth eruption and replacement among eutriconodontans. Replacement of at least one incisor (i2) has been docu-

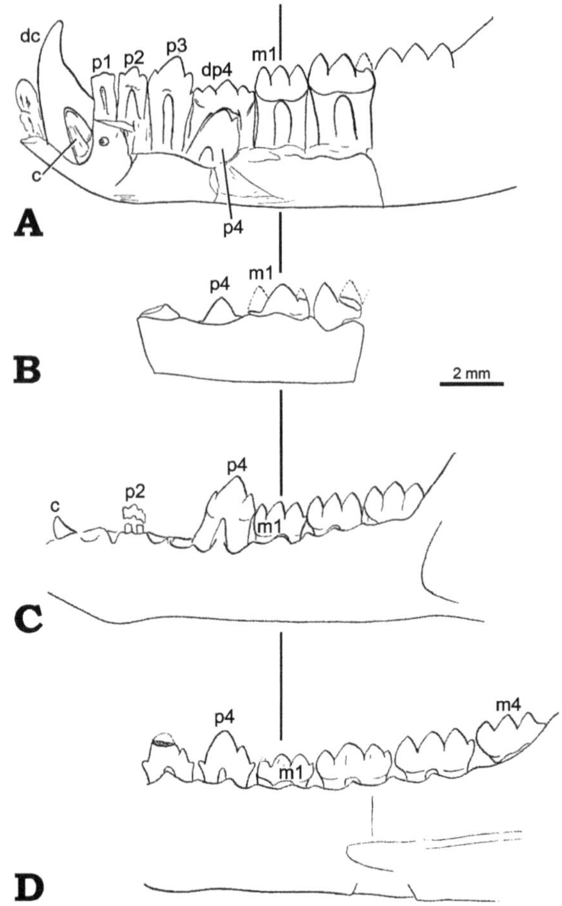

FIGURE 7.7. Tooth eruption and replacement in *Triconodon mordax.* A–D, Labial view of left dentaries, arranged as an ontogenetic series from youngest (A) to oldest (D) individuals. A, The dc and dp4 remain in place, with successor c and p4 visible below; m1–2 are fully erupted and functional. B, The dp4 has been shed, and the successor p4 was in the process of eruption. C, The p4 was fully erupted (partially displaced from alveoli owing to postmortem processes) and C was beginning to erupt; m1–3 were fully erupted and functional, m4 still lay in its crypt. D, All four molars were erupted and in use at time of death. *Source:* modified from Simpson (1928a).

mented for *Jeholodens*, a sister taxon to Triconodontidae, and triconodontids are known to replace at least the canine and last premolar of the lower dentition. By contrast, an essentially cynodont-like replacement of the anterior postcanines is known for Gobiconodontidae, wherein multiple alternate replacement of incisors and canines is known for one genus (*Repenomamus*) and serial replacement of the anterior molariforms has been documented for another (*Gobiconodon*). Whether these are truly primitive conditions or simply atavistic reversals cannot be evaluated with the information that is currently available (see discussion by Jenkins and Schaff, 1988).

Enamel Microstructure. Enamel microstructure of eutriconodontans has thus far been studied only in selected specimens of the gobiconodontid *Gobiconodon ostromi* and the alticonodontines *Astroconodon* sp. and *Jugulator amplissumus* (see Wood et al., 1999, taxon names updated herein). *Gobiconodon* has what has been described as an unusual highly variable type of plesiomorphic prismatic enamel: both sheaths and prisms are present, but they are erratically spaced, with sheaths being of variable size and some seams apparently lacking a visible sheath. *Astroconodon* has enamel with a columnar structure, reminiscent of that of the stem mammal *Sinoconodon;* in *Jugulator*, prism seams are suggested, though they are weaker than in *Gobiconodon*. Wood et al. (1999) concluded that the enamel of eutriconodontans is incipiently prismatic and that it represents a structural transition between synapsid columnar enamel and plesiomorphic prismatic enamel.

POSTCRANIAL SKELETON

Substantial parts of the postcranial skeleton (figures 7.1, 7.8) are known for *Gobiconodon* (Jenkins and Schaff, 1988) and *Jeholodens* (Ji et al., 1999). The holotype of *Priacodon fruitaensis* includes a few associated elements, some of which have been described briefly (Engelmann and Callison, 1998), but add little to knowledge.

The axial skeleton retains some plesiomorphies relative to the conditions in living therians, though most are not unusual in the context of early mammals. The atlas neural arch apparently remained unfused to the centrum, as also seen in cimolodontan multituberculates, the prototribosphenidan *Vincelestes*, and the early eutherian *Asioryctes*. In *Jeholodens*, at least, unfused ribs are present in the entire cervical series, as well as in the lumbar region. *Gobiconodon* is restored as having 13 thoracic vertebrae (Jenkins and Schaff, 1988), as commonly seen among mammals (Slijper, 1946); *Jeholodens* is unusual in having 15, as seen in *Ornithorhychus*. There is no anticlinal vertebra.

The shoulder girdle, best known in *Jeholodens*, is rather similar to that of living therians such as *Didelphis* and contrasts sharply with what is seen in monotremes and nonmammalian cynodonts (Jenkins, 1971b; Jenkins and Parrington, 1976, see our figure 7.8A–C). The scapula is platelike, with a well-developed supraspinous fossa; the spine is strong and possesses a metacromion and a large acromion process. The dorsoposterior part of the supraspinous fossa is somewhat smaller and less fan shaped than in Trechnotheria. The coracoid is represented by a small process that is fused, or at least ankylosed, to the proximal scapula; a procoracoid is lacking. The glenoid is uniformly concave, not saddle shaped as is seen in monotremes and stem mammals. The clavicle is a slender, only slightly curved element; both claviculoacromial and claviculosternal joints were mobile. The interclavicle is present as a separate ossification, as it is in the "symmetrodontan" *Zhangheotherium*, but both the interclavicle and the manubrium of the sternum are small and delicate compared to the massive elements of monotremes and cynodonts. The mobility suggested by the articulations of the clavicle shows that the scapula was capable of fore-aft excursion and that the humerus was able to rotate, with the clavicle acting as a pivotal strut to the interclavicle (this element is fused to the manubrium in adult therians, see Klima, 1987), as in living therians (Jenkins and Weijs, 1979). These features of forelimb locomotion are advanced with respect to what is seen in monotremes (e.g., Jenkins, 1970c).

The humerus retains the primitive condition of a strong degree of proximodistal torsion. Distally, the humerus bears weakly developed ent- and ectepicondyles; weak ulnar and strong radial condyles are present anteriorly, and an incipient ulnar trochlea is present posteroventrally. In these respects, the elbow joint is similar to what is seen in the "symmetrodontan" *Zhangheotherium*, with conditions structurally intermediate between cynodonts, morganucodontids, and monotremes on the one hand and Crown Theria on the other (e.g., Jenkins, 1973; Jenkins and Parrington, 1976). As inferred from the structure of the humerus and elbow joint, in habitual posture the forelimb of *Jeholodens* was probably somewhat abducted, not fully parasagittal. Despite some peculiarities and their great robusticity, the humerus and ulna of *Gobiconodon* (Jenkins and Schaff, 1988) are similar in essential details to those of *Jeholodens*.

In general, the eutriconodontan pelvic girdle and hindlimb contrast with the forelimb in retaining a far greater number of plesiomorphies, more closely resembling those of nonmammalian cynodonts and morganucodontids than those of living therians. The dorsal rim of the acetabulum has an open, V-shaped notch (cotyloid notch of Kühne, 1956) at the contact between ilium and ischium, as in cynodonts, stem mammals, multitubercu-

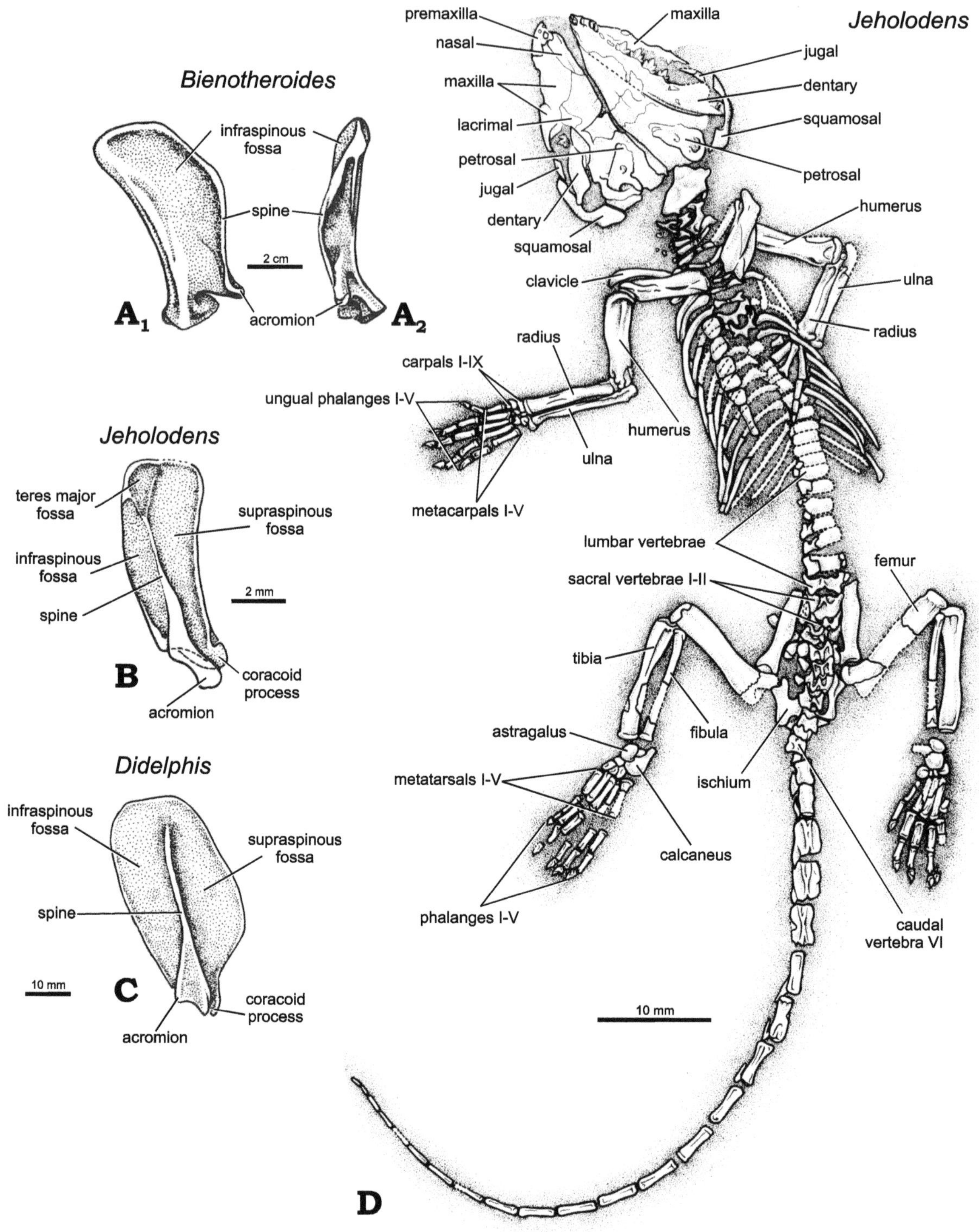

FIGURE 7.8. Postcranial skeleton. A–C, Comparative series of the scapula (all shown as from right side). A, *Bienotheroides wanxianensis* (Cynodontia) in lateral (A_1) and anterior (A_2) views. B, *Jeholodens jenkinsi* (Eutriconodonta, family *incertae sedis;* holotype), lateral view. C, *Didelphis virginiana* (Marsupialia), lateral view. Note that *Jeholodens* possesses a large acromion, as in crown Mammalia, and, like living therians, lacks the procoracoid, has a well-developed supraspinous fossa, and the coracoid is ankylosed to the scapula. D, Holotype skeleton of *Jeholodens jenkinsi* in dorsal view. *Source:* modified from Ji et al. (1999).

lates, and the prototribosphenidan *Vincelestes.* In most trechnotherians, the acetabulum is closed, with a smooth, unbroken, semicircular rim. Within the acetabulum, the sutures of pubis, ilium, and ischium are unfused, contrasting with the fused condition in multituberculates and living mammals. As is the case for living monotremes, marsupials, and most Mesozoic mammals for which the condition is known, an epipubis is present in the pelvis.

The femur lacks a distinct neck. The head is oriented dorsomedially and lacks a fovea for the acetabular ligament; the greater trochanter is oriented dorsolaterally rather than dorsally. The patellar groove is short and broad; in correlation, a cranial crest (cnemial crest of some authors) as such is lacking from the tibia. The tibia possesses a large, hooklike proximolateral tuberosity. These are all plesiomorphic similarities to nonmammalian cynodonts, some stem mammals, and monotremes. The fibula apparently lacked contact with the femur in both *Gobiconodon* and *Jeholodens,* a somewhat unusual condition in view of the fact that this primitive contact is broadly present in early mammals, though lost in some living therians. Both tibia and fibula retained extensive contact with the calcaneus, which is juxtaposed with the astragalus (figure 7.9), rather than lying in partial or full infraposition to the latter element, as in cimolodontan multituberculates (figure 8.14B,D,E) and basal trechnotherians (*Vincelestes* and Crown Theria), respectively. Further, the cuboid is positioned obliquely and medial to the long axis of the calcaneus, metarsal III is oblique to the long axis of the calcaneus, and metatarsal V is separated from the calcaneus by a gap, presumably filled with cartilage in life. These are all similarities to multituberculates and monotremes.

Collectively, features of the hindlimb and pes suggest that the hindlimb was habitually postured in a semi-sprawling position and that the foot was somewhat splayed laterally. In lateral view, terminal (ungual) phalanges of both manus and pes are somewhat curved dorsoventrally. In dorsal view, both sides are convex and mediolateral compression is moderate. Overall, the ungual phalanges are most similar to those of living mammals with generalized terrestrial or fossorial habits (see MacLeod and Rose, 1993; Kielan-Jaworowska and Gambaryan, 1994).

PALEOBIOLOGY

DIET

Although a vaguely triconodont molar pattern is known among some Tertiary therians (see later), there exists no living analog to eutriconodontans in which molars of this pattern occlude. The basis for inference is thus limited. It is clear, however, that all eutriconodontans were strictly or mainly animalivorous, as suggested by the shared presence of long sharp canines; premolars with tall, trenchant main cusps suitable for piercing and grasping prey; limitation of molar function to shearing; strong development of the mandibular adductor musculature (particularly the temporalis muscle, as indicated by the great height and development of the coronoid process, see Maynard Smith and Savage, 1959; Turnbull, 1970); and numerous other features. The size of preferred prey items is generally proportional to the size of the predator and, as noted, eutriconodontans spanned a greater size range than any other group of Mesozoic mammals except for marsupials. Judged by head and body length, an adult *Jeholodens* probably weighed less than 30 g. By analogy with small extant predaceous mammals, the diet of *Jeholodens* likely consisted mainly of insects and other invertebrates. At the large end of the spectrum, an adult *Gobiconodon ostromi,* with bone lengths falling into the range of living *Didelphis virginiana* but with a considerably heavier, more bearlike build (Jenkins and Schaff, 1988), may well have exceeded 2 kg in body mass. At this body size, one would predict greater reliance on vertebrate prey (figure 7.1B).

Somewhat more speculatively, it is reasonable to consider body size, molar function, and other attributes of eutriconodontans in the context of the faunas they lived among. First, eutriconodontans were usually the largest, or among the largest, predaceous mammals of their respective communities (e.g., Osborn, 1907: figure 4). Compared to their contemporaries among docodontans, "symmetrodontans," "eupantotherians," "tribotherians," eutherians, and metatherians, eutriconodontans further differ in having relatively shorter, more robust jaws, better mechanical advantage for jaw adductors, and greater development of the coronoid process (and therefore probably relatively larger temporalis musculature). Molar shearing surfaces are more anteroposteriorly aligned than in these other groups and, for a given jaw length, had a far shorter total length.

Where known, the skull of eutriconodontans is large relative to the body, and the skeleton is robustly built (figure 7.1). Specializations in various lineages include great foreshortening of the jaw and increased depth of the symphysis in *Klamelia* (Chow and Rich, 1984a), increased posterior jaw depth in Triconodontidae and Gobiconodontidae; robust, procumbent incisors designed for piercing (Gobiconodontidae, Jenkins and Schaff, 1988) or grasping (Cifelli and Madsen, 1998); and remarkable molar interlocking patterns (Patterson, 1951; Cifelli et al., 1998: figure 2). Molar shearing surfaces commonly experienced heavy wear, even on newly erupted teeth; replacement of molariforms in Gobiconodontidae may represent an

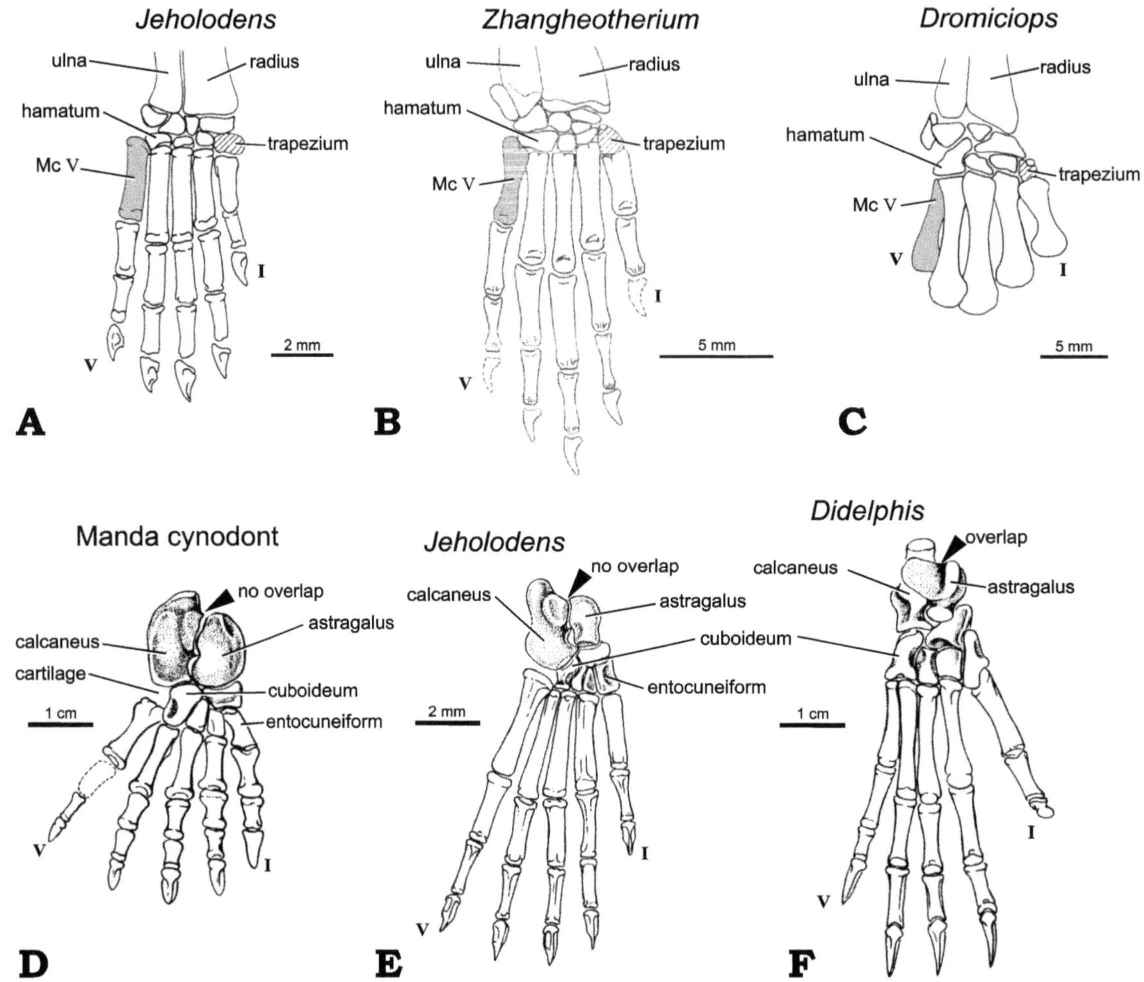

FIGURE 7.9. Comparative series of the hand (A–C) and foot (D–F); all drawn as from right side. A, *Jeholodens jenkinsi* (Eutriconodonta, family *incertae sedis;* holotype). B, *Zhangheotherium quinquecuspidens* (stem Trechnotheria; holotype). C, *Dromiciops australis* (Marsupialia). *Jeholodens* is primitive with respect to trechnotherians in that the hamate is small, the metacarpals are short, and the trapezium is not offset with respect to the scaphoid. D, "Manda cynodont" (see Schaeffer, 1941; Jenkins, 1971b). E, *Jeholodens jenkinsi.* F, *Didelphis virginiana.* In *Jeholodens,* note that the astragalus is not superimposed on the calcaneus, and the cuboid occupies a median position in the tarsus, not supporting the fifth metatarsal (the gap between that element and the calcaneus was presumably occupied by cartilage in life). These features are plesiomorphies relative to the conditions seen in living therians. *Source:* A–C, modified from Ji et al. (2002); D–F, modified from Ji et al. (1999).

adaptation for renewal of shearing surfaces that, unlike the carnassials of living carnivorans, were not self-sharpening (Jenkins and Schaff, 1988). Most other animalivorous early mammals are reasonably interpreted as having relied mainly on insects and other soft-bodied invertebrate prey (see, e.g., chapters 9 and 10). These differences of eutriconodontans lead us to suggest the working hypothesis that, in general, they distinguished themselves among their contemporaries by feeding on prey that tended to be larger relative to their own body sizes (see also Simpson, 1933a) and that they incorporated more vertebrate prey into their diets. The larger eutriconodontans were probably carnivorous and therefore occupied the highest trophic levels, among mammals, in their respective communities. Speculating further, we concur with the suggestion of Jenkins and Schaff (1988) that large-bodied taxa such as *Gobiconodon* (and *Repenomamus*) may well have relied, in part, on scavenging: their enormously robust jaws and dentition presumably conferred some bone-crushing ability (e.g., Werdelin, 1989) and may therefore have provided access to nutrients that were largely unavailable to other Mesozoic mammals.

Posture, Locomotion, and Habitat Preference

Based on the limited data available, both forelimb and hindlimb of eutriconodontans (figure 7.1) were habitually semiabducted, not parasagittal as in most living mammals (Ji et al., 1999). The thoracic and lumbar vertebrae grade into each other and lack evidence for the substantial flexion-extension movement seen in living mammals capable of fast locomotion (Jenkins and Schaff, 1988). These features, together with a robust body build and phalanges similar to those of fossorial or other terrestrial mammals, suggest a terrestrial habitat preference, with an ambulatory form of locomotion for known taxa. The most detailed study of the eutriconodontan skeleton published to date is that by Jenkins and Schaff, who concluded (1988: 23) that the large, robustly built species *Gobiconodon ostromi* "was probably more methodical than agile in its habits, and as a predator had a greater advantage in its size than its speed."

Analogy of molar design to the serially tricuspate pattern found in some cetaceans and phocids led Slaughter (1969) to suggest that the Cretaceous triconodontid *Astroconodon* was piscivorous. He further proposed that *Astroconodon* was semiaquatic, based on faunal associations and occurrence in fluviodeltaic, brackish-water facies. A similar scenario was invoked for the Early Cretaceous ?eutriconodontans *Dyskritodon* and *Ichthyoconodon* from Morocco (Sigogneau-Russell, 1995). As pointed out by Jenkins and Crompton (1979), such analogies in molar design among distantly related taxa of otherwise vastly different body size and general construction can be misleading. This is especially true in view of the fact that the molars of triconodont pattern were used very differently from those of marine mammals: in eutriconodontans, their function was largely or completely restricted to shearing, whereas comparable molars among archaeocete whales and seals do (did) not occlude, but serve (served) a piercing/grasping function during prey capture and manipulation. Similarly, evidence based on faunal association must be interpreted cautiously, and in this instance we find it to be equivocal. With the celebrated exception of certain Mongolian and Chinese assemblages (e.g., Dashzeveg et al., 1995), virtually all occurrences of Mesozoic mammals, including those of all known eutriconodontans, are in sediments laid down by water. Despite variation in facies, the predictable mixing of aquatic and terrestrial taxa is almost universal in the thanatocoenoses represented (Cifelli and Madsen, 1998). The faunal association noted for *Astroconodon* by Slaughter (1969) may well be real, but its significance is not clear. In the fluvially deposited Albian–Cenomanian Cedar Mountain Formation of Utah, for example, Goldberg (2000) found three distinct, contemporaneous faunal assemblages, one predominantly terrestrial and two predominantly aquatic (though dominated by different taxa). Associations for the various terrestrial taxa probably are related to differing habitat preference, including proximity to water, type of water body (e.g., channel or oxbow lake), and other factors. Co-occurrence of mammalian fossils with those belonging to aquatic or semiaquatic taxa does not, in itself, indicate that the mammals were aquatic, despite specific, statistically significant associations of certain taxa. To sum up, the habitat preference for triconodontids remains unknown. There is no convincing evidence to suggest that they were aquatically adapted, although it is entirely possible that some species habituated areas near water or at the margins of bodies of water.

SYSTEMATICS

Marsh's (1887) erection of the family Triconodontidae represents the first recognition of a suprageneric group for taxa now regarded as eutriconodontans. He considered triconodontids to be related to tinodontid "symmetrodonts" (see chapter 9) but was vague on broader affinities. Osborn initially (1888b) regarded triconodontids as a group of primitive carnivorous marsupials. He broadened the scope of the family, later (1907) erecting the order Triconodonta (still treated as probable marsupials and still including mammals now treated as "symmetrodontans") to contain it. Simpson's meticulous and authoritative studies of Mesozoic mammals (1928a, 1929a) created much of the framework for modern understanding of eutriconodontans and provided a morphological basis for distinguishing "triconodonts" from other mammals. He referred seven genera to the Triconodontidae and recognized two major groups within the family: Amphilestinae (*Amphilestes, Aploconodon, Comodon,*[4] *Phascolotherium*) and Triconodontinae (*Priacodon, Triconodon, Trioracodon*). This traditional classification remains in wide use, though the two groups are now generally given family rank (e.g., McKenna and Bell, 1997).

The namesake for "Triconodonta," a serially tricuspate molar pattern, has proven to be problematic as a distinguishing characteristic; this pattern, shared by certain nonmammalian cynodonts, is plesiomorphic for mammals. In the decades that followed Simpson's studies of the

[4] *Phascolodon* of Simpson's usage; see table 7.1.

1920s, the geologic range of mammals was substantially extended by discoveries in rocks of Late Triassic–Early Jurassic age (e.g., Parrington, 1941; Kühne, 1949; Patterson and Olson, 1961; Crompton, 1964b). It came as no surprise that the serially tricuspate (triconodont) molar pattern proved to be common to many of these earliest mammals, such as morganucodontids, megazostrodontids, and sinoconodontids (see chapter 4). Inclusion of some or all of these taxa within "Triconodonta" became generally accepted (e.g., Simpson, 1945; Kermack et al., 1956; Kühne, 1958; Patterson and Olson, 1961; Parrington, 1967; Hopson, 1969), though anatomical differences with respect to geologically younger taxa (e.g., Kermack and Mussett, 1958a) required a considerably broader concept of the order. Recognition of these differences led Kermack et al. (1973) to propose a trifold division of Triconodonta to include (as suborders) Morganucodonta (Morganucodontidae and Sinoconodontidae), Docodonta, and Eutricondonta (the last group equivalent to Triconodonta as conceived by Simpson, 1945; with Amphilestidae and Triconodontidae raised to family rank, following Kühne, 1958).

As noted in the introduction to this chapter and in more detail elsewhere (chapters 4 and 15), there is now compelling reason to abandon this broad-based concept of "Triconodonta." Morganucodontids and structurally similar taxa clearly represent mammalian stem groups, whereas eutriconodontans share common ancestry with more advanced mammals. Monophyly of Eutriconodonta is less clear, as is their position (or positions) on the mammalian tree. Taxonomic sampling of eutriconodontans has been rather limited in most phylogenetic analyses published to date, generally being restricted to one or more of the better-known triconodontids and/or *Gobiconodon.* As detailed in chapter 15, most studies place included eutriconodontans within, at the base of, or just outside Crown Mammalia (Rowe, 1988, 1993; Wible and Hopson, 1993; Rougier, Wible, and Hopson, 1996; Hu et al., 1997; Ji et al., 1999; Luo, Crompton, and Sun, 2001; Luo et al., 2002), though a more distant position (as sister taxon to multituberculates + crown mammals) has been suggested by Wang, Hu, Meng et al. (2001). Results are also mixed with respect to the issue of Eutriconodonta as a natural group, with some studies suggesting paraphyly (Rowe, 1988; Ji et al., 1999; Rougier et al., 1999) and others monophyly (Wang, Hu, Meng et al., 2001). Herein we follow the most comprehensive treatment to date, that of Luo et al. (2002), wherein results indicated weak support for a monophyletic Eutriconodonta. We use this term (coined by Kermack et al., 1973) in order to avoid confusion with "Triconodonta," which are widely understood to include all early mammals with a serially tricuspate molar pattern.

Without doubt, the central problem in interpreting the integrity and phylogenetic placement of eutriconodontans is their heterogeneous contents, and here the major issues concern members of "Amphilestidae." We join Simpson (1928a) and virtually all serious students since in regarding this as a paraphyletic assemblage, united by common possession of primitive characters. "Amphilestids" (nine genera) are herein treated as a structural grade and are included among Eutriconodonta for convenience as much as any other reason. Other taxa of uncertain affinities that are treated here include Austrotriconodontidae (a single genus, represented by two species from the Late Cretaceous of Argentina) and two other somewhat atypical "triconodonts" that presently defy adequate phylogenetic consideration.

The best understood of the eutriconodontans are the Triconodontidae, which are widely recognized as a monophyletic group (e.g., Jenkins and Crompton, 1979). Eight genera can be confidently referred to the family; a ninth, *Jeholodens,* is provisionally excluded from the family, though it appears to be closely related. Also formally included among Eutriconodonta are *Gobiconodon* and allies, here recognized as family Gobiconodontidae. As noted later, there is preliminary evidence to prompt at least the working hypothesis that Gobiconodontidae, *Jeholodens,* and Triconodontidae form a monophyletic group. Continued reliance upon *Gobiconodon* as a representative of (or proxy for) "amphilestids" (with which it shares plesiomorphic structure of the lower molars) seems ill advised based on the data available. We begin systematic descriptions with two *incertae sedis* genera, *Dyskritodon* and *Ichthyoconodon,* and the family Austrotriconodontidae, all of which we assign to Eutriconodonta only tentatively. However, because of their apparently primitive structure, we thought it desirable to discuss them first.

Order ?Eutriconodonta Kermack, Mussett, and Rigney, 1973
Family *incertae sedis*
Genus *Dyskritodon* Sigogneau-Russell, 1995
(figure 7.10A)

Diagnosis (modified from Sigogneau-Russell, 1995). Last molar with high and narrow crown, with cusps a, c, d decreasing regularly in height posteriorly; differs from all Triconodonta . . . by lingual displacement of cusp b and its inclusion in the cingulid; differs from all Eutriconodonta by anterior notch displaced labially with presence of a cuspule f.

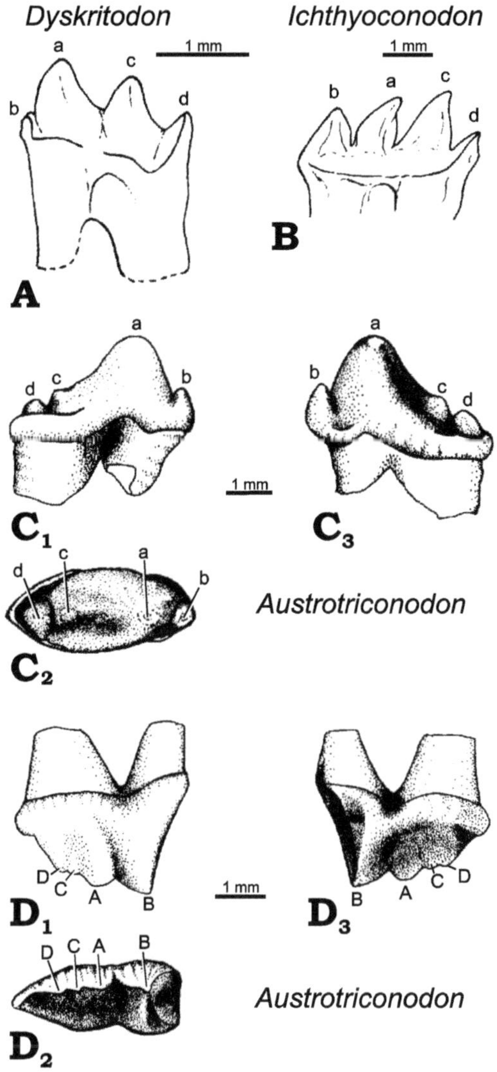

FIGURE 7.10. ?Eutriconodonta, family *incertae sedis.* A, *Dyskritodon amazighi,* right lower molar (holotype) in lingual view. B, *Ichthyoconodon jaworowskorum,* right lower molar (holotype) in lingual view. C, D, Family Austrotriconodontidae, *Austrotriconodon sepulvedai.* C, Right lower cheek tooth (holotype) in labial (C_1), occlusal (C_2), and lingual (C_3) views. D, Ultimate right upper molar in labial (D_1), occlusal (D_2), and lingual (D_3) views. *Source:* A, B, modified from Sigogneau-Russell (1995); C, D, modified from Bonaparte (1992).

Species. *Dyskritodon amazighi* Sigogneau-Russell, 1995, type species; and *D. indicus* Prasad and Manhas, 2002.

Distribution. Early Cretaceous (?Berriasian): Morocco, Talsinnt Province, Synclinal d'Anoual (Séquence B des Couches Rouges); ?Early Jurassic: India, Andhra Pradesh (Kota Formation).

Comments. This odd mammal is known by a complete ?last lower molar belonging to the type species and a less complete lower molar referred to *D. indicus.* The crown bears three major anteroposteriorly aligned cusps, decreasing in height posteriorly; two additional cuspules are present mesially. Sigogneau-Russell (1995) interpreted the main cusps as a, c, and d and the mesial cusps as f (labially) and b (lingually). An alternative possibility that might be considered is that the mesiolingual cusp is e and that the three major cusps are thus b, a, and c (from mesial to distal end of the tooth). Either way, the proportions of the cusps are rather unusual. *Dyskritodon* has a triconodont molar pattern and as it is clearly advanced with respect to morganucodontids, we place it among Eutriconodonta as much for convenience as any other purpose. Occurrence of the genus in the Early Cretaceous of Africa and the Early Jurassic of India seems improbable, but the issue cannot be addressed until both species are better known.

Genus *Ichthyoconodon* Sigogneau-Russell, 1995
(figure 7.10B)

Diagnosis (modified from Sigogneau-Russell, 1995). Molars with very high, narrow, and trenchant crowns, with cusps b, a, and c subequal. Differs from all eutriconodontans, and especially the Triconodontidae, which also have subequal cusps, by their extreme trenchantness and deep separation, as well as by the absence of intermeshing between adjacent molars, possibly replaced by an overlap; differs also by cusp c being slightly dominant. Cingulid faintly developed.

Species. *Ichthyoconodon jaworowskorum* Sigogneau-Russell, 1995, type species by monotypy.

Distribution. Early Cretaceous (?Berriasian): Morocco, Talsinnt Province, Synclinal d'Anoual (Séquence B des Couches Rouges).

Comments. *Ichthyoconodon jaworowskorum* is known by two molariform teeth that are of sufficiently strange design that the author considered and figured several alternative taxonomic allocations, including Chondrichthyes and Pterosauria (Sigogneau-Russell, 1995: figure 6). Based on the illustrations, we concur with Sigogneau-Russell that the specimens in question appear to be mammalian teeth of triconodont design. In general appearance, the teeth look somewhat like those of an advanced triconodontid, but the molar interlocking pattern is quite different, cusp d is present, but different from the homologous cusp in other taxa in overhanging the cingulid; the other cusps are remarkably tall, slender, and sharp.

Family Austrotriconodontidae Bonaparte, 1992

Diagnosis. ?Eutriconodontans with very derived upper molars; cusps arranged "in palisade," little differentiated in external view, with cusp B the largest. Upper molars with distinctive, cirquelike mesial embayment, extending to the apex of cusp B. Lower molars with hypertrophied cusp a and posterior accessory cusp c vestigial; internal cingulid reduced.

Genera. *Austrotriconodon* Bonaparte, 1986a, type genus by monotypy.

Distribution. Late Cretaceous (Campanian–Maastrichtian): South America, Argentina.

Comments. Specimens referred to species of this genus and family are problematic: with one exception, an upper molar, we are uncertain as to what tooth family they represent. That one upper molar, at least (and perhaps the other specimens as well), represents a distinctive mammal, otherwise undocumented, and it is clearly of advanced "triconodont" design. On the basis of this single specimen, we provisionally refer Austrotriconodontidae to Eutriconodonta.

Genus *Austrotriconodon* Bonaparte, 1986a
(figure 7.10C,D)

Diagnosis. As for the family.

Species. *Austrotriconodon mckennai* Bonaparte, 1986a, type species; and *A. sepulvedai* Bonaparte, 1992.

Distribution. Late Cretaceous (Late Campanian–early Maastrichtian): Argentina, Río Negro Province (Los Alamitos Formation).

Comments. The type species, *Austrotriconodon mckennai,* is represented by two cheek teeth, each bearing four mesiodistally cusps: a dominant central cusp, a small but distinct mesial cusp, and two rather small cusps sequentially distal to the principal cusp. The teeth were identified as lower molars (Bonaparte, 1986a, 1992), but they have more of the appearance of premolars, perhaps from the middle of the series, and more likely belonging to the upper than lower dentition. *A. sepulvedai* is represented by two teeth that we tentatively regard as anterior lower premolars, together with an upper molar. The upper molar most probably represents the last locus and therefore is possibly not representative of the series, especially regarding cusp proportions. The great embayment at the mesial end of the tooth appears to have formed part of an interlocking mechanism; if so, the fact that it extends the entire length of cusp B suggests that distal cusps of the preceding tooth may not have been nearly as reduced as they are on this specimen.

Order Eutriconodonta Kermack, Mussett, and Rigney, 1973
Family "Amphilestidae" Osborn, 1888b

Diagnosis. Known range of the dental formula 3–4.1.2–4.5; lower incisors, where known, lacking specializations in crown morphology. Cheek teeth, especially premolars, more or less symmetrical in lateral view; molars with cusps b and c of approximately equivalent height and significantly shorter than cusp a. Lower molars more transversely compressed than in morganucodontids and lack accessory cingulid cuspules commonly seen in morganucodontids and other stem mammals. Occlusal (labial) sides of lower molars tend to be more convex than nonocclusal sides. Adjacent lower molars interlock, with cusp d fitting between cusps e and f of the succeeding molar (except in *Klamelia,* in which the lower molars imbricate). Molars characterized by embrasure occlusion, wherein upper molars occluded between lowers: upper molar cusp A occluded posterior to cusp c of the corresponding lower molar, either wearing against cusp d of that tooth or cusp b of the succeeding molar.

Genera. *Amphilestes* Lydekker, 1887, type genus; *Aploconodon* Simpson, 1925c, *Comodon* Kretzoi and Kretzoi, 2000, *Klamelia* Chow and Rich, 1984a, *Liaotherium* Zhou et al., 1991, ?*Paikasigudodon* Prasad and Manhas, 2002, *Phascolotherium* Owen, 1838, *Tendagurodon* Heinrich, 1998, and *Triconolestes* Engelmann and Callison, 1998.

Distribution. ?Early Jurassic: India; Middle Jurassic: Europe, Britain; Late Jurassic: Asia, east Africa, and North America.

Comments. This suprageneric group was erected as a subfamily by Osborn (1888b) to contain *Amphilestes* and *Amphitylus* (now regarded as a synonym of *Amphitherium,* a "eupantotherian") within Triconodontidae of Marsh (1887), which had previously included only *Triconodon* and *Priacodon.* Osborn (1888b) recognized two other subfamilies of Triconodontidae: Phascolotheriinae (which included *Tinodon,* now recognized as an obtuse-angled "symmetrodont," see chapter 9; as well as *Phascolotherium,* subsequently placed in Amphilestinae) and Spalacotheriinae (including *Menacodon,* now recognized as a junior subjective synonym of *Tinodon;* as well as *Spalacotherium,* another "symmetrodontan"). Simpson (1928a, 1929a) distilled the contents of Triconodontidae, removing several taxa to "Symmetrodonta" and referring those left to Triconodontinae (discussed later) and Amphilestinae (*Amphilestes* and *Phascolotherium,* plus two taxa named by him, *Phascolodon* [now *Comodon*], and *Aploconodon*). Each of these taxa is based on parts of the lower dentition and dentary; with the exclusion of *Gobiconodon* and relatives, all else remains unknown for "amphilestids."

Simpson's (1928a) diagnosis of "Amphilestinae" emphasized three features that contrast with the respective conditions in Triconodontinae: symmetrical premolars, possession of more than four molars, and significant height differential between lower molar cusp a and cusps b and c. Newly discovered fossils have eroded these distinctions, at least in part (Jenkins and Crompton, 1979); the second two characters, at least, are plesiomorphies. Our own attempt at a diagnosis, given earlier, also relies heavily on presumed plesiomorphies. The "amphilestid" molar pattern (three principal serially aligned cusps, with cusp a noticeably taller than cusps b and c), for example, is shared by many chiniquodontid cynodonts, as well as *Sinoconodon* and obtuse-angled "symmetrodontans." Likewise, the possession of at least five molars, as given in the dental formula, is a plesiomorphy shared with some morganucodontids and *Kuehneotherium*. As noted further on, other features (e.g., premolar symmetry, pattern of occlusion, molar interlocking) may be derived within Mammalia, though their significance is presently uncertain.

In establishing a systematic framework that proved to endure for decades, Simpson (1928a: 70) commented: "The subfamily Amphilestinae includes four genera which differ among themselves much more than do the three genera of the Triconodontinae. It is probable that these four genera are not indicative of a single phyletic series, or indeed that they are not very closely related, but they are all characterized by the retention of the three characters listed in the definition of the subfamily. All three of these are clearly primitive characters." Conclusions of later workers (e.g., Patterson and Olson, 1961; Kermack, 1967a; Jenkins and Crompton, 1979; Jenkins and Schaff, 1988; Luo et al., 2002) have been much the same. "Amphilestid" affinities remain poorly understood, though one hypothesis deserves mention.

One of the main themes of Simpson's (1928a, 1929a) revision of Mesozoic mammals was his segregation of major groups along clear morphological lines that he considered as being of fundamental importance. In particular, he differed from all earlier workers in regarding "triconodonts" and "symmetrodonts" as comprising completely different lineages that evolved separately from therapsids. Simpson disagreed with the Cope-Osborn theory of cusp rotation (whereby a triangulated "symmetrodont" configuration of molars cusps evolved from the linear "triconodont" pattern through simple displacement of cusp positions, chapter 9), and took pains to point out that "amphilestids" have a linear configuration of molar cusps (not incipiently triangulated as had been previously supposed, e.g., Simpson, 1928a: 75–76).

A competing hypothesis, suggested by Kermack et al. (1965) and Kermack (1967a) on the basis of unspecified dental similarities, is that "amphilestids" are related to early obtuse-angled "symmetrodontans" ("Welsh Rhaetic pantotheres" of their usage). This hypothesis was further developed by Mills (1971) based mainly on the pattern of molar occlusion. As noted earlier, "amphilestids" are characterized by embrasure occlusion (wherein an upper molar occludes with two lower molars, the distal part of the corresponding lower molar and the proximal part of the succeeding molar). This contrasts with triconodontids (and *Morganucodon*), in which molar opposition is more nearly one to one (see chapter 4). Embrasure occlusion, a necessary precondition to evolution of the triangulated (symmetrodont) cusp pattern, is also seen in *Kuehneotherium*, tinodontids, and other "symmetrodontans" (see also Crompton, 1974; Jenkins and Crompton, 1979). Mills (1971) also noted that "amphilestid" lower molars are not symmetrical in occlusal view: cusp a bulges farther labially than do cusps b and c, implying an incipient stage of cusp triangulation. Indeed, a weakly triangulated cusp pattern has been reported for lower molars of some "amphilestids" (Simpson, 1925c: 335; E. F. Freeman, 1979: 150), and upper molars of gobiconodontids (which have "amphilestid"-like lower molars) are characterized by serially increasing cusp triangulation through the tooth row (Kielan-Jaworowska and Dashzeveg, 1998). "Amphilestid" lower molars also resemble those of obtuse-angled "symmetrodontans" in the manner in which they interlock: cusp d fits between cusps e and f of the succeeding tooth (Crompton, 1974; Kielan-Jaworowska and Dashzeveg, 1998). A final character worth noting is the often-cited symmetry of "amphilestid" premolars, a condition that is also seen in *Tinodon* and in posterior premolars, at least, of *Kuehneotherium* (e.g., Kermack et al., 1968: figure 6). These shared features make it increasingly apparent that the distinction between "amphilestid triconodonts" and "obtuse-angled symmetrodontans" is not quite as clear-cut as Simpson (1928a, 1929a) envisaged.

Gobiconodon and structurally similar taxa have commonly been referred to "Amphilestidae" because of similar (plesiomorphic) structure of the lower molars (e.g., Trofimov, 1978; see review by Jenkins and Schaff, 1988; Kielan-Jaworowska and Dashzeveg, 1998; Godefroit and Guo, 1999). We remove these taxa to Gobiconodontidae in what follows.

Genus *Amphilestes* Lydekker, 1887
(figures 7.6A, 7.11A, 7.12A)

Diagnosis. Dental formula of lower teeth: 3 or 4.1.4.5; differs from *Phascolotherium* in having more premolars. Lower molars are distinctive among "amphilestids" in their short length relative to crown height, gracile construction of cusps, and disproportionately great height of cusp a relative to cusps b and c. Molars differ from otherwise similar *Phascolotherium* in being taller relative to length, less transversely compressed, and lacking enamel ornamentation. Premolars differ from those of most other

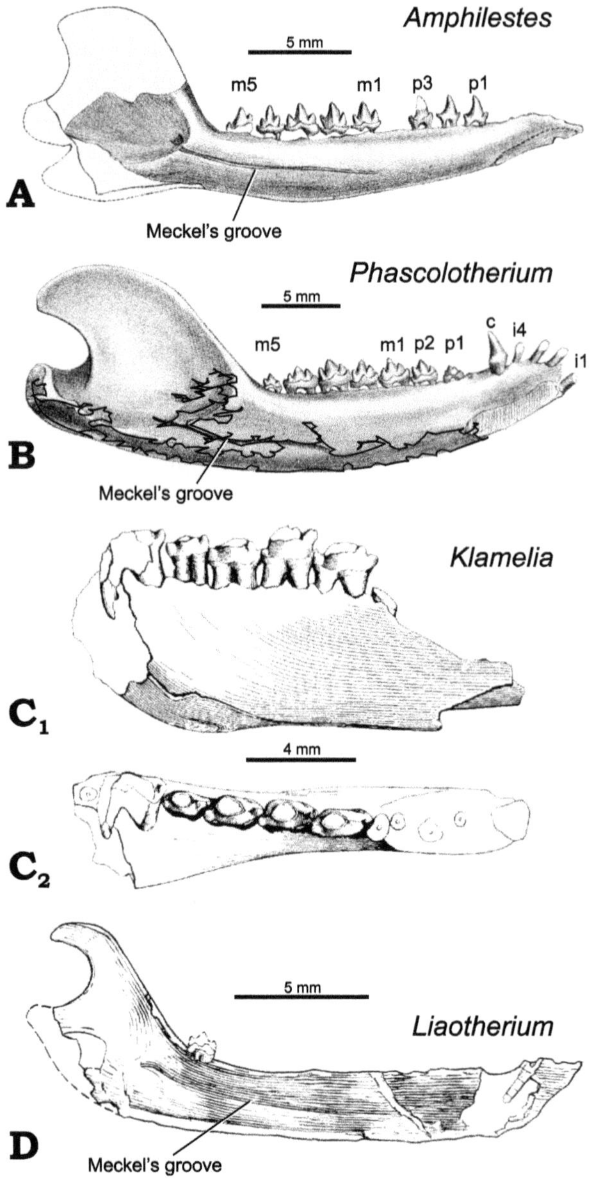

FIGURE 7.11. "Amphilestidae": dentary and lower dentition. A, *Amphilestes broderipii*, left dentary (holotype) in lingual view. B, *Phascolotherium bucklandi*, left dentary (holotype) in lingual view. C, *Klamelia zhaopengi*, right dentary (holotype) in lingual (C_1) and occlusal (C_2) views. D, *Liaotherium gracile*, left dentary (holotype) in lingual view. *Source:* A–B, modified from Osborn (1888b); C, modified from Chow and Rich (1984a); D, from Zhou et al. (1991).

eutriconodontans in lacking a lingual cingulum and in having a symmetrical, nonrecumbent cusp a and symmetrically developed cusps b and c.

Species. *Amphilestes broderipii* (Owen, 1845), type species by monotypy.

Distribution. Middle Jurassic (Bathonian): Britain, Stonesfield Slate (Sharps Hill Formation).

Comments. The basis for this monotypic genus is a species first described as *Amphitherium broderipii* by Owen (1845). The name *Amphilestes* has generally been credited to Owen, who mentioned it conjecturally, and in passing, in his monograph on Mesozoic mammals (Owen, 1871: 16). His clear intent was to retain the species in *Amphitherium;* the first formal referral to *Amphilestes* was by Lydekker (1887; see also Osborn, 1888b). *Amphilestes broderipii* is known by several reasonably complete dentaries.

Genus *Aploconodon* Simpson, 1925c
(figure 7.12C)

Diagnosis. Differs from other "amphilestids" in having short, poorly developed cusps b and c and in completely lacking cusps d–f on penultimate lower molar (cusp e present on last molar).

Species. *Aploconodon comoensis* Simpson, 1925c, type species by monotypy.

Distribution. Late Jurassic (late Kimmeridgian–early Tithonian): United States, Wyoming (Morrison Formation).

Comments. Knowledge of *Aploconodon comoensis* is restricted to the holotype, a dentary fragment preserving the last two molars. Affinities are highly debatable, given structural differences from *Amphilestes* and *Phascolotherium* (e.g., Kermack, 1967a); Patterson and Olson (1961) suggested that *Aploconodon* and *Comodon* may be morganucodontids, a proposal that has not received any support.

Genus *Comodon* Kretzoi and Kretzoi, 2000
(figure 7.12D)

Synonyms: *Phascolodon* Simpson, 1925b, *non Phascolodon* Stein, 1859; *Phascolotheridium* Cifelli and Dykes, 2001.

Diagnosis. Lower molars with weakly triangulated arrangement of principal cusps; molars similar in proportions to those of *Phascolotherium,* differing in lack of enamel ornamentation and in having a single upward curvature (double in *Phascolotherium*) of lingual cingulid.

Species. *Comodon gidleyi* (Simpson, 1925c), type species by monotypy.

Distribution. Late Jurassic (late Kimmeridgian–early Tithonian): United States, Wyoming (Morrison Formation).

Comments. Like *Aploconodon comoensis, Comodon gidleyi* (described as *Phascolodon gidleyi* by Simpson, 1925c, 1929a, see table 7.1) is known by a single specimen (dentary fragments preserving the last four molars) collected in the nineteenth century at Como Quarry 9, Wyoming (see chapter 2). Affinities are highly uncertain, despite a general resemblance in molar structure to *Phascolotherium.*

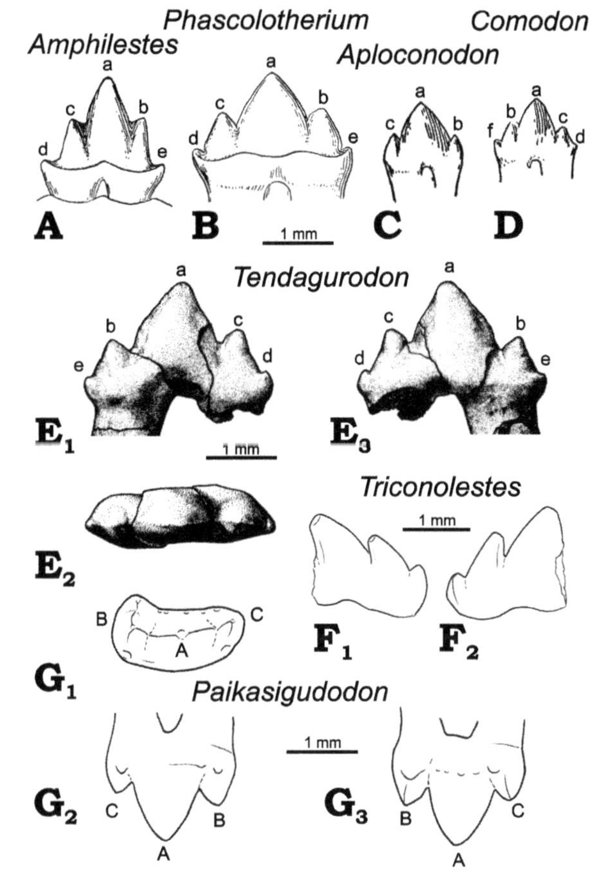

FIGURE 7.12. "Amphilestidae": cheek teeth. A, *Amphilestes broderipii*, left m3 (part of holotype) in lingual view. B, *Phascolotherium bucklandi*, left m2 in lingual view. C, *Aploconodon comoensis*, right penultimate lower molar (part of holotype) in labial view. D, *Comodon gidleyi*, left penultimate lower molar (part of holotype) in labial view. E, *Tendagurodon janenschi*, right lower cheek tooth (holotype) in lingual (E_1), occlusal (E_2), and labial (E_3) views. F, *Triconolestes curvicuspis*, partial right lower cheek tooth (holotype) in oblique labial (F_1) and lingual (F_2) views. G, *Paikasigudodon yadagirii*, left upper molar (holotype) in occlusal (G_1), lingual (G_2), and labial (G_3) views. *Source:* A, B, modified from Simpson (1928a); C, D, modified from Simpson (1929a); F, drawn from photograph in Engelmann and Callison (1998); G, drawn from photographs in Prasad and Manhas (1997).

Genus *Klamelia* Chow and Rich, 1984a
(figure 7.11C)

Diagnosis. Lower molariforms resemble those of Morganucodontidae and differ from those of "amphilestids" in having asymmetrical crowns, with cusp a placed mesial to the center of the tooth, and a very small cusp b developed on the cingulum. Molariforms differ from those of Morganucodontidae in having a weak lingual cingulid, with no development of cuspules such as the kühnecone (cusp g). Resembles Gobiconodontidae and Triconodontidae in the high position of cusp c (with base well above level of the cingulid), and resembles Gobiconodontidae in having a steep mandibular symphysis. Differs from all eutriconodontans in the presence of three extremely large mental foramina and in imbrication of molariforms, wherein cusp b overlaps the lingual side of cusp d on the preceding molar.

Species. *Klamelia zhaopengi*, type species by monotypy.

Distribution. Late Jurassic: China, Xinjiang Region (Shishuguo Formation).

Comments. Chow and Rich (1984a) considered *Klamelia* to be closely related to *Gobiconodon*, placing both in the Gobiconodontinae (herein given family rank). They distinguished these two genera from other eutriconodontans on the basis of a foreshortened mandible and reduction in the number of antemolar teeth to six or fewer. As pointed out by Jenkins and Schaff (1988), however, the sole known dentary of *Klamelia zhaopengi* is incomplete anteriorly and, owing to breakage of tooth crowns, the premolar-molar boundary cannot be determined with any certainty (see also Rougier et al., 2001). Among eutriconodontans, *Klamelia* is similar to *Gobiconodon* in the steepness of the mandibular symphysis and, perhaps, reduction of antemolar teeth. *Klamelia* may be related to Gobiconodontidae, as proposed by Chow and Rich (1984a). Noting their differences in molar structure and the peculiar pattern of imbrication in the molar series of *Klamelia*, however, we regard the genus simply as a eutriconodont of "amphilestid" grade.

Genus *Liaotherium* Zhou, Cheng, and Wang, 1991
(figure 7.11D)

Diagnosis. Lower molars with "amphilestid" crown pattern; differs from all dentally similar mammals in having tall, hooklike coronoid process, deeply incised posteriorly by the supracondylar notch; and by the relatively high position of the mandibular condyle with respect to the tooth row.

Species. *Liaotherium gracile* Zhou, Cheng, and Wang, 1991, type species by monotypy.

Distribution. Middle Jurassic: China, Liaoning Province, Haifanggou Formation.

Comments. The holotype and only specimen of *Liaotherium gracile* is a dentary bearing the last molar. The crown of this tooth is now heavily damaged, but it is said to have had an "amphilestid" crown pattern (Zhou et al., 1991). The dental formula is uncertain, but 2+.1.3.5 is a reasonable interpretation based on available information. The dentary is unusual in several features mentioned in the diagnosis; the hooklike coronoid process, for example, is quite atypical of eutriconodontans, being reminiscent instead of the stem trechnotherian *Zhangheotherium*. The specimen is remarkably small and in this respect is also suggestive of "symmetrodontans." The published illustration suggests that a long, well-defined Meckel's groove was present.

Genus *Paikasigudodon* Prasad and Manhas, 2002 (figure 7.12G)

Diagnosis. Upper molars differ from those of Morganucodontidae in being more transversely compressed, in having slight triangulation of the principal cusps, with cusp B slightly larger than cusp C, and in the presence of a prominent parastylar cusp placed directly labial to cusp B. Upper molars differ from those of Gobiconodontidae in being more transversely compressed, in having taller, less robust cusps and narrower but more cuspidate cingula and in the presence of a prominent parastylar cusp directly labial to cusp B. Upper molars differ from those of Kuehneotheriidae in lacking cusp D, in cusp A being relatively taller and more mesiodistally compressed, in having less triangulation of the principal cusps, and in the labial placement of the parastylar cusp.

Species. *Paikasigudodon yadagirii* (Prasad and Manhas, 1997), type species by monotypy.

Distribution. ?Early Jurassic: India, Andhra Pradesh (Kota Formation).

Comments. *Paikasigudodon yadagirii* is known by a single upper molar initially (Prasad and Manhas, 1997) referred to *Kotatherium* (treated herein as a "symmetrodontan," see chapter 9). Placement of *Paikasigudodon* is highly uncertain. As noted in the diagnosis, the principal molar cusps are arranged in a weakly triangulated configuration. The significance of this pattern is debatable (see earlier comments under "Amphilestidae"), and meaningful comparisons with "Amphilestidae" are precluded by the fact that "amphilestids" are otherwise known solely by the lower dentition. Indeed, upper molar structure of *Paikasigudodon* is comparable in many respects to that of Kuehneotheriidae, so its inclusion in this chapter is somewhat arbitrary.

Genus *Phascolotherium* Owen, 1838 (figures 7.5B, 7.11B, 7.12B)

Diagnosis. Dental formula of lower teeth: 4.1.2.5; differs from *Amphilestes* in fewer number of premolars. Lower molars differ from those of *Amphilestes* in being longer relative to width, with more robust, transversely compressed cusps. Differs from *Comodon* in configuration of lingual cingulum on lower molars (double upward curvature in *Phascolotherium*, single in *Comodon*). Differs from all "amphilestids" in the presence of enamel ornamentation (pitting) on the cheek teeth.

Species. *Phascolotherium bucklandi* (Broderip, 1828), type species by monotypy.

Distribution. Middle Jurassic (Bathonian): Britain, Stonesfield Slate (Sharps Hill Formation).

Comments. The type and only species of *Phascolotherium* was initially referred (Broderip, 1828) to an extant marsupial genus, as *Didelphys* [sic] *bucklandi*. It is known by several relatively complete dentaries that collectively include, or included,[5] representation of all teeth except i1.

Genus *Tendagurodon* Heinrich, 1998 (figure 7.12E)

Diagnosis. ?Lower cheek tooth with principal cusps (a–c) arranged linearly; cusp a is the tallest and cusp b is slightly lower than cusp c. Tooth proportions similar to molars of *Phascolotherium*, differing in more pointed cusp apices, greater development of cusp c relative to cusp b, and lack of cingulids. Small cusps d and e present; cusp f lacking.

Species. *Tendagurodon janenschi* Heinrich, 1998, type species by monotypy.

Distribution. Late Jurassic (Kimmeridgian–Tithonian): Tanzania (Tendaguru Series).

Comments. *Tendagurodon janenschi* is known by a single tooth. Our diagnosis uses molars of *Phascolotherium* as a basis for comparison, but identity of the specimen of *T. janenschi* is highly uncertain. The fact that no occlusal wear facets are present on the tooth, yet some minor wear is seen on the apex of cusp a (Heinrich, 1998), suggests that it may actually be a premolar (which might explain the lack of evidence for contact with preceding and succeeding teeth). A cusp f is lacking, and if the tooth is a molar, it lacks the interlocking mechanism typical of "amphilestids," described earlier.

Genus *Triconolestes* Engelmann and Callison, 1998 (figure 7.12F)

Diagnosis. Lower molars with "amphilestid" cusp pattern differing from those of most other described taxa in having posteriorly curved cusps a and b, and in the presence of a single mesial cusp (identity uncertain) in a median position, rather than cusp e mesiolingually and cusp f mesiolabially.

Species. *Triconolestes curvicuspis* Engelmann and Callison, 1998, type species by monotypy.

Distribution. Late Jurassic (late Kimmeridgian–early Tithonian): United States, Utah (Morrison Formation).

[5] Fossils of Mesozoic mammals are almost invariably both small and extremely fragile; many have suffered damage through the years.

Comments. *Triconolestes curvicuspis* is one of only a few mammals from the Morrison Formation that is not based on remains from Como Quarry 9 (chapter 2). It is by far the poorest known mammal from that formation. The holotype and only specimen consists of a partial cheek tooth, tentatively identified by Engelmann and Callison (1998) as a lower molar. The specimen does resemble "amphilestid" lower molars, yet it remains just possible that it may be a premolar. In either case, it is distinctive.

Family Gobiconodontidae Chow and Rich, 1984a

Synonym: Repenomamidae J.-L. Li, Y. Wang, Y.-Q. Wang, and C.-K. Li, 2000.

Diagnosis. Skull and dentary, where known, of robust construction; dentary deepens markedly toward the distal part of the tooth row (as in Triconodontidae) and bears a steep symphysis (autapomorphy). Facial portion of maxilla well developed, extending well onto zygomatic arch and forming part of orbital margin (shared with Triconodontidae). Anterior dentition autapomorphic: number of incisors reduced; i1, the only lower incisor, is procumbent and enlarged. Canine reduced; anterior premolars, at least, single-rooted, with conical, unicusped crowns. Anterior molariform teeth replaced ontogenetically in a sequential fashion (?autapomorphy). Upper molariforms wide transversely, with well-developed labial and lingual cingula and characterized by increasing triangulation of principal cusps from M1 to M5 (?autapomorphy). Lower molariforms with cusps linearly arranged and generally similar to those of "Amphilestidae," but with relatively taller crowns (height approximately equivalent to length) and with cusp c taller than b (except on the last molar, where the relationship is reversed). Lower molariforms interlock, with cusp d fitting between cusps e and f of the succeeding tooth (shared with most "amphilestids" and obtuse-angled "symmetrodontans," among others).

Genera. *Gobiconodon* Trofimov, 1978, type genus; *Hangjinia* Godefroit and Guo, 1999, and *Repenomamus* Li et al., 2000; unnamed new gobiconodontid, cited by Tang et al. (2001) from the Early Cretaceous Xinminbao Group, Gansu Province, China.

Distribution. Early Cretaceous: Western Europe, Central Asia, East Asia, and North America.

Comments. Chow and Rich (1984a) established Gobiconodontinae as a subfamily of "Amphilestidae," including within it *Klamelia* (from the Late Jurassic of China) as well as *Gobiconodon*. Jenkins and Schaff (1988) raised the group to family level. These authors excluded *Klamelia* from Gobiconodontidae based on reinterpretation of tooth homologies in that genus, also pointing out that *Klamelia* is divergent in its imbricating molar series and in the proportions of its molar cusps. Subsequent workers (e.g., Kielan-Jaworowska and Dashzeveg, 1998; Rougier et al., 2001) have reached the same conclusion; accordingly, *Klamelia* is treated earlier under "Amphilestidae."

Gobiconodontidae have recently been reported from a number of localities, mainly in Central and East Asia (e.g., Maschenko and Lopatin, 1998; Godefroit and Guo, 1999; Rougier et al., 2001; Wang, Hu, Meng et al., 2001). As a result, two additional genera, *Hangjinia* and *Repenomamus*, are known for the family. Though known diversity remains low, these new discoveries hint at a hitherto unappreciated radiation of gobiconodontids in the Early Cretaceous of Asia.

One of the most distinctive features of gobiconodontids is their replacement of teeth that would otherwise be considered molars. However, molars, by definition, are not replaced, and for this reason we refer to these teeth in gobiconodontids as molariforms. Certain stem mammals (such as *Megazostrodon* and *Sinoconodon*, see Gow, 1986a; Zhang et al., 1998) are known to replace molariforms, and in these instances the condition is probably plesiomorphic, inherited from a cynodont pattern of tooth replacement (see Crompton and Luo, 1993; Zhang et al., 1998). Significance of the condition in Gobiconodontidae is less certain (e.g., Jenkins and Schaff, 1988). Gobiconodontids are also unusual in their modifications of the antemolariform dentition (incisors, canines, and premolars). Here, apomorphies of the family are difficult to spell out in detail because *Gobiconodon* itself is the only well-known genus (*Repenomamus* remains to be fully described and illustrated, and *Hangjinia* is known by a single, poorly preserved specimen probably representing a juvenile individual) and tooth homologies remain uncertain (see discussion by Rougier et al., 2001, and accounts for the various genera, in what follows).

Genus *Gobiconodon* Trofimov, 1978
(figures 7.1B, 7.5D, 7.6C, 7.13A,B)

Diagnosis (based on Kielan-Jaworowska and Dashzeveg, 1998 and Rougier et al., 2001). Small to large eutriconodonts. Maxilla with: (1) multiple infraorbital canals connecting orbit to rostrum; and (2) five rounded fossae on the palatal part of the maxilla, situated close and shifted slightly posteriorly with respect to the corresponding upper molariform teeth. Five molariform teeth and five to six antemolariform teeth present in the dentary. Incisors reduced to I2/i1; i1 greatly enlarged and procumbent, c reduced. Incisors, canines, and mesial premolars with simple, conical, sharp crowns. Three or four lower premolars present, all single-rooted (except p4 in *G. hoburensis*); accessory cusps small or absent. Main cusps on M3–5 show incipient triangular pattern, with cusp A placed slightly more lingual than cusps B and C. Interlocking mechanism

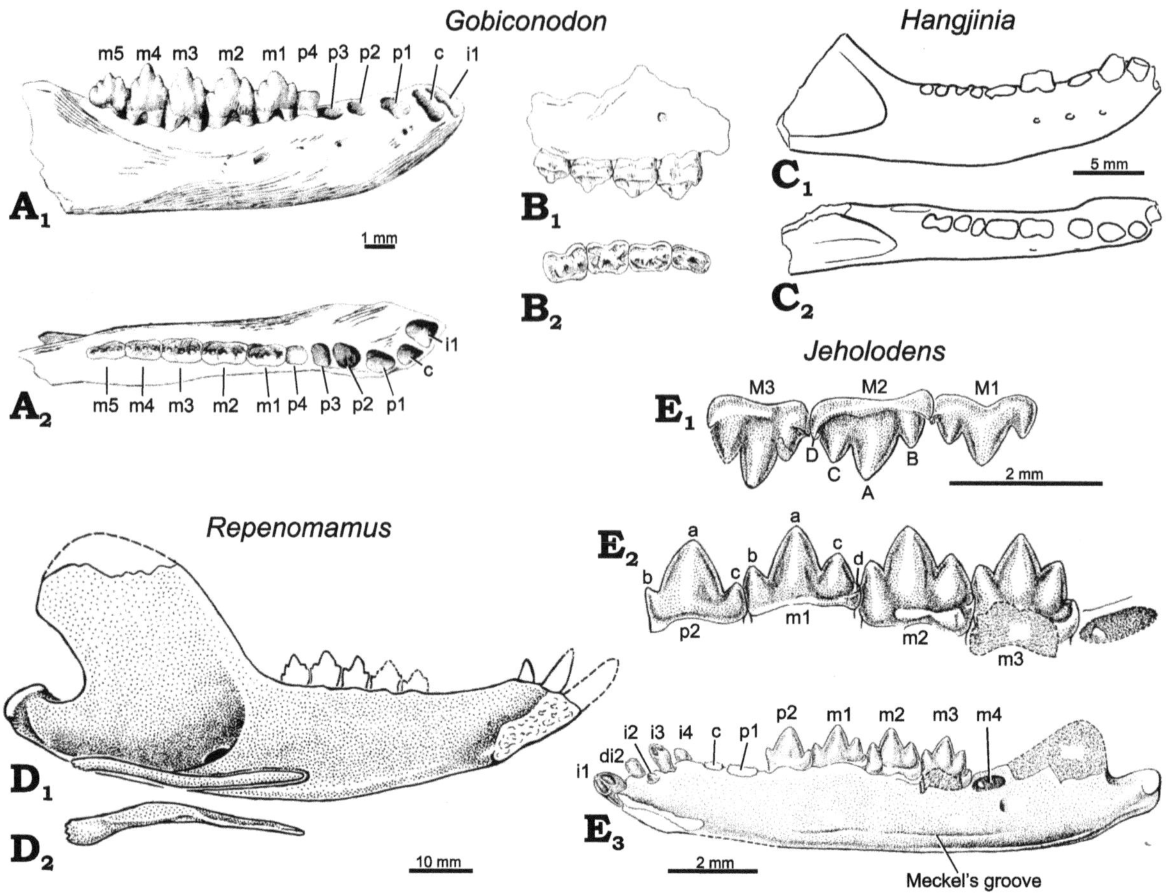

FIGURE 7.13. Gobiconodontidae and family *incertae sedis*. A–B, *Gobiconodon hoburensis*. A, Right dentary (holotype) in labial (A_1) and occlusal (A_2) views. B, Right maxilla with P4, M1–3, in labial (B_1) and occlusal (B_2) views. C. *Hangjinia chowi*, left dentary (holotype, reversed) in labial (C_1) and occlusal (C_2) views. D, *Repenomamus robustus*, left dentary in medial view (D_1), detached Meckel's cartilage of the same in dorsal view (D_2). E, *Jeholodens jenkinsi*, upper dentition, lower dentition, and dentary (parts of holotype): (E_1) right M1–3 in labial view; (E_2) right p2, m1–3 in lingual view; (E_3) right dentary in lingual view. *Source:* A–B, modified from Trofimov (1978); C, redrawn from photographs in Godefroit and Guo (1999, reversed); D, redrawn from Wang, Hu, Meng et al. (2001, D_1 reversed); E, modified from Ji et al. (1999).

of lower molariforms of "amphilestid" type, with cusp d of the anterior tooth fitting into the embayment between small cusps e and f of the succeeding molariform. Anterior molariform teeth undergo sequential replacement from front to back. Molar occlusion characterized by embrasure opposition: main cusp a of lower molariforms occluded immediately mesial to the distal margin of the coresponding upper molariform, rather than between cusps A and B as in Morganucodontidae and Triconodontidae.

Species. *Gobiconodon borissiaki* Trofimov, 1978, type species; *G. hoburensis* (Trofimov, 1978), *G. hopsoni* Rougier, Novacek, McKenna, and Wible, 2001, *G. ostromi* Jenkins and Schaff, 1988, and species left in open nomenclature by Cuenca-Bescós and Canudo (1999), Rougier et al. (2001), and Wang, Hu, Meng et al. (2001).

Distribution. Early Cretaceous (Valanginian–Albian): Europe, Spain, Teruel Province (Artoles Formation); Asia, Mongolia (Oshih Formation and Hövöör beds); China, Liaoning Province (Yixian Formation), Gansu (Xinminbao Group); Russia, Kemerovo Region (Ilek Formation); North America, United States, Montana (Cloverly Formation).

Comments. Fossil representation for *Gobiconodon* is, comparatively speaking, outstanding: much of the skeleton is known (Jenkins and Schaff, 1988) and several complete skulls have been collected, though it remains to be described (Wang, Hu, Meng et al., 2001). Included species vary greatly in size; morphological variation is also significant, and it is possible that comparative study will result in recognition of one or more additional genera.

Genus *Hangjinia* Godefroit and Guo, 1999
(figure 7.13C)

Diagnosis. Differs from *Gobiconodon* in lesser enlargement of i1 and presence of only two lower premolars.

Species. *Hangjinia chowi* Godefroit and Guo, 1999, type species by monotypy.

Distribution. Early Cretaceous (?Barremian): China, Inner Mongolia (Eijinhoro Formation).

Comments. *Hangjinia* is known by a single dentary that preserves parts of two tooth crowns externally, with one more in its crypt. Homologies of the various tooth positions are uncertain. Godefroit and Guo (1999) indicated a dental formula of 3.1.?2.?2, suggesting a greater number of incisors but fewer premolars and molars than in *Gobiconodon.* Morphologic evidence favors homology of the four mesial teeth with those seen in *Gobiconodon,* which would be identified as i1.c.p1–2 following the homologies proposed by Jenkins and Schaff (1988; see also Rougier et al., 2001). A diastema precedes the next tooth, which is two-rooted. Remnants of the tooth crown are not particularly informative (see Godefroit and Guo, 1999: figure 2A), though the morphology is suggestive of a distal premolar. If this were the case, *Hangjinia* would differ from *Gobiconodon* in having a distinctly two-rooted p3 (all premolars of *Gobiconodon* are single-rooted, except p4 of *G. hoburensis,* see Kielan-Jaworowska and Dashzeveg, 1998), with stronger development of accessory cusps. On the other hand, distal premolars were apparently shed ontogenetically in *Gobiconodon,* sometimes leaving a gap between the premolar and molariform series (e.g., Jenkins and Schaff, 1988: figure 3), thus suggesting that the fifth tooth in the holotype of *H. chowi* may be a molariform. Another viable possibility is that the tooth in question is deciduous. If so, identification as to tooth family is highly problematic, given that replacement of molariforms occurred in the closely related *Gobiconodon.* The last tooth of the series is unerupted, suggesting that the individual represented by the specimen was not fully grown. It remains possible that additional molariforms were added to the series in ontogenetically more advanced individuals of *Hangjinia chowi.*

Genus *Repenomamus* J.-L. Li, Y. Wang, Y.-Q. Wang, and C.-K. Li, 2000
(figure 7.13D)

Diagnosis. Differs from closely similar *Gobiconodon* in: presence of three upper incisors, with the lateral incisor subequal to the canine (plesiomorphies; two incisors and reduced canine present in *Gobiconodon*); ?presence of only two upper premolars (apomorphy), and lack of labial cingulum on upper molars (?apomorphy).

Species. *Repenomamus robustus* J.-L. Li, Y. Wang, Y.-Q. Wang, and C.-K. Li, 2000, type species by monotypy.

Distribution. Early Cretaceous (Barremian): China, Liaoning Province (Yixian Formation).

Comments. Li et al. (2000, 2001) referred this genus to its own family, Repenomamidae. Description and illustrations published to date are still limited but suggest close similarity to *Gobiconodon,* and we accordingly place *Repenomamus* in the Gobiconodontidae. Detailed comparisons between the two genera cannot be made on the basis of currently published evidence, which accounts for the preliminary diagnosis given earlier. The distinction in the number of upper premolars must be regarded as provisional. Based on analogy with the lower dentition, three or four upper premolars were probably present in *Gobiconodon* (Kielan-Jaworowska and Dashzeveg, 1998), but the actual number is not yet known.

With a skull length of more than 10 cm, *Repenomamus robustus* is among the largest of known Mesozoic mammals. The sketchiness of available information belies the fact that the species is represented by astonishingly complete remains, including four nearly complete skulls with articulated jaws, as well as postcranial material (Li et al., 2001; Wang, Hu, Meng et al., 2001). When fully described, these fossils promise to provide a virtual treasure trove of new information on the anatomy of Gobiconodontidae.

Family *incertae sedis*
Genus *Jeholodens* Ji et al., 1999
(figures 7.1A, 7.8 BD, 7.9A,E, 7.13E)

Diagnosis (emended from Ji et al., 1999). Dental formula 4.1.2.4/4.1.2.4; molars labiolingually compressed with three main, mesiodistally aligned cusps; uniquely derived among eutriconodontans in possession of spoon-shaped incisors. *Jeholodens* lacks the molar interlocking pattern of Morganucodontidae (wherein cusp d fits between cusps b and e of the succeeding tooth), further differing from morganucodontids in having weak labial cingula on upper molars; in lacking cingulid cuspules e, f, and g on lower molars; and in lacking the angular process and postdentary trough of the dentary (apomorphies). *Jeholodens* resembles Triconodontidae, differing from "Amphilestidae" and Gobiconodontidae in the occlusion of lower molar cusp a into the valley groove between cusps A and B of the corresponding upper molar, rather than into the embrasure between cusps B and D of the corresponding and preceding upper molars, respectively (polarity uncertain). *J. jenkinsi* further resembles Triconodontidae in lacking lower molar cingulid cuspules e and f (present in "Amphilestidae" and Gobiconodontidae) and in having a lower molar interlocking system whereby cusp d fits into the concave mesial margin of cusp b on the succeeding tooth (apomorphies). *Jeholodens* differs from Triconodontidae in retaining primitive proportions of main molar cusps (cusp a much taller than cusps b and c) and in lacking a continuous lingual cingulum on lower molars (apomorphy; also seen in *Alticonodon*).

Species. *Jeholodens jenkinsi* Ji et al., 1999, type species by monotypy.

Distribution. Early Cretaceous (Barremian): China, Liaoning Province (Yixian Formation).

Comments. *Jeholodens jenkinsi* is known by a single specimen—but what a specimen it is! Like many fossils from the Yixian Formation of Liaoning Province, China (see chapter 2), the holotype consists of a virtually complete articulated skull and skeleton. Only a preliminary account has been published to date (Ji et al., 1999). Though the specimen is flattened to the point that it is essentially two dimensional, there is no doubt that it will be a prodigal source of data bearing on the systematics and paleobiology of early mammals.

In the description of *Jeholodens*, Ji et al. (1999) left the genus without familial assignment, noting that it shares certain apomorphies (e.g., patterns of molar occlusion and interlocking of lower molars) with Triconodontidae, yet is primitive in other characters (e.g., incisor count; proportions of molar cusps, with cusps A/a being substantially taller than B/b and C/c) and autapomorphic in others (e.g., spoonlike morphology of incisor crowns). Of the few relevant phylogenetic analyses published since, most show *Jeholodens* as a sister taxon to Triconodontidae (Luo, Cifelli, and Kielan-Jaworowska, 2001; Wang, Hu, Meng et al., 2001, 2002; Rauhut et al., 2002), though Rougier et al. (2001) suggest that it lies outside a grouping of Triconodontidae + Gobiconodontidae. We favor the former interpretation, but formally exclude *Jeholodens* from Triconodontidae in order to permit a diagnosis of the family that is based on a more substantial number of widely accepted and observable apomorphies.

Family Triconodontidae Marsh, 1887

Diagnosis. Monophyletic group of eutriconodontans that share the following apomorphies: petrosal characterized by broad separation between lateral flange and crista parotica. Maxilla dorsally and posteriorly extensive and expanded well onto zygomatic arch, which supports posterior part of molar series; palatal part of maxilla notably incised posteriorly, near alveoli of posterior molars, so that a notch separates posterior molar series from the bony palate proper. Molars with principal cusps A–C/a–c of nearly equal height. Lower molars characterized by reduction to loss of cusp e and by a unique, tongue-in-groove interlocking mechanism that lacks involvement of cingular cuspules: cusp d fits into a basally placed embayment on the mesial face of cusp b on the succeeding molar.

Included Taxa. Subfamilies Triconodontinae Marsh, 1887, and Alticonodontinae Fox, 1976a.

Distribution. Early Cretaceous (Berriasian): Europe, Britain and Portugal; Late Jurassic (Kimmeridgian–Tithonian) through Late Cretaceous (early Campanian); North America.

Comments. Integrity of the Triconodontidae has been well established since Simpson's authoritative revision of the 1920s (Simpson, 1928a, 1929a). Additional apomorphies for Triconodontidae or characters diagnosing the family have been suggested (e.g., Patterson and Olson, 1961; Rougier, Wible, and Hopson, 1996; Wang, Hu, Meng et al., 2001). Most or all of these appear to be primitive, poorly understood, or ambiguous, and they have been omitted for present purposes.

Several genera of Triconodontidae from the Cretaceous of North America have been published since Simpson's (1928a, 1929a) revision of the family. Recent evidence suggests that they form a monophyletic group within the family, herein recognized as the subfamily Alticonodontinae Fox, 1976a. At the same time, there is some support, albeit limited, for a monophyletic grouping of the remaining taxa; we accordingly refer these to the Triconodontinae Marsh, 1887. Remarkably, the contents of the latter subfamily now return to the same roster (*Priacodon, Triconodon,* and *Trioracodon*) recognized by Simpson nearly 80 years ago. Recognition of Triconodontinae as a subfamily within a more inclusive group also prompts the reinstitution of the molar count, a character used by Simpson (1928a, 1929a). His diagnosis of Triconodontinae, which mentions the presence of three or four molars, was intended to distinguish triconodontines from "Amphilestidae" (as used herein), which have five molars. In recent decades, it has been shown that Alticonodontinae (as used herein), where the condition is known, have five molars (Jenkins and Crompton, 1979; Turnbull, 1995; Cifelli et al., 1998)—the probable primitive state. In the present context, then, the presence of fewer than five molars can be regarded as an apomorphy of Triconodontinae.

Subfamily Triconodontinae Marsh, 1887

Diagnosis. Petrosal with small foramen on cochlear housing, anteromedial to fenestra vestibuli; jugular foramen smaller than perilymphatic foramen and fenestra cochleae; and deep pocket present medial to paroccipital process. Four or fewer molars present. Ultimate upper molars, where known, with posterior cusps reduced or lacking.

Genera. *Triconodon* Owen, 1859, type genus; *Priacodon* Marsh, 1887, and *Trioracodon* Simpson, 1928a.

Distribution. Late Jurassic (Kimmeridgian–Tithonian): North America; Early Cretaceous (Berriasian): Europe, Britain, and Portugal.

Comments. Features of the petrosal noted in the diagnosis are based on Rougier, Wible, and Hopson (1996). That study included only one taxon referable to Alticonodontinae, and the characters must be regarded as provisional until their distributions are better understood. Important additions to knowledge of Triconodontinae have been made in recent years, notwithstanding the fact that no genera have been added to this subfamily since Simpson's (1928a, 1929a) review. Some of the fossils from the Purbeck of Britain have undergone additional preparation using new techniques, and a new account of skull structure in *Triconodon* has been published (Kermack, 1963). The most important of the more recent fossil discoveries is a partial skull and associated postcranial remains referable to *Priacodon* from the Morrison Formation of the United States (Rasmussen and Callison, 1981; Rougier, Wible, and Hopson, 1996; Engelmann and Callison, 1998).

Genus *Triconodon* Owen, 1859
(figures 7.3C, 7.7, 7.14C, 7.15E)

Diagnosis. Differs from *Priacodon* in having four (rather than three) premolars and in lacking diastema posterior to the canine. Differs from *Trioracodon* in having four (rather than three) molars. Differs from both *Trioracodon* and *Priacodon* in having molars with cusps A/a more nearly equivalent in height to cusps B/b and C/c and in retaining the full complement of principal cusps (A–C) on last upper molar.

Species. *Triconodon mordax* Owen, 1859, type species by monotypy.

Distribution. Early Cretaceous (Berriasian): Britain, Dorset (Purbeck Limestone Group).

Comments. Specimens of this species are among the most common mammalian fossils from the Purbeck of Britain. Of special significance is the broad ontogenetic range represented in the known sample, which documents replacement at the canine and p4 loci (Simpson, 1928a).

Genus *Priacodon* Marsh, 1887
(figures 7.3A,B, 7.4C, 7.6B, 7.14B, 7.15A–D)

Diagnosis. Differs from *Triconodon* in having only three (rather than four) premolars, molars with greater height differential between cusp A/a and cusps B/b and C/c, and M4 reduced, with only two cusps (A and B present and subequal in development in *Priacodon;* cusps A–C present, serially decreasing in size, in *Triconodon*). Differs from *Trioracodon* in having four (rather than three) molars. Differs from both *Triconodon* and *Trioracodon* in the presence of diastemata separating the canines from the premolar series and in having only three (rather than four) premolars.

Species. *Priacodon ferox* (Marsh, 1880), type species; *P. fruitaensis* Rasmussen and Callison, 1981, *P. grandaevus* Simpson, 1925c, *P. lulli* Simpson, 1925c, *P. robustus* (Marsh, 1879b), and a species left in open nomenclature by Krusat (1989).

Distribution. Late Jurassic (late Kimmeridgian–early Tithonian): United States, Wyoming and Colorado (Morrison Formation); Early Cretaceous (Berriasian): Portugal.

Comments. The four named species of *Priacodon* from Como Quarry 9, Morrison Formation, Wyoming (chapter 2), are similar in size. Considering the small sample size

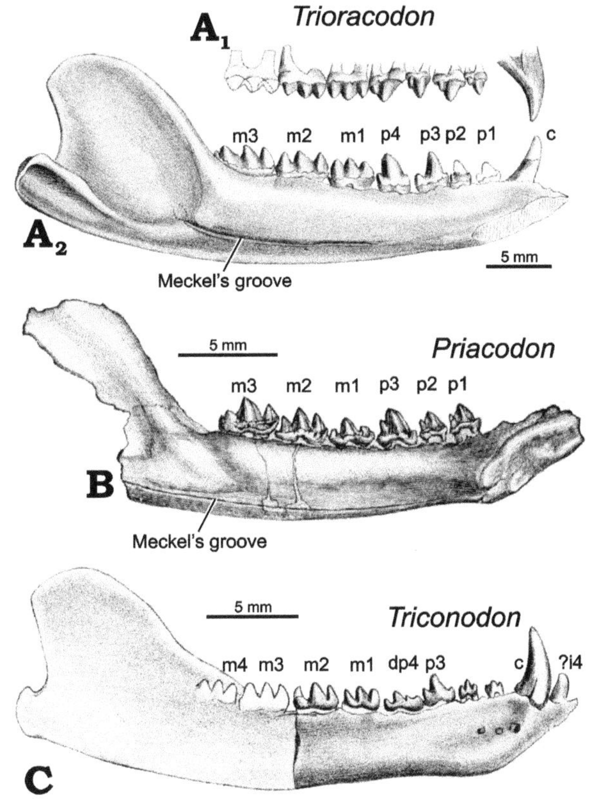

FIGURE 7.14. Triconodontinae: upper tooth series and dentaries, mostly in lingual view. A, *Trioracodon ferox:* (A_1) left upper dentition (C, P1–4, M1–3); (A_2) left dentary (holotype). B, *Priacodon ferox,* left dentary (holotype). C, *Triconodon mordax,* restored dentary showing anterior half as a right dentary in labial view and posterior half as a left dentary in lingual view. *Source:* A, C, modified from Osborn (1888b); B, modified from Marsh (1887).

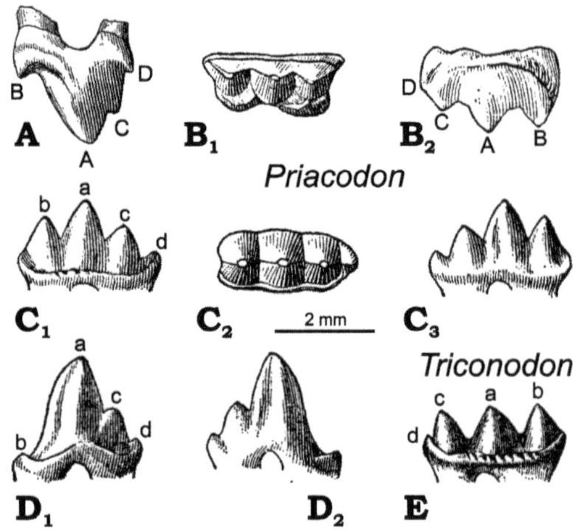

FIGURE 7.15. Triconodontinae: cheek tooth morphology. A–D, *Priacodon*. E, *Triconodon*. A, Left P3, labial view. B, Right upper molar in occlusal (B_1) and labial (B_2) views. C, Right lower molar in lingual (C_1), occlusal (C_2), and labial (C_3) views. D, Right p3 in lingual (D_1) and labial (D_2) views. E, Left lower molar in lingual view. *Source:* modified from Simpson (1929a).

and the possibility of high intraspecific variation (e.g., see Turnbull and Cifelli, 1999), it seems possible that some, or perhaps all, may be conspecific.

Genus *Trioracodon* Simpson, 1928a
(figures 7.2C,F, 7.5C, 7.6D, 7.14A)

Diagnosis. Differs from *Priacodon,* the most closely similar genus, in having four (rather than three) premolars and three (rather than four) molars and in lacking a diastema posterior to the canine. Differs from *Triconodon* in having relatively greater height differential between cusp A/a and cusps B/b and C/c, and last upper molar reduced, bearing only two cusps (A and B present and subequal in development on M3 of *Trioracodon;* cusps A–C present, serially decreasing in size, on M4 of *Triconodon*). Differs from both *Triconodon* and *Priacodon* in having only three molars.

Species. *Trioracodon ferox* (Owen, 1871), type species; *T. bisulcus* (Marsh, 1880), *T. major* (Owen, 1871), and *T. oweni* Simpson, 1928a.

Distribution. Late Jurassic (late Kimmeridgian–early Tithonian): United States, Wyoming (Morrison Formation); Early Cretaceous (Berriasian): Britain, Dorset (Purbeck Limestone Group).

Comments. Simpson's studies of the 1920s (1928a, 1929a) established *Trioracodon* (erected by him) as one of a few mammalian taxa shared at the generic level between the Morrison Formation, United States, and the Purbeck Limestone Group, Britain. The known differences between *Trioracodon* and the other two genera of Triconodontinae are relatively minor, and this is particularly true with respect to *Priacodon.* One of Simpson's primary criteria for differentiating *Trioracodon* from *Priacodon* was the postcanine dental formula: species he referred to *Trioracodon* have four premolars and three molars, whereas those he placed in *Priacodon* have three premolars and four molars. A species described decades later from the Morrison Formation fulfills the role of the proverbial fly in the ointment. Rasmussen and Callison (1981) erected *Priacodon fruitaensis* on the basis of a dentary bearing the canine and entire postcanine dentition, including three premolars and four molars (as in other species referred to *Priacodon*). Later, additional preparation revealed associated elements, including parts of a skull showing that the upper dentition included four premolars and three molars—the formula ascribed to *Trioracodon* (Engelmann and Callison, 1998). We follow these authors in retaining the species in *Priacodon,* based in part on the presence of diastemata separating the canines from P1/p1 in *P. fruitaensis.* However, we point out that each of the characters distinguishing the three genera of Triconodontinae merits scrutiny. The supposed differences in molar cusp proportions, for example, appear to be highly variable, and the distinction between *Trioracodon* and *Triconodon* is not always clear (e.g., Simpson, 1928a: figure 19).

Subfamily Alticonodontinae Fox, 1976a

Diagnosis. Distinguished from Triconodontinae (*Priacodon, Triconodon, Trioracodon*) by the following combination of characters. Dentary with Meckel's groove faint (*Arundelconodon*) or lacking (apomorphy). Greater number of lower molars (five as opposed to three or four in Triconodontinae; plesiomorphy), with cusp d enlarged, subequal in height to cusps a–c, and not overhanging distal tooth root; wear produced by upper molar cusp C bridges adjacent lower molars, resulting in development of a facet on the labial face of cusp d, as well as the labial face of cusp b on the preceding molar. Lower molar cusps posteriorly recumbent and characterized by greater degree of labiolingual asymmetry, with labial faces being gently convex and lingual faces more sharply angulate, with deep reentrants between cusps. Molars with strongly developed tongue-in-groove interlocking system, whereby mesial and distal faces bear deep reentrants and sharp ridges, respectively, that extend along the entire length of the tooth roots. Upper molars lack the lingually bilobed appearance, cusp relief on labial cusp faces, labial cingulum, lingual cingulum, and distinct separation of cusps seen in Triconodontinae (all apomorphies).

Genera. *Alticonodon* Fox, 1969, type genus; *Arundelconodon* Cifelli, Lipka et al., 1999; *Astroconodon* Patterson, 1951; *Corviconodon* Cifelli et al., 1998; *Jugulator* Cifelli and Madsen, 1998.

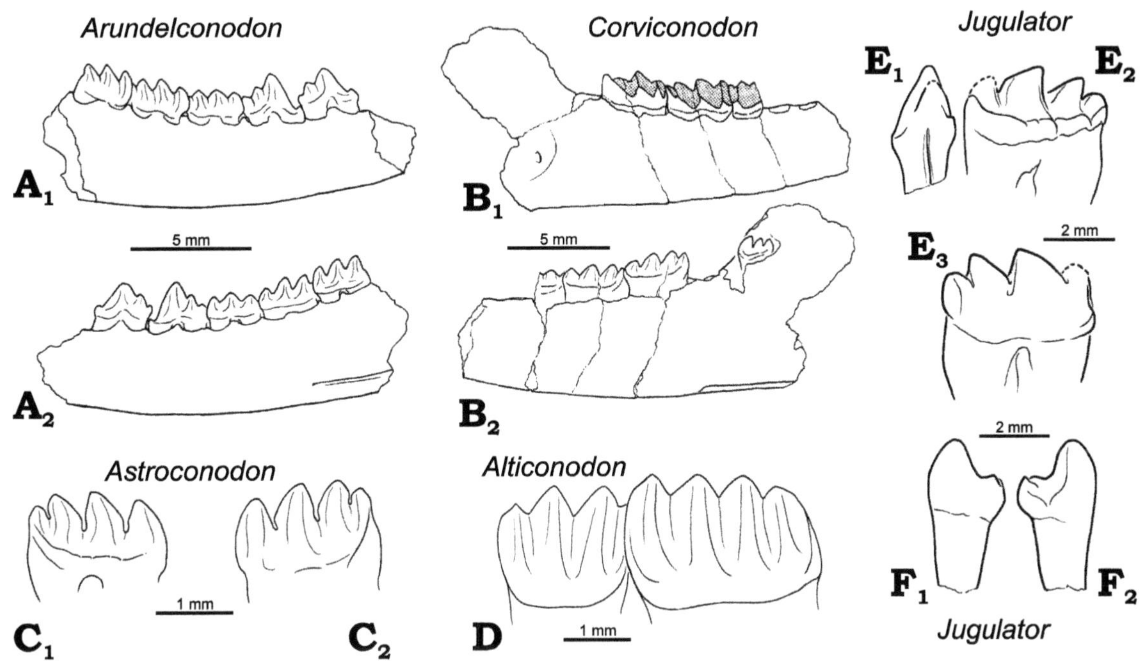

FIGURE 7.16. Alticonodontinae. A, *Arundelconodon hottoni* (holotype), right dentary in labial (A_1) and lingual (A_2) views. B, *Corviconodon montanensis* (holotype), right dentary in labial (B_1) and lingual (B_2) views. C, *Astroconodon denisoni*, left lower molar in lingual (C_1) and labial (C_2) views. D, *Alticonodon lindoei* (holotype), left penultimate and ultimate molars in labial view. E, F, *Jugulator amplissimus*. E, Right lower molar in mesial (E_1), lingual (E_2), and labial (E_3) views. F, Left i1 in labial (F_1) and lingual (F_2) views. *Source:* originals.

Distribution. Early through Late Cretaceous (Aptian through early Campanian): North America.

Comments. Fox (1976a) erected the Alticonodontinae as a monotypic subfamily to recognize the distinctive features seen in Late Cretaceous *Alticonodon.* At that time, the only other Cretaceous triconodontid known from North America was *Astroconodon,* described 25 years earlier by Patterson (1951). *Alticonodon,* the geologically youngest triconodontid, remains unusual in some respects, but additional taxa described in recent years by Cifelli and Madsen (1998), Cifelli et al. (1998), and Cifelli, Lipka et al. (1999) and better understanding of *Astroconodon* (Turnbull and Cifelli, 1999) suggest that *Alticonodon* is a highly derived member of an endemic North American clade. In recognition of these developments, we herein broaden Alticonodontinae to accommodate the remaining taxa from the Cretaceous of North America.

Genus *Alticonodon* Fox, 1969
(figure 7.16D)

Diagnosis. Lower molars differ from those of all other genera in having considerably taller crowns; flatter, nearly planar labial cusp faces (when unworn); last molar considerably longer than preceding molar, with cusp d mesiodistally expanded and equivalent in length to cusps a–c, so that the crown appears serially quadrituberculate.

Species. *Alticonodon lindoei* Fox, 1969, type species by monotypy.

Distribution. Late Cretaceous (early Campanian): Canada, Alberta (Milk River Formation).

Comments. *Alticonodon lindoei* is a highly distinctive taxon known by a dentary fragment bearing the last two molars (Fox, 1969) and an isolated last lower molar (Fox, 1976a). It is also the geologically youngest member of Triconodontidae. The next older North American record of the family is in the Albian–Cenomanian of Utah; it is possible that alticonodontines were restricted to northerly realms during the Late Cretaceous, as discussed by Cifelli and Gordon (1999).

Genus *Arundelconodon* Cifelli, Lipka et al., 1999
(figure 7.16A)

Diagnosis (from Cifelli, Lipka et al. 1999: 199). "Differs from Jurassic and earliest Cretaceous members of the family (*Triconodon, Trioracodon, Priacodon*) in having higher crowned lower molars with primary cusps (a–c) that are posteriorly recumbent and asymmetrical, with slightly curved labial and more angulate lingual faces, in having a cusp d that does not overhang the posterior root, and in having an extensive interlocking system between molars, whereby mesial and distal faces have a groove and ridge, respectively, that extend from the crown along the length of the root. Differs, where known, from the most comparable Cretaceous taxa (*Astroconodon, Corviconodon, Jugulator*) in having a [Meckel's] groove on the medial side of the mandible, a much smaller cusp d on lower molars, and a weaker, occasionally discontinuous lingual cingulid on molars."

Species. *Arundelconodon hottoni* Cifelli, Lipka et al., 1999, type species by monotypy.

Distribution. Early Cretaceous (Aptian): United States, Maryland (Arundel Clay facies of the Potomac Group).

Comments. *Arundelconodon hottoni* is the geologically oldest member of Alticonodontinae. It bears most apomorphies (such as hyperdevelopment of the molar interlocking system) seen in geologically younger taxa, but is characterized by certain plesiomorphies (less development of lower molar cusp d, retention of Meckel's groove) and thus appears to represent the sister taxon to remaining alticonodontines. The species was originally based on a single dentary preserving the last two premolars and first three molars (Cifelli, Lipka et al. 1999). A second specimen described by Rose et al. (2001) represents a juvenile individual probably referable to the same species. This second specimen is of interest in that it suggests early ontogenetic replacement of premolars, by comparison to *Triconodon mordax* (the only species of the family for which an ontogenetically wide-ranging sample has been described, see Simpson, 1928a).

Genus *Astroconodon* Patterson, 1951
(figure 7.16C)

Diagnosis. Differs from the most closely similar genus, *Corviconodon*, in having lower molars with cusp bases more distinctly separated lingually, relatively shorter cusp d (mesial molars), marked increase in crown height on distal molars; last lower molar subequal in size to penultimate molar and bearing well-developed cusp d.

Species. *Astroconodon denisoni* Patterson, 1951, type species; *A. delicatus* Cifelli and Madsen, 1998; and some unnamed species mentioned by by Winkler et al. (1990), Jacobs et al. (1991), and Cifelli, Gardner et al. (1997).

Distribution. Early Cretaceous (Aptian through Albian–Cenomanian): North America.

Comments. The type species, *A. denisoni*, holds the distinction of being the first triconodontid discovered from the Cretaceous (Patterson, 1951). It is rather abundant in the Antlers Formation of Texas (see Slaughter, 1969) and exhibits a wide range of variability, which may be due in part to variation within the molar series and to the fact that most known specimens are isolated teeth that cannot be confidently assigned to a locus (Turnbull and Cifelli, 1999).

Genus *Corviconodon* Cifelli et al., 1998
(figure 7.16B)

Diagnosis (from Cifelli et al., 1998: 237). "Differs from Jurassic Triconodontidae (*Priacodon, Triconodon, Trioracodon*) in having five lower molars; molars deeply interlocking with grooves extending well down mesial roots; cusp d of lower molars strongly developed, subequal to cusps a, b, c, and not overhanging distal root. Differs from *Alticonodon* in having lower crowned molars with well-separated cusps that are oval in cross section, not strongly flattened buccally; differs from *Astroconodon* in having a taller cusp d on mesial molars, less distinct separation of cusp bases lingually, and in lacking a marked increase in crown height on distal molars; differs from both *Alticonodon* and *Astroconodon* in having an extremely small, triscuspate (d cusp lacking) last molar."

Species. *Corviconodon montanensis* Cifelli et al., 1998, type species; and *C. utahensis* Cifelli and Madsen, 1998.

Distribution. Early Cretaceous (Aptian–Albian through Albian–Cenomanian): North America.

Comments. The type species, *Corviconodon montanensis*, is known by a partial dentary that includes most of the molar series; *C. utahensis* is represented by isolated cheek teeth. Lower molars and premolars are rather similar to those of *Astroconodon*, but *Corviconodon* is readily distinguished on the basis of its strongly reduced last molar.

Genus *Jugulator* Cifelli and Madsen, 1998
(figure 7.16E,F)

Diagnosis (from Cifelli and Madsen, 1998: 407). "Largest known member of the Triconodontidae. Lower molars differ from those of most closely similar taxa (*Astroconodon, Corviconodon*) in being relatively wider and in having cusps that are more obtuse and posteriorly recumbent in profile view, in having more broadly separated apices of a, b, and c, and in having more evenly curved labial faces, with only minor concavities separating cusp bases; cusp d developed as an anteroposteriorly ortented ridge rather than a fingerlike projection on posterior molars."

Species. *Jugulator amplissimus* Cifelli and Madsen, 1998, type species by monotypy.

Distribution. Early–Late Cretaceous (Albian–Cenomanian): United States, Utah (Cedar Mountain Formation).

Comments. *Jugulator amplissimus* is known only by isolated teeth and dentary fragments. The species is most distinctive in its relatively large size, with some lower molars exceeding 5 mm in length. The medial lower incisor is greatly enlarged, with a mitten-shaped crown that bears sharp cutting surfaces (Cifelli and Madsen, 1998).

CHAPTER 8

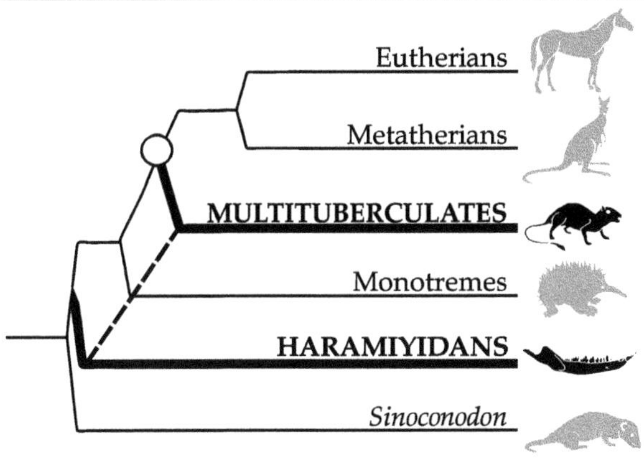

Allotherians

INTRODUCTION

Marsh (1880) proposed the order Allotheria[1] (from Greek *allos*—other, different, *therion*—wild animal) to include two Late Jurassic genera, *Plagiaulax* and *Ctenacodon,* based on well-preserved dentaries with teeth, and he assigned Allotheria to Marsupialia. The dentaries superficially resembled those of rodents, but the teeth designated as molars were different, covered with numerous cusps of subequal height, arranged in longitudinal rows. Moreover, the lower premolars were unusual in being bladelike, covered on their labial and lingual sides by oblique ridges. Four years later Cope (1884) proposed the Multituberculata as a suborder of Marsupialia, including three families: Tritylodontidae, Polymastodontidae (= Taeniolabididae), and Plagiaulacidae. He stated that the Plagiaulacidae are equivalent to Marsh's "order" Allotheria. Tritylodontidae are now referred to Cynodontia (see e.g., Carroll, 1988), while Taeniolabididae and Plagiaulacidae are retained in Multituberculata. Simpson (1929a: 143) raised Allotheria to a subclass, with the single order Multituberculata Cope, whereas McKenna and Bell (1997) regarded Allotheria as an infraclass.

In contrast to the relatively well-known multituberculates were isolated teeth from the Late Triassic–Early Jurassic of Europe, with multiple cusps arranged in longitudinal rows (like those of multituberculates), but of various heights. Simpson (1929a) referred two such single-tooth genera to "?Mammalia of uncertain subclass and order." These teeth were subsequently assigned to the family Haramiyidae Simpson, 1947. Hahn (1973) proposed for the Haramiyidae a suborder Haramiyoidea within Multituberculata and later (Hahn et al., 1989) raised it to ordinal rank as Haramiyida. Butler and MacIntyre (1994) regarded Haramiyida as a sister group of Multituberculata and assigned them to ?Mammalia, Allotheria, but Butler (2000) subsequently assigned Allotheria to Mammalia (= Mammaliaformes of some authors).

Subclass Allotheria Marsh, 1880

Diagnosis (after Butler, 2000, emended). Allotherians are mammals in which the upper and lower molariform teeth have basically two longitudinal rows of cusps, which occlude so that the labial lower row bites into the valley between the upper rows (additional rows may develop on upper teeth). Occlusion is bilateral; the movement is vertical (orthal) or posterior (palinal), or a combination of the two, but not significantly transverse. They differ in the structure and function of molariform teeth from primitive nonallotherian Mammalia, which have one row of cusps that functioned in unilateral occlusion with the labial surfaces of the lower cusps shearing against the lingual surfaces of the upper cusps, involving a transverse jaw movement.

Included Orders. Haramiyida Hahn et al., 1989, tentatively assigned; Multituberculata Cope, 1884.

Distribution. Late Triassic–late Eocene: the world except Australian region and Antarctica.

[1] We tentatively accept in this chapter the idea of Allotheria including the poorly known Haramiyida and Multituberculata (see footnote 3 to table 1.1 and chapter 15 for detailed discussion).

Order Haramiyida Hahn, Sigogneau-Russell, and Wouters, 1989 (tentatively assigned)

HISTORICAL BACKGROUND

The name of the order derives from the generic name *Haramiya,* which in Arabic means trickster, petty thief. Simpson (1947) coined this name to replace the preoccupied *Microcleptes,* which has a similar meaning. Butler and MacIntyre (1994: 434)[2] offered the following, comprehensive survey of the history of research on haramiyids up to 1993:

> Pleininger (1847) described *Microlestes antiquus* from the Rhaetic bonebed of Degerloch, Württemberg (see Clemens, 1980a). The generic name was preoccupied and was replaced by *Thomasia* Poche, 1908. The type specimen was redescribed by Hennig (1922) and Simpson (1928a). Owen (1871) described *Microlestes moorei* from a collection of teeth made by Moore (1867) from a fissure deposit in Holwell Quarry, Somerset, England.

Simpson (1928a), who selected a lectotype, redescribed this species. He made it the type of a new genus *Microcleptes* (in 1947 emended to *Haramiya*), in distinction from *Thomasia,* and from other teeth in the same collection he named two new species: *Microcleptes* (now *Haramiya*) *fissurae* and *Thomasia anglica.* Butler and MacIntyre continued further (1994: 434):

> In 1939 Kühne made a further collection of haramiyid teeth from Holwell (Kühne, 1946), and these were described by Parrington (1947). Huene (1933) and Hahn (1973) described additional specimens from Germany, and Peyer (1956) described 23 teeth, mostly fragmentary, from Hallau, Switzerland. The German and Swiss material was reviewed by Clemens (1980a, 1986). Wouters et al. (1984) reported a tooth of *Haramiya* from Gaume, Belgium. By far the largest sample of haramiyid teeth, consisting of over 200 specimens, is from Saint-Nicolas-de-Port, Lorraine, France (Sigogneau-Russell, 1989b). Both *Haramiya* and *Thomasia* are represented.

Sigogneau-Russell (1989b) recognized in the material from Lorraine not only molars, but also premolars and upper and lower incisors. She proposed a new species, *Haramiya butleri,* subsequently put under synonymy with *Thomasia antiqua* by Butler and MacIntyre (1994). Their study offered a new interpretation of the British haramiyid material, including a meticulous analysis of wear facets, reconstruction of lower jaw movement, and a reconstruction of the upper and lower dentition of *Thomasia antiqua* (see also Sigogneau-Russell and Hahn, 1994).

Originally two genera of haramiyids were recognized: *Thomasia* Poche, 1908, and *Haramiya* Simpson, 1947. Teeth of *Haramiya* and *Thomasia* always occur together, in each sample they are present in equal numbers, and they are of approximately the same length. On this basis, Sigogneau-Russell (1989b) suggested that they are congeneric. As teeth of *Haramiya* are wider than those of *Thomasia,* she regarded those of *Haramiya* as the uppers and those of *Thomasia* as the lowers (in all other mammals, as a rule, the upper molars are wider than the lowers). Accepting her viewpoint, Butler and MacIntyre added (1994: 448):

> Within each "genus" most variations can be ascribed to differences in serial positions (premolars, anterior molars, posterior molars), and no variation attributable to upper-lower differences has been detected. Further, in the incisors, where upper and lower teeth are distinct, it is not possible to distinguish *Thomasia* from *Haramiya.* We therefore support the hypothesis of Sigogneau-Russell (1989b) that *Thomasia* and *Haramiya* are opposing dentitions of the same genus.

Thus *Haramiya* Simpson, 1947, became a junior subjective synonym of *Thomasia* Poche, 1908.

Kermack et al. (1998) created the suborder Eleutherodontida, order *incertae sedis* and assigned it to Allotheria. Eleutherodontids (regarded by Butler, 2000, as a family) are known by isolated teeth from the Bathonian of England. They share with Multituberculata and Haramiyida multicusped teeth with longitudinally arranged cusps, and with multituberculates a backward power stroke of the dentary.

Another group referred to non-multituberculate Allotheria are the Theroteinidae, known from isolated teeth from the Rhaetian (Late Triassic) of Europe. Sigogneau-Russell (1983b) was the first to describe such isolated teeth, which she assigned to Haramiyidae indet. Subsequently Sigogneau-Russell et al. (1986) erected the genus *Theroteinus* and family Theroteinidae for them, while Hahn et al. (1989) placed them in the order Theroteinida. Theroteinidae have multicusped teeth but differ from Multituberculata and Eleutherodontidae in having an orthal power stroke (Sigogneau-Russell et al., 1986; Hahn et al., 1989).

Jenkins et al. (1997) described from the Late Triassic of Greenland *Haramiyavia clemmenseni,* which they assigned to Haramiyidae, represented by dentaries and partial maxillae with teeth and fragments of the postcranial skeleton (undescribed). They argued that *Haramiyavia* had orthal jaw movement, and on this basis excluded Haramiyida from Allotheria. Kermack et al. (1998: 582) stated: "Thus the problem arises whether *Haramiyavia clemmenseni* really belongs to the Haramiyidae, in which case propalinal movement should be then questioned in that family, or whether the taxon is closer to theroteinids

[2] Some dates in the two citations that follow are adjusted following the references in this book.

(also three rows of cusps in the upper molars, against two in haramiyids)."

Butler (2000) came to a different conclusion. He revised all the non-multituberculate allotherians, which he assigned to the order Haramiyida Hahn et al., 1989, and recognized within it two suborders: Theroteinida Hahn et al., 1989 (with a single family Theroteinidae) and Haramiyoidea Hahn, 1973, with three families: Haramiyidae, Eleutherodontidae (assigning family rank to the suborder Eleutherodontida of Kermack et al., 1998), and a new family, Haramiyaviidae Butler, 2000. We tentatively follow assignment of Haramiyida to Allotheria and Butler's division, but see chapter 15 for discussion on the controversy concerning haramiyidan-multituberculate relationship. In the same paper Butler (2000) reconstructed the occlusion pattern in haramiyids based on an analysis of tooth wear.

BRIEF CHARACTERIZATION

Butler (2000) characterized Haramiyida as follows:

> Allotherians, which differ from multituberculates in presence of postdentary-trough (known only in *Haramiyavia*), and differ from multituberculates (except Paulchoffatiidae) in that the cusps of molariform teeth are of unequal height. On upper teeth the labial row has three cusps, the middle one highest, except in Eleutherodontidae where there may be as many as ten cusps, highest near the middle of the series. The number of lingual upper cusps and of labial and lingual lower cusps varies; the highest cusp is towards the distal end in the upper lingual row and towards the mesial end in the lower rows. Molar occlusion ranges from orthal to palinal, but there is always an orthal component that is lacking in multituberculates. A high distal lingual upper cusp occludes in the longitudinal valley of the lower molar, and a high mesial labial lower cusp occludes in the valley of the upper molar. As far as known (*Haramiyavia, Thomasia*) the last upper molar is not displaced lingually as in multituberculates.

Suborders Included. Theroteinida Hahn et al., 1989; Haramiyoidea Hahn, 1973.

Distribution. Late Triassic–Middle Jurassic (Bathonian): Europe; Late Triassic: Greenland; Early Jurassic: North America; Late Jurassic: Africa.

TERMINOLOGY

Butler (2000) named the cusps in haramiyids in accordance with their postulated homologies with *Haramiyavia*, using the system of nomenclature of Jenkins et al. (1997), with minor modifications. This system was adopted from one originally proposed by Hahn (1973) and applied to *Thomasia* and *Haramiya* by Sigogneau-Russell (1989b) and Butler and MacIntyre (1994). We follow Butler's (2000: 319) explanation of cusp nomenclature. The text that follows is taken from Butler's paper, with some shortening.

On lower molars (figure 8.1) the two rows of cusps are termed a (lingual) and b (labial) (= buccal in Butler's terminology); lower-case letters are used for lower teeth, capitals for upper teeth. Within the rows, the cusps are numbered from mesial to distal. The mesial cusps (at the "+ end" of Butler and MacIntyre, 1994) are the largest and most constant, except that (as postulated below) b1 may be reduced or absent. The number of cusps in the rows is variable between taxa, among teeth in different serial positions, and individually; for example, in *Thomasia* additional b cusps develop at the distal end, on the "U ridge" (Butler and MacIntyre, 1994).

The pattern of upper molars is reversed: row A is labial and row B is lingual. Row B occludes in the valley between lower rows a and b, in a manner similar to that in which lower row b occludes between upper rows A and B (Butler, 2000: figure 3). A further complication arises in haramiyids in that opposing teeth are also reversed mesiodistally (figure 8.1). On upper teeth the "+ end," with the largest and most constant cusps, is distal, and the mesial cusps are the most variable. Upper cusps are therefore numbered from distal to mesial. When this system was proposed by Hahn (1973) it was not recognized that *Haramiya* is the upper dentition of *Thomasia*, and the same nomenclature was applied to both "genera"; the distal end of upper teeth was conventionally taken as anterior. Butler (2000) further argued that to renumber upper cusps from mesial to distal would not only create more confusion, but would divorce the nomenclature from homology: for example, the enlarged distal cusp, here called B1 and considered to be homologous in different taxa and on different teeth (figure 8.1A_1–D_1), would have to be numbered variously according to the presence or absence of minor cusps at the mesial end of the tooth.

Butler (2000: 319) further argued:

> The reversed symmetry does not occur, or has been lost, in multituberculates. Hahn and Hahn (1998c) have introduced a cusp nomenclature for paulchoffatiid molars, in which, on upper as well as lower teeth, lingual (L) and labial (B) rows are distinguished, and in each row the cusps are numbered from mesial to distal. This system provides a logical basis for description, though the authors do not discuss its implications for serial homology between adjacent teeth. There is some variation in the number of cusps, especially at the distal end of teeth of both jaws, whereas upper molars of haramiyids vary mainly at the mesial end. In allotherians the row is the functional unit, and only a few cusps are differentiated for individual functions. To apply the system of Hahn and Hahn (1998c) to haramiyids, assumptions would have to be made about the

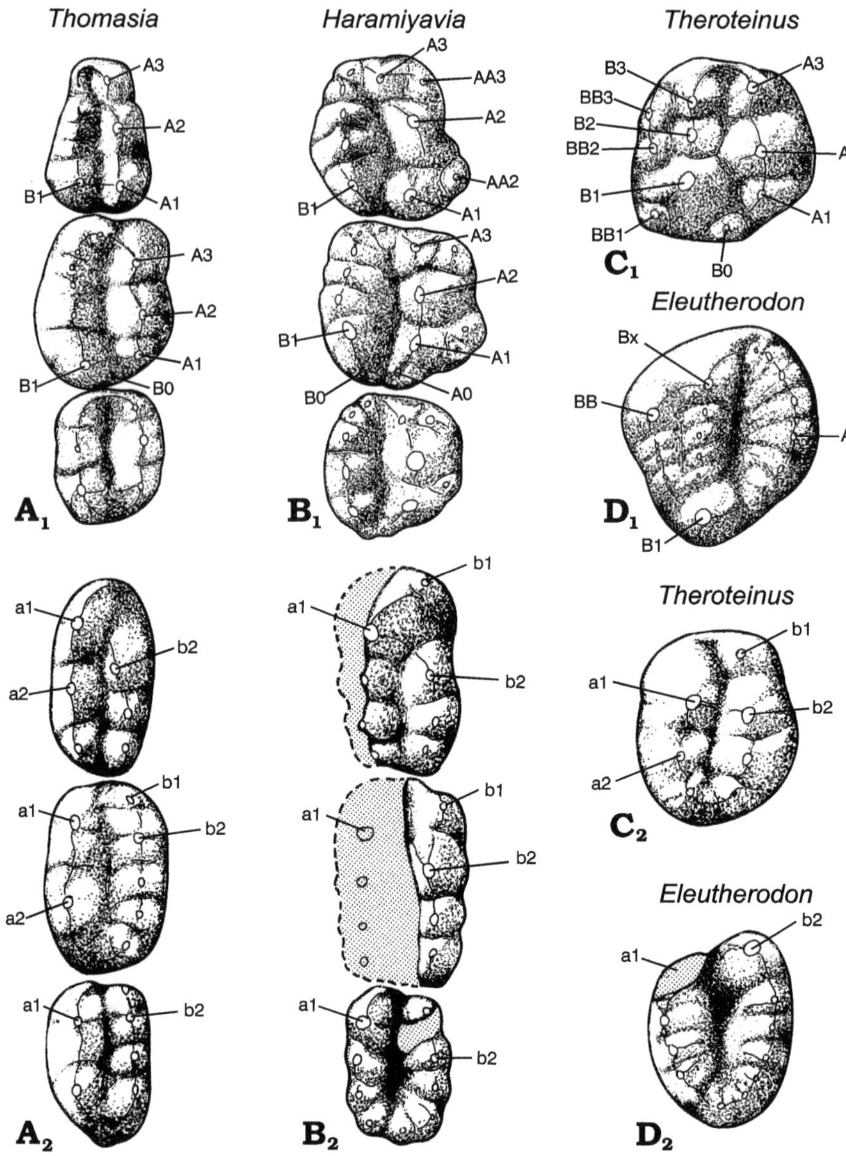

FIGURE 8.1. Comparison of left upper (A_1–D_1) and right lower (A_2–D_2) molariforms of *Thomasia, Haramiyavia, Theroteinus,* and *Eleutherodon* in occlusal views, illustrating cusp homologies postulated by Butler (2000). Labial is to the right and mesial above. On upper molariforms cusps are numbered from distal to mesial and in lowers from mesial to distal, in accordance with their postulated homologies with *Haramiyavia.* On upper molariforms of *Eleutherodon* BB and Bx are enlarged mesial cusps of the lingual (BB) and middle (B) rows, and A indicates the A row. Not to scale. *Source:* modified from Butler (2000).

homologies of their cusps to those of paulchoffatiids, but, on present knowledge, it is only possible to discuss with confidence the homologies of rows of cusps.

SYSTEMATICS

The systematics of the order Haramiyida[3] presented below is based on a recent revision by Butler (2000) and is quoted here from Butler's paper, with a few emendations (see table 8.1).

[3] Although in the systematic parts of this book as a rule we first describe the type family (or the type genus and so on) and then other taxa in alphabetic order, in this section we follow Percy M. Butler's (2000) original arrangement of the high-rank haramiyidan taxa, which reflects his idea on interrelationships of these taxa.

Suborder Theroteinida G. Hahn, Sigogneau-Russell, and Wouters, 1989

Diagnosis. Haramiyids with fully orthal occlusion, in which upper and lower molars alternate, so that each lower molar bites against two upper molars. The highest cusps are more centrally placed on the teeth than in Haramiyoidea, and the longitudinal valleys are short. Molar cusps show a tendency to coalesce.

Families. Theroteinidae Sigogneau-Russell et al., 1986, type family by monotypy.

Distribution. Late Triassic: Europe.

Family Theroteinidae Sigogneau-Russell, Frank, and Hemmerlé, 1986

Diagnosis. Upper molars are short and wide, with an additional lingual row of cusps. Cusps are low and obtuse. Re-

TABLE 8.1. Linnaean Classification of Mesozoic Allotherian Mammals

Subclass Allotheria Marsh, 1880
Order Haramiyida Hahn et al., 1989
Suborder Theroteinida Hahn et al., 1989
Family Theroteinidae Sigogneau-Russell et al., 1986
Theroteinus Sigogneau-Russell et al.,1986
T. nikolai Sigogneau-Russell et al., 1986
Suborder Haramiyoidea Hahn, 1973
Family Haramiyaviidae Butler, 2000
Haramiyavia Jenkins et al., 1997
H. clemmenseni Jenkins et al., 1997
Family Haramiyidae Simpson, 1947
Thomasia Poche, 1908 (synonym: *Haramiya* Simpson, 1947), type genus
T. antiqua Poche, 1908, type species
T. moorei (Owen, 1871)
T. woutersi Butler and MacIntyre, 1994
T. hahni Butler and MacIntyre, 1994
Staffia Heinrich, 1999
S. aenigmatica Heinrich, 1999
Family Eleutherodontidae Kermack et al., 1998
Eleutherodon Kermack et al., 1998
E. oxfordensis Kermack et al., 1998
(Intervening ranks of clades in Table 1.1)
Order Multituberculta Cope, 1884
Suborder "Plagiaulacida" Amghino, 1889
Allodontid line
Family Allodontidae Marsh, 1889b
Ctenacodon Marsh, 1879b (synonym: *Allodon* Marsh, 1881, *partim*), type genus
C. serratus Marsh, 1879b, type species
C. laticeps (Marsh, 1881)
C. scindens Simpson, 1928a
Psalodon Simpson, 1926b (synonym: *Allodon* Marsh, 1881, *partim*)
P. potens (Marsh, 1887), type species
P. fortis (Marsh, 1887)
?*P. marshi* Simpson, 1929a
Family Zofiabaataridae Bakker, 1992
Zofiabaatar Bakker and Carpenter, 1990
Z. pulcher Bakker and Carpenter, 1990
Family *incertae sedis*
Glirodon Engelmann and Callison, 1999
G. grandis Engelmann and Callison, 1999
Paulchoffatiid line
Family Paulchoffatiidae Hahn, 1969
Subfamily Paulchoffatiinae Hahn, 1969, type subfamily
Taxa based on lower dentition
Paulchoffatia Kühne, 1961, type genus
P. delgadoi Kühne, 1961
Guimarotodon Hahn, 1969
G. leiriensis Hahn, 1969
Meketibolodon Hahn, 1993
M. robustus (Hahn, 1978a)
Plesiochoffatia Hahn and Hahn, 1999b
P. thoas (Hahn and Hahn, 1998c)
P. peperethos Hahn and Hahn, 1998c
P. staphylos (Hahn and Hahn, 1998c)
Xenachoffatia Hahn and Hahn, 1998c
X. oinopion Hahn and Hahn, 1998c
Taxa based on upper dentition
Bathmochoffatia Hahn and Hahn, 1998
B. hapax Hahn and Hahn, 1998c
Henkelodon Hahn, 1977a
H. naias Hahn, 1977a
Kielanodon Hahn, 1987b
K. hopsoni Hahn, 1987
Meketichoffatia Hahn, 1993
M. krausei, Hahn, 1993
Pseudobolodon Hahn, 1977a
P. oreas Hahn, 1977a
Renatodon Hahn, 2001
R. amalthea Hahn, 2001
Subfamily Kuehneodontinae Hahn, 1971
Kuehneodon Hahn, 1969
K. dietrichi Hahn, 1969, type species
K. barcasensis Hahn and Hahn, 2001a
K. dryas Hahn, 1977a
K. guimarotensis Hahn, 1969
K. hahni Antunes, 1998
K. simpsoni Hahn, 1969
K. uniradiculatus Hahn, 1978a
Subfamily indet.
Galveodon Hahn and Hahn, 1992
G. nannothus Hahn and Hahn, 1992
Sunnyodon Kielan-Jaworowska and Ensom, 1992
S. notleyi Kielan-Jaworowska and Ensom, 1992, type species
Family and subfamily indet.
Mojo Hahn et al., 1987
M. usuratus Hahn et al., 1987
Family Hahnodontidae Sigogneau-Russell, 1991c
Hahnodon Sigogneau-Russell, 1991
H. taqueti Sigogneau-Russell, 1991
Family Pinheirodontidae Hahn and Hahn, 1999a
Pinheirodon Hahn and Hahn, 1999, type genus
P. pygmaeus Hahn and Hahn, 1999a, type species
P. vastus Hahn and Hahn, 1999a
Bernardodon Hahn and Hahn, 1999
B. atlanticus Hahn and Hahn, 1999a
Ecprepaulax Hahn and Hahn, 1999
E. anomala Hahn and Hahn, 1999a
Gerhardodon Kielan-Jaworowska and Ensom, 1992

(continued)

G. purbeckensis Kielan-Jaworowska and Ensom, 1992
Iberodon Hahn and Hahn, 1999
I. quadrituberculatus Hahn and Hahn, 1999a
Lavocatia Canudo and Cuenca-Bescós, 1996
L. alfambrensis Canudo and Cuenca-Bescós, 1996
Plagiaulacid line
Family Plagiaulacidae Gill, 1872 (synonym: Bolodontidae Osborn, 1887b)
Bolodon Owen, 1871 (synonym: *Plioprion* Cope, 1884)
B. crassidens Owen, 1871, type species
B. falconeri (Owen, 1871)
B. minor (Falconer, 1857)
B. osborni Simpson, 1928a
?non "*B.*" *elongatus* Simpson, 1928a
Plagiaulax Falconer, 1857, type genus
P. becklesii Falconer, 1857
Family Eobaataridae Kielan-Jaworowska et al., 1987
Eobaatar Kielan-Jaworowska et al., 1987, type genus
E. magnus Kielan-Jaworowska et al., 1987, type species
E. hispanicus Hahn and Hahn, 1992
E. minor Kielan-Jaworowska et al. 1987
?*E. pajaronensis* Hahn and Hahn, 2001b
Loxaulax Simpson, 1928a
L. valdensis (Woodward, 1911)
?*Monobaatar* Kielan-Jaworowska et al., 1987
M. mimicus Kielan-Jaworowska et al., 1987
Parendotherium Crusafont-Pairó and Adrover, 1966
P. herreroi Crusafont-Pairó and Adrover, 1966
Sinobaatar Hu and Wang, 2002a
S. lingyuanensis Hu and Wang, 2002a
Family Albionbaataridae Kielan-Jaworowska and Ensom, 1992
Albionbaatar Kielan-Jaworowska and Ensom, 1992
A. denisae Kielan-Jaworowska and Ensom, 1994
Proalbionbaatar Hahn and Hahn, 1998d
P. plagiocyrtus Hahn and Hahn, 1998d
Family *incertae sedis*
Janumys Eaton and Cifelli, 2001
J. erebos Eaton and Cifelli, 2001
Suborder *incertae sedis*
Family Arginbaataridae Hahn and Hahn, 1983
Arginbaatar Trofimov, 1980
A. dmitrievae Trofimov, 1980
Suborder Cimolodonta McKenna, 1975
Superfamily and family *incertae sedis*
Paracimexomys group
Paracimexomys Archibald, 1982, type genus
P. priscus (Lillegraven, 1969), type species
P. magister (Fox, 1971)
P. perplexus Eaton and Cifelli, 2001
?*P. crossi* Cifelli, 1997
?*P. robisoni* Eaton and Nelson, 1991
Bryceomys Eaton, 1995
B. fumosus Eaton, 1995, type species
B. hadrosus Eaton, 1995
B. intermedius Eaton and Cifelli, 2001
Cedaromys Eaton and Cifelli, 2001
C. bestia (Eaton and Nelson, 1991), type species
C. parvus Eaton and Cifelli, 2001
Dakotamys Eaton, 1995
D. malcolmi Eaton, 1995
?*Paracimexomys* group
Ameribaatar Eaton and Cifelli, 2001
A. zofiae Eaton and Cifelli, 2001
Barbatodon Rădulescu and Samson, 1986
B. transylvanicum Rădulescu and Samson, 1986
Cimexomys Sloan and Van Valen, 1965
C. minor Sloan and Van Valen, 1965, type species
C. antiquus Fox, 1971a
C. magnus Sahni, 1972
C. judithae Sahni, 1972
C. gratus (Jepsen, 1930b) (including *C. hausoi* Archibald, 1982)
?*C. gregoryi* Eaton, 1993
Superfamily Djadochtatherioidea Kielan-Jaworowska and Hurum, 2001
Family Djadochtatheriidae Kielan-Jaworowska and Hurum, 1997
Djadochtatherium Simpson, 1925a, type genus
D. matthewi Simpson, 1925a
Catopsbaatar Kielan-Jaworowska, 1994
C. catopsaloides (Kielan-Jaworowska, 1974)
Kryptobaatar Kielan-Jaworowska, 1970
K. dashzevegi Kielan-Jaworowska, 1970a, type species
K. mandahuensis Smith et al., 2001
Tombaatar Rougier et al., 1997
T. sabuli Rougier et al., 1997
Family Sloanbaataridae Kielan-Jaworowska, 1974a
Sloanbaatar Kielan-Jaworowska, 1970a, type genus

TABLE 8.1. Continued

S. mirabilis Kielan-Jaworowska, 1970a
Nessovbaatar Kielan-Jaworowska and Hurum, 1997
N. multicostatus Kielan-Jaworowska and Hurum, 1997
?*Kamptobaatar* Kielan-Jaworowska, 1970a
K. kuczynskii Kielan-Jaworowska, 1970a
Family *incertae sedis*
Bulganbaatar Kielan-Jaworowska, 1974a
B. nemegtbaataroides Kielan-Jaworowska, 1974a
Chulsanbaatar Kielan-Jaworowska, 1974a
Ch. vulgaris Kielan-Jaworowska, 1974a
Nemegtbaatar Kielan-Jaworowska, 1974a
N. gobiensis Kielan-Jaworowska, 1974a
Superfamily *incertae sedis*
Family Cimolomyidae Marsh, 1889a
Cimolomys Marsh, 1889a, type genus
C. gracilis Marsh, 1889a, type species
C. clarki Sahni, 1972
C. major Russell, 1937
C. milliensis Eaton, 1993a
C. trochuus Lillegraven, 1969
Meniscoessus Cope, 1882c
M. conquistus Cope, 1882c, type species
M. borealis Simpson, 1927a
M. ferox Fox, 1971a
M. intermedius Fox, 1976b
M. major (Russell, 1937)
M. robustus (Marsh, 1889a)
?*Buginbaatar* Kielan-Jaworowska and Sochava, 1969
B. transaltaiensis Kielan-Jaworowska and Sochava, 1969
?*Essonodon* Simpson, 1927a
E. browni Simpson, 1927a
Family Eucosmodontidae Jepsen, 1940
Stygimys Sloan and Van Valen, 1965
S. kuszmauli Sloan and Van Valen, 1965, type species
S. cupressus Fox, 1989
?*Clemensodon* Krause, 1992
C. megaloba Krause, 1992
Family Microcosmodontidae Holtzman and Wolberg, 1977
Microcosmodontidae gen. et sp. indet. (Fox, 1989)
Superfamily Taeniolabidoidea Sloan and Van Valen, 1965
Family Taeniolabididae Granger and Simpson, 1929
"*Catopsalis*" Cope, 1882a
"*C.*" *foliatus* Cope, 1882a, type species
"*C.*" *johnstoni* Fox, 1989
"*C.*" cf. *joyneri* (Fox, 1989)
?*Bubodens* Wilson, 1987
B. magnus Wilson, 1987
Superfamily *incertae sedis*
Family Kogaionidae Rădulescu and Samson, 1996
Kogaionon Rădulescu and Samson, 1996, type genus
K. ungureanui Rădulescu and Samson, 1996
Hainina Vianey-Liaud, 1979
H. belgica Vianey-Liaud, 1979, type species
Hainina sp. A, sp. B, sp. (Csiki and Grigorescu 2000)
Family *incertae sedis*
Uzbekbaatar Kielan-Jaworowska and Nessov, 1992
U. kizylkumensis Kielan-Jaworowska and Nessov, 1992, type species
U. wardi Averianov, 1999
Viridomys Fox, 1971a
V. ferox Fox, 1971
Superfamily Ptilodontoidea Sloan and Van Valen, 1965
Family Ptilodontidae Gregory and Simpson, 1926
Kimbetohia Simpson, 1936
K. campi Simpson, 1936
Family Neoplagiaulacidae Ameghino, 1890
Mesodma Jepsen, 1940
M. ambigua Jepsen, 1940, type species
M. formosa (Marsh, 1889b)
M. garfieldensis Archibald, 1982
M. hensleighi Lillegraven, 1969
M. primaeva (Lambe, 1902)
M. senecta Fox, 1971a
M. thompsoni Clemens, 1963b
Parectypodus Jepsen, 1930a
P. simpsoni Jepsen, 1930a, type species
P. foxi Storer, 1991
Family Cimolodontidae Marsh, 1889a
Cimolodon Marsh, 1889a, type genus
C. nitidus Marsh, 1889a, type species
C. electus Fox, 1971a
C. similis Fox, 1971a

In addition a dentary fragment and isolated premolars, assigned to Multituberculata *incertae sedis* have been described from the Campanian or early Maastrichtian of Argentina.

semble Haramiyaviidae in subcircular outline of the upper molars, which have three rows of cusps, but the additional cusps in Haramiyaviidae are labial. Eleutherodontidae have a third row on the lingual side, but they differ from Theroteinidae in tooth shape, cusp form and number, and occlusion. Theroteinidae differ from Haramiyidae (*Thomasia*) in the presence of tubules in the enamel.

Genera. *Theroteinus* Sigogneau-Russell et al., 1986, type genus by monotypy.

Distribution. As for the suborder.

Genus *Theroteinus* Sigogneau-Russell, Frank, and Hemmerlé, 1986 (figure 8.1C)

Diagnosis. As for the family.

Distribution. Late Triassic (lower Rhaetian): France, Lorraine (Saint-Nicolas-de-Port).

Species. *Theroteinus nicolai* Sigogneau-Russell, Frank, and Hemmerlé, 1986, type species and *Theroteinus* sp. (Hahn et al., 1989).

Suborder Haramiyoidea G. Hahn, 1973

Diagnosis. Haramiyidans in which the lower molars are nearly opposite the upper molars, so that there is only transient contact with the more anterior upper molar. The median valley is longer than in Theroteinida, occupying most of the length of the tooth, except on anterior lower molariforms (unknown in Eleutherodontidae), where it is confined to the distal part of the tooth. Palinal occlusal movement developed to various extents; it is incipient in Haramiyaviidae and most extensive in Eleutherodontidae.

Included Families. Haramiyidae Simpson, 1947, type family; Haramiyaviidae Butler, 2000; Eleutherodontidae Kermack et al., 1998.

Distribution. Late Triassic–Middle Jurassic (Bathonian): Europe; Late Triassic: Greenland; Early Jurassic: North America; Late Jurassic: Africa.

Family Haramiyaviidae Butler, 2000

Diagnosis. Upper molariforms wide (subcircular), with additional cusps on the labial side. (In Theroteinidae and Eleutherodontidae the additional cusps are lingual.) On lower molariforms the highest labial cusp is the second, and it is placed more distally than the highest (first) lingual cusp. First upper molariform resembles the second, except that it is narrower mesially; on first lower molariform the longitudinal valley is confined to the distal part of the tooth. Palinal occlusal movement was probably short. Lower premolariforms with a single row of cusps (upper premolariforms unknown). Incisors unspecialized: I1–3 equal in size. The dentary (unknown in other Haramiyida) possesses a groove for postdentary bones; masseteric fossa does not extend forward below the molars as in multituberculates.

Genera. *Haramiyavia* Jenkins et al., 1997, type genus by monotypy.

Distribution. Late Triassic (?Norian–Rhaetic): East Greenland, Jameson Land (Fleming Fjord Formation).

Genus *Haramiyavia* Jenkins, Gatesy, Shubin, and Amaral, 1997 (figure 8.1B)

Diagnosis. As for the family.

Distribution. Late Triassic (?Norian–Rhaetic): East Greenland, Jameson Land (Fleming Fjord Formation).

Species. *Haramiyavia clemmenseni* Jenkins, Gatesy, Shubin, and Amaral, 1997, type species by monotypy.

Family Haramiyidae Simpson, 1947

Diagnosis. Differ from Haramiyaviidae in having upper molariforms longer than wide, with two rows of cusps, the supplementary labial cusps absent or represented by a cingulum. The basin is closed distally by a ridge ("saddle") between the labial and lingual cusps. On lower molars the first labial cusp is rudimentary or absent. Except on the anterior molariform, the highest (second) labial cusp is directly opposite the first lingual cusp and joined to it by a saddle that closes the basin mesially. Palinal chewing, in which cusps moved longitudinally in the basin of the opposing tooth, was well developed. Anterior upper molariform is more narrowed mesially, with lingual cusps confined to the distal part of the tooth. Referred upper incisors differentiated: I2 enlarged, with distal basal heel. The mandible is unknown.

Included Genera. *Thomasia* Poche, 1908 (synonym *Haramiya* Simpson, 1947), type genus; *Staffia* Heinrich, 1999; and ?unnamed genus from the Lower Jurassic Kayenta Formation of North America (Jenkins et al., 1983).

Distribution. Late Triassic–Early Jurassic: Europe; ?Early Jurassic: North America; Late Jurassic: East Africa.

Genus *Staffia* Heinrich, 1999 (figure 8.2C)

Diagnosis. This genus is represented by three isolated teeth of the type species *Staffia aenigmatica*, tentatively identified as a lower posterior premolar or anterior molar (holotype), m3, and M2. Heinrich (2001: 252) defined *Staffia* as follows: "A conspicuous difference [with *Thomasia*] in the preserved lower cheek teeth of *Staffia* is the central position of principal cusp a1 [for cusp numbering see figures 8.1 and 8.3A] at the front of the tooth crown and the presence of a large main notch between cusps a1 and a2. The upper cheek teeth of *Staffia* differ from *Thomasia* and other haramiyids . . . by the presence of four A and four B cusps, and the strong anterolabial cusped cingular ridge. This ridge broadens the width of the tooth crown considerably and increases the number of cusps."

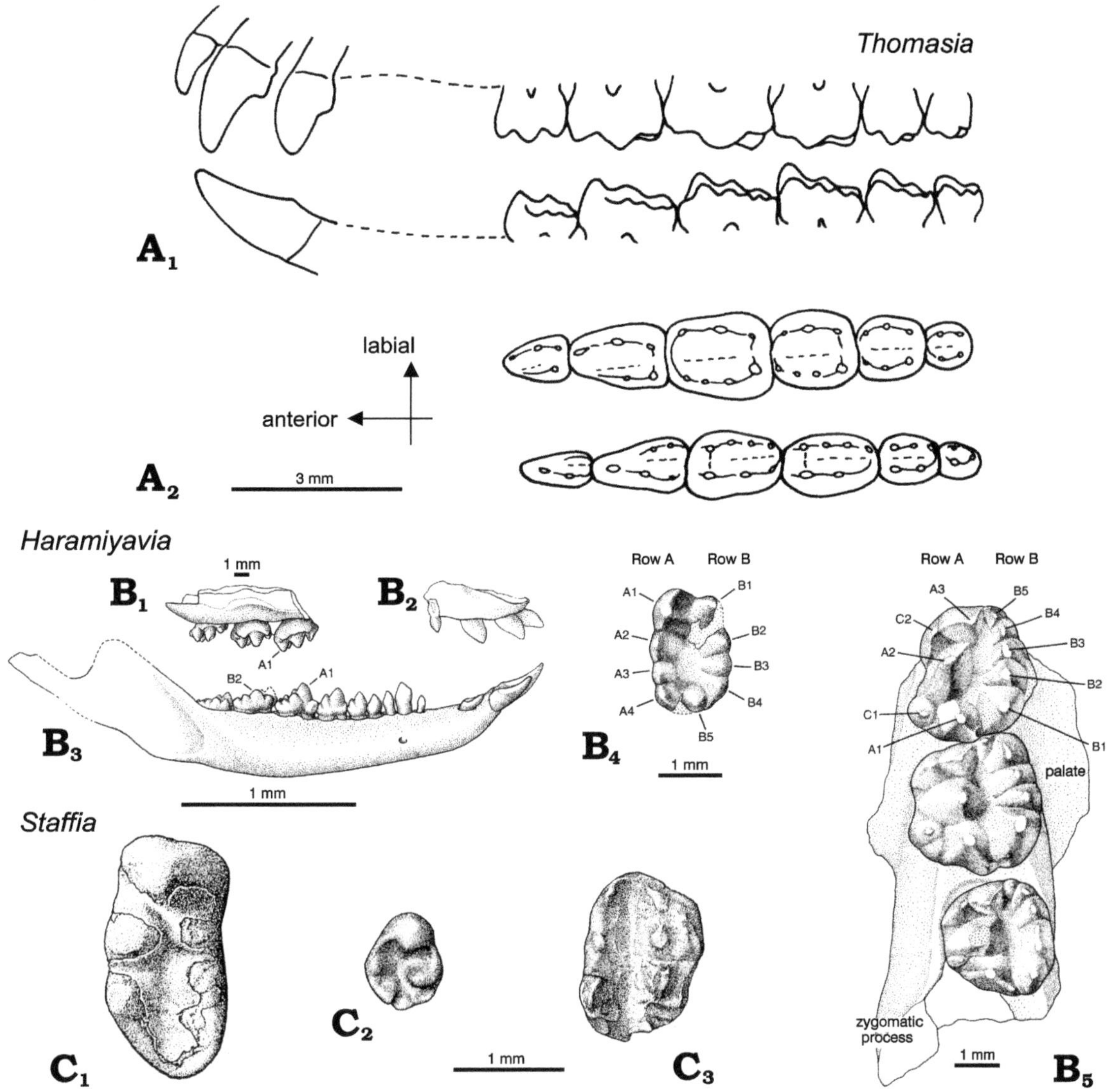

FIGURE 8.2. A. reconstruction of the dentition of *Thomasia antiqua* (= *Haramiya butleri*). The number of cheek teeth is unknown; three upper and one lower incisor (based on Sigogneau-Russell, 1989b), two premolars, two anterior molars, and two posterior molars are reconstructed. A, Left upper and lower dentition seen in labial view (A_1); left upper and right lower cheek teeth in crown view (A_2). B, *Haramiyavia clemmenseni,* right maxilla in lateral view (B_1); left premaxilla in lateral view (B_2); right dentary with teeth in lateral view (B_3); right lower molar in occlusal view (B_4); last upper molars in occlusal view (B_5). C, *Staffia aenigmatica,* ?right lower premolar (C_1), right lower posterior molar (C_2), left upper molar, possibly M2 (C_3), all in occlusal view. *Source:* A, modified from Butler and MacIntyre (1994); B, modified from Jenkins et al. (1997); C, modified from Heinrich (2001).

Species. *Staffia aenigmatica* Heinrich, 1999, type species by monotypy.

Distribution. Late Jurassic (Kimmeridgian–Tithonian): Africa, Tanzania, Tendaguru Beds (Middle Saurian Bed).

Genus *Thomasia* Poche, 1908
(figures 8.1A, 8.2A, 8.3, partly)

Synonyms: *Microlestes* Pleininger, 1847, preoccupied; *Pleiningeria* Krausse, 1919; *Microcleptes* Simpson, 1928a, preoccupied; *Haramiya* Simpson, 1947.

Diagnosis. As for the family.

Distribution. Late Triassic ("Middle Keuper," Norian): Germany (Halberstadt); Late Triassic (?Norian): Britain, Oxfordshire (Holwell Quarry); Late Triassic (early Rhaetian): France, Lorraine (Saint-Nicolas-de Port and Varangéville); Late Triassic (Rhaetian): Belgium, Gaume (Habay-la-Vieille) and Attert, and Luxembourg (Syren); Late Triassic (late Rhaetian): Switzerland (Hallau); Late Triassic ("Upper Keuper," late Rhaetian): Germany (Baden-Württemberg); Early Jurassic ("Liassic"–?Sinemurian): Britain, Wales (Pant Quarry).

Species. *Thomasia antiqua* Poche, 1908, type species; *T. hahni* Butler and MacIntyre, 1994; *T. moorei* (Owen, 1871); *T. woutersi* Butler and MacIntyre, 1994; and some species left in open nomenclature (Clemens, 1980a). For lists of synonyms see Butler and MacIntyre (1994).

Family Eleutherodontidae K. A. Kermack, D. M. Kermack, Lees, and Mills, 1998

Diagnosis. Haramiyoideans with upper molars wide, rhomboidal in outline and possessing three rows of cusps. The additional row is lingual and occludes lingual to the lower molar (a character shared with Theroteinidae, probably by convergence). Lower molars oval, with two rows of cusps that are continuous around the distal end. The largest cusps are at the distal end of the middle row on upper molars and at the mesial end of the labial row on lower molars; the mesial upper lingual cusp is enlarged. Minor cusps are numerous and variable. The longitudinal groove of upper molars, between the labial and middle cusp rows, extends the whole length of the tooth and is not interrupted by a saddle as in Haramiyidae. Palinal occlusion was extensive; an orthal component was retained as in Haramiyidae. Upper molar has a shorter and shallower groove between the middle and lingual cusp rows for occlusion with lower lingual cusps. The sides of the occlusal grooves of upper and lower teeth are covered with numerous minor transverse ridges ("fluting"). Eleutherodontidae differ from all other Haramiyida in having more numerous cusps (e.g., there are up to 10 upper labial cusps); in the anterior position of the large lower labial cusp, which projects mesially beyond the lingual row; and in the fluting.

Genera. *Eleutherodon* K. A. Kermack et al., 1998, type genus by monotypy.

Distribution. Middle Jurassic (late Bathonian): southern England, Oxfordshire, Kirtlington (Forest Marble).

Genus *Eleutherodon* K. A. Kermack, D. M. Kermack, Lees, and Mills, 1998
(figure 8.1 D)

Diagnosis and Distribution. As for the family.

Species. *Eleutherodon oxfordensis* K. A. Kermack, D. M. Kermack, Lees, and Mills, 1998, type species by monotypy.

RELATIONSHIPS WITHIN HARAMIYIDA

As mentioned earlier, Jenkins et al. (1997) argued that *Haramiyavia* had an orthal power stroke and it was on this basis that they excluded Haramiyida from Allotheria. Both the wear surfaces on the teeth and the structure of the dentary in *Haramiyavia* (with a relatively short, posteriorly situated masseteric fossa) are very different from what is found in multituberculates. Gambaryan and Kielan-Jaworowska (1995) argued that in the backward chewing stroke (which among mammals occurs also in Gondwanatheria, see Pascual et al., 1999, and in some haramiyidans, see earlier) in multituberculates the superficial masticatory muscles inserted more anteriorly than in all other mammals, including rodents. Moreover, the masseteric fossa in multituberculates extended far anteriorly, below p4. In rodents, which have longitudinal jaw movement (but with forward power stroke), the masseteric fossa extends relatively far anteriorly (see Gambaryan and Kielan-Jaworowska, 1995: figure 12). Butler (2000) argued that the anterior insertion of the masseter in multituberculates might have had some relation to the evolution of the sectorial premolars. It should be noted that Ride (1959) related the forward insertion of the masseter muscles in the Macropodinae, where they enter the masseteric canal, to the presence of the so-called plagiaulacoid premolars in this group of kangaroos. He stated (Ride, 1959: 49): "there is the direct mechanical advantage that is to be gained by inserting the deep masseter as close as possible to the position at which the shearing surface of the premolar is to be applied." The question to what extent, if any, the anterior insertion of masseter muscles in multituberculates was related to the control their plagiaulacoid premolars requires a special study. On the other hand, the masseter also inserts anteriorly in rodents, which do not have sectorial premolars but share with multituberculates the structure and function of the incisors, used for gnawing. Therefore, it seems more probable that the anterior insertion of the masticatory muscles in multituberculates and rodents is related to the longitudinal jaw movements characteristic of both these groups. However, because of different directions of the power stroke in the two groups, the muscles must have been working differently.

Butler's (2000) conclusion concerning the allotherian nature of *Haramiyavia* differs from that of Jenkins et al. (1997). Butler argued that the molar cusp patterns of *Thomasia* can be compared in detail with those of *Haramiyavia,* although there are differences in the number and relative positions of cusps (see figure 8.1A,B). The number of molars in *Thomasia* is unknown (figure 8.2A), but it seems that that the simplest teeth described by Sigogneau-Russell (1989b) are fourth molars. Butler further argued that in spite of differences between the two genera, which he discussed in detail, there can be little doubt that *Thomasia* and *Haramiyavia* are related. *Thomasia* seems to be more advanced morphologically in the direction of later multituberculates, in the reduction of the b row on m1 (if that tooth is homologous with the multituberculate p4), and in the character of I2. In view of the morphological resemblance between the molars of *Haramiyavia* and *Thomasia*, it is unlikely that their occlusal functions were very different. Butler agreed that in *Haramiyavia* there was no fully developed palinal movement such as in multituberculates, but argued that some degree of palinal movement, at least as much as in the traversodontid *Scalaenodon* (see Crompton, 1972b) might have occurred. The longitudinal cusp rows in *Haramiyavia* would have prevented any significant transverse movement, such as occurred in triconodont mammals with unilateral chewing (see also figure 8.1 and Butler, 2000: figure 3).

Based on these arguments, Butler (2000) assigned *Haramiyavia* (for which he erected a separate family, Haramiyaviidae) to the suborder Haramiyoidea, together with the families Haramiyidae and Eleutherodontidae, the two latter of which had a backward (palinal) power stroke.

The only haramiyidans recognized by Butler as having an orthal power stroke are the Theroteinidae (Sigogneau-Russell et al., 1986; Hahn et al., 1989), which he put in a separate haramiyidan suborder Theroteinida (Butler, 2000).

RELATIONSHIPS BETWEEN HARAMIYIDA AND MULTITUBERCULATA

Owing to the general similarity in gross morphology of multituberculate and haramiyidan teeth, students of the haramiyidans as a rule either assigned them together with Multituberculata to the Allotheria or treated them as a suborder of Multituberculata. The only exception to this rule was the paper by Jenkins et al. (1997), discussed earlier. This contrasts with results of phylogenetic analyses of early mammals of past two decades, which as a rule, because of the incompleteness of haramiyidan material, did not include this group in the analyses (see Kielan-Jaworowska, 1997, for review).

Butler's (2000) paper provides a thorough comparison of multituberculate and haramiyidan teeth and their modes of occlusion. In molars of both Haramiyida and Multituberculata there are two longitudinal rows of cusps, of which the labial row of the lowers occluded in the valley between the labial and lingual upper rows. This type of occlusion precluded transverse jaw movements. In addition, some Haramiyida (*Thomasia* and *Eleutherodon*) had a backward longitudinal (palinal) power stroke as did multituberculates. All other Jurassic mammals had transverse jaw movement, with unilateral chewing (Crompton, 1995).

A difference between Haramiyida and Multituberculata concerns the position of M2. One of the multituberculate apomorphies is the unique position of M2, which is displaced more lingually than M1 in all multituberculates (Krause and Hahn, 1990), so that the lingual cusps of M2 occlude lingual to the lingual row of m2 (see the later section "Multituberculata" and figure 8.8D). In *Haramiyavia*, and probably in *Thomasia*, M2 was situated directly posterior to M1, as in all mammals, except for multituberculates.

Butler (2000) offered a tentative comparison of molar cusps patterns in *Thomasia* and a generalized paulchoffatiid multituberculate, reproduced herein as figure 8.3.

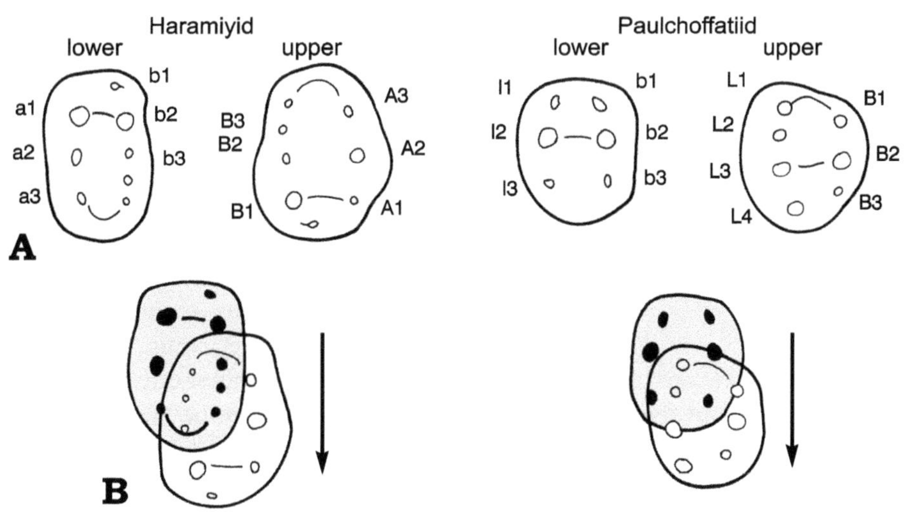

FIGURE 8.3. A, Tentative comparison of molar cusp patterns of a haramiyid (*Thomasia*) and a generalized paulchoffatiid. Paulchoffatiid cusps are labeled following Hahn and Hahn (1998c); labial is to the right, anterior up. B, Postulated occlusal relations at the beginning of the stroke; lower molars stippled; arrows indicate suggested traverse. It is suggested that on lower molars haramiyid a1 = paulchoffatiid l2, and haramiyid b2 = paulchoffatiid b2; on upper molars haramiyid B1 = paulchoffatiid L3, and haramiyid A2 = paulchoffatiid B2. *Source:* modified from Butler (2000).

He suggested that on the first upper molar, the haramiyid B1 may be homologous with the penultimate lingual cusp (L3 in most cases) of paulchoffatiids. In the same paper he argued that although the bladelike (sectorial) p4 is an obvious multituberculate apomorphy, it may be foreshadowed by the so-called *Thomasia II* of Sigogneau-Russell (1989b) and m1 of *Haramiyavia,* in which the labial row is reduced and confined to the distal part of the tooth. If *Thomasia II* is homologous with ml of *Haramiyavia,* as suggested by Butler, the three molars of that genus could represent p4 to m2 of multituberculates.

Butler (2000) further argued that the series *Haramiyavia-Thomasia-Eleutherodon* may indicate how the palinal occlusion of the Multituberculata could have evolved. In *Haramiyavia,* and also in *Theroteinus,* owing to the steplike arrangement of the teeth, the surfaces that meet at the end of the power stroke are tilted to face forward on upper molars and backward on lower molars, so that the force exerted by the lower teeth on the upper ones had a backward component. He noted that a backward force is primitive in synapsids (Barghusen, 1968; Bramble, 1978; Kemp, 1982) and the backward power stroke evolved, as demonstrated by Crompton (1972b), in herbivorous cynodonts (Traversodontidae, Tritylodontidae) independently of Allotheria. Butler admitted that if multituberculates were derived from haramiyids, in view of the dentary structure of *Haramiyavia* (which in the shape of the masseteric fossa resembles those of other primitive mammals such as, e.g., *Morganucodon*), much transformation of the mandible must have taken place.

As long as complete dentitions and dentaries of most non-multituberculate allotherian mammals (*Thomasia, Eleutherodon,* and *Theroteinus*), which might throw light on the multituberculate relationships, are unknown, Butler's hypothesis on the origin of multituberculates from among non-multituberculate allotherians must remain tentative (see also chapter 15 and figures 15.1 and 15.2).

Order Multituberculata Cope, 1884

INTRODUCTION

Multituberculates are the best-known group of Mesozoic mammals, though knowledge about them also comes from studies of Paleocene and Eocene taxa. Numerous upper and lower jaws with teeth and cranial fragments represent the oldest uncontested multituberculates, known from the Kimmeridgian of Portugal. Unfortunately no upper and lower teeth have been found in occlusion and the teeth are, as a rule, strongly worn. Next in stratigraphic sequence are fossils from the Morrison Formation of the United States and the Purbeck of England. The Morrison is conventionally regarded as Upper Jurassic (Simpson, 1929a; Lillegraven, 1979); the Purbeck is now considered to be Lower Cretaceous (see Allen and Wimbledon, 1991; Kielan-Jaworowska and Ensom, 1994, and chapter 2). Teeth, dentaries and maxillae with teeth, and very rarely incomplete skulls likewise represent the Morrison and Purbeck multituberculates. Other Early Cretaceous multituberculates are known by less complete material from Europe, Morocco, and the United States; the largest collection including fragmentary upper and lower jaws with teeth is known from the Early Cretaceous of Mongolia (Kielan-Jaworowska, Dashzeveg, and Trofimov, 1987), and the most complete Early Cretaceous multituberculate specimen (*Sinobaatar*) is known from China (Hu and Wang, 2002b). By far the best multituberculate collection, including hundreds of skulls (often with associated skeletons), comes from the Late Cretaceous of Mongolia. North American multituberculates of comparable age are numerous but less complete, except for a skull of *Meniscoessus* known from the Maastrichtian Hell Creek Formation (Kielan-Jaworowska, 1969a, 1971, 1974a; Novacek et al., 1994; Kielan-Jaworowska and Hurum, 1997; Weil and Tomida, 2001). One Late Cretaceous skull and isolated teeth are known from Europe (Rădulescu and Samson, 1996). Isolated teeth and a dentary fragment with teeth are known from the Late Cretaceous of Argentina. Complete skulls and postcranial skeletons are known from Paleocene and Eocene beds of North America and China, and isolated teeth are known from Europe.

Knowledge of multituberculates dates back to the middle of the ninteenth century. Falconer (1857) described the first multituberculates ever found—dentaries of *Plagiaulax* from the Purbeck of England. Owen (1871), in a monograph on Mesozoic mammals, described *Plagiaulax* and other Purbeck multituberculates. Marsh (1879b) reported the first North American multituberculate, the "plagiaulacidan" *Ctenacodon* from the latest Jurassic Morrison Formation, while Cope (1881b) described the first North American Tertiary multituberculate, *Ptilodus.* During the 1880s Cope and Marsh described numerous multituberculates from the Late Jurassic, Late Cretaceous, and Early Tertiary of North America. During these early times multituberculates were mostly referred to the Marsupialia and, more rarely, to Monotremata (see Hahn, 1983, and Hahn and Hahn, 1983, for comprehensive reviews of multituberculate literature until 1981, including ninteenth century references). At the end of the ninteenth-century, multituberculates were also discovered in the Paleocene of Europe (Lemoine, 1880), represented there by isolated teeth, and purported multituberculates were reported from the Wealden (Lower Cretaceous) of Great Britain (Lydekker,

1893; but see Woodward, 1911, Clemens, 1963a, and Clemens and Lees, 1971, for comments and revision).

Lower and upper multituberculate dentitions were not found in association until the turn of the century. The first multituberculate skull (associated with fragments of the postcranial skeleton) to be discovered, belonging to Paleocene *Ptilodus,* was described by Gidley (1909), who demonstrated that a carnivorous mode of life for multituberculates was not acceptable, but nevertheless allied multituberculates with marsupials. The second skull, described by Broom (1914), belonged to Paleocene *Taeniolabis* (referred to as *Polymastodon*). Broom reviewed multituberculate affinities with monotremes, concluding that the two groups may have originated independently from common Triassic ancestors. Granger (1915), in a short abstract, argued that multituberculates are neither marsupials nor monotremes but belong to a separate group of mammals. This concept was subsequently generally accepted owing mainly to the revisionary work done by Simpson in the 1920s.

The 1920s and 1930s brought several new discoveries. Simpson (1925a) described the first multituberculate skull from the Cretaceous of Asia (*Djadochtatherium*), from the collection assembled in Mongolia by the Central Asiatic Expeditions of the American Museum of Natural History. He subsequently authored (Simpson, 1928a, 1929a) two comprehensive monographs: the first on Mesozoic mammals housed in the Geological Department of the British Museum and the second devoted to American Mesozoic Mammalia. Simpson (1926b) published an attempt at reconstruction of multituberculate cranial musculature and habits, Simpson and Elftman (1928) provided a reconstruction of the hindlimb musculature of Paleocene ?*Eucosmodon,* Granger and Simpson (1929) reconstructed its foot, and Simpson (1937a) gave a detailed description of the multituberculate skull. Multituberculates were also found in Canada in the 1920s (Russell, 1926).

During World War II and in the 1950s only a few papers on multituberculates appeared (e.g., Ride, 1957). The 1960s and 1970s brought a proliferation of scientific publications, when the method of washing and screening sediments came into widespread use, especially in the United States and Canada. The Late Cretaceous of North America yielded numerous multituberculates during this time (e.g., Clemens, 1963b; Fox, 1968, 1971a, 1976b, 1980a, 1989; Lillegraven, 1969; Sahni, 1972; and subsequently Archibald, 1982; Wilson, 1987; Krause, 1992; and others). Later years witnessed the discovery of multituberculates in the early part of the Late Cretaceous (Eaton and Cifelli, 1988) and the Early Cretaceous (Krause et al., 1990; Eaton and Nelson, 1991; Eaton, 1995; Cifelli, 1997a; Eaton and Cifelli, 2001). The adoption of modified washing and screening methods also led to the discovery of a new rich multituberculate fauna of Purbeck age in Britain by Paul C. Ensom (Kielan-Jaworowska and Ensom, 1991, 1992, 1994). The specimens obtained by using washing and screening are mostly isolated teeth or fragments of jaws with teeth; there are no skulls or skeletons.

Beginning in 1964, Late Cretaceous mammals (see Gradziński et al., 1977, Fox, 1978, Averianov, 1997, and chapter 2 for age estimates), including numerous multituberculates, were collected in the Gobi Desert (Mongolia) by Polish-Mongolian Expeditions (1963–1971). In contrast to collections obtained by washing and screening, the mammals collected in Mongolia, at the surface of sandstone and often preserved in sandstone nodules, are as a rule exquisitely preserved and represent entire skulls, frequently associated with skeletons. Description of these materials (see Kielan-Jaworowska, 1969a, 1970a,b, 1971, 1974a,b, 1979, 1980, 1983, 1986, 1989, 1994, 1997, 1998; Kielan-Jaworowska and Dashzeveg, 1978; Kielan-Jaworowska and Sloan, 1979; Kielan-Jaworowska and Gambaryan, 1994; Gambaryan and Kielan-Jaworowska, 1995, 1997, 1998; Hurum et al., 1995, 1996; Kielan-Jaworowska and Hurum, 1997, 2001) increased the knowledge of multituberculate skull and brain structure as well as postcranial anatomy and enabled reconstruction of musculature and habits.

The technique of serial sectioning of multituberculate skulls employed by Kielan-Jaworowska et al. (1986; see also Hurum, 1992, 1994, 1997, 1998a,b) allowed detailed insight into the inner structure of the multituberculate skull and provided new information on, for example, cranial vasculature and other details unavailable otherwise.

The Soviet-Mongolian Paleontological and Soviet-Mongolian Geological Expeditions conducted fieldwork in Mongolia beginning in 1970, and (among other achievements) assembled a rich collection of generally very rare Early Cretaceous multituberculates, using washing and screening (Trofimov, 1980; Kielan-Jaworowska, Dashzeveg, and Trofimov, 1987). They also found multituberculates from the latest Cretaceous (Kielan-Jaworowska and Sochava, 1969; Trofimov, 1975).

Commencing in the 1960s German paleontologists organized fieldwork initiated by Walter Kühne, in the Guimarota Coal Mine in Portugal. They assembled a unique collection of Late Jurassic mammals, among them numerous multituberculate skulls, first reported by Kühne (1961a) and then described in a series of papers by Hahn (1969, 1973, 1977a,b, 1978a,b, 1981, 1985, 1987b, 1988, 1993) and Hahn and Hahn (1994, 1998a,b,c,d, 1999a). These studies cast new light on the evolution of multituberculates. In the 1980s and 1990s multituberculates were also found in Cretaceous rocks on the territory of former Soviet Union by Lev A. Nessov (Nessov, 1987; 1997; Kielan-Jaworowska and Nessov, 1992). Cuenca-Bescós et al. (1995) and Canudo and Cuenca-Bescós (1996) described multituberculate teeth from the Cretaceous of Spain (but see also earlier finds of Crusafont and Gibert, 1976, and Hahn and Hahn, 1992). Grigorescu (1984), Grigorescu et al. (1985), Rădulescu and Samson (1986), and Grigorescu and Hahn (1987) reported Cretaceous multituberculates from Romania. Of great paleogeographic interest are discoveries of multituberculates from Gondwanaland in Argentina (Bonaparte, 1986a,b, 1990, 1994, and references therein; Krause, Wunderlich et al., 1992; Krause, 1993; Krause and Bonaparte, 1993; Kielan-Jaworowska and Bonaparte, 1996, most of which is now assigned to Gondwanatheria, see also Pascual et al., 1999, and chapter 14), and from Morocco (Sigogneau-Russell, 1991c).

The Gobi Desert, which in the 1960s and 1970s yielded the unique multituberculate collections mentioned earlier, is still being explored by paleontologists. A series of Mongolian Academy–American Museum Expeditions that started in 1990 is continuing. This project has produced a spectacular collection of Late Cretaceous mammals from the previously known localities as well as new sites. Of special interest was the discovery in 1993 of the new locality Ukhaa Tolgod in the Nemegt Basin, which yielded several hundred Late Cretaceous mammal skulls, including multituberculates. Only part of this promising material has been described (Novacek et al., 1994; Dashzeveg et al., 1995; Rougier, Wible, and Novacek, 1996a; Rougier et al., 1997, Kielan-Jaworowska, 1998, and Wible and Rougier, 2000), but no doubt further investigation on this collection will lead to a better understanding of the structure and relationships of multituberculates.

Numerous studies on the dentition, skull structure, and postcranial skeleton of Tertiary multituberculates (e.g., Douglass, 1908; Gidley, 1909; Broom, 1914; Matthew and Granger, 1921, 1925; Granger and Simpson, 1928, 1929; Matthew et al., 1928; Simpson and Elftman, 1928; Jepsen, 1930a,b, 1940; Matthew, 1937; Simpson, 1937a; D. E. Russell, 1964; Sloan, 1966, 1979, 1981; Szalay and McKenna, 1971; Novacek and Clemens, 1977; Krause, 1980, 1982a,b, 1984, 1987a,b; Chow and Qi, 1978; Middleton, 1982; Krause and Jenkins, 1983; Miao and Lillegraven, 1986; Miao, 1988; Fox, 1990; Kielan-Jaworowska and Qi, 1990; Krause, Prasad, et al., 1997; Weil, 1998; Montellano et al., 2000; and many others) considerably increased knowledge about Multituberculata as a whole.

Knowledge of enamel microstructure (or ultrastructure) has cast light on interrelationships among multituberculate subgroups (e.g., Fosse et al., 1973, 1978, 1985, 2001; Sahni, 1979; Carlson and Krause, 1985; Krause and Carlson, 1986, 1987; Clemens, 1997; Koenigswald and Sander, 1997b; Wood and Stern, 1997; see also various papers in Koenigswald and Sander, 1997a, and references therein).

Attempts at cladistic analyses of multituberculates have been made since the 1980s (Archibald, 1982; Simmons and Miao, 1986; Krause and Carlson, 1987; Simmons, 1993; Rougier et al., 1997; Kielan-Jaworowska and Hurum, 1997, 2001), but have generally yielded poor resolution. Detailed studies on the cranial (especially braincase) structure of various early mammals and comparison with multituberculates (e.g., Wible, 1991; Rougier et al., 1992, Wible and Hopson, 1993, 1995; Hopson and Rougier, 1993; Lucas and Luo, 1993; Rougier, Wible, and Hopson, 1996; Wible and Rougier, 2000) have provided interesting and, in part, contradictory evidence on multituberculate relationships.

General reviews of multituberculates have been published by Hahn (1978c), Clemens and Kielan-Jaworowska (1979), Sloan (1979), Miao (1991, 1993), and Kielan-Jaworowska (1997), and multituberculate systematics have been proposed by Hahn and Hahn (1983), Stucky and McKenna (1993), McKenna and Bell (1997), and Kielan-Jaworowska and Hurum (2001).

Despite these noteworthy advances and the spectacular completeness of individual specimens, there are numerous gaps in knowledge of the structure and paleogeographic and stratigraphic distribution of multituberculates. Our understanding of this group during 150 years of investigation is gradually increasing and one may hope that the controversial issues concerning multituberculate evolution, and especially the place of multituberculates among mammals, will be soon resolved.

BRIEF CHARACTERIZATION

Multituberculates (Lat. *multum*—much, *multus*—numerous, *tuberculum*—tubercle, in reference to the multicusped molar teeth) are a monophyletic order of Mesozoic and Early Tertiary omnivorous or herbivorous mammals, assigned to the subclass Allotheria Marsh, 1880. They are characterized by multicusped premolars and molars covered with longitudinal rows of low cusps of the same height (except Paulchoffatiidae and Pinheirodontidae, where molar cusps may differ in height). There are three or two upper incisors, the canine is present in some early taxa, five upper premolars in "Plagiaulacida," but one to four in Cimolodonta; two upper molars; a single lower incisor, none to four lower premolars, and two lower molars.

An apomorphic feature of the upper dentition is the lingual shifting of M2 with respect to M1. In the structure of the dentary and the arrangement of the dentition (gnawing incisors and a diastema between them and the premolars) they show superficial similarity to rodents. The characteristic feature of most taxa is the structure of the shearing lower premolar (p4), or several lower premolars in earlier forms, which are bladelike, with a serrate upper margin and oblique ridges that extend along the labial and lingual surfaces.

The skull differs from those of most therian mammals in being wide and dorsoventrally rather than laterally compressed; the postorbital process in some groups, for example, in Djadochtatherioidea, is situated on the parietal, rather than on the frontal (figure 8.4). In relation to the posterior position of the postorbital process, the orbit in these groups is very large. The zygomatic arches are strong and expanded laterally; the glenoid fossa is large and flat, set off from the braincase. The jugal is on the medial side of the zygomatic arch and is not seen in a lateral view. It has been demonstrated (Gingerich, 1977) that multituberculates had a backward power stroke (and share this character with Gondwanatheria, see chapter 14), which resulted in a more anterior insertion of the masticatory muscles than in any other mammal group, including rodents. Different from crown mammals, the cochlea is not coiled but only slightly bent and there are three ossicles in the middle ear as in crown mammals. The epipubic bones are present. The postcranial skeleton shows several apomorphies. The most important post-cranial apomorphies are a very narrow pelvis, with pubes and ischia fused ventrally to form a keel; the platelike structure of the lesser trochanter and the presence of a subtrochanteric tubercle; hooklike proximolateral processes on the tibia and fibula; the calcaneus with a deep peroneal groove. A characteristic feature of the ankle joint is the calcaneo-Mt V contact. It has been argued (Kielan-Jaworowska and Gambaryan, 1994) that multituberculates had a sprawling posture and used a saltatory mode of locomotion; some may have been fossorial, others perhaps scansorial or arboreal, with a squirrel-like reversible hind foot and a prehensile tail (Krause and Jenkins, 1983). The multituberculate brain was probably similar to those of other early mammals in apparently having the midbrain exposed on the dorsal side, but in all available endocasts the midbrain and the vermis are obscured by a cast of a large vessel, called the superior cistern (Kielan-Jaworowska and Lancaster, 2004). A superior cistern also occurs in eutriconodontans (see chapter 7). Most multituberculates were small, of shrew or rat size; exceptions being the Cretaceous *Bubodens* and the Paleocene *Taeniolabis*, which attained the size of a beaver.

DISTRIBUTION

?Middle or Late Jurassic to Early Tertiary (Eocene) of the Northern Hemisphere and Gondwana, except the Australian region and Antarctica. The oldest uncontested multituberculates until recently have been known from the lower Kimmeridgian of Portugal, although purported multituberculates (isolated teeth) have been reported from the Upper Triassic of Belgium (Hahn et al., 1987) and the Middle Jurassic of England (E. F. Freeman, 1976a, 1979). Recently P. M. Butler (pers. comm., May 1999) informed us that in the collection from the Bathonian Forest Marble at Kirtlington, Oxfordshire, housed in the Natural History Museum in London, there are several isolated upper and lower premolars that he considers to belong to Multituberculata (see also Kermack et al., 1998, for description of multituberculate-like teeth from Kirtlington). The only named pre-Kimmeridgian multituberculate is *Mojo*, represented by an incomplete ?upper premolar, and its attribution to Multituberculata remains uncertain (Hahn et al., 1987). Given the scant record and the great hiatus between *Mojo* and the first uncontested multituberculates, we restrict the distribution of Multituberculata to begin with the Late Jurassic, with the possible exception of Middle Jurassic multituberculates just noted. The youngest multituberculate (*Ectypodus*) comes from the late Eocene (referred to as early Oligocene by Krishtalka et al., 1982; but see Swisher and Prothero, 1990) of North America. Most multituberculates are known from the Northern Hemisphere (see Clemens and Kielan-Jaworowska, 1979; Hahn and Hahn, 1983, 1999a; Kielan-Jaworowska and Ensom, 1992; and Kielan-Jaworowska and Hurum, 1997, 2001 for references). They have also been found in Gondwana: in the Early Cretaceous of Morocco (Sigogneau-Russell, 1991c); the Late Cretaceous of Madagascar (Krause and Grine, 1996; Krause, Prasad et al., 1997; Krause, 2000); and—very rare—in the Late Cretaceous of Argentina. We follow Pascual et al. (1999) and regard Argentinean Gondwanatheria, previously assigned to Multituberculata (e.g., Bonaparte, 1986a,b, 1990, 1994; Bonaparte and Rougier, 1987; Krause, Kielan-Jaworowska, and Bonaparte, 1992; Krause, 1993; Krause and Bonaparte, 1993; Kielan-Jaworowska and Bonaparte, 1996), as Mammalia *incertae sedis*. Multituberculates have not been recorded from Australia or Antarctica.

ANATOMY

SKULL

The multituberculate skull has been studied meticulously. In earlier works, summarized and extended by Simpson

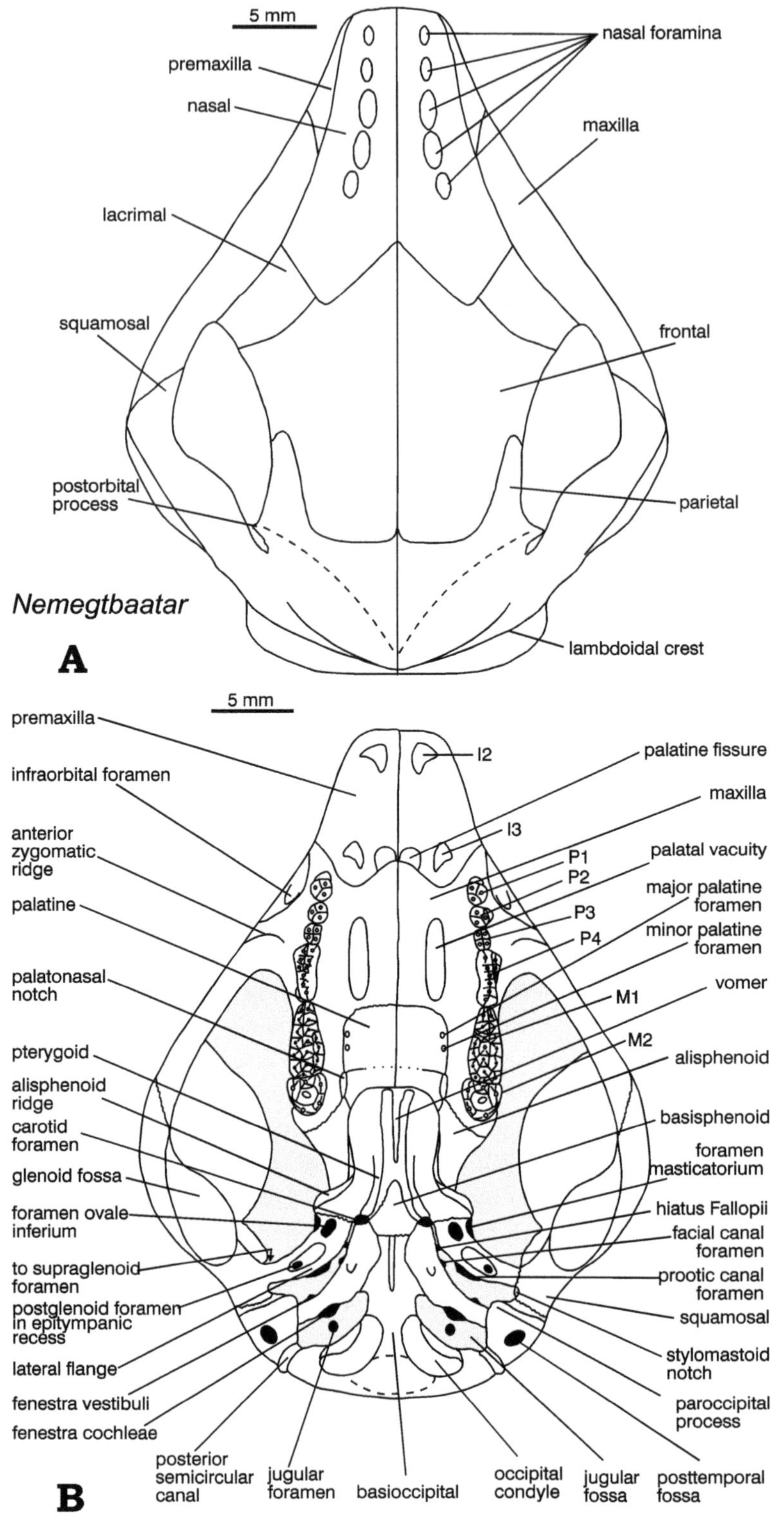

FIGURE 8.4. Reconstructions of multituberculate skull. A–C, *Nemegtbaatar* in dorsal, ventral, and lateral views. Note the structure of the orbit, which has a roof but does not have a floor, the position of pterygoids in the middle of choanal channels, and almost straight (only slightly bent) promontorium. D, Lateral wall of the braincase in Paleocene *Lambdopsalis.* Note the presence of an extensive alisphenoid, absent in other multituberculate taxa for which the braincase is known. E, *Sloanbaatar* in occipital view. *Source:* modified from: A–C, Kielan-Jaworowska et al. (1986); D, Miao (1988); E, Kielan-Jaworowska (1971).

(1937a), the multituberculate braincase was reconstructed on essentially therian lines, with the squamosal and alisphenoid each contributing extensively to the structure of the lateral wall. Kielan-Jaworowska (1970a, 1971, 1974a, see also Kermack and Kielan-Jaworowska, 1971; and Clemens and Kielan-Jaworowska, 1979) demonstrated the presence of a large anterior lamina of the petrosal in the multituberculate braincase, a structure similar to that in the braincase of monotremes but not therians. An exception is Paleocene *Lambdopsalis,* in which there is a large alisphenoid in addition to the anterior lamina of the petrosal (Miao, 1988). The position of the pterygoid bones,

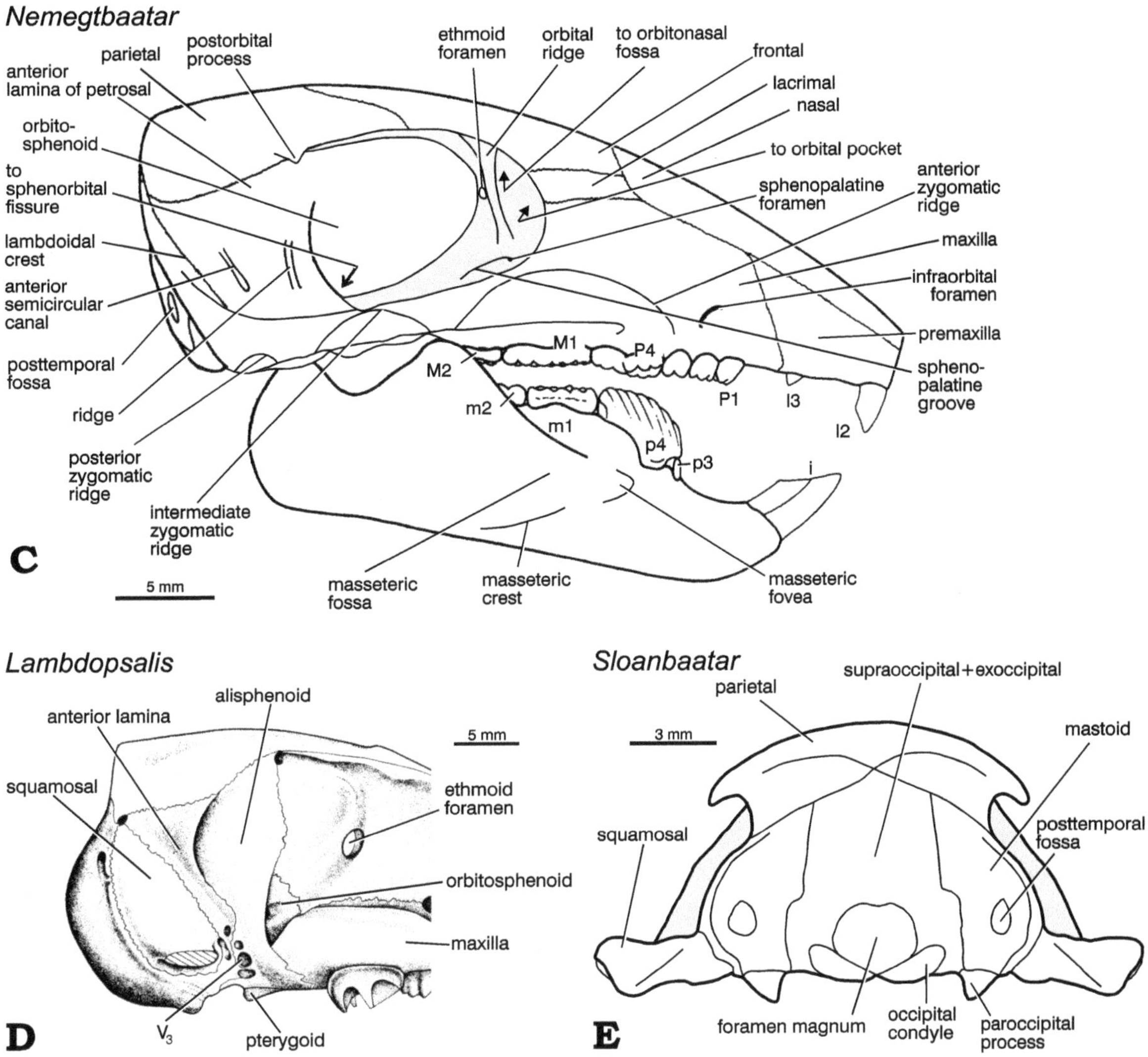

FIGURE 8.4. Continued

with prominent ridges on the roof of the oral pharnyx to form the choanal channels, was found in the Late Cretaceous *Kamptobaatar* (Kielan-Jaworowska, 1970a, 1971) and has been subsequently shown to be characteristic of other multituberculates as well (Hahn, 1981, 1987b; Kielan-Jaworowska et al., 1986). The tabular bone identified by Kielan-Jaworowska (1971) was subsequently recognized, on the basis of serially sectioned skulls, as the mastoid (Kielan-Jaworowska et al., 1986), whereas the ectopterygoid is an artifact (Hurum, 1998a). The description of the multituberculate skull that follows is based on the aforementioned papers, as well as on other important works, such as those of Miao (1988, 1993), Hahn (1993, and references to his earlier papers therein), Krause (1982a, 1987a), Hurum (1994, 1998a,b), Gambaryan and Kielan-Jaworowska (1995), Rougier, Wible, and Novacek (1996a), Rougier et al. (1997), Kielan-Jaworowska and Hurum (1997), Wible and Rougier (2000), and many others cited in those works.

The multituberculate skull is wide and massive, compressed dorsoventrally rather than laterally (as in therians), with a wide and bluntly pointed snout, bent downward.

Snout and Zygoma. A septomaxilla has not been recognized in any multituberculate (Wible et al., 1990). The nasals in Late Jurassic paulchoffatiids are relatively narrow, expanded posteriorly, but narrower than in Late Cretaceous and Tertiary forms (Hahn, 1969, 1977a). In a few cases they bear nasal foramina (figure 8.4A). In Late Cretaceous and Tertiary multituberculates the nasals are very extensive, strongly expanded posteriorly, and bear nasal foramina (referred to in earlier publications as vascular foramina, e.g., Simpson, 1937a; Kielan-Jaworowska, 1971; Kielan-Jaworowska et al., 1986; but see Miao, 1988; and

Hurum, 1994), often asymmetrically arranged. The function of the nasal foramina is not clear; Kielan-Jaworowska (1971) argued that they might transmit the lateral ethmoid nerve as in some lizards. Hahn and Hahn (1994) described the septomaxilla in *Pseudobolodon krebsi* from the Late Jurassic of Portugal; its presence, however, has been challenged by Wible and Rougier (2000). The lacrimal was reconstructed by Hahn (1969, 1978c) as small in Paulchoffatiidae, but it has been very poorly preserved, and one cannot depend on the reliability of this reconstruction. The lacrimal has a large facial exposure, being roughly rectangular in all the Djadochtatherioidea in which this region has been adequately preserved. It is not known whether a large lacrimal is a djadochtatherioid apomorphy or if it occurred in other Cimolodonta as well. The orbital exposure of the lacrimal, situated within the orbital pocket (see later section on "Orbit and Temporal Fossa"), is difficult to recognize in intact skulls, but it has been found in a serially sectioned skull of *Nemegtbaatar* (Hurum, 1994). It forms the medial margin of the infraorbital canal (the lateral one being formed by the maxilla) and part of the anteromedial wall of the orbital pocket. The lacrimal has not been found in any Tertiary multituberculate, including *Ptilodus* (Simpson, 1937a; Krause, 1982a) and *Lambdopsalis*[4] (Miao, 1988). The lacrimal foramen has been reported in *Paulchoffatia* (Hahn, 1969) and the nasolacrimal (lacrimal) canal is shown in the sectioned skulls of *Nemegtbaatar* and *Chulsanbaatar* (see Hurum, 1994).

The facial part of the premaxilla is directed almost vertically and its length varies in proportion to that of the maxilla. The maxilla is extensive; its facial part is almost vertical, with a large infraorbital foramen. Hahn (1985) demonstrated that the infraorbital foramen is double in some Kimmeridgian Paulchoffatiidae, whereas it is single in Berriasian *Bolodon.* He argued that the double infraorbital foramen is a plesiomorphic feature for mammals and regarded the acquisition of a single foramen as an apomorphy of geologically younger multituberculates. However, a double infraorbital foramen also occurs in the allodontid *Ctenacodon* from the Morrison Formation and in the specialized Early Cretaceous *Arginbaatar* (Kielan-Jaworowska, Dashzeveg, and Trofimov, 1987), as well as in the aberrant Late Cretaceous cimolomyid *Meniscoessus* (Archibald, 1982; Clemens, 1973a; Hahn, 1985). Miao (1988) challenged the conclusions of Hahn (1985) on the basis of (among others) a specimen of *Lambdopsalis,* which has a single infraorbital foramen on one side of the skull and a double on the other. He argued that the high degree of variability in the number of the infraorbital foramina in mammals speaks against the taxonomic and phylogenetic utility of this character.

The maxilla contributes extensively to the zygomatic arch, being in contact with the squamosal, which builds the posterior part of the arch. The arch is stout and bears two prominent zygomatic ridges (anterior and intermediate) and a less clear posterior zygomatic ridge, all interpreted by Gambaryan and Kielan-Jaworowska (1995) as places of origin of parts of the masseter muscle (see later section on "Musculature"). The jugal cannot be seen on the lateral aspect of the arch, being placed on its medial side. It is irregular in shape in Paleocene *Ptilodus* (Hopson et al., 1989), but relatively small oval-shaped in *Kryptobaatar* (Wible and Rougier, 2000). In most taxa the zygomatic arches are strongly expanded laterally and are extensive, and the flat glenoid fossa is situated on the expanded posterior part of the arch. The glenoid is situated lateral to the petrosal; it is attached to the braincase by a constricted pedicle rather than being very close to the braincase and anterolateral to the petrosal as in therians and monotremes. It is relatively very large and flat (rather than concave as in most therians), roughly oval in outline in many taxa. Being situated within the arch, the glenoid stands out from the braincase. There is no postglenoid process. Gambaryan and Kielan-Jaworowska (1995) argued that a glenoid fossa set apart from the braincase is a plesiomorphic feature for mammals, but that the multituberculate glenoid is derived as it is positioned much further away from the braincase than the lower jaw joint in other early mammals or cynodonts. This is because of the increase of masticatory musculature and lateral expansion of the zygomatic arches related to it. A unique character of the multituberculate glenoid fossa is that it slopes posteriorly. Gambaryan and Kielan-Jaworowska (1995) explained the backwardly sloping glenoid as being related to the structure of the condylar process, which faces posteriorly or posterodorsally, rather than dorsally (as in therians). These authors argued (p. 84): "Because of such a position of the condylar process, and because of the slightly concave profile of the upper premolars and molars, only at the backward inclination of the glenoid fossa can lower and upper molars and posterior premolars come into occlusion during the power stroke."

Palate. The palatal processes of the premaxilla are extensive and concave (figure 8.4B). Medial to I3 (which in Djadochtatherioidea and Eucosmodontidae are situ-

[4] *Lambdopsalis* has been referred to by various authors as either of Paleocene or Eocene age (see, e.g., Chow and Qi, 1978; Miao, 1988; Kielan-Jaworowska and Qi, 1990). We follow Meng and Wyss (1995), who refer to the Bayan Ulan Formation yielding *Lambdopsalis* as belonging to transitional Paleocene–Eocene beds.

ated on the palatal part of the premaxilla, rather than on its margin; see "Dentition" later) have large incisive foramina (referred to also as palatine fissures, e.g., Kielan-Jaworowska, 1971; Kielan-Jaworowska et al., 1986).

The palatal vacuities are either absent (e.g., Taeniolabididae and Djadochtatheriidae), or present and single (e.g., *Ptilodus* and *Nemegtbaatar*), or double (Sloanbaataridae). The horizontal part of the palatine bone is, as a rule, extensive and roughly quadrangular in Djadochtatherioidea, but subtrapezoidal in, for example, *Lambdopsalis.* There is a pair of large major palatine foramina and one or two pairs of small minor palatine foramina. The postpalatine torus is often present (prominent in Djadochtatherioidea), with a fissurelike palatonasal notch at the side (Kielan-Jaworowska et al., 1986; referred to by Kielan-Jaworowska, 1971 as the palatonasal foramen). The pterygoids are provided with two prominent longitudinal ridges (pterygo-palatine ridges of Barghusen, 1986) situated along the midline of the choanal channel on either side. The choanae are thus divided longitudinally by the vomer and the pterygo-palatine ridges into four channels: two medial, between the vomer and the pterygoid ridge; and two lateral, between the ridge and the lateral wall of the pterygoid, adhering to the alisphenoid (Kielan-Jaworowska, 1971; see also Wible and Rougier 2000). The roof of the choanal channel, above the vomer and the pterygoid, is built of the presphenoid-parasphenoid. The surface lateral to the choanae is formed by the alisphenoid, which anteriorly contacts the maxilla. Posteriorly on the alisphenoid (ventral view) there is an alisphenoid ridge. Kielan-Jaworowska (1971) mistook the anterior part of the alisphenoid in *Kamptobaatar* for an ectopterygoid.

Cranial Roof. The frontals are inserted medially between the nasals (figure 8.4A). The size of the frontals strongly varies among the multituberculate groups and is poorly known in the Late Jurassic Paulchoffatiidae (Hahn, 1969). If correctly reconstructed in paulchoffatiids, the frontals are large and form most of the upper margin of the orbit, each half of the frontoparietal suture being convex posteriorly. In Late Cretaceous Djadochtatherioidea, the frontals are very extensive, forming almost the entire upper rim of the orbit, and the frontoparietal suture as a whole is convex posteriorly. In these multituberculates the parietals form the whole cranial roof posteriorly; they are convex and the sagittal crest is absent or very small. There are prominent lambdoidal crests. In some multituberculates (e.g., in Djadochtatherioidea) the postorbital process is situated on the parietal, rather than on the frontal as in therians (with few exceptions). The position of the process is, however, unknown in most multituberculates.

In Tertiary Ptilodontidae and Taeniolabididae, the frontals are pointed posteriorly and are small in comparison with those in djadochtatherioids. Simpson's (1937a) statement that the frontals are excluded from the orbital rim in *Djadochtatherium* is not valid. A very well-preserved skull of *Djadochtatherium* was found by the Japanese-Mongolian Expedition in 1994 at the Tögrög (previously Toogreeg) locality in the Gobi Desert in beds equivalent to the Djadokhta Formation. Kielan-Jaworowska and Hurum (1997) demonstrated that in that skull, the frontals are extensive as in all other djadochtatherioids and form a large part of the orbital rim. In *Ptilodus* the frontals still contribute to the upper rim of the anterior part of the orbit, but in Taeniolabididae, in particular in *Taeniolabis,* they are reduced in size and are excluded from the orbital rim. Gambaryan and Kielan-Jaworowska (1995) argued that the reduction in the size of the frontals in Taeniolabididae was related to the increase of the cutting function of the incisors and the augmentation of the temporal muscles. The parietals in small Cretaceous forms are rounded and the sagittal crest is small or lacking. In Taeniolabididae there is a prominent sagittal crest and the parietals slope away from it ventrolaterally, forming extensive concave areas for accommodation of the temporal muscles.

Occiput. The occiput forms an extensive plate sloping anteroventrally, with the supraoccipital-exoccipital comprising its central part and the foramen magnum situated ventrally (figure 8.4C,E). Between the supraoccipital-exoccipital and the squamosal there is a mastoid portion of the petrosal bone, pierced in djadochtatherioids by a large posttemporal fossa, which transmitted the arteria diploetica magna (Lillegraven and Hahn, 1993; Wible and Hopson, 1995), referred to by Kielan-Jaworowska et al. (1986) as the posttemporal recess vessel. The mastoid is bounded ventrally by the paroccipital process of the petrosal. The oldest known multituberculate occipital plates are those of Late Cretaceous *Kamptobaatar* and *Sloanbaatar* (Kielan-Jaworowska, 1971). The similarity of the general pattern of this plate to that in therapsids such as tritylodontids (e.g., Kühne, 1956), in particular in the presence of a lateral bone pierced by a large posttemporal fossa, induced Kielan-Jaworowska (1971) to identify this bone as a tabular. Study of serially sectioned skulls of two Late Cretaceous multituberculates, *Nemegtbaatar* and *Chulsanbaatar* (Kielan-Jaworowska et al., 1986), demonstrated that the bone in question houses the semicircular canals and should be identified as the mastoid part of the petrosal. In the same study, it was shown that the posttemporal fenestra in some multituberculates (e.g., *Catopsalis*) is reduced to a small opening. One of the peculiarities of the multituberculate skull is that the semicircular canals are often seen as ridges on the surface of the skull. This holds especially for small taxa, in which the bones of the skull are very thin (e.g., *Kamptobaatar* and *Chulsanbaatar*). The posterior semicircular canal is often visible on the occipital plate as a roughly vertical ridge situated medial to the posttemporal fossa. The occipital plate has been preserved in many djadochtatherioids (Kielan-Jaworowska, 1974a; Kielan-Jaworowska et al., 1986), but has not been completely preserved in the Late Jurassic Paulchoffatiidae. Isolated petrosals of paulchoffatiids, however, allowed Hahn (1988) and Lillegraven and Hahn (1993) to conclude that the squamosal contributed laterally to the structure of the occipital plate; the posttemporal fossa being made by the mastoid and the squamosal, while the pedicles of the occipital condyle are formed in part by the petrosal. Other than djadochtatherioids, this plate has been reconstructed only in Paleocene *Lambdopsalis* (Miao, 1988), shows general similarity to that in djadochtatherioids, except that the lateral portions of the plate are convex posteriorly, owing to an unusual inflation of the vestibule.

Orbit and Temporal Fossa. The orbit is very large and widely open posteriorly, passing gently into the temporal fossa (figure 8.4C). The anterior part of the orbit extends anteriorly, at least in all well-preserved skulls of djadochtatherioids, as a pocketlike structure (orbital pocket of Gambaryan and Kielan-Jaworowska, 1995). The orbital pocket has no floor but is roofed anterolaterally and dorsally by the maxilla and frontal. This contrasts with the condition in Theria, where (with few exceptions, see discussion in Gambaryan and Kielan-Jaworowska, 1995) the anterior part of the orbit has a floor but is not roofed. The orbital wall to the rear of the orbital pocket and part of the temporal fossa anterior to the postorbital process are also roofed by the frontal and parietal, at least in all the djadochtatherioids. The roof is best seen in ventral view of the skulls and in the sectioned skulls (Hurum, 1998a: figure 16). Because of the incomplete state of preservation of non-djadochtatherioid multituberculate skulls, it cannot be stated with any certainty whether the bony roof over the orbit and part of the temporal fossa is characteristic of all multituberculates. It may be a djadochtatherioid apomorphy, as it seems to be absent at least from the Taeniolabididae. Immediately behind the orbital pocket, in the roof (built by the frontal) which overhangs the orbit, there is a rounded fossa described for *Kamptobaatar* and *Sloanbaatar* (two fossae in the latter) and designated by Kielan-Jaworowska (1971) as the orbitonasal fossa. The function of this fossa (found also in other djadochtatherioids, see Kielan-Jaworowska et al., 1986) is unknown; it may be for a gland. Posterior to the orbital pocket on the wall of the orbit there is a vertical ridge (orbital ridge of Kielan-Jaworowska et al., 1986), immediately posterior to which, in *Nemegtbaatar*, is the ethmoid foramen.

The oldest known multituberculate orbital walls are those of Late Jurassic *Kuehneodon* and (less complete) *Henkelodon* from Portugal (Hahn, 1977a). These orbits show a structure similar to very well-preserved orbits of the Late Cretaceous djadochtatherioids from Mongolia, *Kamptobaatar* (Kielan-Jaworowska, 1971), *Nemegtbaatar*, and *Chulsanbaatar* (Kielan-Jaworowska et al., 1986; Hurum, 1994). The bones in the orbital wall of the three djadochtatherioid genera differ in their proportions. In *Kamptobaatar* there is a small orbital process of the maxilla anteriorly, hidden in part in the orbital pocket mentioned earlier; the infraorbital canal opens at the anterior end of the pocket. A small orbital process of the frontal, partly obscured by the overhanging thickened margin of the frontal, extends along the dorsal border of the orbit and tapers ventrally at the sphenopalatine foramen (see later). Ventrally there is a small exposure of the palatine, which tapers at the sphenopalatine foramen. Most of the orbital wall is occupied by a very large, fan-shaped orbitosphenoid. In the anterior part of the orbit, at the junction of four bones (maxilla, frontal, orbitosphenoid, and palatine), is the sphenopalatine foramen. An unnamed groove, which becomes shallower posteriorly, extends from the sphenopalatine foramen along the lower part of the orbitosphenoid. At the middle of the orbitosphenoid, closer to the anterior margin, there is the ethmoid foramen, with a groove that extends from it ventrally. In *Kamptobaatar* the orbitosphenoid contacts the parietal posterodorsally, the anterior lamina of the petrosal posteriorly, and the alisphenoid posteroventrally. In *Nemegtbaatar* and *Chulsanbaatar* (Hurum, 1994; but see emended reconstruction in Hurum, 1998a: figure 14) the frontal has a larger exposure on the orbital wall than in *Kamptobaatar*, and in *Nemegtbaatar* the palatine contributes to the structure of the orbital wall, as in *Kamptobaatar*.

Hurum (1994) found an orbital process of the palatine in *Nemegtbaatar* and argued that one of the diagnostic characters of Multituberculata proposed by Wible (1991)—absence of the orbital process of the palatine—is not valid (see also Rougier et al., 1997; Hurum, 1998a: figure 16). In *Lambdopsalis* from the Paleocene-Eocene boundary (Miao, 1988: figure 17) the arrangement of bones in the orbit is different from that in Djadochtatherioidea. In *Lambdopsalis* the frontal is the largest bone, the sphenopalatine foramen lies at the boundary between the maxilla and the frontal, and the ethmoid foramen is placed within the frontal, while the orbitosphenoid is reduced to a small posteroventral element.

The sphenorbital fissure (synonyms: orbital fissure and anterior lacerate foramen) in therians is a foramen between the orbitosphenoid and alisphenoid. In monotremes it lies between the orbitosphenoid and the anterior lamina of the petrosal. The fissure allows the passage of the contents of the cavum epiptericum (see later) forward to the orbit. In multituberculates it has been described in Late Jurassic *Pseudobolodon* (Hahn, 1981) as a large opening connected to the cavum epiptericum by a long channel. In well-preserved skulls of djadochtatherioids the sphenorbital fissure is scarcely visible in lateral view, being the forward opening of the cavum epiptericum into the orbit. It transmits the ophthalmic and maxillary branches of the trigeminal nerve, V_1 and V_2 (Kielan-Jaworowska et al., 1986; Hopson and Rougier, 1993; Wible and Hopson, 1993). Kielan-Jaworowska (1971) reconstructed the sphenorbital fissure in *Kamptobaatar* as a large area extending anteriorly from the boundary of the orbitosphenoid with the anterior lamina of the petrosal (see later) and occupying a large posteroventral part of the orbitosphenoid. Better-preserved skulls of *Nemegtbaatar* and *Chulsanbaatar* (Kielan-Jaworowska et al., 1986; Hurum, 1998a) confirmed the presence of a sphenorbital fissure, but one that is less extensive than that reconstructed for *Kamptobaatar* (possibly owing to damage of the orbitosphenoid).

The structure of the temporal fossa (the lateral wall of the braincase) is not known in Late Jurassic multituberculates. The earliest forms in which it has been described are Late Cretaceous djadochtatherioids. The structure of this region among the djadochtatherioid genera differs only in details. The anterior lamina of the petrosal has a substantial contribution to the lateral wall of the braincase. In all djadochtatherioids it is in contact with the parietal dorsally and squamosal posteriorly, while the alisphenoid forms a very small ventral element (see reconstructions of the lateral wall of the braincase in various djadochtatherioids by Kielan-Jaworowska, 1971, 1974a; Clemens and Kielan-Jaworowska, 1979; Kielan-Jaworowska et al., 1986; Wible and Hopson, 1993; and Hurum, 1998a). In *Lambdopsalis* the lateral wall of the braincase has a different structure (Miao, 1988, see also chapter 3). There is a large alisphenoid that tapers dorsally at the postorbital foramen; posteriorly it is in contact with the anterior lamina of the petrosal, which is narrow ventrally and widens dorsally. Posteriorly the anterior lamina is in contact with the squamosal (figure 8.4D).

The foramen ovale (for the mandibular branch of the trigeminal nerve, V_3) was first described by Simpson (1937a) for Paleocene *Ptilodus* as being divided into two foramina, designated foramen ovale inferium (facing ventrally) and foramen masticatorium (facing ventrolaterally). Both foramina pierce the petrosal. Such a division of V_3 has also been found in several djadochtatherioid taxa (Kielan-Jaworowska, 1971, 1974a; Kielan-Jaworowska and Dashzeveg, 1978; Kielan-Jaworowska et al., 1986; Hurum, 1998a) and in the Late Jurassic paulchoffatiids (Hahn 1988; Lillegraven and Hahn 1993). The djadochtatherioid *Kamptobaatar* is an exception in this respect, as it has the foramen ovale divided into five foramina, somewhat asymmetrically arranged on either side of the skull (Kielan-Jaworowska, 1971). The division of the foramen ovale, characteristic of multituberculates, is rare in other mammals, but has been described in some rodents (Hill, 1935).

Basicranium and Ear. The basicranium is generally short and very wide. Hahn (1988, and references therein) described it in Late Jurassic Paulchoffatiidae. Kielan-Jaworowska (1971, 1974a) Kielan-Jaworowska and Dashzeveg (1978), Kielan-Jaworowska and Sloan (1979), Kielan-Jaworowska et al. (1986), and Hurum et al. (1996) described and figured basicrania in several taxa of Late Cretaceous djadochtatherioids. Among Tertiary taxa, the basicranium has been described in Paleocene *Ptilodus* (Simpson, 1937a) and *Lambdopsalis* (Miao, 1988).

The differences among these groups are rather minor, except for *Lambdopsalis,* in which the shape of the basicranium has been changed, owing to enormous inflation of the vestibule. The basisphenoid and basioccipital occupy the central part of the basicranium. In Paulchoffatiidae the basisphenoid is separated from the basioccipital by a distinct suture and the carotid foramen (for the internal carotid artery) pierces the basisphenoid. Apparently, the artery entered the pituitary fossa from below, as in primitive mammals—monotremes and some marsupials. In djadochtatherioids, as exemplified by the sectioned skull of *Nemegtbaatar,* the carotid foramen is situated just anterior to the promontorium, at the junction of the petrosal, basisphenoid, alisphenoid, and possibly pterygoid (as is the foramen lacerum medium of extant mammals). The carotid canal entered the pituitary fossa from the side as in some marsupials and in placental mammals. The basioccipital widens posteriorly and is surrounded laterally by the petrosals, the medial, prominent parts of which are the promontoria.

The petrosal as a whole (figure 8.5A) is roughly oval in ventral view (see detailed description of multituberculate petrosal by Luo, 1989). The ascending part of the anterior lamina is not visible in ventral aspect. The promontorium (ventral prominence of the petrosal, which houses the cochlea) in multituberculates is narrow, oriented anteromedially in the basicranium, almost parallel sided, and bent (Fox and Meng, 1997). In Paulchoffatiidae the promontorium widens posteriorly and is less bent than in later multituberculates. The fenestra cochleae and fenestra vestibuli are situated, respectively, at the medial and lateral sides of the posterior margin of the promontorium, rather than on its medial and lateral margins (Hahn, 1969, 1988; Lillegraven and Hahn, 1993). The jugular foramen

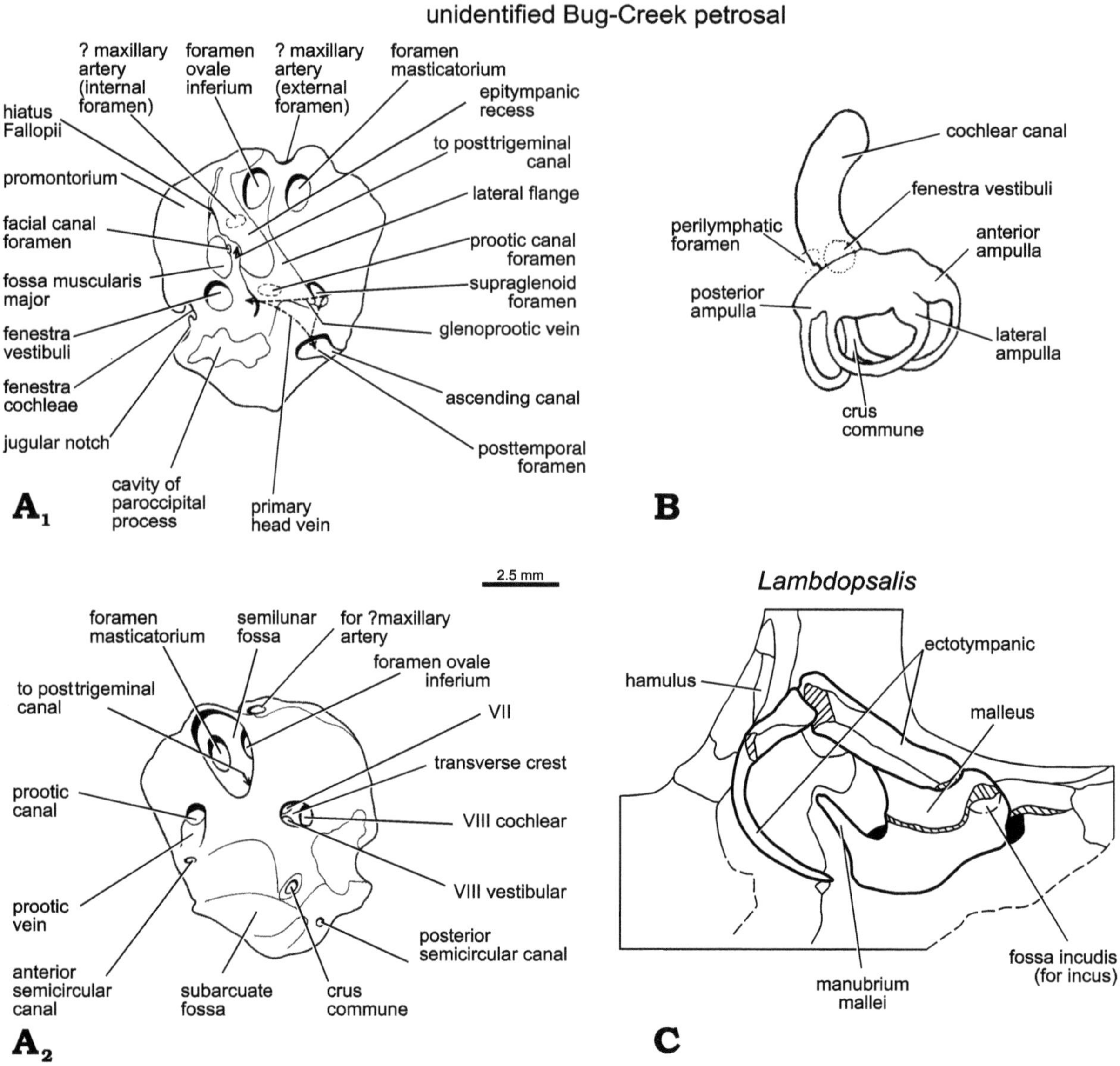

FIGURE 8.5. A, Unidentified Bug Creek petrosal bone in tympanic (A_1) and cerebellar (A_2) side. B, Outlined restoration of the inner ear based on another unidentified Bug Creek petrosal. C, Reconstruction of the left middle ear of *Lambdopsalis,* based on three specimens. The scale bar is for A, the scales for B and C have not been given. *Source:* modified from: A, Kielan-Jaworowska et al. (1986); B, Fox and Meng (1997); C, Meng and Wyss (1995).

is large, placed medially and close to the fenestra cochleae, but there is no jugular fossa; the hypoglossal foramen is double.

In djadochtatherioids the promontorium is less regular and consists of two parts. The anterior part is arranged at about 40–55° to the longitudinal axis of the skull, while the narrow posterior part, which borders the fenestra vestibuli and fenestra cochleae, swings out and is directed more laterally, at about 60–65° to this axis. The posterolateral part of the promontorium forms a bar that divides the cochlear fossula (jugular fossa) and the fossa for the stapedius muscle (posterolateral part of the facial sulcus, see Hurum et al., 1996). A similar prolongation of the promontorium to the rear of the fenestra vestibuli and fenestra cochleae is also present in *Sinoconodon, Morganucodon, Dinnetherium,* triconodontids, and monotremes (Wible, 1991) and has been designated the crista interfenestralis by Wible et al. (1995).

In Djadochtatherioidea the posterior part of the promontorium is flanked anterolaterally by the facial sulcus and posteromedially by the deep and large jugular fossa. In the middle of the roof of the jugular fossa there is a jugular foramen. The hypoglossal foramen has not been recognized in Djadochtatherioidea and may be confluent with the jugular foramen. The facial sulcus in all multituberculates is a deep groove between the promontorium and the lateral flange (crista parotica of embryos of extant mammals). Anterolateral to the facial sulcus lies the hiatus Fallopii (for the greater petrosal nerve), developed either as a separate foramen or as a notch at the most anterior part of the facial sulcus (figure 8.5A$_1$). Anteriorly, the facial sulcus is connected with the posterior end of the semilunar fossa by the posttrigeminal canal. Cranial nerve VII (facial nerve) opens into the facial sulcus (facial canal foramen); its exit is at the posterolateral end of the facial sulcus by the stylomastoid notch, homologous to the foramen stylomatoideum primitivum of extant mammals. The opening of the prootic canal (vascular foramen housing the prootic vein) is posterolateral to the facial canal foramen, at the anterolateral wall of the sulcus. On the posteromedial wall of the facial sulcus is the fenestra vestibuli, an opening from the cochlear capsule forming the site of the footplate of the stapes. Kielan-Jaworowska et al. (1986) demonstrated that in some very well-preserved djadochtatherioids, the facial sulcus is bridged by a delicate strand of bone (often broken) that divides the facial sulcus into anterior and posterolateral parts. In some unidentified petrosals from the Hell Creek Formation, referred to further as Bug Creek petrosals (from the name of the locality), this posterolateral part is enclosed in a bony chamber.

In *Ptilodus* the fenestra vestibuli and fenestra cochleae are situated on the sides of the posterior part of the promontorium, which does not swing laterally as in djadochtatherioids. The jugular foramen is situated within the jugular fossa, and there is a separate hypoglossal foramen that opens into the condylar fossa anterior to the occipital condyle (Simpson, 1937a). In the unidentified Bug Creek petrosals described by Kielan-Jaworowska et al. (1986) and Fox and Meng (1997), the fenestra vestibuli and fenestra cochleae are situated on the sides of the posterior, narrower part of the promontorium, but the promontorium does not swing posterolaterally as in djadochtatherioids.

The facial sulcus in cimolodontans is bounded laterally by a ridge (lateral flange) that extends parallel to the promontorium and overhangs the sulcus ventrally; in djadochtatherioids its anterior part extends parallel to the promontorium while the posterior part turns laterally. The lateral flange has not been recognized in Paulchoffatiidae. Anterolateral to the lateral flange there is the epitympanic recess, which houses the A-shaped incus (figure 8.6). In *Nemegtbaatar* the postglenoid (vascular) foramen opens into the posterior part of the epitympanic recess, whereas in *Kamptobaatar* it is placed in the posterior part of the facial sulcus. Another vascular foramen in this region is the supraglenoid (Simpson, 1937a); the position of this foramen varies among taxa (see the later section on "Vasculature").

Knowledge of the cerebellar side of the basicranial region has been based on several isolated petrosals described from Paulchoffatoidea (Lillegraven and Hahn, 1993), ?Ptilodontoidea (Luo, 1989), Taeniolabididae (Wible and Hopson, 1993, 1995), unidentified Hell Creek Formation

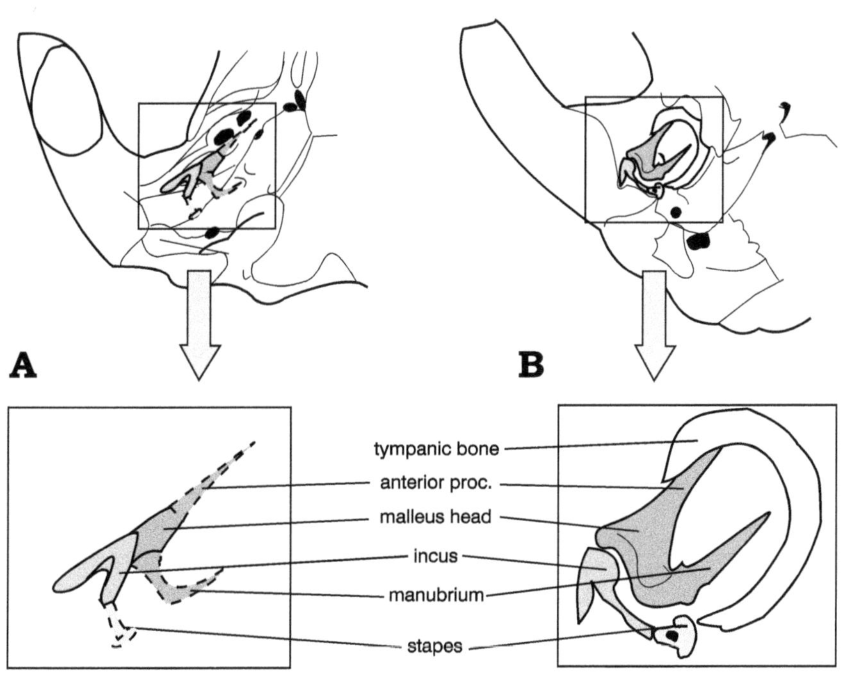

FIGURE 8.6. Right part of basicranium showing arrangement of ear ossicles. A, Cretaceous multituberculate *Chulsanbaatar*. B, Recent marsupial *Didelphis*. *Source:* modified from Hurum et al. (1996).

petrosals that might belong to Ptilodontoidea, Taeniolabididae, or Eucosmodontidae (Kielan-Jaworowska et al., 1986; Fox and Meng, 1997), and on the reconstructions of the dorsal aspect of the brain cavity in the djadochtatherioids *Nemegtbaatar* and *Chulsanbaatar* by Hurum (1998a).

On the cerebellar side the most obvious concavity is the subarcuate fossa, situated in the posterolateral part of the petrosal. In relation to very large paraflocculi (housed by the fossa), the fossa is proportionally larger than in other mammals. The relative size of the subarcuate fossa varies among the multituberculate taxa, probably being comparatively the largest in ?*Catopsalis joyneri* (Kielan-Jaworowska et al., 1986: figure 1D). In Paulchoffatiidae there is no connection between the subarcuate fossa and the posttemporal fossa, whereas in Djadochtatherioidea there is a connection by a canal, which transmitted the arteria diploetica magna. It is also connected with the dorsal part of the cranial cavity by the vascular canal designated by Kielan-Jaworowska et al. (1986) as the ascending canal. In Taeniolabididae the communication between the subarcuate and posttemporal fossae is via a tiny vessel, in front of which is a cranial foramen of the prootic canal (see "Vasculature"). The internal auditory meatus is situated medial to the prootic canal foramen, about in the middle of the bone. In a well-preserved Bug Creek petrosal (Kielan-Jaworowska et al., 1986: figure 8B, see also our figure 8.5A) it is arranged roughly transversely, with a relatively shallow foramen for the cochlear nerve (medially) and a deeper, more laterally situated foramen for the vestibular and facial nerves, separated from the former by a wide crest. A bristle placed through this foramen into the facial nerve canal can be seen through the facial sulcus and through the semilunar fossa. The semilunar fossa is deep, placed anterolateral to the internal auditory meatus and to the two foramina for the mandibular nerve seen in this aspect at the bottom of fossa. There is also a large foramen leading to the posttrigeminal canal, obscured in this view by the vertical medial wall of the fossa.

Hahn (1981) described the dorsal view of the basicranial region in Late Jurassic *Pseudobolodon* and claimed that the cavum epiptericum was incorporated into the braincase. Kielan-Jaworowska et al. (1986) and Miao (1988) have challenged this interpretation. Hurum (1998a) reconstructed the dorsal view of the basicranial region in *Chulsanbaatar* and *Nemegtbaatar.* In both taxa, he recognized the raised tuberculum sellae, where both sides of the taenia clinoorbitalis grow together forming a V-shaped bone. Posterior to the tuberculum there is a shallow, medial groove in the basisphenoid (fossa hypophyseos), limited laterally by the taenia clinoorbitalis. In the posterolateral part of fossa hypophyseos there is the dorsal opening of the carotid canal, which in *Pseudobolodon* is situated in the dorsum sellae (Hurum, 1998a: figure 7).

Ear Ossicles. The ear ossicles, discovered for the first time in Paleocene-Eocene *Lambdopsalis,* were originally interpreted by Miao and Lillegraven (1986) and Miao (1988, the interpretation also accepted by Hurum et al., 1995) as rotated with respect to their position in extant mammals, with the manubrium mallei directed posteriorly (rather than anteromedially). Further discoveries of ear ossicles preserved in *Lambdopsalis* (Meng, 1992; Meng and Wyss, 1995), in *Kryptobaatar* (Rougier, Wible, and Novacek, 1996a), and in *Chulsanbaatar* (Hurum et al., 1996) allowed various authors to challenge this reconstruction. There is now a general consensus that the multituberculate ear ossicles were arranged as in primitive extant mammals (figure 8.6). The best-preserved stapes, known in *Lambdopsalis,* is small and columelliform with a slitlike stapedial foramen (Meng, 1992). The incus, as found in *Chulsanbaatar,* is A-shaped; and the malleus, best preserved in *Kryptobaatar,* has a long anterior process, a robust malleus head, and a shorter manubrium (Hurum et al., 1996; Rougier, Wible, and Novacek, 1996a).

Ectotympanic. Except for uncertain fragments no multituberculate ectotympanic has been found among the numerous exquisitely preserved basicranial regions of the Late Cretaceous multituberculates from Mongolia (e.g., Kielan-Jaworowska, 1971; Kielan-Jaworowska et al., 1986; Hurum, 1998a,b; Wible and Rougier, 2000, and references therein). The only multituberculate ectotympanic reported so far was found by Meng and Wyss (1995) in two specimens of *Lambdopsalis* (recovered from the transitional Paleocene-Eocene beds in China). In *Lambdopsalis* the ectotympanic is a robust bone, completely detached from the lower jaw, lying horizontally against the base of the auditory region and bearing an internal groove for attachment of the tympanic membrane (figure 8.5C).

Inner Ear. The multituberculate inner ear was investigated by Luo and Ketten (1991), Meng and Fox (1995a), and Fox and Meng (1997) on the basis of CT-scan X-radiographic, and SEM study of unidentified Bug Creek petrosals and by Hurum (1998b) on the basis of serial sections and enlarged models of *Nemegtbaatar* and *Chulsanbaatar.* Fox and Meng (1997) compared the inner ear anatomy with that of monotremes and early therians and concluded that the curvature of the multituberculate cochlea is comparable to that in monotremes (figure 8.5B). A conclusion is that the cochlea, at least in some multituberculates, retained a lagena, otherwise known only in monotremes among mammals (but tentatively recognized in the Late Cretaceous eutherian *Daulestes,* see McKenna et al., 2000, and chapter 3). Miao (1988), Luo and Ketten (1991), and Fox and Meng (1997) also showed that the cochlear canal in multituberculates lacks the osseous laminae supporting a short, wide basilar membrane.

According to these authors, the bone-conducted hearing in some multituberculate species might have been inefficient in responding to high-frequency airborne vibrations. Subsequently the issue of high-frequency versus low-frequency hearing in multituberculates was discussed by Meng and Wyss (1995). Hurum (1998b: 65) stated: "A slightly laterally curved, anteromedially directed cochlea, relatively robust ear ossicles, and the estimations of the area of the tympanic membrane and stapedial foot plate in *Chulsanbaatar* suggest high-frequency hearing but a relatively low sensitivity to low-decibel sounds." A different conclusion was reached by Meng and Wyss (1995: 141), who investigated the inner and middle ear structures in Paleocene multituberculate *Lambdopsalis* and stated: "The structure of these new multituberculate auditory ossicles, in conjunction with a greatly inflated vestibule and an uncoiled cochlea, implies an ear inefficient for reception of high-frequency airborne vibrations but well suited for bone-conducted hearing." However, we cannot exclude the possibility that *Lambdopsalis* and most taeniolabidids, which differ from other multituberculates in having unusually expanded vestibules, might also differ from them in their ability to perceive high-frequency airborne vibrations.

The multituberculate vestibule is usually of moderate size (*contra* Luo and Ketten, 1991; see critique by Hurum et al., 1996; and Fox and Meng, 1997), but the vestibule is especially enlarged and inflated in *Lambdopsalis* (Miao, 1988). The multituberculate semicircular canals are comparable in size and proportions to those of extant mammals of similar size (Hurum, 1998b).

Internal Skull Structure. Studies of the sectioned skulls of the djadochtatherioid multituberculates *Nemegtbaatar* and *Chulsanbaatar* showed the presence of ossified turbinals, a cribriform plate (although both structures have been preserved only as broken fragments), and an ossified ethmoid. These studies, augmented by structures found in isolated Bug Creek petrosals, also demonstrate that bones in the multituberculate skull are strongly pneumatized (Kielan-Jaworowska et al., 1986; Hurum, 1994, 1998a). There are large condylar cavities in the basioccipital, cavities in, for example, the basisphenoid, alisphenoid, anterior lamina, and palatine, which are clearly seen in photos of the sectioned skulls published in papers mentioned earlier (Kielan-Jaworowska et al., 1986: figure 12). Hurum (1994) described the sinus frontalis, very extensive in *Nemegtbaatar*, divided into pars nasalis and pars frontalis, and smaller in *Chulsanbaatar*. He interpreted this sinus as related to the lack of a sagittal crest. There is also an extensive sinus maxillaris (figure 8.7A,B). A well-preserved sinus sphenoidalis has been found in *Chulsanbaatar* (figure 8.7C) and a less clear one in *Nemegtbaatar*. Rougier et al. (1997) argued that because the nasal cavity is especially extensive in multituberculates, it is possible that the sphenoid sinus is the posterior expansion of the nasal cavity, as the septum separating these structures has not been found. However, the delicate structures inside the rostral part of the skull, such as the turbinals and cribriform plate (Hurum, 1994), have not been found intact, but only as crushed fragments of bone, and this may have been the fate of the septum separating the nasal cavity and sphenoid sinus in available fossils.

The characteristic feature of the internal structure of the skull is the presence of an extensive cavum epiptericum, separated from the braincase by a strongly ossified pila antotica (figure 8.7D). The cavum epiptericum in reptilian skulls is an extracranial space situated lateral to the primary sidewall of the braincase, incorporated into the brain cavity in mammals (Goodrich, 1930; Beer, 1937; Kuhn and Zeller, 1987b). In all living mammals the cavum epiptericum remains extradural, although it is incorporated into the cranial cavity and the secondary wall of the braincase forms its lateral border. The retention of a separate cavum epiptericum is a plesiomorphic feature, but the ossification of the taenia clinoorbitalis, which in multituberculates is much more extensive than in monotremes, is regarded as probably secondary, related to the general robustness of the cranial bones. The cavum epiptericum is floored by the petrosal and possibly in part by the alisphenoid (Kielan-Jaworowska et al., 1986: figures 28 and 29; Hurum, 1998a: figures 5, 10, 11, see also our figure 8.7D). The semilunar ganglion, as inferred from the size of the semilunar fossa, was very large in multituberculates.

MANDIBLE

The multituberculate mandible is generally rodentlike in appearance (figure 8.4C), with an extensive symphysis. The only "reptilian" bone found in the lower jaw is the coronoid, fused to the dentary, present in the Late Jurassic kuehneodontine *Kuehneodon* (Hahn, 1977b). Otherwise the multituberculate dentary is completely mammalian, with no remnant of Meckel's groove. The condyle primitively is placed below the occlusal level of the molars; it is confluent with the lateral margin of the dentary and faces posteriorly. The angle between the occlusal level of the molars and the lower margin of the dentary primitively is small and the coronoid process is extensive. In later taxa that still retain the low position of the condyle (e.g., Late Cretaceous

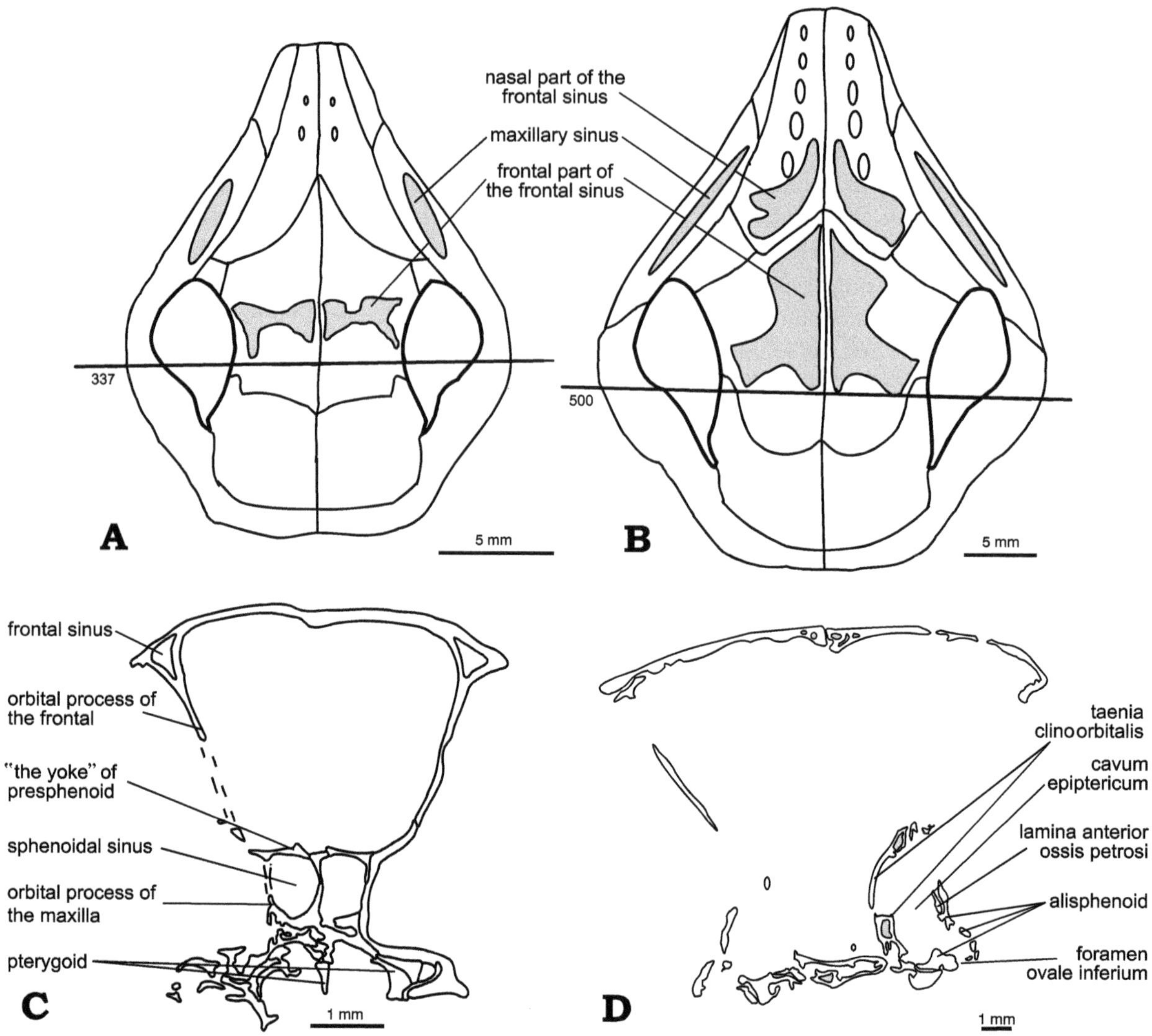

FIGURE 8.7. A,B, Location of frontal and maxillary sinuses (gray) in skulls of *Chulsanbaatar* (A) and *Nemegtbaatar* (B) seen in dorsal view, based on sectioned skulls. Transverse lines indicate the number of section. C, Transverse section through choanal region in *Chulsanbaatar* showing sphenoidal sinus. D, Transverse section through the basicranial region of *Nemegtbaatar,* showing extensive cavum epiptericum. *Source:* modified from: A, B, Hurum (1994); C, D, Hurum (1998a).

djadochtatherioid genera such as *Kryptobaatar* and *Nemegtbaatar*), the coronoid process is small (figure 8.4C). In some advanced forms the condyle is situated above the occlusal level of the molars and faces upward, and the angle between the lower margin and the occlusal level of the molars increases. Gingerich (1977) demonstrated that multituberculates are unique[5] among mammals in having a backward (palinal) power stroke; in other words they chewed posteriorly. In relation to this, in all multituberculates the masseteric muscles inserted more anteriorly and the masseteric fossa and the coronoid process are situated more anteriorly than in all other mammals (Gambaryan and Kielan-Jaworowska, 1995: figure 12). In several taxa there is a masseteric fovea in front of the masseteric fossa (Gambaryan and Kielan-Jaworowska, 1995; Hahn and Hahn, 1998b, see also figures 8.4C, 8.29B,D) for insertion of one part of the masseter muscle. There is no angular process. On the lingual side there is an extensive pterygoid fossa and the whole ventral margin of the dentary is inflected and forms the pterygoideus shelf. The pterygoid muscles were inserted in the pterygoid fossa and on the ptery-

[5] In 1977 when Gingerich published his paper, the backward power stroke characteristic for multituberculates appeared to be unique among mammals. More recently, it has been established that a palinal power stroke also occurred in some haramiyidans (see, e.g., Butler 2000 for summary) and in Gondwanatheria (see Pascual et al., 1999, and chapter 14).

goideus shelf. In early forms (some Paulchoffatiidae) the angle (measured in dorsal view) between the tooth row and the dentary is small (7°); in later forms it increases to 20° or more (figure 8.27A).

DENTITION

Before the tooth replacement was observed multituberculate cheek teeth were identified as premolars and molars on the basis of their morphology. Further studies demonstrated that incisors and at least some premolars are diphyodont (Szalay, 1965; Hahn, 1978b; Kielan-Jaworowska, 1980; Hahn and Hahn, 1998a; see also discussion in Clemens and Kielan-Jaworowska, 1979). On the basis of a juvenile skull of *Taeniolabis taoensis* and other forms and data from the literature, Greenwald (1988) argued that all multituberculates exhibit a pattern of diphyodonty similar to that seen in most placental mammals and that tooth eruption apparently occurred in anteroposterior sequence (see chapter 3 and figure 3.22A). However, Hahn and Hahn (1998a) showed that the pattern is quite different in the paulchoffatiid *Kielanodon* from the Late Jurassic. Here, replacement of the premolars occurred in an alternating mode, in two waves, and replacement ran posteroanteriorly. On this basis, they suggested that the transition from a cynodont to eutherian-like pattern of replacement occurred independently in multituberculates. A further distinction between paulchoffatiids (which are presumably primitive in this regard) and Late Cretaceous multituberculates is that p4 is diphyodont in the former and, apparently, monophyodont in the latter (Simmons Greenwald, 1988; Hahn and Hahn, 1998a).

Early forms ("Plagiaulacida") have three pairs of upper incisors, five or four premolars and two molars (figure 8.8A,B). Of the incisors I1 is relatively small and single-cusped, I2 large and two-cusped, and I3 large and three- or four-cusped in Paulchoffatiidae versus two- or three-cusped in Plagiaulacidae. The canine is absent in advanced multituberculates, but is present in most Late Jurassic Paulchoffatiidae and Pinheirodontidae. The paulchoffatiid canine is similar to the anterior premolars (P1–P3), from which it is separated by a short diastema, but differs from them in being single-rooted. The canines may have from two to five cusps (Hahn, 1993; Hahn and Hahn, 2002). The canine is also present in the Late Jurassic advanced plagiaulacidan *Glirodon* (Engelmann and Callison, 1999), in the allodontid *Ctenacodon* (Engelmann, pers. comm.) and, known from the alveolus, in the highly specialized Early Cretaceous *Arginbaatar* (Kielan-Jaworowska, Dashzeveg, and Trofimov, 1987).

The multituberculate upper premolars and molars carry numerous cusps of subequal height (see later section on "Cusp Formulae"), except for some Paulchoffatiidae and Pinheirodontidae, in which the height of the cusps may vary. In some multituberculates the molars are ornamented with grooves, pits, and ridges, and molar cusps show a tendency to coalesce. Among "Plagiaulacida," Allodontidae have smooth enamel and well-separated cusps (except for poorly known *Psalodon*, the molars of which are unknown, but upper premolars show strong ribbing). In Paulchoffatiidae and Plagiaulacidae the cheek teeth are strongly ornamented. Kielan-Jaworowska and Hurum (2001) argued that the ornamentation and a tendency of molars cusps to coalesce, characteristic for Plagiaulacidae, has been retained in Eobaataridae, and some Cimolodonta, for example, members of the plesiomorphic *Paracimexomys* group and Ptilodontoidea. In Djadochtatherioidea, Eucosmodontidae, Microcosmodontidae, Taeniolabidoidea, and Kogaionidae the enamel is not ornamented (except for the ribbing of the anterior upper premolars). The characteristic feature of the multituberculate upper dentition is the distolingual (rather than distal as in other mammals) position of M2 with respect to M1. Hahn originally believed (1969, 1971, 1977a, 1987b) that M1 and M2 in the Paulchoffatiinae were aligned. Van Valen (1976) and Clemens and Kielan-Jaworowska (1979) questioned this. Finally Krause and Hahn (1990) convincingly demonstrated that the Paulchoffatiinae do not differ from other multituberculates in the position of M2 (figure 8.8D).

The anterior upper premolars in Paulchoffatiidae (Hahn, 1969, 1977a, 1993, and references therein) are roughly quadrangular in occlusal view, with three or four cusps; the posterior premolars are elongated with two or (occasionally) three rows of cusps. In "Plagiaulacida" in which there are five upper premolars, P4 and P5 are subequal in length and as a rule their profile in labial view is more or less horizontal. It has been argued by Clemens (1963b; see also Kielan-Jaworowska and Hurum, 2001) that in Cimolodonta the only retained posterior upper premolar, usually referred to as P4, corresponds to P5 in "Plagiaulacida." The cimolodontan P4 has a different structure than P4 or P5 in "Plagiaulacida," being elongated with respect to the width, asymmetrical, and exhibiting an isosceles triangle in labial view, with the ultimate or penultimate cusp of the medial row being the highest. The molars have two rows of cusps and no posterolingual wing. In Late Jurassic and Early Cretaceous Plagiaulacidae three upper incisors and five premolars are retained; on the posterior premolars an oblique shearing surface develops, while an incipient posterolingual ridge makes its appearance on M1 (also present in a tooth identified as ?P5 of the ?paulchoffatiid *Sunnyodon*, see Kielan-Jaworowska and Ensom, 1992, and our figure 8.31C), and there is an

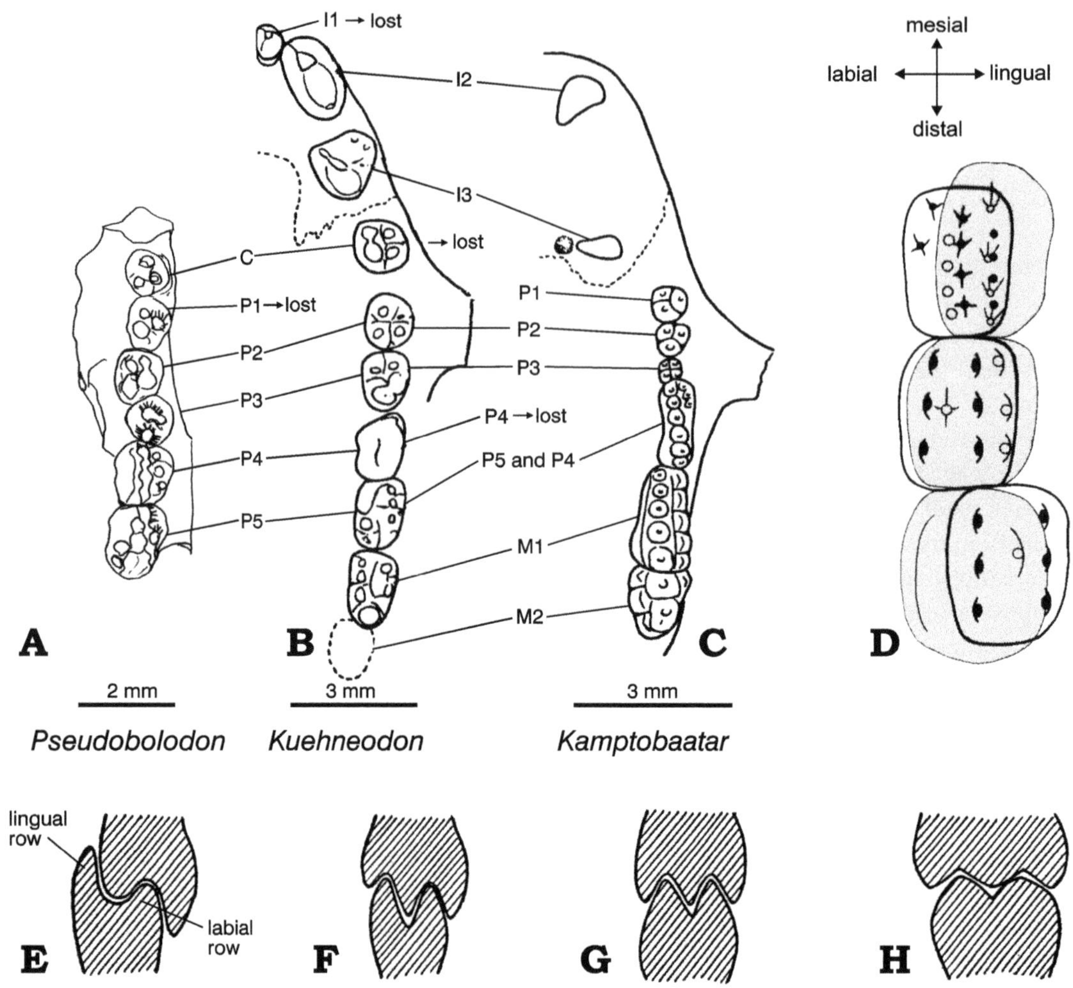

FIGURE 8.8. A–C, comparison of the left upper dentition in palatal view in Late Jurassic *Pseudobolodon* (A), *Kuehneodon* (B), and Late Cretaceous *Kamptobaatar* (C), showing tooth homology according to Clemens's (1963b) hypothesis. P5, present in "plagiaulacidans" (*Pseudobolodon* and *Kuehneodon*), is regarded as homologous to the tooth referred to as P4 in cimolodontans (*Kamptobaatar*), the "plagiaulacidan" P4 being lost. D, diagrammatic representation of occlusal relations between upper and lower last premolars (top), first molars (middle), and second molars (bottom), characteristic of multituberculates. Lower teeth are shaded, with cusp apices represented as circles. Upper teeth are unshaded, with cusp apices represented as solid. E–H, Diagrammatic cross sections through upper and lower molars in Plagiaulacidae (E), Primitive Ptilodontidae (F), Advanced Ptilodontidae (G), and Taeniolabididae (H). *Source:* dentition in A based on Hahn (1977a); B, on Hahn (1969); C, on Kielan-Jaworowska (1971); composition original; D, modified from Krause and Hahn (1990); E–H, modified from Simpson (1929a).

anterolabial ridge on M2. When there are only two rows of cusps on the upper and lower molars, the cusp rows are referred to as labial and lingual. In the case of three rows of cusps (in advanced Cimolodonta) the rows are referred to as labial, medial, and lingual; it follows that the medial row of cusps on the molars in advanced Cimolodonta is homologous to the lingual row in "plagiaulacidans."

In all Late Cretaceous and Tertiary multituberculates (Cimolodonta) there are only two upper incisors, the I1 being lost (figure 8.8C). The I2 in Cretaceous and Tertiary taxa may be two-cusped or single-cusped (but three-cusped in *Meniscoessus*); I3 is single-cusped. In Djadochtatherioidea, Cimolomyidae (*Meniscoessus*), and Eucosmodontidae (*Stygimys*), I3 is situated not at the lateral margin of the premaxilla (as in all other mammals), but medially on the palatal part of the premaxilla. In Cimolodonta the number of upper premolars is reduced to four, but in some specialized lines (e.g., Taeniolabididae) to only one. No Late Cretaceous or Tertiary multituberculate is known to have retained five upper premolars. The upper premolars in Late Cretaceous and Tertiary multituberculates (Cimolodonta) are traditionally designated P1–P4. Clemens (1963b) suggested that they might be homologous to P1–P3 and P5 of "Plagiaulacida." The transitional forms, which would demonstrate reduction of plagiaulacidan P4 in multituberculate evolution, are still to be found. On the other hand, Hahn (1978c: figure 10) suggested that the cimolodontan P4 is homologous to P4 in "plagiaulacidan" Paulchoffatiidae, the anterior "plagiaulacidan" premolar being lost. Given the shearing function of both P4 and P5 in advanced "Plagiaulacida" (in Plagiaulacidae) and retention of the shearing function only by P4 in Cimolodonta, we accept the hypothesis of Clemens (1963b, see our figure 8.8A–C). Multituberculates apparently differ from most mammals (but not all, see, e.g., Cifelli, 2000a) in losing teeth not at the beginning or at the end of tooth series, but in the middle, as demonstrated by djadochtatherioid genera *Catopsbaatar* and *Tombaatar*, which acquired three upper premolars by losing P2 (Kielan-Jaworowska, 1974a; Rougier et al., 1997, see also our figure 8.38C,I).

On M1 (except some Paulchoffatiidae) the posterolingual ridge is present, often developed as a third row of cusps. In Late Cretaceous and Tertiary forms, all of which belong to Cimolodonta, the changes in the dentition occurred along at least three different patterns, paralleled by corresponding changes in the lower dentition (see later).

Only one lower incisor is present in all multituberculates. In earliest forms ("Plagiaulacida") the lower incisor is a stout tooth with a broad base and a curved crown with a convex labial surface. It is completely covered with enamel. The diastema between the incisor and the first premolar is relatively short. In all Paulchoffatiidae and Allodontidae and in most Plagiaulacidae there are four lower premolars, rectangular or oval in labial view (figure 8.9). Multituberculate lower premolars, with the exception of the first one, which is single-rooted and bulbous (see, e.g., Hahn, 1978a), are double-rooted and bladelike. They bear a row of apical cusps associated with ridges on both sides of the crown. The p1 bears a single ridge on each side. An isolated premolar identified by Hahn (1969) as p4 is provided with an additional row of very distinct labial basal cusps, lower than the apical ones. It shows how multituberculate premolars may have developed from teeth with two rows of cusps of equal height (see Hahn, 1969: figure 29, and our figure 8.27B). Such labial cusps occur as a rule on p4s of "Plagiaulacida" and sometimes also on p3s, and even on dp2 (Hahn, 1978b). The labial cusps in older individuals may be strongly abraded by wear and poorly seen. In Pinheirodontidae they may be almost lost or replaced by pits (Hahn and Hahn, 1999a, see also figure 8.33B,D). In Eobaataridae and in Late Cretaceous–Tertiary Cimolodonta, only one (posterior) labial cusp is present. However, in some taxa, for example, Taeniolabididae (Granger and Simpson, 1929), *Arginbaatar* (Kielan-Jaworowska, Dashzeveg, and Trofimov, 1987), *Uzbekbaatar* (Kielan-Jaworowska and Nessov, 1992), and a few other forms (see "Systematics"), there are no labial cusps on p4. Lower molars are provided with two rows of cusps.

In advanced Plagiaulacidae the number of premolars has been reduced to three (p1 being lost), and p4 is strongly enlarged. In Cimolodonta the number of lower premolars is reduced to two or one and the diastema between the incisor and the first existing premolar increased. The p3, if present in cimolodontans, is small and peglike, hidden below the anterior margin of the arcuate enlarged p4.

At the anterior part of p4 in most multituberculates (but not in paulchoffatiids) there is a triangular lobe over the anterior root on the labial side. Krause (1977: 5) wrote: "The term 'exodaenodont lobe' is used throughout the paper in reference to the portion of enamel on p4 that labially overlies and extends ventrally down the anterior root. The term is not currently in use for multituberculate p4s but was suggested to me by R. C. Holtzman and R. E. Sloan (pers. comm., 1975)." However, the term exodaenodont has been used in eutherian literature in a different meaning. For example, Novacek et al. (1985: 10) in the diagnosis of the Erinaceidae stated: "Lower molars semirectangular in occlusal view, with some degree of exodaenodonty (i.e., labial trigonid and talonid cusps are swollen)." In view of different meanings we do not use the term exodaenodont in describing multituberculate p4, and we refer to the lobe in question as the anterior triangular lobe.

The evolutionary changes in the structure of the lower premolars were paralleled by changes in structure and function of the incisors.

In Ptilodontoidea, the lower incisors are long and thin with a narrow base and completely covered with enamel. The p3 is vestigial, while p4 is unusually enlarged—it is arcuate, strongly protruding dorsally over the level of the molars, and provided with numerous ridges (figure 8.9). This p4 served as a shearing blade, whereas the function of the lower incisor was grasping, holding, and piercing (Clemens, 1963b; Krause, 1982a). The ptilodontoids tend to retain four upper premolars, with an increased number of cusps on P4 and on the upper and lower molars.

The Taeniolabididae "chose" a different strategy. Here the lower incisors are very powerful teeth. They are not tapered, but have a relatively uniform diameter throughout their length. The enamel is sharply limited to a distolabial band, producing a self-sharpening tooth, similar to that of rodents and presumably adapted for gnawing. There is only one upper and one lower premolar, both reduced in size; the lower one is triangular in lateral view. The shearing function of the premolars was undoubtedly lost. The upper and lower molars are greatly enlarged in width and length, are provided with numerous cusps, and served as

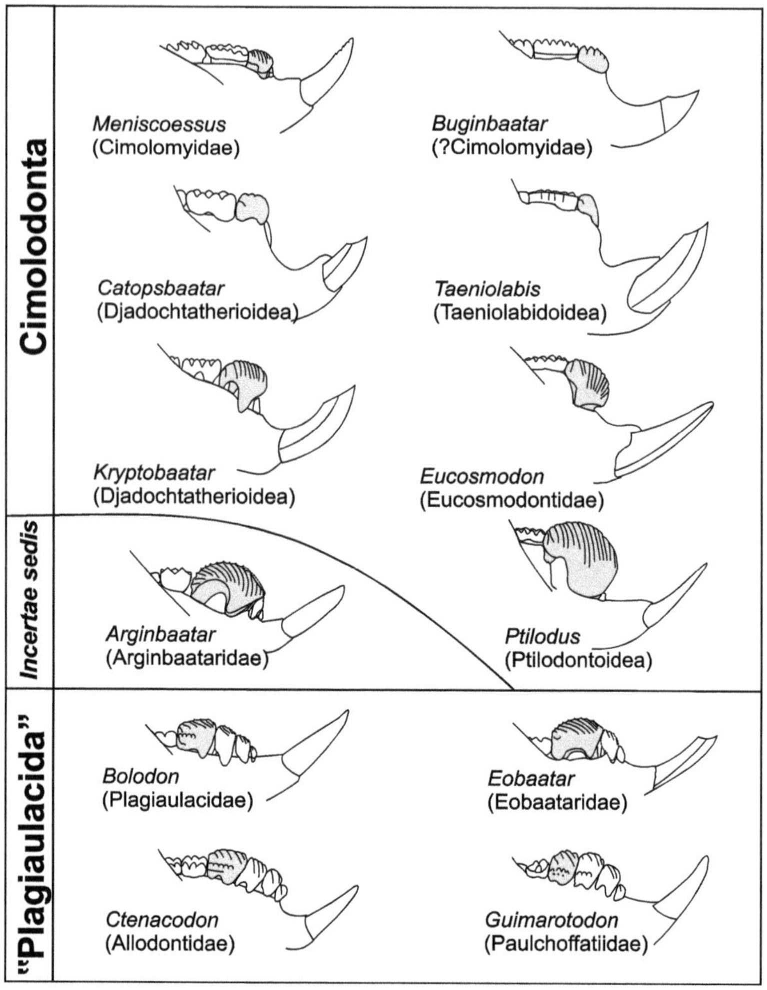

FIGURE 8.9. Comparison of lower dentitions exposed in labial view of selected multituberculates, showing evolution of the dentition. All jaws rendered to approximately the same length, p4 hatched. *Source:* modified from Kielan-Jaworowska and Hurum (2001).

a very effective grinding mechanism (Granger and Simpson, 1929; Miao, 1986, 1988).

The third strategy, intermediate in some respects between those of Ptilodontoidea and Taeniolabidoidea, may be observed in Djadochtatherioidea, Microcosmodontidae, and Eucosmodontidae. Here the enamel on the lower incisor is limited to the distolabial band (not completely limited, but thicker distolabial in some djadochtatherioid genera), p3 is vestigial, and p4 is arcuate but does not protrude dorsally over the level of the molars. Primitively there are four upper premolars, but in some advanced forms such as *Catopsbaatar* and *Tombaatar* (Kielan-Jaworowska, 1974a; Kielan-Jaworowska and Hurum, 1997; Rougier et al., 1997), the number of upper premolars is reduced to three, and p4 (known in *Catopsbaatar*) is reduced in size, although still arcuate (figure 8.9). The lower incisors had a gnawing function, but the shearing function of the premolars was retained, at least in primitive members of the group.

The fourth strategy is that of Cimolomyidae (best known in *Meniscoessus*), in which the lower incisors are robust, but completely covered with enamel. Four upper premolars are retained, but are severely reduced in size, especially in relation to the very large, multicusped upper molars. The p4 is arcuate but relatively small and does not protrude dorsally over the level of the molars (figure 8.9).

Pascual et al. (1999) assigned Gondwanatheria (attributed previously to Multituberculata) to Mammalia *incertae sedis* based on the presence of four molariform teeth in hypsodont representatives of the order (Sudamericidae, see chapter 14). In this book we follow Pascual et al. (1999) in removing Gondwanatheria from Multituberculata, except for a dentary with bladelike lower premolar (p4?) of multituberculate pattern and multituberculate-like anterior upper premolars, all of which we assign to Multituberculata *incertae sedis* (see figure 8.44). The p4? in question retains the rectangular structure of p4 characteristic of "Plagiaulacida," and if future findings demonstrate that it indeed belongs to multituberculates, it would represent a fifth strategy of development in the lower dentition, different from those just listed.

Cusp Formulae. In the cusp formulae used to present the arrangement and number of cusps on particular teeth in multituberculates, the numbers of cusps in consecutive rows are given from labial to lingual, separated by a colon. For example, the cusp formula for M1 of the Late Cretaceous cimolodontan *Catopsbaatar* (figure 8.38I) is 5–6:5–6:4, and for M2 2:2–3:2–3; for M1 of the Early Cretaceous "plagiaulacidan" *Eobaatar* (figure 8.34C) it is 3:4:Ri, and for M2 Ri:3:3 (Ri refers to the noncuspidate ridges, posterolingual on M1 and anterolabial on M2).

Hahn and Hahn (1998c) introduced the numbering of molar cusps in Paulchoffatiidae, discussed by us in detail under "Systematics" (see family Paulchoffatiidae discussed later and figure 8.28).

Occlusion. In contrast to Mesozoic therians, in which, beginning with the classical paper of Crompton (1971, see also chapter 11), the occlusal patterns have been studied in detail, the shearing surfaces numbered, and their homology demonstrated, the details of the evolution of occlusal patterns in multituberculates remain poorly known.

As early as in 1929 Simpson (1929a: figure 3a–d, see also our figure 8.8E–H) observed differences in the mode of occlusion among the Plagiaulacidae, primitive and advanced Ptilodontidae, and Taeniolabididae. In Plagiaulacidae the lingual row of the upper molar occludes in the central furrow of the lower molar. In consequence in the worn "plagiaulacidan" lower molar the lingual cusp row remains high and bears distinct wear facets on its labial side, facing toward the central valley of the tooth; at the same time the labial row becomes much lower and may be completely worn out. In the "plagiaulacidan" upper molars one observes a reversed pattern: the labial row of cusps remains high and worn only along its lingual side, while the lingual row may be more or less worn out. This characteristic "plagiaulacidan" occlusal pattern is seen clearly on SEM micrographs of various representatives of "Plagiaulacida," published among others by Kielan-Jaworowska, Dashzeveg, and Trofimov (1987), Kielan-Jaworowska and Ensom (1992), and Kielan-Jaworowska and Hurum (2001), as well as on the standard photographs and drawings published in numerous papers on Kimmeridgian (Guimarota) and Berriasian (Porto Pinheiro) paulchoffatiid multituberculates from Portugal (see Hahn, 1993, and Hahn and Hahn, 1999a, for references). The same pattern has also been retained in early members of Cimolodonta, from near the Early-Late Cretaceous boundary (Eaton and Cifelli, 2001) or from the earliest Late Cretaceous (Eaton, 1995), assigned by Kielan-Jaworowska and Hurum (2001) and in this book to the informal *Paracimexomys* group.

As shown by Simpson (1929a: figure 3a–d), in multituberculates now assigned to Cimolodonta, the pattern of occlusion is different than in "plagiaulacidans," mostly because of appearance of an additional (third) cusp row on the upper molars, so that in cross section the upper molar embraces the two rows of the lower molar cusps more symmetrically. As a result of this change the molar cusps in cimolodontans are as a rule more horizontally worn, and the wear facets face upward on the lower molars and downward on the uppers. From among cimolodontans the Ptilodontoidea (and *Paracimexomys* group) more resemble the "plagiaulacidan" occlusal pattern than other cimolodontan groups do.

In advanced cimolodontans Krause (1982a) studied the occlusal pattern and jaw movements on the basis of orientation of wear striations and drew conclusions on the diet of the Paleocene *Ptilodus.* Wall and Krause (1992) studied the biomechanics in this genus, while Gambaryan and Kielan-Jaworowska (1995, see also references therein) reconstructed the cranial musculature and jaw movements in Late Cretaceous djadochtatherioidean cimolodontans. None of these studies, however, offered an answer to the problem of the origin of changes in the occlusal pattern through the "plagiaulacidan"-cimolodontan transition.

The question of whether the change from "plagiaulacidan" to cimolodontan occlusal pattern occurred only once or in parallel in various cimolodontan groups is as yet unresolved and requires special investigations.

Enamel Microstructure. The study of mammalian enamel microstructure (also referred to as ultrastructure) dates back to the nineteenth century, but investigation of multituberculate enamel did not begin until the end of the 1960s (see Fosse et al., 1978; Sahni, 1979; and Koenigswald and Clemens, 1992, for references).

In spite of the efforts of Koenigswald and Sander (1997b), the terminology for the types of enamel is confusing, particularly for the most primitive type of mammalian enamel, in which the prisms are not developed. Such enamel occurs in primitive multituberculates assigned to the Paulchoffatiidae (Fosse et al., 1985; Sander, 1997), within the suborder "Plagiaulacida." It is also found in the early mammals *Sinoconodon, Haldanodon, Morganucodon,* and *Kuehneotherium* (Clemens, 1997: figure 2), as well as in Haramiyidae (Frank et al., 1984; Sander, 1997).

Enamel without prisms has been variously referred to as aprismatic, nonprismatic, preprismatic, protoprismatic, or pseudoprismatic. Koenigswald and Sander (1997b) recommended disgarding these terms, suggesting instead the term prismless enamel. But Clemens (1997: 85) stated: "The enamel capping the teeth of vertebrates illustrates intricate patterns of increasing structural complexity ranging from nonprismatic through several types of prismless enamel to prismatic structure." In spite of ambiguity as to whether nonprismatic enamel is synonymous with prismless enamel, we refer to the type of enamel that occurs in the Paulchoffatiidae as prismless. See Koenigswald and Clemens (1992) and Koenigswald and Sander (1997b) for the definition of prisms and other microstructures of mammalian enamel.

The microstructure of enamel in more advanced "plagiaulacidans" (the Plagiaulacidae) is less clear. It is almost unknown in the North American Late Jurassic (Morrison Formation) Allodontidae, except for the statement by Krause and Carlson (1986) that there are no distinct prism boundaries in *Psalodon*. Fosse et al. (1991) studied enamel microstructure in teeth of the Plagiaulacidae from the Early Cretaceous (previously referred to as Late Jurassic) Purbeck Limestone Group of England. We discuss their results after describing the enamel structure in multituberculates which have prismatic enamel.

Fosse et al. (1978) recognized two types of enamel microstructure in Late Cretaceous multituberculates. In *Catopsalis* and *Stygimys*, both assigned at that time to Taeniolabidoidea, they found that prism diameters and their mutual central distances were three times larger than in Ptilodontoidea (two species of *Mesodma*) and in the representatives of Late Cretaceous Eutheria. They designated the taeniolabidoid enamel gigantoprismatic (figure 8.10A). In Ptilodontoidea the size of prisms and their density are similar to what is seen in many recent eutherian mammals (figure 8.10B).

Subsequently, two groups of scientists (Carlson and Krause, 1985; and Fosse et al., 1985) investigated this problem independently. Carlson and Krause (1985) examined the enamel microstructure in 32 multituberculate genera from the Late Cretaceous and Tertiary. They characterized the Ptilodontoidea as having small, circular, and densely spaced prisms and Taeniolabidoidea as having large, arcade-shaped prisms. They stated (Carlson and Krause, 1985: 47): "Surprisingly consistent patterns of enamel ultrastructure were discovered at the subordinal level in multituberculates." Fosse et al. (1985) examined enamel microstructure in numerous teeth belonging to 11 Late Cretaceous djadochtatherioid, eucosmodontid, and ptilodontoid genera, as well as in several other early mammals, and came to the same conclusions as Carlson and Krause (1985). Koenigswald and Sander (1997b) offered their "Glossary of Terms Used for Enamel Microstructures," in which they list prismless enamel (characteristic of most "Plagiaulacida") and gigantoprismatic enamel (characteristic of *Glirodon, Eobaatar,* Arginbaataridae, and Cimolodonta except Ptilodontoidea). The enamel with small, circular prisms, which occurs in Ptilodontoidea, is not mentioned. Simmons (1993) and Clemens (1997) referred to enamel with small circular prisms as small prismatic enamel, and Hahn and Hahn (1999a) introduced the name microprismatic. Craig B. Wood (pers. comm., July 2003) criticized this term, arguing that the "microprisms" in multituberculates are not smaller than the usual range of prism sizes in all other mammals (from Mesozoic mammals to mammoth), and suggested replacing it by "normal prismatic enamel." Fosse (2003) followed this suggestion and we follow it herein.

The enamel microstructure in Plagiaulacidae, known mostly from the Early Cretaceous Purbeck Limestone Group of Britain, is less clear. The only study of this enamel is an extended abstract by Fosse et al. (1991), which presents equivocal results. The authors coined the terms "mixed preprismatic enamel" (see figure 3 in their paper), where there is a combination of preprisms with a mean central distance of 15.5 μm and others with a mean central distance of 4.5 μm. A taxon may display "mixed preprismatic enamel" in one tooth, while all preprisms may be of equal size in another. Fosse et al. (1991: 26) also concluded: "It is probable that in all teeth examined, the structure is different at various levels through the enamel thickness." Because of this ambiguity we use the term "mixed preprismatic enamel," as seen in some plagiaulacid taxa, in quotation marks. Koenigswald and Sander (1997b) did not include "mixed preprismatic enamel" in their " Glossary."

Krause and Carlson (1987) studied the homology and polarity of enamel ultrastructural characters of Late Cretaceous and Tertiary multituberculates in a phylogenetic context. For defining "plagiaulacidan" enamel they used data from the Paulchoffatiidae. Krause and Carlson challenged the previously accepted view (e.g., Boyde and Martin, 1984; Sahni, 1985, and others) that small circular prisms (normal prismatic enamel) are primitive, while large, arcade-shaped prisms (gigantoprismatic enamel) are derived. Krause and Carlson (1987) hypothesized that arcade-shaped prisms (gigantoprismatic enamel of Fosse et al., 1978) represent a primitive type of prismatic multituberculate enamel, whereas small circular prisms (the normal prismatic enamel) constitute a derived type. Normal prismatic enamel also occurs in the enigmatic South American gondwanatherian *Ferugliotherium*, referred previously to Multituberculata (Krause, Kielan-Jaworowska, and Bonaparte, 1992; but see Pascual et al., 1999), whereas now placed in the Gondwanatheria (see chapter 14).

There is general consensus that prismatic enamel evolved from prismless enamel several times in vertebrate

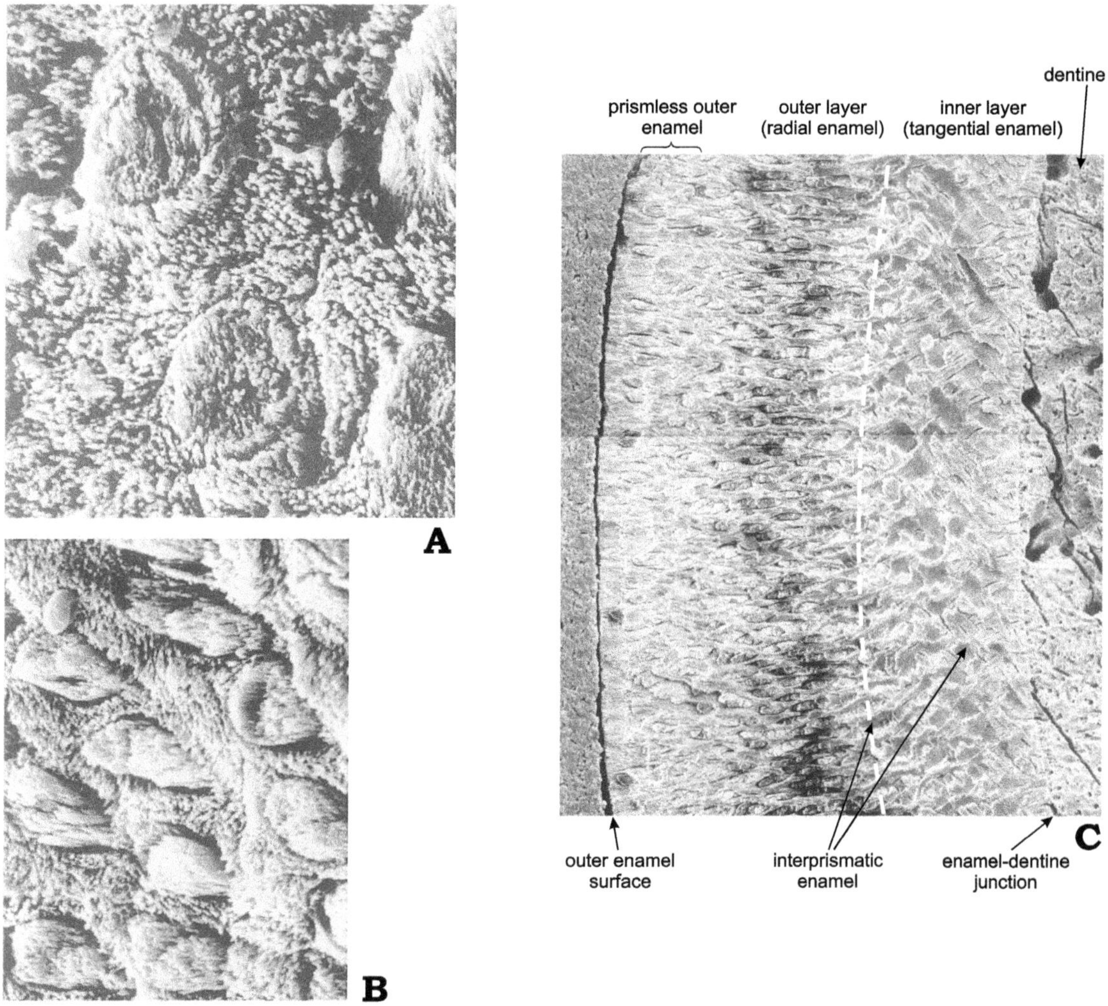

FIGURE 8.10. A,B, SEM micrographs of superficially planed and etched outer enamel surfaces, reproduced with the same magnification ×2900. A, *Kryptobaatar dashzevegi* i1, showing that prisms are widely separated by interprismatic enamel. B, *Mesodma* sp. p4, showing that prisms are smaller and their numerical diversity per unit area considerably higher than in *Kryptobaatar* enamel. C, Cross section of i1 of *Ptilodus* sp., ×350, showing the complexity of the schmeltzmuster in Ptilodontoidea. *Source:* A, B, modified from Fosse et al. (1985). C, original, courtesy of W. v. Koenigswald.

evolution (e.g., Koenigswald and Clemens, 1992; Sander, 1997; Clemens, 1997; see also Wood et al., 1999, for a review). As long as details of enamel microstructure in latest Jurassic and Early Cretaceous Plagiaulacidae are poorly known, it is difficult to demonstrate when exactly prismatic enamel made its appearance in multituberculates and whether it developed only once or several times. The oldest prismatic (gigantoprismatic) enamel was found by Krause, Kielan-Jaworowska, and Bonaparte (1992) in the Late Jurassic "Morrison multituberculate" later named *Glirodon* Engelmann and Callison, 1999 (see also Engelmann et al., 1990). Next in stratigraphic sequence (see Fosse et al., 1985) are *Arginbaatar dmitrievae* and *Eobaatar minor* (the latter identified at that time as family, gen. et sp. indet.) from the Aptian or Albian Höövör (previously Khoboor) locality, Mongolia. Clemens (1997) suggested that normal prismatic enamel evolved from the gigantoprismatic type as many as three and possibly four times, but Kielan-Jaworowska and Hurum (2001) argued that normal prismatic enamel evolved in multituberculate evolution only once, in Ptilodontoidea. Wood and Stern (1997) suggested, contrary to the generally accepted opinion, that the small prisms may be plesiomorphic for multituberculates and the "giant" ones derived, but this conclusion has not been supported by phylogenetic analyses of multituberculates (e.g., Krause and Carlson, 1987; Kielan-Jaworowska and Hurum, 1997, 2001).

Koenigswald (1980, 1997a) and Koenigswald and Clemens (1992) claimed that enamel microstructure could be best understood by distinguishing different levels of complexity. Most of the observations summarized previously are related to the prism level. In prismatic enamel the crystallites may be grouped in various ways, but true prisms, defined by the presence of a prism sheath and interprismatic matrix (Koenigswald and Sander, 1997a), are thus far known only from Late Cretaceous and Tertiary multituberculates. Similar evolution from prismless to prismatic enamel has been observed in therian mammals at about the same time (Koenigswald, 1997a). As Multituberculata and Theria were phylogenetically separated at that time, the prisms evolved independently in these two lineages.

The next level of enamel microstructure is that of enamel types, defined by prism direction. Different prism types, like small prisms and gigantoprismatic prisms, may form the same enamel type. In multituberculates, except for Ptilodontoidea, the prismatic enamel is organized as radial enamel, which means that prisms rise radially from the enamel-dentin junction to the outer surface. In Taeniolabidoidea and some Djadochtatherioidea, a tangential enamel is present. The prisms do not rise but stay almost horizontally and deviate laterally to varying degrees from their radial course. Tangential enamel is also common in marsupials and occurs in a few groups of placental mammals (Koenigswald, 1997a).

The next level of complexity is called the "schmelzmuster level" (enamel pattern). The schmelzmuster can be simple when formed by only a single enamel type, but can also be composed of several different enamel types. The schmelzmuster describes the spatial arrangement of enamel types within one tooth. In Taeniolabidoidea and Djadochtatherioidea, radial enamel is normally combined with outer prismless enamel. In enlarged incisors of Ptilodontoidea, radial and tangential enamels are combined in layers. We figure the cross section of an incisor of *Ptilodus* sp. showing the complexity of the schmelzmuster in Ptilodontoidea (figure 8.10C). The combination of radial and tangential enamels in a schmelzmuster may be explained as a means of strengthening the enamel to hinder crack propagation. Owing to the limited material available from multituberculates, mostly the enlarged incisors were studied, but it is possible that the enamel microstructure of the enlarged incisors may differ fundamentally from that of the cheek teeth (as is the case in therian mammals, see Koenigswald, 1997b). Thus the complete picture of enamel differentiation in multituberculates has not yet been established.

POSTCRANIAL SKELETON

Krause and Jenkins (1983) described the first relatively complete multituberculate skeleton, belonging to Paleocene *Ptilodus,* and reviewed earlier literature and postcranial elements belonging to other North American taxa, especially ?*Eucosmodon,* originally described by Granger and Simpson (1929). Kielan-Jaworowska and Gambaryan (1994) described postcranial elements of Late Cretaceous Mongolian multituberculates, belonging to *Kryptobaatar, Nemegtbaatar,* and *Chulsanbaatar,* and made comparisons with the North American taxa. More recent contributions include those of Sereno and McKenna (1995), Gambaryan and Kielan-Jaworowska (1995), and Kielan-Jaworowska (1998). A mixture of plesiomorphic and derived features characterizes the multituberculate postcranial skeleton.

Vertebral Column. There is no transverse foramen in the atlas; the spinous process in the axis is relatively small in proportion to a large and heavy head (figure 8.11). There are seven cervical vertebrae, with cervical ribs at least in some taxa; possibly 13 thoracic vertebrae; the sternebrae are ossified. The lumbar vertebrae, possibly seven in number (although the diaphragmatic vertebra has not been recognized with any certainty), with very long transverse and spinous processes, are seen in some Late Cretaceous Mongolian multituberculates. The multituberculate lumbars, as preserved in *Nemegtbaatar,* differ from those of monotremes, extant therians, and some cynodonts (but not from tritylodontids) in lacking anapophyses. They also differ from all extant mammals in having poorly developed metapophyses that are not separated from the prezygapophyses. The sacrum consists of four fused vertebrae. In all taxa in which it has been preserved, the sacrum is long, and the second sacral vertebra is longer than the first. The first two sacral vertebrae are fused with the ilium in *Kryptobaatar* and *Chulsanbaatar.* Because of the elongation of S2, the iliosacral contact, as exemplified by *Kryptobaatar,* is relatively very long. This contact, at least in *Kryptobaatar,* is more dorsoventral than mediolateral, as is characteristic of Theria. The number of caudal vertebrae is unknown in Mesozoic multituberculates, and the tail as reconstructed in *Nemegtbaatar* is very long. Krause and Jenkins (1983) estimated that in Paleocene *Ptilodus* there were about 23 caudal vertebrae, the length of the tail was about 180 mm, and the length of the skull was about 40 mm.

Pectoral Girdle and Forelimb. The pectoral girdle consists of scapulocoracoid, interclavicle, and clavicle. None of the described multituberculate scapulacoracoids is complete (Simpson, 1928a; McKenna, 1961; Clemens and Kielan-Jaworowska, 1979; Jenkins and Weijs, 1979; Krause and Jenkins, 1983; Kielan-Jaworowska, 1989; Kielan-Jaworowska and Gambaryan, 1994; Sereno and McKenna, 1995). Best preserved is perhaps that figured by Sereno and McKenna (1995; our figure 8.12), but not described in detail. The multituberculate scapulocoracoid is narrower

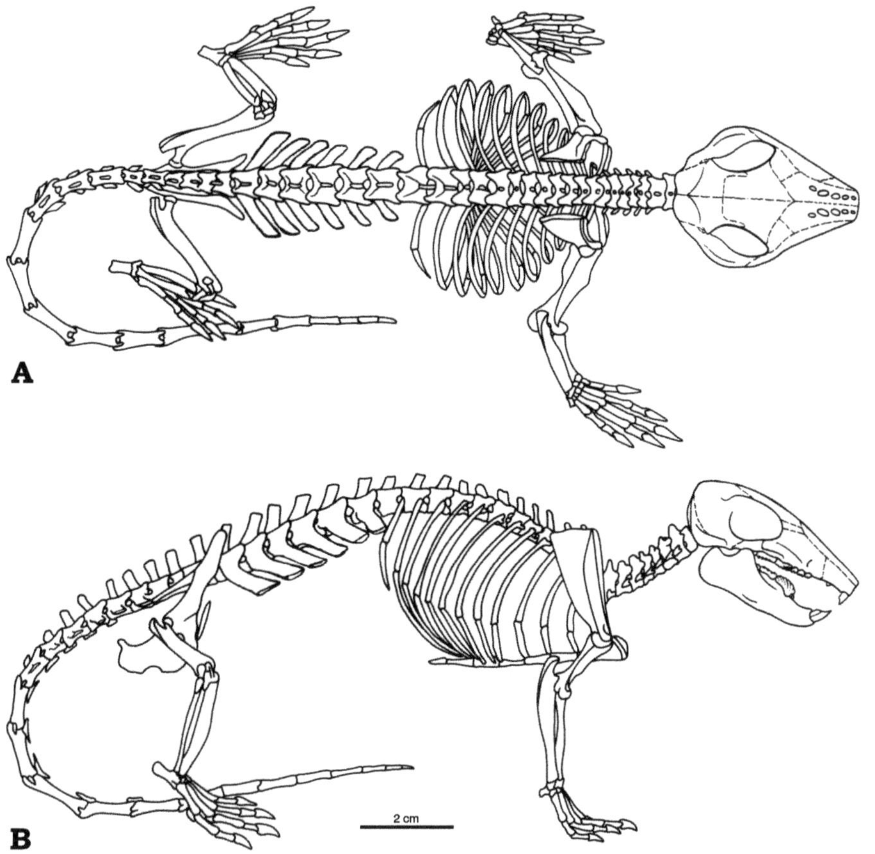

FIGURE 8.11. Reconstruction of the skeleton of *Nemegtbaatar,* based in part on *Kryptobaatar,* in dorsal and lateral views. *Source:* modified from Kielan-Jaworowska and Gambaryan (1994).

than in extant therians, has a deep infraspinous fossa, a prominent spine, and an incipient supraspinous fossa. The coracoid is completely fused with the scapula and forms a small process on the ventral margin of the scapula, which contributes to the glenoid fossa. The acromion is peglike and "embraces" the humeral head craniolaterally (figure 8.12). The clavicle is long and bent and abuts against the roughly heart-shaped interclavicle (Sereno and McKenna, 1995). The manubrium sterni is ossified.

In an abstract McKenna (1996) argued that the multituberculate shoulder girdle is more similar to that of therians than to monotremes. In monotremes the pars dermalis of the T-shaped interclavicle contacts the clavicle, eventually co-ossifying with it. The clavicle is separated from the manubrium by a procoracoid. In embryonic therians the pars chondralis of the interclavicle merges with both the epicoracoid and manubrium. In adult therians, each clavicle contacts the manubrium; in some cases praeclaviculae are present. McKenna introduced the term alloclavicle for a co-ossified praeclavicula. He argued that in multituberculates, as in some embryonic therians, the praeclaviculae are present but are coossified forming alloclavicles, which may or may not fuse to the manubrium. He argued further that this interpretation gives support to the idea that multituberculates and therians are more closely related to each other than either of them is to monotremes, a conclusion not supported by other skeletal parts.

The humerus has a spherical head; the lesser tubercle is only slightly lower than the greater one, but its transverse diameter is larger; the intertubercular groove is wide, the teres tuberosity is crescent shaped; there is a posterior crest that extends in dorsal aspect from the head and continues distally as the ectepicondylar flange. The distal epiphysis is expanded in most known humeri, with a very large, spherical radial condyle, delimited by a deep intercondylar groove from the very prominent ulnar condyle. In all multituberculate humeri there is no trace of even an incipient trochlea, the entepicondyle is large, the ectepicondyle smaller, and the entepicondylar foramen is present. Complete multituberculate humeri have been preserved in only three taxa: *Lambdopsalis* (figure 8.13A), *Kryptobaatar,* and *Bulganbaatar,* but the latter described by Sereno and McKenna (1995) was subsequently identified (Paul C. Sereno, pers. comm., September 2002), as *Kryptobaatar.* The first two show a strong degree of torsion, more than 30° (Kielan-Jaworowska and Gambaryan, 1994; Kielan-Jaworowska 1997), whereas in "*Bulganbaatar*" the torsion is only 15° (Sereno and McKenna, 1995). It follows that the issue of humeral torsion in

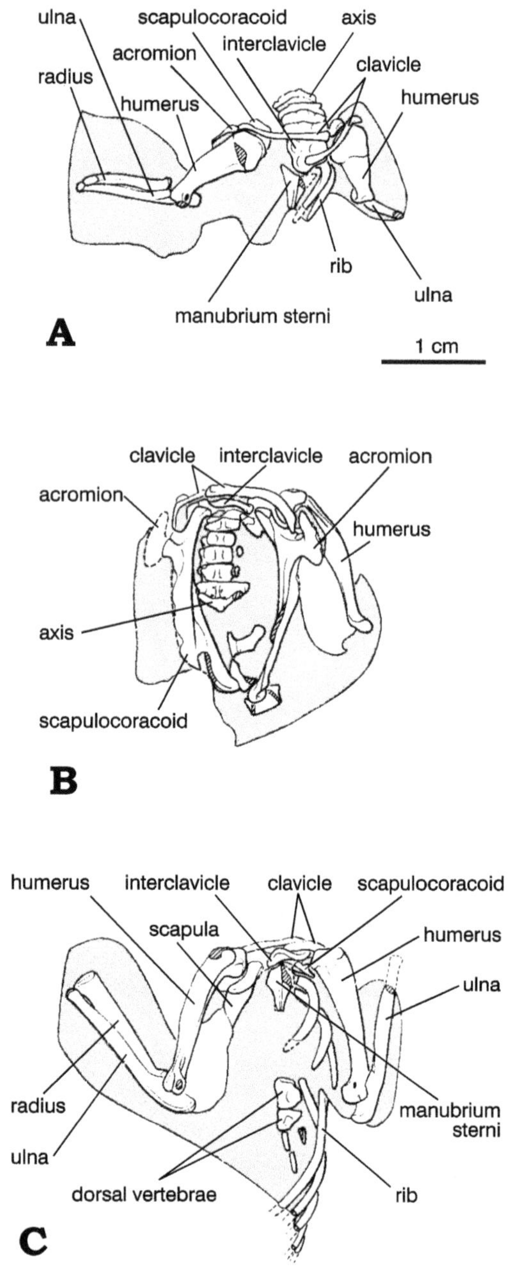

FIGURE 8.12. Pectoral girdle of *Bulganbaatar* sp. from the Late Cretaceous of Mongolia in anterior (A), dorsal (B), and ventral (C) views. *Source:* modified from Sereno and McKenna (1995).

multituberculates requires investigation. In the semilunar notch of the ulna there is a concave, longitudinal facet for the ulnar condyle of the humerus, separated by a prominent ridge from the shallow and concave radial notch. The ridge that separated the ulnar and radial notches in the ulna articulates with the intercondyloid groove on the humerus. The head of the radius is elliptical and bears a concave facet that articulates with the radial condyle. On the proximal side of the radial head there is also a facet for articulation with the ulna. The manus is preserved in *Kryptobaatar* (Minjin, 2001) and in *Sinobaatar* (Hu and Wang, 2002a,b), but only preliminarily described. The fragments of the carpus found in Paleocene *Ptilodus* and Cretaceous *Nemegtbaatar* show that there were three bones in the proximal row of the carpus.

Pelvic Girdle and Hindlimb. The multituberculate pelvis differs from those of extant placental mammals in the presence of epipubic bones, in being deeper, and in having a large (33–36°) iliosacral angle, whereas it is 9–19° in small extant Theria. A large (38–40°) iliosacral angle is also characteristic of monotremes, and this is probably a plesiomorphic feature. The acetabulum is open dorsally, as in morganucodontans, the tritylodontid *Oligokyphus*, and in some gliding marsupials, whereas in monotremes it is only partly covered dorsally. A characteristic feature is the firm fusion of the ventral borders of the pubes and ischia, forming a keel. This fusion shows that the right and left halves of the pubes and ischia could not separate during parturition. Other characteristic features of the pelvis are: a large obturator foramen, a postobturator notch (or foramen), and a parabolic ischial tuber that protrudes dorsally. The ischia meet one another to form an acute angle, and the ischial arc is extremely narrow. The multituberculate pelvis differs from those of therian mammals in being deeper and in this respect resembles that of the monotremes. However, it differs from that in monotremes in having a very small ischial arc, confined to the dorsal part of the ischia, whereas in monotremes the ischial arc, in relation to oviparity, is very wide and U-shaped (Kielan-Jaworowska, 1969a, 1979; Kielan-Jaworowska and Gambaryan, 1994).

The femur is a stout bone and resembles that of therians in having a head with an extensive articular facet, a long neck forming an angle of 50–60° with the shaft, and a prominent greater trochanter extending beyond the head (figure 8.13B). Other features are different, however: the lesser trochanter, placed on the ventral side and at the point of confluence of the greater trochanter with the neck, has a structure unique for mammals. It is prominent, convex lateroproximally, concave mediodistally, and protrudes ventrally beyond the margin of the shaft. Lateral to the lesser trochanter is a fissurelike fossa, referred to by Simpson and Elftman (1928) as a part of the divided trochanteric (digital) fossa, and designated by Kielan-Jaworowska and Gambaryan (1994) the posttrochanteric fossa. The third trochanter is absent. On the dorsal side, at the point of divergence of the neck and the greater trochanter, there is a tubercle, designated by Kielan-Jaworowska and Gambaryan (1994) as the subtrochanteric tubercle, unknown in other mammals. The distal condyles are small, with a shallow trochlea.

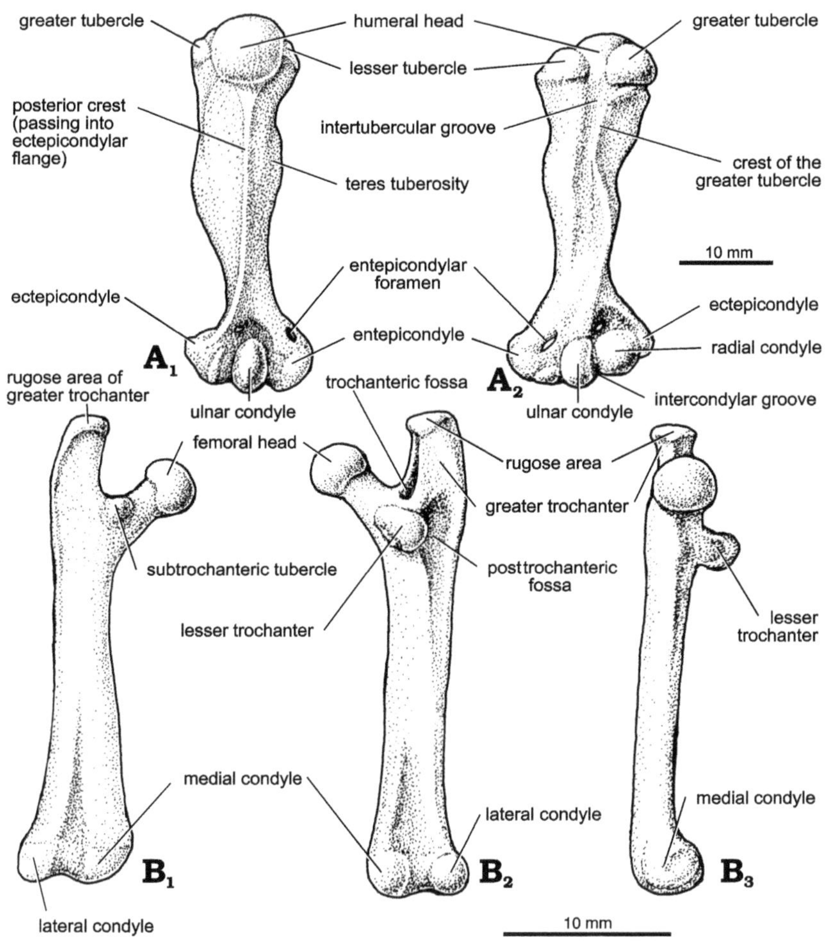

FIGURE 8.13. A, *Lambdopsalis bulla,* from the Paleocene-Eocene boundary of China. Left humerus in posterior (A_1) and anterior (A_2) views. B, *Nemegtbaatar gobiensis* from the Late Cretaceous of Mongolia. Right femur in anterior, posterior, and medial views (B_1–B_3). *Source:* original.

The characteristic feature of the multituberculate tibia is that the mediolateral diameter is relatively large in contrast with the craniocaudal (anteroposterior) diameter, which is the reverse of the condition in therian mammals with parasagittal limbs. In *Kryptobaatar* the craniocaudal diameter is 54% of the mediolateral diameter, in *Nemegtbaatar* it is 60%, and in Paleocene ?*Eucosmodon* it is 63%, whereas in modern rodents of comparable size these values vary between 105 and 150%. These differences are possibly related to the abducted position of multituberculate limbs (see later discussion under "Posture"), as during the propulsive phase the stress on the tibia is directed medially. A markedly asymmetrical proximal end with a small medial facet and a large lateral one characterizes the tibia. A prominent, hooklike triangular process overhangs the shaft laterally and bears a facet for the fibula. The distal end of the multituberculate tibia is similar to that of generalized marsupials such as *Didelphis* in having a prominent medial malleolus and a large, relatively flat lateral condyle. Articular facets for the astragalus are present on the cranial and caudal sides of the medial malleolus.

The fibula is never fused with the tibia. The lateral part of the fibular head bears a hooklike triangular process for articulation with the hooklike triangular process of the tibia. In lateral view these processes are aligned. The multituberculate fibula does not participate in the knee joint and is caudal to the tibia proximally and lateral distally.

The parafibula is a small ossicle in the multituberculate knee joint that articulates with the head of the tibia. An isolated parafibula has been found in Paleocene *Ptilodus* (Krause and Jenkins, 1983). In *Kryptobaatar* and *Chulsanbaatar* it was preserved in articulation with the tibia, but displaced, and its presence is uncertain in *Nemegtbaatar.* Kielan-Jaworowska and Gambaryan (1994) suggested that parafibula might be characteristic of multituberculates as a whole and that the gastrocnemius lateralis muscle possibly originated from it (as in some marsupials), rather than from the femur as in eutherians.

Granger and Simpson (1929) reconstructed the multituberculate pes of Paleocene ?*Eucosmodon* sp., with the longitudinal axis extending along the third ray of the foot. They demonstrated that the cuboid facet on the calcaneus is arranged obliquely (rather than distally as in many other mammals). They reconstructed the calcaneus as not supported distally by any bone. Paleontologists accepted this unusual reconstruction for 65 years, including a thorough description of the multituberculate postcranial skeleton (based on Paleocene taxa) by Krause and Jenkins (1983) and discussion on the multituberculate pes by Szalay (1993b).

Kielan-Jaworowska and Gambaryan (1994) published a revised reconstruction of the multituberculate pes based on two Late Cretaceous taxa, *Kryptobaatar* and *Chulsanbaatar,* and reexamination of the pes of ?*Eucosmodon* on which the original reconstruction of Granger and Simpson (1929) was based (figure 8.14B,D,E). The new reconstruction differs from the previous one in the position of Mt III, which is arranged at 30° with respect to the longitudinal axis of the tuber calcanei. In this reconstruction Mt V articulates with the distal margin of the calcaneus, medial to the peroneal groove. According to this reconstruction, multituberculates differ from other mammals, except for the eutriconodont *Jeholodens* from the Early Cretaceous of China (see Ji et al., 1999, and chapter 7) in having Mt V articulating with the calcaneus. Such an articulation occurs in various extinct tetrapods, for example, in a few therapsids, in primitive archosaurs, in some lizards and some thecodonts, and in extant *Sphenodon* (see Kielan-Jaworowska and Gambaryan, 1994, for references). However, it remains possible that a remnant of cartilage intervened between the calcaneus and Mt V of multituberculates, as apparently was the case in the so-called "Manda cynodont" (figure 8.14A).

Kielan-Jaworowska (1997) speculated that when the fifth tarsal disappeared in the lineage leading to mammals, two different types of mammalian feet developed. In the

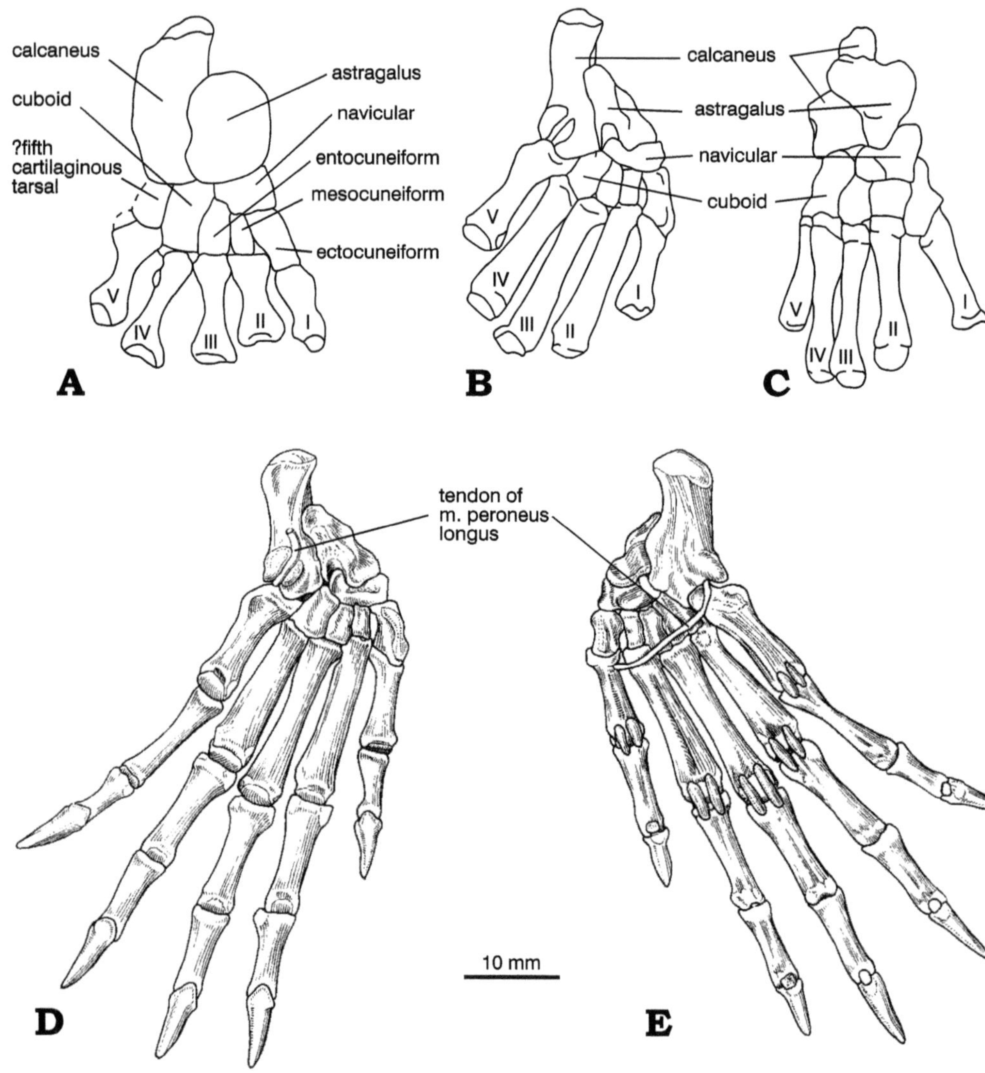

FIGURE 8.14. A–C, Diagrammatic comparison of ankle joints. A, The so-called Manda cynodont. B, Multituberculate ?*Eucosmodon.* C, Recent therian *Didelphis.* D, E, Reconstruction of the right pes of Paleocene ?*Eucosmodon* in dorsal (D) and plantar (E) views. *Source:* modified from: A–C, Kielan-Jaworowska (1997); D, E, Kielan-Jaworowska and Gambaryan (1994).

line leading to Theria the cuboid extended laterally, supporting the distal end of the calcaneus and the cuboid facet on the calcaneus acquired a distal position. In the multituberculate line the calcaneus came into contact with the fifth metatarsal and the cuboid facet on the calcaneus shifted from nearly distal to mediodistal position (figure 8.14A–C).

The multituberculate calcaneus differs from that of Liassic morganucodontans (Jenkins and Parrington, 1976) in having a well-developed and laterally compressed tuber calcanei, similar to that in therians. There are at least two types of multituberculate calcanei. In one type, represented, for example, by Asian Late Cretaceous calcanei, the peroneal groove is wide and the peroneal tubercle is directed laterodistally. In the other type, represented, for example, by ?*Eucosmodon* sp., the peroneal groove is narrow and the peroneal tubercle is directed more distally. However, in both types the peroneal groove is deep and the peroneal tubercle is clearly set off from the calcaneal body. In this respect the multituberculate calcaneus differs from those of other mammals.

Kielan-Jaworowska and Gambaryan (1994) reconstructed the astragalus as situated obliquely with respect to the calcaneus. In dorsal aspect the multituberculate astragalus forms an irregular rectangle. Distally the astragalus articulates with the navicular (central tarsal bone), as was correctly recognized by Szalay (1993b).

Further support for the reconstruction of the multituberculate pes presented by Kielan-Jaworowska and Gambaryan (1994) is found in the grooves, interpreted as for the tendon of musculus peroneus longus, which are preserved on the bones of ?*Eucosmodon* sp. In some extant marsupials and rodents the tendon of musculus peroneus longus lies in the groove on the lateral surface of the calcaneus, and the bones with grooves for the tendon contact one another. The reconstruction of the course of this tendon shows that Mt V articulated with the calcaneus and that other bones of the pes were arranged as shown in figure 8.14D,E.

Among extant mammals the poisonous tarsal spur (os calcaris) occurs only in monotremes. Among fossil mammals it has been described in the eutriconodontan *Gobiconodon* and in the spalacotheriid "symmetrodontan" *Zhangheotherium* (see chapters 3 and 9). Fragments of right and left spurs were also found in multituberculates. Kielan-Jaworowska and Gambaryan (1994: figure 2) figured the partial postcranial skeleton of the Late Cretaceous (Campanian) multituberculate *Kryptobaatar dashzevegi*. Between the distal ends of the tibia, fibula, and astragalus on the left side of the specimen (figure 2A of their paper), a small bone, not completely prepared from the rock, has been preserved; it was not identified by the authors, who regarded it possibly as a displaced fragment of dentary, but Wible and Rougier (2000) suggested that it might represent a displaced tarsal spur.

MUSCULATURE

Cranial Musculature. Multituberculate cranial musculature was reconstructed first by Simpson (1926b) for a Late Jurassic "plagiaulacidan" (based mostly on *Ctenacodon*) and then by Sloan (1979) for Paleocene *Ectypodus.* Wall and Krause (1992) provided a biomechanical analysis of the masticatory apparatus in Paleocene *Ptilodus* and reconstructed the vectors of the principal masticatory muscles. Gambaryan and Kielan-Jaworowska (1995) reconstructed the masticatory musculature in the Late Cretaceous Asian djadochtatherioids *Nemegtbaatar, Chulsanbaatar,* and *Catopsbaatar,* based on surface topography of the bones (on which the muscle scars are evident) and a comparison with extant mammals, especially rodents. The authors concluded that, as in rodents, the masseter muscle was apparently separable into three layers: masseter superficialis, masseter lateralis profundus, and masseter medialis (terminology from Hill, 1937, see our figure 8.15). Although the masseter superficialis is undivided in extant mammals, the presence of two distinct ridges on the external surface of the zygomatic arch, referred to as anterior and intermediate ridges, led Gambaryan and Kielan-Jaworowska (1995) to conclude that the masseter superficialis of multituberculates was separable into anterior and posterior parts. Pars anterior apparently inserted above the masseteric line of the dentary and pars posterior inserted on the flat area at the posteroventral part of the dentary. As the zygomatic ridges (the anterior one being more prominent) are also present in "Plagiaulacida" and Ptilodontoidea, the authors concluded that the division of masseter superficialis into two parts was characteristic for multituberculates as a whole.

The masseter lateralis profundus has been reconstructed as a wide muscle similar to that occurring in most extant mammals. It originated from the whole ventral surface of the zygomatic arch and inserted on a large area on the ventral part of the dentary.

A peculiarity of the multituberculate masticatory system is the place of origin and insertion of the pars anterior of masseter medialis. Gambaryan and Kielan-Jaworowska (1995) reconstructed the masseter medialis as separable into three parts, as in rodents. Pars anterior apparently originated (in *Nemegtbaatar* and other djadochtatherioids) from the medial wall of the orbital space, from a well-developed orbital ridge and a relatively large orbital pocket. It inserted in a masseteric fovea, which is a deepening on the lateral side of the dentary situated in

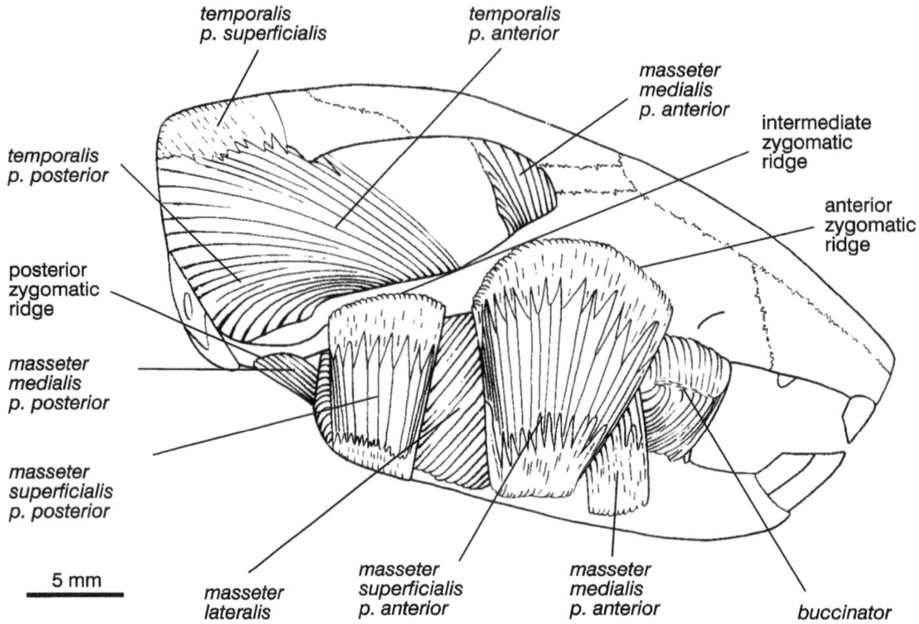

FIGURE 8.15. Reconstruction of the superficial layer of the masticatory musculature in *Nemegtbaatar gobiensis,* based on surface topography of the bones. Masseter lateralis profundus is referred to as masseter lateralis. *Source:* modified from Gambaryan and Kielan-Jaworowska (1995).

front of the masseteric fossa. Sloan (1979) reconstructed a similar position of this muscle for *Ectypodus.* Hahn and Hahn (1998b) described the masseteric fossa, masseteric fovea, and other muscular attachments (e.g., posterior eminences, see figure 8.29B,D in "Systematics"), characteristic of djadochtatherioids, in paulchoffatiid *Meketibolodon* and *Guimarotodon,* which shows that the general pattern of the masticatory system reconstructed by Gambaryan and Kielan-Jaworowska (1995) for the Late Cretaceous multituberculates was already developed by the Late Jurassic. Pars anterior of the masseter medialis rarely originates within the orbital pocket in therian mammals, but it does in some hystricomorph rodents and in extinct South American marsupials of the family Argyrolagidae (Simpson, 1970; see also Gambaryan and Kielan-Jaworowska, 1995, for references).

Musculus temporalis has been reconstructed in *Nemegtbaatar* as consisting of three parts, as in most modern mammals: pars superficialis, pars anterior, and pars posterior. As in extant mammals (see Turnbull, 1970), pars superficialis probably originated as an aponeurosis from the parietal bone below the temporal line and the sagittal crest and passed ventrally into a tendon, which probably inserted in a temporal groove on the dentary. Pars anterior apparently originated from the entire surface of the parietal bone under the aponeurosis of pars superficialis, from the postorbital process, and from the anterior lamina of the petrosal and possibly inserted by a common tendon with pars superficialis in front of the coronoid process. Pars posterior apparently originated, as in modern mammals, from the lambdoidal crest and inserted on the medial and lateral sides of the coronoid process.

Gambaryan and Kielan-Jaworowska (1995) argued that the postorbital process, situated on the parietal in most multituberculates, functionally corresponded to the postorbital process situated on the frontal in therian mammals. The posterior position of the postorbital process in multituberculates resulted in an oblique postero-dorsal-anteroventral position of all the temporal muscles, different from the more vertical orientation in therian mammals.

As in most mammals there were two pterygoideus muscles: pterygoideus lateralis and pterygoideus medialis. The latter apparently originated from the lateral choanal channel (see earlier section "Palate") and inserted in the large pterygoid fossa on the medial side of the dentary and on the pterygoideus shelf. Pterygoideus lateralis apparently originated from the concave area of the horizontal part of the alisphenoid in front of the alisphenoid ridge, from the posterior part of the maxilla to the rear of M2, and from the anteroventral part of the anterior lamina of the petrosal. It inserted in the pterygoid fovea situated on the medial side of the dentary to the rear of the pterygoid fossa. Reconstruction of the origin and insertion of the digastric muscle is uncertain.

Gambaryan and Kielan-Jaworowska (1995) calculated that all the masticatory muscles in multituberculates produced the retractory horizontal components of the resultant force (which is protractory in therians). This difference is related to the backward masticatory power stroke in multituberculates. In order to test the earlier reconstructions of the multituberculate cranial musculature in *Ctenacodon* by Simpson (1926b) and in *Ectypodus* by Sloan (1979), Gambaryan and Kielan-Jaworowska (1995) modified the drawings of these authors with vectors of all the muscles added. The resultant force in *Ctenacodon* gave a large protractory horizontal component, as characteristic of therians but not of multituberculates. It follows that Simpson's (1926b) reconstruction apparently was incorrect. In *Ectypodus* they obtained a retractory horizontal component for this force, as characteristic for the djadochtatherioid multituberculates, as well as for *Ptilodus* (Wall and Krause, 1992).

The facial musculature of multituberculates has not been reconstructed. As pointed out by Gambaryan and Kielan-Jaworowska (1995), the facial musculature may be difficult or even impossible to reconstruct in fossil mammals, as facial muscles, unlike masticatory muscles, do not leave obvious scars on the bones. Of the facial muscles these authors were able to reconstruct tentatively only the buccinator.

Postcranial Musculature. Simpson and Elftman (1928) reconstructed the hindlimb musculature of Paleocene ?*Eucosmodon,* while Kielan-Jaworowska and Gambaryan (1994) reconstructed the postcranial musculature of Asian multituberculates belonging to the Late Cretaceous Djadochtatherioidea (referred to as the Eucosmodontidae) and Cretaceous and Paleocene Taeniolabididae. They based their reconstructions on the surface topography of the partial skeletons of the Late Cretaceous *Krytpobaatar, Nemegtbaatar, Chulsanbaatar,* and *Catopsbaatar* and on isolated bones of Paleocene-Eocene ?*Lambdopsalis.*

Not all the muscles have been reconstructed. Kielan-Jaworowska and Gambaryan (1994; see also Gambaryan and Kielan-Jaworowska, 1995) were unable to reconstruct the occipital musculature because the muscle scars could not be identified on the available neck vertebrae and none of the available occipital plates were adequately preserved. Similarly, the thoracic and caudal vertebrae, ribs, shoulder girdle, carpus, and manus as preserved in both Late Cretaceous (Kielan-Jaworowska, 1989; Kielan-Jaworowska and Gambaryan, 1994; Sereno and McKenna, 1995) and Paleocene (Krause and Jenkins, 1983) multituberculates were incomplete, damaged, or lacking and did not allow for a reconstruction of the relevant musculature. As a result, the musculature of the cervical and thoracic regions, the forelimb, and the tail is incompletely known. Thus Kielan-Jaworowska and Gambaryan (1994) based their reconstruction of the forelimb musculature partly on isolated bones from the Late Cretaceous of North America and from the Paleocene-Eocene boundary of China (Kielan-Jaworowska and Qi, 1990). Such a reconstruction, based on different taxa, is not wholly satisfactory. However, as the multituberculate skeleton appears to be fairly uniform, even among families, we may assume for present purposes (lacking other evidence) that the musculature was also similar.

Of the anterior part of the body, Kielan-Jaworowska and Gambaryan (1994) reconstructed only some muscles of the scapulocoracoid, the humerus, and the forearm. The presence of the incipient supraspinous fossa in *Nemegtbaatar* indicates the presence of a relatively small supraspinatus muscle, although its scars are recognizable only on the cranial surface of the spine. In extant therian mammals this muscle originates on the cranial surface of the spine and in the supraspinous fossa and extends to the lateral surface of the subscapularis muscle. It inserts on the proximal end of the greater tubercle of the humerus (Gambaryan, 1960; Jouffroy, 1971; see also Kielan-Jaworowska and Gambaryan, 1994: figure 28). There is a depression on the greater tubercle of all Late Cretaceous multituberculate humeri and in ?*Lambdopsalis,* which is evidently for the insertion of this muscle. The arrangement of other muscles on multituberculate scapulocoracoids was apparently similar to that in small extant therian mammals.

Of all preserved multituberculate humeri, only five of which are complete (Kielan-Jaworowska, 1998), muscle scars are perhaps best recognizable on two humeri of ?*Lambdopsalis* from the Paleocene-Eocene boundary of China and *Nemegtbaatar* from the Late Cretaceous of Mongolia (figure 8.16A). The arrangement of the muscle scars is generally similar to that in extant small mammals, both placentals (e.g., rodents) and marsupials (e.g., *Antechinus*), but differs in small details. The pectoralis profundus in multituberculates inserted, as in extant mammals, on the wide deltopectoral crest (together with the cutaneus trunci and pectoralis superficialis muscles). In multituberculates its insertion continued up to the greater tubercle and passed directly to the coracoid process, whereas in extant mammals the pectoralis profundus passes along the intertubercular groove to the lesser tubercle and then on to the coracoid process. In ?*Lambdopsalis,* musculus brachialis originated on the proximal one-third of the dorsal surface of the humerus, between the triceps brachii caput laterale and caput mediale. It did not reach the lesser tubercle as in most extant therians (Gambaryan, 1960; Jouffroy, 1971). Muscle scars on the distal part of humeri are less evident and have been reconstructed partly by comparison with extant taxa.

Reconstruction of the arm muscles was based solely on the proximal part of the ulna of an unidentified multi-

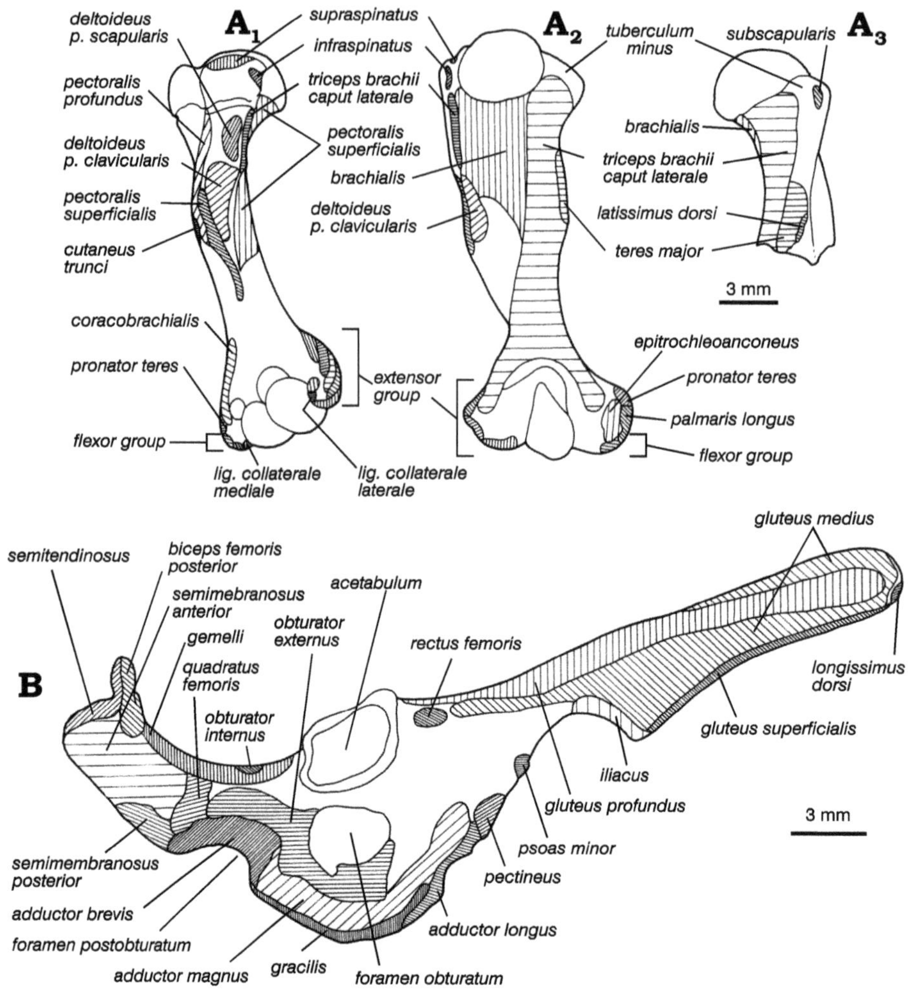

FIGURE 8.16. Reconstruction of muscle attachments based on surface topography of the bone. A, Left humerus of ?*Lambdopsalis bulla* from the Paleocene-Eocene boundary of China, in ventrolateral (A_1), dorsal (A_2), and medial (A_3) views. B, Right pelvis of *Nemegtbaatar gobiensis* from the Late Cretaceous of Mongolia in lateral view. *Source:* modified from Kielan-Jaworowska and Gambaryan (1994).

tuberculate from the Lance Formation. The arrangement of muscle scars resembles those of small marsupials (*Antechinus*) more than rodents (*Mesocricetus*). This concerns especially the digitorum profundus muscle, whose area of origin occupies most of the medial side of the proximal part of the ulna (as in *Antechinus*), whereas in *Mesocricetus* the origin of this muscle is placed more distally (see Kielan-Jaworowska and Gambaryan, 1994: figure 32).

More detailed data have been obtained on the musculature of the posterior part of the multituberculate body. The most important conclusion resulting from study of the muscles of the lumbar region, especially well-preserved in *Nemegtbaatar,* is that because of the unusually long spinous and transverse processes, the epaxial musculature was especially strongly developed. All the muscles of this region, especially semispinalis dorsi, sacrocaudalis dorsalis, longissimus dorsi, quadratus lumborum, and psoas major, were much stronger in *Nemegtbaatar* than they are in the extant rodent *Meriones blackleri,* which is about the same size (Kielan-Jaworowska and Gambaryan, 1994: figure 34). There is a high ventral crest on the ventral surface of L2 in *Nemegtbaatar.* In extant mammals a prominent ventral crest (referred to as the ventral spinous process) occurs only in the Leporidae, in which psoas major is very strongly developed.

When reconstructing the hindlimb musculature in *Nemegtbaatar* (based in part on *Kryptobaatar*), Kielan-Jaworowska and Gambaryan (1994) used Gambaryan's (1974) division of these muscles, which provides a basis for functional analysis. Their reconstruction differs in details from that presented by Simpson and Elftman (1928). Kielan-Jaworowska and Gambaryan (1994) reconstructed the musculature of the pelvic region and hindlimb of *Nemegtbaatar* and *Kryptobaatar* (figures 8.16B, 8.17). They based their reconstructions on well-preserved scars of muscular attachments on the pelvis, femur, tibia, and

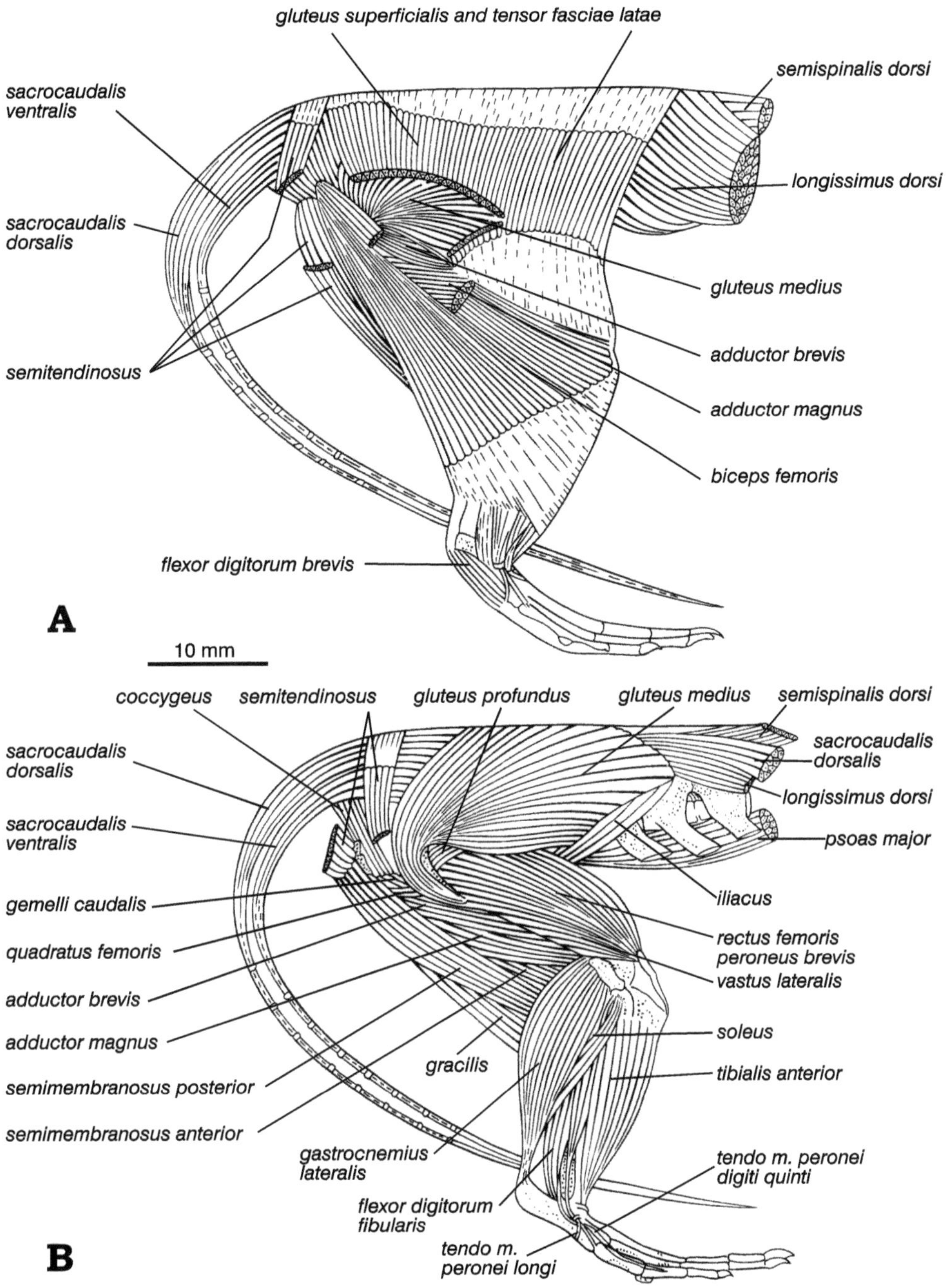

FIGURE 8.17. *Nemegtbaatar gobiensis.* Reconstruction of muscles of right hindlimb in lateral view. A, Superficial layer (*m. gluteus superficialis* and *tensor faciae* latae reconstructed as one muscle). B, Deep layer. *Source:* modified from Kielan-Jaworowska and Gambaryan (1994).

fibula of *Nemegtbaatar* and *Kryptobaatar,* and a comparison with those of extant small rodents (*Meriones blackleri*) and small marsupials (*Antechinus stuartii*).

CRANIAL VASCULATURE

The cranial vascular system was reconstructed by Kielan-Jaworowska et al. (1986) on the basis of the sectioned skulls of *Chulsanbaatar* and *Nemegtbaatar,* other skulls from the Late Cretaceous of Mongolia, and isolated Bug Creek petrosals. Further comparative studies of the cranial structure in multituberculates and other early mammals by Rougier et al. (1992) and Wible and Hopson (1993, 1995) generally confirmed the conclusions of Kielan-Jaworowska et al., with some emendations.

The internal carotid artery entered the pituitary fossa in Late Jurassic Paulchoffatiidae from below, piercing the basisphenoid, whereas in Late Cretaceous forms it entered the fossa laterally at the junction of the pterygoid, basisphenoid, alisphenoid, and possibly pterygoid. A stapedial artery was present, possibly with a meningeal branch that entered the skull through the hiatus Fallopii. (See reconstruction of the endocast and vascular system of *Nemegtbaatar gobiensis,* based on sections, by Kielan-Jaworowska et al., 1986: figure 32.) In addition to these two arteries, Kielan-Jaworowska et al. recognized two other vascular systems: (1) The dural sinus system, which embraces the sagittal, transverse, sigmoid, prootic, and tentorial sinuses. This system received a tributary glenoprootic vein that passed medially through a canal to join the ventral part of the prootic vein. The primary head vein passed from the cavum epiptericum through a posttrigeminal canal into the facial sulcus, received the prootic vein there, and probably left the sulcus as the stylomastoid vein before joining the jugular system. (2) The other system, designated the orbitotemporal system, consists of vessels that run from the posttemporal fossa, through the ascending canal between the anterior lamina of the petrosal and the squamosal, to enter the cranial cavity. As the route of the major vessel that enters the posttemporal fossa shows considerable similarity to arteria diploetica magna of monotremes, we refer to it as arteria diploetica magna.

The course of cranial arteries and veins reconstructed by Wible and Hopson (1995) for the taeniolabidoid *Catopsalis* and the ptilodontoid *Mesodma* (figure 8.18) on the basis of petrosal morphology and comparisons with extant mammals differs in details from that reconstructed for *Nemegtbaatar.*

BRAIN

Simpson (1937a) was the first to reconstruct the multituberculate brain, on the basis of several endocranial casts of Paleocene *Ptilodus.* In his reconstruction the olfactory bulbs widen anteriorly, which is unusual for mammals. Krause and Kielan-Jaworowska (1993) reexamined the material described by Simpson and demonstrated that the olfactory bulbs taper anteriorly in *Ptilodus,* as they do in all other mammals. Hahn (1969) reconstructed the dorsal aspect of the endocranial cast of Late Jurassic *Paulchoffatia* on the basis of skull structure and a comparison with the reconstruction of *Ptilodus.*

Current knowledge of multituberculate brain structure is based mostly on numerous endocranial casts of Late Cretaceous multituberculates from Mongolia. Kielan-Jaworowska (1974a) figured dorsal aspects of several natural endocranial casts preserved in *Chulsanbaatar,* one of which was mistakenly identified as ?*Kamptobaatar.* In 1983 she described the complete, three-dimensional endocranial cast of *Chulsanbaatar,* prepared from the skull by removal of all the bones. Kielan-Jaworowska et al. (1986) published a reconstruction of the brain of *Nemegtbaatar* based on a wax model obtained from study of the serially sectioned skull and provided an emended recon-

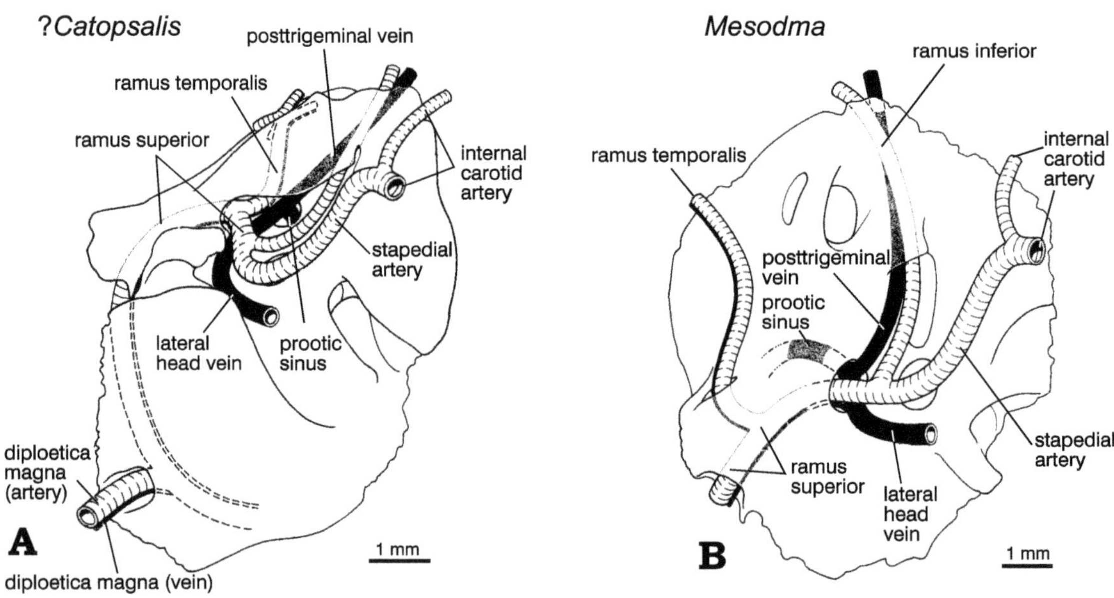

FIGURE 8.18. Reconstruction of the part of the cranial vessels in ?*Catopsalis* (A) and *Mesodma* (B), on the background of petrosal bones in ventral views. Dashed segments of arteries and veins are hidden behind bone; stippled segments are intramural, and dotted and dashed segment of prootic sinus is intracranial. *Source:* modified from Wible and Hopson (1995).

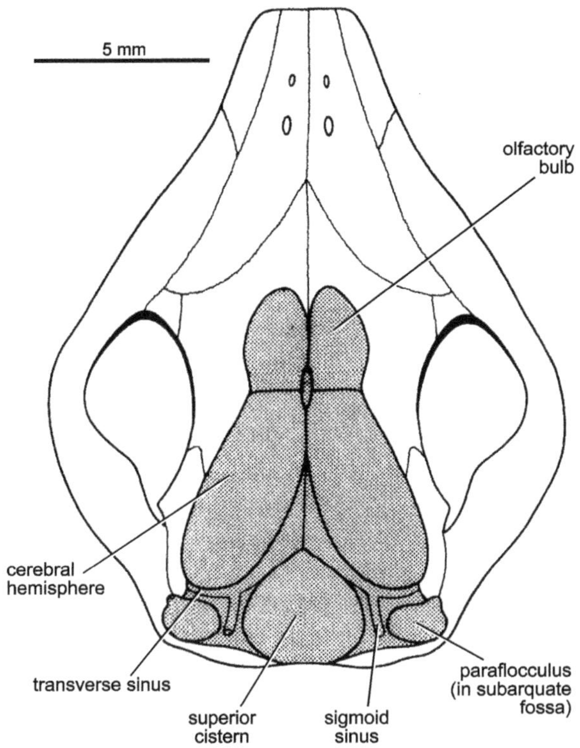

FIGURE 8.19. Reconstruction of the brain in *Chulsanbaatar*, on the outline of the skull. *Source:* modified from Kielan-Jaworowska et al. (1986).

struction of the brain in *Chulsanbaatar* (see figure 8.19, see also the section "Evolution of the Brain" in chapter 3).

Simpson (1937), who was the first to reconstruct the multituberculate endocast (of *Ptilodus*) referred to the triangular bulge inserted between the cerebral hemispheres as the "central lobe of the cerebellum" (= vermis). He has been followed in this respect by the students of multituberculate endocasts (e.g., Jerison, 1973; Hahn, 1969; Kielan-Jaworowska, 1983, 1986, 1997; Kielan-Jaworowska et al., 1986; Krause and Kielan-Jaworowska, 1993).

Using the modern brain terminology for describing endocasts assumes that the endocast provides an accurate picture of the brain. Bauchot and Stephan (1967) demonstrated how the endocranial casts of extant insectivores *sensu lato* differ from their brains. It is well known (e.g., Netter, 1983; Altman and Bayer, 1997) that an endocast is an impression of the brain, the meninges, and vessels that surround it. H. J. Jerison (pers. comm., October 2003) argued that the region described in multituberculates as the "vermis" might instead be the impression of a superior cistern covering the midbrain and vermis. Netter (1983) referred to it as "cistern of the great cerebral vein." H. J. Jerison, in the same personal communication, defined the superior cistern as follows: "it is a relatively large basin of the cerebrospinal fluid in the subarachnoid space in the region of the tectum mesencephali, extending from the anterior vermis of the cerebellum into the posterior space between the cerebral hemispheres. It must have been large enough in multituberculates to press on the internal table of the cranial cavity during development, to create the triangular bulge evident on the endocasts." He also informed us that the only extant species showing a similar "enlarged vermis" known to him is the marsupial koala, in which the enlargement or bulge is due to the impression of the superior cistern, which made an appropriate impression on the skull. The midbrain (tectum mesencephali, which includes anterior and posterior colliculi) is dramatically exposed in the koala brain but is covered on the endocast.

Kielan-Jaworowska and Lancaster (2004) followed the above comments of H. J. Jerison and did not use the term vermis when describing the multituberculate endocasts, replacing it by superior cistern. We follow these authors.

Based on comparison with the brain and endocast of the koala, Kielan-Jaworowska and Lancaster speculated that multituberculates had a brain that was generally similar in structure to those of Cretaceous eutherian mammals, which is well reflected on their endocasts (e.g., Kielan-Jaworowska 1984c, 1997). If so, the cerebellum in multituberculates was probably transversely elongated with a transverse anterior margin, in front of which the space between the diverging cerebral hemispheres might have been occupied by the midbrain, exposed on the dorsal side. There is, however, a difference between multituberculate and early eutherian brains. On the endocasts of Cretaceous eutherian mammals, the vermis is relatively narrow, delimited from the cerebellar hemispheres by a posteriorly directed sigmoid sinus, lateral to which there is an extensive cerebellar hemisphere and then lateral to it a relatively small paraflocculus. In multituberculate endocasts the sigmoid sinus is situated more laterally (lateral to the wide superior cistern) than in eutherian endocasts, and lateral to it there is a large, transversely elongated paraflocculus. It is not clear whether the vermis occupied the whole width enclosed under the superior cistern; if so, it would be relatively wider than in eutherian mammals.

Very large olfactory bulbs and extensive cerebral hemispheres characterize the multituberculate brain; the latter indicates apparent development of the neocortex. The superior cistern is in direct contact with the paraflocculi and there are no cerebellar hemispheres between them, exposed on endocasts, as is characteristic of the therians. A similar type of endocast structure occurs in the eutriconodontan *Triconodon* (figure 7.3C). The endocasts of multituberculate taxa differ in proportions, in the relative length of the olfactory bulbs, and in the size of the superior cistern and paraflocculi.

Jerison (1973) reviewed the literature on endocranial casts of fossil vertebrates and estimated the encephalization quotient (EQ) of the Paleocene multituberculate *Ptilodus* as 0.26. However, as argued by Kielan-Jaworowska (1983) and Krause and Kielan-Jaworowska (1993), Jerison overestimated the body mass of *Ptilodus* and underestimated the brain mass. Krause and Kielan-Jaworowska recalculated the EQ for *Ptilodus montanus* on the basis of new measurements of the diameters of humerus and femur and recalculation of the endocranial volume. They followed Gingerich (1990) and the method suggested by Thewissen and Gingerich (1989) of using two different equations, one of which includes the mass of the olfactory bulbs in the brain mass and the other excludes it. The EQ for *Ptilodus* obtained including the olfactory bulbs is 0.62 and excluding them it is 0.55.

The two other multituberculates for which the EQ has been calculated are the Late Cretaceous *Chulsanbaatar vulgaris* and *Kryptobaatar dashzevegi*, both from Mongolia. Kielan-Jaworowska (1983, 1986) calculated the EQ of *Chulsanbaatar* (including the olfactory bulbs) as 0.54–0.56. Krause and Kielan-Jaworowska (1993) regarded the difference between the EQs of these two multituberculate taxa (owing to various assumptions involved in the calculations) as negligible. Kielan-Jaworowska regarded the EQ of 0.55 for *Chulsanbaatar* as relatively high for a Mesozoic mammal. However, Krause and Kielan-Jaworowska (1993) pointed out that the relative brain size of *Ptilodus montanus* is smaller than that of an average extant mammal. Kielan-Jaworowska and Lancaster (2004) calculated the EQ of *Kryptobaatar* using: (1) Jerison's classical method based on estimation of endocranial volume and body mass, and (2) estimation of body mass based on new equations proposed by them, using measurements of the femur and tibia. In both cases the EQ is about 0.71, which is higher than estimated for *Ptilodus* and *Chulsanbaatar.*

Krause (1986) measured the orbits in Paleocene multituberculate *Ptilodus* and *Ectypodus* based on the reconstructions of Simpson (1937a) and Sloan (1979), respectively, and concluded that the eyes in multituberculates were small. Gambaryan and Kielan-Jaworowska (1995) challenged these reconstructions and Krause's conclusions. They demonstrated that the postorbital process was situated far posteriorly on the parietal in multituberculates, rather than on the frontal as in therians. It follows that the multituberculate orbit was very large and the eyes (as inferred from the size of the orbit) were very big (figure 8.20). It is difficult to evaluate the size of optic and auditory colliculi (*colliculus rostralis* and *colliculus caudalis*) from multituberculate braincasts, owing to the structure of the multituberculate endocasts, as the roof of the midbrain is completely obscured by the superior cistern.

The main multituberculate sense apparently was olfaction. The olfactory bulbs preserved in brain endocasts were unusually large, relatively larger than in any extant mammal brain. The large ear region, as inferred from the structure of the basicranium, also implies acute hearing. The strong development of all these senses points to a nocturnal mode of life of multituberculates, as suggested for all Mesozoic mammals by Jerison (1973) and many others.

PALEOBIOLOGY

Homeothermy. Meng and Wyss (1997) reported the discovery of hair belonging to the multituberculate *Lambdopsalis bulla* from the Bayan Ulan Formation (Paleocene-Eocene boundary) in Chinese Inner Mongolia. The hair was preserved in coprolites of carnivorous mammals. The presence of hair shows that multituberculates, like modern and possibly all fossil mammals, had insulation evidently related to their homeothermy. As discussed in chapter 3, hair may have already appeared in advanced cynodonts.

Dental Function and Diet. Ever since the middle of the nineteenth century, when Falconer described the first multituberculate, the diet of these unusual mammals has been hotly debated. Falconer himself (1857) believed that

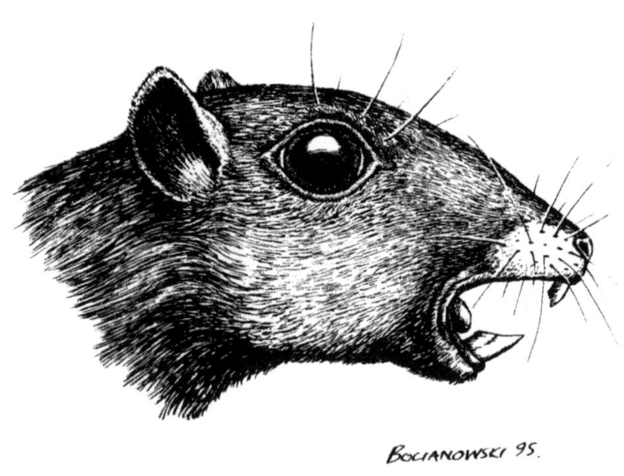

FIGURE 8.20. Reconstruction of the head of *Nemegtbaatar*, approximately ×2. Note big eye situated far posteriorly. *Source:* modified from Gambaryan and Kielan-Jaworowska (1995).

the multituberculate *Plagiaulax* was either herbivorous or frugivorous, an opinion criticized by Owen (1871), who held that *Plagiaulax* was a carnivore preying on small mammals and lizards. Cope (1884) postulated that the Paleocene multituberculate *Ptilodus* was not herbivorous and its diet may have consisted of small eggs picked up by the incisors and cut by the premolars. Gidley (1909) was inclined to believe that both *Plagiaulax* and *Ptilodus* were frugivorous, since the incisors were well fitted for picking small fruits and berries, while the large cutting blades of the lower premolars were adapted for cutting and slicing them. Broom (1910) argued for carnivorous (partly insectivorous) habits of both genera. Four years later he discussed the possibility that early multituberculates might have been mainly herbivorous (and omnivorous) but that some forms such as *Plagiaulax* were carnivorous (Broom, 1914). Hennig (1922) regarded multituberculate dentition as primitive and held that a primitive type of dentition is not adapted to any particular diet. He also expressed doubts that rodentlike (fore-and-aft) jaw movement was characteristic of multituberculates.

Finally Simpson (1926b) summarized these previous opinions and attempted to reconstruct some masticatory muscles in a "plagiaulacidan" multituberculate based mostly on *Ctenacodon* (see the earlier section on "Cranial Musculature" for a critique of Simpson's reconstruction). He compared the multituberculate dentition with that of modern *Bettongia,* a marsupial "rat" kangaroo that feeds on tough, dry vegetable substances. Simpson (1926b: 245) argued that "almost every morphological feature of the skull and dentition is correlated with herbivorous habits either directly or indirectly." In a later paper, Simpson (1933b) referred to the dentition of Jurassic multituberculates as "plagiaulacoid." He enlarged comparisons of the multituberculate dentition with a suite of extant marsupial genera having a plagiaulacoid type of dentition and restated his earlier opinion on the basically herbivorous diet of multituberculates.

The idea of multituberculate herbivory (see also Landry, 1963) has been gradually replaced by their characterization as omnivores, mostly owing to reassessment of food habits of extant "rat" kangaroos, regarded as multituberculate analogues (see Clemens and Kielan-Jaworowska, 1979, for references). The new wave of interpretation of jaw movements, dental function, and in consequence multituberculate diet started with papers by Rensberger (1973) and Greaves (1973). These authors demonstrated that the direction of motion between upper and lower teeth of herbivorous mammals can be inferred from studying wear facets of occluding teeth, by producing them on artificial mammalian teeth and illustrating the wear pattern (see also Costa and Greaves, 1981).

Although horizontal, fore-aft (propalinal) chewing in multituberculates had long been assumed, it remained for Gingerich (1977) to demonstrate, on the basis of analysis of wear facets, that multituberculates had a backward power stroke. Krause (1982a) subsequently corroborated this idea and presented a detailed analysis of jaw movements and dental function in Paleocene *Ptilodus* (figure 8.21). Krause's (1982a) conclusion that ptilodontoids were omnivorous was supported by Wall and Krause (1992).

Gambaryan and Kielan-Jaworowska (1995: figure 9) provided reconstruction of the masticatory musculature in several Late Cretaceous Asian djadochtatherioids (see the earlier section on "Cranial Musculature"), and reconstructed the masticatory cycle in *Nemegtbaatar.* They also produced a diagram (our figure 8.22) showing that the low position of the condyle (characteristic primitively of multituberculates and retained in most taxa) is advantageous for this group. A low condyle provides a greater moment arm for the adductor musculature during the backward power stroke of the dentary. Placental rodents, on the other hand, employ a forward power stroke, and in this case placement of the condyle well above the tooth row gives better mechanical advantage for the adductor musculature.

Wall and Krause (1992) demonstrated that, owing to the structure of the condylar process, Paleocene *Ptilodus*

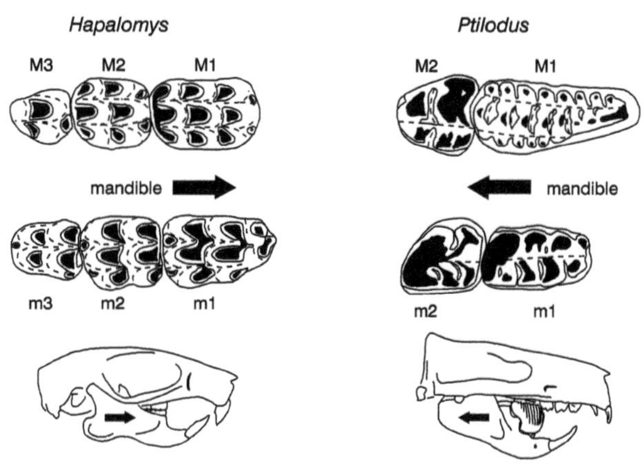

FIGURE 8.21. Comparison of cusp geometry of upper and lower molars of the living murid rodent *Hapalomys* with those of the multituberculate *Ptilodus.* Anterior is to the right. Cusps of *Hapalomys* are concave posteriorly on the upper molars and concave anteriorly on the lowers, indicating a protraction of the mandible during the power stroke of the chewing cycle. Cusps of *Ptilodus* are concave anteriorly on the upper molars and concave posteriorly on the lowers, indicating a retraction of the mandible during the power stroke. Black areas are exposed dentine, gray strips around black areas are unworn enamel, and dashed lines are valleys between the cusps. *Source:* modified from Krause (1982a).

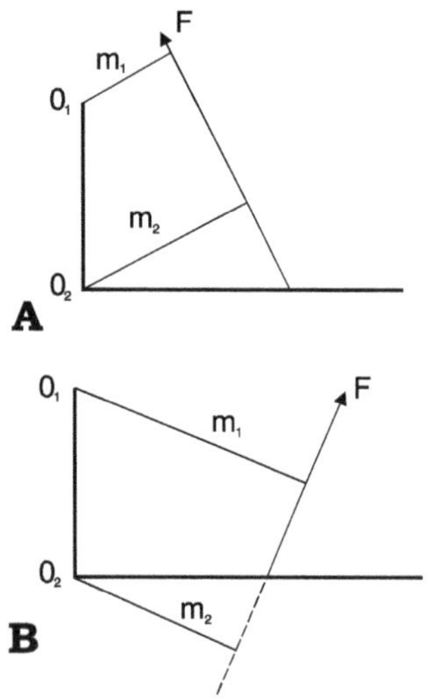

FIGURE 8.22. Diagram illustrating the moment arm of the resultant forces of all the masticatory muscles generated on the teeth during the power stroke. A, Dentary in multituberculates, with retractory horizontal component of the resultant force. B, Dentary in herbivorous therians (e.g., rodents) with protractory horizontal component of the resultant force. Bold lines: horizontal = dentary at the level of the molars, anterior is to the right; vertical = posterior margin of the dentary; F = vector of the resultant force of the masticatory muscles during the power stroke; m_1, m_2 = moment arm of F with the high and low positions of the coronoid process; O_1 = condylar process situated above the level of the molars; O_2 = condylar process situated at the level of the molars. *Source:* from Gambaryan and Kielan-Jaworowska (1995).

was unable to eat 12-mm-diameter *Ginkgo* seeds (*contra* Del Tredici, 1989). Gambaryan and Kielan-Jaworowska (1995: figure 17) speculated as to why the condyle is placed above the level of the molars in some multituberculates (e.g., in *Sloanbaatar, Catopsbaatar,* and in some taeniolabidids). They argued that this may be related to the size of the animals and their different strategies for cutting resistant items. They compared the gape in a Late Jurassic paulchoffatiid multituberculate and in three Late Cretaceous djadochtatherioid genera and attempted an explanation as to why the high position of the condyle and large coronoid process were advantageous in some of them, whereas in other cases the low position of the condyle and low coronoid process were more efficient for eating hard seeds. They concluded (p. 94) that in multituberculates: "the adaptation for crushing hard seeds in extreme cases worked against the benefit of the low position of the condylar process."

Posture. As summarized by Howell (1944), Sukhanov (1974), Gambaryan (1967, 1974), and many others, terrestrial mammals have two basic types of gaits, referred to as symmetrical and asymmetrical. In symmetrical gaits a movement of one forelimb is followed by the movement of either of the hindlimbs, after which the next forelimb and the next hindlimb move; at the same time there are lateral flexures of the body. As a result, the pattern of limb movements on the right side of the body is a mirror image of those on the left. The animals that use symmetrical gaits have a sprawling limb posture (abducted limbs), in which the proximal segments of limbs, when seen from the front or back, are arranged at about 90° (but sometimes less) to the parasagittal (vertical) plane. In asymmetrical gaits the forelimbs move first and then the hindlimbs. Right and left sides of the body do not move symmetrically, there are no lateral flexures, and extension and flexion of the body occurs in a sagittal plane. Among extant tetrapods, urodelans, most reptiles, and monotremes have a sprawling limb posture and use symmetrical gaits, whereas therian mammals mostly use asymmetrical gaits and have a limb posture that is referred to as parasagittal.

However, as demonstrated by Jenkins (1971a), many small mammals have humeri positioned at angles 10–30° from the parasagittal plane and the femora 20–50° to the parasagittal plane. The gait of therians differs from that of animals with a sprawling posture in the lack of adduction and in the movement of the femur, the distal end of which moves in a parasagittal plane during the propulsive phase (Jenkins and Camazine, 1977).

Kielan-Jaworowska and Gambaryan (1994) argued that multituberculates had a sprawling posture. They based their conclusion on the structure of the foot, tibia, and pelvis. They offered a new reconstruction of the multituberculate foot (figure 8.14B,D) in which the main axis, which passes through Mt III, is arranged at an angle of 30° with respect to the longitudinal axis of the tuber calcanei. In mammals with parasagittal limbs, in contrast, the main axis of the foot is parallel to the tuber calcanei and to the parasagittal plane. In mammals with parasagittal limbs, the craniocaudal (anteroposterior) diameter of the tibia is much greater than the mediolateral diameter. But in multituberculates the mediolateral diameter is larger. This may be explained by the abducted position of their limbs, in which the stress on the tibia during the propulsive phase was directed medially. The multituberculate pelvis is deep; Kielan-Jaworowska and Gambaryan proposed that the femoral adductors originated ventral to the acetabulum, as in animals with abducted limbs, rather than posterior to the acetabulum, as in forms with a parasagittal stance. Sereno and McKenna (1995: 146) argued that in *Bulganbaatar* (now *Kryptobaatar* as discussed earlier), "marked ventral (not lateral) orientation of the glenoid, reduction of the size of humeral epicondyles, hinge-like form of the elbow joint (suggested by prominent, narrow ulnar condyle and broad intercondylar groove on the distal end of the humerus, approaching the form of the therian trochlear joint)," and small humeral torsion indicate that the multituberculate forelimb stance was similar to that in *Didelphis* (parasagittal). They also argued that parasagittal posture made its appearance in the evolution of mammals only once, in the common ancestor of therians *sensu stricto* and multituberculates (see also Sereno and McKenna, 1996). Sereno and McKenna's opinion (1995, 1996) has been questioned by Presley (1995), Rougier, Wible, and Novacek (1995b), Gambaryan and Kielan-Jaworowska (1997), Kielan-Jaworowska (1998), and Hu et al. (1997).

Most important are the arguments of Gambaryan and Kielan-Jaworowska (1997), who demonstrated that a lack of torsion is not necessarily indicative of a parasagittal posture, as torsion occurs in terrestrial tetrapods with abducted limbs that use symmetrical diagonal gaits, but not in anurans, which have abducted forelimbs, but use asymmetrical jumps. It does not occur either in fossorial therians with sprawling or semisprawling stance, except the Chrysochloridae. On the other hand, the condylar structure of the elbow joint, characteristic of multituberculates, occurs in tetrapods with primary (rather than secondarily) acquired abducted forelimbs. Therian mammals acquired a trochlea probably during the Late Jurassic. They retained a vestigial radial condyle in Late Cretaceous forms (e.g., in the eutherian *Barunlestes*, see chapter 13) but lost this condyle during the Paleocene. Sereno and McKenna (1995) argued that the multituberculate *Bulganbaatar* (= *Kryptobaatar*) acquired an incipient trochlea. According to Gambaryan and Kielan-Jaworowska (1997), these authors were mistaken in their observations, as multituberculates, which retain prominent radial and ulnar condyles, never acquired a trochlea.

In view of the above arguments we accept that multituberculates had a sprawling stance (figures 8.11, 8.23). Because of the stance the body was held lower to the ground than in mammals with more parasagittally oriented limbs and was similar in proportions to that reconstructed for the amphilestid "triconodont" *Gobiconodon* (Jenkins and Schaff, 1988: figure 1). A characteristic feature of the multituberculate body is the relatively large head and short neck. The multituberculate skull is relatively longer in proportion to body length than in rodents and marsupials of similar size, but shorter than in *Gobiconodon;* it is, however, notably wider than in small extant mammals (Kielan-Jaworowska and Gambaryan, 1994: 57).

Locomotion. Gidley (1909) was the first to describe postcranial bones of the Paleocene multituberculate *Ptilodus* and suggested that it might have been saltatorial. Simpson (1926b) suggested the possibility of a semi-arboreal mode of life for multituberculates, while Simpson and Elftman (1928) went further and argued for an arboreal mode for Paleocene ?*Eucosmodon.* The idea of an arboreal mode for multituberculates flourished in the paleontological literature until 1994, with the main evidence being provided by Krause and Jenkins (1983) on the basis of their analysis of an almost complete skeleton of *Ptilodus* and the hindlimb of ?*Eucosmodon.* These authors argued that both taxa display the high pedal mobility characteristic of arboreal taxa that "descend trees head-first." In addition, the hallux is divergent and the entocuneiform-Mt I joint permitted a considerable hallucal abduction and adduction for prehension in a plane independent of the remaining digits. According to these authors, *Ptilodus* possessed a long tail, a feature that in modern mammals is present in prehensile-tailed forms.

Rowe and Greenwald went further and stated (1987: 25A): "It now appears more likely that arboreality was the ancestral habit for Theriiformes [the taxon comprising the immediate common ancestor of Multituberculata and Theria and all its descendants] and that the origin of Multituberculata involved dietary specializations within that niche." Kielan-Jaworowska and Gambaryan (1994) questioned this conclusion, noting that: (1) questions remain as to how arboreally adapted the North American taxa actually were; and (2) detailed functional analysis of Cretaceous multituberculates from Mongolia suggests terrestrial rather than arboreal life style.

With regard to *Ptilodus* and ?*Eucosmodon,* Kielan-Jaworowska and Gambaryan (1994) noted that the ungual phalanges of the pes are not as distinctly curved or compressed as they tend to be in some arboreal mammals, that Mt I is proportionately shorter than is generally the case in arboreal taxa with an opposable hallux, and that a long, well-muscled tail also occurs in some small terrestrial taxa.

A notable feature of the axial skeleton in Late Cretaceous multituberculates from Mongolia (Kielan-Jaworowska and Gambaryan, 1994) is the presence of very long spinous and transverse processes on the lumbar vertebrae. In urodelans and reptiles, which employ symmetrical locomotion with strong lateral flexures of the body, the transverse processes of the lumbar vertebrae are short, as their elongation would hinder the lateral flexures. The long transverse processes of the lumbar vertebrae in *Nemegtbaatar* thus indicate asymmetrical locomotion, despite the apparent presence of a sprawling posture (which is normally associated with a symmetrical gait; see earlier).

The length of the spinous processes as reconstructed in *Nemegtbaatar* shows that the mass of the erector spinae muscle was strongly developed, as it is in saltatory mammals. Capability for strong, sagittal extension and flexion of the vertebral column in *Nemegtbaatar,* and favorable

FIGURE 8.23. Reconstruction of posture of the Late Cretaceous multituberculate *Nemegtbaatar*. *Source:* modified from Kielan-Jaworowska (1997).

comparison with jumping mammals, is consistent with a saltatory, asymmetrical mode of locomotion. Analysis of the pelvic muscles in *Nemegtbaatar* and a comparison with those of *Rattus, Meriones,* and *Antechinus* allowed Kielan-Jaworowska and Gambaryan (1994) to speculate that the pelvis moved steeply vertically during the propulsive phase in multituberculates, which resulted in a steep jump trajectory (figure 8.24). These authors compared the relative lengths of the tibia and metatarsals in *Kryptobaatar* with those of extant running mammals and found that they were relatively short. Short tibia and short metatarsals indicate slow but powerful locomotion in *Kryptobaatar.* Hence the data obtained from the different parts of the body are contradictory: the axial column indicates strong flexion-extension and a possible saltatory mode of locomotion, whereas limb proportions indicate slow, powerful movement. Kielan-Jaworowska and Gambaryan (1994: 72) suggested that: "This contradiction may be the result of the abducted position of multituberculate limbs, because of which their movements were different from mammals with 'parasagittal' limbs. The direct transposition of data obtained from the analysis of gaits in modern mammals would thus be incorrect." In order to understand the mechanics of multituberculate locomotion these authors compared it with that of frogs, which are the only extant vertebrates that use asymmetrical jumps with abducted limbs. In frogs the trajectory of jumps is very steep: the angle of takeoff is between 30–45° (see Kielan-Jaworowska and Gambaryan, 1994, for references). They speculated that in multituberculates the hindlimbs may have moved rapidly medially during the jump (as in frogs), which resulted in a trajectory that was higher than those of modern therian mammals. In small modern mammals with "parasagittal" limbs, the center of gravity takes off at 3–10° to the horizontal.

Kielan-Jaworowska and Gambaryan (1994: 82) concluded: "On the basis of the functional analysis and comparisons in the preceding sections, we tentatively suggest that in the studied Asian multituberculates the gait, when they were moving fast, was most similar to that of small extant mammals such as, e.g., *Meriones* (Gambaryan, 1974); the important difference was that the trajectory of multituberculate jump was possibly much steeper than in any modern small mammal. Abducted limbs and steep jumps limited multituberculate endurance for prolonged run." These authors also speculated that Asian Late Cretaceous multituberculates inhabited semideserts, like extant *Meriones.* Their conclusions were based in part on sedimentological evidence. Jerzykiewicz et al. (1993) argued that the dinosaur eggs, skeletons of dinosaurs, lizards, mammals, and birds found in the Gobi Desert Djadokhta Formation were buried in eolian sand. Kielan-Jaworowska (1977) claimed that there are no remnants of trees in the Djadokhta and Baruungoyot formations, although tree trunks are common in sandy dinosaur-bearing sediments of the younger Nemegt Formation.

Asian Late Cretaceous forms do not show unequivocal fossorial adaptations, although such a possibility has been suggested for an unidentified "taeniolabidoid" (now possibly djadochtatherioid) from the Djadokhta Formation (Kielan-Jaworowska, 1989). Miao (1988), and Kielan-Jaworowska and Qi (1990) suggested a fossorial mode of life for Paleocene-Eocene *Lambdopsalis* from China.

The conclusion is that most Asian multituberculates for which the living habitat has been reconstructed were terrestrial, some were fossorial, while some later multituberculates might have been arboreal. The combination of an asymmetrical gait with a sprawling posture, as seen in multituberculates, renders the analogy with living taxa difficult.

Reproduction. As described under "Postcranial Skeleton" the multituberculate pelvis differs from those of other mammals in being extremely narrow and in having right and left halves of pubes and ischia firmly fused to form a keel. Kielan-Jaworowska (1979) described the first complete multituberculate pelvis, belonging to the Late

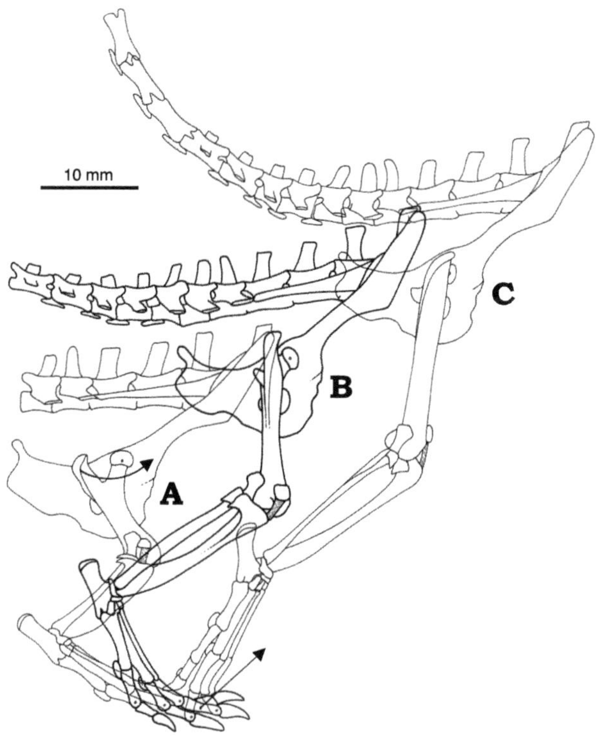

FIGURE 8.24. Diagram illustrating inferred movements of hindlimbs, pelvis, and tail in *Nemegtbaatar* (based in part on *Kryptobaatar*), during the beginning (A), middle (B, bold lines), and end (C) of propulsive phase, in lateral view. Epipubic bones are omitted. Straight arrow denotes direction of force of propulsion; round arrow denotes rotation of femur. *Source:* modified from Kielan-Jaworowska and Gambaryan (1994).

Cretaceous *Kryptobaatar dashzevegi.* In *Kryptobaatar* and other less complete multituberculate pelves from Mongolia, the posterior part of the pelvis (as seen from the side) is bent upward and the ventral keel of the fused ischia continues far dorsally. The length of the pelvic keel and the degree of fusion indicate that the pelvis was rigid and could not have spread ventrally during parturition.

The nature and dimensions of the pelvic girdle bear on the important question of the size of offspring or egg in multituberculates, and so on their mode of reproduction. Kielan-Jaworowska (1979) calculated that if multituberculates were viviparous the maximum width of the head of a newborn *Kryptobaatar* could not be more than 3.4 mm. This is about 0.13 of the width of mother's head, compared with a ratio of 0.33–0.4 in a rat, 0.45–0.5 in a mouse, and 0.1 in the marsupial *Antechinus.* If multituberculates were oviparous the space available for the passage of an egg of circular cross section would be even less than 3.4 mm in *Kryptobaatar* and possibly about 2.2 mm in *Chulsanbaatar,* which would mean an egg smaller than any known cleidoic egg. In monotremes, in correspondence with their oviparity, the ischial arc is very wide, U-shaped, and structurally very different from that in multituberculates. On the basis of the above data Kielan-Jaworowska (1979) concluded that multituberculates were not oviparous, but viviparous with extremely small neonates, similar to the condition in marsupials. She also speculated that the long, immobile symphysis and pelvic ridge, apart from the relevance to the mode of reproduction, might have a bearing on the mode of locomotion in *Kryptobaatar.* Comparison with some jumping ungulates that have an immobile symphysis led her to conclude that some multituberculates were also possibly able to jump. This supposition has been supported by a functional analysis performed by Kielan-Jaworowska and Gambaryan (1994) (see earlier section on "Locomotion").

Hopson (1973) argued that viviparity was one of the last elements in a syndrome of reproductive elements that evolved in mammals. As postcranial skeletons of the oldest uncontested multituberculates from the Kimmeridgian of Portugal are unknown, one cannot speculate when viviparity made its appearance in multituberculates.

Extinction. Multituberculates made their appearance possibly sometime during the Middle Jurassic (P. M. Butler, pers. comm., December 2001) and were certainly important elements of the terrestrial fauna by the Late Jurassic. They survived until the end of the Eocene. Van Valen and Sloan (1966) thoroughly reviewed the problem of their extinction and concluded that first condylarths, then primates, and finally rodents contributed to the gradual extinction of multituberculates. Hopson (1967) discussed the competitive inferiority of multituberculates to placental herbivores. Krause (1986) demonstrated an inverse correlation in generic diversity between multituberculates and rodents in the Paleocene and Eocene of the Western Interior of North America. He argued that the sudden appearance of paramyid rodents in the latest Paleocene of North America might have played an important role in the extinction of multituberculates. Lillegraven (1975, 1979) held that the appearance of the trophoblast and related prolongation of the gestation period enormously increased the adaptative potentiality of eutherians. In view of this, the possible viviparity of multituberculates with an extremely small neonate (Kielan-Jaworowska, 1979) may have contributed to the competitive inferiority of multituberculates to eutherians. Kielan-Jaworowska and Gambaryan (1994) argued that another character implying competitive inferiority of multituberculates to eutherians may be the abducted position of multituberculate limbs, which limited their endurance for prolonged running.

SYSTEMATICS

Simpson (1945) divided the order Multituberculata Cope, 1884, into three families: Plagiaulacidae Gill, 1872; Ptilodontidae Gregory and Simpson, 1926 (= Chirogidae Cope, 1887); and Taeniolabididae Granger and Simpson, 1929. He did not recognize any multituberculate suborder at that time, although he mentioned (Simpson, 1925a) the suborder Plagiaulacoidea earlier and formally diagnosed it three years later (Simpson, 1928a); however, the author of Plagiaulacoidea is not Simpson, but Ameghino (1889). Simpson (1945) treated Plagiaulacoidea as a synonym of Multituberculata as opposed to Tritylodontoidea, which he also regarded as multituberculates. Sloan and Van Valen (1965) raised the family Ptilodontidae to subordinal rank (Ptilodontoidea) and erected the new suborder Taeniolabidoidea.

Hahn (1969) revised, rediagnosed, and assigned a new scope to the Plagiaulacoidea Ameghino, 1889. McKenna (1971) proposed the suborder Plagiaulacida, but subsequently (McKenna, 1975) used the suborder Plagiaulacoidea and proposed within it the new infraorder Cimolodonta. Hahn and Hahn (1983) recognized within Multituberculata (in addition to Haramiyoidea Hahn, 1973, regarded by Hahn et al., 1989, as a separate order, Haramiyida Hahn, 1973), three multituberculate suborders: Plagiaulacoidea Ameghino, 1889; Ptilodontoidea Sloan and Van Valen, 1965; and Taeniolabidoidea Sloan and Van Valen, 1965.

Kielan-Jaworowska and Ensom (1992) proposed a new suborder, Paulchoffatoidea (assigning a new rank to the family Paulchoffatiidae Hahn, 1969). Hahn (1993) rejected it and restored the family status of the Paulchoffatiidae, a course followed by Gambaryan and Kielan-Jaworowska (1995).

Simmons (1993) provided a large-scale cladistic analysis of Multituberculata, using a matrix-based approach. She recognized the paraphyly of Plagiaulacoidea ("Plagiaulacida" in the terminology used herein) and the monophyly of Cimolodonta. She further claimed that Ptilodontoidea are monophyletic if the genera *Boffius* and *Liotomus* are removed to Taeniolabidoidea, whereas the monophyly of Taeniolabidoidea is possible, but not certain. Rougier et al. (1997), using a modified version of Simmons's matrix, came to the conclusion that Asian Late Cretaceous genera (excluding *Buginbaatar*), together with North American Paleocene *Pentacosmodon*, form a monophyletic group. Kielan-Jaworowska and Hurum (1997) reached a similar conclusion on the monophyly of Late Cretaceous Mongolian taxa and erected the suborder Djadochtatheria (= superfamily Djadochtatherioidea in this book) for this group (*Buginbaatar* excluded) and tentatively assigned the North American *Pentacosmodon* and *Paracimexomys* to Djadochtatheria.

McKenna (in Stucky and McKenna, 1993) considered Cimolodonta and Plagiaulacida as orders within the infraclass Multituberculata. McKenna and Bell (1997), however, eliminated the order Plagiaulacida, regarding it as a paraphyletic unit. These authors assigned various paulchoffatiid, plagiaulacid, and allodontid genera (belonging to different well-defined families) to Plagiaulacidae; they placed one plagiaulacid genus, *Bolodon*, in Bolodontidae (regarded by Kielan-Jaworowska and Hurum, 2001, as a synonym of Plagiaulacidae), together with members of the Eobaataridae. This procedure obscured the relationships among early multituberculate subgroups.

Fox (1999) challenged the monophyly of Taeniolabidoidea (*sensu* Sloan and Van Valen, 1965), an idea reached independently by Kielan-Jaworowska and Hurum (2001). Fox (1999) classified Microcosmodontidae (new rank assigned by Fox to previous subfamily), Eucosmodontidae, Cimolomyidae, and Taeniolabididae as families *incertae sedis* within Cimolodonta. Hahn and Hahn (1999a) provided a systematic division of Plagiaulacoidea, within which they recognized eight families. They suggested that Ptilodontoidea might have originated from Allodontidae and Taeniolabidoidea from Plagiaulacidae and argued that if this is true, the suborder Cimolodonta should be abandoned, an idea not accepted herein.

We follow here the revised systematics of Multituberculata proposed by Kielan-Jaworowska and Hurum (2001), in which they divided the order Multituberculata into two suborders, the paraphyletic suborder "Plagiaulacida" Ameghino, 1889, Simpson, 1925a (*nomen correctum* McKenna, 1971, *ex* Plagiaulacoidea), the apparently monophyletic suborder Cimolodonta McKenna, 1975, and one family Arginbaataridae, left in a suborder *incertae sedis*. As McKenna (1971) neither established a new taxon, nor changed the meaning of Ameghino's, but only changed the ending, Ameghino (1889) remains the author of the suborder Plagiaulacida. Kielan-Jaworowska and Hurum (2001) stated that the subordinal name Plagiaulacoidea, commonly used until recently, would be better from the historical point of view. The International Commission on Zoological Nomenclature (ICZN,1999), however, recommends using the suffix "oidea" for superfamily names (Article 29.1). As Kielan-Jaworowska and Hurum (2001) used superfamilies in multituberculate systematics, they chose the name "Plagiaulacida" for the suborder, to avoid confusion with names of the superfamilies. Plagiaulacida are a paraphyletic unit, and therefore this name is used in quotes.

Suborder "Plagiaulacida" Ameghino, 1889

Diagnosis. Plesiomorphic, apparently paraphyletic order of Multituberculata with dental formula: 3.1–0.5–4.2/1.0.4–3.2. Differ from all later multituberculates in having three upper incisors (two in Cimolodonta), I2 enlarged, multicusped with additional cuspules in Paulchoffatiidae and Pinheirodontide, and two-cusped (with exception of Pinheirodontidae, where there are several posterior cuspules or cusps), I3 either enlarged two- to four-cusped, or small single-cusped, with or without a small basal cuspule; canine, if present, premolariform; five or four upper premolars. Differ from advanced Cimolodonta in lingual cusp row of upper molars shearing into the central valley of the lowers, the main wear facets of lower molars facing toward the central valley, rather than upward. The third (lingual) row of cusps on M1 is absent or developed as an incipient, smooth posterolingual wing. There are four or three lower premolars, all except p1 being bladelike. The p4 is rectangular in labial view, with a row of labial cusps (which may be replaced by pits, or almost absent in some Pinheirodontidae, single cusp occurs in Eobaataridae); p3 in labial view primitively rectangular, may be provided with a row of labial cusps, triangular and without labial cusps in later forms, rather than peglike or absent as in Cimolodonta. Lower incisor completely covered with enamel (except *Glirodon* and *Eobaatar*), enamel prismless (except *Glirodon* and *Eobaatar,* which have gigantoprismatic enamel).

Comments. Kielan-Jaworowska and Hurum (2001) grouped the families of "Plagiaulacida" into three informal lineages: the allodontid line, the paulchoffatiid line, and the plagiaulacid line. We follow this division. In addition to the known named "plagiaulacidan" taxa, Takada et al. (2001) reported the finding of a fragmentary dentary with a relatively large, partial p4 with six ridges, alveoli for a double-rooted p3, and a single-rooted p2 in lower Neocomian beds of central Japan, which is one of the oldest multituberculates from Asia.

Families. Allodontid line: Allodontidae Marsh, 1889; Zofiabaataridae Bakker, 1992; and *Glirodon* Engelmann and Callison, 1999, assigned to "Plagiaulacida" *incertae sedis.* Paulchoffatiid line: Paulchoffatiidae Hahn, 1969 (with subfamilies Paulchoffatiinae Hahn, 1969, and Kuehneodontinae Hahn, 1971); Hahnodontidae Sigogneau-Russell, 1991c; Pinheirodontidae Hahn and Hahn, 1999a. Plagiaulacid line: Plagiaulacidae Gill, 1872; Albionbaataridae Kielan-Jaworowska and Ensom, 1994; Eobaataridae Kielan-Jaworowska et al., 1987.

Distribution. ?Middle Jurassic (Bathonian) or Late Jurassic (Kimmeridgian) to Early Cretaceous (Barremian): Europe; Late Jurassic–Late-Early Cretaceous boundary: North America; Early Cretaceous (early Neocomian to Aptian or Albian): Asia; Early Cretaceous (?Berriasian): Morocco. The multituberculate nature of most Bathonian, except perhaps teeth described by Butler and Hooker (in preparation) and pre-Bathonian purported multituberculate (discussed here for the sake of completeness) cannot be demonstrated with any certainty.

ALLODONTID LINE

Brief Characterization. Members of this line differ from paulchoffatiid and plagiaulacid lines by having two-cusped I2 and a small, single-cusped (or with basal cuspule) I3. Differ from plagiaulacid line in having well-separated cusps on lower molars, with smooth enamel, and less elongated p4. Differ from paulchoffatiid line in having m2 with two rows of cusps (rather than basin shaped with a single cusp), p3 lacking labial cusps, M1 with incipient posterolingual wing.

Family Allodontidae Marsh, 1889b

Diagnosis. Dental formula 3.1.5.2/1.0.4.2. Three upper incisors: I1 small, not preserved; I2 large, two-cusped, I3 relatively large with main cusp and basal cuspule. Upper canine present (unknown in *Psalodon*), small, simple, single-rooted. P4 and P5 shorter in relation to M1 than in Plagiaulacidae, with no or single labial cuspule; incipient posterolingual ridge on M1 present (absent in Paulchoffatiidae). Lower incisor uniformly covered with enamel. The p3 lacks labial cusps (present in Paulchoffatiidae, plesiomorphy) and is rectangular in labial view; p4 about 1.5 times longer in labial aspect than p3 (instead of almost twice or twice as in Plagiaulacidae), with five or six serrations and a row of labial cusps. Lower molars with two rows of subequal cusps (plesiomorphy). The m1 is somewhat paulchoffatiid-like in having an enlarged middle cusp in labial row; m2 with two discrete rows of cusps instead of being basin shaped with a single cusp as in Paulchoffatiidae. Lower molar cusps do not coalesce as is characteristic of Plagiaulacidae and Eobaataridae; grooves and pits characteristic of the latter two families are lacking.

Genera. *Ctenacodon* Marsh, 1879b; *Psalodon* Simpson, 1926b; *non Ctenacodon brentbaatar* Bakker, 1998.

Distribution. Late Jurassic: United States, Wyoming (Morrison Formation).

Genus *Ctenacodon* Marsh, 1879b
(figure 8.25A,E,F)

Synonym: *Allodon* Marsh, 1881 (*partim*).

Diagnosis. The genus with characters of the family, differing from *Psalodon* in structure of the upper pre-

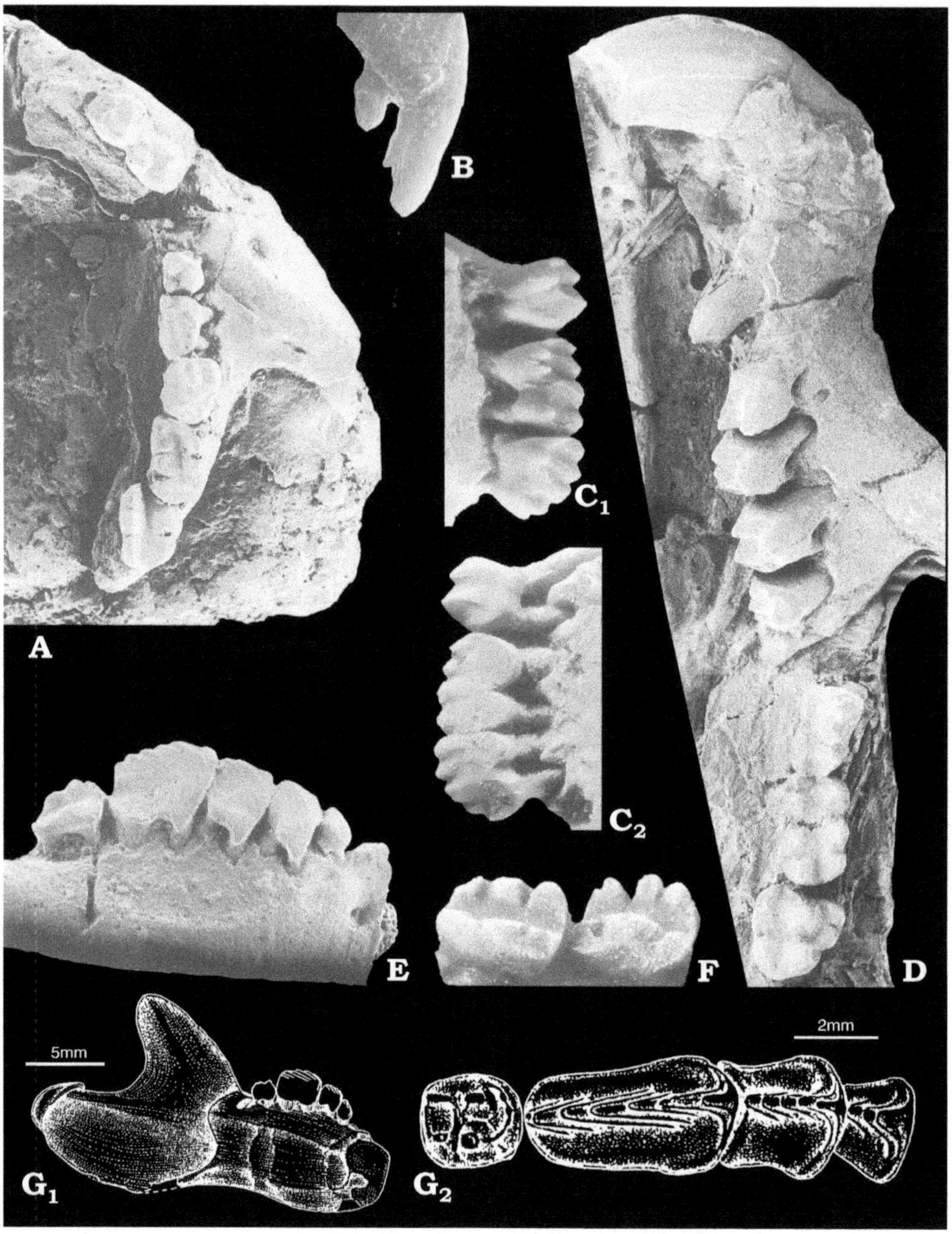

FIGURE 8.25. Members of the allodontid line. A, *Ctenacodon laticeps,* left maxilla with premolars and molars in occlusal view. B, *Psalodon fortis,* I2 (larger, two-cusped) and I3 (smaller single-cusped with a basal cuspule) in lateral view. C, *Psalodon potens,* fragment of right maxilla with P1–P4 in labial (C_1) and lingual (C_2) views. D, *Glirodon grandis,* left side of the palate of the holotype, with I2, alveolus for I3, C, P1–P5, M1, and M2. E, *Ctenacodon serratus,* right dentary with p1–m1 in labial view. F, *Ctenacodon scindens,* right m1 and m2 (anterior is to the right) in labial view. G, *Zofiabaatar pulcher,* left dentary of the holotype with alveoli for incisor, p1 and m2, and p2–m1 in lingual view (G_1); p2–m1 of the same specimen, in occlusal view, anterior is to the right (G_2). A, B, D, E, SEM micrographs. A, F, ×15; B, ×5; C, ×8; D, E, ×10. *Source:* A–F, original photos of epoxy resin casts; G, modified from Bakker and Carpenter (1990).

molars, which are less compressed laterally (less adapted for shearing).

Species. *Ctenacodon serratus* Marsh, 1879b, type species, *C. laticeps* (Marsh, 1881), *C. scindens* Simpson, 1928a, *C.* sp. (Simpson, 1928a), and several teeth left in open nomenclature or identified as *Ctenacodon laticeps/serratus* by Engelmann and Callison (1998).

Distribution. Late Jurassic: United States, Wyoming (Morrison Formation).

Genus *Psalodon* Simpson, 1926b
(figure 8.25B,C)

Synonym: *Allodon* Marsh, 1881 (*partim*).

Diagnosis. Simpson (1929a: 25) characterized *Psalodon* as follows: "P3 with three cusps, but one much the largest and one vestigial. P4–5 more highly developed [than in *Ctenacodon*] for shearing, more compressed laterally, with other cusps reduced in number and vestigial in character. Lower jaws, doubtfully referred, close to those of *Ctenacodon* but considerably larger than any known species of the latter." Differs from *Ctenacodon* in having upper premolar cusps strongly ribbed.

Species. *Psalodon potens* (Marsh, 1887), type species; *P. fortis* (Marsh, 1887); *P.? marshi* Simpson, 1929a, and some species left in open nomenclature.

Distribution. Late Jurassic: North America, United States, Wyoming (Morrison Formation).

Family Zofiabaataridae Bakker, 1992

Diagnosis. Monotypic family including *Zofiabaatar* Bakker and Carpenter, 1990. Only the dentary is known, with dental formula 1.0.4.2. Differ from Allodontidae in having p4 distinctly longer than p3 and share this character and the general structure of the premolars with Plagiaulacidae. Differ from Allodontidae and Plagiaulacidae in having very short m1 with two cusps in two rows, and from Plagiaulacidae and Eobaataridae in having lower molar cusps well separated (not coalesced). *Zofiabaatar* differs from all "plagiaulacidans" in which the dentary is known, in having the condyle facing upward, rather than posteriorly, and shares this character with some advanced Djadochtatherioidea and Taeniolabidoidea. *Zofiabaatar* differs also from other "plagiaulacidan" genera in having a strongly enlarged pterygoid fossa. The two latter characters may be of generic, rather than family distinction.

Genera. *Zofiabaatar* Bakker and Carpenter, 1990, type genus by monotypy.

Distribution. Late Jurassic: North America, United States, Wyoming (Morrison Formation).

Genus *Zofiabaatar* Bakker and Carpenter, 1990
(figure 8.25G)

Diagnosis and Distribution. As for the family.

Species. *Zofiabaatar pulcher* Bakker and Carpenter, 1990; type species by monotypy.

Family *incertae sedis*
Genus *Glirodon* Engelmann and Callison, 1999
(figure 8.25D)

Diagnosis (modified from Engelmann and Callison, 1999). Differs from other "plagiaulacidans" except *Eobaatar* in having restricted enamel band on lower incisor and I2, differs from Paulchoffatiidae and Plagiaulacidae in having single-cusped I3, shares with Allodontidae structure of the lower molars with two rows of separate cusps (not coalesced); differs from Plagiaulacidae and Eobaataridae in lack of "ornamentation" of pits and grooves on the molars. Differs from Cimolodonta in having a complete primitive dental formula of 3.1.5.2/1.0.4.2, differs from most "plagiaulacidans" and shares with *Eobaatar, Arginbaatar,* and cimolodontans (except Ptilodontoidea) gigantoprismatic enamel.

Species. *Glirodon grandis* Engelmann and Callison, 1999, type species by monotypy.

Distribution. Late Jurassic: North America, Utah, Dinosaur National Monument and Colorado, Fruita Paleontological area (Morrison Formation).

PAULCHOFFATIID LINE

Brief Characterization. Members of the paulchoffatiid line differ from those of plagiaulacid and allodontid lines by having (at least in some members) molars with cusps of different heights. Another difference is the presence of very large I2 and I3, the latter four-cusped, lack of even incipient posterolingual ridge on M1 (shared with a few members of the *Paracimexomys* group, assigned to Cimolodonta), presence of labial row of cusps on p3, smaller difference between the length of p3 and p4, and basinlike structure of m2 with only a single cusp.

Family Paulchoffatiidae G. Hahn, 1969

Comments. The classification of the Paulchoffatiidae which we present below is based mostly on the work of Gerhard Hahn, assisted often by his wife Renate Hahn (especially Hahn, 1993; and Hahn and Hahn, 1998b,c, 1999a,b, with some emendations).

The paulchoffatiid material from the Guimarota Coal Mine in Leiria, Portugal, has yielded the most complete paulchoffatiid specimens described so far. The fossils preserved in Guimarota are numerous and consist of isolated teeth, dentaries with teeth, and fragments of skulls with teeth. No postcranial parts have been described. The upper and lower teeth have never been found in occlusion. At first Hahn (1969, 1977a) assigned both upper and lower dentitions to *Paulchoffatia* and *Pseudobolodon* (see also Hahn and Hahn, 1983). In his revision, however, he separated upper and lower dental taxa (Hahn, 1993). For the upper dentition (and skull fragments) previously assigned to *Paulchoffatia,* he created the genus *Meketichoffatia* Hahn, 1993; the lower dentition earlier referred to *Pseudobolodon* was placed in *Meketibolodon* Hahn, 1993. The only paulchoffatiid genus represented by upper and lower dentition is *Kuehneodon* Hahn, 1969 (the only member of the Kuehneodontinae Hahn, 1971), but included species are represented by upper or lower dentitions only. As Hahn (1993) made clear, the number of paulchoffatiid taxa has been artificially enlarged by creating the separate upper and lower dentition taxa. On the other hand, three new genera proposed by Hahn and Hahn (1998c) and numerous unnamed new taxa described in the same paper, based on isolated molars, show that Guimarota multituberculate fauna was more diversified than previously thought.

In addition to the taxa from the Guimarota Coal Mine, several genera attributed (sometimes tentatively) to the Paulchoffatiidae have been described from other localities and strata. All of them are poorly known, represented by isolated teeth, so their subfamilial attribution remains open. We assign them to Paulchoffatiidae *incertae sedis.*

Diagnosis (based on Hahn, 1993, and references therein, emended). Early side branch of "Plagiaulacida" with dental formula 3.1–0.5–4.2/1.0.4–3.2 and nonprismatic enamel. Skull of the same general pattern as in Cimolodonta (figure 8.26, see also figure 8.4 for comparison); nasals relatively narrow, lacrimals small; frontals inserted anteriorly to a pointed end between the nasals and maxillae, apparently rounded posteriorly; postorbital process small, placed on the frontal (Wible and Rougier, 2000). Anterior palatal vacuities small or absent, posterior palatal vacuities always absent; small, thin jugals preserved on the inner flank of the zygomatic process of the maxilla; large carotid foramina piercing basisphenoid; petrosal with "processus promontorii" (Hahn, 1988; Lillegraven and Hahn, 1993); rudimentary coronoid bone (Hahn, 1977b) sometimes present. Angle between tooth row and dentary 7–20° (figure 8.27A).

Differ from other multituberculates (except some Pinheirodontidae) and approach in this respect Haramiyidae in having molars with cusps of different heights (plesiomorphy, figure 8.27D). The paulchoffatiid apomorphies concern the unique structure of the lower molars:

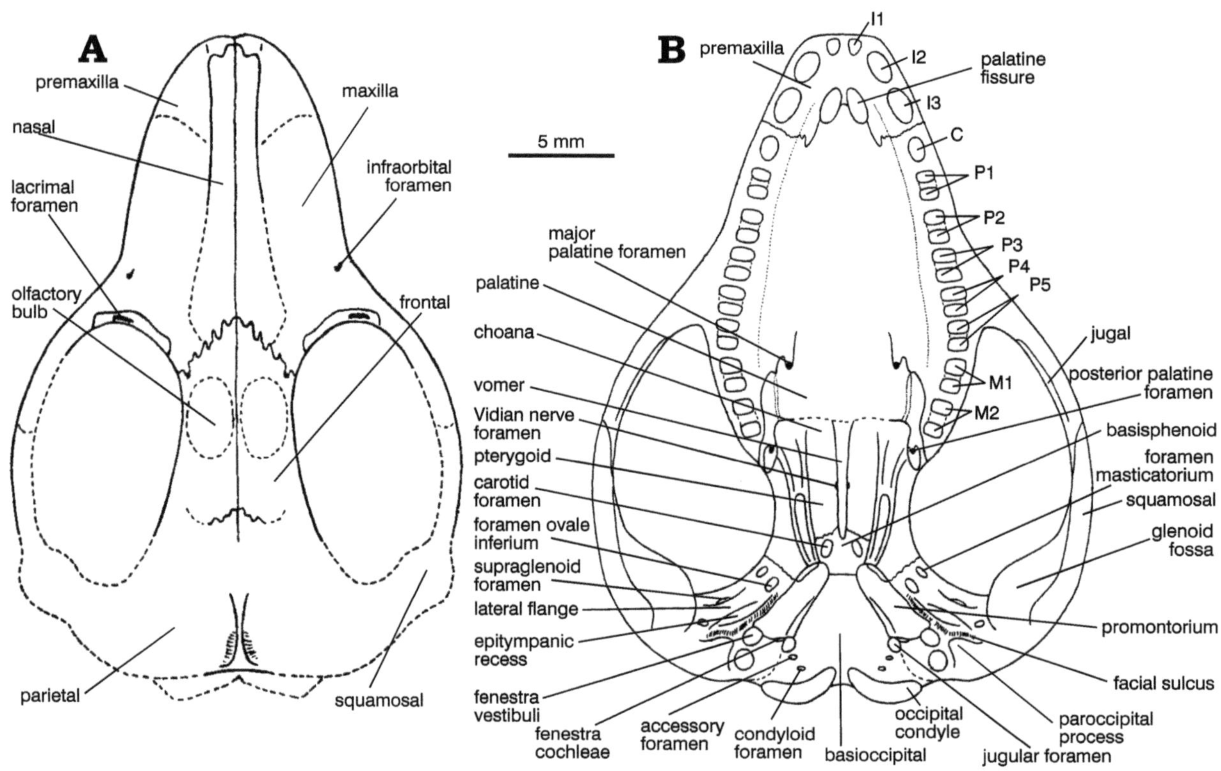

FIGURE 8.26. Reconstruction of the skull of a representative of Paulchoffatiidae in dorsal (A) and palatal (B) views. *Source:* modified A, from Hahn (1969); B, from Hahn (1988).

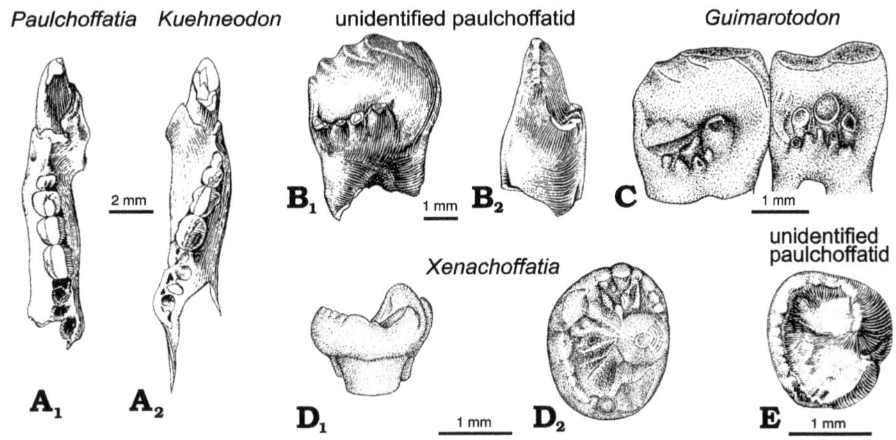

FIGURE 8.27. Morphology of the dentary and teeth in Paulchoffatiidae. A, Comparison of the dentaries of *Paulchoffatia* (A_1) and *Kuehneodon* (A_2), showing differences in angle between the tooth row and dentary, 7° in *Paulchoffatia* and 20° in *Kuehneodon.* B, Right p4 of unidentified paulchoffatiid with well-preserved row of labial cusps, in labial (B_1) and posterior (B_2) views. C, Right p3–p4 of *Guimarotodon leiriensis,* in labial view, showing a row of labial cusps and additional row of cuspules below it. D, Left m2 of *Xenachoffatia oinopion,* in posterior (D_1) and occlusal (D_2) views; note the different heights of the cusps. E, Left m2 of unidentified paulchoffatiid in occlusal view. *Source:* modified from: A, B, E, Hahn (1969); C, G, Hahn and Hahn (1998b).

m1 with anterior cuspidate cingulid and enlarged anterolabial cusp, m2 basin shaped, with cuspules and only one anterolingual cusp, dissimilar to m1, which has two longitudinal rows of cusps (figures 8.27D,E, 8.28m1,m2). Lower premolars show plesiomorphic characters: p1–p3 oval or rectangular, rather than triangular in labial view (as in Plagiaulacidae); p3 with a row of labial cusps (figures 8.27C, 8.29), almost as long as p4; and p4 short with no more than four serrations with ridges. Below a row of labial cusps a row of ornamental cuspules may be present on p3 and p4 in some genera (figure 8.27C, see also Hahn and Hahn, 1998b). Upper dentition (figures 8.28M1, M2, and 8.30): I1 small, single-cusped; I2 large, two-cusped; I3 enlarged, three- to four-cusped, dissimilar to I2, roughly quadrangular or trapezoidal in occlusal view, with main cusp arranged obliquely (anterolingually to posterolabially); canine, if present, premolariform. Anterior premolars (P1–3/4) short, oval to quadrangular, with three or four cusps; posterior premolars (P4–5 or P5 only) not shearing, relatively long, molariform, with two or three longitudinal rows of cusps, sometimes with additional row of cuspules. M1 without posterolingual ridge (plesiomorphy); M2 large with cusp formula 2–3:3–6. Enamel ornamented, especially on m2 and M2; upper premolars often ribbed.

Subfamilies. Paulchoffatiinae G. Hahn, 1969, type subfamily; Kuehneodontinae G. Hahn, 1971; Paulchoffatiidae subfam. indet. (*Galveodon* G. Hahn and R. Hahn, 1992, and ?*Sunnyodon* Kielan-Jaworowska and Ensom, 1992). In addition, Hahn and Hahn (1998c) described several paulchoffatiid taxa identified as Paulchoffatiidae genus A, B, C, D, based on M2s, which we do not describe herein.

Distribution. Late Jurassic (Kimmeridgian): Portugal, Leiria, Guimarota Coal Mine (Guimarota Beds); Late Jurassic (late Kimmeridgian–Tithonian): Portugal, Pai Mongo and Porto das Barcas (Lourinhã Formation); Early Cretaceous (Barremian): Europe. Hahn (1993) referred to the age of Guimarota Beds as "Lusitanian, Oxfordian or lower Kimmeridgian." Because the Lusitanian is a provincial name and, except for listing as a formation does not appear in common geological timetables (Harland et al., 1989; Gradstein et al., 1995), we do not use it here. Upper Triassic (lower Rhaetian) *Mojo,* regarded by Hahn et al. (1987) as a paulchoffatiid, is also described below, but its multituberculate affinity cannot be demonstrated.

Comments. On the lower premolars of "Plagiaulacida" there occurs a row of labial cusps, most prominent in a Paulchoffatiidae gen. et sp. indet. (Hahn, 1969, see also figure 8.27B). On p3–p4 of *Guimarotodon* and p4 of *Meketibolodon* (Hahn, 1987b; Hahn and Hahn, 1998b, and figure 8.27C) there occur small cuspules, situated outside and below the row of labial cusps. Hahn and Hahn (1998b) demonstrated a high degree of variation in the number, shape, and distribution of these cuspules. They introduced a scheme for numbering of molar cusps in Paulchoffatiidae (Hahn and Hahn, 1998c, our figure 8.28). The cusps of the lingual row in upper molars are numbered anteroposteriorly L^{1-x}, of labial row B^{1-x}, and in lower molars, respectively, L_{1-x}, and B_{1-x}. We accept their method, but instead of using lower and upper indexes, we

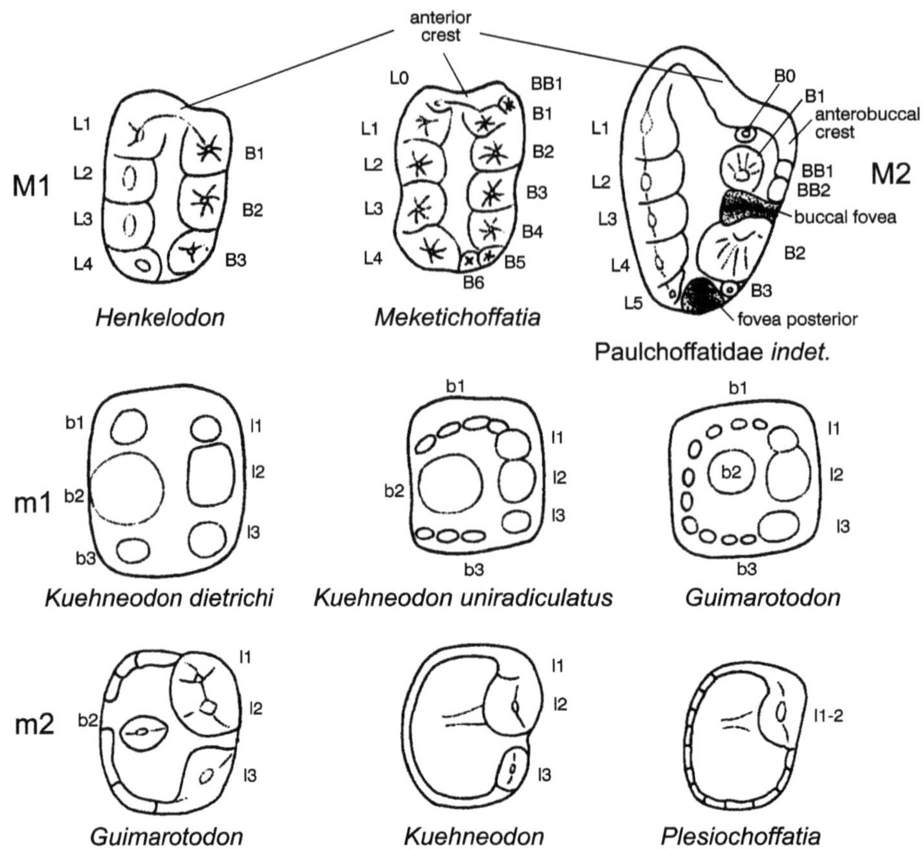

FIGURE 8.28. Schematic comparison of the molars in Paulchoffatiidae showing the numbering of cusps in M1, m1, and m2. *Source:* modified from Hahn and Hahn (1998c).

use our own designation for upper and lower teeth (capital letters for the uppers and lower-case letters for the lowers). So there are, for example, cusps L1 and B1 on upper molars and l1 and b1 on lowers. Other terms introduced by Hahn and Hahn (1998c) to designate the elements of molar structure are shown in figure 8.28.

Subfamily Paulchoffatiinae G. Hahn, 1969

Diagnosis (based on Hahn, 1993). Paulchoffatiidae characterized by tooth formula: 3.1–0.5.2/1.0.4.2. Angle between lower tooth row and dentary from 7 to 16° (figure 8.27A). Lower incisor steeply implanted, crown only slightly curved, root not extended below the molars posteriorly. Posterior upper premolars (P5 always, P4 mostly) with three longitudinal rows of cusps.

Genera. Based on lower dentitions: *Paulchoffatia* Kühne, 1961a, type genus; *Guimarotodon* G. Hahn, 1969; *Meketibolodon* G. Hahn, 1993; *Plesiochoffatia* G. Hahn and R. Hahn, 1999b; *Xenachoffatia* G. Hahn and R. Hahn, 1998c. Based on upper dentitions: *Bathmochoffatia* G. Hahn and R. Hahn, 1998c; *Henkelodon* G. Hahn, 1977a; *Kielanodon* G. Hahn, 1987b; *Meketichoffatia* G. Hahn, 1993; *Pseudobolodon* G. Hahn, 1977a; *Renatodon* G. Hahn, 2001.

Distribution. Late Jurassic (Kimmeridgian): Portugal, Leiria, Guimarota Coal Mine (Guimarota Beds).

GENERA BASED ON LOWER DENTITION

Genus *Paulchoffatia* Kühne, 1961a
(figures 8.27A$_1$, 8.29A)

Diagnosis (based on Hahn, 1993; and Hahn and Hahn, 1998c, emended). Type genus of the Paulchoffatiinae, confined to the lower dentition. Dentary stout with rounded lower margin; angle between tooth row and the dentary 7–10°. Tooth row (p–m) almost horizontal. Incisor stout, steeply implanted, only slightly curved, with root extending posteriorly only below p1–2. The p1 two-rooted; p2–3 of nearly the same length; p4 with one row of labial cusps. Three lingual and three labial cusps on m1. The m2 is roughly rectangular in shape, with large cusp l1–2, joined with l3, subdivided into two cuspules. Crown

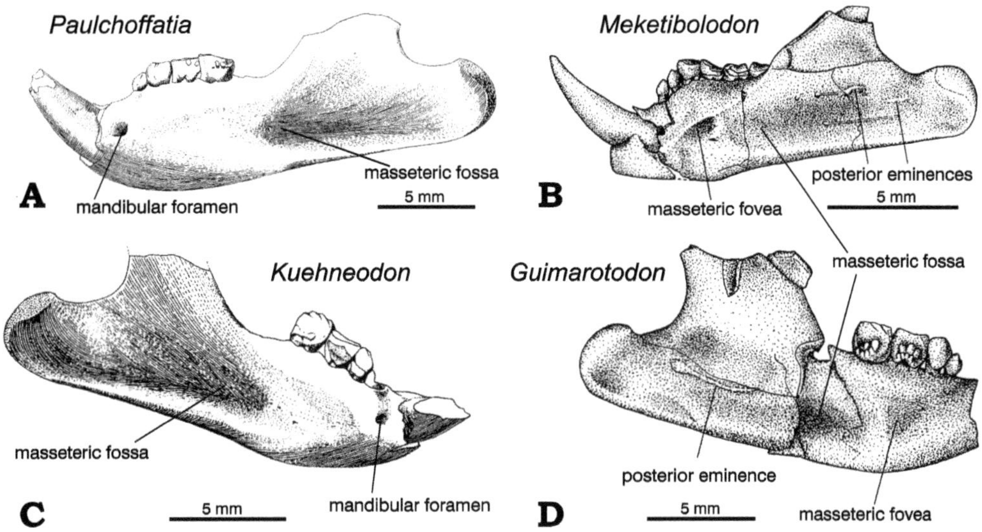

FIGURE 8.29. Dentaries of paulchoffatiid taxa. A, *Paulchoffatia delgadoi,* partial left dentary of the holotype with incisor, abraded p1–p4 and alveoli for m1–m2; reconstruction of labial view. B, *Meketibolodon robustus,* almost complete left dentary with all the teeth and imprints of muscular attachments; note high position of p3; labial view. C, *Kuehneodon dietrichi,* right dentary with incisor, broken p1, and p2–p4, reconstructed after the holotype; labial view. D, *Guimarotodon leiriensis,* incomplete right dentary with p1–p4 and alveoli for incisor and molars, labial view. *Source:* modified from: A, C, Hahn (1969); B, D, Hahn and Hahn (1998b).

basin not divided. Differs from all other paulchoffatiid genera based on lower dentition in having more steeply implanted lower incisor, shorter root of the incisor (shared with *Guimarotodon*), smaller angle between tooth row and the dentary; short p3, not longer than p2. The bipartite root of p1 is autapomorphic and evolved probably independently from other genera.

Species. *Paulchoffatia delgadoi* Kühne, 1961, type species; *Paulchoffatia* sp. A (Hahn, 1978a).

Distribution. As for the subfamily.

Genus *Guimarotodon* G. Hahn, 1969
(figures 8.9, 8.29D)

Diagnosis (based on Hahn and Hahn, 1998b, emended). The genus known only by lower dentition. Dentary slender with rounded lower margin. Incisor relatively slender, steeply implanted, with tips curved somewhat backward. Incisor's root only slightly longer than the crown, extending posteriorly below p1–2. Angle between tooth row and dentary about 15°. The p1 is two-rooted, p3 longer than p2; p4 with additional row of small labial cuspules below the row of labial cusps. The m1 has three labial and three lingual cusps, anterolingual cusp (l1) smaller than mesiolingual (l2), closely attached or confluent with it; mesiolabial cusps (b2) large, shifted to the center of the crown; anterolabial cusp (b1) and posterolabial (b3) subdivided into small cuspules, situated along the labial margin of the crown. The m2 has cusp b2 placed in the center of the crown, and a large posterolingual cusp (l3).

Species. *Guimarotodon leiriensis* G. Hahn, 1969, type species by monotypy.

Distribution. As for the subfamily.

Genus *Meketibolodon* G. Hahn, 1993
(figure 8.29B)

Diagnosis (based on Hahn and Hahn, 1998b, emended). Genus confined to lower dentition, which differs from all other paulchoffatiid genera in having tooth row (p1–m2) bent upward, with p3 being the highest, and in having the lower margin of the dentary bent anterodorsally below p2–3. Incisor steeply implanted. Differs from other paulchoffatine genera based on lower dentition by the relatively longer root of the incisor, extending below p1–4 (apomorphy) and undivided root of p1 (plesiomorphy). Shares with *Guimarotodon* a tendency to develop a row of labial cuspules below the row of labial cusps in p4. The m1 retains relatively plesiomorphic structure with the anterolingual cusp preserved and the enlarged mesiolabial cusp b2 placed close to the labial margin. On m2 anterolabial cusp l1 is completely merged into l2 and the originally broad valley behind this cusp is closed to a narrow notch. The rudimentary posterolabial cusp l3 is preserved. Angle between tooth row and dentary between 12 and 17°.

Species. *Meketibolodon robustus* (G. Hahn, 1978a), type species, originally assigned to *Pseudobolodon?*; *M.* cf. *robustus* (Hahn and Hahn, 1998b).

Distribution. As for the subfamily.

Genus *Plesiochoffatia* G. Hahn and R. Hahn, 1999b
(figures 8.28, 8.31B)

Comment. Hahn and Hahn (1998b) erected genus *Plesiochoffatia*, to replace preoccupied *Parachoffatia* G. Hahn and R. Hahn, 1998c.

Diagnosis (based on Hahn and Hahn, 1998c). Genus based only on m2. Differs from all the paulchoffatiid genera by loss of cusp l3; cusps l1 and l2 are fused into a large cusp. The outer rim, consisting of cuspules that are more or less distinct and vary in number among species, starts at the anterior end of cusp l1–l2, runs around the tooth, and ends at its posterior end.

Species. *Plesiochoffatia thoas* (G. Hahn and R. Hahn, 1998c), type species; *P. peperethos* (G. Hahn and R. Hahn, 1998c); *P. staphylos* (G. Hahn and R. Hahn, 1998c).

Distribution. As for the subfamily.

Genus *Xenachoffatia* G. Hahn and R. Hahn, 1998c
(figure 8.27D)

Diagnosis (based on Hahn and Hahn, 1998c). Genus based only on m2. Of all the lingual cusps only l2 is present. Differs from all the paulchoffatiid genera by position of l2 cusp, which is shifted backward and placed mesiolingually, rather than anterolingually. The outer margin, consisting of varying number of cuspules (up to 17), surrounds the central basin.

Species. *Xenachoffatia oinopion* G. Hahn and R. Hahn, 1998c, type species by monotypy.

Distribution. As for the subfamily.

GENERA BASED ON UPPER DENTITION

Genus *Bathmochoffatia* G. Hahn and R. Hahn, 1998c
(figure 8.31A)

Diagnosis (based on Hahn and Hahn, 1998c, emended). Genus confined to upper dentition based on isolated M1. Differs from all other paulchoffatiid genera by elongation of labial cusp B2 and presence of a narrow shelf designated labial grade (buccal in Hahn and Hahn terminology, gradus buccalis) on the labial margin opposite cusps B1 and B2. Differ from all other paulchoffatiid genera in having five (rather than four) lingual cusps. Shares with *Kuehneodon* closing of the median valley by a broad ridge, extending between L5 and B3.

Species. *Bathmochoffatia hapax* G. Hahn and R. Hahn, 1998c, type species by monotypy.

Distribution. As for the subfamily.

Genus *Henkelodon* G. Hahn, 1977a
(figures 8.28, 8.30A)

Diagnosis (based on Hahn and Hahn, 1998c, emended). Genus based on upper dentition. Tooth formula 3.0.5.2; anterior premolars P1–3 four-cusped; P4 molariform and as long as P5, both with three longitudinal rows of cusps (character shared with *Pseudobolodon*). Differs from other paulchoffatiine genera based on upper dentition by loss of the canine (apomorphy) and by having central valley of M1 open posteriorly (plesiomorphy). Lingual row in M1 with four cusps, labial with three-four cusps, of which cusp B2 is not reduced. Cusps of the labial row pointed, with well-developed enamel ridges (less clear in lingual row). The m2 is unknown.

Species. *Henkelodon naias* G. Hahn, 1977a, type species by monotypy.

Distribution. As for the subfamily.

Genus *Kielanodon* Hahn, 1987b
(figure 8.30D)

Diagnosis (based on Hahn and Hahn, 1998c, emended). Genus based on upper dentition. Alveolus for canine present. P3 longer than wide, with two rows of cusps, distinctly smaller than P4–5; P4–5 molariform, with three longitudinal rows of cusps. Premolar cusps strongly ribbed. M1 with two rows of four cusps each. Cusps B3 and B4 small and partly united, B4 (divided into two parts) situated at the posterior margin of the crown and closes the median valley. Two small cuspules placed between cusps B1 and B2. Lingual row extends further anteriorly than the labial. Differs from other paulchoffatine genera based on upper dentition by having elongated P3.

Species. *Kielanodon hopsoni* Hahn, 1987b, type species by monotypy.

Distribution. As for the subfamily.

Genus *Meketichoffatia* G. Hahn, 1993
(figure 8.30B)

Diagnosis (based Hahn and Hahn, 1998c). Genus based on upper dentition. Tooth formula 3.1.5.2, canine and P1–3 four-cusped, P4 short, similar to P1–3, P5 long, molariform, with three longitudinal rows of cusps. Differs from other paulchoffatine genera based on upper dentition by molarization of only P5 (lengthened and equipped with three rows of cusps). M1 has two rows of cusps and an additional anterolabial BB1 cusp. Lingual row with three to five cusps, situated lower than the labial row. Labial row with four to six cusps. Cusps sharply pointed with ornamentation of radiating ridges.

Species. *Meketichoffatia krausei* G. Hahn, 1993, type species by monotypy; *Meketichoffatia* sp. indet. (Hahn and Hahn, 1998c).

Comment. *Meketichoffatia krausei* includes upper den-

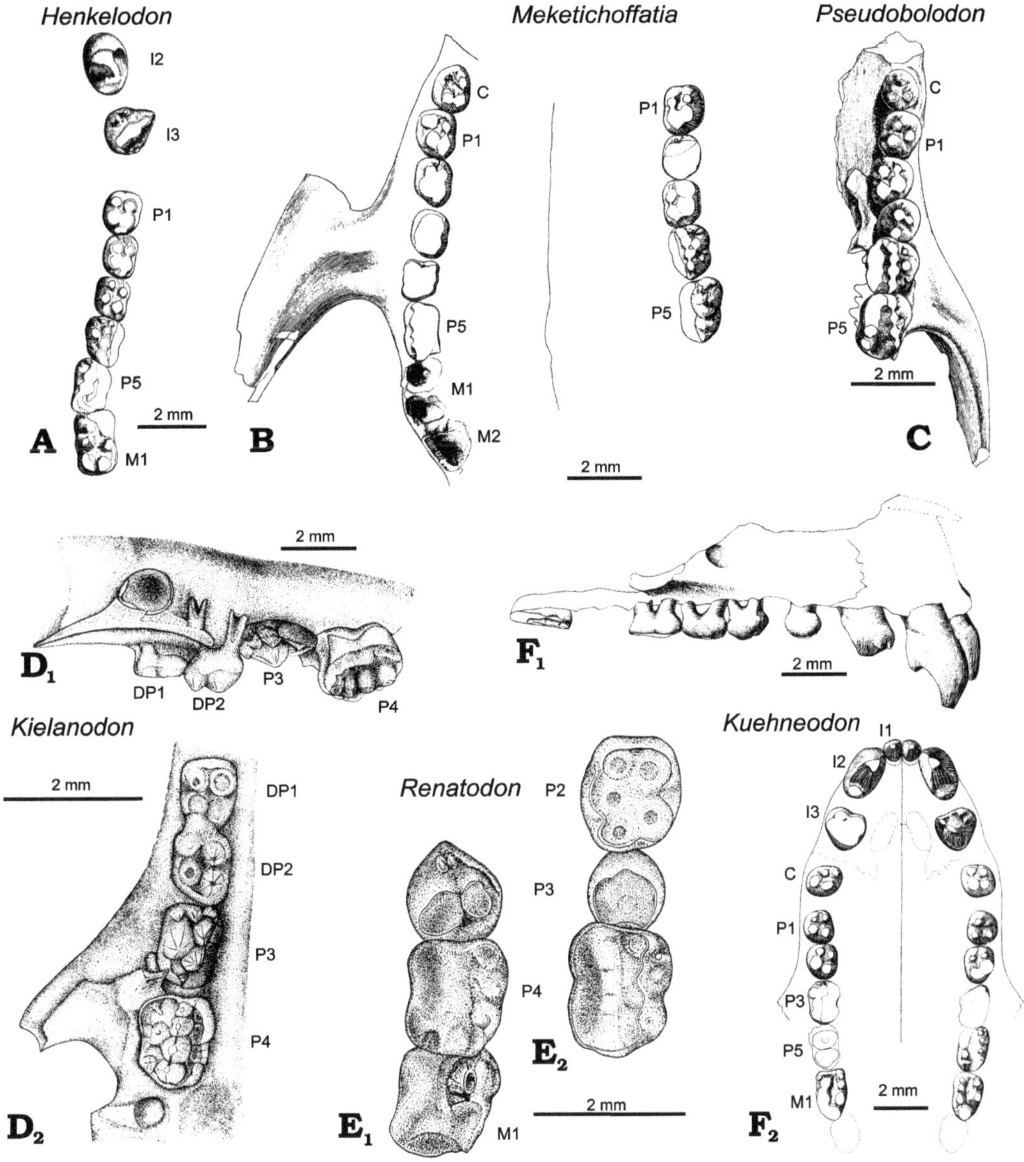

FIGURE 8.30. Skull fragments and upper dentition in paulchoffatiid taxa. A, *Henkelodon naias,* right upper dentition of the holotype, I2, I3, P1–P5, and M1. B, *Meketichoffatia krausei,* ventral view of the holotype specimen, palate with incomplete dentition, originally described as *Paulchoffatia delgadoi.* C, *Pseudobolodon oreas,* left maxilla with C and P1–P5, holotype, ventral view. D, *Kielanodon hopsoni,* right maxilla with DP1, DP2, erupting P3, and P4 in lingual (D_1) and occlusal (D_2) views, showing the tooth replacement. E, *Renatodon amalthea,* left P3, P4, and M1 in occlusal view (E_1); left P2, P3, and P4 (holotype) in occlusal view (E_2). F, *Kuehneodon simpsoni,* rostrum of the holotype (reconstructed) with both premaxillae and maxillae and almost complete dentition (M2s are missing), in labial (F_1) and ventral (F_2) views. *Source:* modified from: A, B, C, Hahn (1977a); D, Hahn and Hahn (1998b); E, Hahn (2001); F, Hahn (1969).

tition and skull fragments assigned by Hahn (1969, 1985, 1988) and Krause and Hahn (1990) to *Paulchoffatia delgadoi,* and by Hahn (1981, 1985, 1987b) to *Pseudobolodon oreas.*

Distribution. As for the subfamily.

Genus ***Pseudobolodon*** **G. Hahn, 1977a**
(figures 8.8A, 8.30C)

Diagnosis (based on Hahn, 1993). Genus based on upper dentition. Tooth formula 3.1.5.2; canine and anterior premolars (P1–3) three- or four-cusped, P4 molariform and as long as P5 with three longitudinal rows of cusps. Differs from other paulchoffatine genera based on upper dentition in having P4 and P5 similarly molariform and canine present.

Species. *Pseudobolodon oreas* G. Hahn, 1977a, type species; *Pseudobolodon* n. sp. (not named) (Hahn, 1993); ?*Pseudobolodon* sp. indet. (Hahn, 1993).

Distribution. As for the subfamily.

Genus *Renatodon* G. Hahn, 2001
(figure 8.30E)

Diagnosis. Genus based on upper dentition, represented by two specimens of the type species with P2–P5 and alveoli for C, P1, and M2. Differs from other paulchoffatine genera by having P3 bicusped, distinctly smaller than P2; the P2 with five cusps, P4 and P5 with three rows of cusps; canine placed close to P1.

Species. *Renatodon amalthea* G. Hahn, 2001, type species by monotypy (the species was figured by Hahn, 1985, as ?*Pseudobolodon* n. sp.; and by Hahn, 1987b, as *Pseudobolodon?* sp.).

Distribution. As for the subfamily.

Subfamily Kuehneodontinae G. Hahn, 1971

Diagnosis (based on Hahn, 1993). A subfamily of the Paulchoffatiidae with tooth formula 3.1.4.2/1.0.4–3.2. Angle between the tooth row and dentary about 20°. Lower incisor procumbent, with distinctly upward curved crown and long root, extended posteriorly below molars. Canine and anterior upper premolars (P1–P3) four-cusped; P5 molariform with two longitudinal rows of cusps; P4 either as anterior premolars or similar to P5; third row of cusps always absent.

Genera. *Kuehneodon* G. Hahn, 1969, type genus by monotypy.

Distribution. Late Jurassic (Kimmeridgian: Portugal, Leiria, Guimarota Coal Mine (Guimarota Beds); Late Jurassic (late Kimmeridgian–Tithonian): Portugal, Pai Mongo and Porto das Barcas (Lourinhã Formation).

Genus *Kuehneodon* G. Hahn, 1969
(figures 8.8B, 8.27A_2, 8.28, 8.29C, 8.30F)

Diagnosis and Distribution. As for the subfamily.

Species. *Kuehneodon dietrichi* G. Hahn, 1969, type species; *K. barcasensis* G. Hahn and R. Hahn, 2001a; *K. dryas* G. Hahn, 1977a; *K. guimarotensis* G. Hahn, 1969; *K. hahni* Antunes, 1998; *K. simpsoni* G. Hahn, 1969; *K. uniradiculatus* G. Hahn, 1978a; *K.?* sp. (Hahn, 1977a).

Family Paulchoffatiidae G. Hahn, 1969
Subfamily indet.
Genus *Galveodon* G. Hahn and R. Hahn, 1992

Diagnosis (based on Hahn, 1993). A paulchoffatiid genus known only from isolated teeth (I2, p3, or p4). Labial cuspule on I2 isolated, not connected to the main cusp by a crest; surface of enamel smooth. Lower premolars eroded by wear down to the base of the crown (figure 8.31D).

Species. *Galveodon nannothus* G. Hahn and R. Hahn, 1992, type species by monotypy.

Distribution. Early Cretaceous (early Barremian): Spain, Teruel Province, Galve (Camarillas Formation) and Cuena Province, Uña.

Genus ?*Sunnyodon* Kielan-Jaworowska and Ensom, 1992 (tentatively assigned)
(figure 8.31C)

Diagnosis (based on Kielan-Jaworowska and Ensom, 1992). Poorly known ?paulchoffatiid genus represented by a single species known from the right ?P5. The ?P5 resembles P5 and P4 of *Meketichoffatia, Kuehneodon,* and *Kielanodon* in having only two rows of cusps, but differs in having two cusps of the labial row arranged symmetrically in the middle of the tooth length and only one anterior and one posterior labial cuspule. Differs from all known paulchoffatiid taxa in the presence of an incipient lingual ridge, with a small cuspule. In number and arrangement of cusps it resembles *Kielanodon,* but differs from P4 and P5 of that genus in being narrower, roughly oval in shape, and in having differently arranged cusps in the labial row. Cusp formula 2:4:Ri. Another tooth, P3 or P4, has been tentatively assigned to *S. notleyi.*

Species. *Sunnyodon notleyi* Kielan-Jaworowska and Ensom, 1992, type species by monotypy.

Distribution. Early Cretaceous (Berriasian): Southern England, Dorset (Purbeck Limestone Group).

Family ?Paulchoffatiidae, subfamily indet.
Genus *Mojo* G. Hahn, Lepage, and Wouters, 1987
(figure 8.31E)

Diagnosis. *M. usuratus* is known from a single incomplete ?upper premolar, with two roots, characterized by the crown with two rows of longitudinal cusps and an ?anterior cingulum-like portion. One row of cusps (?labial) is broader than ?lingual. Crown horizontally eroded near its base.

Species. *Mojo usuratus* G. Hahn, Lepage, and Wouters, 1987, type species by monotypy.

Distribution. Late Triassic (early Rhaetian): Southern Belgium, Gaume (Mortinsart Formation).

Comment. Hahn et al. (1987) based their tentative assignment of the incomplete ?upper premolar to Multituberculata mostly on the manner of wear, different from that in the Haramiyidae and Tritylodontidae, but characteristic of the Paulchoffatiidae. Given the time span of some 32 or 57 Ma that separates *Mojo* (early Rhaetian) from the first uncontested multituberculates (Bathonian or Kimmeridgian) and poor knowledge of this taxon, its attribution to Multituberculata remains open. If a multituberculate, *Mojo* would probably belong to the paulchoffatiid line and therefore we describe it herein.

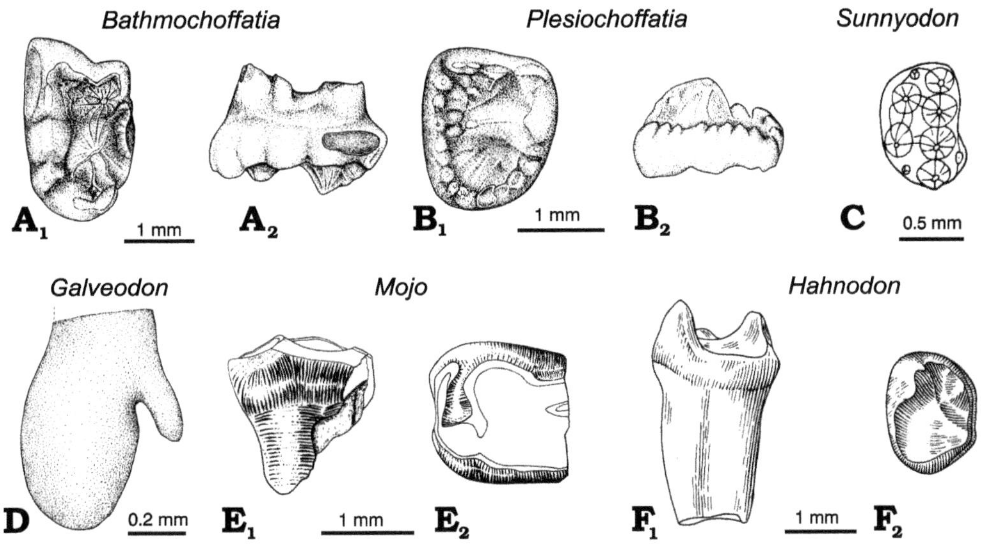

FIGURE 8.31. Poorly known paulchoffatiid and related taxa. A, *Bathmochoffatia hapax*, left M1, holotype, in occlusal (A_1) and lingual (A_2) views. B, *Plesiochoffatia thoas*, left m2, holotype, in occlusal (B_1) and labial (B_2) views. C, *Sunnyodon notleyi*, right ?P5, in occlusal view. D, *Galveodon nannothus*, left I2, holotype, in labial view. E, *Mojo usuratus*, incomplete ?upper premolar, holotype, reconstruction in labial (E_1) and occlusal (E_2) views. F, *Hahnodon taqueti*, left m2 in labial (F_1) and occlusal (F_2) views. *Source:* modified from: A, B, Hahn and Hahn (1998c); C, Kielan-Jaworowska and Ensom (1992); D, Hahn and Hahn (1992); E, Hahn et al. (1987); F, Sigogneau-Russell (1991c).

?Paulchoffatiidae gen. et sp. indet. (E. F. Freeman, 1979)

Comment. E. F. Freeman (1979) described from the upper Bathonian, Oxfordshire, England, an incomplete upper incisor, identified as "multituberculate or tritylodontid incisor." Hahn (1993) assigned it tentatively to the Paulchoffatiidae and noted that it might belong to the same taxon as an eroded molar described by Freeman (1976a) from the same beds. If paulchoffatiid nature of the taxa described by E. F. Freeman (1976a; 1979) and those from the Bathonian Kirtlington Beds studied by Butler and Hooker (in preparation) is demonstrated, then the stratigraphic range of the Paulchoffatiidae would extend to the Bathonian.

Family Hahnodontidae Sigogneau-Russell, 1991c

Diagnosis. Very poorly known family represented by the single genus and species *Hahnodon taqueti* Sigogneau-Russell, 1991, known from a single left m2. Share with Paulchoffatiidae the basin-shaped structure of m2, but differ from them in having two cusps on this molar rather than one, as well as a smooth labial cingulid and cusps of subequal height. Of all the "plagiaulacidan" subgroups, Hahnodontidae appear to be closest to the Paulchoffatiidae.

Genera. *Hahnodon* Sigogneau-Russell, 1991c, type genus by monotypy.

Distribution. Early Cretaceous (?Berriasian): Morocco, Oriental High Atlas, Talsinnt province, synclinal d'Anoual.

Genus *Hahnodon* Sigogneau-Russell, 1991c
(figure 8.31F)

Diagnosis and Distribution. As for the family.

Species. *Hahnodon taqueti* Sigogneau-Russell, 1991c, type species by monotypy.

Family Pinheirodontidae G. Hahn and R. Hahn, 1999a

Diagnosis (based on Hahn and Hahn, 1999a, emended). Family known only from isolated teeth (about 250 in number), assigned to seven species, belonging to six genera. Dental formula unknown; cusps on lower molars of different height. I2 with several small posterior cuspules or cusps (a character known only in one isolated I2 in Paulchoffatiidae). Differ from all multituberculates in structure of I3, which has an obliquely elongated crown, with main ridge bearing up to five cusps and additional posterolabial and anterolabial cusps. Upper canine present, single-rooted. Share with other "plagiaulacidans" P1–P3 three- to five-cusped, P4 with two rows of cusps, the labial shorter than the lingual. Share with Paulchoffatiinae and Albionbaataridae P5 with three rows of cusps.

M1 with two rows of cusps, last cusp in each row enlarged or reduced. M2 with two rows of cusps and a small pit that separates the first and the second labial cusps. Lower incisor completely covered with enamel. The p3 has four serrations with ridges, the second ridge as a rule is elongated; p4 longer than p3 with six serrations. Basal cusps on p4 either present (*Bernardodon*), or hardly discernible, almost absent (*Pinheirodon*), or replaced by a row of pits (*Iberodon*)—a character unique among multituberculates. Share with Paulchoffatiidae and Plagiaulacidae a tendency to enlarge the second labial cusp in m1 and diminish the first and the second. Share with Paulchoffatiidae m2 with central basin and reduced labial cusps, but differ in having three lingual cusps (only one in Paulchoffatiidae). See figure 8.32 for explanation of cusp numbering in Pinheirodontidae.

Genera. *Pinheirodon* G. Hahn and R. Hahn, 1999a, type genus; *Bernardodon* G. Hahn and R. Hahn, 1999a; *Ecprepaulax* G. Hahn and R. Hahn, 1999a; *Gerhardodon* Kielan-Jaworowska and Ensom, 1992; *Iberodon* G. Hahn and R. Hahn, 1999a; *Lavocatia* Canudo and Cuenca-Bescós, 1996.

Distribution. Early Cretaceous (Berriasian): southern England, Dorset (Purbeck Limestone Group); Portugal, Estremadura Province (Porto Pinheiro at Lourinhã); Early Cretaceous (Barremian): Spain, Teruel Province (Galve).

Genus *Pinheirodon* G. Hahn and R. Hahn, 1999a
(figure 8.33A,B)

Diagnosis (based on Hahn and Hahn, 1999a). I3 massive and wide, P5 cusp formula 1–5:3–5:1–4 and cusp row BB (see figure 8.32), extending for the whole length of the crown. M1 with three or four cusps, lingual cusps increasing in size posteriorly; M2 with 2:3 cusps and anterolabial shelf, row of labial cusps shorter than the lingual. The p3 and p4 without labial cusps, but traces are present; p4 considerably longer than m1; m1 with 3:3 cusps, m2 with 0:3 cusps.

Species. *Pinheirodon pygmaeus* G. Hahn and R. Hahn, 1999a, type species; *Pinheirodon vastus* G. Hahn and R. Hahn, 1999a; *Pinheirodon* sp. G. Hahn and R. Hahn, 1999a.

Distribution. Early Cretaceous (Berriasian): Portugal, Estermadura Province (Porto Pinheiro at Lourinhã).

Genus *Bernardodon* G. Hahn and R. Hahn, 1999a
(figure 8.33C)

Diagnosis (based on Hahn and Hahn, 1999a). Differs from *Pinheirodon* in having I3 wider and more robust; on

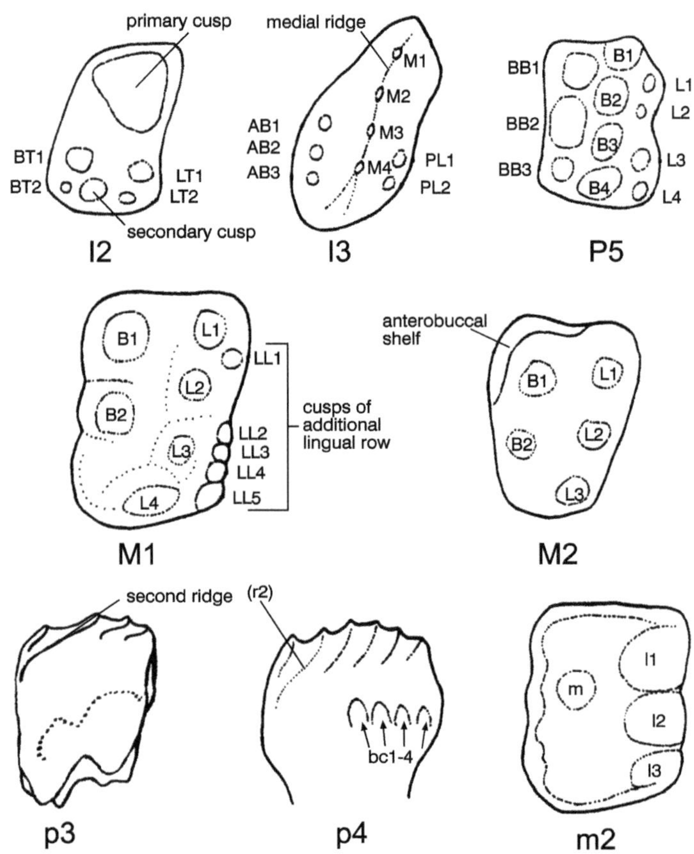

FIGURE 8.32. Schematic drawings showing morphology of teeth and cusp numbering in Pinheirodontidae; for teeth in occlusal view labial is to the left. Abbreviations: B, b, labial; bc1–4 labial basal cusps; L, l, lingual; LL1–5, cusps in additional lingual row; M1–4 cusps in medial ridge in I3; m, medial cusp; r2, prolonged second ridge in p3 and p4. *Source:* modified from Hahn and Hahn (1999a).

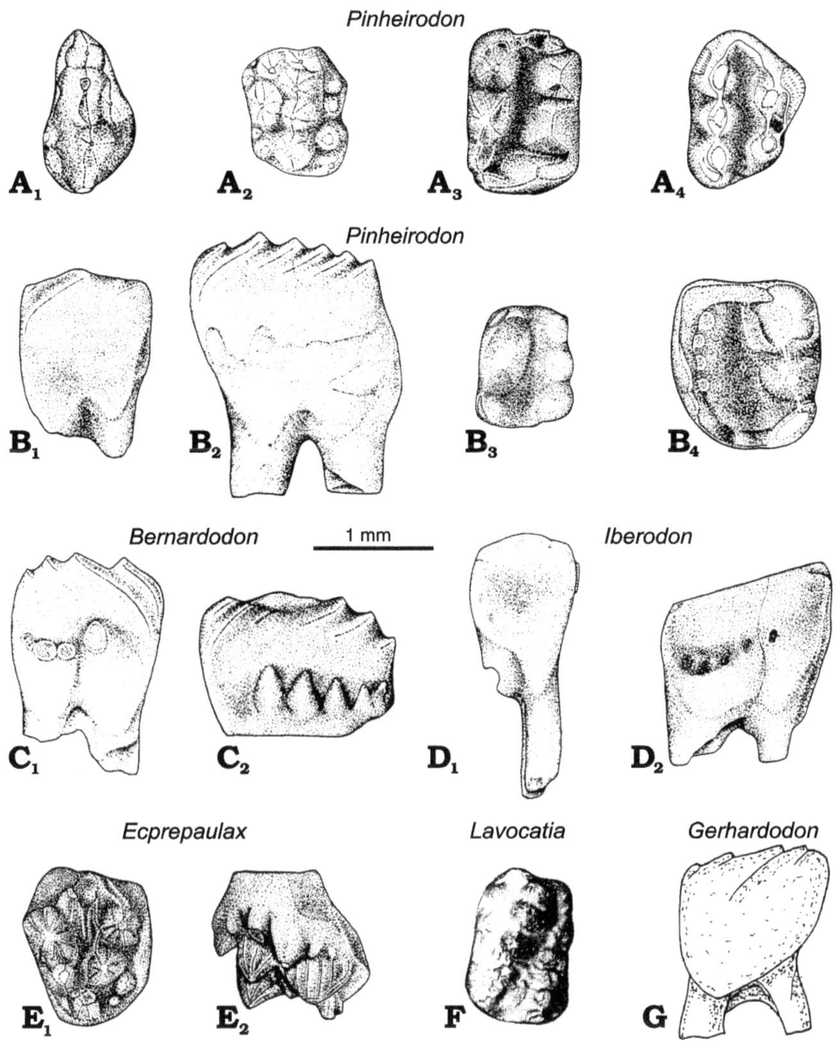

FIGURE 8.33. Teeth of pinheirodontid taxa. A, B, *Pinheirodon pygmaeus.* A, Upper teeth: I2, P3, M1, M2, in occlusal view (A_1–A_4); B, Lower teeth: p3, p4, in labial view m1, m2 (B_1–B_2), and in occlusal view (B_3, B_4). C, *Bernardodon atlanticus,* p3 (C_1); p4 (C_2). D, *Iberodon quadrituberculatus,* p3 (D_1) and p4 (D_2). E, *Ecprepaulax anomala,* I2 in occlusal (E_1) and lingual (E_2) views. F, *Lavocatia alfambrensis,* right P5, holotype, in occlusal view. G. *Gerhardodon purbeckensis,* left ?p3, holotype, in labial view. *Source:* modified from A–E, Hahn and Hahn (1999a); F, Canudo and Cuenca-Bescós (1996); G, Kielan-Jaworowska and Ensom (1992).

P5 the cusps of BB row (see figure 8.32) do not extend for the whole tooth length, cusp formula 3:4:4; M1 with 3:4 cusps and wide anterolabial shelf; p3 with labial cusps, p4 low with very prominent labial cusps, m2 with two lingual cusps.

Species. *Bernardodon atlanticus* G. Hahn and R. Hahn, 1999a, type species, and *B.* sp. indet. Hahn and Hahn (1999a).

Distribution. Early Cretaceous (Berriasian): Portugal, Estermadura Province (Porto Pinheiro at Lourinhã).

Genus *Ecprepaulax* G. Hahn and R. Hahn, 1999a
(figure 8.33E)

Diagnosis. The genus is known only from I2, which differs from I2s of other pinheirodontid taxa by the presence of a longitudinal furrow that extends along the whole length of the crown and is surrounded on both sides by one secondary and three tertiary cusps.

Species. *Ecprepaulax anomala* G. Hahn and R. Hahn, 1999a, type species by monotypy.

Distribution. Early Cretaceous (Berriasian): Portugal, Estermadura Province (Porto Pinheiro at Lourinhã).

Genus *Gerhardodon* Kielan-Jaworowska and Ensom, 1992
(figure 8.33G)

Diagnosis (based on Kielan-Jaworowska and Ensom, 1992). Poorly known pinheirodontid genus represented by a single species known from two isolated premolars, ?p3 and tentatively assigned ?right P4 or P5, and fragments of other teeth. Assigned to Pinheirodontidae, as it resembles *Pinheirodon* (e.g., Hahn and Hahn, 1999a: figures 8, 10, 34) in having p3 with one of the ridges distinctly longer than the others (third in *Gerhardodon* and usually the second in Pinheirodontidae), small number of serrations, and lack of labial cusps. It differs from p3 of paulchoffatiid taxa, for example, *Kuehneodon* and *Guimarotodon,* in having almost transverse (rather than vaulted) upper margin and in being roughly trapezoidal and slightly longer than high, and roughly oval or quadrangular in occlusal view.

Comment. Other teeth, for example, P5 tentatively assigned by Kielan-Jaworowska and Ensom (1992: plate 1, figure 8, and text figure 5D) to *Gerhardodon,* may also belong to the Pinheirodontidae, especially in that ?P5 is characterized by replacement of cusps by pits, which otherwise is known only among Pinheirodontidae, but occurs there in lower premolars.

Species. *Gerhardodon purbeckensis* Kielan-Jaworowska and Ensom, 1992, type species by monotypy.

Distribution. Early Cretaceous (Berriasian): Southern England, Dorset (Purbeck Limestone Group).

Genus *Iberodon* G. Hahn and R. Hahn, 1999a
(figure 8.33D)

Diagnosis (based on Hahn and Hahn, 1999a, emended). Differs from *Pinheirodon* and *Bernardodon* in having long and slender I3, by a tendency for reduction of the number of cusps on P5 and on upper and lower molars, and by replacement of labial cusps on lower premolars by pits. Shares with *Bernardodon* shortening of cusps BB on P5, and with *Pinheirodon* enlargement of cusp L4 on M1, which closes the middle furrow from posterior.

Comment. Replacement of cusps by pits in lower premolars, characteristic of *Iberodon,* is unique among multituberculates and otherwise is known only on P5, tentatively assigned by Kielan-Jaworowska and Ensom (1992: plate 1, figure 8, and text figure 5D) to *Gerhardodon,* a genus that we assign to Pinheirodontidae.

Species. *Iberodon quadrituberculatus* G. Hahn and R. Hahn, 1999a, type species by monotypy.

Distribution. Early Cretaceous (Berriasian): Portugal, Estermadura Province (Porto Pinheiro at Lourinhã).

Genus *Lavocatia* Canudo and Cuenca-Bescós, 1996
(figure 8.33F)

Diagnosis (based on Canudo and Cuenca-Bescós, 1996). Poorly known genus. Cusp formula of P5 (the only tooth known of the type species) is 4:6:5.

Species. *Lavocatia alfambrensis* Canudo and Cuenca-Bescós, 1996, type species by monotypy, represented by single right P5. Some of the isolated teeth (I2, anterior upper premolars, and upper molars) figured and briefly described by Cuenca-Bescós et al. (1995) from the Barremian of Spain and assigned to Eobaatarinae indet. and Paulchoffatinae indet. might also belong to *Lavocatia.*

Comment. We follow Hahn and Hahn (1999a) in assigning *Lavocatia* to Pinheirodontidae, but similarly shaped premolars also occur in Eobaataridae (Kielan-Jaworowska, Dashzeveg, and Trofimov, 1987), and more complete material of this taxon may demonstrate its eobaatarid nature.

Distribution. Early Cretaceous (early Barremian): Spain, Teruel Province, Galve.

PLAGIAULACID LINE

Brief Characterization. Members of this line differ from both paulchoffatiid and allodontid lines by having p4 more elongated with respect to p3, and with a greater number of ridges. Differ from allodontid line by having large I3 (resembling that of paulchoffatiid line, but with fewer cusps), a tendency of molar cusps to coalesce, enamel ornamented with grooves and pits, and asymmetrical lower molars, shorter lingually than labially (the latter characters shared with some members of *Paracimexomys* group). Differ from paulchoffatiid line in lack of basin-shaped m2 and presence of incipient posterolingual ridge (referred to by some authors as the third row of cusps) on M1.

Family Plagiaulacidae Gill, 1872

Synonym: Bolodontidae Osborn, 1887b.

Emended Diagnosis. Dental formula 3.0.5–4.2/1.0.4–3.2. The canine (which occurs in some Paulchoffatiidae) is absent from the Plagiaulacidae. Differ from Allodontidae and Paulchoffatiidae by the following apomorphies: elongated p4 (almost twice as long as p3; also seen in Zofiabaaataridae), with five to seven serrations with ridges; p2 and p3 triangular rather than rectangular in labial view. Share with Eobaataridae and some members of the *Paracimexomys* group the structure of m1 and m2, which are asymmetrical and shorter lingually than labially, with two rows of coalescing cusps. Differ from Eobaataridae and *Glirodon* in having nonprismatic enamel and lower incisors completely covered with enamel. Differ from Eobaataridae in having p4 with a row of labial cusps, rather than a single cusp (mistakenly quoted by Kielan-Jaworowska and Ensom, 1992, as provided with a single cusp in *Bolodon osborni* Simpson, 1928a). Differ from Allodontidae and *Glirodon* in having enlarged, paulchoffatiid-like I3, with two or three cusps, rather than a single-cusp (or with basal cuspule). Differ from Paulchoffatiidae in having I3 roughly triangular (rather than trapezoidal) in occlusal view and in lack of an obliquely arranged, ridgelike main cusp. Differ from Paulchoffatiidae in having p3 without labial cusps, incipient posterolingual ridge on M1, shearing P4–P5, and fewer cusps on M2 (2–3:3). Differ from Allodontidae in having P4 and P5 ornamented with numerous cuspules and share this character with some Paulchoffatiidae, for example, *Kielanodon* Hahn, 1987b. Share with Eobaataridae, some members of the *Paracimexomys* group, and *Ferugliotherium* (Mammalia *incertae sedis,* see chapter 14) "ornamentation" of grooves and pits on the upper and lower molars.

Genera. *Plagiaulax* Falconer, 1857, type genus; *Bolodon* Owen, 1871; ?new genus to be erected for "*Bolodon*" *elon-*

gatus. "*Ctenacodon*" *brentbaatar* Bakker, 1998 (upper P5) from the Morrison Formation may also belong to Plagiaulacidae.

Distribution. Early Cretaceous (Berriasian): southern England, Dorset (Purbeck Limestone Group); possibly Late Jurassic: United States, Wyoming (Morrison Formation).

Genus *Bolodon* Owen, 1871
(figures 8.9, 8.34B, 8.35B, 8.36A)

Synonym: *Plioprion* Cope, 1884.

Diagnosis (based on Kielan-Jaworowska and Ensom, 1992, emended). Smallest and most completely preserved plagiaulacid genus, known from upper and lower dentitions. Differs from *Plagiaulax* in having a smaller lower incisor, four (rather than three) lower premolars (and shares this character with *Ctenacodon*), and fewer labial cusps on p4 (four rather than ?six). Differs from "*Bolodon*" *elongatus* in having P1–P3 without prominent posterior cingulum and P1 of subequal size with P2 rather than distinctly larger. Differs from *Eobaatar* in having smaller (hardly discernible) posterolingual ridges on M1.

Species. *Bolodon crassidens* Owen, 1871, type species; *B. falconeri* (Owen, 1871); *B. minor* (Falconer, 1857); *B. osborni* Simpson, 1928a. "*Bolodon*" *elongatus* Simpson, 1928a, may not belong to this genus.

Comment. Of the species of *Bolodon*, *B. osborni* is most similar to *Eobaatar*, especially in structure of M2 and p4 (but differs from *Eobaatar* in fewer serrations and a row of labial cusps on p4). *B. osborni* appears to be in some respects intermediate between Plagiaulacidae and Eobaataridae. See also Osborn (1887a,b, 1888b).

Distribution. Early Cretaceous (Berriasian): southern England, Dorset (Purbeck Limestone Group).

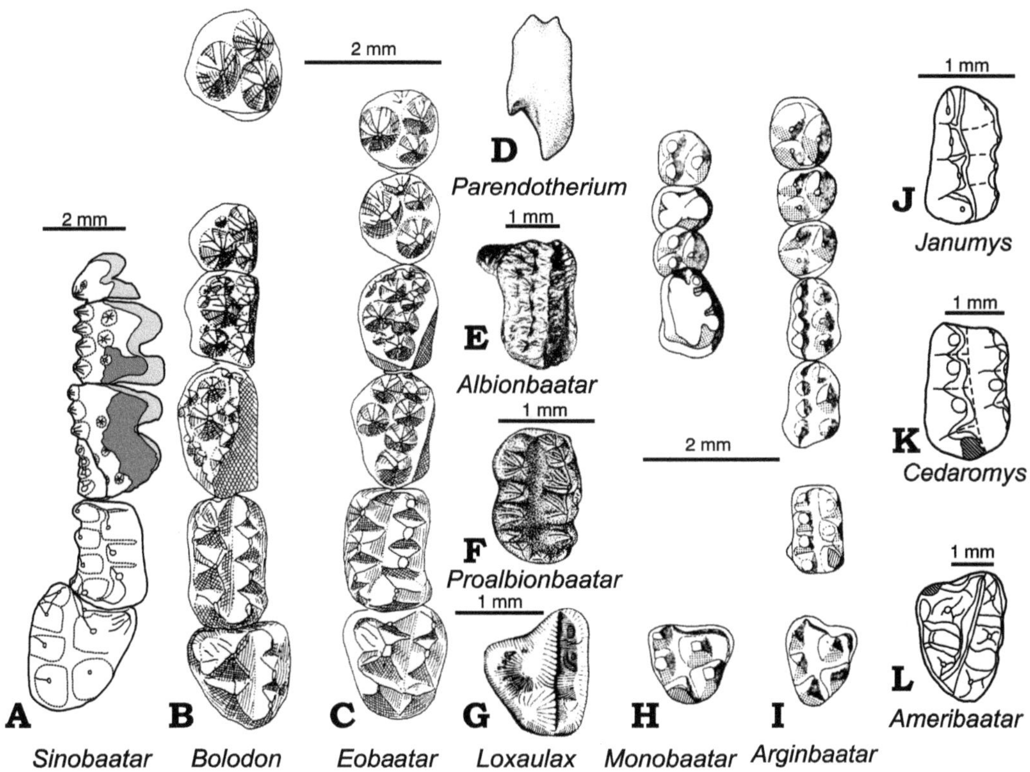

FIGURE 8.34. Upper dentitions in plagiaulacid, eobaatarid, albionbaatarid, arginbaatarid, and family *incertae sedis* taxa. A, *Sinobaatar lingyuanensis,* left P3–M2 of the holotype specimen in occlusal view. B, *Bolodon osborni,* right P1–M2 (P2 missing) in occlusal view. C, *Eobaatar magnus,* reconstruction of right upper premolars and molars based on isolated teeth. D, *Parendotherium herreroi,* I2, holotype, in medial view. E, *Albionbaatar denisae,* right P5, holotype, in occlusal view. F, *Proalbionbaatar plagiocyrtus,* left M1, in occlusal view. G, *Loxaulax valdensis,* right M2 in occlusal view. H, *Monobaatar mimicus,* left upper premolars (preserved together) and isolated left M2 tentatively assigned to the same taxon, all from left side, in occlusal view. I, *Arginbaatar dmitrievae,* upper premolars (three anterior preserved together), M1 and M2, tentatively assigned to the same taxon, in occlusal view. J, *Janumys erebos,* left M1 in occlusal view. K, *Cedaromys parvus,* left M1, holotype, in occlusal view. L, *Ameribaatar zofiae* right M2, holotype, in occlusal view. The scale above *Eobaatar* is for B–D; the scale for H is also for I. *Source:* A, original, courtesy of Yaoming Hu; the remainder modified from: B, C, H, I, Kielan-Jaworowska, Dashzeveg, and Trofimov (1987); D, Crusafont and Gibert (1976); E, Kielan-Jaworowska and Ensom (1994); F, Hahn and Hahn (1998d); G, Clemens (1963a); J–L, Eaton and Cifelli (2001).

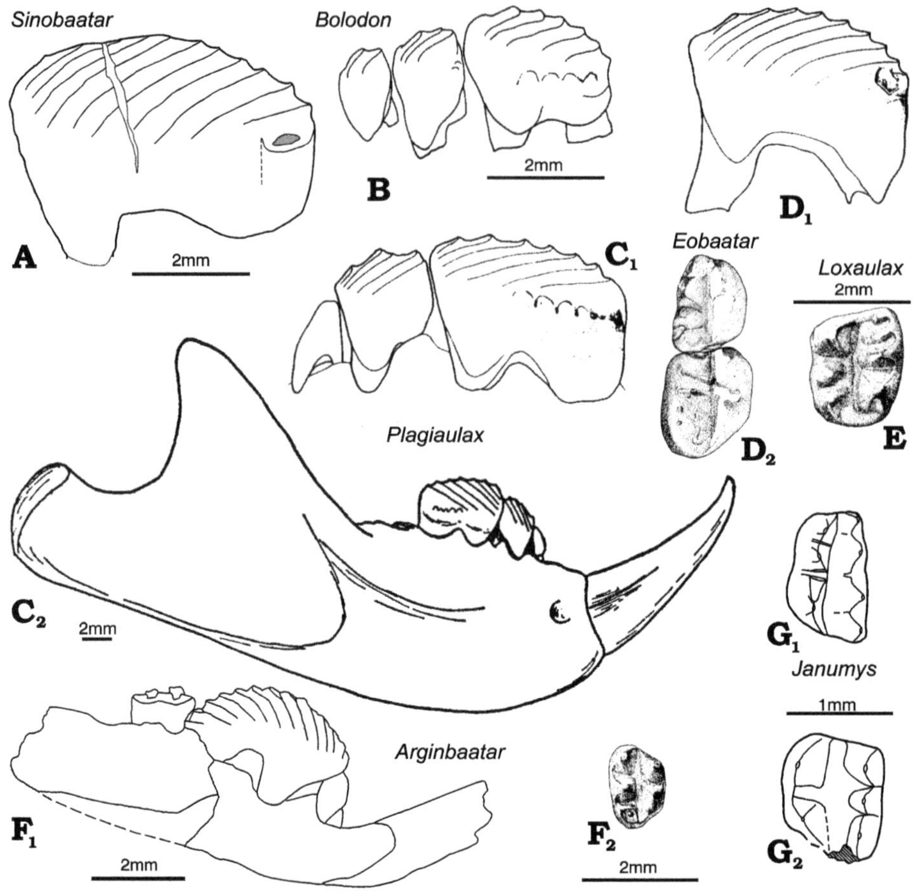

FIGURE 8.35. Lower dentitions in plagiaulacid, eobaatarid, arginbaatarid, and family *incertae sedis* taxa. A, *Sinobaatar lingyuanensis,* left p4 of the holotype in labial view. B, *Bolodon osborni,* left p2–p4 in labial view (p1 not preserved). C, *Plagiaulax becklesi,* details of left p2–p4 of referred specimen in labial view (C_1), right dentary of the holotype with incisor and p2–p3, in labial view (C_2). D, *Eobaatar magnus,* left p4 of the holotype in labial view (D_1), left m1 and m2 of another specimen in occlusal view (D_2). E, *Loxaulax valdensis,* right M2 in occlusal view. F, *Arginbaatar dmitrievae,* right dentary of the holotype with incisor, p2–m1, and alveolus for m2, in labial view (F_1); right m1 of the holotype in occlusal view (F_2). G, *Janumys erebos,* right m1 in occlusal view (G_1); right m2 (tentatively assigned) in occlusal view (G_2). Scale for B is also for C; that or E is also for D. *Source:* A, original, courtesy of Yaoming Hu; the remainder modified from: B, Kielan-Jaworowska and Ensom (1992); C_1, D, E, F_2, Kielan-Jaworowska, Dashzeveg, and Trofimov (1987); F_1 original drawing, based on the cast; C_2, Simpson (1928a); G, Eaton and Cifelli (2001).

Genus *Plagiaulax* Falconer, 1857
(figure 8.35C)

Diagnosis. *Plagiaulax* is known only from the dentary with incisor, premolars, and alveoli for the molars. Differs from *Bolodon* in having three lower premolars (p1 lost) and a very large incisor.

Species. *Plagiaulax becklesii* Falconer, 1857, type species.

Distribution. Early Cretaceous (Berriasian): southern England, Dorset (Purbeck Limestone Group).

Family Eobaataridae Kielan-Jaworowska, Dashzeveg, and Trofimov, 1987

Diagnosis. Estimated dental formula ?5.2/1.0.3.2. Differ from most "plagiaulacidans" (except *Glirodon*) in having gigantoprismatic enamel and lower incisor with limited enamel band. Differ from Plagiaulacidae in having p4 with nine or ten serrations with ridges (rather than five to seven). Differ from all other plagiaualcidans in having single labial cusp on p4. Share with Plagiaulacidae and with some members of the *Paracimexomys* group the structure of m1 and m2, which are asymmetrical, shorter lingually than labially, with coalescing cusps. Share with Plagiaulacidae, some members of *Paracimexomys* group, and *Ferugliotherium* (Mammalia *incertae sedis,* see chapter 14) "ornamentation" of grooves and pits on the molars. Share with Plagiaulacidae generally similar structure of the upper premolars (five in number) and differ in having more prominent incipient lingual ridge of M1. Share with *Bolo-*

don, Plagiaulax, and *Zofiabaatar* similar p3/p4 length ratio and with *Plagiaulax* the presence of only three lower premolars.

Genera. *Eobaatar* Kielan-Jaworowska, Dashzeveg, and Trofimov, 1987, type genus; *Loxaulax* Simpson, 1928a; ?*Monobaatar* Kielan-Jaworowska, Dashzeveg, and Trofimov, 1987; *Parendotherium* Crusafont-Pairó and Adrover, 1966; *Sinobaatar* Hu and Wang, 2002a.

Distribution. Early Cretaceous (Barremian): China, Liaoning Province (Yixian Formation); Spain, Teruel Province, Galve and Cuena Province, Uòa; Early Cretaceous (Aptian or Albian): Mongolia, Gobi Desert, Höövör ("Höövör Beds").

Genus *Eobaatar* Kielan-Jaworowska, Dashzeveg and Trofimov, 1987
(figures 8.9, 8.34C, 8.35D, 8.36B)

Diagnosis (based on Kielan-Jaworowska, Dashzeveg, and Trofimov, 1987). Estimated skull length of *E. magnus,* about 3 cm; of *E. minor,* 2 cm. The p2 is peglike and single-rooted; p3 double-rooted, measuring less than half the length of p4; and p4 with eight or nine serrations with ridges and labial cusp. On lower molars, cusps of the lingual row are crescent shaped, facing toward the middle; cusps of the labial row face posteriorly. On upper teeth (tentatively assigned), P1–P3 three- to four-cusped, P4 and P5 similar to each other, with main row of cusps forming cutting edge. M1 with cusp formula 3:4:Ri. M2 relatively wide, with nearly straight anterior margin.

Species. *Eobaatar magnus* Kielan-Jaworowska, Dashzeveg, and Trofimov, 1987, type species; *E. hispanicus* G. Hahn and R. Hahn, 1992; *E. minor* Kielan-Jaworowska, Dashzeveg, and Trofimov, 1987; ?*E. pajaronensis* G. Hahn and R. Hahn, 2001b; *E.* sp. a, *E.* sp. b. (Kielan-Jaworowska, Dashzeveg, and Trofimov, 1987); ?*Parendotherium vel Eobaatar* (Hahn and Hahn, 1992).

Distribution. Early Cretaceous (early Barremian): Spain, Teruel Province, Galve (Camarillas Formation), and Cuena Province, Pié Pajarón and Uña; Early Cretaceous (Aptian or Albian): Mongolia, Gobi Desert, Höövör ("Höövör Beds").

Genus *Loxaulax* Simpson, 1928a
(figures 8.34G, 8.35E)

Diagnosis. The most characteristic tooth of *Loxaulax,* M2, differs from that in *Eobaatar* in having four (rather than three) cusps on the inner row and a sigmoid anterior margin (outer row shorter with respect to the inner than in *Eobaatar*). It shares with *Eobaatar* two large cusps in the outer row and characteristic "ornamentation" of grooves and pits. Cusps on the lower molars show a tendency to coalesce, as characteristic of the family (see Simpson, 1928a; Clemens, 1963a, Hahn, 1969; Clemens and Lees, 1971).

Species. *Loxaulax valdensis* (Woodward, 1911), type species by monotypy, represented by isolated teeth, including M2, m1 and tentatively assigned anterior upper premolars and I2.

Distribution. Early Cretaceous (Valanginian, "Wealden"): southeastern England, Cliff End (Wadhurst Formation).

Genus ?*Monobaatar* Kielan-Jaworowska, Dashzeveg and Trofimov, 1987
(figure 8.34H)

Diagnosis (based on Kielan-Jaworowska, Dashzeveg, and Trofimov, 1987, emended). Estimated length of the skull 2 cm. Single infraorbital foramen positioned above P3–P4 embrasure. Posteroventral margin of the base of zygomatic arch above P4–P5 embrasure. P1–P3 three-cusped. Cusps 3:4 on P4, the second cusp of the labial row the largest. M2 with Ri:2:3 cusps. Cusps on M2 relatively robust, anterior margin of M2 almost straight. Cusps of premolars and molars ornamented with weak striations; ridges finely granulated. The only M2 (tentatively assigned) differs from that in *Eobaatar* and other plagiaulacid genera in lack of characteristic "ornamentation" of grooves and pits; differs from *Sinobaatar* and *Loxaulax* in having straight anterior margin.

Species. *Monobaatar mimicus* Kielan-Jaworowska, Dashzeveg and Trofimov, 1987, type species by monotypy.

Distribution. Early Cretaceous (Aptian or Albian): Mongolia, Gobi Desert, Höövör ("Höövör Beds").

Comments. The genus is poorly known. *M. mimicus* is based on two premaxillae with premolars, a P4 and tentatively assigned M2. As the lower dentition and especially p4 of *Monobaatar* are not known, its assignment to Eobaataridae cannot be demonstrated with any certainty.

Genus *Parendotherium* Crusafont-Pairó and Adrover, 1966
(figure 8.34D)

Diagnosis. Poorly known genus represented by isolated teeth of the type species, the holotype being a single I2 similar to that referred to *Bolodon.* Isolated premolars assigned to *P. herreroi* show similarity to *Eobaatar.*

Species. *Parendotherium herreroi* Crusafont-Pairó and Adrover, 1966, type species by monotypy.

Distribution. Early Cretaceous (early Barremian): Spain, Teruel Province, Galve (Camarillas Formation).

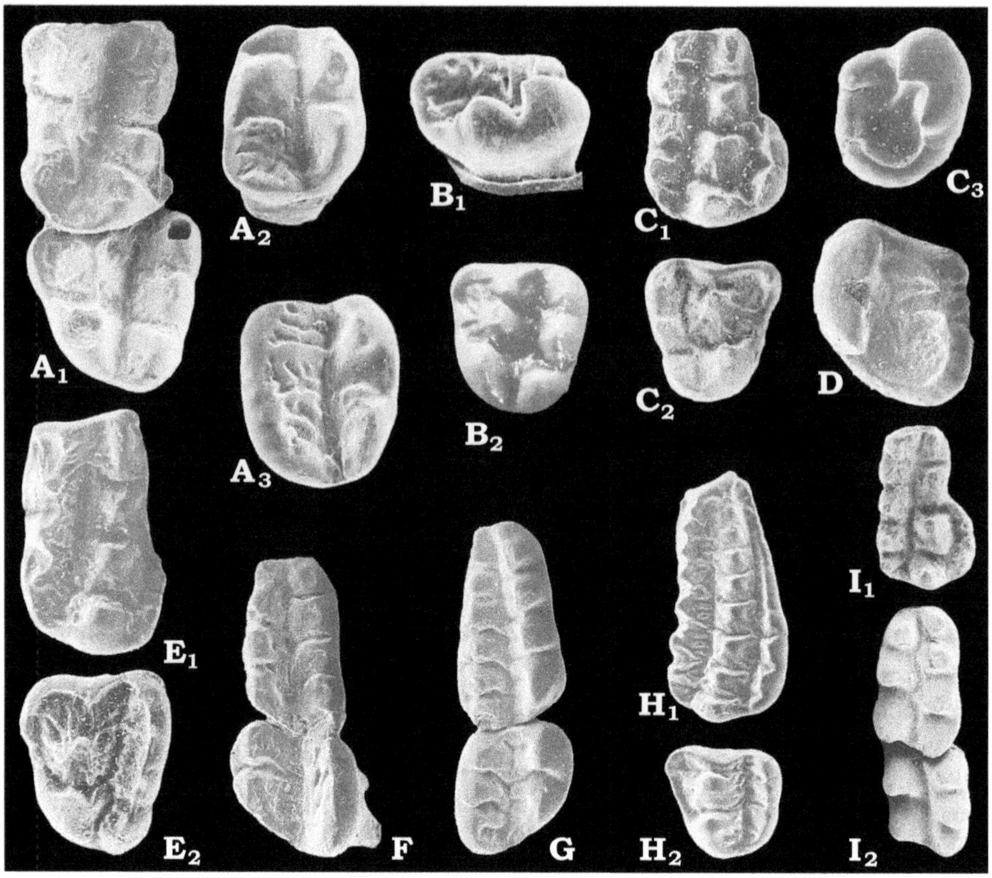

FIGURE 8.36. Comparison of upper and lower molars in plagiaulacid, eobaatarid, *Paracimexomys* group, ptilodontoid, and djadochtatherioid genera, showing ornamentation of grooves and ridges and tendency of cusps to coalesce at least in peripheral aspect in all taxa except djadochtatherioid *Kryptobaatar.* All except B_1 are in occlusal view, with anterior margin up; B_1 is in lingual view, anterior is to the right. All except B_2 are SEM micrographs of epoxy resin casts. A, *Bolodon osborni,* right M1 and M2 (A_1); left m1 and m2 (A_2, A_3). B, *Eobaatar magnus,* left m1 (B_1); right M2 (B_2). C, *Bryceomys fumosus,* right M1 (C_1); left M2 (C_2); left m2 (C_3). D, *Paracimexomys* sp. cf. *P. robisoni,* right m2. E, *Dakotamys malcolmi,* right M1 (E_1); right M2 (E_2). F, *Ptilodus montanus,* left m1 and m2. G, *Parectypodus laytoni,* left m1 and m2. H, *Neoplagiaulax hazeni,* right M1 (H_1) and M2 (H_2). I, *Kryptobaatar dashzevegi,* right M1 (I_1); right m1 and m2 (I_2). All 15×. *Source:* all modified from Kielan-Jaworowska and Hurum (2001).

Genus *Sinobaatar* Hu and Wang, 2002a
(figures 8.34A, 8.35A)

Diagnosis. Skull apparently without postorbital process. Dental formula 3.?.5.2/1.0.3.2. I1, I2, and lower incisor relatively small, lower incisor completely covered with enamel. M1 with rounded posterior margin and cusp formula 3:4:Ri; M2 relatively large in comparison to MI, with sigmoid anterior margin and cusp formula Ri:2:4. The p4 similar to that in *Eobaatar,* but relatively longer with respect to the height, with 10 serrations with ridges and less prominent and placed lower basal cusp.

Species. *Sinobaatar lingyuanensis* Hu and Wang, 2002a, see also Hu and Wang (2002b), type species by monotypy, is the most completely preserved eobaatarid, represented by a skull with dentaries and parts of the postcranial skeleton (only preliminarily described).

Distribution. Early Cretaceous: China, Liaoning Province (Yixian Formation).

Family Albionbaataridae Kielan-Jaworowska and Ensom, 1994

Diagnosis. Shrew-sized multituberculates known from upper premolars of *Albionbaatar* and upper molars of *Proalbionbaatar;* p4 tentatively assigned to *Albionbaatar* by Kielan-Jaworowska and Ensom (1994) may not be long to it. Differ from all multituberculates in having relatively flat, multicusped anterior upper premolars, with 10 to 14 cusps arranged in three rows, rather than three or four, rarely up to nine cusps in two rows; in having lingual slope in all premolars covered by prominent, subparallel ridges; and relatively flat m1 with numerous cusps.

Genera. *Albionbaatar* Kielan-Jaworowska and Ensom, 1994, type genus, and *Proalbionbaatar* G. Hahn and R. Hahn, 1998d; undescribed upper premolars similar to *Albionbaatar* (Wang et al., 1995).

Distribution. Late Jurassic (Kimmeridgian): Portugal, Leiria, Guimarota Coal Mine (Guimarota Beds); Early Cretaceous (Berriasian): southern England, Dorset (Purbeck Limestone Group); Early Cretaceous of northeast China (Wang et al., 1995).

Genus *Albionbaatar* Kielan-Jaworowska and Ensom, 1994
(figure 8.34E)

Diagnosis. Genus known from upper premolars of the type species. ?P1–3 roughly rectangular with three rows of cusps, P5 with cusp formula 5–7:7:7. Differs from known "Plagiaulacida," Djadochtatherioidea, Taeniolabidoidea, and nonspecialized Ptilodontoidea in having P5 with three rows of numerous small cusps. Differs from Paulchoffatiidae in having P5 distinctly longer relative to width. Differs from Plagiaulacidae in having all the cusps on each upper premolar of subequal size. Shares with derived Ptilodontoidea and Taeniolabidoidea numerous cusps on posterior upper premolars.

Species. *Albionbaatar denisae* Kielan-Jaworowska and Ensom, 1994, type species by monotypy.

Distribution. Early Cretaceous (Berriasian): southern England, Dorset (Purbeck Limestone Group).

Genus *Proalbionbaatar* G. Hahn and R. Hahn, 1998d
(figure 8.34F)

Diagnosis. M1s of the type species are 1.5 mm long, with cusp formula 5–6:6–7. Shares with Paulchoffatiidae lack of posterolingual wing. Differs from paulchoffatiid and plagiaulacid M1s in having a more flat crown with very small numerous cusps.

Species. *Proalbionbaatar plagiocyrtus* G. Hahn and R. Hahn, 1998d, type species by monotypy, based on two isolated M1s.

Distribution. Late Jurassic (Kimmeridgian): Portugal, Leiria, Guimarota Coal Mine (Guimarota Beds).

Family *incertae sedis*
Genus *Janumys* Eaton and Cifelli, 2001
(figures 8.34J, 8.35G)

Diagnosis. Upper and lower molar cusps coalesced peripherally. Shares with Eobaataridae m1 and m2 with oblique posterior margins. Shares with *Eobaatar* the structure of cusps in lingual row on m1, which are crescentic, facing medially, with first two partially coalesced. P4 with cusp formula 1–3:4, differs from *Cedaromys, Bryceomys,* and *Dakotamys* in having a very short, poorly developed posterior basin. Differs from *Eobaatar* and all Cedar Mountain Formation genera in lack of the third cusp row on M1 (developed in *Eobaatar* as posterolingual wing) and resembles in this respect *Bolodon,* in which the wing is incipiently developed. M1 is more elongate relative to width than in other Cedar Mountain genera. The m2 (and probably M2) almost as long as m1. Differs from *Bolodon, Eobaatar, Cedaromys, Bryceomys,* and *Dakotamys* in having very weak ornamentation of pits and ribs (as in *Paracimexomys*), discernible only on M1, but shares with them strongly ribbed P4.

Species. *Janumys erebos* Eaton and Cifelli 2001, type species by monotypy, is a very small taxon based on isolated upper and lower molars.

Distribution. Near Early–Late Cretaceous (Albian–Cenomanian) stage boundary: United States, Utah, San Rafael Swell (Cedar Mountain Formation).

Comments. Eaton and Cifelli (2001) assigned *Janumys* to the "plagiaulacid line" because of its similarities to Eobaataridae, especially in structure of m1 (Kielan-Jaworowska, Dashzeveg, and Trofimov, 1987), and because it probably had five upper premolars. However, it differs from *Eobaatar* in having less conspicuous ribs and grooves on the molars, which may be due, at least in part, to the state of preservation. It is primitive in comparison with other Early–Late Cretaceous boundary genera described by Eaton and Cifelli (2001) in lack of a true posterolingual wing on M1, resembling in this respect *Bolodon* (see Kielan-Jaworowska and Ensom, 1992: plate 3: 9). *Janumys* in many respects is primitive, but its molars are elongate relative to width, which is derived relative to the condition in members of "plagiaulacid line."

Suborder *incertae sedis*
Family Arginbaataridae G. Hahn and R. Hahn, 1983

Diagnosis (based on Kielan-Jaworowska and Ensom, 1992, emended). Differ from all multituberculates (and all other mammals) in having a very large p4 with limited enamel, ontogenetically rotating anteroventrally over the worn p3 and p2, which disappear. Share with "Plagiaulacida" (except *Glirodon* and *Eobaatar,* Cimolomyidae, Boffiidae, and Ptilodontoidea) lower incisor completely covered with enamel (plesiomorphy), but differ from Ptilodontoidea in having robust rather than slender incisor. Share with some Paulchoffatiidae, *Ctenacodon,* and *Glirodon* presence of upper canine (plesiomorphy). Differ from Plagiaulacidae and some Eobaataridae and share with Allodontidae P4 and P5 without labial cuspules. Differ from Plagiaulacidae, Eobaataridae, and some members of the *Paracimexomys* group in having distinct conical cusps on lower and upper molars and smooth (not ornamented) enamel. Share with Plagiaulacidae incipient posterolingual ridge on M1. Differ from Plagiaulacidae and Eobaataridae in having molar cusps smooth or weakly striated. Share with Eobaataridae, *Glirodon,* and Cimolodonta (except Ptilodontoidea) gigantoprismatic enamel.

Genera. *Arginbaatar* Trofimov, 1980, type genus by monotypy.

Distribution. Early Cretaceous (Aptian or Albian): Mongolia, Gobi Desert, Höövör ("Höövör Beds").

Genus *Arginbaatar* Trofimov, 1980
(figures 8.9, 8.34I, 8.35F)

Diagnosis and Distribution. The same as for the family.

Species. *Arginbaatar dmitrievae* Trofimov, 1980, type species by monotypy.

Suborder Cimolodonta McKenna, 1975

Diagnosis. Small to large multituberculates with prismatic enamel; dental formula: 2.0.1–4.2/1.0.1–2.2. Differ from "Plagiaulacida" by at least six apomorphies: loss of I1, penultimate upper premolar, p1, and p2; transformation of p3 into a peglike, nonfunctional tooth, change of rectangular p4 into arcuate one (shared with Arginbaataridae). Advanced Cimolodonta (except for members of *Paracimexomys* group and some Ptilodontoidea) differ from "Plagiaulacida" in having wear facets on lower molars facing upward rather than toward the central valley. Two main evolutionary tendencies occur in parallel: (1) increase in number of ridges of p4 and its size, (2) reduction in size of p4 and in premolar number.

Superfamilies. Superfamily and family *incertae sedis:* five genera assigned to the *Paracimexomys* group: Djadochtatherioidea Kielan-Jaworowska and Hurum, 2001; Taeniolabidoidea Sloan and Van Valen, 1965; superfamily *incertae sedis,* families: Cimolomyidae Marsh, 1889; Eucosmodontidae Jepsen, 1940; Microcosmodontidae Holtzman and Wolberg, 1977; Boffiidae G. Hahn and R. Hahn, 1983 (Tertiary family not described herein); Kogaionidae Rădulescu and Samson, 1996; Ptilodontoidea Sloan and Van Valen, 1965; superfamily and family *incertae sedis: Viridomys* Fox, 1971a, and *Uzbekbaatar* Kielan-Jaworowska and Nessov, 1992.

Distribution. Early Cretaceous (Aptian–Albian) to Eocene: Northern Hemisphere.

Superfamily and Family *incertae sedis*
***Paracimexomys* group**

Diagnosis (based on Kielan-Jaworowska and Hurum, 2001). Most plesiomorphic, informal group of Cimolodonta, members of which are known mostly from isolated teeth. Differ from most "Plagiaulacida" (except *Eobaatar* and *Glirodon*) in having gigantoprismatic enamel. Differ from "Plagiaulacida" in having arcuate p4 (cimolodontan apomorphy) and cimolodontan dental formula. Share with Plagiaulacidae and Eobaataridae structure of the molars (especially m2) with coalescing cusps and ornamentation of grooves and ribs on lower and upper molars. At least *Paracimexomys, Dakotamys,* and *Bryceomys* share with Eobaataridae the structure of M2 with straight anterior margin, anterior rim confluent with the first cusp of the medial row, and enlarged second cusp of the medial row. Share with Ptilodontoidea (especially *Cimolodon*) similarly built M2 and a tendency of lower and upper molar cusps to coalesce. Share with some Djadochtatherioidea P4 with a row of main cusps and short anterolabial row of three or four cusps, and M1 with small number of cusps and short posterolabial ridge with few cusps.

Genera. *Paracimexomys* Archibald, 1982; *Bryceomys* Eaton, 1995; *Cedaromys* Eaton and Cifelli, 2001; *Dakotamys* Eaton, 1995.

Distribution. Early Cretaceous (Aptian–Albian) to Late Cretaceous (Maastrichtian): North America; Late Cretaceous (?Maastrichtian): Europe.

Comments. The oldest described record of *Paracimexomys* is that of Cifelli (1997a), based on ?*P. crossi* from the Aptian–Albian of Oklahoma, but the record does not improve until the Albian–Cenomanian, from which Eaton and Nelson (1991) published several species. Early Late Cretaceous species of the *Paracimexomys* group described by Eaton (1995) share with Plagiaulacidae and Eobaataridae the structure of lower molars, with coalescing cusps, and similar structure of M2, but differ in having cimolodontan dental formula and arcuate p4.

Genus *Paracimexomys* Archibald, 1982
(figure 8.36D)

Diagnosis (based on Eaton and Cifelli, 2001, emended). Differs from *Eobaatar* in having molars with generally smooth enamel and lacking complex pitting and ribbing, although occasional pits may be present. M1 cusp formula is 4–5:4–5:1–2, with a short third cusp row (absent on *Cedaromys* or *Janumys*) and not as broadened as in *Bryceomys;* differs from *Dakotamys* in having labial and medial cusp rows of M1 diverging anteriorly. Differs from *Cimexomys* in having cusps in molar rows alternating in position and not crescentic. First lower molars waisted in occlusal view, with cusp formula 4–5:3–4. Lower molars (in particular m2) resemble those of *Eobaatar* in being asymmetrical, shorter lingually than labially, and in having coalesced cusps. The p4 differs from that of *Eobaatar* in being arcuate, rather than rectangular.

Species. *Paracimexomys priscus* (Lillegraven, 1969), type species; *P. magister* (Fox, 1971a); *P. perplexus* Eaton and Cifelli, 2001, ?*P. crossi* Cifelli, 1997; ?*P. robisoni* Eaton and Nelson, 1991, and species left in open nomenclature by Eaton and Cifelli (1998), Eaton and Nelson (1991), and Eaton (1995).

Distribution. Early–Late Cretaceous (Aptian–Albian to Maastrichtian, Lancian): North America. *Paracimexomys? dacicus* from the Upper Cretaceous of Romania (Grigorescu and Hahn, 1987) is an objective junior synonym of *Barbatodon transylvanus* Rădulescu and Samson, 1986, based on the same specimen.

Comments. *Paracimexomys* is a poorly known genus, most species being based on isolated teeth. Two species assigned by Archibald (1982) to *Paracimexomys* (*Cimexomys judithae* Sahni, 1972 and *C. magnus* Sahni, 1972) according to Eaton and Cifelli (2001) do not belong to this genus, and we leave them for the time being in *Cimexomys;* similarly the position of *Paracimexomys robisoni* Eaton and Nelson, 1991, needs revision.

Genus *Bryceomys* Eaton, 1995
(figure 8.36C)

Diagnosis (based on Eaton, 1995, emended). M1 with cusp formula 3–5:4:2–4; differs from *Dakotamys* and *Paracimexomys* in having broader lingual cusp. Cusps of medial and labial rows of M1 not aligned transversely. Cusp formula of m1 is 4–5:3, cusps of labial row conical to weakly crescentic, well divided, separated by deep, wide U-shaped valleys (in side view). The p4 with high arch, which reminds one of a tendency observed in Cimolodontidae; on m2 a continuous ridge to second cusp of labial row connects first cusp of lingual row. M2:M1 length ratio smaller than for *Paracimexomys* and *Dakotamys.* Differs from *Paracimexomys* and *Dakotamys* in having well-separated cusps of lingual row of M2.

Species. *Bryceomys fumosus* Eaton, 1995, type species; *B. hadrosus* Eaton, 1995; *B. intermedius* Eaton and Cifelli, 2001, and at least two species left in open nomenclature by Eaton (1995) and one by Eaton and Cifelli (2001).

Distribution. Late Cretaceous (Albian-Cenomanian through Turonian): United States, southern and central Utah (Cedar Mountain, Dakota, and Straight Cliffs formations).

Genus *Cedaromys* Eaton and Cifelli, 2001
(figure 8.34K)

Diagnosis. Most similar to *Bryceomys.* Lower incisor covered with enamel, oval in cross section, slightly curved. Shares with *Bryceomys* high p4 (resembling those in Cimolodontidae), symmetrical, with strong first serration, well separated from the second. Lower molars broader relative to length than in *Bryceomys.* Cusps of the molars are more robust, proportionally lower, and not as deeply divided as those of *Bryceomys.* P4 with cusp formula 2:4. Shares with *Janumys* lack of the third cusp row (posterolingual wing) on M1 and differs in this respect from *Bryceomys, Paracimexomys,* and *Dakotamys.*

Species Assigned. *Cedaromys bestia* (Eaton and Nelson, 1991), type species; *C. parvus* Eaton and Cifelli, 2001; and two species left in open nomenclature by Eaton and Cifelli (2001).

Distribution. Near Early–Late Cretaceous (Albian-Cenomanian) stage boundary: United States, Utah, San Rafael Swell (Cedar Mountain Formation).

Genus *Dakotamys* Eaton, 1995
(figure 8.36E)

Diagnosis (based on Eaton, 1995). The m1 and M1 with low cusp formulae (m1, 4:3; M1, 4:4:1–2), short lingual row on M1, cusps of labial and medial rows of M1 not aligned transversely. M2/M1 length ratio 0.9. Molar cusps ribbed. Differs from *Paracimexomys* in that M1 has parallel medial and labial cusp rows (not diverging anteriorly) and cusps of labial row are markedly smaller, lower, and more poorly defined posteriorly. Straight central valley (rather than sinuous) on m1. P4 with labially expanded anterolabial platform; p4 with higher arch.

Species. *Dakotamys malcolmi* Eaton, 1995, type species by monotypy.

Distribution. Late Crecateous (Cenomanian): United States, southern Utah (Dakota Formation).

?*Paracimexomys* group

Comment. The three genera described below (*Ameribaatar* Eaton and Cifelli, 2001, *Barbatodon* Rădulescu and Samson, 1986, and *Cimexomys* Sloan and Van Valen, 1965) are only tentatively assigned to the *Paracimexomys* group, and we treat them separately, pending further discoveries.

Genus *Ameribaatar* Eaton and Cifelli, 2001
(figure 8.34L)

Diagnosis. Molar cusps are relatively small and conical, widely spaced. The apomorphy is the unique type of molar wear that leaves wide transverse valleys between the cusps; the valleys are U-shaped in side view, delimited by sharp ridges, and the faces of cusps adjacent to valleys remain concave. On m1 (incompletely known) the central valley is narrow and sigmoid; m2 with cusp formula 3:2, broad with straight anterior and rounded posterior margins, and wide central valley. M1 incompletely known; M2 relatively long with strongly oblique anterior margin (lingual cusp row much longer than the labial), narrow, with deep central valley and cusp formula 2:3. Ornamentation lacking.

Species. *Ameribaatar zofiae* Eaton and Cifelli, 2001, type species by monotypy, based on several isolated upper and lower molars.

Distribution. Near Early–Late Cretaceous (Albian-Cenomanian) stage boundary: United States, Utah, San Rafael Swell (Cedar Mountain Formation).

Comments. The type of wear observed in *Ameribaatar* appears unique among multituberculates. Certain similarity (apparently due to convergence) may be found with m1 of *Arginbaatar* (Kielan-Jaworowska, Dashzeveg, and Trofimov, 1987: pl. 14, figures 1a, 1b) in which m1 when observed in side view shows U-shaped valleys between the widely spaced cusps. However, the same specimen in occlusal view (pl. 14: 1c of the same paper) shows tooth shape and cusp structure different from *Ameribaatar.* The type of wear characteristic of *Ameribaatar* has not been observed in the large collections of Asian multituberculates we have examined. Another genus that invites a comparison is *Essonodon* (see Archibald, 1982: figures 31–34) in which there are also deep (albeit not U-shaped) valleys between the molar cusps in side view. As in *Arginbaatar* the apparent similarity may be due to convergence.

Genus *Barbatodon* Rădulescu and Samson, 1986 (figure 8.37B)

Diagnosis (based on Rădulescu and Samson, 1986). Single left M1 of the type species has a very short posterolingual ridge and cusp formula 3:4:Ri. Length of the tooth 3.38 mm, width 2.06 mm.

Species. *Barbatodon transylvanicum* Rădulescu and Samson, 1986 (synonym: *Paracimexomys? dacicus* Grigorescu and Hahn, 1987), type species by monotypy.

Distribution. Late Cretaceous (Maastrichtian): Europe, Romania, Haţeg Basin (Sînpetru Formation).

Genus *Cimexomys* Sloan and Van Valen, 1965 (figure 8.37A)

Diagnosis (based on Archibald, 1982, emended; but see also Weil, 1999; and Montellano et al., 2000). Small multituberculate with enamel of uniform thickness on lower incisor. Two lower premolars: p3 peglike, p4 arcuate, not protruding dorsally over the level of the molars, with 8–10 serrations. Lateral profile of P4 close to isosceles triangle with the posterior edge shorter. M1 cusp formula 5–7:4–6:Ri, with lingual ridge less than 50% of tooth length. Cusp formula of m1 is 6:4. Molar cusps tending to be subcrescentic to crescentic, molars not waisted. Differs from *Paracimexomys* in having greater number of cusps on the molars. Differs from Ptilodontoidea (except *Cimolodon*) in having gigantoprismatic (rather than normal prismatic) enamel, p4 not protruding dorsally over the level of the molars, P4 lower and smaller number of cusps on the molars, and shorter inner ridge on M1.

Species. *Cimexomys minor* Sloan and Van Valen, 1965, type species; *C. antiquus* Fox, 1971a; *C. gratus* (Jepsen, 1930b) (senior synonym of *C. hausoi* Archibald, 1982, see Lofgren 1995, for correction); *C. judithae* Sahni, 1972; *C. magnus* Sahni, 1972; ?*C. gregoryi* Eaton, 1993; possibly unnamed species of *Cimexomys* (Sloan and Van Valen, 1965; Montellano, 1992).

Distribution. Late Cretaceous to early Paleocene: North America.

Superfamily Djadochtatherioidea Kielan-Jaworowska and Hurum, 2001

Diagnosis (based on Kielan-Jaworowska and Hurum, 2001, emended). A clade of relatively large multituberculates with skull length varying between 18 and 70 mm, dental formula 2.0.3–4.2/1.0.2.2, single-cusped I2 and I3, and double-rooted upper premolars, lower margin of P4 straight in lateral view or forming only incipient isosceles triangle. Synapomorphies: large frontals, deeply inserted between the nasals and pointed anteriorly in the middle; U-shaped frontoparietal suture; a sharp edge between the lateral and palatal walls of premaxilla (rounded in other multituberculates). A large roughly rectangular facial surface of the lacrimal, exposed on the cranial roof and separating the frontal from the maxilla, characteristic for Djadochtatherioidea, may be a plesiomorphy. The postglenoid region of the braincase is apparently longer compared to the skull length than in all other multituberculates. Differ from multituberculates in which this region has been preserved (except *Ptilodus* and *Catopsalis*) in having a large posttemporal fossa (canal). Differ from Taeniolabidoidea, Microcosmodontidae, and Ptilodontoidea in having I3 placed on the palatal part of premaxilla and share this character with Eucosmodontidae (*Stygimys*) and Cimolomyidae (*Meniscoessus*). Differ from "Plagiaulacida" in having characters of Cimolodonta. Djadochtatherioidea have sharply limited enamel band on lower incisor, except Sloanbaataridae in which enamel is thicker on the ventrolateral surface, but not sharply limited. The sharply limited enamel also occurs in the "plagiaulacidans" *Glirodon* and *Eobaatar,* Taeniolabidoidea, Microcosmodontidae, and Eucosmodontidae. Share with *Glirodon, Eobaatar,* and Cimolodonta except Ptilodontoidea gigantoprismatic enamel. Differ from Taeniolabidoidea in having two lower premolars instead of one, arcuate p4 (secondarily subtrapezoidal in *Catopsbaatar*) rather than triangular, three to four upper premolars (rather than one), and smaller number of cusps on the upper and lower molars. Differ from Ptilodontoidea in having a more robust lower incisor,

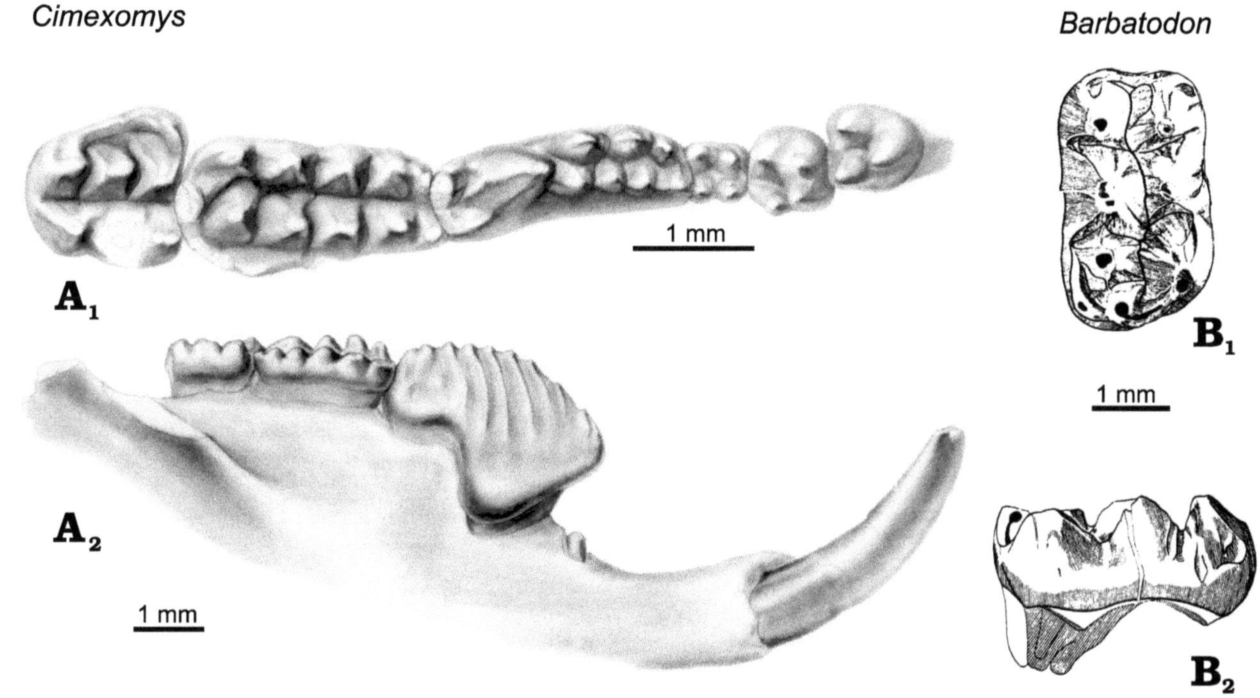

FIGURE 8.37. Late Cretaceous North American (*Cimexomys*) and European (*Barbatodon*) taxa tentatively assigned to *Paracimexomys* group. A, *Cimexomys judithae*, right upper premolars and molars in occlusal view, anterior is to the right (A_1); right dentary of the same with almost complete dentition (p3 broken) in lateral view (A_2). B, *Barbatodon transylvanicum*, left M1 in occlusal (B_1) and lingual (B_2) views. *Source:* modified from: A, Montellano et al. (2000); B, Rădulescu and Samson (1986).

with a limited enamel band, and in having a smaller p4 that does not protrude dorsally over the level of the molars. Differ from *Uzbekbaatar* (Kielan-Jaworowska and Nessov, 1992) and Arginbaataridae in having a labial cuspule on p4, and from Arginbaataridae in having less arcuate crown of p4, completely covered with enamel and not rotating during ontogeny.

Families. Djadochtatheriidae Kielan-Jaworowska and Hurum, 1997, type family; Sloanbaataridae Kielan-Jaworowska, 1974; three Asian genera assigned to family *incertae sedis* (*Chulsanbaatar, Nemegtbaatar,* and *Bulganbaatar*).

Distribution. Late Cretaceous (Campanian): Mongolia and China, Gobi Desert (Djadokhta and Baruungoyot formations and their equivalents in Mongolia and Bayan Mandahu Formation in China), Kazakhstan. Occurrence of this superfamily in the Paleocene of North America is doubtful.

Family Djadochtatheriidae Kielan-Jaworowska and Hurum, 1997

Diagnosis (based on Kielan-Jaworowska and Hurum, 1997, emended). Differ from all other multituberculates (and all other mammals) in having a subtrapezoidal snout in dorsal view, with wide anterior margin and lateral margins confluent with zygomatic arches, rather than incurved in front of the arches. Differ from other members of Djadochtatherioidea in having very long snout, extending for 50% or more of the skull length, and concave areas on lateral sides of parietals to accommodate strong temporal muscles and very long postorbital process. Anterior part of promontorium irregular, with incurvatures on both sides, rather than oval (a character shared with *Lambdopsalis*). They share with *Chulsanbaatar* two pairs of vascular foramina on the nasals (but there are more than two in some specimens of *Catopsbaatar catopsaloides*) and share with *Chulsanbaatar, Kamptobaatar,* and Taeniolabididae lack of palatal vacuities.

Genera. *Djadochtatherium* Simpson, 1925a, type genus; *Catopsbaatar* Kielan-Jaworowska, 1994; *Kryptobaatar* Kielan-Jaworowska, 1970; *Tombaatar* Rougier, Novacek, and Dashzeveg, 1997.

Distribution. Late Cretaceous (Campanian): Mongolia, Gobi Desert (Djadokhta Formation, "Ukhaa Tolgod Beds"; Tögrög Beds; Baruungoyot Formation; Red Beds of Hermiin Tsav); China, Inner Mongolia (Bayan Mandahu Formation).

Genus *Djadochtatherium* Simpson, 1925a
(figure 8.40J)

Diagnosis (based on Kielan-Jaworowska and Hurum, 1997, emended). Skull length about 50–55 mm, m1 cusp formula 4:3. Differs from *Kryptobaatar* in being distinctly larger. Shares with *Kryptobaatar* the presence of four upper premolars and arcuate p4. The p4 is only slightly longer than m1. Differs from *Catopsbaatar* and *Tombaatar* in having four rather than three upper premolars. Shares with *Catopsbaatar* robust lower incisor and with *Kryptobaatar* and *Catopsbaatar* a very long postorbital process. Differs from *Catopsbaatar* in having the orbit placed somewhat more anteriorly and relatively larger and arcuate p4 (subtrapezoidal in *Catopsbaatar*).

Species. *Djadochtatherium matthewi* Simpson, 1925a, type species by monotypy.

Distribution. Late Cretaceous (?early Campanian): Mongolia, Gobi Desert (Djadokhta Formation, "Ukhaa Tolgod Beds"; Tögrög Beds); China, Inner Mongolia (Bayan Mandahu Formation).

Genus *Catopsbaatar* Kielan-Jaworowska, 1994
(figures 8.9, 8.38I, 8.39F, 8.40K)

Revised Diagnosis. Generally similar to *Djadochtatherium* from which it differs in being larger (skull length up to 6 cm), in having only three (instead of four) upper premolars (P2 being lost), and in having smaller and trapezoidal rather than arcuate p4. Cusps 5:1 on P4, 6:5:4 on M1, with inner row extending for 0.75 of the tooth length.

Species. *Catopsbaatar catopsaloides* (Kielan-Jaworowska, 1974), type species by monotypy (see also Kielan-Jaworowska, Hurum, et al., 2002).

Distribution. Late Cretaceous (?late Campanian): Mongolia, Gobi Desert (Baruungoyot Formation and Red Beds of Hermiin Tsav).

Genus *Kryptobaatar* Kielan-Jaworowska, 1970a
(figures 8.10A, 8.38H, 8.39E, 8.40I)

Synonyms: *Gobibaatar* Kielan-Jaworowska, 1970a; *Tugrigbaatar* Kielan-Jaworowska and Dashzeveg, 1978.

Diagnosis (modified from Kielan-Jaworowska and Hurum, 1997, and Wible and Rougier, 2000). Includes the smallest species of Djadochtatheriidae, skull length 25–32 mm; most similar to *Djadochtatherium* with which it shares arcuate p4 (eight serrations, number in *Djadochtatherium* unknown). Differs from *Catopsbaatar* in having a relatively smaller facial surface of the lacrimal. Differs from *Tombaatar* in having alveolus for I3 formed by premaxilla, rather than by both premaxilla and maxilla. Differs from *Catopsbaatar* and *Tombaatar* in having four upper premolars. Differs from *Catopsbaatar* in having shorter inner row of cusps in M1 and shares this character with *Tombaatar.* Differs from *Chulsanbaatar* and *Sloanbaatar* in having cusps on the inner row in M1, rather than a smooth ridge. Differs from *Tombaatar* in having M1 cusp formula 4–5:4:3–5 rather than 5:5:2.

Species. *Kryptobaatar dashzevegi* Kielan-Jaworowska, 1970a, type species (including *Gobibaatar parvus* Kielan-Jaworowska, 1970a, and *Tugrigbaatar saichanensis* Kielan-Jaworowska and Dashzeveg, 1978); *K. mandahuensis* Smith, Guo, and Sun, 2001.

Distribution. Late Cretaceous (Campanian): Mongolia, Gobi Desert (Djadokhta Formation, "Ukhaa Tolgod Beds," Tögrög Beds and Red Beds of Hermiin Tsav; see Kielan-Jaworowska et al. 2003); China, Inner Mongolia (Bayn Mandahu Formation).

Genus *Tombaatar* Rougier, Novacek, and Dashzeveg, 1997
(figure 8.38C)

Diagnosis (based on Rougier et al., 1997, emended). Large djadochtatheriid of *Djadochtatherium/Catopsbaatar* size, known only from rostrum with dentition. Differs from all other Djadochtatherioidea in having the I3 alveolus formed by both the premaxilla and maxilla. Shares with *Catopsbaatar* three upper premolars (P2 absent) but differs from it in having large P1 and P3. Cusp formula 2:5 on P4 and 5:5:2 on M1; inner row of cusps in M1 extending only for half the tooth length.

Species. *Tombaatar sabuli* Rougier, Novacek and Dashzeveg, 1997, type species by monotypy.

Distribution. Late Cretaceous (?early Campanian): Mongolia, Gobi Desert ("Ukhaa Tolgod Beds"); ?China, Inner Mongolia (Bayna Mandahu Formation). See Kielan-Jaworowska et al. (2003).

Family Sloanbaataridae Kielan-Jaworowska, 1974a

Revised Diagnosis. Medium-sized Djadochtatherioidea that differ from Djadochtatheriidae in having anterior part of the snout rectangular (and share this character with *Chulsanbaatar, Nemegtbaatar,* and *Bulganbaatar*) rather than subtrapezoidal, and slender zygomatic arches more expanded laterally. The glenoid fossa is roughly half-oval, oriented anterolaterally. Differ from all other Djadochtatherioidea in having lower incisor with enamel thicker on the ventrolabial side, but not sharply limited, and the coronoid process flared laterally. In all three genera the mandibular condyle is situated high above the level of the molars, and they share this character with *Djadochtatherium* and *Catopsbaatar.*

Genera. *Sloanbaatar* Kielan-Jaworowska, 1970a, type genus; *Nessovbaatar* Kielan-Jaworowska and Hurum, 1997; and tentatively assigned *Kamptobaatar* Kielan-Jaworowska, 1970a.

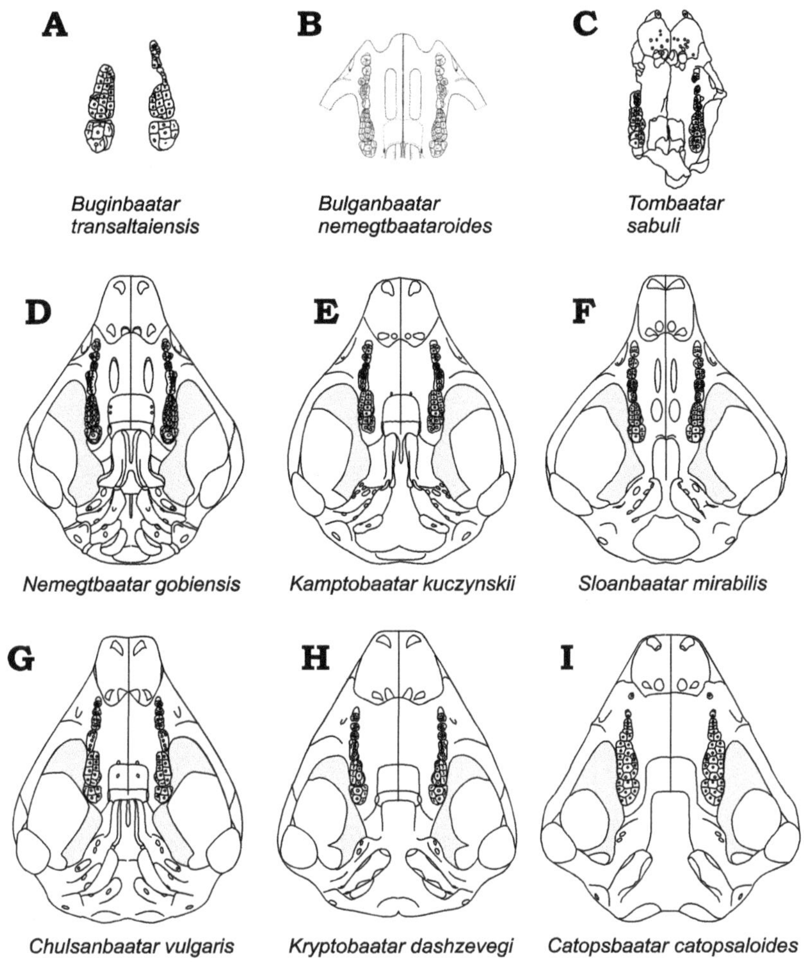

FIGURE 8.38. Comparison of upper dentitions (A–C) and reconstructions of the outline of skulls with dentition in palatal view (D–I) of Late Cretaceous Mongolian multituberculates, rendered to approximately the same length. A, *Buginbaatar* (tentatively assigned to Cimolomyidae); B–I, Representatives of Djadochtatherioidea. A complete skull of *Djadochtatherium* is known (see Kielan-Jaworowska and Hurum, 1997) but it has not been described and we do not figure it here. *Source:* modified from A, Trofimov (1980); B, Kielan-Jaworowska (1974a); C, Rougier et al. (1997); D–I, Kielan-Jaworowska and Hurum (1997).

Distribution. Late Cretaceous (Campanian): Mongolia, Gobi Desert Djadokhta and Barungoyot formations and their equivalents).

Genus *Sloanbaatar* Kielan-Jaworowska, 1970a
(figures 8.4E, 8.38F, 8.39C, 8.40F)

Revised Diagnosis. *Sloanbaatar* (skull length of the type species 25 mm) is characterized by having the narrowest anterior part of the snout among Djadochtatherioidea, the zygomatic arches most strongly expanded laterally, and the largest angle (28°) between the lower margin of the dentary and the occlusal level of the molars. It shares with *Catopsbaatar* the incisive foramen placed entirely within the premaxilla. Differs from all Djadochtatherioidea in having two pairs of palatal vacuities (rather than single or none). Shares with *Kamptobaatar* and *Nessovbaatar* characters of the dentary diagnostic for the family. Shares with *Kamptobaatar, Nessovbaatar, Djadochtatherium,* and *Catopsbaatar* the position of the condyle above the level of the molars and differs in this respect from other Djadochtatherioidea. Cusp formula for P4 is 2:5; M1, 4:4:Ri; M2, 1:2:3; p4, eight serrations and labial cusp; m1, 4:3; m2, 2:2.

Species. *Sloanbaatar mirabilis* Kielan-Jaworowska, 1970a, type species by monotypy.

Distribution. Late Cretaceous (?early Campanian): Mongolia, Gobi Desert (Djadokhta Formation and ?"Ukhaa Tolgod Beds").

Genus *Nessovbaatar* Kielan-Jaworowska and Hurum, 1997
(figure 8.40G)

Diagnosis. Monotypic genus represented by *Nessovbaatar multicostatus,* known only from left and right dentaries with teeth. It differs from *Sloanbaatar* in being smaller, but shares with it and with *Kamptobaatar* a relatively small coronoid process that flares laterally. It shares with *Sloanbaatar, Kamptobaatar, Djadochtatherium,* and *Catopsbaatar* the condyle placed above the occlusal level of the molars and facing dorsally. It differs from all the Djadochtatherioidea in having a larger p4 with ten cusps and nine serrations, eight of which are associated with

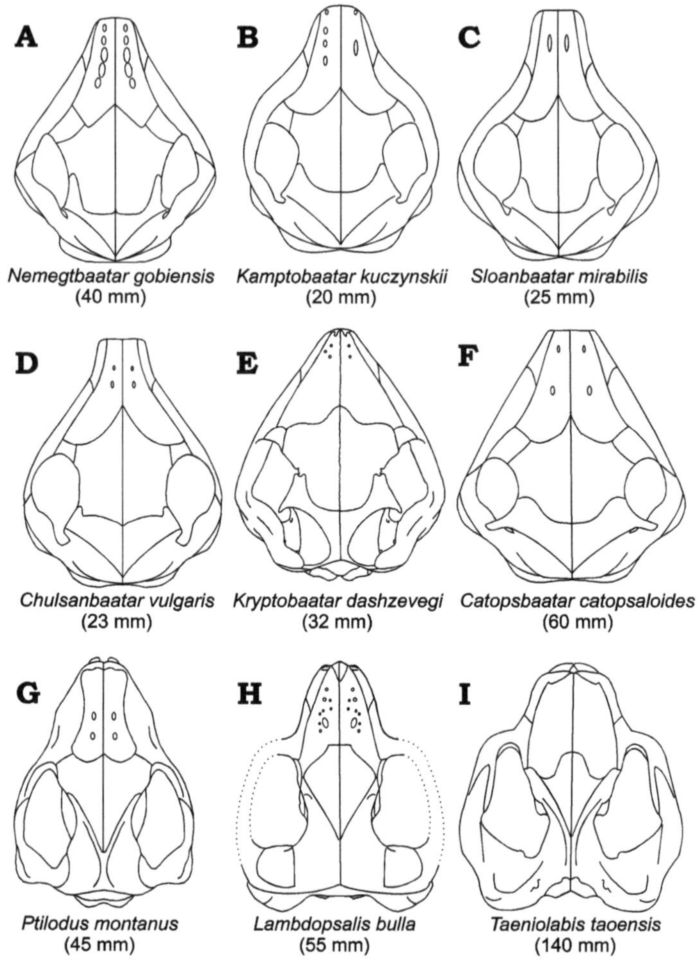

FIGURE 8.39. Diagrammatic reconstructions of skulls in dorsal view, rendered to approximately the same length. A–F, Representatives of Djadochtatherioidea from the Late Cretaceous of Mongolia; G–I, Paleocene genera figured for comparison. G, I, North American taxa; H, Asian taxon. *Ptilodus* belongs to Ptilodontoidea, *Taeniolabis,* and *Lambdopsalis* to Taeniolabidoidea. *Source:* the whole drawing modified from Kielan-Jaworowska and Hurum (1997); G, based on Simpson (1937a); H, on Miao (1988; 1993); I, on Granger and Simpson (1929).

weak ridges. It resembles *Sloanbaatar* in having 4:3 cusps on m1, but differs from it in having an angle of 18° between the lower margin of the dentary and the occlusal level of the molars (28° in *Sloanbaatar*) and 3:2 cusps on m2, rather than 2:2. It differs from *Sloanbaatar* and *Kamptobaatar* in structure of p4, which resembles that of Mongolian Early Cretaceous *Arginbaatar* in being fan shaped, but differs from it in being less vaulted, having fewer serrations with ridges, and being completely covered with enamel. The only specimen of *N. multicostatus* belongs to a juvenile; the length of the dentary, measured from the base of incisor to the posterior end of the condyle, is 13.5 mm.

Species. *Nessovbaatar multicostatus* Kielan-Jaworowska and Hurum, 1997, type species by monotypy.

Distribution. Late Cretaceous (?late Campanian): Mongolia, Gobi Desert (Red Beds of Hermiin Tsav).

Genus *Kamptobaatar* Kielan-Jaworowska, 1970a
(figures 8.8C, 8.38E, 8.39B, 8.40E)

Diagnosis. The snout in *Kamptobaatar* (skull length of the type species between 18 and 20 mm) is incurved in front of the zygomatic arches, with the anterior part of the arches directed roughly transversely or posterolaterally. It has a relatively wide, rectangular, and blunt anterior part of the snout. There are asymmetrical foramina on the nasals, no palatal vacuities, and the foramen ovale is divided into five foramina (two in other multituberculates). The dentary is relatively short; the coronoid process is low and triangular, flaring laterally (as in *Sloanbaatar* and *Nessovbaatar*). The mandibular condyle, which is situated at the top of the rounded posterior edge of the dentary, faces dorsally (not posterodorsally as in most Djadochtatherioidea). Differs from all other Djadochtatherioidea in having the inner row on M1 developed as a smooth ridge, extending along the whole length of the tooth. Cusp formula: P4, 3:5–6; M1, 5:5:Ri; M2, Ri:2:3; p4 with seven serrations with ridges; m1, 4:3; m2 unknown.

Species. *Kamptobaatar kuczynskii* Kielan-Jaworowska, 1970a, type species by monotypy.

Distribution. Late Cretaceous (?early Campanian): Mongolia, Gobi Desert (Djadokhta Formation and ?"Ukhaa Tolgod Beds").

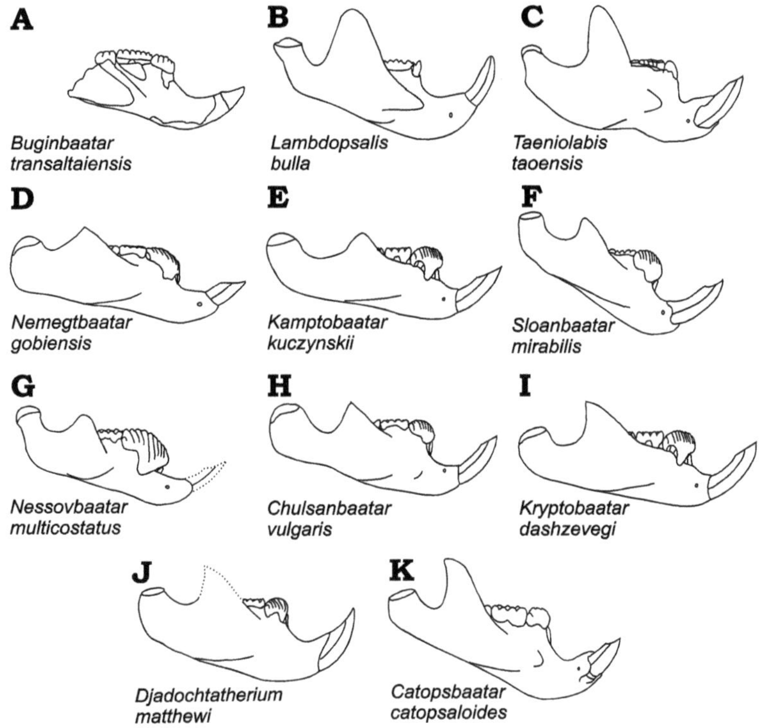

FIGURE 8.40. Comparison of right dentaries in labial view rendered to approximately the same length. A, Late Cretaceous Mongolian *Buginbaatar* (tentatively assigned to Cimolomyidae). B, C, Early Tertiary taeniolabidoids. B, *Lambdopsalis* (from Asia). C, *Taeniolabis* (from North America), figured for comparison. D–K, Late Cretaceous Mongolian Djadochtatherioidea. *Source:* modified from: A, Trofimov (1980); B–K, Kielan-Jaworowska and Hurum (1997); B, original based on Miao (1993); C, based on Granger and Simpson (1929).

Family *incertae sedis*
Genus *Bulganbaatar* Kielan-Jaworowska, 1974a
(figures 8.12, 8.38)

Diagnosis. *Bulganbaatar* is smaller than *Nemegtbaatar* and has fewer cusps on P4 and upper molars. It shares with *Nemegtbaatar* the shape of the anterior part of the snout and one pair of palatal vacuities. The cusp formulae are: P4, 2:5; M1, 5:5:3; M2, Ri:2:2. There are more cusps on M1 than in *Kryptobaatar, Chulsanbaatar,* and *Sloanbaatar,* and the M1 has a long lingual ridge of cusps, in which it resembles *Nemegtbaatar.*

Species. *Bulganbaatar nemegtbaataroides* Kielan-Jaworowska, 1974a, type species, known from a damaged rostrum with teeth (the holotype); ?*Bulganbaatar* sp., a P4 from the early Campanian of Kazakhstan (Averianov, 1997).

Distribution. Late Cretaceous (?early Campanian): Mongolia, Gobi Desert (Djadokhta Formation); Late Cretaceous (early Campanian): Kazakhstan, Grey Mesa (Darbasa Formation).

Genus *Chulsanbaatar* Kielan-Jaworowska, 1974a
(figures 8.6A, 8.7A,C, 8.38G, 8.39D, 8.40H)

Diagnosis. Very small representative of Djadochtatherioidea; skull length varies around 18–22 mm. Like Djadochtatheriidae it lacks the palatal vacuities and has two pairs of foramina on the nasals, but differs from this family in being distinctly smaller, in having a shorter, rectangular rather than trapezoidal snout, small postorbital process, and anterior margins of the skull incurved in front of the zygomatic arches. Also differs from Djadochtatheriidae in cusp formula, which is P4, 2:6; M1, 4:5:Ri; M2, Ri:2:2; p4 has seven ridges and a labial cusp; m1, 4:3; m2, 2:2.

Species. *Chulsanbaatar vulgaris* Kielan-Jaworowska, 1974a, type species by monotypy.

Distribution. Late Cretaceous (Campanian): Mongolia, Gobi Desert ("Ukhaa Tolgod Beds"; Baruungoyot Formation and Red Beds of Hermiin Tsav).

Genus *Nemegtbaatar* Kielan-Jaworowska, 1974a
(figures 8.4A–C, 8.7B,D, 8.11, 8.13B, 8.15, 8.16B, 8.38D, 8.39A, 8.40D)

Diagnosis. Medium-sized djadochtatherioid, with skull length 40 mm. Shares with *Bulganbaatar* one pair of palatal vacuities and differs in this respect from all other Djadochtatherioidea. Differs from all Djadochtatherioidea in having five pairs of vascular foramina on the nasals (a character unknown in *Bulganbaatar*). Differs from Djadochtatheriidae in having the lateral margins of the snout incurved in front of the zygomatic arches and shares this character with other Djadochtatherioidea. The anterior part of the snout is narrow and elongated. Among Djadochtatherioidea, the dentary is characterized by the smallest angle between the lower margin and the horizontal level of the molars and the lowest condyle, confluent with the posterior margin of the dentary and facing posterodorsally. Differs from all Djadochtatherioidea in having the relatively longest upper premolar/molar row and a different cusp formula: P4, 3:6:1; M1, 6:7:4; M2, Ri:3:2; p4 with seven ridges and labial cusp; m1, 5:4; m2, 3:2.

Species. *Nemegtbaatar gobiensis* Kielan-Jaworowska, 1974a, type species by monotypy.

Distribution. Late Cretaceous (?late Campanian): Mongolia, Gobi Desert (Baruungoyot Formation and Red Beds of Hermiin Tsav).

Superfamily *incertae sedis*
Family Cimolomyidae Marsh, 1889b

Emended Diagnosis. Dental formula: 1.0.4.2/1.0.2.2. Share with all Cimolodonta except Ptilodontoidea gigantoprismatic enamel. Apomorphies: differ from other Cimolodonta in having stout lower incisor completely covered with enamel and four upper premolars strongly reduced in width and length in proportion to enlarged molars, P4 triangular in labial view, strongly protruding ventrally over the level of premolars and molars. Share with Eucosmodontidae two-cusped I2 and with Djadochtatherioidea and Eucosmodontidae I3 placed on the palatal part of premaxilla. The p4 is arcuate with 8–11 serrations, shorter than m1, not protruding dorsally over the level of the molars (character shared with Djadochtatherioidea and Eucosmodontidae). Share with Plagiaulacidae, Eobaataridae, and the *Paracimexomys* group ornamentation of grooves and ridges on m1 and weak ribbing on M1, but differ in lack of obvious coalescence of cusps characteristic for these groups. Share with Taeniolabidoidea M1 with three rows of cusps, the inner one as long or almost as long as the others, but differ from them in having four upper premolars. Molar cusps pyramidal to crescentic; upper and lower first molars in *Meniscoessus* and *Cimolomys* develop multiple accessory roots.

Genera. *Cimolomys* Marsh, 1889a, type genus; *Meniscoessus* Cope, 1882c; and two tentatively assigned genera: *Buginbaatar* Kielan-Jaworowska and Sochava, 1969 and *Essonodon* Simpson, 1927b.

Distribution. Late Cretaceous: North America; Late Cretaceous (?Maastrichtian) or ?Paleocene: Mongolia, Gobi Desert, Bügiin Tsav.

Genus *Cimolomys* Marsh, 1889a
(figure 8.41A)

Diagnosis (based on Sahni, 1972, emended). Differs from *Meniscoessus* in having P4 that is long and low with eight or nine cusps, rather than high and short. The lingual row of cusps on M1 terminates near or at the second cusp from the anterior end of the medial row.

Species. *Cimolomys gracilis* Marsh, 1889a, type species, based on isolated upper and lower premolars and molars; *C. clarki* Sahni, 1972; *C. major* Russell, 1937; *C. milliensis* Eaton, 1993a; *C. trochuus* Lillegraven, 1969; and several species left in open nomenclature by Simpson (1929a) and Fox (1971a).

Distribution. Late Cretaceous (?Santonian, Aquilan to Lancian): North America.

Genus *Meniscoessus* Cope, 1882c
(figure 8.41B)

Diagnosis. Large cimolomyid, estimated skull length about 7 cm, with very robust I2 provided with three cusps arranged in a line, lower incisor with denticles on the upper edge. Very large incisive foramina on premaxilla. P4 relatively short and high, p4 with higher crown than in *Cimolomys*, with 8–11 serrations. M2 very large, wider than M1.

Species. *Meniscoessus conquistus* Cope, 1882c, type species; *M. borealis* Simpson, 1927a; *M. ferox* Fox, 1971a; *M. intermedius* Fox, 1976b; *M. major* (L. S. Russell, 1937), *M. robustus* (Marsh, 1889a) (see also Storer, 1991; and Weil and Tomida, 2001).

Distribution. Late Cretaceous (?Santonian, Aquilan, Judithian to Lancian): North America.

Genus ?*Buginbaatar* Kielan-Jaworowska and Sochava, 1969
(figures 8.9, 8.38A, 8.40A)

Diagnosis. Relatively large multituberculate known from fragment of a rostrum with P4–M2 and mandible with incisors and p4–m2. Upper incisors and number of upper premolars unknown. Shares with *Meniscoessus* robust lower incisor completely covered with enamel and p4 not protruding dorsally over the level of the molars. Differs from *Meniscoessus* and from Djadochtatherioidea in lack of p3 and in having vertical, rather than oblique ridges on p4. Shares with *Cimolomys* low p4 (but differs in lack of p3) and relatively small P4 in comparison with enlarged upper molars. Differs from *Cimolomys* and *Meniscoessus* in lack of crescentic cusps on the molars and relatively smaller m2 with smaller number of cusps, similar to those in Djadochtatherioidea.

Species. *Buginbaatar transaltaiensis* Kielan-Jaworowska and Sochava, 1969, type species by monotypy.

Distribution. Late Cretaceous (?Maastrichtian) or ?Paleocene: Mongolia, Gobi Desert, Bügiin Tsav.

Comments. *Buginbaatar* was assigned by various authors to Cimolomyidae, Eucosmodontidae, Buginbaatarinae within Eucosmodontidae, and Taeniolabididae (see Kielan-Jaworowska and Sochava, 1969; Kielan-Jaworowska, 1974a; Trofimov, 1975; Hahn and Hahn, 1983; McKenna

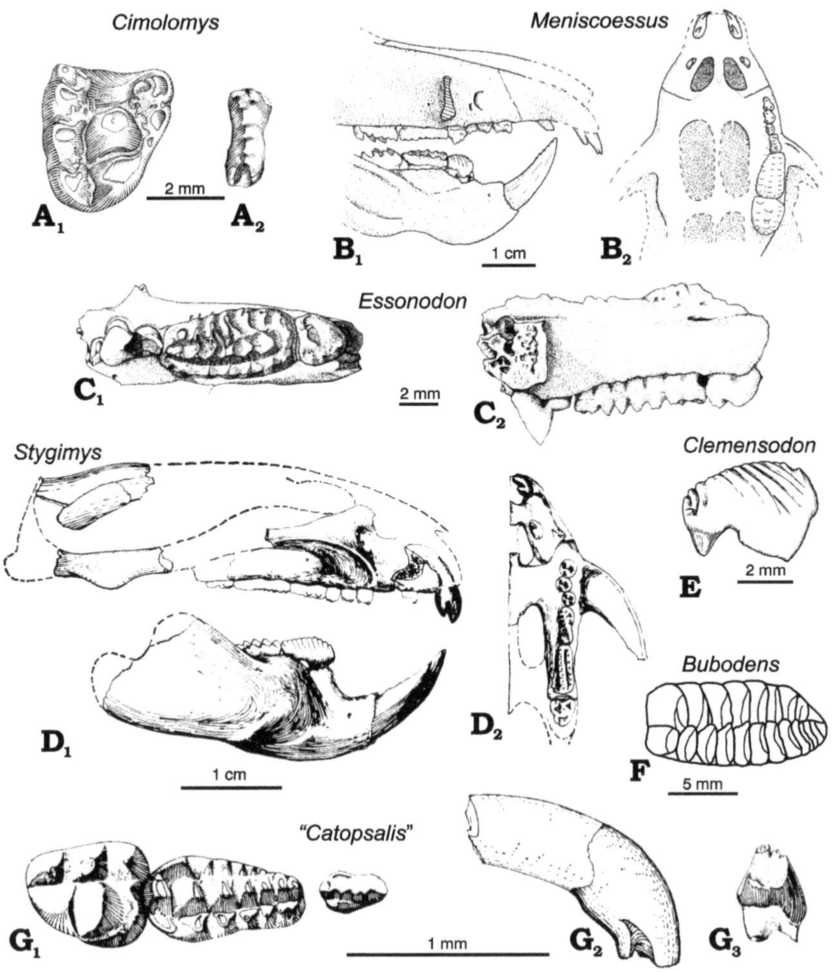

FIGURE 8.41. Comparison of cimolomyid, eucosmodontid, and taeniolabidid genera (all from Late Cretaceous or Paleocene of North America). A, *Cimolomys gracilis,* left M2 in occlusal view (A_1), and left P4 in occlusal view (A_2). B, *Meniscoessus robustus,* reconstruction of the anterior part of the skull and dentary in lateral (B_1) and palatal views (B_2). C, *Essonodon browni,* fragment of maxilla with anterior alveolus and fragment of P3, broken and worn P4, and M1, M2, in occlusal (C_1) and labial views (C_2). D, *Stygimys kuszmauli,* reconstruction of the rostrum and dentary in labial view (D_1) and palatal view of the rostrum (D_2). E, *Clemensodon megaloba,* right p4 in labial view. F, *Bubodens magnus,* left m1 (holotype), anterior is to the left. G, "*Catopsalis*" *joyneri,* right P4 and left M1, M2, in occlusal view, anterior is to the right (G_1); right I2, in labial view (G_2); right p4, labial view (G_3). *Source:* modified from: A–C, Archibald (1982); D, G, Sloan and Van Valen (1965); E, F, drawn on the basis of photographs.

and Bell, 1997; Kielan-Jaworowska and Hurum, 2001). Its systematic position remains unresolved for the time being.

Genus *Essonodon* Simpson, 1927a
(figure 8.41C)

Diagnosis (based on Archibald, 1982, emended). The genus is based on isolated lower and upper molars, and fragments of maxilla, the most complete of which bears a fragment of P3, incomplete P4, and M1–M2. Differs from *Meniscoessus* and *Cimolomys* in lack of well-developed crescentic cusps on the molars, but some molar cusps developed into ridges and numerous accessory ridges, forming a latticework between cusp rows. P4 incompletely known, relatively short and protruding ventrally over the level of the molars, m2 reduced in width and length in relation to m1.

Species. *Essonodon browni* Simpson, 1927a, type species by monotypy, and several species left in open nomenclature.

Distribution. Late Cretaceous (?Edmontonian, Lancian): North America.

Comments. Attribution of *Essonodon* to Cimolomyidae is uncertain. *Essonodon* is highly specialized in having ridges between the cusps of the molars, which otherwise occur only in Ferugliotheriidae (assigned to Mammalia *incertae sedis,* see chapter 14). It resembles *Cimolomys* and *Meniscoessus* in enlargement of the upper molars with respect to the relatively small P4. We follow Archibald (1982) in tentatively placing it in Cimolomyidae.

Family Eucosmodontidae Jepsen, 1940

Diagnosis. Medium-sized multituberculates, with snout (known only in *Stygimys*) incurved in front of zygomatic arches and palatal vacuities present. Share with all Cimolodonta except Ptilodontoidea gigantoprismatic enamel, and with Taeniolabididae, most Djadochtatherioidea, *Eobaatar,* and *Glirodon* lower incisor with limited enamel band. Differ from other Cimolodonta except Cimolomyidae in having two-cusped I2. Differ from Djadochtatherioidea in lack of p3 and in having greater number of cusps (and, on p4, ridges) on the premolars and molars, but share with them (and *Meniscoessus*) I3 placed on the palatal part of premaxilla and four or three upper premolars. P1–P3 single- or double-rooted. Share with Djadochtatherioidea and Cimolomyidae arcuate p4, not protruding over the dorsal level of the molars, and differ in this respect from Ptilodontoidea. Share with Djadochtatherioidea P4 with lower margin almost straight in lateral view and differ in this respect from Cimolomyidae and Ptilodontoidea, which have triangular P4 strongly protruding ventrally. Differ from Taeniolabididae in smaller size, in having four upper premolars, and arcuate rather than triangular p4.

Genera. *Eucosmodon* Matthew and Granger, 1921, type genus (Paleocene genus not described herein), *Stygimys* Sloan and Van Valen, 1965, and tentatively assigned *Clemensodon* Krause, 1992.

Distribution. Late Cretaceous to early Eocene: North America; early Paleocene to early Eocene: Europe.

Genus *Stygimys* Sloan and Van Valen, 1965
(figure 8.41D)

Diagnosis (based on Sloan and Van Valen, 1965, emended). Medium-sized eucosmodontid with single-rooted P1, P2, and P3. The p4 is less arcuate than in *Eucosmodon,* with 11 serrations; posterior portion of its profile relatively straight, anterior portion parabolic or circular. In *S. kuszmauli* P4 is almost as along as M1, p4: m1 length ratio 1.1; P4 cusp formula 3:9:1; M1, 6:7:3; m1, 5:5.

Species. *Stygimys kuszmauli* Sloan and Van Valen, 1965, type species; the only uncontested Late Cretaceous species is *Stygimys cupressus* Fox, 1989, from the Late Cretaceous of Canada. *Stygimys kuszmauli* comes from Hell Creek Formation at Bug Creek, Montana, and is now regarded as one of the early Paleocene elements of that mixed assemblage (Archibald and Lofgren, 1990; Lofgren, 1995). We figure here *Stygimys kuszmauli,* as it is the most complete of *Stygimys* species. See Sloan and Van Valen (1965), Archibald (1982), Johnston and Fox (1984), and Fox (1989) for discussion of Tertiary taxa. The genus has been tentatively recorded from the Campanian of Baja California (Lillegraven, 1972), which (if verified) would be its earliest appearance.

Distribution. Late Cretaceous (?Campanian, Lancian, very rare in Cretaceous strata) to Early Paleocene (more common in Tullock Formation): North America.

Genus ?*Clemensodon* Krause, 1992
(figure 8.41E)

Diagnosis (based on Krause, 1992, emended). Poorly known taxon based on p4, which has gigantoprismatic enamel and low profile. There is a large anterior root and large anterior triangular lobe above it on the labial side of the crown, 11 or 12 serrations, ventrally bifurcating radial ridges (particularly those descending from the sixth and seventh serrations), absence of labial ridges posterior to the eight serration, well-developed and dorsally peaked anterobasal concavity, and short but well-developed posterolabial cusp (ledge). Length of p4 is between 5.8 and 6.4 mm.

Species. *Clemensodon megaloba* Krause, 1992, type species by monotypy.

Distribution. Late Cretaceous (Lancian): United States, Wyoming, Powder River Basin (type Lance Formation).

Comments. Krause (1992) discussed the possible affinities of his new genus and stated that *Clemensodon* is either a derived taeniolabidoid (*sensu lato*) or primitive ptilodontoid. Because of its gigantoprismatic enamel we favor the first alternative and assign it tentatively to the Eucosmodontidae.

Family Microcosmodontidae Holtzman and Wolberg, 1977

Diagnosis (based on Weil, 1998, 1999; and Fox, 1999). Small multituberculates with enlarged lower incisor bearing restricted enamel band; p4 small, shorter than m1, with five or six serrations; p4 crown height subequal to or greater than crown length. Differ from Eucosmodontidae and Djadochtatherioidea in having I3 located at the palatal margin of premaxilla, rather than medially. Enamel microstructure, at least in some species of *Microcosmodon,* contains both normal prisms and gigantoprisms (Carlson and Krause, 1985).

Genera. *Microcosmodon* Jepsen, 1930a, type genus (Paleocene); two other Paleocene genera not described herein (see Jepsen, 1940; and Archibald, 1982), and Microcosmodontidae gen. et sp. indet. (Fox, 1989).

Distribution. Late Cretaceous (very rare) to Paleocene: North America, Canada (Ravenscrag Formation). We base the occurrence of Microcosmodontidae in the Late Cretaceous on Fox (1989), who described a p4 identified as Microcosmodontinae gen. et sp. indet. from the Ravenscrag Formation (Long Fall horizon), Puercan of Canada (see discussion of Lancian-Puercan versus Cretaceous-Tertiary boundaries in Cifelli et al., 2004).

Superfamily Taeniolabidoidea Sloan and Van Valen, 1965

Type Family by Monotypy. Taeniolabididae Granger and Simpson, 1929

Superfamily and Family Diagnosis. Largest multituberculates (skull length in Paleocene *Taeniolabis* reaching 16 cm), with gigantoprismatic enamel (shared with all Cimolodonta except Ptilodontoidea). Dental formula 1.0.1–2.2/1.0.1.2. Apomorphies: snout short and wide with anterior part of zygomatic arches directed transversely, resulting in a square-shaped skull; frontals small, pointed posteriorly, almost or completely excluded from the orbital rim. Share with most Cimolodonta (but not with Cimolomyidae and Eucosmodontidae) single-cusped I2. Differ from all multituberculates in having only one upper premolar (except for one taxon), long diastema between I3 and premolars, P4 and p4 triangular in lateral view, strongly reduced in proportion to enlarged, multicusped molars; p4 without oblique ridges. Share strong, self-sharpening incisors that have enamel limited to the outer surface with most Djadochtatherioidea and Eucosmodontidae. Differ from Djadochtatherioidea, Cimolomyidae, and Eucosmodontidae in position of I3 on the margin of the premaxilla, close behind I2. Differ from Djadochtatherioidea and Cimolomyidae in lack of p3. A tendency toward reduction of the premolars and an increase in the size of the molars and the number of molar cusps also occurs in other multituberculates, for example, in some Djadochtatherioidea and Cimolomyidae. However, in these groups it occurred according to a different pattern, as it is only in Taeniolabidoidea that the anterior upper premolars (P1–P3) are absent, upper and lower fourth premolars so strongly reduced in size, and M1 has three rows of cusps extending for the whole length of the tooth.

Genera. *Taeniolabis* Cope, 1882b, type genus, "*Catopsalis*" Cope, 1882a, and three other Paleocene genera not described herein. Of all the taxa assigned to Taeniolabididae, only two species of "*Catopsalis*" are known from the Late Cretaceous. "*Catopsalis*" has been conventionally regarded as a Cretaceous representative, based on its occurrence at Bug Creek, Montana (Sloan and Van Valen, 1965). This occurrence is now regarded as more probably Paleocene (Archibald and Lofgren, 1990; Lofgren, 1995), but Fox (e.g., 1989, 1997a) records the genus from beds reportedly of latest Cretaceous age, and we herein include "*Catopsalis*" on this basis. Early Tertiary genera *Lambdopsalis* Chow and Qi, 1978 and *Taeniolabis* are figured here for comparison (see figures 8.9, 8.39H,I, 8.40B,C). We also assign tentatively to the Taeniolabididae poorly known Late Cretaceous *Bubodens* Wilson, 1987.

Distribution. Late Cretaceous (rare in Cretaceous strata) to Paleocene: North America; Paleocene: Asia, Mongolia and China, Gobi Desert.

Genus "*Catopsalis*" Cope, 1882a
(figure 8.41G)

Species. "*Catopsalis*" *foliatus* Cope, 1882a, type species, and six other Paleocene species cited by Simmons and Miao (1986); see also Middleton (1982) and references therein for discussion of Paleocene taxa. "*C.*" *johnstoni* Fox, 1989; and "*C.*" cf. *joyneri* are the only two included species with possible occurrences in the Cretaceous (see Fox, 1989).

Distribution. Late Cretaceous (Lancian, represented only by incomplete, poorly identified taxa): Canada; early Paleocene (more common): United States.

Comments. "*Catopsalis*" is a poorly known genus, most species being confined to dentaries with teeth and isolated upper teeth at best. Simmons and Miao (1986) argued that "*Catopsalis*" is a paraphyletic taxon, and therefore we cite it in quotes. These authors recognized within it (in addition to two species now assigned to Djadochtatherioidea, *Djadochtatherium* and *Catopsbaatar*) at least three monophyletic lines "C." *joyneri*, "C." *alexanderi*, and four Paleocene species including the type species. The latter group may be monophyletic, but in the cladogram of Simmons and Miao (1986: figure 1) it forms a polytomy with *Taeniolabis*. None of the relevant "genera" has been defined. Fox (1989; see also Johnston and Fox, 1984) described two "*Catopsalis*" species, "*C.*" cf. *joyneri* and "*C.*" *johnstoni* Fox, 1989 from the Late Cretaceous of Canada. Because of the ambiguity concerning the scope of "*Catopsalis*" we do not diagnose it. We figure "*Catopsalis*" *joyneri* from ?early Paleocene Tullock Formation (figure 8.41G), as it is more complete than Late Cretaceous species from Canada.

Family ?Taeniolabididae Granger and Simpson, 1929
Genus *Bubodens* Wilson, 1987
(figure 8.41F)

Diagnosis. Genus known only from the holotype specimen (m1) of the type species. Differs from all North American Cretaceous multituberculates by larger size (length of m1 is 12.8 mm, width 6.0 mm), by cusp formula 10:8 (or 10:9), and by having strongly crowded cusps. First cusps in both rows (especially lingual) larger than the others, which are lenticular, compressed anteroposteriorly, and slightly decreased in size posteriorly. Anterior margin of the tooth straight, posterior pointed.

Species. *Bubodens magnus* Wilson, 1987, type species by monotypy.

Distribution. Late Cretaceous (Lancian): United States, South Dakota (Fox Hills Formation).

Superfamily *incertae sedis*
Family Kogaionidae Rădulescu and Samson, 1996

Diagnosis. Medium-sized multituberculates with gigantoprismatic enamel, represented by isolated premolars and molars (*Hainina*) and an almost complete skull with dentition, lacking the mandible (*Kogaionon*). Apomorphies: differ from other Cimolodonta in having all upper premolars strongly elongated, of which P3 is the longest; premolar row twice as long as molar row, short and wide

M1 with cusp formula in *Kogaionon* 3:4:3. Share with Ptilodontoidea a short premaxilla (long in Djadochtatherioidea), I2 single-cusped and placed close to I3, which is on the margin of the palatal part of premaxilla. Share with Ptilodontoidea a nearly transverse suture between the nasals and frontals, and vascular foramina on the nasals placed far posteriorly. Another common feature is the shape of the frontals, which are pointed posteriorly, although the suture between the frontals and parietals is straight in *Ptilodus* and interdigitated in *Kogaionon*. Differ from Ptilodontoidea in having I2 with limited enamel band, which also suggests limited enamel band on the lower incisors (unknown). Share with Taeniolabididae roughly rectangular shape of the skull, but differ in having an elongated snout.

Genera. *Kogaionon* Rădulescu and Samson, 1996, type genus; *Hainina* Vianey-Liaud, 1979.

Distribution. Late Cretaceous: Europe, Romania; Paleocene: Europe, France, Spain.

Comments. Attribution of Kogaionidae to a known cimolodontan superfamily poses difficulties. Vianey-Liaud (1979) assigned *Hainina* tentatively to Ptilodontoidea and then (1986) to Cimolomyidae. Peláez-Campomanes et al. (2000) argued that Kogaionidae, to which they assigned *Kogaionon* and *Hainina,* had five upper premolars, and identified the four premolars in *Kogaionon* as P2–P5. They based their conclusion on the assumption that all the multituberculate isolated teeth found in the Paleocene Fontilonga 3 locality in the southern Pyrénées are conspecific, in spite of the fact that they found two different P4s. In our opinion the upper premolars of *Kogaionon* represent the P1–P4, as only the P4 shows shearing structure, whereas the P3, although elongate, has a structure typical of the multituberculate anterior upper premolars, with large piercing cusps, and is not similar to P4s in "Plagiaulacida." We follow Kielan-Jaworowska and Hurum (2001) in regarding Kogaionidae as a separate family that cannot be placed among established multituberculate superfamilies based on present evidence.

Genus *Kogaionon* Rădulescu and Samson, 1996
(figure 8.42A)

Diagnosis. Member of Kogaionidae with family characters and the following cusp formulae: three cusps on P1, four on P2, five on P3, 4:2 on P4, 3:4:3 on M1 and Ri:2:3 on M2. Enamel not ornamented.

Species. *Kogaionon ungureanui* Rădulescu and Samson, 1996, type species by monotypy.

Distribution. Late Cretaceous (late Maastrichtian): Romania, Haţeg Basin (Sînpetru Formation).

Genus *Hainina* Vianey-Liaud, 1979
(figure 8. 42B)

Diagnosis. A kogaionid having m1 (in the type species), with two rows of cusps. Differs from *Kogaionon* in having molar enamel strongly ornamented with grooves. Shares with *Kogaionon* roughly rectangular M1 with three rows of cusps extending for the whole tooth length, but differs in being more elongated with respect to the width and having more (smaller) cusps in inner row, for example, the M1 cusp formula of the type species (*H. belgica* Vianey-Liaud, 1979) is 4:4:7. M2 cusp formula (in *H. godfriauxi* Vianey-Liaud, 1979 is 2:3. P4 distinctly narrower and shorter than M1, with 1:5 cusps in the type species. P3 longer than P2 and P1, but the difference in length is not as dramatic as in *Kogaionon.* The p4 attributed to *H. belgica* is arcuate, has nine serrations, eight ridges, and no posterolabial cusp; it protrudes dorsally over the level of the molars as in Ptilodontoidea.

Species. *Hainina belgica* Vianey-Liaud, 1979, type species (Paleocene), represented by isolated lower and upper premolars and molars; three Paleocene species from Europe (Vianey-Liaud, 1979 and Peláez-Compomanes et al., 2000); and *Hainina* sp. A, *Hainina* sp. B, and *Hainina* sp. (two specimens), all described by Csiki and Grigorescu (2000) from the late Maastrichtian of Romania.

Distribution. Late Cretaceous (late Maastrichtian): Romania, Haţeg Basin (Sînpetru Formation); Paleocene: Belgium and Spain.

Superfamily and Family *incertae sedis*
Genus *Uzbekbaatar* Kielan-Jaworowska and Nessov, 1992
(figure 8.42C)

Diagnosis (based on Kielan-Jaworowska and Nessov, 1992, emended). The genus probably had two lower premolars. The p4 is arcuate but relatively low, with nine serrations and eight ridges. Shares with Djadochtatherioidea arcuate p4, but differs from them and from most Eucosmodontidae and Ptilodontoidea in lack of the posterolabial cusp and in having p3 possibly less reduced in relation to the size of p4. Shares with Arginbaataridae lack of the posterolabial cusp on p4, but differs from them in having the crown less arcuate and completely covered with enamel and in having fewer serrations.

Species. *Uzbekbaatar kizylkumensis* Kielan-Jaworowska and Nessov, 1992, type species, based on isolated p4 (unidentified incisor, fragment of a dentary, and postcranial fragments have been found at the same locality); *U. wardi* Averianov, 1999.

Distribution. Late Cretaceous (Turonian–Coniacian): Uzbekistan, Kyzylkum Desert, Dzharakuduk (Bissekty and Aitym formations).

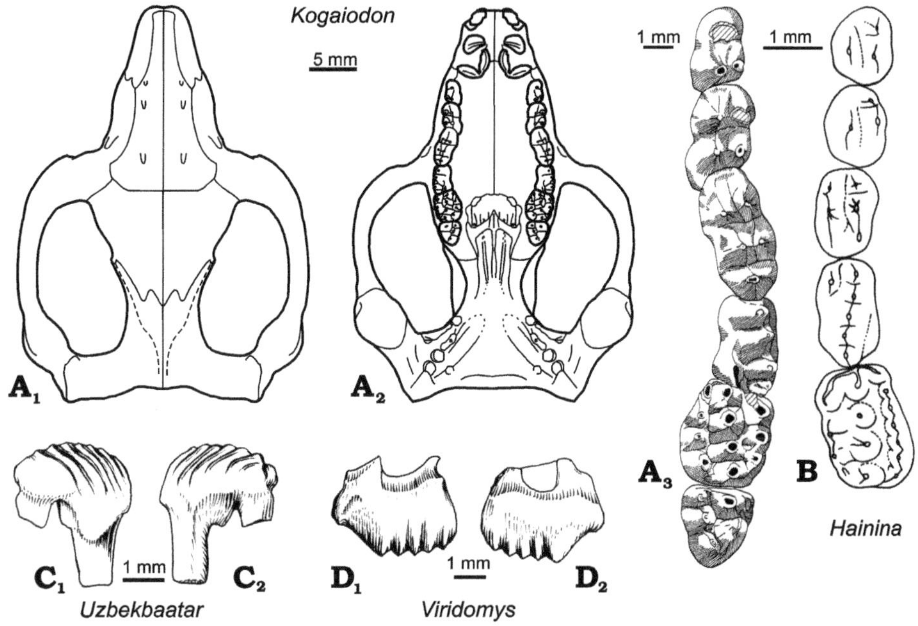

FIGURE 8.42. A, *Kogaionon ungureanui,* skull in dorsal (A_1) and ventral (A_2) views; left upper premolars and molars in occlusal view (A_3). B, Paleocene *Hainina belgica,* right P1–M1 in occlusal view. C, *Uzbekbaatar kizylkumensis,* right p4 in lingual (C_1) and labial (C_2) views. D, *Viridomys ferox,* P4 in labial (D_1) and lingual (D_2). *Source:* A_1, A_2, courtesy of Costin Rădulescu; A_3, modified from Rădulescu and Samson (1996); B, modified from Vianey-Liaud (1979); C, based on photographs in Kielan-Jaworowska and Nessov (1992); D, based on photographs in Fox (1971a).

Genus *Viridomys* Fox, 1971a
(figure 8.42D)

Diagnosis (based on Fox, 1971a, emended). Poorly known genus based on several isolated P4s of the type species, characterized by a high and narrow crown with a single row of cusps. Cusp formula is zero (two specimens) and five (one specimen).

Species. *Viridomys ferox* Fox, 1971, type species by monotypy.

Distribution. Late Cretaceous (early Campanian, Aquilan): Canada, Alberta, east of Milk River Village (upper part of Milk River Formation).

Comments. Fox (1971a) assigned *Viridomys* to Taeniolabidoidea (then a suborder) *incertae sedis.* Given the superfamily status of the group adopted herein, we leave *Viridomys* in a superfamily *incertae sedis.*

Superfamily Ptilodontoidea Sloan and Van Valen, 1965

Diagnosis. Dental formula 2.0.4.2/1.0.2.2. Apomorphies: normal prismatic enamel (except *Cimolodon,* which has gigantoprismatic enamel); slender lower incisor completely covered with enamel. In tentatively assigned *Neoliotomus* the lower incisor is robust, with limited enamel band. The p4 is very large, arcuate, strongly protruding dorsally over the level of the molars (apomorphy). Snout is wide, gently incurved in front of zygomatic arches. Share with Taeniolabididae small frontals pointed posteriorly (convergence) and differ in this respect from Djadochtatherioidea. I2 probably single-cusped, I3 placed on the margin of premaxilla. P4 elongated, in lateral view shaped as an isosceles triangle, protruding ventrally over the level of the anterior premolars and molars. Share with Plagiaulacidae, Eobaataridae, and the *Paracimexomys* group ornamentation of grooves and ridges on the molars and tendency of molar cusps to coalesce.

Families. Ptilodontidae Gregory and Simpson, 1926, type family; Neoplagiaulacidae Ameghino, 1890, Cimolodontidae Marsh, 1889b, and *Neoliotomus* Jepsen, 1930a (tentatively assigned to Ptilodontoidea, family *incertae sedis*).

Distribution. Late Cretaceous: North America and Europe; Paleocene-Eocene: North America, Europe, and Asia.

Family Ptilodontidae Gregory and Simpson, 1926

Synonym: Chirogidae Cope, 1887.

Diagnosis (based on Hahn and Hahn, 1983, emended). Differ from Neoplagiaulacidae and Cimolodontidae in structure of P4, which does not show a tendency for disappearance of the labial row of cusps. The shearing margin of p4 is moderately vaulted, less protruding dorsally over the level of the molars than in Cimolodontidae. Two pairs of vascular nasal foramina are present.

Genera. *Ptilodus* Cope, 1881b, type genus (Paleocene), and three other Paleocene genera not discussed herein. *Kimbetohia,* also a Paleocene genus, has been reported from the Late Cretaceous of New Mexico (Clemens, 1973b; Lillegraven and McKenna, 1986). We figure Paleocene *Ptilodus* for comparison in figures 8.9 and 8.39G.

Distribution. ?Late Cretaceous (?Judithian–late Paleocene): North America.

Genus *Kimbetohia* Simpson, 1936
(figure 8.43E)

Diagnosis (based on Simpson, 1936, emended). Genus known only from the type species, of very small size and represented by upper premolars. Differs from *Ptilodus* in having three-cusped P2, P3 slightly less reduced than in *Ptilodus,* and P4 shorter and broader, with second row long but with only about half as many cusps as the inner row. P4 cusp formula in *K. campi* 7:4:1.

Species. *Kimbetohia campi* Simpson, 1936, type species and "*Kimbetohia* n. sp." cited but not described by Van Valen and Sloan (1965) from the Hell Creek Formation (we follow other compendia in omitting this record from our distribution listing).

Distribution. Late Cretaceous and Paleocene: United States, New Mexico; early Paleocene: North America.

Family Neoplagiaulacidae Ameghino, 1890

Synonyms: Ectypodidae Sloan and Van Valen, 1965, modified to Ectypodontidae by Van Valen and Sloan, 1965.

Diagnosis (based on diagnoses of Sloan and Van Valen, 1965; and Hahn and Hahn, 1983). Small Ptilodontoidea, with p4 lower than in Cimolodontidae and Ptilodontidae; ratio between lengths of p4 and m1 ranging from 1.4 to 2.0. P4 shows a tendency to disappearance of the labial row of cusps.

Genera. *Neoplagiaulax* Lemoine, 1882, type genus (Paleocene genus not described herein); *Mesodma* Jepsen, 1940, *Parectypodus* Jepsen, 1930a, and eight Tertiary genera, not listed herein.

Distribution. Late Cretaceous–late Eocene: North America, late Paleocene–early Eocene: Europe; early Eocene: Asia.

Genus *Mesodma* Jepsen, 1940
(figure 8.43C,D)

Diagnosis (based on Hahn and Hahn, 1983, emended). Most primitive member of the Neoplagiaulacidae, with p3 present. Differs from the Tertiary representatives of the family in having p4 lower crowned. In P4 the posterior

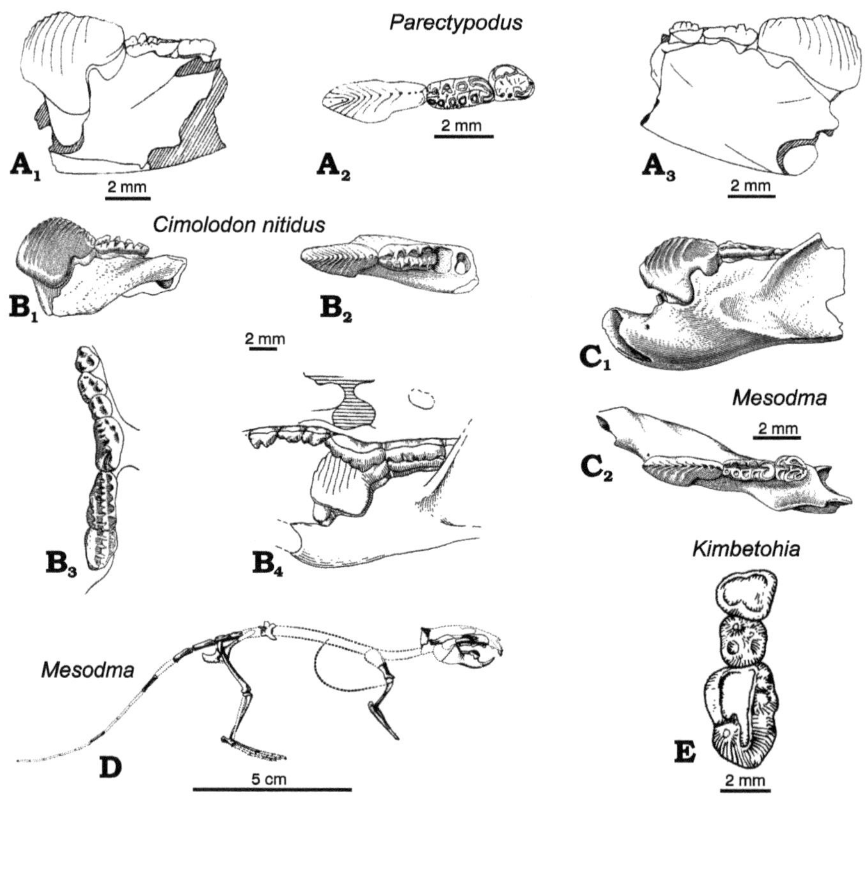

FIGURE 8.43. North American Late Cretaceous ptilodontoidean genera. A, *Parectypodus foxi,* fragment of left dentary in labial (A_1), occlusal (A_2), and lingual (A_3) views. B, *Cimolodon nitidus,* fragment of left dentary containing p4 and m1 in labial (B_1) and occlusal (B_2) views; reconstruction of part of left maxilla with premolars and molars in occlusal view, anterior is up (B_3); reconstruction of parts of left dentition in occlusion in labial view (B_4). C, *Mesodma formosa,* left dentary with root of p3, complete p4, m1–m2, in labial (C_1) and occlusal (C_2) views. D, Reconstruction of the skeleton of *Mesodma thompsoni.* E, *Kimbetohia campi,* right P2–P4, in occlusal view. *Source:* modified from: A, Storer (1991); B, C, Clemens (1963b); D, Sloan and Van Valen (1965); E, Simpson (1936b).

cusps of the medial row are the highest, but the shearing specializations of this tooth are less developed than in later genera (e.g., *Ectypodus*). The p4 has 9–15 serrations; length varies between 2.7 and 4.7 mm.

Species. *Mesodma ambigua* Jepsen, 1940, type species; *M. formosa* (Marsh, 1889b), *M. garfieldensis* Archibald, 1982, *M. hensleighi* Lillegraven, 1969, *M. primaeva* (Lambe, 1902), *M. senecta* Fox, 1971a, *M. thompsoni* Clemens, 1963b, several taxa left in open nomenclature by Clemens (1963b), Fox (1971a), Novacek and Clemens (1977), Armstrong-Ziegler (1978), Archibald (1982), and Storer (1991), and other species confined to the Tertiary.

Distribution. Late Cretaceous (Aquilan-Lancian)–late Paleocene: North America.

Genus *Parectypodus* Jepsen, 1930a
(figure 8.43A)

Diagnosis (based on Sloan, 1981, emended). Neoplagiaulacid with height of the first serration in p4 above 0.45 of tooth length (see Sloan, 1981: figure 6.1 for additional explanation). Labial height of enamel at the anterior triangular lobe above the anterior root is approximately equal to or greater than the tooth length; occlusal surface of the molars and anterior face of p4 meet at a right or usually acute angle. The p4–m1 length ratio is 1.8–2.0 (high for Neoplagiaulacidae). P4 cusp formula 2–5:7:9 (additional labial cusps occur in advanced species); length of P4 80–85% of p4 length.

Species. *Parectypodus simpsoni* Jepsen, 1930a, type species; *P. foxi* Storer, 1991, at least seven Tertiary species, and several others left in open nomenclature.

Distribution. Late Cretaceous (Lancian): Canada; Paleocene–early Eocene: North America.

Family Cimolodontidae Marsh, 1889a

Diagnosis (based on Hahn and Hahn, 1983, emended). Medium-size Ptilodontoidea that differ from Neoplagiaulacidae and Ptilodontidae by structure of p4, with extremely high crown, strongly vaulted upper margin, highest in the middle of the tooth length; and by having gigantoprismatic enamel. Number of serrations on p4 reaches 15. P4 forms a high isosceles triangle in lateral view, P4 and p4 together forming a strong shearing apparatus.

Genera. *Cimolodon* Marsh, 1889a, type genus; Cimolodontidae gen. and sp. undetermined (Eaton, 1995); and one (Jepsen, 1940) Paleocene genus not described herein.

Distribution. Late Cretaceous (Cenomanian-Lancian)–late Paleocene: North America.

Genus *Cimolodon* Marsh, 1889a
(figure 8.43B)

Diagnosis. Cimolodontid with molar cusp formula M1, 5–7:7–8:3–7; M2, 1–5:3–6:4–6; m1 4–8:4–6; m2, 4–6:2–3; and 12–15 serrations on p4.

Species. *Cimolodon nitidus* Marsh, 1889a, type species; *C. electus* Fox, 1971a, *C. similis* Fox, 1971a, and several taxa left in open nomenclature by Simpson (1927b), Szalay (1965), Sahni (1972), Armstrong-Ziegler (1978), Eaton and Cifelli (1988), and Eaton (1995).

Distribution. Late Cretaceous (Cenomanian–Lancian): North America.

Multituberculata *incertae sedis*
(figure 8.44)

We assign to Multituberculata *incertae sedis* several upper premolars and an incomplete dentary with a premolar from the Campanian (or lower Maastrichtian) Los Alamitos Formation of Argentina. Krause, Kielan-Jaworowska, and Bonaparte (1992: figure 2C,D) illustrated upper premolars, assigned tentatively to the gondwanatherian genus *Ferugliotherium*, believed at that time to belong to multituberculates. One of these teeth, illustrated here in figure 8.44A, has structure characteristic of the anterior upper premolars of multituberculates, with four cusps in the lingual row and two in the labial row. The dentary with a premolar also tentatively assigned by Kielan-Jaworowska and Bonaparte (1996) to *Ferugliotherium* shows the alveolus for an incisor, a diastema, and a single bladelike premolar of multituberculate pattern

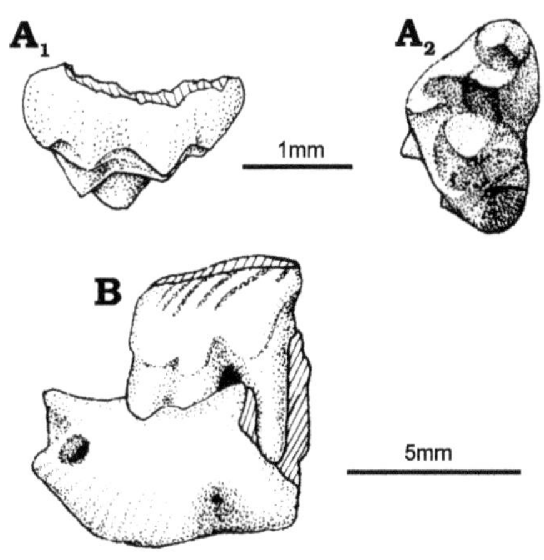

FIGURE 8.44. Multituberculata *incertae sedis* from the Late Cretaceous of Argentina. A, ?Right upper premolar (P3?) in labial (A_1) and occlusal (A_2) views. B, Fragment of a left dentary with alveolus for incisor and bladelike premolar (p4?) in labial view. *Source:* B, modified from Kielan-Jaworowska and Bonaparte (1996).

covered on labial and lingual sides with oblique ridges (figure 8.44B). The p4 differs from that of Cimolodonta in apparently being rectangular rather than arcuate in labial view and shares this character with "Plagiaulacida." As Pascual et al. (1999) demonstrated that Gondwanatheria have four molarized lower teeth (see chapter 14) and no bladelike premolars, the dentary in question cannot belong to that suborder. Therefore, we remove this dentary and tentatively also upper premolars from Gondwanatheria and assign them to Multituberculata *incertae sedis.* This suggests that multituberculates might have lived during the Late Cretaceous in South America, but their radiation there is as yet poorly known.

Distribution. Late Cretaceous (Campanian or early Maastrichtian, Alamitian): Argentina, Patagonia, southeastern Río Negro Province, Los Alamitos (Los Alamitos Formation).

RELATIONSHIPS WITHIN MULTITUBERCULATA (figure 8.45)

Although Multituberculata are the best-known group of Mesozoic mammals, there are many unresolved problems concerning their origin (see chapter 15) and phylogeny. Kielan-Jaworowska and Hurum (2001) discussed the algorithm-based cladistic analyses of Multituberculata (e.g., Simmons, 1993; Rougier et al., 1997) as well as their own attempts at analysis (Kielan-Jaworowska and Hurum, 1997, 2001) and demonstrated that these analyses did not yield results that were robust or well resolved. They offered a taxon-character matrix, choosing Haramiyoidea (see also Hahn, 1973; Sigogneau-Russell, 1989b; Butler and MacIntyre, 1994; Butler, 2000) as an outgroup, and constructed a handmade cladogram, which we show here as figure 8.45. We base our considerations of multituberculate interrelationships on the conclusions of Kielan-Jaworowska and Hurum (2001); the text that follows (which should be read in conjunction with figure 8.45) is a summary of their paper.

One of the most important unresolved problems of multituberculate systematics involves the interrelationships among the families of "Plagiaulacida." Kielan-Jaworowska and Hurum (2001) recognized within "Plagiaulacida" three informal lines, referred to as allodontid, paulchoffatiid, and plagiaulacid (see earlier section on "Systematics").

Figure 8.45 indicates that "Plagiaulacida" are a grade characterized by three upper incisors, lower incisor completely covered with enamel (except *Glirodon* and *Eobaatar*), five upper premolars (with the exception of two paulchoffatiid genera), and four (three in advanced forms) bladelike lower premolars, rectangular or triangular in lateral view (figure 8.9). Bladelike lower premolars are an apomorphy of Multituberculata and differentiate them from other mammals (including the Haramiyidae), except for a few marsupials (homoplasy).

"Plagiaulacida" include eight families and *Glirodon,* placed by Engelmann and Callison (1999) in "Plagiaulacida" *incertae sedis.* It is difficult to decide whether the allodontid or paulchoffatiid line is more plesiomorphic, as different mixtures of plesiomorphic and derived characters characterize both lines.

The allodontid line is characterized by characters 2^1, 3^1, 4^1, 5^1, 29^1, 30^1 in figure 8.45. We follow Kielan-Jaworowska and Hurum (2001) in tentatively accepting that members of the allodontid line might be the most plesiomorphic multituberculates, as they retain plesiomorphic structure of the lower molars with two rows of well-separated cusps, smooth enamel (lack of grooves and ribbing on the molars), and a small I3 (see Simpson, 1929a; Bakker and Carpenter, 1990; Carpenter, 1998; Engelmann and Callison, 1999, and figure 8.25A–F). They all retain five upper and four lower premolars. They are, however, more advanced than members of the paulchoffatiid line in the structure of the lower premolars and in having a larger angle between the tooth row and the longitudinal axis of the dentary (as seen in dorsal view), which is very low (7°) in at least one paulchoffatiid genus, *Paulchoffatia,* which in this respect is the most plesiomorphic among the multituberculates (figure 8.27A_1). The Allodontidae also have a slightly younger stratigraphic range occurrence than the Paulchoffatiidae (Simpson, 1928a, 1929a; Hahn, 1969, 1993; Kielan-Jaworowska and Ensom, 1992; Carpenter et al., 1998; Engelmann and Callison, 1998, and chapter 2).

Upper premolars are smooth in *Ctenacodon* and *Glirodon,* but ornamented in *Psalodon* (Simpson, 1929a: pl. 5, figure 2). The labial cusps on m1 in *Ctenacodon serratus* (figure 8.25E), and on m1 and m2 of *Ctenacodon scidens* (figure 8.25F) are strongly worn and it cannot be demonstrated whether they were coalesced or not. However, in *Ctenacodon* sp. figured by Simpson (1929a: pl. 1, figure 3) both labial and lingual cusps are well preserved and separated.

Upper incisors are unknown in *Ctenacodon.* In *Psalodon,* I1 (not preserved) is small, I2 is enlarged and two-cusped, I3 is small with a strong main cusp and a minute basal cuspule (figure 8.25B). In *Glirodon* I3 is single-cusped (Kielan-Jaworowska and Hurum, 2001: plate 1, figure 2). Simmons (1993) and Rougier et al. (1997) scored I3 of *Psalodon* together with *Paulchoffatia* and *Bolodon* as multicusped. This is misleading as the small I3 in *Psalodon* is very different from the enlarged, three- or four-cusped I3 in Paulchoffatiidae and Plagiaulacidae. Rougier et al.

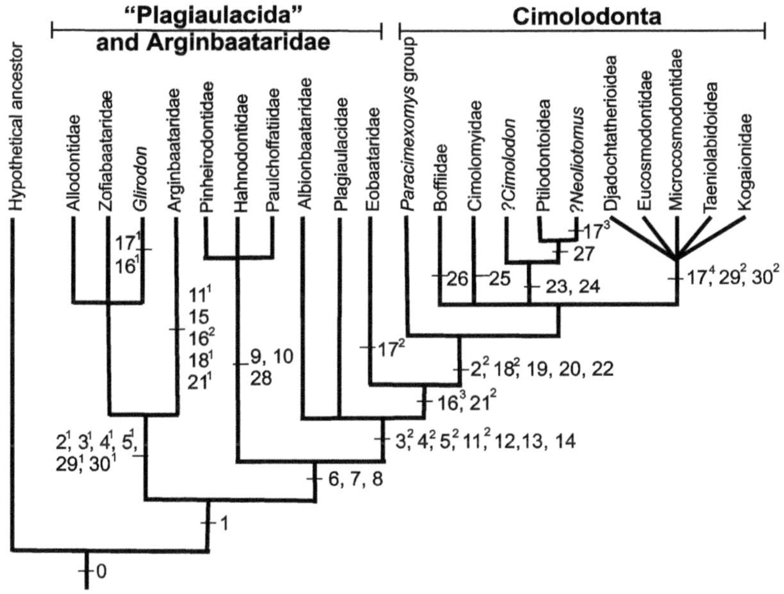

FIGURE 8.45. Cladogram showing hypothetical interrelationships of Multituberculata. The numbers show distribution of selected characters. If a character appears twice or several times, the same number is repeated, marked by different indices, for example, 3^1, 3^2, and so on. The Arginbaataridae, which are a very specialized side branch of Multituberculata, have been placed together with "Plagiaulacida" based on technical reasons. For characters defining all of the groups, see "Systematics." *Source:* from Kielan-Jaworowska and Hurum (2001).

0. Hypothetical ancestor at allotherian non-multituberculate grade of evolution. Three upper incisors; lower molars with two parallel rows of cusps of different height; lower canine present, four lower premolars not modified as "blades," molar cusps separate; enamel surface smooth, more than one lower incisor. Propalinal jaw movement not developed or incipient.
1. Containing all except ancestor.
2^1, 2^2. I3 small and single-cusped, or with main cusp and basal cuspule (allodontid line and Cimolodonta, homoplasy or common ancestry?).
3^1, 3^2. Loss of labial cusps on p3 (plagiaulacid and allodontid lines, ?homoplasy).
4^1, 4^2. The p4 increased in size with respect to p3 (plagiaulacid and allodontid lines, ?homoplasy, more strongly expressed in plagiaulacid than in allodontid line).
5^1, 5^2. Incipient lingual ridge on M1 (plagiaulacid and allodontid lines, ?homoplasy).
6. I3 enlarged, with three or four cusps (paulchoffatiid and plagiaulacid lines).
7. The m2 with coalesced labial cusps (paulchoffatiid and plagiaulacid lines).
8. Incipient ornamentation of molar enamel (paulchoffatiid and plagiaulacid lines).
9. I3 roughly quadrangular or trapezoid in occlusal view, three- or four-cusped, with main cusps arranged obliquely forming a ridge (paulchoffatiid line).
10. The m2 basinlike (paulchoffatiid line) differing in details in three families (see diagnoses of Paulchoffatiidae, Hahnodontidae, and Pinheirodontidae in "Systematics").
11^1, 11^2. Anterior lower premolars roughly triangular in labial view (plagiaulacid line and Arginbaataridae, homoplasy).
12. Enlarged I3 (smaller than in Paulchoffatiidae), roughly triangular in occlusal view, with three cusps (plagiaulacid line).
13. Tendency of molar cusps to coalesce (retaining two-row structure of m2), more strongly expressed in the lowers than in the uppers (Plagiaulacidae, Eobaataridae, *Paracimexomys* group; less strongly expressed in Ptilodontoidea)
14. Ornamentation of grooves and pits on the molars (Plagiaulacidae, Eobaataridae, *Paracimexomys* group; less strongly expressed in Ptilodontoidea, Cimolomyidae, and Boffiidae).
15. Unusually large p4, with limited enamel, rotating during the ontogeny over p2 and p3, which are ultimately shed (arginbaatarid autapomorphy).
16^1, 16^2, 16^3. Gigantoprismatic enamel (appearing ?independently in *Glirodon, Arginbaatar, Eobaatar* and characteristic also of Cimolodonta, except for the Ptilodontoidea).
17^1, 17^2, 17^3, 17^4. Limited enamel band on lower incisor (appearing independently at least four times, in *Glirodon, Eobaatar, Neoliotomus,* and five groups of nonptilodontoidean Cimolodonta, possibly separately in each of them).
18^1, 18^2. Arcuate p4 (appearing independently at least twice, homoplasy, in Cimolodonta and Arginbaataridae).
19. Loss of I1 (apomorphy of Cimolodonta).
20. Loss of P4 (apomorphy of Cimolodonta).
21^1, 21^2. Loss of p1 (Arginbaataridae, Eobaataridae, and a few other plagiaulacid genera, the latter not shown in the cladogram, and Cimolodonta, homoplasy; between Eobaataridae and Cimolodonta possibly due to common ancestry).
22. Transformation of p3 into a peglike nonfunctional tooth (apomorphy of Cimolodonta).
23. Increase in size of p4, which becomes strongly arcuate with numerous ridges and protrudes dorsally over the level of the molars (apomorphy of Ptilodontoidea).
24. Lower incisor very gracile (apomorphy of Ptilodontoidea, except *Neoliotomus,* tentatively assigned).
25. Molar cusps pyramidal to crescentic, correlated with reduction of upper premolars and enlargement of the molars, enamel grooves only on m1 (apomorphies of Cimolomyidae).
26. Upper molars strongly enlarged with three rows of cusps; strong ribbing of upper premolars and molars, weak on m1, grooves absent (apomorphies of Boffiidae).
27. Acquisition of normal prismatic enamel (apomorphy of Ptilodontoidea, which appears once after the separation of *Cimolodon,* tentatively assigned, from the main ptilodontoidean line).

We also recognize three plesiomorphic characters:

28. Plesiomorphic structure of p3, with labial cusps present (paulchoffatiid line).
29^1, 29^2, 29^3. Lower molar cusps separated (not coalesced) (allodontid line, Arginbaataridae, and five cimolodontan groups in polytomy).
30^1, 30^2, 30^3. Molar enamel smooth (allodontid line, Arginbaataridae, and five cimolodontan families in polytomy).

(1997) also mistakenly scored I3 in *Kuehneodon* (figure 8.8B), *Henkelodon* (figure 8.30A), and *Bolodon* (figure 8.34A) as single-cusped.

Allodontidae share with Plagiaulacidae the loss of the labial cusps on p3 (homoplasy, number 4 in figure 8.45), an increase in the length of p4 (figure 8.9), and an incipient posterolingual ridge on M1 (figure 8.25A).

Zofiabaatar Bakker and Carpenter, 1990, assigned by Bakker (1992) to the monotypic family Zofiabaataridae (see also Bakker, 1998) is a specialized genus from the Morrison Formation, known only from a dentary with p1–p4 and m1 (figure 8.25G). Carpenter (1998) assigned *Zofiabaatar* to Plagiaulacidae. However, lower molars with coalescing cusps characterize Plagiaulacidae, whereas in *Zofiabaatar* cusps are well separated, as in Allodontidae. Therefore we follow Kielan-Jaworowska and Hurum (2001) and assign Zofiabaataridae to the allodontid line.

Engelmann and Callison (1999) assigned *Glirodon* (characters 16^1, 17^1 in figure 8.45) to Plagiaulacoidea *incertae sedis*. *Glirodon* retains the plesiomorphic "plagiaulacidan" dental formula and shares with Allodontidae the structure of the upper premolars and molars (figure 8.25). It differs from Paulchoffatiidae and Plagiaulacidae in having a single-cusped I3. It is more advanced than *Ctenacodon* in the structure of M2 (with three rows of cusps), but less advanced in lack of incipient posterolingual ridges on M1. The gigantoprismatic enamel (unique for multituberculates among mammals, see Fosse et al., 1985, and Carlson and Krause, 1985) and lower and upper incisors with a limited enamel band appear in *Glirodon* for the first time in multituberculate evolution. *Glirodon* shares these characters with Early Cretaceous Mongolian *Eobaatar*, which we refer to the plagiaulacid line, and with numerous Cimolodonta (see later).

The paulchoffatiid and plagiaulacid lines appear to be more closely related to one another (numbers 6–8 in figure 8.45) than either is to the allodontid line. The shared derived characters of the two lines are enlarged I3 with three or four cusps, the structure of m2 with no labial cusps, and ornamentation of at least the lower molars, but less constant and only incipient in the paulchoffatiid than the plagiaulacid line. These three character states differentiate both lines from the allodontid line. However, in spite of the general similarity, the details of I3, p3, and m2 structure, as well as ornamentation, are different in paulchoffatiid and plagiaulacid lines (characters 9, 10, and 28 in figure 8.45).

Members of the paulchoffatiid line (as exemplified by Paulchoffatiidae) are more plesiomorphic in the structure of the dentary and lower premolars than those of the allodontid line. They retain plesiomorphic molar cusps of different heights, but at the same time show specialized features in the structure of the upper incisors and lower molars. The autapomorphies of this line include the almost complete coalescence of cusps on m2, producing a basinlike tooth with only a single cusp in Paulchoffatiidae, and enlarged three- or four-cusped I3, roughly quadrangular or trapezoidal in occlusal view. Both m2 and I3 differ in details from other families of this line and from Plagiaulacidae (see "Systematics" for details). Hahn (1971) accepted that small I3 is plesiomorphic for multituberculates, whereas an enlarged I3 is apomorphic for the plagiaulacid and paulchoffatiid lines, which we accept. However, this view has been challenged by Van Valen (1976), who argued on the basis of isolated, presumably haramiyid incisors (assumed to be the plesiomorphic sister taxon of Multituberculata) described by Peyer (1956) and Parrington (1947) that multicusped I3 may be plesiomorphic for Multituberculata. Among isolated incisors attributed to Haramiyidae, Sigogneau-Russell (1989b) described two types considered as possible uppers, one of which was two-cusped and the other three-cusped, but of simpler structure than in the Paulchoffatiidae. On the other hand, the upper incisors in *Haramiyavia* (assigned by Jenkins et al., 1997, to Haramiyidae and by Butler, 2000, to his new family Haramiyaviidae within the suborder Haramiyoidea Hahn, 1973) are all single-cusped. Therefore, it seems to us more probable that I3 single-cusped or with a basal cuspule is plesiomorphic for Multituberculata.

Paulchoffatiidae are plesiomorphic in having a p3 with a row of labial cusps (absent in other families) and a crown almost as long as that of p4 (figure 8.9) and, as discussed above, a small angle (in *Paulchoffatia*) between the tooth row and the longitudinal axis of the dentary. Molars and premolars may show complicated ornamentation. Hahn (1969, 1971) argued that the basinlike structure of m2 is apomorphic for Paulchoffatiidae (the whole paulchoffatiid line in our division) and excludes Paulchoffatiidae from the ancestry of Cimolodonta. We follow this opinion (see also Hahn, 1993) and regard Paulchoffatiidae as a monophyletic dead-end of "Plagiaulacida." (See "Systematics" for a description of the two other families of the paulchoffatiid line.)

Members of the plagiaulacid line (characters 3^2, 4^2, 5^2, 11^2, 12, 13, 14 in figure 8.45), Plagiaulacidae, Eobaataridae (figure 8.9), and tentatively assigned Albionbaataridae, are more advanced than those of the paulchoffatiid line. Plagiaulacidae share with Paulchoffatiidae the basic similarities in structure of I3 and m2, ornamentation of the enamel, incipiently developed in some Paulchoffatiidae (Simpson, 1928a, 1929a; Hahn, 1969, 1993; Kielan-Jaworowska, Dashzeveg, and Trofimov, 1987; Kielan-Jaworowska and Ensom, 1992), but differ in having several apomorphies (see "Systematics"). The plagiaulacid I3 is roughly triangular in occlusal view, with three cusps and the absence of the oblique ridge characteristic of the Paulchoffatiidae. They share with the paulchoffatiid line the coalescence of labial cusps on m2, but differ in having separated lingual cusps and a longitudinal valley along the tooth (figures 8.35D_2, 8.36A_3,B_1), rather than having a basinlike m2. A tendency of molar cusps to coalesce and distinct ornamentation of molar enamel with grooves, pits, and ridges, incipient in Paulchoffatiidae, is well expressed in Plagiaulacidae, and we regard it as characteristic of the plagiaulacid line (figure 8.36A,B). Another characteristic feature is the strong ribbing of the upper premolar cusps and cuspules (see, e.g., Kielan-Jaworowska and Ensom, 1992: plate 4, figures 8, 9), which is shared with some Paulchoffatiidae (e.g., *Kielanodon,* see Hahn, 1987b). Plagiaulacidae also share with Paulchoffatiidae a robust lower incisor and a low position of the dentary condyle.

Eobaataridae (figure 8.9 and characters 16^3, 17^2 in figure 8.45) are closely allied with Plagiaulacidae as indicated by the structure of molars, which have coalescing cusps and ornamentation in the form of grooves and ribbing (figure 8.35D_2, 8.36B); upper premolars are similar in shape and in the presence of strong ribbing (Kielan-Jaworowska, Dashzeveg, and Trofimov, 1987: plates 4, 5). They differ from Plagiaulacidae by several apomorphies (see "Systematics" for a description of Eobaataridae and Albionbaataridae).

We leave the strongly specialized Arginbaataridae (figure 8.9 and characters 11^1, 15, 16^2, 18^1, 21^1 in figure 8.45), endemic to the Early Cretaceous of Mongolia, as suborder *incertae sedis* because they show a mixture of characters of the "Plagiaulacida" and Cimolodonta (Trofimov, 1980; Kielan-Jaworowska, Dashzeveg, and Trofimov, 1987, see also earlier section on "Systematics"). *Arginbaatar* as seen in figure 8.45 shares several characters with members of the allodontid line, and may have had its origin within this group.

Cimolodonta (characters 2^2, 18^2, 19, 20, 22 in figure 8.45), which embrace some Early Cretaceous (Aptian–Albian) and all Late Cretaceous and Tertiary multituberculates, have prismatic enamel. They might be monophyletic, as argued by Simmons (1993), but see Archibald (1982) and Hahn and Hahn (1999a) for alternative opinions. They are well defined by at least five apomorphies (see "Systematics"). Their origin remains mysterious. It has not been demonstrated with any certainty which of the five upper premolars characteristic of "Plagiaulacida" disappeared in the line leading to Cimolodonta. Clemens (1963b) speculated that it was P4, while Hahn (1978c: figure 10) suggested that it was rather P1. In favor of Hahn's hypothesis, one may argue that the shortening of the cheek tooth row in mammals usually begins either from the anterior or from the posterior end. Hahn (1977a) demonstrated that in *Kuehneodon,* a member of the Paulchoffatiidae, P1 has been lost, as there is a diastema between C and P2 (figure 8.8). However, Paulchoffatiidae, as argued by Hahn (1993) and accepted herein, are not in the evolutionary line leading to Cimolodonta. On the other hand, there are examples in various mammal groups of loss of a tooth in the middle of the premolar series. Cifelli (2000a) argued that in early eutherian mammals, which had five premolars, it was the middle premolar, referred to by him as px (figure 1 in his paper), that was lost. Similarly in some Djadochtatherioidea (e.g., *Catopsbaatar* and *Tombaatar*) with three upper premolars, it was P2 and not P1 that disappeared, leaving a short diastema between P1 and P3. This loss of a tooth in the middle of the tooth row in these taxa gives additional support to Clemens's (1963b) hypothesis, which is accepted here (figure 8.8). Advanced Cimolodonta differ from "Plagiaulacida" in the mode of occlusion (see "Occlusion" earlier), but members of the *Paracimexomys* group (discussed later) and at least some Ptilodontoidea retained the "plagiaulacidan" mode of occlusion.

The most plesiomorphic of Cimolodonta (and apparently paraphyletic) is a group of poorly known Aptian–Albian to Maastrichtian, mostly North American genera, referred to by Archibald (1982) and Eaton and Cifelli (2001) as the *Paracimexomys* grade, and which Kielan-Jaworowska and Hurum (2001) informally called the *Paracimexomys* group. Members of this group, *Paracimexomys* Archibald, 1982, and related forms, are known almost exclusively from isolated teeth (Lillegraven, 1969; Sahni, 1972; Archibald, 1982; Eaton and Cifelli, 1988, 2001; Eaton and Nelson, 1991; Eaton, 1995; Cifelli, 1997a). However, *Cimexomys,* which we only tentatively assign to this group, is known from almost complete upper and lower dentitions, fragments of the skull, and the dentary (Montellano et al., 2000). It has four upper premolars of cimolodontan pattern, a slender lower incisor completely covered with enamel, p4 not protruding dorsally over the level of the molars, molar cusps (especially in m2) showing a tendency to coalesce, and cusps covered with grooves. *Paracimexomys* and *Dakotamys,* assigned here, resemble Eobaataridae in the structure of the upper and lower molars, with cusps showing a tendency to coalesce and with ornamentation of grooves and ribs on the molars (figure 8.36D,E). Another genus, *Bryceomys,* shares with the two former genera the characteristic ornamentation of the molars and coalescence of cusps on m2, but differs in having the cusps of M2 separated (figure 8.36C).

Members of this group apparently differ from Eobaataridae in having a slender lower incisor uniformly covered with enamel or with enamel thinner on the lingual side but not sharply limited (Eaton, 1995), and share with the Eobaataridae gigantoprismatic enamel (where known). The similarities of isolated molars (especially M2) of *Paracimexomys, Bryceomys,* and *Dakotamys* to *Eobaatar,* are striking (figure 8.36). However, the genera of the *Paracimexomys* group differ from *Eobaatar* and other members of the plagiaulacid line in having an arcuate rather than a rectangular p4 and a cimolodontan dental formula. Ornamentation of molars and the molar cusp arrangement (especially on M2) of the members of *Paracimexomys* group are also similar to those of Ptilodontoidea, discussed in the next section.

On the other hand, the genera of the *Paracimexomys* group share a small number and similar arrangement of cusps on M1 with some Djadochtatherioidea, for example, with *Kryptobaatar, Sloanbaatar,* and *Chulsanbaatar* (see, e.g., figures 8.36C_1, I_1). Also P4s of members of this group, figured by Eaton (1995), are strikingly similar to P4s in Djadochtatherioidea (e.g., Kielan-Jaworowska, 1970a, 1974a; Kielan-Jaworowska and Hurum, 1997). It follows that the *Paracimexomys* group embraces forms intermediate in some respects between the plagiaulacid line and at least some Cimolodonta.

Origin of Ptilodontoidea, Cimolomyidae, and Boffiidae from members of the "Plagiaulacida" (via the *Paracimexomys* group) appears highly probable. Ptilodontoidea are the only Cimolodonta that share with the *Paracimexomys* group the coalescence of molar cusps associated with grooves, which is also characteristic of the plagiaulacid line (see figures in Fox, 1971a; Krause, 1977; Vianey-Liaud, 1986; and many others; discussion in Krause, Kielan-Jaworowska, and Bonaparte, 1992, and our figure 8.36F–H). *Cimolodon* seems to be more closely related to the *Paracimexomys* group than are other ptilodontoideans (compare, e.g., M1 of *Cimolodon electus* in Fox, 1971a: figure 5 with our figure 8.36E). Cimolodontidae, to which Hahn and Hahn (1983) refer *Cimolodon, Anconodon,* and *Liotomus,* are assigned to Ptilodontoidea (but see comments on *Cimolodon* later), with which they share a very high p4, protruding dorsally over the level of the molars, and a lower incisor completely covered with enamel but less slender than in other Ptilodontoidea (Clemens, 1963b; Lillegraven, 1969). *Cimolodon* differs from other cimolodontid genera in retaining gigantoprismatic enamel, which is a more plesiomorphic character than the normal prismatic type (Carlson and Krause, 1985; Kielan-Jaworowska and Hurum, 2001, *contra* Wood and Stern, 1997). Kielan-Jaworowska and Hurum (2001) suggested that Ptilodontoidea acquired normal prismatic enamel (number 27 in figure 8.45) after separation of *Cimolodon* (which we only tentatively assign to Ptilodontoidea) from the main ptilodontoidean line. Eaton and Cifelli (2001) went further and suggested that Cimolodontidae may have arisen independently of ptilodontoids from some form within the *Paracimexomys* group close to *Bryceomys* and *Cedaromys,* which have a relatively high p4 (see figures 9A,B and 14 in their paper).

In all "Plagiaulacida" the cutting edge of p4 is above the level of the molars (figure 8.9). Among Cimolodonta, only in the Ptilodontoidea does the p4 protrude dorsally over the level of the molars; however, it differs considerably from that in "Plagiaulacida," where it is rectangular in lateral profile. The arcuate, strongly enlarged p4, with numerous ridges and serrations is an apomorphy of Ptilodontoidea (character 23 in figure 8.45). The presence of uniform enamel on the lower incisor is a plesiomorphic character, albeit the gracile incisor is an apomorphy of Ptilodontoidea (character 24 in figure 8.45).

On the basis of the foregoing comparisons, we suggest that Ptilodontoidea originated from among the plagiaulacid line, the intermediate links being forms close to Eobaataridae and *Paracimexomys* group. *Eobaatar* apparently had a limited enamel band on the lower incisor and therefore cannot be a direct ancestor of Ptilodontoidea. However, as a limited enamel band made its appearance several times in multituberculate evolution, one can imagine the existence of forms close to eobaatarids with their incisors completely covered with enamel, especially insofar as such incisors are characteristic of the *Paracimexomys* group (Eaton, 1995; Montellano et al., 2000).

The similarities of the *Paracimexomys* group to *Eobaatar,* on the one hand, and Ptilodontoidea (*Cimolodon*), on the other, have also been recognized by Eaton (1995), who noted that members of this group share an alternating arrangement of cusps on M1 with *Eobaatar.* With respect to similarity to Ptilodontoidea he stated p. 782): "*Dakotamys* could be considered morphologically transitional between *Paracimexomys* and *Cimolodon.* Both *Dakotamys* and *Cimolodon* have a pocket labial to the anteriormost cusps of the labial row of M1, *Dakotamys* shows some tendency toward the ribbed pyramidal cusps and pitted valleys found in some species of *Cimolodon,* and both taxa have straight central valleys on M1s."

Neoliotomus Jepsen, 1930a, previously assigned to Eucosmodontidae (see Krause, 1982b), shares with Ptilodontoidea normal prismatic enamel and p4 protruding dorsally over the level of the molars. It differs from other Ptilodontoidea in having a very robust lower incisor with a limited enamel band (character 17^3 in figure 8.45). But as such a band made its appearance several times in multituberculate evolution, it is possible that it also appeared in a ptilodontoid. On this basis we assign *Neoliotomus* tentatively to Ptilodontoidea.

Another genus with normal prismatic enamel, which was originally assigned to Eucosmodontidae is *Xyronomys*

Rigby, 1980. It is incompletely known, represented only by p4 and P4. Its eucosmodontid attribution was based on the presence of a low, roughly triangular p4. Because of its possession of normal prismatic enamel we assign it to Neoplagiaulacidae within Ptilodontoidea.

Cimolomyidae (character 25 in figure 8.45), of which the best known is *Meniscoessus*, approach Ptilodontoidea in having the lower incisor completely covered with enamel, albeit in Cimolomyidae the incisor is robust (which is a plesiomorphic character) rather than slender as in Ptilodontoidea. Another similarity is ornamentation; for example, the upper and lower molars of *Meniscoessus* (Sahni, 1972; Fox, 1976b, 1980a) show heavily ribbed cusps, with groove-like valleys between the ribs. The same applies to *Cimolomys* (Archibald, 1982). The tentatively assigned *Essonodon* differs from other multituberculates in having transverse ridges between the molar cusps and shares this character only with a gondwanatherian *Ferugliotherium* (Krause, Kielan-Jaworowska, and Bonaparte, 1992). Cimolomyidae differ from Ptilodontoidea in having differently built upper dentition with strongly reduced upper premolar row and crescentic molar cusps. The p4 is small and low, very different from the enlarged p4 in Ptilodontoidea, and instead resembles that in Djadochtatherioidea (figure 8.9). We tentatively assign the incompletely known Asian *Buginbaatar* to Cimolomyidae (Kielan-Jaworowska and Sochava, 1969; Trofimov, 1975, see also our figure 8.9). As the Cimolomyidae are generally poorly known, their origin from members of the *Paracimexomys* group (with which they share ornamentation) remains uncertain.

Boffius (character 26 in figure 8.45), a relatively large taxon with upper molars bearing three rows of cusps, was originally assigned to the Ptilodontoidea. Hahn and Hahn (1983) erected the family Boffiidae within Ptilodontoidea. Carlson and Krause (1985) demonstrated that *Boffius* has gigantoprismatic enamel. McKenna and Bell (1997) assigned the rank of a tribe Boffiini to the Boffiidae; they attributed to Boffiini four genera, *Boffius, Essonodon, Liotomus,* and *Neoliotomus.* We regard Boffiidae as a monotypic family of uncertain affinities and place it among Cimolodonta as superfamily *incertae sedis.* Boffiidae resemble Cimolomyidae in an increase in the number of molar cusps, grooves, and ridges on P4 and M1. Vianey-Liaud (1986: figure 56) derived *Boffius* from Cimolomyidae. As in the case of Cimolomyidae the origin of Boffiidae from the *Paracimexomys* group (with which they share ornamentation) is uncertain.

The origin of Djadochtatherioidea, Eucosmodontidae, Microcosmodontidae, Taeniolabidoidea, and Kogaionidae (characters 17^4, 29^2, 30^2 in figure 8.45) appears more enigmatic than those of the groups previously discussed. Sloan and Van Valen (1965) assigned the Eucosmodontidae, Taeniolabididae, and Cimolomyidae to their suborder Taeniolabidoidea (now a superfamily). Kielan-Jaworowska and Hurum (2001) agreed with Fox (1999) that there are no apomorphies uniting Taeniolabidoidea *sensu* Sloan and Van Valen (1965) and restricted Taeniolabidoidea to Taeniolabididae, which we follow. These families share a limited enamel band on the lower incisor (except some Djadochtatherioidea), a character that also occurs in some "Plagiaulacida" and the ptilodontoidean *Neoliotomus.* Other characters uniting them are shared with all other cimolodontans. Members of these subgroups differ from the *Paracimexomys* group, Ptilodontoidea, Cimolomyidae, and Boffiidae in the lack of enamel ornamentation (except for upper incisors in some cases and anterior upper premolars, which might be ribbed) and in having well-separated molar cusps. The most plesiomorphic of these five groups are perhaps Djadochtatherioidea, some members of which (e.g., *Kryptobaatar*) share the small number of cusps and the general shape of m1 and of upper premolars and molars with members of the *Paracimexomys* group (compare figure 8.36C_1,E_1,I). They differ, as noted earlier, in lacking ornamentation and in having the molar cusps well separated.

Djadochtatherioidea and Eucosmodontidae approach one another in some characters of the dentary and skull, which in Eucosmodontidae is known only from a single, incomplete rostrum of *Stygimys* (Sloan and Van Valen, 1965). Resolving the problem of Djadochtatherioidea-Eucosmodontidae relations is not possible as long as the details of the upper dentition and skull structure of eucosmodontids remain largely unknown.

The superfamily assignment of the aberrant Late Cretaceous–Paleocene European monotypic family Kogaionidae Rădulescu and Samson, 1996, of which the dentary and lower dentition remain unknown, cannot be resolved. The skull shape is similar to that of Taeniolabidoidea in being roughly rectangular (figure 8.42A), but the upper dentition differentiates them from all other multituberculates. The Paleocene genus *Hainina*, known from Belgium (Vianey-Liaud, 1979), France (Vianey-Liaud, 1986), and Romania (Csiki and Grigorescu, 2000), shows some resemblance to *Kogaionon* in the general morphology of the molars, as suggested by Csiki and Grigorescu. *Hainina* was originally assigned by Vianey-Liaud (1979) to ?Cimolomyidae, and Carlson and Krause (1985) showed that it has gigantoprismatic enamel. Subsequently Fosse et al. (2001) demonstrated that *Kogaionon* also has gigantoprismatic enamel. We assign *Hainina* to the Kogaionidae (superfamily *incertae sedis*) tentatively only, as it differs from *Kogaionon* in having ornamented enamel.

If Cimolodonta are monophyletic (figure 8.45), one should search for the closest relatives of these five subgroups within the plagiaulacid line, perhaps via the *Paracimexomys* group. It would be difficult to visualize the origin of forms with well-separated molar cusps and smooth enamel from such forms as *Paracimexomys* and *Dakotamys,* in which discrete labial cusps, for example, on m2, disappeared. However, a separation of the cusps is observed in some teeth of members of the *Paracimexomys* group (e.g., M2 but not m2 of *Bryceomys,* figure 8.36C). Another possible candidate close to ancestral forms of the cimolodontans discussed might be the poorly known Mongolian Early Cretaceous *Monobaatar,* which has well-separated molar cusps (Kielan-Jaworowska, Dashzeveg, and Trofimov, 1987), and has been only tentatively assigned to Eobaataridae.

CHAPTER 9

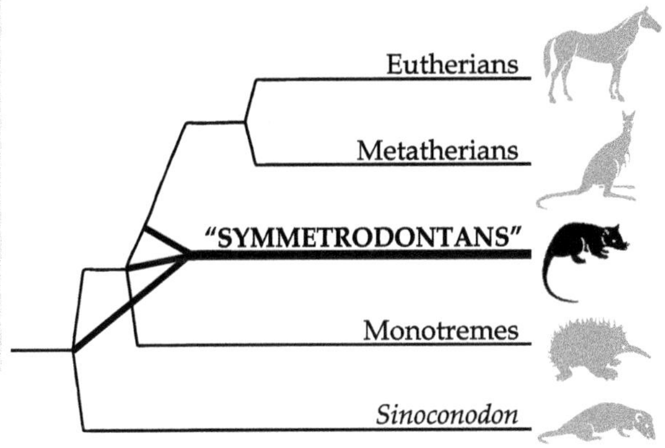

"Symmetrodontans"

INTRODUCTION

"Symmetrodontans" are poorly known, rare, and mainly small mammals, characterized by a simple reversed-triangle molar pattern. Functionally, the pattern represents a fundamental step in the evolution of mammalian molar design (e.g., Crompton and Sita-Lumsden, 1970): it is structurally intermediate between the cusp-in-line "triconodont" molar (chapters 4, 7) and the more elaborate molars of tribosphenic mammals and their precursors among "eupantotheres" (Patterson, 1956; Crompton and Jenkins, 1967, see chapters 10, 11). Formal recognition of "symmetrodontans," in concept as well as name, stems from the work of Simpson (1925e, 1928a, 1929a), who placed then-known taxa (all from the Late Jurassic and earliest Cretaceous and all with molars that are vaguely symmetrical in occlusal view) in the order Symmetrodonta. Simpson (e.g., 1928a) viewed "symmetrodontans" as uniquely specialized, distant relatives of "eupantotherians" (Pantotheria in his usage), though he remained skeptical as to certain cusp homologies between these two kinds of early mammals.

In subsequent years, the concept of Symmetrodonta was considerably broadened by the addition of geologically older and more primitive taxa, particularly those from the Late Triassic–Early Jurassic (e.g., Kermack et al., 1968; Sigogneau-Russell and Hahn, 1995). Morphologically, the least common denominator shared by all is the angulated cusp pattern. As this design has long been considered diagnostic of the more inclusive Theria *sensu lato* (e.g., Hopson and Crompton, 1969) or "Holotheria" (Hopson, 1994; Wible et al., 1995, see subsequent discussion in "Systematics"), the symmetrodont molar pattern represents a structural grade, not an apomorphy unique to "Symmetrodonta." To further complicate matters, it now appears that the reversed-triangle molar pattern probably evolved iteratively among mammals, so that this basic molar design is not, in itself, uniformly reliable for interpreting relationships (e.g., Rougier, Wible, and Novacek, 1996a; Pascual et al., 2002).

Cladistic analysis by Prothero (1981) yielded weak support for a restricted Symmetrodonta, as conceived by Simpson (1945, Amphidontidae + Spalacotheriidae + Tinodontidae as used herein), but most subsequent classifications continued to recognize the group (e.g., Fox, 1985; McKenna and Bell, 1997). Uniqueness of the few supporting characters has been challenged, however (Sigogneau-Russell and Ensom, 1998; Cifelli and Madsen, 1999), and a more comprehensive recent analysis suggests a nonexclusive common ancestry for Spalacotheriidae and Tinodontidae (Luo et al., 2002). Amphidontidae are very poorly known and, given the current state of knowledge, practically indeterminate (Averianov, 2002). Similar uncertainty surrounds the relationships of a number of other "symmetrodontans" described over the past quarter-century (Trofimov, 1980; Sigogneau-Russell, 1989a, 1991b; Bonaparte, 1990; Prasad and Manhas, 1997; Sigogneau-Russell and Ensom, 1998).

For organizational purposes, we find it convenient to treat these poorly understood mammals in a single chapter. We emphasize, however, that no implication of monophyly is intended: we use the term "symmetrodontan" in a descriptive, not taxonomic, sense. Present evidence supports monophyly and phylogenetic placement of only one of the seven families treated herein: Spalacotheriidae, which are a stem clade of Trechnotheria (Hu et al., 1997;

Cifelli and Madsen, 1999; Luo et al., 2002; see the taxon-based definition in the systematics section). The profound uncertainties surrounding the remaining taxa are reflected in our classification (tables 9.1, 9.2); we regard the remaining six families of "symmetrodontans" (Amphidontidae, Bondesiidae, Kuehneotheriidae, Tinodontidae, Thereuodontidae, Woutersiidae, all of which are either monotypic or not demonstrably monophyletic) as Mammalia *incertae sedis,* and an additional four genera are left in family *incertae sedis.*

BRIEF CHARACTERIZATION

"Symmetrodontans" are mostly small mammals similar in their common possession of the reversed-triangle molar pattern, with little development of a talonid on the lower molars (plesiomorphy). Beyond this plesiomorphous character, great variability is the norm for known features of the taxa assigned to "symmetrodontans" by various workers. Known dental formula ranges as follows. Upper dentition: I3 (known for *Zhangheotherium* only), C1, P2 to 5 or 6? (inferred from lower dentition), M4 or 5? (inferred from lower dentition) to 7. Lower dentition: 3.1.2– 5 or 6?.4–7. Canines are generally small; these and the anterior premolars may be single- or double-rooted; teeth at premolar loci may be retained through life or shed ontogenetically, with "plugging" of alveoli. Premolars have diphyodont replacement in *Zhangheotherium,* although in alternating sequence as in *Dryolestes* and different from the sequential premolar replacement as in extant placentals. Replacement at least at the last premolar locus is known for some taxa, whereas the entire premolar series appears to have been monophyodont in others. Angulation of molar cusps varies from negligible to acute; within the series, angulation generally becomes progressively more acute posteriorly, with the trend modestly reversed between the penultimate and ultimate molar. Molar cingula/cingulids may be complete, incomplete, or lacking entirely. Lower molar cusps d and e are usually present, with cusp f present in some taxa. Except for the invariably tall paracone, cusps on upper molars (B′, parastyle, stylocone = B, metastyle, cusp "C")

TABLE 9.1. Cladistic Classification of Mammals with a Symmetrodontan Molar Pattern above the Species Level[1]

Class Mammalia Linnaeus, 1758
  Unnamed clade (*Morganucodon* + Crown Mammalia)
    Family Kuehneotheriidae D. M. Kermack et al., 1968, (*sedis mutabilis*)
      Kuehneotherium D. M. Kermack et al., 1968
      Kotatherium Datta, 1981
      Kuehneon Kretzoi, 1960, *nomen dubium*
    Family *incertae sedis*
      Trishulotherium Yadagiri, 1984
    Family Woutersiidae Sigogneau-Russell and Hahn, 1995
      Woutersia Sigogneau-Russell, 1983
  Unnamed clade (Docodonta + Crown Mammalia)
(Intervening ranks of clades in Table 1.1)
  Unnamed clade (Eutricondonta + Allotheria + Trechnotheria)
    Family Amphidontidae Simpson, 1925 (*sedis mutabilis*)
      Amphidon Simpson, 1925
      Manchurodon Yabe and Shikama, 1938
      Nakunodon Yadagiri, 1985
    Family Bondesiidae Bonaparte, 1990 (*sedis mutabilis*)
      Bondesius Bonaparte, 1990
    Family Thereuodontidae Sigogneau-Russell and Ensom, 1998 (*sedis mutabilis*)
      Thereuodon Sigogneau-Russell, 1989
    Family Tinodontidae Marsh, 1887 (*sedis mutabilis*)
      Tinodon Marsh, 1879
    Family *incertae sedis*
      Atlasodon Sigongeau-Russell, 1991
      Gobiotheriodon Trofimov, 1997
      Microderson Sigogneau-Russell, 1991
(Intervening ranks of clades in table 1.1)
  Clade Trechnotheria McKenna, 1975 (Spalacotheriidae + Crown Theria)
    Family Spalacotheriidae Marsh, 1887
    Subfamily *incertae sedis*
      Spalacotherium Owen, 1854
      Zhangheotherium Hu et al., 1997
    Subfamily Spalacolestinae Cifelli and Madsen, 1999
      ?*Shalbaatar* Nessov, 1997
    Spalacolestes Cifelli and Madsen, 1999
      Spalacotheridium Cifelli, 1990
      Spalacotheroides Patterson, 1955
      Symmetrodontoides Fox, 1976

[1] For practical purposes we summarize information about mammals with symmetrodontan molar patterns (traditionally known as "symmetrodonts") in a single chapter "Symmetrodontans." Kuehneotheriids and tinodontids were traditionally assigned to the grouping of "symmetrodonts" (see table 9.2). In the best available phylogenetic trees (see figures 1.1, 15.1, 15.2), however, kuehneotheriids are a stem group outside the Crown Mammalia, and tinodontids are more distant to trechnotherians ("symmetrodontans" *sensu stricto*) than to eutriconodontans and multituberculates. It has also been proposed that the "symmetrodontan" *Woutersia* may be related to docodontans. In the comprehensive phylogeny (figures 15.1, 15.2) and complete cladistic classification (table 1.1) these groups would occupy different positions in the classification scheme, separated by intervening ranks of clades (shown here). For information about type genera, type species and synonyms see table 9.2 and the text.

TABLE 9.2. Linnaean Classification of Mammals with a Symmetrodontan Molar Pattern

- Mammalia, *incertae sedis*
  - Family *incertae sedis*
    - *Atlasodon* Sigogneau-Russell, 1991
      - *A. monbaroni* Sigogneau-Russell, 1991
    - *Gobiotheriodon* Trofimov, 1997
      - *G. infinitus* (Trofimov, 1980)
    - *Microderson* Sigogneau-Russell, 1991
      - *M. laaroussii* Sigogneau-Russell, 1991
    - *Trishulotherium* Yadagiri, 1984
      - *T. kotaensis* Yadagiri, 1984
  - Family Amphidontidae Simpson, 1925
    - *Amphidon* Simpson, 1925, type genus
      - *A. superstes* Simpson, 1925
    - ?*Manchurodon* Yabe and Shikama, 1938
      - *M. simplicidens* Yabe and Shikama, 1938
    - ?*Nakunodon* Yadagiri, 1985
      - *N. paikasiensis* Yadagiri, 1985
  - Family Bondesiidae Bonaparte, 1990
    - *Bondesius* Bonaparte, 1990, type genus
      - *B. ferox* Bonaparte, 1990
  - Family Kuehneotheriidae[1] D. M. Kermack et al., 1968
    - *Kuehneotherium* D. M. Kermack et al., 1968, type genus
      - *K. praecursoris* D. M. Kermack et al., 1968
    - *Kotatherium* Datta, 1981
      - *K. haldanei* Datta, 1981
    - *Kuehneon* Kretzoi, 1960, *nomen dubium*
      - *K. duchyense* Kretzoi, 1960, *nomen dubium*
  - Family Thereuodontidae[2] Sigogneau-Russell and Ensom, 1998
    - *Thereuodon* Sigogneau-Russell, 1989
      - *T. dahmanii* Sigogneau-Russell, 1989, type species
      - *T. taraktes* Sigogneau-Russell and Ensom, 1998
  - Family Tinodontidae Marsh, 1887
    - *Tinodon*[3] Marsh, 1879
      - *T. bellus* Marsh, 1879, type species (including *T. lepidus* Marsh, 1879)
      - *T. micron* Ensom and Sigogneau-Russell, 2000
  - Family Woutersiidae Sigogneau-Russell and Hahn, 1995
    - *Woutersia* Sigogneau-Russell, 1983
      - *W. mirabilis* Sigogneau-Russell, 1983, type species
      - *W. butleri* Sigogneau-Russell and Hahn, 1995
- Superlegion Trechnotheria McKenna, 1975
  - Family Spalacotheriidae Marsh, 1887
  - Subfamily *incertae sedis*
    - *Spalacotherium*[4] Owen, 1854, type genus
      - *S. tricuspidens* Owen, 1854, type species
      - *S. evansae* Ensom and Sigogneau-Russell, 2000
      - *S. henkeli* Krebs, 1985
      - *S. taylori* Clemens and Lees, 1971
      - *Zhangheotherium* Hu et al., 1997
      - *Z. quinquecuspidens* Hu et al., 1997
  - Subfamily Spalacolestinae Cifelli and Madsen, 1999
    - *Spalacolestes* Cifelli and Madsen, 1999, type genus
      - *S. cretulablatta* Cifelli and Madsen, 1999, type species
      - *S. inconcinnus* Cifelli and Madsen, 1999
    - ?*Shalbaatar* Nessov, 1997
      - *S. bakht* Nessov, 1997
    - *Spalacotheridium* Cifelli, 1990
      - *S. mckennai* Cifelli, 1990, type species
      - *S. noblei* Cifelli and Madsen, 1999
    - *Spalacotheroides* Patterson, 1955
      - *S. bridwelli* Patterson, 1955
    - *Symmetrodontoides* Fox, 1976[5]
      - *S. canadensis* Fox, 1976, type species
      - *S. foxi* Cifelli and Madsen, 1986
      - *S. oligodontos* Cifelli, 1990

[1] *Cyrtlatherium* E. F. Freeman, 1979, originally described as a kuehneotheriid, has recently been transferred to Docodonta (see Sigogneau-Russell, 2001, and chapter 5).
[2] Included taxa thought by Martin (2002) and Averianov (2002) to be represented by milk teeth of stem zatherians.
[3] Includes *Eurylambda* Simpson, 1929.
[4] Includes *Peralestes* Owen, 1871.
[5] *Mictodon* Fox, 1984, known by a single tooth representing the monotypic species *M. simpsoni* Fox, 1984, is omitted; the specimen appears to be a deciduous premolar of *Symmetrodontoides canadensis* (see Cifelli, 1999).

are variable in presence, development, and position (cusp terminology is shown in figure 9.1). The dentary is long and slender, with an unfused, subhorizontal symphysis. An angular process is uniformly lacking; a pterygoid crest (medially) and flaring of the inferior margin of the masseteric fossa (laterally) are commonly but not universally present. A postdentary trough, overlying ridge, and scars indicating attached postdentary elements may be present; Meckel's groove and scar(s) for the coronoid and/or splenial are also variable in their development and expression, or may be altogether absent in some taxa. The cochlea of known forms was probably uncoiled. The skeleton, known only for *Zhangheotherium*, presents a mosaic of primitive and derived features, including: unfused ribs on postaxial cervical vertebrae, interclavicle present but with mobile clavicular joint, strong supraspinous fossa on scapula; humerus with at least 30° of torsion, lesser tubercle large relative to greater tubercle, incipient ulnar trochlea and vestigial ulnar condyle; presence of epipubic in pelvis; tarsal spur present (see "Postcranial Skeleton," later).

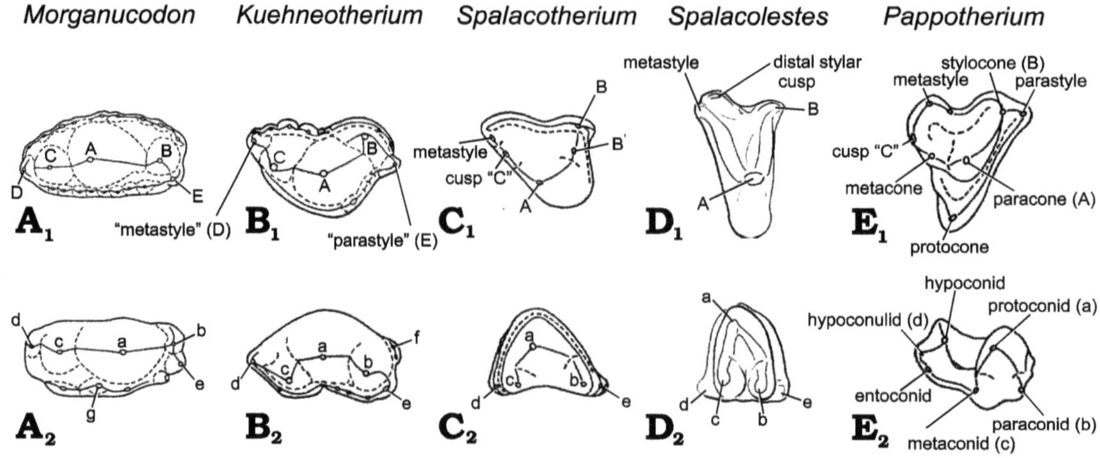

FIGURE 9.1. The "symmetrodontan" pattern in the context of mammalian molar evolution. A_1–E_1, Right upper molars. A_2–E_2, Left lower molars. A, *Morganucodon* sp.; B, *Kuehneotherium* sp.; C, *Spalacotherium tricuspidens;* D, *Spalacolestes cretulablatta;* E, *Pappotherium pattersoni. Source:* A–C, E, modified from Crompton and Jenkins (1968); D, modified from Cifelli and Madsen (1999). (Not to scale.)

DISTRIBUTION

"Symmetrodontans" include some of the geologically oldest mammals known. Woutersiidae (Late Triassic), represented by the single genus *Woutersia,* are known only by isolated upper and lower molariforms from two localities in France (Sigogneau-Russell and Hahn, 1994, 1995). Kuehneotheriidae (Late Triassic–Early Jurassic), known from scattered localities in continental Europe, Britain, Greenland, and possibly also in India (*Kotatherium,* see Prasad and Manhas, 1997), are also generally scarce. *Kuehneotherium,* which has played a prominent role in the interpretation of early mammal radiations (e.g., Cassiliano and Clemens, 1979), is by far the best known: fissure-fill deposits in Wales have yielded large samples of isolated teeth, edentulous dentary and maxilla fragments, and poorly preserved petrosals (Kermack et al., 1968; Parrington, 1971). Preliminary study has permitted the reconstruction of most of the dentary, molar occlusion, and (more tentatively) both upper and lower molar series, together with identification of premolars and probable deciduous premolars (Crompton and Jenkins, 1967, 1968; Kermack et al., 1968; Parrington, 1971; Gill, 1974; Mills, 1984). Fossils described to date come from two quarries. The samples are said to be taxonomically distinct, perhaps at the generic level; for comparative purposes herein, we follow Mills (1984) in referring to them collectively as *Kuehneotherium.* This genus is also represented by a relatively large sample of isolated teeth from Saint Nicolas-de-Port (Late Triassic), France (Godefroit and Sigogneau-Russell, 1999).

Of the five mammalian genera (all known by isolated teeth, some incomplete) from the Lower Jurassic Kota Formation of India, two merit passing mention here. *Kotatherium* is tentatively referred to Kuehneotheriidae. We follow Prasad and Manhas (1997) in provisionally allocating *Nakunodon* to Amphidontidae. Amphidontids are otherwise known by *Amphidon superstes* (represented by a single, somewhat equivocal dentary fragment from the Late Jurassic of North America, Simpson, 1929a) and, tentatively, by *Manchurodon simplicidens.* This species is based on a dentary bearing eight postcanine teeth, from the Middle or Late Jurassic of China. Although some uncertainties remain as to the molar pattern in *Manchurodon* (see Averianov, 2002), its traditional placement in Amphidontidae (Yabe and Shikama, 1938; Patterson, 1956; Cassiliano and Clemens, 1979) is supported by several apomorphies.

Late Jurassic–earliest Cretaceous "symmetrodontans" are known from the western United States, Britain, and Morocco. Faunal similarity between the Morrison Formation (Upper Jurassic), United States, and the Purbeck Limestone Group (now regarded as earliest Cretaceous), Britain, is well established (e.g., Simpson, 1928a, see chapter 2). The "symmetrodontan" *Tinodon,* long known by dentaries and (tentatively) an upper molar from the Morrison, has recently been reported from the Purbeck based on isolated teeth (Ensom and Sigogneau-Russell, 2000). A notable difference between the faunas of the two units is the presence of the distinctive spalacotheriid *Spalacotherium* (represented by dentaries and a maxilla) in the Purbeck. *Spalacotherium* persisted into the Valanginian of

Britain (Clemens and Lees, 1971) and the Barremian of Spain (Krebs, 1985). Spalacotheriidae are the best represented and most broadly distributed "symmetrodontans." *Zhangheotherium,* from the Barremian of China, is known by most of the skull and skeleton (Hu et al., 1997, 1998). Greatest diversity for the family is recorded from the Cretaceous of North America, from which nine species (placed in four genera) have been described (Patterson, 1955; Fox, 1976a; Cifelli and Madsen, 1986, 1999; Cifelli, 1990c; Cifelli and Gordon, 1999). North American spalacotheriids (known by dentaries and isolated teeth) form a monophyletic subfamily (Spalacolestinae) and survived later in time than elsewhere, ranging from the Aptian–Albian to the beginning of the Campanian. A notable exception to the general pattern of modest abundance and diversity among "symmetrodontans" is the fauna of the Cedar Mountain Formation, Utah, where Spalacotheriidae are represented by at least four species and, in terms of individuals, far outnumber all other nonmultituberculate mammals (Cifelli and Madsen, 1999). *Shalbaatar,* known by an edentulous dentary fragment from the Turonian of Uzbekistan, may be related to the otherwise North American Spalacolestinae (Averianov, 2002). The potential for such a faunal link is suggested by several other mammalian taxa that are possibly common to the Late Cretaceous of Uzbekistan and North America (Cifelli, 2000c; Archibald and Averianov, 2001b).

Four "symmetrodontans," each represented by isolated teeth, are known from the Berriasian of Morocco. One (*Thereuodon dahmanii*) is congeneric with a species from the Purbeck; the ultimate affinities of *Thereuodon* and other Moroccan "symmetrodontans" are debatable (Sigogneau-Russell and Ensom, 1998). A final record of a "symmetrodontan" mammal grade comes from the Late Cretaceous (Campanian or Maastrichtian) Los Alamitos fauna of Argentina. Initially, five genera (placed in three families) from this fauna were referred to "Symmetrodonta" (Bonaparte, 1990). We follow recent accounts (Bonaparte, 1994; Sigogneau-Russell and Ensom, 1998) in regarding only one, *Bondesius,* as a "symmetrodontan." *Bondesius,* referred to the monotypic family Bondesiidae, is represented by a single lower molar.

ANATOMY

SKULL

Aside from undescribed (and reportedly not well preserved) petrosals referred to *Kuehneotherium* (see Kermack et al., 1968), knowledge of "symmetrodontan" skulls is confined to the holotype of the spalacotheriid *Zhangheotherium quinquecuspidens.*[1] The specimen includes a fairly complete but badly crushed skull; we confine our account to a few salient features of the auditory region (figure 9.2A), based on the brief description given by Hu et al. (1997). Posterior to the glenoid fossa, which is flanked posterolaterally by a weak postglenoid crest, the squamosal is mediolaterally constricted where it forms a lateral wall to the epitympanic recess. The presence of such a squamosal wall is an apomorphy shared by other trechnotherians, to the exclusion of multituberculates, *Ornithorhynchus* (but not *Tachyglossus*), and stem mammal lineages (Luo et al., 2002). More posteriorly, the cranial moiety of the squamosal appears to be broad, a derived condition otherwise seen mainly among "eupantotherians" and living therians. The postglenoid region of the squamosal further resembles that of other trechnotherians in having a posterolateral depression, or external auditory meatus.

The petrosal of *Zhangheotherium* (figure 9.2B) is characterized by a combination of primitive and derived features. Most obvious among the plesiomorphies is the fact that the promontorium is uninflated and subcylindrical, as it is in morganucodontans, eutriconodontans, and multituberculates. This contrasts with the condition in living therians, where the promontorium is more bulbous and oval-shaped. The promontorium houses the cochlea, which is coiled at least 360° in living therians; hence, it may be reasonably inferred that the cochlea of *Zhangheotherium* was either straight or only slightly curved. (The petrosal fragments ascribed to *Kuehneotherium,* never described and mentioned only in passing, are said to lack any indication of a spiral cochlea [Kermack et al., 1968].) The paroccipital process differs from that of living therians and resembles those in monotremes, multituberculates, and certain stem mammal groups in projecting ventrally

[1] A possible but highly uncertain addition is an isolated petrosal from the Aptian–Albian of Mongolia, described by Wible et al. (1995). Cladistic analysis by these authors placed the fossil where one might expect a Cretaceous "symmetrodontan" to lie on the mammalian tree: nested within Crown Mammalia in an unresolved trichotomy with eutriconodontans and *Vincelestes* + living therians. The specimen appears to be of appropriate size to belong to the "symmetrodont" *Gobiotheriodon infinitus,* known from the same locality (Trofimov, 1980, 1997). This petrosal shows many similarities to that of *Zhangheotherium* described later. However, Rougier, Wible, and Hopson (1996) subsequently reported another petrosal, of similar evolutionary grade but representing a different taxon, from the same locality. At present, both specimens must be regarded as Mammalia *incertae sedis.*

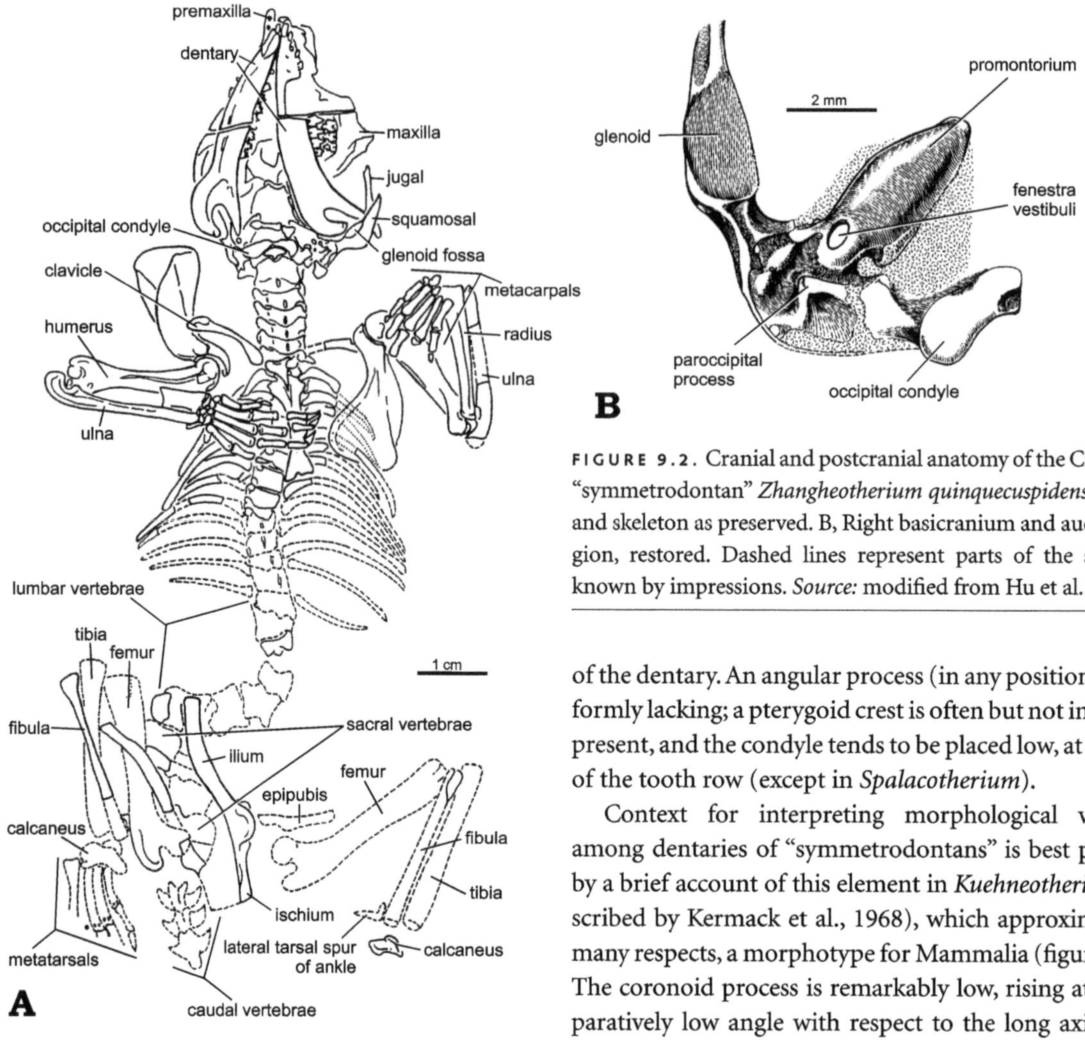

FIGURE 9.2. Cranial and postcranial anatomy of the Cretaceous "symmetrodontan" *Zhangheotherium quinquecuspidens*. A, Skull and skeleton as preserved. B, Right basicranium and auditory region, restored. Dashed lines represent parts of the specimen known by impressions. *Source:* modified from Hu et al. (1997).

beyond surrounding structures. Two other features of the petrosal, however, are derived attributes shared with other trechnotherians: the presence of a large posterior tympanic recess and a caudal tympanic process.

MANDIBLE

Of any element known, the dentary (figure 9.3) best exemplifies morphological heterogeneity among "symmetrodontans," thereby serving to underscore the problem of regarding them as a phylogenetic unit, or natural group. Reasonably informative dentaries are known for some 10 genera; structurally, these span much of the mammalian range from entirely plesiomorphic to rather modern in most characters; some appear to bear unique specializations. A few vague, nonexclusive generalizations deserve mention. The dentary of "symmetrodontans" tends to be long, slender, and shallow. The symphyseal region is quite shallow, and the symphysis has an elongate oval shape, inclined at a rather low angle with respect to the long axis of the dentary. An angular process (in any position) is uniformly lacking; a pterygoid crest is often but not invariably present, and the condyle tends to be placed low, at the level of the tooth row (except in *Spalacotherium*).

Context for interpreting morphological variation among dentaries of "symmetrodontans" is best provided by a brief account of this element in *Kuehneotherium* (described by Kermack et al., 1968), which approximates, in many respects, a morphotype for Mammalia (figure 9.3A). The coronoid process is remarkably low, rising at a comparatively low angle with respect to the long axis of the dentary. The masseteric fossa is shallow and its borders are weakly defined. The presence of a condyle on the dentary shows that at least part of the temporomandibular joint was formed by the dentary and squamosal. As in morganucodontids (Kermack et al., 1973), docodontans (Lillegraven and Krusat, 1991), and other lineages of stem mammals, the inferior border of the dentary is emarginated posteriorly for the reflected lamina of the angular, a homologue of the mammalian tympanic (Allin and Hopson, 1992). Medially, a well-marked trough and overlying ridge extend posteriorly from the mandibular foramen to the condyle. These features reflect the presence of postdentary elements (articular, prearticular, angular, surangular) that were attached to the medial surface of the dentary. Association of these postdentary elements with the dentary is a plesiomorphic condition inherited from nonmammalian cynodonts (e.g., Allin, 1975, 1986). Hence, despite the fact that the postdentary elements presumably served in the reception and transmission of airborne sound, the characteristic configuration of the mammalian middle ear had not been achieved. A scar for one of the paradentary elements, the coronoid, is present. Meckel's groove is well de-

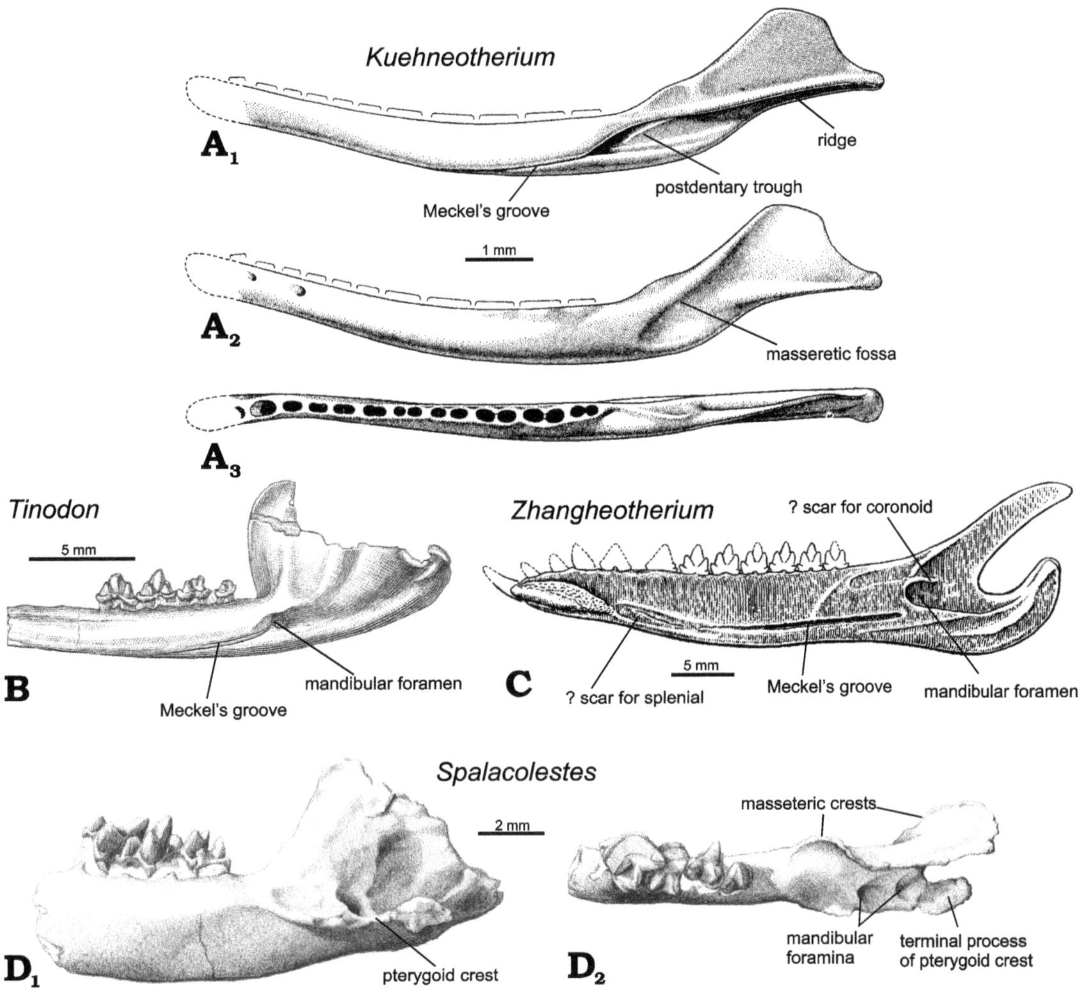

FIGURE 9.3. Dentary structure in "symmetrodontans." A, *Kuehneotherium praecursoris,* in medial (A_1), lateral (A_2), and occlusal (A_3) views (brackets above the dentary in A_{1-2} denote tooth positions). B, *Tinodon bellus,* medial view. C, *Zhangheotherium quinquecuspidens,* medial view. D, *Spalacolestes cretulablatta,* in medial (D_1) and occlusal (D_2) views. *Source:* A, modified from Kermack et al. (1968); B, modified from Marsh (1887); C, modified from Hu et al. (1997); D, modified from Cifelli and Madsen (1999).

fined, extending anteriorly from the mandibular foramen near the inferior margin of the dentary.

In all other "symmetrodontans" for which the relevant part of the dentary is preserved (*Amphidon, Tinodon, Gobiotheriodon, Spalacolestes, Spalacotherium, Zhangheotherium*), there is no trace of the medial trough, its overhanging ridge, or possible scars for the postdentary elements,[2] and Meckel's groove (where present) is well separated from the mandibular foramen (figure 9.3B–D). Clearly, the postdentary elements were detached mediolaterally from the angular region of the mandible. We cannot exclude the possibility that the middle ear elements were still linked anterior to the horizontal ramus of the mandible via Meckel's cartilage (as evidenced by a similar arrangement of Meckel's cartilage in *Gobiconodon*). The only "symmetrodontan" auditory region known is that of *Zhangheotherium;* the presence of a therian-like epitympanic recess and other features suggests that the postdentary elements in this taxon, at least, probably achieved

[2] Allin and Hopson (1992) identified scars for the angular and articular + prearticular in the Middle Jurassic "eupantotherian" *Amphitherium.* As the postdentary trough and overlying ridge are lacking, these authors interpreted *Amphitherium* as having had partially detached postdentary elements that retained a mobile contact with the dentary at their anterior extremity. The scars on which this hypothesis rests are indistinct, however, and alternate interpretations are possible. For example, the presence of an ossified Meckel's cartilage has recently been reported for two eutriconodontans (Wang, Hu, Meng et al., 2001).

some derived characteristics of the middle ear in a typically mammalian fashion (see Crompton and Sun, 1985).

Other plesiomorphies of the dentary have somewhat varied distributions among "symmetrodontans." A scar indicating the presence of the coronoid bone is present in most taxa (*Zhangheotherium* also has a scar suggesting retention of the splenial), as is a trace (at least) of Meckel's groove; absence of both in *Spalacolestes* suggests loss of these elements within the family Spalacotheriidae (Cifelli and Madsen, 1999). The coronoid process is rather weak and recumbent posteriorly (plesiomorphies, see Prothero, 1981) in the basal spalacotheriid *Zhangheotherium* (figure 9.3C), but more erect and anteroposteriorly expanded in other Spalacotheriidae and in *Tinodon* (figure 9.3B). A pterygoid crest, lacking in *Kuehneotherium*, is present in all other "symmetrodontans," though it is rather variable in position and development. Possession of a pterygoid crest was cited as a synapomorphy of Symmetrodonta by Prothero (1981), but it is present in all mammals save a few stem groups (Miao, 1988; Rowe, 1988; Cifelli and Madsen, 1999; Luo et al., 2002). Specialization within Spalacotheriidae involves anterior expansion of the pterygoid fossa and elaboration of the pterygoid crest; in advanced genera, the crest originates anterodorsally near the back of the tooth row and is developed into a large, curving process near the mandibular foramen (Cifelli and Madsen, 1999, see our figure 9.3D). An associated feature is the fact that the posteroventral margin of the dentary in Spalacotheriidae is strongly reflected laterally. These unusual characteristics appear to be related to integrated action of the pterygoid (medially) and masseter (laterally) muscles in rotation and lateral translation of the mandible during the masticatory cycle (Oron and Crompton, 1985; Crompton, 1995).

DENTITION

Incisors, Canines, and Premolars. Knowledge of the incisors among "symmetrodontans" is limited mainly to the single specimen of *Zhangheotherium* and to alveoli and/or broken teeth in *Spalacotherium* and *Gobiotheriodon*. *Spalacotherium* is said to have had three or four well-spaced lower incisors (Simpson, 1928a). Three closely spaced lower incisors were present in *Gobiotheriodon;* i3 was markedly larger than i1–2 and, judged by the alveoli, was subequal in size to the canine (Averianov, 2002). *Zhangheotherium* has three upper and lower incisors, with i1 enlarged and procumbent (Hu et al., 1997, 1998). The canine of most "symmetrodontans" differs from that of eutriconodontans and many other early mammals in being rather small and noncaniniform. In most genera it is single-rooted; the canine root is furrowed but not surely divided in *Tinodon* (Simpson, 1925d: figure 2), whereas the canine is reportedly two-rooted in *Spalacotherium* (e.g., Owen, 1871; Osborn, 1888b).

The highest premolar count among "symmetrodontans" is that of *Kuehneotherium*, wherein five or six (the first four of which were single-rooted) were present (Kermack et al., 1968; Gill, 1974). The premolar count of most other "symmetrodontans" is, by comparison, reduced, though at present there is no discernible pattern regarding taxonomic distribution: *Amphidon* had four (Simpson, 1925d; but see Averianov, 2002, for an alternative interpretation), *Gobiotheriodon* had four (Trofimov, 1980) or three (Averianov, 2002), *Tinodon* and *Manchurodon* had three (Simpson, 1925d; Yabe and Shikama, 1938), and the number in Spalacotheriidae ranged from three in *Zhangheotherium* (see Hu et al., 1997) to five in an unnamed taxon from North America (Cifelli et al., 2000). In all taxa, the premolars are simple and trenchant, dominated by cusp A/a, with cusps B/b and C/c (where present) generally small and basally placed.

Molars. Triangulation of the principal molar cusps has long been interpreted as an important evolutionary innovation among mammals. The first widely accepted model to explain the origin of this pattern from the simpler, linear cusp arrangement seen in nonmammalian cynodonts is the Cope-Osborn theory (e.g., Osborn, 1907), which posits transformation via rotation of auxiliary cusps (displacement of mesial and distal cusps B/b and C/c labially on upper molars and lingually on lower molars). An opposing and influential interpretation was proposed by Simpson (e.g., 1925d, 1928a, 1929a), who recognized "symmetrodontans" as distinct from other early mammals based on the pattern and inferred nonhomology of the molar cusps. He viewed accessory (mesial and distal) cusps of the typical three-cusped molar pattern as having originated independently, in place, in separate lineages (see discussions by Butler, 1939, and Patterson, 1956). There is now little doubt that the three principal cusps of "symmetrodont" molars are homologous with those of "triconodonts," and that the reversed-triangle pattern is a result of cusp rotation from a cusp-in-line configuration (see review by Crompton and Jenkins, 1968). With one possible exception, each of the three primary cusps on upper and lower molars of "symmetrodontans" is readily homologized with corresponding cusps on molars of "eupantotherians" and tribosphenic mammals. The possible exception concerns the distolabial cusp, "C," of the upper molars. Crompton (1971) referred to this as cusp "c" (a designation prone to confusion with a similarly termed cusp, homologous to the metaconid, on lower molars) and hypothesized the metacone to be a neomorph that developed among eupantotherians. Others (e.g., Kermack et al., 1968; Butler, 1972b) have suggested the simpler explanation that cusp "C" is homologous with the metacone (see also Clemens and Mills, 1971; Hopson, 1997). We tentatively follow the homologies of Crompton (1971, see chapters 10 and 11), but as Crompton's cusp "c" belongs to upper molars, we replace it with the upper case letter (cusp "C") following the terminology accepted in this book (see Preface).

The molar count in "symmetrodontans" correlates roughly with the acuteness of cusp angulation (Patterson, 1956): "obtuse-angled" taxa such as *Tinodon* and *Amphidon* have (or are interpreted to have had) four molars, whereas the count is five (upper series of *Zhangheotherium* subadult) to seven (lower series, at least, of *Spalacotherium*) among the "acute-angled" Spalacotheriidae. An exception is "obtuse-angled" *Kuehneotherium,* which may have had up to six molars, perhaps owing to an ontogenetic posterior shifting in the postcanine tooth row (involving the addition of a molar to the back of the tooth series late in life).

Upper molars of *Kuehneotherium* generally have a complete cingulum that differs from that of morganucodontids in lacking consistent cuspules (figure 9.1A,B), an apomorphy according to Sigogneau-Russell and Ensom (1998). The upper molars are similar in width to the lowers (presumed plesiomorphy). In addition to the three primary cusps (paracone or cusp A, stylocone or cusp B, cusp "C"), parastyle and metastyle are present mesiolabially and distolabially, respectively. The metastyle fits into an embayment between the parastyle and stylocone (B) of the succeeding molar, providing some degree of molar interlocking (apomorphy relative to the condition in morganucodontids).

Lower molars of *Kuehneotherium* have strong lingual and mesial cingulids, with the number of cingular cusps reduced in comparison to *Morganucodon.* The mesial cingulid bears two cusps, e (lingual) and f (labial). Distally, the lingual cingulid terminates in a talonid cusp (d, or hypoconulid). Cusp d fits into an embayment between cusps e and f of the succeeding molar, which provides for interlocking of the molar series (an apomorphy relative to most stem mammals but present in *Dinnetherium*). Like morganucodontids and in contrast to crown mammals, extensive wear was needed in order to produce matching shearing surfaces on corresponding parts of the upper and lower molars in *Kuehneotherium.*

The principal shearing surfaces on molars of *Kuehneotherium* are the prevallum/postvallid surfaces (surfaces 1 of Crompton, 1971, see our figure 11.3A). Shearing surfaces 2 (postvallum/prevallid) are limited to a small area between cusp "C" and the metastyle (on the upper molars) and the mesolabial face of the protoconid (on the lowers). Small accessory wear facets (A and B of Crompton, 1971) on adjacent surfaces of the paracone and cusp "C" occluded with counterparts on the labial face of the paraconid. Upper molar facets A and B of *Kuehneotherium* are analogous to, or homologous with, those on the centrocrista (facets 3 and 4 of Crompton, 1971) of tribosphenic mammals and proximal relatives. However, their occlusal relationships are different: in tribosphenic mammals, the corresponding lower molar facets 3 and 4 lie on the labial faces of a neomorphic cusp, the hypoconid, on a greatly expanded talonid (see discussions by Butler, 1972b; Prothero, 1981).

Of the many variations in "symmetrodont" molar design (see review by Sigogneau-Russell and Ensom, 1998), those of Spalacotheriidae are the best understood, and we conclude this section with a brief account of them (figure 9.1C,D). The most significant difference compared to the pattern seen in *Kuehneotherium* involves the shearing surfaces, which are developed as continuous vertical faces on the mesial and distal surfaces of both upper and lower molars; cusps are developed as topographic features on the crowns only, with no expression of relief on the embrasure surfaces. The postvallum/prevallid surfaces are considerably expanded in comparison to *Kuehneotherium* and are composed of single, continuous facets. In contrast to the condition in *Kuehneotherium,* the upper molars are noticeably wider (more transversely developed) than the lowers. Upper molars are dominated by the tall paracone; other cusps are variable in presence and expression, but their surface relief is generally slight. An accessory, presumably neomorphic cusp (B^1 of Hu et al., 1997, and subsequent authors) and cusp "C" are present on the preparacrista and postparacrista, respectively, of primitive forms, and a stylar cusp is present on the distal moiety of the stylar shelf in some advanced taxa (see Fox, 1985; Cifelli and Madsen, 1999). Lower molars are characterized by great crown height, particularly labially; the two roots are subequal in development and are notably compressed anteroposteriorly. Cingulids (lacking on upper molars) may be complete, discontinuous, or lacking altogether. The talonid cusp d (or hypoconulid) is greatly reduced and lingually situated. Mesially, cusp f is lacking, and cusp e is placed at the lingual margin of the tooth. An apparently unique apomorphy is the manner in which adjacent molars interlock: cusp d lies lingual to the abutting cusp e of the succeeding molar. Deep V-shaped notches are present in the paracristid and protocristid. Occlusal relations with the upper molars (dominated by a very tall paracone and considerably broader transversely than the lowers), together with the orientation of wear striations (more oblique on upper than lower molars), suggest that the masticatory cycle in spalacotheriids involved considerable lateral translation and rotation of the mandible (Patterson, 1956; Crompton, 1971; Fox, 1976a; Cifelli and Madsen, 1999).

Various authors (e.g., Patterson, 1956; Fox, 1976a; Cifelli and Madsen, 1986) have noted the presence of flat, straplike wear features on the principal cristids (paracristid, protocristid) of spalacotheriid lower molars; similar wear surfaces are present on the paracristae of spalcotheriid upper molars (Cifelli and Madsen, 1999). Patterson (1956) supposed that these resulted from direct apposition of corresponding upper and lower molar crests. However, these are nonocclusal surfaces in spalacotheriids and structurally similar mammals: owing to the reversed-triangle configuration, lower molars passed between uppers, and their contact was restricted mainly to the vertical embrasures (Crompton and Sita-Lumsden, 1970). Apical wear on spalacotheriid molar crests resulted not from tooth-tooth contact, but from tooth-food contact. By contrast, very little wear occurred on the vertically oriented, occlusal surfaces of the molars—the embrasures. Apical wear surfaces like those of spalacotheriids are developed on the structurally similar molars of dryolestoid "eupantotherians." Crompton et al. (1994) have shown that differential apical and occlusal wear on dryolestoid molars results from differing patterns of crystallite orientation, within the tooth enamel, with respect to direction of force by the eroding agent. The apical wear facets of molar crests are suborthogonal to the embrasure faces. By analogy with a paper cutter, this relationship is responsible for effective shearing function. Crompton et al. (1994) have proposed that differential molar wear, with apical wear taking place much more rapidly than occlusal wear, provides a mechanism for maintenance of sharp leading edges to the shearing surfaces.

Tooth Eruption, Replacement, and Loss. Some data are now available for the developmental aspects of the dentition in "symmetrodontans." Direct evidence for tooth replacement is known in *Zhangheotherium*. Replacement of premolars was diphyodont and occurred in an alternating sequence, as in *Dryolestes*. The ultimate premolar was replaced just before or around the time of eruption of m5. There are also limited to sketchy data for *Tinodon* and some Spalacotheriidae. In *Tinodon*, the last premolar (at least) underwent normal eruption after m1 was in place (Cifelli, 1999a), and the same appears to have been true of *Spalacotherium* (see Ensom and Sigogneau-Russell, 2000). Somewhat more tenuously, isolated teeth of *Kuehneotherium* suggest that a premolariform successor may have replaced a molariform primary tooth at least at the last premolar locus (Mills, 1984). In North American spalacotheriids (Spalacolestinae), the deciduous canine and all premolars were retained late in life (Cifelli, 1999a) and may never have been replaced; the canines and postcanine dentition (incisors remain unknown) appear to have been monophyodont (Cifelli et al., 2000). Deciduous premolars of "symmetrodontans," inferred for *Kuehneotherium* and Spalacotheriidae (Mills, 1984; Cifelli, 1999a; Ensom and Sigogneau-Russell, 2000), have attributes analogous to those of most other mammals: the enamel tends to be thinner than on the replacement premolars or molars; they are low crowned and anteroposteriorly elongate; the angle formed by the principal cusps tends to be more obtuse than on the molars, with cusps B/b and C/c tending to be lower and more poorly developed; and molariform appearance increases posteriorly through the series.

There is as yet no evidence for replacement of molariforms among "symmetrodontans," as there is for some lineages of stem mammals (e.g., Gow, 1986a). However, at least one molar may have been added ontogenetically at the posterior end of the series in *Kuehneotherium* (Mills, 1984). Additionally, anterior premolars of *Kuehneotherium* were shed, without replacement, in aged individuals (Gill, 1974). Both features are associated with an ontogenetic pattern of posterior shifting in the postcanine dentition, thought to be primitive among mammals (Butler and Clemens, 2001). The uniqueness of this pattern to *Kuehneotherium* among "symmetrodontans" is consistent with other anatomical data that distinguish *Kuehneotherium* by its numerous plesiomorphies from more derived spalacotheriids and highlights the problems inherent in recognition of a broad-based "Symmetrodonta" that include Late Triassic–Early Jurassic forms.

Enamel Microstructure. To date, study of enamel microstructure among "symmetrodontans" has been limited to three taxa: the kuehneotheriid *Kuehneotherium* (Sigogneau-Russell et al., 1984; Wood, 2000) and the spalacotheriids *Spalacotheridium* and *Spalacolestes* (Wood and Stern, 1997; Wood et al., 1999; Wood, 2000). As reported by Sigogneau-Russell et al. (1984), the enamel of *Kuehneotherium* has a columnar, prismless structure, in a pattern later termed synapsid columnar enamel (Sander, 1997). Enamel of a single specimen of *Kuehneotherium* examined by Wood (2000) proved to be virtually structureless, perhaps because of its extreme thinness. The pattern seen in both spalacotheriids, on the other hand, is the more derived plesiomorphic prismatic enamel (Wood et al., 1999; Wood, 2000). The significance of this distinction is unclear, but it is worth pointing out that it is consistent with the morphological differences between *Kuehneotherium* and Spalacotheriidae and their respective positions on the mammalian tree.

POSTCRANIAL SKELETON

Information on the postcranium of "symmetrodontans" is limited to the partial skeleton of the spalacotheriid *Zhangheotherium* (Hu et al., 1997, 1998, see our figures 9.2A, 9.4), which has not been fully described. The most notable feature of the axial column is the presence of unfused postaxial cervical ribs. This is a plesiomorphy also seen in morganucodontids, multituberculates, and the eutriconodont *Jeholodens* (see Ji et al., 1999); in living therian mammals (variable in monotremes), all postaxial cervical ribs are fused to their respective vertebrae. A V-shaped interclavicle (figure 9.4A) is present in the shoulder girdle of *Zhangheotherium*. This element is ontogenetically incorporated into the sternal manubrium in living therians (Klima, 1987); its retention as a separate element in *Zhangheotherium* is a plesiomorphy shared with morganucodontids, monotremes, multituberculates, and eutriconodontans. However, the lateral process of the interclavicle is reduced in comparison to the condition in *Ornithorhynchus*, and it has a mobile articulation (presumed to be a derived condition) with the clavicle. The scapula is triangle-shaped, with a tall spine and a well-developed supraspinous fossa, as in other trechnotherians. The procoracoid is lacking; presumably it was fused to the sternal apparatus, as in living therians, in contrast to the condition in morganucodontids and *Ornithorhynchus*. The humerus (figure 9.4B,C) retains several plesiomorphies, such as a large lesser tubercle (relative to the greater tubercle), some degree of torsion between proximal and distal ends, and a weakly developed ulnar condyle; however, an incipient ulnar trochlea (apomorphy) is also present. Similarly, the anterior part of the radial articulation is developed as a bulbous condyle (plesiomorphy), whereas its posterior aspect is cylindrical (apomorphy), as in other trechnotherians. The elbow joint is thus structurally intermediate between that of stem mammals (e.g., morganucodontans, docodontans), monotremes, and multituberculates, on the one hand, and living therians on the other. A generally similar elbow joint is found in eutriconodontans and "eupantotherians."

The pelvis includes the epipubis (figure 9.2A), the presence of which has now been documented for a number of groups of early mammals, including eutriconodontans, "eupantotherians," multituberculates, and primitive eutherians (e.g., Kielan-Jaworowska, 1969a, 1975a; Krebs, 1991; Novacek et al., 1997; Ji et al., 1999). In contrast to various stem mammal lineages and eutriconodontans, the femur is derived in having a medially directed head, distinct neck, dorsally directed greater trochanter, and lesser trochanter oriented ventromedially (rather than medially as in stem mammals). As in multituberculates (Kielan-

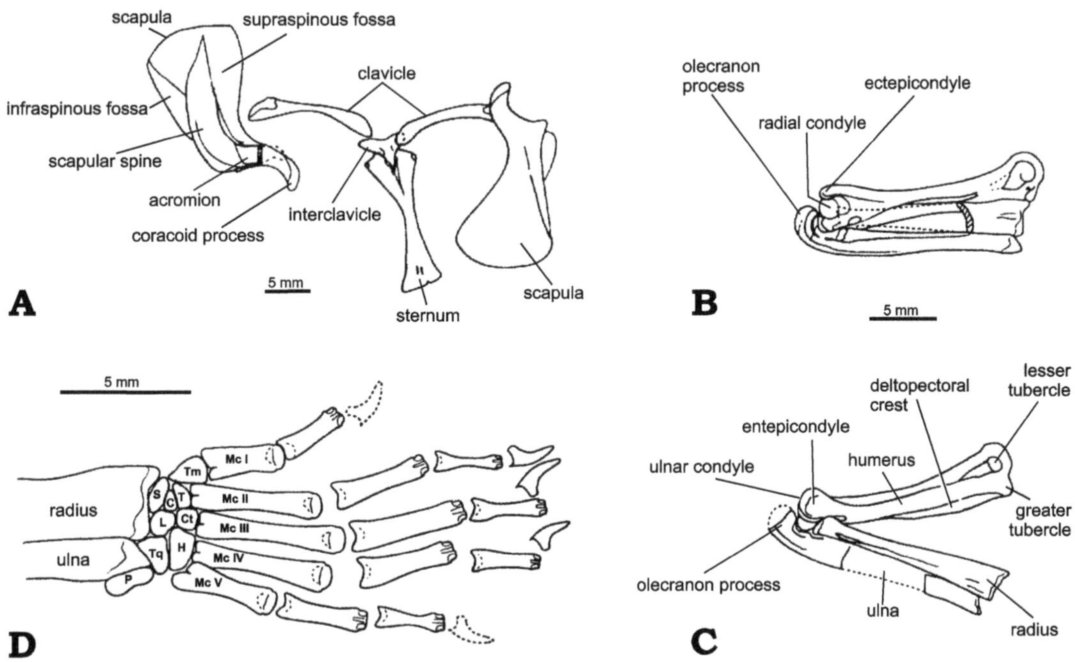

FIGURE 9.4. Shoulder girdle and forelimb of *Zhangheotherium quinquecuspidens*. A (left to right): Right scapula in lateral view, sternal complex, and left scapula in medial view. B, Right humerus, radius, and ulna, lateral view. C, Left humerus, radius, and ulna, medial view. D, Right hand, dorsal view. Abbreviations: C, centrale; Ct, capitum; H, hamatum; L, lunatum; Mc, metacarpals I through V, respectively; P, pisiform; S, scaphoideum; T, trapezoideum; Tm, trapezium; Tq, triquetrum. *Source:* modified from Hu et al. (1998).

Jaworowska and Gambaryan, 1994), the ankle shows partial superposition of the astragalus on the calcaneus, an intermediate condition between the minimally or nonsuperimposed proximal tarsals of morganucodontids, monotremes, and eutriconodontans, on the one hand, and the more fully superimposed elements of living therians, on the other. A distinctive feature is the presence of an external tarsal spur. A similar spur is found among living monotremes, mainly in males, where it is associated with a venom delivery system (Griffiths, 1978). Among Mesozoic mammals, tarsal spurs are also known for *Gobiconodon* (Jenkins and Schaff, 1988), multituberculates (Wible and Rougier, 2000), and the "eupantotherian" *Henkelotherium* (G. W. Rougier, pers. comm.).

PALEOBIOLOGY

Diet. With the exception of poorly known Amphidontidae (described by Simpson, 1925d: 469 as "functionally monocuspid," but subject to alternative interpretation; see Crompton and Jenkins, 1968; Cassiliano and Clemens, 1979), shearing was the primary function of "symmetrodont" molars. Shearing was largely or completely restricted to the opposing surfaces of reversed upper and lower molars (embrasure shear, Patterson, 1956). Embrasure shear in primitive taxa such as *Kuehneotherium* has been described as "rather inefficient" (Crompton et al., 1994: 341, see later). "Symmetrodont" molars also have tall, pointed cusps, suggesting some degree of adaptation for puncturing or "puncture crushing" (e.g., Crompton and Hiiemae, 1970; Osborn and Sita-Lumsden, 1978), particularly in primitive taxa such as *Kuehneotherium* (Mills, 1984). "Symmetrodontans" were almost uniformly small mammals; most were extremely small. With the caveat that the appropriateness of models among living mammals remains to be established, rough comparisons of jaw size suggest that one of the smaller "symmetrodontans," *Kuehneotherium praecursoris,* may have been about the size of the living shrew *Sorex trowbridgii,* which weighs between 4 and 5.5 g as an adult. At the other end of the spectrum, *Spalacotherium tricuspidens,* one of the larger "symmetrodontans," may have been comparable in size to *Monodelphis domestica,* which ranges from 70 to 155 g (body masses from Silva and Downing, 1995).

Simpson (1933a) compared "symmetrodontans" favorably with eutriconodontans and suggested carnivorous fare for both groups of early mammals. That hypothesis remains probable for eutriconodontans (chapter 7), but now appears unlikely for "symmetrodontans," wherein molar design for puncturing and embrasure shearing, coupled with small body size (e.g., Kay, 1975), argues convincingly for an insectivorous diet.

As with "eupantotherians," a notable feature of the "symmetrodont" record is the proliferation of "acute-angled" taxa and their relative success (as judged by diversity and abundance) compared to "obtuse-angled symmetrodontans." The acute-angled cusp pattern, which resulted in more transversely oriented embrasures, allowed for increased shearing capacity (as judged by total length of shearing surfaces) for a given jaw length; as noted, "acute-angled symmetrodontans" (e.g., Spalacotheriidae) generally have a higher molar count than do "obtuse-angled" taxa (Patterson, 1956). Several further lines of evidence indicate that shearing capacity and effectiveness was greater in Spalacotheriidae than in Tinodontidae or Kuehneotheriidae (see Crompton et al., 1994). Postvallum/prevallid shearing, restricted to limited surfaces of the upper and lower molars in "obtuse-angled symmetrodontans" (e.g., Mills, 1984), took place along the entire respective molar faces in Spalacotheriidae. Expansion of postvallum/prevallid shearing and consolidation of the primitively multiple occlusal features into a single set of shearing surfaces probably increased both capacity and effectiveness of shearing function. Corresponding shear surfaces of unworn (or newly erupted) teeth match each other more closely than in *Kuehneotherium* or *Tinodon,* so that precise fit was obtained without the need for a prolonged period of occlusal wear. As described above, cutting edges of the well-developed, closely matching shear surfaces in spalacotheriids appear to have been maintained through differential wear of vertical and apical surfaces. A correlative feature is the fact that the upper molars of Spalacotheriidae are considerably wider than the lowers; full use of shearing surfaces on upper molars was dependent on lateral translation of the dentary during occlusion. The masticatory cycle (as inferred from tooth occlusion, wear striations, and relative development of the relevant jaw adductors, the pterygoid, and masseter muscles) of spalacotheriids thus appears to have incorporated significantly greater lateral translation and rotation of the dentary than in *Kuehneotherium* or *Tinodon.*

Ongoing studies of the relationship between diet and dentition suggest that a refined interpretation of food preference in "symmetrodontans" and dentally similar mammals may soon be possible. In particular, attention to the importance of food texture within dietary category (Lucas, 1979) and to specializations for insectivory provide useful direction for such investigations (e.g., P. W. Freeman, 1979). Prey for insectivorous mammals varies considerably in hardness. Studies of extant insectivores (bats, primates, lipotyphlans) show that attributes of the cranium, mandible, and dentition vary in predictable ways, depending on prey preference. Mammalian insectivores with a predilection for soft-bodied prey (e.g., larvae, earthworms, lepidopterans) are characterized by greater molar shearing capacity, with more transversely oriented shearing surfaces, smaller but more numerous molars; smaller canines; and a more gracile mandible, among other features (P. W. Freeman, 1979, 1981, 1984; Strait, 1993b; see also Evans and Sanson, 1998). The reverse conditions apply to insectivores with a preference for hard-bodied prey (e.g., coleopterans), which have a thick, chitinous exoskeleton. Hard-object feeders may feed opportunistically on soft-bodied prey, whereas molar design of soft-object feeders is poorly suited to processing hard-prey items (Strait, 1993b). In this sense, soft-object feeders appear to be more specialized, with a narrower feeding niche (P. W. Freeman, 1984).

Insectivory, long inferred for "symmetrodontans" and many other Mesozoic mammals (see various chapters in Lillegraven et al., 1979), is commonly viewed as a generalized, primitive feeding strategy (e.g., Eisenberg, 1981). The foregoing consideration of food texture suggests that this perspective is oversimplified and may be misleading. Analogy with extant mammals suggests the working hypothesis that most Spalacotheriidae, at least, were probably dietary specialists: they appear to have been adapted specifically to eating soft-bodied prey and thus were confined to a narrow range of feeding opportunities (a possible exception is *Spalacotherium* itself, the largest spalacotheriid). Such a specialized feeding strategy could well have influenced other aspects of spalacotheriid ecology, such as the degree and nature of specificity in habitat preference. Hypotheses regarding dietary preference in "symmetrodontans" might be tested by detailed morphologic comparisons to extant forms; dental microwear shows special promise for specific food preferences among insectivorous mammals (Strait, 1993a).

Locomotion and Substrate Preference. The single known "symmetrodontan" skeleton, that of *Zhangheotherium,* shows no striking, habitat-specific modifications. The small body size of *Zhangheotherium* and other "symmetrodontans" implies similar functional requirements for terrestrial and arboreal habitats (Jenkins, 1974a). Like living therians, the shoulder girdle was fully mobile; in habitual posture, however, the forelimb was noticeably abducted and in this respect was intermediate between the parasagittal or slightly abducted forelimb of most living therians (Jenkins, 1971a) and the more sprawling posture of monotremes (Jenkins, 1970c; Pridmore, 1985). The articulations of the wrist suggest considerable flexion-extension capability; the hand is of the "spreizhand" pattern (Altner, 1971, see full account in chapter 13) common to many small mammals, both arboreal and terrestrial. Hu et al. (1998) suggested that, like many small generalized mammals, *Zhangheotherium* was capable of both climbing and walking, but may have had a preference for a terrestrial habitat.

Preliminary studies show a correlation between dental microwear features and substrate preference in small, insectivorous mammals (Cifelli et al., 1988), suggesting the possibility of inference based on nonskeletal data. Dental microwear patterns of "symmetrodontans" have not yet been studied. However, it is worth pointing out that the only published image in which microwear can be seen clearly (Cifelli and Madsen, 1999: figure 18) suggests a terrestrial habitat preference for one individual of the species *Spalacolestes cretulablatta.*

SYSTEMATICS

Owen's (1854) description of *Spalacotherium* represents the first account of a mammal later considered a "symmetrodont." He subsequently (1871) placed *Spalacotherium* among the Marsupialia, to which Mesozoic mammals were commonly referred until early in the twentieth century. Marsh, on the other hand, placed *Spalacotherium* and all other nonallotherian Mesozoic mammals in his order Pantotheria, which he alternately viewed as ancestral to all living therians (Marsh, 1880) or allied more specifically with eutherians (Marsh, 1887). Marsh did not attribute great significance to differences between symmetrodont and triconodont molar patterns, resulting in some mixed and confusing taxonomic groupings (Spalacotheriidae and Tinodontidae, see later). Subsequent workers continued to ally "symmetrodontans" with "triconodonts" (e.g., Osborn, 1888b, 1907; Gregory, 1910), until the comprehensive and authoritative work of Simpson (1925d, 1928a, 1929a). Simpson's grouping of Spalacotheriidae (Marsh, 1887) and Amphidontidae (Simpson, 1925d) into the order Symmetrodonta reflected his belief that "triconodont" and "symmetrodont" molar types are fundamentally dissimilar and that each originated separately from a simpler (single-cusped) design among nonmammalian synapsids. This view was founded, in turn, on Simpson's conviction that mammals arose in a polyphyletic fashion. Mammalian polyphyly was to remain a dominant hypothesis until the 1960s (see review by Hopson and Crompton, 1969, and chapter 3).

Simpson (e.g., 1936a, 1945) tentatively recognized "symmetrodontans" as distant relatives of "eupantotherians" and living therians. Detailed study led Patterson (1956; see also Crompton, 1971) to recognize the symmetrodont molar pattern as structurally antecedent to that of "eupantotherians," with the acquisition of reversed-triangle cusp design representing a significant initial step ("the symmetrodont stage") in the evolution of therian molars. In turn, description of some of the geologically oldest (Late Triassic–Early Jurassic) mammals later provided convincing evidence for derivation of the symmetrodont molar pattern from a triconodont configuration (Crompton and Jenkins, 1968). With general acceptance of mammalian monophyly, the reversed-triangle cusp pattern was identified as a fundamental specialization of Theria. Accordingly, "symmetrodontans" came to be regarded as basal therians (Crompton and Jenkins, 1968; Hopson, 1969; McKenna, 1975).

Later authors have advocated a taxon- and crown-based definition of Theria, restricting the group to include the last common ancestor of marsupials + placentals and all of its descendants (e.g., Rowe, 1988, 1993). Hopson (1994) informally introduced the name Holotheria to replace the traditional, broad-based concept of "Theria"—that is, a major clade of Mammalia diagnosed by the possession of a reversed-triangle molar pattern. The term Holotheria was formalized by Wible et al. (1995) as a clade (of unspecified rank) including the last common ancestor of *Kuehneotherium* and living therians plus all of its descendants. Wible et al. (1995) and, later, McKenna and Bell (1997) excluded both monotremes and multituberculates from Holotheria. The definition and practical utilization of Holotheria are problematic (see Luo et al., 2002). For example, *Kuehneotherium* itself (a taxon critical to the definition of the group) is poorly known. The phylogenetic position of *Kuehneotherium* is unstable, leading to similar instability in contents of Holotheria. Significantly, however, recent studies place Monotremata within Holotheria, as defined on the basis of *Kuehneotherium* (see review by Cifelli, 2001). Finally, recent analyses also suggest that the diagnosing character for Holotheria, molars of reversed-triangle pattern, appears to have evolved iteratively within the group; the pattern is lacking in some apparent members of Holotheria, such as Eutriconodonta (Luo et al., 2002, see chapter 7). With these fundamental issues on the table, at present we see no practical utility in the use of the term Holotheria: its contents are uncertain, it does not serve the purpose for which it was intended, and it is diagnosed by a single character that is of dubious value.

Problems in addressing the relationships of "symmetrodontans" to each other and their placements on the mammalian tree stem partly from the fact that we know little about them. Common possession of triangulated molars—all that is known for many taxa—does not represent a unique apomorphy; it simply defines a structural grade. Prothero (1981) tentatively recognized a restricted Symmetrodonta (Amphidontidae + Tinodontidae + Spalacotheriidae as used herein) as monophyletic, based on the presence of a strong pterygoid crest, reduction of lower molar talonids, and reduction to four premolars. The diagnostic value of each of these characters is questionable, however (Sigogneau-Russell and Ensom, 1998; Cifelli and Madsen, 1999). A more comprehensive survey of taxa and characters found no support for a monophyletic grouping of Spalacotheriidae + Tinodontidae (Luo et al., 2002), and remaining "symmetrodontans" are so incompletely known that their relationships are largely speculative.

A second problem relating to the affinities of "symmetrodontans" and their possession of triangulated molars concerns the uniqueness of the pattern itself. Recognition of a monophyletic grouping on the basis of the triangulated molar pattern implies homoplasies in other characters and character complexes. Most notable among these are features of the jaw and auditory system; *Kuehneotherium* and a number of other early mammals retained attached postdentary elements, implying multiple origin (or reversal) of the mammalian middle ear (e.g., Hopson, 1966; Allin, 1986; Allin and Hopson, 1992; McKenna and Bell, 1997). The problem is highlighted by results of analyses by Luo et al. (2002), which place taxa with nontriangulated molars within Holotheria. Given these uncertainties, we do not recognize any suprafamilial grouping for "symmetrodontans."

We start systematic descriptions with genera assigned to family *incertae sedis* because of their primitive structure.

Class Mammalia Linnaeus, 1758
Order and Family *incertae sedis*
Genus *Atlasodon* Sigogneau-Russell, 1991b
(figure 9.5A)

Diagnosis (based on Sigogneau-Russell, 1991b; Sigogneau-Russell and Ensom, 1998). Upper molars primitive with respect to Spalacotheriidae in being obtuse-angled and strongly asymmetrical; derived with respect to Tinodontidae in presence of cusp on distal moiety of stylar shelf, incorporation of metacone into distal trigon wall, and presence of a single, extensive postvallum shearing surface.

Species. *Atlasodon monbaroni* Sigogneau-Russell, 1991b, type species by monotypy.

Distribution. Early Cretaceous (?Berriasian): Morocco, Talsinnt Province, Synclinal d'Anoual (Séquence B des Couches Rouges).

Comments. This genus is poorly known, being founded on a single upper molariform tooth of the type species. Averianov (2002) has suggested that the tooth may be a deciduous premolar.

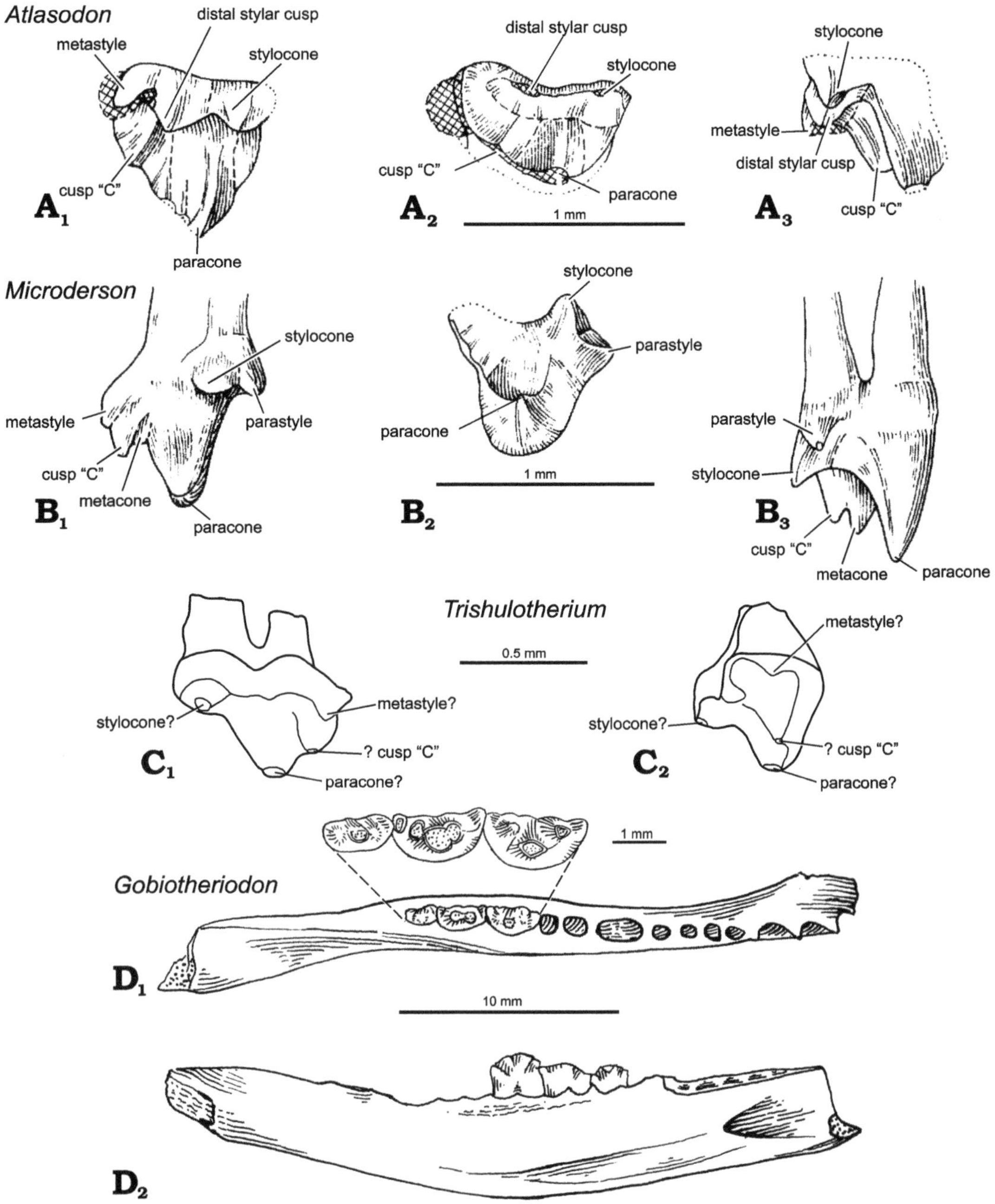

FIGURE 9.5. "Symmetrodontans" of uncertain affinities. A, *Atlasodon monbaroni*, right upper molar (holotype) in labial (A_1), occlusal (A_2), and mesial (A_3) views. B, *Microderson laaroussii*, right upper molar (holotype) in labial (B_1), occlusal (B_2), and mesial (B_3) views. C, *Trishulotherium kotaensis*, ?left upper molar (holotype) in labial (C_1) and oblique distal (C_2) views. D, *Gobiotheriodon infinitus*, right dentary and last three molars in occlusal (D_1) and lateral (D_2) views. *Source:* A, B, modified from Sigogneau-Russell and Ensom (1998); C, modified from Prasad and Manhas (1997); D, modified from Trofimov (1980).

Genus *Gobiotheriodon* Trofimov, 1997
(figure 9.5D)

Synonym: *Gobiodon* Trofimov, 1980 (preoccupied).

Diagnosis. Resembles Kuehneotheriidae and differs from Tinodontidae, Amphidontidae, and Spalacotheriidae in presence of posteriorly broad Meckel's groove that is confluent with the mandibular foramen (plesiomorphy). Resembles some or all Spalacotheriidae and differs from Kuehneotheriidae and Tinodontidae in reduction or loss of cusp f on lower molars (note variability in *Kuehneotherium;* however, see Godefroit and Sigogneau-Russell, 1999), in development of strong, continuous prevallid shearing surface, and in lateral deflection of posteroinferior margin of dentary (apomorphies). Uniquely derived among "symmetrodontans" in: hypertrophy of cusp e into tall, anteriorly positioned cusp, hyperdevelopment of prevallid shearing surface and its extension onto labial face of cusp e, and drastic reduction of postvallid shearing surface.

Species. *Gobiotheriodon infinitus* (Trofimov, 1980), type species by monotypy.

Distribution. Early Cretaceous (Aptian or Albian): Mongolia, Gobi Desert, Höövör.

Comments. This taxon was first mentioned in the literature, as *Gobiodon infinitus,* by Belyaeva et al. (1974), but as the species was not described, it remained a *nomen nudum* until it was formally published by Trofimov (1980). However, the name *Gobiodon* is preoccupied by a genus of osteichthyan, which prompted Trofimov (1997) to transfer this species to the new genus *Gobiotheriodon.*

The holotype of *G. infinitus* consists of a nearly complete horizontal ramus of a dentary bearing the last three molars and alveoli for preceding teeth. The postcanine dental formula has been alternately interpreted as p4, m5 (assuming single-rooted premolars, Trofimov, 1980) or p3, m4 (assuming double-rooted premolars and different placement of the premolar-molar boundary, Averianov, 2002). Trofimov (1980, 1997) also referred two M3s to the species; Fox (1984a) questioned this referral, and Averianov (2002) removed one M3 (the other cannot be located) to *Gobiconodon,* thereby restricting *G. infinitus* to the holotype alone.

Trofimov (1980) placed *Gobiotheriodon* in Amphidontidae, a referral accepted without restudy by some authors (e.g., McKenna and Bell, 1997; Cifelli and Madsen, 1999; Kielan-Jaworowska et al., 2000), but questioned by others. Fox (1984a) considered *Gobiotheriodon* of uncertain affinities, whereas Sigogneau-Russell and Ensom (1998) and Averianov (2002) considered it to be a possible late-surviving member of Tinodontidae. Many interpretations are viable, given the poor state of knowledge of *Gobiotheriodon* and structurally similar mammals. The cusp pattern is strikingly obtuse only on the last molar, which is rather reduced; otherwise, cusp angulation is similar to that of either Tinodontidae or certain Spalacotheriidae, depending on tooth locus being compared (see Sigogneau-Russell and Ensom, 1998). Meckel's groove is broadly open posteriorly, but the significance of this is unclear: is it possible that postdentary elements remained attached, at least anteriorly, or, perhaps, was an ossified Meckel's cartilage present in *Gobiotheriodon,* as in the eutriconodontans *Gobiconodon* and *Repenomamus* (see Wang, Hu, Meng et al., 2001)? Most striking about the dentition of *Gobiotheriodon* are the hyperdevelopment of the prevallid shearing surface (including its extension onto a hypertrophied cusp e) and the reduction of the primary, postvallid shearing surface: these features sharply distinguish the genus from all other mammals with a symmetrodont molar pattern.

Genus *Microderson* Sigogneau-Russell, 1991b
(figure 9.5B)

Diagnosis (based on Sigogneau-Russell and Ensom, 1998). Upper molars similar to Spalacotheriidae in acuteness of cusp angle, but differ in narrower trigon, presence of three lobes on postparacrista, presence of two postvallum wear facets and a median ridge, labial projection of stylocone, and presence of three roots.

Species. *Microderson laaroussii* Sigogneau-Russell, 1991b, type species by monotypy.

Distribution. Early Cretaceous (?Berriasian): Morocco, Talsinnt Province, Synclinal d'Anoual (Séquence B des Couches Rouges).

Comments. This genus, based on a single upper molar of the type species, was initially referred to Spalacotheriidae (Sigogneau-Russell, 1991b), but later Sigogneau-Russell and Ensom (1998) assigned it to family *incertae sedis* because of autapomorphies mentioned in the foregoing diagnosis.

Genus *Trishulotherium* Yadagiri, 1984
(figure 9.5C)

Diagnosis. Not diagnosable based on published data.

Species. *Trishulotherium kotaensis* Yadagiri, 1984, type species by monotypy.

Distribution. Early Jurassic: India, Andhra Pradesh (Kota Formation).

Comments. Knowledge of this species is limited to the holotype, a heavily worn molariform tooth. The original diagnosis of *T. kotaensis* was predicated on identification of the type as a lower molar (Yadagiri, 1984; Prasad and Manhas, 1997). We concur with Sigogneau-Russell and Ensom (1998) in considering the specimen to be an upper molar with obtuse cusp angulation, but cannot attempt a revised diagnosis based on available illustrations.

Family Amphidontidae Simpson, 1925d

Diagnosis. Lower molars with obtuse angulation of principal cusps; apparently differ from other "symmetrodontans" in the extreme weakness of paraconid and metaconid (apomorphy), which are developed as minor undulations on the tooth crests rather than distinct cusps. Postcanine tooth formula p4, m4 (apomorphic for both premolars and molars). Differ from both Tinodontidae and Spalacotheriidae in strong development of cusp c on p4, presence of a marked diastema between lower canine and p1; and lack of lingual cingulid, mesial cuspules (e and f), and interlocking of lower molars (apomorphies).

Genera. *Amphidon* Simpson, 1925d, type genus; ?*Manchurodon* Yabe and Shikama, 1938; and ?*Nakunodon* Yadagiri, 1985.

Distribution. Late Jurassic (late Kimmeridgian–early Tithonian): United States, western states (Morrison Formation); possibly Early Jurassic: India; and ?Middle or ?Late Jurassic: China.

Comments. The diagnosis is based almost wholly on *Amphidon,* and even in this respect there remains margin for alternative interpretation (see *Amphidon,* later). Status of the family and affinities of included genera are uncertain. We have followed the traditional interpretation of postcanine dental formula in *Amphidon* (four premolars and four molars, see Simpson, 1929a). Averianov (2002), however, suggested that the fourth tooth in the holotype and only specimen of *Amphidon superstes* is a molar, thus yielding a formula of p3 m5 and removing the supposed difference from Tinodontidae and Spalacotheriidae in structure of the last premolar.

Simpson (1925d,e) believed the molar morphology of Amphidontidae to approximate the primitive pattern for "symmetrodontans." Patterson (1956), with the benefit of newly described fossils for comparison, tentatively suggested the alternative view that amphidontids are specialized in having reduced (rather than incipient) paraconid and metaconid. This view has been upheld by subsequent studies (Crompton and Jenkins, 1968; Prothero, 1981). Taken at face value, the unusual molar morphology indicates reduction of shearing function and emphasis on puncturing, in turn suggesting a diet favoring engorged ticks or similar fare.

Genus *Amphidon* Simpson, 1925d
(figure 9.6A)

Diagnosis. As for the family.

Species. *Amphidon superstes* Simpson, 1925d, type species by monotypy.

Distribution. Late Jurassic (late Kimmeridgian–early Tithonian): United States, Wyoming (Morrison Formation).

Comments. *A. superstes* is known only by the holotype, a dentary fragment with the last premolar (p4) and molar series (m1–4, each damaged). Study of the specimen by Crompton and Jenkins (1968) indicated that the molars are considerably worn, so that the original heights of paraconid and metaconid are difficult to judge. Simpson (1925d) based a second species, *A. aequicrurius,* on an upper molar; he later (1929a) transferred this to its own genus, *Eurylambda* (now considered as representing the upper dentition of *Tinodon;* see later).

Genus ?*Manchurodon* Yabe and Shikama, 1938
(figure 9.6B)

Diagnosis. Not diagnosable based on available data.

Species. *Manchurodon simplicidens* Yabe and Shikama, 1938, type species by monotypy.

Distribution. ?Middle or Late Jurassic: China, Liaoning Province (Wa-fang-dian Formation).

Comments. The holotype of *M. simplicidens,* consisting of a dentary and possibly associated partial scapula, was reportedly lost during World War II (Cassiliano and Clemens, 1979). Neither the original description nor illustrations (Yabe and Shikama, 1938) are particularly informative, and the morphology shown in the latter appears somewhat stylized and of questionable fidelity (see figure 9B). As in *Amphidon,* a diastema separated the canine and first premolar. The presence of such a diastema is unusual among mammals with a symmetrodont molar pattern and is a possible apomorphy uniting the two genera. Eight postcanine teeth were present; we tentatively favor the interpretation that the dental formula was p4, m4 (Patterson, 1956), as in *Amphidon,* rather than p3, m5 (Yabe and Shikama, 1938; Averianov, 2002). Assuming this dental formula, both it and the unusually well-developed cusp c on p4 appear to be further apomorphies shared with *Amphidon.* The molars, illustrated diagrammatically and only in labial view, appear to have been similar to those of *Amphidon,* with reduced paraconid and metaconid.

Genus ?*Nakunodon* Yadagiri, 1984
(not figured)

Diagnosis. Not diagnosable based on available data.

Species. *Nakunodon paikasiensis* Yadagiri, 1984, type species by monotypy.

Distribution. Early Jurassic: India, Andhra Pradesh (Kota Formation).

Comments. The holotype and only specimen of *N. paikasiensis,* an incomplete tooth fragment said to represent an upper molar, has not yet been described and illustrated in detail, so meaningful comparisons cannot be made at present (see Sigogneau-Russell and Ensom, 1998). The specimen is said to be "monodont" (Yadagiri, 1985: 415), with strongly reduced parastylar and metastylar cusps (Prasad and Manhas, 1997). Referral to Amphidontidae, not otherwise known by upper molars, must be regarded as highly tentative, as it is based on analogy rather than direct comparison.

Family Bondesiidae Bonaparte, 1990

Diagnosis. Lower molars of moderately "obtuse-angled symmetrodont" pattern, differing from *Kuehneo-*

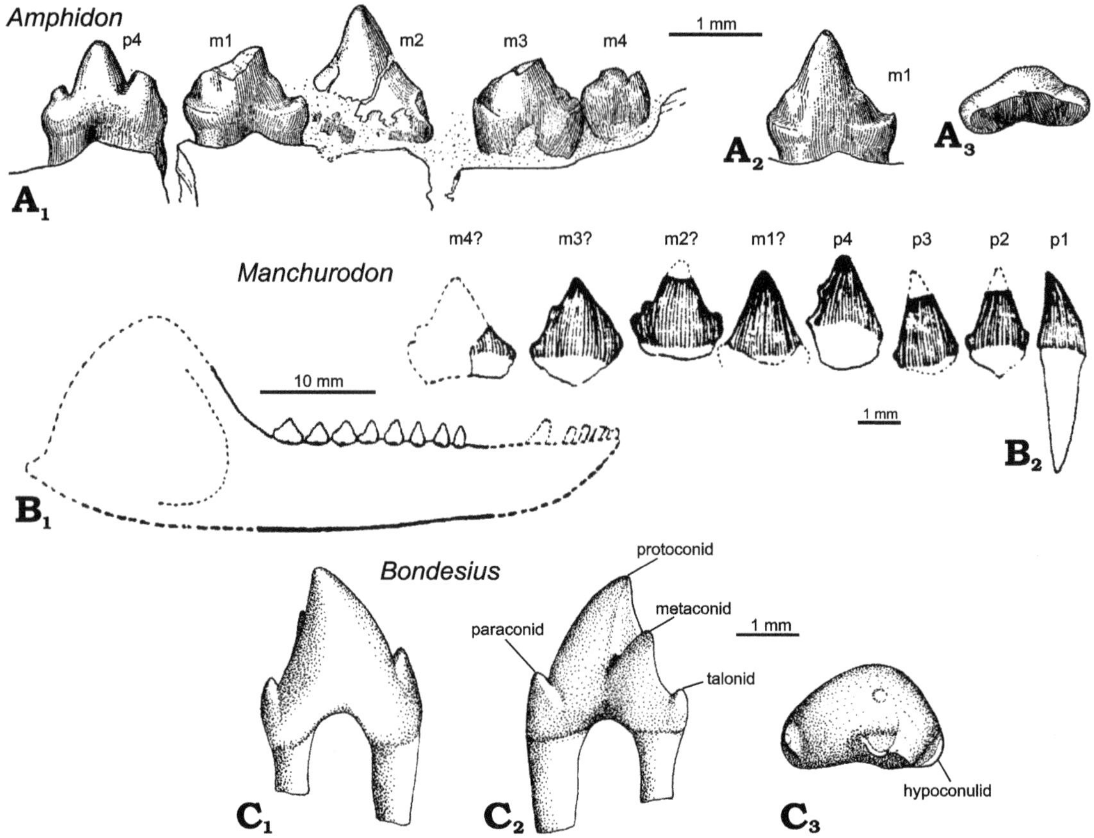

FIGURE 9.6. Amphidontidae and Bondesiidae. A, *Amphidon superstes:* right p4 and m1–4 (holotype) in lingual view (A_1); lingual (A_2) and occlusal views of restored m1 (A_3). B, *Manchurodon simplicidens,* holotype. Lateral view of restored dentary (B_1); labial view of cheek teeth, tentatively identified as p1–4, m1–4 (B_2). C, *Bondesius ferox,* right lower molar (holotype) in labial (C_1), lingual (C_2), and occlusal (C_3) views. *Source:* A, modified from Simpson (1929a); B, modified from Yabe and Shikama (1938); C, modified from Bonaparte (1990).

therium and *Tinodon* in having inflated protoconid (apomorphy); a relatively taller metaconid that is more lingually placed and almost transversely aligned with protoconid; smaller paraconid (developed as a low cuspule just lingual to the center of the tooth); lack of paracristid and associated prevallid shearing surface(s); lack of cusps e, f, and of basal cingulid (all presumed apomorphies).

Genera. *Bondesius* Bonaparte, 1990, type genus by monotypy.

Distribution. Late Cretaceous (Campanian or Maastrichtian): South America, Argentina, Río Negro Province.

Comments. *Bondesius* is one of five genera originally described as a "symmetrodontan" from the Los Alamitos local fauna, Argentina (Bonaparte, 1990). The other four (*Casamiquelia, Brandonia, Barberenia, Quirogatherium*) were later referred to Dryolestida among "eupantotherians" (Bonaparte, 1994; see also Sigogneau-Russell and Ensom, 1998, and chapter 10). The fauna from Los Alamitos is clearly an endemic one, and interpretation of tooth morphology is further complicated by the probability that at least two genera (*Barberenia, Quirogatherium*) appear to be represented by deciduous premolars (Martin, 1999a; Averianov, 2002, see chapter 10).

Genus *Bondesius* Bonaparte, 1990
(figure 9.6C)

Diagnosis. As for the family.

Species. *Bondesius ferox* Bonaparte, 1990, type species by monotypy.

Distribution. Late Cretaceous (Campanian or Maastrichtian): Argentina, Río Negro Province (Los Alamitos Formation).

Comments. *Bondesius ferox* was originally described on the basis of the holotype lower molar and a referred specimen thought to be an upper molar (Bonaparte, 1990). The referred specimen, of baffling identity (see Bonaparte, 1990: figure 3G–L), was later implicitly removed from the species (Bonaparte, 1994). Comparatively speaking, at least, *B. ferox* was a large mammal; at more than 2.5 mm long, the holotype is by far the largest molar of "symmetrodont" pattern (Sigogneau-Russell and Ensom, 1998). In overall appearance, *Bondesius* somewhat resembles primitive "symmetrodontans" such as *Kuehneotherium,* though it is clearly distinguished by apomorphies listed in the previous diagnosis. Wear facets and shearing structures have not been described; identity of the holotype as a molar seems most probable, but the possibility that it represents a molarized premolar cannot be dismissed based on published evidence. Averianov (2002) suggested that the holotype of *B. ferox* is a deciduous premolar.

Family Kuehneotheriidae D. M. Kermack, K. A. Kermack, and Mussett, 1968

Diagnosis. Small, plesiomorphic mammals with "obtuse-angled symmetrodont" molar pattern. Dentary differs from otherwise similar Morganucodontidae in being relatively longer and more gracile, with lower coronoid process; and lacking angular process (apomorphies?). Dentary resembles Morganucodontidae and differs from that of Tinodontidae, Spalacotheriidae, and other "symmetrodontans" in continuity of Meckel's groove with mandibular foramen, presence of postdentary trough and overlying ridge, and emargination of posteroventral border (plesiomorphies). Dentition differs from that of all other mammals with symmetrodont molar pattern in greater number of premolars (up to six; probable plesiomorphy), ontogenetic posterior shift in tooth row (progressive loss of anterior premolars and plugging of alveoli, addition of molar at back of series; plesiomorphies), and more extensive upper molar cingula (plesiomorphy). Resembles Tinodontidae and differs from Spalacotheriidae and other Trechnotheria in imprecise "fit" of shearing surfaces between newly erupted upper and lower molars and in lesser and discontinuous development of postvallum/prevallid shearing, represented by three distinct facets instead of one (plesiomorphy).

Genera. *Kuehneotherium* D. M. Kermack et al., 1968, type genus; *Kotatherium* Datta, 1981; and *Kuehneon* Kretzoi, 1960, *nomen dubium.*

Distribution. Late Triassic–Early Jurassic: India, Greenland, and Western Europe.

Comments. The discovery of a "symmetrodontan" molar in one of the geologically oldest mammal assemblages (a Welsh fissure fill now regarded as of Early Jurassic age) was first reported by Kühne (1950) and widely discussed in following years (see summary by Parrington, 1971). The specimen went unnamed for some time (see account of *Kuehneon,* later). In the meantime, similar fossils were being collected, also from fissure fill deposits in Britain; these, too, were cited for some years without formal treatment (Kermack et al., 1956; Kermack, 1965, 1967b; Parrington, 1967). The foregoing diagnosis is based solely on *Kuehneotherium,* the only genus that is reasonably well represented. Ironically, despite the comparative wealth of fossils available, existing reports on *Kuehneotherium* (Kermack et al., 1968 ; Parrington, 1971; Mills, 1984) are still preliminary in some respects; for example, it remains uncertain how many species are represented, or even if all are congeneric.

K. A. Kermack et al. (1965), K. A. Kermack (1967a,b), and D. M. Kermack et al. (1968) considered Kuehneotheriidae to be specially allied to "eupantotherians" (as used herein) and, ultimately, tribosphenic mammals. *Kuehneotherium* possesses a lingual cingulum on upper molars and, compared to most "symmetrodontans," a large hypoconulid on lowers (note variability reported by Godefroit and Sigogneau-Russell, 1999); these structures were viewed as predecessors to the protocone and expanded talonid, respectively, of more advanced mammals. Detailed study of tooth occlusion by Crompton and Jenkins (1967, 1968) shows that the molars of *Kuehneotherium* are fundamentally similar to those of morganucodontids, on the one hand, and primitive "symmetrodontans" (*Tinodon*), on the other, and that they have no special resemblance to "eupantotherans." This view has been upheld by cladistic studies (Prothero, 1981; Hopson, 1994; Luo et al., 2002).

Passing mention should be made of *Cyrtlatherium,* from the Middle Jurassic (Bathonian) of England. *Cyrtlatherium* was described as a kuehneotheriid (E. F. Freeman, 1979) and later referred to Tinodontidae by McKenna and Bell (1997). We follow Sigogneau-Russell (2001) in recognizing *Cyrtlatherium* as a docodont (chapter 6).

Genus *Kuehneotherium* D. M. Kermack, K. A. Kermack, and Mussett, 1968
(figures 9.1B, 9.3A, 9.7A)

Diagnosis. "Obtuse-angled symmetrodontan" differing from *Kotatherium* in having upper molars with complete lingual cingulum; stronger paracone, cusp "C," stylocone, and parastyle; cusp "C" more labially placed and in line with paracone and metastyle; metastyle connected to cusp "C" by postmetacrista. Differs from dentally similar *Tinodon* in retention of plesiomorphies on the dentary: presence of postdentary trough and overlying ridge, continuity of Meckel's groove with mandibular foramen, and feeble development of coronoid process. Dentition differs from *Tinodon* in canine and p1–4 being single-rooted; greater number of postcanine teeth (p5–6, m5–6, as opposed to p3, m4); ontogenetic posterior shift in dentition (loss of anterior premolars and plugging of their alveoli, addition of molar at back of series); upper molars less labiolingually compressed, lingual cingulum complete and labial cingulum cuspidate, well-developed parastyle present, metastyle more hooklike and better separated from cusp "C"; lower molars with cusps e and f better sep-

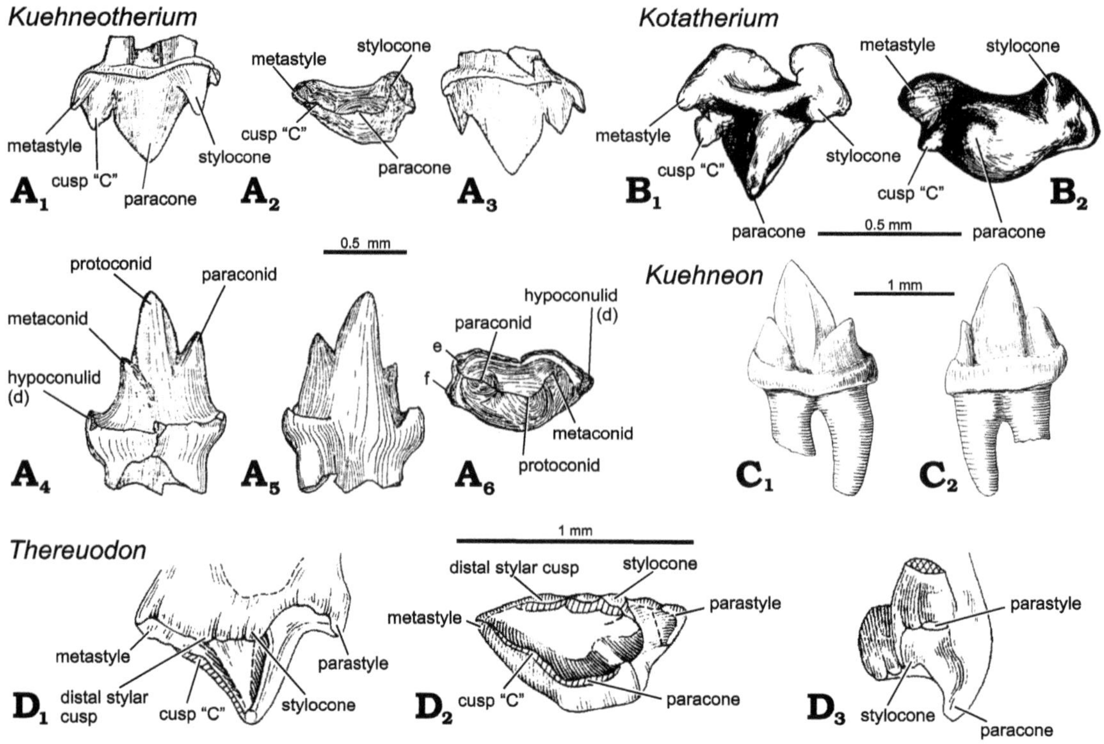

FIGURE 9.7. Kuehneotheriidae and Thereuodontidae. A, *Kuehneotherium praecursoris.* A_{1-3}, right upper molar (holotype) in labial (A_1), occlusal (A_2), and lingual (A_3) views. A_{4-6}, left lower molar in lingual (A_4), labial (A_5), and occlusal (A_6) views. B, *Kotatherium haldanei,* right upper molar (holotype) in labial (B_1) and occlusal (B_2) views. C, *Kuehneon duchyense, nomen dubium,* lower molar (holotype) in lingual (C_1) and labial (C_2) views. D, *Thereuodon taraktes,* right upper molar (holotype) in labial (D_1), occlusal (D_2), and mesial (D_3) views. *Source:* A, modified from Kermack et al. (1968); B, modified from Datta (1981); C, from Kühne (1958); D, modified from Sigogneau-Russell and Ensom (1998).

arated and with hypoconulid larger, less lingually situated, and more posteriorly projecting.

Species. *Kuehneotherium praecursoris* D. M. Kermack et al., 1968, type species by monotypy, and species left in open nomenclature (Mills, 1984; Fraser et al., 1985; Jenkins et al., 1994; Sigogneau-Russell and Hahn, 1994; Godefroit et al., 1998; Godefroit and Sigogneau-Russell, 1999).

Distribution. Late Triassic (Norian)–Early Jurassic (?Sinemurian): Britain; Late Triassic (Rhaetian): France and Luxembourg; Late Triassic (Norian): Greenland, Jameson Land.

Comments. The holotype (an upper molar) and a limited hypodigm (nine additional teeth and four dentary fragments) of *Kuehneotherium praecursoris* were described by Kermack et al. (1968). These specimens, some illustrated and/or described earlier (Kermack et al., 1956; Kermack and Mussett, 1958b, Kermack et al., 1965), are reportedly part of a much larger sample taken from a single pocket in Pontalun Quarry, near Brigend, Wales. Additional fossils (reportedly including more than 100 teeth and several jaw fragments) of *K. praecursoris* from Pontalun Quarry (although evidently not the same pocket) were described by Parrington (1971).

Kermack et al. (1968) reported the presence of another taxon, allied but distinct (at the generic level, according to the authors), based on a large sample of fossils from the nearby Pant Quarry site. This sample has never been studied or described, but one of two additional samples from the quarry was reported by Mills (1984: 190), who simply referred to the "symmetrodont" as "the Pant *Kuehneotherium.*"

Kuehneotherium has been reported from the Triassic of France, based on a comparatively large sample of isolated teeth. The specimens show considerable variability and include the range of variation known for *K. praecursoris* (see Godefroit and Sigogneau-Russell, 1999).

Genus *Kotatherium* Datta, 1981
(figure 9.7B)

Diagnosis. Upper molars with asymmetrical, "obtuse-angled symmetrodont" cusp pattern, differing from *Kuehneotherium* in the lingual cingulum being weak or absent;

paracone, cusp "C," stylocone, and parastyle smaller; stylocone placed lower on crown and smaller than cusp "C"; cusp "C" more lingually placed and separated from metastyle by deeper valley. Differs from *Tinodon* in presence of parastyle and stronger development of metastyle and stylocone.

Species. *Kotatherium haldanei* Datta, 1981, type species by monotypy.

Distribution. Early Jurassic: India, Andhra Pradesh (Kota Formation).

Comments. Datta (1981) did not allocate *Kotatherium* to family, but clearly regarded the genus as structurally intermediate between the "obtuse-angled symmetrodontans" *Kuehneotherium* and *Tinodon.* Prasad and Manhas (1997) followed Fox's (1985) concept of Tinodontidae, wherein *Kuehneotherium* is included. We restrict Tinodontidae to the type genus and place *Kotatherium* with the structurally similar *Kuehneotherium.* A second species was originally referred to *Kotatherium,* as *K. yadagirii* Prasad and Manhas, 1997. Further study has revealed that this species is probably not a symmetrodont, and it has been referred to the eutriconodontan genus *Paikasigudodon* (see Prasad and Manhas, 2002; chapter 7).

Genus *Kuehneon* Kretzoi, 1960, *nomen dubium*
(figure 9.7C)

Diagnosis. Not diagnosable based on available data.

Species. *Kuehneon duchyense* Kretzoi, 1960, *nomen dubium,* type species by monotypy.

Distribution. Early Jurassic (?Sinemurian): Britain, Wales, fissure fillings.

Comments. The basis for *Kuehneon duchyense* is a single molariform tooth, described and illustrated by Kühne (1950, 1958), but formally named by Kretzoi (1960; see Cifelli and Madsen, 1986, for comments on spelling of the generic name). Early authors (e.g., Kühne, 1950; Patterson, 1956) believed the specimen to be a lower molar, in which case the presence of a complete cingulid purportedly distinguishes the taxon from *Kuehneotherium* (Kermack et al., 1968). Partly on this basis, Kermack et al. (1968) suggested the possibility that the specimen may be an upper molar, in which case it is distinct from *Kuehneotherium* in lacking the parastyle and metastyle and in having a taller paracone. These authors also noted that the tooth appears to be damaged and that it is not diagnostic. Unfortunately, collection of further specimens at the type locality (Duchy Quarry, which has been inactive for decades) is unlikely, and the single specimen of *Kuehneon duchyense* has been lost or mislaid (Cassiliano and Clemens, 1979). As the systematics of British Kuehneotheriidae remain to be addressed, however, it is possible that some or many existing specimens are referable to the species. For example, a complete labial cingulid is variably present in *Kuehneotherium praecursoris* (see Parrington, 1971: figure 12c).

Family Thereuodontidae Sigogneau-Russell and Ensom, 1998

Diagnosis (modified from Sigogneau-Russell and Ensom, 1998). Similar to Spalacotheriidae in acute angulation of principal cusps on upper molars and great height of paracone; resembles certain Spalacolestinae (e.g., *Spalacolestes*) in presence of stylar cusp on distal moiety of crown. Differs from Spalacotheriidae in transversely narrower, more asymmetrical profile of crowns (in occlusal view), with postparacrista much longer than preparacrista; better development of stylocone; occasional presence of three roots (invariably two where known for Spalacotheriidae); lack of continuous postvallum shear facet; and presence of stronger parastylar lobe.

Genera. *Thereuodon* Sigogneau-Russell, 1989a, type genus by monotypy.

Distribution. Early Cretaceous (Berriasian): Western Europe and North Africa.

Comments. Martin (1999a) suggested that *Thereuodon* may be based on deciduous premolars of a "eupantothere." As with other intriguing possibilities for the affinities of Berriasian mammals from Morocco, judgment is best withheld until the fauna is better known.

Genus *Thereuodon* Sigogneau-Russell, 1989a
(figure 9.7D)

Diagnosis. As for the family.

Species. *Thereuodon dahmanii* Sigogneau-Russell, 1989a, type species; *T. taraktes* Sigogneau-Russell and Ensom, 1998.

Distribution. Early Cretaceous (?Berriasian): Morocco; Early Cretaceous (Berriasian): Britain.

Comments. Perhaps most noteworthy of the enigmas surrounding *Thereuodon* is its geographic distribution. Terrestrial realms of Britain and northern Africa were separated by marine habitat during the Late Jurassic–Early Cretaceous. It is presently uncertain whether the species of *Thereuodon* were Berriasian relicts of a more broadly distributed ancestor dating back to the Middle Jurassic, or whether their distributions resulted from more recent migration via ephemeral land bridges and/or successive island colonization (Sigogneau-Russell and Ensom, 1998).

Family Tinodontidae Marsh, 1887

Diagnosis. "Obtuse-angled symmetrodontans" with molar pattern generally similar to that of Kuehneotheriidae. Dentition differs from Kuehneotheriidae in ?canine and all premolars being double-rooted (polarity uncertain; canine and p1–4 single-rooted in Kuehneotheriidae); fewer premolars and molars (apomorphy; p3, m4 as opposed to p5–6, m5–6 in Kuehneotheriidae); upper molars more labiolingually compressed, with reduced stylocone, parastyle weak or absent, lingual cingulum weak or absent and labial cingulum noncuspidate, metastyle less hooklike and better joined to cusp "C"; lower molars with more closely approximated cusps e and f and with cusp d (hypoconulid) smaller, more lingually placed, and less posteriorly projecting (presumed apomorphies). Dentary differs from that of Kuehneotheriidae in having taller, more anteroposteriorly expanded coronoid process with an anterior edge that arises almost vertically (apomorphy; coronoid process low, anteroposteriorly short, and posteriorly recumbent in Kuehneotheriidae); Meckel's groove well separated from mandibular foramen; postdentary trough, overlying ridge, and posteroventral emargination of dentary lacking; pterygoid crest present lingually (apomorphies).

Genera. *Tinodon* Marsh, 1879a, type genus by monotypy.

Distribution. Late Jurassic: North America; Early Cretaceous: Western Europe.

Comments. The family Tinodontidae was erected by Marsh (1887) to include *Tinodon,* from the Morrison Formation, United States, and *Phascolotherium,* from the Stonesfield Slate, Britain. As noted above, Marsh seems to have recognized little distinction between the "reversed-triangle" and "cusp-in-line" molar patterns. This resulted in a rather confused taxonomy, as reflected in the grouping of *Tinodon* (a "symmetrodont") with *Phascolotherium* (a "triconodont"). *Tinodon* itself at one time included not only the two species (dubiously distinct) now regarded as "symmetrodontans" (*T. bellus* and *T. lepidus*), but also *T. ferox* (Marsh, 1880; removed by Marsh, 1887, to *Priacodon*) and *T. robustus* (Marsh, 1879b; removed by Simpson, 1925b to *Priacodon*), which are triconodontids. At the same time, Marsh (1887) referred *Menacodon rarus* Marsh, 1887 (later found by Simpson, 1925d, to be a junior synonym of *Tinodon lepidus* Marsh, 1879) to the Spalacotheriidae. Simpson (1925b,d, 1928a, 1929a) provided a clear morphologic and systematic basis for segregating "triconodonts" from "symmetrodontans." He referred *Tinodon* to Spalacotheriidae, adding his then-new *Eurylambda* to the family (Simpson, 1929a). *Eurylambda* is now regarded as probably representing the upper dentition of *Tinodon,* otherwise known only by the dentary and lower dentition (see later).

For many years following Simpson's (1928a, 1929a) comprehensive revision of Mesozoic Mammalia, the concept of Tinodontidae was largely ignored or forgotten (the name is not even listed as a synonym in, e.g., Simpson, 1945). Detailed study by Crompton and Jenkins (1967) showed that molar occlusion in *Tinodon* differed in important respects from that of *Spalacotherium,* retaining primitive features reminiscent of *Kuehneotherium* (named as such a year later by Kermack et al., 1968, and until then commonly referred to as a "Rhaetic pantothere"). Crompton and Jenkins (1968) formalized this distinction by placing *Tinodon* (and *Eurylambda*) together with *Amphidon* in Amphidontidae Simpson, 1925d, apparently unaware of Marsh's prior family group named based on *Tinodon.* Prothero (1981) hypothesized *Tinodon* to represent the sister taxon to *Spalacotherium,* a relationship he recognized taxonomically by placing each in its own infraclass within his sublegion Spalacotherioidea. Fox (1985) revised Marsh's Tinodontidae and included *Kuehneotherium* within the family because of shared plesiomorphies in molar structure. This placement is unsatisfactory because of the great differences between *Kuehneotherium* and *Tinodon* in other aspects of the dentition, and (especially) features of the dentary. Our concept of Tinodontidae is based on *Tinodon* alone.

Genus *Tinodon* Marsh, 1879a
(figures 9.3B, 9.8)

Diagnosis. As for the family.

Species. *Tinodon bellus* Marsh, 1879a, type species; *T. micron* Ensom and Sigogneau-Russell, 2000; and a species left in open nomenclature by Engelmann and Callison (1998).

Distribution. Late Jurassic (late Kimmeridgian–early Tithonian): North America, Wyoming (Morrison Formation); Early Cretaceous (Berriasian): Britain, Dorset (Purbeck Limestone Group); and, with some doubt, Early Cretaceous (Berriasian): Portugal, Porto Pinheiro (Lourinhã Formation).

Comments. North American species of *Tinodon* are based on lower dentitions. As noted by Simpson (1925d, 1929a), differences between *T. bellus* and *T. lepidus* are negligible; we consider them herein to be synonymous. *Eurylambda aequicruris,* based on an upper molar from the same site (Como Quarry 9) as known specimens of *T. bellus* and *T. lepidus,* is now considered as belonging to *Tinodon* (e.g., Crompton and Jenkins, 1967; Prothero, 1981; Fox, 1985; McKenna and Bell, 1997). Hence, it is probable that all described specimens of *Tinodon* from the Morrison Formation belong to the single species *T. bellus.*

The diminutive *T. micron,* recently described from the Purbeck of Britain, is known by both upper and lower molars. The upper molar is strikingly similar to that from the Morrison Formation; however, comparisons call into question the homologies of cusps mesiolabial to the paracone and suggest the possibility that accessory cusp B^1

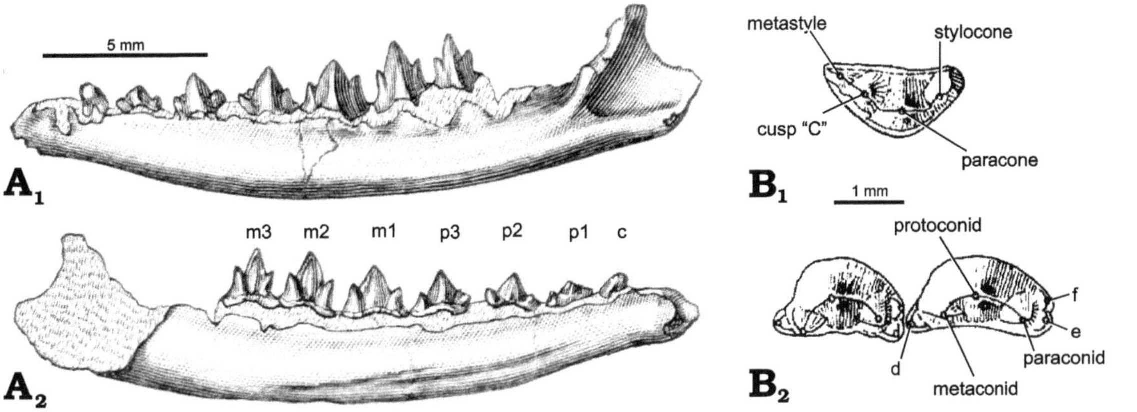

FIGURE 9.8. Tinodontidae. A, *Tinodon bellus*, left dentary with c, p1–3, m1–3 (holotype of *Menacodon rarus*), in lateral (A_1) and medial (A_2) views. B, *Tinodon* sp., molars in occlusal view: right upper molar (holotype of *Eurylambda aequicruris*) (B_1); two left lower molars (B_2). *Source:* A, modified from Marsh (1887); B, modified from Prothero (1981).

(otherwise documented only in Spalacotheriidae) may have been present in Tinodontidae (Ensom and Sigogneau-Russell, 2000).

Tinodon has also tentatively been recorded from Porto Pinheiro (?Berriasian, Portugal), based on a single lower molar, now quite incomplete (Krusat, 1989). Referral was largely based on lack of an external cingulum, a character that is not particularly diagnostic; other features, such as acute angulation of cusps, suggest the possibility that the specimen may represent a spalacotheriid instead.

Family Woutersiidae Sigogneau-Russell and Hahn, 1995

Diagnosis (after D. Sigogneau-Russell, pers. comm.). "Obtuse-angled symmetrodontans" generally similar to Kuehneotheriidae, differing in having lower molars that are wider labiolingually with more robust cusps. Lower molars differ in having a more obtuse angle, stronger cusp g, anterolabial cingulid rather than cuspule f, cuspule e situated more distally, and smaller cusp d. Upper molars differ in the more consistent presence of a lingual cusp.

Genera. *Woutersia* Sigogneau-Russell, 1983a, type genus by monotypy.

Distribution. Late Triassic (Rhaetian): Western Europe, France.

Comments. *Woutersia* was initially (Sigogneau-Russell, 1983a) referred to Kuehneotheriidae, and later (Sigogneau-Russell and Hahn, 1995) placed in its own, monotypic family. These authors noted a strong resemblance of *Woutersia* to Docodonta, suggesting the possibility that docodontans may have originated from early "therians." Later, Sigogneau-Russell and Godefroit (1997) described *Delsatia rhupotopi* as a true docodont. Molar structure of *Delsatia* is similar to that of *Woutersia* (known from the same locality, Saint-Nicolas-de-Port, France), the main difference being a more acute angulation of molar cusps. Partly for this reason, Butler (1997) suggested that *Delsatia* should be regarded as a "symmetrodont" and *Woutersia* as a docodont, while at the same time agreeing with Sigogneau-Russell and Godefroit (1997) that Docodonta may have arisen from an early mammal with a triangulated cusp pattern. We follow Sigogneau-Russell and Godefroit (1997) in tentatively referring *Delsatia* to Docodonta (chapter 5) and in categorizing *Woutersia* simply as a "symmetrodontan." Given the similarities and co-occurrence of the two genera, however, we suggest that it might be worth considering, at least as a null hypothesis, the possibility that they are based on teeth of the same taxon and that observed differences may be due to position in the molar series.

Genus *Woutersia* Sigogneau-Russell, 1983a (figure 9.9)

Diagnosis. As for the family.

Species. *Woutersia mirabilis* Sigogneau-Russell, 1983a, type species; and *W. butleri* Sigogneau-Russell and Hahn, 1995.

Distribution. Late Triassic (Rhaetian): France, Lorraine (Keuper Formation).

Comments. The holotypes of both species are lower molariforms; upper molariforms and possible premolars have also been referred to both *W. butleri* and *W. mirabilis.* Most fossils of both species are from the single locality of Saint-Nicolas-de-Port, but *W. mirabilis* has also been reported from Varangéville, also in Lorraine, France (Godefroit, 1997).

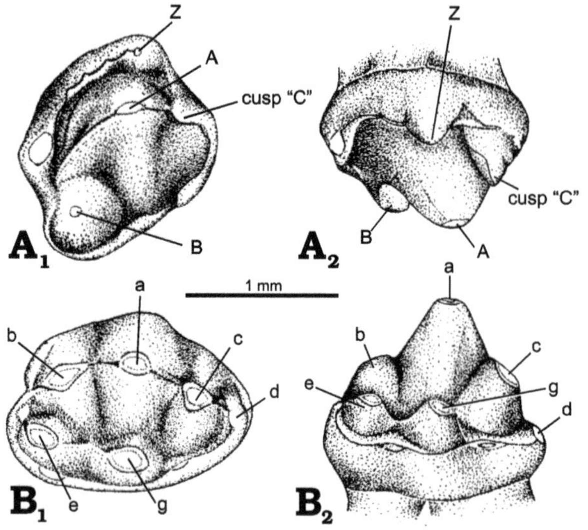

FIGURE 9.9. Woutersiidae: *Woutersia butleri*. A, Right upper molar in occlusal (A_1) and lingual (A_2) views. B, Right lower molar (holotype) in occlusal (B_1) and lingual (B_2) views. *Source:* modified from Sigogneau-Russell and Hahn (1995, orientation of original maintained to preserve shading).

Superlegion Trechnotheria McKenna, 1975

Comments. We recognize Trechnotheria as the clade that includes the last common ancestor of *Zhangheotherium* and living therians plus all of its descendants. An extensive list of apomorphies diagnosing this clade was given by Luo et al. (2002). The geologically oldest trechnotherians are Bathonian (Middle Jurassic) *Palaeoxonodon* and *Amphitherium;* Spalacotheriidae (in which we place *Zhangheotherium*), which range from Berriasian to Campanian, constitute the stem taxon of Trechnotheria.

Family Spalacotheriidae Marsh, 1887

Diagnosis. Mammals with "symmetrodont" molar pattern and acute angulation of principal cusps; primitive molar count generally higher (five or more) than in other "symmetrodontans"; molar roots anteroposteriorly compressed. Cusps developed as high points on occlusal surface but lack expression on crown faces. Postvallum/prevallid shearing surfaces complete, occupying entire faces of respective upper and lower molars. Lower molars high crowned, with well-developed primary cusps (protoconid, metaconid, paraconid) and reduced talonid. Sole talonid cusp (hypoconulid) small and placed at lingual margin of tooth; distinct, lingually placed e cusp present as the only mesial cingular cusp on lower molars. Interlocking pattern of lower molars unique with hypoconulid placed labial to cusp e of the succeeding tooth. Upper molars primitively with cusp B^1 (lost in advanced taxa) on preparacrista between paracone and stylocone.

Included taxa. *Spalacotherium* Owen, 1854, type genus; *Zhangheotherium* Hu et al., 1997 (subfamily *incertae sedis*); and Spalacolestinae Cifelli and Madsen, 1999.

Distribution. Early Cretaceous (Berriasian–Barremian): Western Europe, Britain and Spain; Early Cretaceous (Barremian): Asia, China; Late Cretaceous (Turonian): Asia, Uzbekistan; Early–Late Cretaceous (Aptian–Albian to early Campanian): North America, United States and Canada.

Comments. Marsh (1887) erected the Spalacotheriidae to recognize the similarity of *Spalacotherium,* from the Purbeck, with *Menacodon* (= *Tinodon*) from the Morrison. Osborn (1888b) ranked the group as a subfamily, placing it in the Triconodontidae. In bringing order to a very confused taxonomy (see Tinodontidae, earlier), Simpson (1925d, 1928a) distanced "symmetrodontans" from "triconodonts." However different in concept, the net result is that contents of Spalacotheriidae (*Spalacotherium, Tinodon*) remained similar. Patterson (1956), noting the difference in angulation of cusps and number of molars, more cautiously considered *Tinodon* as ?Spalacotheriidae *incertae sedis* (see also Cassiliano and Clemens, 1979). Crompton and Jenkins (1967) showed that *Tinodon* shares with *Kuehneotherium* many primitive attributes of molar structure and occlusal relationships. These authors later (Crompton and Jenkins, 1968) recognized the distinction of these genera from *Spalacotherium* by placing them in Amphidontidae Simpson, 1925d.[3] Taxonomic separation of *Tinodon* from *Spalacotherium* was widely followed in subsequent years (e.g., Prothero, 1981; Fox, 1985; Cifelli and Madsen, 1999; Averianov, 2002, see Tinodontidae, earlier), though McKenna and Bell (1997) implicitly maintain them as sister taxa.

Discovery and description of new taxa in the past two decades has permitted recognition of Spalacotheriidae as a robustly supported, moderately diverse clade, allied with remaining Trechnotheria on the basis of a number of shared apomorphies (Hu et al., 1997, 1998; Cifelli and Madsen, 1999; Luo et al., 2002). Recent studies further suggested that North American Cretaceous spalacotheriids (*Spalacolestes, Spalacotheridium, Spalacotheroides, Symmetrodontoides*) form a monophyletic clade, Spalacolestinae, within the family (Cifelli and Madsen, 1999). Existing evidence suggests that the two remaining genera of Spalacotheriidae, *Spalacotherium* and *Zhangheotherium,* do not share a unique common ancestor, and we see no useful purpose in designating a subfamily to contain them.

[3] To which Tinodontidae (Marsh, 1887) are clearly antecedent, regardless of the merits of the proposal.

Genus *Spalacotherium* Owen, 1854
(figures 9.1C, 9.10A)

Diagnosis. Differs from Spalacolestinae in: presence of Meckel's groove, smaller pterygoid fossa, pterygoid crest lacking process and not extending anterodorsally toward tooth row (plesiomorphies); presence of one or two roots on lower canine (polarity uncertain), fewer premolars (apomorphy); molars less-acutely angled (plesiomorphy); upper molars lacking parastyle (polarity uncertain), paracristae of subequal height at anterior loci, lack of cusp on distal moiety of stylar shelf (plesiomorphies); m7 less reduced. Differs from *Zhangheotherium* in: lack of facet for splenial; coronoid process better developed and less

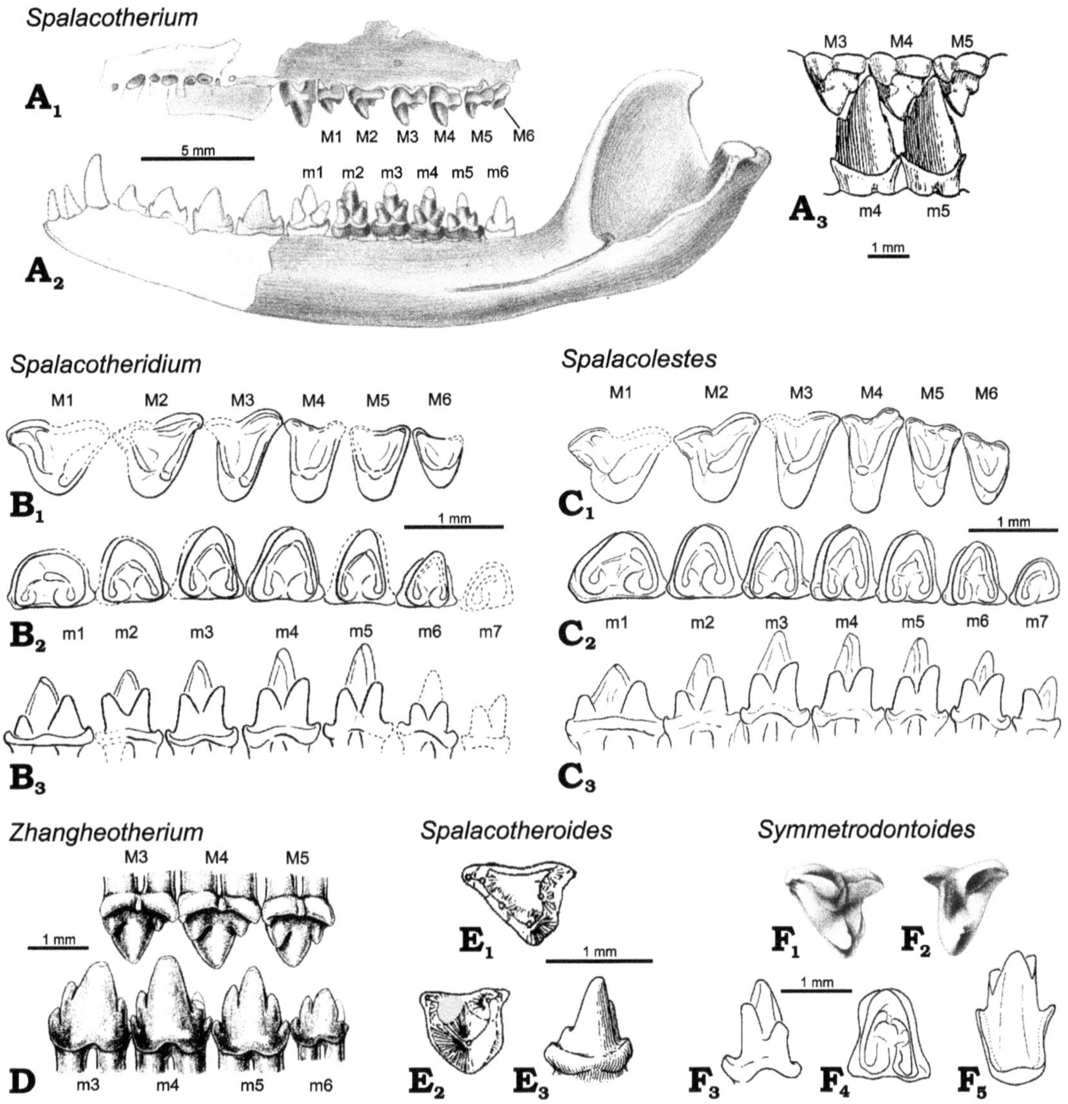

FIGURE 9.10. Spalacotheriidae. A, *Spalacotherium tricuspidens:* (A_1) maxilla with dentition (holotype of *Peralestes longirostris,* reversed) in lateral view; (A_2) right dentary in medial view. (Note: the basis for this reconstruction, published by Osborn, 1888b, is not clear, but it includes at least one important error: *Spalacotherium* has only three lower premolars (and perhaps seven molars), whereas four premolars are sketched in here.); (A_3) series of left upper and lower molars coming into occlusion. B, *Spalacotheridium noblei,* reconstructed molar series: (B_1) left upper molars in occlusal view; right lower molars in occlusal (B_2) and lingual (B_3) views. C, *Spalacolestes cretulablatta,* reconstructed molar series: (C_1) left upper molars in occlusal view; right lower molars in occlusal (C_2) and lingual (C_3) views. D, *Zhangheotherium quinquecuspidens,* left upper and lower molar series (holotype) in labial view. E, *Spalacotheroides bridwelli:* (E_1) right upper molar in occlusal view; left lower molar (holotype) in occlusal (E_2) and labial (E_3) views. F, *Symmetrodontoides* spp.; upper molars of *Symmetrodontoides canadensis,* occlusal view right (F_1) and left (F_2); left lower molar of *S. oligodontos* in lingual (F_3), occlusal (F_4), and labial (F_5) views. *Source:* A_{1-2}, modified from Osborn (1888b); A_3, modified from Simpson (1928a); B, C, modified from Cifelli and Madsen (1999); D, modified from Hu et al. (1997); E, modified from Prothero (1981); F_{1-2}, from Fox (1985).

posteriorly recumbent; condyle more dorsally situated (apomorphies); canine double-rooted (polarity uncertain); greater numbers of premolars (plesiomorphy) and molars (apomorphy); molar cusps more robust and less conical in shape; upper molars with smaller cusp B^1 (polarities uncertain); lower molars with stronger development of cingulid (plesiomorphy).

Species. *Spalacotherium tricuspidens* Owen, 1854, type species; *S. evansae* Ensom and Sigogneau-Russell, 2000; *S. henkeli* Krebs, 1985; and *S. taylori* Clemens and Lees, 1971.

Distribution. Early Cretaceous (Berriasian–Valanginian): Britain (Purbeck Limestone Group and Wealden Supergroup); Early Cretaceous (Barremian): Spain, Teruel Province (Camarillas Formation).

Comments. *Spalacotherium tricuspidens* (including *Peralestes longirostris*, see Hu et al., 1997) is comparatively well known, being represented by the dentary and most of the dentition (Simpson, 1928a). Except for *Zhangheotherium*, remaining species are known by dentary fragments and isolated teeth (Clemens and Lees, 1971; Krebs, 1985; Ensom and Sigogneau-Russell, 2000).

Genus *Zhangheotherium* Hu et al., 1997
(figures 9.2, 9.3C, 9.4, 9.10D)

Diagnosis (modified and shortened after Hu et al., 1997). A spalacotheriid "symmetrodont" with the dental formula 3.1.2.5/3.1.2.6; differing from all other known "symmetrodontans" in having a hypertrophied cusp (B^1) between the main cusp A and the stylocone (cusp B); distinguishable from other spalacotheriids in the conical shape of the main cusps and the lack of lingual and labial cingulids on the lower molars; distinctive from *Spalacotherium* in having more robust and rounded main cusps that lack connecting cristae; unique among early mammals in having fused sternebrae and a posteriorly expanded xiphoid process.

Species. *Zhangheotherium quinquecuspidens* Hu et al., 1997, type species by monotypy.

Distribution. Early Cretaceous (Barremian): China, Liaoning Province (Yixian Formation).

Comments. *Zhangheotherium quinquecuspidens* is the most completely known of all "symmetrodontans." The holotype includes the dentition and much of the skull and skeleton (Hu et al., 1997, 1998). Current evidence suggests that *Zhangheotherium* represents the sister taxon to all other Spalacotheriidae (Cifelli and Madsen, 1999).

Subfamily Spalacolestinae Cifelli and Madsen, 1999

Diagnosis. Differ from most closely similar and related taxon, *Spalacotherium*, in: dentary lacks Meckel's groove, pterygoid crest extends anterodorsally toward tooth row and developed into process near mandibular foramen, pterygoid fossa more expanded anteriorly (apomorphies); canine single-rooted (known only for *Spalacotheridium* and an unnamed taxon; polarity uncertain); retention late in life of deciduous canine and premolars (apomorphy); greater number of premolar loci; molars higher crowned and more acutely angled; anterior upper molars with parastyle and preparacrista distinctly lower than postparacrista, cusp present on distal moiety of stylar shelf (apomorphies).

Genera. *Spalacolestes* Cifelli and Madsen, 1999, type genus; ?*Shalbaatar* Nessov, 1997; *Spalacotheridium* Cifelli, 1990c; *Spalacotheroides* Patterson, 1955; and *Symmetrodontoides* Fox, 1976a.

Distribution. Early–Late Cretaceous (Aptian or Albian through early Campanian): North America; ?Late Cretaceous (Turonian): Uzbekistan.

Comments. Spalacolestinae are best known by the type genus. By comparison to *Spalacotherium*, they are derived in a number of respects, a notable exception being the retention of at least four (five in an unnamed taxon) premolars. An unusual feature is the apparent retention and nonreplacement of the deciduous dentition (Cifelli and Madsen, 1999; Cifelli et al., 2000). *Shalbaatar*, the only non–North American genus, is included with some doubt.

Genus *Spalacolestes* Cifelli and Madsen, 1999
(figures 9.1D, 9.3D, 9.10C)

Diagnosis (from Cifelli and Madsen, 1999: 180). "Differs from *Spalacotherium* in having more acutely angled trigonids on posterior molars and in having lower molar paraconid much lower than metaconid. Differs from *Symmetrodontoides* in having proportionately narrower posterior lower molars with more obtusely angled trigonids and lesser height differential between paraconid and metaconid; m1 differs from that of *Symmetrodontoides* in having lower, more conical paraconid and lower paracristid. Lower molars differ from those of *Spalacotheridium* in having a more pronounced height differential between paraconid and metaconid. Upper molars similar, where known, to those of *Symmetrodontoides*, except that M1–2 have a more bulbous-based paracone with a gently curving (not tightly arced or folded) lingual face. Upper molars differ from those of *Spalacotheroides*, *Spalacotherium*, and *Zhangheotherium* in reduction of the stylocone, lack of cusps B^1 and "C," presence of an extremely low preparacrista (anterior loci only), and presence of an enlarged distal stylar cusp. Differs from the otherwise similar *Spalacotheridium* in having deeper trigon basins and, on posterior upper molars, parastyle reduced."

Species. *Spalacolestes cretulablatta* Cifelli and Madsen, 1999, type species; and *S. inconcinnus* Cifelli and Madsen, 1999.

Distribution. Early–Late Cretaceous (Albian–Cenomanian): United States, Utah (Cedar Mountain Formation).

Comments. The type species, *S. cretulablatta,* is by far the best known of North American spalacotheriids, being represented by several dentaries, isolated teeth including redundant specimens at all upper and lower molar loci (Cifelli and Madsen, 1999), and isolated teeth interpreted as being deciduous premolars (Cifelli, 1999a). The dental formula is unknown, though a count of p4, m7 for the lower dentition is a reasonable approximation based on available data (Cifelli and Madsen, 1999; Cifelli, 1999a). The dentary has a strongly efflected (laterally reflected) posteroventral border, a probable synapomorphy within Spalacotheriidae (shared by *Spalacotherium* but lacking in *Zhangheotherium*). More unusual still is the hyperdevelopment of the pterygoid crest, which forms a large process at its termination near the mandibular foramen; condition of this character among other Spalacotheriidae is unknown.

Genus ?*Shalbaatar* Nessov, 1997
(not figured)

Diagnosis. Not presently diagnosable.

Species. *Shalbaatar bakht* Nessov, 1997, type species by monotypy.

Distribution. Late Cretaceous (Turonian): Uzbekistan, Kyzylkum Desert (Bissekty Formation).

Comments. *Shalbaatar bakht* was named but not illustrated in the posthumous monograph of Nessov (1997) on the basis of an edentulous dentary. Nessov tentatively assigned *Shalbaatar* to ?Plagiaulacoidea within Multituberculata because the coronoid process is anteriorly placed (derived feature thought to be shared with multituberculates), but p4 is not enlarged and the masseteric fossa is posteriorly placed (features that are plesiomorphic with respect to all known multituberculates). Restudy of the specimen by Averianov (2002) suggests that *Shalbaatar* may be a spalacotheriid, perhaps referable to the otherwise North American subfamily Spalacolestinae. We tentatively follow Averianov's assessment, noting that judgment must be deferred until the holotype of *S. bakht* is fully described and illustrated.

Genus *Spalacotheridium* Cifelli, 1990c
(figure 9.10B)

Diagnosis (from Cifelli and Madsen, 1999: 195). "Lower molars differ from those of *Spalacotherium* in being more nearly symmetrical and acutely angled; from *Spalacotheroides* in having a complete labial cingulid; and from *Symmetrodontoides* and *Spalacolestes* in being lower crowned, with paraconid and metaconid subequal in development and of approximately equal height, only slightly lower than the protoconid, and in having posterior molars that are proportionately narrower and have more obtuse trigonid angles. Upper molars distinct from *Spalacotheroides* in the presence of a larger distally placed stylar cusp and more prominent parastylar hook (anterior loci), the lack of cusps B^1 and C, and the extremely low placement of the preparacrista (anterior loci). Upper molars differ from those of *Spalacolestes* in having shallower trigon basin, and in the presence of a prominent parastyle on M6."

Species. *Spalacotheridium mckennai* Cifelli, 1990c, type species; and *S. noblei* Cifelli and Madsen, 1999.

Distribution. Early–Late Cretaceous (Albian-Cenomanian through Turonian): United States, Utah (Cedar Mountain and Straight Cliffs formations).

Comments. Each of the included species is based on isolated teeth. However, moderately large samples available for *Spalacotheridium noblei* permit reconstruction of most of the molar series (see also Cifelli and Gordon, 1999; Cifelli and Madsen, 1999), together with the ?dc and deciduous lower premolar series (Cifelli, 1999a).

Genus *Spalacotheroides* Patterson, 1955
(figure 9.10E)

Diagnosis. Lower molars differ from most other Spalacotheriidae (except *Spalacotherium evansae* and *Zhangheotherium quinquicuspidens*) in incompleteness of labial cingulid. Upper molars differ from those of otherwise similar *Spalacotherium* in presence of a strong parastyle, presence of a cusp on the distal moiety of the stylar shelf, and (for anterior loci) preparacrista lower than postparacrista. Upper molars differ from those of *Spalacotheridium, Spalacolestes,* and *Symmetrodontoides* in presence of cusps B^1 and C, lack of prominent parastylar hook, smaller cusp on distal moiety of stylar shelf, and stronger stylocone.

Species. *Spalacotheroides bridwelli* Patterson, 1955, type species by monotypy.

Distribution. Early Cretaceous (Aptian–Albian): United States, Texas (Antlers Formation).

Comments. *Spalacotheroides bridwelli* was named by Patterson (1955) on the basis of a dentary fragment with a single lower molar, to which he later added one complete upper molar and two fragments thereof (Patterson, 1956). Another edentulous dentary fragment is probably referable to the species (Cifelli and Madsen, 1999), and a lower molar vaguely similar to the type is known from the Cloverly Formation, Montana (Jacobs et al., 1991). *Spalacotheroides* has justifiably figured prominently in the literature (e.g., Patterson, 1956; Fox, 1976a; Cassiliano and Clemens, 1979; Prothero, 1981; Sigogneau-Russell and Ensom, 1998), as it was the first uncontested spalacotheriid to be reported from North America and the first undoubted Early Cretaceous "symmetrodont" in general. It is thus an irony that, in the face of newly discovered taxa, *Spalacotheroides* is practically indeterminate, at least insofar as the holotype is concerned. The locus of the single tooth preserved in this specimen has been the subject of wide speculation: based on analogy with *Spalacotherium,* Patterson (1955) suggested m5, Fox (1976a; see also Cifelli and Madsen, 1986) proposed m3 or 4, and Cifelli and Madsen (1999) hypothesized it to be m2. Uncertainty is due partly to the fact that the paraconid is missing (*contra* Prothero, 1981: figure 7E), which also precludes other important comparisons. Much currency has been placed on the incompleteness of the labial cingulid on the holotype of *S. bridwelli* (e.g., Fox, 1976a; Cifelli and Madsen, 1986). However, it is possible that this is an artifact of preservation (Cifelli and Madsen, 1999). Regardless, this feature is not as unique among Spalacotheriidae as once envisaged (Ensom and Sigogneau-Russell, 2000). Fortunately, the referred upper molars are more distinctive (at least at the present state of knowledge) and hence form the basis for much of our revised diagnosis.

Genus *Symmetrodontoides* Fox, 1976a
(figure 9.10F)

Synonym: *Mictodon* Fox, 1984a.

Diagnosis. Differs from most closely similar genus, *Spalacolestes,* in having posterior lower molars that are broader transversely, with greater height differential between paraconid and metaconid; m1 with taller paraconid and paracristid; M1–2 more acutely angled, with less bulbous, more tightly folded lingual face to paracone.

Species. *Symmetrodontoides canadensis* Fox, 1976a, type species; *S. foxi* Cifelli and Madsen, 1986; *S. oligodontos* Cifelli, 1990c; and species left in open nomenclature (Eaton, Cifelli et al., 1999; Eaton, Diem et al., 1999).

Distribution. Late Cretaceous (Turonian–early Campanian): Canada, Alberta (Milk River Formation); United States, Utah (Straight Cliffs and Wahweap formations).

Comments. Collectively, the species of *Symmetrodontoides* are known by several fragments of the dentary and a number of isolated teeth (e.g., Fox, 1976a; 1985; Cifelli and Madsen, 1986; Cifelli, 1990c). Comparison with the closely similar *Spalacolestes* provides a basis for identification of tooth locus; known specimens of *Symmetrodontoides* include parts of upper and lower molar series, together with presumed deciduous premolars (Cifelli, 1999a; Cifelli and Gordon, 1999). A rare but geographically widespread taxon, *Symmetrodontoides* is the last of North America's Spalacotheriidae, which presumably became extinct before onset of the Judithian land-mammal age.

CHAPTER 10

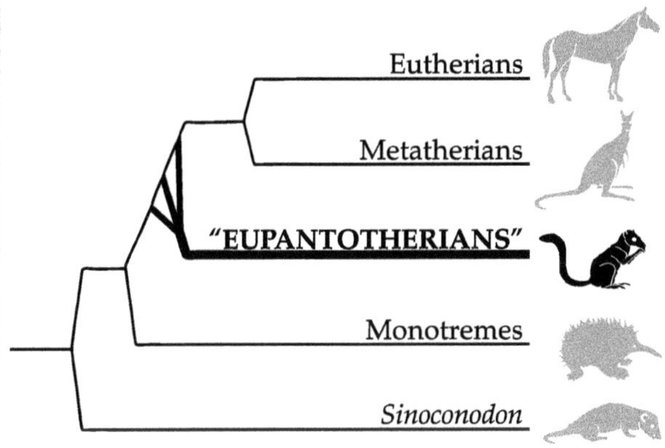

"Eupantotherians" (Stem Cladotherians)

INTRODUCTION

"Eupantotherians" (= "eupantotheres"; "Eupantotheria" Kermack and Mussett, 1958) represent a very important assemblage of Mesozoic mammal clades that are placed between the plesiomorphic "symmetrodontans" (chapter 9) and the more advanced boreosphenidan mammals with tribosphenic molars (chapter 11). The three main groups traditionally assigned to "eupantotherians" are peramurids and their kin, amphitheriids, and dryolestoids, including paurodontids and dryolestids.

Cladotheria (McKenna, 1975) is a stem-based clade defined by the common ancestor of boreosphenidans (including marsupials and placentals) and the fossil taxa more closely related to Crown Theria than to the more plesiomorphous Spalacotheriidae. By current prevailing hypotheses of cladistic relationships from the best available evidence, peramurids, dryolestoids, and amphitheriids are stem clades within the monophyletic Cladotheria. These are related to boreosphenidans in a successively more distant order, but do not form a natural group by themselves.

Since Kermack and Musset (1958a) proposed the group "Eupantotheria," its membership has been stable in many taxonomic studies of recent decades (see the review by Kraus, 1979). However, study of the group had a long and varied history before the 1950s. Marsh (1880) proposed and defined the order Pantotheria and in a further study (1887), he included four families: Dryolestidae and Paurodontidae (assigned to Dryolestoidea in the present chapter), Diplocynodontidae (now assigned to Docodonta, see chapter 5), and Dromatheriidae (now considered to be nonmammalian cynodonts).

In a well-known textbook, Zittel (1893) placed dryolestids with some polyprotodont marsupials in the group of Trituberculata, a term borrowed from Henry Fairfield Osborn. Osborn (1887a,b) coined the descriptive term "trituberculate molars," and later (1893) used "trituberculates" as an informal unit in reference to some Late Cretaceous eutherians and marsupials with true tribosphenic molars (the latter term having been established long after his time by Simpson, 1936a, see chapter 11). The term "Trituberculata" (Zittel, 1893) was used by some authors to replace Pantotheria. After a thorough critical review by Simpson (1928a), the taxon Trituberculata was abandoned by all experts in Mesozoic mammals, but was still used in some textbooks, for example, by Romer (1966), as an infraclass encompassing Symmetrodonta and Pantotheria, and by Carroll (1988), as an infraclass including Symmetrodonta and Eupantotheria.

Simpson (1925b) erected the order Symmetrodonta for part of Marsh's Pantotheria. Later Simpson (1928a, 1929a) used the two separate orders, Symmetrodonta and Pantotheria, and restricted the order Pantotheria to *Amphitherium*, paurodonts (including *Peramus*), dryolestids, and docodonts (Simpson, 1929a). Butler (1939) offered an analysis of non-multituberculate Jurassic mammals, and within Pantotheria recognized the Docodontoidea (see chapter 5) and Dryolestoidea (see later). In this respect he followed Simpson (1929a), who treated Docodontidae as specialized pantotherians.

In 1931 Simpson formally proposed the infraclass Pantotheria to include two orders: Symmetrodonta Simpson, 1925, and Pantotheria Marsh, 1880. In his *Classification of Mammals*, Simpson (1945) noted that use of this confusing classification of these groups was inappropriate. To avoid the distraction caused by giving the same name to two different taxa, Kermack and Mussett (1958a) proposed the order Eupantotheria to replace the ordinal name Pantotheria, while retaining the term Pantotheria as the name of the infraclass. Simpson (1971) attempted to address the same problem by using infraclass Patriotheria, as a replacement for Pantotheria *sensu lato,* including the orders Symmetrodonta and Pantotheria; but his scheme was not followed by subsequent authors.

BRIEF CHARACTERIZATION

"Eupantotherians" differ from "symmetrodontans" in the presence of an angular process of the dentary and share this character with Boreosphenida (figure 10.1B). The upper and lower molars form a series of reversed and interlocking triangles (as in "symmetrodontans"), but differ from most of them in having the upper molars wider than the lowers, having a third (lingual) root, and more strongly developed labial stylar cusps, with elaborate structure in the anterolabial region (figure 10.2). On the lower molars, the talonid is clearly differentiated from the trigonid; the talonid structures may range from a single cusp (cusp d, or hypoconulid) in dryolestoids, to a cusp plus a small basin in *Vincelestes* and *Amphitherium,* or to a wider basin with two cusps on the rim, as in *Peramus.* But being more primitive than Boreosphenida, none of the "eupantotherians" has an entoconid (separated from the hypoconulid and hypoconid), or a talonid basin with grinding function on the lower molars, or a protocone on the upper molar. The groups traditionally designated as "eupantotheres" differ from one another in dental formulae and in the structure of the molars.

Elements of the skull are known for *Dryolestes* (Martin, 1999a, our figure 10.1A), and for *Henkelotherium* (Krebs, 1991). Details of the braincase structure and the petrosal are known only in *Vincelestes* ("stem-lineage of Zatheria"). Only two postcranial skeletons are known, one of which belongs to *Henkelotherium* (Dryolestida, Paurodontidae) and was well described by Krebs (1991) and Vázquez-Molinero et al. (2001). *Vincelestes* is also known by numerous skeletal elements, described by Rougier (1993) in his Ph.D. thesis. Because many dryolestoids and "peramurans" are represented by isolated teeth or disassociated upper and lower dentitions, some genera were based either on the upper or on the lower teeth. After several thorough studies of dental replacement in taxa with more extensive fossil materials (e.g., Martin, 1997, 1999a), it has also been recognized in recent years that taxonomic distinctions of some "dental genera" were wrongly based on deciduous premolars that had been mistaken for permanent molars. This problem could be more widespread than currently known, but it is difficult to solve until materials better than isolated teeth are known to prove otherwise.

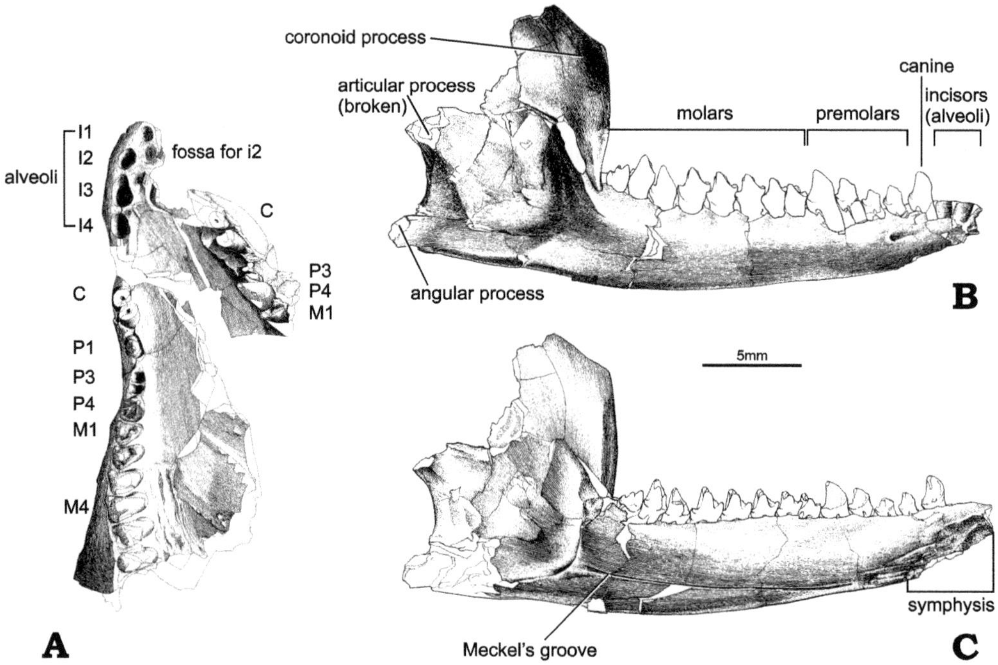

FIGURE 10.1. Anatomy of the rostrum and dentary in Dryolestoidea, exemplified by *Dryolestes leiriensis.* Fragment of a rostrum in occlusal view (A); dentary in labial (B) and lingual (C) views. *Source:* modified from Martin (1999a).

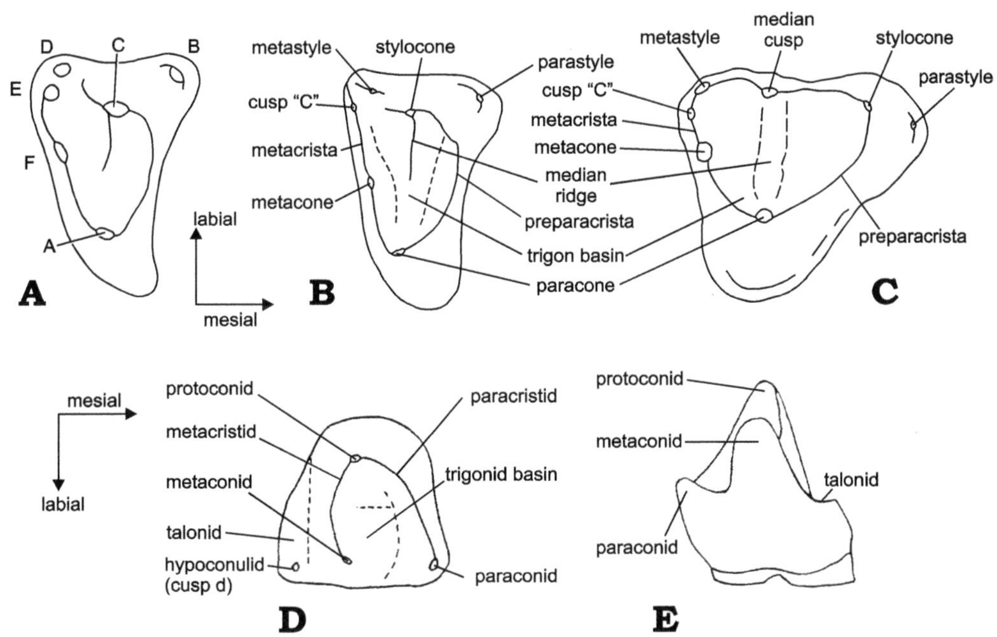

FIGURE 10.2. Molar structure and terminology in Dryolestidae. A, Generalized dryolestoid right upper molar in occlusal view, with cusp terminology proposed by Simpson (1961). B, The same with cusp terminology proposed by Martin (1999a), emended here. C, Cusp terminology of the dryolestoid *Donodon,* emended here. D, Left lower molar of generalized dryolestoid in occlusal view. E, Right lower molar of same in lingual view. Median stylar cusp is labeled median cusp. Labial is up and mesial to the right. *Source:* A, modified from Simpson (1961); B, D, E, modified from Martin (1999a); C, courtesy of D. Sigogneau-Russell.

DISTRIBUTION

"Eupantotherian" or stem cladotherian groups are relatively common in Late Jurassic rocks and in some earliest Cretaceous formations. For example, dryolestoideans make up 49% of all the mammal teeth in the Guimarota Coal Mine (Kimmeridgian) in Portugal (Martin, 2001). Dryolestoids are also common and taxonomically diverse in the Late Jurassic Morrison Formation of North America (Simpson, 1929a; Prothero, 1981; Martin, 1999a). In the Early Cretaceous dryolestoids declined in diversity, but peramurans had a modest diversity in the earliest Cretaceous sites of Britain and Morocco and late Early Cretaceous sites of Mongolia. Of these three groups, according to the currently available evidence, dryolestoids survived to the Late Cretaceous only in South America. The best-preserved fossils of "eupantotherians" are: (1) most of a skeleton and some skull elements of the paurodontid *Henkelotherium* (Krebs, 1991); and (2) numerous skulls and skeletal elements of *Vincelestes* (Bonaparte and Rougier, 1987; Rougier, 1993). The distribution of the groups discussed in this chapter includes: Middle Jurassic to Early Cretaceous of Europe; Late Jurassic to ?Late Cretaceous of North America; Late Jurassic to Early Cretaceous of Africa; Early Cretaceous (Hauterivian or Barremian) to Paleocene of South America (Gelfo and Pascual, 2001); Aptian or Albian of Mongolia; and a possible record from the Albian of Australia (Clemens et al., 2003).

TERMINOLOGY

Molar cusps of "dryolestoids" require explanation (figure 10.2). Authors of the nineteenth and the early twentieth centuries, including Simpson (1928a, 1929a), avoided using terminology of tribosphenic (tritubercular) molars for describing "eupantothere" molar cusps. This was related to ambiguity concerning homology, in particular the designation of the lingual cusp (paracone versus protocone) on "eupantothere" upper molars. Butler (1939) demonstrated that the protocone is absent in "eupantotherians." He regarded the lingual cusp in dryolestid upper molars as the paracone and the cusp situated near the middle of the posterior crista as the metacone. The designations of cusp homology were further expanded by Patterson (1956) in a more comprehensive theory for the evolution of the tribosphenic molar.

Butler (1939) referred to the three labial cusps (styles) as the anterior, buccal (the middle one), and posterior cusps. Patterson (1956) went further and referred to the anterior stylar cusp in dryolestoideans as the parastyle, the medial one as the stylocone, and the posterior one as the metastyle. He named the cusp situated on the posterior crista between the metacone and metastyle the posteroexternal cingulum cusp. According to Patterson (1956), the stylocone in "eupantotherians" is situated more posteriorly than its homologue on the tribosphenic molar of derived mammals, even attaining a central position in some derived taxa within dryolestids.

Simpson (1961: figure 1) offered a different cusp terminology, using capital letters for the designation of cusps in "eupantoptherians." Simpson's scheme is exemplified by a dryolestoid upper molar (figure 10.2A): the paracone corresponds to cusp A; four stylar cusps correspond to cusps B, C, D, and E; the metacone is designated cusp F. Thus B is the parastyle; C, the stylocone of Butler (1939) and Patterson (1956); D, the metastyle; and E, the cusp between the metastyle and metacone. Simpson's (1961) terminology has not been widely used and is not accepted by students who engaged in further study of dryolestoids (e.g., Clemens and Mills, 1971; Butler, 1978b; Prothero, 1981; Bonaparte, 1986a, 1990; Krebs, 1991; Martin, 1999a; Sigogneau-Russell, 1999; and many others). Different designations for dryolestid molar cusps were also proposed by Vandebroek (1961) and Hershkovitz (1971), but these were not generally accepted.

Simpson's (1961) cusp E of the eupantotherian molar was subsequently referred to by Crompton (1971) as cusp "c" (referred to by us as cusp "C"). Crompton (1971) argued that this cusp is homologous to one of three main cusps (cusp C) in morganucodontans and early "symmetrodontans" such as kuehneotheriids (e.g., Crompton and Jenkins, 1968; Cassiliano and Clemens, 1979, see also chapters 4, 9, 11), and hence it was given the notation "C." This designation of cusp "C" received some acceptance because it is consistent with the widely accepted theory of tribosphenic molar evolution, as initially proposed by Patterson (1956) and supported by a systematic analysis of wear facets by Crompton (1971).

This pattern is relatively easy to follow in "eupantotherian" taxa in which there are three stylar cusps and, sometimes, cusp "C" of Crompton (1971: figure 7, see also our figures 10.10B_1,C, 10.13C_1). Crompton (1971) suggested that the metacone of "eupantotherians" is not homologous to cusp C of "symmetrodonts" (see chapter 9, and also Clemens and Mills, 1971). He argued that metacone was a neomorph that made its appearance in some of the more derived "eupantotherians," in conjunction with the hypocone on the lower molar, to form a new shearing surface (see the section "Origin of Tribosphenic Molar" in chapter 11 and figure 11.2).

We follow the designation of cusp homology for dryolestids by Martin (1999a: figure 7A; figure 10.2B herein), except that we do not designate metastyle(s) to two cusps simultaneously. If there are two cusps in the metastylar region, we designate the more lingual of the two as cusp "C" and the more labial as the metastyle. Cusp "C" so defined is located between the metastyle and the metacone. In this scheme the paracrista (preparacrista of mammals with tribosphenic molars) joins the stylocone with the paracone, which may be variable in its location in different taxa depending on the curvature of the paracrista. The metacrista joins the metacone with the metastyle and the triangular basin embraced by these two cristae is called the "trigon basin." The trigon basin associated with the paracone in "eupantotherians" is the "primary trigon," not homologous to the "secondary trigon," associated with the true protocone in tribosphenic molars (Patterson, 1956; Butler, 1978b: figure 1).

On the upper molars of some dryolestoideans there is a median ridge that extends between the centrally placed median stylar cusp and the paracone, or that extends from the paracone labially and does not reach the stylar cusp. It divides the trigon basin into two valleys, and it has been referred to by Simpson (1929a) as "median transverse crest," by Bonaparte (1990) as median ridge, and by Martin (1999a) as "medianer Grat." In some genera such as the donodontid *Donodon* and the dryolestid *Krebsotherium,* this structure is less sharp. Herein we use Bonaparte's term, the median ridge (figure 10.2B).

The nomenclature for the stylar cusps is more complicated in some dryolestoidean genera such as *Donodon* (figure 10.2C). Most dryolestoids have three stylar cusps: the parastyle in the anterior position; the stylocone is connected with the paracrista and placed more centrally; and the metastyle is in a posterior position (figure 10.2B). In *Donodon* (Sigogneau-Russell, 1991b), there are four cusps on the stylar shelf; an additional median cusp is present between the metastyle and the stylocone (the latter situated more anteriorly in *Donodon* than in other dryolestoideans) and is connected to the median ridge in the "primary trigon." Sigogneau-Russell (1991b) referred to it as cusp D, although cusp D in Simpson's scheme is usually the metastyle. Bonaparte (1990) termed the centrally placed cusp the stylocone in the dryolestoid *Leonardus* from Argentina and then in a further paper renamed it "the centrocone" (Bonaparte, 1994).

In tribosphenic molars the centrally placed cusp on the stylar shelf is commonly known as the mesostyle in basal eutherians (e.g., Van Valen, 1966) or stylar cusp C in marsupials. But the mesostyle (or cusp C) of tribosphenidans is obviously not homologous with the centrally placed median cusp of dryolestoids. This cusp is clearly a separate structure from the stylocone in at least one dryolestid; yet neither "cusp D" nor "centrocone" is an adequate designation.

Upon evaluating all alternatives, we decided to term this cusp the median stylar cusp, which, if present, is also connected to median ridge (although not all median ridges are connected to a median cusp, such as in *Laolestes*); the median stylar cusp is abbreviated in the drawings as median cusp (figures 10.2, 10.11). We restrict the term stylocone to the cusp that is fully connected to the paracrista.

For dental terminology of peramurids and related forms we follow Sigogneau-Russell (1999). For the new term introduced by her (in French) "sillon creusant le stylocone" we use "stylocone groove." Other French terms introduced by that author only have different suffixes in English and do not require explanation; they are illustrated in figure 10.16A.

ANATOMY OF DRYOLESTOIDEA

SKULL AND DENTARY

The dryolestoidean skull is hardly known, as no braincase or petrosal has yet been described. In *Dryolestes* the rostrum is elongate and narrow anteriorly, expanding laterally at the level of M1 (figure 10.1A), but the rostrum is shorter and probably more robust in Paurodontidae than in Dryolestidae and Amphitheriidae. The lacrimal forms the orbital margin and extends on the dorsal side of the skull; it is small but its shape could not be established. The maxilla participates in the structure of the anterior part of the zygomatic arch, most of which was built mainly by the jugal. The mandible is known from several fairly complete dentaries, of which the horizontal ramus is very gracile in Dryolestidae, similar to that of Amphitheriidae, but more robust than that of Paurodontidae. The vertical ramus of the mandible in all groups is extensive and high, with a steeply directed coronoid process (figure 10.1B,C). The articular process is often directed obliquely in the posterodorsal direction and is situated above the occlusal level of the lower molars; it has a well-defined and transversely formed condyle. The angular process is prominent, directed posteriorly, and leveled with the ventral border of the mandible. There is a deep emargination of the posterior margin of the dentary between the articular and angular processes, as well as a long symphysis that may extend posteriorly up to the level of the penultimate premolar in Dryolestidae, but it is shorter and vertically oriented in Paurodontidae. The masseteric fossa on the external aspect of mandible is deep and well defined. The pterygoid fossa on the medial side is also deep with a sharp ventral border, known as the medial pterygoid shelf. As demonstrated by Krebs (1969, 1971) and Martin (1995), the coronoid, the splenial, and Meckel's groove were present in adult dryolestoideans of the Late Jurassic (figure 10.1C), but the coronoid and splenial disappeared in the Early Cretaceous *Crusafontia*.

DENTITION

There are four lower and five upper incisors, canine, four premolars, and usually eight or nine molars in Dryolestidae (figure 10.1); the dental formula for lower teeth in Amphitheriidae is 4.1.5.6–7, but there are only two to six upper molars and up to seven lowers in Paurodontidae (*Henkelotherium* has four premolars and six or seven molars). The number of lower molars may exceed that of the uppers, but probably not more than by one. The lower incisors are usually peglike with a modest procumbency; the canine is double-rooted and of various heights. The premolars differ in morphology from the molars; the ultimate lower premolar is the largest of the premolars and has a dominant main cusp and a continuous lingual cingulid, without any triangulating crown pattern.

The lower molars (figure 10.2D,E) have a well-defined trigonid with a paraconid that is more or less procumbent; the talonid is well defined, but as a rule small with a single cusp. In Dryolestidae the posterior lower molar root is smaller than the anterior root; the difference in root size is less dramatic in Amphitheriidae and hardly discernible or absent in most Paurodontidae. There is no anterolingual cingulid on lower molars, but there is an anterolabial or labial cingulid.

The upper molars (figure 10.2A,C) in Dryolestidae are strongly compressed mesiodistally and widened labiolingually, more rectangular than triangular in occlusal view, whereas in Paurodontidae they tend to be shorter labiolingually, sometimes roughly rectangular in occlusal view. The paracrista and metacrista encircle the crown from the anterior and posterior sides, respectively. The paracone is situated lingually and the metacone in about the middle of the metacrista. The parastylar region is often enlarged, developed as a prominent wing, and usually has only one cusp (parastyle), but sometimes includes a second cusp, the stylocone. There is often a central cusp, the median stylar cusp (labeled as the median cusp in the drawings). The metastylar region is smaller. In many dryolestoideans there is a median ridge that extends from the median stylar cusp to the paracone (or from the stylocone to the paracone). The median ridge does not generally occur in Paurodontidae, but might be present, albeit indistinct, in *Henkelotherium*.

The milk dentition in dryolestids was first recognized by Butler (1939), who concluded that two dryolestid genera "*Malthacolestes*" (synonym of *Melanodon* = *Laolestes*, see "Systematics") and *Asthenodon* (synonym of *Dryolestes*) were based unknowingly on the milk dentition. Tooth replacement has been studied by Martin (1997) in three dryolestid genera: *Krebsotherium*, cf. *Guimarotodus*, and *Dryolestes*. Butler and Krebs (1973) first described purported milk teeth in dryolestoids, but as argued by Martin (1997) these teeth belong to the permanent dentition. Martin demonstrated that all antemolar teeth are replaced and that tooth replacement, at least in the lower jaw, took place in two waves. The first replacing wave consists of i2, i4, p1, p3; the second wave of i1, i3, c, p2, p4; the p4 was the last premolar to erupt and the milk tooth in this position was present when the sixth molar (m6) started to erupt. Enamel ultrastructure has been investigated in Late Jurassic dryolestoideans from Europe by Lester and Koenigswald (1989), Sander (1997), and Martin (1999a), and in Late Cretaceous *Groebertherium* and *Mesungulatum* from South America by Crompton et al. (1994). All these authors observed the presence of prismatic enamel in the dryolestoids studied.

Crompton et al. (1994) reconstructed the occlusal pattern and shearing surfaces in the molars of *Groebertherium* and *Mesungulatum*. Molars of *Groebertherium* have the simple shearing function of their mesial and distal surfaces, but the puncturing function of the large median cusp (stylocone of Bonaparte, 1990). Crompton et al. (1994) concluded that shearing had more importance than puncturing. The occlusion pattern and wear of *Mesungulatum*, as reconstructed, were more complicated. The crowns in *Mesungulatum* are bulbous, but the primary trigon and trigonid, with near-vertical shearing surfaces, are dominant features of crown morphology. The unique features of this form are large cingula at the base of shearing facets. The authors concluded that *Mesungulatum* molars combined crushing and shearing functions.

POSTCRANIAL SKELETON

The postcranial skeleton is known from a single incomplete skeleton, the holotype of *Henkelotherium guimarotae* Krebs, 1991, preserved in articulation (figures 10.3, 10.4) on a slab from the Guimarota Coal Mine. The estimated length of the holotype specimen, without a tail, is 7 cm (Vázquez-Molinero et al., 2001). The specimen is slightly flattened and it was impossible to prepare the individual bones from the surrounding rock so as to examine them from all sides. Thus, even in the case of completely preserved bones, some structural details remain obscure. The description below is based on Krebs's (1991) monograph, as emended by Vázquez-Molinero et al. (2001).

Vertebral Column. Only fragments of cervical vertebrae have been preserved. There is a series of two last thoracic, five lumbar, and two sacral vertebrae preserved together and exposed in ventral view. The transverse processes of the lumbar vertebrae are short. A number of caudal vertebrae have been preserved. The first five or six caudals have proportions similar to those of presacrals. Beginning with the sixth or the seventh caudal vertebra and continuing poteriorly, the centrum becomes increasingly elongate. The ninth caudal has the length of three presacrals. In these long caudals the spinous and transverse processes are reduced; posteriorly their length increases slightly and they become more sticklike. Distal to the fifteenth caudal vertebra some vertebrae are missing but the last eight are preserved; these are very thin and long. Because of the elongation of the caudal vertebrae the tail of the animal was quite long (Krebs, 1991).

Pectoral Girdle and Forelimb. The pectoral girdle in *Henkelotherium* has only a preserved scapula and clavicle. Both bones are very much like those of extant therian mammals. The coracoid process is large, sutured to the scapula, and contributes to the glenoid cavity. There is no interclavicle or procoracoid. Along the middle of the scapula there extends a scapular spine, in front of which there is a supraspinous fossa, larger than the infraspinous fossa.

The proximal extremity of the humerus is not preserved (figure 10.4A). The humerus is a stout bone. In his earlier description of the skeleton, Krebs (1987: 138) stated: "The two distal condyles of the humerus are located side by side and clearly separated." The condylar structure of the elbow joint has been confirmed by Vázquez-Molinero et al. (2001: 210), who stated: "The distal extremity of the left humerus of *Henkelotherium* exhibits two condyles (figure 5). Both condyles are clearly separated by a wide, intercondylar groove." The degree of twisting of the humerus has not been measured. The radius and ulna, preserved in articulation with the humerus, are long stout bones. More interesting is the manus (incompletely preserved), which shows elongation of the phalanges.

Pelvic Girdle and Hindlimb. The pelvis is built as in Cretaceous eutherian mammals (e.g., *Barunlestes*, see chapter 13), with small differences. The ischium and pubis surround the large obturator foramen, but the pubis does not contribute to the acetabulum; the ilium has the shape of an elongated shaft, the inner surface of which articulates with two sacral vertebrae. The epipubic ("marsupial") bones are present (figure 10.3). The femur is a stout bone with the head offset from the shaft, the femoral neck being arranged at 30° to the shaft. There is a strong greater trochanter and wide lesser trochanter. The two femoral condyles articulate with the tibia and fibula.

Vázquez-Molinero et al. (2001) regarded the structure of the knee joint in *Henkelotherium* as similar to those in *Morganucodon* (see chapter 4) and *Vincelestes* (this chapter). The tarsus is incomplete, but the foot is almost completely preserved, including ungual phalanges; it is relatively long (figure 10.4B). Krebs (1987) suggested that the

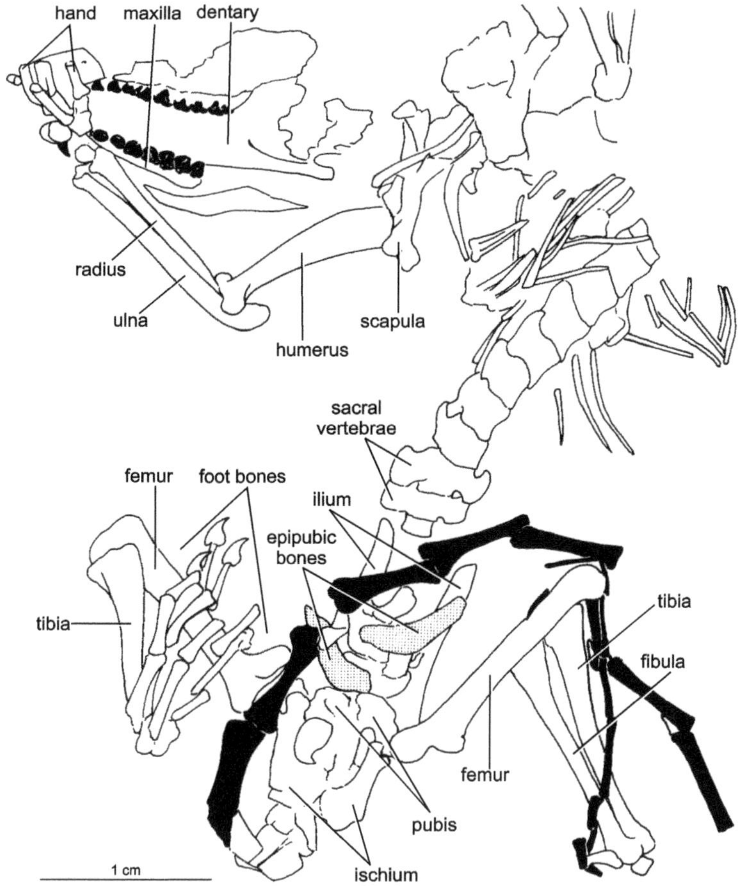

FIGURE 10.3. Schematic drawing of the skeleton, associated with dentary and skull fragments of *Henkelotherium guimarotae*. The epipubic bones are hatched and teeth and caudal vertebrae are black. *Source:* modified from Henkel and Krebs (1977).

increased length of the penultimate phalanges and the shape of the distal phalanges were indicative of climbing adaptations in *Henkelotherium* (see "Paleobiology," which follows).

PALEOBIOLOGY

The dental structure of *Henkelotherium* and other dryolestoideans, with a full complement of five upper and four lower incisors, stout canines, and high-cusped premolars and molars, points to an insectivorous diet. The high molar count and acute angulation of cusps, typical of Dryolestidae, suggests the possibility that some "eupantherians," like spalacotheriid "symmetrodontans," may have had a specialized diet including mainly soft-bodied invertebrates (see chapter 9). Krebs (1987, 1991) argued, on the basis of the postcranial skeleton, that *Henkelotherium* was an arboreal mammal, climbing with its claws ("Krallenkletterer") and leaping with the help of a tail that may have served as a dynamic stabilizer. He based his conclusion on the significant elongation of the caudal vertebrae, clawlike distal phalanges of the foot, and the considerable length of the penultimate phalanges of the hand and foot. Climbing proclivities for *Henkelotherium* were supported by the study of Vázquez-Molinero et al. (2001).

Elongation of the *Henkelotherium* tail is not sufficient to indicate arboreality, as tails are just as elongated in some extant terrestrial mammals (see, e.g., Aristov et al., 1980). The distal ungual phalanges of extant "Krallenkletterer" are strongly compressed laterally (see, e.g., Kielan-Jaworowska and Gambaryan, 1994: figure 60). In order to evaluate the extent of their compression, the unguals must be seen in cross section or dorsal view, but in *Henkelotherium* they have only been shown in lateral view (Krebs, 1991: figure 11, and plate 4). We agree with Vazquez-Molinero et al. (2001) that *Henkelotherium* had some limb and foot features characteristic of climbing mammals. However, for small mammals, climbing ability is a sufficient but not a necessary condition for a mode of life that is obligatorily arboreal (Jenkins, 1974a). We believe that *Henkelotherium* was capable of climbing, but its arboreality has not been convincingly demonstrated.

SYSTEMATICS

McKenna (1975) was the first to argue that the order Pantotheria Marsh, 1880, and the identically named infraclass erected by Simpson are both paraphyletic taxa. In their place, he erected the legion Cladotheria with two sublegions, Dryolestoidea and Zatheria, the latter with two in-

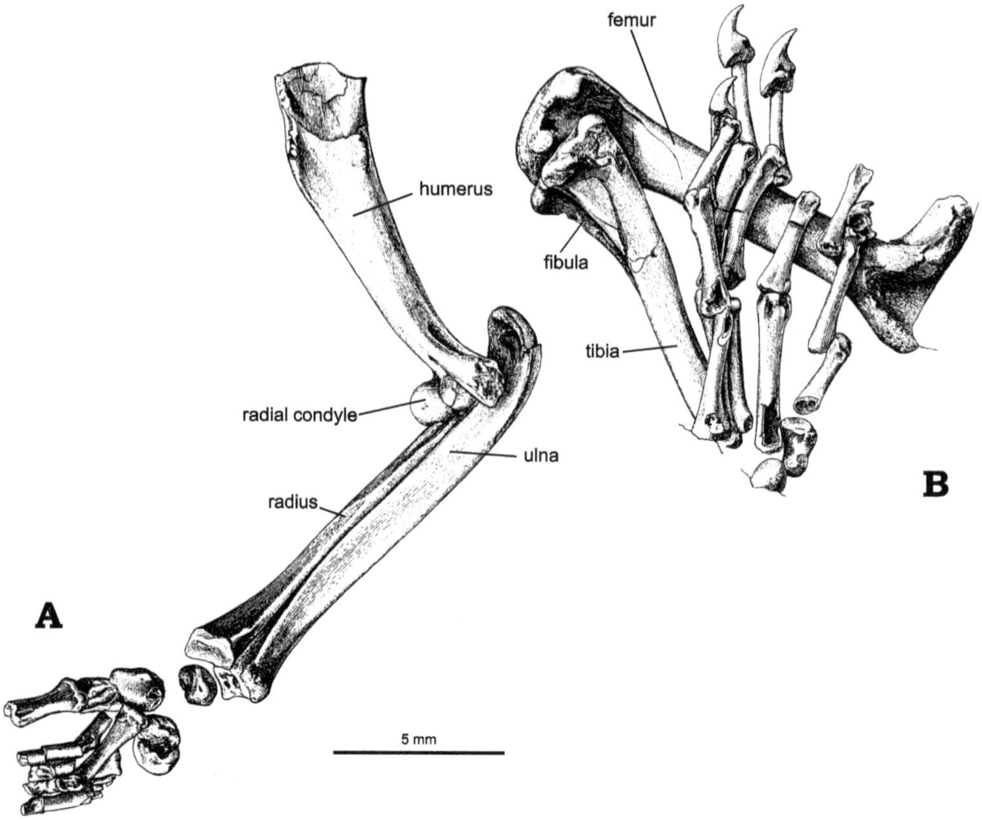

FIGURE 10.4. *Henkelotherium guimarotae.* Right forelimb, including partial carpus and metacarpals (A); hindlimb, including two cuneiforms, metatarsalia, and phalanges (B), both as preserved. *Source:* modified from Krebs (1991).

fraclasses, Peramura and Tribosphenida. In other words, members of the former Pantotheria were attributed to two different taxa of very high taxonomic rank (sublegions) (see also McKenna and Bell, 1997). McKenna's phylogenetic idea received full support in a comprehensive review of the nontribosphenic Theria by Prothero (1981), who explicitly stated (p. 323): "The use of a wastebasket term like 'pantothere' is pernicious, not only because it obscures relationships, but also because it gives a false impression that the whole group can be lumped together without further thought or discussion."

In spite of McKenna's (1975) and Prothero's (1981) criticism, the terms Pantotheria or Eupantotheria are deeply entrenched in the literature (e.g., Kraus, 1979; Krebs, 1987, 1991, 1993, 1998; Carroll, 1988; Bakker and Carpenter, 1990; Bonaparte, 1990; Canudo and Cuenca-Bescós, 1996; Martin, 1997; Rich et al., 1999; and many others). Some authors have accepted McKenna's idea that Eupantotheria are a paraphyletic group and have simply cited "eupantotherians" in quotation marks (e.g., Clemens, 1997; Hu et al., 1997; Kielan-Jaworowska, 1997; Kielan-Jaworowska et al., 1998; and many others). Gradually, more authors in the 1990s who dealt with "eupantotherians" in detail explicitly accepted McKenna's and Prothero's taxonomic framework, with some emendations (e.g., Sigogneau-Russell, 1991b, 1994a, 1999; Bonaparte, 1994; Hopson, 1994; Carpenter, 1998; Engelmann and Callison, 1998; Ensom and Sigogneau-Russell, 1998; Martin, 1999a; Butler and Clemens, 2001).

In this chapter (as in other chapters dealing with paraphyletic groups) we present two alternative classifications, one cladistic (table 10.1) and the other Linnaean (table 10.2). Compelled by the need to summarize the stem clades of the Cladotheria and cognizant of the fact that "eupantotherians" is a concept used by many Mesozoic mammal workers since the 1950s, we organized the three stem clades, dryolestoids, amphitheriids, and peramurids, into one chapter, which coincides with the traditional and widely used concept of "eupantotherians." The arrangement of dryolestoideans, amphitheridans, and "peramurans" in table 10.1 is derived from a phylogeny by Luo et al. (2002). Phylogenetic arrangement for families is emended from Prothero (1981) and McKenna and Bell (1997) and reflects the cladistic study of Martin (1999a). The phylogenetic details we adopted are based mostly on previous taxonomic studies of Prothero (1981), McKenna

TABLE 10.1. Cladistic Classification of the Mammalian Clade Cladotheria McKenna, 1975, above the Species Level[1]

- Clade Cladotheria (*Dryolestes* + *Amphitherium* + *Vincelestes* + *Peramus* + Boreosphenida) McKenna, 1975
  - Superorder Dryolestoidea Butler, 1939
    - Order Dryolestida Prothero, 1981
      - Family Dryolestidae Marsh, 1879b
        - *Dryolestes* Marsh, 1878, type genus
        - *Amblotherium* Owen, 1871
        - *Crusafontia* Henkel and Krebs, 1969
        - *Groebertherium* Bonaparte, 1986a
        - *Guimarotodus* T. Martin, 1999a
        - *Krebsotherium* T. Martin, 1999a
        - *Laolestes* Simpson, 1927c
        - *Leonardus* Bonaparte, 1990
        - *Peraspalax* Owen, 1871
        - *Phascolestes* Owen, 1871
        - *Portopinheirodon* T. Martin, 1999a
      - Family Paurodontidae Marsh, 1887
        - *Paurodon* Marsh, 1887, type genus
        - *Araeodon* Simpson, 1937b
        - *Archaeotrigon* Simpson, 1927c
        - ?*Brancatherulum* Dietrich,1927
        - *Comotherium* Prothero, 1981
        - *Dorsetodon* Ensom and Sigogneau-Russell, 1998
        - *Drescheratherium* Krebs, 1998
        - *Euthlastus* Simpson, 1927c
        - *Foxraptor* Bakker and Carpenter, 1990
        - *Henkelotherium* Krebs, 1991
        - *Tathiodon* Simpson, 1927c
      - Family Donodontidae Sigogneau-Russell, 1991 (*sedis mutabilis*)
        - *Donodon* Sigogneau-Russell, 1991
      - Family Mesungulatidae Bonaparte, 1986a (*sedis mutabilis*)
        - *Mesungulatum* Bonaparte and Soria, 1985
      - Family Brandoniidae Bonaparte, 1992 (*sedis mutabilis*)
        - *Brandonia* Bonaparte, 1990
        - ?*Casamiquelia* Bonaparte, 1990
- Clade (*Amphitherium* + *Vincelestes* + *Peramus* + Boreosphenida)
  - Order Amphitheriida Prothero, 1981
    - Family Amphitheriidae Owen, 1846
      - *Amphitherium* de Blainville, 1838
  - Order and family *incertae sedis* (*sedis mutabilis*)
    - *Chunnelodon* Ensom and Sigogneau-Russell, 1998
- Clade (*Vincelestes* + *Peramus* + Boreosphenida)
  - Family Vincelestidae Bonaparte, 1986a
    - *Vincelestes* Bonaparte, 1986a
- Clade Zatheria (*Peramus* + Boreosphenida) McKenna, 1975
  - Stem-lineages of Zatheria (Martin, 2002)
    - *Afriquiamus* Sigogneau-Russell, 1999
    - *Arguimus* Dashzeveg, 1979
    - *Arguitherium* Dashzeveg, 1994
    - *Nanolestes* T. Martin, 2002 (*sedis mutabilis*)
  - Family Peramuridae Kretzoi, 1946
    - *Peramus* Owen, 1871
    - ?*Tendagurutherium* Heinrich, 1998
    - *Palaeoxonodon* E. F. Freeman, 1979
    - *Abelodon* Brunet et al., 1990
    - *Magnimus* Sigogneau-Russell, 1999
    - *Minimus* Sigogneau-Russell, 1999
- Clade Boreosphenida Luo, Cifelli, and Kielan-Jaworowska, 2001 (including Tribosphenida McKenna, 1975)

[1] The arrangement of dryolestoideans, amphitheridans, and "peramurans" is derived from a phylogeny by Luo et al. (2002). Phylogenetic arrangement for families is emended from Prothero (1981) and McKenna and Bell (1997). For the clade-ranks above the subclass and infraclass, we simply use the "clade" with its attendant taxic definition, instead of the clade ranks used by other authors, e.g., Prothero (1981) and McKenna and Bell (1997). For the taxonomic ranks below the infraclass, we use the traditionally accepted Linnaean units such as superorders, orders, and families and subfamilies. In a cladistic phylogeny it is also unavoidable that some stem taxa will cluster in polytomies in various types of consensus trees. These taxa are listed as *sedis mutabilis*, following the procedures recommended by Wiley et al. (1991). The families and/or genera within monophyletic suborders and families are as a rule listed in alphabetic order, and their arrangement does not reflect the phylogenetic relationships. In other places in this book, e.g., table 10.2 and in the text, we made exceptions to this rule with respect to the order Dryolestida, in which the families are arranged following Martin's (1999a) cladistic relationships.

and Bell (1997), Ensom and Sigogneau-Russell (1998), Sigogneau-Russell (1999), and Martin (1999a, 2002).

Superorder Dryolestoidea Butler, 1939

BRIEF CHARACTERIZATION

Dryolestoideans are very small mammals. With the single exception of *Henkelotherium*, they are represented mostly by isolated teeth, dentaries with dentitions, and fragments of the rostrum with teeth. The lower jaw is long and slender, with an extensive horizontal ramus of the mandible. Familiar and primitive features of the Late Jurassic (Kimmeridgian) taxa are the retention of coronoid and splenial bones, as well as the presence of Meckel's groove in specimens representing adult individuals. The splenial and coronoid bones are, however, absent from the Cretaceous forms, although a reduced Meckel's groove is still present. All dryolestoideans have a typical dentary-squamosal "mammalian" jaw joint (see chapter 3). They differ from "symmetrodotans," amphitheriids, peramurids, and boreosphrenidan mammals in that the mandibular angle ex-

TABLE 10.2. Linnaean Classification of "Eupantotherian" Mammals (Stem Cladotherians)

- "Eupantotheria" Kermack and Mussett, 1958
  - Superorder Dryolestoidea Butler, 1939, emended
    - Order Dryolestida Prothero, 1981
      - Family Dryolestidae Marsh, 1879
        - *Dryolestes* Marsh, 1878, type genus
          - *D. priscus* Marsh, 1878, type species
          - *D. leiriensis* T. Martin, 1999a
        - *Amblotherium* Owen, 1871
          - *A. soricinum* Owen, 1871, type species
          - *A. gracile*[1] (Marsh, 1879c)
          - *A. nanum* (Owen, 1871)
          - *A. pusillum* (Owen, 1866)
          - ?*A. minimum*[1] (Simpson, 1927c)
        - *Crusafontia* Henkel and Krebs, 1969
          - *C. cuencana* Henkel and Krebs, 1969
        - *Groebertherium* Bonaparte, 1986a
          - *G. stipanicici* Bonaparte, 1986a, type species
          - *G. novasi* Bonaparte, 1986a
        - *Guimarotodus* T. Martin, 1999a
          - *G. inflatus* T. Martin, 1999a
        - *Krebsotherium* T. Martin, 1999a
          - *K. lusitanicum* T. Martin, 1999a
        - *Laolestes* Simpson, 1927c
          - *L. eminens* Simpson, 1927c, type species
          - *L. andresi* T. Martin, 1997
          - *L. goodrichi* (Simpson, 1929a)
          - *L. hodsoni* (Clemens and Lees, 1971)
          - *L. oweni* (Simpson, 1927c)
        - *Leonardus* Bonaparte, 1990
          - *L. cuspidatus* Bonaparte, 1990
        - *Peraspalax* Owen, 1871
          - *P. talpoides* Owen, 1871
        - *Phascolestes* Owen, 1871
          - *P. mustelulus* (Owen, 1871)
        - *Portopinheirodon* T. Martin, 1999a
          - *P. asymmetricus* T. Martin, 1999a
      - Family Paurodontidae Marsh, 1887
        - *Paurodon* Marsh, 1887, type genus
          - *P. valens* Marsh, 1887
        - *Araeodon* Simpson, 1937b
          - *A. intermissus* Simpson, 1937b
        - *Archaeotrigon* Simpson, 1927c
          - *A. brevimaxillus* Simpson, 1927c
        - ?*Brancatherulum* Dietrich, 1927
          - *B. tendagurense* Dietrich, 1927
        - *Comotherium* Prothero, 1981
          - *C. richi* Prothero, 1981
        - *Dorsetodon* Ensom and Sigogneau-Russell, 1998
          - *D. haysomi* Ensom and Sigogneau-Russell, 1998
        - *Drescheratherium* Krebs, 1998
          - *D. acutum* Krebs, 1998
        - *Euthlastus* Simpson, 1927c
          - *E. cordiformis* Simpson, 1927c
        - *Foxraptor* Bakker and Carpenter, 1990
          - *F. atrox* Bakker and Carpenter, 1990
        - *Henkelotherium* Krebs, 1991
          - *H. guimarotae* Krebs, 1991
        - *Tathiodon* Simpson, 1927c
          - *T. agilis* Simpson, 1927c
      - Family Donodontidae Sigogneau-Russell, 1991
        - *Donodon* Sigogneau-Russell, 1991
          - *D. prescriptoris* Sigogneau-Russell, 1991
      - Family Mesungulatidae Bonaparte, 1986a
        - *Mesungulatum* Bonaparte and Soria, 1985
          - *M. houssayi* Bonaparte and Soria, 1985
      - Family Brandoniidae Bonaparte, 1992
        - *Brandonia* Bonaparte, 1990
          - *B. intermedia* Bonaparte, 1990
        - ?*Casamiquelia* Bonaparte, 1990
          - ?*C. rionegrina* Bonaparte, 1990
    - Order Amphitheriida Prothero, 1981
      - Family Amphitheriidae Owen, 1846
        - *Amphitherium* de Blainville, 1838
          - *A. prevostii* (Mayer, 1832), type species
          - *A. rixoni* Butler and Clemens, 2001
    - Order and family *incertae sedis*
        - *Chunnelodon* Ensom and Sigogneau-Russell, 1998
          - *C. alopekodes* Ensom and Sigogneau-Russell, 1998
  - Superorder Zatheria McKenna, 1975 (referred to as stem-lineage Zatheria by Martin, 2002)
      - Family *incertae sedis:*
        - *Afriquiamus* Sigogneau-Russell, 1999
          - *A. nessovi* Sigogneau-Russell, 1999
        - *Arguimus* Dashzeveg, 1979
          - *A. khosbajari* Dashzeveg, 1979
        - *Arguitherium* Dashzeveg, 1994
          - *A. cromptoni* Dashzeveg, 1994
        - *Nanolestes* T. Martin, 2002
          - *N. drescherae* T. Martin, 2002
    - "Peramurans" (former order Peramura McKenna, 1975)
      - Family Peramuridae Kretzoi, 1946
        - *Peramus* Owen, 1871
          - *P. tenuirostris* Owen, 1871
      - Family ?Peramuridae Kretzoi, 1946
        - *Tendagurutherium* Heinrich, 1998
          - *T. dietrichi* Heinrich, 1998
      - Family *incertae sedis*
        - *Palaeoxonodon* E. F. Freeman, 1979
          - *P. ooliticus* E. F. Freeman, 1979
        - *Abelodon* Brunet et al., 1990
          - *A. abeli* Brunet et al., 1990
        - *Magnimus* Sigogneau-Russell, 1999
          - *M. ensomi* Sigogneau-Russell, 1999
        - *Minimus* Sigogneau-Russell, 1999
          - *M. richardfoxi* Sigogneau-Russell, 1999

[1] Emended to reflect the gender (neuter) of the genus.

tends horizontally from the ventral border of the mandibular horizontal ramus. "Symmetrodontans" (chapter 9) have no projecting mandibular angle. Amphitheriids and peramurids have a slightly downturned angular process of the mandible (Owen, 1871; Simpson, 1928a), as do eutherians.

The postcranial skeleton (known only in *Henkelotherium*) is advanced by comprison to that of stem mammals such as morganucodontans (chapter 4) and eutriconodontans such as *Jeholodens* (chapter 7). The scapula has a supraspinous fossa and the pelvis is similar to those of modern boreosphenidans. Dental formulae, especially the number of molars, can vary among families and genera. Except for Paurodontidae the number of molars is greater than five, distinguishable from Zatheria with fewer than five. Upper molars (unknown in Amphitheriida) are specialized in various ways. Stylar cusps are well developed as a rule, and in some taxa a median ridge joins the paracone to the median stylar cusp, whereas in a few others it joins the paracone to the stylcone. The metacone is present. Lower molars have a well-developed trigonid and a small talonid, with one cusp in some taxa and with a cingulid-like structure without cusps in others.

Comment. The transversely wide and mesiodistally short upper molars of dryolestoids have a superficial resemblance to those of docodontans (chapter 5, see also Simpson, 1928a; Butler, 1939; Patterson, 1956). Asher and Sánchez-Villagra (2000) compared dryolestoidean teeth with those of zalambdodont eutherians, noting that in dryolestoids, as in zalambdodont molars of tenrecs, the main lingual cusp of the upper molar does not occlude in the lower talonid basin, but rather on the labial margin of the lower tooth, just posterior to the trigonid. The authors correctly stated that the broadly convergent molar structures of unrelated taxa such as tenrecs and dryolestoids have similar functions, but this does not mean that these structures are phylogentically homologous.

Distribution. Middle Jurassic to Early Cretaceous: Europe; Late Jurassic: North America; Late Jurassic to Early Cretaceous: Africa; ?Late Cretaceous: North America; Late Cretaceous and Paleocene: South America.

Orders Assigned. Dryolestida Prothero, 1981; Amphitheriida Prothero, 1981.

Order Dryolestida Prothero, 1981

Synonym: Quirogatheria Bonaparte, 1992.

Comments. Bonaparte (1992) erected the dryolestoid order Quirogatheria to include the family Barbereniidae. In 1994 he also included Brandoniidae Bonaparte, 1992, in Quirogatheria. Martin (1999a, pers. comm.) suggested that *Barberenia* was based on upper deciduous premolars of the dryolestoid *Bradonia*, thus invalidating "Barbereniidae," so there is no need to recognize Quirogatheria as a distinct order. Here we include the family Brandoniidae in the order Dryolestida.

Diagnosis. Dryolestida differ from Amphitheriida in the structure of the lower molars, which are shorter mesiodistally and wider labiolingually. Upper molars (unknown in Amphitheriidae) are as for the superorder. Differ from Zatheria (*Peramus*) in having the power stroke of the dentary less steeply inclined (Butler, 1972b).

Families.[1] Dryolestidae Marsh, 1879b, type family; Paurodontidae Marsh, 1887; Donodontidae Sigogneau-Russell, 1991; Mesungulatidae Bonaparte, 1986a; Brandoniidae Bonaparte, 1992.

Family Dryolestidae Marsh, 1879b

Diagnosis. Abundant and diverse family of Dryolestida, known mostly from dentaries and maxillae with dentition. Dryolestids differ from other dryolestidan families by the following apomorphies: upper and lower molars strongly shortened mesiodistally and widened labiolingually; anterior root of lower molars very robust and enlarged, in comparison with a small and thin posterior root. Differ from Paurodontidae and Amphitheriidae in having unequal height of the alveolar border on contralateral sides of the molars. The molars have markedly higher crowns on the labial side. The median ridge is present in some genera, as is the median stylar cusp. By contrast, these are always absent in Paurodontidae (figures 10.9, 10.10). The metacone and stylar cusps, especially the stylocone, are more prominent in dryolestids than in paurodontids (figure 10.2A). Lower molars differ from those in Paurodontidae in having a smaller paraconid. Dryolestidae differ further from Paurodontidae and Amphitheriidae in having a greater number of molars, as many as eight or nine, in contrast to a maximum of five in North American paurodontids (*Foxraptor*), six upper and seven lower in the European paurodontid *Henkelotherium*, six or seven lowers in Amphitheriidae, and only three in peramurids.

Distribution. Middle Jurassic (late Bathonian): Britain, Oxfordshire; Late Jurassic (Kimmeridgian): Portugal, Leiria, Guimarota Coal Mine (Guimarota Beds); Late Jurassic (late Kimmeridgian–early Tithonian): United States, western states (Morrison Formation); Jurassic-Cretaceous boundary: Portugal, Porto Pinheiro; Early Cretaceous (Berriasian): Britain (Purbeck Limestone group); Early Cretaceous (Valanginian): Britain, Isle of Oxney (Wealden Supergroup); Early Cretaceous (Barremian): Spain, Uña and Galve; Late Cretaceous (Campanian):

[1] In Dryolestida we arrange the families in the phylogenetic order proposed by Martin (1999a), rather than alphabetically.

Argentina (Los Alamitos Formation); ?Late Cretaceous: United States, Wyoming ("Mesaverde" Formation).

Genera. *Dryolestes* Marsh, 1878, type genus (= *Herpetairus* Simpson, 1927c); *Amblotherium* Owen, 1871 (synonyms: *Kepolestes* Simpson, 1927c, possibly also *Miccylotyrans* Simpson, 1927c, and *Kurtodon* Osborn, 1887); *Crusafontia* Henkel and Krebs, 1969; *Groebertherium* Bonaparte, 1986a; *Guimarotodus* T. Martin, 1999a; *Krebsotherium* T. Martin, 1999a; *Laolestes* Simpson, 1927c (synonyms: *Melanodon* Simpson, 1927c, and *Malthacolestes* Simpson, 1927c); *Leonardus* Bonaparte, 1990; *Peraspalax* Owen, 1871; *Phascolestes* Owen, 1871; *Portopinheirodon* T. Martin, 1999a; and gen. et sp. indet. (E. F. Freeman, 1976a).

Comment. The systematics of the dryolestid genera and justification for synonymies are based on the recent revision of this family by Martin (1999a). More complete lists of synonyms are presented in McKenna and Bell (1997: 46), emended by Martin (1999a: table 1) and noted under the following descriptions of the genera.

Genus *Dryolestes* Marsh, 1878
(figures 10.1, 10.5A)

Synonym: *Pocamus* Canudo and Cuenca-Bescós, 1996.

Comment. See Martin (1999a) for justifications for synonymy.

Diagnosis. *Dryolestes* is a large-size genus (e.g., about 35% larger than *Krebsotherium*), characterized by a robust mandibular horizontal ramus and a very large coronoid process oriented between 80 and 90° to the tooth row. It differs in this respect from other dryolestid genera (except *Crusafontia*), in particular from *Amblotherium*, which have a much lower coronoid angle. Dental formula: 4.1.1.8/4.1.4.8–9. Differs from other dryolestids (except *Krebsotherium*) in lacking a conspicuous ectoflexus on upper molars. Differs from *Laolestes* in lacking the median ridge on upper molars; instead, there is an approximately centrally placed stylocone connected via a convex paracrista to the paracone. The most characteristic feature that differentiates *Dryolestes* from all remaining dryolestid genera is the strongly enlarged metacone that projects into the trigon basin. The lower dentition is characterized by a strongly enlarged, semiprocumbent i1 (at least in *D. leiriensis*). The lower molars differ from *Krebsotherium* in having the paraconid more procumbent, main cusps less sharply pointed and lower, and talonid less extended lingually. Differs from *Laolestes* in that the latter has a bifid metaconid.

Species. *Dryolestes priscus* Marsh, 1878, type species (synonyms: *Herpetairus arcuatus* Marsh, 1879b, *Laolestes grandis* Simpson, 1927c, and *Herpetairus humilis* Simpson, 1929a); *D. leiriensis* T. Martin, 1999. *D. leiriensis* is one of the best-known dryolestids, represented by dentaries and rostra with the dentition.

Distribution. Late Jurassic (Oxfordian or Kimmeridgian): Portugal, Leiria, Guimarota Coal Mine (Guimarota beds); Late Jurassic (late Kimmeridgian–early Tithonian): United States, western states (Morrison Formation).

Genus *Amblotherium* Owen, 1871
(figure 10.5B)

Synonyms: *Kepolestes* Simpson, 1927c, possibly also *Miccylotyrans* Simpson, 1927c, and *Kurtodon* Osborn, 1887.

Comment. One of the species of *Amblotherium*, *A. pusillum*, was cited by Simpson (1928a: 133) as: "one of the three species of Jurassic mammals in which the upper and lower molars are found in actual possession." Still the diagnosis of the genus given by Simpson (1928a, 1929a) was limited to the dentary and lower teeth. Martin (1999a) argued that *Miccylotyrans*, known from one upper jaw fragment with molars might belong to *Amblotherium*. We assign *Miccylotyrans* herein tentatively, with the caveat that its upper molars figured by Simpson (1929a: figure 33) differ considerably from those of *A. pusillum* figured by Simpson (1928a: figure 42).

Diagnosis. *Amblotherium* is small, with an estimated length of the dentary of about 18 mm, in contrast to the dentary of *Dryolestes leiriensis*, which is about 35 mm in length. Dental formula for lower teeth is: 4.1.4.7–9. Simpson (1929a: 69) diagnosed it as follows: "Premolars similar to those of *Laolestes*, slender, recurved, with no anterior accessory cusp. Metaconid of molars simple, pointed; paraconid erect, nearly or quite as high as metaconid on posterior molars, separated from metaconid by a deep, V-shaped notch; paraconid, metaconid and talonid cusp in a straight antero-posterior line. External cingulum (cingulid) on molars (faint in one English species). Small forms with slender jaw with pointed, recurved, hook-like coronoid process." We extend the diagnosis as follows: differs from *Dryolestes*, *Krebsotherium*, *Laolestes*, *Phascolestes*, and *Crusafontia* in having a relatively low angle (45°) of coronoid process in relation to the tooth row (see Prothero, 1981: table 4). Differs from other dryolestid genera in having the roots of m1 and m2 of subequal size (see Martin, 1999a: 60). Root size difference in *Amblotherium* occurs only in posterior molars, beginning with m3. The number of upper molars is unknown; M2 is as broad as long; more posterior molars are broader than long. The crowns are obliquely triangular. There is a small ectoflexus; parastyle forms a hooklike projection, the metastyle less prominent than the parastyle. There is no median ridge. On the metacrista labial to the metacone cusp "C" is present.

Species. *Amblotherium soricinum* Owen, 1871, type species; *A. gracile* (Marsh, 1879c, including *A. debile* Simpson, 1927c, see Martin 1999a: 60); *A. nanum* (Owen, 1871);

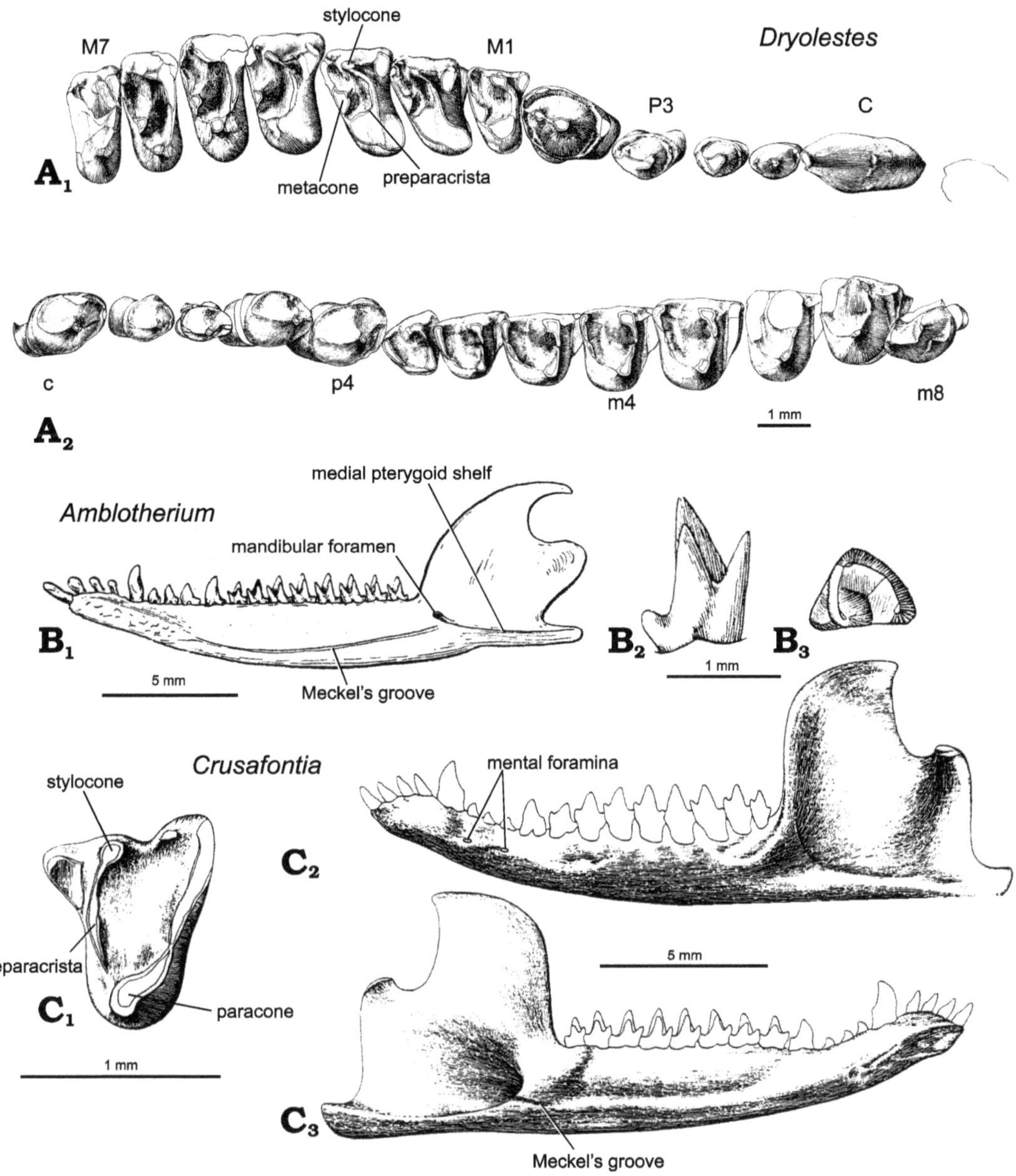

FIGURE 10.5. Three dryolestid genera from Europe. A, *Dryolestes leiriensis,* right upper canine, premolars, and molars in occlusal view (A_1); left lower canine, premolars, and molars in occlusal view (A_2). B, *Amblotherium pusillum,* reconstruction of the right dentary in lingual view (B_1); left lower molar in lingual view (B_2); the same in occlusal view (B_3). C, *Crusafontia cuencana,* left upper molar in occlusal view (C_1); left dentary in labial (C_2), and lingual (C_3) views. *Source:* modified from: A, Martin (1999a); B, Simpson (1928a); C, Krebs (1993).

A. pusillum (Owen, 1866); *A.* sp. indet. (=*Kepolestes coloradensis* Simpson, 1927c); *A.* sp. indet. Simpson 1928a; ?*A. minimum* (Simpson, 1927c), described as *Miccylotyrans.*

Distribution. Late Jurassic (late Kimmeridgian–early Tithonian): United States, western states (Morrison Formation); Early Cretaceous (Berriasian): southern England, Dorset (Purbeck Limestone Group).

Genus *Crusafontia* Henkel and Krebs, 1969
(figure 10.5C)

Synonym: *Pocamus* Canudo and Cuenca-Bescós, 1996.

Diagnosis (based on Krebs, 1993, emended). Differs from all Jurassic dryolestid genera by the absence of the coronoid and splenial bones on the mandible and a more reduced Meckel's groove. Differs from *Dryolestes* by having the condyle oriented dorsally, rather than posterodorsally. Differs from most dryolestid genera by simplified structure of the upper molars, which are characterized by lack of the median stylar cusp and median ridge, reduction of metastyle and parastyle, and lack of a metacone, and presence of a stylocone. Differ from most dryolestids (although similar to *Leonardus* and *Groebertherium*) in the absence of the metacone. Posterior upper molars have only two roots. Upper and lowers premolars have no cingula; p1 and p2 are single-rooted. Differ from other genera by presence of a strong labial cingulid on lower molars, which extends mesially to end as an anterior cuspule below notch of paracristid. Talonid small, developed as a triangular lingual wing, which is obscured in the tooth row by the paraconid of the preceding molar.

Species. *Crusafontia cuencana* Henkel and Krebs, 1969 (= *Pocamus pepelui* Canudo and Cuenca-Bescós, 1996, P4 of *C. cuencana* see Martin, 1998), type species by monotypy, based on the dentaries with dentition and isolated upper premolars and molars.

Distribution. Early Cretaceous (Barremian): Spain, Uña and Galve.

Genus *Groebertherium* J. F. Bonaparte, 1986a
(figure 10.6A)

Diagnosis. The genus is known from isolated upper and lower molars assigned to two species (but see later). Upper molars are triangular in occlusal view, with ectoflexus. *Groebertherium* shares with *Leonardus* and *Crusafontia* lack of the metacone and differs in this respect from the Late Jurassic genera. In the stylar region there is a large stylocone, placed more lingually than the prominent metastyle and less prominent parastyle. There is a deep furrow separating the median cusp from the metastyle. The paracone is narrow mesiodistally and high. The trigon basin is moderately concave without an obvious median ridge. Differs from *Leonardus* in being more expanded mesiodistally on labial side and relatively shorter labiolingually and in the lack of characteristic oval-like convexity in the middle of the trigon. In occlusal outline it resembles *Brandonia,* from which it differs in being shorter mesiodistally. Lower molars have almost equilateral trigonid; protoconid is very high in respect to lower paraconid and metaconid; the metaconid larger than the paraconid. Differs from other dryolestid genera in having prominent precingulid, which is almost as wide as the very narrow talonid, limited to the lingual part of the tooth.

Species. *Groebertherium stipanicici* J. F. Bonaparte, 1986a, type species; *G. novasi* J. F. Bonaparte, 1986a. As noted by Bonaparte (1994), the second species may represent a variation of the type species.

Distribution. Late Cretaceous (Campanian or Maastrichtian): Argentina (Los Alamitos Formation).

Genus *Guimarotodus* T. Martin, 1999a
(figure 10.7A)

Diagnosis. The dentary is robust. Dental formula for lower teeth is ?.1.4.8–9. Differs from all known dryolestid genera by extremely enlarged and swollen metaconid, which is almost as high as the protoconid. Shares with *Krebsotherium* a low and strongly procumbent paraconid. The paraconid is separated from the metaconid by a narrow incision. Talonid is small, very narrow, and less extended labially than in *Krebsotherium.* On the middle molars there is labial cingulid and weak lingual cingulid between the paraconid and talonid. The canine and premolars are stout; the premolars very low.

Species. *Guimarotodus inflatus* T. Martin, 1999a, type species by monotypy, represented by lower dentition of large size, known from almost complete postincisor lower dentition and parts of a dentary.

Distribution. Late Jurassic (Kimmeridgian): Portugal, Leiria, Guimarota Coal Mine (Guimarota Beds).

Genus *Krebsotherium* T. Martin, 1999a
(figure 10.7B)

Diagnosis. Small animals with eight or nine lower molars, the last two of which may be very reduced in size. The main lower molar cusps are tall and sharply pointed. The metaconid is columnar, almost as high as the protoconid. The paraconid is procumbent and is much smaller than the metaconid. The talonid is relatively narrow mesiodistally, triangular, strongly narrowing labially. There is a tiny lingual edge and a weak labial cingulid. The trigonid angle of the anterior lower molars is 40°, posterior 35°. The P4 has a wide lingual cingulum with anterior, medial, and posterior cuspules. There is a weak median ridge associated with a centrally placed stylocone. On last upper molars the median ridge is vestigial. The metacone is large, labiolingually extended. Trigon angle of the anterior upper molars is 50°, posterior 45°.

Species. *Krebsotherium lusitanicum* T. Martin, 1999a, type species by monotypy.

Distribution. Late Jurassic (Kimmeridgian): Portugal, Leiria, Guimarota Coal Mine (Guimarota Beds).

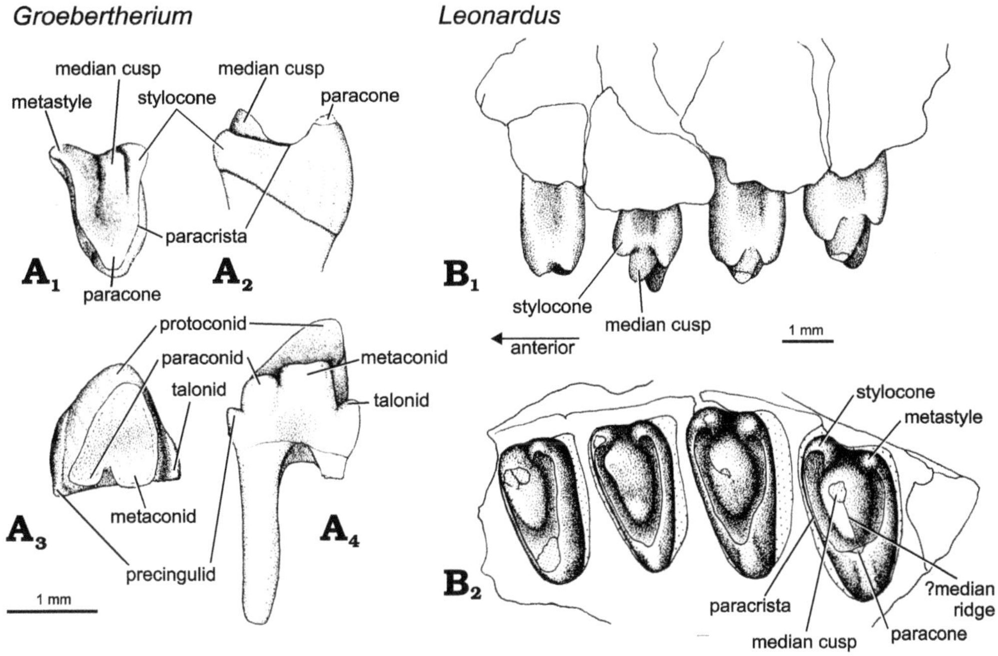

FIGURE 10.6. Dryolestid genera from Argentina. A, *Groebertherium stipanicici,* right upper molar in occlusal (A_1), and distal (A_2) views; right lower molar in occlusal (A_3) and lingual (A_4) views. B, *Leonardus cuspidatus,* fragment of left maxilla with four molars in labial (B_1) and occlusal (B_2) views. *Source:* modified from: A, Bonaparte (1986a); B, Bonaparte (1990).

Genus *Laolestes* Simpson, 1927c
(figure 10.7C)

Synonym: *Melanodon* Simpson, 1927c.

Comment. See Martin (1999a) for argument on synonymy of *Melanodon* Simpson, 1927c, and *Laolestes* Simpson, 1927c, a possibility suggested earlier by Simpson (1929a) and Prothero (1981).

Diagnosis. Differs from all dryolestid genera in having bifid metaconid. Simpson (1929a: 61) diagnosed *Laolestes* as follows: "Premolars with continuous internal cingulum, without anterior accessory cusp. Metaconid of molars columnar, bifid at apex; paraconid slightly procumbent, lower than metaconid; paraconid and talonid cusp distinctly internal relative to metaconid. External cingula on molars." We further note that *Laolestes* differs from *Dryolestes* and *Krebsotherium* in having a very large stylocone, placed centrally. Parastyle and metastyle are placed more labially and much lower than the stylocone. Differs from *Dryolestes* in having a very prominent median ridge. In the middle of the metacrista there is a small metacone. In some species, for example, *L. andresi,* there is a precingulum on the paracone slope. The body of the mandible is more slender than in most dryolestid genera. The coronoid process is arranged at 70° to the tooth row and ends as a hooklike process. Lower dentition dental formula is: 4.1.4.8.

Species. *Laolestes eminens* Simpson, 1927c, type species; *L. andresi* T. Martin, 1997; *L. goodrichi* (Simpson, 1929a); *L. hodsoni* (Clemens and Lees, 1971); *L. oweni* (Simpson, 1927c); *non L. grandis* Simpson, 1929a (= *Dryolestes*).

Distribution. Late Jurassic (late Kimmeridgian–early Tithonian): United States, western states (Morrison Formation); Jurassic-Cretaceous boundary: Portugal, Porto Pinheiro; Early Cretaceous (Valanginian): England, Isle of Oxney (Kent and Cliff End bonebeds of Hastings).

Genus *Leonardus* J. F. Bonaparte, 1990
(figure 10.6B)

Diagnosis (based on Bonaparte, 1990, emended). Genus represented by only partial upper dentition, known from the last four molars preserved in a fragmentary left maxilla of the type species. Bonaparte (1990: 74) diagnosed *Leonardus* as follows: "A dryolestid with the upper molars anteroposteriorly narrower than in two known species of *Groebertherium,* without parastylar hook and with subequal parastylar and metastylar cusps. Large 'stylocone' placed in the labial half of the crown in front of the middle of the smooth ectoflexus. Molars widely separated

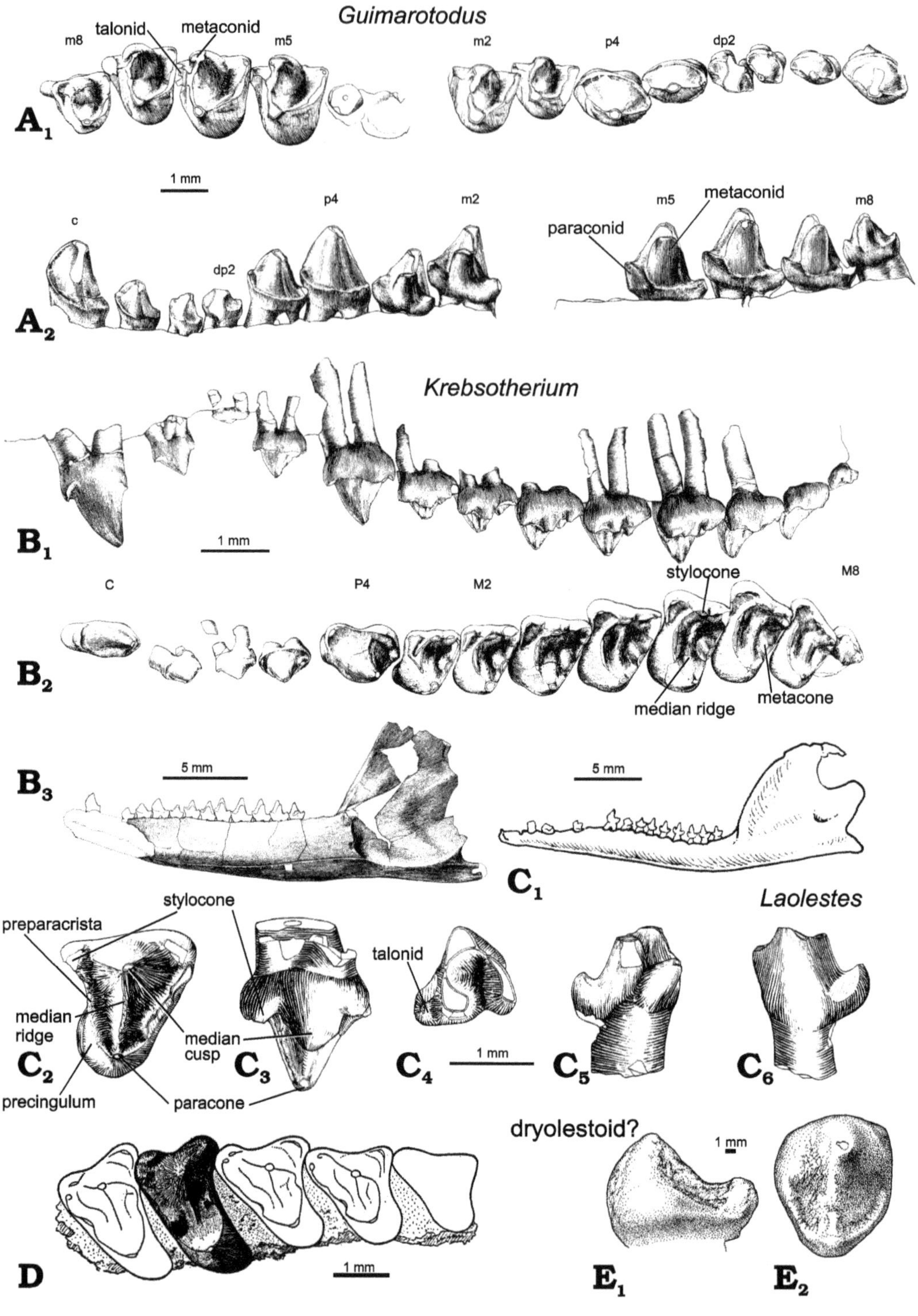

FIGURE 10.7. Four dryolestid genera. A, *Guimarotodus inflatus,* right lower dentition in occlusal (A_1) and lingual (A_2) views. B, *Krebsotherium lusitanicum,* left upper dentition in labial (B_1) and occlusal (B_2) views; reconstruction of the dentary in lingual view (B_3). C, *Laolestes,* reconstruction of the dentary in lingual view, based on original specimen of *L. eminens* (C_1). *L. andresi,* left upper molar in occlusal (C_2) and labial (C_3) views; left lower molar in occlusal (C_4), lingual (C_5), and labial (C_6) views. D, *Laolestes andresi,* reconstruction of upper molars in occlusal view. E, An unidentified ?dryolestoid tooth from the Early Cretaceous of New South Wales, Australia, in anterior (E_1) and occlusal (E_2) views. *Source:* all except C_1, D, and E, modified from Martin (1999a); C_1, from Prothero (1981); D, from Simpson (1929a); E, from Clemens et al. (2003).

from one another. Paracone anteroposteriorly narrow." We further note that *Leonardus* differs from many dryolestids (except *Groebertherium* and *Crusafontia*) in lacking a metacone; differs from *Krebsotherium, Laolestes,* and *Portopinheirodon* by lack of a median ridge (except perhaps a weak crest on the ultimate molar) and by having the central part of the crown developed as an oval convexity and encircled by the paracrista and the metacrista. The median stylar cusp (labeled by Bonaparte as the stylocone) is near the central part of the crown, while this cusp is more labially positioned in *Groebertherium.*

Species. *Leonardus cuspidatus* J. F. Bonaparte, 1990, type species by monotypy.

Distribution. Late Cretaceous (Campanian): Argentina (Los Alamitos Formation).

Genus *Peraspalax* Owen, 1871
(figure 10.8A)

Comment. Simpson (1928a: 141) stated: "This genus is very close to *Phascolestes* and more material may prove these two to be identical, although the evidence in hand does not permit this conclusion."

Diagnosis. Lower dentition genus, similar to *Phascolestes,* known only from the type species represented by an incomplete dentary with apparently very large canine, four premolars and seven molars (see Owen, 1871: plate 2, figure 9). Fourth lower premolar with a well-developed and cuspidate heel and with a distinct anterior cingulid cusp and lingual cingulid. Molars differ from those of *Phascolestes* by having trigonid basin less compressed anteroposteriorly.

Species. *Peraspalax talpoides* Owen, 1871, type species by monotypy.

Distribution. Early Cretaceous (Berriasian): southern England, Dorset (Purbeck Limestone Group).

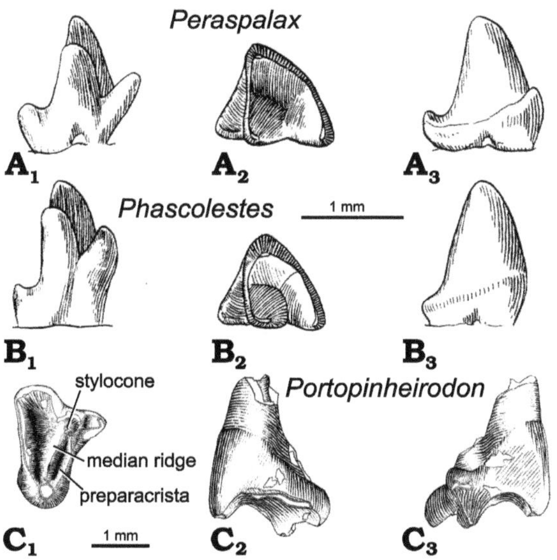

FIGURE 10.8. Three dryolestid genera. A, *Peraspalax talpoides,* left lower molar in lingual (A_1) and occlusal (A_2) views, last premolar in lingual view (A_3). B, *Phascolestes mustelulus,* right lower molar in lingual (B_1), and occlusal (B_2) views, last premolar in lingual view (B_3). C, *Portopinheirodon asymmetricus,* right upper molar in occlusal (C_1), distal (C_2), and mesial (C_3) views. *Source:* A, B, modified from Simpson (1928a); C, modified from Martin (1999a).

Genus *Phascolestes* Owen, 1871
(Erected as a subgenus of *Peralestes*)
(figure 10.8B)

Diagnosis. Genus based on lower dentition. Dental formula (after Simpson, 1928a) is: 3–4.1.4.7–8. The dentary is slender with a steep coronoid process, similar to that in *Amblotherium.* Differs from all dryolestid genera in having the trigonid very short, compressed mesiodistally, with broad blunt cusps, and somewhat recurved or spatulate paraconid, which is shorter than metaconid.

Species. *Phascolestes mustelulus* Owen, 1871, type species by monotypy.

Distribution. Early Cretaceous (Berriasian): southern England, Dorset (Purbeck Limestone Group).

Genus *Portopinheirodon* T. Martin, 1999a
(figure 10.8C)

Diagnosis. The upper molar is strongly asymmetrical, with ectoflexus. There is an enlarged, rounded metastylar region that protrudes labially, but no obvious metastyle is developed. There is a large and inflated stylocone, placed more lingually than the metastylar region, with paracrista extending from it, first anterolingually and then bent and continuing lingually. The paracone is very large, but only slightly higher than the stylocone. The median ridge is strong; it extends between the stylocone and paracone, close to the paracrista, and divides the trigon into two unequal concavities, the anterior one being much shorter than the posterior. The metacrista is weak; the metacone is small, situated close to the paracone. The parastylar region forms a roughly triangular plate that protrudes anteriorly; the parastyle is small, hardly discernible.

Species. *Portopinheirodon asymmetricus* T. Martin, 1999a, type species by monotypy, based on one isolated upper molar.

Distribution. Jurassic-Cretaceous boundary: Portugal, Porto Pinheiro.

Family Paurodontidae Marsh, 1887

Synonym: Henkelotheriidae Krebs, 1991.

Diagnosis. Dryolestoidean family, known from upper and lower jaws with dentition and fragments of a skull associated with postcranial skeleton (in *Henkelotherium*). Differ from Dryolestidae and Amphitheriidae in having a robust dentary that is high anteriorly, with vertical symphysis. Differ from Dryolestidae in having smaller number of molars (two to six upper and up to seven lower), rather than eight or nine; in most genera (except *Henkelotherium*) the number of the lower molars is smaller than in Amphitheriidae (where there are six or seven lower molars, upper count unknown). The upper molars are slightly narrower labiolingually in Paurodontidae than in Dryolestidae; for example, in *Henkelotherium* the transverse and longitudinal dimensions of the upper molars approach one another. The median stylar cusp and median ridge are lacking, but there may be a swelling in the middle of the trigon basin. The lower molars differ from those in Dryolestidae in having a shelf-like paraconid and reduced height of the metaconid relative to the paraconid. In some paurodontids the hypconulid is in a median position, rather than lingual as in dryolestids. The difference in the root size of the lower molars is very small, sometimes negligible, never so dramatic as in Dryolestidae.

Comment. In his revision of the Purbeck mammals, Butler (1939: 334) stated: "The shortening of the cheek teeth must have accompanied considerable changes in the form of the head, and, in spite of the similarity of molar pattern, the Paurodontidae must have been animals of different habits and structure from *Amphitherium*." The difference is even greater between the Paurodontidae and Dryolestidae, as in the latter family there are more cheek teeth than in Paurodontidae.

Paurodontidae still require revision even after the earlier work by Prothero (1981) and the more recent study by Ensom and Sigogneau-Russell (1998), especially for the taxa from the Morrison Formation. A key issue is that there are only minute differences among several "lower teeth" genera established on single type specimens by Simpson (1927c, 1929a, 1937b; see also Bakker and Carpenter, 1990; Carpenter, 1998; Engelmann and Callison, 1998). Some of these taxa may turn out to be congeneric when known by more complete material (see the more detailed discussion by Ensom and Sigogneau-Russell, 1998).

Genera. *Paurodon* Marsh, 1887, type genus; *Araeodon* Simpson, 1937b; *Archaeotrigon* Simpson, 1927c; ?*Brancatherulum* Dietrich, 1927; *Comotherium* Prothero, 1981; *Dorsetodon* Ensom and Sigogneau-Russell, 1998; *Drescheratherium* Krebs, 1998; *Euthlastus* Simpson, 1927c; *Foxraptor* Bakker and Carpenter, 1990; *Henkelotherium* Krebs, 1991; *Pelicopsis* Simpson, 1927c; *Tathiodon* Simpson, 1927c.

Distribution. Late Jurassic to Early Cretaceous: Europe; Late Jurassic: North America; Late Jurassic: Africa.

Genus *Paurodon* Marsh, 1887
(figure 10.9A)

Comment. *Paurodon* is based on the lower dentition. Prothero (1981: figure 7) figured it in association with upper molars of *Pelicopsis.*

Diagnosis. Simpson (1929a: 49) diagnosed *Paurodon* as follows: "Dental formula i?, c, p2, m4. Paraconid and talonid shelf-like, not forming true cusps, metaconid very low. Talonid semicircular in plan. With postcanine diastema, molars spaced. Horizontal ramus stout. Symphysis short. Coronoid arising some distance posterior to the last molar."

Species. *Paurodon valens* Marsh, 1887, type species by monotypy.

Distribution. Late Jurassic (late Kimmeridgian–early Tithonian): United States, western states (Morrison Formation).

Genus *Araeodon* Simpson, 1937b
(figure 10.9B)

Diagnosis. Poorly known genus based on lower dentition, diagnosed by Simpson (1937b: 2) as follows: "Cheek teeth seven or eight, probably p3, m4. p?1 minute, following tooth large, typically premolariform. m?1 (fourth tooth counting from posterior end of series) molariform, with small and low but distinctly cuspidate paraconid and metaconid subequal in height, paraconid projecting anteriorly and shelf-like, talonid very small, subtriangular, internal, unbasined. Two roots of each molar subequal. m4 somewhat reduced in size."

Species. *Araeodon intermissus* Simpson, 1937b, type species by monotypy.

Distribution. Late Jurassic (late Kimmeridgian–early Tithonian): United States, western states (Morrison Formation).

Genus *Archaeotrigon* Simpson, 1927c
(figure 10.9C)

Diagnosis. Genus based on the lower dentition, diagnosed by Simpson (1929a: 51) as follows: "Dental formula i4, c1, p2, m3–4. Paraconid low but distinct, talonid with low distinct cusp, metaconid intermediate in height. Talonid semicircular in plan. No postcanine diastema, molars not spaced. Horizontal ramus stout. Symphysis short. Coronoid arising immediately posterior to last molar."

Species. *Archaeotrigon brevimaxillus* Simpson, 1927c, type species; *A.* sp. indet. (Simpson, 1929a).

Distribution. Late Jurassic (late Kimmeridgian–early Tithonian): United States, western states (Morrison Formation).

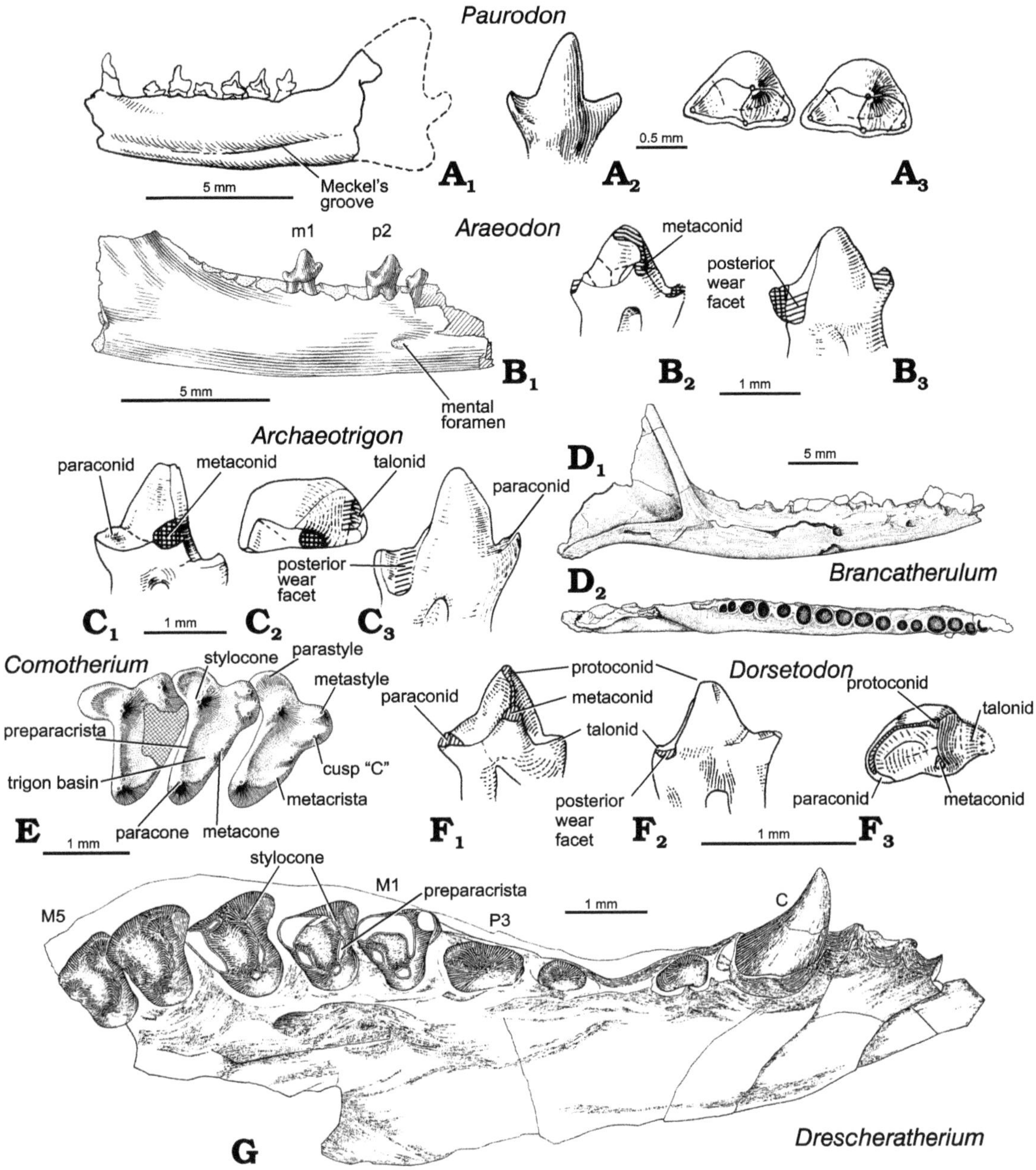

FIGURE 10.9. Seven paurodontid genera. A, *Paurodon valens,* reconstruction of incomplete right dentary in lingual view (A_1); right lower molar in labial view (A_2); left lower molars in occlusal view (A_3). B, *Araeodon intermissus,* right dentary with p1, p2, and m1 in labial view (B_1); right lower molar in lingual (B_2), and labial (B_3) views. C, *Archaeotrigon brevimaxillus,* right lower molar in lingual (C_1), occlusal (C_2), and labial (C_3) views. D, *Brancatherulum tendagurense,* right edentulous dentary in labial (D_1) and occlusal (D_2) views. E, *Comotherium richi,* upper left molars in occlusal view. F, *Dorsetodon haysomi,* right lower molar in lingual (F_1), labial (F_2), and occlusal (F_3) views. G, *Drescheratherium acutum,* right upper dentition in occlusal view, lingual side is up. *Source:* A, E, modified from Prothero (1981); B_1, modified from Simpson (1937b); B_2, B_3, C, F, modified from Ensom and Sigogneau-Russell (1998); D, modified from Heinrich (1991); G, modified from Krebs (1998).

Genus ?*Brancatherulum* Dietrich, 1927
(figure 10.9D)

Comment. *Brancatherulum* is known from a single edentulous dentary. As the teeth are not known, we assign it to the Paurodontidae tentatively only (see also Simpson, 1928a; Kraus, 1979; Prothero, 1981; Heinrich, 1991), especially because it differs from paurodontid dentaries in lack of a vertical symphysis, as best evidenced by *Foxraptor* (see Bakker and Carpenter, 1990).

Diagnosis. Poorly known genus. The single specimen of the type species (edentulous dentary) shows 12 post-canine alveoli, indicating the presence of six or seven post-canine teeth. Differs from members of the Dryolestidae and Amphitheriidae in the smaller number of the post-canine teeth. Differs from paurodontid genera in which the dentary is known (e.g., *Paurodon, Henkelotherium,* and *Foxraptor*) and from some "stem-lineage of Zatheria" (e.g., *Arguitherium* and *Arguimus*) in having a less robust mandibular ramus, which is reduced in height anteriorly. The coronoid process forms an angle of about 70° with the tooth row (measured on the figures in Heinrich, 1991), the masseteric fossa is deep and sharply limited, and the angular process directed posteroventrally.

Species. *Brancatherulum tendagurense* Dietrich, 1927, type species by monotypy.

Distribution. Late Jurassic (Kimmeridgian–Tithonian): Tanzania (Tendaguru Beds).

Genus *Comotherium* Prothero, 1981
(figure 10.9E)

Comment. Prothero (1981) assigned *Comotherium* to Dryolestidae. We follow Martin (1999a) in considering *Comotherium* to be a paurodontid.

Diagnosis. Genus based on upper dentition. Upper molars generally similar to *Euthlastus* in having a strong ectoflexus, a prominent stylocone joined with the paracone by a prominent paracrista, and a shelflike, enlarged parastyle. Other similarities concern lack of ridges or cusps in the trigon basin. Differs from *Euthlastus* in having molars narrower mesiodistally and more elongated labiolingually. Another difference is a relatively smaller metacone, situated at the midlength of the metacrista, rather than a large metacone adhering to another metacrista cusp labial to it. In *Comotherium* the metastylar region is more expanded distally, with two distinct cusps, whereas in *Euthlastus* the additional metastylar cusp is placed more lingually on the metacrista. Anterior interdental embrasures between the trigons are narrower than in any other dryolestoid.

Species. *Comotherium richi* Prothero, 1981, type species by monotypy.

Distribution. Late Jurassic (late Kimmeridgian–early Tithonian): United States, Wyoming (Morrison Formation).

Genus *Dorsetodon* Ensom and Sigogneau-Russell, 1998
(figure 10.9F)

Diagnosis. Genus represented by only lower dentition; one of the smallest known dryolestoids. The diagnosis that follows is adopted with minor emendations from Ensom and Sigogneau-Russell (1998: 37). Protoconid moderately high, less so than in *Paurodon.* Differ from other Paurodontidae (except *Archaeotrigon* and *Tathiodon*) in having a distinct, shelflike, but not spatulate paraconid, which is narrower than the metaconid. Metaconid of moderate height. Posterior face of trigonid more strongly concave than in other Paurodontidae. Talonid relatively long, notably longer than in *Paurodon* and *Araeodon;* subtriangular as in *Araeodon,* low and mediolingually situated without a well-defined cusp. Lower molars closest to *Araeodon* in labial "convexity" (flatter in other genera including *Henkelotherium*). Closest to *Tathiodon* in general proportions, but metaconid less stout and talonid less sharply triangular. Differs from *Henkelotherium* by posterior trigonid face more concave, less compressed trigonid, paraconid relatively more gracile, talonid less wide labiolingually.

Species. *Dorsetodon haysomi* Ensom and Sigogneau-Russell, 1998, type species by monotypy, represented by isolated lower molars.

Distribution. Early Cretaceous (Berriasian): southern England, Dorset (Purbeck Limestone Group).

Genus *Drescheratherium* Krebs, 1998
(figure 10.9G)

Diagnosis. Genus based on upper dentition; dental formula ?5.1.3.5. Differs from *Henkelotherium* in having a very large and sharp canine and only three premolars (P2 is missing and there is a short diastema between P1 and P3), whereas in *Henkelotherium* P2 is reduced in size, but present. Further differences concern the structure of the upper molars, which are more triangular in occlusal view with a narrower lingual part. Upper molars have a small ectoflexus plus a convex labial cingulum and prominent parastylar wing. In the stylar region the prominent cusp is the stylocone, situated lingual to the labial cingulum, while other styles are hardly discernible. The prominent paracrista joins the stylocone with the paracone, which is the highest cusp of the tooth. The metacrista is less prominent and there is a small metacone situated at the lingual part of the metacrista. There is a swelling in the middle of the trigon basin, but the median ridge is not developed. In the triangular shape of the upper molars *Drescheratherium* is similar to *Groebertherium* from Argentina, from which it differs in the presence of a metacone and in being more expanded mesiodistally. It differs from *Comotherium* and *Euthlastus* in being less elongated labiolingually and having less prominent cusps in the stylar region, and from *Tathiodon* in being more pointed lingually and in having a smaller metacone.

Species. *Drescheratherium acutum* Krebs, 1998, type species by monotypy, represented by fairly complete upper jaws with teeth.

Distribution. Late Jurassic (Kimmeridgian): Portugal, Leiria, Guimarota Coal Mine (Guimarota Beds).

Genus *Euthlastus* Simpson, 1927c
(figure 10.10C)

Comment. Simpson (1929a) and Prothero (implicitly, 1981) referred *Euthlastus* to Dryolestidae. Engelmann and Callison (1998) and independently Martin (1999a) argued that *Euthlastus* is a paurodontid; we follow the latter opinion.

Diagnosis. Shares with *Comotherium* strong ectoflexus, prominent stylocone, and lack of median stylar cusp and median ridge. Similar to *Comotherium* in having a large platelike parastylar region and prominent paracrista joining the stylocone with paracone. Differs from *Comotherium* in having molars longer mesiodistally and shorter labiolingually, a larger metacone, and a more labial position of cusp "C" (*sensu* Crompton, 1971) on the metacrista, whereas this cusp is less distinctive and placed closer to the metacone in *Comotherium.* Differs from the latter by being 50% smaller and in having a lower paracone.

Species. *Euthlastus cordiformis* Simpson, 1927c, type species by monotypy, represented by only five upper molars.

Distribution. Late Jurassic (late Kimmeridgian–early Tithonian): United States, western states (Morrison Formation).

Genus *Foxraptor* Bakker and Carpenter, 1990
(figure 10.10A)

Diagnosis (based on Bakker and Carpenter, 1990; Carpenter, 1998, emended). Genus based on lower dentition; dental formula 4.1.3.5. Differs from *Paurodon, Araeodon,*

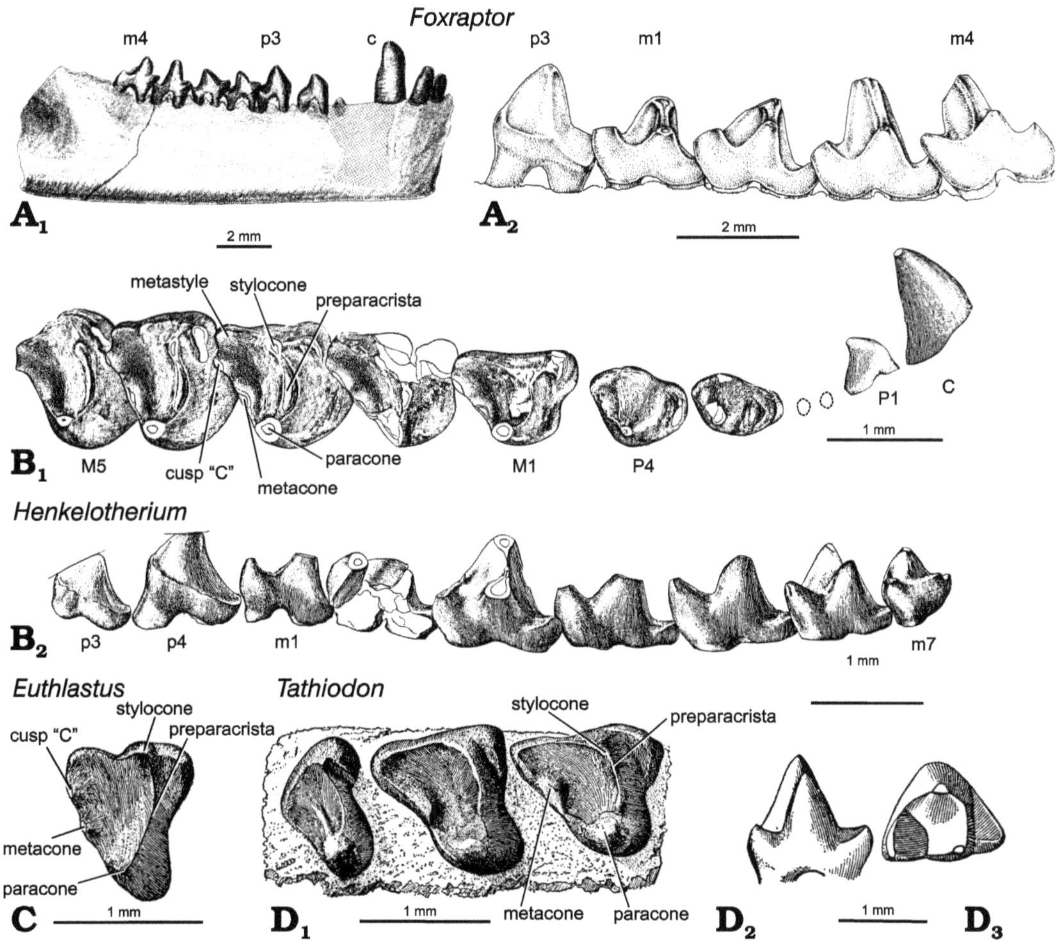

FIGURE 10.10. Five paurodontid genera. A, *Foxraptor atrox,* partial right dentary in labial view (A_1); the p3–m4 of the same in lingual view (A_2). B, *Henkelotherium guimarotae,* right upper dentition in occlusal view (B_1); right lower dentition in lingual view (B_2), P2 alveolus reconstructed according to the original description. C, *Euthlastus cordiformis,* reconstruction of the right upper molar in occlusal view. D, *Tathiodon agilis,* right upper molars in occlusal view (type of "*Pelicopsis dubius*") (D_1); right lower molar in lingual (D_2) and occlusal (D_3) views. *Source:* A_1, modified from Carpenter (1998); A_2, modified from Bakker and Carpenter (1990); B, modified from Krebs (1991); C, D, modified from Simpson (1929a).

Archaeotrigon, ?Brancatherulum, and *Tathiodon,* but not from *Henkelotherium,* in having a greater number of post-canine teeth. Incisors and canine are spatulate, with tips curved lingually. The molar trigonids are not as elongated as in other paurodontids, being about as long as wide, with procumbent, shelflike paraconid, and metaconid about half the height of the protoconid. Talonids of m2–3 are broad, semicircular, with a hypoconulid; talonid of m4 compressed labiolingually; m5 not preserved, but apparently smaller than m4.

Species. *Foxraptor atrox* Bakker and Carpenter, 1990, type species by monotypy.

Distribution. Late Jurassic (late Kimmeridgian–early Tithonian): United States, Wyoming (Morrison Formation).

Genus *Henkelotherium* Krebs, 1991
(figure 10.10B)

Comment. *Henkelotherium* is the most completely known genus of all dryolestoids, represented by large parts of the postcranial skeleton as well as a fragmentary skull and the dentition. In order to make the diagnosis compatible with those of other genera, we limit it to the dentition and dentary; for a description of the postcranial skeleton see earlier section on "Anatomy: Postcranial Skeleton."

Diagnosis. Differs from other paurodontids, such as *Paurodon* and *Foxraptor,* in that the dentary is more gracile, although still more robust than those of dryolestids and *Amphitherium.* Krebs (1991) estimated the dental formula as 4–5.1.4.6/4.1.4.7. *Henkelotherium* differs from most paurodontid genera in having more post-canine teeth. There is a sharp morphological boundary between the premolars and molars. The upper molars differ from one another slightly in proportions, but are more quadrangular than rectangular in occlusal view and in this respect differ from *Paurodon, Comotherium, Euthlastus, Drescheratherium,* and *Tathiodon.* There is a labial cingulum and the stylocone is situated lingual to it. The parastyle and metastyle are moderately prominent. The paracone is very high, the paracrista and metacrista prominent. There are two cusps on the metacrista (as in *Euthlastus,* from which it differs in the shape of the crown), the metacone placed close to the paracone and a prominent more labial cusp "C" between the metacone and metastyle. The lower dentition has been figured only in lingual view. The m7 is strongly reduced in size with respect to all the others. The protoconid (as preserved in m6) is highest of all the trigonid cusps, the paraconid lower than the metaconid. There is also a very low and small talonid with a single cusp.

Species. *Henkelotherium guimarotae* Krebs, 1991, type species by monotypy.

Distribution. Late Jurassic (Kimmeridgian): Portugal, Leiria, Guimarota Coal Mine (Guimarota Beds).

Genus *Tathiodon* Simpson, 1927c
(figure 10.10D)

Synonym: *Pelicopsis* Simpson, 1927c.

Comment. We follow Krebs (1991: 96) and Martin (1999a: 63) and regard *Pelicopsis,* based on the upper dentition, as a synonym of *Tathiodon,* based on the lower dentition.

Diagnosis. Dental formula of the upper teeth is not known but Simpson (1929a) suggested that there were only four or five molars. Upper molars without ectoflexus, roughly triangular in occlusal view. There is a prominent semicircular parastylar region, stylocone very week, metastyle not developed as a cusp. In the middle of the metacrista length is a large metacone, which enters the concave trigon basin. Lower dentition is known from a single fragment of a dentary with two molars of the type species. The dentary is robust, as characteristic of the Paurodontidae, but less so than in *Paurodon* and *Foxraptor.* Trigonid is wider than long in occlusal view; the protoconid the highest of the trigonid cusps, the paraconid very low, strongly procumbent, the metaconid intermediate in height. The talonid is small with a single cusp, limited to the labial side of the teeth. Differs from *Euthlastus* in lack of ectoflexus and cusp "C." Differs from *Henkelotherium* in having upper molars triangular rather than rectangular in occlusal view, fewer cusps in the stylar region, and lack of cusp "C" ("metastyle" of Prothero 1981). Differs from *Comotherium* and *Euthlastus* in having upper molars longer mesiodistally and shorter labiolingually and stylar cusps not developed as high cones. Apparently similar to *Archaeotrigon* in lower molar proportions.

Species. *Tathiodon agilis* Simpson, 1927c (including *Pelicopsis dubius* Simpson, 1927c), type species by monotypy.

Distribution. Late Jurassic (late Kimmeridgian–early Tithonian): United States, western states (Morrison Formation).

Family Donodontidae Sigogneau-Russell, 1991

Diagnosis. Differ from Dryolestidae in having upper molars not compressed mesiodistally, only slightly wider than long. In the presence of a stylocone and parastyle in the occlusal view, the two known upper molars (*Donodon*) resemble Argentinean Late Cretaceous *Brandonia,* from which it differs in lack of ectoflexus and broader and lower median ridge. Differ from Dryolestidae and Paurodontidae in having a high labial cingulum with two prominent cusps, median stylar cusp and stylocone (referred to by

Sigogneau-Russell, 1991b as cusps D and B, respectively). Parastylar region enlarged, forming a protruding wing, with a small parastyle; median stylar cusp more prominent than stylocone. Paracrista sharp and prominent. A small metacone is situated near the middle point of the metacrista and enters the primary trigon basin. A wide and low median ridge extends from the median stylar cusp to the paracone, dividing the trigon basin into two shallow concavities. The lower molars found in the same beds as the upper teeth are roughly circular in occlusal view. They consist of a compact trigonid, with a high protoconid, and lower paraconid and metaconid; the paraconid is arranged vertically, not procumbent. The talonid is extremely narrow mesiodistally, crescent shaped, without an obvious cusp.

Genera. *Donodon* Sigogneau-Russell, 1991, type genus by monotypy.

Distribution. Early Cretaceous (?Berriasian): Morocco, Talsinnt Province, Synclinal d'Anoual (Séquence B des Couches Rouges).

Genus *Donodon* Sigogneau-Russell, 1991
(figures 10.2C, 10.11A)

Diagnosis and Distribution. As for the family.

Species. *Donodon prescriptoris* Sigogneau-Russell, 1991, type species by monotypy.

Family Mesungulatidae J. F. Bonaparte, 1986a

?Synonym: Barbereniidae J. F. Bonaparte, 1992 (*partim*).

Diagnosis. Monotypic family of relatively large forms, known from taxa represented by isolated upper and lower molars and tentatively assigned deciduous upper premolars. Differ from other dryolestoid families by having strong precingulum and postcingulum on the upper molars, which are extended lingually but do not meet around the paracone. There are three cusps on the stylar shelf (parastyle, stylocone, and metastyle). Stylocone largest of all the tooth cusps, extending the whole length of the labial margin of the trigon. Metacone absent. Median ridge joins the paracone with stylocone and divides trigon basin into two concave areas. Lower molars are roughly rectangular in occlusal view, with metaconid larger and higher than paraconid, the protoconid the largest of the trigonid cusps. Lower molars differ from those in Dryolestidae in having an anterior cingulid. The talonid is not much wider than the anterior cingulid and extends along the whole posterior margin of the trigonid.

Genera. *Mesungulatum* Bonaparte and Soria, 1985, type genus; two very large, but undescribed genera from the Late Cretaceous La Colonia Formation of Patagonia, reported by Rougier et al. (2000).

Distribution. Late Cretaceous (Campanian–Maastrichtian): Argentina (Los Alamitos and La Colonia formations).

Genus *Mesungulatum* J. F. Bonaparte and Soria, 1985
(figure 10.11B)

?Synonym: *Quirogatherium* J. F. Bonaparte, 1990.

Comments. Bonaparte (1990) erected the monotypic genus *Quirogatherium* and assigned it to his new symmetrodont family, Barbereniidae (which also included *Barberenia* Bonaparte, 1990). He subsequently (Bonaparte, 1992) erected the order Quirogatheria for Barbereniidae, and in 1994 also included Brandoniidae in Quirogatheria. We agree with T. Martin (pers. comm. June 2000) that *Quirogatherium,* which belongs to the same size category as *Mesungulatum,* might represent its deciduous upper premolars.

Diagnosis and Distribution. As for the family.

Species. *Mesungulatum houssayi* J. F. Bonaparte and Soria, 1985, including *Quirogatherium major* J. F. Bonaparte, 1990.

Family Brandoniidae J. F. Bonaparte, 1992

?Synonym: Barbereniidae J. F. Bonaparte, 1990 (partim).

Comments. As discussed by Martin (1999a: 82) *Barberenia* Bonaparte, 1990 (based on upper deciduous molars), the type genus of Barbereniidae, might be a junior synonym of *Brandonia* Bonaparte, 1990. As Barbereniidae in part are a synonym of Mesungulatidae (see earlier), we choose for *Brandonia* the family name Brandoniidae.

Diagnosis. Poorly known family, confined to upper molars and possibly deciduous upper premolars. Differ from Dryolestidae, Mesungulatidae, and Reigitheriidae in having upper molars less elongated labiolingually and less constricted mesiodistally, triangular rather than roughly rectangular in occlusal view. Differ from most Dryolestidae, Donodontidae, Mesungulatidae, and Reigitheriidae in having a distinct ectoflexus. Differ from Mesungulatidae in lack of prominent pre- and postcingula and share with them lack of a metacone.

Genera. *Brandonia* Bonaparte, 1990, type genus (possibly including *Barberenia* Bonaparte, 1990); and tentatively assigned *Casamiquelia* Bonaparte, 1990.

Distribution. Late Cretaceous (Campanian): Argentina (Los Alamitos Formation).

Genus *Brandonia* J. F. Bonaparte, 1990
(figure 10.11D)

?Synonym: *Barberenia* J. F. Bonaparte, 1990.

Comments. We follow Martin (1999a: 82), who argued that *Barberenia araujoae* may represent deciduous pre-

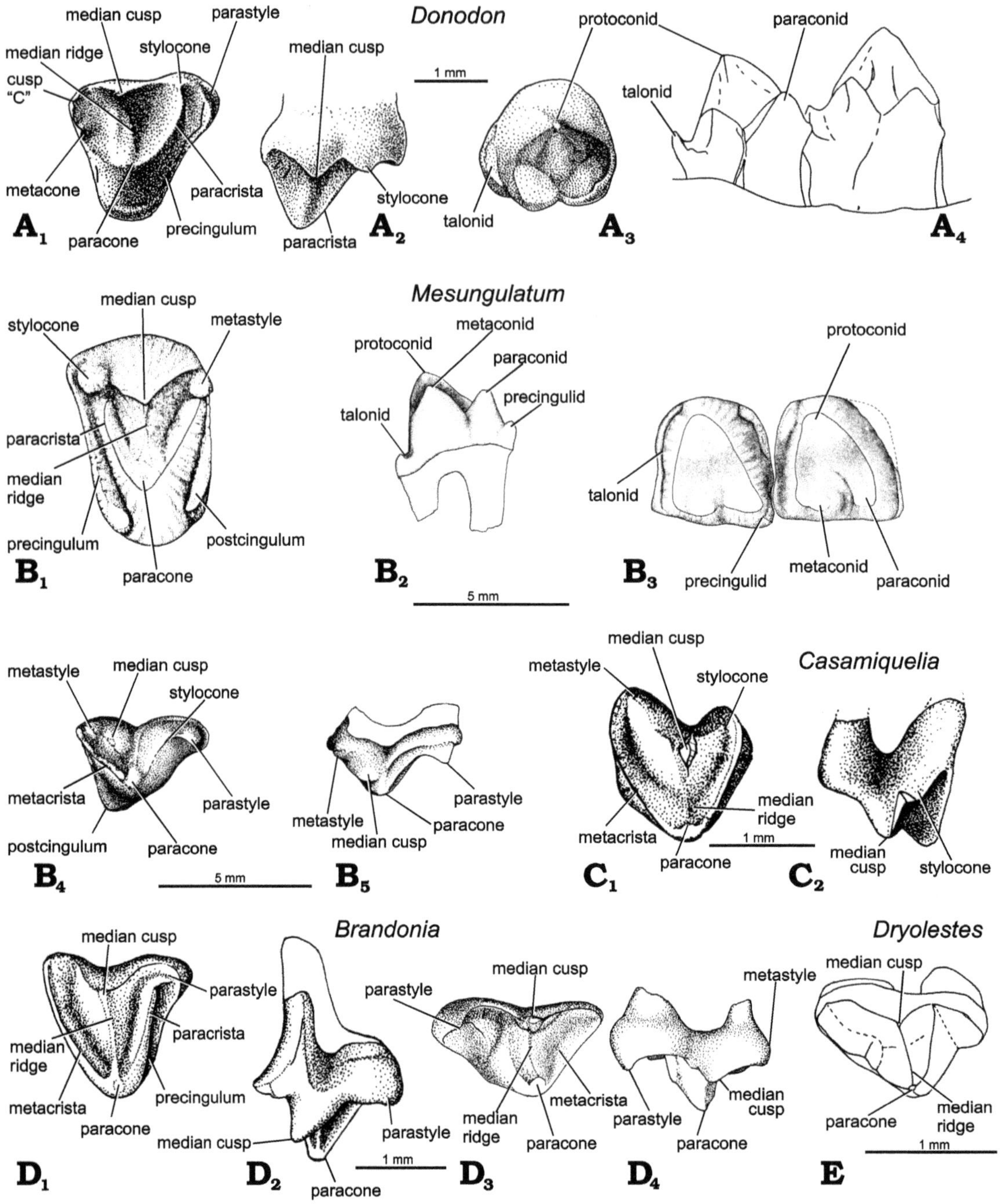

FIGURE 10.11. Dryolestoids from the Cretaceous of Gondwana (except E, shown here for comparison). A, *Donodon perscriptoris,* right upper molar in occlusal (A_1) and labial (A_2) views; left lower molar in occlusal view (A_3); two left lower molars in lingual view (A_4). B, *Mesungulatum houssayi* left upper molar in occlusal view (B_1); left lower molars in occlusal view (B_3), one of the molars in B_3 in lingual view (B_2); right deciduous premolar ?DP3, tentatively assigned, in occlusal (B_4) and labial (B_5) views (B_4 and B_5 are two views of holotype of *Quirogatherium major*). C, *Casamiquelia rionegrina,* right upper molar in occlusal (C_1) and labial (C_2) views. D, *Brandonia intermedia,* right upper molar in occlusal (D_1) and labial (D_2) views; left deciduous premolar, tentatively assigned, possibly DP3 (holotype of "*Barberenia araujoae*") in occlusal (D_3) and labial (D_4) views. E, *Dryolestes leiriensis,* left DP3 in occlusal view, shown for comparison with B_4, B_5, D_3, and D_4, which are probably also deciduous premolars; median stylar cusp is referred to as median cusp. *Source:* A, courtesy of Denise Sigogneau-Russell. Remaining figures modified from: B_1–B_3, Bonaparte (1986a); B_4, B_5, C, D, Bonaparte (1990); E, Martin (1999a).

molars of *Brandonia intermedia.* If so, *Barberenia* would be a synonym of *Brandonia.*

Diagnosis. A brandoniid with relatively weak ectoflexus; the stylocone is placed somewhat lingually from the labial margin, parastyle prominent, protruding anteriorly, while metastyle is less prominent and protrudes labially. Paracone higher than stylocone; there is a narrow, well-defined median ridge between them. Differs from *Casamiquelia* in having shallower ectoflexus, metastylar lobe less protruding labially, and smaller and more labially placed stylocone.

Species. *Brandonia intermedia* J. F. Bonaparte, 1990, type species by monotypy, possibly including *Barberenia araujoae* J. F. Bonaparte, 1990.

Distribution. Late Cretaceous (Campanian): Argentina (Los Alamitos Formation).

Genus ?*Casamiquelia* J. F. Bonaparte, 1990
(figure 10.11C)

Diagnosis. A poorly known genus confined to upper dentition, which differs from *Brandonia* by included species being distinctly smaller, having much deeper, asymmetrical ectoflexus, enlarged metastyle that strongly protrudes labially, and very large stylocone.

Species. *Casamiquelia rionegrina* J. F. Bonaparte, 1990, type species by monotypy, represented only by the holotype specimen (right upper molar).

Distribution. Late Cretaceous (Campanian or Maastrichtian): Argentina (Los Alamitos Formation).

RELATIONSHIPS OF DRYOLESTOIDEANS

Prothero (1981) and Martin (1999a) proposed a close relationship between dryolestids and paurodontids, and Martin (1999a: figure 41) presented a cladogram depicting hypothesized relationships among dryolestid genera. There is general consensus that Dryolestida, including the two families (sister groups) Dryolestidae and Paurodontidae, are a monophyletic, well-defined group, characterized by simplification of the talonid and representing a blind line of "eupantotherian" evolution.

Order Amphitheriida Prothero, 1981

Diagnosis. The order is monotypic, including a single family and a single genus, known only from the lower dentition. Differ from Dryolestida in having five premolars (three or four in Dryolestida). Number of molars (six or seven) is higher than those known in the Zatheria (maximum five in *Nanolestes*), fewer than in Dryolestidae (eight or nine), and about equal to that in *Henkelotherium* (seven). Talonid on lower molars is larger than in Dryolestida, overlapping the following trigonid labially. The talonid cusp (cusp d, or hypoconulid) is labial (in line with the protoconid), whereas it is median in paurodontids or lingual in dryolestids. There is a small difference between the sizes of the roots, the posterior one (on the posterior molars only) being smaller, but this difference is much smaller than in Dryolestidae (although not paurodontids). Premolars have continuous lingual cingulids and last premolar is not enlarged. Amphitheriida are also characterized by a downturned angular process (mandibular angle), similar to that of *Peramus* (Owen, 1871), and differing from the leveled angular process of dryolestoideans. Butler and Clemens (2001) suggested that *Amphitherium* probably had reached the grade of modern therians in the division between diphyodont premolars and monophyodont molars.

Distribution. Middle Jurassic (middle Bathonian): England, Oxfordshire, Stonesfield (Sharps Hill Formation); Middle Jurassic (late Bathonian): England, Oxfordshire, Kirtlington (Forest Marble).

Family Amphitheriidae Owen, 1846
Genus *Amphitherium* de Blainville, 1838
(figure 10.12A)

Diagnoses and Distribution. As for the order for both the family and the genus. Dental formula (lower dentition) of *Amphitherium* is 4.1.5.6–7 (see Butler and Clemens, 2001).

Species. *Amphitherium prevostii* (Mayer, 1832), type species; *A. rixoni* Butler and Clemens, 2001; *Amphitherium* sp. (E. F. Freeman, 1979).

Order and Family *incertae sedis*
Genus *Chunnelodon* Ensom and Sigogneau-Russell, 1998
(figure 10.12B)

Diagnosis. Ensom and Sigogneau-Russell (1998) diagnosed *Chunnelodon* as follows: Lower molars with trigonid very flattened transversely. Cusps sharp. Protoconid moderately high; small paraconid not inclined anteriorly but recurved, and not shelflike; metaconid high, slightly visible in labial view; strong backward inclination of the posterior wall of the trigonid. Talonid reduced to a sharp, lingual, and relatively high cusp. Roots slightly unequal, with a preeminence of the anterior one, but labially the two roots are nearly aligned anteroposteriorly.

Species. *Chunnelodon alopekodes* Ensom and Sigogneau-Russell, 1998, type species by monotypy, represented by isolated lower molars.

Distribution. Early Cretaceous (Berriasian): Southern England, Dorset (Purbeck Limestone Group).

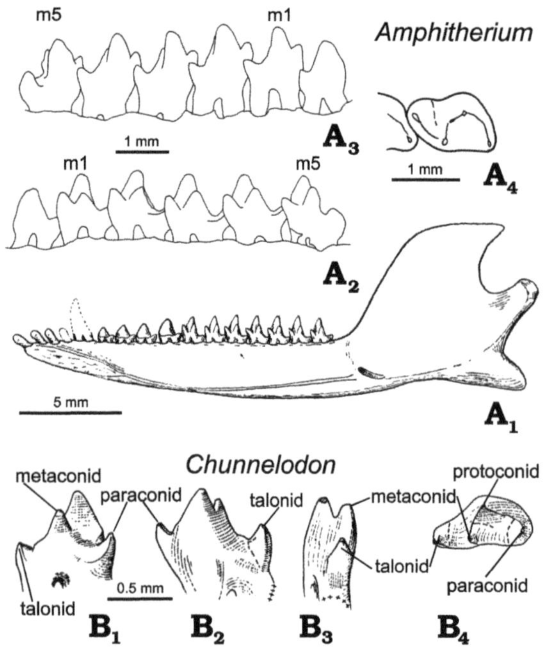

FIGURE 10.12. *Amphitherium* (A) and *Chunnelodon* (B). A, Reconstruction of the dentary of *Amphitherium prevostii,* in lingual view (A_1); enlarged p5 (?p4) and m1–m5 of the same in lingual (A_2) and labial (A_3) views; selected molar of the same in occlusal view (A_4). The specimen in A_1 belongs to a juvenile individual and has only five molars. Number of premolars in reconstruction by Simpson (1928a) in A_1 was four, but Butler and Clemens (2001) argued that there are five premolars in *Amphitherium.* B, *Chunnelodon alopekodes,* left lower molar (holotype) in lingual (B_1), labial (B_2), posterior (B_3), and occlusal (B_4) views. Source: modified from: A_1, Simpson (1928a); A_2, A_3, Fox (1975a); A_4, Butler and Clemens (2001); B, Ensom and Sigogneau-Russell (1998).

RELATIONSHIPS OF AMPHITHERIIDA

There are two alternative placements for Amphitheriida: as relatives of Dryolestida, on the one hand, and of "peramurans" and boreosphenidans, on the other. Prothero (1981: 319) provided a balanced review of these alternatives and favored the former, uniting amphitheriidans with the order Dryolestida within the superorder Dryolestoidea (Prothero, 1981). Here *Amphitherium* is more closely related to dryolestids than to zatherians (*Peramus* + boreosphenidans). Traditionally, however, *Amphitherium* has been considered to be more similar in their derived molar structures to *Peramus* and tribosphenic mammals than to dryolestoids (Butler, 1939; Mills, 1964; Crompton, 1971; Sigogneau-Russell, 1999; Butler and Clemens, 2001). Our reevaluation of the relevant evidence for both alternatives shows that not all the characters supporting a relationship of dryolestidans and *Amphitherium* to the exclusion of zatherians are valid. The evidence currently available from the lower jaw and dentition of *Amphitherium* still favors its closer placement to the clade of *Peramus* and boreosphenidan mammals (Luo et al., 2002), which will hold true unless and until this latter hypothesis is disproven by contradicting evidence of some better materials of this taxon in the future.

Superorder Zatheria McKenna, 1975

Comments. McKenna (1975) erected the sublegion Zatheria within his legion Cladotheria. According to his classification, Zatheria include two infraclasses Peramura McKenna, 1975, and Tribosphenida McKenna, 1975 (the latter taxon replaced by Boreosphenida Luo, Cifelli and Kielan-Jaworowska, 2001a, in the classification accepted herein). The high-rank taxa erected by McKenna (1975) have been defined only by taxonomic content and have not been diagnosed. In a revised and improved phylogeny for the non-tribosphenic therians (*sensu lato*), Prothero (1981: figure 12, node 17, table 9) gave several diagnostic characters for Zatheria, including reduction to three molars, basined talonid, presence of hypoconulid and hypoconid, and reduced stylocone. He included the sole genus *Peramus* in Peramura and approximately diagnosed it as follows (Prothero, 1981: figure 12, node 18, table 9): "*Peramus:* reduction of the last molar, reduce stylar region, reduce stylocone, reduce posterior cingulum on molars." By contrast, Tribosphenida (node 19, table 9) was diagnosed as: "add protocone, paracone anterior to metacone."

McKenna and Bell (1997) assigned to Zatheria *incertae sedis* three monotypic families: Arguitheriidae Dashzeveg, 1994, Arguimuridae Dashzeveg, 1994, and Vincelestidae Bonaparte, 1986a. They limited Peramura to the single family Peramuridae Kretzoi, 1946, with four genera: *Peramus* Owen, 1871; *Palaeoxonodon* E. F. Freeman, 1976; *Pocamus* Canudo and Cuenca-Bescós, 1996 (junior synonym of *Crusafontia* Henkel and Krebs, 1969, see earlier), and *Abelodon* Brunet et al., 1990. Sigogneau-Russell (1999) described three new peramurid-like genera based on isolated molars from the Early Cretaceous of England and Morocco and assigned them to Zatheria *incertae sedis.* She discussed the contents and validity of Peramura McKenna and argued that:

> It appears that this group of therian mammals [Peramura McKenna, 1975], whose molars establish a morphological link between those of primitive Symmetrodonta and those of tribosphenic mammals, does not in fact show any exclusive synapomorphy, at least in our present state of knowledge; we are confronted, rather, with a few representatives of an evolutionary grade whose vast distribution in space and time suggests an even wider diversification, as well as an early origin of the pretribosphenid line. (Sigogneau-Russell, 1999: 90).

In spite of this statement, Sigogneau-Russell (1999: 112) still used the taxa Peramura McKenna and Peramuridae Kretzoi as formal taxonomic units (albeit including only *Peramus*).

Martin (2002) confirmed the lack of synapomorphies for several taxa formerly assigned to "peramurans." He further noted that Arguimuridae, as erected by Dashzeveg (1994), do not show derived diagnostic characters either. He provided an informal grouping of "stem-lineage of Zatheria" to include the taxa that are more closely related to Zatheria than dryolestoids and amphitheriids but lack the apomorphic features of the Zatheria as diagnosed by Prothero (1981: 321). In this informal "stem-lineage of Zatheria" group he placed *Arguitherium*, *Arguimus*, and *Nanolestes*, all of which lack the basined talonid of *Peramus* and tribosphenic mammals. By Martin's grouping, *Vincelestes* from the Early Cretaceous of Argentina should also be a "stem-lineage zatherian." Lacking better evidence from these taxa, which are represented by incomplete fossils, we follow Martin (2002) in this expedient treatment of the stem taxa. Because McKenna and Bell (1997) assigned some of these taxa to Peramura, we treat the former Peramura McKenna, 1975 as an informal group referred to them as "peramurans." We should also make it clear that when speaking about Zatheria in this chapter, we refer only to non-tribosphenic zatherians or stem taxa of the monophyletic group Zatheria.

"Stem-Lineage of Zatheria" (Martin, 2002)

Brief Characterization. According to Martin (2002) the Zatheria comprise all taxa that are more closely related to Tribosphenida than to Dryolestida, including the stem-lineages between the last common ancestor ("Stammart") of Cladotheria and the last common ancestor ("Stammart") of Zatheria. In other words, he included non-dryolestidan "eupantotherians" that have more than three molars and/or lack the fully basined talonid.

Distribution. Late Jurassic (Kimmeridgian): Portugal, Leiria, Guimarota Coal Mine (Guimarota Beds); Jurassic-Cretaceous boundary: Portugal, Porto Pinheiro; Early Cretaceous (late Hauterivian or early Barremian): Argentina (La Amarga Formation); Early Cretaceous (Aptian or Albian): Mongolia, Gobi Desert ("Höövör Beds").

Taxa Included. *Arguimus* Dashzeveg, 1979; *Arguitherium* Dashzeveg, 1994; *Nanolestes* T. Martin, 2002; family Vincelestidae Bonaparte, 1986a.

Comments. As Martin noted that "Arguimuridae" lack any autapomorphies, we assign the sole genus of "Arguimuridae," *Arguimus*, to "stem-lineage of Zatheria," family *incertae sedis*. We also place in family *incertae sedis* another poorly known genus *Arguitherium*, which was placed in its own family, "Arguitheriidae," by Dashzeveg (1994), but see Sigogneau-Russell (1999) and Martin (2002). This assignment agrees with the scheme of McKenna and Bell (1997), in which these taxa and Vincelestidae are Zatheria *incertae sedis*.

The best-preserved taxon of the "stem-lineage of Zatheria" is *Vincelestes*, the only member of the family Vincelestidae represented by several complete skulls, dentaries (Bonaparte, 1986a; Bonaparte and Rougier, 1987; Rougier et al., 1992), and numerous elements of the postcranial skeleton (Rougier, 1993). With its shortened rostrum and reduced number of postcanine teeth, *Vincelestes* differs from other members of "stem-lineage zatherians," such as *Arguimus* and *Nanolestes*, that had an elongated snout as evidenced by the preserved anterior portion of the dentary or rostrum. We do not characterize the "stem-lineage zatherians" by highlighting *Vincelestes* as a representative. Instead, we describe it under the Vincelestidae below.

Family *incertae sedis*
Genus *Arguimus* Dashzeveg, 1979
(figure 10.13A)

Comment. Dashzeveg (1979, 1994) interpreted the teeth in the holotype specimen as three last premolars and two molars. However, Butler and Clemens (2001) pointed out that the supposed p5 is more worn than the supposed m1. This is an unlikely scenario because in a typical diphyodont replacement, m1 is always more worn than p5. Butler and Clemens (2001) reinterpreted the preserved teeth as two last premolars and three anterior molars (figure 10.13A). Dashzeveg (1979) initially reconstructed the molar talonid with two cusps (entoconid absent) but he later recognized three talonid cusps: hypoconid, tiny entoconid, and a questionable hypoconulid (Dashzeveg, 1994). This interpretation has been questioned by Sigogneau-Russell (1999), who determined that there was only a single talonid cusp in *Arguimus*, which she referred to as the hypoconulid.

Diagnosis. *Arguimus* is a poorly known genus, confined to lower dentition; it probably had a large canine, as indicated by the canine alveolus preserved in the type specimen of *A. khosbajari*, four double-rooted premolars, and four molars, three of which (m1–m3) have been preserved and the last indicated by an alveolus. Paraconid and metaconid on m2 and m3 are much lower than the protoconid.

Species Assigned. *Arguimus khosbajari* Dashzeveg, 1979, type species by monotypy, represented by a fragment of the left dentary with five teeth, plus a posterior molar alveolus.

Distribution. Early Cretaceous (Aptian or Albian): Mongolia, Gobi Desert ("Höövör Beds").

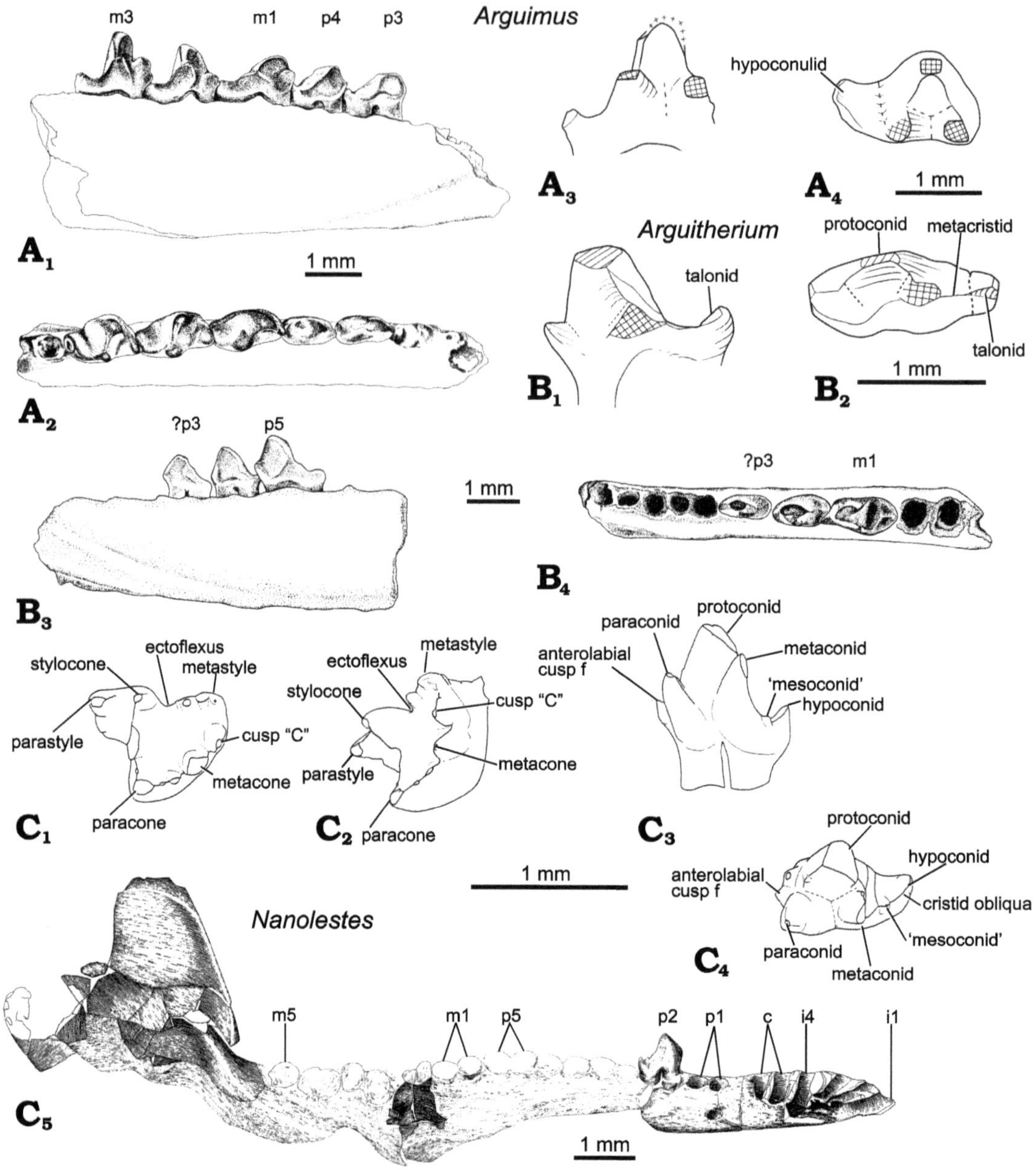

FIGURE 10.13. "Stem-lineage of Zatheria." A, *Arguimus khosbajari*, left fragmentary dentary (holotype) with p3, p4, and m1–m3, in lingual (A_1) and occlusal (A_2) views; m2 of the same in lingual (A_3) and occlusal (A_4) views. B, *Arguitherium cromptoni*, right fragmentary dentary (holotype) with ?p3–p5 in lingual (B_3), and occlusal (B_4) views; p5 of the same in lingual (B_1) and occlusal (B_2) views. C, *Nanolestes drescherae*, left upper molar in occlusal (C_1) and occlusodistal (C_2) views; right lower molar in lingual (C_3) and occlusal (C_4) views; right dentary in labial view (C_5). *Source:* modified from: A_{1-2}, B_{3-4}, Dashzeveg (1994); A_{3-4}, B_{1-2}, Sigogneau-Russell (1999); C, Martin (2002).

Genus *Arguitherium* Dashzeveg, 1994
(figure 10.13B)

Comments. Dashzeveg (1994) erected the monotypic genus *Arguitherium,* on the basis of the holotype specimen of the type species, which is a fragmentary right dentary with several alveoli and three teeth, interpreted by Dashzeveg as p4, p5, and m1. Butler and Clemens (2001) reinterpreted the specimen as having a molar (m1) and the two last premolars preserved, but the number of the premolars remains uncertain. However, Martin (2002) argued that the preserved teeth are p3–p5, following an earlier suggestion by Sigogneau-Russell (1999). The m1 (or p5, see Martin, 2002) has the talonid with a single cusp and a poorly defined basin (as shown by Sigogneau-Russell, 1999; *contra* Dashzeveg, 1994).

Diagnosis. Structure of the molars is unknown (following the interpretation of Martin, 2002). The ?p4 consists of the main cusp and posterior heel; the trigonid is not developed. The p5 has a small paraconid and metaconid, set far away from one another, and a talonid with a single cusp and without an definite basin.

Species Assigned. *Arguitherium cromptoni* Dashzeveg, 1994, type species by monotypy, represented by incomplete lower dentition.

Distribution. Early Cretaceous (Aptian or Albian): Mongolia, Gobi Desert ("Höövör Beds").

Genus *Nanolestes* T. Martin, 2002
(figure 10.13C)

Diagnosis (based on Martin, 2002, emended). The dentary is slender and strongly elongated, with distinct Meckel's groove and well-developed angular process. Dental formula for lower teeth: 4.1.5.5. All lower teeth except incisors and m5 are double-rooted. Lower molars with enlarged, unicusped talonid (no talonid basin) and additional cuspule on cristid obliqua. A cuspule in this position is very uncommon among "symmotrodontans," "eupantotherians," and tribosphenic mammals, so this cusp is not referable to any of the standard talonid cusps of these groups. But a similar cuspule was identified as the "mesoconid" by Butler (1990a) in some (not all) specimens of *Palaeoxonodon* and *Peramus* (as it also occasionally occurs in tribosphenic molars, see, for example, Van Valen 1966: figure 1). Upper molars with a small ectoflexus, parastyle, stylocone, and metastyle, paracone lower than stylar cusps, cusp "C" present on metacrista labial to metacone; there are also additional cusps on the paracrista. Martin's (2002) SEM micrographs show that the stylocone groove is well developed.

Species. *Nanolestes drescherae* T. Martin, 2002, type species, represented by fragmentary dentaries and numerous isolated upper and lower teeth. *N. krusati* T. Martin, 2002 (erected for the so-called "Porto Pinheiro molar" of Krusat, 1969; see also Kraus, 1979).

Distribution. Late Jurassic (Kimmeridgian): Portugal, Leiria, Guimarota Coal Mine (Guimarota Beds); Jurassic-Cretaceous boundary: Portugal, Porto Pinheiro.

Family Vincelestidae J. F. Bonaparte, 1986a

Diagnosis. The only known genus, *Vincelestes* Bonaparte, 1986a, is a relatively large mammal by Mesozoic standards (skull length about 7 cm). Skull is strongly shortened, with robust zygomatic arch; septomaxilla present; lateral wall of the braincase is built of alisphenoid and anterior lamina of the petrosal, both structures being of approximately similar size (Hopson and Rougier, 1993). Mandible is robust with a very high coronoid process. Dental formula: 4.1.2.3/1.1.2.3. Canines are very large. First premolars P1/p1 and last molars M3/m3 are reduced in size; P2/p2 enlarged. Upper molars are longer than wide, roughly triangular in occlusal view; there is no ectoflexus, stylar region is very wide with four stylar cusps, paracone and metacone are small, placed parallel to one another, there is a small lingual cusp, sometimes regarded as an incipient protocone. Lower molars are strongly elongated mesiodistally, with low trigonid cusps, paraconid and metaconid placed far away from one another, forming an obtuse triangle; talonid small, unbasined, with a single cusp. Vincelestidae differ from dryolestoids and "peramurans" by having strongly shortened rostrum and reduced number of postcanine teeth and in the shape of the upper molars, which are mesiodistally elongated and labiolingually shortened, and lack of obvious cristae.

Genera. *Vincelestes* Bonaparte, 1986a, type genus by monotypy.

Distribution. Early Cretaceous (late Hauterivian or early Barremian): Argentina (La Amarga Formation).

Genus *Vincelestes* J. F. Bonaparte, 1986a
(figures 10.14, 10.15)

Diagnosis and Distribution. As for the family.

Species Assigned. *Vincelestes neuquenianus* J. F. Bonaparte, 1986a, type species by monotypy.

Comments. *V. neuquenianus* is one of the best-preserved representatives of the Early Cretaceous cladotheres. Its skull anatomy has been partially described or discussed in numerous papers, prime among which are those of Bonaparte (1986a), Bonaparte and Rougier (1987), Rougier and Bonaparte (1988), Rougier et al. (1992), Hopson and Rougier (1993), Wible and Hopson (1993), Wible et al. (1995), and Sigogneau-Russell (1999). Of special value is Rougier's (1993) Ph.D. thesis, which summarized the previously published information on skull and dentition and provided a detailed description of the postcranial skeleton. Below we briefly describe the anatomy of the skull and dentition of *Vincelestes* from the published papers just mentioned, but defer the discussion of the postcranial skeleton to the formal publication by the original author.

ANATOMY OF *VINCELESTES*

Skull

The skull of *Vincelestes* (figure 10.14) is massive and robust, with a strongly shortened snout. A plesiomorphic feature is the presence of a septomaxilla, which among extant mammals occurs in monotremes and xenarthrans (regarded in the latter group as a covergent neomorph, see

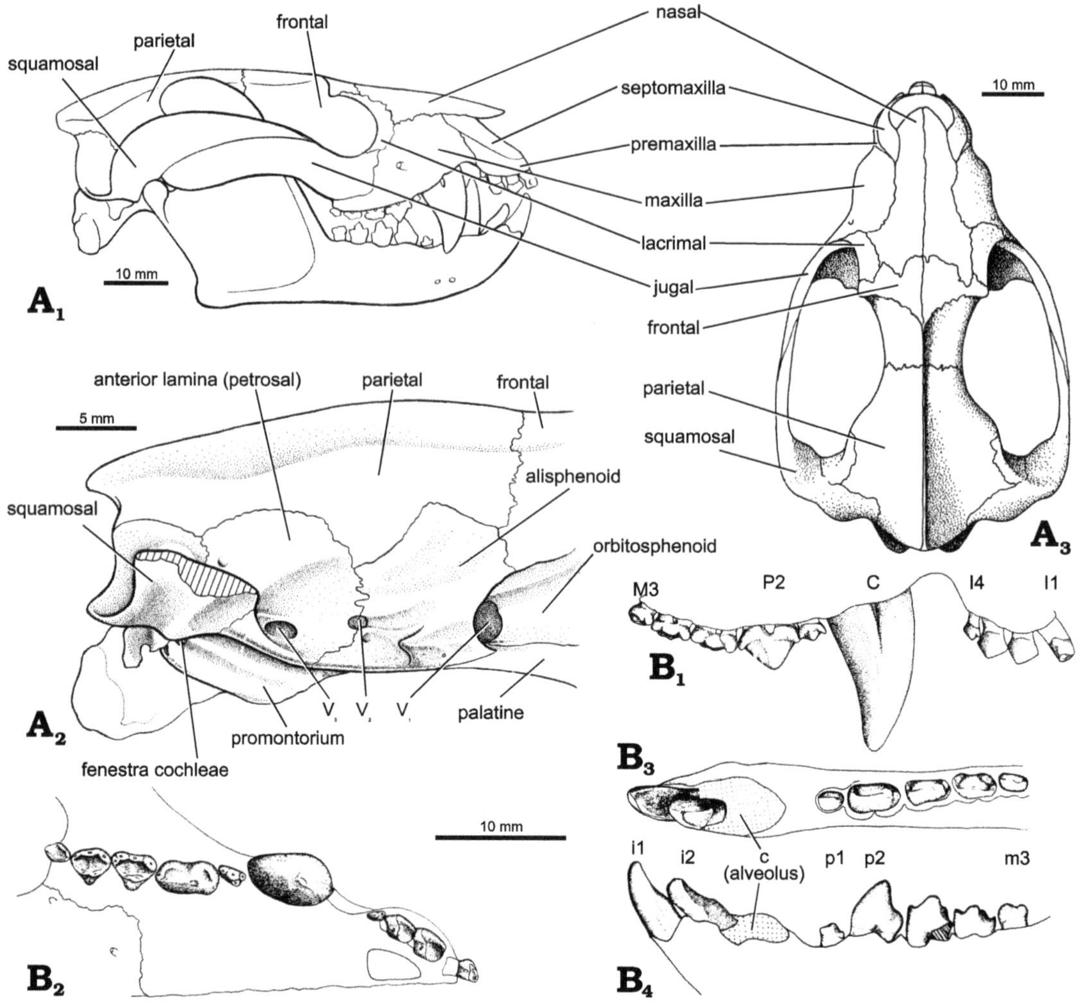

FIGURE 10.14. *Vincelestes neuquenianus.* A, Reconstruction of the skull in lateral (A_1) and dorsal (A_3) views; reconstruction of the lateral wall of the braincase of the same (A_2). B, Reconstruction of the dentition, right upper dentition in labial (B_1) and occlusal (B_2) views; left lower dentition in occlusal (B_3) and labial (B_4) views. *Source:* A_1, A_3, and B, modified from Bonaparte and Rougier (1987); A_2, modified from Hopson and Rougier (1993); A_1 and A_2 reversed.

Wible et al., 1990). Among Mesozoic mammals it has been described in *Sinoconodon, Morganucodon,* and *Haldanodon.* The septomaxilla in *Vincelestes* is most similar to that in *Sinoconodon* (Wible et al., 1990). The arrangement of the bones of the cranial roof reminds one vaguely of that in extant *Didelphis.* There is a sagittal crest formed by the parietal and a very robust, deep zygomatic arch. The robust and shortened skull, its deep zygomatic arch, and very large canines are unequivocal indications of a carnivorous mode of life.

Vincelestes differs from modern Theria and Monotremata in the structure of the lateral wall of the braincase, with a large anterior lamina of the petrosal (lamina obturans) present behind the large alisphenoid. The foramina of the maxillary (V_2) and mandibular (V_3) branches of the trigeminal nerve lie within the anterior lamina, while the ophthalmic branch leaves the braincase by a sphenorbital fissure between the orbitosphenoid and alisphenoid. Similar structure of the lateral wall of the braincase among mammals occurs only in *Morganucodon* (Kermack et al., 1981). In monotremes (chapter 6) and multituberculates (chapter 8) the lateral wall is built mostly of the anterior lamina, with the exception of Paleocene *Lambdopsalis* (Miao, 1988), where there is a large alisphenoid lying in front of the anterior lamina, resembling to some extent the condition in *Vincelestes* and *Morganucodon.*

Hopson and Rougier (1993), who described the lateral wall of the braincase in *Vincelestes,* argued that: "Because it has both a large alisphenoid and a well-developed anterior lamina (lamina obturans) separated by a distinct interdigitating suture, *Vincelestes* supports the non-homology of the sheet-like ossification of the alisphenoid and lam-

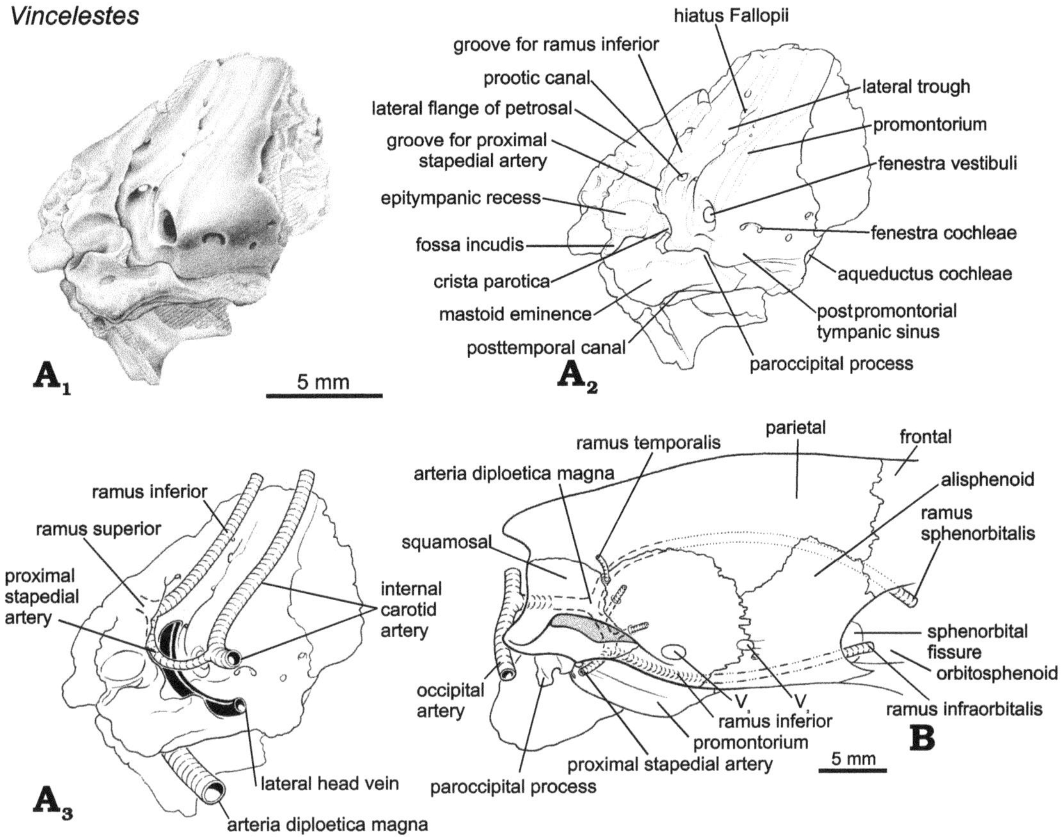

FIGURE 10.15. *Vincelestes neuquenianus*. A, Isolated right petrosal in tympanic view (A_1) and explanatory drawing of the same (A_2); reconstruction of the vessels (A_3). B, Braincase in right lateral view with reconstructed vessels. *Source:* modified and simplified after Rougier et al. (1992); B, reversed.

ina obturans. Likewise, because it shows the primitive tetrapod relation of the alisphenoid (processus ascendens) to the maxillary branch of the trigeminal nerve (V_2), as also seen in Early Jurassic mammals, it indicates that this is the outgroup condition for modern therians, retained in pouch young of didelphid marsupials."

Cranial Vasculature

Rougier et al. (1992) described the basicranial region of *Vincelestes.* On the basis of analysis of vascular grooves, canals, and foramina and a comparison with arteries and veins occurring in extant mammals they reconstructed the cranial vasculature in *Vincelestes* in great detail (figure 10.15). They concluded (p. 213):

> Regarding the basicranial veins, *Vincelestes* exhibited essentially the same pattern as in *Tachyglossus* and *Ornithorhynchus* with a well-developed lateral head vein that drained the prootic sinus through the prootic canal and a sigmoid sinus that likely drained through the foramen magnum. Regarding the cranial arteries, the pattern reconstructed for *Vincelestes* is a composite of the pattern in *Tachyglossus* and *Ornithorhynchus.* As in the echidna, *Vincelestes* had a well-developed, intramural arteria diploetica magna extending from the occiput to the orbit. As in the platypus, *Vincelestes* had a proximal stapedial artery, a ramus inferior that ran ventral to the roof of the lateral trough, a ramus superior enclosed within the anterior lamina of the petrosal (= lamina obturans), and a ramus temporalis that penetrated the bone. *Vincelestes* also had a connection between the arteria diploetica magna and stapedial system through ramus superior, a condition that occurs in the juvenile platypus. The overall pattern of arterial distribution reconstructed for *Vincelestes* is also very similar to that proposed for a hypothetical ancestral placental by Wible (1987).

Reconstruction of the main basicranial arteries and veins in *Vincelestes* is important as it serves as a comparative model to reconstruct the homologous vascular channels in other stem mammals.

RELATIONSHIPS OF *VINCELESTES*

Bonaparte and Rougier (1987) initially identified an incipient protocone on upper molars of *Vincelestes.* However, in the SEM stereophotographs of upper teeth of

Vincelestes provided by Sigogneau-Russell (1999: figure 7C) the protocone is not discernible. For these reasons, she concluded that molars of this taxon are of a "pre-tribosphenic" design that is consistent with the single-cusped talonid of the lower molar. However, Guillermo Rougier informed us (pers. comm., November 2003) that the molars figured by Sigogneau-Russell were worn, and in unworn specimens a distinct lingual cusp—an incipient protocone—is present in *Vincelestes.* On the basis of petrosal features, Rougier, Wible, and Hopson (1996) placed *Vincelestes* and extant therians in a prototribosphenidan clade, to the exclusion of eutriconodontans, multituberculates, and "symmetrodontans." This conclusion was corroborated by comparative studies of postcranial features of the relevant mammal taxa, such as the dryolestoidean *Henkelotherium* and the "symmetrodontan" *Zhangheotherium* (Krebs, 1991; Hu et al., 1997). On the basis of the petrosal and skeletal characters, we consider *Vincelestes* to be more closely related to extant therians than are the dryolestids and "symmetrodontans." Upon incorporating the dental characters into parsimony analysis of all morphological evidence, we conclude that *Peramus* is more closely related to the boreosphenidans than is *Vincelestes* (see review by Luo et al., 2002).

"Peramurans"

Comments. For reasons discussed earlier under "Comments" to Zatheria, we treat "peramurans" as an informal, paraphyletic unit.

Brief Characterization. Differ from all other "eupantotherians" by having lower molars with at least incipiently basined talonid, one or two cusps present; in upper molars of Peramuridae the paracone and metacone are placed lingually, side by side in a line subparallel to the sagittal plane, but obliquely in genera assigned to "peramurans" *incertae sedis.* The upper molar lingual cingulum is very small in Peramuridae and absent in other genera. Differ from all other "eupantotherians" in having dental formula of the postcanine teeth (known only for *Peramus*) P5/5, M3/3 and in having last lower premolar partly molarized. Differ from Vincelestidae in lack of an incipient protocone and presence of a talonid basin. Differ from tribosphenic mammals in lack of a protocone (figure 10.16A).

Taxa Included. Family Peramuridae Kretzoi, 1946, and five *incertae sedis* genera (see later).

Distribution. Middle Jurassic (late Bathonian): England, Oxfordshire, Kirtlington (Forest Marble); Late Jurassic (Kimmeridgian–Tithonian): Tanzania (Tendaguru beds); Early Cretaceous (Berriasian): southern England, Dorset (Purbeck Limestone Group); Early Cretaceous (?Berriasian): Morocco, Talsinnt Province.

Family Peramuridae Kretzoi, 1946

Diagnosis. Advanced monotypic family of "peramurans," with postcanine dental formula P5/5, M3/3, and semimolariform P5 (dental formula in other "peramurans" remains unknown). M1 and M2 widened labiolingually, with deep ectoflexus, enlarged parastylar region with small stylocone and parastyle, cusp "C" present on the metacrista between the metacone and metastyle. M3 with completely reduced metastylar region. Differ from Dryolestida in having power stroke of mandible more steeply inclined (Butler, 1972b). The upper molars differ from those of other "peramurans" by presence of a short lingual cingulum and position of the paracone and metacone in a line subparallel to the sagittal plane, whereas in other "peramurans" the metacone is placed more labially than the paracone. The presence of a stylocone groove ("sillon creusant de stylocone" of Sigogneau-Russell, 1999) is uncertain. Differ from other "peramurans" by presence of a talonid basin with two cusps (in other "peramurans" talonid is also basined, but less deep and provided with a single cusp).

Genus Assigned. *Peramus* Owen, 1871, type genus by monotypy (following Sigogneau-Russell, 1999).

Distribution. Early Cretaceous (Berriasian): southern England, Dorset (Purbeck Limestone Group); Early Cretaceous (?Berriasian): Morocco, Talsinnt Province.

Genus *Peramus* Owen, 1871
(figure 10.16B)

Comments. The upper and lower dentitions were not found in association, but were regarded by Clemens and Mills (1971) as conspecific. These authors followed Simpson (1928a) and Mills (1964) in accepting the presence of four molars in *Peramus* and reconstructed its dental formula as ?.1.4.4/4.1.4.4 (also see Kraus, 1979). However, the tooth identified as M1 is narrower than that identified as M2, and further differs in having only two roots and in being only partly molariform; likewise, purported m1 is only partly molariform. Because of these morphological differences, McKenna (1975) reinterpreted the cheek tooth formula of *Peramus* as including five premolars and three molars. This interpretation has been adopted by most subsequent workers (e.g., Prothero, 1981; Dashzeveg and Kielan-Jaworowska, 1984; Novacek, 1986b; Dashzeveg, 1994; Canudo and Cuenca-Bescós, 1996; Sigogneau-Russell, 1999; Butler and Clemens, 2001; and others). We follow this interpretation herein.

Diagnosis. As for the family.

Species Assigned. *Peramus tenuirostris* Owen, 1871, type species, represented by one incomplete maxilla and

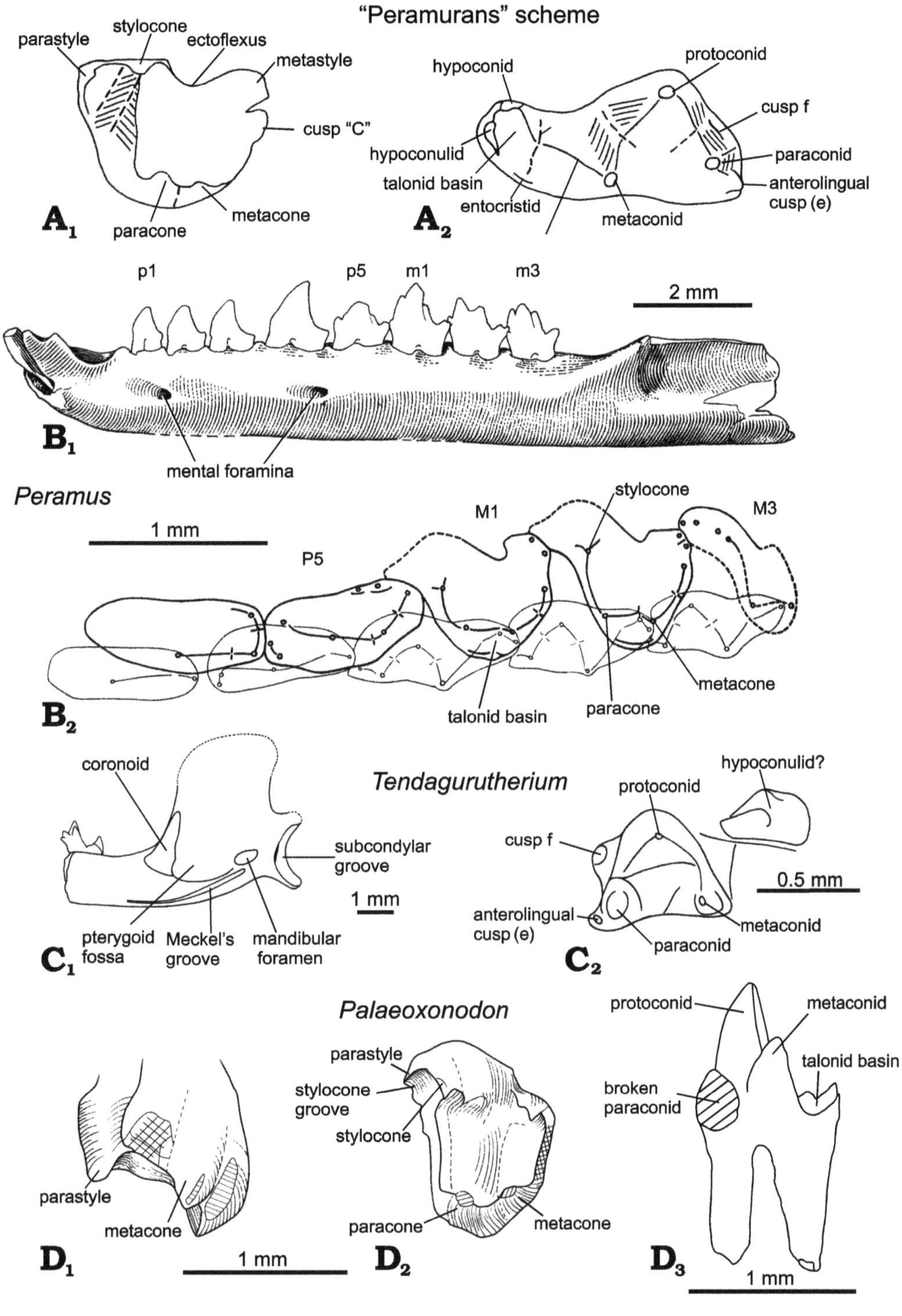

FIGURE 10.16. Dentition of "peramurans" and other stem zatherians. A, Schematic drawings of left upper (A_1) and lower (A_2) molars. B, *Peramus tenuirostris,* reconstruction of the left dentary in labial view (B_1); reconstruction of part of the upper and lower postcanine teeth, superimposed, in occlusal view (B_2). C, *Tendagurutherium dietrichi,* posterior part of an incomplete dentary, right side in medial view (C_1); ultimate molar, broken, in occlusal view (C_2). D, *Palaeoxonodon ooliticus,* left upper molar in anterior (D_1) and occlusal (D_2) views; right lower molar in lingual view (D_3). *Source:* A, D_{1-2}, modified from Sigogneau-Russell (1999); B, modified from Clemens and Mills (1971), note different identification of molars than in the original paper; C, modified from Heinrich (1998); D_3, original drawing based on the stereophotograph in E. F. Freeman (1979).

several dentaries with teeth; and ?*Peramus* sp. (Sigogneau-Russell, 1999).

Distribution. As for the family.

Family ?Peramuridae Kretzoi, 1946
Genus *Tendagurutherium* Heinrich, 1998
(figure 10.16C)

Comments. Heinrich (1998) described from the Late Jurassic of Tendaguru (Tanzania) a partial dentary with damaged last molar designated *Tendagurutherium dietrichi* and assigned it to the ?Peramuridae. As the talonid on the molar is damaged, it is impossible to state whether it was basined or not, and therefore the systematic position of the new taxon cannot be established with any certainty. We follow Heinrich (1998) in tentatively assigning *Tendagurutherium* to the Peramuridae.

Diagnosis (modified from Heinrich, 1998). Dentary with a sharp angular process, projecting backward and directed slightly medially. Below the condylar process there is a distinct subcondylar groove with a central concavity. Meckel's groove and the coronoid bone are present. The mandibular foramen is large, oval, and positioned far posteriorly. Trigonid of ultimate molar high, compressed anteroposteriorly, with three principal cusps and two anterior basal cusps. Trigonid dominated by protoconid, paraconid similar or equal in height to metaconid. Metaconid placed farther lingually than paraconid; protoconid slightly anterior to metaconid.

Species Assigned. *Tendagurutherium dietrichi* Heinrich, 1998, type species by monotypy.

Distribution. Late Jurassic (Kimmeridgian–Tithonian): Tanzania (Tendaguru Beds).

Zatheria and "Peramurans" *incertae sedis*

Comments. We follow Sigogneau-Russell (1999) by placing five genera as Zatheria *incertae sedis: Palaeoxonodon* E. F. Freeman, 1976b; *Abelodon* Brunet et al., 1990 (assigned to Peramuridae by McKenna and Bell, 1997); *Magnimus, Minimus,* and *Afriquiamus* Sigogneau-Russell, 1999. McKenna and Bell (1997) also assigned *Pocamus* Canudo and Cuenca-Bescós, 1996, to Peramuridae; we regard *Pocamus* as a synonym of the dryolestid *Crusafontia* (see earlier). Although compared by various authors to the Peramuridae, these taxa nonetheless lack the apomorphic features diagnostic of *Peramus.* In a few characters, some of the taxa included under this heading are more derived than *Peramus* and in these respects resemble stem tribosphenic mammals. Until we have a further parsimony analysis of all the relevant features of these taxa, the most expedient way to treat them is as Zatheria and "peramurans" *incertae sedis.* We begin the description of this group with *Palaeoxonodon,* which is most similar to *Peramus;* remaining genera follow in alphabetic order.

Genus *Palaeoxonodon* E. F. Freeman, 1979
(figure 10.16D)

Diagnosis. Lower molar with high and sharply pointed trigonid cusps, of which the protoconid is the highest, and small anterolabial cingulid. There is a short and very narrow, but distinctly basined talonid with one cusp. The upper molars attributed by E. F. Freeman (1979) herein are characterized by a shallow ectoflexus, large parastylar region with a stylocone and parastyle, stylocone groove present, small metastylar region, prominent paracone situated more lingually than smaller metacone, lack of lingual cingulum, presence of cusp "C." Differ from *Peramus* in lack of lingual cingulum, in having paracone and metacone arranged obliquely with respect to the longitudinal axis of the tooth, rather than subparallel, and in having a shallower ectoflexus. Sigogneau-Russell (1999) illustrated the metacone placed at almost the same level as the paracone (see figure 10.16D_2), but on the upper molars of *P. oolíticus* the paracone is larger than the metacone and placed distinctly more lingually (see E. F. Freeman, 1979: pl. 18, figures 1, 5).

Species Assigned. *Palaeoxonodon ooliticus* E. F. Freeman, 1976b, type species by monotypy, based on isolated upper and lower molars.

Distribution. Middle Jurassic (late Bathonian): England, Oxfordshire, Kirtlington (Forest Marble).

Genus *Abelodon* Brunet et al., 1990[2]
(figure 10.17A)

Diagnosis. The upper molar of *Abelodon* is characterized by a very deep ectoflexus, large parastylar area with stylocone and parastyle, and stylocone groove; the paracone is a dominant cusp, situated more lingually than the relatively small metacone. There is also a cusp "C" on the metacrista and no obvious cusps in metastylar region. Differs from *Peramus* and resembles *Palaeoxonodon* in having a deeper ectoflexus, more symmetrical metastylar and parastylar areas, presence of stylocone groove, lack of a lingual cingulum. Differs from *Palaeoxonodon* in having a deeper and narrower ectoflexus, more prominent metastylar region, and different proportions (being more expanded longitudinally).

Species Assigned. *Abelodon abeli* Brunet et al., 1990, type species by monotypy, based on a single left upper mo-

[2] There are nine authors of the genus *Abelodon.* We refer the reader to the References to see all the names.

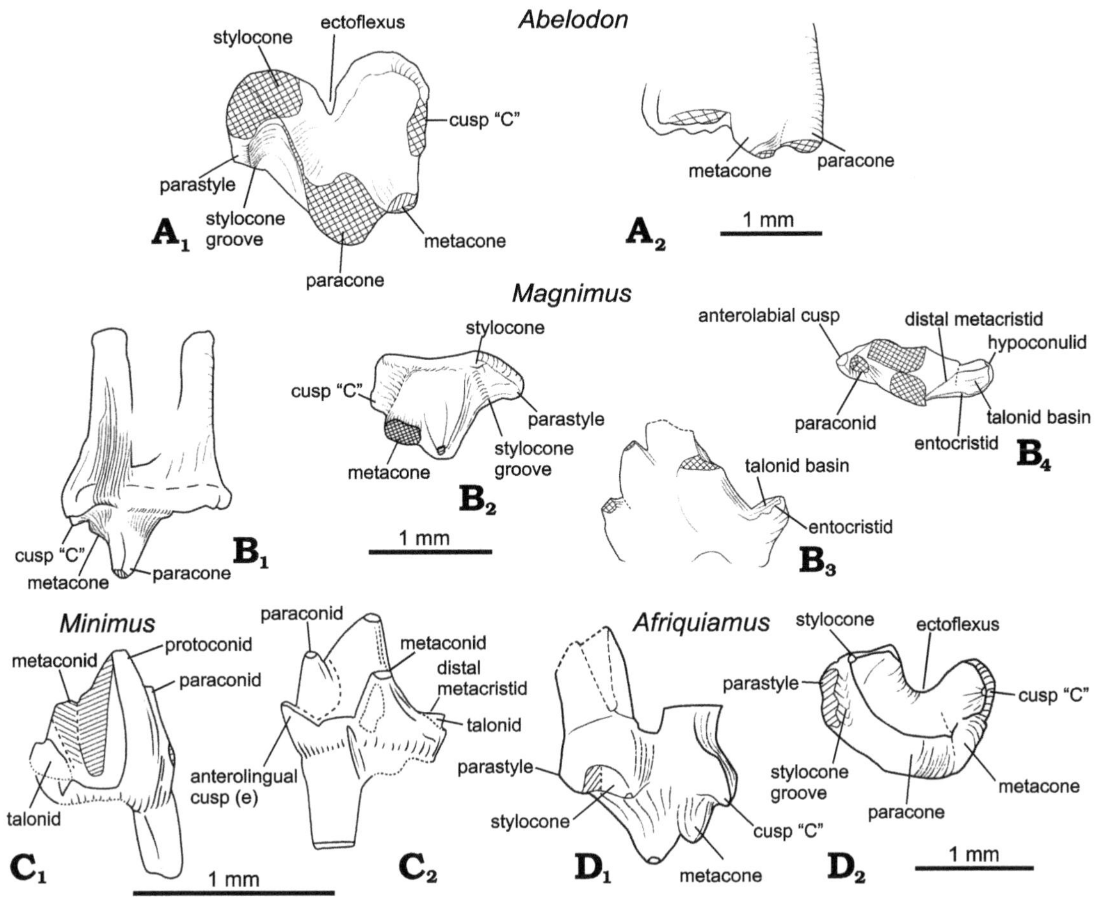

FIGURE 10.17. "Peramuran" taxa and *Afriquiamus*. A, *Abelodon abeli*, left upper molar of the holotype in occlusal (A_1) and distal (A_2) views. B, *Magnimus ensomi*, right upper molar, the holotype in labial (B_1) and occlusal (B_2) views; right lower molar in lingual (B_3) and occlusal (B_4) views. C, *Minimus richardfoxi*, right lower molar in labial (C_1) and lingual (C_2) views. D, *Afriquiamus nessovi*, left upper molar (holotype) in labial (D_1) and occlusal (D_2) views. *Source:* all modified and simplified after Sigogneau-Russell (1999).

lar, the holotype of the type species, and attributed right lower molar.

Distribution. Early Cretaceous (Barremian–Aptian): Cameroon, Bassin de Koum.

Genus *Magnimus* Sigogneau-Russell, 1999 (figure 10.17 B)

Diagnosis. Upper molars relatively short labiolingually and extended mesiodistally, without ectoflexus. Differs from *Peramus* by lack of lingual cingulum, by having enlarged parastylar region with parastyle and stylocone, and in the presence of a stylocone groove extending lingually from the stylocone. By contrast, this groove is uncertain in *Peramus*, as the parastyle and stylocone in the parastylar wing are too tightly packed to allow contact with the lower molar protoconid (Clemens and Mills, 1971). Another difference concerns a poorly developed paracone, situated slightly more lingually than the small and very low metacone. Cusp "C" on the metacrista is weakly developed. No styles are discernible in metastylar region. Lower molars attributed to the same taxon are characterized by a relatively stout trigonid, strongly compressed labiolingually, with a basal, prominent anterolingual cusp. The talonid is small in comparison with large trigonid and is distinctly basined, with a single cusp from which a ridge (distal metacristid) extends onto the metaconid; another ridge (entocristid) surrounds the talonid lingually. This is different from (and more primitive than) the talonid of *Peramus* in that the hypoconid is not separated from the hypconulid.

Species Assigned. *Magnimus ensomi* Sigogneau-Russell, 1999, type species by monotypy, based on isolated, relatively large upper and lower molars of the type species.

Distribution. Early Cretaceous (Berriasian): southern England, Dorset (Purbeck Limestone Group).

Genus *Minimus* Sigogneau-Russell, 1999
(figure 10.17C)

Diagnosis. *Minimus* differs from *Magnimus* by having the trigonid relatively wider labiolingually and a smaller talonid, with a single cusp and incipient basin sloping lingually. Shares with *Magnimus* the presence of a distinct anterolabial cusp, which may be prolonged into a cingulid, and differs in this respect from *Peramus* and *Palaeoxonodon.*

Species Assigned. *Minimus richardfoxi* Sigogneau-Russell, 1999, type species by monotypy, based on isolated lower molars of the type species, which is a very small taxon.

Distribution. Early Cretaceous (?Berriasian): Morocco, Talsinnt Province, Synclinal d'Anoual (Séquence B des Couches Rouges).

Genus *Afriquiamus* Sigogneau-Russell, 1999
(figure 10.17D)

Diagnosis. Genus based on upper dentition (isolated upper molar, holotype of the type species). Upper molar is relatively long but narrow transversely, almost symmetrical, with very deep ectoflexus. The stylocone and parastyle are present, but do not form a prominent anterior lobe. The paracone is higher than the metacone, which is placed relatively lingually. Differs from *Peramus* by lack of the lingual cingulum, from *Peramus* and *Magnimus* by lack of the third root, from *Magnimus* by having a deep ectoflexus and smaller parastylar region, and from *Abelodon* by having the metacone placed more lingually.

Species. *Afriquiamus nessovi* Sigogneau-Russell, 1999, type species by monotypy.

Distribution. Early Cretaceous (?Berriasian): Morocco, Talsinnt Province, Synclinal d'Anoual (Séquence B des Couches Rouges).

INTERRELATIONSHIPS OF "PERAMURAN" GENERA

Among the "peramuran" taxa under consideration, discovery of *Peramus* itself long antedates that of remaining genera. Furthermore, *Peramus* is represented by far more complete remains, and it has been thoroughly studied in a series of investigations dating back well over 100 years (e.g., Owen, 1871; Simpson, 1928a; Clemens and Mills, 1971). Historically, *Peramus* has figured prominently in comparative studies of mammalian dental evolution, and it has long been considered to be close to the origin of tribosphenic mammals (e.g., Crompton, 1971). The description of additional "peramuran" taxa, listed above, begs the question as to their respective proximity of relationship to boreosphenidan mammals.

Sigogneau-Russell (1999) suggested the possibility that *Peramus* may be more distantly related to tribosphenic mammals than are the recently discovered Moroccan taxa that are "*Peramus*-like." *Peramus* apparently lacks the stylocone groove ("sillon creusant le stylocone" of Sigogneau-Russell, 1999) that bears Crompton's shearing surface 1a on the vertical and lingual surface of the stylocone. This facet would contact the protoconid of the lower molar (Crompton, 1971; Crompton and Kielan-Jaworowska, 1978). It appears that the cusps of the parastylar wing are too tightly packed to allow any contact with the lower molar protoconid in *Peramus* (Clemens and Mills, 1971). However, the parastyle-stylocone region is poorly preserved in the only maxillary specimen of *Peramus.* Sigogneau-Russell (1999: 115) herself noted that: "l'absence apparente de sillon sur la face lingual du stylocone n'est pas non plus assurée."

By comparison, Sigogneau-Russell (1999) recognized that the derived feature of stylocone groove is present in several genera that are otherwise fairly similar to *Peramus* morphologically, including *Palaeoxonodon, Abelodon, Magnimus,* and *Afriquiamus.* To this assemblage of taxa with this derived feature Martin (2002) further added *Nanolestes.* The stylocone groove of these "peramurans" is homologous to the groove for the protoconid as identified in tribosphenic molars (Crompton and Kielan-Jaworowska, 1978). This is a derived character complex that is widespread in stem boreosphenidans with true tribosphenic molars, as surveyed by Cifelli (1993b). Moreover, it also is a structural precursor for further development of a series of characters for elaborated prevallum/postvallid shearing for tribosphenic molars (Fox, 1975; Cifelli, 1993b). This stylocone groove, as pointed out by Sigogneau-Russell (1999), could indicate a close relationship between the *Peramus*-like taxa from Morocco and tribosphenic mammals. She also argued that the lingual position of the metacone, a characteristic of *Peramus,* might be a plesiomorphic rather than a derived character, as it also occurs in the "symmetrodont" *Tinodon* (see chapter 9). However, one can argue that even if this position is primitive, it was nevertheless retained in tribosphenic mammals, in which the development of the protocone pushed the paracone and the metacone labially.

While the Moroccan *Peramus*-like taxa are more similar to tribosphenic mammals in possessing the stylocone groove, these taxa unfortunately lack such derived talonid features as a clear separation of hypoconid from hypoconulid, which, according to the currently available evidence, is shared only by *Peramus* and tribosphenic mammals (see Sigogneau-Russell, 1999: table 2). It should be noted that other characters of the "peramurans" also foreshadow tribosphenic molars; that is, metacone distal to paracone and, perhaps, lingual cingulum on upper molars. These characters have a mosaic distribution among the "peramuran-like" taxa. It remains to be seen whether *Peramus* and boreosphenidans can be united by the derived talonid characters to the exclusion of *Magnimus* and *Nanolestes*, or if the other "peramuran" taxa and boreosphenidans can be united by the stylocone groove features to the exclusion of *Peramus* itself. These conflicting characters need to be sorted out by a parsimony analysis including all relevant taxa, which is beyond the scope of our present work. Here we adopt the position that *Peramus* is the sister taxon to boreosphenidans and closer to the latter group than other less-well-preserved "peramurans" (see figure 15.1).

INTERRELATIONSHIPS OF "EUPANTOTHERIANS"

Ignoring for present purposes other stem zatherians (for which similar comments apply), the most advanced clade among all "eupantotherians" are the Peramuridae, which bear strong resemblance in derived features of the molar talonid to the tribosphenic mammal group Boreosphenida. Peramurids and boreosphenidans form a monophyletic group of Zatheria (McKenna, 1975; Prothero, 1981; Ensom and Sigogneau-Russell, 1998; Sigogneau-Russell, 1999; Martin, 2002). Zatherians, vincelestids, amphitheriids, and dryolestidans form the more inclusive monophyletic group Cladotheria, as established by McKenna (1975) and studied in detail by Prothero (1981), Krebs (1991), Rougier (1993), Martin (1999a), and Butler and Clemens (2001). The phylogenetic hierarchy of the major taxonomic groups formerly assigned to "Eupantotheria," as first proposed by McKenna (1975), is now corroborated by additional, detailed studies of these groups by many workers (Prothero, 1981; Rougier, 1993; McKenna and Bell, 1997; Sigogneau-Russell, 1999; Martin, 1999a; Luo et al., 2002). However, there is still an unresolved issue concerning the phylogenetic placement of the family Amphitheriidae. Owing to the conflicting characters shown on its very limited fossils (lower dentition and jaw), amphitheriids could be associated either with dryolestoids (Prothero, 1981; Martin, 1999a) or with a clade of peramurids and boreosphenidans (Butler and Clemens, 2001; Luo et al., 2002). The latter relationship is supported by several derived features of the molar talonid shared by amphitheriids, peramuids, and tribosphenic mammals (Crompton, 1971; Butler and Clemens, 2001). This placement of amphitheriids is supported by characters on the premolar and molar cingulids (see balanced review of both lines of evidence by Prothero, 1981). There is also an issue as to how to place the incompletely known peramurid-like taxa, represented by isolated teeth discovered in the 1990s from the Lower Cretaceous of Morocco and England, in relation to earliest Cretaceous *Peramus* and boreosphenidans. These taxa are more derived than *Peramus* and more similar to stem boreosphenidan mammals in parastylar features of the upper molars (with the caveat that the condition in *Peramus* cannot be established with confidence), but they are more primitive than *Peramus* and stem boreosphenidans in the talonid features (Sigogneau-Russell, 1999, see also earlier systematic comments on "peramurans").

CHAPTER 11

"Tribotherians" (Stem Boreosphenidans)

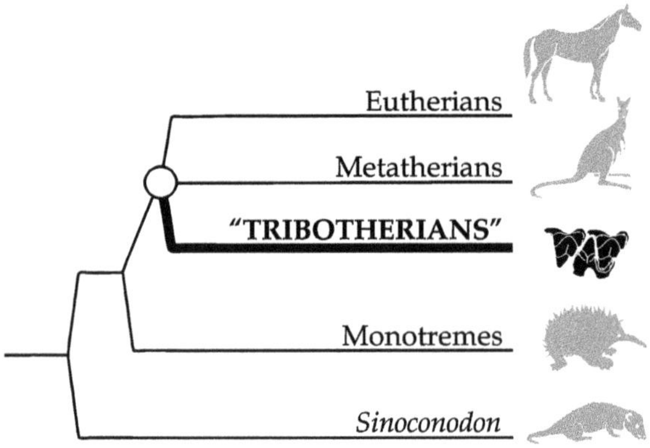

INTRODUCTION

In 1936 Simpson introduced the term "tribosphenic molars" in reference to a pattern common to both basal eutherians and metatherians. These teeth had previously been referred to as tritubercular (upper) molar and tritubercular sectorial or tuberculosectorial (lower) molars (Osborn, 1907). The tribosphenic molar (from Greek *tribein,* to grind; and *sphen,* wedge) is, as described by Simpson (1936a: 797): "suggestive of the mortar and pestle, opposing action of protocone and talonid of the wedge-like, alternating and shearing action of trigon and trigonid." During occlusion the upper tribosphenic molar shears posterior to its lower counterpart (figure 11.1C). Simpson's term tribosphenic molar succinctly highlighted the talonid, the most crucial and derived structural element of the tribosphenic molar—and distinguished it from the more primitive trituberculate molar. Although it had long been recognized that the tritubercular (triangulated, without talonid) molar is more primitive than the molar with a talonid (e.g., Osborn, 1907), Simpson's term helped to distill this important distinction.

McKenna (1975) classified mammals with tribosphenic molars in the infraclass Tribosphenida, whereas McKenna and Bell (1997) assigned the rank of superlegion to Tribosphenida. As discussed in chapter 6, it has long been accepted that tribosphenic mammals made their appearance only once, on the Laurasian continents, and all mammals with tribosphenic molars and their descendants originated from a common ancestor and were monophyletic. Discoveries of mammals with tribosphenic teeth from the Jurassic and Cretaceous of Gondwanan continents (see chapter 6) inspired Luo, Cifelli, and Kielan-Jaworowska (2001) to suggest that the tribosphenic molars evolved not once, as previously believed, but twice: first among australosphenidans, appearing in the Middle Jurassic on southern continents; and subsequently (probably during the latest Jurassic or the earliest Cretaceous), in boreosphenidans on Laurasian continents. The boreosphenidan clade proposed by Luo, Cifelli, and Kielan-Jaworowska (2001) is equivalent to Tribosphenida McKenna, 1975, and is intended to avoid confusion that would result from application of the name Tribosphenida to only one of the groups with tribosphenic molars.

In this chapter we discuss the structural evolution of tribosphenic molars and describe several Early and Late Cretaceous boreosphenidans that cannot be confidently allocated to either Metatheria or Eutheria. Most of these are represented by isolated teeth. Their molar structure has enough derived characters for them to be placed in the character-based clade Boreosphenida. However, their dental formulae cannot be established owing to their fragmentary nature, and their molars lack the diagnostic features necessary for them to be assigned unequivocally to either Eutheria or Metatheria. As an expedient way to describe these important yet very incomplete taxa, we group them together as stem taxa of boreosphenidans, a paraphyletic assemblage traditionally known as "tribotherians" (Butler, 1978a; Clemens and Lillegraven, 1986) or "therians of metatherian-eutherian grade" (Patterson, 1956; Kielan-Jaworowska, Eaton, and Bown, 1979).

HISTORICAL BACKGROUND

Bryan Patterson (1956) was the first to describe structurally primitive tribosphenic molars from what was then known as the Trinity Sandstone (Lower Cretaceous) of Texas and referred to them (p. 13) as "Therian mammals of uncertain infraclass affinities, but of metatherian-eutherian grade." In subsequent literature (e.g., Kielan-Jaworowska, Eaton, and Bown, 1979; Carroll, 1988; Cifelli, 1993b, and many others) the cumbersome term "Theria of metatherian-eutherian grade" has been used as an informal grouping of these tribosphenid mammals. Though the morphological informativeness of the Texas specimens collected by Patterson (and, later, by B. H. Slaughter, D. D. Gillette, L. L. Jacobs, D. A. Winkler, and associates) has since been partly eclipsed by more recent discoveries elsewhere, these fossils remain of great importance. Widely known as the Trinity therians, the taxa represented by these materials formed the basis for our modern understanding of molar evolution among mammals (e.g., Patterson, 1956; Crompton, 1971; Crompton and Kielan-Jaworowska, 1978; Butler, 1990a).

Kermack et al. (1965) described a tiny lower molar from the Wealden (Valanginian) of England as *Aegialodon dawsoni,* which has an undoubted tribosphenic pattern and long reigned as the oldest known tribosphenic molar. However Sigogneau-Russell (1991a, 1992, 1995a) described isolated tribosphenic molars from the Berriasian of Morocco, older but structurally more derived than *Aegialodon.* Currently the oldest tribosphenic mammal from Laurasia is *Tribactonodon* Sigogneau-Russell et al., 2001, from the Purbeck Limestone Group (Berriasian) in Dorset (southern England), and also a fragment described earlier from the same stratigraphic level by Sigogneau-Russell and Ensom (1994).

Beginning in 1965 Slaughter started to describe important material of the therians from the Trinity Group (Aptian–Albian) of Texas, based on isolated teeth and fragments of jaws with teeth housed at Southern Methodist University, Dallas (Slaughter, 1965, 1968a,b, 1971, 1981). He erected three genera: *Pappotherium, Holoclemensia,* and *Kermackia. Pappotherium* and *Holoclemensia* are both based on upper molars, and differ mostly in the arrangement of cusps on the stylar shelf. A major difference is the presence in *Holoclemensia* of a very large medial stylar cusp, a mesostyle (or cusp C of Bensley, 1906). Slaughter (1968a) regarded *Holoclemensia* as a marsupial because its large mesostyle and (purportedly) relatively large metacone are similar to those in many marsupials. Subsequently, Slaughter (1971) attributed to *Pappotherium* a partial dentary with one newly erupted and another erupting premolar and worn molars, showing premolar diphyodonty, as characteristic of placental mammals. On this basis he assigned *Pappotherium* to Eutheria. Butler (1978a) erected for this dentary the new genus *Slaughteria,* which has been restudied by Kobayashi et al. (2002).

Turnbull (1971) described the collection of Trinity therians housed in the Field Museum in Chicago and previously studied by Patterson (1956), erecting for them (and also for *Peramus*), the new cohort Tribosphenata and order Tribosphena, assigned to the infraclass Eutheria. He included some mammals with nontribosphenic molars in addition to tribosphenic mammals in the Tribosphenata. Turnbull's taxonomic arrangement was criticized by Fox (1975) and has since not been accepted by subsequent authors. On the basis of the coronal structure of penultimate and ultimate upper molars and a comparison with mammals that have either three or four molars (early eutherians and deltatheroids), Fox (1975) argued that *Pappotherium* had three, whereas *Holoclemensia* had four molars, and he supported Slaughter's systematic allocations for these mammals (accepted also by Crompton, 1971).

Almost simultaneously, primitive mammals with tribosphenic molars were also discovered in Asia. Dashzeveg (1975) described a tiny lower molar from the Aptian or Albian Höövör locality (previously Khoboor) in Mongolia that shows similarities to *Aegialodon.* Subsequently, Dashzeveg and Kielan-Jaworowska (1984) described a dentary of *Kielantherium* with four molars and alveoli showing the presence of four or five premolars.

Butler (1978a) offered a new interpretation of the Trinity therians. He reexamined collections previously described by Patterson, by Slaughter, and by Turnbull and reconstructed the molar and ultimate premolar series of both *Pappotherium* and *Holoclemensia* on the basis of fragments of upper and lower molars. In his reconstruction, both genera have four molars; Butler also argued that the differences between these genera are small. His reconstruction of the dentition of *Pappotherium,* based on fragmentary teeth (except for ultimate and penultimate upper molars of the holotype), was criticized by Slaughter (1981). The latter held that the poorly preserved tooth, recognized by Butler as M1 of *Pappotherium,* is more probably a molariform P4. Thus the existence of four molars in *Pappotherium* was far from certain. Subsequent students have sometimes regarded *Pappotherium* to be a eutherian.

Butler's reconstruction of the dentition of *Holoclemensia* is perhaps more reliable, as all the teeth involved have a recognizable mesostyle and match one another in shape and size better than those used for the reconstruction of *Pappotherium.* In addition, as pointed out by Fox (1975), the structure of the penultimate and ultimate molars strongly resembles that in Deltatheroida, which have four molars. However, even in *Holoclemensia,* the number of molars must be regarded as uncertain.

Butler (1978a) erected the infraclass Tribotheria with two orders: (1) Aegialodontia, with the families Aegialodontidae and Kermackiidae (erected by him), Deltatheridiidae, and *Potamotelses* Fox, 1972; and (2) Pappotherida, which include the family Pappotheriidae with *Pappotherium* and *Holoclemensia.* Butler (1978a) did not provide a formal diagnosis of his infraclass, but indicated that the Tribotheria have tribosphenic molars and stand between the Pantotheria, on the one hand, and the Metatheria and Eutheria on the other. He identified the origin of the infraclasses Metatheria and Eutheria from among of the Tribotheria.

Fox (1980b) accepted the infraclass Tribotheria, providing a revision of the group context. He changed the scope of Aegialodontia by excluding the Kermackiidae and including the Picopsidae and *Potamotelses.* In addition, he followed Slaughter (1981) and assigned *Pappotherium* and *Slaughteria* to Eutheria, and *Holoclemensia* to Marsupialia rather than to Tribotheria.

Aplin and Archer (1987) erected the family Holoclemensiidae, which they assigned to Marsupialia. Kielan-Jaworowska and Nessov (1990) argued that Deltatheroida belong to Metatheria. Owing to this and other arguments, Butler (1990a: 549) withdrew his infraclass Tribotheria, stating: "If the infraclass Metatheria is extended to include the Deltatheroida (?with *Potamotelses* and *Picopsis*), and the Pappotheriidae and Kermackiidae are included in the Eutheria, the need for an infraclass Tribotheria (Butler, 1978a; Fox, 1980b) disappears, and the order Aegialodontia (Tribosphenida *incertae sedis*) is restricted to *Aegialodon* and *Kielantherium.*" Cifelli (1993b), however, argued against the metatherian nature of Deltatheroida. Muizon (1994, 1998) demonstrated that the alisphenoid bulla could not be regarded as an apomorphy of Metatheria, which challenges, to some extent, the conclusions of Kielan-Jaworowska and Nessov (1990). The subsequent findings of Asian deltatheroids and marsupials has supported recognition of Deltatheroida as basal metatherians (Rougier et al., 1998; Averianov and Kielan-Jaworowska, 1999; but see Luo et al., 2002; see also chapter 12). The infraclass Tribotheria Butler, 1978, although withdrawn by its author, has still been used as a formal taxon, for example, by Sigogneau-Russell (1995b). An important discovery was made by Kobayashi et al. (2002), who, using high-resolution CT analysis, demonstrated a previously unrecognized alternating pattern of premolar replacement in *Slaughteria.* This is similar to the primitive pattern of dental replacements of "symmetrodontan" *Zhangheotherium* and dryolestids, and shared by stem eutherians (see chapter 3 and figure 3.24C).

In addition to the forementioned taxa, several isolated molars, which we regard as "tribotherians" of Early and Late Cretaceous age have also been described from North America by Fox (1972b, 1980b), Clemens and Lillegraven (1986), Jacobs et al. (1989), Cifelli (1990d, 1994), and Eaton (1993b).

Given that the infraclass Tribotheria can no longer be maintained, we refer the mammals formerly assigned to the assemblage of "tribotherians" (*sensu* Clemens and Lillegraven, 1986), as stem boreosphenidans. Because of the ambiguity concerning the eutherian versus metatherian nature of *Pappotherium, Slaughteria,* and *Holoclemensia,* we describe them in this chapter, pending new discoveries. Several families and one order within tribotherians have been erected. We use them in the "Systematics" section below, assigning other taxa to order and families *incertae sedis* (table 11.1).

STRUCTURE AND EVOLUTION OF THE TRIBOSPHENIC MOLAR

Structure. The upper tribosphenic molar (figure 11.1A) consists of a large trigon, with three main cusps: protocone (neomorph) lingual, paracone (cusp A of stem mammals) anterolabial, and metacone posterolabial. Two cingula (absent in the earliest forms) surrounding the protocone, precingulum (anterior) and postcingulum (posterior) may be present. In advanced forms a new cusp, the hypocone, may be developed on the postcingulum, and if so, the shape of the tooth may assume a roughly rectangular or trapezoidal rather than triangular shape. The area labial to the paracone and metacone is called the stylar shelf. Additional, smaller cusps known as styles may be present on the shelf. In early forms the stylar shelf may be very wide; its labial margin may be concave to form an ectoflexus. The main styles are called, from mesial to distal, parastyle, stylocone (also known as cusp B and, fortunately, homologous with cusp B of stem mammals), mesostyle, and metastyle. In a few forms (e.g., *Prokennalestes*), an additional style known as the preparastyle occurs lingual to the parastyle, on the anterior margin of the tooth (Kielan-Jaworowska and Dashzeveg, 1989). The mesial (anterior) stylar area is called the parastylar wing; in early forms it often appears as a lobe protruding mesially and supports the parastyle (and sometimes preparastyle). The parastylar wing commonly overlaps the metastylar lobe of the preceding molar, providing some degree of interlocking for the molar series. The labial slope of the parastylar wing may develop a stylocone groove between the crest that joins the paracone with the stylocone (preparacrista) and the mesial margin of the tooth. This transverse groove is for the protoconid of the lower molar (Crompton and Kielan-Jaworowska, 1978: figure 1B "groove for the protoconid") and is homologous to the "sillon creusant le stylocone" of Sigogneau-

TABLE 11.1. Linnaean Classification of "Tribotherian" (Stem Boreosphenidan) Mammals[1]

- Subclass Boreosphenida Luo, Cifelli, and Kielan-Jaworowska, 2001, new rank
  - Early Cretaceous genera
    - Order Aegialodontia Butler, 1978a
      - Family Aegialodontidae Kermack, 1967a
        - *Aegialodon* Kermack et al., 1965
          - *A. dawsoni* Kermack et al., 1965
        - *Kielantherium* Dashzeveg, 1975
          - *K. gobiensis* Dashzeveg, 1975
    - Order ?Aegialodontia Butler, 1978a
      - Family *incertae sedis*
        - *Tribactonodon* Sigogneau-Russell et al., 2001
          - *T. bonfieldi* Sigogneau-Russell et al., 2001
    - Order *incertae sedis*
      - Family Kermackiidae Butler, 1978a
        - *Kermackia* Slaughter, 1971
          - *K. texana* Slaughter, 1971
        - *Trinititherium* Butler, 1978a
          - *T. slaughteri* Butler, 1978a
      - Family Pappotheriidae Slaughter, 1965
        - *Pappotherium* Slaughter, 1965
          - *P. pattersoni* Slaughter, 1965
      - Family Holoclemensiidae Aplin and Archer, 1987
        - *Holoclemensia* Slaughter, 1968c
          - *H. texana* (Slaughter, 1968a)
      - Family *incertae sedis*
        - *Hypomylos* Sigogneau-Russell, 1992
          - *H. phelizoni* Sigogneau-Russell, 1992, type species
          - *H. micros* Sigogneau-Russell, 1995
        - *Slaughteria* Butler, 1978a
          - *S. eruptens* Butler, 1978a
        - *Tribotherium* Sigogneau-Russell, 1991
          - *T. africanum* Sigogneau-Russell, 1991
  - Late Cretaceous genera (+ *Comanchea*)
      - Family Picopsidae Fox, 1980
        - *Picopsis* Fox, 1980
          - *P. pattersoni* Fox, 1980
        - *Comanchea* Jacobs et al., 1989
          - *C. hilli* Jacobs et al., 1989
      - Family *incertae sedis*
        - *Falepetrus* Clemens and Lillegraven, 1986
          - *F. barwini* Clemens and Lillegraven, 1986
        - *Palaeomolops* Cifelli, 1994
          - *P. langstoni* Cifelli, 1994
        - *Potamotelses* Fox, 1972
          - *P. aquilensis* Fox, 1972
        - *Zygiocuspis* Cifelli, 1990
          - *Z. goldingi* Cifelli, 1990

[1] In contrast to other chapters confined to the paraphyletic taxa (chapters 4, 9, and 10) in which we provide both cladistic and Linnaean classification tables, here we give only the Linnaean classification. The reason for this is that the taxa described in this chapter are as a rule very poorly known (represented by single teeth) and establishing cladistic relationships between them, in our judgment, is not warranted on the basis of available data. In addition to taxa described herein, Hershkovitz (1995) erected the new genus and species *Adinodon pattersoni,* based on an edentulous dentary fragment from the Aptian or Albian of Texas. Based on an apparently "staggered" incisor series, he placed *Adinodon* in the Marsupialia (family "Marmosidae"). Regardless of the condition of the incisor series, the specimen is unidentifiable beyond the fact that it represents some type of a boreosphenidan. We follow Cifelli and Muizon (1997) in regarding *Adinodon pattersoni* Hershkovitz, 1995, as a *nomen dubium.* Further, we assign Genera *Dakotadens* Eaton, 1993, and *Bistius* Clemens and Lillegraven, 1986, classified by their authors as Theria *incertae sedis,* to Marsupialia *incertae sedis* and describe them in chapter 12.

Russell (1999), which occurs in some "peramurans" (see chapter 10).

The cusps are connected with crests, which are called cristae. The names of particular cristae are shown in figure 11.1A. From the protocone the two crests extending anterolabially and posterolabially are called, respectively, the preprotocrista and the postprotocrista. Primitively, these terminate near the bases of the paracone and metacone, respectively, but in more advanced taxa they extend to the labial margin of the tooth. Small additional cusps, called conules, may be present on these crests, the mesial one the paraconule and the distal one the metaconule. In earliest forms the conules are poorly developed and variably absent; primitively they are positioned closer to the protocone, whereas in more advanced forms they tend to assume positions near the base of paracone and metacone, respectively. Conular crests, lacking in primitive forms, are common to most tribosphenic mammals. Such crests bear the formal and cumbersome names preparaconular crista, postparaconular crista, premetaconular crista, and postmetaconular crista (figure 11.1A); these cristae are winglike in appearance and conules bearing them are more simply termed "winged."

In many taxa, there is an additional cusp on the crest that joins the metacone to the metastyle (postmetacrista), which Crompton (1971) designated cusp "c." As discussed in chapter 10, we refer to this as cusp "C," in order to avoid confusion with cusps of the lower dentition. Cusp "C" also occurs in many stem cladotherians (= "eupantotherians," see chapter 10).

The paracone and metacone support the primary shearing crests of upper molars. The preparacrista extends

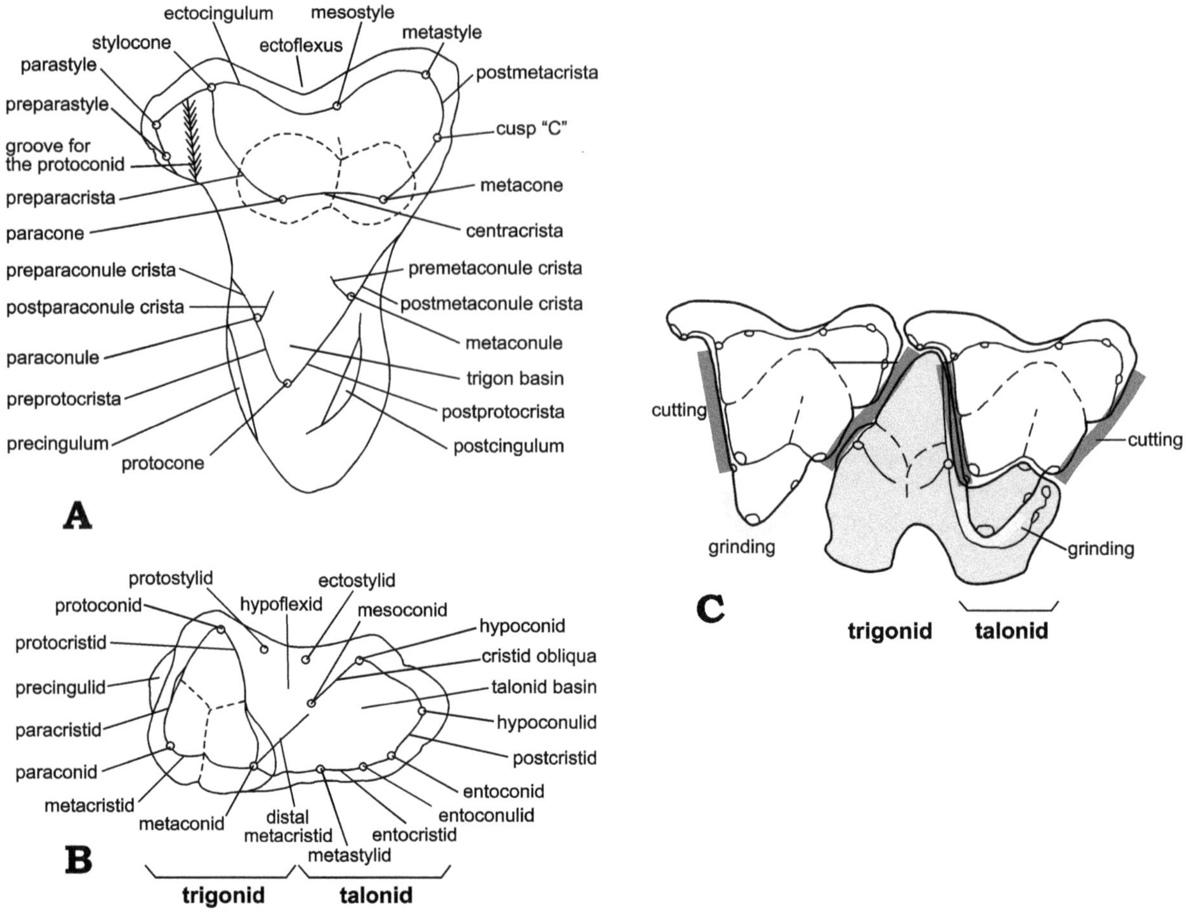

FIGURE 11.1. Diagrammatic drawings of upper and lower tribosphenic molars showing terminology of cusps and crests. A, Upper molar (left M2) in coronal view. B, Lower molar (right m2) in coronal view; for convenience the left lower molar was drawn as transparent in view from the root side (and looks like the right one). C, Teeth shown in A and B, in occlusion; the morphology of the teeth in C has been simplified and not all of the cusps are shown. Dark shading indicates cutting surfaces, light shading is the grinding surface of the protocone in the talonid basin.

labially or mesiolabially from the paracone to the stylocone (cusps A and B, respectively, of stem mammals). Between paracone and metacone is a V-shaped crest, the centrocrista, comprised of the postparacrista (ascending[1] distally from the apex of the paracone) and the premetacrista (ascending mesially from the apex of the metacone). Finally, the postmetacrista ascends distolabially from the apex of the metacone to the metastylar region of the tooth.

The names of cusps and other structures on the lower molars bear the suffix "id." The lower tribosphenic molar (figure 11.1B) consists of the anterior trigonid, with three cusps, protoconid (situated labially; homologous to cusp a of stem mammals), paraconid (mesiolingual; homologous to cusp b of stem mammals), and metaconid (distolingual; homologous to cusp c of stem mammals), forming a triangle, and the lower, distally positioned talonid. During the initial phase of occlusion the trigonid enters the triangular embrasure between the trigons of the corresponding and more anterior upper molar (figure 11.1C). The talonid is basined, which contrasts with the condition in "eupantotherians," where an incipient basin occurs only in *Peramus* and *Palaeoxonodon* (chapter 10). During a later phase of occlusion the protocone of the upper molar enters the talonid basin of its lower counterpart (figure 11.1C). Three main cusps border the talonid: lingual entoconid, labial hypoconid, and distal hypoconulid (presumed to be homologous to cusp d of stem mammals). A strong crest (cristid obliqua) runs from the hypoconid to the posterior wall of the trigonid. Several cusps may be present in variable conditions in some (but not most) taxa; an additional cusp, the mesoconid, may be present

[1] We use the terms "ascend" and "descend" with reference to anatomical orientation, irrespective of the orientation the teeth are viewed in.

on the cristid obliqua. The labial border of the tooth is often incurved in the middle, and the concave area lingual to it (between the trigonid and cristid obliqua) is called the hypoflexid. The protostylid and the ectostylid, two small, variable cusps, may be present in the hypoflexid area between the cristid obliqua and the labial margin of the tooth. On the lingual side of the lower molar, two small and variable cusps, the metastylid and the entoconulid, may be present on the crest that joins the metaconid with entoconid. In early forms a crest may extend distally from the metaconid, called the distal metacristid.

A cingulid is often present near the base on the anterolateral wall of the trigonid, which is referred to as precingulid. The precingulid often bears a small cusp f, which we consider to be homologous to the mesiolabial cingulid cusp f in Early Jurassic mammals (e.g., *Kuehneotherium, Morganucodon,* and *Dinnetherium,* Luo, 1994). There is a mesiolingual cingulid cusp e in many boreosphenidan mammals, which is homologous to the cingulid cusp e in *Kuehneotherium* and *Dinnetherium* (Luo, 1994) and in several "peramurans." The notch between the cingulid cusps e and f very often receives the distal cusp d (hypoconulid) of the preceding molar for the interlocking mechanism between the adjacent lower molars. This kind of interlocking mechanism (distal d of preceding molar between cusps e and f of the succeeding molar) is developed in many "peramurans" and boreosphenidans, but never in the australosphenidans.

Patterson's Homology on Protocone. Simpson (1936a) introduced the term tribosphenic molar as a way to recognize the grinding structure of this type of molar, absent in "symmetrodontans," which was clearly an advance from the Cope-Osborn theory (see Osborn, 1907). However, he retained the fundamental principle of the Cope-Osborn theory: the apical cusps in the triangular arrangements of upper and lower molars (protocone on uppers and protoconid on lowers) are homologues to the single cusps of basic reptilian teeth.

In the context of his study of the Trinity therians, Patterson (1956) reevaluated the Cope-Osborn theory of mammalian molars and challenged its most important assumption of tooth cusp homology. According to Patterson, the protocone of tribosphenic mammals is an evolutionary neomorph with no homologous structure in their phylogenetic relative dryolestoid "eupantotherians." He suggested that the primary lingual cusp in the dryolestoid molar is the homologue of the paracone rather than of the protocone, as had been assumed by Cope, Osborn, Simpson, and Butler. The triangulate structure or the trigon of the dryolestoid molar is the primary trigon (enclosed within the paracrista linking the paracone with the stylocone, and the metacrista linking the metacone with the metastyle, see chapter 10 and figure 10.2). This trigon is equivalent to the part of tribosphenic molars that is labial to the paracone and metacone; that is, the homologue of the dryolestid (or "symmetrodontan") trigon is the stylar shelf of tribosphenic mammals. Patterson proposed that, in addition to the primary trigon, the advanced tribosphenic molar has a secondary trigon, with paracone, metacone, and the new cusp protocone at its apices. This new trigon came into existence concomitantly with the development of the neomorphic protocone, and it was to have profound adaptive consequences for the mammalian dentition. An important contribution of Patterson's then-new theory was that the derived function of grinding by the "secondary trigon" of the tribosphenic molar was correlated with an evolutionary neomorphic feature—the protocone.

When it became evident that the protocone is not a primary cusp but a neomorphic feature of tribosphenic mammals, Vandebroek (1961) proposed a new terminology for the elements of the tribosphenic molars, replacing that of Cope and Osborn. Etymologically, the Cope-Osborn terminology was only partly incorrect, but as it had already been in use for about 80 years, Vandebroek's terminology was generally ignored, as use of two different terminologies for the familiar elements of dental anatomy would have led to confusion. It is noteworthy that while the paleontological world generally accepted Patterson's (1956) conclusion, Simpson was cautious about it for several years (e.g., Simpson, 1961) and only accepted the new idea in 1971.

Functional Studies of Molar Occlusion. Detailed studies on morphology and function of molar occlusion in extant and fossil tribosphenic mammals were carried out by, among others, Butler (1939, 1961), Mills (1966), Crompton and Hiiemae (1970), Crompton and Sita-Lumsden (1970), Crompton (1971), Kay and Hiiemae (1974), and Crompton and Kielan-Jaworowska (1978). These studies were accompanied by cineradiographic investigations of jaw movements in the American opossum (*Didelphis marsupialis*), which retains tribosphenic molars of primitive design (Crompton and Hiiemae, 1969, 1970). A crucial observation was that each lower jaw undergoes lateral translation and rotates slightly along a triangular trajectory during occlusion, which is possible owing to the mobile mandibular symphysis of primitive tribosphenic mammals. With this "triangular" rotation during occlusion, the trigonid of the lower molar first shears past the labial part of the upper molar, followed by the talonid grinding against the protocone in the lingual part of the upper molar (Crompton and Hiiemae, 1969; Kay and Hiiemae, 1974).

The labial parts of upper molars and trigonids of lower molars of basic tribosphenic design, when brought into occlusion, form a series of alternating reversed triangles. Although the cusp arrangement is slightly different owing to the strong development and displacement of the metacone in tribosphenic mammals, this pattern is homologous to that of more primitive mammals with the reversed-triangle cusp pattern, such as "symmetrodontans" (see chapter 9). The lower teeth are correspondingly reversed and fit into the embrasures between the uppers (figure 11.2A). Crompton and Sita-Lumsden (1970) explained why these alternating triangles (and embrasures between them) are not equilateral. Were the triangles isosceles, and the shearing surfaces vertical, the leading edges of these surfaces would shear against each other only if the lower jaw moved orthally (vertically). However, the characteristic feature of mammalian mastication is that the lower jaw moves both vertically and medially in a triangular trajectory during occlusion. The transverse component of the mandibular movement would separate the cutting edges of such teeth as they come into occlusion. For the upper and lower shearing surfaces to make contact during the transverse phase of occlusion, they remain nearly vertical and the anterior shearing edge of the upper molar is transverse with respect to the longitudinal axis of the tooth row, so that the anterior edge of upper trigon (prevallum) and the posterior edge of the lower trigonid (postvallid) can maintain the contact. Crompton and Sita-Lumsden (1970) provided a model showing how the teeth constructed this way work during occlusion (figure 11.2B,C).

Crompton's Transformation Series (figure 11.3). Our current understanding is based on a landmark study by Crompton (1971) that provided an elegant outline of the main morphological and functional stages involved in the origin of tribosphenic molars.

When Patterson (1956) proposed that the protocone is a neomorph that is absent in pretribosphenic mammals, the fossils showing intermediate conditions between the dryolestoids and the typical tribosphenic molar of crown therians were still poorly understood. Patterson's main thesis on tribosphenic molar cusp homology offered no explanation for the morphological transformation that might have led from the ancestral taxon with a primary trigon (such as a dryolestoid, see chapter 10) to the true tribosphenic molar with a neomorphic protocone.

The subsequent descriptions of the dentition of *Amphitherium* and *Peramus* (Mills, 1964; Clemens and Mills, 1971) led to the conclusion that *Peramus* and related forms lay closer to the ancestry of mammals with tribosphenic dentition than drolestoideans do. In a very detailed study of *Aegialodon*, Kermack et al. (1965) showed that the lower molar talonid basin has at least two wear facets as a result of its inferred contact with the upper molar protocone. These studies provided important background for a synthetic theory of molar evolution and occlusal patterns in mammals.

Crompton (1971) established such a transformational series and showed how the complex form and function of the tribosphenic molars could have been assembled in a stepwise and incremental fashion. In a major insight he noted that derived wear facet features in tribosphenic molar can be explained by functional evolution.

Crompton's transformation series (figure 11.3) includes the "symmetrodontan" *Kuehneotherium* (chapter 9),

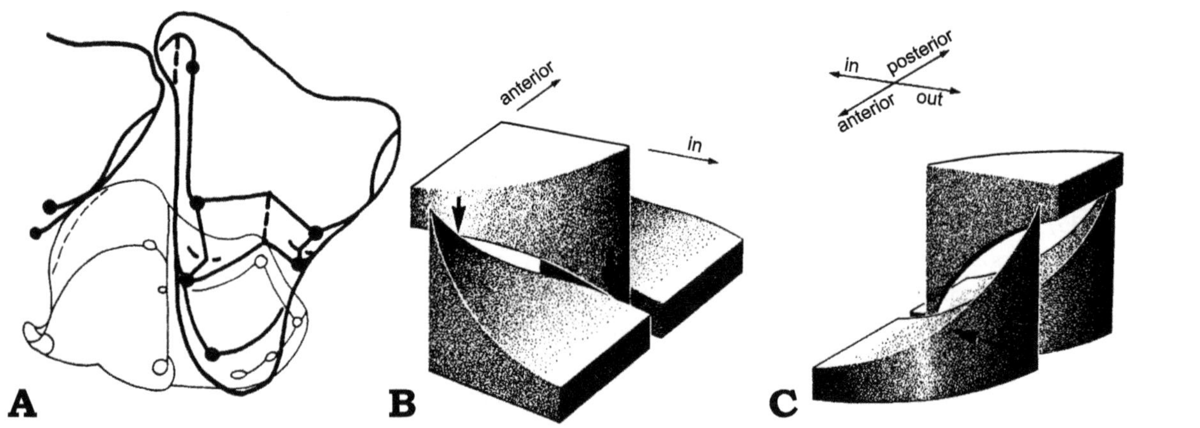

FIGURE 11.2. A, diagrammatic drawing of the left upper and left lower (drawn as transparent in view from the root side, and looking like the right one) tribosphenic molars in the palaeoryctid eutherian *Didelphodus* sp., showing asymmetrical shape of the upper molars and of the trigonid in lowers. B, Diagrammatic model of reversed-triangle shearing system in tribosphenic molars (talonids omitted) in which the transverse shearing edges are straight and the diagonal edges curved; the curved diagonal edge is seen at the beginning of occlusion. C, The same model showing transversely oriented shearing edges at the beginning of occlusion. *Source:* modified from Crompton and Sita-Lumsden (1970).

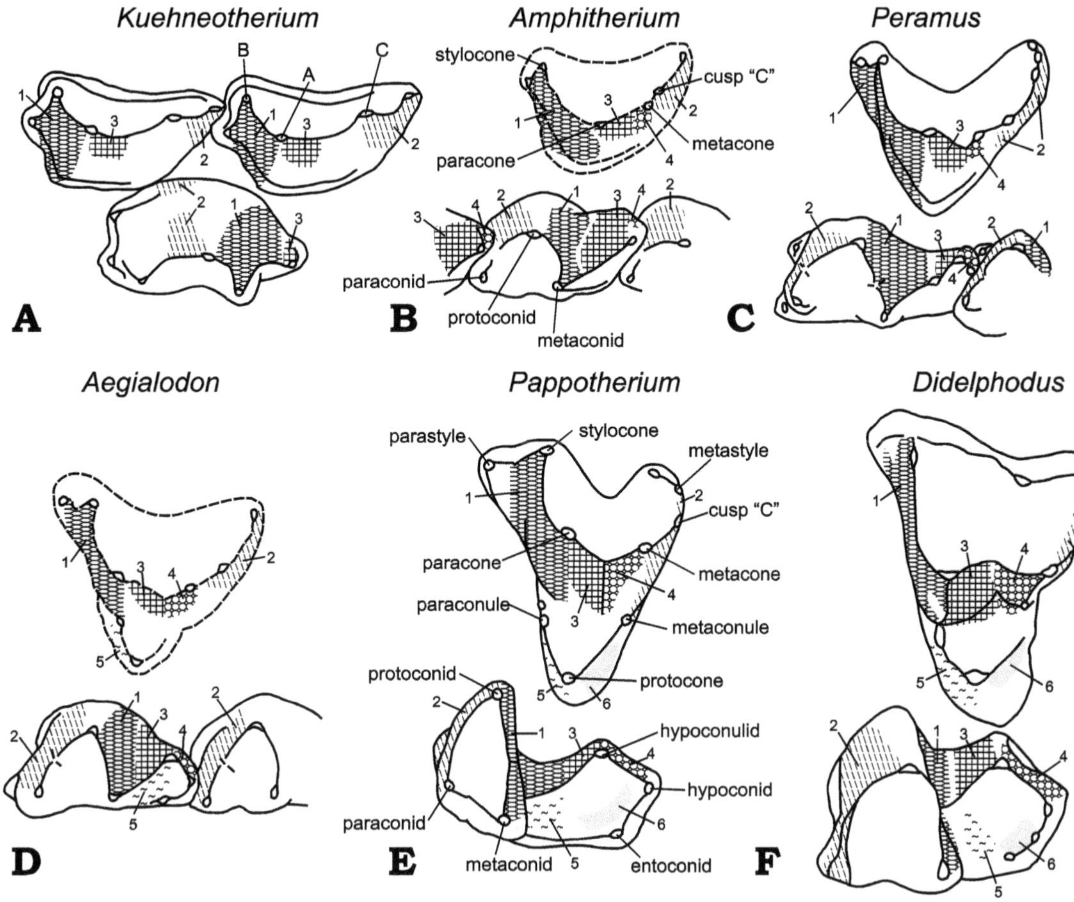

FIGURE 11.3. Diagrammatic drawings showing important stages in the evolution of the tribosphenic molar. The molars are shown in coronal view; for each pair of drawings the upper teeth are above and lowers below; anterior is to the left. Designation of wear facets (1–6) as introduced by Crompton (1971). A, *Kuehneotherium* (a Rhaeto-Liassic "symmetrodontan"). B, *Amphitherium* (a Middle Jurassic "eupantotherian"), upper molar reconstructed. C, *Peramus* (Early Cretaceous "eupantotherian"). D, *Aegialodon* (Early Cretaceous "tribotherian"). E, *Pappotherium* (Early Cretaceous "tribotherian"). F, *Didelphodus* (Paleocene palaeoryctid eutherian). E and F are tribosphenic molars. Not to scale. *Source:* modified from Crompton (1971).

"eupantotherians" *Amphitherium* (upper molar reconstructed) and *Peramus* (chapter 10), then *Aegialodon,* the oldest known at that time lower tribosphenic molar (upper molar reconstructed), the primitive "tribotherian" (see later) *Pappotherium,* and finally the Paleocene palaeoryctid eutherian *Didelphodus.* Of the six principal shearing surfaces in tribosphenic molars (such as in *Didelphodus*) only three are present in the Late Triassic–Early Jurassic *Kuehneotherium* (figure 11.3A). Most important of these is shearing surface 1, which is developed on the mesial side of upper molars (on the paracrista between cusps A and B) and on the distal side of lowers (protocristid between cusps a and c). These are homologous to the shearing surfaces on preparacrista (between paracone and stylocone) and protocristid (between protoconid and metaconid) of tribosphenic mammals (figure 11.3D–F). Shearing surfaces 2 (figure 11.3A) lie at the opposite ends of the respective teeth: between the distal side of upper molars (distolabial to cusp C) and mesial side of lowers (between cusps a and b). These are homologous to the shearing surfaces on postmetacrista (between metacone and metastyle) and paracristid (between protoconid and paraconid) of tribosphenic molars (figure 11.3D–F). Shearing surfaces 3 are small in *Kuehneotherium* (figure 11.3A); they are developed on the distolingual face of cusp A on the uppers and the mesiolabial face f cusp d on the lowers.

In *Kuehneotherium* upper and lower teeth are of approximately similar width, and the lower jaw moved dorsomedially during occlusion. Evolution toward the tribosphenic molar from the *Kuehneotherium* type is accompanied by an increase in the transverse component of the lower jaw movement, adding new shearing surfaces on the lingual side of the upper molar and in the talonid part of the lower. This resulted in an increase in the width of the upper molar, with new cusps appearing in the lingual region, and with the enlargement of the talonid from a single cingulid cusp (hypoconulid or cusp d) toward a broadened talonid bearing a grinding basin rimmed with neomorphic cusps.

In *Peramus,* and presumably also in *Amphitherium* (figure 11.3B,C), the upper molar is transversely wider than the matching lower. In these "eupantotherians," a new cusp, the metacone, appeared between the paracone and cusp "C" (see chapter 10). Crompton (1971) believed that cusp "C" of *Kuehneotherium* is not a homologue to the true metacone in *Peramus* and that the metacone in "eupantotherians" is a neomorph that developed in conjunction with the hypoconid of the lower molar to form a new shearing surface (facet 4, figure 11.3B,C, but see contrasting opinion by Hopson, 1997). Although the upper molars of some "eupantotherians," especially *Peramus,* are triangular, resembling those of earliest boreosphenidans (albeit being narrower transversely), the lingual cusp, the protocone, is not developed; only a lingual cingulum is present. The talonid basin, however, is far more developed than a mere cingulid cusp (d) of *Kuehneotherium;* it has a broader form in two "eupantotherian" genera (*Peramus* and *Palaeoxonodon*) and preceded the development of the protocone. In *Aegialodon* the strongly worn talonid is still very small but was probably basined, suggesting contact with a protocone. Because of the poor preservation of the single known molar of *Aegialodon,* it cannot be stated with any certainty whether three talonid cusps were present. The closely related *Kielantherium* (see later) has only two talonid cusps, the entoconid not being developed. Wear facet 4 in *Aegialodon* is larger than in "peramurans," and a new surface (wear facet 5) made its appearance (figure 11.3C,D). It is of interest that in the oldest known tribosphenic molar of the boreosphenidan *Tribactonodon* (Sigogneau-Russell et al., 2001), the talonid is extensive, with three cusps (including entoconid) well developed.

In *Pappotherium* the protocone is well developed, and the new wear facet (6) makes its appearance on its posterior wall (figure 11.3E). On the lower molar the talonid basin is enlarged, surrounded by the hypoconid, hypoconulid, and entoconid cusps. Wear facet 6 develops within the talonid basin, opposite the entoconid. In *Didelphodus* (figure 11.3F), as in most genera of stem boreosphenidans, the structure of the molars and shearing surfaces show basically the same structure as in *Pappotherium.* The difference, however, is that with the development of the winged conules, shearing surfaces 3 and 4 are duplicated, and the postparaconule and premetaconule cristae form the leading edges of a second rank of shearing surfaces (Fox, 1975).

Crompton and Kielan-Jaworowska (1978) studied the mode of occlusion and shearing surfaces in a number of Cretaceous boreosphenidans. In the earliest stem taxa of boreosphenidans, the pre- and postcingula did not develop. By contrast, with appearance of pre- and postcingula in more derived boreosphenidans, the second and third ranks of the shearing surfaces 1 (on the anterior wall of the upper molar) and surface 2 (on the posterior wall of the upper molar) may develop in correlation with these cingula, an observation corroborated by Fox (1975).

SYSTEMATICS

(table 11.1)

Subclass Boreosphenida Luo, Cifelli, and Kielan-Jaworowska, 2001, new rank

INTRODUCTION

Boreosphenida (Latin *boreas,* northern wind, and Greek *sphen,* wedge, referring to their origin in the Northern Hemisphere and to tribosphenic molars) are a subclass (new rank) erected by Luo, Cifelli, and Kielan-Jaworowska (2001) to include mammals with primarily tribosphenic molars that originated in the Northern Hemisphere during the latest Jurassic or Early Cretaceous. Boreosphenida include "tribotherians," Metatheria, and Eutheria, and they correspond to Tribosphenida *sensu* McKenna (1975) and McKenna and Bell (1997). The characterization given below is based on Luo, Cifelli, and Kielan-Jaworowska (2001), with some emendations. Nothing is known of the cranial and postcranial anatomy of stem boreosphenidans ("tribotherians"), except for an incomplete dentary of *Kielantherium.* So for the stem taxa of northern tribosphenic mammals, we do not have a brief characterization of skull and skeletal anatomy as in other chapters. The dental and mandibular features are presented in systematic sections under the genera.

BRIEF CHARACTERIZATION

Boreosphenidans differ from: (1) other mammals by the posterior placement of the mandibular angle, more elevated and near or above level of the postcanine alveoli (below the dentary condyle); (2) kuehneotheriids, some "eupantotherians," and australosphenidans by the absence of the primitive postdentary trough for the postdentary bones in the dentary (evidence of a paradentary element, the coronoid, is present in some taxa); (3) all mammals except Australosphenida by the presence of tribosphenic molars; (4) *Shuotherium* in the lack of the postdentary trough and in having talonid placed posterior to the trigonid on the lower molars, but convergent to the latter's pseudoprotocone of the uppers; (5) Australosphenida by having distinctive cingulid cusps (e, f) for molar interlocking, but lacking the continuous mesial cingulid on the molars of the latter, and in lacking the triangulated trigonid on the ultimate lower premolar in the Early Cretaceous taxa.

Taxa Included. "Tribotherians"; Metatheria Huxley, 1880; Eutheria Gill, 1872 (as modified by Huxley, 1880).

Distribution. Restricted to the Northern Hemisphere during the Early Cretaceous; present in the latest Cretaceous of South America and northern continents; Tertiary–Recent of the world.

Comment. In the present chapter we describe only the stem boreosphenidans, informally grouped for convenience as "tribotherians." For descriptions of Metatheria and Eutheria see chapters 12 and 13, respectively.

STEM BOREOSPHENIDANS ("TRIBOTHERIAN" GRADE)

We refer several poorly known Early and Late Cretaceous taxa, represented mainly by isolated tribosphenic molars to a taxonomic nether world, the informal group of "tribotherians" (= "tribotheres"). These taxa clearly share the synapomorphies of boreosphenidans. However, their further assignment within boreosphenidans, either to Metatheria or to Eutheria, cannot be established. Although one order and five families have been erected, the systematics of "tribotherians" pose difficulties. We are also aware that when they are better known, some of the genera may become synonyms. In addition, we describe several genera assigned to family *incertae sedis.* We divide the latter into two groups, describing Early Cretaceous *incertae sedis* genera separately, to enable comparison with established Early Cretaceous taxa and then Late Cretaceous *incertae sedis* genera (see also table 11.1).

EARLY CRETACEOUS GENERA

Order Aegialodontia Butler, 1978a
Family Aegialodontidae K. A. Kermack, 1967a

Order and Family Diagnosis. Poorly known order and family of very small forms. Differ from Metatheria in having four (or five) premolars; differ from Eutheria in having four molars. Share with Deltatheroida the structure of the trigonids, in which the paraconid is higher than the metaconid, and relatively small talonids, limited to the lingual side of the teeth; but the talonids in Aegialodontia are smaller than in the Deltatheroida and have (at least in *Kielantherium*) only two cusps, the entoconid not being developed (and entoconid is variably present in Deltatheroida). Share with many early mammals, among others with "eupantotherians" (e.g., *Peramus,* see chapter 10) a deep Meckel's groove (plesiomorphy); share with numerous Rhaeto-Liassic and Jurassic mammals (chapters 4–6, and 9), but also with Early Cretaceous eutherian *Prokennalestes* (chapter 13) the presence of an articular facet for the coronoid bone (plesiomorphy); share with the eutherian Otlestidae and with a few marsupials (see Dashzeveg and Kielan-Jaworowska, 1984) the presence of a labial mandibular foramen in the masseteric fossa (Kielan-Jaworowska and Dashzeveg, 1989).

Genera Included. *Aegialodon* K. A. Kermack et al., 1965; *Kielantherium* Dashzeveg, 1975.

Distribution. Early Cretaceous (Valanginian to Aptian or Albian): Eurasia, England; Mongolia, Gobi Desert.

Genus *Aegialodon* K. A. Kermack, Lees, and Mussett, 1965
(figure 11.4B)

Diagnosis. Poorly known genus represented by a single molar. Lower molar differs from that of *Kielantherium* in having the semiprocumbent paraconid oriented obliquely anterodorsally rather than dorsally and the protoconid relatively more robust with respect to the paraconid and metaconid; further differences concern the structure of the talonid, which is roughly triangular in occlusal view (possibly at least partly owing to wear) and shorter anteroposteriorly, rather than elongated and roughly rectangular. Because of the poor state of preservation, the number of talonid cusps cannot be established with any certainty; in particular the presence of the entoconid is uncertain.

Species. *Aegialodon dawsoni* K. A. Kermack, Lees, and Mussett, 1965, type species by monotypy, based on the holotype (single lower molar) of the type species.

Distribution. Early Cretaceous (Valanginian): Britain, East Sussex (Cliff End, Wadhurst Formation).

Genus *Kielantherium* Dashzeveg, 1975
(figure 11.4A)

Diagnosis. Aegialodontid genus with four molars and possibly four double-rooted premolars, but the presence of five premolars cannot be excluded. Differs from *Aegialodon* in having a higher trigonid, with cusps (especially the paraconid) directed vertically upward, and protoconid (in coronal view) relatively smaller with respect to the paraconid and metaconid. The other difference concerns the structure of the talonid, which is more elongated than in *Aegialodon,* being rectangular rather than triangular, and has only two cusps. The dentary is slender with a gently curved lower margin. The coronoid process has been broken, but its base is preserved and shows that it slopes

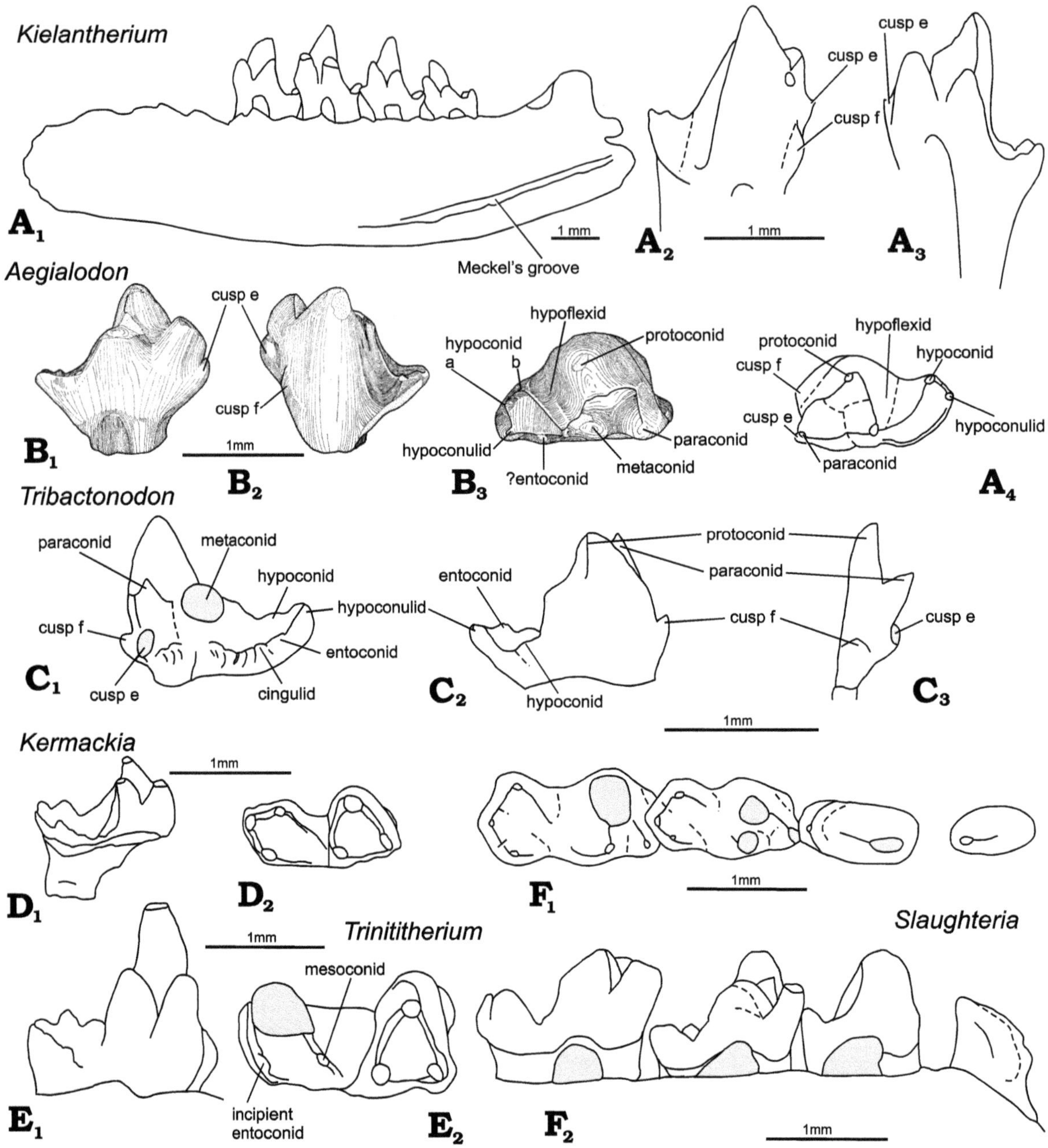

FIGURE 11.4. Some Early Cretaceous "tribotherian" taxa based on lower dentitions. A, *Kielantherium gobiensis,* right dentary with four molars in lingual view (A_1); right lower molar m2 or m3 (holotype) in labial (A_2), lingual (A_3), and occlusal (A_4) views. B, *Aegialodon dawsoni,* left lower molar (holotype) in lingual (B_1), labial (B_2), and occlusal (B_3) views. C, *Tribactonodon bonfieldi,* right lower molar (holotype) in lingual (C_1), labial (C_2), and mesial (C_3) views. D, *Kermackia texana,* left lower molar (holotype) in lingual (D_1) and occlusal (D_2) views. E, *Trinititherium slaughteri,* left lower molar (holotype) in lingual (E_1) and occlusal (E_2) views. F, *Slaughteria eruptens* left dentary fragment (holotype) with erupting penultimate premolar, ultimate premolar and two first molars, in occlusal (F_1) and lingual (F_2) views. *Source:* A_1, original, based on cast; A_{2-4}, modified from Crompton and Kielan-Jaworowska (1978); B, modified from Kermack et al. (1965); C, original, based on sketches provided by Denise Sigogneau-Russell; D–F modified from Slaughter (1971).

very gently upward. Behind the base of the coronoid process on the labial side, there is a small labial mandibular foramen (Kielan-Jaworowska and Dashzeveg, 1989), which enters the mandibular canal. On the lingual side there is a deep, sharply limited Meckel's groove and a facet for the coronoid bone.

Species. *Kielantherium gobiensis* Dashzeveg, 1975, type species by monotypy, based on an isolated lower molar (holotype), and a dentary with four molars and possibly four double-rooted alveoli for premolars.

Distribution. Early Cretaceous (Aptian or Albian): Mongolia, Gobi Desert ("Höövör Beds").

Order ?Aegialodontia Butler, 1978a
Family *incertae sedis*
Genus *Tribactonodon* Sigogneau-Russell, Hooker, and Ensom, 2001
(figure 11.4C)

Diagnosis (based on Sigogneau-Russell et al., 2001, emended). The lower molar is characterized by the prominent protoconid with a wide base, the paraconid much lower, and the metaconid (broken) with the base set off from the paraconid. A lingual cingulid extends from the mesial border, where it underlies a large mesiolingual cusp (cusp e),[2] to the entocristid distally, with a possible interruption under the metaconid. Mesiolabial cusp (cusp f) sharp, well detached, and extending into a mesiolabial cingulid. Paraconid vertically directed. Relatively long talonid with three cusps; hypoconid and entoconid distinct and well separated from the higher hypoconulid.

Species. *Tribactonodon bonfieldi* Sigogneau-Russell, Hooker, and Ensom, 2001, type species by monotypy, based on a single lower molar.

Distribution. Early Cretaceous (Berriasian): southern England, Dorset (Purbeck Limestone Group).

Order *incertae sedis*
Family Kermackiidae Butler, 1978a

Diagnosis. Very small "tribotherians" that differ from Aegialodontidae in having relatively larger talonid with three cusps, of which the entoconid is small or incipient, and the paraconid and metaconid are of subequal height, rather than the metaconid being higher. Differ from *Slaughteria* in having the paraconids directed vertically upward, rather than slightly procumbent, and lower talonids slightly narrower than the trigonids, rather than of subequal width. From lower molars tentatively assigned to *Pappotherium* and *Holoclemensia* they differ in smaller dimensions, smaller talonids, with less distinct entoconid, paraconid, and metaconid subequal, rather than paraconid lower, and by the presence of a distal metacristid.

Genera. *Kermackia* Slaughter, 1971; *Trinititherium* Butler, 1978.

Distribution. Early Cretaceous (Aptian–Albian): United States, north-central Texas (Trinity Group).

Genus *Kermackia* Slaughter, 1971
(figure 11.4D)

Diagnosis. Lower molar differs from that of *Trinititherium* in having relatively narrower trigonid, lower protoconid (not differing so conspicuously in height from the paraconid and metaconid), and in lack of the mesoconid.

Species. *Kermackia texana* Slaughter, 1971, type species by monotypy, represented by a single lower molar, the holotype of the type species.

Distribution. As for the family.

Genus *Trinititherium* Butler, 1978a
(figure 11.4E)

Diagnosis. Differs from *Kermackia* by having the protoconid much higher with respect to the height of the paraconid and metaconid (which, however, are strongly worn), a relatively wider trigonid, and the presence of the mesoconid.

Comment. The differences between the holotypes of *Trinititherium slaughteri* and *Kermackia texana* are slight and may simply reflect positional variation within the tooth row of the same species.

Species. *Trinititherium slaughteri* Butler, 1978a, type species by monotypy, represented by a single lower molar, the holotype of the type species.

Distribution. As for the family.

Family Pappotheriidae Slaughter, 1965

Comment. The family is monotypic, the diagnosis is as of the type genus *Pappotherium*. Although it was previously considered to be eutherian (Slaughter, 1968b), several recent phylogenetic studies have established that it is a basal stem taxon that cannot be placed within the crown therians (Cifelli, 1993b; Rougier et al., 1998; Luo et al., 2002).

Genus *Pappotherium* Slaughter, 1965
(figure 11.5D)

Family and Generic Diagnosis. The stylar shelf is slightly narrower than in *Deltatheroides* and *Atokatheridium* (see chapter 12). Shares with these genera deep ectoflexus, strong paracrista, large parastylar region, and unwinged conules, but differs in having a slightly wider pro-

[2] Sigogneau-Russell et al. (2001) named two mesial cusps on the lower molar in *Tribactonodon* as cusp e (mesiolingual) and cusp f (mesiolabial), following Godefroit and Sigogneau-Russell (1999: figure 1), who used these terms for *Kuehneotherium*. We follow these designations.

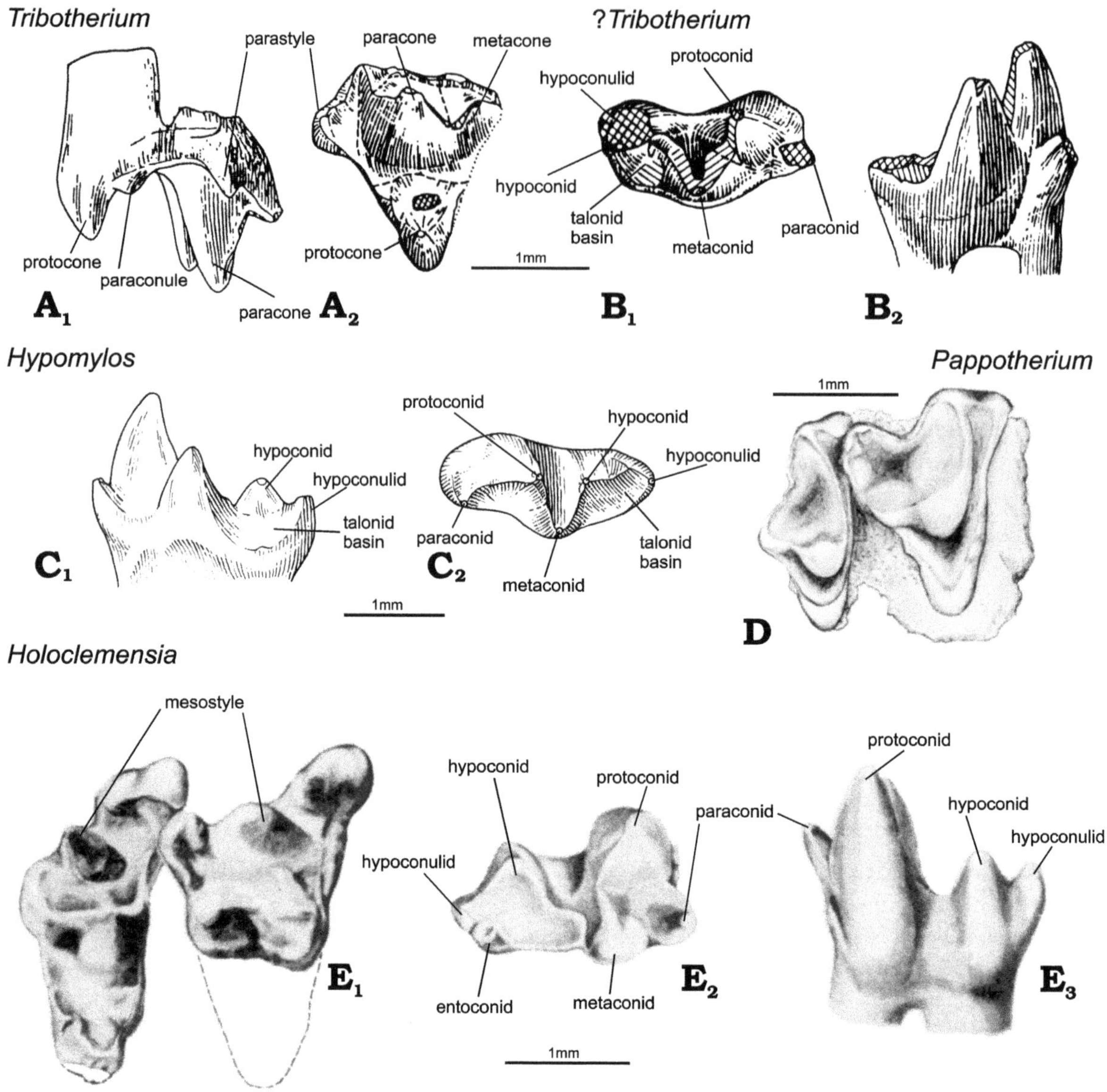

FIGURE 11.5. Selected "tribotherian" Early Cretaceous upper and lower molars. A, *Tribotherium africanum* left upper molar in mesial (A_1) and occlusal (A_2) views. B, ?*Tribotherium africanum* attributed left lower molar in occlusal (B_1) and lingual (B_2) views. C, *Hypomylos phelizoni,* right lower molar, holotype, in lingual (C_1) and occlusal (C_2) views. D, *Pappotherium pattersoni,* penultimate and ultimate upper molars (holotype) in occlusal view. E, *Holoclemensia texana,* penultimate and ultimate upper molars (holotype) in occlusal view (E_1); referred left lower molar in occlusal (E_2) and labial (E_3) views. *Source:* modified from: A, B, Sigogneau-Russell (1995a); C, Sigogneau-Russell (1992); D, Slaughter (1965); E, Slaughter (1968a).

tocone. Differs from *Holoclemensia* in the lack of a strong mesostyle, but shares with it the general shape of the upper molars, with the parastylar region strongly protruding labially and anteriorly. The lower molar attributed to *Pappotherium* by Butler (1978a), differs from that of Deltatheroida in having subequal paraconid and metaconid and a wider talonid; however, the attribution of the lower molar is not certain.

Species. *Pappotherium pattersoni* Slaughter, 1965, type species by monotypy, based on penultimate and ultimate upper molars preserved together (holotype), and tentatively attributed lower molar.

Distribution. Early Cretaceous (Aptian–Albian): United States, north-central Texas (Trinity Group).

Family Holoclemensiidae Aplin and Archer, 1987

Comment. Aplin and Archer (1987) assigned Holoclemensiidae to Didelphia, but because of the ambiguity concerning its dental formula and other details we prefer to treat it informally as a "tribotherian." The family is monotypic, the diagnosis and distribution are the same as for the type genus: *Holoclemensia.* Cifelli (1993b) and Rougier et al. (1998) showed by parsimony analysis that this taxon cannot be placed with the crown therians.

Genus *Holoclemensia* Slaughter, 1968c
(figure 11.5E)

Family and Generic Diagnosis. Largest boreosphenidan genus among the Trinity therians, which shares with *Pappotherium* the shape of the two last upper molars, with the parastylar region strongly protruding labially and anteriorly. Differs from *Pappotherium* in the presence of an enlarged mesostyle and lack of ectoflexus, and shares a large mesostyle with several Cretaceous marsupial genera, for example, *Albertatherium,* some species of *Alphadon,* and *Glasbius* (see chapter 12). Shares with *Pappotherium* wide stylar shelf, which is, however, narrower than in Deltatheroida and *Atokatheridium* (chapter 12). The protocone has been preserved only on the last molar and appears relatively wider than in Deltatheroida. The lower molar attributed to *Holoclemensia* (e.g., by Butler, 1978a) differs from those of Deltatheroida and *Kielantherium* in having the paraconid shorter than the metaconid and larger talonid.

Species. *Holoclemensia texana* (Slaughter, 1968a), type species by monotypy.

Distribution. Early Cretaceous (Aptian–Albian): United States, north-central Texas (Trinity Group).

Family *incertae sedis*
Genus *Hypomylos* Sigogneau-Russell, 1992
(figure 11.5C)

Diagnosis. Genus confined to lower dentition; lower molars characterized by a high protoconid with a flat lingual surface, very low (strongly reduced) paraconid, and metaconid strongly inclined lingually; bases of paraconid and metaconid well separated. The talonid is long and very narrow, inclined lingually, with only two cusps (hypoconid and hypoconulid), separated at the bases; the entoconid is lacking in type species (*H. phelizoni*), but recognized in *H. micros.*

Species. *Hypomylos phelizoni* Sigogneau-Russell, 1992, type species; *H. micros* Sigogneau-Russell, 1995; ?*Hypomylos?* sp. (Sigogneau-Russell, 1995a).

Distribution. Early Cretaceous (?Berriasian): Morocco, Talsinnt Province, Synclinal d'Anoual (Séquence B des Couches Rouges).

Genus *Slaughteria* Butler, 1978a
(figures 3.24C, 11.4F)

Comment. Slaughter (1971) interpreted the lower teeth of *Slaughteria* as three premolars and first molar, in which case the ultimate premolar would be completely molariform, as in advanced eutherians. Butler (1978a) interpreted them as two premolars and two molars, which appeared more probable. Slaughter (1971) assigned this dentary to *Pappotherium,* but Butler (1978a) argued that it is too small to belong to this genus and erected the genus *Slaughteria* for it. However, after examining this specimen by CT scan, Kobayashi et al. (2002: 371) suggested that *Slaughteria* may be based on a juvenile of *Pappotherium.* These authors also found that the first molariform tooth in the holotype of *S. eruptens* is underlain by a premolariform successor, and thus the tooth identified by Slaughter (1971) as the ultimate premolar and by Butler (1978a) as the first molar is the deciduous last premolar. We elect to recognize *Slaughteria* as a valid taxon until further materials become available to validate potential synonymy of these two taxa.

Diagnosis. Differs from *Kielantherium* and members of Kermackiidae in having semiprocumbent (rather than vertical) paraconids on lower molars and shares this character with *Aegialodon;* differs from Aegialodontidae and Kermackiidae in having the talonids as wide as the trigonids rather than distinctly shorter, with three well-developed cusps.

Species. *Slaughteria eruptens* Butler, 1978a, type species by monotypy, based on the holotype, which is a dentary fragment with the penultimate premolar in eruption, the deciduous last premolar underlain by its successor, and the first molar.

Distribution. Early Cretaceous (Aptian–Albian): United States, north-central Texas (Trinity Group).

Genus *Tribotherium* Sigogneau-Russell, 1991
(figure 11.5A,B)

Diagnosis. A "tribotherian" known from upper and tentatively assigned lower molars. Differs from most Early Cretaceous "tribotherians" in the lack of an ectoflexus and in having a narrow stylar area. Shares straight labial margin with *Picopsis, Comanchea,* and more advanced *Falepetrus* and *Zygiocuspis,* but differs from the first two genera in the lack of a mesostyle. Parastylar wing large, protruding anteriorly, situated lingual to the paracone, with parastyle and high stylocone; there is a small metastyle and cusp "C." Paracone and metacone situated far labially, paracone much larger than metacone. Protocone narrow, lower than the parastyle; conules present. Attributed lower molar differs from *Hypomylos* in having protoconid situated directly labial to the metaconid, with bases of paraconid and metaconid joined, but shares the same proportions between the two cusps, of which the paraconid is much lower. The talonid is small and less inclined lingually than in *Hypomylos.* As in *Hypomylos,*

Kielantherium, ?*Atokatheridium* (see chapter 12), and *Trinititherium* only two talonid cusps are present.

Species. *Tribotherium africanum* Sigogneau-Russell, 1991, type species, based on upper molars; ?*Tribotherium africanum*, lower molar tentatively assigned (Sigogneau-Russell, 1995a).

Distribution. Early Cretaceous (?Berriasian): Morocco, Talsinnt Province, Synclinal d'Anoual (Séquence B des Couches Rouges).

LATE CRETACEOUS GENERA[3]

Family Picopsidae Fox, 1980

Comment. Fox (1980b) erected the monotypic family Picopsidae for *Picopsis*, based on upper molars and the trigonid of a lower molar. Upper molars of *Picopsis* are triangular, very short transversely, and resemble the deciduous premolars of Marsupialia (see, e.g., Clemens, 1966: figure 30). Because of uncertainty we retain the family Picopsidae, and assign to it also *Comanchea*, which is similar to *Picopsis* in structure of the stylar shelf, but differs in being more elongated transversely. It is also interesting that deciduous premolars of Dryolestoidea are similar in shape to those of *Picopsis* (see chapter 10).

Diagnosis. Poorly known family of small mammals that differ from Early Cretaceous "tribotherian" genera (except for *Tribotherium*) in having a straight labial margin of upper molars, with ectoflexus lacking, a very reduced anterior stylar shelf, stylocone, and a parastyle. Individually, some of these characters may be present in some other Late Cretaceous boreosphenidan taxa, but not in the same combination. The partial or complete reduction of the anterior stylar shelf together with a relatively large metastylar region are apomorphic. Stylar cusps are variously developed, mesostyle present, paracone larger than metacone; protocone low and small; cingula very weak or absent (plesiomorphy).

Genera. *Picopsis* Fox, 1980, type genus; *Comanchea* Jacobs et al., 1989; upper molars described by Eaton (1993b: 107–109, figure 3C–G), as "Theria gen. et sp. undetermined" resemble *Picopsis* and may be congeneric with it, or at least belong to the Picopsidae; *Falepetrus* Clemens and Lillegraven, 1986.

Distribution. Early–Late Cretaceous (Aptian–Campanian): United States and Canada.

[3] We describe in this informal section Late Cretaceous genera, except for *Comanchea*, which is of Early Cretaceous age, and which we assign to the Picopsidae, because of its similarity to *Picopsis*.

Genus *Picopsis* Fox, 1980
(figure 11.6A)

Diagnosis. Upper molars longer anteroposteriorly than transversely, without anterior stylar shelf and stylocone, very little or no ectoflexus, mesostyle present, paracone higher than metacone, small protocone, no conules and no precingulum (weak postcingulum present). Attributed lower molar is represented only by the trigonid, with bladelike anteriorly placed paraconid, and reduced posterior metaconid.

Species. *Picopsis pattersoni* Fox, 1980; cf. *Picopsis* sp. (Fox, 1980b).

Distribution. Canada, Alberta (upper part of the Milk River Formation); Late Cretaceous (Santonian): United States, Utah (unnamed unit).

Genus *Comanchea* Jacobs, Winkler, and Murray, 1989
(figure 11.6B)

Diagnosis. Poorly known genus represented only by the holotype of the type species (incomplete left upper molar). Shares family characters with *Picopsis*, but differs in being more elongate transversely than anteroposteriorly, in having the protocone elongated anteroposteriorly, and in the presence of a small paraconule (not seen on the photograph published by Jacobs et al., 1989). There are five stylar cusps of which parastyle, stylocone, and mesostyle are reduced in size, whereas two medial cusps mesostyle (cusp C) and cusp D are enlarged.

Species. *Comanchea hilli* Jacobs, Winkler, and Murray, 1989, type species by monotypy.

Distribution. Early Cretaceous (Albian): United States, north-central Texas (Paluxy Formation).

Genus *Falepetrus* Clemens and Lillegraven, 1986
(figure 11.6C)

Diagnosis. Upper dentition genus, characterized by a straight labial margin, with minute stylocone, mesostyle, and metastyle (referred to by Clemens and Lillegraven as cusps B, C, and D). Stylar shelf narrow, large parastylar wing (broken), extending lingual to the paracone, which is slightly higher than the metacone. Small, weakly winged conules are placed close to the paracone and metacone. The protocone is large but low, and there are small, but distinct pre- and postcingula.

Species. *Falepetrus barwini* Clemens and Lillegraven, 1986, type species by monotypy, represented by two upper molars.

Distribution. Late Cretaceous (middle Campanian, Judithian): United States, Wyoming ("Mesaverde" Formation), Montana (Judith River Formation).

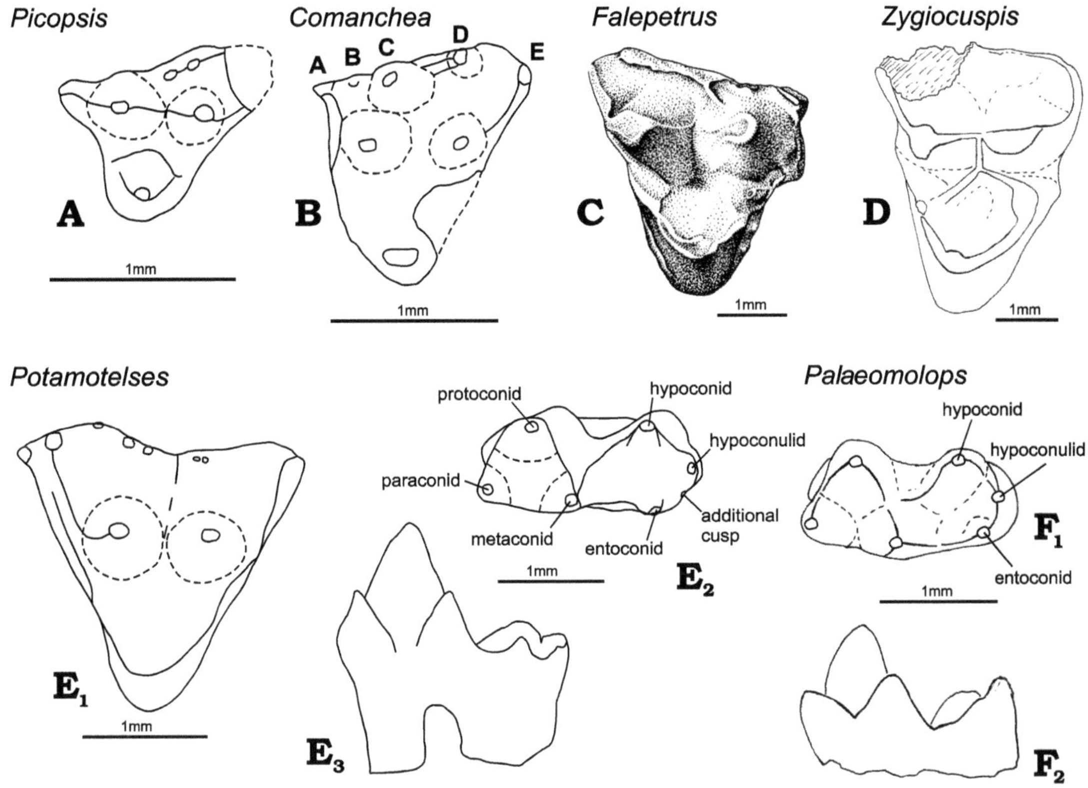

FIGURE 11.6. Albian *Comanchea* and Late Cretaceous "tribotherian" genera. A, *Picopsis pattersoni,* left upper molar (holotype) in occlusal view. B, *Comanchea hilli,* left upper molar (holotype) in occlusal view (stylar cusp designations after Jacobs et al., 1989). C, *Falepetrus barwini,* right upper molar in occlusal view. D, *Zygiocuspis goldingi,* left upper molar (holotype) in occlusal view. E, *Potamotelses aquilensis,* left upper molar, holotype (reversed), in occlusal view, associated with a ?penultimate upper molar (E_1); referred right lower molar in occlusal (E_2) and lingual (E_3) views. F, *Palaeomolops langstoni,* right lower molar in occlusal (F_1), and lingual (F_2) views. *Source:* A, original based on stereophotograph in Fox (1980b), courtesy of P. M. Butler; B, original based on the SEM photograph in Jacobs et al. (1989); C, modified from Clemens and Lillegraven (1986); D, original based on the photograph in Cifelli (1990d); E_1, modified from Fox (1975), E_2, E_3, original, based on stereophotographs in Fox (1972b); F, original, based on SEM photographs in Cifelli (1994).

Family *incertae sedis*
Genus *Palaeomolops* Cifelli, 1994
(figure 11.6F)

Diagnosis. Labial margin of the lower molar concave in occlusal view, protoconid high, paraconid and metaconid subequal in height, more approximated than in *Potamotelses;* talonid three cusped (rather than four cusped as in *Potamotelses*). Shares with Cretaceous marsupial taxa (of which the most similar is *Iugomortiferum*) slightly inflated cusps, but differs in having paraconid and hypoconulid in more median positions, in hypoconulid not being twinned with entoconid, and in having weak or absent labial postcingulid.

Species. *Palaeomolops langstoni* Cifelli, 1994, type species by monotypy, represented by several lower molars, and tentatively assigned two last upper molars.

Distribution. Late Cretaceous (Campanian, Judithian): United States, southern Texas (Aguja Formation).

Genus *Potamotelses* Fox, 1972
(figure 11.6E)

Diagnosis. A "tribothere" represented by isolated upper and lower molars. Upper molar crown triangular in occlusal view with long labial border and relatively wide stylar shelf, but narrower and shorter than in *Pappotherium, Atokatheridium,* and Deltatheroida. Small parastyle, large stylocone, and small marginal labial cuspules. Protocone low, but wider transversely than in above-cited genera, with which it shares the lack of pre- and postcingula; conules lacking. Trigonid of lower molars elongated anteroposteriorly; talonid short and narrow, four-cusped. Protoconid high, paraconid and metaconid low, of similar height, separated at bases; distal metacristid extending from apex of the metaconid to cristid obliqua; hypoconulid posterolabial, nearer to hypoconid than to entoconid; additional small talonid cusp between the hypoconulid and entoconid.

Species. *Potamotelses aquilensis* Fox, 1972, type species by monotypy.

Distribution. Late Cretaceous (early Campanian): Canada, Alberta (upper part of the Milk River Formation).

Comment. Nessov (1987: figure 1) used the family name Potamotelsidae, not mentioned in his text. However, according to ICZN (1999: Art. 13, Exclusion 13.6.1.), this name is not available.

Genus *Zygiocuspis* Cifelli, 1990
(figure 11.6D)

Diagnosis. Relatively large animal compared to other known stem boreosphenidans ("tribotherians"); upper molar characterized by straight labial margin (no ectoflexus) and short stylar area (but broken parastylar wing apparently very large); no stylar cusps posterior to the paracone. Paracone and metacone subequal; large winged paraconule and metaconule with a strong yoke uniting internal conular cristae (autapomorphy); protocone large, no pre- and postcingula.

Species. *Zygiocuspis goldingi* Cifelli, 1990, type species by monotypy, represented only by the holotype (incomplete upper molar, with broken parastylar wing).

Distribution. Late Cretaceous (early Campanian): United States, Utah (middle part of Wahweap Formation).

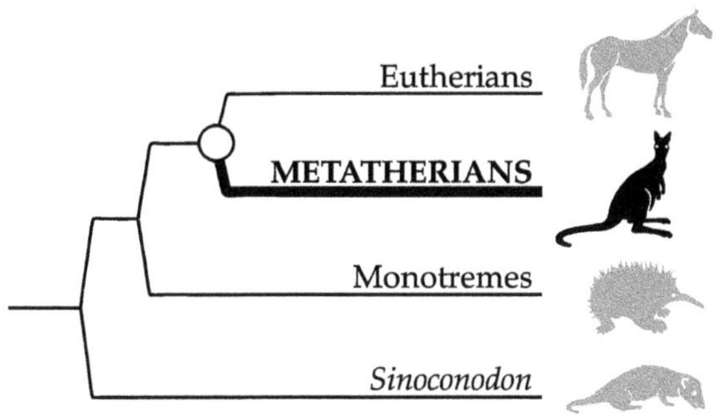

CHAPTER 12

Metatherians

INTRODUCTION

Marsupials and their putative fossil relatives, collectively termed Metatheria, have long been of general interest among evolutionary biologists. Like eutherians—and in contrast to monotremes—living marsupials give live birth; however, their reproductive biology is distinctly different (e.g., Tyndale-Biscoe, 1973; Renfree, 1981, 1983, 1993; Tyndale-Biscoe and Renfree, 1987; Zeller and Freyer, 2001). Living marsupials are mainly restricted to southern landmasses (though this is not the case for fossil relatives), and the origin and biogeographic deployment of the group have been widely discussed (e.g., Simpson, 1953; Hoffstetter, 1972; Lillegraven, 1974; Marshall, 1980, and references therein).

There are nearly 80 genera of living marsupials, placed in some 16–18 families and grouped variably at higher taxonomic levels (see Aplin and Archer, 1987; Nowak, 1991). We recognize 19 marsupial genera from the Mesozoic, distributed among six families (Asiatheriidae, "Alphadontidae," Glasbiidae, Peradectidae, "Pediomyidae," and Stagodontidae; three genera are left *incertae sedis*). Recent evidence suggests that the Deltatheroida, including mainly Asiatic taxa formerly thought to be eutherians (Gregory and Simpson, 1926; Van Valen, 1966) or "tribotheres" (Butler, 1978a), may be related to marsupials (Kielan-Jaworowska and Nessov, 1990; Marshall and Kielan-Jaworowska, 1992; Rougier et al., 1998). Although the issue remains open to alternative interpretation (Cifelli, 1993b; Luo et al., 2002), we tentatively recognize Deltatheroida as a basal group of Metatheria.

Inclusion of Deltatheroida within Metatheria has significant consequences for interpretation of the marsupial-placental dichotomy. As recognized by Gregory and Simpson (1926) and all authors since, the molar structure of deltatheroidans appears to be remarkably primitive, suggesting a temporally remote common ancestry for Metatheria and Eutheria (Kielan-Jaworowska, 1992; Cifelli, 1993b). Clearly, the hypothesis of an inclusive Metatheria is ripe for testing via new discoveries from the fossil record.

With these caveats, we tentatively recognize Metatheria as a clade defined by the common ancestor of all extant marsupials plus all extinct mammals that are more closely related to extant marsupials (such as kangaroos) than to extant placentals (such as hedgehogs). For reasons given later, we employ a stem-based definition of Marsupialia, recognized as the clade including all mammals more closely related to living marsupials than to Deltatheroida or Eutheria. We provisionally regard marsupials as sister taxon to Deltatheroida, under Metatheria.

BRIEF CHARACTERIZATION

The most significant differences between living Metatheria and Eutheria (see chapter 13) concern the reproductive system and other aspects of "soft anatomy," thus involving features that cannot be directly determined (or, in general, reasonably inferred) from fossils. Marsupials lack the trophoblast seen in eutherians; furthermore, most marsupials are characterized by a choriovitelline placenta and short-lived corpus luteum. As a result, maternal and developing fetal tissues are poorly suited to prolonged intrauterine development, owing to immunological recognition and rejection of the young by the mother. The young are born in an altricial condition following a very short gestation period (Lillegraven, 1975, 1979; Lillegraven et al., 1987, and references therein) and develop during a prolonged period of maternal lactation, typically (but not invariably) in a pouch. These aspects of marsupial reproduction and development appear to pose constraints on many aspects of their biology. Relative to placentals, marsupials typically are characterized by lower metabolic rates, fewer chromosomes, smaller brains, and lesser degrees of specialization of the mouth and forelimb, among other features (Lillegraven, 1984). The extent to which these characteristics (and many other aspects of marsupial soft anatomy) are primitive retentions from a common ancestor shared with eutherians is debatable (see Renfree, 1983, 1993; Lillegraven et al., 1987; Tyndale-Biscoe and Renfree, 1987; Szalay, 1994; and references cited earlier). Other marsupial characteristics, such as the presence of a pseudovaginal canal and the end arteries on the surface of the brain, are quite probably derived (Lillegraven, 1984).

The skeleton of marsupials is substantially similar to that of eutherians; an often-cited difference among living taxa, the presence of epipubic bones in marsupials, is primitive: epipubics are found in early eutherians (Kielan-Jaworowska, 1975a; Novacek et al., 1997; Ji et al., 2002, see chapter 13), monotremes, and a number of extinct Mesozoic groups. The marsupial ankle is probably not much modified from a common therian ancestor (Szalay, 1994); marsupials lack the placental condition, in which lateral mobility of the astragalus is restricted by the medial and lateral malleoli at the upper ankle joint (Szalay, 1984). In the carpus, the apparent incorporation of the centrale into the scaphoid may be an apomorphy of (or within) marsupials. The condition is unknown in Deltatheroida and uncertain in the ?stem marsupial *Asiatherium,* though it lacks the prominent distolateral process of the scaphoid seen in remaining "ameridelphians" (Szalay, 1994; Szalay and Trofimov, 1996).

In the skull, marsupials commonly have an alisphenoid bulla; this, however, appears to be derived within Marsupialia rather than an attribute of their common ancestor. The most obvious possible apomorphy of the ear region in Metatheria is the lack of a groove on the promontorium for the stapedial artery (present in eutherians and *Vincelestes,* but lacking in all other outgroups to Boreosphenida); additional apomorphies listed by Rougier et al. (1998) include the lack of a foramen for the superior ramus of the stapedial artery and the absence of the related ascending canal, the separation of the jugular foramen from the inferior petrosal opening, and the posterolateral placement of the transverse sinus with respect to the subarcuate fossa.

Like eutherians, the mandible of Metatheria lacks any trace of postdentary bones; known forms also lack a separate coronoid bone (or facet indicating its presence, a primitive feature that is retained in several Cretaceous eutherians (Kielan-Jaworowska and Dashzeveg, 1989; Cifelli, 1999b). Undoubtedly the most distinctive feature of the metatherian dentary is the inflected shelflike angular process. The angular process is slightly inflected in a number of early eutherians (Kielan-Jaworowska, Bown, and Lillegraven, 1979), but does not have the shelflike appearance seen in Deltatheroida and some living marsupials (see discussion in Averianov and Kielan-Jaworowska, 1999: 78).

With the inclusion of Deltatheroida, the dentition of the ancestral metatherian is characterized mainly by features presumed to be primitive for Boreosphenida as a whole. The molars have the basic tribosphenic pattern (chapter 11) also seen in Eutheria and "tribotherians," and there is a sharp morphological break in the premolar-molar series (in eutherians, the posterior premolars tend toward molarization); both conditions are plesiomorphic. The ancestral dental formula, as represented by the primitive condition in marsupials, is 5.1.3.4/4.1.3.4. The only apomorphy in this series is the count of three premolars, which represents a reduction from four or five in the ancestral boreosphenidans. Another apomorphy of marsupials, and possibly present in deltatheroidans as well, is reduction of diphydonty, in which postnatal tooth replacement is limited to the third premolar (Luckett, 1993). Dental microstructure of most living and fossil marsupials differs from that of most living placentals in having tubules; owing to uncertainties in homology, the significance of this is not clear (see later).

DISTRIBUTION

MARSUPIALIA

True Cretaceous Marsupialia (in the sense used herein) were first described from what is now known as the Lance Formation, Wyoming; and the Hell Creek Formation, South Dakota (Marsh, 1889a; Cope, 1892). The record of marsupials in North America now extends through the entire Late Cretaceous, though marsupials are best known from Campanian and Maastrichtian faunas, with most known fossils coming from the northern and central parts of the Western Interior (see chapter 2). We recognize 19 genera of Cretaceous marsupials from North America. Of these, four are present in the Albian–Cenomanian, three in the Cenomanian, three in the Turonian, nine in the Aquilan (early Campanian), seven in the Judithian (late

Campanian), and six in the Lancian (late Maastrichtian). Some of this variation is clearly due to vastly greater sampling for the younger faunas, but it is notable that the greatest diversity appears to be in the Aquilan, which is represented by only two local faunas. Bearing in mind that the Cretaceous record leaves much to be desired, there is nonetheless some basis for the impression that marsupials declined in generic diversity during the Campanian–Maastrichtian of North America.

While it is clear that marsupial radiations during the Mesozoic were mainly restricted to North America, there is considerable uncertainty as to the origin of Marsupialia, both geographically and temporally. Estimates based on molecular data (e.g., Kumar and Hedges, 1998; Penny et al., 1999) place the marsupial-placental dichotomy as far back as the Middle Jurassic. The most notorious complaint about the fossil record—its incompleteness—is especially well deserved when it comes to Mesozoic mammals (chapter 2). Nonetheless, such results are perplexing in the context of what little is known from the fossil record. As noted previously, the earliest marsupials that are recognizable on the basis of known fossils are from North American rocks that approximate the Early–Late Cretaceous boundary in age. At this first appearance in the fossil record, marsupials are rare, poorly diversified (both morphologically and taxonomically), and plesiomorphic in almost all known characters. Whatever the ultimate origin of Marsupialia, current evidence suggests that these Albian–Cenomanian taxa lie near the base of North America's marsupial radiation (Cifelli, 2004).

North American Cretaceous marsupials are known almost entirely by jaws and isolated teeth; of these, only a single specimen preserves upper and lower dentitions in occlusion (Lillegraven, 1969). Except for highly distinctive taxa (such as the stagodontids *Eodelphis* and *Didelphodon*), this lack of associated fossils presents the obvious problem of identifying upper and lower teeth belonging to the same species. Simpson (1929a) dealt with the upper and lower dentitions under separate systematic headings. With a vastly improved sample, Clemens (1966) was able to propose reasonable associations for marsupials of the type Lance Formation. Subsequent studies (e.g., Lillegraven, 1969) have corroborated these associations, and taxonomic allocation of upper and lower dentitions is now routine, at least in cases in which sampling is reasonably good and morphology is comparable to known taxa. Aside from teeth and jaws, the North American record of Cretaceous marsupials includes petrosals and skull fragments, mostly representing Stagodontidae (Matthew, 1916; Clemens, 1966; Wible, 1990; Meng and Fox, 1995c), and isolated tarsals ascribed to Stagodontidae and "Pediomyidae" (Szalay, 1994).

A notable development in recent years has been recognition of Marsupialia in the Cretaceous of Asia, although there is not universal agreement on this point. Three taxa are tentatively included. Geologically oldest of these is Coniacian *Marsasia*, known by fragmentary specimens (Nessov, 1997; Averianov and Kielan-Jaworowska, 1999). The most completely known is *Asiatherium*, represented by a skull and skeleton from the Campanian of Mongolia (see Trofimov and Szalay, 1994; Szalay and Trofimov, 1996). Finally, an unnamed, possible marsupial is known as the "Guriliin Tsav skull" (see comments under Deltatheroida, later), from the Maastrichtian of Mongolia.

There are several possible occurrences of Cretaceous marsupials in South America (chapter 2). Noteworthy among these is cf. *Peradectes*, represented by fragmentary teeth of debatable age from Peru. If the Cretaceous age and generic referral are correct, the occurrence is highly notable. *Peradectes* is otherwise known only from North America and Europe (Crochet, 1980), with a first appearance in the early Paleocene (Archibald et al., 1987; Cifelli et al., 2004).

DELTATHEROIDA

The geologically oldest taxa that can be confidently referred to Deltatheroida are *Sulestes* Nessov, 1985b and *Deltatherus* Nessov, 1997, both placed in the Deltatherididae. Both of these are from the Coniacian of Uzbekistan and each is known by a few teeth and jaw fragments. Two somewhat older Asiatic mammals may be deltatheroidans; we refer them to the group based on their large size (deltatheroidans tend to be large relative to their contemporaries). *Khuduklestes* Nessov, Sigogneau-Russell, and Russell, 1994 is known by an axis vertebra from the ?Cenomanian of Gansu Province, China; *Oxlestes* Nessov, 1982, is represented by an axis and possibly associated parietal and canine fragment (see Nessov, Sigogneau-Russell, and Russell, 1994, and references cited therein).

Far more informative specimens have been collected from Mongolia. Jaws and partial skulls have been described for *Deltatheroides* (poorly known) and *Deltatheridium* (Gregory and Simpson, 1926; Kielan-Jaworowska, 1975c; Rougier et al., 1998). Additional skulls and skeletons of *Deltatheridium* have reportedly been collected by the American Museum–Mongolian Expeditions, but thus far only a calcaneus has been mentioned briefly in the literature (Horovitz, 2000). Both *Deltatheroides* and *Deltatheridium* are known from the Campanian of Mongolia; *Deltatheridium*, represented by *D. nessovi* Averianov, 1997, is also known from the early Campanian of Kazakhstan.

An unnamed mammal from the latest Cretaceous of Mongolia is represented by a cranium referred to here, as in other literature, as the "Guriliin Tsav skull." The specimen was illustrated (as *Deltatheridium*) in a brochure describing the activities of the Paleontological Institute, Russian Academy of Sciences, Moscow (Anonymous, 1983). Kielan-Jaworowska and Nessov (1990) commented on the specimen, considering it deltatheroidan (but not *Deltatheridium;* also see comments in table 12.1). The skull was subsequently illustrated by Szalay and Trofimov (1996). Pending a full description, a reasoned appraisal of the affinities of this important specimen cannot be given. However, analyses presented by Rougier et al. (1998) suggest that it may represent a marsupial, and we tentatively place it there.

North American deltatheroidans are known only from isolated teeth. The geologically oldest of these is tentatively-referred *Atokatheridium* (upper and possible lower molar), from the Aptian–Albian of Oklahoma (Kielan-Jaworowska and Cifelli, 2001). A fragmentary lower molar from the Turonian of Utah was described by Cifelli (1990c) as an indeterminate deltatheridiid. Upper molars similar to those of *Deltatheroides* were reported by Fox (1974a) from the Judithian of Alberta and the Lancian of Wyoming. Fox and Naylor (1995) mentioned a new deltatheridiid from the Late Cretaceous of North America, but this has not yet been described.

ANATOMY

SKULL

There are only two reasonably complete skulls from the Cretaceous that may belong to marsupials. Data available for these are rather limited and not very useful for present purposes. The skull of *Asiatherium* (described by Szalay and Trofimov, 1996) is nearly complete but badly flattened, with a poorly preserved basicranium. In the following account, we draw from anatomy of *Mayulestes* (figures 12.1A, 12.2) and *Pucadelphys,* well-known taxa from the early Paleocene of Bolivia (Marshall and Muizon, 1995; Muizon, 1998), together with extant *Didelphis* (figure 12.1B), generally considered to be plesiomorphic in skull structure. We focus mainly on characters that appear to differentiate the skull of basal marsupials from that of basal eutherians.

In general shape, the skull of early marsupials appears to have approximated that of later omnivorous-carnivorous forms such as *Didelphis,* with strong sagittal and lambdoidal crests, high zygomatic arch, and substantial postorbital constriction. Judged from the dentition, small taxa such as *Aenigmadelphys* might have had a total skull length of less than 25 mm. Skull length is about 30 mm in *Asiatherium* and 65 mm in the Guriliin Tsav specimen; species of *Didelphodon,* the largest of Cretaceous marsupials, may have had a skull 100 mm or more in length.

Among therians, the presumed plesiomorphic orientation of the occipital plate is posteroventrally sloping from the lambdoidal crests, as seen in early eutherians (see figure 13.2). In *Asiatherium,* the occipital plate (which is badly damaged) has been reconstructed as roughly vertical; in the Guriliin Tsav skull and the Paleocene marsupials *Pucadelphys* (see Marshall and Muizon, 1995) and *Mayulestes* (figure 12.2B), it appears to slope posteroventrally, though less so than in early eutherians. In living *Didelphis,* the occipital plate slopes anteroventrally.

The palatal part of the premaxilla bears incisive foramina that are strongly elongated, reaching the boundary with the maxilla. This is in apparent contrast to early eutherians, where the incisive foramina (known only in *Asioryctes*) are small openings lying in the premaxilla-maxilla suture. Posterior to this suture in marsupials is a well-developed diastema to accommodate the lower canine. Palatine vacuities or fenestrae are variable: they are large in *Didelphis* but lacking in *Pucadelphys* and *Mayulestes.* Similar variability is encountered among early Eutheria. Unlike eutherians, the pterygoids of marsupials are not fused to the alisphenoids. The foramen ovale lies lateral to the carotid foramen; a foramen rotundum is also present. A transverse canal variably occurs between the foramen rotundum and the foramen ovale (Archer, 1976): it is present in *Didelphis* but lacking in *Mayulestes* and *Pucadelphys* (see Marshall and Muizon, 1995; Muizon, 1998).

The alisphenoid is relatively larger than in eutherians, and in some genera (e.g., *Mayulestes, Didelphis*) the posterolateral process of the alisphenoid contributes to the glenoid fossa, which is never the case in Eutheria. The jugal contributes to the lateral side of the glenoid. Another condition of the alisphenoid, widespread among Marsupialia but not encountered among Eutheria, is the presence of a posterior extension (tympanic process) that reaches the auditory region. This is commonly termed an alisphenoid bulla, though it does not fully conceal the petrosal. The horseshoe-shaped tympanic fits into the incurvature of the bulla, but does not fuse to it. The presence of an alisphenoid bulla has long been considered a diagnostic marsupial character (e.g., Clemens, 1979a). However, it is lacking in *Pucadelphys* and *Mayulestes* and appears to have evolved several times within Marsupialia (Muizon, 1994). Among Cretaceous taxa, an alisphenoid bulla is present in both the Guriliin Tsav skull (Kielan-Jaworowska and Nessov, 1990) and *Asiatherium* (see Szalay and Trofimov, 1996), though in the latter it is poorly preserved and was apparently very small. In *Asiatherium,* another process is present, identified

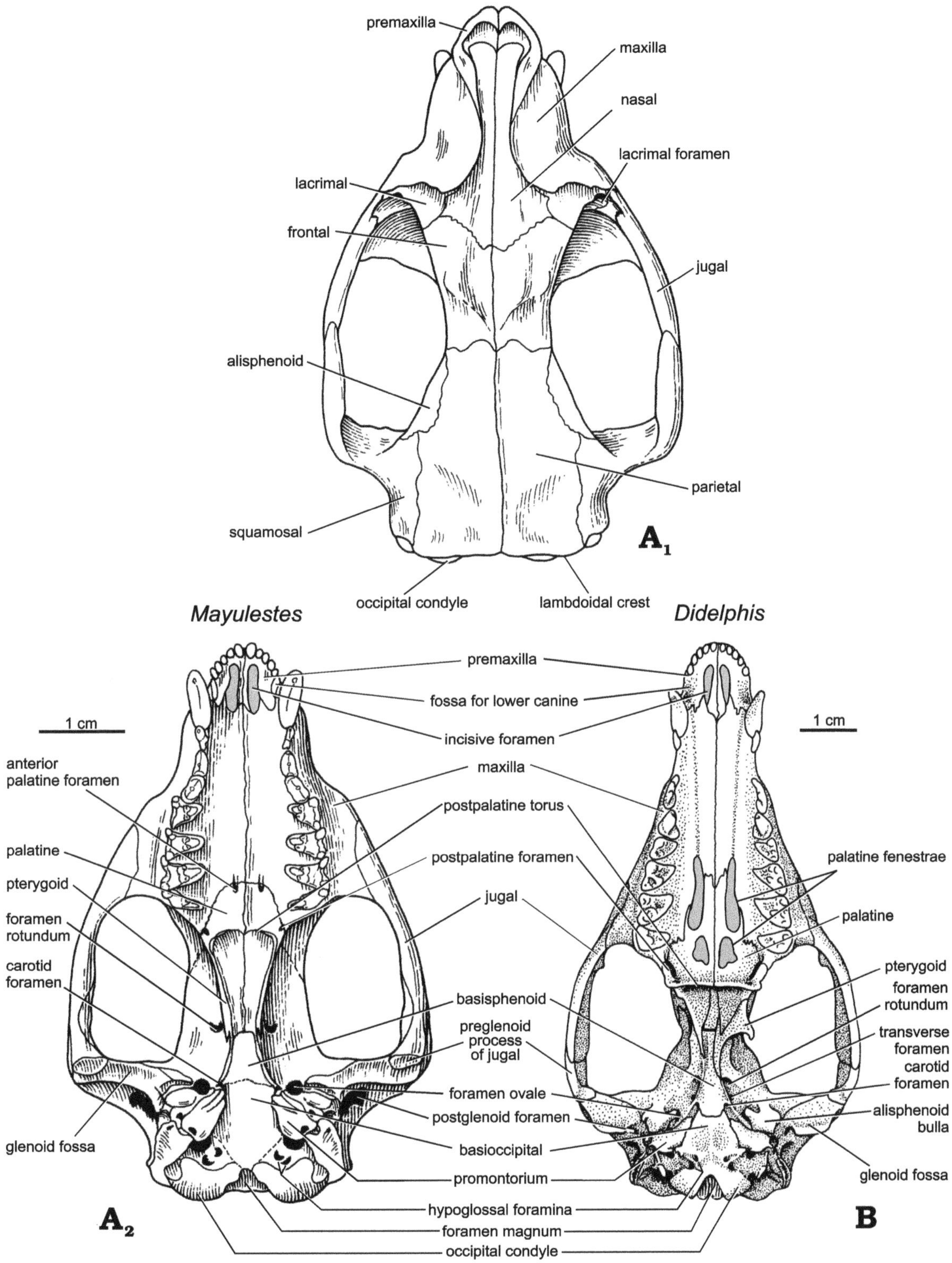

FIGURE 12.1. Marsupial skulls. A, Early Paleocene *Mayulestes ferox*, skull in dorsal (A_1) and ventral views (A_2). B, Extant *Didelphis virginiana*, skull in ventral view. *Source:* A, modified and simplified from Muizon (1998); B, modified from Jollie (1962).

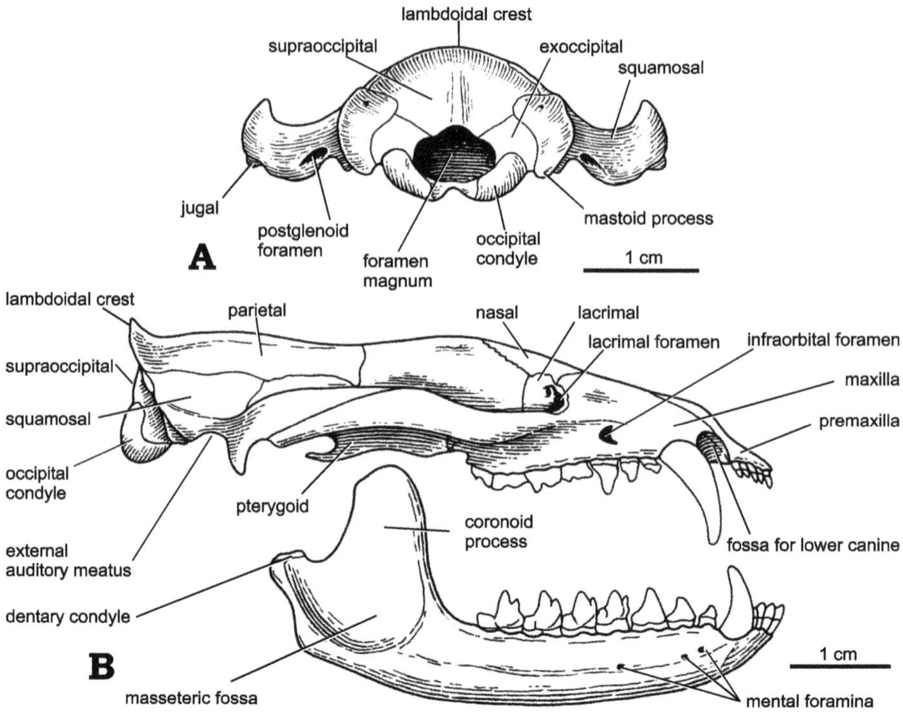

FIGURE 12.2. Early Paleocene *Mayulestes ferox*. A, Skull in occipital view. B, Skull and dentary in lateral view. *Source:* modified and simplified from Muizon (1998).

by Szalay and Trofimov (1996) as a rostral tympanic process of the petrosal. Medioventrally, this process partly overlaps the tympanic process of the alisphenoid.

Petrosals are known for several Cretaceous marsupials, including *Eodelphis, Didelphodon,* and several undetermined taxa (Matthew, 1916; Clemens, 1966; Archibald, 1979; Wible, 1990; Meng and Fox, 1995c, see figure 12.3). Some features of the basicranium and auditory region have also been described for *Deltatheridium* (see Rougier et al., 1998). These works, particularly the cladistic analyses of Wible (1990) and Rougier et al. (1998), identify some distinctive differences between marsupials and eutherians in the structure of the petrosal. One such difference is the lack of a groove for the stapedial artery on the promontorium of marsupials and *Deltatheridium.* In extant marsupials, the stapedial artery is lost ontogenetically (Wible, 1990). Polarity for this character is uncertain. A stapedial sulcus is present in *Vincelestes* (see Rougier et al., 1992), but lacking in the stem trechnotherian *Zhangheotherium,* triconodontids, and most stem mammals, at least (Luo et al., 2002). Living marsupials also have the internal carotid artery placed in a median position; this primitive condition was apparently present in Cretaceous marsupials as well, inferred from the lack of a groove for this vessel on the promontorium. A sulcus on the anterior pole of the promontorium was identified as part of the course for the internal carotid in early Paleocene *Pucadelphys* and *Mayulestes* (Marshall and Muizon, 1995; Muizon, 1998). However, the internal carotid sulcus is not present on the main part of the promontorium in these or other fossil marsupials, as in many eutherians (chapter 13). Other characteristic features of the petrosal in early marsupials include a prominent rostral tympanic process, a reduced prootic canal (which transmits the lateral head vein) with intramural opening, the indicated presence of a sphenoparietal emissary vein, and the placement of the prootic sinus within a deep sulcus between the petrosal and squamosal (figure 12.3).

The facial part of the rostrum is dominated by very extensive maxillae, which in some fossil taxa tend to cover the lateral parts of the nasals (figure 12.1A_1). The nasals are thus narrow anteriorly but strongly expanded posteriorly. The lacrimal is large, as are the parietals, which constitute the entire posterior part of the skull roof. A postparietal bone is usually lacking in marsupials, though it is present and fused with the supraoccipital in *Didelphis.* The braincase of *Didelphis* is relatively smaller than in living eutherians of similar size, though early Paleocene marsupials have a braincase of comparable size to that of Late Cretaceous eutherians (see, e.g., figures 12.1A and 13.7 for comparisons).

Not much has been published on the skull structure of Deltatheroida. The material described before 1998 consists of incomplete rostra and dentaries (Gregory and Simpson, 1926; Kielan-Jaworowska, 1975c). Additional, more complete specimens are now available; illustrations and preliminary description of the rostrum, dentary, and petrosal are given by Rougier et al. (1998, and figure

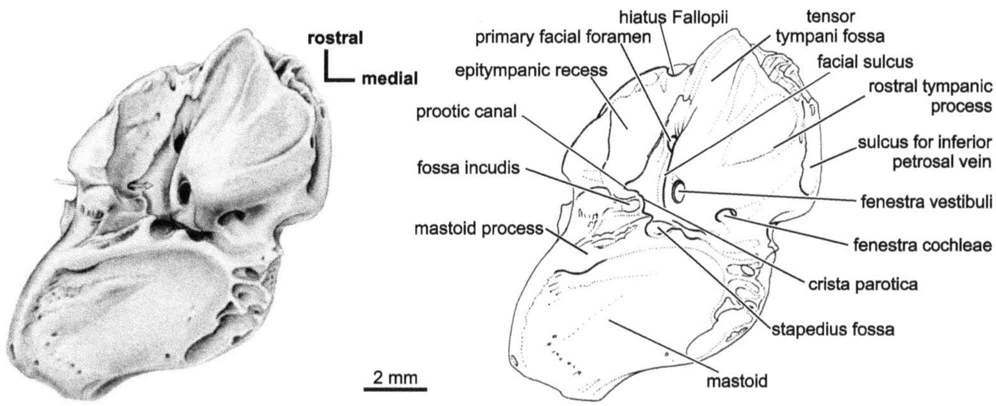

FIGURE 12.3. Right marsupial petrosal from the Late Cretaceous Lance Formation of Wyoming, in tympanic view (type A of Wible, 1990). *Source:* modified and simplified from Wible (1990).

12.4B,C herein). The account given below is based on published cranial material of *Deltatheridium.*

The skull (figure 12.4A,B) has a relatively short snout that is narrow and parallel sided anteriorly, widening above P3. The premaxilla has a posteriorly directed process that reaches the alveolus of the canine. The anterior margin of the orbit is placed above the P3–M1 embrasure. The infraorbital foramen, similarly placed, is relatively large. The nasals are parallel sided anteriorly and strongly expanded posteriorly, contacting the lacrimals. The nasofrontal suture is convex posteriorly. The lacrimal has an extensive exposure on the dorsolateral part of the face, entering with a pointed end between the nasal and the maxilla. The lacrimal foramen is situated at the emargination of the lacrimal bone. The zygomatic arches are strong and notably expanded laterally. The jugal extends anteriorly to the level of the anterior margin of the lacrimal. There are no palatal vacuities.

The deltatheroidan braincase has not been figured. It appears from the preliminary report by Rougier et al. (1998) that there is an extensive squamosal, while the anterior lamina of the petrosal does not contribute to the lateral wall of the braincase. The petrosal (figure 12.4C) resembles those of early marsupials and differs from those of Cretaceous eutherians (Wible, 1990; Wible and Hopson, 1993; Meng and Fox, 1995b, see also description of eutherian petrosal in chapter 13). Most noteworthy in this respect are the absence of vascular sulci on the promontorium, indicating reduction or complete absence of the stapedial arterial system (polarity uncertain), and the presence of a small, horizontally oriented prootic canal connected to the postglenoid venous system.

MANDIBLE

The dentary of extant and fossil marsupials differs from that of most eutherians in having an inflected angular process. The dentary of *Deltatheridium* (Rougier et al., 1998) bears an angular process that forms a shelflike, inflected projection, as in some marsupials and in contrast to the general condition in eutherians. The significance of this deltatheroidan-marsupial similarity is open to differing interpretations. It may be a synapomorphy of the two groups (Rougier et al., 1998). On the other hand, early eutherians, where known, often bear a rodlike angular process that is inflected (e.g., Lillegraven, 1969; Kielan-Jaworowska, Bown, and Lillegraven, 1979; Cifelli, 1999b; Cifelli and Madsen, 1999), according to the definition of Sánchez-Villagra and Smith (1997: 120). The angular process of extant marsupials is variably developed, ranging from rodlike to shelflike. The plesiomorphic condition cannot be inferred by distribution of states among living taxa. Given that the rodlike angle is correlated with an animalivorous diet (Sánchez-Villagra and Smith, 1997), however, this condition may well be the ancestral state for marsupials. Averianov and Kielan-Jaworowska (1999) distinguished the in-turned angle of early eutherians as projecting ventrally and hence being visible in labial view, but this condition is also widespread among early marsupials (e.g., Simpson, 1929a: figure 48; Clemens, 1966: figures 60c, 65b; Lillegraven, 1969: figure 23, 1b; Szalay and Trofimov, 1996: figure 3; Muizon et al., 1997: figure 1d).[1]

[1] It should be noted that the angular process is also inflected in multituberculates and the posteroventral corner of the bone is completely rounded in lateral view. This type of inflection is, however, very different from that in metatherians and some early eutherians. In multituberculates the process is inflected together with the whole ventral margin of the dentary to form the extensive shelf for insertion of the pterygoideus muscles (see Gambaryan and Kielan-Jaworowska, 1995, and chapter 8). It is thus possible that the inflection of the angular process took place independently three times among Mesozoic mammals.

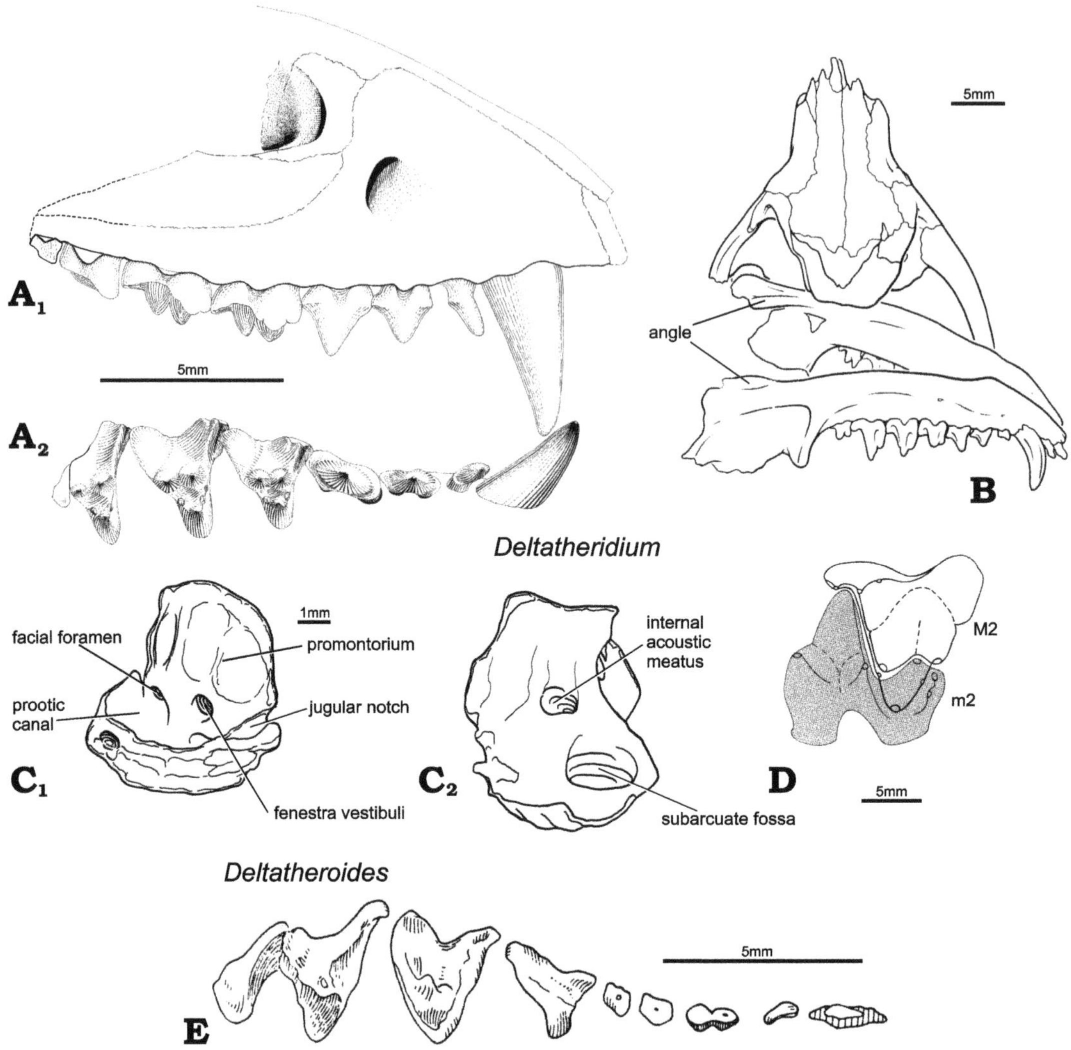

FIGURE 12.4. Skulls and teeth of selected deltatheridiids. A–D, *Deltatheridium pretrituberculare,* reconstruction of the rostrum in lateral view (A_1) and left upper dentition of the same (incisors not preserved) in occlusal view (A_2). B, Anterior part of the skull in dorsal view, associated with both dentaries, overturned. C, Diagrammatic drawings of the right petrosal in ventral view (C_1) and inverted left petrosal in endocranial view (C_2). D, Oblique view of M2 and m2 in occlusion. E, *Deltatheroides cretacicus,* right dentition in occlusal view (holotype). *Source:* A, modified from Kielan-Jaworowska (1975c), using data from Rougier et al. (1998). B, modified from Rougier et al. (1998). C, original, based on stereophotographs published by Rougier et al. (1998). D, modified from Kielan-Jaworowska (1975c). E, modified from Gregory and Simpson (1926).

The oldest marsupial dentary is that of Albian–Cenomanian *Kokopellia* (Cifelli, 1993a; Cifelli and Muizon, 1997), and this element is reasonably well represented among Late Cretaceous marsupials, being known for Asiatheriidae (Szalay and Trofimov, 1996), Stagodontidae (Clemens, 1968a; Fox and Naylor, 1986), "Alphadontidae," and "Pediomyidae" (Clemens, 1966; Lillegraven, 1969). In all essential respects, the dentary of early marsupials (figures 12.5C, 12.9B,C, 12.11D, 12.13G, 12.14C) is fully modern in appearance. The symphysis was unfused; in *Kokopellia,* it is more horizontally oriented and elongate than is typical of living didelphids of comparable size. Mental foramina are variable, but are often found below p2 and m1. A labial mandibular foramen, absent in most marsupials as it is in Deltatheroida (Marshall and Kielan-Jaworowska, 1992), is present in *Kokopellia,* as may be the case in Cretaceous *Alphadon* and Miocene *Microbiotherium* (Sinclair, 1906; Marshall, 1982; Cifelli and Muizon, 1998a). The phylogenetic history of this feature, which is variably present elsewhere among early mammals, is poorly understood (Cifelli and Muizon, 1997). In all early marsupials for which the ascending ramus of the dentary is known, it ascends subvertically, and the dentary condyle resembles that of modern "didelphoids."

Lingually, a faint trace of a Meckel's groove may be present in *Kokopellia* (figure 12.8A), as it is in Paleocene marsupials from Bolivia (Marshall and Muizon, 1995; Muizon, 1998). As shown elsewhere in this book, Meckel's groove is a primitive character that is widely distributed among groups of early mammals and was lost independently numerous times (e.g., Cifelli and Madsen, 1999). Among relevant comparative taxa, a somewhat stronger Meckel's groove is present in *Kielantherium* (Dashzeveg and Kielan-Jaworowska, 1984) and in the early eutherian *Prokennalestes* (Kielan-Jaworowska and Dashzeveg, 1989), but is lacking in Deltatheroida (Kielan-Jaworowska, 1975c). There are no traces of scars for attached postdentary elements, from which we assume that the middle ear was of fully modern aspect. Similarly, there is no trace of the coronoid bone, which was apparently retained in some early eutherians (e.g., Kielan-Jaworowska, 1981; Kielan-Jaworowska and Dashzeveg, 1989; Nessov, Sigogneau-Russell, and Russell, 1994) and the stem boreosphenidan *Kielantherium* (Dashzeveg and Kielan-Jaworowska, 1984). Like Meckel's groove, a coronoid (or scar indicating its presence) is widespread among early mammal groups and was apparently lost independently in various lineages (e.g., Cifelli and Madsen, 1999). The pterygoid fossa is well developed, and the angle of the dentary (where known) forms a prominent, inflected shelf.

The deltatheroidan dentary (figure 12.4B) is structurally similar to those of early marsupials and is typical of carnivorous mammals (e.g., Maynard Smith and Savage, 1959). The coronoid process is tall and the condyle is placed low, near the level of the tooth row. The masseteric fossa is deep and bordered by salient masseteric crests; the anterior margin of the masseteric fossa is placed just posterior to the last molar. A notable feature on the lingual side of the dentary is the presence of an inflected and shelflike angular process, as in marsupials (see earlier). The dentary is considerably foreshortened mesial to the large canine.

DENTITION

Dental Formula: Incisors, Canines, and Premolars. Living didelphoids have five upper and four lower incisors, presumed to represent the primitive condition (Clemens and Lillegraven, 1986). Five upper and four lower incisors are also seen in early Paleocene didelphoids such as *Pucadelphys* (see Marshall and Muizon, 1995). The upper incisor series is not known for Cretaceous marsupials. Four lower incisors are known to be present in Albian–Cenomanian *Kokopellia* (see Cifelli and Muizon, 1997) and some specimens, at least, of *Alphadon* (Clemens and Lillegraven, 1986). Hershkovitz (1982, 1995) suggested that the ancestral count was five lower incisors, with Marsupialia being characterized by reduction of the anterior part of the jaw, resulting in loss of i1 and "staggering" at the second (interpreted as i3) of the remaining tooth positions. He further observed an apparently staggered pattern in an edentulous dentary from the Early Cretaceous, thus implying great antiquity for this incisor configuration and for Marsupialia in general. The few Cretaceous taxa in which the condition can be judged—*Alphadon, Kokopellia,* and *Eodelphis*—lack staggering of the incisor series. The presence of staggered incisors in early Paleocene taxa from South America suggests that the condition may be a derived feature within (rather than of) Marsupialia (Cifelli and Muizon, 1997). Judged by the alveoli, the first three lower incisors of *Kokopellia* were subequal in size, with the fourth perhaps somewhat reduced (Cifelli and Muizon, 1997). It is possible that at least one species of *Alphadon* had a (reduced) count of three incisors, with the tooth at the second position considerably enlarged (Cifelli and Muizon, 1998a, see figure 12.5C). Among Cretaceous marsupials for which incisors are known, Stagodontidae appear to be most variable and to have the most notable specializations. Three incisors, with the second enlarged, are present in Judithian *Eodelphis* (see Matthew, 1916; Simpson, 1929a). Three subequal incisors appear to have been present in Lancian *Didelphodon vorax* (see Clemens, 1966), whereas its "Edmontonian" congener, *D. coyi,* had only one lower incisor (Fox and Naylor, 1986).

Little can be said about the canine of Cretaceous marsupials except that, where known, it is usually well developed in both upper and lower jaws. Notably, the canine is invariably single-rooted. This contrasts with the condition in early Eutheria (chapter 13) and most other groups of early mammals, in which the canine is variably or usually double-rooted, sometimes varying in root count between the deciduous and replacement tooth (see Clemens and Lillegraven, 1986; Cifelli, 1999a).

Three upper and lower premolars, and four upper and lower molars, are present in all Mesozoic marsupials for which the postcanine tooth formula is known. We take these counts to represent the primitive metatherian conditions. Establishing the homologies of these teeth with respect to those of other mammals (eutherians, stem boreosphenidans, and proximal relatives of Boreosphenida) and specifying exactly what (if anything) is derived about the marsupial postcanine dental formula is hampered by a paucity of ontogenetic data for fossil taxa, among other things (see detailed reviews by Clemens and Lillegraven, 1986; Luckett, 1993, and the section "Tooth Replacement" that follows). Based on comparison with *Kielantherium* (the most dentally primitive boreo-

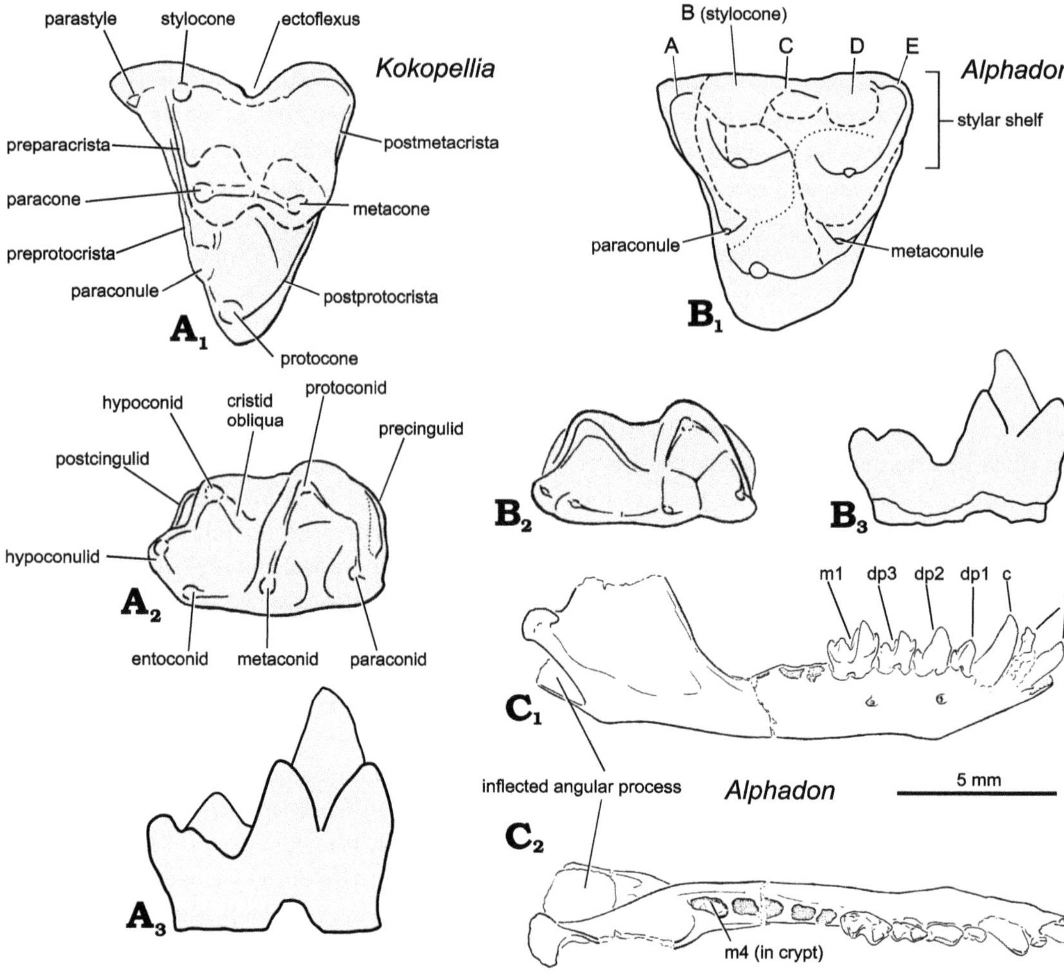

FIGURE 12.5. Structure of the molars and dentary in early marsupials. A, Molars of *Kokopellia juddi*, illustrating presumed basal conditions of Marsupialia: (A_1) left upper molar in occlusal view, (A_{2-3}) left lower molar in occlusal (A_2) and lingual (A_3) views. B, Molars of *Alphadon* sp., illustrating apomorphies typical of many latest Cretaceous marsupials: (B_1) left upper molar in occlusal view, showing the cusps (labeled A–E) of the stylar shelf, (B_{2-3}), left lower molar in occlusal (B_2) and lingual (B_3) views. C, *Alphadon eatoni* (holotype), right dentary with two incisors, c, dp1–3, m1, illustrating features of the dentary and tooth replacement: (C_1) labial view, (C_2) occlusal view. *Source:* A, modified, with additions, from Cifelli and Muizon (1997). B_2, modified from Cifelli (1993a). C, modified from Cifelli et al. (1996).

sphenidan that is reasonably well known, see Dashzeveg and Kielan-Jaworowska, 1984) and proximal relatives with nontribosphenic molars (chapter 10), the primitive count for Boreosphenida appears to include four or more premolars and either four or (less probably) three molars (see supplementary information to Rougier et al., 1998; Butler and Clemens, 2001). This suggests that Metatheria may retain a primitive molar count and are probably apomorphic in having reduced the number of premolars from four (or more) to three. We follow convention in referring to these teeth as P/p 1–3 and M/m 1–4, while again emphasizing the fact that problems remain in establishing homologies (e.g., Archer, 1978; Luckett, 1993).

With few exceptions, the premolars of Mesozoic marsupials resemble those of living didelphids in having rather simple, lanceolate crowns dominated or formed exclusively by a single cusp (paracone or cusp A on uppers; protoconid or cusp a on lowers), though a small anterobasal cuspule (B/b) and posterobasal cuspule or heel (C/c) are usually present. A short diastema commonly (but not invariably) separates p1 from the preceding canine and succeeding p2. The premolars increase in size, most notably in height, from first to third. The last premolar is less anteriorly recumbent than its predecessors, and the fact that its crown is markedly taller than those of the succeeding molars suggests that it may be serially homologous to the last premolariform tooth of nontribosphenic *Peramus* (chapter 10) and to the penultimate premolar of eutherians (McKenna, 1975; Luckett, 1993). Premolars of certain large "alphadontid" taxa such as *Turgidodon* have moderately inflated crowns, but the most significant variance from the primitive pattern (seen, e.g., in *Kokopellia*) is encountered among Stagodontidae (Clemens, 1966, 1968a). Here, the first two premolars tend to reduction (P/p1 is double-rooted in some *Eodelphis*, but becomes single-rooted in other *Eodelphis* and in *Didelphodon*), whereas P/p3 is enlarged, robust, inflated, and bears a somewhat more complex crown—specializations presumably related to durophagy (Fox and Naylor, 1986).

Fossil incompleteness long hampered interpretation of the dental formula in deltatheroidans. Newly collected specimens establish the formula in *Deltatheridium* as 4.1.3.4/3.1.3.4 (Rougier et al., 1998); the same postcanine dental formula is present in *Deltatheroides* (see Kielan-Jaworowska, 1975c). Both the incisor and premolar counts are presumed to represent reductions from the ancestral condition in Boreosphenida; reduction to three premolars is an apomorphy shared with Marsupialia. The four upper incisors of *Deltatheridium* are followed by a very strong, single-rooted canine. The presence of a single canine root is a similarity to marsupials. The three upper premolars of *Deltatheridium* increase in size posteriorly. P1 is peglike and single-rooted; P2–3 are triangular in labial view, double-rooted, single-cusped, and bear anterior and posterior cingula (more extensive on P3 than P2). The first lower incisor of deltatheroidans, but presumed to be i2 of *Deltatheridium*, was clearly the largest and has been described as "staggered" (Rougier et al., 1998), a condition known in many advanced marsupials (Hershkovitz, 1995), though not in stem taxa (Cifelli and Muizon, 1997, 1998a, and above). Both the second and third lower incisors are spatulate. The lower canine, like the upper, is very strong and single-rooted. All three lower premolars are double-rooted. They increase in size through the series; p1 is very small and is oriented obliquely with respect to the longitudinal axis of the dentary. Both p2 and p3 bear tall main cusps and small distal cuspules.

Molars. Molars of latest Cretaceous marsupials are generally distinguishable from those of contemporary eutherians and the more primitive boreosphenidans in a number of characteristics (Clemens, 1979a). The situation is somewhat less clear for earlier, more primitive marsupials. As represented by the Albian–Cenomanian marsupial *Kokopellia* (figure 12.5A_1), the upper molars differ from those of most "tribotheres" and Deltatheroida in that the protocone is more transversely and anteroposteriorly expanded, the conules somewhat better defined and placed somewhat closer to the bases of paracone and metacone, respectively; and the postprotocrista extends labially past the base of the metacone. A presumed deltatheroidan specialization, the hyperdevelopment of the distal stylar shelf (correlated with emphasis on postvallum/prevallid shearing, see Kielan-Jaworowska and Cifelli, 2001), is lacking. Although these features serve to distinguish early marsupials from other presumed metatherians (Deltatheroida) and from most "tribotherians," they are not particularly diagnostic, especially insofar as early Eutheria (such as *Eomaia, Prokennalestes,* and *Murtoilestes*) are concerned.

Geologically younger taxa, for which *Alphadon* (figure 12.5B_1) serves as a useful example, are more similar to living "didelphoids" in the structure of the upper molars. The paracone (primitively taller than metacone) is subequal to the metacone, or (more commonly) the latter is the taller of the two. Compared to *Kokopellia,* the stylar shelf is proportionately narrower and the protocone more transverse; the protocone is notably more anteroposteriorly developed as well. In Late Cretaceous taxa such as *Alphadon,* the conules are typically placed very close to the bases of paracone and metacone, are well developed, and generally bear strong cristae (features that, except for position, tend to be reversed in later didelphoids, in which conules are variably lacking).

A feature that has attracted broad attention among students of marsupial evolution is the presence of a series of cusps distributed along (or near) the margin of the stylar shelf and termed stylar cusps (figure 12.5B_1). An extended discussion of these cusps was presented by Clemens (1979a), so we give only a basic account of them here, integrating new information made available in the past two decades. The number of stylar cusps is variable, but five major positions are generally recognized. Following the Bensley-Simpson system (see Bensley, 1903, 1906; Simpson, 1929a) as redefined by Clemens (1966), these are labeled as stylar cusps A–E, from the mesial to distal part of the shelf (figure 12.5B_1). Unfortunately, other lettering schemes for molar cusps exist (see the section "Terminology" in chapter 10 and figure 10.2). As noted elsewhere (chapters 4, 7, 9), the same alphabetic designations are currently employed for the primary cusps of the upper molar (in which cusp A, e.g., is the paracone). For this reason, we recommend that usage of the Bensley-Simpson system include clear indication that reference is being made to cusps of the stylar shelf.

We emphasize our deliberate use of the term "positions" implicit in the definitions of these cusps: this labeling system was developed mainly for descriptive purposes, not necessarily to imply homology (Bensley, 1903, 1906; Simpson, 1929a; Clemens, 1966). Only one of the five, stylar cusp B (stylocone), is readily recognized as having a homologous counterpart on molars of other groups of Mesozoic mammals, including basal taxa. The stylocone is believed to be one of the three primary cusps of the earliest mammals (Patterson, 1956); it is connected to the paracone by a primary shearing crest, the preparacrista. By fortunate coincidence, this cusp bears the same letter designation, B, in "symmetrodontans" and structurally similar mammals. Stylar cusps A and E, lying, respectively, at the mesiolabial and distolabial corners of the tooth, may also be, respectively, termed "parastyle" and "metastyle." Their homologies with similarly placed cusps of other mammals are debatable (e.g., Cifelli and Madsen, 1999): in early eutherians, for example, at least two cusps may be present in the parastylar region (chapter 13). In marsupials, parastylar and metastylar projections typically overlap between adjacent molars and serve as a mechanical interlocking mechanism for the upper molar series.

There are varied interpretations to explain the frequent presence of cusps in the C (between paracone and metacone) and D (labial to the metacone) positions (Clemens, 1979a). The hypothesis that an *Alphadon*-style configuration of the stylar shelf (with cusps present in both the C and D positions) was primitive for Marsupialia (Clemens, 1966, 1968b) received support with the description of Aptian–Albian *Holoclemensia* (see chapter 11) and early Campanian *Albertatherium*. *Holoclemensia* was originally regarded as a marsupial (Slaughter, 1968a,c, 1971), but it does not share any obvious marsupial apomorphies and is generally regarded as a boreosphenidan of uncertain affinities (chapter 11). *Holoclemensia* has a well-developed cusp in the C position, as does *Albertatherium*, which was originally interpreted as a possible descendant of *Holoclemensia* (see Fox, 1971b). This interpretation was subsequently challenged by Fox (1987a,b; see also Clemens, 1979a), who hypothesized that stylar cusp C was ancestrally lacking in marsupials. This view has since received corroborative support (Cifelli, 1990b, 1993a). A cusp in the D position is widespread in marsupials, and a similarly placed cusp is present in certain early eutherians such as *Paranyctoides* and *Prokennalestes* (chapter 13). Fox (1987b) interpreted the respective marsupial and eutherian cusps as homologous and suggested that a D cusp was present in their common ancestor (Crown Theria). However, both C and D cusps, as such, are lacking in a number of early and otherwise primitive marsupials (such as *Kokopellia*), as they are in deltatheroidans. Thus, present evidence suggests that cusps in these positions were acquired later in marsupial evolution than were other apomorphic features, such as dental formula and characteristics of the lower molars (Cifelli, 1993a,b). Even where cusps C and D are present, their homologies within Marsupialia cannot always be presupposed (e.g., Fox, 1987b). Scenarios explaining models of evolution among cusps of the stylar shelf (e.g., Marshall et al., 1990) cannot be adequately tested with the existing fossil record. A thorough, recent treatment of stylar cusps in Cretaceous marsupials is given by Johanson (1996b).

Among Cretaceous marsupials, one of the most significant variations in morphology of upper molars is reduction to loss of the anterior part of the stylar shelf. This appears to have occurred independently, at least twice, among taxa referred to "Pediomyidae" (Fox, 1987a,b). Unlike eutherians, where development of protoconal cingula and attendant cusps (such as the hypocone) is common, protoconal cingula are rare in early marsupials. A small preprotoconal cingulum is present in *Glasbius;* both cingula, including a hypertrophied postprotoconal cingulum, are seen in the enigmatic Mongolian genus *Asiatherium*. In stagodontids, the distal stylar shelf is expanded and the paracone somewhat reduced, in correlation with emphasis on postvallum/prevallid shearing. However, a large stylocone is retained.

Upper molars of Deltatheroida (figures 12.4A,E, 12.7A,B) bear a protocone that is small, both transversely and mesiodistally. Conules are faint and generally placed closer to the protocone than to the respective bases of paracone and metacone; the postprotoconal cingulum does not extend labially past the base of the metacone. The stylar shelf is transversely broad, especially the distal lobe, which bears a prominent elongate postmetacrista. Except for a well-developed stylocone (cusp B), stylar cusps as such are lacking, though small variable cuspules are present along the rim of the stylar shelf in *Sulestes* (figure 12.7B). The strong elongate postmetacrista and the transverse development of the distal stylar shelf appear to be apomorphies (Kielan-Jaworowska and Cifelli, 2001), presumably related to carnivory (e.g., Muizon and Lange-Badré, 1997). This may also be the case for other features of deltatheroidan upper molars, just noted, though in most respects the conditions approximate what might be expected in the ancestral boreosphenidan.

Kielan-Jaworowska (1975c), Fox (1975), and Crompton and Kielan-Jaworowska (1978) discussed the molar shearing surfaces of *Deltatheridium*. Of those recognized by Crompton (1971) for primitive tribosphenic molars (see chapter 11 and figure 11.3), surface 1a (on the preparacrista) is well developed, a condition that is presumably primitive. Notably, however, surface 2a (on the postmetacrista) is strongly enlarged. This is an advanced condition, correlated with the great width of the distal part of the stylar shelf in deltatheroidans.

Lower molars, though less complex than the uppers, paradoxically appear to be somewhat more useful for recognition of marsupial affinities among the geologically older taxa. As represented by *Kokopellia*, for example (figure 12.5A_{2-3}), the paraconid and metaconid in early marsupials tend toward subequal development (probable plesiomorphy), differing on the one hand from Deltatheroida, in which the paraconid is hypertrophied, and on the other from Eutheria, where (except for *Montanalestes*), there is an early trend toward reduction of the paraconid, which commonly becomes appressed to the metaconid (chapter 13). The paraconid of marsupials bears a noticeable keel (?apomorphy) on its mesiolingual edge, rather than being gently rounded, as is generally true of other boreosphenidans (with the exception of *Murtoilestes*). This feature may be related to interlocking of the molar series in correlation with changes in the configuration of the talonid. The distal metacristid (see figure 11.1B), characteristic of stem boreosphenidans and Deltatheroida (Fox, 1975), is lacking; the cristid obliqua is fully developed and attaches to the trigonid below the notch between the paraconid and the metaconid. Unlike most stem Boreosphenida and Deltatheroida, the three talonid cusps are well developed, with a large talonid basin, and the talonid is generally as wide or wider than the trigonid. The hypoconulid and entoconid are joined by a crest, but are not "twinned" (see later) in primitive taxa. Perhaps most diagnostic of the talonid among otherwise plesiomorphic taxa such as *Kokopellia* is the presence of a labial postcingulid, extending mesiolabially from the apex of the hypoconulid to a point near the base of the hypoconid.

Lower molars of "typical" Late Cretaceous marsupials (such as *Alphadon*, figure 12.5B_{2-3}) bear additional specializations of the crown group that are absent in basal taxa (e.g., *Kokopellia*). There is a lesser height differential between trigonid and talonid cusps. The paraconid achieves an extremely lingual position, so that it is in line with the metaconid and entoconid (Clemens, 1979a). Most characteristic is the fact that the hypoconulid is lingually placed and closely integrated with (sometimes almost indistinguishable from) the entoconid. The two cusps are prominent, often projecting above the hypoconid (primitively the tallest cusp of the talonid) in labial view.

Among Late Cretaceous marsupials, a common deviation from the presumed primitive pattern is the labial displacement of the attachment of the cristid obliqua to the trigonid. This modification, characteristic of "pediomyids," is also seen in *Glasbius* and advanced Stagodontidae (*Didelphodon*). Stagodontids show another specialization of the lower molars: hypertrophy of the paraconid, so that it is noticeably taller than the metaconid.

Lower molars of Deltatheroida (figures 12.4D, 12.7A_2, A_3,B_2,B_3,C) are dominated by their trigonids, the respective talonids being much lower and narrower. Cingulids as such are generally lacking, though anteromesial cuspule e and anterolabial cuspule f are variably present. The most notable features of the lower molars are the relatively tall paraconid and short metaconid (a presumed apomorphy associated with emphasis on postvallum/prevallid shearing), the poorly developed talonid basin, the weak (variably lacking) entoconid, and the presence of a distal metacristid (presumed plesiomorphies). The first lower molar is considerably smaller than m2–3; m4 is strongly reduced and is much lower than the preceding molars.

Tooth Replacement. The pattern of tooth replacement is poorly known in early mammals (e.g., Cifelli, 1999a). The ancestral pattern for Boreosphenida is assumed to include replacement at all incisor, canine, and premolar loci, as seen in the "eupantothere" *Dryolestes* (Butler and Krebs, 1973; Martin, 1997; see also chapter 10). Living eutherians are rather variable in terms of tooth replacement, but most retain the pattern of diphyodonty at all antemolar loci except for the first premolar, where the deciduous tooth is generally retained (e.g., Luckett, 1993). By contrast, living marsupials are more stereotyped, and tooth replacement is largely suppressed. Deciduous precursors appear from the primary dental lamina at incisor and canine positions, but these do not erupt and are rapidly resorbed; the teeth that eventually erupt at these loci are successor or replacement teeth (Luckett, 1993). Teeth at the first and second premolar positions derive from the primary dental lamina and thus are deciduous teeth that are retained throughout life. Homology at the succeeding tooth locus has been contended. Archer (1974, 1978), for example, suggested that P/p3 arises from the primary dental lamina anterior to the tooth it generally displaces. Under this interpretation, the adult tooth at the third premolar locus would be a late-appearing primary (deciduous) premolar, and the succeeding tooth is considered a displaced molar. The four molars retained by adult marsupials would thus be M/m2–5, resulting in a confusing terminology adopted ephemerally by some authors (e.g., Marshall, 1987). Detailed study by Luckett (1993), however, shows that the late-erupting tooth at the third premolar locus is a true successor to the tooth it displaces, so that they should be interpreted as P/p3 and dP/dp3, respectively. Hence, marsupials postnatally replace only one tooth in each jaw—the third premolar. Luckett (1993) further suggested that the seven postcanine teeth seen in most marsupials and eutherians may be serially homologous (an idea dating back to Owen, 1868), and that the marsupial first molar is actually a retained deciduous premolar (P/p4), as suggested by its unusual morphology with respect to succeeding molars (see discussion by Butler and Clemens, 2001). As yet, no convincing data have been brought forward to test this intriguing hypothesis.

The stereotyped and advanced suppression of diphyodonty in living marsupials has been linked to reproductive specializations related to birth of extremely altricial young, in particular, the period of fixation that occurs during extended maternal lactation (Luckett, 1993). A survey of South American fossil marsupials, including a reasonably well-represented ontogenetic series for the Paleocene didelphoid *Pucadelphys,* suggests a similar pattern of suppressed tooth replacement (Cifelli and Muizon, 1998b). North American Cretaceous taxa appear to share this pattern of tooth replacement, in turn implying that marsupial reproductive specializations may extend back at least to the Late Cretaceous (Cifelli et al., 1996).

A juvenile specimen of *Deltatheridium* described by Rougier et al. (1998) has p3 in eruption, showing that replacement occurred at this locus. Rougier et al. (1998) argued that the preceding two teeth represent retained deciduous premolars and that replacement occurred only at the P/p3 locus—an advanced condition seen in marsupials (Luckett, 1993, see earlier). Given the relatively advanced ontogenetic age of the specimen, homologies of the teeth (deciduous or replaced) at the p1–2 loci are uncertain. In this context, however, it is worth noting that fossil and living marsupials are distinctive in having replacement at the p3 locus that is retarded with respect to eruption of the canine and molar series (Cifelli et al., 1996; Cifelli and Muizon, 1998b). This pattern is lacking in *Deltatheridium.*

Enamel Microstructure. Traditionally, a primary focus of debate on dental microstructure in marsupials has been the presence and significance of enamel tubules (see review by Clemens, 1979a). It has long been noted that these are present in teeth of marsupials, whereas they are lacking from those of most placentals (Tomes, 1849). Structurally, tubules are uncalcified or secondarily filled areas within the enamel matrix. Enamel tubules are not limited to marsupials, but are found in a number of extinct mammalian clades, as well as in a variety of nonmammalian amniotes (see Lester et al., 1987; Sahni and Lester, 1988, and references therein). This suggests that the presence of enamel tubules may be primitive for Mammalia (Koenigswald and Clemens, 1992). However, tubules are variable in their relationships to other tissues and structures (such as enamel prisms, where present), and their morphogenesis and phylogenetic significance are incompletely understood. The situation is compounded by the fact that tubules, like other enamel structures, are affected by variations in preparation technique (e.g., Carlson and Krause, 1985). Living marsupials studied by Lester et al. (1987) showed the presence of tubules restricted to individual enamel prisms, thus suggesting that they are formed during amelogenesis. However, an association with odontoblasts is indicated in at least some cases, as suggested by the alignment of the enamel with dentinal tubules, and their apparent continuity across the dentinoenamel junction (Lester et al., 1987). Wood (1992) noted that these separate origins suggest the possibility, at least, of nonhomology between tubules generated by the respective pathways. He further observed that, unlike eutherians and various extinct mammalian groups, marsupials appear to have enamel tubules that are consistently ameloblast related (confined to individual prisms). On this basis, he hypothesized the condition in marsupials as a synapomorphy within the group. The question remains an open one.

Among Cretaceous marsupials (see Moss, 1969; Sahni, 1979; Wood, 1992, 2000), enamel microstructure has been studied in *Alphadon* ("Alphadontidae"), *Glasbius* (Glasbiidae), "*Pediomys*" ("Pediomyidae"), *Didelphodon, Eodelphis,* and *Pariadens* (Stagodontidae). Results of some studies have suggested that enamel tubules are lacking in *Alphadon* and *Didelphodon* (e.g., Sahni, 1979; Wood, 1992). Tubules are now known to be present in the enamel of both genera, though they are relatively scarce and difficult to observe in *Didelphodon* (see Wood, 2000). Tubules are reported to be absent in the other two stagodontids (Wood, 1992), though the recent results just cited suggest that reexamination is warranted. Enamel tubules are found in teeth of all of the remaining taxa.

All of the Cretaceous marsupials for which the enamel structure has been studied to date (listed earlier) have radial enamel (Koenigswald and Sander, 1997b) presenting variations of the plesiomorphic prismatic pattern, as defined by Wood and Stern (1997) and Wood et al. (1999). The following brief account is based on Wood (1992, 2000). The enamel structure of *Alphadon* is rather variable but does not depart drastically from that seen in outgroups to Boreosphenida: prisms tend to be somewhat sparse and erratically spaced, with sheaths that are small in diameter, and a rather thick outer zone of aprismatic enamel is often present. Enamel structure of "*Pediomys*" is rather similar to that of *Glasbius,* despite their divergent molar morphology. Structurally, the enamel of these two genera is somewhat more advanced than in *Alphadon:* the outer aprismatic layer is generally thinner and prisms range to larger size, have greater density, and are more regularly spaced. The stagodontids present a consistent and unique pattern, combining a thick outer aprismatic layer and prism sheaths that are less arc-shaped (plesiomorphies), with very regularly spaced, large prisms (apomorphies).

Enamel microstructure of *Deltatheridium* has been described by Wood (2000). The pattern is plesiomorphic prismatic enamel in plesiomorphic condition, similar to what is seen "eupantotherians" and spalacotheriid "symmetrodontans."

POSTCRANIAL SKELETON

The published record of the postcranium in Mesozoic marsupials is restricted to a single, fairly complete skeleton of Mongolian *Asiatherium* (Szalay and Trofimov, 1996, our figure 12.6A) and to isolated proximal tarsals from the Late Cretaceous of North America, described by Szalay (1993b, 1994; our figure 12.6B,C). These are complementary, insofar as the tarsals are unknown for *Asiatherium.* Otherwise, the skeleton of early marsupials is known from exquisitely preserved specimens from the early Paleocene of Bolivia (Marshall and Sigogneau-Russell, 1995; Muizon, 1998; Argot, 2001; Muizon and Argot, 2003) and a great number of isolated elements from the late Paleocene of Brazil (Szalay, 1994). These cannot be treated in detail here, though we make passing reference to them for the purposes of comparison and inference on paleobiology of early marsupials. The following account of the postcranium in Mesozoic marsupials is based on the descriptions of Szalay and Trofimov (1996) and Szalay (1994).

Little can be said regarding the vertebral column in *Asiatherium,* except that contact between pre- and postzygapophyses is nearly vertical in the posterior thoracic region, where anapophyses brace metapophyses tightly; significantly, no anticlinal vertebra is present. There are apparently two fused sacral vertebrae, as is also the case for *Pucadelphys.* The shoulder girdle is poorly preserved; the acromion process appears to have been long and the clavicle robust. The humerus has a well-defined intertubercular groove and teres major tuberosity; the deltoid tuberosity is faint and located about one-third of the distance from proximal to distal end. The capitulum is somewhat spindle shaped and bears a large lateral facet for the radius, which has an oval-shaped head. A medial facet for the humerus appears to be lacking, so that the central region of the distal humerus articulated with the ulna rather than with the radius. Contact between radius and ulna appears to have been flat, or nearly so. On the proximal ulna, the proximal trochlear crest is long relative to the crest that borders the distal part of the trochlear notch; the olecranon process is medially inturned. Combined features of the distal humerus and proximal ulna indicate lateral stability of the elbow joint. Of the carpals, the scaphoid is long proximally, suggesting that it dominated carporadial contact, but a lateral process is lacking from the scaphoid. It is possible that a fused prepollex was present, but this cannot be determined with certainty. The metacarpals are broadly expanded distally and lack grooving lateral to the central area of articulation with the phalanges.

In the pelvic girdle, there is no clear separation on the lateral side of the ilium distinguishing attachments for the gluteus maximus and iliacus muscles, nor do the distal tips of the ilia flare much laterally. An epipubic bone is preserved on one side of the specimen. The femur has a spherical head, and the relative distance from greater to lesser trochanter is similar to what is seen in living *Caenolestes.* There is no trace of an osseus patella, nor even a patellar groove. The lateral condyle is considerably broader than the medial one, suggesting the presence of a parafibula, upon which the calf muscles (gastrocnemius and soleus) are thought to have attached. The fibula is slender proximally, suggesting that it assumed less of a load-bearing function than in a therian morphotype. The fibula appears to have articulated with the femur, unlike the condition commonly encountered in eutherians.

Szalay (1994) described isolated proximal tarsals (astragalus, calcaneus) from the Cretaceous Judith River Group and Frenchman Formation, Canada, and the Lance Formation, Wyoming. The tarsals are similar in important details, and we cover only one set (thought to represent a single species of "Pediomyidae," figure 12.6B,C) following Szalay's description. The calcaneus bears a peroneal process that is large, though smaller than that of stem mammals such as *Morganucodon.* The sustentacular facet lies almost below the calcaneoastragalar facet, indicating that the astragalus lay more medial to (rather than superimposed upon) the calcaneus than in living didelphids. The calcaneocuboid facet forms an evenly concave surface and lacks the depression (formed by a conical protuberance on the cuboid) seen in the calcaneocuboid facet of Didelphidae. This facet suggests the ability for dorsoplantar rotational movement in the calcaneocuboid joint, in turn implying considerable pronation-supination capability of the foot. The calcaneofibular facet is large and suggests that, unlike didelphids, the fibula distributed load directly through the calcaneus.

The astragalus is sharply distinct from that of "didelphoids" in having a neck. It bears a fibular facet that is rather vertical compared to that seen in fossil or Recent marsupials of South America. The proximal surface of the element is very slightly trochleated, with a tibial facet that extends onto the neck. On the plantar side, the medial tuberosity occupies the proximal third of the astragalus. The narrow sustentacular facet is indistinguishable from the navicular facet distally; the astragalus may have contacted the cuboid, but there is no distinction of facets to indicate this.

PALEOBIOLOGY

Given that the record of Mesozoic marsupials consists largely of teeth and jaws, direct interpretation of their lifestyles is mainly limited to data bearing on body size and dietary preferences. Living Didelphidae, commonly regarded as "living fossils" (but see Gardner, 1982), fortunately provide good analogues in this instance. Dental morphology between the living and fossil taxa is generally comparable (Clemens, 1979a), as are details of molar oc-

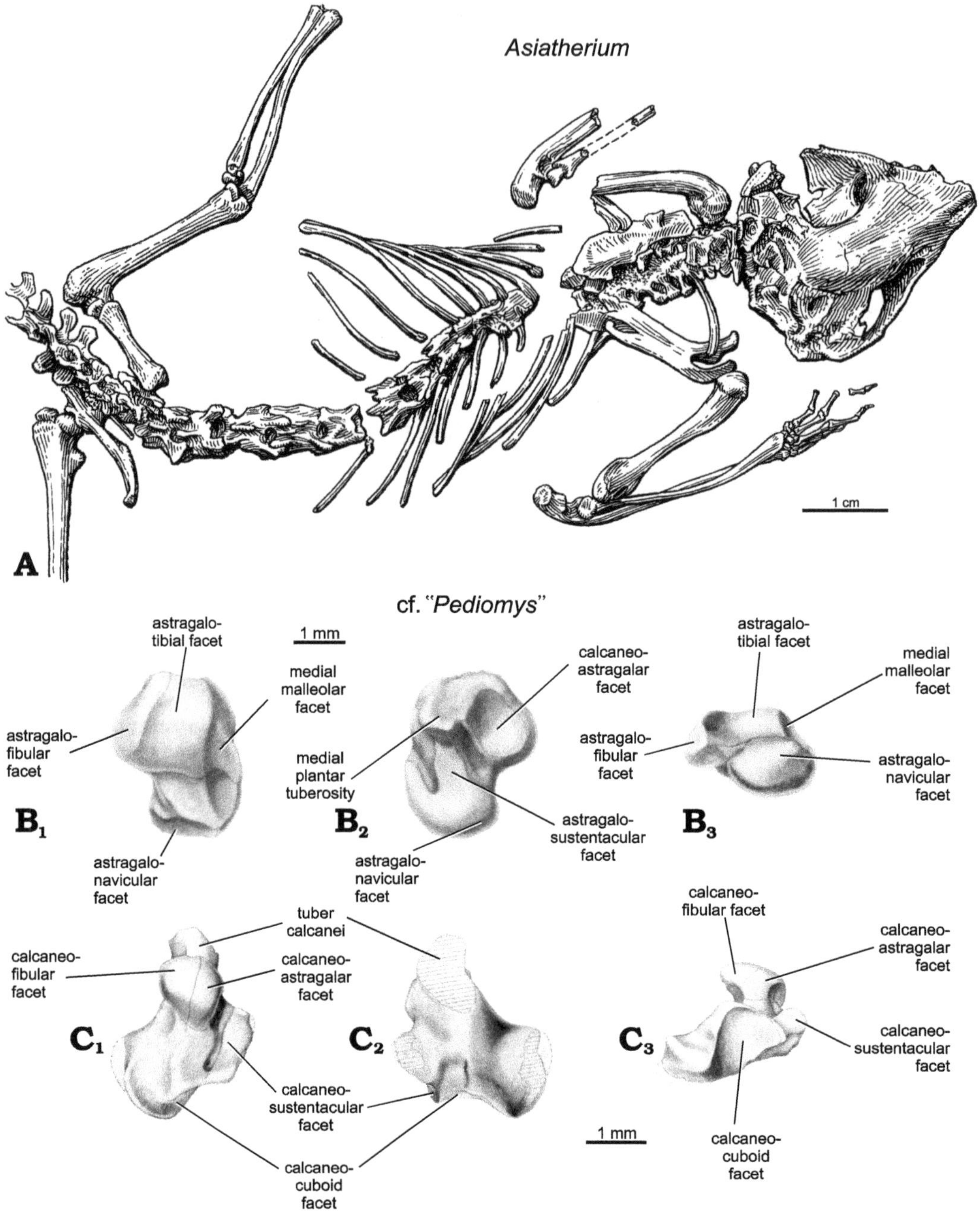

FIGURE 12.6. Postcranial skeleton in Cretaceous Marsupialia. A, Skeleton of *Asiatherium reshetovi* (holotype); B, C, Isolated tarsals referred to cf. "*Pediomys*": B, Right astragalus in dorsal (B_1), plantar (B_2), and distal (B_3) views. C, Incomplete right calcaneus in dorsal (C_1), plantar (C_2), and distal (C_3) views. *Source:* A, modified from Szalay and Trofimov (1996). B, C, modified from Szalay (1994).

clusion (Crompton and Kielan-Jaworowska, 1978). Inference as to posture and locomotion is more problematic. Lacking reasonable sampling from the fossil record, interpretation is based on reconstruction of hypothetical ancestral conditions. Recent work has shown that commonly used living models (e.g., Didelphidae) are not uniformly appropriate and have resulted in misleading conclusions.

Dentition and Diet. The postcranial skeleton of deltatheroidans remains to be described, and paleobiological interpretation is thus limited to what may be inferred from craniodental morphology. The tall coronoid process and low mandibular condyle (reflecting emphasis on the temporalis muscle), short snout, large canines, emphasis on postvallum/prevallid molar shear, and relatively large body size suggest that deltatheroidans incorporated vertebrate prey in their diet.

Evidence from the dentition suggests a reasonably high level of ecological diversity for Cretaceous marsupials, at least by comparison to other groups of Mesozoic mammals. The 19 genera recognized herein include species of widely varying body mass. Judged by dental measurements, known Cretaceous marsupials ranged from the size of the smallest living marsupial "mice," *Planigale* (5 g or less) to that of *Didelphis* (2.0–5.5 kg, these and other body masses from Nowak, 1991; Silva and Downing, 1995). This range spans some three orders of magnitude and presumably resulted in differing life histories for included species (Eisenberg, 1981). For present purposes, discussion of diet in Cretaceous marsupials can be treated under three categories: "Alphadontidae" plus "Pediomyidae" (together with certain taxa left *incertae sedis*), Stagodontidae, and Glasbiidae.

Of the three categories, "Alphadontidae," "Pediomyidae," and unassigned genera such as *Kokopellia* are the most conservative in terms of molar structure. The most important differences from basal Metatheria and outgroups are associated with an increase in crushing function. Diet may be reasonably approximated by reference to a spectrum of dentally similar living marsupials among the Didelphidae and Dasyuridae. Smaller species of these families are mainly insectivorous, though they may take some vertebrate prey. Larger didelphids, in particular, are opportunistic feeders, with a varied diet including insects, vertebrate prey, carrion, fruit, and other items (Nowak, 1991). Species in this fossil group ranged upward in size to *Turgidodon rhaister,* which may have been about the size of living *Philander* (ca. 500 g). Most of these Cretaceous marsupials thus probably fed mainly on insects and arthropods, with the dietary regime broadening at larger sizes. Species of *Turgidodon,* for example, are characterized by large size and low, bulbous molar cusps (Cifelli, 1990a), suggesting a more omnivorous fare.

Stagodontidae include a series of taxa characterized by increasing body size and dental specialization (Clemens, 1966; Fox, 1981; Fox and Naylor, 1986), culminating in species of *Didelphodon.* Emphasis on postvallum/prevallid shearing in the molar series suggests a carnivorous diet, particularly in the largest species (e.g., Butler, 1990b; Muizon and Lange-Badré, 1997). The hypertrophied, blunt posterior premolars are suggestive of enhanced crushing ability, a supposition supported by features of the dentary and skull, indicating good mechanical advantage for the mandibular adductors and strong development of the temporalis muscle. There is agreement that these specializations reflect durophagy, but it remains uncertain as to what types of hard objects were most routinely sought by stagodontids. One possible explanation for this modification is bone crushing (Clemens, 1968a,b), as seen in living *Sarcophilus* and various eutherian carnivores (e.g., Werdelin, 1986, 1989). An alternative hypothesis, a diet of shelled invertebrates (molluscs), is based on similarity of premolar structure to the sea otter, *Enhydra* (Clemens, 1979a). This possibility is particularly intriguing in light of postcranial specializations suggestive of an aquatic habitat preference for *Didelphodon* (see later).

The molar dentition of *Glasbius,* characterized by reduced crown height and low, bulbous cusps, reflects a reduction of shearing function (compared to, e.g., "alphadontids") and a concomitant increase in crushing and grinding function. Given that body mass was probably comparable to living species of "mouse opossums" (*Marmosa* and allies, perhaps 50 g or less), tooth morphology and energy needs suggest specialization toward feeding on high-calorie plant material (rather than foliage). Based on analogy with Recent mammals (Kay, 1975), a reasonable working hypothesis is that *Glasbius* may have represented an early evolutionary experiment in frugivory. Angiosperm plants radiated widely during the last half of the Cretaceous, and a number of major types of fruiting bodies made their first appearance in the Campanian and Maastrichtian (Friis and Crepet, 1987). It is conceivable that the lineage leading to *Glasbius* developed its distinctive dental specializations to take advantage of such resources.

Habitat Preference and Locomotion. The hypothesis that marsupials had an arboreal ancestry has had broad appeal since the suggestion was first made (Huxley, 1880; see also Bensley, 1901a; Haines, 1958; Lewis, 1980). This hypothesis was based on a traditional reliance on didelphids as proxy for the ancestral morphotype for all marsupials, with the implication that the climbing specializations of many (although not all) didelphid taxa would be ancestral for marsupials. A nagging concern with this approach is that specializations of didelphid skeletons may not represent the generalized condition (Szalay, 1984, 1994; Muizon and Argot, 2003). A striking example lies in the structure of the didelphid upper ankle joint, which is modified for extreme inversion capabilities associated with reversal of the foot (Jenkins and McLearn, 1984). Another issue is the fact that extant didelphids have a wide range of locomotory habits, from the largely terrestrial *Metachirus* to the fully arboreal *Caluromys.* Despite these caveats, some level of analogous comparison between extant marsupials and fossils of Late Cretaceous and Early Tertiary age can be achieved (Szalay, 1994; Muizon, 1998; Argot, 2001; Szalay and Sargis, 2001).

The issue of an arboreal versus a terrestrial ancestry for marsupials may, in any case, be largely moot. As indicated in the foregoing sections, early marsupials, like most other Mesozoic mammals, were small. At such small body sizes, climbing capability is required for a mammal to move on both terrestrial and arboreal substrates; therefore, skeletal differences between tiny scansorial and arboreal mammals tend to be minor (Jenkins, 1974a). Discussion on the ancestral habits of marsupials are most productively focused on the extent to which scansorial features are developed (or not) among early fossil marsupials (see Szalay, 1994).

Asiatherium is the only Cretaceous marsupial known by associated skeletal remains. The most notable features of *Asiatherium* include the lack of an anticlinal vertebra, as seen in leaping and cursorial mammals, and the transversely stabilized elbow joint and slender proximal fibula, which are common to small, mainly terrestrial marsupials (Szalay and Trofimov, 1996). Tarsals assigned to the North American Cretaceous "*Pediomys*" reportedly lack specializations seen in arboreal didelphids, differing from these modern forms in showing better capabilities for flexion-extension and in lacking the capability for extreme inversion of the foot of the extant didelphids (Szalay, 1994). Tarsals assigned to the stagodontid *Didelphodon* are uniquely apomorphic, apparently permitting a wide range of rotational mobility, and suggestive of aquatic adaptation (Szalay, 1994).

The oldest complete skeletal remains of marsupials are from the early Paleocene of Bolivia (Muizon, 1998; Argot, 2001; Muizon and Argot, 2003). Additional information on early marsupial skeletal anatomy is provided by abundant but isolated postcranial elements of diverse taxa from the Paleocene Itaboraí fauna, Brazil (Szalay, 1994; Szalay and Sargis, 2001). These South American Paleocene taxa considerably postdate the marsupial-placental dichotomy and the earliest known fossil marsupials, but nonetheless provide some of the only data for fossil-based inference on early locomotor patterns within Marsupialia.

The three best-known taxa (*Pucadelphys, Andinodelphys, Mayulestes*) have some skeletal features associated with climbing in living didelphids (Muizon, 1998; Argot, 2001; Muizon and Argot, 2003). Marshall and Sigogneau-Russell (1995) argued that *Pucadelphys* had a terrestrial mode of life and was probably capable of some digging. However, more detailed study led Muizon (1998) to conclude that *Pucadelphys* was an agile, leaping animal with many features suitable for climbing. Scansorial adaptations for *Pucadelphys* were also suggested by Argot (2001). Collectively, the morphology of *Pucaldelphys* suggests that it was an agile, partly terrestrial animal, with good capabilities for leaping and climbing, similar to small living dasyurids or to some of the terrestrial didelphids such as *Metachirus. Mayulestes* has a greater number of features associated with climbing, and it has been reconstructed as a largely arboreal mammal (Muizon, 1998; Argot, 2001; Szalay and Sargis, 2001). The diverse marsupial assemblage from Itaboraí, Brazil, includes both arboreal and terrestrial forms (Szalay, 1994; Szalay and Sargis, 2001).

SYSTEMATICS

Infraclass Metatheria Huxley, 1880
(table 12.1)

INTRODUCTION

In the nineteenth century through the early twentieth, most Mesozoic mammals were referred without issue to Marsupialia (e.g., Owen, 1871; Marsh, 1889a; Osborn, 1887a).[2] As with so many aspects of knowledge on early mammals, the first major advances were made by Simpson, summarized in his two monographs (Simpson, 1928a, 1929a). The truly extraneous taxa were relegated to other major groups and a basis for distinguishing between early marsupials and eutherians was recognized, although problems remained.[3] His subsequent classification (Simpson, 1945) placed all then-known Cretaceous marsupials within the living family Didelphidae, recognizing two major groups for the fossils (Pediomyinae and Thlaeodontinae, see also Simpson, 1927d). With the advantage of many decades of hindsight and infinitely more fossils than Simpson had at his disposal, these groups now appear to be mixed assemblages. The basis for our current systematic arrangement was established by Clemens (1966), who made general order among a bewildering array of poorly based taxonomic names created in the nineteenth century. Intensive field investigations over the past four decades, coupled with widespread use of microfossil recovery techniques, have resulted in dramatically expanded knowledge of the taxonomic and morphologic diversity of early marsupials and related mammals. Simpson's (1945) classification included five of the currently recognized genera of metatherians (two deltatheroidans, then placed among Eutheria, and three marsupials); the current compendium

[2] This is one point upon which there is at least general agreement between Marsh and Osborn, whose heated exchanges (late 1880s and early 1890s) on the subject of Mesozoic mammals bring a breath of entertainment to the subject.

[3] *Cimolestes,* now universally recognized as a eutherian, remained in Marsupialia until much later (see Simpson, 1951; Lillegraven, 1969; Clemens, 1973a).

TABLE 12.1. Linnaean Classification of Mesozoic Metatherian Mammals

- Infraclass Metatheria Huxley, 1880
  - Cohort Deltatheroida Kielan-Jaworowska, 1982
    - Family Deltatheridiidae Gregory and Simpson, 1926
      - *Deltatheridium* Gregory and Simpson, 1926, type genus
        - *D. pretrituberculare* Gregory and Simpson, 1926, type species
        - *D. nessovi* Averianov, 1997
      - *Deltatheroides* Gregory and Simpson, 1926
        - *D. cretacicus* Gregory and Simpson, 1926
      - *Deltatherus* Nessov, 1997
        - *D. kizylkumensis* (Nessov, 1993)
      - *Sulestes* Nessov, 1985b
        - *S. karakshi* Nessov, 1985b
    - Family *incertae sedis*
      - ?*Atokatheridium* Kielan-Jaworowska and Cifelli, 2001
        - *A. boreni* Kielan-Jaworowska and Cifelli, 2001
      - *Khuduklestes* Nessov, Sigogneau-Russell, and Russell, 1994
        - *K. bohlini* Nessov, Sigogneau-Russell, and Russell, 1994
      - *Oxlestes* Nessov, 1982
        - *O. grandis* Nessov, 1982
  - Cohort Marsupialia Illiger, 1811[1,2]
  - Superorder "Ameridelphia" Szalay, 1982, new rank
  - Order and family *incertae sedis*
      - *Anchistodelphys* Cifelli, 1990b
        - *A. archibaldi* Cifelli, 1990b
        - ?*A. delicatus* Cifelli, 1990c
      - *Iugomortiferum* Cifelli, 1990b
        - *I. thoringtoni* Cifelli, 1990b
      - *Kokopellia* Cifelli, 1993a
        - *K. juddi* Cifelli, 1993a
  - Order Asiadelphia Trofimov and Szalay, 1994
    - Family Asiatheriidae Trofimov and Szalay, 1994
      - *Asiatherium* Trofimov and Szalay, 1994, type genus
        - *A. reshetovi* Trofimov and Szalay, 1994
      - ?*Marsasia* Nessov, 1997
        - *M. aenigma* Nessov, 1997
  - Order "Didelphimorphia" Gill, 1872
    - Family "Alphadontidae" Marshall et al., 1990
      - *Alphadon* Simpson, 1927d, type genus
        - *A. marshi* Simpson, 1927d, type species
        - *A. attaragos* Lillegraven and McKenna, 1986
        - *A. clemensi* Eaton, 1993b
        - *A. eatoni* Cifelli and Muizon, 1998a
        - *A. halleyi* Sahni, 1972
        - *A. jasoni* Storer, 1991
        - *A. lillegraveni* Eaton, 1993a
        - *A. perexiguus* Cifelli, 1994
        - *A. sahnii* Lillegraven and McKenna, 1986
        - *A. wilsoni* Lillegraven, 1969
      - *Aenigmadelphys* Cifelli and Johanson, 1994
        - *A. archeri* Cifelli and Johanson, 1994
      - *Albertatherium* Fox, 1971b
        - *A. primum*[3] Fox, 1971b, type species
        - *A. secundum*[3] Johanson, 1994
      - *Protalphadon* Cifelli, 1990a
        - *P. lulli* (Clemens, 1966), type species
        - *P. foxi* Johanson, 1996b
      - *Turgidodon* Cifelli, 1990a
        - *T. praesagus* (Russell, 1952), type species
        - *T. lillegraveni* Cifelli, 1990a
        - *T. madseni* Cifelli, 1990a
        - *T. parapraesagus* (Rigby and Wolberg, 1987)
        - *T. petaminis* (Storer, 1991)
        - *T. rhaister* (Clemens, 1966)
        - *T. russelli* (Fox, 1979b)
      - *Varalphadon* Johanson, 1996b
        - *V. wahweapensis* (Cifelli, 1990a), type species
        - *V. creber* (Fox, 1971b)
        - *V. crebreforme* (Cifelli, 1990b)
    - Family "Pediomyidae" Simpson, 1927d
      - "*Pediomys*" Marsh, 1889a, type genus
        - "*P.*" *elegans* Marsh, 1889a, type species
        - "*P.*" *clemensi* Sahni, 1972
        - "*P.*" *cooki* Clemens, 1966
        - "*P.*" *exiguus* Fox, 1971b
        - "*P.*" *fassetti* Rigby and Wolberg, 1987
        - "*P.*" *florencae* Clemens, 1966
        - "*P.*" *hatcheri* (Osborn, 1898)
        - "*P.*" *krejcii* Clemens, 1966
        - "*P.*" *prokrejcii* Fox, 1979c
      - *Aquiladelphis* Fox, 1971b
        - *A. incus* Fox, 1971b, type species
        - *A. minor* Fox, 1971b
        - *A. paraminor* Rigby and Wolberg, 1987
      - *Iqualadelphis* Fox, 1987a
        - *I. lactea* Fox, 1987a
    - ?Family Peradectidae Crochet, 1979a
      - cf. *Peradectes* Matthew and Granger, 1921, type genus
        - cf. *P. austrinus* (Sigé, 1971)
    - Family Stagodontidae Marsh, 1889b
      - *Didelphodon* Marsh, 1889a, type genus
        - *D. vorax* Marsh, 1889a, type species
        - *D. coyi* Fox and Naylor, 1986
        - *D. padanicus* (Cope, 1892)
      - *Eodelphis* Matthew, 1916
        - *E. cutleri* (Woodward, 1916), type species
        - *E. browni* Matthew, 1916
      - ?*Pariadens* Cifelli and Eaton, 1987
        - *P. kirklandi* Cifelli and Eaton, 1987
        - *P. mckennai* Cifelli, 2004
  - Order Paucituberculata Ameghino, 1894
    - Family Glasbiidae Clemens, 1966
      - *Glasbius* Clemens, 1966, type genus
        - *G. intricatus* Clemens, 1966, type species
        - *G. twitchelli* Archibald, 1982

[1] As noted in the text, we tentatively follow Rougier et al. (1998) in assigning the unnamed Guriliin Tsav skull to Marsupialia. In addition to incorrect referral to *Deltatheridium* (see Anonymous, 1983), this specimen has been referred to "*Eodeltatheridium kursanovi*" and "*Neodeltatheridium*" (no species designation) by Gambaryan (1989). Both of these names were apparently intended to apply to this specimen alone; each name is a *nomen nudum.*

[2] We view *Adinodon pattersoni* Hershkovitz, 1995 (placed by its author within "Adinodontinae," considered as a subfamily of "Marmosidae") as a *nomen dubium* (see Cifelli and Muizon, 1997).

[3] Emended to reflect the gender (neuter) of the genus.

includes 28 genera (7 deltatheroidans and 21 marsupials; see table 12.1). As noted in chapter 11, we have excluded certain genera of debatable affinities, such as *Holoclemensia* (see Slaughter, 1968a,c, 1971; Butler, 1978a; Kielan-Jaworowska, Bown, and Lillegraven, 1979).

Until recently, the terms Marsupialia Illiger, 1811, and Metatheria Huxley, 1880, were widely regarded as equivalent (see Aplin and Archer, 1987; McKenna and Bell, 1997). With recognition of Deltatheroida as possible relatives of Marsupialia, a practical distinction in usage would be to recognize Marsupialia as a crown-based group (i.e., the last common ancestor of all living marsupials, plus all of its descendants) within a more inclusive Metatheria that also contains deltatheroidans (e.g., Rougier et al., 1998). This would be equivalent to the use of the terms Eutheria and Placentalia, adopted herein (Novacek et al., 1997, see chapter 13). In the case of Metatheria, however, there are a number of fossil taxa that are sharply distinct from Deltatheroida and resemble crown marsupials. It is virtually certain that most of these are more closely related to Crown Marsupialia than are Deltatheroida, highly probable that they form a hierarchy of nested clades approaching Marsupialia, and conceivable that some may lie within the crown group (as reflected in our classification, table 12.1). Under these circumstances, we find it neither practical nor desirable to restrict Marsupialia to a definition based on living taxa and employ a stem-based definition of the group to include all metatherians more closely related to living marsupials than to Deltatheroida. This definition permits us to avoid the confusion that would be introduced by employing new or cumbersome terminology for fossils that have long and universally been placed within Marsupialia.

As the fossil record and knowledge of the profound differences among living taxa have improved, an increasing representation of marsupial diversity has become apparent in higher-level classifications. Simpson's (1945) influential classification included only six superfamilies of marsupials, all placed within a single order. More recently, taxonomists have followed Ride (1964) in distributing fossil and Recent marsupials among a number of orders. Most classifications published in the past two decades have followed the influential work of Szalay (1982; see also Szalay and Sargis, 2001) in recognizing a fundamental division between Australidelphia (Australian taxa + South American microbiotheres) and Ameridelphia (e.g., Archer, 1984; Aplin and Archer, 1987; Marshall, 1987; Reig et al., 1987; McKenna and Bell, 1997). We follow this division herein, noting later that there are significant difficulties in characterizing "Ameridelphia" and defending it as a monophyletic group. Szalay (1994) united Asiadelphia Trofimov and Szalay, 1994 (then represented by the single genus *Asiatherium*), with the Deltatheroida Kielan-Jaworowska, 1982, into a cohort Holarctidelphia. We tentatively recognize Asiadelphia as being more closely related to Marsupialia than to Deltatheroida. As "Ameridelphia" are implicitly recognized as paraphyletic (including presumed ancestry of Australidelphia), we simply expand its concept herein to include Asiadelphia and stem marsupials of North America.

Cohort Deltatheroida Kielan-Jaworowska, 1982, New Rank

INTRODUCTION

Among the many significant discoveries of the American Museum of Natural History's Central Asiatic Expeditions (1920s) were skulls of three relatively large (by Mesozoic standards) carnivorous mammals. These were briefly described by Gregory and Simpson (1926), who named them as *Deltatheridium, Deltatheroides,* and *Hyotheridium.* Noting that the molars were of rather primitive design, Gregory and Simpson (1926) placed these genera in a new family, Deltatheridiidae, which they referred to the eutherian order Insectivora. Of the three, *Hyotheridium* may be dismissed from further consideration in this chapter, as is represented by a single, poorly preserved specimen that we tentatively place among Eutheria, as a *nomen dubium* (chapter 13).

Gregory and Simpson (1926) regarded deltatheridiids as structurally suitable ancestors of zalambdadont insectivores and creodonts (then placed within Carnivora and now regarded as close relatives of carnivorans, e.g., Lillegraven, 1969; McKenna and Bell, 1997). Gregory and Simpson (1926) also noted some ambiguous similarity of *Deltatheridium* to Eocene *Didelphodus* (later widely regarded as a "palaeoryctid"; see chapter 13). These views became generally accepted (Matthew, 1928; Simpson, 1928c, 1945) and, as a result, Deltatheridiidae subsequently played an important role in interpretation of the early phylogeny of Eutheria. Van Valen (1966) made them the namesake of his order Deltatheridia, including "palaeoryctids" and other "insectivores," together with creodonts (but not Carnivora, see McKenna et al., 1971; Szalay and McKenna, 1971; McKenna, 1975, for somewhat different viewpoints).

The proposed affinities of Deltatheridiidae to (or within) Eutheria were based on interpretation of cheek tooth homologies in addition to minor modifications for carnivory and retention of numerous plesiomorphies in the dentition. The tooth series was interpreted (see McKenna et al., 1971; McKenna, 1975) as having no more than three molars, with the first complex cheek tooth sometimes considered as a semimolariform premolar. Both the molar count and the elaboration of the last premolar are characteristics generally associated with Eutheria (chapter 13). On the basis of newly collected and better-preserved specimens, Butler and Kielan-Jaworowska (1973) reinterpreted the cheek tooth formula of *Deltatheridium* as including three premolars and (as many as) four molars. This combination is a similarity to marsupials; however, in view of various uncertainties, these authors considered *Deltatheridium* to be a therian of metatherian-eutherian grade, not referable to either living clade of therian mammals. This view was widely adopted in subsequent years (e.g., Kielan-Jaworowska, 1975c; Butler, 1978a; Kielan-Jaworowska, Bown, and Lillegraven, 1979). In recognition of their distinctive morphology, *Deltatheridium* and allies were eventually placed in their own order, Deltatheroida Kielan-Jaworowska, 1982.

A relationship of Deltatheroida to Marsupialia was forwarded by Kielan-Jaworowska and Nessov (1990) and Marshall and Kielan-Jaworowska (1992), in part based on similarities of the former to Stagodontidae, an early-diverging clade of marsupials from the Late Cretaceous of North America. This view was challenged for a number of reasons (Cifelli, 1993b; Muizon, 1994, 1998; Fox and Naylor, 1995; Muizon et al., 1997). Additional data bearing on the issue were presented by Rougier et al. (1998), based on preliminary study of newly collected specimens of *Deltatheridium*. These authors provided several apomorphies for a comprehensive Metatheria (Deltatheroida + Marsupialia). At present, the most compelling character supporting a relationship between Deltatheroida and Marsupialia is the shared presence of only three premolars. We provisionally recognize Deltatheroida as sister taxon to Marsupialia and give them a stem-based definition: all Metatheria more closely related to *Deltatheridium* than to Marsupialia.

Diagnosis. Dentally primitive Metatheria, plesiomorphic with respect to Marsupialia in the following features. Upper molars with transversely and mesiodistally small protocone; conules poorly developed and not placed near bases of paracone and metacone; little or no development of auxiliary cusps on the stylar shelf; postprotocrista not extending labially past metacone; paracone taller than metacone. Lower molars with great height differential between trigonid and talonid; talonid markedly narrower than trigonid; lingual cuspule e often present, as is labial cuspule f, which may be associated with small mesiolabial precingulid; labial postcingulid absent; entoconid either lacking or poorly developed and, when present, not lingually situated; distal metacristid present. Eruption of last premolar not delayed with respect to eruption of molar series. Apomorphic with respect to basal Eutheria and Marsupialia in having the snout foreshortened, with a tendency toward reduction of the last molar; incisors reduced to I4, i3; upper molars with distal stylar shelf more transversely developed and with more salient postmetacrista; lower molars with relatively taller paraconid and shorter metaconid and relatively stronger paracristid compared to protocristid.

Included Taxa. Family Deltatheridiidae Gregory and Simpson, 1926, and tentatively assigned *Atokatheridium* Kielan-Jaworowska and Cifelli, 2001; *Oxlestes* Nessov, 1982; and *Khuduklestes* Nessov, Sigogneau-Russell, and Russell, 1994.

Distribution. ?Early–Late Cretaceous (?Aptian or Albian through latest Maastrichtian, Lancian): North America; Late Cretaceous (?Turonian–?late Campanian): Asia.

Family Deltatheridiidae Gregory and Simpson, 1926

Synonyms: Sulestinae Nessov, 1985b; Deltatheroididae Kielan-Jaworowska and Nessov, 1990.

Diagnosis. As for the Deltatheroida.

Included Genera. *Deltatheridium* Gregory and Simpson, 1926; *Deltatheroides* Gregory and Simpson, 1926; *Deltatherus* Nessov, 1997; *Sulestes* Nessov, 1985b; and several taxa left in open nomenclature (Fox, 1974a; Cifelli, 1990c).

Distribution. Late Cretaceous (Coniacian–Campanian): Asia, Uzbekistan and Mongolia. Late Cretaceous (Turonian-Lancian): North America, Canada and United States.

Comments. Nessov (1985b) erected within Deltatheridiidae the monotypic subfamily Sulestinae. This distinction was followed by Kielan-Jaworowska and Nessov (1990), who divided Deltatheridiidae into Deltatheridiinae and Sulestinae. McKenna and Bell (1997) regarded Sulestinae as a synonym of Deltatheridiidae. As Sulestinae are represented only by the type genus, of which only a fragment of maxilla with two upper molars (*Sulestes karakshi*) and a dentary fragment with ?m1 and broken p3 (*Sulestes* sp., Kielan-Jaworowska and Nessov, 1990) are known, there is insufficient basis for evaluating the taxonomic significance of differences between *Sulestes* and *Deltatheridium*. Pending recovery of fossils informative enough to substantiate distinction of Sulestinae, we simply assign *Sulestes* to Deltatheridiidae.

Genus *Deltatheridium* Gregory and Simpson, 1926
(figures 12.4A–D, 12.7A)

Diagnosis. Primitive member of Deltatheridiidae, resembling poorly known *Deltatheroides*, from which it differs in having an asymmetrical, relatively smaller M3; differs from *Sulestes* and from *Deltatheroides*-like mammals

(Fox, 1974a) in having a deeper ectoflexus on the upper molars, lacking marginal stylar cusps posterior to the stylocone, and having a less anteroposteriorly expanded protocone. Lower molars of *Deltatheridium* differ from those of *Sulestes* sp. (Kielan-Jaworowska and Nessov, 1990) in having relatively higher trigonid and considerably narrower talonid. Among the species assigned to *Deltatheridium*, imperfectly known *D. nessovi* is probably more advanced than *D. pretrituberculare* in having a shallower ectoflexus and narrower stylar shelf (Averianov, 1997).

Species. *Deltatheridium pretrituberculare* Gregory and Simpson, 1926, type species; and *Deltatheridium nessovi* Averianov, 1997.

Distribution. Late Cretaceous (Campanian): Mongolia, Gobi Desert (Djadokhta Formation, "Ukhaa Tolgod Beds," Baruungoyot Formation and Red Beds of Hermiin Tsav); ?China, Inner Mongolia (Bayn Mandahu Formation, see Kielan-Jaworowska et al. 2003); Kazakhstan (Darbasa Formation).

Comments. The last molars of *Deltatheridium* are extremely reduced. Where visible, they were mistakenly interpreted as parts of the preceding molar (Gregory and Simpson, 1926), or thought to be absent (Kielan-Jaworowska, 1975c). Similarly, the lower incisor series was initially interpreted as including only one or two teeth; two subspecies of *D. pretrituberculare* were recognized on the basis of incisor count and skull length (Kielan-Jaworowska, 1975c). Newly collected specimens show that these differences are artifactual and that distinction of subspecies is not warranted (Rougier et al., 1998).

Genus *Deltatheroides* Gregory and Simpson, 1926
(figure 12.4E)

Diagnosis. Poorly known genus represented by an incomplete skull (without a dentary) of the type species from the Djadokhta Formation and an attributed dentary from the Baruungoyot Formation. Differs from *Deltatheridium* in shape of the upper molars; M1 being more reduced in width, M2 relatively larger with a deeper ectoflexus, M3 more symmetrical with prominent parastylar region and existing metastylar region (the latter completely reduced in *Deltatheridium*), and relatively larger M4.

Species. *Deltatheroides cretacicus* Gregory and Simpson, 1926, type species by monotypy.

Distribution. Late Cretaceous (?early Campanian): Mongolia, Gobi Desert (Djadokhta Formation).

Comments. The poorly known upper molars of *Deltatheroides* approach more those of the Guriliin Tsav skull than those of *Deltatheridium;* on the other hand, reduction in the size of M4 allies *Deltatheroides* with Deltatheroida, rather than marsupials (in which we tentatively place the Guriliin Tsav taxon).

Genus *Deltatherus* Nessov, 1997
(figure 12.7C)

Diagnosis. This poorly known, monotypic genus is represented by a few specimens of the type species (*Deltatherus kizylkumensis,* referred by Nessov, 1993, to *Deltatheroides*): two lower molars and a large, edentulous maxillary fragment bearing alveoli. It approaches *Deltatheridium* in size of the maxilla, but differs in having a canine alveolus that is almost twice as large, double-rooted P1 (single-rooted in *Deltatheridium*), lower position of the infraorbital foramen and higher position of the suture between maxilla and jugal. The lower molars differ from those in *Deltatheridium* and *Sulestes* in having a lower talonid and relatively sharper and higher trigonid cusps, resembling in this respect *Kielantherium* (see Dashzeveg and Kielan-Jaworowska, 1984). *Deltatherus kizylkumensis* is about twice as large as *Sulestes karakshi.*

Species. *Deltatherus kizylkumensis* (Nessov, 1993), type species by monotypy.

Distribution. Late Cretaceous (Coniacian): Uzbekistan (Bissekty Formation).

Genus *Sulestes* Nessov, 1985b
(figure 12.7B)

Diagnosis. More advanced genus than *Deltatheridium* (although its Coniacian occurrence is some 7 Ma older), from which it differs in having lower crowns of the upper molars, weaker ectoflexus, smaller stylar shelf, more anteroposteriorly expanded protocone, presence of a row of marginal cuspules posterior to the stylocone, and a row of cuspules labial to two paraconules, along the preparaconule crista. Lower molars differ in having lower crowns, lesser height differential between trigonid and talonid, and distinctly wider talonid. In all these characters *Sulestes* resembles *Potamotelses* (see Fox, 1972b).

Species. *Sulestes karakshi* Nessov, 1985b, type species by monotypy; and a species left in open taxonomy (Kielan-Jaworowska and Nessov, 1990).

Distribution. Late Cretaceous (Coniacian): Uzbekistan (Bissekty Formation).

Family *incertae sedis*
Genus *Atokatheridium* Kielan-Jaworowska and Cifelli, 2001
(figures 12.7D–E)

Diagnosis. Monotypic genus based on the holotype of the type species (single upper molar). The upper molar is similar to Deltatheroida and differs from *Pappotherium,*

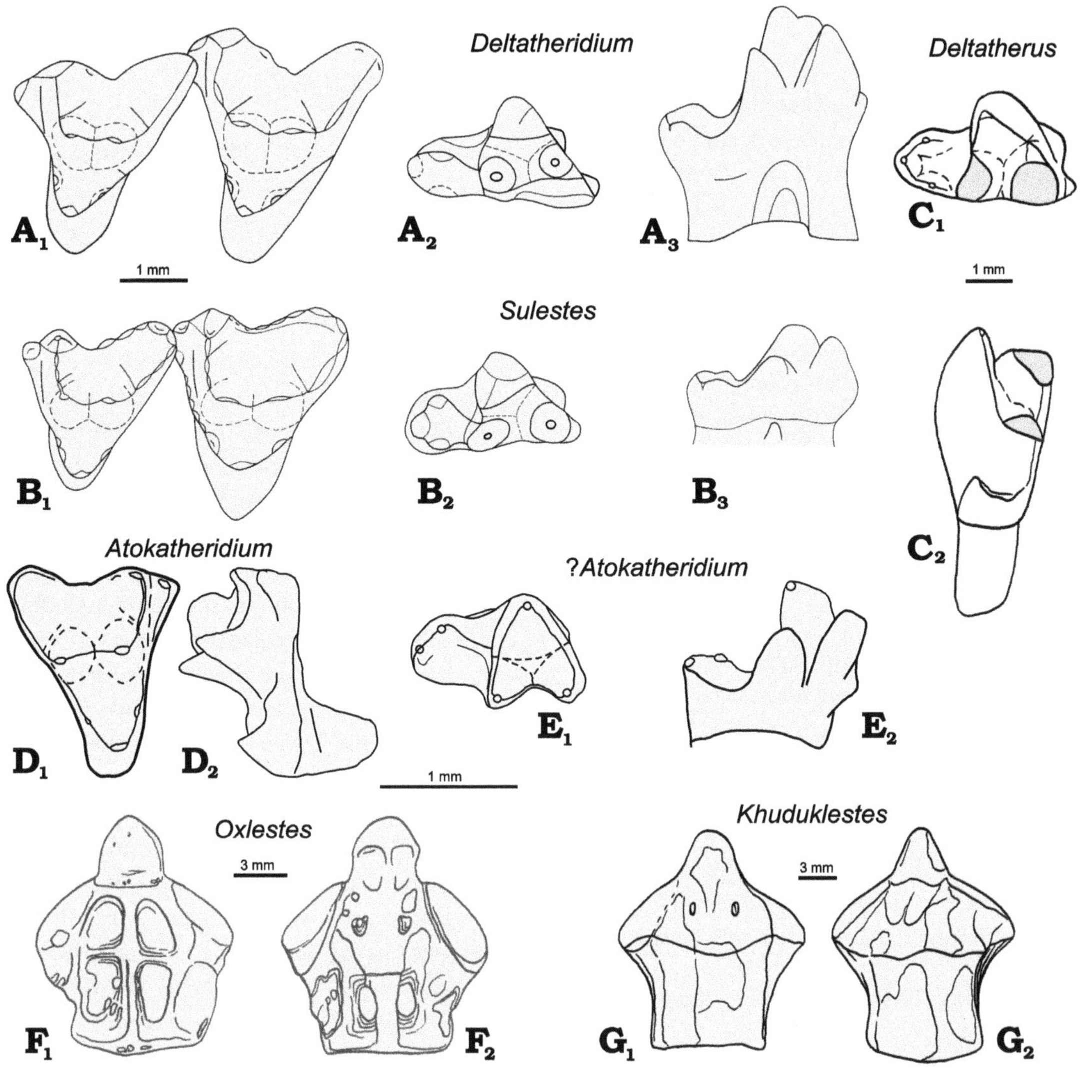

FIGURE 12.7. Deltatheridiidae (A–C) and ?Deltatheroida, family *incertae sedis* (D–G). A, *Deltatheridium pretrituberculare:* left M1–2 in occlusal view (A_1); left m1 in occlusal (A_2) and lingual (A_3) views. B, *Sulestes:* left M1–2 of *Sulestes karakshi* (holotype) in occlusal view (B_1); left m1 of *Sulestes* sp. in occlusal (B_2) and lingual (B_3) views. C, *Deltatherus kizylkumensis:* left ?m2 (holotype) in occlusal (C_1) and posterior (C_2) views. D, *Atokatheridium boreni:* right upper molar (holotype) in occlusal (D_1) and mesial (D_2) views. E, ?*Atokatheridium boreni:* left lower molar in occlusal (E_1) and lingual (E_2) views. F, *Oxlestes grandis:* partial axis vertebra (holotype) in ventral (F_1) and dorsal (F_2) views. G, *Khuduklestes bohlini:* axis vertebra (holotype) in dorsal (G_1) and ventral (G_2) views. *Source:* A and B, modified from Kielan-Jaworowska and Nessov (1990); D, E, modified from Kielan-Jaworowska and Cifelli (2001); C, F, originals; G, original based on Bohlin (1953).

Holoclemensia, Tribotherium, and early eutherians and marsupials in extreme development of the distal stylar shelf, which projects labially and lacks cusps. Differs from all early boreosphenidans except *Tribotherium* (chapter 11) in the lesser development of conules, particularly the metaconule. Differs from early eutherians, marsupials, *Pappotherium,* and *Holoclemensia* in less mesiodistal expansion of the protocone. A lower molar of uncertain position, tentatively referred to the species, has the following characteristics: paraconid and metaconid are well separated, with paraconid much taller than metaconid, and talonid much lower and narrower than trigonid, with a very small, shallow basin that is open lingually. Two cusps, hypoconid and hypoconulid, are present; despite the presence of some wear on the rim of the talonid, it is clear that no entoconid was ever present.

Species. *Atokatheridium boreni* Kielan-Jaworowska and Cifelli, 2001, type species by monotypy.

Distribution. Early Cretaceous (Aptian or Albian): United States, Oklahoma (Antlers Formation).

Genus *Oxlestes* Nessov, 1982
(figure 12.7F)

Diagnosis. Poorly known genus, based on the holotype of the type species, which consists of a partial axis that represents one of the largest mammals known from the early Late Cretaceous. The type species may include a large parietal with a strong sagittal crest, and a large upper canine (see figures in Nessov, 1981; Nessov and Kielan-Jaworowska, 1991; Nessov, Sigogneau-Russell, and Russell, 1994). The axis is relatively narrow, with a long, pointed, anteriorly directed process. Anterior paired articular surfaces are directed strongly obliquely. Length of the axis with dens is about 19.5 mm.

Species. *Oxlestes grandis* Nessov, 1982, type species by monotypy.

Distribution. Late Cretaceous (?early Cenomanian): Uzbekistan, Karapalkaplian Autonomous Republic (Khodzhakul Formation).

Comments. *Oxlestes* and *Khuduklestes* (described later) are very poorly known. The type materials are, for practical purposes, indeterminate, though they are notable in documenting the presence of rather large mammals in their respective faunas. Nessov (1982) originally assigned *Oxlestes* to Eutheria ?Palaeoryctidae, but later (Nessov, Sigogneau-Russell, and Russell, 1994) suggested that it might be a large deltatheroidan. We tentatively follow the latter referral, cautioning that it is based only on size.

Genus *Khuduklestes* Nessov, Sigogneau-Russell, and Russell, 1994
(figure 12.7G)

Diagnosis. This monotypic genus was established by Nessov, Sigogneau-Russell, and Russell (1994) for the holotype and only known specimen of *Khuduklestes bohlini*. The holotype is a relatively large axis, described and figured by Bohlin (1953: figure 20). As diagnosed by Nessov et al. (1994), the axis differs from that of *Oxlestes* in that the base of the odontoid process lacks a pronounced necklike area and is more strongly pointed anteriorly; the anterior part of the ventral ridge is not as strongly marked as in *Oxlestes*. The axis narrows behind the paired articular surfaces more than in *Oxlestes*, so that the lateral ridges are much closer to the edges of the posterior half of the centrum. Bohlin (1953) did not provide measurements of the axis, but judging from the magnification of his drawing, the axis with dens was about 20.7 mm, suggesting that it was subequal in size to *Oxlestes*.

Species. *Khuduklestes bohlini* Nessov, Sigogneau-Russell, and Russell, 1994, type species by monotypy.

Distribution. Late Cretaceous (?Cenomanian): China, Gansu Province, Tsondolein Khuduk.

Cohort Marsupialia Illiger, 1811

INTRODUCTION

As noted, we place all Cretaceous marsupials in superorder "Ameridelphia" Szalay, 1982 (see table 12.1). Among North American taxa, we recognize five genera as basal "ameridelphians," left *incertae sedis*, and four families. Three of these ("Alphadontidae," "Pediomyidae," and Stagodontidae) are placed in "Didelphimorphia" Gill, 1872; the fourth, Glasbiidae, is tentatively allocated to the otherwise South American Paucituberculata Ameghino, 1894. Marsupials did not fare well in the extinctions that occurred at or near the Cretaceous-Tertiary boundary (Archibald, 1996a): three of the four families are restricted to the Cretaceous, and the fourth is a mainly Tertiary family with only a doubtful occurrence in the Cretaceous. The "Alphadontidae" are the most diverse, including six genera. "Alphadontids" are dentally conservative and so have often been placed in the Didelphidae (which include living opossums) or Peradectidae (known mainly or exclusively from the Tertiary). The smallest species (e.g., *Alphadon perexiguus, Aenigmadelphys archeri*) were shrew sized and presumably insectivorous. Larger taxa (species of *Turgidodon*) ranged up to approximately squirrel size and generally have teeth with more bulbous cusps, suggesting more omnivorous dietary fare. "Alphadontidae" first appear in the Cenomanian (Eaton, 1993b).

"Pediomyidae" (three genera) are distinguished on molar specializations, such as reduction or loss of the anterior stylar shelf. Partly on this basis, they have been compared favorably or allied with various South American and Australian taxa (see, e.g., Marshall, 1977, 1987; Reig et al., 1987). We adhere to the consensus view that "Pediomyidae" represent a monophyletic group restricted to the Late Cretaceous of North America (e.g., Clemens, 1979a; Fox, 1979a; Aplin and Archer, 1987). "Pediomyids" have a range of size variation similar to what is seen in "Alphadontidae." A possible unnamed "pediomyid" has been reported from the Santonian (Eaton, 1999a), but the oldest fossils identified to the generic level are of early Campanian age (Fox, 1971b, 1987a). "Pediomyids" are common elements of Judithian through Lancian faunas in the central and northern parts of the Western Interior, but they are rare or absent in more southerly assemblages (e.g., Cifelli, 1994).

Stagodontidae (three genera, one tentatively included), also endemic to the Late Cretaceous of North America, were large relative to other mammals of their respective faunas. The largest species, placed in *Didelphodon*, reached an adult size close to that of the living Virginia opossum. Stagodontidae are sometimes viewed as an early-diverging lineage of North American Marsupialia, though there is no consensus and published opinions are varied (e.g., Marshall and Kielan-Jaworowska, 1992; Fox and Naylor, 1995; Rougier et al., 1998). The dentition bears some dental features associated with carnivory. This has led to the suggestion of a special relationship to the similarly specialized Borhyaenoidea of South America (Marshall et al., 1990), among others. Herein we adopt what we take to be the consensus view (e.g., Aplin and Archer, 1987; Szalay, 1994; Fox and Naylor, 1995; McKenna and Bell, 1997; Muizon and Lange-Badré, 1997; Muizon, 1998) in considering the molar similarities to borhyaenoids as having been acquired independently. The oldest possible stagodontid is *Pariadens*, from the Albian–Cenomanian (see Cifelli and Eaton, 1987; Fox and Naylor, 1995 for contrasting viewpoints on affinities of *Pariadens*). Like "pediomyids," stagodontids are notably scarce in southerly assemblages.

Glasbiidae (often considered as a subfamily of Didelphidae, e.g., Clemens, 1966) are represented by the single genus *Glasbius*. The two known species are small and are characterized by low-crowned molars that are morphologically distinctive among North American forms. *Glasbius* is known only from the Lancian and it is restricted to a few local faunas in Wyoming, Montana, and Saskatchewan.

A notable development in recent years has been the recognition of Marsupialia in the Cretaceous of Asia, though there is not universal agreement on this point. Three taxa (one unnamed) are tentatively included. *Asiatherium*, known by a skull and skeleton, is morphologically divergent relative to North American taxa, and is placed in its own order, Asiadelphia (see Trofimov and Szalay, 1994; Szalay and Trofimov, 1996). The other named taxon is *Marsasia*, also possibly referable to Asiadelphia, but based on more fragmentary specimens (Nessov, 1997; Averianov and Kielan-Jaworowska, 1999). Finally, an unnamed taxon represented by the Guriliin Tsav skull (see earlier comments under Deltatheroida) is tentatively recognized as a marsupial of uncertain affinities and is not treated in Systematics.

One named marsupial species from South America may be of Cretaceous age. We regard its generic affinities as uncertain and tentatively place it in the "didelphimorphian" family Peradectidae Crochet, 1979a, which are otherwise restricted to the Tertiary.

Diagnosis. Metatherians with the following derived features that are absent in Deltatheroida: eruption of last premolar (P/p3) delayed with respect to eruption of the molar series. Upper molars with protocone more strongly developed, both transversely and anteroposteriorly; postprotocrista extends labially past base of metacone; conules more strongly developed and placed closer to bases of paracone and metacone; tendency for development of stylar cusps and enlargement of metacone with respect to the paracone. Lower molars lacking distal metacristid and lingual cuspule on precingulid, and with more broadly expanded talonid, with lesser height differential relative to trigonid, entoconid much more strongly developed, labial postcingulid present; tendency for lingual placement of hypoconulid and its twinning with entoconid. Derived dental specializations of Deltatheroida, including upper molars with expanded postmetacrista and distal stylar shelf, and lower molars with enlarged paraconid and paracristid, are primitively lacking, and their presence in certain groups (Stagodontidae, Borhyaenoidea, Dasyuroidea) is assumed to represent independent acquisition (Muizon and Lange-Badré, 1997).

Included Taxa. Superorders "Ameridelphia" Szalay, 1982, new rank; and Australidelphia (Szalay, 1982), new rank (Australidelphia are not known from the Mesozoic and are not treated herein).

Distribution. (Mesozoic distribution only) Early–Late Cretaceous (Albian-Cenomanian through Maastrichtian): North America; ?Late Cretaceous (?Maastrichtian): South America; Late Cretaceous (Coniacian–?late Campanian): Asia.

Superorder "Ameridelphia" Szalay, 1982, New Rank

Included Taxa. Orders Asiadelphia Trofimov and Szalay, 1994; "Didelphimorphia" Gill, 1872, Paucituberculata Ameghino, 1894, and Sparassodonta Ameghino, 1894; and the following taxa of uncertain ordinal affinities: *Anchistodelphys* Cifelli, 1990b; *Iugomortiferum* Cifelli, 1990b; and *Kokopellia* Cifelli, 1993a.

Distribution. (Mesozoic distribution only) Early–Late Cretaceous (Albian-Cenomanian through late Maastrichtian, Lancian): North America; ?latest Cretaceous: South America; Late Cretaceous (Coniacian through ?late Campanian): Asia.

Comments. An extended characterization for "Ameridelphia" was given by Szalay (1994: 305–308). As he pointed out, most or all of the characters cited are presumably primitive for Marsupialia as a whole and do not serve to exclude—where known—North American Cretaceous taxa. A further complication is that Australidelphia are conceived as having originated from among "Ameridelphia" (e.g., Szalay and Trofimov, 1996). From this it is apparent that providing definition and diagnosis is not a straightforward matter and, as the issue is peripheral to the current work, it is not attempted here.

North American Cretaceous marsupials have traditionally been placed within (or allied closely with) Didelphidae (e.g., Simpson, 1945) and, for some taxa at least, this practice continues (McKenna and Bell, 1997). Accumulating data in the form of derived dental, cranial, and postcranial characters (e.g., Szalay, 1994; Muizon et al., 1997; Muizon and Cifelli, 2001) support placement of

Didelphidae within the context of an endemic South American radiation and sharpen their distinction from the Cretaceous taxa of North America. On this basis, we distinguish the North American taxa by placing them in their own families. Nonetheless, current evidence continues to support the hypothesis that South American marsupials took origin from within the North American radiation (e.g., Muizon and Cifelli, 2001). We recognize this by placing most of the Cretaceous taxa within "Didelphimorphia," conceived as a paraphyletic taxon (which it is in any case, using any of the other published definitions). *Glasbius* is excluded, being tentatively placed in the otherwise South American Paucituberculata. *Iugomortiferum* is inadequately known to permit ordinal placement, while *Kokopellia* and *Anchistodelphys* are structurally suitable antecedents to Asiadelphia (see Cifelli and Muizon, 1997); these three are excluded from "Didelphimorphia."

Order and Family *incertae sedis*
Genus *Anchistodelphys* Cifelli, 1990b
(figure 12.8C)

Diagnosis (modified from Cifelli, 1990b). Small, primitive marsupials, resembling *Kokopellia, Aenigmadelphys,* and *Albertatherium* in having an anteroposteriorly compressed protocone; differs from *Kokopellia* in having small, variable cuspules developed on the stylar shelf, better-developed, more labially placed conules, and (on lower molars) lingually placed hypoconulid (twinned with entoconid); differs from *Aenigmadelphys, Albertatherium,* and other Cretaceous marsupials in lacking consistent, well-developed stylar cusps C and D.

Species. *Anchistodelphys archibaldi* Cifelli, 1990b, type species; and tentatively referred *A. delicatus* Cifelli, 1990c.

Distribution. Late Cretaceous (?Turonian and early Campanian): United States, Utah (?Straight Cliffs Formation and Wahweap Formation).

Comment. Revision by Johanson (1996b) transferred a specimen described as *A. archibaldi* (see Cifelli, 1990b) to *Varalphadon wahweapensis.*

Genus *Iugomortiferum* Cifelli, 1990b
(figure 12.8B)

Diagnosis (modified from Cifelli, 1990b). Molar cusps similar to those of *Turgidodon* in being inflated, but differ in being lower crowned. Upper molars differ from those of *Turgidodon* and other "Alphadontidae" in having a more

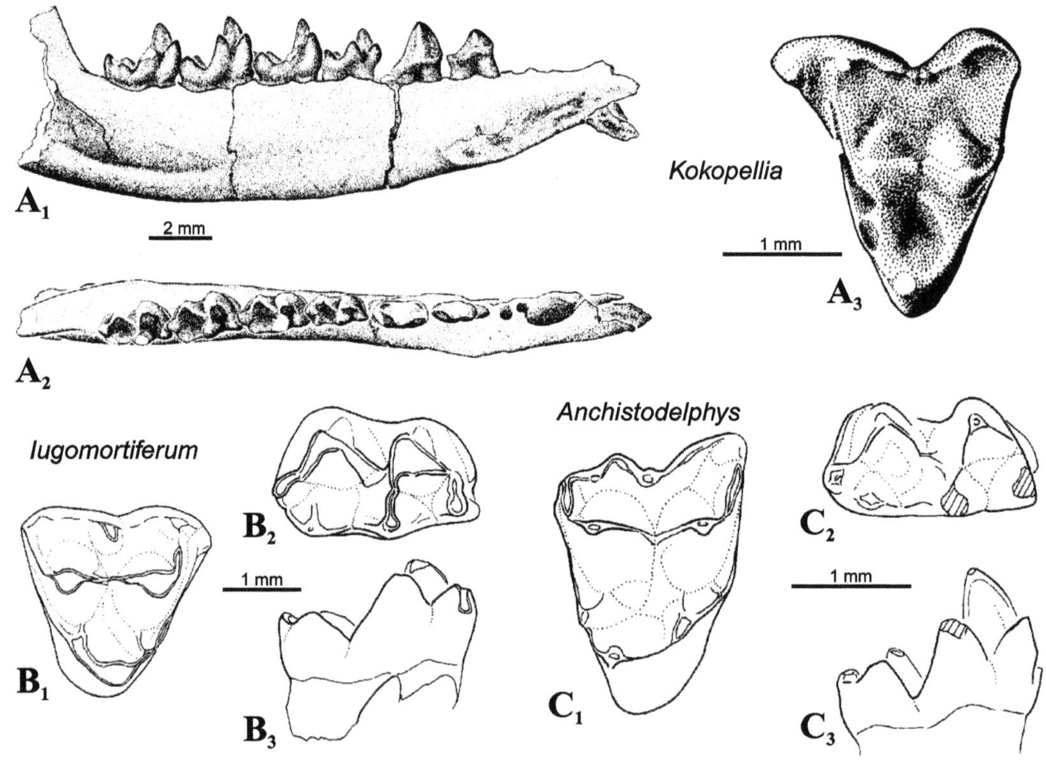

FIGURE 12.8. "Didelphimorphia" *incertae sedis.* A, *Kokopellia juddi:* left dentary with p2–3, m1–4 (holotype) in lingual (A_1) and occlusal (A_2) views; left M3 in occlusal view (A_3). B, *Iugomortiferum:* B_1, right upper molar (holotype of *I. thoringtoni*) in occlusal view; left lower molar of the same species in occlusal (B_2) and lingual (B_3) views. C, *Anchistodelphys archibaldi:* C_1, left M3 (holotype) in occlusal view; C_{2-3}, left lower molar in occlusal (C_2) and lingual (C_3) views. *Source:* A_{1-2} from Cifelli (1993a); A_3, B, C, originals.

anteroposteriorly developed, centrally placed protocone, with outline of ?M1 approximating an equilateral triangle in occlusal view. Conules differ from those of most Cretaceous Marsupialia in being more weakly developed. Stylar cusps A and D lacking; stylar cusp C low but broad based.

Species. *Iugomortiferum thoringtoni* Cifelli, 1990b, type species by monotypy, and a possibly referable species left in open taxonomy by Cifelli (1990b).

Distribution. Late Cretaceous (Aquilan): United States, Utah (Wahweap Formation).

Comment. *Iugomortiferum* is obviously specialized in some respects, such as the inflated, low, blunt molar cusps. In other features, such as the symmetry of upper molars and the presence of a stylar cusp C but not D, it is atypical of North American Cretaceous marsupials, so that its referral remains uncertain.

Genus *Kokopellia* Cifelli, 1993a
(figures 12.5A, 12.8A)

Diagnosis (modified from Cifelli, 1993a). Differs from metatherian morphotype in having the following presumed apomorphies of Marsupialia: loss of distal metacristid, lower molars with keeled paraconid, broad talonid with lesser trigonid-talonid height differential and labial postcingulid; upper molars with protocone more anteroposteriorly and transversely developed, postprotocrista extending labially past base of metacone and somewhat stronger, more labially placed conules. Plesiomorphic with respect to most Cretaceous marsupials (e.g., *Alphadon*) in having lower molars with hypoconulid somewhat more centrally placed and not fully twinned with the entoconid, less development of entoconid; upper molars with proportionately broader stylar self, cusps C and D lacking, more anteroposteriorly compressed protocone, and conules more weakly developed and not placed so close to bases of paracone and metacone.

Species. *Kokopellia juddi* Cifelli, 1993a, type species by monotypy.

Distribution. Early–Late Cretaceous (Albian-Cenomanian boundary): United States, Utah (Cedar Mountain Formation). [Note: *Kokopellia* is the best represented of pre-Campanian marsupials in North America, being known by most of the dentary, lower dentition, and upper molar series (Cifelli and Muizon, 1997). At present, *Kokopellia* is the oldest generally accepted marsupial (e.g., McKenna and Bell, 1997).]

Order Asiadelphia Trofimov and Szalay, 1994

Diagnosis. As for the family Asiatheriidae.

Included Taxa. Family Asiatheriidae Trofimov and Szalay, 1994.

Distribution. As for the family Asiatheriidae.

Family Asiatheriidae Trofimov and Szalay, 1994

Diagnosis (adapted from Trofimov and Szalay, 1994; Cifelli and Muizon, 1997; Averianov and Kielan-Jaworowska, 1999). Primitive, where known, with respect to other "Ameridelphia" in lacking prominent distolateral process on the scaphoid and in having a fibula that is extremely slender proximally. Dentary differs from that of most other Marsupialia in being of subequal depth along the tooth row and in having a very deep masseteric fossa. Upper molars, where known, differ from those of most other early marsupials in the trenchant protocone, with strong development of pre- and (especially) postprotoconal cingula; stylar shelf reduced, especially anteriorly, with preparacrista and stylocone lacking.

Included Taxa. *Asiatherium* Trofimov and Szalay, 1994, type genus; and *Marsasia* Nessov, 1997.

Distribution. Late Cretaceous (Coniacian–?late Campanian): Asia.

Comments. Diagnosis of the family is based, perforce, mainly on the single known specimen of *Asiatherium reshetovi*, described in detail by Szalay and Trofimov (1996). *Asiatherium* has been summarily excluded from Marsupialia in some recent accounts (Fox, 1997a; McKenna and Bell, 1997). We tentatively retain Asiatheriidae in Marsupialia based on dental evidence (Cifelli and Muizon, 1997). Allocation of *Marsasia* to Asiatheriidae is suggested by derived features of the dentary (Averianov and Kielan-Jaworowska, 1999).

Genus *Asiatherium* Trofimov and Szalay, 1994
(figures 12.6A, 12.9A)

Diagnosis. Differs from *Marsasia* in having condylar process of the mandible situated much higher above the tooth row, less deeply excavated mandibular fossa, and much taller entoconid on lower molars.

Included Species. *Asiatherium reshetovi* Trofimov and Szalay, 1994, type species by monotypy.

Distribution. Late Cretaceous (?late Campanian): Mongolia (?Baruungoyot Formation).

Genus *Marsasia* Nessov, 1997
(figure 12.9B)

Diagnosis (abbreviated from Averianov and Kielan-Jaworowska, 1999: 72–73). "Shares with *Asiatherium* the shape of the dentary, having the horizontal ramus of almost equal depth along the tooth row (subparallel alveolar and ventral borders), and similar shape of a large and very deep masseteric fossa (known also in other groups, e.g., in zhelestids). Differs from *Asiatherium* in having the

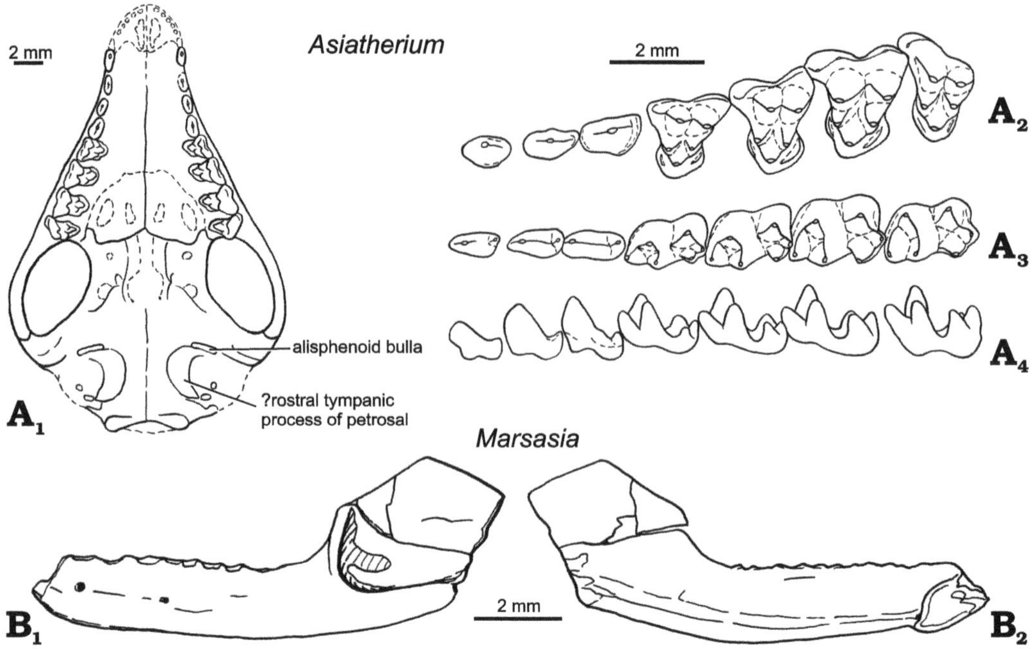

FIGURE 12.9. Asiadelphia. A, *Asiatherium reshetovi:* skull of the holotype specimen in palatal view (A_1); postcanine dentition of the same, reconstructed from right and left sides; left upper teeth in occlusal view (A_2); right lower teeth in occlusal (A_3) and lingual (A_4) views. B, *Marsasia aenigma:* incomplete left dentary (holotype) with alveoli for seven double-rooted postcanine teeth in lateral (B_1) and medial (B_2) views. *Source:* originals.

condylar process situated very low, only slightly above the level of the molars, and deeper masseteric fossa, probably the deepest among known Cretaceous marsupials. Shares with . . . *Asiatherium, Pariadens, Glasbius,* and several other Cretaceous marsupials relatively low cusps of the molars, and a small difference between the height of the trigonid and talonid. Differs from *Asiatherium* in having the talonid (at least in m4) slightly longer than the trigonid, and shares this character with *Kokopellia,* as well as with some Cretaceous eutherians, such as zhelestids. Shares with *Kokopellia* and *Asiatherium,* and with many Cretaceous and Paleogene eutherians, the posterior (rather than lingual) position of the hypoconulid, and the entoconid and hypoconulid slightly approximated (but not strongly twinned as in Late Cretaceous marsupials). The characteristic feature of *Marsasia* is its extremely small and low entoconid, probably the lowest among the known Cretaceous marsupials; it is also small in *Kokopellia,* while high in *Asiatherium.*"

Included Species. *Marsasia aenigma* Nessov, 1997, type species by monotypy, and a species left in open nomenclature by Averianov and Kielan-Jaworowska (1999).

Distribution. Late Cretaceous (Coniacian): Uzbekistan (middle part of the Bissekty Formation).

Comments. When first described (Nessov, 1997), *Marsasia aenigma* was known only by a partial, edentulous dentary, leaving some doubt as to its affinities. Additional specimens described by Averianov and Kielan-Jaworowska (1999), including another edentulous dentary and (as *Marsasia* sp.) a dentary fragment with one molar, have improved knowledge of the genus. Although many details remain unknown, characteristics of the dentary (cited in the diagnosis of the family) are sufficient, at present, to group *Marsasia* with *Asiatherium.*

Order "Didelphimorphia" Gill, 1872

Diagnosis. Generally plesiomorphic marsupials; differ from Asiadelphia and basal "ameridelphians" such as *Kokopellia* in presence of stylar cusp D and twinning of hypoconulid with entoconid (apomorphies). Differ from Asiadelphia in having a shallower masseteric fossa, less trenchant protocone, stylocone, and preparacrista and lacking protoconal cingula (plesiomorphies).

Included Taxa. Superfamily "Didelphoidea" Gray, 1821, and families "Alphadontidae" Marshall, Case, and Woodburne, 1990; "Pediomyidae" Simpson, 1927d, Peradectidae Crochet, 1979a, and Stagodontidae Marsh, 1889a.

Distribution. (Mesozoic distribution only) Late Cretaceous (Cenomanian–Lancian): North America; Late Cretaceous (?Maastrichtian): South America.

Comments. In addition to the Cretaceous taxa noted above, "didelphoids" are taken here to include Didelphidae Gray, 1821, Sparassocynidae Reig, 1958, and didelphid-like genera (currently lacking secure family placement) from the Early Tertiary of South America (Muizon and Cifelli, 2001). Didelphidae are readily recognizable as a monophyletic group (e.g., Szalay, 1982, 1994; Reig et al., 1987) but comprehensive definition and diagnosis of the more inclusive "Didelphoidea" is more problematic (Muizon and Cifelli, 2001). Treatment of the issue is beyond our present scope, but we tentatively recognize "Didelphoidea" as distinct from the stem taxa of didelphimorphians of the North American Cretaceous in sharing (or in having independently acquired) the following derived features (see Muizon et al., 1997; Muizon and Cifelli, 2001): upper molars with more robust protocone bearing a posteriorly swollen base, metacone taller than paracone, centrocrista V-shaped, postprotocrista not extending labially past the base of metacone; lower incisors with i1 enlarged and i2 staggered; lower molars with taller metaconid (relative to both protoconid and paraconid) and entoconid, with cristid obliqua attaching to back of trigonid in labial position (behind the protoconid).

Family "Alphadontidae" Marshall, Case, and Woodburne, 1990

Diagnosis. Primitive didelphimorphians, lacking asiadelphian, "didelphoid," "pediomyid," peradectid, and stagodontid specializations (listed in other diagnoses and not repeated here), and differentiated from *Kokopellia* and structurally similar genera in having the following apomorphies: upper molars with well-developed conules, placed near bases of paracone and metacone; anteroposteriorly longer protocone, proportionately narrower stylar shelf; stylar shelf with at least stylar cusp D and often stylar cusp C present; stylar cusp A placed lower with respect to occlusal surface than cusp B and separated from that cusp by a deep notch; lower molars with paraconid placed at lingual margin of tooth, in line with metaconid and entoconid.

Included Genera. *Alphadon* Simpson, 1927d, type genus; *Aenigmadelphys* Cifelli and Johanson, 1994; *Albertatherium* Fox, 1971b, *Protalphadon* Cifelli, 1990a, *Turgidodon* Cifelli, 1990a, and *Varalphadon* Johanson, 1996b.

Distribution. Late Cretaceous (Cenomanian–Lancian): North America.

Comments. The rationale for removing these genera from Didelphidae is discussed under "Ameridelphia" and "Didelphomorphia," earlier. Additional brief comments are called for here regarding an alternative arrangement that has been employed in recent years. With recognition that modern didelphids share a number of advanced characters not seen in putative members of the family known from the Cretaceous, some workers adopted the taxonomy of Crochet (1979a, 1980), grouping *Alphadon* and structurally similar taxa in Peradectidae (or lower taxonomic categories based on the same name). Peradectidae, which otherwise include Tertiary marsupials, have also traditionally been placed in Didelphidae. The concept of an expanded Peradectidae has drawn criticism, in part because of uncertainties surrounding the origin of Didelphidae *sensu stricto.* Tertiary peradectids may be distinguished from modern didelphids on much the same basis as for the Cretaceous taxa. However, these more advanced peradectids have specializations not seen in *Alphadon* and others from the Cretaceous (e.g., Krishtalka and Stucky, 1983a; Montellano, 1992). In some respects, certain of the Tertiary peradectids resemble modern didelphids more than do *Alphadon* and others from the Cretaceous. Did modern didelphids originate from a *Peradectes*-like taxon (Crochet, 1979b) or from a structurally more primitive ancestor (e.g., Muizon and Cifelli, 2001)? In the latter case, similarities of peradectids (*sensu stricto*) to modern didelphids would be regarded as homoplasies. Resolution of these taxonomic conundra cannot be attempted here, and it is in any case debatable whether data from the existing fossil record are adequate (see Szalay, 1993a). Analysis by Johanson (1996b) supports the integrity of a monophyletic Peradectidae (see Krishtalka and Stucky, 1983a) and highlights the problems regarding classification of more primitive marsupials from the Cretaceous. Our interim solution, pending additional fossils and analysis, is to group the Cretaceous taxa into a paraphyletic family, "Alphadontidae."

For many years, *Alphadon* was one of only a few marsupials with a generalized molar pattern recognized from the Cretaceous of North America (see summary by Clemens, 1979a). Recent work has resulted in the splitting of *Alphadon* species into several genera (Cifelli, 1990a; Johanson, 1996b) and the recognition of new taxa, some from the earlier part of the Late Cretaceous (Cifelli, 1990c; Eaton, 1993b). Notably, some of these are referable to *Alphadon* itself, which as presently known has the remarkably long geologic range of Cenomanian–late Maastrichtian.

Genus *Aenigmadelphys* Cifelli and Johanson, 1994 (figure 12.10B,C)

Diagnosis (modified after Cifelli and Johanson, 1994). Differs from *Alphadon* and structurally similar genera in lacking a stylar cusp C and in having a taller paracone with more lingually extended base relative to metacone, more anteroposteriorly compressed protocone, and lower molars with greater height differential between trigonid and talonid. Differs from *Kokopellia* in having lower molars with more strongly developed entoconid and lingually placed hypoconulid; differs from *Kokopellia* and other

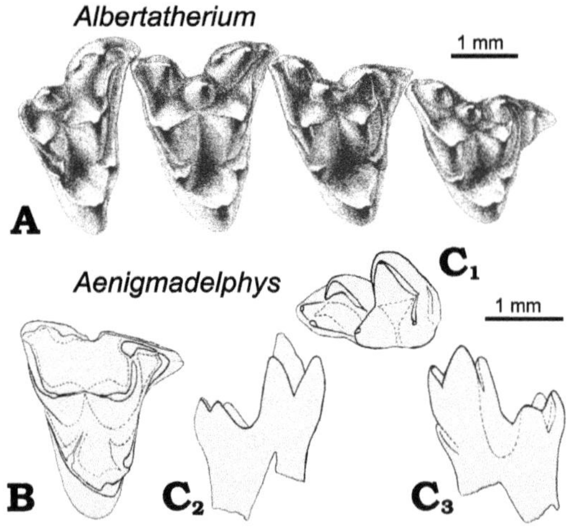

FIGURE 12.10. "Alphadontidae." A, *Albertatherium primum*, right M1–4 (composite) in occlusal view. B, C, *Aenigmadelphys archeri:* B, right M2 in occlusal view; C, left m4 in occlusal (C_1), lingual (C_2), and labial (C_3) views. *Source:* A, from Fox (1987b); B, C, originals.

basal didelphimorphians in having lingually placed paraconid in line with metaconid and entoconid. Similar to *Anchistodelphys* in anteroposterior compression of protocone and differing in having stylar cusp D consistently present. Differs from *Iqualadelphis* in having a relatively larger stylar cusp B that is positioned labial rather than anterior to the paracone, preparacrista joining stylar cusp B and paracone, paracone relatively taller than metacone, and anterior stylar shelf broader.

Included Species. *Aenigmadelphys archeri* Cifelli and Johanson, 1994, type species by monotypy; and a species left in open nomenclature by Diem (1999).

Distribution. Late Cretaceous (Judithian–"Edmontonian"): North America.

Comment. Fossils of *Aenigmadelphys archeri* were erroneously referred to *Iqualadelphis lactea* by Cifelli (1990a).

Genus *Albertatherium* Fox, 1971b
(figure 12.10A)

Diagnosis (after Johanson, 1994: 595). "Upper molars similar to *Aenigmadelphys* and *Protalphadon*, differing from *Alphadon* and *Turgidodon* in being generally short anteroposteriorly and wide transversely; protocone anteroposteriorly compressed, no inflation of protoconal sides; similar to *Aenigmadelphys* in stylar shelf wide on M2 and M3; base of protocone extending farther lingually than base of the metacone; differing from *Protalphadon* in the consistent presence or absence of stylar cusp C at the species level; similar to *Alphadon* and *Turgidodon*, differing from *Aenigmadelphys* and *Protalphadon*, in paracone and metacone subequal in height or paracone slightly taller; stylar cusps A, B, and D strongly developed; upper molars differing from *Aenigmadelphys, Protalphadon, Alphadon*, and *Turgidodon* in: ectoflexus deep, rounded, giving stylar shelf bilobed appearance on all upper molars; posterior stylar shelf strongly developed on M1–M3; conules strongly developed, preparaconular crista and postmetaconular crista appear to 'overhang' the pre- and postprotocrista, internal conular cristae high, sharp; trigon basin deep; molars overall appearing slender, narrow, with sharp cusps and crests."

Included Species. *Albertatherium primum* Fox, 1971b, type species; and *A. secundum* Johanson, 1994.

Distribution. Late Cretaceous (Aquilan): Canada, Alberta (Milk River Formation).

Genus *Alphadon* Simpson, 1927d
(figures 12.5B,C, 12.11A)

Diagnosis (after Johanson, 1996b). Differs from more primitive "alphadontids" (*Varalphadon, Protalphadon*) in: stylar cusp C consistently present; paracone and metacone more rounded, conical, and widely separated at their bases; deeper ectoflexus on upper molars; protocone more anteroposteriorly elongate; reduced height differential of trigonid to talonid. Differs from *Turgidodon* in having taller cusps, with stronger height discrepancy between paracone and cusp B; lower, more crestlike conular cristae, premolars less bulbous and robust. Differs from Peradectidae in having cusp A taller and better separated from cusp B; taller cusp B relative to cusps C and D; deeper ectoflexus; paracone and metacone subequal in height (metacone taller in Peradectidae); postprotocrista extending labially past base of protocone; cristid obliqua terminates below notch between metaconid and protoconid, rather than labial to the notch.

Included Species. *Alphadon marshi* Simpson, 1927d, type species; *A. attaragos* Lillegraven and McKenna, 1986; *A. clemensi* Eaton, 1993b; *A. eatoni* Cifelli and Muizon, 1998a; *A. halleyi* Sahni, 1972; *A. jasoni* Storer, 1991; *A. lillegraveni* Eaton, 1993b; *A. perexiguus* Cifelli, 1994; *A. sahnii* Lillegraven and McKenna, 1986; *A. wilsoni* Lillegraven, 1969; and numerous species left in open nomenclature (Fox, 1971b, 1976b, 1989; Breithaupt, 1982; Wilson, 1983; Standhardt, 1986; Fiorillo, 1989; Eaton, 1993b, 1999a; Hoganson et al., 1994; Sankey, 1998; Cifelli, Nydam, Eaton et al., 1999; Eaton, Cifelli et al., 1999).

Distribution. Late Cretaceous (Cenomanian–Lancian): North America.

Comments. *Alphadon* has traditionally been a broadly conceived genus (e.g., Clemens, 1966; Lillegraven, 1969; Sahni, 1972). Cifelli (1990a,b) removed several (presumably primitive) species to *Protalphadon* and placed others

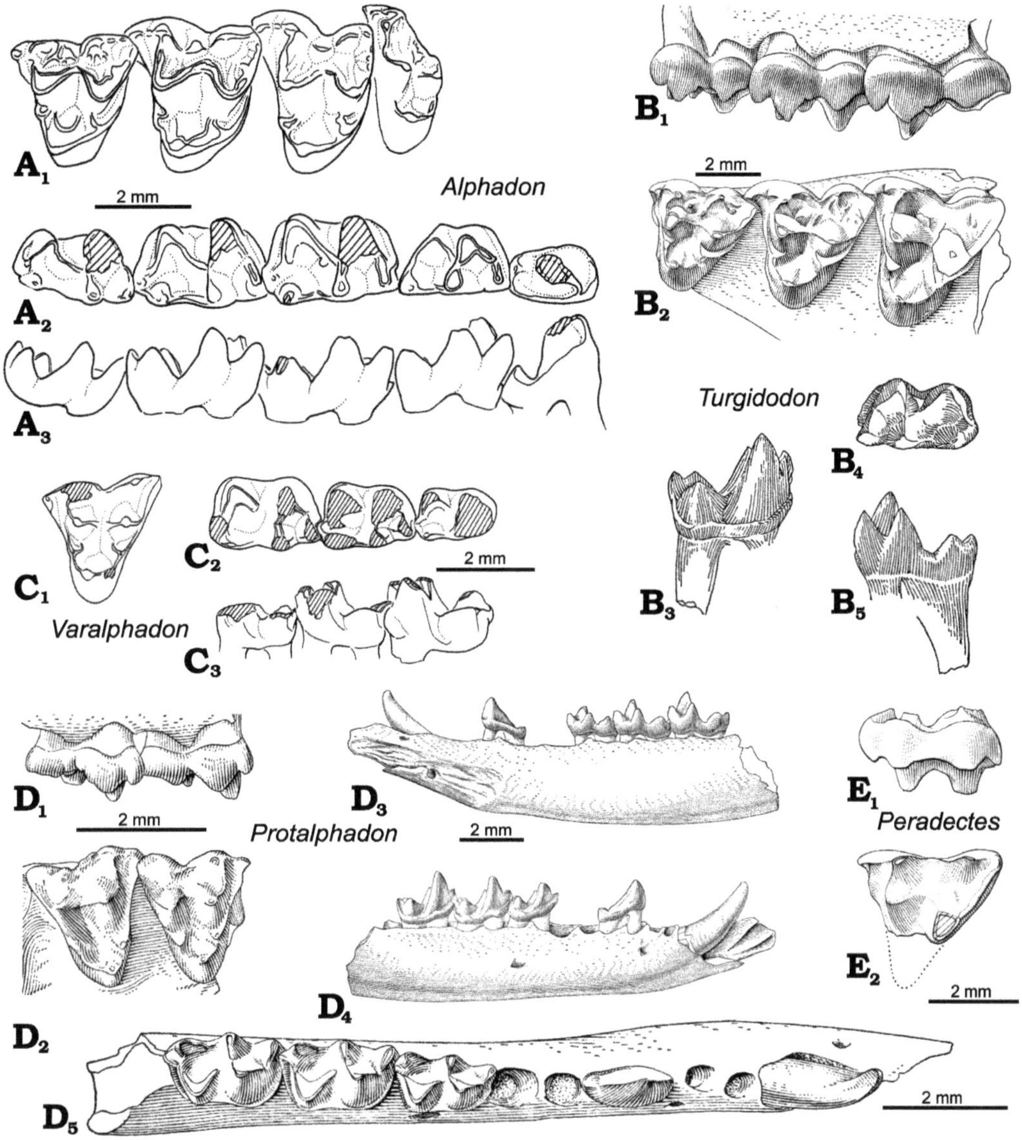

FIGURE 12.11. "Alphadontidae" and Peradectidae. A, *Alphadon jasoni:* A_1, left M1–4 in occlusal view; A_{2-3}, left p3, m1–4 in occlusal (A_2) and lingual (A_3) views. B, *Turgidodon rhaister:* B_{1-2}, left M1–3 (holotype) in labial (B_1) and occlusal (B_2) views; B_{3-5}, right m4 in labial (B_3), occlusal (B_4) and lingual (B_5) views. C, *Varalphadon wahweapensis:* C_1, left M3 (holotype) in occlusal view; C_{2-3}, left m1–3 in occlusal (C_2) and labial (C_3) views. D, *Protalphadon lulli:* D_{1-2}, right M2–3 in labial (D_1) and occlusal (D_2) views; D_{3-5}, right dentary with c, p2, m1–3 in lingual (D_3), labial (D_4) and occlusal (D_5) views. E, cf. *Peradectes austrinum:* Left upper molar (holotype) in labial (E_1) and occlusal (E_2) views. *Source:* A, C, originals; B, D, from Clemens (1966); E, modified from Sigé (1972).

(thought to share apomorphies) in *Turgidodon.* Our systematic arrangement of *Alphadon* and similar genera largely follows Johanson (1996b), the most recent and comprehensive account of the subject. *A. sahnii* was considered a junior synonym of *A. halleyi* by Montellano (1988, 1992; see also Johanson, 1996b); we consider the species as distinct, based on the consistent presence of both morphs at Judithian localities in the southern and central part of North America (Lillegraven and McKenna, 1986; Cifelli, 1990a, 1994). An *Alphadon*-like upper molar has been reported from the Late Cretaceous of Portugal (Antunes et al., 1986).

Genus *Protalphadon* Cifelli, 1990a
(figure 12.11D)

Diagnosis (modified after Johanson, 1996b). Differs from *Varalphadon, Alphadon,* and *Turgidodon* in having paracone taller than metacone, the cusps being of subequal width when viewed labially; postmetacingulum absent; stylar cusp C, if present, anterior to ectoflexus; stylar cusp D small, especially on M3. Resembles *Varalphadon* and differs from the remaining genera in variable presence of stylar cusp C, close approximation of paracone and metacone bases, protocone more anteroposteriorly constricted, and greater height differential between trigonid and talonid.

Included Species. *Protalphadon lulli* (Clemens, 1966), type species; *P. foxi* Johanson, 1996b; and species left in open nomenclature (Eaton, 1993b; Eaton, Cifelli et al., 1999).

Distribution. Late Cretaceous (Cenomanian–Lancian): North America.

Comments. *Protalphadon* was erected by Cifelli (1990a) to include species previously placed in *Alphadon* but differing from *A. marshi* and similar species in variably lacking a stylar cusp C and several other features. Several species were later transferred from *Protalphadon* to *Varalphadon* by Johanson (1996b). *Protalphadon lulli* is, thus far, the only mammal shared at the species level between Late Cretaceous faunas of eastern and western North America (Grandstaff et al., 1992; Johanson, 1996b).

Genus *Turgidodon* Cifelli, 1990a
(figure 12.11B)

Diagnosis (after Johanson, 1996b). Differs from most closely similar genus, *Alphadon*, in having relatively lower, more rounded cusps on upper molars, with cusp B more nearly equal to paracone in height; internal conular cristae taller but less sharp and more wall-like; premolars more robust and bulbous.

Included Species. *Turgidodon praesagus* (Russell, 1952), type species; *T. lillegraveni* Cifelli, 1990a; *T. madseni* Cifelli, 1990a; *T. parapraesagus* (Rigby and Wolberg, 1987); *T. petaminis* Storer, 1991; *T. rhaister* (Clemens, 1966); *T. russelli* (Fox, 1979b); and species left in open nomenclature (Cifelli, 1990a; Eaton, 1993a).

Distribution. Late Cretaceous (Judithian–Lancian): North America.

Comments. At present, these large, somewhat distinctive marsupials are not known prior to the Judithian, and *Turgidodon* thus may prove useful in recognizing the onset of this land-mammal age (Cifelli, 1994; Cifelli et al., 2004).

Genus *Varalphadon* Johanson, 1996b
(figure 12.11C)

Diagnosis (modified after Johanson, 1996b). Closely similar to *Protalphadon*, differing in having paracone and metacone of subequal height and metacone broader than paracone in labial view. Differs from *Alphadon* and *Turgidodon* in variable presence of stylar cusp C, paracone and metacone uninflated and with more closely approximated bases, and greater height differential between trigonid and talonid.

Species. *Varalphadon wahweapensis* (Cifelli, 1990a), type species; *V. creber* (Fox, 1971b); *V. crebreforme* (Cifelli, 1990b).

Distribution. Late Cretaceous (Aquilan–Judithian): North America.

Family "Pediomyidae" Simpson, 1927d

Diagnosis. Cretaceous marsupials distinguished from the otherwise similar "Alphadontidae" in having upper molars with anterior part of stylar shelf somewhat reduced to lacking altogether; paracone placed more labially with respect to metacone; stylocone somewhat reduced to lacking; preparactista, where present, shorter, terminating labially anterior to stylocone, rather than at that cusp; lower molars with anterior attachment of cristid obliqua labial to notch between metaconid and protoconid (all presumed apomorphies).

Included Genera. "*Pediomys*" Simpson, 1927d, type genus; *Aquiladelphis* Fox, 1971b; and *Iqualadelphis* Fox, 1987a.

Distribution. Late Cretaceous (Aquilan–Lancian): North America.

Comments. This suprageneric group was proposed by Simpson (1927d) as a subfamily and elevated to family rank by Clemens (1966; see also Clemens, 1973a). Szalay (e.g., 1982, 1994) has repeatedly advocated a broad concept of "Pediomyidae," to include all nonstagodontid marsupials from the North American Cretaceous, but this view has found little support (e.g., Fox, 1983). On the other hand, dental specializations of "Pediomyidae," in particular, the reduction of the anterior part of the stylar shelf, have led others to suggest a special relationship to certain groups endemic to South America and/or Australia, such as borhyaenoids and other carnivorous marsupials (Marshall, 1977) or microbiotheriids (Marshall, 1987; Reig et al., 1987). Other evidence, insofar as known (Szalay, 1982, 1994), does not lend support to these views, and they have not been generally accepted (see summary by Aplin and Archer, 1987).

Fox (1987a,b) described *Iqualadelphis lactea* from the early Campanian of Canada and, noting that it possesses a stylar cusp D (but not C) and an anterior stylar shelf that is only moderately reduced, suggested that the strong reduction of the stylar shelf seen in advanced species of "*Pediomys*" probably took place independently in two or more lineages. If correct, this indicates that the genus and the family that contains it are mixed assemblages (see also Johanson, 1993). Resolution of the problem, which depends in part on undescribed fossils from the Milk River Formation, cannot be attempted here. Because *Iqualadelphis* shares dental specializations (admittedly subtle) that are probably synapomorphies with some (and perhaps all) "Pediomyidae," we choose to tentatively place it in this family, *contra* Fox (1987a) and Johanson (1993). Given the uncertainties surrounding the integrity of both the family and the type genus (Fox, 1987a), we place both names in quotes, except when referring to *Pediomys elegans*, the type species.

The range given previously under "Distribution" does not take into account unnamed taxa possibly referable to "Pediomyidae" (Sigé, 1971; Antunes et al., 1986; Eaton et al., 1998; Eaton, 1999a).

Genus "*Pediomys*" Marsh, 1889a
(figure 12.12C)

Diagnosis. Upper molars with anterior stylar shelf, stylocone, and preparacrista reduced or absent; stylar cusp D and, variably, stylar cusp C present. Differs from *Iqualadelphis* in the greater reduction of the anterior stylar shelf and stylocone (generally absent), greater length proportional to width of upper molars, protocone more anteroposteriorly expanded with posterobasal expansion, paracone more labially placed with respect to metacone, stronger development of conular cristae. Differs from *Aquiladelphis* in having taller, less massive cusps; stylar cusp C smaller or lacking; and single (or, more commonly, no) development of cusp B.

Included Species. "*Pediomys*" *elegans* Marsh, 1889a, type species; "*Pediomys*" *clemensi* Sahni, 1972; "*P.*" *cooki* Clemens, 1966; "*P.*" *exiguus* Fox, 1971b; "*P.*" *fassetti* Rigby and Wolberg, 1987; "*P.*" *florencae* Clemens, 1966; "*P.*" *hatcheri* (Osborn, 1898); "*P.*" *krejcii* Clemens, 1966; "*P.*" *prokrejcii* Fox, 1979c; and species left in open nomenclature (Lillegraven, 1972; Clemens et al., 1979; Fox, 1979c, 1989; Lillegraven and McKenna, 1986; Hunter et al., 1997; Peng, 1997; Hunter and Archibald, 2002).

Distribution. Late Cretaceous (Aquilan–Lancian): North America.

Comment. A premolar described as being similar to that of "*Pediomys*" has been reported from the Late Cretaceous of Portugal (Antunes et al., 1986).

Genus *Aquiladelphis* Fox, 1971b
(figure 12.12A)

Diagnosis (modified after Fox, 1971b). Stylar shelf moderately reduced anteriorly, with two small B cusps present; stylar cusp D larger than B cusps but smaller than C; stylar cusp C is the tallest cusp of the stylar shelf. Principal molar cusps low and robust; paracone and metacone labiolingually compressed and divergent; talonid basin of lower molars broad and shallow.

Included Species. *Aquiladelphis incus* Fox, 1971b, type species; *A. minor* Fox, 1971b; and *A. paraminor* Rigby and Wolberg, 1987.

Distribution. Late Cretaceous (Aquilan–Judithian, "?Edmontonian"): North America.

Comment. The possible presence of *Aquiladelphis* in the "Edmontonian" of the Williams Fork Formation, Colorado, is based on Diem (1999).

Genus *Iqualadelphis* Fox, 1987a
(figure 12.12B)

Diagnosis (after Johanson, 1993: 375). "Upper molars similar to '*Pediomys*' *prokrejcii* and '*P.*' *krejcii* in: anterior stylar shelf reduced labial to paracone, relative to posterior stylar shelf; ectoflexus shallow, broad, rimmed by thin edge; paracone, metacone subequal in height, conical; paracone shifted slightly labially relative to the metacone; differs from "*Pediomys*" *prokrejcii* and "*P.*" *krejcii* in: upper molars smaller with greater transverse width to anteroposterior length ratio; anteroposteriorly compressed protocone, lacking posterobasal expansion; anterior shelf labial to paracone relatively wider; small stylar cusp B positioned anterolabially relative to paracone; cusp A closely appressed to B on shelf; stylar cusp D approximately same size as or larger than B; paracone and metacone sharp, uninflated; paracone only slightly shifted labially relative to the metacone, internal conular cristae present but low and weakly developed."

Included Species. *Iqualadelphis lactea* Fox, 1987a, type species by monotypy.

Distribution. Late Cretaceous (Aquilan): Canada, Alberta (Milk River Formation).

Comment. Cifelli (1990a) erroneously reported the presence of this species in the Kaiparowits Formation (Judithian) of Utah; the fossils in question were later transferred to *Aenigmadelphys archeri* (see Cifelli and Johanson, 1994; Johanson, 1993). Fossils possibly referable to *Iqualadelphis*, not included in the distribution cited above, were reported from an unnamed unit by Eaton (1999a).

Family Peradectidae Crochet, 1979a

Diagnosis (modified after Johanson, 1996a,b, and Szalay, 1994). Primitive "didelphimorphians" distinguished from "Alphadontidae" in having the following derived features of the dentition. Upper molars with relatively smaller stylar cusp A placed close to stylar cusp B and about on the same level on the stylar shelf; cusp B reduced in size and subequal to stylar cusp C; ectoflexus shallow on all upper molars, not becoming deeper posteriorly; metacone

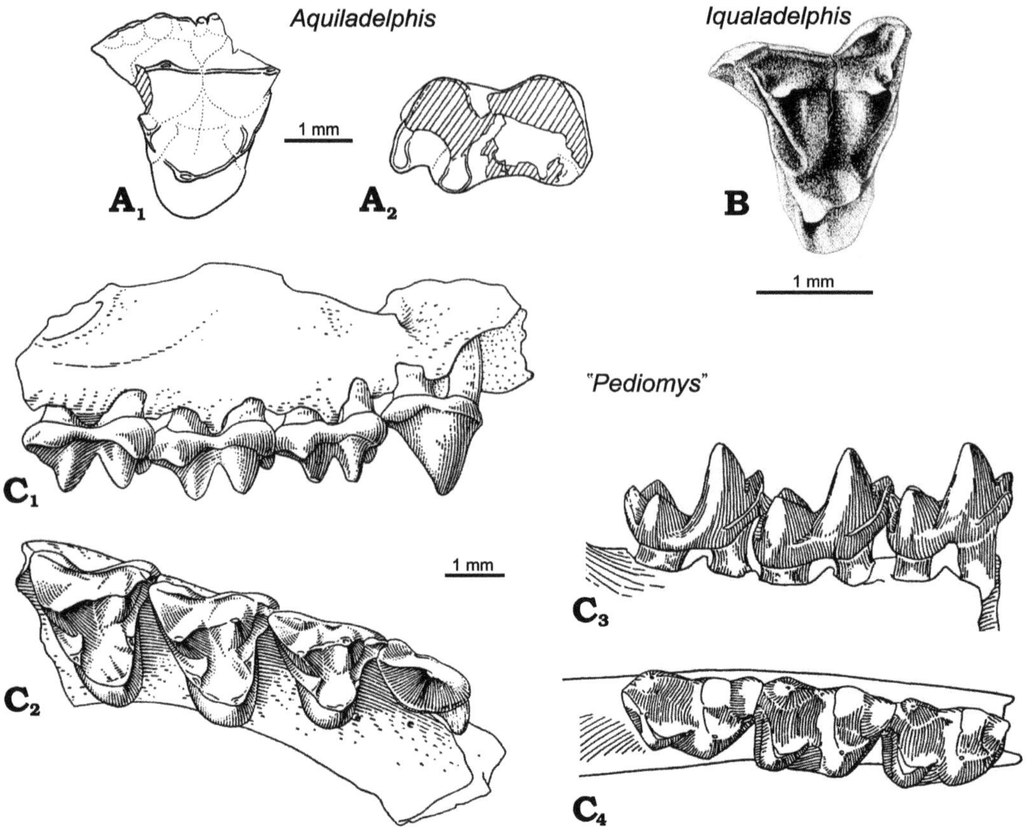

FIGURE 12.12. "Pediomyidae." A, *Aquiladelphis incus:* left M3 (holotype) in occlusal view (A_1); right mx in occlusal view (A_2). B, *Iqualadelphis lactea,* left M2 (holotype) in occlusal view. C, "*Pediomys*" *cooki* and "*P.*" ?*cooki:* C_{1-2}, "*P.*" *cooki,* right maxilla with P3, M1–3 (holotype) in labial (C_1) and occlusal (C_2) views. C_{3-4}, "*P.*" ?*cooki,* right dentary with m2–4 in labial (C_3) and occlusal (C_4) views. *Source:* A, original; B, modified from Fox (1987b); C, modified from Clemens (1966).

taller than paracone; postmetacingulum lost; preparacrista extends from paracone to a point anterior to cusp B, rather than terminating at that cusp; conules and conular cristae reduced, with postmetaconular crista lost. Lower molars with cristid obliqua attaching to trigonid labial to notch between metaconid and protoconid. Where known, ankle differs from that of other stem "didelphimorphians" and resembles Didelphidae in lacking articulation between the fibula and calcaneus. Differ from Didelphidae in lacking dilambdodonty and a posteriorly expanded protoconal base; hypoconulid taller and not situated so far lingually.

Included Genera. *Peradectes* Matthew and Granger, 1921, type genus; and various Tertiary genera not treated herein.

Distribution. (Mesozoic occurrence only) ?Late Cretaceous (?late Maastrichtian): Peru (Vilquechico Group).

Comments. The position of Peradectidae with respect to *Alphadon* and structurally similar taxa, on the one hand, and Didelphidae (*sensu stricto*) on the other is discussed above. The presence of Peradectidae in the Cretaceous is highly tenuous. The genus *Peradectes* has, at times, been tentatively identified from the Late Cretaceous of North America (e.g., Lillegraven and McKenna, 1986). We follow Cifelli et al. (2004) in regarding these as probable cases of mistaken taxonomic or stratigraphic identity. The remaining occurrence is discussed later.

Genus *Peradectes* Matthew and Granger, 1921
(figure 12.11E)

Diagnosis (based on Johanson, 1996a,b). Upper molars differ from those of Didelphidae *sensu stricto* and Herpetotheriinae in being predilambdodont and in having more weakly developed stylar cusps. Lower molars differ in having a shorter entoconid and in lacking a strong notch separating that cusp from the hypoconulid. Distinct from *Peratherium* in having paracone and metacone uninflated and less rounded; hypoconulid not as closely approximated to entoconid.

Included Species. *Peradectes elegans* Matthew and Granger, 1921, type species (of Paleocene age and not treated herein); cf. *Peradectes austrinus* (Sigé, 1971); and various species of Tertiary age, not treated herein.

Distribution. ?Late Cretaceous (?late Maastrichtian): Peru (Vilquechilco Group); and various occurrences of Tertiary age, not treated herein.

Comments. Recent systematic treatment of *Peradectes* and structurally similar marsupials from the Late Cretaceous and Early Tertiary include works by Crochet (1979a, 1980), Krishtalka and Stucky (1983a,b), and Johanson (1996a,b). The species in question (from Laguna Umayo, Perú) is based on fragmentary teeth initially described as *Alphadon austrinum* by Sigé (1971) and subsequently transferred to *Peradectes* by Crochet (1980). There seems to be agreement that the species is not referable to *Alphadon,* but placement in *Peradectes* is also problematic (Clemens, 1982; Montellano, 1992), and uncertainties exist as to the age of the rock unit from which the fossils were collected (chapter 2). All that can be said at present about the specimens is that they represent a marsupial, that they are comparable to teeth of *Peradectes,* and that they may be of Cretaceous age.

Family Stagodontidae Marsh, 1889b

Diagnosis. Relatively large marsupials, by Cretaceous standards, characterized by several plesiomorphies, including a broad stylar shelf and lack of stylar cusp C. Apomorphies include the presence of large epitympanic sinuses in the squamosal; upper molars with deep ectoflexus and strongly developed metastylar region, resulting in a bilobed appearance; upper and lower molars characterized by emphasis on postvallum/prevallid shearing, reflected in strong development of distal stylar shelf and postmetacrista on the upper molars, and strong development of paraconid and paracristid (the latter bearing a well-developed carnassial notch), with metaconid reduced, on lower molars. Lower molars differ from those of "Alphadontidae" in having anterior attachment of cristid obliqua labial to, rather than below, notch between protoconid and metaconid. Anterior premolars reduced; P/p3 strongly enlarged, robust, somewhat complex; three or fewer incisors present in the lower dentition. Tarsals, where known, autapomorphic in: calcaneus with peroneal tubercle terminating adjacent to distal end of calcaneoastragalar facet, with prominent groove for peroneus longus tendon; calcaneocuboid facet on calcaneus ovoid in shape; extensive articulation present between astragalus and cuboid; calcaneoastragalar facet on astragalus clearly separated by a sulcus from anteromedial plantar tubercle.

Included Genera. *Didelphodon* Marsh, 1889a, type genus; *Eodelphis* Matthew, 1916; and *Pariadens* Cifelli and Eaton, 1987.

Distribution. Late Cretaceous (Cenomanian–Lancian): North America.

Comments. Most of the limited information available for cranial anatomy of Cretaceous marsupials is based on Stagodontidae (Matthew, 1916; Clemens, 1966). Isolated tarsals belonging to *Eodelphis* and *Didelphodon* have been described by Szalay (1994); otherwise, knowledge of the group is based on teeth and jaws, with several relatively complete dentaries being known (Matthew, 1916; Clemens, 1968a; Fox and Naylor, 1986). Relationships of and within Stagodontidae were discussed by Fox and Naylor (1986, 1995). Analysis by Rougier et al. (1998) places an unnamed taxon from Mongolia (represented by the Guriliin Tsav skull) with Stagodontidae, suggesting the possibility that the group may not be exclusively North American in distribution.

Genus *Didelphodon* Marsh, 1889a
(figure 12.13)

Diagnosis (based on Fox, 1981). Includes the largest known species of Stagodontidae and of Cretaceous Marsupialia in general. Differs from the closely similar genus, *Eodelphis,* in being larger; having more enlarged, inflated P/p3 with large lingual lobe on P3; stylocone on upper molars larger; paracone and metacone more flattened labiolingually, with paracone proportionately smaller; lobes of stylar shelf more symmetrical and rounded; conules smaller with more rounded, lower cristae; stylar cusp D present; lower molars more robust, with proportionately smaller metaconid and taller paraconid, precingulid more strongly developed; trigonid wider, with more elongate paracristid, and decreased trigonid angle.

Included Species. *Didelphodon vorax* Marsh, 1889a, type species; *D. coyi* Fox and Naylor, 1986; *D. padanicus* (Cope, 1892); and species left in open nomenclature (Fox and Naylor, 1986; Hunter and Pearson, 1996).

Distribution. Late Cretaceous ("Edmontonian"–Lancian): North America.

Comments. Aside from some fragments assigned to *Eodelphis* (Clemens, 1979a; Fox, 1981), *Didelphodon* is the only North American Cretaceous marsupial known from cranial remains (Clemens, 1966). The genus first appears in the "Edmontonian," where it is represented by *D. coyi* and an unnamed species. Of antecedent taxa, *Didelphodon* seems more likely to have evolved from a clade including *Eodelphis cutleri* than from another that includes its congener, *E. browni* (Fox and Naylor, 1986).

Genus *Eodelphis* Matthew, 1916
(figure 12.14B,C)

Diagnosis (based on Fox, 1981). Differs from the most similar genus, *Didelphodon,* in the following. Included species of smaller size. Distal premolars (P/p3) not as in-

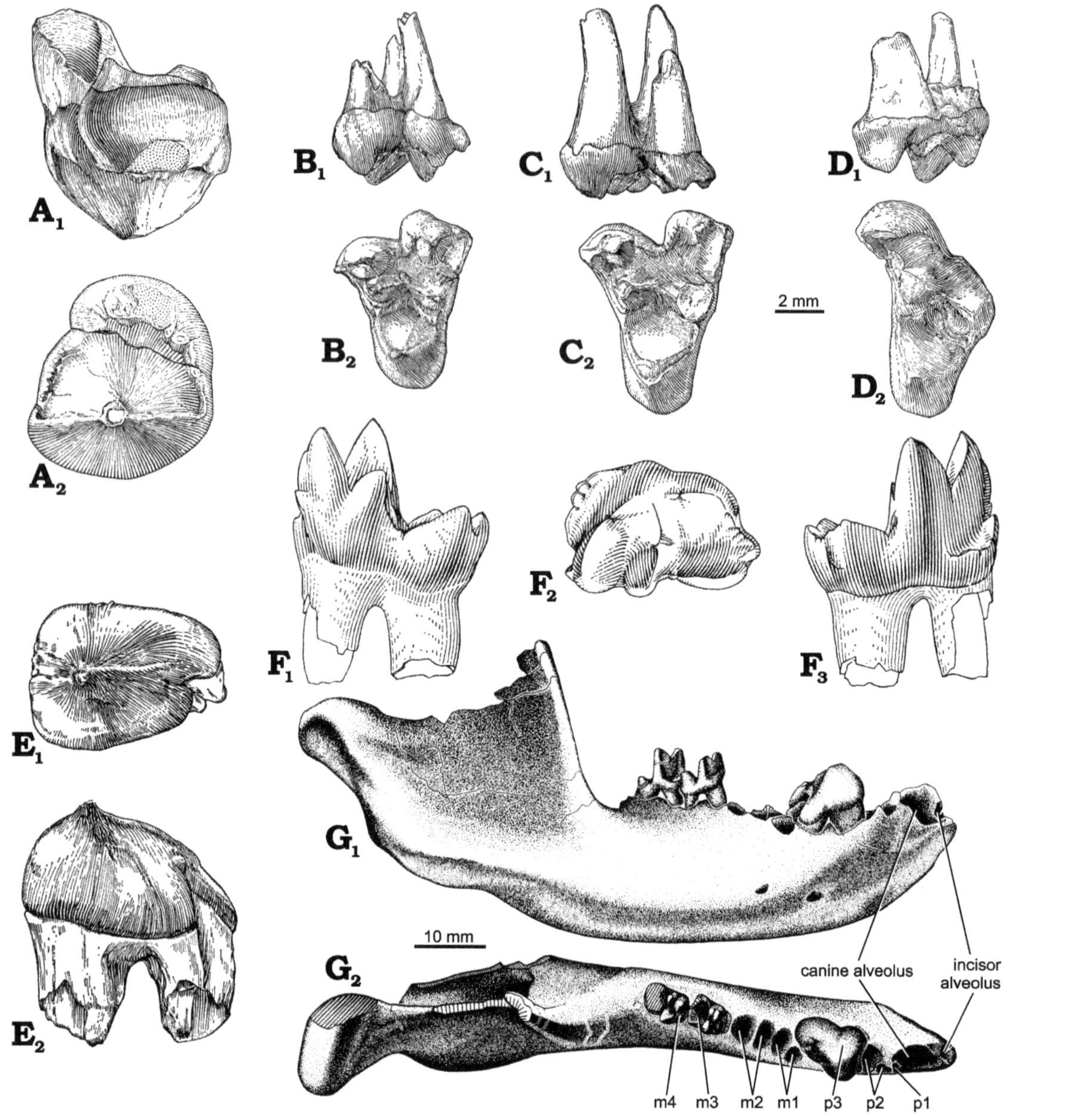

FIGURE 12.13. Representatives of *Didelphodon*. A–F, *Didelphodon vorax:* A, ?Left P3 in labial (A_1) and occlusal (A_2) views. B, Left M1 in labial (B_1) and occlusal (B_2) views. C, Left M2 in labial (C_1) and occlusal (C_2) views. D, Left M4 in labial (D_1) and occlusal (D_2) views. E, Left p3 in occlusal (E_1) and labial (E_2) views. F, Right M4 in lingual (F_1), occlusal (F_2), and labial (F_3) views. G, *Didelphodon coyi:* right dentary with p3, m3–4 (holotype) in labial (G_1) and occlusal (G_2) views. *Source:* A–F, from Clemens (1966); G, modified from Fox and Naylor (1986).

flated or robust; accessory labial lobe on P3 lacking. Upper molars more gracile, with stylocone smaller, lower than paracone; paracone and metacone more conical, with paracone proportionately larger; lobes of stylar shelf less rounded and symmetrical; conules stronger, with more trenchant cristae; stylar cusp D lacking. Lower molars with proportionately smaller paraconid and larger metaconid; trigonid less mesiodistally compressed; attachment of cristid obliqua to trigonid more lingual; precingulid narrower, not forming as prominent a shelf.

Included Species. *Eodelphis cutleri* (Woodward, 1916), type species; *E. browni* Matthew, 1916; and species left in open nomenclature (Fox, 1971b; Lillegraven and McKenna, 1986; Fiorillo, 1989; Montellano, 1992; Peng, 1997; Diem, 1999).

Distribution. Late Cretaceous (Aquilan–"Edmontonian"): North America.

Comments. The Aquilan record for *Eodelphis* is based on the presence of an unidentified species in the mammalian fauna of the upper Milk River Formation, Alberta

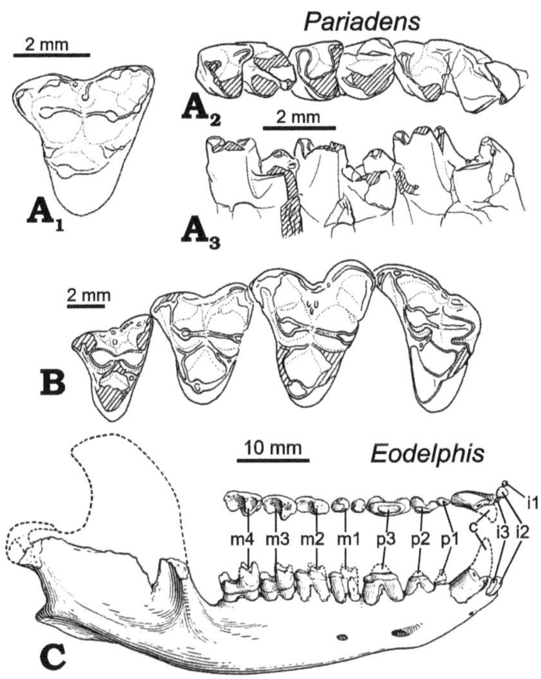

FIGURE 12.14. Stagodontidae. A, *Pariadens kirklandi:* A_1, left upper molar in occlusal view; A_{2-3}, left m2–4 (holotype) in occlusal (A_2) and labial (A_3) views. B, C, *Eodelphis:* B, *Eodelphis* sp., left M1–4 (composite) in occlusal view. C, *Eodelphis browni*, right dentary with i1–3, c, p1–3, m1–4 (part of holotype) in occlusal and labial views. *Source:* A, B, originals, C, modified from Matthew (1916).

(Fox, 1971b). In the Judithian, *Eodelphis* is represented by two named species, both from northerly faunas; Stagodontidae are apparently not present in Judithian or Lancian assemblages from more southerly parts of North America (Cifelli, Nydam, Eaton et al., 1999).

Genus *Pariadens* Cifelli and Eaton, 1987
(figure 12.14A)

Diagnosis (after Eaton, 1993b: 118–119). "Paraconid equal to or larger than metaconid on m2–4 as in stagodontids and unlike other marsupial families, but lacking the extreme enlargement of the paraconid seen in *Eodelphis* or *Didelphodon.* Paracristid and protocristid notches well developed on unworn specimens. Cristid obliqua originates medially on m1, m3, and m4 like alphadontids but unlike stagodontids or pediomyids, and labially on m2, unlike alphadontids. Cusps of m1–3 talonids close to subequal in height, in unworn specimens entoconid slightly taller than hypoconulid, hypoconulid taller than hypoconid. On m4, entoconid tallest cusp of the talonid. Molars increase in size posteriorly as in stagodontids, but unlike alphadontids and pediomyids."

Included Species. *Pariadens kirklandi* Cifelli and Eaton, 1987, type species by monotypy.

Distribution. Late Cretaceous (Cenomanian): United States, Utah (Dakota Formation).

Comments. *Pariadens kirklandi* was originally based on a dentary fragment with m2–4 (Cifelli and Eaton, 1987); additional lower and upper molars were referred to the species by Eaton (1993b). Some distinctive features of Stagodontidae are present, such as the proportions of paraconid and metaconid and the sequential increase in molar length through the series. However, *Pariadens* is plesiomorphic with respect to *Eodelphis* and *Didelphodon* in other features, such as the proportionately taller paracone with respect to metacone, lack of distinct wings on conules, cristid obliqua not generally with a labial attachment to the trigonid, less discrepancy in height between paraconid and metaconid, and lack of mesiodistal compression of the trigonid (Eaton, 1993b). These features are consistent with the older geological age of *Pariadens.* The basis for referral to Stagodontidae is not strong and has been challenged by Fox and Naylor (1995).

Order Paucituberculata Ameghino, 1894

Diagnosis (modified after Marshall, 1987). Molars primitively bunodont with respect to those of stem "Didelphimorphia." Lower molars with little height differential between trigonid and talonid, paraconid and metaconid closely approximated; metaconid large relative to protoconid; hypoconid and entoconid large and subequal. Upper molars with enlarged cusps B and D. Anterior lower dentition (incisors, canines) tend toward procumbency.

Included Taxa. Superfamilies Caroloameghinoidea Ameghino, 1901; Caenolestoidea Trouessart, 1898; Argyrolagoidea Ameghino, 1904; Polydolopoidea Ameghino, 1904; and Family Glasbiidae Clemens, 1966, Superfamily *incertae sedis.*

Distribution. Late Cretaceous (Lancian): North America; and Cenozoic, South America and Antarctica.

Comments. Our concept of Paucituberculata follows that of Marshall (1987), Aplin and Archer (1987), and McKenna and Bell (1997), differing from other recent treatments (e.g., Marshall et al., 1990; Szalay, 1994) in being more inclusive. As such, a great diversity of taxa is included (see discussion by Aplin and Archer, 1987), and some apparent similarities, such as enlargement of posterior premolars and anterior molars, may have developed independently among various included groups (see Szalay, 1994). *Glasbius* is often grouped within Caroloa meghiniidae Ameghino, 1901, an otherwise strictly South American family (e.g., Marshall et al., 1990). Given the primitive morphology of *Glasbius* with respect to the remarkable variety of specializations seen among the various Paucituberculata, we more conservatively retain it in its own family, of uncertain suprafamiliar affinities.

Family Glasbiidae Clemens, 1966

Diagnosis. As for the type and only genus.

Included Genera. *Glasbius* Clemens, 1966, type genus by monotypy.

Distribution. As for the type and only genus.

Comments. Glasbiinae were erected by Clemens (1966) as a subfamily of Didelphidae. *Glasbius,* the only included genus, is highly distinctive among North American Cretaceous marsupials and appears on the continent in the latest Maastrichtian (Lancian). As *Glasbius* lacks obvious structural antecedents in earlier faunas, its origin has been somewhat puzzling, and it has been considered to be an "alien" in the Lancian (Weil and Clemens, 1998; Clemens, 2002; Cifelli et al., 2004).

Genus *Glasbius* Clemens, 1966
(figure 12.15)

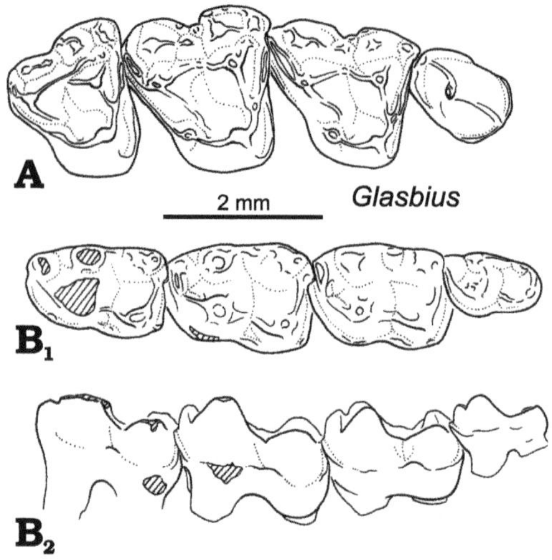

FIGURE 12.15. Glasbiidae. A, B, *Glasbius intricatus:* A, right P3, M1–3 in occlusal view; B_{1-2}, left m1–4 in occlusal (B_1) and labial (B_2) views. *Source:* original.

Diagnosis (after Archibald, 1982: 137). "Principal cusps on upper molars low relative to protofossa; metacone higher than paracone and metaconule larger than paraconule; stylar shelf broad; B large on M1–4; D higher than B on M1–2, smaller than B on M3, absent on M4; A and E small; C small or absent; m2 longer than other molars; talonid (including labial cingulum) wider than trigonid, except on m4; width of m3 > m2 > m1; cristid obliqua contacts trigonid on back of protoconid; trigonid short anteroposteriorly, paraconid and metaconid closely approximated on m3–4; difference in height of trigonid over talonid not great and decreases from m1 to m4; lingual side of crown higher than labial, except on m1–2, where protoconid is subequal to metaconid; basal cingulum on anterior, labial, and posterior sides of m1–4 with variously developed cusps on some molars; M/m 4 much smaller than preceding molars; p1–3 double-rooted, increasing in size posteriorly; anteroposterior axis of p1 rotated at 30–45° angle to long axis of dentary."

Included Species. *Glasbius intricatus* Clemens, 1966, type species; and *G. twitchelli* Archibald, 1982.

Distribution. Late Cretaceous (Lancian): North America.

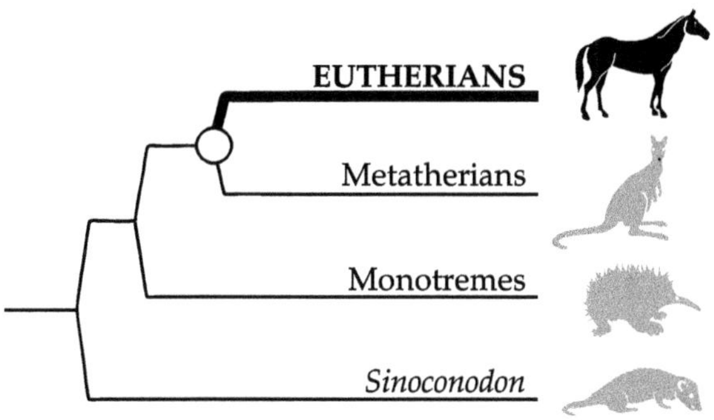

CHAPTER 13

Eutherians

INTRODUCTION

Eutheria are the most successful group of mammals and are of special interest in that they have dominated Tertiary mammal faunas of all major continents except Australia and South America. Eutherians are a clade consisting of all extant placentals plus all extinct mammals that are more closely related to extant placentals (such as humans) than to extant marsupials (such as kangaroos). The extant placentals are a subgroup of the Eutheria. Placentals have some 18 extant orders, comprise the vast majority of about 4,600 modern living mammal species (see chapter 1), and many more fossil representatives have been described.

Within the context of mammalian phylogeny, Eutheria are clearly related to their tribosphenic cousins, Marsupialia, but the proximity of this relationship, the timing of the marsupial-placental dichotomy (Lillegraven et al., 1987), and the affinities of various "tribotherians" (Butler, 1978a; Cifelli, 1993b)—to eutherians on the one hand or marsupials on the other—remain elusive. Indeed, though the earliest known eutherian *Eomaia* (Ji et al., 2002) is of Early Cretaceous age (middle Barremian, about 125 Ma), some authors have advocated a far earlier split with marsupials (e.g., Kielan-Jaworowska, 1992). A large part of this problem stems from the fact that these two major groups of living mammals are distinguished primarily on the basis of soft anatomy, particularly the reproductive system, which is not preserved in the fossil record and usually cannot be inferred from fossils. A second problem stems from the fossil record itself: only a few Cretaceous taxa are known by anything resembling complete material (skull and/or skeleton), with most being represented by jaws or isolated teeth. Apparently, early eutherians were dentally conservative as a general rule, and some of the established specializations seem to represent general trends, or iterative patterns, that may have been acquired independently in various lineages. As currently known, there are some 42 genera of Cretaceous eutherians, collectively spanning the last 60 Ma or so of the Mesozoic. With few exceptions, though, the relationships of these taxa to one another—and, perhaps more importantly, to mammalian groups that rose to prominence in the Cenozoic—remain poorly understood. For these reasons, systematic arrangement is arbitrary and unsatisfactory in many cases, and the general adequacy of the Mesozoic record to either calibrate or test models of mammalian evolution based on molecular data (e.g., Foote et al., 1999) is highly suspect. Overall phylogenies should be taken for what they are: hypotheses rather than definitive statements of relationships.

BRIEF CHARACTERIZATION

Members of Eutheria (including living placental mammals) are similar to Metatheria (chapter 12) in skeletal anatomy. The most fundamental differences between the two groups lie in the structure of the reproductive tract and the pattern of reproduction. Both marsupials and placentals are viviparous; however, placental mammals differ from marsupials in possessing the unique trophoblastic tissue and a chorioallantoic placenta. The trophoblast forms an active barrier between the maternal and embryonic tissues, preventing rejection of the developing embryo. As a result, eutherians undergo a prolonged period of development and morphogenesis in the uterine environment. This permits higher levels of metabolism, more extensive brain development, greater karyotypic variability, and a far greater range of morphological divergence as adults (Lillegraven et al., 1987).

The skeletal anatomy of extant eutherian mammals has been described in numerous anatomy textbooks (e.g., Starck, 1967; Getty, 1975; Moore, 1981; Schaller, 1992; Evans, 1995) and its characterization is not repeated herein. The skull of known early eutherians generally has an elongate, narrow snout with a large infraorbital foramen, suggesting that vibrissae were well developed (Kay and Cartmill, 1977). Brain size was on a par with small extant insectivores and marsupials, with variable development of the neocortex and large olfactory bulbs, the latter suggesting that olfaction played an important role in the habits of early eutherians (Kielan-Jaworowska, 1984c). The ear region is large (suggesting acute hearing), with a horseshoe-shaped ectotympanic. In many later forms, the ectotympanic is transformed into a tympanic bulla, or part thereof, which is different from the condition in Metatheria (see chapter 12). There is no evidence of attached postdentary bones even in the earliest Eutheria; however, a coronoid (a primitive feature) is retained in some taxa, as is a remnant of Meckel's groove. In relation to the elongated snout, the mandible is slender and elongate. The angular process is peg- or hooklike, sometimes slightly bent ventromedially. In the postcranial skeleton, the best-known difference between extant placentals and marsupials is the former's lack of marsupial (epipubic) bones, but it has been shown that epipubic bones were retained in early (Late Cretaceous) Eutheria from Mongolia (Kielan-Jaworowska, 1975a; Novacek et al., 1997). In association with an early trend toward terrestriality, most known eutherians are characterized by an advanced condition of the proximal ankle joint, whereby the astragalus is restricted from mediolateral movement by the medial malleolus of the tibia (medially) and the distal fibula (Szalay, 1984). However, this condition is poorly developed in asioryctitherians (Horovitz, 2000).

Eutheria share with Metatheria (and "tribotherians") a tribosphenic molar pattern (see chapter 11), but differ in dental formulae and mode of tooth replacement. It is generally accepted that the primitive dental formula of extant placental mammals is 3.1.4.3/3.1.4.3, but this is not the case with the earliest eutherians, which might have as many as five upper and four lower incisors and five premolars (Kielan-Jaworowska, 1981; Nessov, 1985; Kielan-Jaworowska and Dashzeveg, 1989; Sigogneau-Russell et al., 1992; Novacek et al., 1997; Cifelli, 2000a; Ji et al., 2002). The high incisor count is a plesiomorphy shared with Metatheria. A high number of premolars (four or more), as found in eutherians, is a presumed plesiomorphy, whereas the premolar count is reduced to three in metatherians. The presence of only three molars in Eutheria, however, may well be a derived condition distinguishing them from Metatheria, which have four. Eutherians also differ from metatherians in lacking a sharp morphological break between the premolar and molar series; the premolars in Eutheria (especially the uppers) are gradually molarized posteriorly (the presumed derived condition), which is not the case in Metatheria. Most of the antemolar dentition of eutherians (excepting perhaps the first premolar) undergo single replacement (diphyodonty), which is presumed to be the primitive condition (Martin, 1997). Eutherians differ in this respect from marsupials, in which only one tooth—the last premolar—is replaced, which is a derived condition (Luckett, 1993; Kobayashi et al., 2002; see also chapter 3 and the discussion on tooth replacement in early boreosphenidans therein).

DISTRIBUTION

The oldest generally accepted eutherian to be described is *Eomaia*, from the middle Barremian Yixian Formation of Liaoning Province, China. *Eomaia* is known from an exceptionally complete skeleton, with a fur halo preserved around it (Ji et al., 2002). Other Early Cretaceous eutherian taxa from Asia are *Prokennalestes* and *Murtoilestes*. *Prokennalestes* has been known informally since the early 1970s (Belyaeva et al., 1974) but was not described until later (Kielan-Jaworowska and Dashzeveg, 1989; Sigogneau-Russell et al., 1992). *Prokennalestes* is known from the presumed Aptian or Albian of Mongolia and a similar taxon (*Murtoilestes*) is known from somewhat ?earlier beds in Russia (Averianov and Skutschas, 2000a, 2001).

Pappotherium and undescribed taxa from the approximately equivalent Trinity Group of Texas have been proffered as eutherians (Slaughter, 1971, 1981), but they are poorly known and this referral has been viewed with general skepticism (e.g., Butler, 1978; Kielan-Jaworowska, Eaton, and Bown, 1979; Kobayashi et al., 2002, see also chapter 11). Similarly, eutherian affinities have been suggested for the Australian Early Cretaceous mammal *Ausktribosphenos* (Rich et al., 1997, 1999), but this view has not achieved wide acceptance (Kielan-Jaworowska et al., 1998; Musser and Archer, 1998; Archer et al., 1999; Luo, Cifelli, and Kielan-Jaworowska, 2001; Luo et al., 2002, see chapter 6). An early presence of Eutheria in Asia, coupled with the fact that eutherians are the dominant therians in the Late Cretaceous faunas of that continent, has led to the widespread view that they arose in Asia, later spreading to North America and elsewhere (Kielan-Jaworowska, 1982). The question remains an open one, however: a probable eutherian, *Montanalestes*, has recently been described from the Early Cretaceous of North America. *Montanalestes* is from the Aptian–Albian of Montana, and thus is slightly younger than *Eomaia* (Cifelli, 1999b).

A variety of eutherians is known from the early Late Cretaceous of middle Asia (Nessov, 1997). Most of these are "zhelestids" or archaic hoofed mammals (Nessov, Archibald, and Kielan-Jaworowska, 1998), but also included are a few taxa that appear to be related to groups known from younger rocks of central Asia: Kennalestidae and Zalambdalestidae. To the west, in Western Europe, several eutherians are known from the Campanian and Maastrichtian, where they are represented by rare, fragmentary fossils. The best known of these, *Lainodon* and *Labes,* appear to represent a distinctive group related to "zhelestids," suggesting: (1) endemism in the latest Cretaceous of Western Europe; and (2) probable earlier continuity with faunas of Asia (e.g., Gheerbrant and Astibia, 1994). The earliest occurrence of "Zhelestidae" is in the Cenomanian–Turonian of Japan (Setoguchi, Tsubamoto et al., 1999), showing that the group was widely distributed. To the east, eutherians are well known from the Campanian of Mongolia, where they are represented by five genera and three families (Asioryctidae, Kennalestidae, Zalambdalestidae), and one genus assigned to a family *incertae sedis.* The marked compositional difference between the eutherians of Mongolia and those of middle Asia is probably due to differences in habitat and environment (Nessov, 1992; Nessov, Archibald, and Kielan-Jaworowska, 1998).

With one possible exception (represented by a molar talonid from the Santonian of Mississippi), eutherians are curiously lacking from the North American record between the Aptian–Albian and the early Campanian, leaving a "eutherian hiatus" of more than 30 Ma. Possibly, this is an artifact of the fossil record, which is notoriously poor for the early Late Cretaceous. If so, eutherians may have been present all along in North America and, perhaps, some of the groups that appeared later may have evolved in situ, rather than arriving as immigrants from Asia. On the other hand, eutherians may have fared poorly or may even have become extinct during this interval in North America. Interestingly, the early Late Cretaceous was a time of major diversification for marsupials on the North American continent (Eaton, 1993a; Cifelli, Nydam, Weil et al., 1999). Because of differences in their modes of reproduction, eutherians enjoyed a number of adaptive advantages over marsupials in some environments (Lillegraven et al., 1987). Taken at face value, the fossil record of North America thus suggests the intriguing but untestable possibility that eutherians did not evolve their distinctive reproductive system until later in the Cretaceous. Whatever the case, eutherians appear in the early Campanian of North America, represented by the nyctitheriid insectivore *Paranyctoides* (see Fox, 1984b; Cifelli, 1990e). Interestingly, Nessov (1993) reported the same genus from a slightly older (Coniacian) horizon in middle Asia, suggesting dispersal from that continent (see also Archibald and Averianov, 2001b). Other groups, including primitive ungulatomorphs (e.g., *Gallolestes, Avitotherium,* see Nessov, Archibald, and Kielan-Jaworowska, 1998), leptictid insectivores (*Gypsonictops*), and, possibly, cimolestids (Lillegraven and McKenna, 1986) appear in North America by the late Campanian. Nonetheless, eutherians were minor elements in North American faunas until the latest Cretaceous, when there is some diversification among Palaeoryctidae and, with some doubt, when ungulates (Arctocyonidae, Periptychidae) make their first appearance (Fox, 1989, 1997b).

The little that is known of South America's Mesozoic mammals suggests that eutherians did not arrive on that continent until the end of the Cretaceous (Bonaparte, 1994). *Perutherium,* the only named Cretaceous eutherian from South America (Grambast et al., 1967), is of debatable affinities but is generally believed to be a primitive ungulate (Cifelli, 1983; Marshall et al., 1983). The link for South America's tribosphenic mammals is probably North America, but that of another southern landmass, India, is not so obvious. *Deccanolestes,* a eutherian from the Maastrichtian of India, is primitive in a number of respects; an Asiatic origin has been suggested (Prasad et al., 1994).

ANATOMY

SKULL

The skull known in Cretaceous Eutheria differs in some details from those of basic extant representatives of the group and in our description we confine ourselves mostly to pointing out these differences. The oldest known eutherian skull is that of *Eomaia* from the Barremian of China. The cranial roof of the skull is badly damaged; the skull has been figured (Ji et al., 2002: figure 1), but not described. The Coniacian of Uzbekistan yielded a more complete skull, which is of a juvenile individual of *Daulestes nessovi,* tentatively recognized as belonging to Asioryctitheria (McKenna et al., 2000). The remaining complete skulls of Mesozoic eutherian mammals are from the Late Cretaceous of Mongolia. These include the zalambdalestids *Zalambdalestes* and *Barunlestes* (see Kielan-Jaworowska and Trofimov, 1980, 1986; Kielan-Jaworowska, 1984b), the kennalestid *Kennalestes,* and the asioryctitherians *Asioryctes* and *Ukhaatherium* (Kielan-Jaworowska, 1981, 1984c; Novacek et al., 1997). The cranial anatomy of *Ukhaatherium* has not yet been published in detail. An exquisitely preserved skull of *Zalambdalestes* collected by Mongolian–American Museum of Natural History Expeditions has been studied by Timothy Rowe at the University of Texas in Austin, using computed topography. Images of this skull appeared in a film produced by *Nova* in the United States, but the scientific results have not yet been published. We base our description of the skull in Mesozoic Eutheria mostly on the papers mentioned earlier. All six Late Cretaceous Asian eutherian genera for which the skulls are known share several characters and are described here jointly.

Skull as a Whole. Known Cretaceous eutherians were small mammals, the length of the adult skull of *Daulestes* (estimated by comparison with *Kennalestes*) might have been between 19 and 20 mm. Skull length of other taxa is between 26 and 30 mm, except for the Zalambdalestidae, where it varies between 40 and 50 mm. The skull is slender (figure 13.1), with a narrow snout and a relatively slender dentary (most robust in *Barunlestes*). There is a conspicuous interorbital constriction, no postorbital process, and no paroccipital process. The promontorium is large. The skull is distinguished from those of extant eutherians by a long mesocranial region (relatively longer in Asioryctitheria than in Zalambdalestidae), associated with a large alisphenoid. The ectotympanic forms about three-quarters of a ring and is inclined (in *Asioryctes*) at about 45° to the horizontal plane; the occipital plate (uncertain in *Daulestes* and *Ukhaatherium*) is inclined forward from the condyles that protrude behind. The skulls of *Kennalestes, Asioryctes,* and *Daulestes* (and probably also *Ukhaatherium*) are more similar to each other than any one of them is to the representatives of the Zalambdalestidae.

Snout and Zygoma. The snout (figure 13.1) is more elongated in Zalambdalestidae (particularly in *Zalambdalestes*) than in Asioryctitheria. The nasals appear very long and expanded posteriorly, but the nasofrontal suture has not been recognized with any certainty in any of the genera. The infraorbital foramen is relatively large in all the genera. The lacrimal is placed near the edge of the margin of the orbit and there is a small facial wing. The zygomatic arch is relatively the deepest in Asioryctidae and much more slender in Zalambdalestidae, especially so in *Zalambdalestes,* where the zygomatic arches are strongly expanded laterally. The jugal is very extensive and forms almost the entire zygomatic arch, posteriorly reaching the anterior margin of the glenoid fossa. The maxilla does not contribute to the zygomatic arch, at least in *Asioryctes* and *Daulestes.*

Palate. The palate as preserved in Asioryctitheria apparently does not differ from those in extant primitive eutherian mammals. The incisive foramina, preserved only in *Asioryctes,* are very small and placed within the premaxilla-maxillary suture, in contrast to the strongly elongated foramina in basal marsupials (see chapter 12). There is a weak postpalatine torus and the posterior palatine foramen is probably developed as a notch in *Kennalestes* and *Asioryctes,* but as a foramen in *Daulestes.* In Zalambdalestidae the postpalatine torus is more prominent and the posterior palatine foramen is developed as a very large oval opening.

Cranial Roof. The sutures between the bones of the cranial roof are usually poorly seen. The parietals form the whole cranial roof posteriorly and are pointed anteriorly in *Kennalestes.* A distinct sagittal crest and lambdoidal crests, most prominent in the Zalambdalestidae, are present in all the genera (but not preserved in *Daulestes*). The squamosal is extensive, the posterior root of the zygomatic arch being placed far posteriorly. When seen from the side, the squamosal is roughly triangular and is embraced by two crests: the temporal crest and, posteriorly, the lateral mastoid flange. Dorsally the two crests meet and continue as the lambdoidal crest. The surface of the squamosal between the postglenoid and posttympanic processes forms a convex recess for the external auditory meatus. In the ventral part of *Asioryctes,* a very large, roughly rectangular tympanohyal is superimposed on the squamosal. A large, fissurelike subsquamosal foramen opens on the lateral surface of the zygomatic arch, above the postglenoid process in *Daulestes, Asioryctes,* and *Kennalestes,* but not in Zalambdalestidae.

Occiput. The occiput (where preserved) is inclined forward from the condyles (figure 13.2). It is best preserved in *Kennalestes* and *Asioryctes* and consists of a large supraoccipital, exoccipitals, basioccipitals, and extensive mastoids, which differ in shape between the two genera. There is a mastoid process, more prominent in *Kennalestes* than in *Asioryctes,* in which the mastoid is pierced by two foramina: mastoid foramen and lower mastoid foramen.

Orbit and Temporal Fossa. The orbit is confluent with the temporal fossa in all the genera. It has proven impossible to determine the arrangement of sutures in the orbital region in most of the genera, and they have been only tentatively recognized in *Barunlestes.* The lacrimal foramen opens into the orbit and the maxillary foramen is placed medial to and below it. A groove extends from the maxillary foramen to the sphenopalatine foramen (the latter uncertain in *Daulestes*) through the floor of the orbit in all the genera, but the exposure of the particular bones within the orbit differs in details. The sphenopalatine foramen lies in a recess in the palatine bone in *Asioryctes* and *Kennalestes,* but it pierces the maxilla in Zalambdalestidae (figure 13.2). Kielan-Jaworowska (1984b) pointed out that the orbit in *Zalambdalestes* resembles that in *Didelphis* in many details.

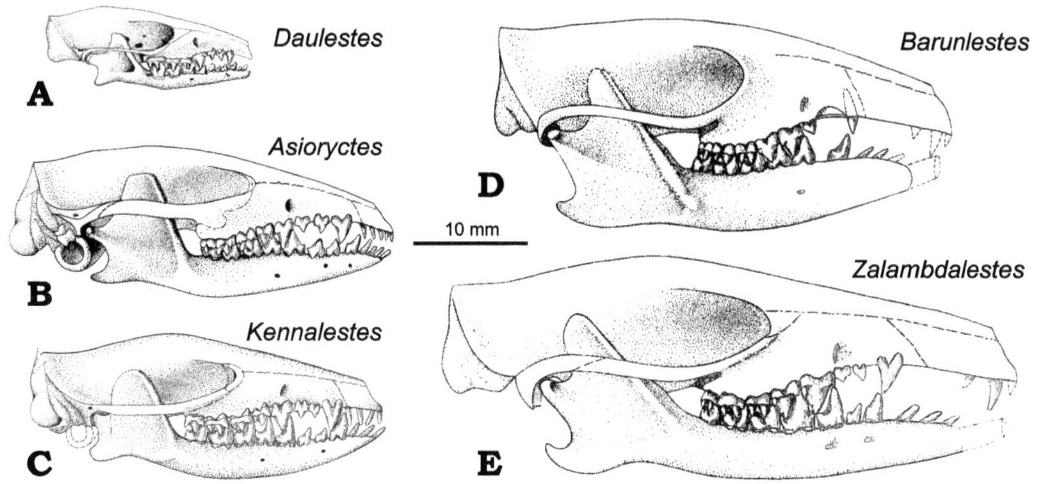

FIGURE 13.1. Reconstructions of Asian Late Cretaceous eutherian skulls in lateral view. A, *Daulestes nessovi.* B, *Asioryctes nemegetensis.* C, *Kennalestes gobiensis.* D, *Barunlestes butleri.* E, *Zalambdalestes lechei.* All in the same scale. *Source:* B–E, modified from Kielan-Jaworowska (1975b).

The temporal fossa (lateral wall of the braincase) is almost identical in *Kennalestes* and *Asioryctes,* but differs from that in Zalambdalestidae (figure 13.2). In *Daulestes* the sutures between the bones are fused and reconstruction of their shape by McKenna et al. (2000) is conjectural (figure 13.3). In Asioryctitheria the orbitosphenoid is roughly trapezoidal in lateral view and is relatively small in comparison with the large alisphenoid in lateral view (the shape of orbitosphenoid is uncertain in *Daulestes*). The orbitosphenoid contacts the palatine anteriorly, the parietal dorsally, and the alisphenoid posteriorly. The alisphenoid bears four foramina: the ethmoidal foramen, the sinus canal foramen, the optic foramen (two latter tentatively identified in *Daulestes*), and the sphenorbital fissure. The last is united with the foramen rotundum and situated at the posteroventral corner of the orbitosphenoid, at the boundary with the alisphenoid. The anterior margin of the alisphenoid projects slightly over the margin of the orbitosphenoid, which has a different appearance than the other bones, being smoother and whiter (best preserved in *Asioryctes*). It is thus possible that, in life, a thin sheet of bone covered the posterior part of the orbitosphenoid. The optic foramen is situated relatively low, close to its counterpart, with which it probably communicates, as, for example, in some Tenrecidae. Ventrally the orbitosphenoid in *Asioryctes* is bounded by the pterygoid, which is prolonged ventrally as a hamulus and posteriorly as a pterygoid process bounding the alisphenoid. The alisphenoid is extensive and roughly triangular in lateral view, surrounded dorsally by the parietal and posterodorsally by the squamosal. At its posteroventral corner there is a large oval foramen ovale. The alisphenoid canal is absent; instead there is a groove (best preserved in *Asioryctes*) along the ventral margin of the alisphenoid, interpreted as the groove for the internal maxillary artery (figure 13.2).

In Zalambdalestidae (as described for *Barunlestes*) the perpendicular part of the palatine is relatively larger than in *Asioryctes* and extends more posteriorly, being bounded ventrally by the pterygoid (figures 13.4, 13.5). The hamulus is relatively smaller than in *Asioryctes.* The size and shape of the orbitosphenoid and alisphenoid, as well as the contact between them, have only tentatively been recognized. Both bones are bounded ventrally by an extensive pterygoid process of the basisphenoid. The orbitosphenoid is apparently fan-shaped; the alisphenoid is roughly triangular. In contrast to *Asioryctes,* the foramen rotundum is separated from the sphenorbital fissure; other foramina are similarly distributed.

Choanae and Basicranium. The zalambdalestid skulls differ from those of other Cretaceous eutherians in having a shorter mesocranial region. We start our description with the skull of *Asioryctes,* in which the basicranial region is best preserved. This region in *Daulestes* is generally similar to that of *Asioryctes,* but the mesocranial region is relatively shorter in *Daulestes,* which may be related to the juvenile nature of the only skull preserved in this taxon. In *Asioryctes* the vomer has been only tentatively recognized. The presphenoid, as preserved, appears to be a paired bone, which is probably due to the state of preservation (break along the midline). It extends between the optic and Vidian foramina to the rear of the bone tentatively

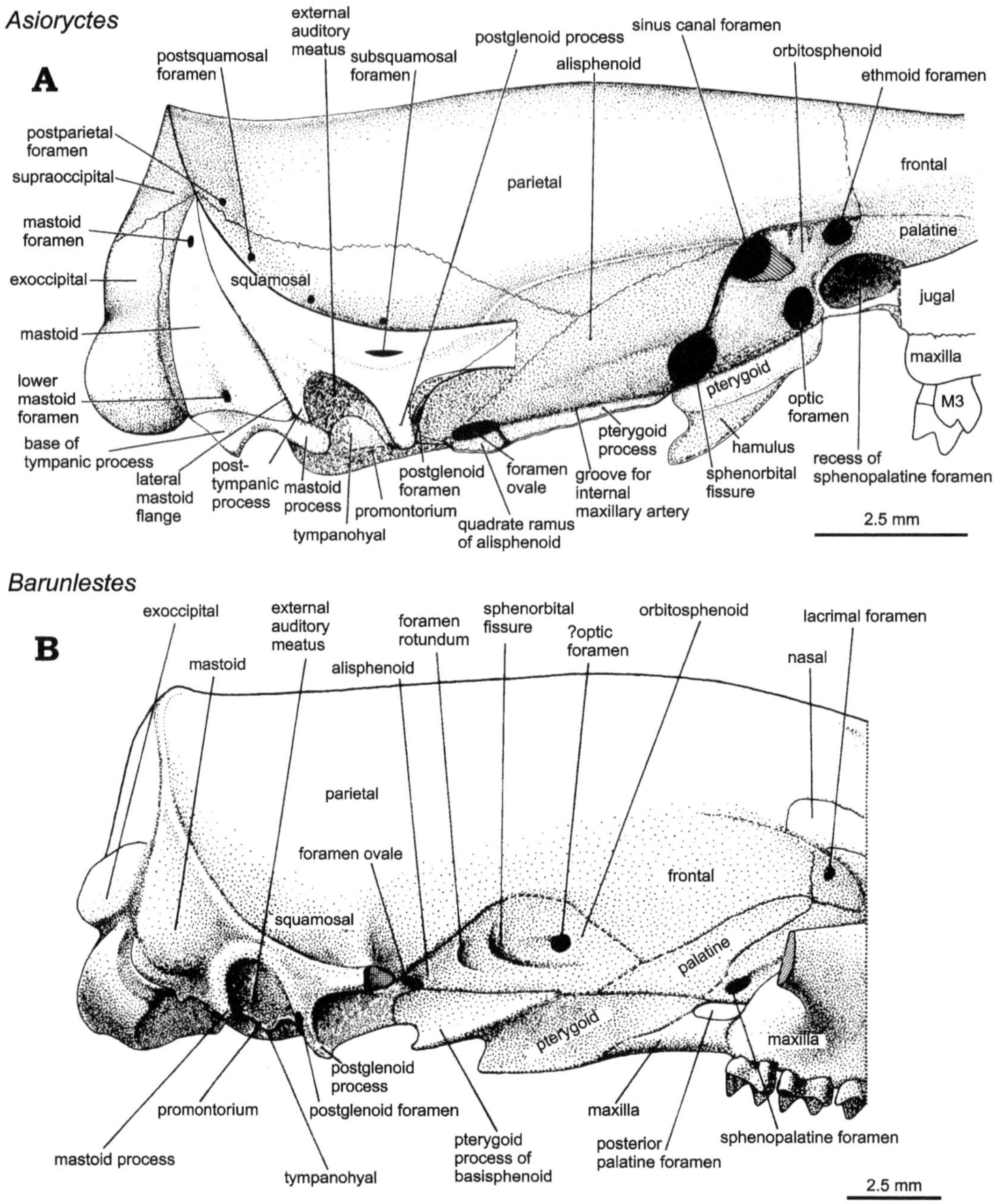

FIGURE 13.2. Lateral views of the posterior part of the skull in *Asioryctes nemegetensis* (A) and *Barunlestes butleri* (B). The opening identified in *Barunlestes* as the sphenorbital fissure resembles more the anterior lacerate foramen of extant mammals than the sphenorbital fissure of most plesiomorphic Mesozoic mammals. We call it the sphenorbital fissure to enable comparison with, for example, *Asioryctes*, and other Mesozoic genera. *Source:* A, modified from Kielan-Jaworowska (1981); B, from Kielan-Jaworowska and Trofimov (1980).

identified as a vomer (figure 13.4). The basisphenoid on each side consists of the medial part and two lateral flanges designated the basisphenoid wings. Extending along the middle of the basisphenoid is a narrow ridge. The medial part is arranged horizontally; the basisphenoid wings are arranged obliquely with regard to the median part and project ventrolaterally. The anteromedial corner of the basisphenoid wing embraces the Vidian foramen (also present in *Daulestes*). The posterior margin of both parts of the basisphenoid is strongly emarginated, embracing the anterior part of the promontorium. The medial part of the basisphenoid extends some distance between the promontoria. The lateral margin of the basisphenoid wing and the medial margin of the alisphenoid between the optic and

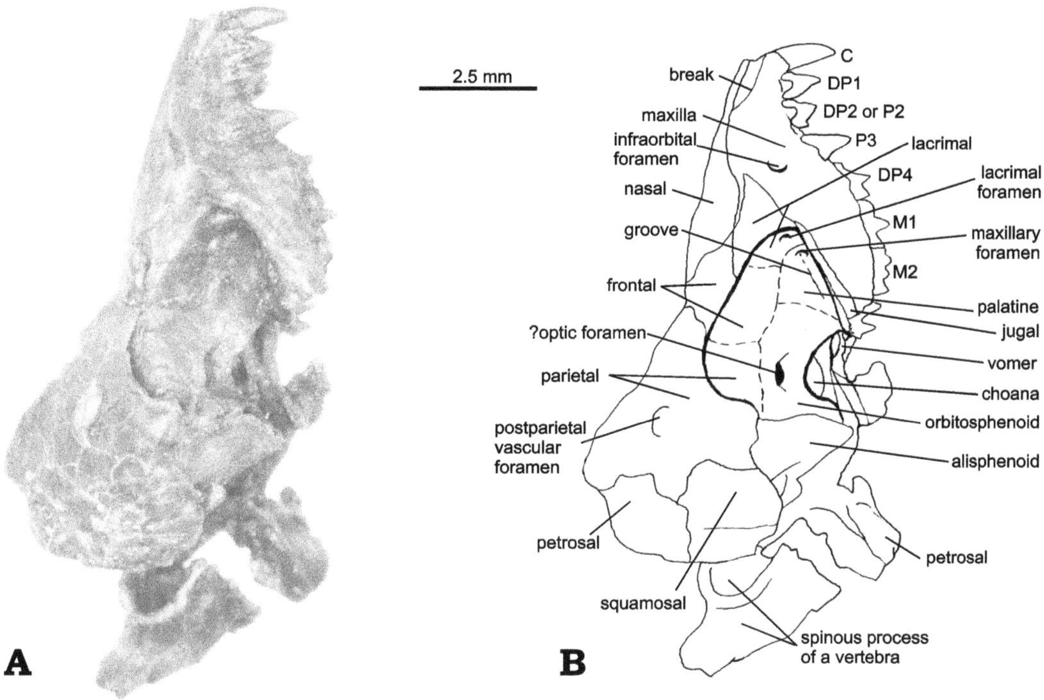

FIGURE 13.3. Skull of *Daulestes nessovi* as preserved in lateral view, photograph (A) and explanatory drawing of the same (B). *Source:* modified from McKenna et al. (2000).

Vidian foramina together form a prominent pterygoid ridge, which is part of a continuous flange that extends anteromedially from the medial border of the glenoid fossa as a prolongation of the postglenoid process. Part of the flange, medial to the foramen ovale, recalls the quadrate ramus of the epipterygoid of therapsids (see Kemp, 1972, and references therein) and morganucodontids (Kermack et al., 1981, see also chapter 4), but in *Asioryctes* it is more inflated.

Crompton and Jenkins (1979) and Crompton and Sun (1985) demonstrated that a principal innovation in the structure of the basicranial region in mammals, in comparison with that of cynodonts, is the formation of the bony floor to the cavum epiptericum. This floor is formed by the petrosal in most mammals outside the Crown Theria (Rougier, Wible, and Novacek, 1996a), but in eutherians it is built of the ventral part of the alisphenoid, which corresponds to the quadrate ramus of the epipterygoid in the basal mammals. The alisphenoid floor of the cavum epiptericum in *Asioryctes* is built by the large quadrate ramus of alisphenoid (see figure 13.4).

The basioccipital in *Asioryctes* is relatively flat and wide, and the sulcus jugularis and foramen nervi hypoglossi are very large. The alisphenoid is bounded posterolaterally by the squamosal. The glenoid fossa is situated lateral to the anterior part of the promontorium. It is gently concave, with a weak postglenoid process; lying posteromedial to the postglenoid process is a large, transversely elongated postglenoid foramen. To the rear of the postglenoid process there is a concave area of the external auditory meatus.

The basicranial region in *Kennalestes* differs in small details from that in *Asioryctes*. Figures 13.1–13.6 demonstrate conspicuous differences in the basicranial region of *Asioryctes, Daulestes,* and *Kennalestes,* on the one hand, from that of the Zalambdalestidae (*Barumlestes*), on the other. In the latter the mesocranial region is much shorter. The pterygoid bone has a pointed end inserted between the posterior prolongations of the maxilla and palatine. The posterior end of the pterygoid is pointed, forming a small hamulus. The anterior part of the presphenoid cannot be defined. The presphenoid has a prominent median process, which protrudes ventrally and posteriorly opposite the basisphenoid. The boundary between the presphenoid and basisphenoid is not very clear and there is no Vidian foramen, which is present at that boundary in *Asioryctes* and *Daulestes.* The basisphenoid is wide, gently concave and provided with a very extensive, roughly triangular pterygoid process. Between the posterior margin of the pterygoid process and the postglenoid process is a deep groove for chorda tympani. Across the middle of the basisphenoid there extends a rounded ridge, which ends

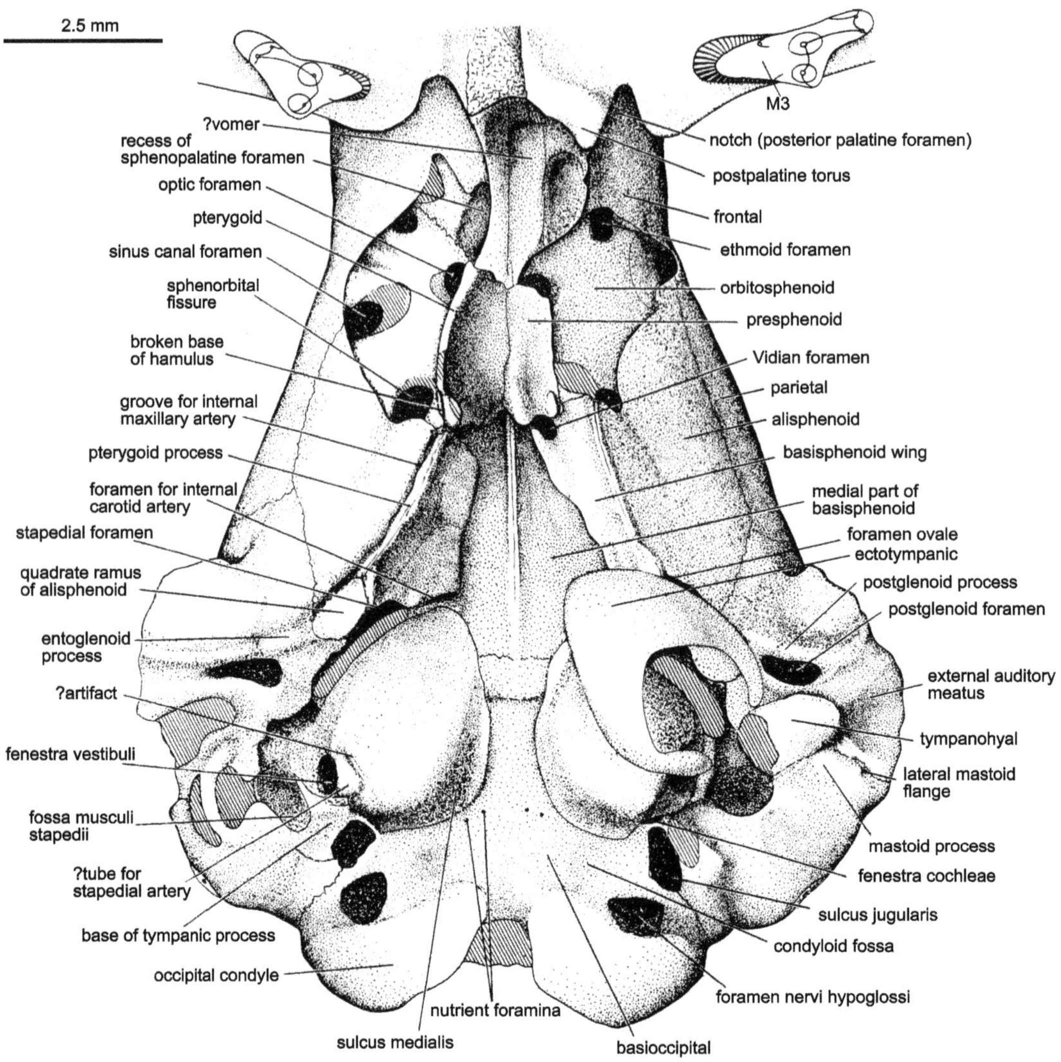

FIGURE 13.4. Ventral view of the braincase in *Asioryctes nemegetensis. Source:* modified from Kielan-Jaworowska (1981).

as a rounded knob where the basisphenoid-basioccipital suture should lie.

The basioccipital is not distinguishable from the exoccipitals. The foramen nervi hypoglossi is divided into two foramina situated to the rear of and medial to the jugular foramen. Lateral to the basioccipital there is a mastoid, separated from it by a distinct suture.

The glenoid fossa in Asioryctitheria and Zalambdalestidae is situated lateral to the anterior part of the promontorium. It is slightly concave in the former, but nearly flat in Zalambdalestidae, with a short postglenoid process (extending only medially) and a tentatively recognized postglenoid foramen.

Ear Region. In addition to the ear region preserved in Late Cretaceous skulls from Asia, isolated eutherian petrosals have been described from the Bug Creek Anthills, Montana (MacIntyre, 1972; Archibald, 1979; Wible, 1990; Meng and Fox, 1995b). As noted in chapter 2, these assemblages are now interpreted as mixed and time averaged, including fossils of both Cretaceous and Paleocene age. Regardless, we include discussion of these petrosals for comparison. MacIntyre (1972) recognized two types of eutherian petrosals from the Hell Creek Formation, referred to as "ferungulate" and "unguiculate," a division not recognized by subsequent authors. Finally, a petrosal referred to Early Cretaceous *Prokennalestes* has recently been described. We begin treatment of the petrosal with its structure in *Prokennalestes* based on the account of Wible et al. (2001).

The petrosal of *Prokennalestes* (figure 13.6B) is apparently primitive in a number of respects, resembling nontribosphenic *Vincelestes* (chapter 10) rather than the Late Cretaceous and younger eutherians and metatherians. It differs from those of Metatheria in having grooves for

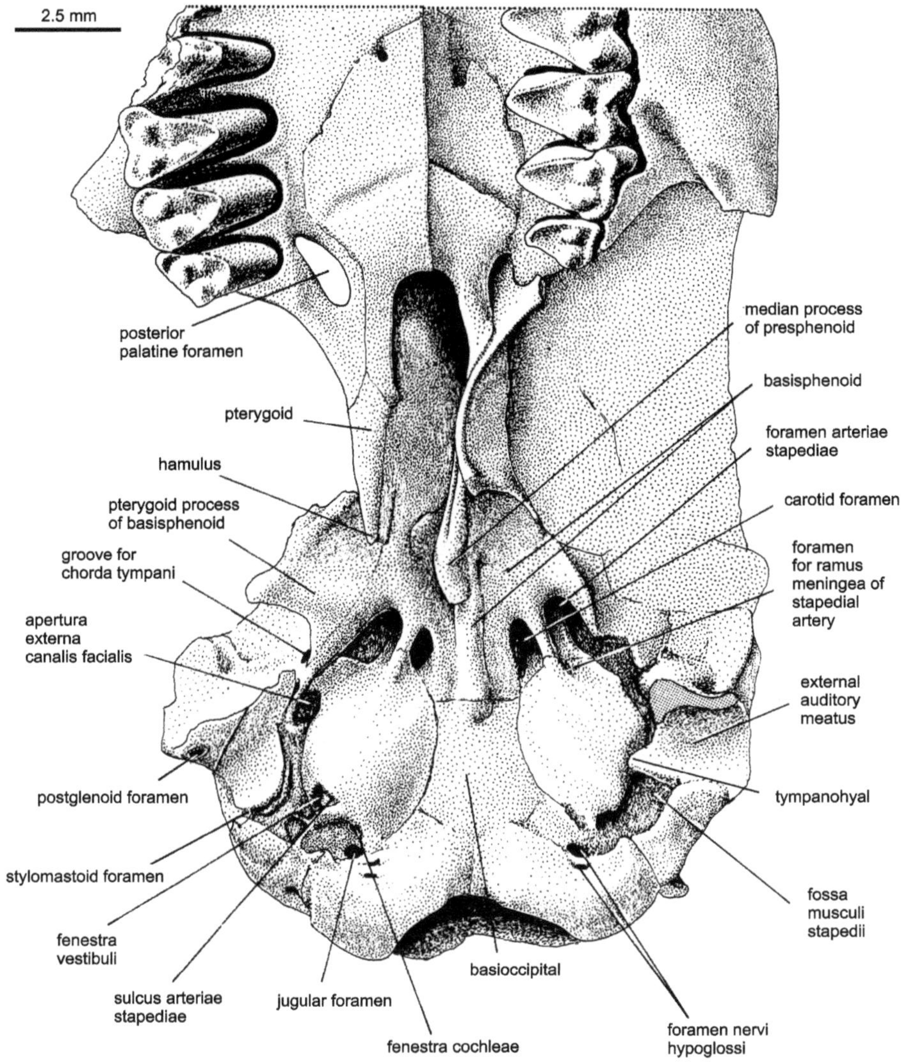

FIGURE 13.5. Ventral view of the braincase in *Barunlestes butleri. Source:* modified from Kielan-Jaworowska and Trofimov (1980).

stapedial and internal carotid arteries. It has been accepted for many years in paleontological literature that possession of two main vessels of the internal carotid artery, referred to as the medial internal carotid and promontory arteries, is characteristic of a eutherian morphotype (McDowell, 1958; McKenna, 1966; Van Valen, 1966; MacIntyre, 1972; Szalay, 1975; Archibald, 1977; and many others). Presley (1979) challenged this hypothesis on embryological evidence, arguing that the internal carotid artery may move medially or laterally during the growth of the promontorium along its medial border, or cross the middle or lateral side of the promontorium and lie within the wall of the tympanic cavity. Presley's idea has been supported by, among others, Kielan-Jaworowska and Trofimov (1980), Kielan-Jaworowska (1981), and especially Wible (1986, 1990), who discussed this issue extensively. In extant mammals the medial position of the internal carotid is retained (among others) in marsupials. In view of these data we do not use the term promontory artery, referring to the artery that leaves the groove on the promontorium (in *Prokennalestes* and *Daulestes,* figure 13.6B) as the internal carotid artery. In other Late Cretaceous eutherians (see later) only the groove for the stapedial artery has been preserved.

Plesiomorphies seen in *Prokennalestes* include the presence of a prootic canal, an intrapetrosal inferior petrosal sinus, a vertical paroccipital process, a fenestra semilunaris, and the lack of complete separation between the cavum epiptericum and the cavum supracochlaeare. In *Vincelestes,* the anterior lamina and lateral flange of the petrosal are well developed; these are lacking in Late Cretaceous and younger eutherians and metatherians and are reduced in *Prokennalestes,* which therefore is morphologically intermediate for these characters. A therian-like characteristic is the coiling of the cochlea through a minimum of 360°. *Prokennalestes* especially resembles Late Cretaceous eutherians in having a shallow internal acoustic meatus with a thin prefacial commissure.

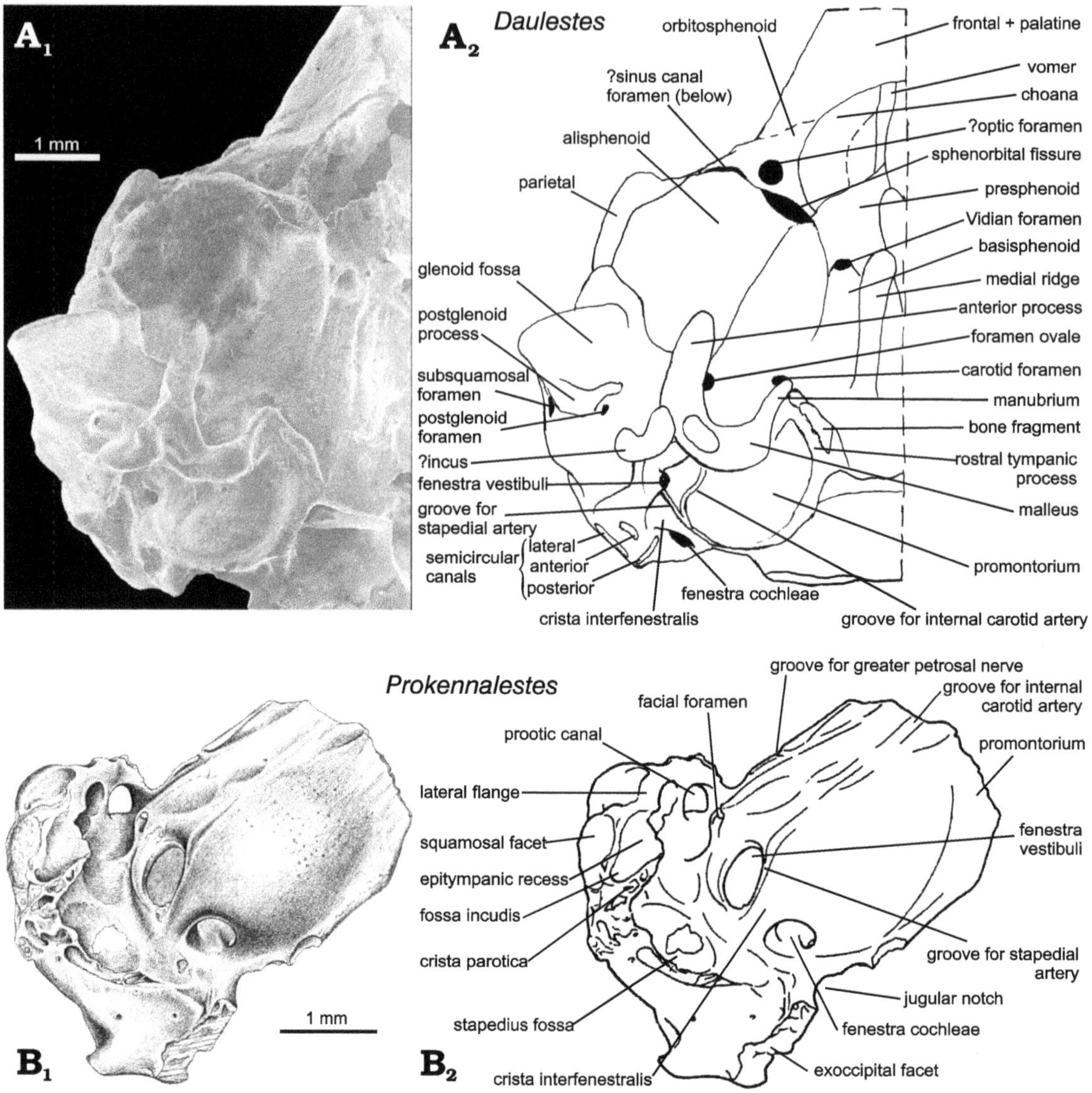

FIGURE 13.6. Petrosal bones in *Daulestes* (A) and *Prokennalestes* (B). A, SEM micrograph of the ventral view of the right basicranial region of *Daulestes nessovi* (A_1) and explanatory drawing of the same (A_2). B, Right petrosal of *Prokennalestes trofimovi.* Shaded drawing (B_1), and explanatory drawing of the same (B_2). *Source:* A, modified from McKenna et al. (2000); B, modified from Wible et al. (2001).

The promontorium is very large in all Late Cretaceous Asian genera, relatively larger than in extant eutherian mammals (figures 13.4B, 13.6). In *Asioryctes* it is pear-shaped and strongly convex. The rostral apex of the promontorium abuts against the basisphenoid. In *Daulestes* it is similarly shaped, but its long axis is more transversely arranged (figure 13.6). Lateral to the rostral apex there is a sulcus at the posterior margin of the basisphenoid wing, interpreted as a foramen for the internal carotid artery. On the lateral side of the promontorium in *Asioryctes* there is a large opening in the roof of the middle ear, which resembles the piriform fenestra in the middle ear roof of the Solenodontidae and Soricidae (McDowell, 1958). However, the piriform fenestra was not recognized in *Daulestes* (McKenna et al., 2000) and because of the state of preservation, it is possible that the vacuity in *Asioryctes* is due to distortion. The fenestra vestibuli is a comparatively large, oval opening on the lateral face of the promontorium. The fenestra cochleae is probably situated in a large, ovoid, funnel-like structure called the recessus fenestrae cochleae. Along the medial margin of the promontorium there extends a shallow concavity, which may be interpreted as a sulcus medialis, tentatively identified as being for the medial internal carotid artery. This sulcus is, however, less distinct than in the eutherian petrosals from Bug Creek (MacIntyre, 1972; Archibald, 1977). The grooves for the stapedial and internal carotid arteries are absent in *Asioryctes,* but there is a short tube medial to the fenestra vestibuli, which might have encased the stapedial artery. In *Kennalestes,* only a groove for the stapedial artery has been preserved (Kielan-Jaworowska, 1981), while in *Daulestes* both sulci are preserved (figure 13.6A).

In *Barunlestes,* at the posterior margin of the basisphenoid adjacent to the boundary with the promontorium, there are two distinct foramina. The medial of the two is recognized as a carotid foramen; the lateral is the foramen arteriae stapediae for the ramus inferior of the stapedial artery. On the bridge of bone between the two foramina there is a small foramen, interpreted as a foramen for the ramus meningea of the stapedial artery (figure 13.5). The fenestra vestibuli is a comparatively large, oval opening on the lateral face of the promontorium (only partly exposed in ventral view). Extending medially from the fenestra is a very short, triangular sulcus for the stapedial artery. The fenestra cochleae faces posteriorly, toward the sulcus that housed the jugular foramen. The apertura externa canalis facialis is discernible on the lateral side of the petrosal, from which a groove for the facial canal (sulcus facialis of MacIntyre, 1972) extends toward the fossa musculi stapedii. The latter is partly covered ventrally by the tympanohyal. The foramen stylomastoideum (developed as a sulcus) is placed between the tympanohyal and the mastoid process.

The cerebellar side of the petrosal could be observed only in a juvenile skull of *Kennalestes* (see Kielan-Jaworowska, 1981) and two isolated petrosals. In the mastoid portion of the petrosal there is a very large and deep subarcuate fossa, encircled laterally and dorsally by the anterior semicircular canal and medially by crus commune. The sharp petrosal crest divides the petrosal into cerebellar and squamosal sides. On the petrosal portion there are two foramina, the smaller of which, situated just under the petrosal crest, is for cranial nerve VII and continues in the petrosal ventrolaterally as the facial canal. The larger foramen of nerve VIII is subdivided into vestibular and cochlear parts. All three foramina are situated together in a common fossa of the internal auditory meatus. On the squamosal side of the petrosal, just below the petrosal crest, there is an incurvature, possibly for a semilunar ganglion of trigeminal nerve V, but there is no clearly defined fossa.

The X-ray photograph of the cochlea of *Daulestes* (McKenna et al., 2000: figure 10) demonstrates that it has one full turn, with an expanded apex, which suggests that a lagena might have been present (but see Wible et al., 2001). The coiling in *Daulestes* is slightly less extensive than that of a therian from the Late Cretaceous Milk River Formation of Canada, described by Meng and Fox (1995c). In the Milk River specimen, the apex of the cochlear canal overlaps the basal canal, but in *Daulestes* there is no overlap. In isolated eutherian petrosals described by MacIntyre (1972), Archibald (1979), Wible (1990), Meng and Fox (1995a,c), and Wible et al. (1995), the cochlea is fully coiled; Wible et al. (2001) estimate a minimum of about 450°. The cast of a cochlea is preserved in one specimen of *Zalambdalestes,* where it possibly had about one full coil of rotation (Kielan-Jaworowska, 1984c), but owing to the lack of an X-ray photograph, this cannot be regarded as a certainty. The presence of a lagena, recognized in *Daulestes,* would be unique among therian mammals. Among mammals, monotremes and probably multituberculates have a lagena, which is indicated osteologically by an expanded apex of the cochlear canal (Griffiths, 1968; 1978; Fox and Meng, 1997, and references therein). The cochlea of *Prokennalestes* is of uniform diameter to its tip, but the significance of this is uncertain (Wible et al., 2001).

In *Daulestes* the three semicircular canals are preserved and do not differ from those of other mammals.

Ear Ossicles. Archibald (1979) described a eutherian stapes, associated with a petrosal, from the Bug Creek Anthills. Whether of Cretaceous or Paleocene age, it is the oldest eutherian stapes that is completely preserved. It has an elongated footplate, gracile crura set near either end to the footplate, and large stapedial foramen. The only Cretaceous eutherian malleus and ?incus have been preserved in the Coniacian *Daulestes* (see McKenna et al., 2000). The malleus is nearly complete on the right side of the skull (figure 13.6A) and is partly preserved on the left side. The right malleus has been moved from its original position anteriorly and is rotated slightly within the horizontal plane, with its anterior end shifted more laterally, whereas the posterior end is shifted more medially. Consequently, the anterior process (goniale) now points anteriorly. In life, it would have pointed more anteromedially. The anterior process is long, although its tip may be incomplete. McKenna et al. (2000) were unable to determine whether the foramen for the chorda tympani is present as in monotremes (Doran, 1878; Fleischer, 1973) and multituberculates (Meng and Wyss, 1995), but the homology of the anterior process (goniale) with the nonmammalian synapsid prearticular was assumed. The lamina (body) of the malleus is in all aspects typically therian. Its ventral surface is concave, and laterally the body is bounded by a distinct ridge, which represents part of the thickening at what appears to be the incus-malleus articulation. The ridged structure suggests that the incus was placed posterior to the malleus, as in living therians. The manubrium is a very thin process, extending from its rounded base anteriorly. The anterior end of the manubrium abuts against the carotid foramen.

The ?incus, as interpreted by McKenna et al. (2000), preserved on the right side of the skull (figure 13.6), adheres to and ventrally partly covers the posterolateral border of the malleus. If this bone is indeed an incus, it has apparently been overturned posterolaterally. The ?stapes has not been identified with any certainty in *Daulestes.*

The ectotympanic has been preserved in two specimens of *Asioryctes,* in one of *Kennalestes,* and in one of *Daulestes,* but not in Zalambdalestidae. It is horseshoe-shaped in the first two genera and U-shaped in *Daulestes,* which might be a juvenile character. In specimens of *Asioryctes,* preserved with the dentaries in occlusion (Kielan-Jaworowska, 1975b: plate 1), the angular process of the dentary conceals the anteromedial part of the ectotympanic bone, so that the anterior part of the ectotympanic fits into the emargination in the posterior part of the dentary. This clearly shows that the bone in question is homologous to the angular of nonmammalian synapsids, being the ectotympanic and not entotympanic.

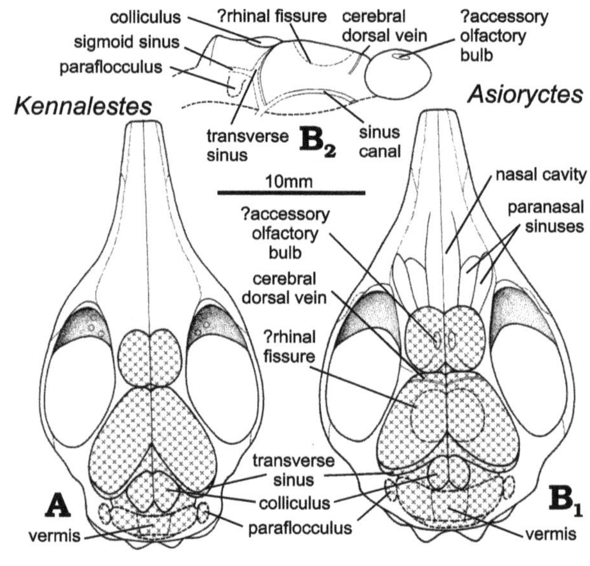

FIGURE 13.7. Reconstruction of the skull and endocast of *Kennalestes gobiensis* (A) and *Asioryctes nemegetensis* (B). B_1, the skull and endocast in dorsal view, B_2, the endocast in lateral view. Sigmoid sinus and a part of transverse sinus reconstructed in B_2 have not been preserved. *Source:* modified from Kielan-Jaworowska (1984c).

BRAIN

Ariëns Kappers et al. (1960) indicated that the mammalian cerebellum differs from those of all other vertebrates by its greater development of the transverse diameter (presence of cerebellar hemispheres and paraflocculi) and regarded monotremes as an exception in this respect. Studies of Holst (1986), however, demonstrated that the monotreme (especially tachyglossid) brain is generally therian-like. Ariëns Kappers et al. were speaking only about the brains of extant mammals.

Partial or complete endocranial casts are known for only four Mesozoic eutherians (figure 13.7), all from the Late Cretaceous of Mongolia: *Kennalestes, Asioryctes, Zalambdalestes,* and *Barunlestes* (see Kielan-Jaworowska, 1984b, 1986, 1997, and references therein; Kielan-Jaworowska and Trofimov, 1986). They indicate a primitive therian lissencephalic brain with very large olfactory bulbs and cerebral hemispheres widely separated posteriorly. They have a comparatively short and wide cerebellum with well-developed cerebellar hemispheres. The brain cavity in numerous therian mammals is separated into cerebral and cerebellar cavities by the tentorium osseum. As inferred from endocranial casts there apparently was a midbrain exposure on the dorsal side in Cretaceous eutherians, as characteristic of some extant primitive placentals (see Starck, 1962).

Kielan-Jaworowska (1984c) tentatively estimated encephalization quotients of 0.36 for *Kennalestes gobiensis,* 0.56 for *Asioryctes nemegetensis,* and 0.70 for *Zalambdalestes lechei.*

DENTITION

As already noted, the fossil record for early tribosphenic mammals consists mainly of teeth and jaws, and for this reason the dentition has assumed a preeminent role in interpreting the early history of both eutherians and marsupials. Given the often subtle or equivocal nature of the differences, however, the dental evidence is best interpreted cautiously and with a healthy degree of skepticism. Whereas the skull and skeleton merit individual attention because they are known for only a few Cretaceous eutherians, virtually all taxa are represented by at least part of the dentition. In this section, we review the basic morphology of the dentition in early eutherians, identifying, where possible, the probable primitive condition. We also provide a brief overview of the major trends and variations in the dentitions of Cretaceous eutherians, but defer discussion of individual specializations and their biological and systematic interpretation to the "Systematics" section of this chapter. Similarly, most of the illustrations cited in this section are grouped at the back of the chapter, among the accounts of individual genera.

Dental Formula. The presence of three upper and lower incisors is almost universal among at least the early members of modern and fossil eutherians of the Cenozoic and distinguishes them from primitive marsupials (which have I5/i4). Cretaceous eutherians for which the count is known paint a different picture. The oldest eutherian, *Eomaia,* has five upper and four lower incisors (see Ji et al., 2002). The incisor count for *Prokennalestes* (see Sigogneau-Russell et al., 1992) and other Early Cretaceous to early Late Cretaceous taxa is not known. Campanian asioryctids *Asioryctes* and *Ukhaatherium* (figure 13.22B) have five upper and four lower incisors, and the kennalestid *Kennalestes* has four upper and three lower incisors (Kielan-Jaworowska, 1981; Novacek et al., 1997). This suggests that eutherians primitively had a similar incisor count to that of marsupials (I5/i4), with a later reduction to three in each jaw (Clemens and Lillegraven, 1986). Intriguingly, the last upper incisor in asioryctitherians is located within or behind the suture between the maxilla and premaxilla (figures 13.1B,C, 13.21B). *Cimolestes incisus* had three lower incisors (Clemens, 1973a), and at least two were present in *Gypsonictops* (Fox, 1979d). Zalambdalestids have ?three upper and three lower incisors (Kielan-Jaworowska, 1984b). The count is unknown for most other Cretaceous Eutheria.

Interpretation of the primitive premolar count is also problematic, though for different reasons. Many latest Cretaceous eutherians and virtually all eutherian groups of the Cenozoic are thought to have had four premolars ancestrally. But a growing number of taxa from the Cretaceous, including the three oldest known by relatively complete jaws, *Eomaia, Prokennalestes,* and *Otlestes,* have five—suggesting that this may be the primitive number (McKenna, 1975; Clemens and Lillegraven, 1986; Novacek, 1986a; Kielan-Jaworowska and Dashzeveg, 1989; Nessov, Sigogneau-Russell, and Russell, 1994; Nessov, Archibald, and Kielan-Jaworowska, 1998, see figures 13.8A, 13.19B_3). Five antemolar cheek teeth are also known in the lower dentition of the leptictoid *Gypsonictops* (Lillegraven, 1969; Clemens, 1973a; Fox, 1977, 1979), the upper dentition of the juvenile (but not adult) asioryctitherian *Kennalestes* (Kielan-Jaworowska, 1981), and one or more of the ungulatomorph "Zhelestidae" (Nessov, Archibald, and Kielan-Jaworowska, 1998; Archibald and Averianov, 2001a). The last tooth of the series is usually semimolariform and the penultimate trenchant. The homologies of these, and the first two in the series, are not disputed: at issue is the tooth in the middle. In *Gypsonictops* and "Zhelestidae," this tooth tends to be considerably smaller than adjacent premolars, and it also appears to be relatively small in *Kennalestes* and *Prokennalestes.* McKenna (1975) interpreted the small size of this tooth to indicate that it was a replacement premolar undergoing phylogenetic reduction, thus suggesting that five premolars were present in the ancestral eutherian. Luckett (1993) noted that premolar loss among mammals generally proceeds backward from the front of the series, not in the middle.[1] Luckett (1993) forwarded an alternative hypothesis, suggested earlier by Clemens (1973a), that the tooth in the middle of the series is a retained deciduous tooth, probably dP2, that was not displaced by the eruption of the successor P2. In addition to the discrepancy in size, this hypothesis is supported by the fact that the variable presence of the middle premolar at least in *Kennalestes,* and probably *Gypsonictops* as well, is ontogenetically related: it is lost in older individuals (Clemens, 1973a; Kielan-Jaworowska, 1981). The adult dentition of the Coniacian eutherian *Daulestes* is not known, but existing data are consistent with what is

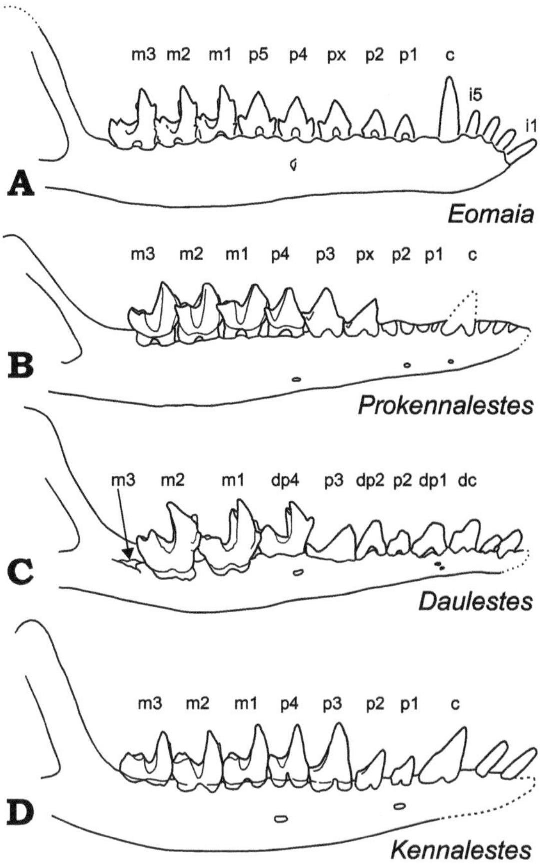

FIGURE 13.8. Dentaries of Cretaceous eutherians in lateral view, showing variation in the premolar series. A, *Eomaia.* B, *Prokennalestes.* In both *Eomaia* and *Prokennalestes* five premolars are preserved, the third labeled px, following the system advocated in the text (but see figure 13.19G, for alternative numbering of premolars in *Eomaia*). C, *Daulestes* juvenile, showing the retention of dp2 after the eruption of the successor p2. D, *Kennalestes* adult, showing the typical primitive eutherian complement of p1–4, although juveniles retain dp2 after eruption of p2, as in *Daulestes.* Not to scale; reversed as necessary. *Source:* A, modified from Ji et al. (2002); B–D, modified from Cifelli (2000a).

[1] This rule does not hold for Multituberculata, however. In the djadochtatheriid genera *Catopsbaatar* and *Tombaatar,* for example, the formula of three upper premolars was acquired through the loss of P2 (see Kielan-Jaworowska, 1974a; Rougier et al., 1997; chapter 8 and figure 8.38C,I).

known for *Kennalestes,* showing that dp2 and p2 were present at the same time in a juvenile individual (figure 13.21A), whereas the adult probably had four premolars. In this case, the traditional interpretation, that eutherians primitively had four premolars, could be retained. Unfortunately, existing fossils are inadequate to address the ontogenetic hypothesis, and outgroup comparisons are not of much help either. Early Cretaceous *Kielantherium,* a "tribotherian" (see chapter 11) had at least four premolars (Dashzeveg and Kielan-Jaworowska, 1984), and nontribosphenic *Peramus* can be alternatively interpreted as having four or five (see Clemens and Lillegraven, 1986, and chapter 10). Similarly, the "eupantotherian" *Amphitherium* may have four (Simpson 1928a) or five (Butler and Clemens, 2001) premolars, as is also the case for the *Peramus*-like *Arguitherium* and *Arguimus* (see Dashzeveg, 1994; Butler and Clemens, 2001, and chapter 10). We can reasonably conclude that eutherians are primitive in retaining at least four premolars, and that this distinguishes them from the more advanced marsupials and deltatheroidans, which have only three. Whether eutherians primitively had five premolars or early taxa tended to retain a deciduous tooth into adulthood is uncertain (see discussion in Nessov, Archibald, and Kielan-Jaworowska, 1998). Also left open is the larger question of whether, in either case, the presence of five antemolar cheek teeth in some early eutherians is: (1) a plesiomorphy inherited from a noneutherian ancestor; (2) an apomorphy of Eutheria, later reversed; or (3) an apomorphy of a restricted subset of Cretaceous Eutheria. Given these uncertainties and the confusion stemming from alternatively designating the teeth P1–5 (in taxa with five premolars) or P1–4 (in taxa with only four), despite the unquestioned homology of the posterior two teeth, we prefer to use the more neutral designations of P1, P2, Px, and P3, P4 (see Cifelli, 2000a).

There is general agreement that the primitive molar count for eutherians is three. Marsupials, deltatheroidans, *Kielantherium,* and another primitive tribosphenic mammal (Wang et al., 1995) have four molars. We provisionally regard this as the primitive number for Boreosphenida, so that the presence of three in eutherians is a reduction. Nontribosphenic *Peramus* (see chapter 10) has either three or four molars, but phylogenetic analysis suggests that the count of three in eutherians is derived, regardless of how *Peramus* is interpreted (Rougier et al., 1998).

Incisors, Canines, and Premolars. In *Eomaia* there are five upper and five lower peglike incisors, the I1 being semiprocumbent and the following uppers nearly vertical. The lower incisors show decreasing procumbency, the i1 lying in prolongation of the oblique lower margin of the dentary and successive teeth being gradually more vertical. Some of the incisor alveoli are known for *Prokennalestes* (see Sigogneau-Russell et al., 1992), *Otlestes* (see Nessov, Sigogneau-Russell, and Russell, 1994), *Gypsonictops* (Fox, 1979d), and *Cimolestes* (Clemens, 1973a). These are not very informative, other than suggesting that some degree of procumbency is common among Cretaceous Eutheria. Most of our knowledge of these teeth comes from Mongolian taxa. In asioryctids, most upper and lower incisors are simple and peglike, with i1 being semiprocumbent and procumbency decreasing from i2 to i4. I3–4 are larger than I1–2 and 5 in both *Asioryctes* (Kielan-Jaworowska, 1981) and *Ukhaatherium* (Novacek et al., 1997); notably, I3 and I4 may be bicusped in the latter (figure 13.21B). In zalambdalestids (Kielan-Jaworowska, Bown, and Lillegraven, 1979), i1 is greatly enlarged and procumbent, with i2 and i3 being much smaller and semiprocumbent, an obviously apomorphic condition related to the elongated snout.

Judged from the condition in most Mesozoic mammals and eutherian outgroups, the canine presumably was moderately large in the ancestral eutherian, as it is in *Otlestes,* asioryctids, *Gypsonictops, Cimolestes,* and ungulatomorphs. Despite difficulty in interpreting alveoli on available fossils, *Prokennalestes* appears to have had a rather small canine (Sigogneau-Russell et al., 1992). The lower canine of zalambdalestids (Kielan-Jaworowska, Bown, and Lillegraven, 1979) is derived in being small, semiprocumbent, and functionally integrated with the lateral incisors (figure 13.23A).

Most living and Cenozoic tribosphenic mammals, both eutherians and marsupials, have single-rooted canines. However, double-rooted canines have a broad and puzzling distribution among Mesozoic mammal groups (Clemens and Lillegraven, 1986; Cifelli and Madsen, 1999) and are found in some early eutherians. At present, the primitive condition in eutherians, as well as the significance of the observed distribution, defies interpretation. In *Eomaia* both lower and upper canines are single-rooted. Sigogneau-Russell et al. (1992) reconstructed the lower canine in *Prokennalestes* as single-rooted, but one of us examined a dentary of this taxon with a double-rooted canine in the collection of the Paleontological Institute in Moscow.

Otlestes probably had a single canine root (Nessov, Sigogneau-Russell, and Russell, 1994). Variability is the norm among asioryctitherians and zalambdalestids: *Kennalestes* has a single-rooted deciduous and a double-rooted replacement canine (figure 13.22); *Asioryctes* has double-rooted canines (Kielan-Jaworowska, 1981), whereas they are single-rooted in closely related *Ukhaatherium* (Novacek et al., 1997, our figure 13.21B). Similarly, the canines are double-rooted in *Zalambdalestes* but single-rooted in its close relative, *Barunlestes* (Kielan-Jaworowska and Trofimov, 1980, see also our figure 13.1D,E). Remaining taxa for which the condition can be determined have a single-rooted canine: "zhelestids," *Protungulatum, Gypsonictops,* and *Cimolestes* (Lillegraven, 1969; Clemens, 1973a; Fox, 1979d; Nessov, Archibald, and Kielan-Jaworowska, 1998).

With the exception of Px, noted earlier, premolars of Cretaceous eutherians generally increase in size from first to last. In *Eomaia* both upper and lower premolars are double-rooted, and the lowers increase in size distally. The upper premolars show a different pattern (Ji et al., 2002: figure 2): the first four premolars decrease in size distally and there are short diastemata between them, but the P5 is by far the largest of all. The premolars are all double-rooted in *Prokennalestes, Otlestes,* and most other taxa, which probably represents the primitive condition. However, P1 or p1 occasionally has only a single root, such as in certain species of *Cimolestes* (Lillegraven, 1969; Clemens, 1973a), where the tooth is presumably undergoing reduction. Interestingly, both two-rooted (*Sorlestes*) and single-rooted (*Zhelestes*) first premolars are known for "zhelestids," whereas a single-rooted first premolar characterizes *Protungulatum* and other early ungulates (figure 13.28C), suggesting that the condition may be a synapomorphy at some level within Ungulatomorpha. The presence of small anterobasal and posterobasal cuspules on each of the premolars (e.g., figures 13.21, 13.22) is also probably primitive for eutherians, though in a few cases some of these are lacking, as in *Asioryctes,* where the premolars are anteroposteriorly compressed. The posterior cuspule (talonid) on lower premolars often, though not invariably, increases in height and length through the series.

Without doubt, the most important characteristics of the premolars in early eutherians are found on the last, or occasionally last and next to last, in the series. In eutherians, these generally are somewhat molariform, so that the premolar and molar series intergrade to some extent. Deltatheroidans and marsupials, on the other hand, resemble most Mesozoic mammal groups in having simple, premolariform last premolars with a tall primary cusp; there is a sharp morphological break between the premolars and the molars. We recognize the intergrading of premolar and molar series as a derived feature of Eutheria. However, it should be noted that the appearance of certain cusps on the last premolar is a tendency achieved independently in various lineages. Furthermore, there are problems in recognizing the serial homology of cheek teeth in tribosphenic mammals and their close relatives (Clemens and Lillegraven, 1986; Luckett, 1993; Butler and Clemens, 2001). For example, if the cheek tooth formula of nontribosphenic *Peramus* is interpreted as P5 M3 (rather than P4 M4), then its last premolar would be somewhat molariform (see chapter 10). This would raise the question as to whether: (1) this unusual condition was achieved independently; (2) the simple last premolar of marsupials and deltatheroidans represents a reversal (or the penultimate premolar and the first "molar" in these taxa is a retained deciduous premolar, Luckett, 1993); or (3) the major groups of higher mammals arose from different, nontribosphenic ancestors. Lacking definitive evidence, we favor the first hypothesis as the simplest explanation and thus regard molarization of the last premolar as an apomorphy of eutherians. Interestingly, the last upper premolar of the early eutherian *Prokennalestes* is more complex and molariform than the last lower (Kielan-Jaworowska and Dashzeveg, 1989; Sigogneau-Russell et al., 1992), as is generally the case for other Cretaceous eutherians, showing that the upper dentition is more advanced in this respect than the lower.

Important molariform characteristics of P4 in *Prokennalestes* (figure 13.19A) are the presence of three (rather than two) roots, a well-developed protocone, parastyle, and metastyle, and a slight stylar shelf. A metacone is lacking, which is probably the primitive condition for Eutheria. The p4 of *Prokennalestes* is simple. We take the morphology of the last premolars in *Prokennalestes* to approximate the ancestral pattern for Eutheria. Interestingly, in the approximately contemporaneous *Montanalestes* from North America (figure 13.19D), a distinct paraconid and metaconid (the former lingually placed) are present on p4, as is an incipient talonid basin (Cifelli, 1999b). These features show it to be more advanced; unfortunately, upper premolars of *Montanalestes* are unknown. The p4 of Cenomanian *Otlestes* (figure 13.19B) is more advanced yet, with a tall, lingually placed metaconid and better developed paraconid, also lingually placed.

We note in passing that premolars referred to the Trinity therians *Pappotherium* and *Holoclemensia* (chapter 11) are also somewhat molariform. The significance of these features and the referral of the specimens themselves remain uncertain (Butler, 1978a), but as discussed later, it does suggest the possibility, at least, that one or both may eventually prove to be eutherian, as suggested originally (Slaughter, 1971, 1981).

Major modifications of what we interpret as the ancestral pattern for eutherian posterior premolars occur in various groups. Whether features were acquired through a common ancestor or independently is generally difficult or impossible to assess. The most common modification is increased molarization of P4/p4. P4 of many taxa bears a metacone, and in certain instances conules and pre- and postcingula are present as well (figures 13.22B, 13.24B,C, 13.25C, 13.28A,B). Lower p4 commonly has more molariform development of the trigonid cusps and expansion of the small talonid basin. In a few groups, such as ungulatomorphs and the leptictoid *Gypsonictops,* molarization also extends to P3/p3 (figures 13.24C, 13.28C). The conditions in the Lancian cimolestidan *Cimolestes* (figure 13.25C) are enigmatic. P4 bears a protocone, but the p4 is remarkably simple for such a late-occurring taxon, so that its structure may be genuinely primitive or, possibly, secondarily simplified.

Molars. Study of latest Cretaceous mammal faunas, particularly those of North America, has led to the recognition of morphological guidelines that serve to distinguish between the included marsupials and eutherians on the basis of molar structure (Lillegraven et al., 1979). However, the discovery of more primitive taxa from the medial Cretaceous has eroded the reliability of many such features for recognizing the split between these major clades, though they remain useful as general trends. Upper molars of the oldest eutherian, *Eomaia,* have not been preserved, except for M3, but that has only been figured for the schematic labial view (Ji et al., 2002: figure 2c). The authors characterized the upper molars of *Eomaia* as having larger metastylar and metaconal regions than *Prokennalestes* and less developed conules than both *Prokennalestes* and *Murtoilestes.* Lower molars in *Eomaia* are well preserved; they differ from *Prokennalestes* and *Murtoilestes* in having a shorter trigonid and a longer talonid basin (also see the diagnosis of *Eomaia* under "Systematics" in this chapter). Molars of *Prokennalestes* (figure 13.19A) conform, in general, to what might be expected in an ancestor of either Eutheria or Marsupialia. On the upper molars, a broad stylar shelf is retained, the conules are small, unwinged, and not closely approximated to paracone and metacone, respectively; the paracone is significantly taller than the metacone, and there are no pre- and postcingula. The absence of stylar cusps (presumably a primitive condition) distinguishes many Cretaceous eutherians from certain Late Cretaceous marsupials (Kielan-Jaworowska, Bown, and Lillegraven, 1979), though stylar cusps are known in some eutherians (Fox, 1984b), and primitive marsupials appear to have lacked them (Cifelli and Muizon, 1997).

Common, early tendencies in Eutheria include reduction of the width of the stylar shelf and reduction to loss of the stylocone, strengthening of the conules and development of strong conular cristae, and development of pre- and postcingula, particularly posteriorly, where a hypocone often develops in advanced taxa. The protocone is expanded lingually in some groups (e.g., Cimolestidae, figure 13.25B), so that the upper molars are very broad transversely. In other groups, such as ungulatomorphs (figures 13.26B, 13.28), the protocone becomes enlarged anteroposteriorly and cusp height is reduced, in apparent correlation with a shift to omnivory.

Lower molars, like the uppers, are rather primitive in early forms, on the basis of both overall shape and coronal morphology (Fox, 1984b; Kielan-Jaworowska and Dashzeveg, 1989; Cifelli, 1999b). Indeed, they are distinguished from marsupials primarily because they lack marsupial specializations, including a keeled, lingually placed paraconid, labial postcingulid, and hypoconulid medially shifted, or "twinned" with the entoconid (Clemens, 1979a; Clemens and Lillegraven, 1986; Cifelli, 1993b). Perhaps the most diagnostic feature of lower molars belonging to most Cretaceous Eutheria is the reduced paraconid, which is distinctly lower than the metaconid and tends to be closely appressed to that cusp (figure 13.19A,B). However, the paraconid and metaconid of *Montanalestes* (figure 13.19D), an Early Cretaceous eutherian from North America, are subequal in height and well separated (Cifelli, 1999b). Tendencies in the early evolution of eutherian lower molars include the increase (e.g., Cimolestidae, figure 13.25C) or decrease (e.g., ungulatomorphs, figures 13.26B, 13.28) in the height differential between trigonid and talonid, widening of the talonid (certain insectivores, ungulatomorphs), and further reduction of the paraconid, with its appression to the metaconid (many groups).

Tooth Replacement. Most (excepting perhaps the first premolar) of the antemolar dentition of eutherians undergoes single replacement (diphyodonty), which is presumed to be the primitive condition (Martin, 1997). Marsupials, on the other hand, are derived in that only one tooth—the last premolar—is replaced postnatally in each jaw (Luckett, 1993, see also chapter 12). The condition in the tribotherian *Slaughteria* has been recently reported by Kobayashi et al. (2002, see chapter 11). The replacement mode is unknown for most Mesozoic mammals, though it is clear that a wide variety of patterns is represented (Cifelli, 1999a).

POSTCRANIAL SKELETON

In the oldest known eutherian mammal, *Eomaia* (assigned to Eutheria, order and family *incertae sedis,* see Ji et al., 2002), the complete (but flattened) postcranial skeleton has been preserved (figure 13.9), but has not yet been described in detail. Other postcranial material of early eutherians includes isolated skeletal elements of *Protungulatum* and *Deccanolestes* and large fragments of the skeleton of the Late Cretaceous Mongolian eutherians, classified as Asioryctitheria and Zalambdalestidae. As these skeletal parts differ considerably from one another, we describe

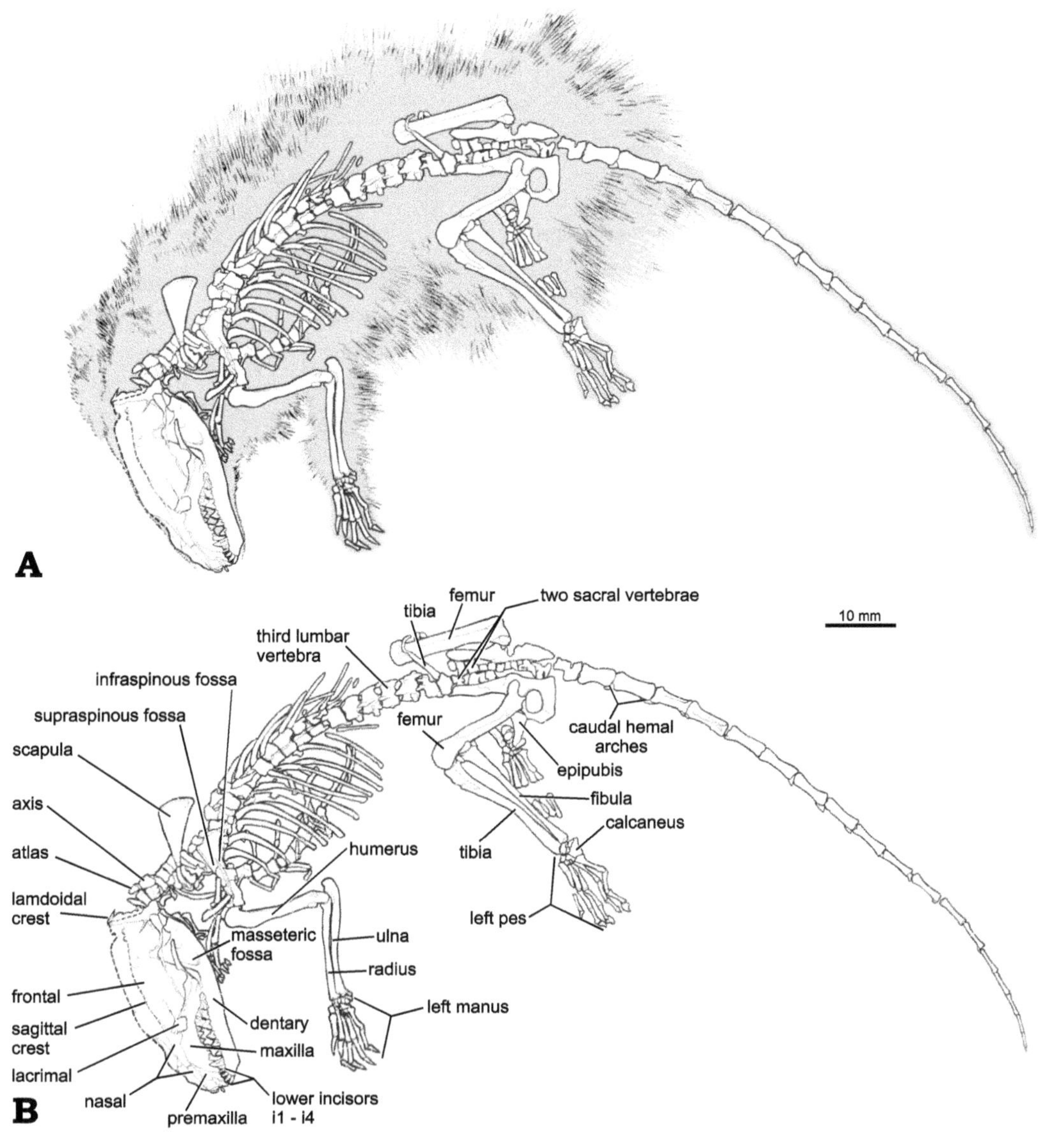

FIGURE 13.9. *Eomaia scansoria.* Skeleton as preserved, with fur halo (A); the same with identification of skeletal structures (B). Note that there are seven cervical vertebrae, 13 thoracic (with ribs), six lumbar, two sacral and 25 caudal. *Source:* modified from Ji et al. (2002).

the postcranial skeletons in these groups separately, beginning with the most complete skeleton of *Eomaia*, followed by the isolated tarsal bones of *Protungulatum* (see Szalay and Decker, 1974) and *Deccanolestes* (see Godinot and Prasad, 1994; Prasad and Godinot, 1994), and concluding with Late Cretaceous Mongolian taxa (Asioryctitheria, Zalambdalestidae).

POSTCRANIAL SKELETON IN *EOMAIA*

The following description of the postcranial skeleton in *Eomaia* is a summary of the account given by Ji et al. (2002). The *Eomaia* postcranial skeleton differs from those of other Cretaceous eutherians (in particular those from the Late Cretaceous of the Gobi Desert), earliest known metatherian and nontribosphenic therians by several apomorphies. We figure herein the skeleton of *Eomaia* as preserved (figure 13.9) and details of its limb structure (figure 13.10), all based on the preliminary study of Ji et al. (2002).

Forelimb. The scapula is slender, with a prominent coracoid process and a relatively large acromion process on a tall scapular spine. The clavicle is robust and curved, with the proximal end abutting the lateral process of the clover-shaped manubrium. *Eomaia* differs from the Late Cretaceous eutherians *Asioryctes* and *Zalambdalestes* in having an enlarged and elongated trapezium in the carpus (figure 13.10A). The hamatum is large; the trapezoid and

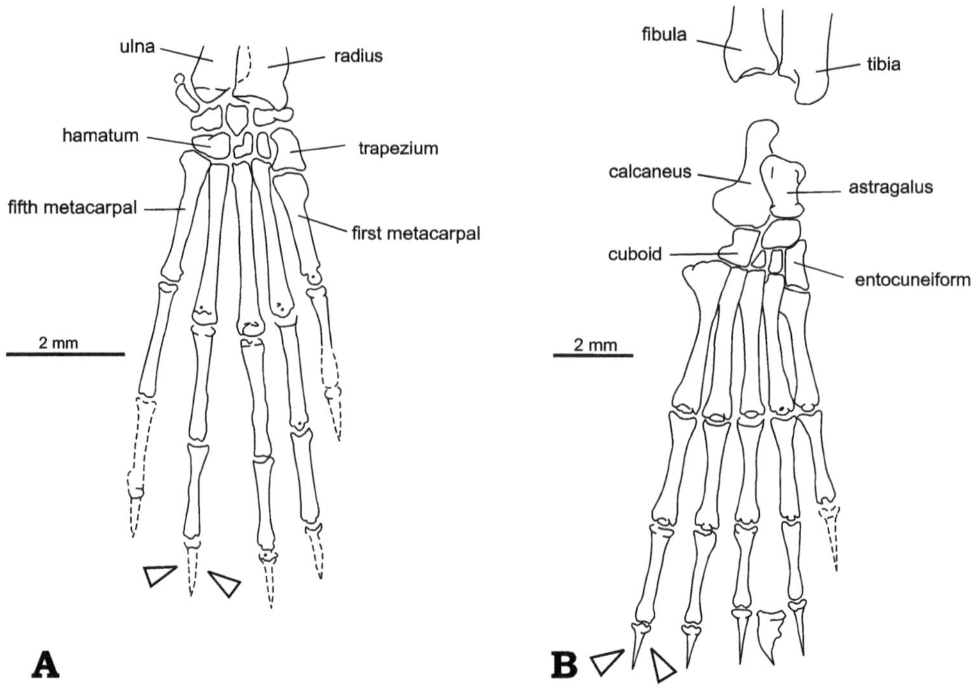

FIGURE 13.10. *Eomaia scansoria.* A, Reconstruction of the right manus in dorsal view. B, Reconstruction of the right foot in dorsal view. In both reconstructions the claws are arched and laterally compressed (arrows), as is characteristic of scansorial or arboreal mammals. *Source:* modified from Ji et al. (2002).

capitatum are small and their proportions to the hamatum and trapezium are comparable to the condition in the grasping hand of living scansorial and arboreal mammals (see figure 13.18B). *Eomaia* and other eutherians retain the primitive mammalian condition in which the scaphoid and triquetrum are small relative to other carpals, in contrast to the hypertrophied scaphoid and triquetrum seen in marsupials (see chapter 12). Metacarpal V is aligned with the anterior edge of the hamatum (a synapomorphy of crown therians), in contrast to the eutriconodont *Jeholodens* (chapter 7) and the "symmetrodontan" *Zhangheotherium* (chapter 9), in which the proximal end of metacarpal V overhangs and is offset from the hamatum.

Pelvic Girdle and Hindlimb. The ilium, ischium, and pubis are fused, and the epipubic bone is present. The patella is present, which is a derived placental feature that is lacking in most basal metatherians. The ankle bears a strong resemblance to those of Late Cretaceous eutherians (figure 13.10B). As diagnostic for Eutheria, the medial astragalotibial facet is well developed, vertical, and separated by a sharp crest from the lateral astragalotibial facet. The navicular facet is distinctive from the astragalar neck, and does not extend to the medial side of the neck as in metatherians. The entocuneiform is elongate, and its joint with metatarsal I is offset anteriorly from the intermediate cuneiform and metatarsal II. The calcaneus is similar to those of *Asioryctes, Ukhaatherium,* and *Zalambdalestes* in retaining many primitive features of crown therians.

TARSAL BONES OF *PROTUNGULATUM* AND *DECCANOLESTES*

Outside of Mongolia (see later), proximal tarsals assigned to *Protungulatum* and *Procerberus* have been described by Szalay and Decker (1974). The specimens in question, from the Bug Creek Anthills, are probably of Paleocene age (Lofgren, 1995), but we include the description of *Protungulatum* herein, as its morphology has played a central role in understanding the locomotion of early eutherians (e.g., Szalay, 1984). *Protungulatum* is tentatively known from the Cretaceous of Saskatchewan (Fox, 1989, 1997b, see chapter 2). Calcanei and astragali referred to *Deccanolestes* have been reported from India (Godinot and Prasad, 1994; Prasad and Godinot, 1994).

Szalay and Decker (1974) were the first to describe the tarsal bones (calcaneus and astragalus) in mammals of presumed Late Cretaceous age. They provided a detailed description of the tarsus of the ?Late Cretaceous–Paleocene ungulate *Protungulatum donnae,* reconstructed the movements of the tarsus, and compared it with tarsal bones of the ?Late Cretaceous–early Paleocene cimolestid *Procerberus formicarum* and the late Paleocene plesiadapid primate *Plesiadapis* cf. *P. gidleyi*. They also mentioned the presence of the tarsal bones of *Cimolestes* in the collection they studied, but described them only as being virtually identical to those of *Procerberus*. *Procerberus* differs in some respects from *Protungulatum* (notably in the highly modified tibial trochlea of the astragalus, which permitted strong plantar flexion), but for present purposes can be omitted from further discussion.

Szalay and Decker (1974) presented a list of characters presumed to be primitive for the eutherian astragalocalcaneal complex (see figure 13.11 for terminology): (1) distally placed peroneal tubercle; (2) cuboid facet oblique to the long axis of the calcaneus; (3) posterior astragalocalcaneal articulation forming a relatively large angle (35–40°) with the long axis of the calcaneus; (4) oblique axis of rotation of the astragalocalcaneal articulation; (5) short calcaneal body anterior to the astragalocalcaneal facet; (6) plantar anterior tubercle on the calcaneus; (7) low, slightly grooved tibial trochlea on the astragalus; (8) large astragalar canal piercing the body; (9) possibly cuboid contact with the astragalus (alternating tarsus); (10) astragalar sustentacular facet not continuous with other facets; and (11) wide astragalar head, laterally thicker and rotated slightly dorsolaterally. They demonstrated that these features are present in the tarsus of *Protungulatum* (figure 13.11A,C) and concluded that this taxon closely approximates the primitive condition for Eutheria.

Godinot and Prasad (1994) and Prasad and Godinot (1994) described isolated calcanei and astragali referred to the two species of the eutherian *Deccanolestes* (figure 13.11B,D) from the Maastrichtian of India (see Prasad et al., 1994). As may be seen in figure 13.11, the astragalocalcaneal complex of *Deccanolestes* differs conspicuously from that in *Protungulatum*. The distal part of the calcaneus is more expanded laterally (more prominent peroneal tubercle) and the distal margin is more transversely oriented. Hence the cuboid facet is placed at almost a right angle with respect to the long axis of the calcaneal body, rather than being oblique to it, and the calcaneal sustentacular facet is relatively larger, rectangular in shape, and extends further distally. The astragalar head is smaller (narrower), as is the astragalar neck, which gives the astragalus

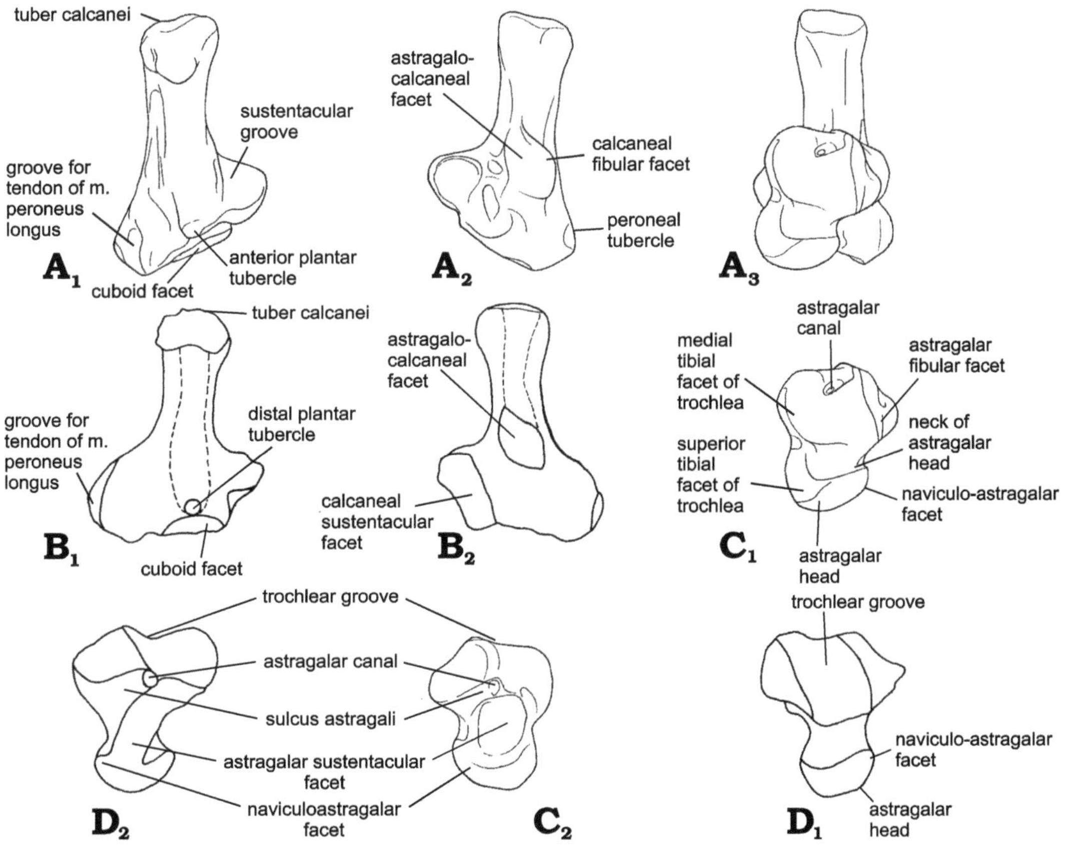

FIGURE 13.11. Comparison of the tarsal bones in cf. *Protungulatum* sp. (A, C) and *Deccanolestes hislopi* (B, D). Left calcaneus in plantar view (A_1, B_1); the same in dorsal view (A_2, B_2). Astragalocalcaneal complex of cf. *Protungulatum* in dorsal view (A_3). Left astragalus in dorsal view (C_1, D_1); the same in plantar view (C_2, D_2). D_1 and D_2 represent right astragalus, reversed. Not to scale. *Source:* A, C, modified from Szalay and Decker (1974); B, D, from Prasad and Godinot (1994).

a more constricted appearance. The trochlea is broad and shallow and there is a small astragalar foramen. The lateral trochlear crest is prominent, curving distomedially and slightly higher than the medial crest. In these and other respects, the implied movements of the foot in *Deccanolestes* differ from those of *Protungulatum,* instead resembling those of early primates (see "Paleobiology" later).

POSTCRANIAL SKELETON IN ASIORYCTITHERIA

Vertebral Column. Fragments of an atlas are known for *Daulestes* (left half of the arch), *Kennalestes* (damaged dorsal arch), and *Asioryctes* (well-preserved dorsal arch, figure 13.12A). In all of them the transverse processes lack a transverse foramen. The sulcus arteriae vertebralis is deep and well preserved in *Asioryctes* (Kielan-Jaworowska, 1977). McKenna et al. (2000) concluded that the two parts of the arch were not fused in *Daulestes,* owing either to the juvenile age of the individual or to its possessing of a more primitive condition than in *Asioryctes.* Kielan-Jaworowska (1977) speculated that as the intercentrum is not preserved in known specimens of *Kennalestes* and *Asioryctes,* it is possible that it was not fused to the arches, as is the case for known Late Triassic–Early Jurassic mammals. The transverse foramen is absent in Monotremata, Marsupi-

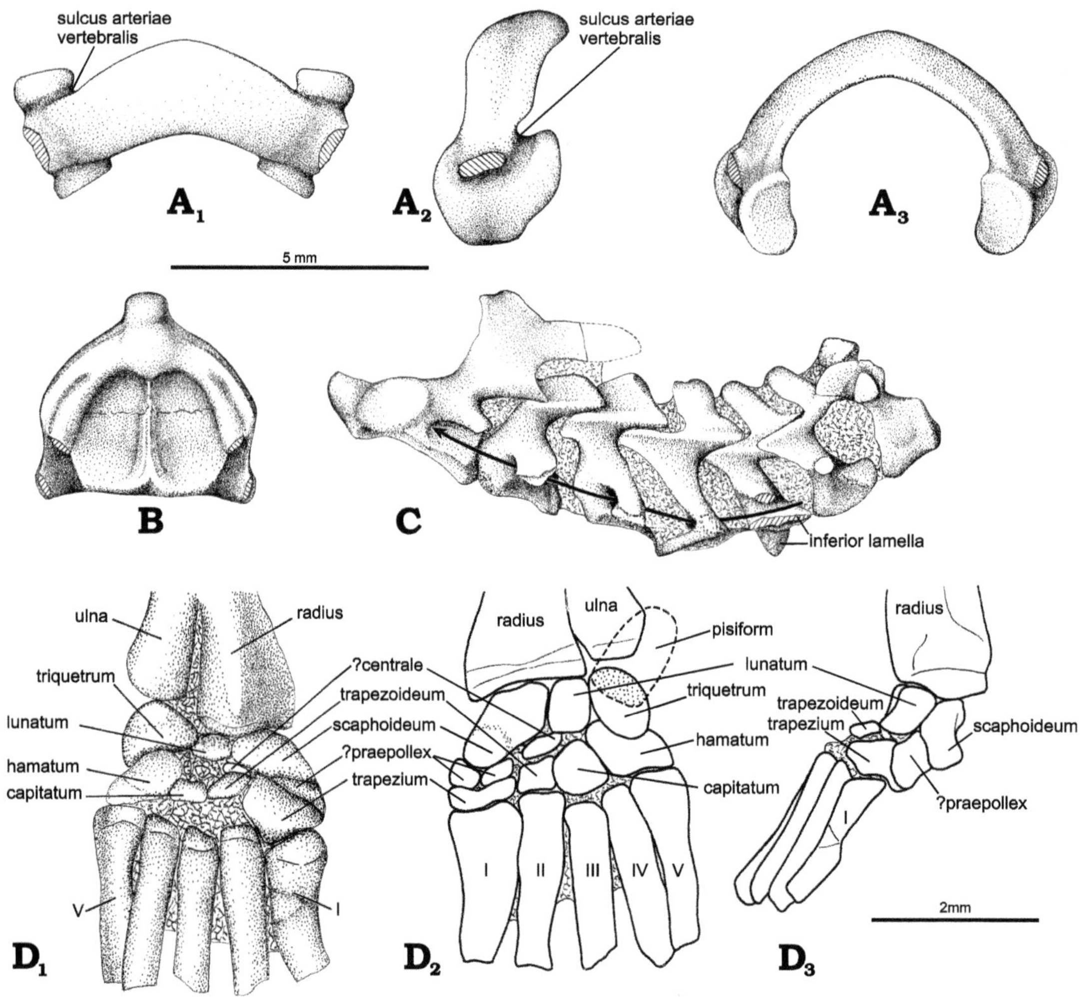

FIGURE 13.12. Cervical vertebrae and carpus of *Asioryctes nemegetensis.* A, Reconstruction of the atlas (without intercentrum) in dorsal (A_1), right lateral (A_2), and anterior (A_3) views. B, Reconstruction of the axis in ventral view. The hatched areas in A and B denote the broken bases of the transverse processes. C, Second cervical to first thoracic vertebrae as preserved in one of the specimens, not reconstructed, in left lateral view. The arrow denotes the course of the arteria vertebralis. D, Right carpus, in dorsal (D_1), ventral (D_2), and medial (D_3) views. The shadowed area on the triquetrum denotes the articulating surface for the pisiform, which is reconstructed; I–V, metacarpals. *Source:* modified from Kielan-Jaworowska (1977).

alia, Cetacea, Sirenia, Ruminantia (except the Tylopoda), and Rhinocerotidae (Lessertisseur and Saban, 1967, and references therein). Kielan-Jaworowska (1977) argued that the absence of this foramen is a primitive feature for Monotremata, Marsupialia, *Kennalestes,* and *Asioryctes* and that the lack of the foramen in living eutherian groups represents secondary loss.

Most of the neck is known for *Asioryctes* (figure 13.12). In ventral view, the axis shows a weak transverse joint, dividing the body into atlantal and axial parts. The transverse process (both preserved on one specimen) is not pierced by a foramen, with the arteria vertebralis running in a groove below the process. The arch is strongly expanded dorsally into a long, anvil-shaped spinous process. The arch of the third vertebra is low and the transverse process is directed obliquely, being more posteriorly than laterally oriented. It arises by two roots, one from the body and the other from the axis; between these is the large transverse foramen. On the fourth through sixth vertebrae the transverse foramina are distinct. The transverse processes are long, but the free cervical ribs have not been recognized (Kielan-Jaworowska, 1977).

The characteristic feature of the sixth cervical vertebra in most living mammals is the presence of the so-called inferior lamella (Howell, 1926; see also Lessertisseur and Saban, 1967). It originates as a ventral branch of the transverse process that has lost contact with the rest of the transverse process and forms a longitudinally arranged lamella, projecting strongly downward. The size and shape of the lamella depend on the development of the longus coli muscle that extends from it in both directions. In *Asioryctes* (figure 13.12C) the inferior lamella is only incipiently developed, less prominent than in extant mammals. As usual among mammals, the transverse process of the seventh cervical vertebra in *Asioryctes* is not perforated by a transverse foramen, and the costal facet for the first rib is discernible on the body. The first thoracic vertebra is generally similar to the seventh cervical.

The neural spines are lacking on the cervical vertebrae of *Asioryctes,* which is probably related to its small body size (see Slijper, 1946, and Kielan-Jaworowska, 1977, for discussions on the role of neural spines in relation to the size). Other parts of the vertebral column are unknown in Asioryctitheria described thus far; they are preserved (e.g., lumbar vertebrae) in *Ukhaatherium* (Novacek et al., 1997: figure 2), but still await description.

Forelimb. An incomplete forelimb known for *Asioryctes* preserves the distal radius and ulna, together with a nearly complete carpus (figure 13.12D). Kielan-Jaworowska (1977) reconstructed the carpus as having 11 bones. The proximal row consists of the scaphoideum (naviculare), the lunar (lunatum), and the cuneiform (triquetrum). The pisiform is not preserved, but there is an articular surface for it on the ventral side of the cuneiform. The proximal row forms an arch that embraced the bones of the distal row, which includes trapezium, trapezoideum, magnum (capitatum), and hamatum. In the middle there is a small element tentatively identified as the centrale. The metacarpals are subparallel to one another.

Kielan-Jaworowska (1977) demonstrated that the carpus of *Asioryctes* is very different from that in *Didelphis,* which has the structure characteristic of grasping hands in the terminology of Altner (1971) and Yalden (1972) (see figure 13.18). It more closely resembles the carpus of *Tenrec,* with the so-called convergent hand, characteristic of terrestrial insectivores.

Hindlimb. Kielan-Jaworowska (1977: figure $4A_2$) reconstructed the tarsus of *Asioryctes* on the basis of a single incomplete specimen. The calcaneus is elongated anteroposteriorly and compressed dorsoventrally. The peroneal tubercle is extensive and projects posterolaterally. On the lateral margin above the peroneal tubercle there is another, crescent-shaped, process. The cuboid facet is oblique to the long axis of the calcaneus. The cuboid facet is confluent with a small facet on the astragalar head. The tuber of the calcaneus is comparatively short and slightly bent medially. The sustentacular facet, exposed in plantar aspect, is roughly rectangular, with a foramen in the middle. The astragalus is anteroposteriorly elongate, with an inflated proximal body and a narrower, well-defined head. On the proximal body there is a medial facet for articulation with the tibia. The tibial trochlea is not developed. There is probably a plantar astragalar foramen (Szalay, 1966). The cuboid is elongate and slightly constricted in the middle. The medial wall is concave to fit the lateral cuneiform. The latter is elongated anteroposteriorly. The lateral cuneiform is short and transversely elongate. The medial cuneiform is very large, projecting far distally beyond the distal margin of the other cuneiforms. Its distal margin is strongly oblique.

Of the five metatarsals, the first is the shortest and the third the longest. The fourth and fifth articulate with the distal margin of the cuboideum. In all metatarsals the epiphyses are very distinct. Two phalanges of the first digit and the first phalanges of the second and third digits are also preserved.

Kielan-Jaworowska (1977) characterized the tarsus of *Asioryctes* by two features: (1) The proximal body of the astragalus (talus) does not override the calcaneus but is situated medial to it. (2) The sustentacular facet of the calcaneus is well developed, but extends only beneath two-thirds of the width of the astragalar head.

It is well known that the astragalus is situated medial to the calcaneus in reptiles, a primitive condition retained by morganucodonts (Jenkins and Parrington, 1976). In Eutheria, including ?Late Cretaceous *Protungulatum* and *Procerberus* (Szalay and Decker, 1974), the astragalus is supported by the calcaneus and has no plantar contact. Kielan-Jaworowska (1977) compared the tarsus of *Asioryctes* with that of *Didelphis* and argued that the calcaneoastragalar contact in *Didelphis* is more extensive than in *Asioryctes;* in *Didelphis* a large part of the proximal astragalar body, as well as the head, is superimposed on the calcaneus. She concluded that the tarsus of *Asioryctes* represents the most primitive type found in therian mammals and is in this respect in an intermediate stage of development between Triassic mammals and later eutherians.

The latter conclusion was challenged by Szalay (1984), who argued that the most important modifications of the ankle that distinguish eutherians from marsupials are associated with eutherian terrestriality. In Eutheria the upper ankle joint develops a mortise-and-tenon configuration, which is an interlocking joint in which the mortise (in engineering: a recess cut in a piece of timber) receives a tenon (a wooden peg). The astragalar body is sandwiched between a well-developed medial malleolus of the tibia and (laterally) the fibular-calcaneal contact. The result is restriction of lateral mobility (inversion-eversion) at the upper ankle joint. Szalay (1984) stated that the astragalocalcaneal contact cannot be reconstructed in *Asioryctes* because of the poor state of preservation of the specimen. Nonetheless, the medial and lateral malleoli of the crus of *Asioryctes* appear to be well developed, suggesting a well-established mortise-and-tenon upper tarsal joint, as in all eutherians. Horovitz (2000) described the ankle joint of *Ukhaatherium* (figure 13.13) and argued (*contra* Kielan-Jaworowska, 1977) that the astragalus was completely superimposed over the calcaneus in Asioryctitheria. Otherwise, she supported Kielan-Jaworowska's conclusion concerning the primitive nature of the postcranial skeleton in *Asioryctes.* Comparisons suggest that asioryctitherians are primitive with respect to all other known eutherians for which the hindlimb is known (including Zalambdalestidae) in: (1) mortise-tenon articulation of leg and foot being only incipiently developed; (2) thick fibula and, possibly, femoro-fibular articulation; (3) lack of pulley-shaped astragalar trochlea; (4) lack of anterior plantar tubercle on the calcaneus; (5) ventral curvature in tuber calcanei; (6) larger peroneal process; and (7) more medially oriented sustentacular facet. However, asioryctitherians may be unique among mammals in possessing a pedicillate, anteroposteriorly oriented posterior calcaneoastragalar facet; strong dorsal orientation of the sustentacular facet; and hypertrophy of the astragalar medial plantar tuberosity (Horovitz 2000).

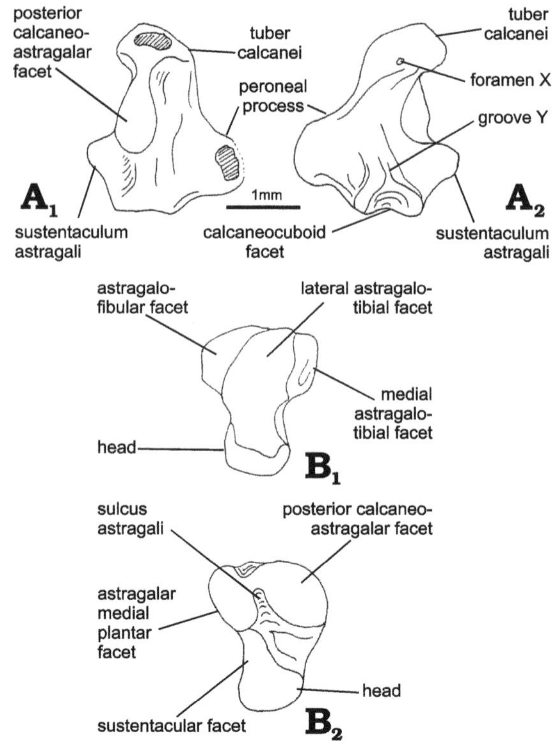

FIGURE 13.13. *Ukhaatherium nessovi.* A, Left calcaneus in dorsal (A_1) and plantar (A_2) views. B, Right astragalus in dorsal (B_1) and plantar (B_2) views. *Source:* modified from Horovitz (2000).

POSTCRANIAL SKELETON IN ZALAMBDALESTIDAE

The two Mongolian Late Cretaceous zalambdalestid genera *Zalambdalestes* and *Barunlestes* differ from one another in skull proportions, dental formula, and few details of the dentition, but the postcranial skeletons appear to be quite similar. Kielan-Jaworowska (1978) reconstructed the postcranial skeleton of *Zalambdalestes* partly on the basis of elements preserved in *Barunlestes.* We base the description that follows on her paper. The postcranial skeleton in Zalambdalestidae approaches that of living Eutheria (more so than *Asioryctes*), being at the same time specialized.

Vertebral Column. The atlas in Zalambdalestidae (preserved in *Barunlestes*) differs from that in Asioryctitheria in the presence of a transverse foramen. The ventral arch (intercentrum) is synostosed with the arches, but the sutures are present (figure 13.14A). The sulcus arteriae vertebralis is seen clearly on both sides of the transverse canal. The suture between the atlantal and axial parts of the body

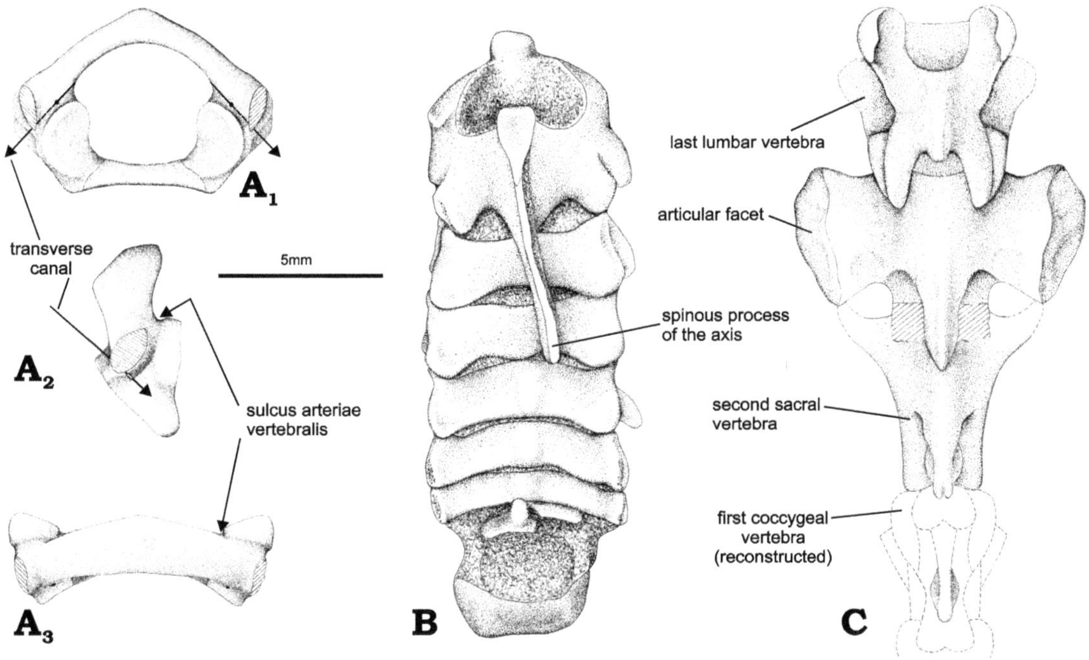

FIGURE 13.14. Selected vertebrae of Zalambdalestidae. A, Atlas of *Barunlestes butleri* in posterior (A_1), lateral (A_2), and dorsal (A_3) views. The hatched areas denote the broken bases of the transverse processes. The arrows in A_1 and A_2 denote the course of the transverse canal. Compare the atlas of *Asioryctes nemegetensis* (figure 13.12A), in which the transverse canal is not present. B, Second to seventh cervical vertebrae of *Zalambdalestes lechei;* note very long spinous process of the axis. C, Last lumbar vertebra, sacrum (composed of two vertebrae), and first coccygeal vertebra (partly reconstructed) of *Barunlestes butleri* in dorsal view. Note that only the first sacral vertebra articulates with the ilium. *Source:* modified from Kielan-Jaworowska (1978).

of the axis is present (as in *Asioryctes*). The most characteristic feature of the neck in Zalambdalestidae is the unusually long spinous process of the axis, which projects posteriorly, roughly horizontally above the third and fourth vertebrae (figure 13.13B). The anterior part of the dorsal edge of the spinous process is dorsoventrally flattened for attachment of the rectus capitis dorsalis. The succeeding cervical vertebrae do not bear spinous processes. The inferior lamella on the sixth vertebra, characteristic of most mammals, including *Asioryctes,* is weakly developed. The free cervical ribs have not been recognized. The lumbar vertebrae (preserved in *Barunlestes*) have short transverse processes. The sacrum (also preserved in *Barunlestes*) consists of two vertebrae, only the first of which articulates with the ilium, which is a primitive character (figure 13.14C).

Pectoral Girdle and Forelimb. Part of the scapula and fragments of the forelimb are known for *Barunlestes.* The scapula (figure 13.15A) differs from most modern claviculate mammals in the presence of a very extensive coracoid process, which together with the small tuber scapulae strongly projects ventrally beyond the glenoid fossa. In this respect, it is reminiscent of some small claviculate eutherians, such as *Tenrec, Solenodon, Elephantulus,* and *Tupaia,* as well the marsupial *Didelphis.* Although the acromion and clavicle are not preserved, Kielan-Jaworowska (1978) argued that the clavicle might have been present in *Barunlestes.*

The distal humerus of *Barunlestes* (figure 13.15B) bears an entepicondylar foramen and large fossa olecrani. The trochlea is developed, but the vestige of the radial condyle is still visible as a spherical structure (capitulum), which is a primitive character. The proximal part of the radius is placed fully anterior to the ulna, not anterolaterally. Carpal structure in *Barunlestes* is characteristic of a convergent hand (Altner, 1971; Yalden, 1972); its characteristic feature is fusion of the scaphoideum and lunatum into a large, crescent-shaped scapholunatum (figure 13.14C).

Pelvic Girdle and Hindlimb. Kielan-Jaworowska (1975a) demonstrated the presence of a triangular fossa on the lower part of the anterior margin of pubic bone, which she interpreted as an articular surface for an epipubic bone. This has been confirmed by Novacek et al. (1997), who found the epipubic bone preserved in *Zalambdalestes.*

A characteristic feature of the postcranial skeleton of zalambdalestids is the elongation of the hindlimbs and the

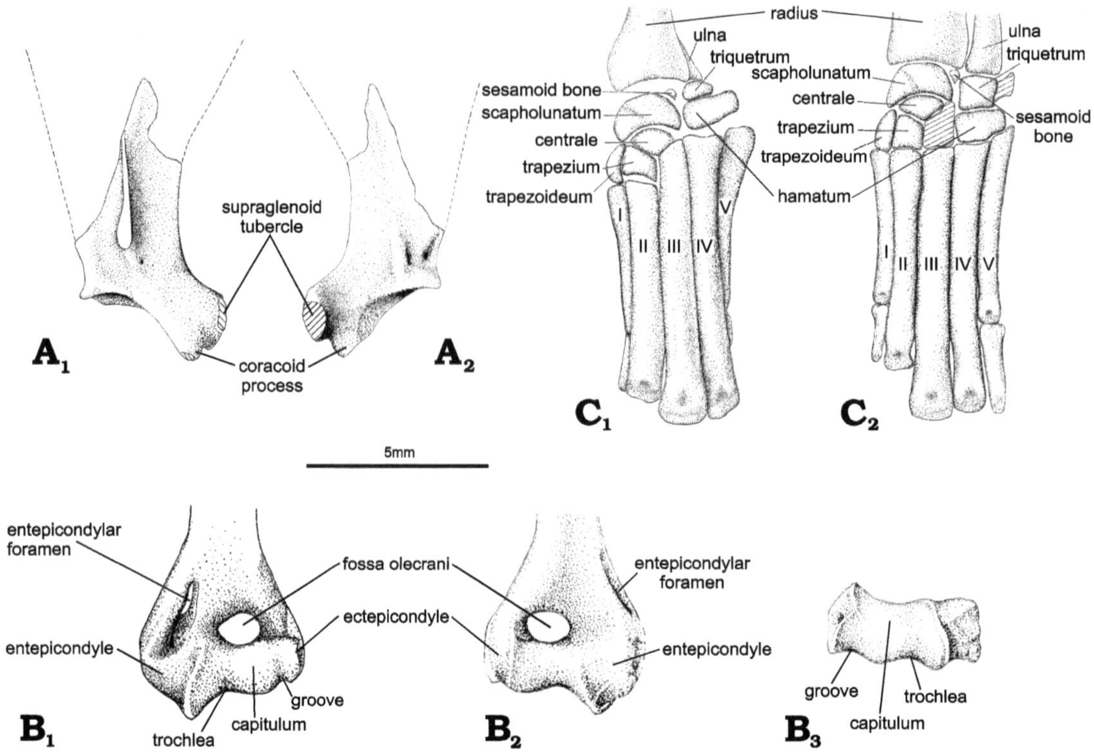

FIGURE 13.15. Selected fragments of the pectoral girdle and forelimb of *Barunlestes butleri*. A, Partial right scapula in lateral (A_1) and medial (A_2) views. B, Distal part of the left humerus in anterior (B_1), posterior (B_2), and distal (B_3) views. C, Distal part of the left forearm, carpus and metacarpals in dorsal view, as preserved (C_1), and reconstruction (C_2); I–V, metacarpals. *Source:* modified from Kielan-Jaworowska (1978).

feet. Kielan-Jaworowska (1978) demonstrated that the proportions of the hindlimb segments and the proportions of the hindlimb to forelimb resemble those in extant elephant shrews. The femur is long and slender; the head forms more than a hemisphere and has a flattened upper (proximal) surface. The neck is stouter than in Macroscelididae and the greater trochanter is rather short.

The fibula is fused with tibia along about two-thirds of its distal extent and consists only of a slender proximal part, fused again at its upper extremity to the tuberosity of the tibia (figure 13.17B).

The tarsus, preserved only in *Zalambdalestes* (figure 13.16), resembles that of other Eutheria except asioryctitherians in having a pulleylike astragalar trochlea, full development of a mortise-tenon upper ankle joint, and an anterior plantar tubercle on the calcaneus (Horovitz, 2000). The calcaneus is strong, compressed laterally, and provided with a prominent tuber calcanei. The astragalar head is separated from the trochlea by a short neck; the trochlea is asymmetrical and provided with a deep groove, which might be an effect of the state of preservation or preparation. The navicular is a small bone that articulates with the astragalar head and the tuber tibialis. The cuboideum is relatively large and is wider proximally, opposite the contact with the navicular, than distally. Of the three cuneiforms, the medial is strongly elongated, projecting distally over the distal margin of the other cuneiforms and the cuboideum.

The metatarsals, almost completely preserved in *Zalambdalestes* and partially in *Barunlestes*, are extremely elongated. The first is the shortest and the second the widest; the fifth is pointed proximally and protrudes proximally along the cuboideum. The phalanges are incompletely preserved and have been partly reconstructed by Kielan-Jaworowska (1978).

POSTURE AND GENERAL PROPORTIONS OF THE BODY

Reconstruction of the skeleton has been published for only two Cretaceous eutherians: *Eomaia* (Jiet al., 2002: figure 1c, see also our figure 13.17A) and *Zalambdalestes* (Kielan-Jaworowska, 1978: figure 17, see also our figure 13.17B). These two reconstructions differ considerably in details, owing to the different lifestyles of the two genera (discussed in the following section). One notable point of

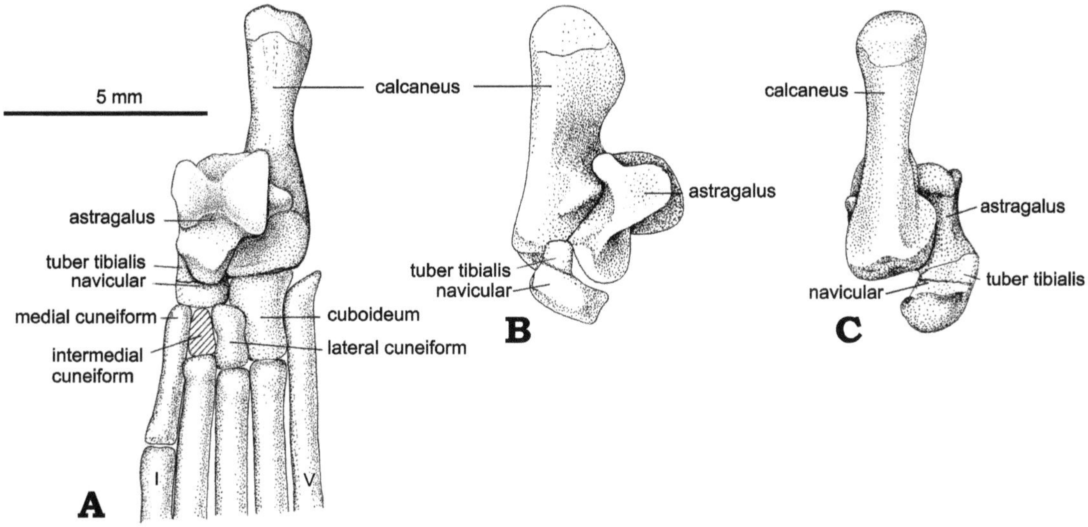

FIGURE 13.16. Tarsus and foot of *Zalambdalestes lechei.* Reconstruction of the left tarsus and proximal ends of metatarsals in dorsal view (A). Calcaneus, astragalus, and navicular of the same specimen, as preserved, in medial view (B); the same in plantar view (C). *Source:* modified from Kielan-Jaworowska (1978).

similarity, however, is the small, lightly built head, which distinguishes these early eutherians from other groups of early mammals (e.g., docodontans, eutriconodontans, morganucodontans, multituberculates), as well as living didelphids (see, e.g., figures 5.6, 7.1, 8.11, 8.23, and 12.6A).

PALEOBIOLOGY

Both marsupials and placentals were long thought to have been primitively arboreal (e.g., Dollo, 1899; Bensley, 1901a,b; Matthew, 1904, 1909, 1937), although dissenting

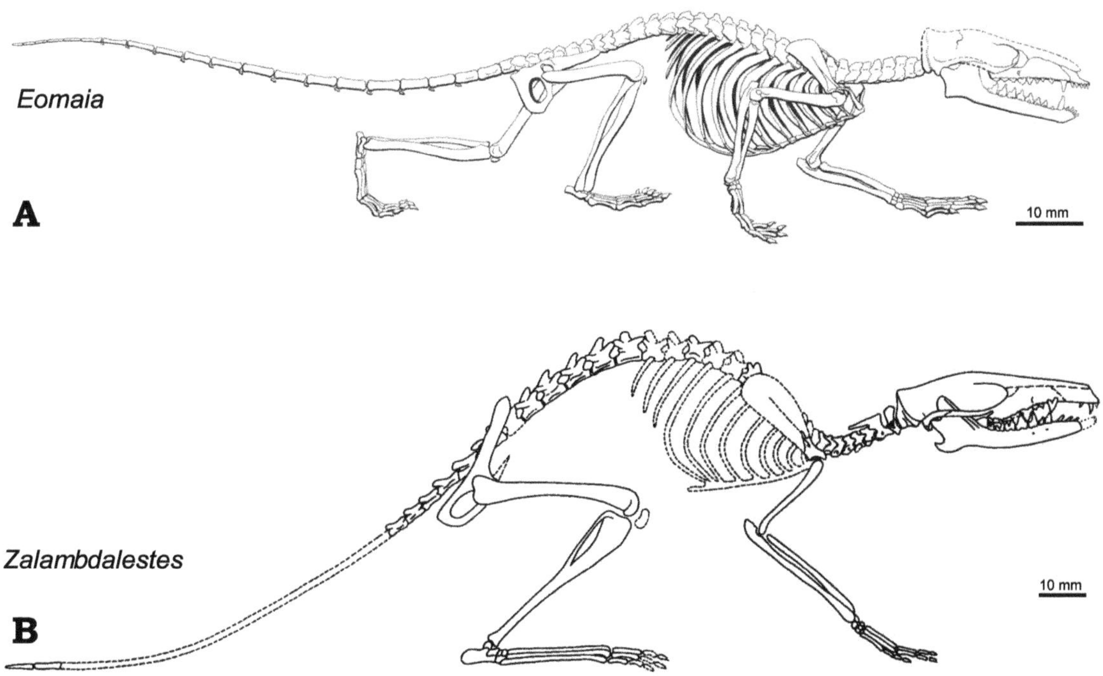

FIGURE 13.17. Reconstructed skeletons of Cretaceous eutherian mammals. A, *Eomaia scansoria* (scansorial or arboreal mammal). B, *Zalambdalestes lechei* (terrestrial, probably ricochetal mammal). *Source:* A, modified from Ji et al. (2002); B, modified from Kielan-Jaworowska (1978).

views were expressed by Gidley (1919) and Haines (1958). Until the postcranium of Cretaceous eutherians, including well-preserved astragalocalcaneal complexes (Szalay and Decker, 1974), and associated skeletons of Late Cretaceous eutherians from Mongolia (Kielan-Jaworowska, 1977, 1978) were described, the discussion could only be theoretical.

Szalay and Decker (1974), although having only astragalocalcaneal complexes of a few eutherian genera from the ?early Paleocene of Montana at their disposal, convincingly demonstrated that the oldest known ungulate *Protungulatum* (see Archibald, 1996b; Nessov, Archibald, and Kielan-Jaworowska et al., 1998) displayed terrestrial adaptations. Their work was based on study of well-preserved facets on the bones and reconstruction of the axes of rotation of various joints. However, the hallmark features of terrestrialism in eutherians, the mortise-tenon interlocking of the lower leg and astragalus together with increased capacity for flexion-extension at the upper ankle joint (Szalay, 1984) are poorly developed in asioryctitherians (Horovitz, 2000).

Although the postcranial skeletons of Mongolian Late Cretaceous eutherians (especially *Asioryctes* and Zalambdalestidae, see figure 13.17B and Kielan-Jaworowska, 1977, 1978) are more complete than the isolated elements referred to *Protungulatum*, the state of preservation of particular bones—especially the astragalocalcaneal complex—is imperfect, as the bones are fused together and cannot be separated. Therefore, the conclusions concerning the lifestyles of these Mongolian eutherians have been based on other elements, rather than on studies of the details of movement in the astragalocalcaneal complex.

Traditionally, the structure and habits of extant *Tupaia* figured prominently in the debate on arborealism versus terrestrialism in early eutherians. Especially important was the conclusion of Jenkins (1974a), who argued that for small primitive mammals, terrestrialism and arborealism are not discrete phenomena. He demonstrated that most of the tree shrew species move freely between ground and trees. A similar idea was expressed earlier by Altner (1971). More recent work has shown that most tupaiids, including virtually all species of *Tupaia,* are secondarily modified for terrestriality. Highly arboreal *Ptilocercus* more closely approximates a morphotype for Archonta, which in turn appear to have arisen from a terrestrially adapted ancestor (Sargis, 2000).

Altner (1971) showed that in semiarboreal *Sciurus* and terrestrial *Xerus, Erinaceus,* and *Nesogale,* in spite of different lifestyles, the structure of the hand is essentially the same. He designated this type of the hand as "Spreizhand," which corresponds to the convergent hand of Haines (1958) and Napier (1961) and contrasts with the grasping hand (figure 13.18). The "Spreizhand" as defined by Altner (1971) may be characterized as follows: (1) The proximal row of carpal bones is concave distally and embraces the distal row. (2) The centrale may be incorporated either in the proximal row (Insectivora) or in the distal row (Rodentia). (3) The trapezoideum and capitatum are small in comparison with the trapezium and the hamatum. (4) The fifth digit projects laterally beyond the hamatum and may contact the triquetrum. (5) The trapezium is longitudinally elongated and projects distally beyond the distal margin of the remaining carpal bones. (6) The carpometacarpal joint of the pollex is of a hinge type. The first digit is not opposable.

Kielan-Jaworowska (1977) demonstrated that the hand of *Asioryctes* shows most of these characteristics. It differs from the hand of small modern mammals in having the trapezoideum, which is not elongated longitudinally, although it projects somewhat distally, and also in having

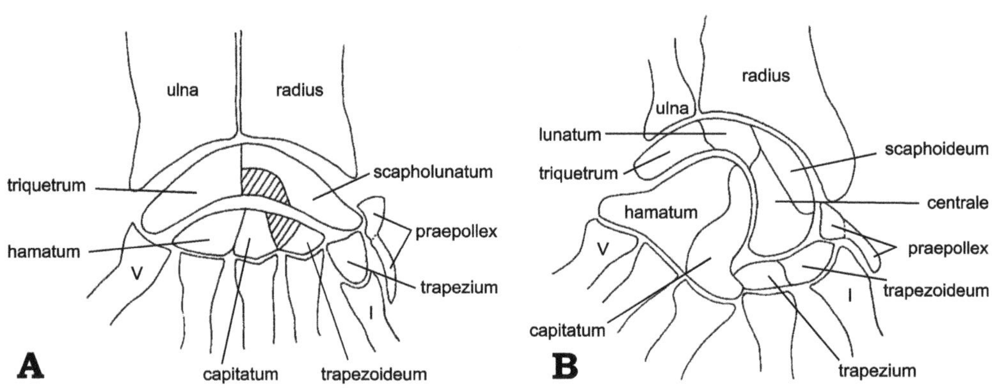

FIGURE 13.18. Diagrammatic comparison of the structure of the carpus in dorsal view in a convergent hand (A) and grasping hand (B). The hatched areas denote the centrale, which may be incorporated into either the proximal row (as in Insectivora), or distal row (as in Rodentia); I–V, metacarpals. *Source:* modified from Altner (1971).

the fifth metacarpal not projecting laterally. As in modern mammals with convergent hands, the pollex in *Asioryctes* was not opposable.

The structure of the pes has been less extensively discussed with respect to arborealism versus terrestrialism. Kielan-Jaworowska (1977) demonstrated that the hallux was certainly not opposable in the foot of *Asioryctes* and that the foot is moderately elongated. Despite the fact that *Asioryctes* had a nonprehensile convergent hand and a foot with a nonopposable hallux, it remains possible that it was semiarboreal or facultatively arboreal (like *Tupaia* or *Sciurus*), especially in light of the study by Jenkins (1974a). Additional information bearing on this question comes not from skeletal morphology, but from studies of the environment in which skeletons of *Asioryctes* were deposited. Sedimentological investigations in the Djadokhta and Baruungoyot formations of Mongolia (Lefeld, 1971; Gradziński and Jerzykiewicz, 1974; see also Jerzykiewicz and Russell, 1991; Jerzykiewicz et al., 1993, for summaries) were deposited in inland areas with semiarid and arid climates. No remnants of trees are found in these sediments, although tree trunks are found in the sandy, dinosaur-bearing sediments of the overlying Nemegt Formation (Gradziński, 1970), which has yielded only two mammals. On the basis of these data Kielan-Jaworowska (1977) concluded that Late Cretaceous eutherian mammals from Mongolia, such as *Asioryctes* and possibly also *Kennalestes,* lived in a semidesert environment and were probably not tree-runners.

The lifestyle of another eutherian group that flourished during the Late Cretaceous of the Gobi Desert is much clearer than those of *Asioryctes* and *Kennalestes.* Kielan-Jaworowska (1978) described fairly complete postcranial skeletons of two zalambdalestid genera (*Zalambdalestes* and *Barunlestes*) and compared the proportions of their fore- and hindlimbs with those of extant Macroscelididae. The elongated horizontal spinous process on the atlas of *Zalambdalestes* (figure 13.14B) indicates that the anterior part of the neck had limited mobility. Such immobility occurs in mammals of various habits, including aquatic, fossorial, and ricochetal adaptations. Kielan-Jaworowska (1978) concluded that it is possible that partial immobility of the anterior part of the neck is indicative of a tendency to ricochetal behavior, although Zalambdalestidae presumably did not leap bipedally (see later). The intermembral index (lengths of humerus + radius × 100 to lengths of femur + tibia) and metatarsal/femoral index (length of Mt III × 100 to length of femur) of zalambdalestids compare favorably with those of Macroscelididae, except that the metatarsals of Zalambdalestidae are relatively longer than in most macroscelidid genera. Kielan-Jaworowska (1978: 31) concluded: "On the basis of intermembral and metatarsal/femoral indices, and other proportions of particular segments of the limbs, I conclude that locomotion by the zalambdalestids was similar to that of present-day macroscelidids. While it seems reasonable to conclude that they were not bipedal leapers, it is perhaps too much to expect a mode of progression identical with that of any other living or extinct group of mammals."

The year 1994 brought an important discovery on the variety of lifestyles among early eutherians. Godinot and Prasad (1994) and Prasad and Godinot (1994) described the astragalocalcaneal complexes of *Deccanolestes* from the Maastrichtian of India, and provided meticulous functional analyses of their movements, showing undoubted adaptation to arboreal life. Prasad and Godinot (1994: 901) concluded: "What appears now, and in fact might have been suspected, is that the early history of eutherian locomotion must have been a very complex one. Several eutherian radiations on different continents during the Cretaceous must have led to a variety of arboreal, terrestrial, and other adaptations." Fossorial adaptations have not yet been found among Cretaceous eutherians, but existed among contemporary multituberculates (see chapter 8), and it may be expected that they might also be discovered sometime in the future.

Arborealism in early Eutheria has been convincingly demonstrated for *Eomaia* by Ji et al. (2002: 816), who concluded that: "The new eutherian has limb and foot features that are known only from scansorial (climbing) and arboreal (tree-living) extant mammals, in contrast to terrestrial or cursorial (running) features of other Cretaceous eutherians." The scansorial adaptations of *Eomaia* concern the phalangeal proportions and curvature, which are similar to those of extant arboreal mammals, such as the didelphid *Caluromys.* The authors argued that (p. 818): "In phalangeal features *Eomaia* is more similar to arboreal mammals than to such scansorial taxa as the tree shrew and opossum," but later stated (p. 818): "Both manual and pedal claws are more similar to scansorial mammals . . . than to fully arboreal taxa." Finally they concluded (p. 820): "The available evidence is insufficient to determine whether *Eomaia* was scansorial (as in some species of *Tupaia* or *Glis*) or fully arboreal (e.g., as the marsupial *Caluromys* and the tupaiid *Ptilocercus*). Because the basal metatherians are scansorial . . . scansorial skeletal features appear to be primitive for the earliest known eutherians. But the evidence for an ancestral scansorial adaptation for the crown group therians as a whole is less clear."

As discussed earlier, Late Cretaceous mammals of Mongolia lived in inland areas with semiarid and arid climates. In contrast, mammal-bearing rocks of middle Asia (especially the best-known Coniacian Bissekty Formation) were deposited on low coastal plains in semihumid subtropical conditions (Nessov and Kielan-Jaworowska, 1991, and references therein). A similar inferred environment was characteristic for most Upper Cretaceous formations of North America (see also Lehman, 1987, 1997; Archibald, 1996a). In addition to mammals, Late Cretaceous vertebrate assemblages from both middle Asia and the Western Interior of North America include sharks, bony fishes, amphibians, turtles, and crocodilians, all of which are much less common or absent in Mongolian sites. Notably, though multituberculates were very common during the Late Cretaceous of Mongolia and western North America, they were very rare in middle Asia (Kielan-Jaworowska and Nessov, 1992). Nessov, Archibald, and Kielan-Jaworowska (1998) hypothesized that "zhelestid" eutherians and multituberculates were ecological competitors. According to these authors the rarity of multituberculates in Upper Cretaceous deposits of middle Asia may be explained by the diversification of "zhelestids," the first eutherians adapted to herbivorous niches, which ecologically replaced the multituberculates.

Other aspects of early eutherian lifestyles are based on study of endocranial casts (see the earlier section "Brain"). Jerison (1973: 200) developed the widely accepted hypothesis that the brains of Mesozoic mammals: "evolved to accommodate life in nocturnal niches in which hearing and smell, rather than the normal reptilian vision could be the characteristic sense modalities for information about events at a distance." Discoveries of numerous endocranial casts of Late Cretaceous Mongolian mammals, both multituberculates and eutherians (see Kielan-Jaworowska, 1986, for summary), support this view. At least the few eutherians known by endocranial casts were more dependent on smell than most Tertiary and Recent mammals, and favored nocturnal niches in which olfaction and hearing played an important role.

SYSTEMATICS

Infraclass Eutheria Gill, 1872[2]

INTRODUCTION

A scholarly treatment for taxonomic reference to this group is given by McKenna and Bell (1997), who prefer the term Placentalia. This name might well be applied to a node-based definition of the group, that is, the most recent common ancestor of all living placentals, plus all of its descendants. In this case, the term Eutheria could be employed to include fossil taxa that are suspected of being more closely related to placentals than to other living mammals. The terms Placentalia and Eutheria would thus be analogous to those employed for the other major group of tribosphenic mammals, Marsupialia, and the more inclusive Metatheria (see Rougier et al., 1998). As shall be seen, this distinction is difficult to implement in practice, owing to the great uncertainty in placement of Mesozoic with respect to living taxa. For present purposes, use of the term Eutheria, as defined earlier and without further qualification, appears to be the most useful.

Mesozoic mammals were traditionally referred to Marsupialia without issue, and the presence of Eutheria in the Cretaceous was problematic (see, e.g., Osborn, 1893). It was not until the 1920s, with the discovery of relatively complete specimens from Mongolia, together with careful study of Lancian fossils, that eutherians were recognized unambiguously from Mesozoic strata (Gregory and Simpson, 1926; Simpson, 1929a, 1951). Given the small size, relatively primitive dental structure, and presumed insectivorous diet of these early mammals, it is not surprising that they were referred to the most similar living group, Insectivora, without issue. This became established as common taxonomic practice in subsequent years, resulting in the concept of Insectivora as a taxonomic wastebasket for generally small, dentally primitive mammals. As noted later, recent years have witnessed major strides in defining monophyletic groups for many Cenozoic (and certain Cretaceous) "insectivores," and in conceptual advances permitting hypothesized relationships of certain Cretaceous eutherians to other, more derived groups. However, given the fragmentary nature of many fossils, together with the paucity of compelling derived features, the relationships of many Cretaceous eutherians remain questionable or altogether speculative.

Simpson's (1945) classification of mammals included only five Cretaceous genera referred to the Eutheria. Of these, *Deltatheridium* and *Deltatheroides* are now assigned to the suborder Deltatheroida within Metatheria (chapter 12), while we regard *Hyotheridium*, which might be eutherian, as *nomen dubium*. The remaining genera, *Zalambdalestes* and *Gypsonictops*, were placed in the Zalambdalestidae and Leptictidae, respectively, within the erinaceoid Insectivora, cohort Unguiculata (insectivores, dermopterans, primates, edentates, and certain extinct groups). *Cimolestes*, then known only by isolated teeth, was retained in the Marsupialia, until more complete specimens clearly demonstrated its eutherian affinities (see Clemens, 1973a). Detailed study of the skull in Ter-

[2] Gill (1872) included in Eutheria both marsupials and placentals. Huxley (1880) modified the scope of Eutheria, limiting it to Placentalia only. We follow Huxley's (1980) modification of Eutheria.

tiary and Recent taxa led Butler (1956) and McDowell (1958) to suggest that leptictids were only remotely related to lipotyphlans. This issue remained unsettled for approximately three decades (see, e.g., Van Valen, 1967b; Szalay, 1968).

Additional fossils brought inevitable taxonomic complexity, and Romer (1966) created the suborder Proteutheria to include most Cretaceous eutherians with some problematic taxa from the early Tertiary and placed the remaining insectivores in other suborders, including Macroscelidea, Dermoptera, Lipotyphla, and ?Zalambdodonta. This had the effect of defining more-or-less monophyletic groups within the "Insectivora," thereby relegating wastebasket status to the subgroup "Proteutheria." Additional Cretaceous and Early Tertiary genera of problematic affinities were assigned to "Proteutheria" in subsequent years, often placed in the "Palaeoryctidae," a family known mainly from the Tertiary. In observing the fact that Proteutheria is not a natural group, Butler (1972a: 264) noted: "further research should reduce the size of the Proteutheria, by discovering the relationships of some of its groups to other orders, and possibly by making new orders out of other groups."

Two major conceptual shifts occurred in the late 1960s and early 1970s. One involved the recognition of two distinct groups among then-known Mesozoic eutherians. Van Valen (1966) proposed the now-abandoned Deltatheridia to include possible Cretaceous relatives of carnivores and "palaeoryctids," among others. This was expanded upon by Lillegraven (1969), who, based on new and much better fossils of *Cimolestes,* provided detailed hypotheses of relationships for Tertiary Carnivora, Creodonta, and "Palaeoryctidae" among Cretaceous species of this genus (see also McKenna, 1975). In Lillegraven's (1969) scheme, the other main group consisted of North American *Gypsonictops,* derived from a *Kennalestes*-like taxon from Asia. In turn, *Gypsonictops* was postulated as having given rise to later radiations of insectivores and several other groups of mammals (see also, e.g., McKenna, 1969). *Kennalestes,* the most primitive eutherian known at that time (Kielan-Jaworowska, 1969b), was commonly accorded an ancestral role in eutherian evolution (see, e.g., Fox, 1979d).

A second important concept was introduced by Szalay and McKenna (1971: 301), who proposed the Anagalida as a new order, to include Zalambdalestidae, together with various extinct groups, regarding it as: "an endemic Cretaceous and early Tertiary Asian radiation, whose closest living relatives are the Lagomorpha." Although some of the shared similarities now appear to be convergent (Kielan-Jaworowska, 1975b, 1978; also see discussion by Novacek, 1986b), this view has persisted (McKenna, 1975; Szalay, 1977) and is tentatively followed herein. Study of newly collected fossils from Uzbekistan support a relationship of Zalambdalestidae to Glires, which include rodents as well as lagomorphs (Archibald et al., 2001, but see Fostowicz-Frelik and Kielan-Jaworowska, 2002).

The first comprehensive attempt to place Cretaceous eutherians within the context of a cladistically based taxonomy is that of McKenna (1975), who created a number of higher categories, a few of which currently remain in use. Nonedentate eutherians were placed in cohort Epitheria, with fundamental divisions into two magnorders, Ernotheria and Preptotheria. Late Cretaceous Mongolian *Asioryctes* and *Kennalestes,* together with palaeoryctines, were regarded as sister taxa to remaining ernotherians, including leptictids, anagalids, and macrosceledids, and, within Lagomorpha, zalambdalestids and modern lagomorphs. Preptotheres constituted "Deltatheridia" (now abandoned) as a sister taxon to remaining eutherians, including Ferae (Cimolesta, as an order, encompassing several Tertiary groups as well as carnivorans and creodonts), Insectivora (with contents equivalent to Lipotyphla), Archonta (tupaiids, dermopterans, chiropterans, and primates), and Ungulata (extinct and modern ungulate groups) as grandorders. For our purposes, the following are the essential features of this classification: (1) Placement of Mongolian *Asioryctes* and *Kennalestes* within the same group as leptictids and Tertiary palaeoryctines (omitting the Cretaceous taxa sometimes referred to Palaeoryctidae). Within the same group, a close relationship was posited for living macrosceledids and lagomorphs, together with Cretaceous Zalambdalestidae. (2) Cretaceous "palaeoryctids" (e.g., *Cimolestes*) were placed in another major eutherian clade, closely allied with carnivores and creodonts (Ferae), lipotyphlan insectivores, and remaining eutherians, including ungulates.

A second comprehensive classification, departing from most other recent schemes in emphasizing functional-adaptive systems and particularly the tarsal complex appeared two years later (Szalay, 1977). In fundamental points it is similar to that of McKenna (1975). Three major divisions of Eutheria are recognized, with many of the major groupings more or less the same: contents of Archonta, and their placement with ungulates and Ferae; segregation of edentates; and placement of anagalids, lagomorphs (including zalambdalestids), and macrosceledids within one group (in this case called the cohort Glires, also including rodents and other groups not relevant herein). Here, however, Cretaceous *Cimolestes* and relatives were placed not with the Ferae, but rather with "Leptictidomorpha." Asioryctidae were regarded as lying somewhere within Glires, but neither their postcranium nor that of *Kennalestes* (not included in the classification or cladogram) had yet been described.

The review of Kielan-Jaworowska, Bown, and Lillegraven (1979) included 14 genera of Cretaceous eutherians, two cited as *nomina nuda* and two unnamed. The taxa were distributed among three orders, as follows: (1) Proteutheria, including superfamilies Leptictoidea (for *Gypsonictops* and, tentatively, *Kennalestes*) and Palaeoryctoidea (*Cimolestes, Batodon, Procerberus, Asioryctes, Telacodon*), with the Zalambdalestidae (*Zalambdalestes* and *Barunlestes*) listed *incertae sedis.* (2) Condylarthra, with superfamily Arctocyonoidea (*Protungulatum*), and *Perutherium* included *incertae sedis.* (3) Primates, infraorder Plesiadapiformes (*Purgatorius*). *Endotherium* (regarded by us as *nomen dubium*), together with several undescribed taxa, were assigned to Eutheria *incertae sedis.* Of these, *Purgatorius* is no longer recognized from the Cretaceous, and the inclusion of *Protungulatum* and *Procerberus* in the Cretaceous is tentative (chapter 2). As we attempt to place taxa to family level where possible, we follow McKenna and Bell (1997) in allocating *Gypsonictops* to Gypsonictopsidae within Leptictoidea.

Synthetic phylogenetic studies were published during the 1980s by M. J. Novacek, best summarized in his influential study of leptictids (Novacek, 1986b), which resulted in a widely used classification (e.g., Carroll, 1988; Benton, 1990b). Unfortunately, the classification is restricted to higher-level categories and is not of much help for Mesozoic Eutheria, though some are discussed peripherally in the text. Epitheria were recognized as including all eutherians except Edentata. Most important in the current context is recognition of a monophyletic group, superorder Insectivora, to include, as orders, Lipotyphla and Leptictida. This removes leptictoids (including Cretaceous *Gypsonictops*) from ancestry of other "insectivores" and many additional mammalian groups, as variously postulated earlier. Although uncertainty remains (e.g., MacPhee and Novacek, 1993), this welcome corroborated and therefore significant hypothesis is followed here.

A veritable flood of new taxa resulting from the prodigious collecting efforts of Lev Nessov in middle Asia (chapter 2) produced the inevitable further complications, but, felicitously, also resulted in at least one additional conceptual advance. Various surprisingly advanced (considering their age) eutherians were initially (Nessov, 1985a) grouped into the proteutherian suborder Mixotheridia because they seemed to possess features characteristic of marsupials on the one hand and eutherians on the other. The core of the group consisted of predominantly Asiatic taxa later recognized to be related to ungulates, but also included (Nessov, Sigogneau-Russell, and Russell, 1994) the aforementioned Zalambdalestidae. Nessov (1997) relegated more primitive taxa, the Kennalestidae and Otlestidae, to Proteutheria. *Prokennalestes,* the oldest well-known eutherian, was referred to the latter family (Kielan-Jaworowska and Dashzeveg, 1989; see also Sigogneau-Russell et al., 1992), a course tentatively followed here (see also Averianov and Skutschas, 2000a).

The conceptual advance just noted was the creation of a superordinal taxon, Ungulatomorpha, to include not only living ungulates and various early Tertiary taxa ("condylarths"), but also presumed relatives from the Cretaceous (Archibald, 1996b). Most of the latter are "Zhelestidae" described by Nessov from Uzbekistan, but a few North American and, perhaps, European taxa appear to be related (Nessov, Archibald, and Kielan-Jaworowska, 1998; Gheerbrant and Astibia, 1999). Our systematic treatment with respect to Ungulatomorpha follows these authors. Relationships among placental orders are currently undergoing dramatic reinterpretation based on molecular data. Most significant in the present context is the distribution of living ungulates and "ungulates" among two major clades, Laurasiatheria and Afrotheria (Liu and Miyamoto, 1999; Springer et al., 1999; Liu et al., 2001). Fortunately, except for a possible change in the name of the supraordinal grouping, the issue is presently moot with respect to the relationship of Cretaceous ungulatomorphs to Tertiary taxa and at least some of the living ungulates among Laurasiatheria.

The classification of McKenna and Bell (1997), in press at the time Ungulatomorpha was erected, is similar in overall structure to that of McKenna (1975), a major difference being that the Anagalida (macroscelidids, lagomorphs, rodents, and zalambdalestids) were removed from a close relationship to Leptictida and placed instead within Preptotheria. This had the effect of recognizing leptictoids as belonging to a different clade than all other Cretaceous and later mammals. The distribution of genera among higher taxa is, in many cases, unorthodox (see Szalay, 1999).

Almost simultaneous with the publication of McKenna and Bell (1997), a posthumous compendium by Nessov (1997) appeared, which included three new genera of Cretaceous Eutheria. Of these, *Ortalestes* may be synonymous with previously described *Aspanlestes,* the single species of *Sazlestes* has not been figured and its status is difficult to evaluate, and the affinities of *Eozhelestes* are unclear (see note added in proof to Nessov, Archibald, and Kielan-Jaworowska, 1998, written by the second and third authors after Nessov's death).

Novacek et al. (1997) described a new asioryctid, *Ukhaatherium,* from the Late Cretaceous of Mongolia.

Cranial comparisons led them to recognize a relationship between Asioryctidae and Kennalestidae, and they grouped the two together in Asioryctitheria, a higher category (of unspecified rank). Distant relationship of asioryctitherians to other eutherians is corroborated by a study of the tarsus (Horovitz, 2000). This view was accepted by Kielan-Jaworowska et al. (2000), who placed the Zalambdalestidae in Anagalida, following earlier work cited previously. Three other recent additions include *Montanalestes*, described from the Early Cretaceous of North America (Cifelli, 1999b); *Murtoilestes*, from the Early Cretaceous of Siberia (Averianov and Skutschas, 2000a); and *Eomaia*, from the Early Cretaceous of China (Ji et al., 2002).

Thirty-two genera of Cretaceous Eutheria were included in McKenna and Bell (1997). With various additions and emendations, we recognize 42, one of which is included as *nomen dubium*. In structuring our own classification (table 13.1), we have tried to extract the common threads, including points of general similarity and at least a modicum of agreement from the foregoing works, as well as from others not mentioned. We have also incorporated some of the recent, synthetic hypotheses of relationships among Cretaceous eutherians, where basic supporting evidence has been presented. The following are some explanatory comments for the course we have adopted.

A basic, fundamental split of eutherians between Xenarthra (with or without other "edentates") and all other taxa (Epitheria), as proposed by McKenna (1975), has been largely upheld by subsequent morphology-based studies (e.g., Novacek et al., 1988; Rose and Emry, 1993). On the other hand, most molecular studies (e.g., Liu et al., 2001) recognized Afrotheria as sister taxon to Xenarthra + remaining placentals. In practice, neither distinction can yet be recognized for Cretaceous eutherians based on strictly morphological criteria; rather it must be generally inferred, with comparisons to possible (mainly Tertiary) relatives for which good cranioskeletal material is known. Accordingly, we have left some of the oldest and most primitive Eutheria unassigned to magnorder and have simply assumed (often without adequate data) that the remainder are probably Epitheria (or Laurasiatheria, which as far as the Mesozoic fossils are concerned can be regarded as equivalent at the current level of resolution).

Within Epitheria, several taxa are either poorly known, largely plesiomorphic, or both, and these have been left *incertae sedis*. A recent hypothesis forwarded by Novacek et al. (1997; see alternate view by Archibald et al., 2001) suggests the possibility that two families of taxa from the Cretaceous of Mongolia (Kennalestidae, Asioryctidae) may represent an endemic radiation of mammals unrelated to contemporary or later taxa, and we follow these authors in placing these groups in the Asioryctitheria (recognized by us as a superorder), with affinities within Epitheria uncertain. Similarly, a recent study has provided evidence supporting a relationship of "zhelestids" to later ungulates (Archibald, 1996b; Archibald and Averianov, 2001a; Nessov, Archibald, and Kielan-Jaworowska, 1998; Archibald et al., 2001), and we accordingly place these in the Ungulatomorpha, together with more apomorphic taxa of the latest Cretaceous "condylarths." Anagalida are widely recognized as a divergent group of Epitheria (e.g., Szalay, 1977; Novacek, 1986b) and inclusion of Zalambdalestidae among them is accepted here on the basis of recent work by Archibald et al. (2001; but see Meng and Wyss, 2001, and Fostowicz-Frelik and Kielan-Jaworowska, 2002).

The concept of Archonta, a supraordinal grouping of primates, dermopterans, chiropterans, and tree shrews (Gregory, 1910), has enjoyed revival in recent years (Szalay, 1977; Novacek and Wyss, 1986; Sargis, 2000, 2002). *Deccanolestes*, described as a "palaeoryctid" from the Maastrichtian of India (Prasad and Sahni, 1988; Prasad et al., 1994), has been shown to share archontan specializations of the tarsus (Godinot and Prasad, 1994; Prasad and Godinot, 1994; but see Szalay and Lucas, 1996). We tentatively recognize Archonta from the Cretaceous on this basis.

Remaining Cretaceous eutherians fall into three groups. *Gypsonictops* has consistently been placed with Tertiary leptictoids, a course we follow. Szalay (1977) has shown that tarsals assigned to ?Late Cretaceous–early Paleocene *Procerberus* (variably placed among mammalian groups) are quite similar to those of Tertiary Leptictidae. In contrast, the dentition of *Procerberus* is similar to that of Cimolestidae (Lillegraven, 1969), and as these taxa are based on teeth, we place *Procerberus* in this family. Tarsals assigned to *Cimolestes* are said to be quite similar to those referred to *Procerberus* (Szalay and Decker, 1974), as well as to Tertiary Leptictidae (Szalay, 1977). Given the extreme scarcity of dental specimens of *Cimolestes* at the Bug Creek Anthills (0.1% of the mammalian fauna), together with the fact that species-level identification of these specimens has yet to be made (Lofgren, 1995), we find the case for association of tarsals to this genus problematic. Cimolestidae are referred to Ferae for reasons given later.

A good deal of uncertainty surrounds recognition of true lipotyphlans in the Cretaceous. The only taxon for which recent and detailed comments have been made is *Paranyctoides*, which Fox (1979d, 1984b; see also Cifelli, 1990e) placed in the soricoid (see Krishtalka, 1976; Bown and Schankler, 1982; Novacek et al., 1985; Rose and Gingerich, 1987) family "Nyctitheriidae." We find McKenna and Bell's (1997) relegation of Leptictida to its own superorder unrelated to other insectivores to be perplexing,

TABLE 13.1. Linnaean Classification of Mesozoic Eutherian Mammals

- Infraclass Eutheria Gill, 1872
  - Order *incertae sedis*
    - Family *incertae sedis*
      - *Eomaia* Ji et al., 2002
        - *E. scansoria* Ji et al., 2002
      - *Montanalestes* Cifelli, 1999b
        - *M. keeblerorum* Cifelli, 1999b
    - Family Otlestidae Nessov, 1985a
      - *Otlestes* Nessov, 1985a, type genus
        - *O. meiman* Nessov, 1985a
      - *Murtoilestes* Averianov and Skutschas, 2001
        - *M. abramovi* (Averianov and Skutschas, 2000a)
      - *Prokennalestes* Kielan-Jaworowska and Dashzeveg, 1989
        - *P. trofimovi* Kielan-Jaworowska and Dashzeveg, 1989, type species
        - *P. minor* Kielan-Jaworowska and Dashzeveg, 1989
  - Magnorder Epitheria McKenna, 1975
  - Superorder and order *incertae sedis*
    - Family Bobolestidae Nessov, 1989
      - *Bobolestes* Nessov, 1985a
        - *B. zenge* Nessov, 1985a
  - Order *incertae sedis*
    - Family *incertae sedis*
      - *Eozhelestes* Nessov, 1997
        - *E. mangit* Nessov, 1997
  - Superorder Asioryctitheria Novacek et al., 1997
    - Family Asioryctidae Kielan-Jaworowska, 1981, type family
      - *Asioryctes* Kielan-Jaworowska, 1975b, type genus
        - *A. nemegetensis* Kielan-Jaworowska, 1975b
      - *Ukhaatherium* Novacek et al., 1997
        - *U. nessovi* Novacek et al., 1997
      - ?*Bulaklestes* Nessov, 1985a
        - *B. kezbe* Nessov, 1985a
    - Family Kennalestidae Kielan-Jaworowska, 1981
      - *Kennalestes* Kielan-Jaworowska, 1969
        - *K. gobiensis* Kielan-Jaworowska, 1969, type species
        - *K. uzbekistanensis* Nessov, 1997
      - ?*Sailestes* Nessov, 1982
        - *S. quadrans* Nessov, 1982
    - Family *incertae sedis*
      - *Daulestes* Trofimov and Nessov, 1979 in Nessov and Trofimov, 1979
        - *D. kulbeckensis* Trofimov and Nessov, 1979, in Nessov and Trofimov, 1979, type species
        - *D. inobservabilis* (Nessov, 1982)
        - *D. nessovi* McKenna et al., 2000
  - Superorder Anagalida Szalay and McKenna, 1971
    - Family Zalambdalestidae Gregory and Simpson, 1926
      - *Zalambdalestes* Gregory and Simpson, 1926, type genus
        - *Z. lechei* Gregory and Simpson, 1926
      - *Alymlestes* Averianov and Nessov, 1995
        - *A. kielanae* Averianov and Nessov, 1995
      - *Barunlestes* Kielan-Jaworowska, 1975b
        - *B. butleri* Kielan-Jaworowska, 1975b
      - *Kulbeckia* Nessov, 1993
        - *K. kulbecke* Nessov, 1993, type species
        - *K. rara* Nessov, 1993
        - *K. kansaica* Nessov, 1993
      - ?*Beleutinus* Bazhanov, 1972
      - *Beleutinus orlovi* Bazhanov, 1972
  - Superorder Archonta Gregory, 1910
    - Order *incertae sedis*
      - Family *incertae sedis*
        - *Deccanolestes* Prasad and Sahni, 1988
          - *D. hislopi* Prasad and Sahni, 1988, type species
          - *D. robustus* Prasad et al., 1994
  - Superorder Insectivora Bowdich, 1821
    - Order Leptictida McKenna, 1975
      - Family Gypsonictopsidae Van Valen, 1967b
        - *Gypsonictops* Simpson, 1927b
          - *G. hypoconus* Simpson, 1927b, type species
          - *G. illuminatus* Lillegraven, 1969
          - *G. lewisi* Sahni, 1972
          - *G. clemensi* Rigby and Wolberg, 1987
    - Order Lipotyphla Haeckel, 1866
    - Suborder Soricomorpha Gregory, 1910
      - Family "Nyctitheriidae" Simpson, 1928f
        - *Paranyctoides* Fox, 1979d
          - *P. sternbergi* Fox, 1979d, type species
          - *P. aralensis* Nessov, 1993
          - *P. maleficus* Fox, 1984b
          - *P. megakeros* Lillegraven and McKenna, 1986
  - Superorder Ferae Linnaeus, 1758
    - Order Cimolesta McKenna, 1975
      - Family Cimolestidae Marsh, 1889a
        - *Cimolestes* Marsh, 1889a, type genus
          - *C. incisus* Marsh, 1889a, type species
          - *C. cerberoides* Lillegraven, 1969
          - *C. magnus* Clemens and Russell, 1965
          - *C. propalaeoryctes* Lillegraven, 1969
          - *C. stirtoni* Clemens, 1973a
        - *Batodon* Marsh, 1892
          - *B. tenuis* Marsh, 1892
        - *Procerberus* Sloan and Van Valen, 1965
          - *P. formicarum* Sloan and Van Valen, 1965
        - *Telacodon* Marsh, 1892
          - *T. laevis* Marsh, 1892
  - Superorder Ungulatomorpha Archibald, 1996b
    - Order *incertae sedis*
      - Family "Zhelestidae" Nessov, 1985a
        - *Zhelestes* Nessov, 1985a, type genus

TABLE 13.1. Continued

Z. temirkazyk Nessov, 1985a
Alostera Fox, 1989
A. saskatchewanensis Fox, 1989
Aspanlestes Nessov, 1985a
A. aptap Nessov, 1985a
Avitotherium Cifelli, 1990e
A. utahensis Cifelli, 1990e
Eoungulatum Nessov, Archibald, and Kielan-Jaworowska,1998
E. kudukensis Nessov, Archibald, and Kielan-Jaworowska,1998
Gallolestes Lillegraven, 1976
G. pachymandibularis Lillegraven, 1976, type species
G. agujaensis Cifelli, 1994
Labes Sigé *in* Pol et al., 1992
L. quintanillensis Sigé in Pol et al., 1992, type species
L. garimondi Sigé in Pol et al., 1992
Lainodon Gheerbrant and Astibia, 1994
L. orueetxebarriai Gheerbrant and Astibia, 1994
Parazhelestes Nessov, 1993
P. robustus Nessov, 1993, type species
?*P. minor* Nessov, Archibald, and Kielan-Jaworowska, 1998
Sorlestes Nessov, 1985a
S. budan Nessov, 1985a, type species
S. kara Nessov, 1993
S. mifunensis Setoguchi, Tsubamoto et al., 1999
Order "Condylarthra" Cope, 1881a
Family "Arctocyonidae" Giebel, 1855
Baioconodon Gazin, 1941a
B. denverensis Gazin, 1941a, type species (Tertiary)
Baioconodon sp.
Oxyprimus Sloan and Van Valen, 1978
O. erikseni Van Valen, 1978, type species (Paleocene)
O. cf. *erikseni*
Protungulatum Sloan and Van Valen, 1965
P. donnae Sloan and Van Valen, 1965
Family Periptychidae Cope, 1882d
"*Mimatuta*" Van Valen, 1978
"*M.*" *morgoth* Van Valen, 1978, type species
?Order Notoungulata Roth, 1903
Family ?Perutheriidae Van Valen, 1978
Perutherium Grambast et al., 1967
P. altiplanense Grambast et al., 1967

given the extensive recent comparisons suggesting otherwise; we follow what we believe to be current consensus in recognizing Leptictida and Lipotyphla as divisions of Insectivora (Novacek, 1986b; MacPhee and Novacek, 1993). The third category (a term we prefer over "group," given the nature of the evidence) is the most problematic: this includes taxa commonly referred to, and compared with, Tertiary "Palaeoryctidae." We tentatively recognize shared dental specializations (noted later) in *Cimolestes, Batodon,* and *Procerberus* and place them together in the Cimolestidae. Many arguments can be, and have been, made regarding the affinities of these taxa: most compelling in our view are those that suggest that they represent a basal branch of Ferae (Van Valen, 1966; Lillegraven, 1969). Historically, one or more of these three genera have often been compared favorably with leptictoids (e.g., Van Valen, 1967b) or lipotyphlans (McKenna and Bell, 1997), herein placed in a separate superorder. In this context, we note that recent studies suggest a possible relationship of Ferae to Insectivora (MacPhee and Novacek, 1993; Wyss and Flynn, 1993), and this view is reflected in our summary phylogram of Cretaceous Eutheria. Origin of the Tertiary Taeniodonta from Cimolestidae has been suggested on the basis of recent investigations (Eberle, 1999).

Order and Family *incertae sedis*
Genus *Eomaia* Ji, Luo, Yuan, Wible, Zhang, and Georgi, 2002
(figures 13.8A, 13.9, 13.10, 13.17A, 13.19G)

Diagnosis (shortened and emended after Ji et al., 2002). Among eutherians previously known from the late Early Cretaceous, *Eomaia scansoria* differs from *Prokennalestes* in lacking a labial mandibular foramen in the masseteric fossa and in having a larger metastylar and metaconal region on upper molar M3; differs from *Murtoilestes* in having less developed conules on the upper molars, and from both *Murtoilestes* and *Prokennalestes* in having an anteroposteriorly shorter trigonid and a longer talonid basin; differs from *Montanalestes* in having a paraconid lower than the metaconid. Differs from *Montanalestes* and all Late Cretaceous eutherians in retaining the Meckel's groove, and from most eutherians (but not *Prokennalestes, Montanalestes,* and several asioryctitherians) in having a slightly in-turned angular process. Differs from *Deltatheridium* and other metatherians in having a eutherian dental formula, 5.1.5.3/4.1.5.3; differs from *Ausktribosphenos* and *Bishops* in lacking a shelflike mesial cingulid on the molars and in having laterally compressed ul-

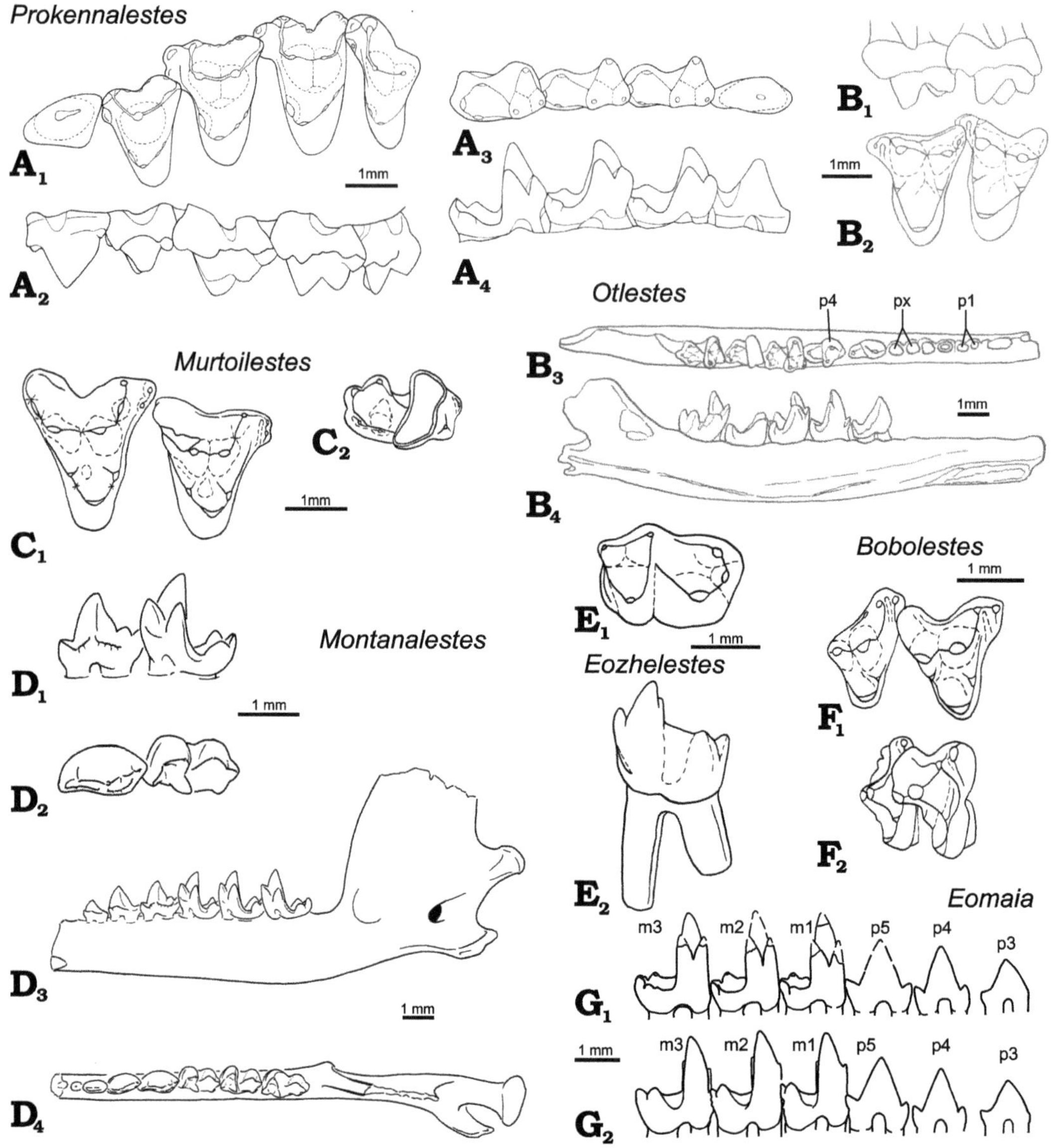

FIGURE 13.19. Dentition in Otlestidae and selected Eutheria of uncertain affinities. A, *Prokennalestes trofimovi:* reconstruction of posterior cheek teeth; left P3, P4 and M1–3 in occlusal view (A_1); the same in labial view (A_2); left p4–m3 in occlusal view (A_3); the same in lingual view (A_4). B, *Otlestes meiman,* left M1 and M2 in labial view (B_1); the same in occlusal view (B_2); left dentary with alveoli for c, p1–px, and preserved teeth p3, p4, and m1–3, in occlusal view (B_3); the same in lingual view (B_4). C, *Murtoilestes abramovi,* right M1–2 in occlusal view (C_1); left m3 in occlusal view (C_2). D, *Montanalestes keeblerorum:* right p4–m1 in lingual (D_1) and occlusal (D_2) views; restored dentary in lingual (D_3) and occlusal (D_4) views, holotype. E, *Eozhelestes mangit,* left m1 in occlusal (E_1) and labial (E_2) views. F, *Bobolestes zenge,* right M2, M3 in occlusal (F_1) and anteroocclusal (F_2) views. G, *Eomaia scansoria,* partial right lower dentition (p3–m3), reversed, in lingual view (G_1). The same in labial view (G_2). Note that in Figure 13.8B one of the premolars bears the notation px. *Source:* A, modified from Kielan-Jaworowska and Dashzeveg (1989); B–F, originals based on specimens; G, modified from Ji et al. (2002).

timate and penultimate lower premolars without full cusp triangulation; differs from *Ausktribosphenos, Bishops,* and most stem mammaliaforms in lacking the primitive postdentary trough on the mandible.

Species. *Eomaia scansoria* Ji, Luo, Yuan, Wible, Zhang, and Georgi, 2002, type species by monotypy.

Distribution. Early Cretaceous (middle Barremian): China, Liaoning Province (Yixian Formation).

Genus *Montanalestes* Cifelli, 1999b
(figure 13.19D)

Diagnosis. Generally plesiomorphic eutherian, differing from *Prokennalestes* in having a semimolariform p4; from *Otlestes* in having an unreduced paraconid on lower molars and a lower, more posteriorly placed metaconid on p4; and from both in lacking a Meckel's groove on the dentary.

Species. *Montanalestes keeblerorum* Cifelli, 1999b, emended (Cifelli, 2000b), type species by monotypy.

Distribution. Early Cretaceous (Aptian–Albian): United States, Montana (Cloverly Formation).

Comments. An expanded diagnosis, including comparison to noneutherian tribosphenic mammals, is given elsewhere (Cifelli, 1999b). *M. keeblerorum* is known by comparatively good material: both dentaries, with the entire back of the jaw collectively represented, together with the last three premolars (and a pair of alveoli anterior to the first) and all three molars. *Montanalestes* was initially compared to Eutheria but not formally referred to them (Cifelli, 1999b). Our referral is based on the presence of only three molars and the molarization of the last premolar. In the second feature, *Montanalestes* is more advanced than *Prokennalestes* (see Kielan-Jaworowska and Dashzeveg, 1989; Sigogneau-Russell et al., 1992), in which the last lower premolar is simple. Similarly, *Montanalestes* lacks the Meckel's groove seen in *Prokennalestes* and certain other early Eutheria, but its molar structure is exceedingly primitive; the paraconid, for example, is unreduced. This genus represents the oldest occurrence of Eutheria in North America, the next oldest being early Campanian in age (Fox, 1984b; Cifelli, 1990e).

Family Otlestidae Nessov, 1985a

Diagnosis. Family of Eutheria characterized by the following plesiomorphies: remnants of the coronoid bone and Meckel's groove present; single labial mandibular foramen in the masseteric fossa, apparently four lower incisors (reconstructed only in *Prokennalestes*), five premolars, pre- and postcingula absent from upper molars (a character shared, among others, with *Daulestes,* Asioryctidae, Zalambdalestidae, *Bobolestes,* some species of *Cimolestes*), large stylar area, cusp present on postmetacrista. Apomorphies: ultimate upper premolar (P4) semimolariform, but without metacone, ultimate lower premolar (p4) not molariform in *Prokennalestes,* but semimolariform in *Otlestes,* penultimate upper premolar (P3) strong and piercing, but relatively lower than in Asioryctitheria, preparastyle present on all the molars. Differ from Asioryctitheria in having parastylar and metastylar regions equally protruding labially, shallow ectoflexus on M1, deep on M2, conules unwinged or with incipient wings, talonid distinctly narrower than the trigonid.

Genera. *Otlestes* Nessov, 1985a, type genus; *Murtoilestes* Averianov and Skutschas, 2001; *Prokennalestes* Kielan-Jaworowska and Dashzeveg, 1989.

Distribution. Early–Late Cretaceous (?late Barremian to early Cenomanian): Asia, Russia, Mongolia, and Uzbekistan.

Genus *Otlestes* Nessov, 1985a
(figure 13.19B)

Diagnosis. Apparently differs from *Prokennalestes* in having a single-rooted lower canine. Differs from *Prokennalestes* in having relatively lower talonids on the lower molars and semimolariform ultimate premolar (p4) with an incipient three-cusped trigonid, whereas in *Prokennalestes* p4 has only one anterior cusp and posterior basal cuspule. Differs from *Prokennalestes* in having winged conules (a character shared with *Murtoilestes*) and in having a deeper ectoflexus on M2.

Species. *Otlestes meiman* Nessov, 1985a, type species by monotypy.

Distribution. Late Cretaceous (early Cenomanian): Uzbekistan, Kyzylkum Desert (Khodzhakul Formation).

Genus *Murtoilestes* Averianov and Skutschas, 2001
(figure 13.19C)

Diagnosis. Poorly known genus based on several isolated teeth of the type species. Generally similar to *Prokennalestes,* from which it differs in having incipiently winged conules, smaller parastylar lobe and smaller parastylar cusps, greater size difference between M1 and M2; M1 narrower transversely and relatively smaller in proportion to M2 than in *Prokennalestes,* with more robust (longer) protoconal region. Cristid obliqua attaches more labially to the trigonid; talonid wider in proportion to the trigonid and talonid cusps bigger than in *Prokennalestes.*

Species. *Murtoilestes abramovi* (Averianov and Skutschas, 2000a), type species by monotypy.

Distribution. Early Cretaceous (?late Barremian–middle Aptian): Russia, Transbaikalia (Murtoi Formation).

Genus *Prokennalestes* Kielan-Jaworowska and Dashzeveg, 1989
(figures 13.8B, 13.19A)

Diagnosis. Differs from *Otlestes* in having a double-rooted lower canine. Differs from *Kennalestes* and *Daulestes* in having five premolars (rather than four); shares with *Daulestes* (and many other genera) lack of pre- and postcingula on the upper molars and differs in this respect from *Kennalestes* and *Sailestes.* Differs from *Daulestes* in having ultimate upper premolar semimolariform and ultimate lower premolar nonmolariform. Differs from *Daulestes* in having three distinct cusps in parastylar region and less reduced metastylar region; differs from *Kennalestes* in having a well-developed preparastyle (hardly discernible in *Kennalestes*) and shares three-cusped parastylar region with *Sailestes.* Differs from *Sailestes* in lack of cusp "C" on the ectoflexus.

Species. *Prokennalestes trofimovi* Kielan-Jaworowska and Dashzeveg, 1989, type species; and *P. minor* Kielan-Jaworowska and Dashzeveg, 1989.

Distribution. Early Cretaceous (Aptian–Albian): Mongolia, Gobi Desert ("Höövör Beds").

Comment. As argued by Kielan-Jaworowska and Dashzeveg (1989, see also Sigogneau-Russell et al., 1992), *P. trofimovi* and *P. minor* differ only in size and may be sexual morphs within the same species.

Magnorder Epitheria McKenna, 1975
Superorder and Order *incertae sedis*
Family Bobolestidae Nessov, 1989 (Nessov, 1992)

Diagnosis (based on Nessov, Sigogneau-Russell, and Russell, 1994, emended). Paracone of M2 and M3 noticeably larger than the metacone; both cusps situated slightly labial to the middle of the distance between protocone and ectoflexus and separated by a deep furrow situated between postparaconule and premetaconule cristae. Postparacrista and premetacrista considerably reduced. Stylocone and parastyle on M2 and M3 moderately developed. Preparaconule crista long, extending labially beyond the paracone. Protocone not large; preprotocrista and postprotocrista placed relatively widely apart. Conules well developed, with strong wings. Pre- and postcingula absent.

Genera. *Bobolestes* Nessov, 1985a, type genus by monotypy.

Distribution. Early Cretaceous (late Albian): Uzbekistan, Karakalpakian Autonomous Republic (Khodzhakul Formation).

Comments. Originally Nessov (1985a) attributed *Bobolestes* to Theria, Pappotheriidae *incertae sedis,* but he subsequently (Nessov, 1989; see also Nessov, Sigogneau-Russell, and Russell, 1994) erected for it a separate eutherian family, within suborder Proteutheria Romer, 1966. The plesiomorphic feature of *Bobolestes* is the lack of pre- and postcingula (shared with numerous Cretaceous eutherians such as Otlestidae, *Daulestes, Asioryctes, Ukhaatherium,* Zalambdalestidae, and many others), but it is advanced in having well-developed, winged conules. Because of this, eutherian affinities of *Bobolestes,* which seems to be more advanced than *Prokennalestes* (Kielan-Jaworowska and Dashzeveg, 1989), appear probable (see also Nessov, Sigogneau-Russell, and Russell, 1994; Nessov, 1997), but McKenna and Bell (1997) regarded Bobolestidae as a junior synonym of Pappotheriidae Slaughter, 1965, and assigned *Bobolestes* to Pappotheriidae.

Genus *Bobolestes* Nessov, 1985a
(figure 13.19F)

Diagnosis and Distribution. As for the family.

Species. *Bobolestes zenge* Nessov, 1985a, type species by monotypy.

Order and Family *incertae sedis*
Genus *Eozhelestes* Nessov, 1997
(figure 13.19E)

Diagnosis (translated from Nessov, 1997). Size as in *Aspanlestes.* Protoconid and metaconid relatively high. Paraconid not strongly reduced and slightly approaching (close to) the metaconid. Talonid of m1 narrower than trigonid. Cristid obliqua reaches the trigonid lingual to the notch between protoconid and metaconid, then passes up toward the tip of the metaconid. Precingulid situated relatively low with respect to the notch between protoconid and metaconid. Entoconid and hypoconulid closely approximated.

Species. *Eozhelestes mangit* Nessov, 1997, type species by monotypy.

Comments. Nessov (1997) compared *Eozhelestes* to "zhelestids," noting similarity in protoconid and metaconid with bulbous swollen bases; some reduction of paraconid, and entoconid and hypoconulid closely approximated. He noted that it differs from Turonian–Coniacian "zhelestids" in having a higher trigonid, relatively low talonid (in comparison to *Aspanlestes* and *Sorlestes*), cristid obliqua with longer anterior ridge attached lingual to (rather than directly below) the notch in the protocristid, and a low precingulid.

Nessov (1997) assigned this poorly known monotypic genus, represented by a single molar (identified as m1) of the type species *E. mangit,* to the family "of the ?grade Zhelestidae" (translated from Russian). In a "note added in proof" to Nessov, Archibald, and Kielan-Jaworowska (1998: 87), written after Nessov's death, the two remaining authors of that paper stated: "The type of *Eozhelestes mangit* does appear either m1 or m2. Although there is nothing precluding a mammal with this kind of lower molar and of this earlier age from being ancestral to 'zhelestids', it lacks the expanded talonid of 'zhelestids' and later ungulatomorphs. Accordingly, it is excluded from Ungulatomorpha (including 'Zhelestidae')." As far as can be judged from the poor illustration in Nessov's (1997: plate 43, figure 4) monograph, in proportion and distribution of the trigonid and talonid cusps the *E. mangit* specimens resemble the molars of *Prokennalestes* (Kielan-Jaworowska and Dashzeveg, 1989), from which they differ in having a somewhat shorter and wider talonid. Because of the poor knowledge of *Eozhelestes,* we leave its familial identification unresolved.

Distribution. Late Cretaceous (early Cenomanian): Uzbekistan (Kodzhakul Formation).

Superorder Asioryctitheria Novacek, Rougier, Wible, McKenna, Dashzeveg, and Horovitz, 1997

Diagnosis. Small eutherians with long mesocranial region, a tall, curved crista interfenestralis that links the paroccipital process to the promontorium, a strong entoglenoid process, a postglenoid foramen that opens inside the glenoid area, subsquamosal foramen, horseshoe-shaped free ectotympanic, no entotympanic. Vidian foramen present, large sphenorbital fissure, no foramen rotundum, large alisphenoid. Up to five upper incisors present; four premolars present, of which the penultimate is the tallest of the series, strong and piercing; parastylar region on P4 and upper molars more prominent than the strongly reduced metastylar region. On lower molars the entoconid and hypoconulid are not twinned. Hindlimb plesiomorphic with respect to that of other eutherians in ?retaining femoro-fibular contact, in having only incipient mortise-tenon interlocking of astragalus, and in lacking a pulley-shaped astragalar trochlea and an anterior plantar tubercle on the calcaneus. Proximal tarsals differ from those of other eutherians in having a pedicillate, anteroposteriorly oriented posterior astragalocalcaneal facet, strong dorsal orientation of sustentacular facet, and hypertrophied astragalar medial plantar tuberosity.

Distribution. Late Cretaceous: Asia.

Families. Asioryctidae Kielan-Jaworowska, 1981, type family; Kennalestidae Kielan-Jaworowska, 1981; and tentatively assigned *Daulestes* Trofimov and Nessov, 1979 (in Nessov and Trofimov, 1979), family *incertae sedis,* see McKenna et al., 2000.

Family Asioryctidae Kielan-Jaworowska, 1981

Diagnosis. Differ from Kennalestidae and *Daulestes* in having longer mesocranial region, P2 smaller than P1, and very slender upper molars, strongly elongated transversely, with parastylar region more expanded labially, lack of pre- and postcingula (a character shared with Otlestidae, *Daulestes,* and many other Cretaceous eutherian taxa), and noticeably reduced metastylar region. Differ from Kennalestidae in incisor count five/four rather than four/three and in having lower molars with smaller paraconids and more compressed trigonids. Differ from *Prokennalestes* in having two cusps in parastylar region (rather than three).

Genera. *Asioryctes* Kielan-Jaworowska, 1975b, type genus; *Ukhaatherium* Novacek et al., 1997; and tentatively assigned ?*Bulaklestes* Nessov, 1985a.

Distribution. Late Cretaceous (Coniacian to Campanian): Asia, Uzbekistan, and Mongolia.

Genus *Asioryctes* Kielan-Jaworowska, 1975a
(figures 13.1C, 13.2A, 13.4, 13.7B, 13.8D, 13.12, 13.20D)

Diagnosis. Differs from *Ukhaatherium* in having peg-like upper incisors, double-rooted upper and lower canines, more expanded parastylar and more reduced metastylar regions on the upper molars, which are very slender and more elongated transversely, and lack of metaconule on M2 (see Crompton and Kielan-Jaworowska, 1978). Differs from *Bulaklestes* in lack of an incipient precingulum.

Species. *Asioryctes nemegetensis* Kielan-Jaworowska, 1975a, type species by monotypy.

Distribution. Late Cretaceous (?late Campanian): Mongolia, Gobi Desert (Baruungoyot Formation and Red Beds of Hermiin Tsav).

Genus *Ukhaatherium* Novacek, Rougier, Wible, McKenna, Dashzeveg, and Horovitz, 1997
(figures 13.13, 13.21B)

Diagnosis. Differs from *Asioryctes* in having two- or three-cusped I3 and I4, enlarged, single-rooted upper and lower canines, long diastema between I4 and upper canine to accommodate lower canine, less robust P3 (shorter in lateral view), upper molars less elongated transversely and less slender, only two mental foramina in the dentary (up to six in *Asioryctes*), smaller facial process of the lacrimal (so that nasal-maxillary contact in facial region is broader), and only one foramen in the mastoid.

Species. *Ukhaatherium nessovi* Novacek, Rougier, Wible, McKenna, Dashzeveg, and Horovitz, 1997, type species by monotypy.

Distribution. Late Cretaceous (?early Campanian): Mongolia, Gobi Desert ("Ukhaa Tolgod Beds").

Genus ?*Bulaklestes* Nessov, 1985a
(figure 13.20C)

Diagnosis. Monotypic genus known from a single M3 of the type species. This M3 (see Nessov, Sigogneau-

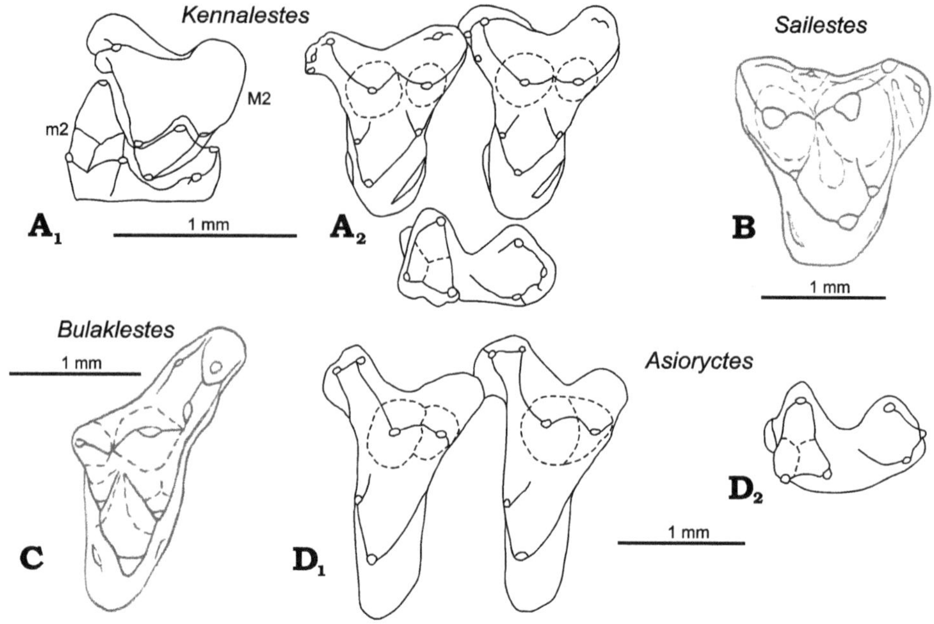

FIGURE 13.20. Dentition of selected Asioryctitheria. A, *Kennalestes gobiensis,* oblique view of left M2 and right m2 showing the teeth in occlusion (A_1); occlusal view of left M1, M2 and right m2 (A_2). B, *Sailestes quadrans,* right M1 or M2, occlusal view. C, *Bulaklestes kezbe,* right M3, occlusal view. D, *Asioryctes nemegetensis,* occlusal view of left M1, M2 (D_1); occlusal view of right m2 (D_2). *Source:* original drawings based on specimens or casts.

Russell, and Russell, 1994: plate 4, figure 2) resembles that of *Asioryctes* in being very slender and having very long (anterolabially) parastylar region with two cusps, and winged conules, but differs from it in having small and short (incipient) precingulum (absent in other asioryctids).

Species. *Bulaklestes kezbe* Nessov, 1985a, type species by monotypy.

Distribution. Late Cretaceous (Coniacian): Uzbekistan, Kyzylkum Desert (middle part of the Bissekty Formation).

Family Kennalestidae Kielan-Jaworowska, 1981

Diagnosis. Differ from Asioryctidae in having a relatively shorter mesocranial region, smaller number of incisors (four/three rather than five/four), and upper molars relatively wider and less elongated transversely, with less reduced metastylar region.

Genera. *Kennalestes* Kielan-Jaworowska, 1969, type genus, and tentatively assigned *Sailestes* Nessov, 1982.

Distribution. Late Cretaceous (Coniacian–Campanian): China and Uzbekistan.

Genus *Kennalestes* Kielan-Jaworowska, 1969
(figures 13.1B, 13.7A, 13.8C, 13.20A, 13.22)

Diagnosis. Differs from *Prokennalestes* in having four (rather than five) premolars in the adult and shares this character with *Daulestes.* Differs from *Prokennalestes* and *Daulestes* in having pre- and postcingula, from *Prokennalestes* in lacking an obvious preparastyle (albeit a remnant may be discernible), and from *Sailestes* in having M1 more elongated transversely, with a slender protocone and lack of a strong cusp "C" in the ectoflexus.

Species. *Kennalestes gobiensis* Kielan-Jaworowska, 1969, type species; *K. uzbekistanensis* Nessov, 1997.

Distribution. Late Cretaceous (Coniacian–?early Campanian): Mongolia, Gobi Desert (Djadokhta Formation and "Ukhaa Tolgod Beds"); China (Bayan Mandahu Formation); Uzbekistan, Kyzylkum Desert (Bissekty Formation).

Genus ?*Sailestes* Nessov, 1982
(figure 13.20B)

Diagnosis. Poorly known genus, represented by a single M1 or M2. Differs from *Kennalestes* in having molars less elongated transversely, having weaker cingula, and presence of a distinct cusp "C" on the ectoflexus (in *Kennalestes* an incipient cusp "C" occurs only on M2), but shares with it winged conules and with *Prokennalestes* the structure of the enlarged, three-cusped parastylar region. Differs from *Prokennalestes* in presence of weak cingula, winged conules, and cusp "C."

Distribution. Late Cretaceous (Coniacian): Uzbekistan, Kyzylkum Desert (middle part of Bissekty Formation).

Species. *Sailestes quadrans* Nessov, 1982, type species by monotypy.

Comment. Nessov (1982) assigned *Sailestes* to Kennalestidae; Nessov, Sigogneau-Russell, and Russell (1994) followed this assignment, pointing out its similarities to European late Paleocene *Bustylus* Gheerbrant and Russell, 1991. Nessov (1997) erected for *Sailestes* the subfamily Sailestinae within Kennalestidae. We agree that *Sailestes* differs noticeably from the typical kennalestid (*Kennalestes*) and we assign it to Kennalestidae only tentatively. Affinities of *Sailestes* are highly uncertain: Archibald and

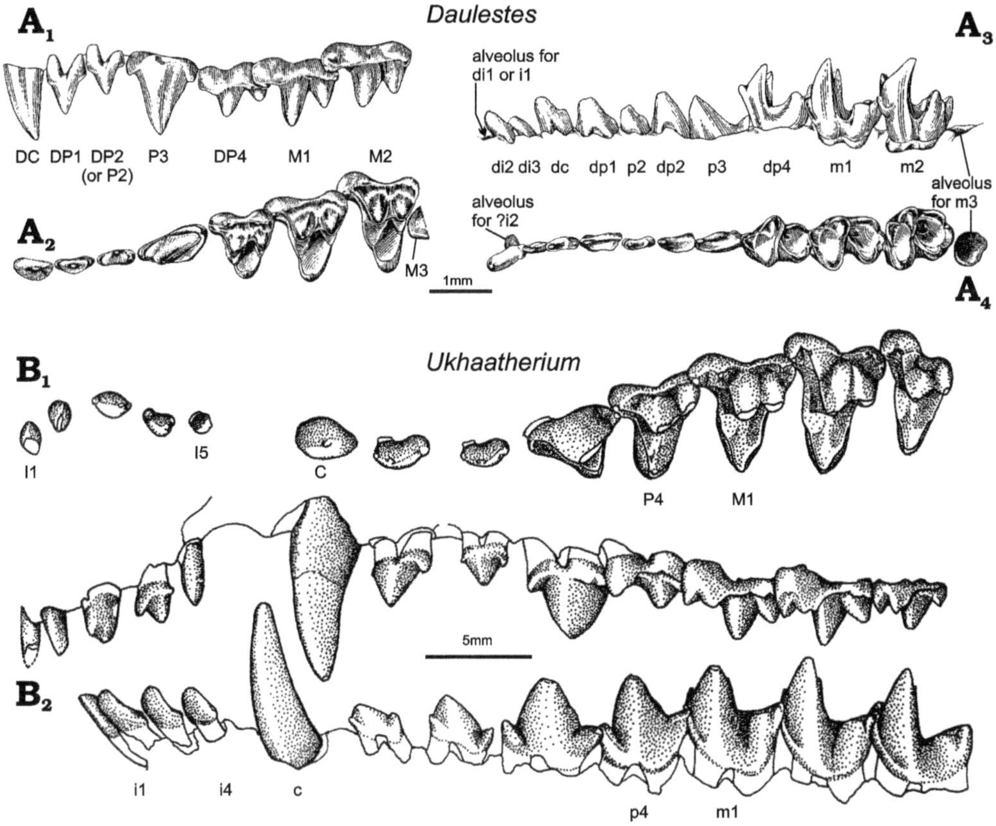

FIGURE 13.21. Dentition of *Daulestes* and *Ukhaatherium*. A, Slightly diagrammatic drawing of the left upper and lower dentitions of *Daulestes nessovi*. Upper dentition in labial (A_1) and occlusal (A_2) views (incisors and M3 are not preserved). Lower dentition in labial (A_3) and occlusal (A_4) views. B, Dentition of *Ukhaatherium nessovi*. Left upper teeth in occlusal view (B_1) and left upper and lower teeth in labial view (B_2). *Source:* A, modified from McKenna et al. (2000); B, drawing, courtesy of M. J. Novacek.

Averianov (2001a) suggested that it may represent the upper molar of *Paranyctoides* or even a metatherian.

Family *incertae sedis*
Genus *Daulestes* Trofimov and Nessov, 1979 in Nessov and Trofimov, 1979
(figures 13.1A, 13.3, 13.6, 13.8C, 13.21A)

Synonym: *Taslestes* Nessov, 1982; *Kumlestes* Nessor, 1985a.

Diagnosis. Differs from other Asioryctitheria in lack of a piriform fenestra (at least in the only available skull of a juvenile individual of *D. nessovi* McKenna et al., 2000), a character regarded by Novacek et al. (1997) as diagnostic for Asioryctitheria. Shares with members of Asioryctidae a strongly asymmetrical stylar shelf on the upper molars, with a large parastylar and small metastylar regions. Differs from *Prokennalestes* and *Otlestes* in lack of a preparastyle and single labial mandibular foramen (multiple, small foramina present in *Prokennalestes* and *Otlestes*), and in having the talonid almost as wide as the trigonid (rather than markedly narrower); differs from *Kennalestes* in lack of pre- and postcingula, and shares this character with Otlestidae, Asioryctidae, and many other Cretaceous eutherians. Shares with *Asioryctes* generally similar proportions of the orbitosphenoid and alisphenoid and arrangement of alisphenoid/orbitosphenoid foramina.

Species. *Daulestes kulbeckensis* Trofimov and Nessov, 1979, in Nessov and Trofimov, 1979, type species (synonym *Kumlestes olzha* Nessor, 1985a); *D. inobservabilis* (Nessov, 1982), and *D. nessovi* McKenna, Kielan-Jaworowska, and Meng, 2000.

Distribution. Late Cretaceous (late Turonian–Coniacian): Uzbekistan, Kyzylkum Desert (Bissekty Formation).

Superorder Anagalida Szalay and McKenna, 1971

Diagnosis (based on Szalay and McKenna, 1971). Posterior premolars molariform, trigonids compressed anteroposteriorly, a tendency toward prismatic or unilaterally hypsodont upper cheek teeth with crown pattern tending to be obliterated early in wear, a tendency toward procumbent, ever-growing incisors (variously expressed), and a skeleton (particularly the foot), where known, that is somewhat lagomorph-like.

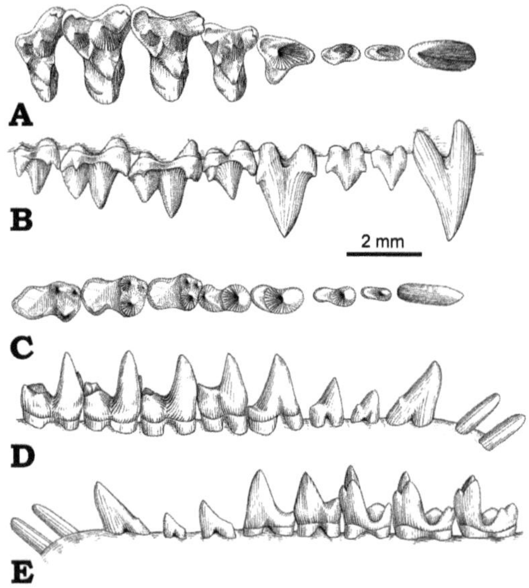

FIGURE 13.22. Dentition of *Kennalestes gobiensis.* Right C, P1–4, M1–3 in occlusal (A) and labial (B) views; right i2–3, c, p1–4, m1–3 in occlusal (C), labial (D), and lingual (E) views. *Source:* modified from Kielan-Jaworowska (1969b).

Families. From among numerous high-rank taxa assigned by McKenna and Bell (1997) to Anagalida, regarded by them as a grandorder within superorder Preptotheria McKenna, 1975, only Zalambdalestidae Gregory and Simpson, 1926 occur in the Mesozoic and we limit our discussion on the Anagalida to that family.

Distribution. Mesozoic distribution as for the Zalambdalestidae (see later).

Family Zalambdalestidae Gregory and Simpson, 1926

Synonym: Kulbeckiidae Nessov, 1993.

Comment. Nessov (1993) erected the family Kulbeckiidae and assigned it to his order Mixotheridia Nessov, 1985a (see also Nessov, 1989). In the monograph published posthumously, Nessov (1997) assigned Kulbeckiidae to superfamily Zalambdalestoidea within Mixotheridia, but McKenna and Bell (1997) assigned Kulbeckiidae to the superorder Leptictida McKenna, 1975. Archibald et al. (2001) demonstrated that *Kulbeckia* is a zalambdalestid, an assignment followed herein.

Diagnosis (based on Kielan-Jaworowska, 1984a,b, emended). Differ from Otlestidae, Kennalestidae, and Asioryctidae in being larger and in having a very long, narrow snout. Skull constricted in front of P1. Differ from Kennalestidae, Asioryctidae, and *Daulestes* in having a more inflated braincase (Kielan-Jaworowska and Trofimov, 1980; Kielan-Jaworowska, 1984a,b) and share with them position of the occipital plate, which is inclined forward from the condyles. Differ from the eutherians mentioned previously in having very large oval posterior palatine foramen, maxilla that extends backward along the choanae, presence of a foramen rotundum, a prominent median process of the presphenoid, very large and differently shaped pterygoid process of the basisphenoid, and postglenoid process extending only along the medial part of the cupola-like glenoid fossa. The promontorium is flattened and lacks definite grooves for transpromontorial arteries, the carotid foramen is placed medially, and there is a foramen arteriae stapediae and sulcus arteriae stapediae, but no sulcus arteriae promontorii. This arrangement shows that the carotid arteries, the main channels supplying blood to the brain, entered the skull along the midline rather than at the sides, as they do in most living mammals. Piriform fenestra absent. Dental formula is 3.1.3-4.4/3.1.3–4.3. I2 is enlarged, caniniform; long diastema present between small I3 and C, which is large and placed well behind the premaxilla-maxillary suture. P1 small or absent, P2 small, P3 tallest of all teeth (a character shared with Kennalestidae, Asioryctidae, and *Daulestes*), P3 and P4 submolariform, but without metacone, upper molars without pre- and postcingula, small stylar shelves, and paracone and metacone of similar height; M3 very small. Of lower incisors i1 is the largest, procumbent, open-rooted (Archibald et al., 2001; Fostowicz-Frelik and Kielan-Jaworowska, 2002); i2, i3 small; p4 with three-cusped trigonid and unbasined talonid. Lower molars with very small trigonids, paraconid and metaconid connate at the bases, and large, high talonids that are distinctly wider and longer than the trigonids. Skeleton shows a mosaic of primitive and advanced characters. The axis has a very long horizontal spinous process, but the thoracic vertebrae bear only short spinous processes. The tibia and fibula are strongly fused, a calcaneal fibular facet is lacking, and the tibial trochlea on the astragalus is well developed. The hindlimbs, especially the metatarsals, are very long.

Genera. *Zalambdalestes* Gregory and Simpson, 1926, type genus; *Alymlestes* Averianov and Nessov, 1995; *Barunlestes* Kielan-Jaworowska, 1975b; *Kulbeckia* Nessov, 1993, and tentatively assigned *Beleutinus* Bazhanov, 1972. Attribution of Paleocene *Praolestes* Matthew, Granger, and Simpson, 1929, to Zalambdalestidae is uncertain.

Distribution. Late Cretaceous (late Turonian–?late Campanian): Uzbekistan, Tadjikistan, Kazakhstan, and Mongolia.

Genus *Zalambdalestes* Gregory and Simpson, 1926 (figures 13.1E, 13.14B, 13.16, 13.17B, 13.23A)

Diagnosis. Differs from *Barunlestes* in having a somewhat longer (about 50 mm long) skull of more gracile

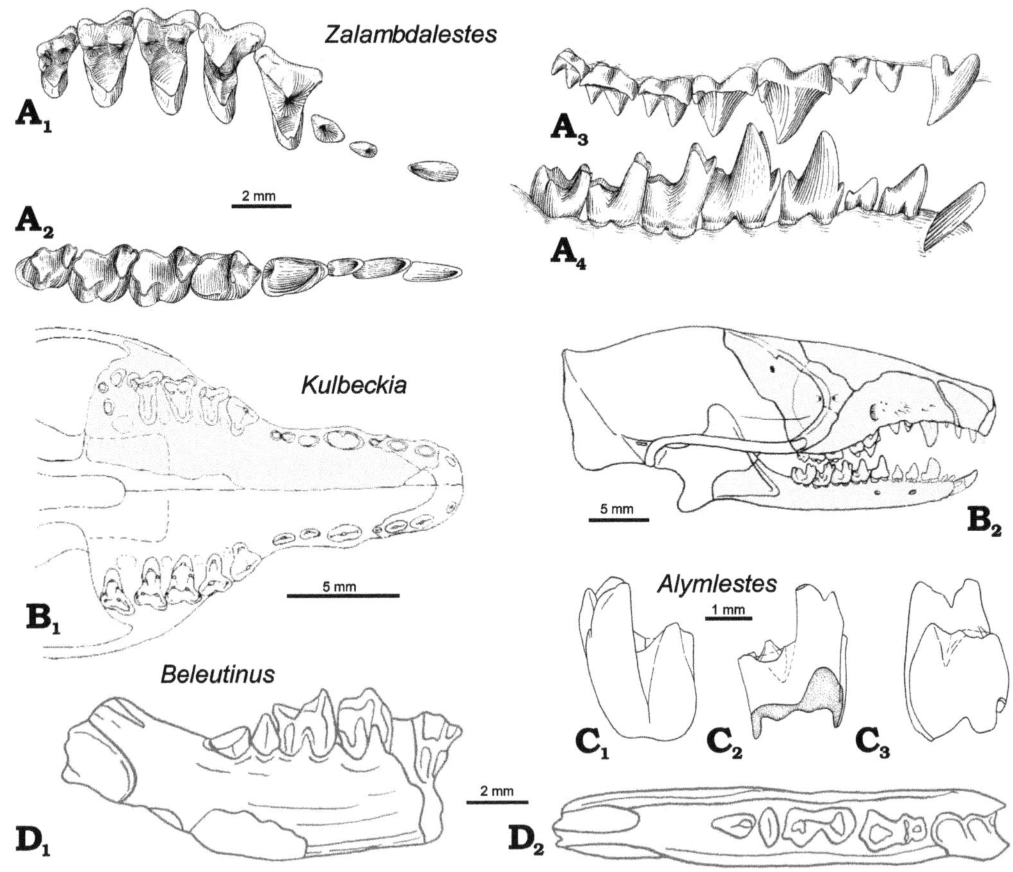

FIGURE 13.23. Dentition of selected Zalambdalestidae. A, *Zalambdalestes lechei,* right C, P1–4, M1–3 in occlusal (A_1) and labial (A_3) views; right c, p1–4, m1–3 in occlusal (A_2) and labial (A_4) views. B, *Kulbeckia kulbecke,* reconstruction of the snout in palatal (B_1) and lateral (B_2) views (B_2 reversed). C, *Alymlestes kielanae,* left m1 in labial (C_1), lingual (C_2), and posterior (C_3) views. D, *Beleutinus orlovi,* posterior part of right dentary with alveoli for penultimate and ultimate premolars, strongly worn m1 and m2 and roots of m3, in labial (D_1) and occlusal (D_2) views. *Source:* A, modified from Kielan-Jaworowska (1969b); B_1, B_2, modified from Archibald et al. (2001); C, D, original drawings based on specimens or casts.

structure, expressed by slender zygomatic arches and a more slender dentary provided with a coronoid crest without basal swelling. Further differences concern presence of four upper and lower premolars (rather than three) and structure of the upper canine, which is large, double-rooted, and situated far behind the premaxilla-maxillary suture. Differs from *Kulbeckia* in being larger, having a more elongated snout, and having three lower incisors rather than four. Further difference concerns the structure of the conules, which are hardly expressed and placed close to the paracone and metacone, whereas in *Kulbeckia* the conules are more prominent and placed more lingually. Differs from *Alymlestes* in smaller dimensions, in having lower crowns of the molars, and not showing the incipient unilateral hypsodonty characteristic of that taxon.

Species. *Zalambdalestes lechei* Gregory and Simpson, 1926, type species by monotypy; *non Zalambdalestes mynbulakensis* Nessov, 1985b (probably = *Sorlestes budan* Nessov, 1985a, see Nessov, Archibald, and Kielan-Jaworowska, 1998: 62).

Distribution. Late Cretaceous (?early Campanian): Mongolia, Gobi Desert (Djadokhta Formation, "Ukhaa Tologod Beds" and Tögrög Beds); ?China, Inner Mongolia (Bayn Mandahu Formation) (see Kielan-Jaworowska and Trofimov, 1981, and Kielan-Jaworowska et al., 2003).

Genus *Barunlestes* Kielan-Jaworowska, 1975a
(figures 13.1D, 13.2B, 13.5, 13.14A,C, 13.15)

Diagnosis. Differs from *Zalambdalestes* in having a shorter and somewhat more robust skull (less than 40 mm long), a deeper dentary, higher coronoid process provided with a knoblike projection and a medial prominence. Further differences concern the presence of only three premolars (P2 and p2 are lacking) and a single-rooted upper canine.

Species. *Barunlestes butleri* Kielan-Jaworowska, 1975b, type species by monotypy.

Comment. Li and Ting (1985) noted that, among the specimens assigned by Kielan-Jaworowska and Trofimov (1980) to *Barunlestes butleri,* the dentary figured by these authors (pl. 7, figure 2) differs from other specimens in having a strongly enlarged incisor, extending posteriorly as far as m2, and the trigonids strongly compressed anteroposteriorly. According to Li and Ting this specimen may belong to a different genus, related to Eurymylidae.

Distribution. Late Cretaceous (?late Campanian): Mongolia, Gobi Desert (Baruungoyot Formation and Red Beds of Hermiin Tsav).

Genus *Kulbeckia* Nessov, 1993
(figure 13.23B)

Diagnosis (based on Archibald et al., 2001; and Nessov, 1997). Differs from *Zalambdalestes* in being distinctly smaller, but shares with it gracile structure of skull and dentary and elongated snout, which, however, is relatively shorter than in *Zalambdalestes.* Differs from *Barunlestes* in being slightly smaller and having a much less robust skull. Differs from both *Zalambdalestes* and *Barunlestes* in having four lower incisors (three in other genera), and shares with *Zalambdalestes* three upper incisors (number uncertain in *Barunlestes*). Shares with *Zalambdalestes* number of premolars (4/4) and differs in this respect from *Barunlestes* which has 3/3 premolar count. The molars are of zalambdalestid type, but *Kulbeckia* differs from *Zalambdalestes* in having more prominent and more lingually placed conules (hardly discernible in *Zalambdalestes* and unknown in *Barunlestes* because of the strong wear of all known specimens). Differs from *Zalambdalestes* in having ultimate molars less reduced in size with respect to the penultimates.

Distribution. Late Cretaceous (late Turonian–early Santonian): Uzbekistan, Kyzylkum Desert (Bissekty Formation), and Tadjikistan, Fergana Valley (Yalovach Formation).

Species. *Kulbeckia kulbecke* Nessov, 1993, type species; *K. rara* Nessov, 1993; *K. kansaica* Nessov, 1993; and a species left in open nomenclature by Nessov (1993).

Genus *Alymlestes* Averianov and Nessov, 1995
(figure 13.23C)

Diagnosis (based on Averianov and Nessov, 1995, emended). Poorly known genus represented by a single m1 of the type species. Differs from *Zalambdalestes* and *Barunlestes* in being distinctly larger and showing incipient unilateral hypsodonty of the lower molars, in which the labial side of the talonid basin is high. Trigonid and talonid basins are reduced in size and relatively small. The differences between heights of the talonid and trigonid are smaller than in other genera.

Species. *Alymlestes kielanae* Averianov and Nessov, 1995, type species by monotypy.

Distribution. Late Cretaceous (early Campanian): southern Kazakhstan, Chimkent Province (Darbasa Formation).

Genus ?*Beleutinus* Bazhanov, 1972
(figure 13.23D)

Species. *Beleutinus orlovi* Bazhanov, 1972, type species by monotypy.

Distribution. Late Cretaceous (Santonian–early Campanian): Kazakhstan, Kyzyl-Orda Province (Bostobe Formation).

Comments. Because of the poor knowledge of this taxon we refrain from attempting to diagnose it. The type species is represented only by the holotype specimen (posterior part of a right dentary with alveoli for penultimate and ultimate premolars, strongly worn m1 and m2, and roots of m3). The value of this taxon is mostly historical, as it was the first species of a Mesozoic mammal found in the territory of the former Soviet Union. As recognized by Clemens et al. (1979: 37), Bazhanov (1972) misinterpreted the specimen, inverting it anteroposteriorly and left to right. Nessov (1987) tentatively suggested possible attribution to Zalambdalestidae on the basis of the position of teeth with respect to the masseteric fossa, an idea supported by Nessov, Sigogneau-Russell, and Russell (1994). Although the specimen is poorly preserved, the type of molar wear, as seen in photographs published by Nessov, Sigogneau-Russell, and Russell (1994: plate 1, figure 1a–c) suggests zalambdalestid affinities, and we tentatively refer it to Zalambdalestidae.

Superorder Archonta Gregory, 1910
Order and family *incertae sedis*
Genus *Deccanolestes* Prasad and Sahni, 1988
(figures 13.11B,D, 13.24A)

Diagnosis. Molars generally primitive; uppers with wide stylar shelf, pre- and postcingula lacking. Upper molars differ from those of *Prokennalestes* in being less transversely developed, with a smaller parastyle, and in having a more lingually developed paracone; differs from *Cimolestes* in having a more lingually developed paracone and in lacking hypertrophied conules and crests, with carnassial notches lacking from secondary crests. Lower molars similar to those of *Otlestes* and differing from those of *Prokennalestes* and *Cimolestes* in having a smaller height and width differential between trigonid and talonid. Assigned

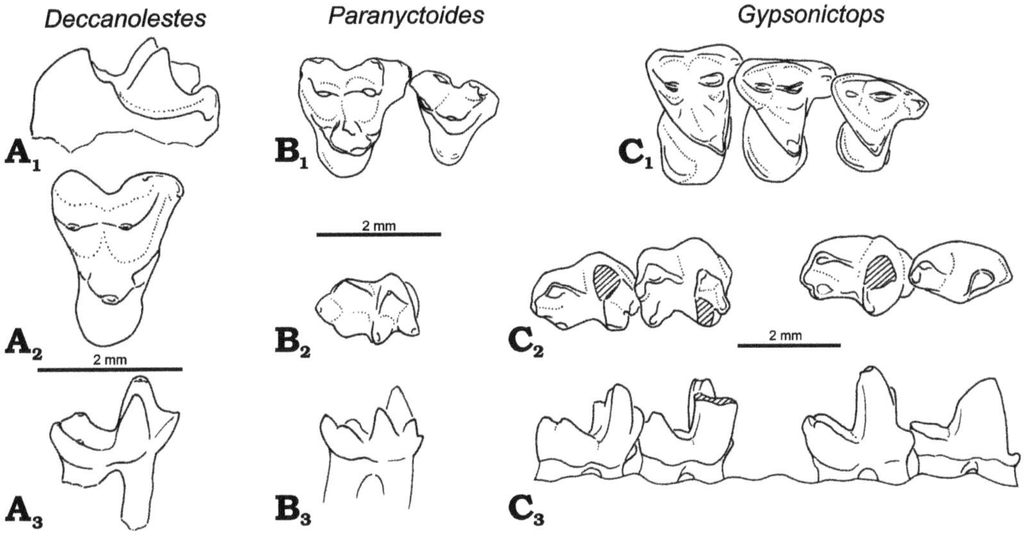

FIGURE 13.24. Archonta and Insectivora. A, *Deccanolestes robustus*, right M2 in mesial (A_1) and occlusal (A_2) views; left m1 in lingual (A_3) view. B, Dentition of *Paranyctoides*, right P4–M1 (*Paranyctoides* sp.) in occlusal view (B_1); left m3 (*P. maleficus*) in occlusal (B_2) and lingual (B_3) views. C, *Gypsonictops hypoconus*, right P3, P4, and M1 in occlusal view (C_1); left p3–4, m2–3 in occlusal (C_2) and lingual (C_3) views. *Source:* A, from Prasad et al. (1994); B, C, original drawings based on specimens or casts.

tarsals depart from presumed primitive eutherian condition (Szalay and Decker, 1974; Szalay, 1984) and resemble Archonta in lacking calcaneofibular contact and in having an ovoid, concave, transversely oriented calcaneal cuboid facet, a well-developed astragalar neck, and mediolaterally rounded astragalar head strongly dorsoplantarly developed on its lateral side, with extended navicular facet.

Species. *D. hislopi* Prasad and Sahni, 1988, type species; *D. robustus* Prasad, Jaeger, Sahni, Gheerbrant, and Khajuria, 1994.

Distribution. Late Cretaceous (Maastrichtian): India, Andhra Pradesh (Intertrappean Beds).

Comments. The two species of *Deccanolestes* are rather similar, differing in size and a few morphological details (Prasad et al., 1994). *D. hislopi* is the better known of the two, being represented by dentulous jaw fragments in addition to isolated teeth. Unfortunately, the tooth surfaces are abraded and corroded (see Khajuria and Prasad, 1998), so that interpreting the details of coronal morphology is problematic. *Deccanolestes* was originally (Prasad and Sahni, 1988; Prasad et al., 1994) placed in the "Palaeoryctidae," based presumably on the concept of this family as a generally primitive group including ancestral "insectivores" of the Cretaceous and Early Tertiary. Molar morphology of *Deccanolestes* differs substantially from that of *Cimolestes* and similar taxa (herein removed from the Palaeoryctidae and placed in the Cimolestidae), mainly in the presumed retention of plesiomorphies. *Deccanolestes* is tentatively placed in the Archonta based on features of the referred ankle bones (Godinot and Prasad, 1994; Prasad and Godinot, 1994).

Superorder Insectivora Bowdich, 1821
Order Leptictida (McKenna, 1975)

Comment. McKenna (1975) erected Leptictida as a superorder. The ordinal rank was assigned to it by Novacek (1986b).

Family Gypsonictopsidae Van Valen, 1967b

Diagnosis (based on Van Valen, 1967b; Fox, 1979d). Differ from Leptictidae in having p4 with reduced paraconid but tall metaconid, subequal to protoconid; lower molars with taller trigonid; m3 entoconid and hypoconulid closely approximated; upper molars with paracone and metacone somewhat connate.

Genera. *Gypsonictops* Simpson, 1927b, type genus by monotypy.

Distribution. Late Cretaceous (Judithian–Lancian): North America.

Comments. The family Gypsonictopsidae was originally erected as a subfamily, Gypsonictopsinae, by Van Valen (1967b) and raised to family rank by Stucky and McKenna (1993). *Gypsonictops* has been consistently placed in the Leptictidae since it was first described (e.g., Simpson, 1927b, 1945). Van Valen (1967b) was the first to identify significant differences in its dentition from other members of the family; Novacek (1977) agreed, noting that it was autapomorphic with respect to Tertiary taxa, and suggested that it be placed in its own family, as was done later by Stucky and McKenna (1993).

Genus *Gypsonictops* Simpson, 1927b
(figure 13.24C)

Diagnosis. As for the family.

Species. *Gypsonictops hypoconus* Simpson, 1927b, type species; *G. illuminatus* Lillegraven, 1969; *G. lewisi* Sahni, 1972; *G. clemensi* Rigby and Wolberg, 1987; and taxa left in open nomenclature (Clemens, 1973b, 1995; Sloan and Russell, 1974; Carpenter, 1979; Clemens et al., 1979; Fox, 1989; Cifelli, 1990e; Eberle and Lillegraven, 1998b; Hunter and Archibald, 2002).

Distribution. Late Cretaceous (Judithian–Lancian): United States, New Mexico (Fruitland-Kirtland formations); Canada, Scabby Butte; Alberta, and numerous Lancian (and its equivalents) localities and formations in United States and Canada (see chapter 2 and table 2.27).

Comments. *Gypsonictops* was the first unambiguous eutherian to be described from the Cretaceous of North America (Simpson, 1927b, 1951). The dentition is well known (e.g., Lillegraven, 1969; Clemens, 1973a; Fox, 1979d): it is rather distinctive, being characterized by well-molarized premolars (including the penultimate) and upper molars with strong cingulae, the postcingulum being expanded into a hypocone; five premolars are variably present, of which the third (Px) is small and is sometimes lost ontogenetically (Clemens, 1973a). Because of its distinctiveness, relative abundance, and broad distribution, *Gypsonictops* has been proposed as a first appearance taxon for the Judithian (Cifelli, 1994; Cifelli et al., 2004).

Order Lipotyphla Haeckel, 1866
Suborder Soricomorpha Gregory, 1910
Family "Nyctitheriidae" Simpson, 1928f

Diagnosis (based on Butler, 1988b). Paraphyletic family of soricomorphs generally characterized by plesiomorphies. Advanced characters occurring among some Tertiary genera include shortened infraorbital canal, reduction of the canine, development of multicusped incisors, and broadening of the upper molar postcingulum and mandibular condyle.

Genera. *Paranyctoides* Fox, 1979d, and a number of Early Tertiary genera not treated herein.

Distribution. (Mesozoic distribution only) Late Cretaceous (Coniacian–Maastrichtian): Asia, Uzbekistan; North America, Canada and United States.

Comments. "Nyctitheriid" Lipotyphla are tentatively recognized from the Cretaceous, based on arguments presented by Fox (1979d, 1984b). Subordinal placement of the family is uncertain; we follow Krishtalka (1976), Novacek et al. (1985), and Rose and Gingerich (1987) in recognizing probable soricomorph affinities for "Nyctitheriidae," which have been suggested to be the most plesiomorphic of soricomorphs (Butler, 1988b).

Genus *Paranyctoides* Fox, 1979d
(figure 13.24B)

Diagnosis (based on Fox, 1979d, and Cifelli, 1990e). Small eutherians with dentition closely similar to that of Tertiary "Nyctitheriidae"; differ from presumed primitive eutherian condition in having upper molars with more anteroposteriorly expanded, less transverse protocone, lower cusps; strong, winged conules; pre- and postcingula and incipient hypocone present; stylar shelf broad on P4; stylar cusp D present on both P4 and upper molars; lower molars with relatively low cusps, smaller height differential between trigonid and talonid, broad talonid; paraconid arising low on the trigonid (and not appressed to metaconid as in many derived taxa); entoconid tall and anteroposteriorly developed; p4 resembles the primitive condition, lacking molarized trigonid found in many other taxa, but talonid tall and bladelike.

Species. *Paranyctoides sternbergi* Fox, 1979d, type species; *P. maleficus* Fox, 1984b; *P. megakeros* Lillegraven and McKenna, 1986; *P. aralensis* Nessov, 1993; and several taxa left in open nomenclature (Rigby and Wolberg, 1987; Cifelli, 1990e; Montellano, 1992; Archibald and Averianov, 2001b).

Distribution. Late Cretaceous (Coniacian–?Santonian): Asia, Uzbekistan; Late Cretaceous (Aquilan-Edmontonian): Canada, Alberta, and United States, Utah.

Comments. This genus is a rather rare element in Cretaceous faunas and is only known by posterior premolars and molars (most of which are isolated teeth) of the upper and lower dentitions (Fox, 1979d, 1984b; Lillegraven and McKenna, 1986; Cifelli, 1990e). Nessov (1993; see also Archibald and Averianov, 2001b) referred a species from the Coniacian of Uzbekistan to *Paranyctoides*—a placement that, if upheld by further study, indicates a Late Cretaceous dispersal of this genus between North America and Asia.

Superorder Ferae Linnaeus, 1758
Order Cimolesta McKenna, 1975
Family Cimolestidae Marsh, 1889a

Diagnosis. Primitive cimolestans characterized by anteroposteriorly short, transversely broad upper molars with anteroposteriorly compressed protocone, paracone much taller than metacone, and pre- and postcingula weak or lacking; upper premolars with preparacrista lacking but postparacrista strong and trenchant; lower premolars lack (p1–3) or variably have (p4) a small paraconid; anterior lower premolars anteriorly inclined, tending toward reduction, with p1 (and sometimes p2) single-rooted, and diastemata usually separating c, p1–2; lower molars generally with extreme height and width differential between the tall, broad trigonid and the low, narrow talonid.

Genera. *Cimolestes* Marsh, 1889a, type genus; *Batodon* Marsh, 1892; *Telacodon* Marsh, 1892; *Procerberus* Sloan and Van Valen, 1965; and several Cretaceous taxa left in open nomenclature from North America (Clemens et al., 1979; Fox, 1979d; Whitmore, 1985; Eberle and Lillegraven, 1998b; Eaton, Cifelli et al., 1999; Hunter and Archibald, 2002), South America (Crochet and Sigé, 1993), and Europe (Pol et al., 1992), and a number of Tertiary genera not treated here.

Distribution. (Mesozoic distribution only) Late Cretaceous (Judithian–Lancian): North America; possibly Late Cretaceous (Maastrichtian): Europe; Late Cretaceous (Maastrichtian)–early Paleocene: South America.

Comments. Cimolestidae are well known from the Lancian of North America (Lillegraven, 1969; Clemens, 1973a); their occurrence in rocks as old as Judithian is based on fragmentary remains (Fox, 1979d; Eaton, Cifelli et al., 1999). Similarly, specimens from the Cretaceous of South America and Europe (references cited earlier) are inadequate for definitive assignment.

As noted previously, it has been common practice to refer small, dentally primitive Cretaceous eutherians to "Palaeoryctidae" (see, e.g., Kielan-Jaworowska, Bown, and Lillegraven, 1979), a practice that continues (e.g., Prasad et al., 1994). The foregoing diagnosis is intended as a step toward redressing this situation. Although character polarities are not well established, we view Cimolestidae as apomorphic in a number of respects. Thus, for example, eutherians primitively had a noticeable discrepancy in height between trigonid and talonid, but it is substantially greater in cimolestids, which appear to be divergently specialized in regard to this and other aspects of molar shape (Butler, 1990a; Cifelli, 1999b). The specializations appear to be related to carnivory, particularly in the larger species (Lillegraven, 1969).

Genus *Cimolestes* Marsh, 1889a
(figure 13.25C)

Diagnosis. Cimolestids characterized by taller cusps, with greater height differential between paracone and metacone (upper molars) and trigonid and talonid (lower molars), than other Cretaceous members of the family; carnassial notches developed on accessory shearing surfaces of upper molars; p4 with unicusped talonid, metaconid lacking; cusp anterior to entoconid commonly present on lower molars.

Comment. Fox (1989) noted some variation within the genus and suggested that it may be a "grade."

Species. *Cimolestes incisus* Marsh, 1889a, type species; *C. magnus* Clemens and Russell, 1965; *C. cerberoides* Lillegraven, 1969; *C. propalaeoryctes* Lillegraven, 1969; *C. stirtoni* Clemens, 1973a; and a Puercan species not treated here; species left in open nomenclature (Clemens et al., 1979; Johnston and Fox, 1984; Fox, 1989; Eaton, Cifelli et al., 1999). "*Cimolestes*" *lucasi* Rigby and Wolberg, 1987, is based on a lower molar probably belonging to a marsupial (Cifelli, 1990e) and is not included here.

Distribution. Late Cretaceous (?Judithian–"Endmontonian"): Canada, Saskatchewan (Ravenscrag and Frenchman formations); Alberta (Scollard Formation); United States, Montana (Hell Creek Formation); Wyoming (Lance Formation).

Comments. *Cimolestes* represents one extreme within the Cretaceous Cimolestidae (in many of the characters cited in the familial diagnosis), the other being *Procerberus.* Species of *Cimolestes* are characterized by strong development of molar shearing crests and large canines (known in few other taxa and hence omitted from the diagnosis), presumably associated with a carnivorous diet. Lillegraven (1969; see also Van Valen, 1966) hypothesized origin of Creodonta, Carnivora, and Taeniodonta from species of *Cimolestes.* Little more than the dentition and a few fragments of the skull are known, but most species of *Cimolestes* are represented by relatively good series of specimens, including well-preserved jaws (Lillegraven, 1969; Clemens, 1973a).

Genus *Batodon* Marsh, 1892
(figure 13.25A)

Diagnosis. Extremely small cimolestid; differs from *Cimolestes* in having stronger pre- and postcingula on upper molars and a metaconid on p4, the latter cusp lower than in *Procerberus;* p4 talonid unicuspid and not transversely expanded, as in *Procerberus.*

Species. *Batodon tenuis* Marsh, 1892, type species by monotypy.

Distribution. Late Cretaceous (Lancian): Canada, Alberta (upper part of Edmonton Formation); United States, Wyoming (Lance Formation), Montana (Hell Creek Formation).

Comments. *Batodon* is a rare and poorly known component of Lancian faunas. In addition to the type, a dentary bearing the last three premolars (Marsh, 1892) and another jaw fragment with parts of the same teeth plus the first two molars (both from the type Lance, Clemens, 1973a), there are a few specimens, tentatively referred, from Trochu, Gryde, and several sites in the Hell Creek Formation (Lillegraven, 1969; Archibald, 1982; Storer, 1991; Hunter and Archibald, 2002). As noted by Clemens (1973a), scarcity of specimens belonging to *B. tenuis* may

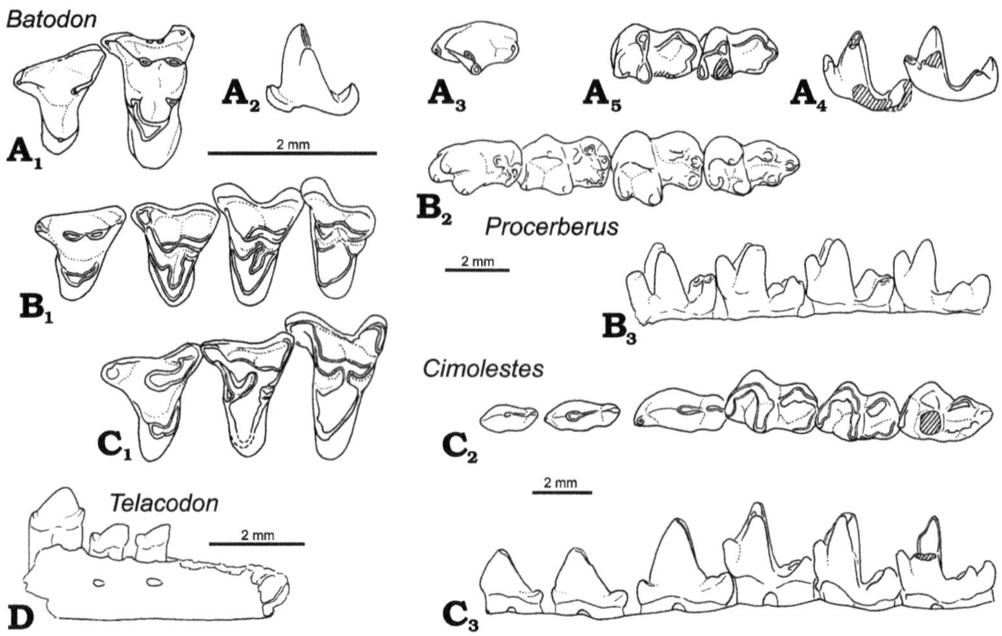

FIGURE 13.25. Ferae. A, *Batodon tenuis,* left P4 and M1 in occlusal view (A_1); right p4 in lingual (A_2) and occlusal (A_3) views; right m2, m3 in lingual (A_4) and occlusal (A_5) views. B, *Procerberus formicarum,* left P4, M1–3 in occlusal view (B_1); right p4, m1–3 in occlusal (B_2) and lingual (B_3) views. C, *Cimolestes incisus,* left P4, M1–2 in occlusal view (C_1); right p2–4, m1–3 in occlusal (C_2) and lingual (C_3) views. D, *Telacodon laevis,* fragment of right dentary with p2–4 in labial view. *Source:* original drawings based on specimens or casts.

well be due to its small size. Analysis of molar function and estimate of body size were recently presented by Wood and Clemens (2001), who concluded that *Batodon tenuis,* at an estimated 5 g adult body mass, is the smallest known Cretaceous eutherian.

Genus *Telacodon* Marsh, 1892
(figure 13.25D)

Diagnosis. Small eutherians, about the size of *Batodon* and apparently differing from that genus in having p1 double-rooted, proportionately longer p3 (compared to p2), and in having the mental foramina situated below p2 and p3, rather than p2 and p4.

Species. *Telacodon laevis* Marsh, 1892, type species by monotypy.

Distribution. Late Cretaceous (Lancian): United States, Wyoming (Lance Formation).

Comments. This tiny eutherian is known only by the type specimen of the type species, a dentary fragment that includes the last three premolars and some alveoli for more anterior teeth. Marsh (1892) speculated that five premolars were present; Simpson (1929a) disagreed, suggesting instead that the enlarged alveolus identified by Marsh as for the canine had housed an enlarged and procumbent incisor. If so, *Telacodon* would be rather atypical among Cimolestidae and, for that matter, among most other Cretaceous Eutheria. The type might profitably be reexamined in light of controversies surrounding the premolar count in early eutherians, noted previously, and in light of the additional specimens that have since been referred to *Batodon.* The latter is closely similar in size, and the significance (if any) of the distinctions noted in the diagnosis cannot be evaluated (Clemens, 1973a).

Genus *Procerberus* Sloan and Van Valen, 1965
(figure 13.25B)

Diagnosis. Molars lower crowned than in other Cimolestidae, with smaller height differential between paracone and metacone (upper molars) and trigonid and talonid (lower molars). Molar shearing surfaces less bladelike than in *Cimolestes* and other Cimolestidae; upper molar protocone more anteroposteriorly and less transversely developed. Premolars differ from those of other taxa in having less trenchant crests, with posterior premolars being more molarized: P3–4 with metacone consistently present; p4 with well-developed lingually placed paraconid and metaconid and basined multicusped talonid.

Species. *Procerberus formicarum* Sloan and Van Valen, 1965, type species by monotypy, and a species left in open nomenclature (Johnston and Fox, 1984).

Distribution. Early Paleocene (Puercan): Canada, Saskatchewan (Ravenscrag Formation); United States, Montana (Tullock Formation).

Comments. *Procerberus* is known by numerous specimens from the Bug Creek Anthills, Montana, and is now believed to be an early Paleocene element of that mixed assemblage (Lofgren, 1995, see chapter 2). Its tentative inclusion here is based on the occurrence of *Procerberus* cf. *P. formicarum* in what is interpreted as being the Cretaceous part of the Ravenscrag Formation, Saskatchewan (Fox, 1997b). As noted, the dentition of *Procerberus* represents an extreme in Cimolestidae, differing from *Cimolestes* in a number of features listed in the diagnoses. Indeed, *Procerberus* was first placed in the Leptictidae (see Sloan and Van Valen, 1965) and accorded an ancestral position within the family (Van Valen, 1967b). Its affinities have been debated ever since; our placement in Cimolestidae follows detailed point-by-point comparisons made by Lillegraven (1969). The differing degree of premolar molarization among genera included here in the Cimolestidae deserves some comment. The posterior premolars of *Procerberus* are rather strongly molarized (though less so than in *Gypsonictops*), but much simpler in *Cimolestes,* which, for example, has a simple p4. Given the relatively young geological age of these taxa, either condition could be ancestral within the group—hence Cimolestidae could have derived from an extremely primitive taxon (with a simpler premolar than Cenomanian *Otlestes*) or, alternatively, the condition in *Cimolestes* could represent secondary simplification, associated with what we interpret to be carnivorous specializations. Existing evidence is inadequate to judge between these alternatives, but perhaps most plausible is an intermediate interpretation: that the ancestral condition in the family may have resembled what is seen in *Batodon* and that *Procerberus,* on the one hand, and *Cimolestes,* on the other, independently acquired different apomorphic conditions.

Tarsals ascribed to *Procerberus* have been described by Szalay and Decker (1974). According to Szalay (1977), they are strikingly similar to those of Tertiary Leptictidae, showing great flexion-extension mobility in the upper ankle joint and great inversion-eversion capability in the lower ankle joint. These features indicate arboreal specializations and suggest that *Procerberus* may have been capable of reversing its foot, as seen in recent Sciuridae, Tupaiidae, and some Carnivora (see Jenkins and McLearn, 1984).

Superorder Ungulatomorpha Archibald, 1996b
Order *incertae sedis*

Diagnosis (based on Nessov, Archibald, and Kielan-Jaworowska, 1998, emended). Differ from other Late Cretaceous eutherians (including earliest anagalidans) and earliest primatomorphs (i.e., *Purgatorius*) in having the protocone anteroposteriorly expanded, upper molar crown in occlusal view only slightly constricted in conular region; parastylar region with two cuspules in earliest diverging taxa (also in *Kennalestes* and *Sailestes*), reduced to one or none in later taxa; paracone/metacone and protocone relatively far apart in earlier diverging taxa (in later diverging taxa protocone migrates labially), with conules closer to paracone/metacone. Paraconid of lower molars lingual to sublingual, with some appression to metaconid; entoconid and hypoconulid twinned (also in some zalambdalestids). Differ from other Cretaceous eutherians, but share with earliest primatomorphs upper molar crown shape (trapezoidal, subrectangular, rectangular, but not triangular in occlusal view; also seen in *Paranyctoides*); trigonid height lower relative to talonid height, talonid as wide or wider than trigonid (also in zalambdalestids). Differ from most Late Cretaceous eutherians, but share with *Gypsonictops* and earliest primatomorphs and some later anagalidans a narrow stylar shelf. Similar to *Kennalestes, Batodon, Gypsonictops,* and *Paranyctoides* in consistent presence of pre- and postcingula, but differ from these taxa and are (convergently?) like *Purgatorius* and *Tribosphenomys* (early rodent) in having pre- and postcingula that extend below the conules. Share with *Sailestes,* zalambdalestids, *Gypsonictops, Purgatorius,* and *Tribosphenomys* paracone and metacone of similar height. Share with *Kennalestes, Zalambdalestes, Gypsonictops, Purgatorius,* and *Tribosphenomys* ultimate upper premolar with metacone or swelling (metaconid or swelling on lowers). Share with *Paranyctoides, Otlestes, Sailestes, Zalambdalestes, Gypsonictops, Purgatorius,* and *Tribosphenomys* paracone and metacone with separate bases.

Taxa Assigned. Family "Zhelestidae" Nessov, 1985a, and grandorder Ungulata Linnaeus, 1766, not discussed herein except for *Protungulatum* and a few other taxa possibly occurring in the latest Cretaceous (see Nessov, Archibald, and Kielan-Jaworowska, 1998: 72, for definition and discussion of Ungulata).

Distribution. (Mesozoic distribution only) Late Cretaceous (Cenomanian–Maastrichtian): Asia, Uzbekistan, and Japan; Europe, France and Spain; North America, Canada, United States, Mexico; and possibly South America, Peru.

Family "Zhelestidae" Nessov, 1985a

Diagnosis. Paraphyletic family (see Nessov, Archibald, and Kielan-Jaworowska, 1998), based on isolated upper and lower teeth, fragments of maxillae and dentaries with teeth, and edentulous dentaries, as a rule not found in occlusion. Most of the diagnosis of Ungulatomorpha given earlier pertains to the "Zhelestidae" and is not repeated here.

Genera Assigned. *Zhelestes* Nessov, 1985a, type genus; *Alostera* Fox, 1989; *Aspanlestes* Nessov, 1985a; *Avitotherium* Cifelli, 1990e; *Eoungulatum* Nessov, Archibald, and Kielan-Jaworowska, 1998; *Gallolestes* Lillegraven, 1976; *Labes* Sigé in Pol et al., 1992; *Lainodon* Gheerbrant and Astibia, 1994; *Parazhelestes* Nessov, 1993; *Sorlestes* Nessov, 1985a.

Distribution. Late Cretaceous (Campanian–Maastrichtian): Asia, Uzbekistan and Japan; Europe, France and Spain; and North America, Canada, United States, and Mexico.

Genus *Zhelestes* Nessov, 1985a
(figure 13.26A)

Diagnosis. Five upper premolars. Very large single-rooted canine followed by a diastema, small, single-rooted P1 and Px, large double-rooted P2, P3 strong, piercing, P4 semimolariform, but without metacone. M1 and M2 narrow, with small stylar area and very weak ectoflexus, especially on M1.

Species. *Zhelestes temirkazyk* Nessov, 1985a, type species by monotypy. *Non: Z.* cf. *temirkazyk* Nessov, 1985a; *non: Z. bezelgen* Nessov, 1987, which possibly belong to *Aspanlestes aptap* (see later). Nessov, Archibald, and Kielan-Jaworowska (1998: figure 10) reconstructed the maxilla with teeth of *Zhelestes temirkazyk* together with the dentary of *Sorlestes budan,* suggesting that they might be conspecific.

Distribution. Late Cretaceous (late Turonian–Coniacian): Uzbekistan, Kyzylkum Desert (Bissekty Formation).

Genus *Aspanlestes* Nessov, 1985a
(figure 13.26F)

Synonym: *Ortalestes* Nessov, 1997 (see note added in proof in Nessov, Archibald, and Kielan-Jaworowska, 1998).

Diagnosis. Relatively well-known zhelestid, including only type species represented by numerous specimens: three fragmentary dentaries with molars or premolars, two maxillae with P4 and molars, and some isolated teeth. Differs from most "zhelestids" in having M2 with deeper ectoflexus and metastylar region more expanded labially; shares with *Avitotherium* incurvature on the anterior margin of the parastylar region. The following part of the diagnosis is based on that of *Aspanlestes aptap* Nessov, 1985a, by Nessov, Archibald, and Kielan-Jaworowska (1998: 50), with only some editorial modification and shortening. Compared with other "zhelestids": upper molar crowns slightly constricted through conular region, but similar to *Parazhelestes minor;* ectoflexus moderately deep; protocone with slight anteroposterior expansion, protocone with little or no labial shift of apex; smallest in size. Of these five characters only the last might be an autapomorphy. The first four are shared with some or all other Late Cretaceous *non*-"zhelestids." Upper dentition averages 83% of size of next larger "zhelestid," *Parazhelestes minor.* As in lower dentition of other "zhelestids": trigonid low relative to talonid, slight appression of paraconid and metaconid, pre- and postcingulids present, labial and lingual cingulids absent, entoconid and hypoconulid with some twinning, unlike in cf. *Eoungulatum kudukensis* and *Sorlestes budan,* m3 hypoconulid not posteriorly shifted and no suggestion of apical closure of cusps.

Species. *Aspanlestes aptap* Nessov, 1985a, type species by monotypy; see also synonym list given by Nessov, Archibald, and Kielan-Jaworowska (1998: 49). Upper dentition of *Zhelestes bezelgen* Nessov, 1987, probably belongs to *Aspanlestes aptap* (see Nessov, Archibald, and Kielan-Jaworowska, 1998: figure 8).

Distribution. Late Cretaceous (late Turonian–Coniacian): Uzbekistan, Kyzylkum Desert (Bissekty Formation).

Genus *Eoungulatum* Nessov, Archibald and Kielan-Jaworowska, 1998
(figure 13.26B,C)

Diagnosis (based on Nessov, Archibald, and Kielan-Jaworowska, 1998). Poorly known genus, based on the holotype (M2) of the type species and tentatively assigned lower molars. Compared with other "zhelestids": upper molar crown without ectoflexus; protocone with greatest anteroposterior expansion; protocone with greatest labial shift of apex; largest in size. Upper dentition is comparable in length, but about 6% wider than the slightly smaller "zhelestid," *Parazhelestes robustus.* Nessov, Archibald, and Kielan-Jaworowska (1998) tentatively attributed several isolated lower molars to *E. kudukensis* and commented that if the lower teeth are correctly referred to *E. kudukensis,* the lower dentition averages 8% longer and 20% wider than the slightly smaller *Sorlestes budan.* Nessov et al. (1998) also tentatively attributed a single P4 to *E. kudukensis.*

Species. *Eoungulatum kudukensis* Nessov, Archibald, and Kielan-Jaworowska, 1998, type species (M2), and tentatively assigned as cf. *E. kudukensis* several lower molars and P4 (figure 13.26C).

Distribution. Late Cretaceous (Coniacian): Uzbekistan, Kyzylkum Desert (middle part of the Bissekty Formation).

Genus *Parazhelestes* Nessov, 1993
(figure 13.26D)

Diagnosis (based in part on Nessov, Archibald, and Kielan-Jaworowska, 1998). Upper molar crowns only slightly (*Parazhelestes minor*) or not constricted through

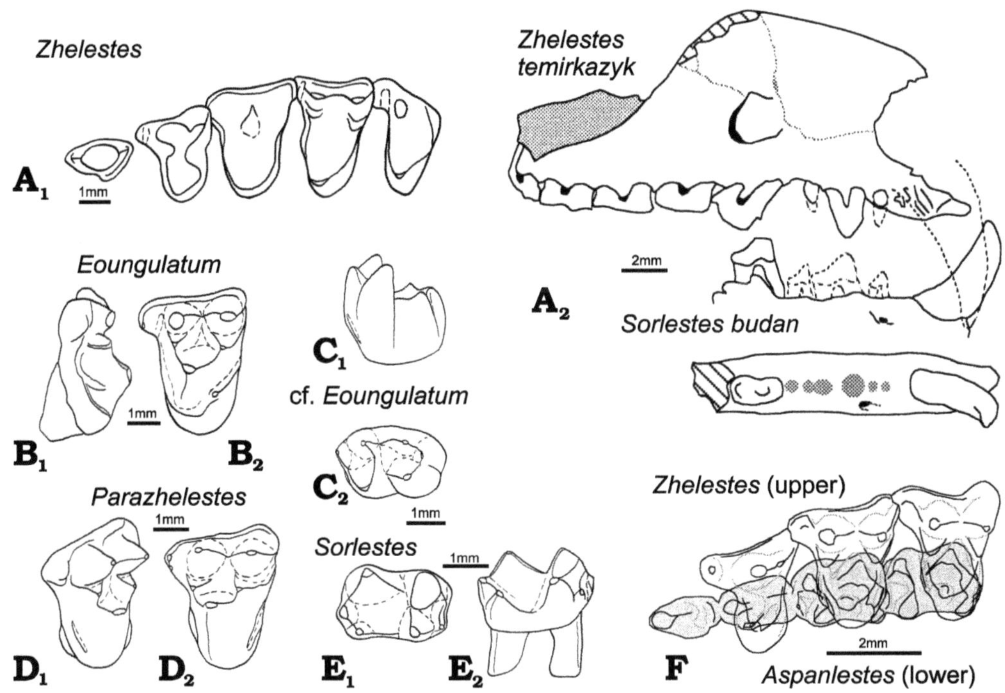

FIGURE 13.26. Selected teeth of Asian "Zhelestidae." A, *Zhelestes,* schematic drawing of the left upper teeth of *Zhelestes temirkazyk* (A_1), and lateral view of the left maxilla of *Z. temirkazyk* compared to the dentary of *Sorlestes budan* (A_2); A_2 reversed. Reconstructed teeth are shown by dotted lines, shadings are broken areas. B, *Eoungulatum kudukensis,* left M2 in mesial (B_1) and occlusal (B_2) views. C, Cf. *Eoungulatum kudukensis,* left m3 in labial (C_1) and occlusal (C_2) views. D, *Parazhelestes robustus,* left M1 in mesial oblique (D_1) and occlusal (D_2) views. E, *Sorlestes budan,* left m2 in occlusal (E_1) and oblique-lingual (E_2) views. F, Superimposed occlusal outlines of the maxillary dentition of *Zhelestes bezelgen* and the mandibular dentition (reversed) of the holotype of *Aspanlestes aptap.* The two specimens probably belong to the same species. *Source:* A_2 and B modified from Nessov, Archibald, and Kielan-Jaworowska (1998); all others original drawings based on specimens or casts.

conular region (*P. robustus*) as is also the case in other Asian "zhelestids." Shares with *Eoungulatum* a well-defined, roughly semicircular parastylar region and with *Aspanlestes* the structure of metastylar region, which strongly protrudes labially. Protocone intermediate among "zhelestids" in degree of both anteroposterior expansion and labial shift of apex; intermediate in size among "zhelestids." M3 markedly linguolabially narrower compared to M1–2 in *P. robustus* (condition unknown in *P. minor*).

Species. *Parazhelestes robustus* Nessov, 1993, type species, and assigned *P. minor* Nessov, Archibald, and Kielan-Jaworowska, 1998.

Distribution. Late Cretaceous (Coniacian): Uzbekistan, Kyzylkum Desert (Bissekty Formation).

Genus *Sorlestes* Nessov, 1985a
(figure 13.26A_2,E)

Diagnosis (based on Nessov, Archibald, and Kielan-Jaworowska, 1998). Lower dentition genus. As in lower dentition of other "zhelestids": trigonid low relative to talonid, slight appression of paraconid and metaconid, pre- and postcingulids present, labial and lingual cingulids absent, entoconid and hypoconulid with some twinning. Lower dentition some 22% larger than the smallest "zhelestid," *Aspanlestes aptap,* and about 20% smaller than cf. *Eoungulatum kudukensis.* Nessov, Archibald, and Kielan-Jaworowska (1998: 63) concluded with respect to the type species: "The only possible autapomorphy is the intermediate size of this species, but as many as three species could be included."

Species. *Sorlestes budan* Nessov, 1985a, type species (probably including *Zalambdalestes mynbulakensis* Nessov, 1985b); *S. kara* Nessov, 1993 (see also Nessov, Sigogneau-Russell, and Russell, 1994: plate 7, figure 3); *S. mifunensis* Setoguchi, Tsubamoto, et al., 1999. See also an earlier note under *Zhelestes* regarding possible conspecificity of *Sorlestes budan* and *Zhelestes temirkazyk.*

Distribution. Late Cretaceous (late Cenomanian–early Turonian): Japan, Kumamoto Prefecture, Mifune town ("Upper Formation," Mifune Group); Late Cretaceous (early Turonian): Kazakhstan, Chimkent Province (unnamed beds); Late Cretaceous (late Turonian–Coniacian): Uzbekistan, Kyzylkum Desert (Bissekty Formation).

Genus *Alostera* Fox, 1989
(figure 13.27A)

Diagnosis. Poorly known monotypic genus, represented by several isolated M1 and M2 of the type species, and tentatively assigned m1. M1 resembles those of the "zhelestid" genera *Aspanlestes, Zhelestes, Parazhelestes,* and *Avitotherium* in almost complete lack of ectoflexus, but differs from them in having more symmetrical stylar shelf, with parastylar region less strongly protruding anteriorly. M2 differs from *Aspanlestes, Zhelestes, Parazhelestes,* and *Eoungulatum* in having metastylar region less strongly protruding labially and from *Eoungulatum* in being more constricted in conular region, a character shared with *Aspanlestes* and *Avitotherium.* Differs from *Aspanlestes* in having relatively more extensive postcingulum. Differs from most "zhelestids" in presence of a mesostyle (at least on some molars) and shares this character with *Paranyctoides.*

Species. *Alostera saskatchewanensis* Fox, 1989, type species by monotypy.

Distribution. Late Cretaceous (Lancian): Canada, Saskatchewan (Frenchman Formation), Alberta (Scollard Formation); United States, Montana (Hell Creek Formation).

Genus *Avitotherium* Cifelli, 1990e
(figure 13.27B)

Diagnosis. Monotypic genus represented by several isolated upper and lower molars and fragment of a mandible with broken premolars and roots of other teeth of the type species. Upper molars (M1 or M2) share with M1 of *Aspanlestes, Parazhelestes, Zhelestes, Alostera,* and *Eoungulatum* almost complete lack of ectoflexus and mesostylar region not protruding labially. Differ from them in having more symmetrical stylar region, owing to the relatively smaller parastylar lobe. An apomorphy is an incurved anterior margin of the parastylar lobe between the two cusps (see also incurvature in this region in *Aspanlestes*).

Species. *Avitotherium utahensis* Cifelli, 1990e, type species by monotypy.

Distribution. Late Cretaceous (Judithian): United States, Utah (lower part of Kaiparowits Formation).

Genus *Gallolestes* Lillegraven, 1976
(figure 13.27C)

Diagnosis. Lower dentition genus, but uncertain fragments of upper molars are also known. Height and width differential between trigonid and talonid less than in most other eutherians from the Late Cretaceous of North America, from which it also differs in lingual placement of hypoconulid and better development of hypoflexid.

Species. *Gallolestes pachymandibularis* Lillegraven, 1976, type species; *G. agujaensis* Cifelli, 1994 (see also Lillegraven, 1972; Clemens, 1980b; Nessov, Archibald, and Kielan-Jaworowska, 1998).

Distribution. Late Cretaceous (Campanian): Mexico, Baja California del Norte ("El Gallo Formation"); Late Cretaceous (Judithian): United States, Texas (Aguja Formation).

Comments. The affinities of *Gallolestes* have been debated, in part because of uncertain tooth homologies in

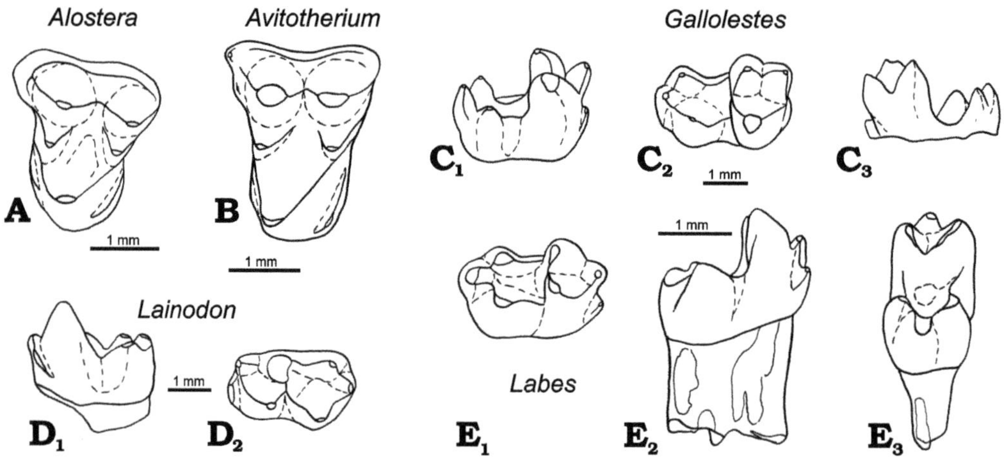

FIGURE 13.27. Selected teeth of non-Asian "Zhelestidae." A, *Alostera saskatchewanensis,* left M1 (holotype) in occlusal view. B, *Avitotherium utahensis,* left M1 or M2 (holotype) in occlusal view. C, *Gallolestes pachymandibularis;* right m2 in oblique labial (C_1), occlusal (C_2), and lingual (C_3) views. D, *Lainodon orueetxebarriai,* left m1 (holotype) in labial (D_1) and occlusal (D_2) views. E, *Labes quintanillensis,* right m3 (holotype) in occlusal (E_1), labial (E_2), and posterior (E_3) views. *Source:* original drawings based on casts or photographs.

the type specimen of *G. pachymandibularis* (see Lillegraven, 1976; Clemens, 1980b). We follow Butler (1977, 1990a) in considering the tooth preceding the first unquestioned molar to be a molariform deciduous tooth. Upper molars referred to *G. agujaensis* bear strong pre- and postcingula, supporting eutherian affinities for the genus, which we place in Ungulatomorpha following arguments presented by Nessov, Archibald, and Kielan-Jaworowska (1998).

Genus *Labes* Sigé in Pol et al., 1992
(figure 13.27E)

Diagnosis (based on Pol et al., 1992). Lower dentition genus. Trigonid moderately high and moderately mesiodistally constricted; paraconid conical, mesially projecting, and submedian. Talonid long, wide, and high, with massive and bulbous cusps; hypoconid important and median; precingulid isolated and prominent; m3 not very reduced.

Species. *Labes quintanillensis* Sigé in Pol et al., 1992, type species; *Labes garimondi* Sigé in Pol et al., 1992, known previously as "Champ-Garimond tooth" (Ledoux et al., 1966).

Distribution. Late Cretaceous (Maastrichtian): Spain, Burges Province, Laño; Late Cretaceous (Campanian) France, Gard (Champ-Garimond Beds).

Genus *Lainodon* Gheerbrant and Astibia, 1994
(figure 13.27D)

Diagnosis. Lower dentition genus similar to *Labes*, from which it differs in having generally more bunodont structure of lower molars (only ?m1 and ?m2 of the type species are known), trigonid incurved mesially, paraconid placed more lingually, protoconid distinctly bigger than paraconid and metaconid, stronger postmetacristid, shorter talonid, precingulid less incurved.

Species. *Lainodon orueetxebarriai* Gheerbrant and Astibia, 1994, type species by monotypy.

Distribution. Late Cretaceous (Maastrichtian): Spain, Burges Province, Laño.

Order "Condylarthra" Cope, 1881a

Diagnosis (based on Nessov, Archibald, and Kielan-Jaworowska, 1998). Dentally primitive ungulates distinguished from "zhelestids" in having upper molar conules closer to protocone than to midpoint of the tooth, with internal cristae weak or absent; parastylar region, groove, and lobe reduced, with a single cusp present; metacingulum formed by postmetaconular crista continuing onto the metastylar region; upper molars more rectangular in occlusal view; premolars reduced to four; mandibular condyle dorsal to the occlusal surface of the tooth row.

Distribution. ?Late Cretaceous: North America; Early Tertiary: North America, South America, Europe, and Asia.

Families. "Arctocyonidae" Giebel, 1855; Periptychidae Cope, 1882d; and families restricted to the Tertiary and not treated herein.

Comment. It has been widely recognized that "Condylarthra," as traditionally conceived (see review by Cifelli, 1983) are a paraphyletic assemblage of generally primitive ungulates, and for this reason we refer to this group in quotes. Despite problems inherent in usage of the term (e.g., Prothero et al., 1988; Archibald, 1998), most "condylarths" have continued to defy well-supported placement in monophyletic groups (Muizon and Cifelli, 2000), a problem that cannot be resolved herein. For a discussion of other characters used to diagnose Ungulata, see Archibald (1998).

Protungulatum and other "condylarths" occur at the Bug Creek Anthills and nearby sites in Montana. Once thought to be of Cretaceous age (see review by Clemens et al., 1979), these fossils are now interpreted as being Paleocene elements of time-averaged fluvial deposits (Lofgren, 1995). The presence of *Protungulatum* and other Puercan ungulates in the Cretaceous is now based on their presence at two sites in Saskatchewan, where they co-occur with other species that are typical of the Lancian (Fox, 1997b). The ages of these sites have been contended; we provisionally regard them to be latest Cretaceous (see discussion by Cifelli et al., 2004).

Family "Arctocyonidae" Giebel, 1855

Diagnosis. Dentally primitive ungulates sharing basic apomorphies of Ungulata, differing from remaining "condylarths" in lacking specializations of the respective families, including lesser degree of premolar molarization, lack of marked cusp inflation, lesser degree of paraconid reduction and appression to metaconid, and lesser development of hypocone.

Distribution. ?Late Cretaceous: North America; Early Tertiary: North America, Europe, and Asia.

Genera. *Baioconodon* Gazin, 1941a; *Oxyprimus* Van Valen, 1978; *Protungulatum* Sloan and Van Valen, 1965; and various Tertiary genera not treated herein.

Comment. We herein recognize "Arctocyonidae" as including unspecialized "condylarths"; as such, the family cannot be properly diagnosed. Characterization of "Arctocyonidae" is given by Matthew (1937), but this is not useful in the present context.

Genus *Baioconodon* Gazin, 1941a
(figure 13.28A)

Diagnosis (cited after Eberle and Lillegraven, 1998b: 60). "Relatively large compared with most contemporary condylarths (see Middleton, 1983); cheek teeth low-crowned and bulbous, but not to the extent in *Loxolophus;* enamel rugose; upper molars and P4 relatively transverse (Van Valen, 1978), compared with other loxolophines; stylar cingulum strong; well-developed parastylar lobe on M1–3; metastylar lobe best developed on M2 labial to metacone; mesostyle commonly present on M2; strong para- and metacristae on M1 and M2; well-developed pre- and postcingula; protoconal apex central; hypocone variable in size on M1 and M2, lacking on M3; internal conular wings usually weak (Middleton, 1983); inflated molar trigonids; molar paraconid intermediate in position between that of *Loxolophus* and *Protungulatum;* molar paraconid lingual to groove separating metaconid and protoconid; molar talonids relatively small, intermediate in size between *P. donnae* and *Loxolophus* (see Middleton, 1983); unlike *Loxolophus,* m2 trigonid longer and usually wider than talonid."

Species. Six species, including the type species *B. denverensis* Gazin, 1941a, all of Tertiary age and not treated herein. The single possible Cretaceous record of the genus is that of *Baioconodon* sp., evidently representing a new but currently undiagnosable species, from Saskatchewan (Johnston and Fox, 1984; Fox, 1989).

Distribution. Early Paleocene (Puercan): Canada, Saskatchewan (Ravenscrag Formation).

Comments. *Baioconodon* has been linked to Cete by Prothero et al. (1988). The only dental character supporting this allocation, inflation of molar cusps, is of somewhat dubious utility. However, phylogenetic affinities of major ungulate clades are now supported by a host of cranial characters (Prothero et al., 1988; Archibald, 1998). A reasonably complete but unpublished skull is known for Puercan *Baioconodon nordicus,* offering the potential to test the hypothesis of Prothero et al. (1988).

Genus *Oxyprimus* Van Valen, 1978
(figure 13.28E)

Diagnosis (based on Archibald, 1982, Middleton, 1983, and Luo, 1991). Differs from other "arctocyonids" in having a more vertical protocone, with shorter lingual slope, on upper molars; lower molars with greater height differential between trigonid and talonid. Closely similar to *Protungulatum,* with similar cusp height, differing in having a narrower p4 with large and divergent paraconid and metaconid; p4 longer relative to m1; relatively longer and wider talonid on m1–2; lower molars with more closely appressed paraconid and metaconid, tendency toward closure of talonid notch, and more weakly developed pre- and postcingulids; m3 talonid relatively shorter; upper molars transversely developed, with internal conular cristae and nearly straight pre- and postprotocristae.

Species. *Oxyprimus erikseni* Van Valen, 1978, type species, and several other Paleocene species not treated herein. The only reported Cretaceous occurrence of the genus is that of *Oxyprimus* cf. *erikseni.*

Distribution. Early Paleocene (Puercan): Canada, Saskatchewan (Ravenscrag Formation).

Comment. *Oxyprimus* possesses a mixture of primitive and advanced dental characters, and its relationships are uncertain. Comparative study by Luo (1991) suggests that *Oxyprimus* retains a number of plesiomorphies (e.g., relative height of trigonid to talonid on lower molars, short lingual slope on upper molars) compared to *Protungulatum.* Archibald (1982, 1998), on the other hand, identified at least one apomorphy (relative length of p4 to m1) indicating affinities to Hyopsodontidae.

Genus *Protungulatum* Sloan and Van Valen, 1965
(figure 13.28B)

Diagnosis (based on Luo, 1991, and Archibald, 1998, emended). Differs from closely similar taxa (e.g., *Oxyprimus*) in having m2–3 with more lingually placed paraconid, upper molars with larger, more rounded conules, lacking internal cristae, and more procumbent hypoconulid on m3. Protocone of upper molars more labially inclined, with longer lingual slope, than in *Oxyprimus,* but less than in *Baioconodon* or Periptychidae. Differs from most "Arctocyonidae" in lacking a complete lingual cingulum on upper molars, from Hyopsodontidae in having a lesser degree of premolar molarization, and from *Baioconodon,* Mioclaenidae, and Periptychidae in lacking notable inflation of molar cusps.

Species. *P. donnae* Sloan and Van Valen, 1965, type species, and various species restricted to the Paleocene and not treated herein. The only ?Cretaceous occurrence of the genus is that of *P.* cf. *donnae.*

Distribution. Early Paleocene (Puercan): Canada, Saskatchewan (Ravenscrag and Frenchman formations).

Comments. The dentition of *Protungulatum* is well known by abundant specimens from the Bug Creek assemblages and elsewhere (e.g., Sloan and Van Valen, 1965; Archibald, 1982; Luo, 1991; Lofgren, 1995; Eberle and Lillegraven, 1998b). In addition, isolated tarsals (Szalay and Decker, 1974) and petrosals (MacIntyre, 1972) have been referred to the genus. The type species, *P. donnae,* is the first appearance datum for the Puercan land-mammal age (Archibald and Lofgren, 1990; Cifelli et al., 2004).

Family Periptychidae Cope, 1882d

Diagnosis (based on Cifelli, 1983, and Archibald, 1998). Upper molars with base of protocone somewhat to significantly expanded lingually and apex shifted labially, producing a long lingual slope on this cusp; hypocone de-

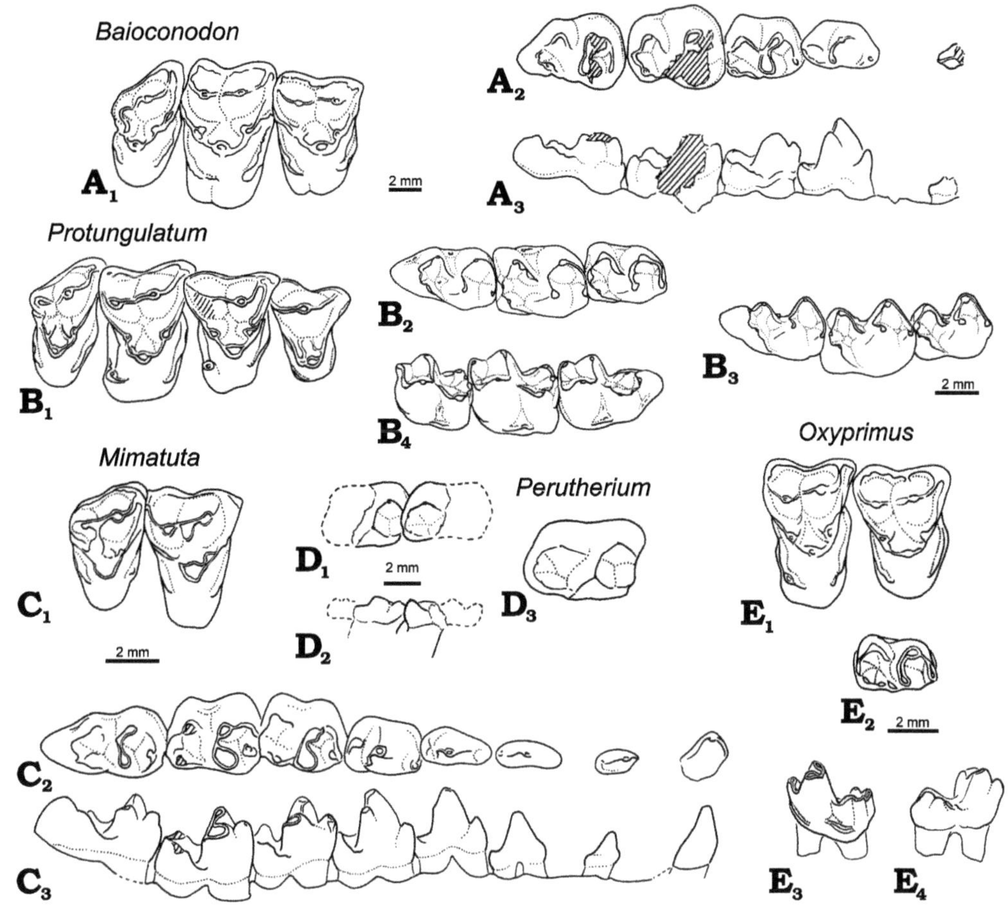

FIGURE 13.28. "Condylarthra." A, *Baioconodon nordicus,* right M1–3 in occlusal view (A_1); fragment of left p2, p4, and m1–3 in occlusal (A_2) and lingual (A_3) views. B, *Protungulatum donnae,* right P4, M1–3 in occlusal view (B_1); left m1–3 in occlusal (B_2), oblique lingual (B_3), and oblique labial (B_4) views. C, "*Mimatuta*" *morgoth,* right M2–3 in occlusal view (C_1); left c, p1–4, m1–3 in occlusal (C_2) and lingual (C_3) views. D, *Perutherium altiplanense.* Holotype, left m1, m2 in occlusal (D_1), lingual (D_2), and restored occlusal (D_3) views. E, *Oxyprimus erikseni,* right M1, M2 in occlusal view (E_1); left m1 in occlusal (E_2), labial (E_3), and lingual (E_4) views. *Source:* original drawings, based on specimens or casts.

velops in extremely lingual position; internal conule cristae salient and buccally directed. P4 with enlarged protocone; posterior upper and lower premolars slightly to greatly inflated. Lower molars with lingual and, variably, labial cingulids; paraconid in a median position and often lost; lower molar cusps tend to converge apically.

Genera. *Periptychus* Cope, 1881a, type genus, and other Tertiary genera not treated herein; the only named genus possibly occurring in the Cretaceous is "*Mimatuta*" Van Valen, 1978.

Distribution. Early Paleocene (Puercan): Canada, Saskatchewan (Ravenscrag Formation); Paleocene: ?Asia and ?South America.

Comments. Periptychids were among the first ungulate groups to develop strong dental specializations in the Paleocene, and they are both abundant and characteristic elements of North American faunas of the Puercan (e.g., Matthew, 1937; Archibald et al., 1987). In addition to "*Mimatuta,*" noted above, an unidentified genus of ?Periptychidae has been reported from the ?Cretaceous of Saskatchewan (Fox, 1989).

Genus "*Mimatuta*" Van Valen, 1978
(figure 13.28C)

Diagnosis (based on Archibald et al., 1983, Cifelli, 1983, Luo, 1991, and Archibald, 1998). Possibly paraphyletic, primitive periptychids differing from other basal ungulates (e.g., *Protungulatum, Oxyprimus*) in presence of a transverse crest and well-formed basin on the talonid of p4; upper molar protocone labially shifted and with longer lingual slope, protocristae nearly straight; lower molar cusps somewhat appressed, with talonid short relative to width. Differs from other Periptychidae except *Oxyacodon* in retention of stronger stylar shelves, styles, and distinct conules on upper molars and lesser appression of cusps on lower molars. Differs from *Oxyacodon* in having smaller, less lingually situated hypocone on upper molars; in retaining a protocone on P3; and in having lower molars with more strongly developed paraconid that is less appressed to metaconid.

Species. "*Mimatuta*" *morgoth* Sloan and Van Valen, 1978, type species, and one or more other species, all Paleocene in age and not treated herein.

Distribution. Early Paleocene (Puercan): Canada, Saskatchewan (Ravenscrag Formation); Paleocene: ?South America.

Comment. The possible presence of this genus in South America is based on a maxilla fragment with p3–4, referred to cf. *Mimatuta?*, from the Santa Lucía Formation (early Paleocene) of Tiupampa, Bolivia (Muizon, 1991). The presence at Tiupampa of pantodont and mioclaenid condylarths closely similar to those of the North American Puercan suggests a proximal North American origin for many eutherians of the fauna (Van Valen, 1988; Muizon and Marshall, 1992; Muizon and Cifelli, 2000).

?Order Notoungulata Roth, 1903

Diagnosis. Cifelli (1993c: 203) characterized the order as Ungulata with the following synapomorphies: "upper molars with distinctive coronal pattern (protoloph, metaloph, ectoloph, and, most importantly, crochet extending anterolabially from the metaloph); lower molars lacking paraconid and with paracristid short, extending anteriorly from the protoconid; lower molars with entoconid transversely expanded into entolophid; posterior root of zygomatic arch originating high on the side of the skull; epitympanic sinus developed in squamosal; prominent vagina processus hyoidei; tubular ectotympanic with crista meatus; canal of Hugier opening externally at the posterior end of the fissura Glaseri; and astragalus with medial protuberance, long, constricted neck, and sulcus extending laterally from the superior astragalar foramen."

Families. ?Perutheriidae Van Valen, 1978, and a number of Cenozoic families not treated herein.

Distribution. ?Late Cretaceous (Maastrichtian): Peru; Cenozoic: South America.

Comment. With exclusion of Holarctic Arctostylopidae (Cifelli et al., 1989), Notoungulata are entirely restricted to South America and, with one possible and rather dubious exception noted later, are known only from the Paleocene–Pleistocene.

Family Perutheriidae Van Valen, 1978
Genus *Perutherium* Grambast, Martinez, Mattauer, and Thaler, 1967
(figure 13.28D)

Diagnosis (quoted portion translated from Marshall et al., 1983: 147). Poorly known dental genus, represented by a dentary fragment with the talonid of m1 and the trigonid of m2, plus a referred protocone of an upper molar. "Molars brachyodont; premetastylid and postmetastylid small but distinct, conical; anterolophid obliquely oriented and constituted of protoconid, metaconid, and metastylid, equally spaced, aligned, and united by a protocristid and postmetacristid, respectively; postmetastylid lingual and posterior to metaconid; talonid unspecialized, with weakly developed cristid obliqua and postcristid."

Species. *P. altiplanense* Grambast et al., 1967, type species by monotypy.

Distribution. Late Cretaceous (?Maastrichtian): Peru, Department of Puno (Vilquechico Group). The age of the site, Laguna Umayo, is disputed; herein we tentatively accept a Late Cretaceous (Maastrichtian) age (see chapter 2).

Comments. Ungulate affinities for this poorly known genus are generally (though not universally, see Hoffstetter, 1981) recognized, but there is little agreement beyond this point. *Perutherium* has been suggested to be an "arctocyonid" (Grambast et al., 1967; Sigé, 1972), didolodontid (Tedford, 1974), periptychid (Van Valen, 1978), or indeterminate (Cifelli, 1983) "condylarth." The most detailed study thus far, that of Marshall et al. (1983), proposes that *Perutherium* is a notoungulate, a view adopted herein, partly for lack of a more satisfactory alternative. The basis for this referral is the presence of pre- and postmetastylids on lower molars. Perutheriinae were originally proposed (Van Valen, 1978) as a subfamily of Periptychidae, later raised to familial status with transferral to Notoungulata (Marshall et al., 1983).

CHAPTER 14

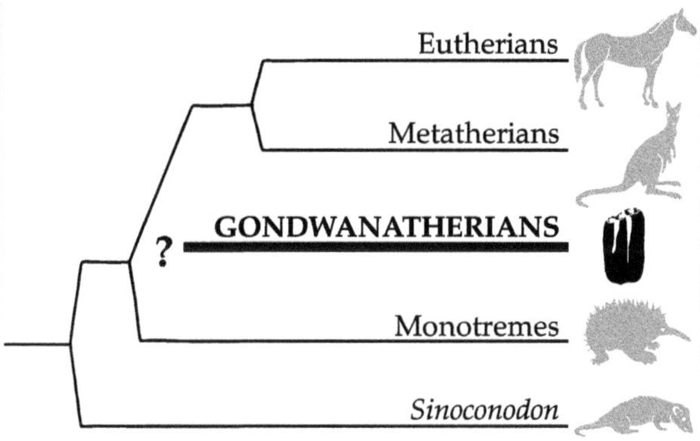

Gondwanatherians

INTRODUCTION

Gondwanatheria are an enigmatic group of Late Cretaceous and early Paleocene mammals from Gondwana landmasses (figures 14.1, 14.2). The two South American gondwanatherian genera, characterized by hypsodont molars and very thick enamel (the Late Cretaceous *Gondwanatherium* and the early Paleocene *Sudamerica*) were initially assigned to Edentata (e.g., Scillato-Yané and Pascual, 1984, 1985; Bonaparte, 1986a,b, 1990). Bonaparte (1986a) described m2 of a brachyodont Late Cretaceous taxon, *Ferugliotherium*, which he assigned to Multituberculata. Mones (1987) proposed an edentate order Gondwanatheria for *Gondwanatherium* and *Sudamerica*, but Krause and Bonaparte (1990) erected suborder Gondwanatheria within Multituberculata, to include hypsodont Sudamericidae (*Sudamerica* and *Gondwanatherium*) and brachyodont Ferugliotheriidae (*Ferugliotherium*). Krause, Kielan-Jaworowska, and Bonaparte (1992) described all isolated teeth attributed to *Ferugliotherium* then known, and argued that the genus belonged to Multituberculata. Krause and Bonaparte (1993) replaced Gondwanatheria by a superfamily Gondwanatherioidea, which they assigned tentatively to the suborder ?Plagiaulacida within Multituberculata. Kielan-Jaworowska and Bonaparte (1996) described a fragmentary dentary with p4 of multituberculate pattern, which they tentatively assigned to *Ferugliotherium*. Krause, Prasad et al. (1997) described the poorly known gondwanatherian genus *Lavanify* from the Late Cretaceous of Madagascar, which they assigned to the subclass ?Allotheria, tacitly implying that Gondwanatheria might be a sister group of multituberculates. Bonaparte (1988) studied the hypsodonty of *Gondwanatherium*, and Sigogneau-Russell et al. (1991), Krause, Prasad et al. (1997), Koenigswald et al. (1999), and Patnaik et al. (2001) investigated the enamel microstructure of Gondwanatheria. Koenigswald et al. (1999) concluded that hypsodonty and enamel microstructure had been acquired by Gondwanatheria independently from that of therian mammals.

The multituberculate affinities of gondwanatherians were challenged by the study of Pascual et al. (1999), who described a dentary with two molariform teeth and two more molar loci posterior to them (figure 14.2) belonging to *Sudamerica*, from the Paleocene of Argentina. *Sudamerica* apparently had four molariform teeth, a condition unknown for multituberculates. Therefore, Pascual et al. (1999) placed Gondwanatheria in Mammalia *incertae sedis*. Relationships of Gondwanatheria to other mammals are unresolved, although it is evident that this group is monophyletic with several highly distinctive dental autapomorphies.

We accept the assignment of Pascual et al. (1999), but exclude from the Ferugliotheriidae the dentary with a bladelike ?p4 and upper premolars of multituberculate pattern, tentatively assigned, respectively, by Kielan-Jaworowska and Bonaparte (1996) and by Krause, Kielan-Jaworowska, and Bonaparte (1992: figure 2C–F) to *Ferugliotherium*. We describe these fossils under Multituberculata *incertae sedis* (see our figure 8.44). It cannot be unequivocally demonstrated whether these fragments and the molars assigned here to Ferugliotheriidae belong to the same taxonomic unit, pending discovery of more complete material of *Ferugliotherium*.

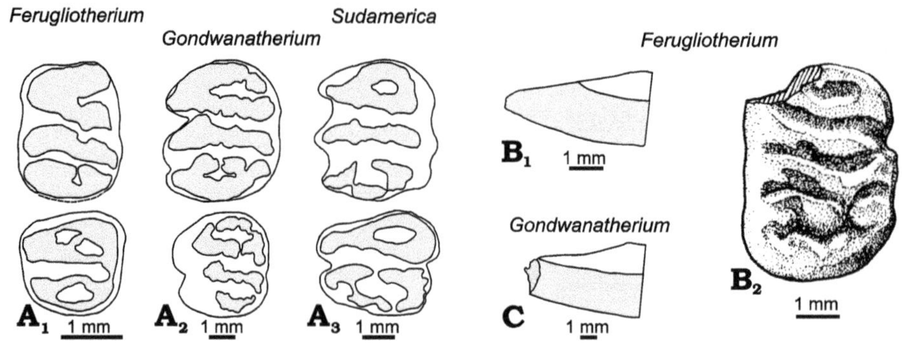

FIGURE 14.1. A, comparison of occlusal morphology of right m1 and m2 in *Ferugliotherium windhauseni* (A_1), *Gondwanatherium patagonicum* (A_2), and *Sudamerica ameghinoi* (A_3), showing similar structure and wear pattern. Anterior is up. White areas indicate unworn enamel and, if present, cementum. Gray areas represent outlines of worn enamel and, if present, exposed dentine. B, *Ferugliotherium windhauseni.* C, *Gondwanatherium patagonicum.* B_1, C, comparison of lower incisor morphology in *Ferugliotherium* (B) and *Gondwanatherium* (C). B_2, right M1 of *Ferugliotherium* in occlusal view. *Source:* A, B_1, C, modified from Krause and Bonaparte (1993); B_2, original.

SYSTEMATICS

(table 14.1)

Mammalia *incertae sedis*
Suborder Gondwanatheria Mones, 1987 (Krause and Bonaparte, 1990)

Diagnosis. Gondwanatheria are poorly known, small mammals with a rodentlike dentary. They share with multituberculates propalinal jaw movement with backward power stroke and with some multituberculates (Ptilodontoidea) normal prismatic enamel. They differ from multituberculates (except for the Late Cretaceous North American ?cimolomyid *Essonodon*) in having transverse ridges connecting the cusps of adjacent rows and transverse furrows between the ridges. The most important difference concerns the number of lower molariform teeth (four in Gondwanatheria and two in Multituberculata) and in absence of bladelike lower premolars in Gondwanatheria (the latter are an apomorphy of Multituberculata).

Families. Ferugliotheriidae Bonaparte, 1986a and Sudamericidae Scillato-Yané and Pascual, 1984.

Distribution. Late Cretaceous (Campanian)–early Paleocene: Argentina; Late Cretaceous: Madagascar and India.

Family Ferugliotheriidae J. F. Bonaparte, 1986a

Diagnosis. Poorly known family of Gondwanatheria, represented by isolated fragmentary upper and lower in-

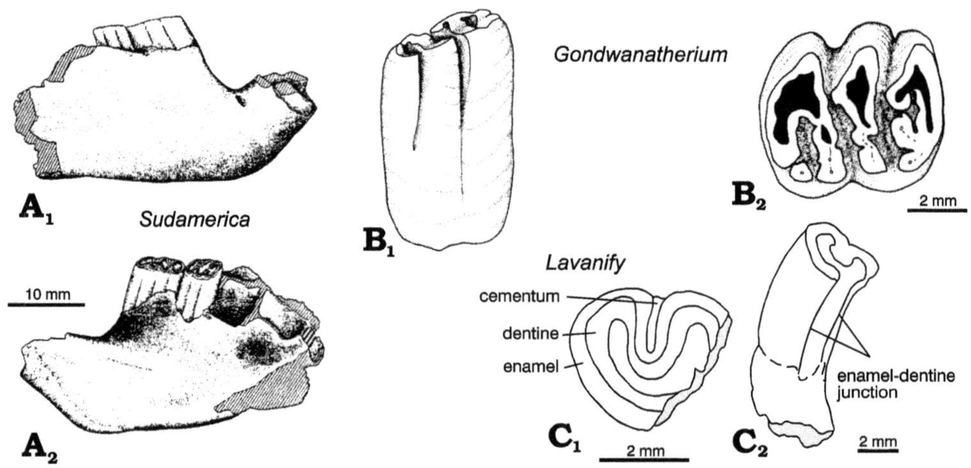

FIGURE 14.2. Representatives of Sudamericidae. A, *Sudamerica ameghinoi,* right dentary bearing root of the incisor, separated by a large diastema from two molariform cheek teeth, and alveoli for two cheek teeth distally, in labial (A_1), and lingual (A_2) views. B, *Gondwanatherium patagonicum,* postcanine tooth (holotype), presumed labial view (B_1), occlusal view of the same (presumed labial side down), (B_2). C, *Lavanify miolaka,* cheek tooth, holotype, in occlusal (C_1), and left-side (C_2) views. *Source:* modified from: A, Pascual et al. (1999); B, Bonaparte (1986b); C, Krause, Prasad et al. (1997).

TABLE 14.1. Linnaean Classification of Gondwanatherian Mammals

Subclass *incertae sedis*
  Suborder Gondwanatheria Mones, 1987 (Krause and Bonaparte, 1990)
    Family Ferugliotheriidae Bonaparte, 1986a
      Ferugliotherium Bonaparte, 1986a
    Family Sudamericidae Scillato-Yané and Pascual, 1984
      Sudamerica Scillato-Yané and Pascual, 1984, type genus (Paleocene; not treated herein)
      Gondwanatherium Bonaparte, 1986b
        G. patagonicum Bonaparte, 1986b
      Lavanify Krause et al., 1997
        L. miolaka Krause et al., 1997
  In addition, unnamed isolated teeth from the Late Cretaceous (Maastrichtian) Deccan Intertrappean sequence of India, were referred by Krause, Prasad et al. (1997) to Sudamericidae.

cisors (tentatively assigned), lower molars, and M1. The m1 and m2 show similar occlusal morphology and are apparently congeneric. Share with some Multituberculata (e.g., Plagiaulacidae, Eobaataridae, and *Paracimexomys* group) a tendency of molar cusps to coalesce in peripheral aspect and ornamentation of grooves and pits on the molars. Share with numerous multituberculates, but also with several eutherian mammals lower incisor with limited enamel band. Differ from Sudamericidae in having brachyodont rather than hypsodont molars, but share with them the occlusal morphology of the molars and other characters of the suborder.

Genera. *Ferugliotherium* Bonaparte, 1986a, type genus by monotypy.

Distribution. Late Cretaceous (Campanian–Maastrichtian): Argentina, Río Negro Province, Los Alamitos (Los Alamitos Formation).

Genus *Ferugliotherium* J. F. Bonaparte, 1986a
(figure 14.1A_1,B_1,C)

Synonym: *Vucetichia* J. F. Bonaparte, 1990 (see Krause, 1993).

Species. *Ferugliotherium windhauseni* J. F. Bonaparte, 1986a, type species by monotypy.

Diagnosis and Distribution. As for the family.

Family Sudamericidae Scillato-Yané and Pascual, 1984

Synonym: Gondwanatheriidae J. F. Bonaparte, 1986b.

Diagnosis (based on Krause and Bonaparte, 1993, emended). Differ from Ferugliotheriidae in being much larger and in possessing large, hypsodont molars, each supported by massive root covered with cement. Coronal pattern of ridges and furrows on molars is generally more complex than in Ferugliotheriidae. Life style of *Sudamerica* might have been semiaquatic and possibly burrowing, similar to that of living beavers (Koenigswald et al., 1999).

Distribution. Late Cretaceous–early Paleocene: South America; Late Cretaceous: Madagascar and India.

Genera. *Sudamerica* Scillato-Yané and Pascual, 1984, type genus; *Gondwanatherium* Bonaparte, 1986b; *Lavanify* Krause, Prasad et al., 1997. *Sudamerica* is Paleocene in age and is not described in this book, but as it is the only gondwanatherian in which a fairly complete dentary has been preserved, it is figured herein (figure 14.1A_3, 14.2A). In addition to the two formally named Late Cretaceous genera from Argentina and Madagascar, isolated teeth from the Late Cretaceous (Maastrichtian) of Deccan Intertrappean sequence of India were referred by Krause, Prasad et al.(1997) to Sudamericidae.

Genus *Gondwanatherium* J. F. Bonaparte, 1986b
(figures 14.1A_2,C, 14.2B)

Diagnosis (based on Bonaparte, 1986b). Differs from *Sudamerica* (not described herein) in being smaller, in having occlusal surface of the molars made of three asymmetrical lophs (rather than four symmetrical as in *Sudamerica*) and in having enamel strip around occlusal surface divided into parts, rather than continuous.

Species. *Gondwanatherium patagonicum* J. F. Bonaparte, 1986b, type species by monotypy.

Distribution. Late Cretaceous (Campanian–Maastrichtian): Argentina, Río Negro Province, Los Alamitos (Los Alamitos Formation).

Genus *Lavanify* Krause, Prasad, Koenigswald, Sahni, and Grine, 1997
(figure 14.2C)

Diagnosis (based on Krause, Prasad et al., 1997). The cheek teeth of *Lavanify* differ from those of *Gondwanatherium* and *Sudamerica* in possessing prominent and continuous interrow sheets of prismatic matrix in dental enamel and at least one cheek tooth position that has a single V-shaped dentine island and lacks enamel on one side of the crown. Differs from *Gondwanatherium* in having cheek teeth with vertical furrows that extend to the base of the crown and onto the root.

Species. *Lavanify miolaka* Krause, Prasad, Koenigswald, Sahni, and Grine, 1997, type species by monotypy.

Distribution. Late Cretaceous (Maastrichtian): northwestern Madagascar, Mahajanga Basin, Berivotra (Anembalema Member of the Maevarano Formation).

CHAPTER 15

Interrelationships of Mesozoic Mammals

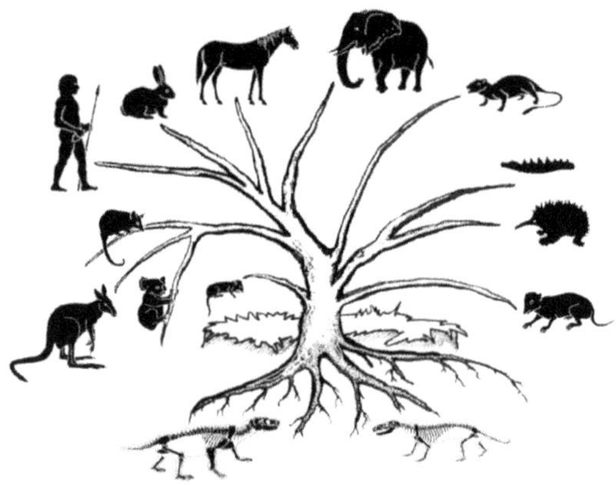

INTRODUCTION

Phylogenetic relationships are the foundation for understanding the diversification of Mesozoic mammals. A phylogeny with extensive sampling of taxa and characters is absolutely crucial for establishing the framework in which to clarify the taxic evolution of Mesozoic mammals. Mesozoic mammals also underwent great anatomical evolution in dentition, skull, and skeleton, as best reflected by the vast differences between the basal taxa from the Late Triassic or Early Jurassic and the more derived taxa of the Late Cretaceous. The most basic approach to the understanding of the Mesozoic mammalian evolution is to map the pattern of anatomical evolution within a well-supported hypothesis of relationships among these Mesozoic mammal groups. For these purposes, we present a comprehensive hypothesis on the cladistic relationships of all major clades of Mesozoic mammals, based on a morphological dataset that includes all characteristics known to us, including features of the dentition, crania, and postcrania. Our principal hypothesis on the placements of the major Mesozoic mammal groups is illustrated in figure 15.1; possible alternative placements of some key clades of Mesozoic mammals are illustrated in figure 15.2.

If this book is to serve the role of a general summary of the current knowledge of Mesozoic mammals, a critical question we have to answer is whether the currently available morphological data are adequate for us to infer a comprehensive cladistic phylogeny of all Mesozoic mammals. As a prelude to a synthetic review of these relationships, we discuss the quality of the morphological data from which our Mesozoic mammalian phylogeny must be derived. We also discuss our views of the potential corroboration and conflicts between the morphological phylogenies based on fossils and the deep phylogenies of extant mammalian clades from molecular evolutionary studies. In this chapter, we hope to demonstrate that the cladistic relationship for most (if not all) Mesozoic mammals can be resolved by parsimony methods on the basis of currently available morphological data, despite the fact that most Mesozoic mammals are still quite incomplete.

CURRENT ISSUES OF TAXON AND CHARACTER SAMPLING

Beginning with several pioneer cladistic studies, such as McKenna (1975), Prothero (1981), Kemp (1983), and Rowe (1988), relationships among early mammals have been discussed mostly in cladistic terms. By the 1990s, parsimony phylogenies of early mammal clades were already established on a growing body of morphological data that has been described in detail (Wible, 1991; Cifelli, 1993b; Wible and Hopson, 1993; Luo, 1994; Szalay, 1994; Wible et al., 1995; Rougier, Wible, and Hopson, 1996; Hu et al., 1997; Springer, 1997; Rougier et al, 1998; Shoshani and McKenna, 1998; Ji et al., 1999; Luo, Cifelli, and Kielan-Jaworowska, 2001; Luo, Crompton, and Sun, 2001; Luo et al., 2002).

Although much of the discussion has focused on alternative placements of clades, several studies also addressed the more general systematic issues about the sampling of morphological data that would affect the estimate of large-scale phylogenies (Gauthier et al., 1988; Novacek, 1992a; Rowe and Gauthier, 1992). Most fossils are incomplete by nature of preservation. As a consequence, paleontologists who rely on morphological data to esti-

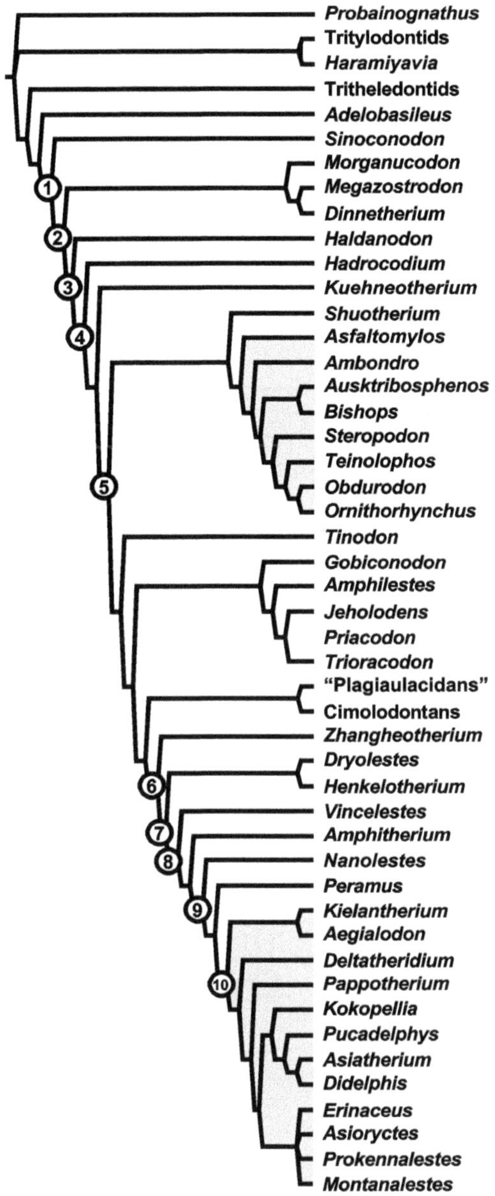

FIGURE 15.1. Phylogenetic relationships of all major Mesozoic mammal lineages. Numbers in circles (1 through 10) denote the nodes of the stable monophyletic clades, discussed in the text. Node (1) denotes Mammalia (*sensu lato*). Node (5) denotes the mammalian crown group. Shadowed areas denote Australosphenida (upper shading) and Boreosphenida (lower shading: node 10). This Mesozoic mammalian family tree is based on the strict consensus tree of 50 equally parsimonious trees from 1,000 heuristic searches by PAUP 4.0b (Swofford, 2000) of a data matrix of 275 morphological characters on 48 clades of Mesozoic mammal and premammalian cynodont taxa (Appendix). Each of the 50 equally parsimonious trees has: L = 943 steps, CI = 0.495; RI = 0.764. All multistate characters are unordered. No topological constraints were enforced. See the text for further explanations. *Source:* modified from Luo et al. (2002) with recently discovered taxa (Martin 2002; Rauhut et al. 2002).

mate phylogenetic relationships must face several vexing technical difficulties related to sampling of incomplete taxa and missing data inherent in the fossil record.

First, the well-resolved phylogenies by earlier pioneer studies tended to have relatively few (although more complete) taxa, lacking sufficient taxonomic coverage. The vast evolutionary bush of Mesozoic mammals has more than 25 independent lineages and there is a need to include most of them in a parsimony phylogeny. Second, the best anatomical studies tended to be selective in character complexes, lacking comprehensive sampling of characters. Many conflicting placements of controversial taxa in higher-level Mesozoic mammal phylogeny are attributable to the difference in character sampling and/or different emphasis on those data that are more frequently preserved on the incomplete taxa under study. While many investigators were cognizant of this recurrent problem, they were nonetheless limited to dealing with those characters that happen to be preserved on their incomplete fossils. Third, conflicting tree topologies resulting from different studies were often not comparable in taxonomic sampling, making it difficult to resolve these alternative topologies in parsimony terms. For example, Prothero (1981) developed the best and most extensive cladistic analysis of dental characters of the Mesozoic nontribosphenic therian clades (*sensu lato*); Wible et al. (1995) and Rougier, Wible, and Hopson (1996) provided the best summaries of basicranial characters. Although excellent in analyses of their respective character complexes, these studies sampled different taxa and their resultant trees cannot be compared directly.

To address these problems it is necessary to combine all dental, cranial, and postcranial features known to be informative of the Mesozoic mammal relationships in a single dataset that also covers all major early mammal clades for parsimony study. This is a prerequisite for the students of Mesozoic mammals to evaluate all conflicting hypotheses on the basis of the total morphological evidence and on comparable selection of taxa. The cladistic dataset recently published by Luo et al. (2002), as slightly expanded here (Appendix), represents a first step toward the inclusion of all relevant morphological data. This dataset samples 45 Mesozoic mammals, including the recently published *Asfaltomylos* and *Nanolestes* (Rauhut et al., 2002; Martin, 2002), plus three nonmammalian cynodont groups as outgroups. It includes nearly all published morphological (dental, cranial, and postcranial) characters known to be informative in regard to Mesozoic mammal relationships from many previous studies.

Individually, cranial, dental, and postcranial character complexes have all been shown to be homoplastic to some extent by empirical studies (Kemp, 1983; Rowe, 1988; Ji

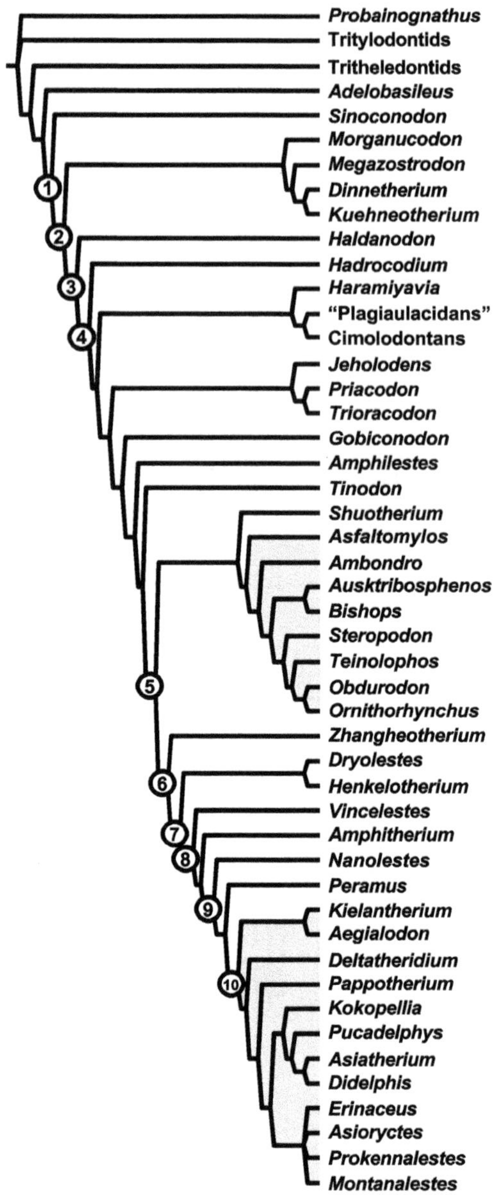

FIGURE 15.2. Alternative interpretation of phylogenetic relationships of all major Mesozoic mammal lineages. This alternative topology is from searches that were constrained to retain the trees compatible with an allotherian clade (*Haramiyavia* + multituberculates) outside the mammalian crown group. Numbers in circles (1 through 10) denote the nodes of the stable monophyletic clades, discussed in the text. Node (1) denotes Mammalia (*sensu lato*). Node (5) denotes the mammalian crown group. Shadowed areas denote Australosphenida (upper shading) and Boreosphenida (lower shading: node 10). Tree topology is the strict consensus tree of seven equally parsimonious trees from 1,000 heuristic runs of PAUP 4.0b (Swofford, 2000) of the matrix in Appendix, with unordered multistate characters. Each of the seven equally parsimonious trees has: L = 951 steps; CI = 0.491; RI = 0.760. The enforced topological constraints have excluded the monophyletic allotherians (*Haramiyavia*, "plagiaulacidans," cimolodontants) from the crown mammalian group to reflect the preference of several dental specialists (e.g., Butler, 2000) that allotherians are likely a highly specialized and early divergent lineage of mammals, even though this topology is not strictly parsimonious for all the morphological evidence that is currently available. See the text for further explanations. *Source:* modified from Luo et al. (2002) with recently discovered taxa (Martin, 2002; Rauhut et al., 2002).

et al., 1999; Luo, Cifelli, and Kielan-Jaworowska, 2001). Yet these character complexes are all known to be phylogenetically informative in combined analyses of all available morphological evidence. In the most ideal situation, dental, cranial, and postcranial characters must all be included (e.g., Rowe, 1988; Luo et al., 2002). Because all morphological datasets (teeth, skull, and skeleton) of Mesozoic mammals must have had a single underlying history, combining different character complexes into one matrix is consistent with the premise of the combined phylogenetic analysis, as discussed thoroughly in the recent literature on systematic methods (Kluge, 1989; Sanderson and Donoghue, 1989; Novacek, 1992b; Wheeler, 1992; Queiroz, 1993; Hillis, 1995; Miyamoto and Fitch, 1995; Queiroz et al., 1995; Wiens and Reeder, 1995; Huelsenbeck et al., 1996, but see also the precaution of Bull et al., 1993, and Wiens, 1998a).

Traditionally, Mesozoic mammal studies were based on molar features. But in the past two decades, much fresh data from the anterior dentition and from cranial and skeletal features have become available. It is no longer acceptable, in our view, to continue to insist on using molar characters alone for phylogenetic inference. When all morphological characters are incorporated into the phylogenetic studies, different, yet more informative, estimates of phylogeny may emerge.

For example, the exciting discoveries of mammals with tribosphenic molars in the Cretaceous and Jurassic sediments on the Gondwanan landmasses by Rich et al. (1997, 1999), Flynn et al. (1999), and Rauhut et al. (2002) challenged the century-old idea that tribosphenic mammals originated from Laurasian landmasses and secondarily spread to the southern continents. Rich and colleagues regarded Australian Early Cretaceous mammals as placentals (see also the most recent review on this topic by Woodburne et al., 2003). These findings initiated a vigorous discussion on the relationships among early mammals. However, the initial assessment of the affinities of the ausktribosphenids was based on characteristics of only the talonid of molars. Our team (Kielan-Jaworowska et al., 1998) pointed out that if the mandibular features are also considered, the first Australian mammal with tribosphenic molars *Ausktribosphenos* would have to be excluded from placentals. In our next paper on this subject (Luo, Cifelli, and Kielan-Jaworowska, 2001), we showed that the Jurassic tribosphenic mammal from Madagascar (Flynn et al., 1999) and the Early Cretaceous ausktribosphenids (Rich et al., 1997, 1999) share with *Ausktribosphenos* several derived features of premolars and nonfunctional features of the molars. These tribosphenic mammals are best interpreted as being a group of mammals endemic to southern continents, probably related to toothed monotremes, in a southern clade designated Australosphenida. This clade has several derived features that are not present in the boreosphenidans. Our idea that tribosphenic molars evolved independently in southern and northern Jurassic–Cretaceous mammals was corroborated by the merger of dental and mandibular characters observable on the southern tribosphenic mammals and the cranial and postcranial characters observable on the more complete Mesozoic mammals that are known to be informative for the reliable phylogenetic estimates of all mammals.

MISSING DATA AND CONFLICTS OF CHARACTERS

A long-standing obstacle for phylogenetic studies of fossils is that many fossil taxa are represented by different, yet incomplete anatomical parts. It is a daunting challenge to encompass all relevant clades with varying degrees of completeness in a combined dataset so that we can combine all the character complexes that were used separately for supporting alternative phylogenetic hypotheses. Apart from the sampling, choice of the anatomical characteristics of Mesozoic mammals can also be a complicated issue (see the critique by Wible, 1991, on Rowe, 1988). Nonetheless, we hope to demonstrate here that a parsimony analysis with extensive sampling of incomplete Mesozoic mammals can achieve good resolution of their relationships, provided that the phylogenetically informative characters for the relevant taxa are sampled, regardless of the fact that they may not be complete and some significant amount of their anatomical information may still be missing (figures 15.1, 15.2).

Merging and combining the previously separate character complexes (e.g., dentition versus mandible) would result in a varying amount of missing data for the majority of fossil taxa (see Appendix). This is inevitable when we combine the previously separate datasets into a comprehensive taxon-character matrix (Wiens and Reeder, 1995). Earlier studies suggested that missing characters were the root cause of the ambiguities in resolving the relationship of many incomplete early mammals (e.g., Rowe, 1988; Novacek, 1992b; Simmons, 1993). More recent studies, however, show that incomplete preservation of fossils is only one of several causes for lack of resolution of cladistic relationships among fossil taxa. The ambiguities in resolution of the relationships could be caused by missing data (such as incomplete preservation), by character conflicts (homoplasies), or a mixture thereof (Wilkinson, 1995; Kearney, 2001). Missing data can result from either incomplete taxa with fragmentary preservation or incomplete characters that are so transformed that they can be only coded for some taxa but have to be coded as non-applicable for others (Novacek, 1992b; Wiens, 2003). The dataset on which we based our hypothesis of Mesozoic mammal phylogeny is no exception in this regard, containing as it does both types of missing data, and our proposed phylogenetic relationship (figure 15.1) is also affected by a conflict of characters (figure 15.2).

Missing data are not desirable. Nonetheless, recent simulation models show that the requisite missing data of the incomplete taxa in the combined dataset, by themselves, are not misleading (Wiens, 1998b). More important for the resolution of phylogenetic reconstruction is that enough informative characters be coded for the incomplete taxa. Under some conditions, the addition of highly transformed (and incomplete) characters can improve the phylogenetic resolution (Wiens, 2003). Conclusions of these simulation studies are consistent with our recent study of Mesozoic mammals, in which we demonstrated empirically that if appropriate and informative dental characters are included, phylogenetic relationships of the incomplete dental taxa of Mesozoic mammals can be resolved in the combined analysis of both incomplete and the more complete Mesozoic mammals (e.g., Luo, Cifelli and Kielan-Jaworowska, 2001; Luo et al., 2002). Ultimately, the ambiguities in phylogenetic relationships owing to missing data must be improved by further discoveries of more complete fossils.

Character conflict is more likely to contribute to ambiguities in phylogeny than the lack of character evidence (Kearney, 2001). As in any phylogenetic study, resolution of Mesozoic mammalian family trees is more seriously affected by character conflicts or homoplasies. Traditional studies on relationships of Mesozoic mammals were almost entirely reliant on dental features. As anatomical features of the mandible, cranium, and postcranium became available with discoveries of more complete fossils, an issue arose in the combined analyses of their relationships as to which character complexes would be the best source of characters for inference of phylogeny if there are conflicts among various such complexes. Several studies of the 1980s and 1990s debated the relative merits of dental *versus* postcranial characters in phylogenetic studies of nonmammalian cynodonts and mammals (Kemp, 1983, 1988; Sues, 1985b; Hopson, 1991, 1994; Rowe, 1993).

One of the most prominent cases of character conflicts resulting in uncertainty in phylogenetic placement concerns multituberculates. Molars of multituberculates resemble those of haramiyidans and nonmammalian tritylodontids. Based on molar characteristics alone, multituberculates are certainly more comparable to haramiyidans. However, multituberculates are far more derived than haramiyidans and tritylodontids in mandibular characters, and in these characters they are more closely comparable to the more derived mammalian clades in the mammalian crown group. Obviously, there is only one phylogenetic history for these seemingly conflicting characteristics of multituberculates. A hypothesis of the multituberculate relationship, if based only on either dental or mandibular features, would be incomplete at best and misleading at worst. The best way to assess these character conflicts would be not only to combine the dental and mandibular characters, but also to combine the dental and mandibular characters with those from crania and postcranial skeletons, so that all character complexes that share a common phylogenetic history can be used for phylogenetic reconstruction.

MORPHOLOGICAL AND MOLECULAR PHYLOGENIES OF MESOZOIC MAMMALS

Most of the Mesozoic groups such as *Sinoconodon*, morganucodontans, haramiyidans, eutriconodontans, multituberculates, docodontans, and kuehneotheriids have no living descendants. Most of the families nested within Trechnotheria and Australosphenida are also extinct. As noted by two molecular evolutionists, Springer and de Jong (2001: 1710), "Beyond living taxa, the primacy of morphological data remains unchallenged." Therefore conclusions on interrelationships among Mesozoic mammals must be based on analyses of anatomical characters.

Molecular evolutionary studies have made tremendous strides in estimating the sequence and timing of extant mammalian clades that have a long history into the Mesozoic. Based on certain assumptions of molecular evolutionary rates, molecular studies have now postulated explicit hypotheses on the sequence and timing of the major events in the Mesozoic history of major extant mammalian lineages, such as the splits of the monotremes, placentals, and marsupials, as well as the split of superorders within placentals and marsupials.

There is good concordance between the morphological phylogeny of monotremes based on fossil evidence and the protein and nuclear DNA sequence studies on monotremes. Recent phylogenetic studies using DNA sequences of diverse nuclear genes supported the traditional therian monophyly to the exclusion of monotremes (Retief et al., 1993; Kullander et al., 1997; Lee et al., 1999; Gilbert and Labuda, 2000; Killian et al., 2000, 2001). The crown therian clade is also supported by studies of protein sequences (Messer et al., 1998; Belov et al., 2002). The study of molecular evolutionary rates by some protein sequences suggests that modern monotremes split from crown therians about 170 to 168 Ma ago (Belov et al., 2002). This is consistent with the hypothesis that toothed monotremes are closely related to the southern tribosphenic mammals (or australosphenidans), whose earliest representatives are from the Bathonian, around 165 Ma ago. However, the studies of mitochondrial genomes tend to support the now-abandoned morphological hypothesis of "Marsupionta" in which monotremes are nested within marsupials to the exclusion of placentals (Janke et al., 1996, 1997, 2002; Kirsch et al., 1997; Penny and Hasegawa, 1997). This hypothesis gains no support from any morphological characters (Zeller, 1999a; Szalay and Sargis, 2001; Luo et al., 2002). Comparative studies on the nuclear versus mitochondrial genomes by Springer et al. (2001) also suggest that the nuclear exon genes are more informative than the protein-coding mitochondrial genes for resolving deep mammal phylogeny, weakening the relative strength of the mitochondrial evidence for recovering higher-level phylogenetic history in mammals.

There are some significant differences between the molecular phylogenies of placental lineages and morphological phylogenies based on fossil evidence from the Mesozoic. A series of molecular studies in the late 1990s suggested that major groups of placental and marsupial orders originated in the Cretaceous, much earlier than the previously accepted timeline for mammal evolution as established by paleontologists (e.g., Hedges et al., 1996; Kumar and Hedges, 1998; Esteal, 1999; Gee, 1999). One study even suggested that the split of marsupials and placentals occurred in the Late Jurassic (Kumar and Hedges, 1998), postulating an enormous gap between the supposed time of origination and the then earliest eutherian *Prokennalestes* (Aptian or Albian, about 110 Ma ago).

The gulf between the molecular estimate of therian mammal evolution and the fossil record narrowed in the early 2000s. The recently revised molecular estimate of the divergence time for the superorders with the crown placental group is around 104 ± 8 Ma ago (Eizirik et al., 2001; Madsen et al., 2001; Murphy, Eizirik, Johnson et al., 2001; Murphy, Eizirik, O'Brien et al., 2001). For the fossil records, several eutherians have also been discovered in recent years: *Montanalestes* (Aptian or Albian, Cifelli, 1999b), *Murtoilestes* (Barremian or Aptian, Averianov and

Skutschas, 2001) and *Eomaia* (Barremian, Ji et al., 2002), adding three new taxa to a growing list of Early Cretaceous eutherians. The earliest known eutherians (ca. 125–110 Ma ago) are now certainly older than the estimated time window of the molecular evolution of the placental crown group.

Despite these advances, there are still some seemingly unbridgeable gaps between the molecular estimates of divergence between extant placental superorders and the currently known fossil taxa that can be placed in the extant placental superordinal lineages with confidence. Recent molecular studies postulated that ordinal diversification of living placental groups ranged from 104 to 60 Ma (Eizirik et al., 2001; Murphy, Eizirik, Johnson et al., 2001; Murphy, Eizirik, O'Brien, et al., 2001; but see Kumar and Hedges, 1998), whereas recognizable fossil members of most living placental orders do not appear until the late Paleocene or early Eocene. Molecular evolutionists offered two interpretations for the lack of Cretaceous fossils that can be assigned to the crown group of placentals. They suggest that either the phyletic lineage splitting occurred without much morphological divergence (Kumar and Hedges, 1998; Esteal, 1999; Gee, 1999; Waddell et al., 1999) or that the fossil record is too incomplete. During the Early Cretaceous, the geological time period in which the splits of major groups of crown therians were supposed to have occurred, the fossil records from the Gondwanan continents are very poor. Therefore, we cannot rule out the possibility that the earliest diversification of crown placentals occurred on the southern continents and are only waiting to be discovered (Foote et al., 1999; Murphy, Eizirik, O'Brien et al., 2001; Rich et al., 2002). Conversely, several morphological studies based on the fossil records have questioned the crucial assumption of constancy of rate in molecular evolution, and the calibration for the molecular clock by the spurious evidence on which the earlier time estimates for divergence were established (Allard et al., 1999; Benton, 1999; Foote et al., 1999).

For the corroboration or falsification of molecular versus morphological estimates of phylogeny, the most crucial issue centers on whether any eutherian fossil taxa from the Cretaceous can be placed within the placental crown group. Currently there is disagreement in three areas: (1) The most prominent phylogenetic issue is the early evolution of a clade of African placental mammals. Molecular studies suggest that elephants, manatees, hyraxes, aardvarks, elephant shrews, tenrecs, and golden moles belong to a monophyletic group (Afrotheria) that originated in Africa in the Early Cretaceous. This is drastically different from morphological phylogenies for these mammals (e.g., Novacek, 1992a; Prothero, 1993; Fischer and Tassy, 1993; Shoshani and McKenna, 1998; Asher, 1999; Allard et al., 1999). Paleontologist could concede, however, that the morphological data are questionable in some cases and that fossil evidence bearing directly on the question (i.e., early Tertiary of Sub-Saharan Africa) is virtually nonexistent. (2) Rich et al. (1997, 1999, 2000) argued that the Early Cretaceous ausktribosphenids are closely related to modern erinaceomorphs or are associated with some basal placentals (Woodburne et al., 2003). In our view, it is far more parsimonious to postulate that these southern tribosphenic mammals are part of an endemic mammalian radiation and are likely to be related to toothed monotremes (Luo, Cifelli, and Kielan-Jaworowska, 2001; Luo et al., 2002). (3) Archibald (1996b; also Archibald et al., 2001) proposed that zhelestids and zalambdalestids are nested within the laurasiatherian clade of crown placentals, although this has been questioned on other characters (Novacek et al., 1997, 2000). In a recent study that has a more extensive sampling of taxa and characters (Meng and Wyss, 2001, see also Fostowicz-Frelik and Kielan-Jaworowska, 2002), zalambdalestids were not placed in gliriforms. However, zalambdalestids and gliriforms together are nested within crown placentals.

There is less discordance between molecular-based time estimates of marsupial diversification and the appearance of relevant clades of crown Marsupialia in the fossil record. Springer (1997) estimated a divergence of Ameridelphia from Australodelphia around 78 to 70 Ma ago, and within Ameridelphia a split of Paucituberculata from Didelphimorphia between 65 and 61 Ma ago. The earliest putative member of Paucituberculata (*Glasbius*) is about 70 Ma old. The earliest known South American didelphimorphs are from the early Paleocene, about 63 Ma old (Marshall and Muizon, 1995; Muizon et al., 1997; Muizon, 1998).

A key to the resolution of the conflicts between molecules and fossils for the timing of the splits of extant clades is to obtain reliable paleontological data for testing possible placement of the earliest eutherian and metatherian fossils in one of the groups of crown marsupials or placentals. Field paleontologists will have to continue their searches for ever-earlier fossils of modern lineages. At the same time the molecular evolutionists have to reconsider the assumptions and the datasets for their time estimates, especially in regard to the constancy and calibration of the molecular clock.

MAJOR MAMMALIAN CLADES AND THEIR INTERRELATIONSHIPS

The interrelationships of Mesozoic mammals, as presented in this chapter (figures 15.1, 15.2), are based mostly on a recently published comprehensive paper by our team, "In

Quest for a Phylogeny of Mesozoic Mammals" (Luo et al., 2002). After the basic cladistic framework was established for the clades with relatively complete preservation (e.g., Kemp, 1983; Rowe, 1988), the big challenge was to encompass most of the lineages that have only teeth and mandibles, so that the relationships of dental taxa can be estimated in the context of total morphological evidence. Further, there is an urgent need to evaluate phylogenetic affinities of the so-called dental taxa of Mesozoic mammals in the larger framework of an all-mammal phylogeny.

In the following pages, we discuss: (1) the well-supported monophyletic groups of Mesozoic mammals in the sequence of their nested hierarchies (see figure 15.1); (2) the alternative placements of eutriconodontans, multituberculates, and kuehneotheriids (some of these alternative placements are illustrated in figures 15.1 and 15.2). Lastly we briefly outline why several traditional groupings of early mammals must be abandoned because they are shown to be paraphyletic by the currently available and extensive evidence.

Among the 48 taxa chosen for parsimony analysis (figure 15.1, Appendix), *Probainognathus*, tritylodontids, and tritheledontids belong to nonmammalian cynodonts and have been used as outgroups. As noted in chapter 3, we used an inclusive, stem-based definition of Mammalia, embracing *Sinoconodon* through the crown group of mammals and all the intervening clades placed between them.

Mammalia. Defined as the common ancestor of *Sinoconodon* and crown mammals (figures 15.1, 15.2: node 1), Mammalia are diagnosed by a jaw hinge formed by the dentary condyle and the squamosal glenoid and a large number of petrosal features. The dentary and squamosal contact is absent in most nonmammalian cynodonts (Crompton, 1972a). Tritheledontids, generally considered to be the sister taxon to Mammalia (Hopson and Barghusen, 1986; McKenna, 1987; Shubin et al., 1991; Crompton and Luo, 1993; Miao, 1993; Luo, 1994, and this book), have a dentary-squamosal contact; but there is no dentary condyle, nor is there a clearly defined glenoid area on the squamosal (Crompton, 1972a). Thus tritheledontids have an intermediate character state to the typical mammalian craniomandibular joint and still differ from Mammalia in this feature.

Mammalia are also diagnosed by the presence of a petrosal promontorium or the external eminence of the pars cochlearis (MacIntyre, 1972; Crompton and Sun, 1985; Gow, 1985, 1986b; Hopson and Barghusen, 1986; Rowe, 1988; Wible, 1991; Luo, 1994; Luo et al., 1995). The pars cochlearis is the bony housing of the hearing organ, the cochlea; enlargement and emergence of the pars cochlearis on the ventral surface of the skull provides more internal space for a longer cochlea within the petrosal, probably for better hearing (Rosowski and Graybeal, 1991; Rosowski, 1992). Other diagnostic features of mammals include: the extensive development of a petrosal floor for the cavum epipterícum in the braincase for the trigeminal cranial nerve ganglion (Crompton and Sun, 1985), the presence of a separate tympanic aperture for the prootic canal for the prootic vein (Wible and Hopson, 1995), the loss of the thickened rim of the fenestra vestibuli (Lucas and Luo, 1993), and the separation of the hypoglossal foramen (cranial nerve XII) from the jugular foramen (cranial nerves IX, X, XI) (Lucas and Luo, 1993). Other, equivocal features in the palate and the orbital wall can also diagnose Mammalia, depending on whether tritylodontids or tritheledontids are considered to be the sister taxon and the immediate outgroup to mammals (Luo, 1994). Within Mammalia, *Sinoconodon* (figure 15.1) is the sister taxon to a clade that includes *Morganucodon* (and related forms) and the living mammals (Crompton and Sun, 1985; Crompton and Luo, 1993; Rowe, 1993; Wible and Hopson, 1993; Luo, 1994; Rougier, Wible, and Hopson, 1996).

Adelobasileus is a putative taxon of stem-defined Mammalia. One of the oldest Late Triassic (Carnian) fossils described as a mammal (see chapter 4), *Adelobasileus* is incompletely known, being represented by a partial braincase (Lucas and Hunt, 1990; Lucas and Luo, 1993). It possesses an incipient petrosal promontorium, together with several other mammalian apomorphies (see table 3.1 and the discussion on the diagnosis of stem-defined Mammalia in chapter 3). Initial analyses including *Adelobasileus* nested this genus within Mammalia (Lucas and Hunt, 1990; Lucas and Luo, 1993; Hopson, 1994), but more recent analyses placed *Adelobasileus* outside the clade of *Sinoconodon* and living mammals (figures 15.1, 15.2; see also Rougier, Wible, and Hopson, 1996; Luo, Cifelli, and Kielan-Jaworowska, 2001; Luo et al., 2002). The fossil of *Adelobasileus* is incomplete. In our classification (table 3.1), we consider *Adelobasileus* to have a position of *incertae sedis* within stem-defined Mammalia, noting that a well-supported appraisal of its status ultimately depends on discovery of additional fossils.

Clade (*Morganucodon* + Crown Mammalia). Morganucodonta (morganucodontids and megazostrodontids) are more closely related to the crown group of mammals than are *Sinoconodon* and *Adelobasileus*. This clade is defined by the common ancestor of *Morganucodon*, living mammals, and all fossil taxa nested within this clade (figures 15.1, 15.2: node 2). This clade is diagnosed by a very derived dental replacement pattern and many molar features associated with the presence of precise occlusion between upper and lower molars. These include one-to-one opposition of upper to lower molars and the development of matching upper and lower facets associated with individual molar cusps as the result of occlusal wear (K. A. Kermack et al., 1965; Crompton and Jenkins, 1968; D. M. Kermack et al., 1968; Crompton, 1971, 1974; Mills, 1971). Most of the taxa in this clade share an apomorphic condition of tooth replacement, wherein the anterior postcanines are diphyodont, but molars (posterior postcanines) are not replaced (see chapter 4 for details). Known exceptions are *Gobiconodon* (see Jenkins and Schaff, 1988) and *Megazostrodon* (see Gow, 1986a), where replacement occurs at one or more anterior "molariform" positions.

The large samples of fossil jaws and crania of *Morganucodon* show that it had achieved a derived pattern of determinate growth (Gow, 1985; Luo, 1994; Luo, Crompton, and Sun, 2001). By comparison, *Sinoconodon* retains the primitive pattern of indeterminate skull growth, coupled with successive replacement of molariforms at posterior loci while the skull continued to grow in size, as seen in cynodonts and most other nonmammalian amniotes (Crompton and Luo, 1993; Luo, 1994; Zhang et al., 1998). In *Sinoconodon,* as in nonherbivorous cynodonts, the molariform postcanines lack one-to-one opposition between respective upper and lower teeth.

The clade of *Morganucodon* and living mammals is also characterized by a large suite of derived basicranial features. One most notable feature is a crista parotica on the anterior paroccipital process of the petrosal (Luo and Crompton, 1994). In all derived mammals (including extant species), the incus is articulated directly on the crista parotica of the petrosal. The incus, reduced in size and suspended by the petrosal, has a greater degree of mobility, which presumably facilitated sound transmission in the middle ear. By contrast, *Sinoconodon* and tritheledontids retain the primitive condition, wherein the incus (quadrate) is entirely suspended by the squamosal, as in all nonmammalian cynodonts except for tritylodontids (Crompton, 1964a; Hopson, 1964; Sues, 1985b, 1986; Luo and Crompton, 1994).

Morganucodon is known by relatively complete, abundant fossils representing the dentition, skull, and postcranial skeleton (see chapter 4). Because it is the most completely known Late Triassic–Early Jurassic mammal and because its characteristics have received thorough coverage in previous phylogenetic analyses, the phylogenetic position of *Morganucodon* among basal mammals has remained very stable.

Clade (Docodontans + Crown Mammalia). In our phylogeny, docodontans (chapter 5) are more derived and placed closer to crown mammals than morganucodontans. The common ancestor of docodontans and crown mammals is indicated by node 3 in figures 15.1 and 15.2. This clade is diagnosed by several derived basicranial features, such as the anterior placement of the pterygo-paroccipital foramen and a more extensive ventral floor of the cavum epiptericum (Rougier, Wible, and Hopson, 1996).

Although docodontans are characterized by typical mammalian diphyodont dental replacement, as are morganucodontans, their molar occlusal features are far more derived. The molars of docodontans are capable of some grinding function (Jenkins, 1969b; Gingerich, 1973), resembling in some aspects therians. Because their molars are very derived (at least more derived than the features of their mandibles), for a long time after their initial discovery, docodontans were classified as "pantotherians" (e.g., Simpson, 1928a, 1929a). After Patterson (1956) pointed out that the most dominant crown features of docodontans resemble those of morganucodontids, it became more widely accepted that docodontans represent a stem lineage of basal mammals, and are not closely associated with therians (Hopson and Crompton, 1969). Detailed study of the skull of the docodontan *Haldanodon* by Lillegraven and Krusat (1991) reinforced this view. These authors even went further to conclude that docodontans are more primitive than morganucodontans and regarded them as the plesiomorphic sister group of all other mammals.

Recent parsimony analyses placed Docodonta closer to Crown Mammalia than morganucodontids and *Sinoconodon* (Hopson, 1994; Luo, 1994; Rougier, Wible, and Hopson, 1996; Luo et al., 2002), but more basal than *Hadrocodium* (Luo, Crompton, and Sun, 2001), triconodontids, and multituberculates. Our placement of docodontans on the Mesozoic mammalian tree reflects this latest view (figures 15.1, 15.2).

Clade (*Hadrocodium* + Crown Mammalia). This clade is defined as the common ancestor of *Hadrocodium* (described in chapter 4) and living mammals plus all its descendants (figures 15.1, 15.2: node 4). *Hadrocodium* is more derived than docodontans, morganucodontans, and *Sinoconodon* in several cranial and mandibular features. In the latter three groups, the braincase is smaller relative to the size of the skull (see figure 3.7), whereas the cranial cavity of *Hadrocodium* is more enlarged. The enlargement of the cranial cavity is a shared derived character of *Hadrocodium,* Triconodontidae, multituberculates, and the mammalian crown group, more so than in cynodonts, *Sinoconodon, Morganucodon,* and *Haldanodon* (Simpson, 1927a, 1928a; Patterson and Olson, 1961; Quiroga, 1979c; Kermack et al., 1981; Crompton and Sun, 1985; Kielan-Jaworowska, 1986, 1997; Lillegraven and Krusat, 1991; Crompton and Luo, 1993; Luo, 1994; Rowe, 1996a,b). Related to the enlargement of the cranial cavity, the craniomandibular joint of *Hadrocodium* is more anteriorly placed than in morganucodontans and *Sinoconodon* (see figure 3.15). *Hadrocodium* and all extant clades of the mammalian crown group lack an ossified pila antotica inside the cranial cavity that separates the cavum epiptericum from the braincase, with the notable exception of multituberculates, which might be regarded as a reversal (Hurum, 1998a).

The second suite of diagnostic apomorphies concentrates on the medial side of the dentary. Taxa of this clade either have a much reduced postdentary trough without the overhanging medial ridge, as in *Shuotherium* and some stem taxa of australosphenidans, or the postdentary trough is absent altogether, as in *Hadrocodium*. *Hadrocodium* lacks the postdentary trough and its associated medial ridge for buttressing the surangular, the articular and the prearticular, and the medial concavity of the angular process for accommodating the reflected lamina of the angular bone. Based on these features, Luo, Crompton, and Sun (2001) suggested that *Hadrocodium* achieved the separation of the articular (malleus) and angular (ectotympanic) from the mandible, a derived feature otherwise only known in the very derived mammals. They suggested that detachment of the middle ear elements (malleus and ectotympanic) might be a synapomorphy of the clade of *Hadrocodium* and crown mammals.

However, subsequent analyses showed that this total separation of the middle ear elements is more complicated (Luo et al., 2002). Several taxa nested within the clade of *Hadrocodium* and Crown Mammalia, such as *Shuotherium*, *Ausktribosphenos*, and possibly also the toothed monotreme *Steropodon* (Luo et al., 2002), retained the postdentary trough, although without the associated medial ridge and medial angular concavity (Kielan-Jaworowska et al., 1998; Kielan-Jaworowska, Cifelli, and Luo, 2002). This suggests that the postdentary trough may have been lost independently in at least two separate evolutionary lineages.

The postdentary trough is also well developed in *Haramiyavia* (Jenkins et al., 1997; described in chapter 8) and in *Kuehneotherium* (described in chapter 9). If *Haramiyavia* is proven to be related to multituberculates in the allotherian clade (as hypothesized in figure 15.2) and if *Kuehneotherium* is nested within the clade of *Hadrocodium* and crown mammals or if *Kuehneotherium* is a basal "therian" (*sensu lato*), then the loss of the postdentary trough is likely to have happened four times in mammalian evolution. Therefore, the current weight of the evidence appears to support the classic hypothesis that separation of the middle ear from the mandible occurred several times (Crompton and Parker, 1978; Allin and Hopson, 1992; McKenna and Bell, 1997), but not a monophyletic origin of the mammalian middle ear as initially proposed by Rowe (1996a,b) and supported by Luo, Crompton, and Sun (2001). Most recently, Wang, Hu, Li et al. (2001) discovered that gobiconodontids have preserved the ossified Meckel's cartilage that remains connected to Meckel's groove on the mandible. The postdentary trough is no longer present on the mandible and the articular-prearticular complex and the ectotympanic were separated mediolaterally from the dentary, as can be inferred from the preserved part of the mandible (Wang, Hu, Li et al., 2001). This suggests that the mediolateral separation of the articular (malleus) and the angular (ectotympanic) is not always correlated with the severance of Meckel's cartilage. The malleus and the ectotympanic can be separated laterally from the dentary, but still connected anteriorly to the dentary via an ossified Meckel's cartilage.

In an alternative scheme of character evolution, the absence of the postdentary trough can still be considered as a synapomorphy of the clade of *Hadrocodium* and crown mammals by optimizing the distribution of this character. If so, the presence of the postdentary trough in *Haramiyavia*, *Kuehneotherium*, *Shuotherium*, and *Ausktribosphenos* would have to be considered secondary, atavistic reversals within this clade. Because of the transformations involved, we consider this possibility to be highly improbable.

The Crown Group Mammalia (= Mammalia of Rowe, 1988; Rowe and Gauthier, 1992; McKenna and Bell, 1997). This clade is defined as the common ancestor of all living mammals and all its descendants (figures 15.1, 15.2: node 5). A dental synapomorphy shared by all fossil and living members of this clade is the presence of occlusal surfaces that match precisely between upper and lower molars upon eruption. By comparison, precise molar occlusion in the stem mammals outside this clade required the development of extensive wear facets, wherein a significant part of the tooth crowns must be worn away in order to achieve matching occlusal surfaces between upper and lower molars (Crompton, 1971, 1974; Mills, 1971, 1984; Godefroit and Sigogneau-Russell, 1999). Docodontans represent an exception among the stem mammals in having a complex pattern of occlusion (Jenkins, 1969b; Gingerich, 1973; Krusat, 1980; Butler, 1988a; Pascual et al., 2000). A mandibular synapomorphy of this clade is the presence of a distinctive masseteric fossa with a well-defined ventral margin. This fossa occupies the entire mandibular angle region and is far more expanded in crown mammals than in such stem taxa as *Sinoconodon*, morganucodontids, *Kuehneotherium*, or *Hadrocodium*. This character is also absent in *Haramiyavia*, which can be placed outside the crown group if the searches are ordered.

The cochlear canal is more elongate and curved in many cases (although not always coiled) in all members of the mammalian crown group in which this feature is known (Luo and Ketten, 1991; Meng and Wyss, 1995; Fox and Meng, 1997; Hurum, 1998b), at least in comparison to the relatively straight cochlear canals of *Sinoconodon* (Luo et al., 1995), *Morganucodon* (Graybeal et al., 1989), *Haldanodon* (Lillegraven and Krusat, 1991), and cynodonts (Luo, 2001). The astragalus and calcaneus are in partial superposition in the majority of the taxa for which the tarsals are known (except *Ornithorhynchus;* also *Jeholodens* if the latter proves to lie among crown mammals). All living orders of this group have greatly enlarged gyrencephalic cerebral hemispheres (with external gyri and sulci on the surface of endocast).

Within the crown mammals, there are two major divisions, one leading to extant monotremes and the other to extant clades of marsupials and placentals. The first of these clades includes, in a successively more distant order, extant monotremes, extinct toothed monotremes, ausktribosphenids, *Ambondro, Asfaltomylos,* and shuotheriids.

Clade of *Shuotherium* + Australosphenida. In several recent phylogenetic analyses, *Shuotherium,* stem australosphenidans, and monotremes appear to be a monophyletic group (Luo, Cifelli, and Kielan-Jaworowska, 2001; Luo et al., 2002; Rauhut et al., 2002; Kielan-Jaworowska, Cifelli, and Luo, 2002). Within this clade, *Shuotherium* and Australosphenida are sister taxa. The Middle Jurassic *Shuotherium* has lower molars that possess the trigonid and "talonid," but are unique in having the "talonid" situated in front of the trigonid, rather than behind it as in molars of usual tribosphenic pattern. *Shuotherium* and Australosphenida share several highly derived characters on the ultimate lower premolar (see chapter 6). The ultimate premolar of *Shuotherium* and *Ambondro* and penultimate and ultimate premolars of *Ausktribosphenos* and *Bishops* are very derived. These taxa have a fully triangulated trigonid on the ultimate lower premolar and even on the penultimate premolar. The ultimate premolar is a posteriorly wide premolar with transverse distal cingulid. These derived premolar features are unseen in any other mammals of the Jurassic–Early Cretaceous. This feature is not preserved in the earliest monotreme *Steropodon* and secondarily lost in living monotremes. The ultimate lower premolar of *Obdurodon* is transversely wide in its posterior part, as in *Ambondro.* By comparison, pretribosphenic therians *sensu lato*—"symmetrodontans" and "eupantotherians"—have an ultimate premolar with a laterally compressed crown. Most of the Cretaceous eutherians and metatherians lack a triangulated trigonid on the ultimate premolar, and none has such a premolar that also lacks a molarized talonid. None of the Cretaceous eutherians and metatherians has a triangulated trigonid on the penultimate premolar.

Other dental features, albeit equivocal, may also be used to support the grouping of *Shuotherium* and Australosphenida. Kielan-Jaworowska, Cifelli, and Luo (2002) argued that on the basis of comparisons with australosphenidans *Ausktribosphenos* and *Bishops,* the pseudotalonid in molars of *Shuotherium* is homologous to the shelflike mesial cingulid of most of the australosphenidans. *Shuotherium* is more derived than stem australosphenidans in this regard in that its pseudotalonid appears to have had occlusal contact with the upper molar (Sigogneau-Russell, 1998; Wang, Clemens et al., 1998) whereas the mesial cingulid of australosphenidans did not. It is conceivable that the pseudotalonid is derived in *Shuotherium* after its split from australosphenidans. But the autapomorphic characters of the pseudotalonid do not preclude the possibility of *Shuotherium* and australosphenidans having shared a common ancestor. Occlusal structures of the pseudotalonids anterior to the trigonids are not unique to shuotheriids. Docodontans have a similar pattern of an anterior grinding basin in the lower molar. This feature is best developed in the docodontan *Simpsonodon,* although variable in other docodontans, which led to an earlier speculation that *Shuotherium* may be also related to docodontans (Kermack et al., 1987). Regardless of how the pseudotalonid is interpreted in Mesozoic mammal phylogeny, it is clear that the anteriorly positioned grinding structure on the lower molar shows significant homoplasy (Averianov, 2002). A corollary of the hypothesis that *Shuotherium* and Australosphenida are sister taxa is that the ancestral lineage inclusive of *Shuotherium* and australosphenidans had a global distribution no younger than early Middle Jurassic and that the respective clades diverged prior to full separation of Gondwanan and Laurasian landmasses.

Australosphenida. This clade consists of living monotremes and all extinct mammals more closely related to monotremes than to *Shuotherium* (chapter 6). This stem-based clade includes *Asfaltomylos, Ambondro,* ausktribosphenids, toothed and toothless fossil monotremes, and living taxa. The significant findings of Mesozoic mammals with tribosphenic molars on the southern continents in the late 1990s helped to open up new frontiers in the studies of Mesozoic mammals (Rich et al., 1997; Flynn et al., 1999; Rauhut et al., 2002). There are two contrasting views about the phylogenetic relationships of these southern tribosphenic mammals. Thomas Rich and his collaborators have argued that the Australian *Ausktribosphenos nyktos* and *Bishops whitmorei* were close to extant Erinaceidae (Rich et al., 1997, 1999; Rich, Flannery, and Vickers-Rich, 1998; Rich and Vickers-Rich, 1999; Rich, Flannery, Trusler, and Vickers-Rich, 2001; Rich, Flannery, Trusler, Kool, van Klaveren, and Vickers-Rich, 2002). Woodburne, Rich, and their colleagues further suggested that *Ambondro* and *Asfaltomylos* are also placentals (Woodburne et al., 2003). An alternative view is that the Gondwanan tribosphenic mammals of Middle Jurassic to Early Cretaceous age represent an endemic radiation of Mesozoic mammals; some of these tribosphenic mammals (Sigogneau-Russell et al., 2001) or all of them (Luo, Cifelli, and Kielan-Jaworowska, 2001; Luo et al., 2002; Rauhut et al., 2002) are related to monotremes.

Owing to their highly specialized skull and dental features, but very primitive postcranial skeleton, monotreme relationships have been controversial ever since modern monotremes became known to science. The discovery of the toothed monotreme *Steropodon* from the Early Cretaceous of Australia in 1985 showed that early monotremes had complex molars that are structurally closer to the tribosphenic pattern than had been previously expected. According to the interpretation of Archer et al. (1985), the molar of *Steropodon* represented a highly modified tribosphenic pattern, implying an origin of monotremes from a relatively advanced tribosphenic stage of "therian" evolution. Kielan-Jaworowska, Crompton, and Jenkins (1987; see also Jenkins 1990) pointed out that the molars of *Steropodon* lack wear function within the talonid basin and are not fully tribosphenic. Kielan-Jaworowska, Crompton, and Jenkins (1987) proposed an alternative origin for monotremes, among "pretribosphenic therians" (e.g., *Peramus*). Bonaparte (1990) emphasized the dental similarities between *Steropodon* and dryolestid "eupantotherians," an observation endorsed by Pascual et al. (1992a) and Archer et al. (1993).

After the initial report of *Ausktribosphenos* (Rich et al., 1997), Kielan-Jaworowska et al. (1998) noted that *Ausktribosphenos* has the postdentary trough for the attachment of the postdentary bones, a primitive feature common to the plesiomorphic Late Triassic–Early Jurassic mammals but entirely absent from the currently known eutherians. This precludes the possibility that *Ausktribosphenos* is a placental. Subsequently, Luo, Cifelli, and Kielan-Jaworowska (2001) attempted to place Australian Cretaceous monotremes, *Ausktribosphenos* and related forms, plus *Ambondro* into the overall scheme of mammalian phylogeny using parsimony analysis, including representatives of all relevant Mesozoic and living clades and all morphological features. A major hypothesis of this work is that *Ausktribosphenos* and *Ambondro* are endemic mammals of the southern Mesozoic fauna. They are not related to the Cretaceous tribosphenic mammals (marsupials, placentals, and their close kin) on the northern Laurasian continents, and are more likely to be related to monotremes. Based on this phylogenetic interpretation, Luo, Cifelli, and Kielan-Jaworowska (2001) erected two new infraclasses (subclasses in the book) of mammals: the Boreosphenida, which include marsupials and placentals and their close relatives with tribosphenic dentition that arose on the northern continents, and Australosphenida, which arose on the southern Gondwanan lands. Taking into account the broad geographic distribution of known fossils, these authors concluded that australosphenidans had diversified and spread throughout Gondwana before the end of the Jurassic and that the mammalian faunas of the Southern and Northern hemispheres were already distinct by the Middle–Late Jurassic. The notion of Australosphenida received support in the paper by Rauhut et al. (2002), who discovered *Asfaltomylos patagonicus,* the first Jurassic mammal from Argentina.

Crucial to the hypothesis of a dual origin of mammals with tribosphenic molars is the strength of the separate placements for australosphenidan and boreosphenidan mammals (see Luo et al., 2002). Within the hypothetical framework of dual evolution of the tribosphenic mammals, several issues have yet to be settled, especially regarding the possible transformation between the tribosphenic morphotype of the stem australosphenidan taxa (as currently defined) and the more specialized molar pattern of toothed monotremes. Luo et al. (2002) did not regard the toothed monotremes as having typical tribosphenic molars (see also Kielan-Jaworowska, Crompton, and Jenkins, 1987). This is not, however, incompatible with the hypothesis that monotremes are closely related to taxa in which a recognizable tribosphenic pattern is present, such as ausktribosphenids (see the phylogenetic argument by Sigogneau-Russell et al., 2001). Given the latest evidence that *Asfaltomylos, Ambondro,* and ausktribosphenids are three ranks of clades that are successively closer to the crown monotremes (Rauhut et al., 2002, also see figure 15.1), we favor a scenario that toothed monotremes, such as *Steropodon, Monotrematum,* and *Obdurodon,* have a secondarily modified molar pattern evolved from a tribosphenic ancestry; as independently proposed by Archer et al. (1985), molars of toothed monotremes are not typical tribosphenic teeth but they bear some significant resemblance to the "typical" tribosphenic molars. Here it is relevant to note that most Cenozoic clades of Marsupialia and Placentalia also lack tribosphenic molars, though they undoubtedly evolved from ancestors in which that pattern was present.

Trechnotheria (modified from McKenna, 1975). This clade is defined as the common ancestor of Crown Theria and *Zhangheotherium* (and by extrapolation the monophyletic group of Spalacotheriidae, see Cifelli and Madsen, 1999) and all of its descendants (figures 15.1, 15.2: node 6, see also chapter 9). This node-based definition corresponds to Trechnotheria of McKenna (1975). The most prominent diagnostic character of this group is the presence of a hypertrophied shearing mechanism between the posterior side of the upper molar trigon (postvallum) and the anterior side of the lower molar trigonid (prevallid) (Crompton and Jenkins, 1968; Crompton and Sita-Lumsden, 1970; Krebs, 1971; Fox, 1975; Hu et al., 1997, 1998; Cifelli and Madsen, 1999). With strong development of the postvallum crest, the upper molar cusps form a more acute triangulation in spalacotheriids, dryolestids, paurodontids, *Amphitherium* (Mills, 1964; Butler and Clemens, 2001), and the therian crown group. The upper molar triangulation is less developed in "peramurids." In some spalacotheriids, there is a slight gradient of triangulation from the anterior to the more posterior tooth loci (Fox, 1976a; Cifelli and Madsen, 1986, 1999); but nonetheless the triangulation is a distinctive feature of the entire molar series, far better developed than the variable triangulation of "obtuse-triangle symmetrodonts."

A trechnotherian clade is also supported by several derived characters of petrosals, such as a large posttympanic recess, a caudal tympanic process, and a squamosal wall to the epitympanic recess (Rougier, Wible, and Hopson, 1996; Hu et al., 1997). The skeleton of trechnotherians has several synapomorphies. One synapomorphy is an expanded supraspinous fossa of the scapula so that the fossa is fan-shaped or triangular in outline and the supraspinous fossa is wider than the infraspinous fossa (Krebs, 1991; Rougier, 1993; Szalay, 1994; Marshall and Sigogneau-Russell, 1995; Hu et al., 1997; Muizon, 1998). Another synapomorphy is on the humerus; the posterior aspect of its distal articulation for the radius is cylindrical (Hu et al., 1997, 1998), although its anterior aspect retains a primitive bulbous condyle. In the hindlimb, the tibia has no proximolateral process, a primitive feature that is strongly developed and hooklike in eutriconodontans, multituberculates, monotremes, morganucodontids. A large proximolateral tibial process is also present in *Gobiconodon* and *Jeholodens.*

The basalmost group of trechnotherians are spalacotheriids. According the latest systematic revision (Cifelli and Madsen, 1999), spalacotheriids are a monophyletic group with a Laurasian distribution during the Cretaceous (see chapter 2).

Cladotheria (*sensu* McKenna, 1975). This clade is defined as the common ancestor of dryolestoids and living Theria plus all its descendants (figures 15.1, 15.2: node 7, see also chapter 10). Spalacotheriidae (represented by *Zhangheotherium* in figures 15.1 and 15.2) are the plesiomorphic sister taxon to Cladotheria. The most reliable dental apomorphy for Cladotheria is the elevation of the talonid above the level of the cingulid on lower molars. The hypoconulid is developed as an elevated version of cuspule d. A distinctive wear facet is developed on the hypoconulid of the lower molar as it occludes against the paracone of the upper molar in dryolestids and paurodontids, as it does in living therians (Crompton, 1971; Prothero, 1981; Krebs, 1991; Martin, 1999a). In contrast, in such noncladotherian mammals as spalacotheriids and kuehneotheriids, "talonid" cuspule d (where present) is usually a part of the cingulid and does not have occlusal function with the upper molar (Hu et al., 1998; Cifelli and Madsen, 1999).

The second suite of apomorphies is on the angular process of the mandible. The angular process is posteriorly positioned, directly below the dentary condyle. The angle is not elevated as in australosphenidans (see Luo et al., 2002: figure 5). The crest of the pterygoid shelf on the medial side of the mandible extends posteriorly along the ventral border of the mandible to reach the posterior apex of the angular process. In contrast, in spalacotheriids, eutriconodontans, and multituberculates, a distinctive angle is absent and the posteroventral mandibular border is rounded. In these noncladotherian groups, the pterygoid shelf on the medial side of the mandible extends continuously along the posteroventral mandibular border toward the dentary condyle. In the stem taxa of mammals, such as *Sinoconodon, Morganucodon, Kuehneotherium,* and *Haramiyavia,* the angle is anteriorly positioned, below the anterior part of the coronoid process (not below the dentary condyle), and the pterygoid shelf is absent.

Prototribosphenida (*sensu* Rougier, Wible, and Hopson, 1996a). This clade is defined as the common ancestor of *Vincelestes* (described in chapter 10) and living therians, plus all of its descendants (figures 15.1, 15.2: node 8). This grouping corresponds to Prototribosphenida as defined by Rougier, Wible, and Hopson (1996). The most important diagnostic features are in the inner ear. The bony cochlear canal contains a primary bony lamina (and even secondary bony lamina among the earliest-known metatherian and eutherian petrosals) that provides internal support for the coiled cochlear duct (*Vincelestes,* Rougier, 1993; Rougier, Wible, and Hopson, 1996; early marsupials, Meng and Fox, 1995a; early eutherians, Wible et al., 2001). By contrast, the primary and secondary bony laminae are absent in outgroups of Prototribosphenida.

Another character is the greater degree of coiling of the bony cochlear canal, which in Prototribosphenida achieved about 270°. In monotremes (herein considered to be part of Australosphenida), the bony cochlear canal is coiled less than 270° and lacks internal bony laminae for support of the membranous cochlear duct. Other outgroups do not have a fully coiled cochlear canal (although the canal may be curved to some degree). The perilymphatic duct of prototribosphenidans is completely enclosed within bone, and its endocranial foramen is separated from the fenestra cochleae (Zeller, 1987; Wible, 1990; Rougier et al., 1992; Wible and Hopson, 1993). Outside the Prototribosphenida, there is no separation of the perilymphatic foramen and the fenestra cochleae in the immediate outgroups (*Zhangheotherium*, Hu et al., 1997, and some putative "symmetrodontan" petrosals, Wible et al., 1995) or in the successively more distant outgroups.

Zatheria (*sensu* McKenna, 1975). The clade, defined as the common ancestor of *Peramus* and living therians, plus all its descendants was designated Zatheria by McKenna (1975; figures 15.1, 15.2: node 9, see also chapter 10). *Peramus*, like other "peramurids," is known only by the dentition and mandible (Clemens and Mills, 1971; Sigogneau-Russell, 1999), and for this reason the diagnosis is restricted to dental features. It is diagnosed by the presence of wear within the talonid. In the earliest and most primitive members of this clade, the wear facet in the talonid, equivalent to Crompton's (1971) facet 5, extends onto the posterior face of the metaconid, so a well-defined distal metacristid is present (Crompton, 1971; Clemens and Mills, 1971; Fox, 1975; Sigogneau-Russell, 1999). Another important feature is the stylocone groove on the upper molar that received the protoconid of the lower molar trigonid. This feature is well developed in *Nanolestes* (Martin, 2002) and several other *Peramus*-like taxa (Sigogneau-Russell, 1999), although less developed in *Peramus* itself. This derived character is absent in the immediate and the successively more distant outgroups of Zatheria, including: *Vincelestes* (Rougier, 1993; Sigogneau-Russell, 1999), *Amphitherium* (Mills, 1964; Butler and Clemens, 2001), dryolestids (Krebs, 1991; Martin, 1999a), and spalacotheriids (Hu et al., 1998; Cifelli and Madsen, 1999). Basal taxa within the monophyletic group, including stem boreosphenidans, have developed a deep groove in the parastylar region of the upper molar. This is an apomorphy for Zatheria.

The best preserved taxon among stem Zatheria is *Peramus*, known only by the dentition and mandible (Clemens and Mills, 1971; Sigogneau-Russell, 1999). The recently described *Nanolestes* also provides significant anatomical information for basal zatherians (Martin 2002). *Peramus, Palaeoxonodon, Magnimus, Minimus, Afriquiamus,* and *Nanolestes* are the basal member of this clade (Sigogneau-Russell, 1999; Martin, 2002, see chapter 10).

Boreosphenida (Luo, Cifelli, and Kielan-Jaworowska, 2001, modified from Tribosphenida McKenna, 1975). This clade is defined as the common ancestor of *Kielantherium* and living marsupials and placentals, plus all of its descendants (shaded area in figures 15.1 and 15.2: node 10, see also chapters 11, 12, and 13). The most important apomorphy of this group is the presence of grinding features within the talonid basin that are the true derived characters of tribosphenic molars. Other features of the basal boreosphenidans are unknown because most of the stem boreosphenidan taxa are represented only by dental specimens.

Basal members of Boreosphenida include a host of taxa represented by isolated teeth from the Aptian–Albian and Late Cretaceous of North America (e.g., Fox, 1972b, 1976a, 1980b, 1982; Butler, 1978a; Slaughter, 1981; Clemens and Lillegraven, 1986; Jacobs et al., 1989; Cifelli, 1990c,d, 1993b). The earliest-known taxa belonging to this clade, from the Berriasian (Sigogneau-Russell and Ensom, 1994; Sigogneau-Russell et al. 2001) and Valanginian (Kermack et al., 1965) of Britain and the Berriasian of Morocco (Sigogneau-Russell, 1991a, 1992, 1994a), are known by one or a few isolated teeth. *Kielantherium* is somewhat better represented, being known by a dentary and the lower molar series (Dashzeveg and Kielan-Jaworowska, 1984). There are 20 published Early Cretaceous boreosphenidan genera (including eutherians and metatherians), distributed among Europe, Asia, and North America. Of these, stem boreosphenidans (11 genera) have been reported from all three major landmasses. Eutherians (5 genera) are known from Asia and North America for this time interval; thus far, the uncontested marsupials (4 genera) of this age are restricted to North America (Kielan-Jaworowska and Cifelli, 2001). Putative stem metatherians are known from both Asia and North America. The total number of Early Cretaceous boreosphenidans would be considerably higher (about 30 taxa) if unnamed taxa are included. Boreosphenidans of the Early Cretaceous are only known from the northern continents (chapter 11).

ALTERNATIVE PLACEMENTS OF EUTRICONODONTANS AND MULTITUBERCULATES

Eutriconodontans. In the cladogram in figure 15.1 the eutriconodontans (see chapter 7) are placed within the crown group Mammalia. *Gobiconodon* and *Triconodon* were placed within the mammalian crown group by Rowe (1988), and this was corroborated by additional petrosal apomorphies from the study of Rougier, Wible, and Hopson (1996). In these previous cladistic analyses of the cranial and mandibular characters, eutriconodontans, multituberculates, and trechnotherians belong to the same monophyletic group. However, in recent phylogenetic analyses that emphasized the postcranial characters, the eutriconodontans *Gobiconodon* and *Jeholodens* were placed outside the mammalian crown group (see Hu et al., 1997; Ji et al. 1999).

In the most recent review on the alternative placements of eutriconodontans, Luo et al. (2002) concluded that the different placements of eutriconodontans with regard to the mammalian crown group is caused by different sampling of cranial characters, especially the derived petrosal characters. When all available postcranial characters are combined with all available cranial characters, the derived petrosal characteristics dominate the outcome of the parsimony analysis in favor of the placement of eutriconodontans within the mammalian crown group, albeit only weakly (Luo et al., 2002). It is also shown that the monophyly of Eutriconodonta is weakly supported by strict parsimony analysis (figure 15.1). In the constrained search whereby allotherians were forced outside of the mammalian crown group, if the characters were treated as unordered, eutriconodontans collapsed as a clade and became a paraphyletic group (figure 15.2). Given these results, we only tentatively regarded Eutriconodonta as a clade (see chapter 7).

Multituberculates. As discussed earlier, the alternative placements of multituberculates with either haramiyidans or within the mammalian crown group are primarily influenced by the conflicts of different character complexes (or data partitions, see chapter 8). The putative phylogenetic affinities, as indicated by the highly specialized dental features, are often in conflict with the phylogenetic signals from the skull and postcranial characters. These homoplasies and character conflicts are the primary reasons for the long-standing difficulty in achieving a consensus concerning the relationships of multituberculates to other major Mesozoic mammal groups. All hypotheses of multituberculates, as presently proposed, would have to invoke some significant homoplasies in some character complexes.

The multicuspate rows of multituberculates are not unlike those of nonmammalian tritylodontids and are very similar to the features of haramiyidans of the Late Triassic to Late Jurassic. Around the turn of twentieth century, when tritylodontids and multituberculates were known only by teeth and jaw fragments, they were considered to be related to each other. The tritylodontid-multituberculate relationship was part of the reason for the now-abandoned idea for multiple origins of mammals (Owen, 1884; Simpson, 1928a, 1929a). This early idea was abandoned upon the discovery of complete skulls of tritylodontids (Young, 1940, 1947) and multituberculates (Simpson, 1937a). Watson (1942), and in particular, Young (1947) pointed out that tritylodontids are "reptilian" in lacking the dentary-squamosal jaw joint and in retaining the postdentary bones on the mandible, whereas multituberculates are clearly mammalian in these features. Subsequent and exhaustive studies of the cranium and skeleton have further reinforced the view that tritylodontids and multituberculates are not related despite their similar dental features. The multituberculate-tritylodontid hypothesis has been abandoned by almost all students of cynodonts and multituberculates for the past 60 years, including Simpson (1971), with the exception of Tatarinov (1985) and Butler and MacIntyre (1994), who reemphasized the dental similarities of multituberculates and tritylodontids.

Since their initial discovery in the late nineteenth century, haramiyidans were regarded as being as possible relatives of multituberculates on the basis of their dental similarities (reviewed by Simpson, 1928a, 1929a). From the 1960s through the 1980s, as more and better dental fossils of haramiyidans became known, there was a growing recognition that haramiyidans are a distinctive group of the Late Triassic to the Early Jurassic (recently described *Staffia* extends this record to the Late Jurassic, at least in Africa; Heinrich, 1999, 2000). Haramiyidans are separated by a large stratigraphic gap from multituberculates, which make their first undoubted appearance in the Late Jurassic (e.g., Hahn, 1973; Clemens, 1980a; Hahn et al., 1989; Sigogneau-Russell, 1989b). Nonetheless, several specialists who have studied the dentition of the Triassic-Jurassic haramiyidans advocated placement of haramiyidans with multituberculates in an allotherian group (e.g., Hahn et al., 1989; Sigogneau-Russell, 1989b; Butler and MacIntyre, 1994). This is essentially the same as Simpson's (1928a) view of their relationship (though Simpson was very tentative as to relationships of haramiyidans), but much better documented by more exhaustive morphological and functional analyses of haramiyidan teeth, as most recently summarized by Butler (2000, see chapter 8). As haramiyidans are among the earliest-known mammals, this hypothesis implied that the allotherian clade represents an early, divergent lineage of mammals, a sister taxon to all other mammals (as argued by McKenna, 1987, and Miao, 1993).

This view is not without its critics. Rowe (1988, 1993) offered different placements of haramiyidans and multituberculates on the cladogram. Although not discussed in detail, Rowe's (1988) mammalian tree implies that multituberculates are part of the mammalian crown group, whereas haramiyidans are not. Discovery by Jenkins et al. (1997) of a Late Triassic haramiyidan, *Haramiyavia*, cast doubts on the haramiyidan-multituberculate relationship. *Haramiyavia* revealed that the mandibular structure of haramiyidans is different from that of uncontested multituberculates in several key features. *Haramiyavia* has a primitive masseteric fossa that does not extend below the last premolar as in multituberculates. This indicates that the anterior insertion of the masseter muscle of multituberculates (Gambaryan and Kielan-Jaworowska, 1995) and the inferred power stroke did not occur in *Haramiyavia* (Jenkins et al., 1997). These authors showed that

Haramiyavia has the primitive postdentary trough for accommodating the postdentary bones. By contrast, these structures are entirely absent in uncontested multituberculates, wherein the postdentary elements are incorporated into the middle ear, as in Crown Mammalia. Jenkins et al. (1997) concluded that *Haramiyavia* had orthal jaw movement during the power stroke, in contrast to the backward power stroke of multituberculates (but see Butler, 2000). On the basis of these differences, Jenkins et al. (1997) excluded haramiyidans from Allotheria (chapter 8).

Our parsimony analyses produce results that counter the opinion of most specialists that haramiyidans might be relatives of multituberculates. In the cladogram in figure 15.1, the haramiyidans (*Haramiyavia*) lie outside Mammalia and are not related to multituberculates, which is similar to the position of Rowe (1988) and Jenkins et al. (1997). However, even if there is enough evidence to conclude that haramiyidans do not belong within the mammalian crown group, we do not have enough information to propose an explicit alternative sister-group relationship of haramiyidans with any other stem mammals or nonmammalian cynodonts. We cannot rule out the possibility that haramiyidans are related to tritylodontids (figure 15.1).

The Sister-Group Relationship of Multituberculates and Monotremes. This idea was first developed by Wible and Hopson (1993) on the basis of derived braincase and petrosal structures. Similarities in braincase structure between multituberculates and monotremes have been noted since they were pointed out by Kielan-Jaworowska (1971). The only subsequent support for this hypothesis was from features of the ear ossicles, provided by Meng and Wyss (1995). As discussed by Rougier, Wible, and Novacek (1996a), extensive sampling and comparison of other cranial and postcranial features reveal that the shared similarities between multituberculates and monotremes in some basicranial and middle ear features are outnumbered by conflicting features. The systematic characters supporting a multituberculate-monotreme relationship were all included in the dataset of Luo et al. (2002), but the hypothesis is not supported by their parsimony analysis, as most recently advocated by Meng and Wyss (1995).

Sister-Group Relationships of Multituberculates and Eutriconodontans. Kielan-Jaworowska (1986) argued that multituberculates and eutriconodontans shared a close relationship on the basis of some brain endocast features. She suggested that multituberculate and eutriconodontan endocasts are characterized by their lack of the evident exposure of the midbrain on the dorsal side and differ in this respect from endocasts of Cretaceous eutherians and from brains of primitive extant placentals and marsupials, in which the midbrain is largely exposed dorsally. Recent studies (Kielan-Jaworowska and Lancaster, 2004), however, suggested that the difference between brain endocasts of multituberculates and eutriconodontans, on the one hand, and therians *sensu stricto*, on the other, can be interpreted as the presence in the former of a cast of a large sinus (the superior cistern) that obscures the vermis and the midbrain (see also chapters 3, 7, and 8). Thus the difference between multituberculates plus eutriconodontans and therians concerns the cranial vasculature, rather than the structure of the brain itself.

In both multituberculates (Kielan-Jaworowska and Gambaryan, 1994) and the eutriconodontan *Jeholodens* (Ji et al., 1999), Mt III is abducted from the longitudinal axis of the tuber calcanei (as in monotremes), rather than parallel to it as in crown therians. In both groups Mt V is offset from the cuboid and separated by a short distance from the peroneal end of the calcaneus (see figures 7.9E and 8.14). By contrast, Mt V, the cuboid, and the calcaneus are in alignment in morganucodontids (Jenkins and Parrington, 1976) and in Crown Theria (Kielan-Jaworowska, 1977, 1978; Horovitz, 2000; Ji et al., 2002). In multituberculates and *Jeholodens,* there are no obvious facets on the distal margin of the calcaneus and proximal margin of Mt V, so we cannot exclude the possibility that a small cartilage might also have been retained in the space between the offset Mt V and the cuboid, as has been recognized in the so-called "Manda" cynodont (Schaeffer, 1941; Jenkins, 1971b; Szalay, 1993b). Support for a multituberculate-eutriconodont relationship from these ankle features is equivocal. Another character common to multituberculates and *Jeholodens* is the medially placed jugal on the zygoma. Again, this feature is also present in monotremes (Kuhn, 1971; Zeller, 1989b).

Sister-Group Relationship of Multituberculates and Trechnotherians (modified from Theriiformes of Rowe, 1988). This hypothesis has been based on several derived features of the postcranial skeleton (see Rowe, 1986, 1988, 1993; Rowe and Greenwald, 1987; Sereno and McKenna, 1995; Hu et al., 1997). Several unambiguous derived features are from the hindlimb. Multituberculates and trechnotherians both have a spherical femoral head on a constricted neck, with a prominent greater trochanter that is vertically directed and separated from the shaft by a deep incisure. In the ankle joint both groups have a high, laterally compressed tuber calcanei and partial superposition of the astragalus and the calcaneus. On the other hand, the forelimb and pectoral girdle features used to support a multituberculate-trechnotherian relationship appear to be ambiguous and variable (Rougier, Wible, and Novacek, 1996b; Gambaryan and Kielan-Jaworowska, 1997) or turn out to have a wider distribution outside multituberculates and trechnotherians (Ji et al., 1999). A possible argument against multituberculate-trechnotherian affinities may be the unique structure of the multituberculate dentition, which is similar to that of haramiyidans (Butler, 2000). However, in the evolution of mammals during the Tertiary one may cite several groups of therians *sensu stricto* that developed a comparatively "unique" type of dentition, very different from the basal tribosphenic type.

A variant of the Theriiformes hypothesis is that multituberculates (not allotherians) may be the sister taxon to "holotherians" (i.e., the clade of *Kuehneotherium* + monotremes + living therians) (see Miao, 1991: figure 1). Several studies have placed multituberculates in an unresolved trichotomy with living monotremes and living therians (Kemp, 1983; Wible, 1991; Rougier et al., 1992; Lillegraven and Hahn, 1993). This is somewhat similar to the theriiform hypothesis, inasmuch as multituberculates are placed with Crown Mammalia, but differs in that the position of multituberculates is not further resolved.

The multituberculate-trechnotherian hypothesis has been supported by the parsimony studies of Luo et al. (2002, see our figure 15.1). The cladogram represents the strict consensus tree from unconstrained searches, based on 50 equally parsimonious trees. Multituberculates (or allotherians *sensu stricto* excluding haramiyidans) in this cladogram are placed very close to the trechnotherian clade, well within the mammalian crown group. Such placement of multituberculates is in general agreement with previous cladistic analyses of early mammals, for example, by Rowe (1988, 1993), Wible (1991), Rougier, Wible, and Hopson (1996), and Rougier, Wible, and Novacek (1996a), although the detailed placements of Multituberculata relative to eutriconodontans in these analyses vary due to the different sampling of taxa (see Kielan-Jaworowska, 1997, and Luo et al., 2002, for reviews).

From this review, it is clear that the most important disagreement is whether multituberculates should be placed outside the mammalian crown group (e.g., in close relation to haramiyidans, as argued by several specialists on the dentition) or within it (e.g., in close relation to trechnotherians). To test these different placements, Luo et al. (2002) imposed a topological constraint excluding multituberculates from the crown mammalian clade and obtained a second, suboptimal cladogram (our figure 15.2), for comparison to the most parsimonious tree in which multituberculates are separated from haramiyidans and are sister taxon to trechnotherians (our figure 15.1). In this second cladogram, allotherians occupy a position outside the mammalian crown group, between *Hadrocodium* and eutriconodonts (*Jeholodens*), which would be consistent with the traditional view of an early divergence of multituberculates from remaining mammals based on the uniqueness of the multituberculate dentition. Luo et al. (2002) stressed that the two alternative topologies that they obtained (figures 15.1, 15.2) differ by only about 1% of the total tree length (based on a search with either ordered or unordered characters) and that statistical tests show that these positions do not differ significantly from one another. The position of allotherians among early mammals remains as one of the most controversial and unresolved problems of early mammalian evolution.

ALTERNATIVE PLACEMENTS OF KUEHNEOTHERIIDS

Kuehneotherium has principal molar cusps arranged in an obtuse-triangle pattern (when seen in occlusal view), with upper and lower triangles reversed with respect to each other. This pattern has been regarded as representing a great advance in the functional evolution of mammalian molar occlusion (chapter 9), as linking "obtuse-angle symmetrodontans" to "eupantotherians" and, finally, to Theria *sensu stricto*. Since its discovery in the 1950s (although not formally erected as a taxon until 1968), it has been assumed by all specialists on mammalian dentition that all triangulated molar cusp patterns of the later and more derived "therian" groups are derived from a precursor condition similar to that seen in *Kuehneotherium* (e.g., Butler, 1939; Patterson, 1956; Crompton and Jenkins, 1967, 1968; Kermack et al., 1968; Crompton, 1971, 1995; Parrington, 1973; McKenna, 1975).

Kuehneotherium, however, is very primitive in having a postdentary trough and a medial ridge on the dentary, indicating that the postdentary elements remained associated with the mandible, as in cynodonts, rather than being incorporated into the middle ear, as in living mammals (Kermack et al., 1968; Allin and Hopson, 1992; Rowe, 1993; Hopson, 1994; Rougier, Wible, and Novacek, 1996a; Kielan-Jaworowska et al., 1998). The combination of molars of advanced structure together with a mandible showing primitive "reptilian" features results in character conflict for any placement of *Kuehneotherium* in mammalian phylogeny: either a triangulated molar cusp pattern evolved more than once or the separation of the middle ear from the mandible occurred more than once (e.g., Miao and Lillegraven, 1986; Allin and Hopson, 1992; Rowe, 1993; Rougier, Wible, and Novacek, 1996a; Luo et al., 2002). A third possibility is that both the molar cusp triangulation and the separation of the middle ear from the dentary are homoplastic and evolved independently in more than one lineage.

Most of the early-mammal studies in the 1970s clearly considered the reversed triangulation of molar cusps to be supremely important and to be homologous among all "therian" mammals, whereas the conflicting, primitive mandibular features have been regarded as homoplasies. From the 1950s (Kermack and Mussett, 1958a) through the early 1990s (e.g., Allin and Hopson, 1992; Hopson, 1994), the loss of the postdentary trough and the separation of middle ear elements from the mandible were interpreted as convergent features. This line of thinking was consistent with the long-held interpretation that the mammalian jaw joint and middle ear evolved independently in multiple cynodont lineages, resulting in polyphyletic origins of the Mammalia (e.g., Olson, 1959; Simpson, 1959).

More recent large-scale phylogenetic analyses have incorporated an increasing number of cranial and skeletal features, the importance of which has been emphasized throughout this chapter. All of these studies supported an interpretation that the reversed-triangle pattern of molar cusps evolved more than once (see reviews by Rowe, 1993; Rougier, Wible, and Hopson, 1996; Rougier, Wible, and Novacek, 1996a; Cifelli and Madsen, 1999; Pascual et al., 1999, 2000, 2002; and Luo et al., 2002). Our most recent studies clearly suggest that *Kuehneotherium* must be placed outside the mammalian crown group (figures 15.1, 15.2). Because kuehneotheriids are very incomplete and the little morphology that is preserved on the limited fossils shows very significant homoplasies, the position of *Kuehneotherium* is inherently unstable.

PARAPHYLETIC AND ABANDONED HIGH-RANK TAXONOMIC UNITS

During the past two centuries of scientific studies of extant and fossil mammals, several high-rank taxonomic units have been erected and later abandoned, many of which are synonymous, others homonymous, and some either polyphyletic or paraphyletic. Following the principles of phylogenetic systematics (e.g., Hennig, 1966), in this book we restricted formal usage of taxonomical names to those high-level groups for which monophyly is supported. This criterion eliminates paraphyletic taxa (see later). However, in such a review book as ours, use of informal but widely understood names and traditional vernacular names that are not monophyletic and are not formally recognized as valid cannot be avoided.

As noted in chapter 1, we identify paraphyletic names by citing them in quotes. Some paraphyletic taxa often must be mentioned for historical reasons. In other cases, traditional groupings were established on and named after a specific molar pattern that is now considered to be either primitive or homoplastic, but nonetheless the name must be mentioned in the discussion of dental morphology. For example, the term "triconodont" must be used in a morphological sense, referring to mammals with molars bearing the three main cusps arranged in a longitudinal row; "triconodonts" however, cannot be regarded as a clade, as taxa with "triconodont" molars are assigned to widely separated groups, treated in different chapters: *Sinoconodon* and morganucodontans (chapter 4) and "eutriconodontans" (chapter 7). "Symmetrodontans" (or "symmetrodonts") refer to mammals with a primitive and (in our judgment) homoplastic reversed-triangle molar pattern. We use "symmetrodontans" as a heading for chapter 9, treating them, however, as a grade but not a clade; similar are the cases of "Eupantotheria" (or "eupantotherians," see chapter 10) and "tribotherians" (or tribotherians, see chapter 11). Another problematic group is "Holotheria" (see discussion in chapter 9). We advocate abandonment of these taxonomic names for the purposes of formal classification.

"TRICONODONTS"

When "Triconodonta" as an order were first introduced in the literature, the taxon only included amphilestids and triconodontids (*sensu stricto*) (e.g., Osborn, 1907; Simpson, 1928a, 1929a). On the basis of dental features, Kermack et al. (1973) included Morganucodonta (*Morganucodon* and dentally similar taxa from the Rhaeto-Liassic), Docodonta, and Eutriconodonta (= "amphilestids" + triconodontids) as suborders within the order Triconodonta (see chapters 4, 5, and 7). The placement of morganucodonts in "Triconodonta" was widely accepted for more than two decades because they all have similar and "triconodont-like" teeth (see reviews by Jenkins and Crompton, 1979; Miao and Lillegraven, 1986; and Hopson, 1994: figure 8).

Kemp (1983) and Rowe (1988) showed that morganucodontans are far more primitive than eutriconodontans in mandibular characters. Further studies helped to demonstrate that morganucodontans are much more primitive than eutriconodontans in features of the basicranium (Wible and Hopson, 1993; Rougier, Wible, and Hopson, 1996) and skeleton (Ji et al. 1999), despite their similarities in some (plesiomorphic) dental characters. Taking into account all of the dental, mandibular, and postcranial features, morganucodontans are no more closely related to eutriconodontans than they are to other members of the mammalian crown group. The dental similarities of morganucodontans and eutriconodontans would have to be overruled as homoplasies on parsimony grounds. Subsequently, it also became clear that *Sinoconodon* (see Patterson and Olson, 1961) has a more primitive dentition than any other mammal (Crompton and Sun, 1985; Crompton and Luo, 1993; Zhang et al., 1998), and it was removed from the "Triconodonta." The study of Luo et al. (2002), with a comprehensive sampling of characters, reaffirmed the paraphyletic status of "Triconodonta" *sensu lato* (as used by Kermack et al., 1973, and followed by many in the 1970s and 1980s).

"SYMMETRODONTANS"

In the traditional vernacular use of terms, spalacotheriids, tinodontids, and kuehneotheriids were all referred to as "symmetrodonts" (reviewed in chapter 9) because these groups all appear to have a reversed-triangle pattern of molar cusps. The Spalacotheriidae, with "acute-angle symmetrodont" molars, are clearly a monophyletic family (Cifelli and Madsen, 1999, also see chapter 9), and it belongs to the trechnotherian clade (McKenna, 1975; Prothero, 1981; Hu et al., 1998). However, the affinities of *Kuehneotherium,* which has "obtuse-triangle" molars, to the more derived trechnotherian clade have been in doubt (see Rougier, Wible, and Hopson, 1996). The recent analysis by Luo et al. (2002), which included all known dental and mandibular features of *Kuehneotherium* and *Tinodon,* demonstrated that there are no synapomorphies to support a broadly conceived "Symmetrodonta" that include these kuehneotheriids, tinodontids, and spalacotheriids. The "obtuse-angle symmetrodonts" themselves are not closely related to each other. These authors argued that the archaic "obtuse-angle symmetrodonts" *Kuehneotherium* and *Tinodon* represent a heterogeneous evolutionary grade that lacks reliable diagnostic features and are probably unrelated to each other or to "acute-angle symmetrodonts," as reflected in figures 15.1 and 15.2, and occupy some basal positions in mammalian phylogeny that are not well resolved by the limited anatomical data on the basis of so far known incomplete fossils.

Molar cusp triangulation is a variable feature in the so-called "obtuse-triangle symmetrodonts." Triangulation of cusps can vary between the anterior and the posterior molars. The posterior lower molars have better-developed triangulation than the anterior lower molars, as has been observed for *Kuehneotherium* (Parrington, 1978), *Tinodon* (Crompton and Jenkins, 1967), and *Gobiotheriodon* (Averianov, 2002). The variability of molar cusp triangulation observed in "obtuse-angle symmetrodonts" also occurs in several non-"symmetrodontan" mammals. A gradient of increasing triangulation toward the posterior upper molars is seen in *Gobiconodon* (Jenkins and Schaff, 1988; Kielan-Jaworowska and Dashzeveg, 1998) and, to a lesser degree, in *Megazostrodon* as well (Crompton, 1974; Parrington, 1978). It should be noted that in *Gobiconodon* and *Megazostrodon* the variable triangulation is known for the upper but not the lower molars. These observations call into question whether the obtuse-triangle of molar cusps, with known variability, would be a reliable apomorphy for defining a wide range of early mammal groups. This undermines the "symmetrodontans" as an apomorphy-based taxonomic unit.

"THERIANS" *SENSU LATO* AND "HOLOTHERIANS"

Theria Parker and Haswell, 1897, were originally proposed to distinguish two of the living mammalian groups, Metatheria and Eutheria, from the third, Monotremata. In a widely adopted approach to defining a major taxon by its crown group today, Rowe (1988, 1993) advocated that the name Theria be restricted to the clade including the common ancestor of extant marsupials and placentals, plus all of its descendants. However, for almost a century since it was first coined, the term "therians" has been used with a number of different meanings. Simpson (1945) recognized certain Mesozoic mammals as more closely related to living therians than to monotremes and formalized an expanded concept of "Theria" by adding the infraclass "Pantotheria" (including "eupantotheres" of Kermack and Mussett, 1958a, see chapter 10). This concept was further refined and expanded in subsequent years, ultimately leading to a widespread acceptance of the reversed-triangle molar as an apomorphy-based definition for "therians" (e.g., Kermack, 1967b; Hopson and Crompton, 1969; McKenna, 1975). Wherever feasible in this book, we use the name crown therians (for living groups of therians), which is equivalent to Rowe's (1988) Theria. We also use boreosphenidans, zatherians, cladotherians, and trechnotherians for the successive ranks of clades in place of "Therian" *sensu lato.*

Hopson (1994) erected the informal taxon Holotheria as a character-based unit, diagnosed by the reversed triangle of molar cusps. "Holotherians" were designated for one of two clades representing an early, fundamental dichotomy in mammalian history. The need for replacing Theria *sensu lato* by Holotheria was to distinguish the more inclusive group of "holotherians" from the more restricted (crown group) definition for Theria proposed by Rowe (1988, 1993; see also Rowe and Gauthier, 1992, and McKenna and Bell, 1997) which excludes "symmetrodontans" and "eupantotherians." However, in Hopson's original proposal, monotremes are nested within the Holotheria, whereas triconodontids and multituberculates are excluded. Wible et al. (1995: 11) followed up in using the "holotherians," but with an important difference: monotremes were excluded from "holotherians." McKenna and Bell (1997: 43) followed the definition of Holotheria given by Wible et al. (*Kuehneotherium* + crown therian group) and conceived of it as a replacement name for "therians" *sensu lato,* excluding monotremes, eutriconodontans, and multituberculates.

As discussed earlier and also pointed out in several previous papers (Rougier, Wible, and Hopson, 1996; Luo et al., 2002), the position of *Kuehneotherium* in mammalian phylogeny is not stable. It is currently incomplete so there are few diagnostic features. The limited features known for this taxon have demonstrated some tremendous conflicts of very primitive mandibular characters and relatively derived molar cusp triangulation. The main diagnostic feature of molar triangulation is also variable in the intended basal taxa. The Holotheria, as a taxonomic unit defined by *Kuehneotherium* (*sensu* Wible et al., 1995, and McKenna and Bell, 1997) as well as other archaic taxa with a reversed-triangle molar pattern, up to extant mammals, are not monophyletic by the best available evidence (figures 15.1, 15.2). Because *Kuehneotherium* is very incomplete and its phylogenetic position is unstable, any group defined on the basis of it is not a stable one. For this reason we do not use Holotheria in this book as a formal taxonomic name. We use the term "therians *sensu lato*" (= "holotherians" of Hopson, 1994), to convey the historically important (and still widely used) concept of a group including *Kuehneotherium* and other "symmetrodontans," "eupantotherians," Metatheria, and Eutheria. A number of traditionally used taxa nested within "therians" now appear to be paraphyletic. In this book these are all cited in quotation marks.

CONCLUDING REMARKS

The vast taxonomic diversity of Mesozoic mammal groups and their enormous scope of anatomical evolution present a great challenge for building a taxonomically comprehensive phylogeny from parsimonious analyses of a single "supermatrix" that combines characters from different anatomical areas (dentitions, mandibles, crania, and postcranial skeletons). Definition of highly transformed characters and their character states can be controversial. The empirical assessment of the often distorted and partially preserved fossils can also be difficult. Choice for characters for resolving certain phylogenetic problems can also be debated. The proposed phylogeny for Mesozoic mammals (figures 15.1, 15.2) represents the best available morphological evidence that we have managed to gather thus far. We expect that future discoveries of new fossils will help to fill in the blank areas of the dataset and lead to new questions in Mesozoic mammal evolution. We offer this dataset (Appendix) in the hope and expectation that it will be critically reevaluated, corrected, and added to by fellow aficionados of early mammals, present and future.

APPENDIX

MORPHOLOGICAL CHARACTERS OF MESOZOIC MAMMALS AND THEIR SYSTEMATIC DISTRIBUTIONS AMONG MAJOR CLADES

The following character list is adopted from Luo et al. (2002) and represents a summary of the morphological characters of early mammals by previous studies from the 1980s and 1990s. The reader should refer to Luo et al. (2002) for the original sources of the morphological characters, detailed rationale, and justification for the definition of characters and division of character states.

CHARACTER LIST

1. Postdentary trough (behind the tooth row): (0) Present; (1) Absent.
2. Separate scars for surangular/prearticular in the postdentary trough: (0) Present; (1) Absent.
3. Overhanging medial ridge above the postdentary trough (behind the tooth row): (0) Present; (1) Absent.
4. Curvature of Meckel's groove (under the tooth row) in adults: (0) Present, parallel to ventral border of mandible; (1) Present, convergent with ventral border of mandible.
5. Degree of development of Meckel's groove in adults: (0) Well developed; (1) Weakly developed; (2) Vestigial or absent.
6. Mandibular symphysis: (0) Fused; (1) Unfused; (2) Unfused and further reduced.
7. Groove for the replacement dental lamina: (0) Present; (1) Absent.
8. Angular process of dentary—presence versus absence: (0) Present; (1) Reduced or weakly developed; (2) Present, transversely flaring; (3) Present, slightly inflected; (4) Present, strongly inflected and anteriorly continuing as a mandibular shelf; (5) Absent.
9. Angular process of dentary—anteroposterior position: (0) Anterior position (angular process below main body of coronoid process, widely separated from dentary condyle); (1) Posterior position (angular process positioned at same level of posterior end of coronoid process; either close to or directly under dentary condyle).
10. Angular process of dentary—vertical elevation: (0) Low position, at or near level of ventral border of mandibular horizontal ramus; (1) High position, at or near level of molar alveolar line (and far above ventral border of mandibular horizontal ramus).
11. Coronoid (or its attachment scar) in adults: (0) Present; (1) Absent; (A) (0/1 polymorphic).
12. Mandibular foramen (posterior opening of the mandibular canal) for the inferior alveolar nerve and vessels: (0) Located within postdentary trough (depression around foramen part of Meckel's groove and postdentary trough); (1) Foramen not associated with either postdentary trough or Meckel's groove.
13. Medial concavity (excavated fossa) for the reflected lamina of the angular bone on the medial side of the dentary angular process: (0) Present; (1) Angular region has no fossa for angular bone.
14. Splenial as a separate element (as indicated by its scar on the dentary) in adults: (0) Present; (1) Absent.
15. Relationship of the surangular to the craniomandibular joint (CMJ): (0) Participate in CMJ either as

separate bone or fused with articular; (1) Surangular as bony element lost in adult.

16. Pterygoid muscle fossa on the medial side of the mandible: (0) Absent; (1) Present.

17. Medial pterygoid ridge (shelf) along the medial aspect of the ventral border of the coronoid part of the mandible: (0) Absent; (1) Present; (2) Pterygoid.

18. Ventral border of masseteric fossa on the lateral aspect of the mandible: (0) Absent; (1) Present as low and broad crest; (2) Present as well-defined and thin crest.

19. Crest of the masseteric fossa along the anterior border of the coronoid process: (0) Absent or weakly developed; (1) Present; (2) Present and laterally flaring.

20. Anteroventral extension of the masseteric muscle fossa: (0) Below or posterior to ultimate molar; (1) Anterior extension below ultimate premolar.

21. Orientation of the dentary peduncle and condyle: (0) Peduncle posteriorly directed; (1) Dentary condyle continuous with rounded posterior margin of dentary; condyle facing up owing to upturning of posteriormost part of dentary; (2) Dentary articulation extends vertically for entire depth of horizontal ramus of mandible; confluent with horizontal ramus and without a peduncle; dentary articulation posteriorly directed; (3) Vertically directed dentary peduncle; (B) (1/2 polymorphic).

22. Shape and relative size of the dentary articulation: (0) Small and dorsoventrally compressed: tritheledontids (represented by lateral ridge); (1) Condyle massive and bulbous, transversely broad in dorsal aspect; (2) Condyle mediolaterally narrow and vertically deep, forming broad arc in lateral outline, either ovoid or triangular in posterior view; (A) (0/1 polymorphic).

23. Ventral (inferior) border of dentary peduncle: (0) Posteriorly tapering; (1) Columnar, with lateral ridge; (2) Ventrally flaring; (3) Robust and short; (4) Ventral part of peduncle and condyle continuous with ventral border of mandible.

24. Position of the dentary condyle relative to the vertical level of postcanine alveoli: (0) Below or at about same level as postcanine alveoli; (1) Above level of postcanine alveoli; (A) (0/1 polymorphic).

25. Tilting of the coronoid process of the dentary (measured as the angle between the imaginary line of the anterior border of the coronoid process and the horizontal alveolar line of all molars): (0) Coronoid process strongly reclined and coronoid angle obtuse (≥150°); (1) Coronoid process less reclined (135–145°); (2) Coronoid process less than vertical (115–125°); (3) Coronoid process near vertical and coronoid angle small (95–105°); (B) (1/2 polymorphic: i.e., with range of variation of 130~150°).

26. Alignment of the erupting ultimate molar to the anterior margin of the dentary coronoid process (and near the coronoid scar if the scar is present): (0) Ultimate functional molar erupts medial to coronoid process; (1) Ultimate functional molar erupts in alignment with coronoid process.

27. Ultimate lower premolar symmetry of the main cusp a (protoconid) (as measured by the length ratio of the anterior and posterior cutting edges extending from the cusp: (0) Asymmetrical (anterior edge of cusp a more convex in outline than posterior edge); (1) Symmetrical (anterior and posterior cutting edges equal or subequal in length; neither edge more convex or concave in lateral profile).

28. The ultimate lower premolar—anterior cusp b (= paraconid). (Absence or reduction of anterior cusp b tends to make the premolar appear more asymmetrical; this character is an indication of the asymmetry of the premolar): (0) Present (at least subequal to cusp c of same tooth); (1) Small (much smaller than cusp c of same tooth) or vestigial to absent.

29. Ultimate lower premolar—arrangement of principal cusp a, cusp b (if present), and cusp c (assuming that the cusp is c if there is only one cusp behind the main cusp a): (0) Aligned in single straight line or at slight angle; (1) Distinct triangulation; (2) Premolar (or anterior postcanines) with multicusps in longitudinal row(s): (3) Premolar multicuspate and bladelike.

30. Ultimate lower premolar—posterior (distal) cingulid or cingular cuspule d (in addition to the main cusp c): (0) Absent; (1) Posterior cingular cusp present; (2) Presence of continuous posterior (distal) cingulid.

31. Ultimate lower premolar—outline: (0) Laterally compressed (or slightly angled); (1) Transversely wide (coding for *Obdurodon* follows Archer et al., 1993).

32. Labial cingulid of ultimate lower premolar: (0) Absent or vestigial; (1) Present (at least along length of more than half the crown).

33. Lower premolar lingual cingulid: (0) Absent or vestigial: (tritylodontids coded from anterior postcanines); (1) Present.

34. Relative height of primary cusp a to main cusps b and c of the ultimate lower premolar (measured as the height ratio of a and c from the bottom of the valley between the two adjacent cusps): (0) Posterior cusp c indistinct; (1) Posterior cusp c distinct but less than 30% of the primary cusp a; (2) Posterior cusp c and primary cusp a equal or subequal in height (c 40–100% of a).

35. Ultimate upper premolar—functional protocone (character applicable to mammals with reversed triangulation of molar cusps): (0) Absent; (1) Present.

36. The penultimate lower premolar—anterior cusp b (= paraconid): (0) Vestigial (much smaller than cusp c of

same tooth) to absent; (1) Distinct; (2) Multicuspate row(s).

37. Penultimate lower premolar—arrangement of principal cusps a, b (if present), and c (we assume the cusp to be c if there is only one cusp behind the main cusp a): (0) Cusps in straight alignment (for tooth with single cusp, anterior and posterior crests from main cusp in alignment); (1) Cusps in reversed triangulation; (2) Premolar with multicusps in longitudinal row(s).

38. Alignment of main cusps of posterior lower molar(s) (m3 or more posterior if present) (Note: this character differs from character 39 because in *Tinodon, Gobiotheriodon,* and *Kuehneotherium,* the cusps on m1 (or anterior molars) are hardly triangulated but the posterior molar cusps are distinctively triangulated. There clearly is a gradient of increasing triangulation of molar cusps. Split of characters 38 and 39 reflects these morphological differences and may influence the placement of the taxa with the gradient): (0) Single longitudinal row; (1) Reversed triangle—obtuse (≥95°); (2) Reversed triangle—acute (≤90°); (3) Multiple longitudinal multicuspate rows.

39. Alignment of main cusps of anterior lower molar (m1): (0) Single longitudinal row; (1) Acute reversed triangle; (2) Multiple longitudinal rows.

40. Postvallum/prevallid shearing (angle of the main trigonid shear facets): (0) Absent; (1) Present, weakly developed and slightly oblique; (2) Present, strongly developed and more transverse.

41. Development of postprotocrista on upper molar for double rank postvallum shear (applicable to molars with reversed triangulation of molar cusps): (0) Postprotocrista short, not extending labially beyond metacone; (1) Postprotocrista long, extending labially beyond metacone.

42. Precise opposition of upper and lower molars (either one-to-one or occluding at the opposite embrasure or talonid): (0) Absence of precise opposition of upper and lower molars; (1) Present (either one-to-one or occluding at opposite embrasure or talonid); (2) Present (one lower molar contacts sequentially more than one upper molar).

43. Relationships between the cusps of the opposing upper and lower molars: (0) Absent; (1) Present, lower primary cusp a occludes in groove between upper cusps A and B; (2) Present, lower main cusp a occludes in front of upper cusp B and into embrasure between opposite and preceding upper tooth; (3) Present, parts of talonid occluding with lingual face (or any part) of upper molar; (4) Lower multicuspate rows alternately occluding between upper multicuspate rows.

44. Relative height of primary cusp a (protoconid) to cusp c (metaconid) of the anterior lower molars (measured as the height ratio of a and c from the bottom of the valley between the two adjacent cusps, on m1): (0) Posterior cusp c less than 40% of primary cusp a (protoconid); (1) Posterior cusp c and primary cusp a equal or subequal in height (c 50–100% of a).

45. Relative size/height of anterior cusp b (paraconid) to posterior cusp c (metaconid) (based on m2): (0) c taller than b; (1) b taller than c; (2) b and c more or less equal in height.

46. Elevation of the cingulid base of the paraconid (cusp b) relative to the cingulid base of the metaconid (cusp c) on the lower molars: (0) Absent; (1) Present.

47. The cristid obliqua (oblique crest anterior to and connected with the labialmost cusp on the talonid heel, the leading edge of facet 3; applicable only to molars with at least a hypoconid on the talonid or a distal cingulid cuspule): (0) Absent; (1) Present, pointed lingual of the metaconid-protoconid notch; (2) Present, hypertrophied and directed to posterior (distal) of metaconid; (3) Present, short and pointed anteriorly between metaconid-protoconid notch and protoconid (labial of notch).

48. Lower molar—medial and longitudinal crest (="preentocristid" or "prehypoconulid") on the talonid heel (only applicable to taxa with talonid or at least a cusp d): (0) Talonid (or cusp d) lacks medial and longitudinal crests; (1) Medialmost cristid ("preentoconid cristid") of talonid in alignment with metaconid or with postmetacristid if the latter is present (postmetacristid defined as posterior crest of metaconid parallel to lingual border of crown), but widely separated from the latter; (2) Medialmost cristid of talonid ("prehypoconulid" cristid) hypertrophied and in alignment with postmetacristid and abuts the latter by a V-notch; (3) "Preentocristid" crest offset from metaconid (and postmetacristid if present), "preentocristid" extending anterolingually past base of metaconid.

49. Labial curvature of the primary cusp a of the lower molars (at the base level) relative to those of cusps b and c: (0) Cusp a and cusps b and c have same degree of bulging; (1) Cusp a bulges far more than cusps b and c.

50. Labial curvature of main cusps A, B, and C at the level of the cusp valley on the penultimate and ultimate upper molars; (0) Cusps A, B, and C with about the same degree of curvature; (1) cusp A slightly more concave (or far less convex) than either cusp B or C.

51. Lingulolabial compression of the primary functional cusps of the lower molar (at the level of the cusp base but above the cingulid): (0) Absent; (1) Present.

52. Posterior lingual cingulid of the lower molar: (0) Absent or weak; (1) Distinctive; (2) Strongly developed, crenulated with distinctive cuspules (such as kuhneoconecusp g).

53. Anterior internal (mesiolingual) cingular cuspule (e) on the lower molar: (0) Present; (1) Absent; (A) (0/1 polymorphic).

54. Anterior and labial (mesiobuccal) cingular cuspule (f): (0) Absent; (1) Present.

55. Mesial transverse cingulid above the gum: (0) Absent; (1) Present below trigonid but weak and discontinuous (as individual cuspule e, or f, or both, but e and f not connected); (2) Present in continuous shelf below trigonid (with no relation to protoconid and paraconid), without occlusal function; (3) Cingular area present, has occlusal contact with upper molar.

56. Cingulid shelf wrapping around the anterointernal corner of the molar to extend to the lingual side of the trigonid below the paraconid: (0) Absent; (1) Present, weakly developed, its lingual part without occlusal function to upper molars; (2) Present, strongly developed, without occlusal function to upper molars; (3) Present, weakly developed, its lingual part with occlusal function to upper molars.

57. Postcingulid (distal transverse cingulid) on the lower molar: (0) Absent; (1) Present, oblique and connected to hypoconulid; (2) Present, horizontal above gum level: *Shuotherium* (assuming the positional homology of the distal "cingulid" of this taxon to the postcingulid of other therians).

58. Interlocking mechanism between two adjacent lower molars: (0) Absent; (1) Present, posterior cingular cuspule d (or base of hypoconulid) of preceding molar fits between cingular cuspules e and f of succeeding molar or base of hypoconulid interlocks with concave area between irregular mesial cingulid cuspules in front of paraconid; (2) Present, posterior cingular cuspule d fits between cingular cuspule e and cusp b of succeeding molar; (3) Present, posterior cingular cuspule d of preceding molar fits into embayment or vertical groove of anterior aspect of cusp b of succeeding molar (without any involvement of distinctive cingular cuspules in interlocking).

59. Size ratio of posterior lower molars: (0) Last three postcanines form series of posteriorly decreasing size: penultimate molar larger than ultimate molar but smaller than preceding molar (for total of five molars: m3 ≥ m4 ≥ m5, for total of four molars: m2 ≥ m3 ≥ m4, or for total of three molars, m1 ≥ m2 ≥ m3); (1) Penultimate molar the largest (m1 ≤ m2 ≤ m3 >>>> m4); (2) Ultimate molar the largest.

60. Trigonid configuration on the lower molar: (0) Paraconid in anterolingual position, paraconid-protoconid line forming more oblique angle to longitudinal axis of tooth; (1) Paraconid lingually positioned, paraconid-protoconid line forming more transverse triangle to longitudinal axis of tooth; (2) Paraconid lingually positioned, appressed to metaconid ("twinned").

61. Orientation of paracristid (crest between cusps a and b) relative to the longitudinal axis of the molar (from Hu et al., 1998) (this is separated from character 60 because of a different distribution of the a–b crest among mammals with a nontriangulated cusp pattern): (0) Longitudinal orientation; (1) Oblique; (2) Nearly transverse.

62. Mesiolingual surface of the paraconid on lower molars (applicable only to taxa with reverse triangulation of molar cusps: (0) Rounded; (1) Forming a keel; (A); (0/1 polymorphic).

63. Molar (m2) trigonid/talonid heel width ratio: (0) Narrow (talonid ≤40% of trigonid width); (1) Wide (talonid 40–70% of trigonid width); (2) Talonid as wide or wider than trigonid.

64. Lower molar hypoflexid (the labial embayment between trigonid part and the talonid part of the molar located between the proconid and hypoconid above the cingulid level): (0) Absent or shallow; (1) Deep (but 50% of talonid width); (2) Very deep (≥60% of talonid width).

65. Morphology of the talonid (or the posterior heel) of molar: (0) Absent; (1) Present as incipient heel, cingulid, or cingular cuspule (d); (2) Present as heel (with at least one functional cusp); (3) Present as transverse "V-shaped" basin with two functional cusps; (4) Present as functional basin, rimmed with three functional cusps (there is a functional crest to define medial rim of basin in the case that entoconid is vestigial).

66. Hypoconulid (we designate the distal cingulid cuspule d as the homologue to the hypoconulid in the teeth with linear alignment of the main cusps; we assume that cusp to be the hypoconulid if there is only a single cusp on the talonid in the teeth with reversed triangulation): (0) Present, but not elevated above cingulid level; (1) Present (as distal cusp d, *sensu* Crompton 1971), elevated above cingulid level, labially positioned (or tilted in lingual direction).

67. Hypoconid (if there are only two functional cusps on the talonid, it is assumed that the second and more labial of the two is the hypoconid): (0) Absent; (1) Present, posterior, median, and equidistant to entoconid and hypoconid; (2) Present, placed on lingual rim of talonid basin.

68. Hypoconulid anteroposterior orientation: procumbent versus reclined (applicable to taxa with at least two cusps on the talonid): (0) Cusp tip reclined and posterior wall of hypoconulid slanted and overhanging root; (1) Cusp tip procumbent and posterior wall of cusp vertical; (2) Cusp tip procumbent and posterior wall gibbous.

69. Postcingulid (shelf) labial to hypoconulid on lower molars (this is distinctive from the postcingulid because of different relationship to talonid cusps; only applicable to taxa with identifiable hypoconid and hypoconulid): (0) Absent; (1) Present as crest descending mesiolabially from apex of hypoconulid to base of hypoconid.

70. Entoconid (if there are three functional cusps on the talonid, we assume that the third and the lingualmost functional cusp on the talonid is the entoconid): (0) Absent; (1) Present; (2) Present and twinned with hypoconulid; (A) (0/1 polymorphic).

71. Height of entoconid as compared to other cusps of the talonid (applicable only to taxa with reversed triangulation and a talonid heel or cusp d): (0) Absent on talonid heel; (1) Lower than hypoconulid (or even vestigial); (2) Equal in height to hypoconulid; (A) (0/1 polymorphic).

72. Alignment of paraconid, metaconid, and entoconid on lower molars (applicable only to taxa with triangulation of trigonid cusps and entoconid present on talonid): (0) Cusps not aligned; (1) Cusps aligned.

73. Aspect ratio in occlusal view (length versus width) of the functional talonid basin at the cingulid level (based on m2): (0) Longer than wide (or narrows posteriorly); (1) Length equals width; (2) Wider than long.

74. Elevation of the talonid (measured as the height of the hypoconid from the cingulid on the labial side of the crown) relative to the trigonid (measured as the height of the protoconid from the cingulid) (applicable only to the teeth with reversed triangle): (0) Hypoconid/protoconid height ratio less than 20% (hypoconid or cusp d on the cingulid); (1) Between 25 and 35% (talonid cusp elevated to about cingulid level); (2) Between 40 and 50%; (3) Between 50 and 60%; (4) Between 60 and 70%.

75. Width of upper molar labial stylar shelf (the area labial to the paracone/metacone): (0) Absent; (1) Present and narrow; (2) Present and broad; (3) Present and broad, with a hyptertrophied ectoflexus.

76. Morphology of labial cingulum of the upper molars: (0) Absent or weak; (1) Distinctive cingulum, straight; (2) Distinctive cingulum with strong ectoflexus (but without hypertrophied stylar cusps); (3) Wide cingulum with ectoflexus, plus individualized and hypertrophied stylar cusps; (4) Cingulum crenulated with distinctive and even-sized multiple cuspules; (A) (0/1 polymorphic).

77. Upper molar with a lingual functional protocone or pseudoprotocone that grinds against a basin on the lower: (0) Absent; (1) Present.

78. Transverse width of protocone on upper molars (applicable only to taxa with protocone present; M2 measured where possible): (0) Narrow (distance from protocone apex to paracone apex less than 60% of total tooth width); (1) Strongly transverse (distance from protocone apex to paracone apex greater than 60% of total tooth width).

79. Anteroposterior development of the lingual region on upper molars (applicable only to taxa with reversed triangulation and an occluding lingual portion of the upper molar; for taxa with and without conules, this is measured between the paraconule and metaconule or as the length of the tooth medial to the base of paracone in the absence of conules, respectively; where possible M2 is measured): (0) Narrow (anteroposterior distance medial to paracone and metacone less than 30% of total tooth length); (1) Moderate development (distance between position of conules 31–50% of total tooth length); (2) Long (distance between conules greater than 51% of total tooth length).

80. Conules on upper molars (character applicable to mammals with reversed triangulation of molar cusps and a lingual functional cusp): (0) Absent; (1) Present but weak, without cristae; (2) Conules distinctive, with cristae.

81. Relative height and size of the paracone (cusp B) and metacone of upper molars: (0) Paracone higher and larger than metacone; (1) Metacone higher and larger than paracone.

82. Centrocrista between the paracone and the metacone of upper molars (applicable only to taxa with a well-developed metacone and distinctive wear facets 3 and 4): (0) Straight; (1) V-shaped, with labially directed postparacrista and prematacrista.

83. Upper molar cuspule E (an enlarged version of E would be the parastyle): (0) Present; (1) Absent.

84. Upper molar interlock: (0) Absent; (1) Tongue-in-groove interlock.

85. Upper M1—Number of cusps within the main functional straight cusp row (for teeth with multiple rows, the labial row is designated): (0) Three main functioning cusps (or fewer) within a row; (1) Four main functioning cusps or more; (A) (0/1 polymorphic).

86. Lower multicuspate m1—Number of cusps within the main functional straight multicusp row (if there are multiple rows, the labial row is designated): (0) Three or fewer main functioning cusps within a row; (1) Four or more main functioning cusps within a row: *Haramiyavia*, plagiaulacidans, cimolodontans; (A) (0/1 polymorphic).

87. Outline of the lower m1: (0) Oval-shaped; (1) Laterally compressed; (2) Oblong with slight labial bulge; (3) Oblong with a strong labial bulge; (4) Triangular or teardrop-shaped; (5) Rectangular (or slightly rhombdoidal).

88. Aspect ratio and outline of the upper M1: (0) Laterally compressed; (1) Longer than transversely wide (oval or spindle-shaped); (2) Transversely wider than long (tri-

angular outline); (3) Transversely wide (dumbbell-shaped); (4) Rectangular or nearly so.

89. Multicuspate row in upper molar: cusp height gradient within the individual longitudinal rows of cusps: (0) Cusps in a row more or less equal in height; (1) Distal cusp highest, with gradient of anteriorly decreasing height; (A) (0/1 polymorphic).

90. Lower molars with multicuspate rows—U-ridge: (0) Absence of U-shaped ridge (anterior crest) at mesial end of lower molar with open valley/basin between longitudinal cusp rows; (1) Presence of U-shaped ridge (anterior crest) at mesial end of lower molar enclosing valley/basin between longitudinal cusp rows.

91. Multicuspate upper M2 with longitudinal multicuspate rows—lingual offset with M1: (0) Upper M2 lingually offset from M1 so that lower m2 lingual row occludes lingual side of M2 upper labial row; (1) m2 lower labial row occludes labial side of M2 upper labial row.

92. Multicuspate lower molar: cusp height ratio within the labial longitudinal row: (0) First cusp (b1 by Butler's, 2000, designation) highest so that labial cusp row forms series of decreasing height posteriorly; (1) Second cusp (b2 by same designation) highest.

93. Functional development of occlusal facets on individual molar cusps: (0) Absent for lifetime: *Probainognathus*, tritheledontids, *Sinoconodon;* (1) Absent at eruption but developed later by wearing of crown; (2) Wear facets match upon eruption of teeth (inferred from flat contact surface upon eruption).

94. Topographic relationship of wear facets to the main cusps: (0) Wear facet absent or a simple longitudinal facet extending entire length of crown; (1) Lower cusps a and c support two different wear facets (facets 1 and 4) that contact the upper primary cusp A; (2) Lower cusps a and c support single wear facet (4) that contacts upper primary cusp B (it extends onto cusp A as wear continues, but 1 and 4 do not develop simultaneously in these taxa); (3) Multicuspate series, each cusp may support two wear facets.

95. Development and orientation of prevallum/postvallid shearing: (0) Absent; (1) Present, obtuse; (2) Present, hypertrophied and transverse.

96. Wear facet 1 (a single facet supported by cusps a and c) and facet 2 (a single facet supported by cusps a and b): (0) Absent; (1) Present.

97. Upper molar—development of wear facet 1 and preprotocrista on upper molar (applicable to molars with reversed triangulation of molar cusps): (0) Facet 1 (prevallum crest) short, not extending to stylocone area; (1) Facet 1 extends into hooklike area near stylocone; (2) Long preprotocrista (below paracone-stylocone crest) added to prevallum shear, extending labially beyond paracone.

98. Differentiation of wear facet 3 (on the anterolabial aspect of talonid) and facet 4 (on the posterolabial aspect of the talonid): (0) Absent; (1) Present; (2) Facets 3 and 4 hypertrophied on flanks of strongly V-shaped talonid.

99. Orientation of facet 4 (on the posterior aspect of the hypoconid): (0) Present, oblique to long axis of tooth; (1) Present, forming a more transverse angle to long axis of tooth.

100. Morphology of the posterolateral aspect of the talonid (the labial face of the hypoconid, applicable to the taxa with a fully basined talonid): (0) Gently rounded; (1) Angular.

101. Wear pattern within the talonid basin: (0) Absent; (1) Present.

102. Development of the distal metacristid (applicable only to taxa with reversed triangulation cusp pattern): (0) Distal metacristid present; (1) Distal metacristid absent.

103. Differentiation of wear facets 5 and 6 on the labial face of the entoconid: (0) Absent; (1) Present.

104. Surficial features on the occluding surfaces of facets 5 and 6 on the talonid (only applicable to taxa with reversed triangulation): (0) Smooth surface on talonid heel (or on cusp d); (1) Multiple ridges within talonid basin.

105. Number of lower incisors: (0) Three or more; (1) Two or fewer.

106. Upper canine size, presence versus absence: (0) Upper canine present and enlarged; (1) Present and small; (2) Absent; (B) (1/2 polymorphic).

107. Lower canine size, presence versus absence: (0) Present and enlarged; (1) Present and small; (2) Absent.

108. Total number of lower premolars: (0) Five or more; (1) Four; (2) Three; (B) (1/2 polymorphic); (3) Two or fewer.

109. Diastema separating the procumbent upper P1 from P2: (0) Absent; (1) Present.

110. Number of lower molars or molariform postcanines: (0) Six or more; (1) four or five; (2) Three or fewer; (3) Two; (A) (0/1 polymorphic).

111. Lower postcanine roots: (0) Root division incipient or incomplete; (1) Roots divided; (2) Multiple roots (more than three); (A) (0/1 polymorphic); (B) (1/2 polymorphic).

112. Replacement of incisors and canines: (0) Alternating and multiple replacement; (1) Diphyodont replacement or none.

113. Replacement of premolariform: (0) Multiple replacement; (1) One replacement or none.

114. Replacement of at least some functional molariforms: (0) Present; (1) Absent.

115. Procumbency and enlargement of the anteriormost lower incisor: (0) Absent; (1) Both procumbent

and enlarged (at least 50% longer than second functional incisors).

116. Bicuspate second upper incisor: (0) Absent; (1) Present; A (0/1 polymorphic).

117. Enlarged diastema in the lower incisor-canine region (better developed in older individuals, see Crompton and Luo, 1993): (0) Absent; (1) Present behind the canine; (2) Present behind the posterior incisor.

118. Enamel microstructures: (0) Synapsida columnar enamel prismless; (1) "Transitional" character state (sheath indistinct, "prismatic" crystallites inclined at less than 45° to "interprismatic" matrix); (2) Plesiomorphic prismatic enamel.

119. Proatlas neural arch as a separate ossification in adults: (0) Present; (1) Absent.

120. Fusion of atlas neural arch and intercentrum in adults: (0) Unfused; (1) Fused.

121. Atlas rib in adults (this feature is equivalent to presence/absence of transverse foramen in the atlas): (0) Present; (1) Absent.

122. Prezygapophysis on axis: (0) Present; (1) Absent.

123. Fusion of dens to axis: (0) Unfused; (1) Fused.

124. Rib of axis in adults: (0) Present; (1) Absent (rib fused to become transverse process).

125. Postaxial cervical ribs in adults: (0) Present; (1) Absent.

126. Thoracic vertebrae: (0) 13; (1) 15 or more.

127. Lumbar ribs: (0) Unfused to the vertebra; (1) Synostosed to the vertebra to form transverse process.

128. Interclavicle in adults: (0) Present; (1) Absent.

129. Contact relationships in adults between the interclavicle (embryonic membranous element) and the sternal manubrium (embryonic endochondral element) (assuming the homologies of these elements by Klima, 1973, 1987): (0) Two elements distinct from each other, posterior end of interclavicle abuts anterior border of manubrium; (1) Two elements distinct from each other, interclavicle broadly overlaps ventral side of manubrium; (2) Complete fusion of embryonic membranous and endochondral elements.

130. Cranial margin of the interclavicle: (0) Anterior border emarginated or flat; (1) With median process (assuming interclavicle fused to sternal manubrium in living therians, Klima, 1987).

131. Clavicle-sternal apparatus joint (assuming that homologous elements of the interclavicle and the manubrium are fused to each other in therians, Klima, 1973, 1987): (0) Immobile; (1) Mobile.

132. Acromioclavicular joint: (0) Extensive articulation; (1) Limited articulation (either pointed acromion, or pointed distal end of clavicle, or both).

133. Curvature of clavicle: (0) Boomerang-shaped; (1) Slightly curved.

134. Scapula—supraspinous fossa: (0) Absent (acromion extending from dorsal border of scapula, positioned anterior to glenoid); (1) Weakly developed (present only along part of scapula and acromion positioned lateral to glenoid); (2) Fully developed, present along entire dorsal border of scapula.

135. Scapula—acromion process: (0) Short stump (level with or behind glenoid); (1) Hooklike, extending below glenoid.

136. Scapula—a distinctive fossa for the teres major muscle on the lateral aspect of the scapular plate: (0) Absent; (1) Present.

137. Procoracoid (as a separate element in adults): (0) Present; (1) Fused to sternal apparatus (Klima, 1973); (2) absent in adult.

138. Procoracoid foramen: (0) Present; (1) Absent (assuming the procoracoid fused to sternal apparatus in living therians, Klima, 1973).

139. Coracoid: (0) Large, with posterior process; (1) Small, without posterior process.

140. Size of the anteriormost element (manubrium) relative to the subsequent sternebrae in the sternal apparatus in adults: (0) Large; (1) Small.

141. Orientation ("facing" of articular surface) of the glenoid (relative to the plane or the axis of the scapula): (0) Nearly parallel to long axis of scapula and facing posterolaterally; (1) Oblique to long axis of scapula and facing more posteriorly; (2) Articular surface of glenoid perpendicular to main plane of scapular plate.

142. Shape and curvature of the glenoid: (0) Saddle-shaped, oval, and elongate; (1) Uniformly concave, more rounded in outline.

143. Medial surface of scapula: (0) Convex; (1) Flat.

144. Humeral head: (0) Subspherical, weakly inflected; (1) Spherical, strongly inflected.

145. Intertubercular groove that separates the deltopectoral crest from the lesser tubercle: (0) Shallow and broad; (1) Narrow and deep.

146. Size of the lesser tubercle of humerus (relative to the greater tubercle): (0) Wider than the greater tubercle; (1) Narrower than the greater tubercle.

147. Torsion between the proximal and distal ends of the humerus: (0) Strong (≥30°); (1) Moderate (30–15°); (2) Weak; (B) (1/2 polymorphic).

148. Ventral extension of deltopectoral crest or position of deltoid tuberosity: (0) Not extending beyond midpoint of humeral shaft; (1) Extending ventrally (distally) past midpoint of humeral shaft; (A) (0/1 polymorphic).

149. Ulnar articulation on the distal humerus: (0) Bulbous ulnar condyle; (1) Cylindrical trochlea (in posterior view) with vestigial ulnar condyle in anterior view); (2) Cylindrical trochlea without ulnar condyle (cylindrical trochlea extended to anteroventral side).

150. Radial articulation on the distal humerus: (0) Distinct and rounded condyle in both anterior (ventral) and posterior (dorsal) aspects of the structure (that does not form a continuous synovial surface with the ulnar articulation in the anteroventral view of the humerus); (1) Radial articulation forms rounded condyle anteriorly but posterior surface near cylindrical; (2) Capitulum (radial articulating structure that forms continuous synovial surface with ulnar trochlea, surface cylindrical in both anterior and posterior aspects).

151. Entepicondyle and ectepicondyle of the humerus: (0) Robust; (1) Weak; (A) (0/1 polymorphic).

152. Rectangular shelf for the supinator ridge extended from ectepicondyle: (0) Absent; (1) Present.

153. Styloid process of radius: (0) Weak; (1) Strong.

154. Enlargement of the scaphoid with a distomedial projection: (0) Absent; (1) Present.

155. Size and shape of hamatum (unciform) in the wrist: (0) Anteroposteriorly compressed (wider than long in dorsal view); (1) Mediolaterally compressed (longer than wide).

156. Pelvic acetabular dorsal emargination: (0) Open (emarginated); (1) Closed (with complete rim).

157. Sutures of the ilium, the ischium, and the pubis within the acetabulum in adults: (0) Unfused; (1) Fused.

158. Ischiatic dorsal margin and tuberosity: (0) Dorsal margin concave (emarginated), ishchiatic tuberosity present; (1) Dorsal margin concave, ischiatic tubercle hypertrophied; (2) Dorsal margin straight, ishchiatic tubercle small.

159. Inflected head of the femur set off from the shaft by a neck: (0) Neck absent (head oriented dorsally); (1) Neck present (head inflected medially); head spherical, inflected.

160. Fovea for acetabular ligament on the femoral head: (0) Absent; (1) Present.

161. Greater trochanter: (0) Directed dorsolaterally; (1) Directed dorsally.

162. Orientation of lesser trochanter: (0) On medial side of shaft; (1) On ventromedial or ventral side of shaft.

163. Size of lesser trochanter: (0) Large; (1) Small.

164. Patellar facet ("groove") of the femur: (0) Absent; (1) Shallow, weakly developed; (2) Well developed.

165. Proximolateral tubercle or tuberosity of tibia: (0) Large, hooklike; (1) Indistinctive.

166. Distal tibial malleolus: (0) Weak; (1) Distinct.

167. Fibula contacting distal end of the femur: (0) Present; (1) Absent.

168. Distal fibular styloid process: (0) Weak or absent; (1) Distinct.

169. Fibula contacting the calcaneus (= tricontact in upper ankle joint): (0) Extensive contact; (1) Reduced: *Deltatheridium;* (2) Mortise and tenon contact of fibula to the ankle.

170. Superposition (overlap) of the astragalus over the calcaneus (lower ankle joint): (0) Little or absent; (1) Weakly developed; (2) Present.

171. Orientation of sustentacular facet of the calcaneus with regard to the horizontal plane: (0) Nearly vertical orientation, with the facet nearly perpendicular to the horizontal plane; (1) Oblique (=70°) or nearly horizontal orientation of the facet to the horizontal plane of the calcaneus.

172. Astragalar neck: (0) Absent; (1) Weakly developed; (2) Present.

173. Astragalar trochlea: (0) Absent; (1) Present.

174. Calcaneal tubercle: (0) Short, without terminal swelling; (1) Elongate, vertically deep, and mediolaterally compressed, with terminal swelling.

175. Peroneal process and groove of calcaneus: (0) Forming laterally directed shelf, without a distinct process; (1) Weakly developed with shallow groove on lateral side of process; (2) With a distinct peroneal process.

176. Contact of the cuboid on the calcaneus: (0) On anterior (distal) end of calcaneus (cuboid aligned with long axis of calcaneus); (1) On anteromedial aspect of calcaneus (cuboid is skewed to medial side of long axis of calcaneus).

177. Sustentacular facet on the calcaneus: (0) Vertically oriented on medial edge of calcaneus; (1) Sustentacular facet on dorsal aspect of calcaneus, positioned medial or medioanterior to astragalar facet on calcaneus; (2) Sustentacular facet on dorsal aspect of calcaneus, positioned anterior to astragalar facet on calcaneus.

178. Relationships of the proximal end of metatarsal Mt V to the cuboid: (0) Mt V offset to cuboid; (1) Mt V far offset to cuboid, contacting calcaneus; (2) Mt V aligned with cuboid.

179. Angle of metatarsal Mt III to the calcaneus (indicating the degree to which the sole of the foot is "bent" from the long axis of the ankle): (0) Mt III aligned with (or parallel to) imaginary line of long axis of calcaneus; (1) Mt III arranged obliquely from imaginary line of long axis of calcaneus.

180. Sesamoid bones in flexor tendons: (0) Absent; (1) Present and unpaired; (2) Present and paired.

181. External pedal (tarsal) spur present in male: (0) Absent; (1) Present.

182. External size of the cranial moiety of the squamosal: (0) Narrow; (1) Broad.

183. Contribution of the cranial moiety of the squamosal to the braincase: (0) Does not contribute to endocranial wall of braincase; (1) Contributes to endocranial wall of the braincase.

184. Entoglenoid constriction (neck) between the craniomandibular joint (or glenoid) and the cranial moiety of the squamosal (only applicable to taxa with the dentary-squamosal joint; this character is best seen in the ventral view): (0) Absent; (1) Present.

185. Postglenoid depression on the squamosal (= external auditory meatus): (0) Present as postcraniomandibular joint sulcus (external auditory meatus); (1) Absent; (2) Present.

186. Position of craniomandibular joint: (0) Posterior or lateral to the level to the fenestra vestibuli; (1) Anterior to the level of fenestra vestibuli; (A) (0/1 polymorphic).

187. Orientation of the glenoid on the squamosal for the craniomandibular joint: (0) On inner side of zygoma, facing ventromedially; (1) On platform of zygoma, facing ventrally; (B) (1/2 polymorphic).

188. Postglenoid process: (0) Absent; (1) Present as distinctive process; (B) (1/2 polymorphic).

189. Postglenoid foramen within the squamosal bone: (0) Absent; (1) Present.

190. The basisphenoid wing on the ventral aspect of the skull ("the parasphenoid ala"): (0) Present, overlapping part of or whole petrosal cochlear housing; (1) Absent, bone does not overlap petrosal cochlear housing.

191. Relationship of the pars cochlearis to the lateral lappet of the basioccipital: (0) Pars cochlearis entirely covered by basioccipital; (1) Pars cochlearis partially covered by basioccipital; (2) Pars cochlearis fully exposed as promontorium.

192. Medial flat facet of the promontorium of the pars cochlearis: (0) Medial flat facet present on pars cochlearis; (1) Medial aspect of promontorium inflated and convex.

193. External outline and morphology of promontorium: (0) Triangular, with steep and slightly concave lateral wall; (1) Elongate and cylindrical petrosal cochlear housing; (2) Bulbous and oval-shaped promontorium.

194. Cochlea: (0) Short and uncoiled; (1) Elongate and partly coiled; (2) Elongate and coiled to about 360°.

195. Morphology of internal acoustic meatus: (0) Region for VII and VIII cranial nerves poorly ossified and no clearly developed internal auditory meatus; (1) Floor well ossified and meatus forming a deep tube; (2) Present as shallow depression; (3) Floor forms cribriform foramina for auditory nerve; (B) (1/2 polymorphic).

196. Primary bony lamina within the cochlear canal: (0) Absent; (1) Present.

197. Secondary bony lamina for the basilar membrane within the cochlear canal: (0) Absent; (1) Present.

198. Crista interfenestralis: (0) Horizontal, extending to base of paroccipital process; (1) Vertical, delimiting back of promontorium.

199. Posttympanic recess: (0) Absent; (1) Present.

200. Caudal tympanic process of petrosal: (0) Absent; (1) Present; (2) Caudal tympanic process notched; (B) (0/1) Polymorphic;

201. Prootic canal: (0) Prootic canal absent; (1) Prootic canal present, its tympanic aperture a distinct foramen (and separated from pterygoparoccipital foramen = foramen for ramus superior of stapedial artery, if the latter is present); (2) Prootic canal present, its tympanic aperture confluent with pterygoparoccipital foramen; (B) (1/2 polymorphic).

202. Lateral trough floor anterior to the tympanic aperture of the prootic canal and/or the primary facial foramen: (0) Open lateral trough but not bony floor; (1) Present; (2) Lateral trough absent.

203. Enclosure of the geniculate ganglion by the bony floor of the petrosal: (0) Absent; (1) Present.

204. Anteroventral opening of the cavum epiptericum: (0) Present; (1) Present with reduced size (owing to anterior expansion of lateral trough floor); (2) Present, partially enclosed by petrosal; (3) Enclosed by both alisphenoid and petrosal.

205. Anterior lamina of the petrosal and ascending process of the alisphenoid and their relationship to the exit of mandibular branch (V_3) of the trigeminal nerve; (0) V_3 trigeminal foramen at suture of alisphenoid ascending process and anterior lamina of petrosal; (1) V_3 trigeminal foramen within enlarged anterior lamina of petrosal; (2) Double trigeminal foramina within anterior lamina (foramen masticatorium versus inferium) in addition to trigeminal foramen at anterior lamina border with alisphenoid; (3) V_3 trigeminal foramen within ascending process of alisphenoid; (D) (2/3 polymorphic).

206. Quadrate ramus of alisphenoid: (0) Forms rod overlapping anterior part of lateral flange; (1) Absent.

207. Orientation of the anterior part of the lateral flange: (0) Horizontal shelf; (1) Ventrally directed; (2) Medially directed, contacting promontorium; (3) Vestigial or absent.

208. Vertical component of lateral flange (L-shaped and forming a vertical wall to pterygoparoccipital foramen): (0) Present; (1) Absent.

209. Vascular foramen in the posterior part of the lateral flange (and anterior to the pterygoparoccipital or the ramus superior foramen): (0) Present; (1) Absent.

210. Relationship of the petrosal lateral flange to the crista parotica (or the anterior paroccipital process that

bears the crista): (1) Narrowly separated; (2) Continuous bone formed by petrosal.

211. Morphology of the pterygoparoccipital foramen (for the ramus superior of the stapedial artery): (0) Laterally open notch (laterally open pterygoparoccipital foramen); (1) Foramen enclosed by petrosal or squamosal or both.

212. Position of pterygoparoccipital foramen relative to the fenestra vestibuli: (0) Foramen posterior or lateral to level of fenestra vestibuli; (1) Foramen anterior to level of fenestra vestibuli; (A) (0/1 polymorphic).

213. Bifurcation of paroccipital process—presence versus absence (modified from the character used in several previous studies): (0) Bifurcation absent; (1) Bifurcation present.

214. Posterior paroccipital process of the petrosal: (0) No ventral projection below level of surrounding structures; (1) Ventral projection below level of surrounding structures.

215. Morphological differentiation of anterior paroccipital region: (0) Anterior paroccipital region indistinct from surrounding structures; (1) Anterior paroccipital region bulbous and distinct from surrounding structures; (2) Anterior paroccipital region has a distinct crista parotica.

216. Epitympanic recess lateral to the crista parotica: (0) Absent; (1) Present (as large depression on crista parotica); (A) (0/1 polymorphic).

217. Relationship of the squamosal on the paroccipital process: (0) Squamosal covers entire paroccipital region; (1) No squamosal cover on anterior paroccipital region; (2) Squamosal covers part of paroccipital region, but not crista parotica (squamosal wall and crista parotica separated by epitympanic recess).

218. Medial process of squamosal reaching toward the foramen ovale: (0) Absent; (1) Present.

219. Stapedial artery sulcus on the petrosal: (0) Absent; (1) Present; (A) (0/1 polymorphic).

220. Transpromontorial sulcus for the internal carotid artery on the pars cochlearis: (0) Absent; (1) Present.

221. Site for the attachment of the tensor tympani muscle on the petrosal: (0) Absent. (1) Present on shallow anterior embayment on lateral trough; (2) Present as longitudinal groove on lateral trough. (3) Present on oval-shaped fossa (although position of fossa may be variable).

222. Bullar process of alisphenoid: (0) Absent; (1) Present.

223. Hypotympanic recess in the junction of alisphenoid, squamosal, and petrosal: (0) Absent; (1) Present.

224. Separation of the fenestra cochleae from the jugular foramen: (0) Fenestra cochleae and jugular foramen within same depression; (1) Separate (do not share same depression).

225. Bony channel of perilymphatic duct: (0) Open channel and sulcus; (1) Channel partially or fully enclosed; (A) (0/1 polymorphic).

226. Stapedial muscle fossa: (0) Absent; (1) Present, in alignment with crista interfenestralis; (2) Present, lateral to crista interfenestralis.

227. Hypoglossal foramen: (0) Indistinct, either confluent with or sharing depression with jugular foramen; (1) Separated from the jugular foramen.

228. Geometry (shape) of the incudomallear contact: (0) Trochlear (convex and cylindrical) surface of incus; (1) Trough or saddle-shaped contact on incus; (2) Flat surface.

229. Alignment of incus and malleus by the "center of mass": (0) Posteroanterior; (1) Posterolateral to anteromedial; (2) Dorsoventral.

230. Presence of a quadrate/incus neck (separation of the dorsal plate and the trochlea; this represents the differentiation between the "body" and crus brevis of the incus): (0) Absent; (1) Present.

231. The stapedial process (crus longum) of the incus/quadrate: (0) Absent; (1) Present.

232. Dorsal plate (= crus brevis) of the quadrate/incus: (0) Broad plate; (1) Pointed triangle; (2) Reduced.

233. Incus: angle of the crus brevis to crus longum of the incus (this is the equivalent of the angle between the dorsal plate and the stapedial process of the quadrate): (0) Alignment of stapedial process (crus longum) and dorsal plate (crus brevis) (or obtuse angle between the two structures); (1) Perpendicular; (2) Acute angle of crus brevis and crus longum (A-shaped incus).

234. Primary suspension of the incus/quadrate on the basicranium: (0) By squamosal and quadratojugal; (1) By squamosal only; (2) By petrosal (either by preserved direct contact of incus or by inference from presence of well-defined crista parotica).

235. Quadratojugal notch on squamosal: (0) Present as independent element in adult; (1) Absent.

236. Morphology of stapes: (0) Columniform—macroperforate; (1) Columelliform—imperforate (or microperforate); (2) Bicrurate—perforate.

237. Bony secondary palate: (0) Ending anterior to posterior end of tooth row; (1) Level with posterior end of tooth row; (2) Extending posterior to tooth row.

238. Relationship of the maxilla to the subtemporal margin or the orbit: (0) Participating in rounded subtemporal margin of orbit; (1) Forming well-defined edge along subtemporal margin.

239. Pterygopalatine ridges: (0) Present; (1) Absent.

240. Transverse process of the pterygoid: (0) Present and massive; (1) Present but reduced (as hamulus); (2) Greatly reduced or absent; (A) (0/1 polymorphic); (B) (1/2 polymorphic).

241. Basisphenoid constricture (= palatal width anterior to the basisphenoid): (0) Strongly developed (very narrow anterior to basisphenoid); (1) Intermediate (wide anterior to basisphenoid); (2) No constricture (palatal width as broad at basisphenoid as internal choate).

242. Vault of the naso-oral pharyngeal passage near the pterygoid-basisphenoid junction: (0) Roof of pharynx V-shaped in transverse section, narrowing toward basisphenoid; (1) Roof of pharynx U-shaped in transverse section.

243. Complete ossification of the orbital floor: (0) Absent; (1) Present.

244. Pattern of orbital mosaic as exposed externally: (0) Alisphenoid contacts frontal and parietal, thereby separating petrosal anterior lamina from orbitosphenoid in external view of orbit; (1) Petrosal anterior lamina contacts orbitosphenoid, thereby separating alisphenoid from frontal and parietal.

245. Overhanging roof of the orbit: (0) Formed by prefrontal; (1) Absent; (2) Formed by frontal.

246. Outline of the facial part of the lacrimal: (0) Large, triangular and pointed anteriorly; (1) Small, rectangular or crescentic; (2) Excluded from facial (and preorbital) part of skull.

247. Pila antotica: (0) Present; (1) Absent (in adult).

248. Frontal/parietal suture on alisphenoid: (0) Dorsal plate of alisphenoid contacting frontal by anterior corner; (1) Dorsal plate of alisphenoid has more extensive contact to frontal (~50% of dorsal border).

249. Jugal on the zygoma: (0) Anterior part of jugal extending to facial part of maxilla, forming part of anterior orbit; (1) Anterior part of jugal not extending to facial part of maxilla and excluded from anterior rim of orbit.

250. Maximum vertical depth of the zygomatic arch relative to the length of the skull (this character is designed to indicate the robust versus gracile nature of the zygomatic arch): (0) Between 10 and 20%; (1) Between 5 and 7%.

251. Posterior opening of the posttemporal canal: (0) At junction of petrosal, squamosal, and tabular; (1) Between petrosal and squamosal.

252. Anterior ascending vascular channel (for the arteria diploetica magna) in the temporal region: (0) Open groove; (1) Partially enclosed in a canal; (2) Completely enclosed in a canal or endocranial.

253. Lambdoidal crest: (0) Crest overhanging concave or straight supraoccipital; (1) Weak crest with convexing supraoccipital; (A) (0/1 polymorphic).

254. Sagittal crest: (0) Prominently developed; (1) Weakly developed; (2) Absent; (B) (1/2 polymorphic).

255. Tabular bone: (0) Present; (1) Absent.

256. Shape of the occipital condyles: (0) Bulbous; (1) Ovoid; (2) Cylindrical; (B) (1/2 polymorphic).

257. Occiput slope: (0) Occiput sloping posterodorsally (or vertically oriented) from occipital condyles; (1) Occiput sloping anterodorsally from occipital condyles (such that lambdoidal crest level anterior to occipital condyle and condyle fully visible in dorsal view of skull).

258. Foramina on the dorsal surface of nasal bones: (0) Absent; (1) Present.

259. Septomaxilla: (0) Present, with ventromedial shelf; (1) Present, without septomaxillary shelf; (2) Absent.

260. Premaxillary internarial process: (0) Present; (1) Absent.

261. Facial part of premaxilla borders on the nasal: (0) Absent; (1) Present; (A) (0/1 polymorphic).

262. Ossified ethmoidal cribriform plate of the nasal cavity: (0) Absent; (1) Present.

263. Posterior excavation of nasal cavity into the bony sphenoid complex: (0) Absent; (1) Present; (2) Present, partitioned from nasal cavity.

264. External bulging of the braincase in the parietal region: (0) Absent; (1) Expanded (parietal part of cranial vault wider than frontal part, but expansion not extending to lambdoidal region); (2) Greatly expanded (extending to lambdoidal region); (B) (1/2 polymorphic).

265. Interparietal: (0) Present as separated element in adult; (1) Absent.

266. Bony tentorium septum: (0) Present; (1) Absent.

267. Presence of the vascular superior cistern on the midbrain (Kielan-Jaworowska and Lancaster, 2004): (0) Absent; (1) Present, covers and conceals mesencephalon (midbrain) on endocast.

268. Overall size of vermis: (0) Small; (1) Enlarged.

269. Lateral cerebellar hemisphere (excluding the paraflocculus): (0) Absent; (1) Present.

270. Lateral extension of the paraflocculus: (0) For less than 30% of total cerebellar width; (1) For more than 30% of the cerebellar width.

271. External division on the endocast between the olfactory lobe and the cerebral hemisphere (presence versus absence of the circular sulcus that defines the external aspect of the olfactory lobe): (0) Absent; (1) Present.

272. Anterior expansion of the cerebral hemisphere: (0) Absent; (1) Developed.

273. Expansion of the posterior cerebral hemisphere (for each hemisphere, not the combined width of the posterior hemispheres): (0) Absent; (1) Present.

274. Direction of the lower jaw movement during oc-

clusion: (0) Dorsomedial movement (as inferred from teeth); (1) Dorsomedial movement with significant medial component; (2) Dorsoposterior movement.

275. Mode of occlusion as inferred from the mandibular symphysis: (0) Bilateral (with more or less rigid symphysis); (1) Unilateral (with mobile symphysis).

Data Matrix. The data matrix has 48 taxa and 275 characters. Character-states are scored: 012345. Missing and inapplicable data identified by "?". Polymorphic codings are: A = 0/1, B = 1/2, C = 0/2, D = 2/3.

Parsimony Searches. Parsimony searches were obtained by PAUP 4.0b (Swofford, 2000). All characters are of type "unordered." All characters have equal weight. Number of parsimony-informative characters = 272. Multistate taxa interpreted as polymorphism.

Figure 15.1. The data matrix below is the basis for the tree shown in figure 15.1. Tree topology is the strict consensus tree of 50 equally parsimonious trees, each of which has: tree length = 943, consistency index = 0.495; retention index = 0.764. All multistate characters were unordered. No topological constraints were enforced for the figure. (For further details about the searches see figure 15.1.)

Figure 15.2. Tree topology is the strict consensus tree of seven equally parsimonious trees, each of which has: tree length = 951; consistency index = 0.491; retention index = 0.760. All multistate characters were unordered. Topological constraints were enforced for the figure. The enforced constraints excluded the monophyletic allotherians (*Haramiyavia*, plagiaulacidans, cimolodontants) from the crown mammalian group to account for the preferred hypothesis that, on the basis of dental evidence, allotherians are likely a highly specialized and early divergent lineage of mammals, even though this topology is not strictly parsimonious for all of the morphological evidence that is currently available. (For further details about the searches see figure 15.2.)

TAXON CHARACTER MATRIX

The scores in the following matrix correspond to states for the characters described above, and are cited in horizontal rows.

Probainognathus

0000000000	0000000000	0?0?00????	???????00?	?00???????	00??0?00??	??????????
????000???	??00??01??	??00??????	????000?0?	0000000??0	0??00?0???	?0?000000?
0000000000	0000?0??00	0000?00000	???00????0	?00000??00	0??0000000	1000001110
10000?000?	0000000000	0000000000	0000000000	0000000000	000000??00	00001

Tritylodontids

0000001000	0000000000	0?0?30??2?	??0??2232?	?24???????	00??0?00??	??????????
?????00???	??10AA54A0	0023??????	????122??A	B?11102000	00000?0000	0000000000
0000000A00	0000?00000	0000000000	0000000??0	000000??00	0??0100000	0000100000
00101?100?	0000010000	100000000A	0010100000	0000000000	A010100?00	00020

Tritheledontids

0000010000	0000000000	001?20????	??0????00?	?00???????	00??0?00??	0?????????
????040???	??000001??	??00??????	????000?0?	00000A02?0	0???0?00?1	101000000?
0000000000	0000?0??0?	000???0???	??????????	?000100?00	0??0?00000	0000000110
00000?000?	0000000001	0001101000	0000100000	0000010?00	1?001?????	???00

Adelobasileus

??????????	??????????	??????????	????????0?	????????0?	00????????	??????????
??????????	??????00??	??????????	??????????	0?????????	??????????	??????????
??????????	??????????	??????????	??????????	?10??????0	1?0????000	1100201101
10000?0000	0001001???	????????0?	1110??00??	1101120???	???01?????	?????

Sinoconodon

0000010000	0000000000	0110001000	0000?0000?	?00000??00	10A000001?	0???000??0
????000???	??000010??	??0000?0??	0?0?000311	A0100011?0	0???0??000	0000000000
0000000000	0000????00	0000??0???	??????????	?000100001	1000100000	1111200000
00101?0000	0001011100	0001101101	1010100001	1000000000	0010100000	00000

Morganucodon

0001010000 0000000000 0A10000000 0000?10000 ?11000??00 120000020? 0???000??0
????040??? ??000010?? ??1100?0?? 0?0?000001 1111001?0 01000?10?1 ?0010?000?
0000000000 0000?00001 00010?0?00 000000??? ?00110AA01 2110200000 1111201101
0011201000 1001011001 1002101101 1010100001 1001110110 0010100000 00001

Megazostrodon

0001010100 0000000000 0020100100 0000?10000 ?12000??00 120000020? 0???000??0
????040??? ??000010?? ??1100?0?? 0?0?000001 1??00012?0 01000?1??? ?0?100000?
0000000000 00?0000001 000?000000 00000?0000 00?110??01 201020?000 1111011?1
0011201000 ?001011001 ?0?21?11?1 1?101??001 ?001?10??? ????1????? ???01

Dinnetherium

0001011100 0000?00000 0220100000 0000?10000 ?11010??00 120100010? 0???000??0
????040??? ??000010?? ??1100?0?? 0?0?000001 1??10010?? ?????????? ??????????
?????????? ?????????? ?????????? ?????????? ?????????1 2110???000 11111?11?1
0011201000 10010110?? 10?21??1?? 1????????? ?????????? ????1????? ???01

Haldanodon

0000011000 0000010000 0110100000 0010?10000 ?11?10??00 1?0033000? 0???000??0
????011??? ??000053?? ??1100?0?? 0?0?000201 11?10000?? ?????????? ???100????
00?00?0100 000?????0? 000?0????? ?????????? ?001201101 2000200000 1112001???
?010201000 200111100? ???2?02112 2110100000 ?101111110 000000???? 10?01

Hadrocodium

111?12?300 ?11??00000 00102?1100 00?0???000 ?12000??00 1?0000000? 0???000??0
????000??? ??000010?? ??1100?0?? 0?0?0003?2 1???001??? ?????????? ??????????
?????????? ?????????? ?????????? ?????????? ?100210101 212????000 1111?11102
1000211000 1001011??? ???2??2112 2110101001 111212101? 00?21???11 11101

Kuehneotherium

0001011100 0000?00000 0010000100 00000??101 0120100011 0101000100 1000100??0
0??0110??0 0?00??21?? ??121100?? 0000???0?0 A?11???0?? ?????????? ??????????
?????????? ?????????? ?????????? ?????????? ?????????? ?????????? ??????????
?????????? ?????????? ?????????? ?????????? ?????????? ?????????? ???01

Shuotherium

00110?15?? 0011?000?0 ???011001 0110?00211 01?0200011 0101332001 1000110??0
0??0221??0 0000??32?? ??221100?? 0100?????1 1????????? ?????????? ??????????
?????????? ?????????? ?????????? ?????????? ?????????? ?????????? ??????????
?????????? ?????????? ?????????? ?????????? ?????????? ?????????? ???01

Asfaltomylos

001???1211 001??00220 311121??11 ??1????212 ?13??13?1? 00110000?1 1001411001
1113?0???? ?????????? ??2211?100 1010?????2 1????????? ?????????? ??????????
?????????? ?????????? ?????????? ?????????? ?????????? ?????????? ??????????
?????????? ?????????? ?????????? ?????????? ?????????? ?????????? ?????

Ambondro

??????1??? ???1?????? ????????12 2111???212 ?131?1331? 00012200?1 1022411201
1113?????? ??????3??? ??2211?110 1010?????? 1????????? ?????????? ??????????
?????????? ?????????? ?????????? ?????????? ?????????? ?????????? ??????????
?????????? ?????????? ?????????? ?????????? ?????????? ?????????? ???01

Ausktribosphenos

```
00110?1201 0011?00220 ??11211012 2111?11212 ?13101331? 0001220002 2021412202
2124?????? ??????5??? ??2221?111 1111?????2 1????????? ?????????? ??????????
?????????? ?????????? ?????????? ?????????? ?????????? ?????????? ??????????
?????????? ?????????? ?????????? ?????????? ?????????? ?????????? ???01
```

Bishops

```
0???1?1201 ?011?00220 3111211012 2?11?11212 ?13101331? 0001220012 2021412202
2124?????? ??????5??? ??2221?111 1111???0?2 1????????? ?????????? ??????????
?????????? ?????????? ?????????? ?????????? ?????????? ?????????? ??????????
?????????? ?????????? ?????????? ?????????? ?????????? ?????????? ???00
```

Steropodon

```
0?100?1??? ?0?1?????? ?????1???? ???????212 ?13121221? 0001212002 2022312100
0?24?????? ??????5??? ??2221?211 0100?????2 1????????? ?????????? ??????????
?????????? ?????????? ?????????? ?????????? ?????????? ?????????? ??????????
?????????? ?????????? ?????????? ?????????? ?????????? ?????????? ???01
```

Teinolophos

```
111?2?1201 0111?00220 311121???? ???????212 ?13121221? 00012120?2 2022312100
0?23?????? ??????5??? ??2221?211 0100?????? 1????????? ?????????? ??????????
?????????? ?????????? ?????????? ?????????? ?????????? ?????????? ??????????
?????????? ?????????? ?????????? ?????????? ?????????? ?????????? ???01
```

Obdurodon

```
111?221101 1111111220 3111211012 211??00212 ?13121221? 0001212002 2022312100
0?2421???? ???0??54?? ??2221?211 0100122?2 2????????? ?????????? ??????????
?????????? ?????????? ?????????? ?????????? ?000111001 211????000 2111111112
1001201000 2000000??? ???2??2112 211112??11 1?12121011 0??21????? ???01
```

Ornithorhynchus

```
111?2215?? 1111111220 311121???? ?????????? ?1???1???? 0???2?20?2 202231210?
??2421???? ???0??54?? ???????2?? ?1??1223?2 2??1???011 1110111000 0000010100
0000000100 0000?11100 0002010000 0000210111 1000101001 2111300000 2111111112
1001201000 200000022? 1212112112 2111121?11 1012121011 0002110110 111C1
```

Gobiconodon

```
11101115?? 0111?12210 1140101101 0101?00000 ?12020??01 110100010? 0???000??0
????010??? ??000010?? ??1201?0?? 0?0?111101 1?101002?? ?????00??? ?1?21?1?1?
1110000110 000??00200 000100100? ?????????? 1001201101 2?10?0?000 ?1?2?111??
?????0?0?? 200??????? ??????21?? ????10??10 12001?0?1? 0??01????? ???01
```

Amphilestes

```
11111115?? 1111?122?? 1140201101 0001?00000 ?12020??0? 110100010? 0???000??0
????0????? ????0010?? ??1201?0?? 0?0?0?01?1 1?11??0??? ?????????? ??????????
?????????? ?????????? ?????????? ?????????? ?????????? ?????????? ??????????
?????????? ?????????? ?????????? ????????1? ?????????? ?????????? ???01
```

Jeholodens

```
11101115?? ?111?12?10 1240100001 0001?00000 ?11020??00 101000031? 0???000??0
????0A0??? ??000010?? ??1100?0?? 0?0?011311 1??1101??0 0100010011 1112111?10
1110??0110 1000000200 0001001000 0000010010 000120110? ??102????? 1?????????
???????0?? ?????????? 122????1?? ????10??11 ??011??01? 0??11????? ???01
```

Priacodon

11101115?? 1111?12210 1240100101 0002?00000 ?11120??00 111000031? 0???000??0
????010??? ??110010?? ??1100?0?? 0?0?000201 1??10011?? ????????? ?????????
???????110 100??????? ?????????? ?????????? ?????????? 21102??000 11121?1111
00??20??00 2??1021??? ???2??21?? ?????????? 11???2???? ????????? ???01

Trioracodon

11101115?? 1111?12210 1240100101 0002?00000 ?11120??00 11A000031? 0???000??0
????010??? ??110010?? ??1100?0?? 0?0??00102 11110?0??? ????????? ?????????
?????????? ?????????? ?????????? ????0????? ?????????1 21102??000 1112111111
0011201?00 2??102???? ???2???1?? ?????????? 11???????? ???1?11101 11001

Haramiyavia

0?0?0?1100 ?0?????000 0?1110??2? 000??2232? ?24??????? 00??0?000? ?????????
?????00??? ??10115411 0123?????? ????0111?2 1???1100?? ????????? ?????????
????????00 ?????????? ?????????? ?????????? ?????????? ????????? ?????????
?????????? ?????????? ?????????? ?????????? ?????????? ????????? ???20

"Plagiaulacidans"

111?2115?? A111110211 224010??3? 100??2232? ?24??????? 00??0?000? ?????????
?????00??? ??10115411 1123?????? ????1B2B?3 11111120?? ????????? ?????????
?????????? ?????????? ?????????? ?????????? ??01111001 21102??000 111221211?
??012110?0 3001021??? ???2??0102 21??21??11 12??120??? ??2?1????? ???20

Cimolodontans

111?2115?? 1111110211 224AB0??3? 100??2232? ?24??????? 00??0?000? ?????????
?????00??? ??10115400 1023?????? ????1223?3 11111A2210 0110001000 1111101110
110100B000 A010001111 1102010001 0001211112 10011A1001 2110B0010A B112D12112
1A012A10A0 3001A2122? 122211010B 211A200?11 12AB1B0121 102B111101 11020

Tinodon

11111?15?? 011??121?0 0141311001 0000?00101 ?120200011 0101000101 1000100??0
0??0110??0 0?00??21?? ??1211?0?? 0000??12?1 1????????? ????????? ?????????
?????????? ?????????? ?????????? ?????????? ?????????? ????????? ?????????
?????????? ?????????? ?????????? ?????????? ?????????? ????????? ???01

Zhangheotherium

11101215?? 0110?12210 1141011001 0000000212 0120200011 0100000001 1000100??0
0??0210??0 0?00??42?? ??222100?? 0000112?0 111110021? 1110001010 1112111110
1111111111 100001021? 1102101001 100110???2 1??1211101 2110???111 11?????12
1001212000 ?001?2???? ???2???1?? ?????????1 ?????????? ????????? ???11

Dryolestes

1110111010 0111?11210 1131210000 0010000211 0131000011 0010000001 1000220??0
0??1100??0 0?00??42?? ??221100?0 0100000100 11110?02?? ????????? ?????????
???????111 100??????? ?????????? ?????????? ?????????? ????????? ?????????
?????????? ?????????? ?????????? ?????????? ?????????? ????????? ???11

Henkelotherium

1110111010 0111?11210 1131210000 0?00000211 0131000011 001?000001 100?220??0
0??1100??0 0?00??42?? ??221100?0 0100?00110 1?????0??? ??????1??? ?112111?1?
11111?1011 1010??0211 110211010? ???1?????? 11?12?110? ?1???????? ?????????
???????0?? ?????????? ???????1?? ?????1??01 ?????????? ????????? ???11

Vincelestes

```
111?211010  0111?11110  1131211100  0000000211  0130100011  0010000000  1000210??0
0?01210?00  0?00??31??  ??221100?0  0000100302  1???000?10  1111101121  11?2101110
1111010011  1000?01211  1002110102  101110120?  0110211101  21211?1111  1112111112
1101212011  2001121???  ???2??1111  1110111000  1200110111  0??11?????  ???11
```

Amphitherium

```
1111111010  0111?11210  1131210100  0010?00211  0130001011  0011000100  1000210??0
0??1110??0  0?00??41??  ??22111100  00000?00?0  10????0???  ??????????  ??????????
??????????  ??????????  ??????????  ??????????  ??????????  ??????????  ??????????
??????????  ??????????  ??????????  ??????????  ??????????  ??????????  ???11
```

Nanolestes

```
111011?010  1111?112?0  1131210100  0000000212  0130001011  0011000100  1000210000
0?01310??0  0000???1??  ??221110?0  00000?0001  1?????02??  ??????????  ??????????
??????????  ??????????  ??????????  ??????????  ??????????  ??????????  ??????????
??????????  ??????????  ??????????  ??????????  ??????????  ??????????  ?????
```

Peramus

```
1111111010  0111?11210  1131210101  0001000212  0130101011  0001000100  1A00211000
0?01310?00  0000??31??  ??22111100  00100?0002  1?????0???  ??????????  ??????????
??????????  ??????????  ??????????  ??????????  ??????????  ??????????  ??????????
??????????  ??????????  ??????????  ??????????  ??????????  ??????????  ???11
```

Kielantherium

```
1110111010  0111?11110  ???1310100  0000?00211  0130101111  00A1100101  100141100A
1001??????  ??????3???  ??2221?100  1000??01?1  1?????0???  ??????????  ??????????
??????????  ??????????  ??????????  ??????????  ??????????  ??????????  ??????????
??????????  ??????????  ??????????  ??????????  ??????????  ??????????  ???11
```

Aegialodon

```
??????????  ??????????  ??????????  ???????211  ?13??0111?  00011001?1  1001411001
1002??????  ??????3???  ??2221?100  1000??????  ??????????  ??????????  ??????????
??????????  ??????????  ??????????  ??????????  ??????????  ??????????  ??????????
??????????  ??????????  ??????????  ??????????  ??????????  ??????????  ?????
```

Deltatheridium

```
111?211410  1111?11110  0131310101  0001000212  0131101111  00A1100111  101141100A
A001321001  0000??32??  ??22212100  1000000211  1??10?02??  ??????????  ??????????
??????????  ??????????  ????????12  1??1??1???  ?1112111?1  2122???111  1213?13?1?
??00212?00  31?1121???  ???2??2111  101?11??01  1??????02?  ????1?????  ???11
```

Asiatherium

```
111?1?1410  1111111110  0??1310101  0001000212  1131203111  0011101111  1121412011
2113211022  1100??32??  ??22212100  1110?0?211  1?????????  ????101???  ?11210111?
2111112022  1111?0121?  111111101?  ?????????2  0111211111  2122???111  ?2?3313???
??00?121??  ?1111?1???  ???21?2111  10101???01  ???11200??  ???21?????  ???11
```

Kokopellia

```
?1??1114??  1??1?1?110  ????310101  0001?00212  1130203111  0011101110  1121411011
2103321012  0000??32??  ??22212100  1110002?1  1??10?02??  ??????????  ??????????
??????????  ??????????  ??????????  ??????????  ??????????  ??????????  ??????????
??????????  ??????????  ??????????  ??????????  ??????????  ??????????  ???11
```

Pucadelphys

111?111410 1111111110 0131310101 0001000212 1131003111 0011101110 1121412012
2113331022 1100??32?? ??22212100 1110000211 1?11000211 1111101121 1112101111
2111112022 1111?11211 1111111012 111110120? 0111211111 21223??111 1213313?1?
??00212100 3011121??? ???21?2111 1010111101 1201120021 1?211????? ???11

Didelphis

111?211410 1111111110 0131310101 0001000212 1131103111 0011101111 1121412012
2113331020 1100??32?? ??22212100 1110000211 1111000211 1111101121 1112101111
2111112022 1111111211 1111111012 1111101202 0111211111 2122311111 1213313?1?
??00212100 3111121111 1112122111 1010111101 1200120021 1121100110 11111

Pappotherium

??????1??? ?????????? ??????0100 0000???212 0131001111 00111001?0 1011411001
2002331001 0000??32?? ??22212100 1110?????? ?????????? ?????????? ??????????
?????????? ?????????? ?????????? ?????????? ?????????? ?????????? ??????????
?????????? ?????????? ?????????? ?????????? ?????????? ?????????? ?????

Erinaceus

111?211010 1111111110 0131311011 2000100212 1130003111 0011100101 1021411101
2013211122 0000??52?? ??22212111 1110111302 1111000211 1111101121 1112101111
2111112022 1010111211 1112110122 1211102202 0111211111 2122311112 0213313?1?
1100212011 3001121111 1112122111 1010111101 1201121021 1122100110 11111

Asioryctes

111?211310 1111111120 0131311101 0000100212 1130001111 0011100120 1011411001
2003321112 0000??32?? ??22212100 1110000102 1?11000?10 11111????? ??????????
?????????? ?0101????? ?????10122 ?21110220? 0111211111 2122???112 0213313?1?
1100212010 3001121??? ???21?2111 1010111101 1201121021 1??1100110 11111

Prokennalestes

1110111010 0111?11120 0131211101 0000100212 1130001111 0011100120 1011411001
2003331111 0000??32?? ??22212100 11100?00?2 1??1??0??? ?????????? ??????????
?????????? ?????????? ?????????? ?????????? ?????????1 2122311112 2213??3?1?
110021??11 3??1121??? ???2?????? ??????1??? ?????????? ?????????? ???11

Montanalestes

111?2?1310 0111?11120 0131311101 0010?00212 ?130201111 0011100120 1011411001
2003?????? ??????3??? ??2221?100 1110???0?2 1????????? ?????????? ??????????
?????????? ?????????? ?????????? ?????????? ?????????? ?????????? ??????????
?????????? ?????????? ?????????? ?????????? ?????????? ?????????? ???11

REFERENCES

Agadjanian, A. K. 1999. The first docodont of Asia. *In* V. N. Orlov (ed.), *VI Congress of Theriological Society:* 7. Theriological Society, Moscow (in Russian).

Aiello, L. C., and Wheeler, P. 1995. The expensive-tissue hypothesis: the brain and the digestive system in human and primate evolution. *Current Anthropology* 36: 199–221.

Alifanov, V. R., Efimov, M. B., Novikov, I. V., and Morales, M. 1999. A new psittacosaur complex of tetrapods from the Lower Cretaceous locality Shestakovo (southern Siberia). *Doklady Akademii Nauk* 369: 491–493 (in Russian).

Allard, M. W., Honeycutt, R. L., and Novacek, M. J. 1999. Introduction: advances in high level mammalian relationships. *Cladistics* 15: 213–219.

Allen, P., and Wimbledon, W. A. 1991. Correlation of NW European Purbeck-Wealden (nonmarine Lower Cretaceous) as seen from the English type-areas. *Cretaceous Research* 12: 511–526.

Allin, E. F. 1975. Evolution of the mammalian middle ear. *Journal of Morphology* 147: 403–438.

Allin, E. F. 1986. The auditory apparatus of advanced mammal-like reptiles and early mammals. *In* N. Hotton, P. D. MacLean, J. J. Roth, and E. C. Roth (eds.), *The Ecology and Biology of Mammal-like Reptiles,* 283–294. Smithsonian Institution Press, Washington, D. C.

Allin, E. F., and Hopson, J. A. 1992. Evolution of the auditory system in Synapsida ("mammal-like reptiles" and primitive mammals) as seen in the fossil record. *In* D. B. Webster, R. R. Fay, and A. N. Popper (eds.), *The Evolutionary Biology of Hearing,* 587–614. Springer-Verlag, New York.

Allman, J. 1990. The origin of the neocortex. *The Neurosciences* 2: 257–262.

Alroy, J. 1998. Cope's rule and the dynamics of body mass evolution in North American fossil mammals. *Science* 280: 731–734.

Altman, J., and Bayer, S. A. 1997. *Development of the Cerebellar System: Its Relation to Its Evolution, Structure, and Functions.* viii + 783 pp. CRC Press, Boca Raton/New York.

Altner, G. 1971. Histologische und vergleichend-anatomische Untersuchungen zur Ontogenie und Phylogenie des Handskellets von *Tupaia glis* (Diard 1820) und *Microcebus murinus* (J. H. Miller 1777). *Folia Primatologica* Supplement 14: 1–106.

Ameghino, F. 1889. Contribución al conocimiento de los mamíferos fósiles de la República Argentina. *Actas de la Academía Nacional de Ciencias de Córdoba* 6: 1–1027.

Ameghino, F. 1890. Los plagiaulácidos Argentinos y sus relaciones zoológicas, geológicas y geográficas. *Boletín del Instituto Geográfico Argentino* 11: 143–208.

Ameghino, F. 1894. Enumeration synoptique des espèces de mammifères fossiles des formations éocènes de Patagonie. *Boletín de la Academia de Ciencias de Córdoba* 13: 259–452.

Ameghino, F. 1901. Notices préliminaires sur des ongulés nouveaux de terrains crétacé de Patagonie. *Boletín de la Academia de Ciencias de Córdoba* 16: 329–426.

Ameghino, F. 1904. Nuevos especies de mamíferos cretáceos y terciarios de la República Argentina. *Anales de la Sociedad Científica Argentina* 58: 35–71, 182–192, 225–291.

Andreis, R. R. 1987. The Late Cretaceous fauna of Los Alamitos, Patagonia, Argentina. Part 1—Stratigraphy and paleoenvironment. *Revista del Museo Argentino de Ciencias Naturales "Bernardino Rivadavia," Paleontología* 3: 103–110.

Andrews, R. C. 1932. *The New Conquest of Central Asia.* 678 pp. American Museum of Natural History, New York.

Anonymous. 1983. *Palaeontological Institute of the USSR Academy of Sciences.* 35 pp. Vneshtorgizdat. Izdatelstvo No. 3496, Moscow (in Russian).

Antunes, M. T. 1998. A new Upper Jurassic paulchoffatiid multituberculate (Mammalia) from Pai Mogo, Portugal. *Memórias da Academia das Ciências de Lisboa* 37: 125–153.

Antunes, M. T., Sigogneau-Russell, D., and Russell, D. E. 1986. Sur quelques dents de Mammifères du Crétacé supérieur de Taveiro, Portugal (Note préliminaire). *Comptes Rendus de l'Académie des Sciences, Paris, sér. II,* 303: 1247–1250.

Aplin, K. P., and Archer, M. 1987. Recent advances in marsupial systematics with a syncretic classification. *In* M. Archer (ed.), *Possums and Opossums: Studies in Evolution*, xv–lxxii. Surrey Beatty & Sons and the Royal Zoological Society of New South Wales, Sydney.

Archer, M. 1974. The development of the cheek-teeth in *Antechinus flavipes* (Marsupialia, Dasyuridae). *Journal of the Royal Society of Western Australia* 57: 54–63.

Archer, M. 1976. The basicranial region of marsupicarnivores (Marsupialia), relationships of carnivorous marsupials, and affinities of the insectivorous peramelids. *Zoological Journal of the Linnean Society* 59: 217–322.

Archer, M. 1978. The nature of the premolar-molar boundary in marsupials and a reinterpretation of the homology of marsupial cheekteeth. *Memoirs of the Queensland Museum* 18: 157–164.

Archer, M. 1981. A review of the origin and radiation of Australian mammals. *In* A. Keast (ed.), *Ecological Biography of Australia*, 1437–1488. The Hague, Junk.

Archer, M. 1984. The Australian marsupial radiation. *In* M. Archer and G. Clayton (eds.), *Vertebrate Zoogeography and Evolution in Australasia*, 633–808. Hesperian Press, Carlisle.

Archer, M., and Bartholomai, A. 1978. Tertiary mammals of Australia: a synoptic review. *Alcheringa* 2: 1–19.

Archer, M., Plane, M. D., and Pledge, N. S. 1978. Additional evidence for interpreting the Miocene *Obdurodon insignis* Woodburne and Tedford, 1975, to be a fossil platypus (Ornithorhynchidae: Monotremata) and a reconsideration of the status of *Ornithorhynchus agilis* De Vis, 1995. *The Australian Zoologist* 20: 9–27.

Archer, M., Flannery, T. F., Ritchie, A., and Molnar, R. 1985. First Mesozoic mammal from Australia—an Early Cretaceous monotreme. *Nature* 318: 363–366.

Archer, M., Jenkins, F. A., Jr., Hand, S. J., Murray, P., and Godthelp, H. 1992. Description of the skull and non-vestigial dentition of a Miocene platypus (*Obdurodon dicksoni* n. sp.) from Riversleigh, Australia, and the problem of monotreme origins. *In* M. L. Augee (ed.), *Platypus and Echidnas*, 15–27. The Royal Zoological Society of New South Wales, Sydney.

Archer, M., Murray, P., Hand, S. J., and Godthelp, H. 1993. Reconsideration of monotreme relationships based on the skull and dentition of the Miocene *Obdurodon dicksoni*. *In* F. S. Szalay, M. J. Novacek, and M. C. McKenna (eds.), *Mammal Phylogeny: Mesozoic Differentiation, Multituberculates, Monotremes, Early Therians, and Marsupials*, 75–94. Springer-Verlag, New York.

Archer, M., Arena, R., Bassarova, M., Black, K., Brammal, J., Cooke, B., Creaser, P., Crosby, K., Gillespie, A., Godthelp, H., Gott, M., Hand, S. J., Kear, B., Krikmann, A., Mackness, B., Muirhead, J., Musser, A., Myers, T., Pledge, N., Wang, Y.-Q., and Wroe, S. 1999. The evolutionary history and diversity of Australian mammals. *Australian Mammalogy* 21: 1–45.

Archibald, J. D. 1977. Ectotympanic bone and internal carotid circulation of eutherians in reference to anthropoid origins. *Journal of Human Evolution* 6: 609–622.

Archibald, J. D. 1979. Oldest known eutherian stapes and a marsupial petrosal bone from the Late Cretaceous of North America. *Nature* 281: 669–670.

Archibald, J. D. 1982. A study of Mammalia and geology across the Cretaceous–Tertiary boundary in Garfield County, Montana. *University of California Publications in Geological Sciences* 122: 1–286.

Archibald, J. D. 1987. Late Cretaceous (Judithian and Edmontonian) vertebrates and geology of the Williams Fork Formation, N.W. Colorado. *In* P. J. Currie and E. H. Koster (eds.), *Fourth Symposium on Mesozoic Terrestrial Ecosystems*, 7–11. Tyrrell Museum of Palaeontology, Drumheller.

Archibald, J. D. 1996a. *Dinosaur Extinction and the End of an Era: What the Fossils Say.* 237 pp. Columbia University Press, New York.

Archibald, J. D. 1996b. Fossil evidence for a Late Cretaceous origin of "hoofed" mammals. *Science* 272: 1150–1153.

Archibald, J. D. 1998. Archaic ungulates ("Condylarthra"). *In* C. M. Janis, K. M. Scott, and L. L. Jacobs (eds.), *Evolution of Tertiary Mammals of North America. Vol. 1: Terrestrial Carnivores, Ungulates, and Ungulatelike Mammals*, 292–331. Cambridge University Press, Cambridge.

Archibald, J. D. 1999. Divergence times of eutherian mammals. *Science* 285: 2031.

Archibald, J. D., and Averianov, A. O. 2001a. Zhelestids, archaic ungulates, and Placentalia. *Journal of Vertebrate Paleontology* 21, supplement to no. 3: 29A.

Archibald, J. D., and Averianov, A. O. 2001b. *Paranyctoides* and allies from the Late Cretaceous of North America and Asia. *Acta Palaeontologica Polonica* 46: 533–551.

Archibald, J. D., and Lofgren, D. L. 1990. Mammalian zonation near the Cretaceous–Tertiary boundary. *Geological Society of America, Special Paper* 243: 31–50.

Archibald, J. D., Schoch, R. M., and Rigby, J. K., Jr. 1983. A new subfamily, Conacodontinae, and a new species, *Conacodon kohlbergeri*, of the Periptychidae (Condylarthra, Mammalia). *Postilla* 191: 1–24.

Archibald, J. D., Clemens, W. A., Gingerich, P. D., Krause, D. W., Lindsay, E. H., and Rose, K. D. 1987. First North American land mammal ages of the Cenozoic. *In* M. O. Woodburne (ed.), *Cenozoic Mammals of North America*, 24–76. University of California Press, Berkeley.

Archibald, J. D., Averianov, A. O., and Ekdale, E. G. 2001. Late Cretaceous relatives of rabbits, rodents, and other extant eutherian mammals. *Nature* 414: 62–65.

Argot, C. 2001. Functional-adaptive anatomy of the forelimb in the Didelphidae, and the paleobiology of the Paleocene marsupials *Mayulestes ferox* and *Pucadelphys andinus*. *Journal of Morphology* 247: 51–79.

Ariëns Kappers, C. U., Huber, C. C., and Crosby, E. C. 1960. *The Comparative Anatomy of the Nervous System in Vertebrates, Including Man.* 1845 pp. 3 Vols. Hafner Publishing Company, New York.

Aristov, A. A., Krasts, I. V., and Gambaryan, P. P. 1980. Biomechanics of the turning of a large gerboa *Allactaga jaculus* Pall. *Trudy Zoologicheskogo Instituta Akademii Nauk SSSR* 91: 56–62 (in Russian).

Armstrong-Ziegler, J. G. 1978. An aniliid snake and associated vertebrates from the Campanian of New Mexico. *Journal of Paleontology* 52: 480–483.

Asher, R. 1999. A morphological basis for assessing the phylogeny of the Tenrecoidea (Mammalia, Lipotyphla). *Cladistics* 15: 231–252.

Asher, R. 2001. Cranial anatomy in tenrecid insectivores: character evolution across competing phylogenies. *American Museum Novitates* 3352: 1–54.

Asher, R., and Sánchez-Villagra, M. 2000. Jurassic molars in modern mammals: 150 million years of zalambdodont teeth. *Journal of Vertebrate Paleontology* 20, supplement to no. 3: 37A.

Astibia, H., Buffetaut, E., Buscalioni, A. D., Cappetta, H., Corral, C., Estes, R., Garcia-Garmilla, F., Jaeger, J.-J., Jimenez-Fuentes, E., Le Loeff, J., Mazin, J. M., Orue-Etxebarria, X., Pereda-Suberbiola, J., Powell, J. E., Rage, J.-C., Rodriguez-Lazaro, J., Sanz, J. L., and Tong, H. 1991. The fossil vertebrates from Laño (Basque Country, Spain); new evidence on the composition and affinities of the Late Cretaceous of continental Europe. *Terra Research* 2: 460–466.

Averianov, A. O. 1997. New Late Cretaceous mammals of southern Kazakhstan. *Acta Paleontologica Polonica* 46: 243–256.

Averianov, A. O. 1999. A new species of multituberculate mammal *Uzbekbaatar* from the Late Cretaceous of Uzbekistan. *Acta Palaeontologica Polonica* 44: 301–304.

Averianov, A. O. 2000. Mammals from the Mesozoic of Kirghizia, Uzbekistan, Kazakhstan and Tadzhikistan. *In* M. J. Benton, M. A. Shishkin, E. N. Kurochkin, and D. M. Unwin (eds.), *The Age of Dinosaurs in Russia and Mongolia,* 627–652. Cambridge University Press, Cambridge.

Averianov, A. O. 2002. Early Cretaceous "symmetrodont" mammal *Gobiotheriodon* from Mongolia and the classification of "Symmetrodonta." *Acta Palaeontologica Polonica* 47: 705–716.

Averianov, A. O., and Kielan-Jaworowska, Z. 1999. Marsupials from the Late Cretaceous of Uzbekistan. *Acta Palaeontologica Polonica* 44: 71–81.

Averianov, A. O., and Nessov, L. A. 1995. A new Cretaceous mammal from the Campanian of Kazakhstan. *Neues Jahrbuch für Geologie und Paläontologie* 1995: 65–74.

Averianov, A. O., and Skutschas, P. P. 1999a. The oldest placental mammal. *In* V. N. Orlov (ed.), *VI Congress of Theriological Society,* 6. Theriological Society, Moscow (in Russian).

Averianov, A. O., and Skutschas, P. P. 1999b. Phylogenetic relationships within basal tribosphenic mammals. *Proceedings of the Zoological Institute, Russian Academy of Sciences* 281: 55–60.

Averianov, A. O., and Skutschas, P. P. 2000a. A eutherian mammal from the Early Cretaceous of Russia and biostratigraphy of the Asian Early Cretaceous vertebrate assemblages. *Lethaia* 33: 330–340.

Averianov, A. O., and Skutschas, P. P. 2000b. The oldest placental mammal. *Doklady Akademii Nauk* 374: 130–132 (in Russian).

Averianov, A. O., and Skutschas, P. P. 2001. A new genus of eutherian mammal from the Early Cretaceous of Transbaikalia, Russia. *Acta Palaeontologica Polonica* 46: 431–436.

Averianov, A. O., Archibald, J. D., and Martin, T. 2003. Placental nature of the alleged marsupial from the Cretaceous of Madagascar. *Acta Palaeontologica Polonica* 48: 149–151.

Bakker, R. T. 1971. Dinosaur physiology and the origin of mammals. *Evolution* 25: 636–658.

Bakker, R. T. 1975. Experimental and fossil evidence for the evolution of tetrapod bioenergetics. *In* D. Gates and R. Schmerl (eds.), *Perspectives on Biophysical Ecology,* 365–399. Springer-Verlag, Berlin.

Bakker, R. T. 1992. Zofiabaataridae, a new family of multituberculate mammals from the Breakfast Bench fauna at Como Bluff. *Hunteria* 2: 24.

Bakker, R. T. 1998. Dinosaur mid-life crisis: the Jurassic–Cretaceous transition in Wyoming and Colorado. *New Mexico Museum of Natural History and Science Bulletin* 14: 67–77.

Bakker, R. T., and Carpenter, K. 1990. A new latest Jurassic vertebrate fauna from the highest levels of the Morrison Formation at Como Bluff, Wyoming, with comments on Morrison biochronology. Part III. The mammals: a new multituberculate and a new paurodont. *Hunteria* 2: 2–8.

Barghusen, H. 1968. The lower jaw of cynodonts (Reptilia, Therapsida) and the evolutionary origin of mammal-like adductor jaw musculature. *Postilla* 116: 1–49.

Barghusen, H. 1986. On the evolutionary origin of the therian tensor veli palatini and tensor tympani muscles. *In* N. Hotton, P. D. MacLean, J. J. Roth, and E. C. Roth (eds.), *The Ecology and Biology of Mammal-like Reptiles,* 253–262. Smithsonian Institution Press, Washington, D. C.

Barghusen, H., and Hopson, J. A. 1970. Dentary-squamosal joint and the origin of mammals. *Science* 168: 573–575.

Bauchot, R., and Stephan, H. 1966. Donées nouvelles sur l'encephalisation des Insectivores et des Prosimiens. *Mammalia* 30: 160–196.

Bauchot, R., and Stephan, H. 1967. Encéphales et moulages endocraniens de quelques insectivores et primates endocraniens. *In* Problèmes Actuels de Paléontologie (Evolution des Vertébrés). Colloques Internationaux du CNRS 163. Paris, Editions du CNRS: 575–586.

Bazhanov, V. S. 1972. First Mesozoic mammal (*Beleutinus orlovi* Bazhanov) from the Soviet Union. *Teriologiya* 1: 74–80 (in Russian).

Beer, G. R. de. 1926. Studies on the vertebrate head. II: The orbito-temporal region of the skull. *Quarterly Journal of Microscopic Science* 70: 263–370.

Beer, G. R. de. 1937. *The Development of the Vertebrate Skull.* 554 pp. Clarendon Press, Oxford.

Belov, K., Hellman, L., and Cooper, D. W. 2002. Characterization of echidna IgM provides insights into the time of divergence of extant mammals. *Developmental and Comparative Immunology* 26: 831–839.

Belyaeva, E. I., Trofimov, B. A., and Reshetov, V. Y. 1974. General stages in evolution of late Mesozoic and early Tertiary mammalian faunas in central Asia. *Trudy Sovmestnoi Sovetsko-Mongol'skoi Paleontologicheskoi Ekspeditsii* 1: 19–45 (in Russian).

Bennett, A. F., and Ruben, J. A. 1986. The metabolic and thermoregulatory status of therapsids. *In* N. Hotton, P. D. MacLean,

J. J. Roth, and E. C. Roth (eds.), *The Ecology and Biology of Mammal-like Reptiles,* 207–218. Smithsonian Institution Press, Washington, D. C.

Bensley, B. A. 1901a. On the question of arboreal ancestry of the Marsupialia and the interrelationships of the mammalian subclasses. *American Naturalist* 35: 117–138.

Bensley, B. A. 1901b. A theory of the origin and evolution of the Australian Marsupialia. *American Naturalist* 35: 245–269.

Bensley, B. A. 1903. On the evolution of the Australian Marsupialia: with remarks on the relationships of the marsupials in general. *Transactions of the Linnean Society of London* 9: 83–217.

Bensley, B. A. 1906. The homologies of the stylar cusps in the upper molars of the Didelphidae. *University of Toronto Studies in Biology* 5: 1–13.

Benton, M. J. (ed.) 1988. *The Phylogeny and Classification of the Tetrapods, Vol. 2.* 329 pp. The Systematics Association, Special Volume No. 35B. Clarendon Press, Oxford.

Benton, M. J. 1990a. *Vertebrate Palaeontology.* 377 pp. Unwin Hyman, London.

Benton, M. J. 1990b. Reptiles. *In* K. J. McNamara (ed.), *Evolutionary Trends,* 279–300. Belhaven Press, London.

Benton, M. J. 1999. Early origins of modern birds and mammals: molecules vs. morphology. *BioEssays* 21: 1043–1051.

Benton, M. J. 2000a. Conventions in Russian and Mongolian paleontological literature. *In* M. J. Benton, M. A. Shishkin, E. N. Kurochkin, and D. M. Unwin (eds.), *The Age of Dinosaurs in Russia and Mongolia,* xvi–xxxix. Cambridge University Press, Cambridge.

Benton, M. J. 2000b. Stems, nodes, crown clades, and rank-free lists: is Linnaeus dead? *Biological Reviews* 75: 633–648.

Berkowitz, B. K. 2000. Tooth replacement patterns in non-mammalian vertebrates. *In* M. F. Teaford, M. M. Smith, and M. W. J. Ferguson (eds.), *Development, Function, and Evolution of Teeth,* 186–200. Cambridge University Press, Cambridge.

Bertini, R. J., Marshall, L. G., Gayet, M., and Brito, P. 1993. Vertebrate faunas from the Adamantina and Marilia formations (upper Bauru Group, Late Cretaceous, Brazil) in their stratigraphic and paleobiogeographic context. *Neues Jahrbuch für Geologie und Paläontologie, Abhandlungen* 188: 71–101.

Bien, M. N. 1941. "Red Beds" of Yunnan. *Bulletin of Geological Society of China* 21: 159–198.

Bilbey, S. A. 1998. Cleveland-Lloyd Dinosaur Quarry—age, stratigraphy and depositional environments. *Modern Geology* 23: 87–120.

Blainville, H. M. D. de. 1834. "Cours de la faculté des sciences, 1834." *Fide* Palmer, T. S. *Index generum mammalium,* 1904, 780 pp.

Blainville, H. M. D. de. 1838. Doutes sur le prétendu Didelphe de Stonesfield. *Comptes Rendus de l'Académie des Sciences* 7: 402–418.

Bloch, J. I., Rose, K. D., and Gingerich, P. D. 1998. New species of *Batodonoides* (Lipotyphla, Geolabididae): smallest known mammal? *Journal of Mammalogy* 79: 804–827.

Bloch, J. I., Boyer, D. M., Gingerich, P. D., and Gunnell, G. F. 2002. New primitive paromomyid from the Clarforkian of Wyoming and dental eruption in Plesiadapiformes. *Journal of Vertebrate Paleontology* 22: 366–379.

Bohlin, B. 1953. *Fossil Reptiles from Mongolia and Kansu.* 113 pp. Statens Etnografiska Museum, Stockholm.

Bonaparte, C. L. 1837. New systematic arrangement of vertebrated animals. *Transactions of the Linnean Society of London* 18: 247–304.

Bonaparte, J. F. 1962. Descripción del cráneo y mandíbula de *Exaertodon frenguellii* Cabrera, y su comparación con Diademodontidae, Tritylodontidae y los cinodontes sudamericanos. *Publicaciones del Museo Municipal de Ciencias Naturales y Tradicional de Mar del Plata* 1: 135–202.

Bonaparte, J. F. 1963. Descripción del esqueleto poscraneano de *Exaertodon* (Cynodontia–Traversodontidae). *Acta Geológica Lilloana* 4: 5–51.

Bonaparte, J. F. 1966. Sobre las cavidades cerebral, nasal y otras estructuras del cráneo de *Exaertodon* sp. (Cynodontia–Traversodontidae). *Acta Geológica Lilloana* 1966: 5–29.

Bonaparte, J. F. 1980. El primer ictidosaurio (Reptilia–Therapsida) de América del Sur, *Chalminia musteloides,* del Tríasico Superior del La Rioja, República Argentina. *Actas del II Congreso Argentino del Paleontología y Bioestratigrafía, y I Congreso Latinamericano de Paleontología* 1: 123–133.

Bonaparte, J. F. 1986a. Sobre *Mesungulatum houssayi* y nuevos mamíferos cretácicos de Patagonia. *Actas IV Congreso Argentino de Paleontología y Bioestratigrafía* 2: 48–61.

Bonaparte, J. F. 1986b. A new and unusual Late Cretaceous mammal from Patagonia. *Journal of Vertebrate Paleontology* 6: 264–270.

Bonaparte, J. F. 1988. Specialized dentition in two Cretaceous mammals from Patagonia. *Mémoirs du Museum national d'Histoire naturelle, Paris, sér. C,* 53: 207–215.

Bonaparte, J. F. 1990. New Late Cretaceous mammals from the Los Alamitos Formation, northern Patagonia. *National Geographic Research* 6: 63–93.

Bonaparte, J. F. 1992. Una nueva especie de Triconodonta (Mammalia) de la Formación Los Alamitos, Provincia de Río Negro y comentarios sobre su fauna de mamíferos. *Ameghiniana* 29: 99–110.

Bonaparte, J. F. 1994. Approach to the significance of the Late Cretaceous mammals of South America. *Berliner geowissenschaftliche, Abhandlungen E* 13: 31–44.

Bonaparte, J. F. 1999. New Dryolestida (Theria) from the Late Cretaceous Los Alamitos Formation (Argentina) and paleobiogeographical comments. *In* H. A. Leanza (ed.), *VII International Symposium on Mesozoic Terrestrial Ecosystems, Abstracts* 10. Museo Argentino de Ciencias Naturales "Bernardino Rivadavia," Buenos Aires.

Bonaparte, J. F., and Barberena, M. C. 1975. A possible mammalian ancestor from the Middle Triassic of Brazil (Therapsida–Cynodontia). *Journal of Paleontology* 49: 931–936.

Bonaparte, J. F., and Barberena, M. C. 2001. On two advanced carnivorous cynodonts from the Late Triassic of southern Brazil. *In* F. A. Jenkins, Jr., T. Owerkowicz, and M. D. Shapiro (eds.), *Studies in Organismic and Evolutionary Biology in*

Honor of A. W. Crompton, 59–80. *Bulletin of the Museum of Comparative Zoology* 156, Cambridge.

Bonaparte, J. F., and Rougier, G. W. 1987. Mamíferos del Cretácico Inferior de Patagonia. *IV Congreso Latinomericano de Paleontología* 1: 343–359.

Bonaparte, J. F., and Soria, M. F. 1985. Nota sobre el primer mamífero del Cretácico Argentino, Campaniano–Maastrichtiano (Condylarthra). *Ameghiniana* 21: 178–183.

Bonaparte, J. F., Baez, A. M., Cione, A. L., Andreis, R., Broin, F. de, Powell, J. E., and Albino, A. 1987. The Late Cretaceous fauna of Los Alamitos, Patagonia, Argentina. Part IX—Resume. *Revista del Museo Argentino de Ciencias Naturales "Bernardino Rivadavia," Paleontología* 3: 171–178.

Bowdich, T. E. 1821. *An Analysis of the Natural Classifications of Mammalia for the Use of Students and Travellers.* 115 pp. J. Smith, Paris.

Bown, T. M., and Schankler, D. 1982. A review of the Proteutheria and Insectivora of the Willwood Formation (lower Eocene), Bighorn Basin, Wyoming. *Bulletin, United States Geological Survey* 1523: 1–79.

Boyde, A., and Martin, L. B. 1984. A non-destructive survey of prism packing patterns in primate enamel. *In* R. W. Fearnhead and S. Suga (eds.), *Tooth Enamel IV,* 417–421. Elsevier Science Publishers, Amsterdam.

Braman, D. R. 2001. Terrestrial palynomorphs of the upper Santonian–?lowest Campanian Milk River Formation, southern Alberta, Canada. *Palynology* 25: 57–107.

Braman, D. R., Johnston, P. A., and Haglung, W. M. 1995. *Canadian Paleontology Conference Field Trip Guidebook No. 4. Upper Cretaceous Paleontology, Stratigraphy and Depositional Environments at Dinosaur Park and Drumheller, Alberta.* 119 pp. Geological Association of Canada, St. John's.

Bramble, D. M. 1978. Origin of the mammalian feeding complex: models and mechanisms. *Paleobiology* 4: 271–301.

Breithaupt, B. H. 1982. Paleontology and paleoecology of the Lance Formation (Maastrichtian), east flank of Rock Springs uplift, Sweetwater County, Wyoming. *Contributions to Geology, University of Wyoming* 21: 123–151.

Brink, A. A. 1956. Speculations on some advanced mammalian characteristics in higher mammal-like reptiles. *Palaeontologica Africana* 4: 77–95.

Brink, A. A. 1980. The road to endothermy—a review. *Mémoires de la Société géologique de France N. S.* 139: 29–38.

Brinkman, D. B. 1990. Paleoecology of the Judith River Formation (Campanian of Dinosaur Provincial Park, Alberta, Canada): evidence from vertebrate microfossil localities. *Palaeogeography, Palaeoclimatology, Palaeoecology* 78: 37–54.

Brinkman, D. L., Cifelli, R. L., and Czaplewski, N. J. 1998. First occurrence of *Deinonychus antirrhopus* (Dinosauria: Theropoda) from the Antlers Formation (Lower Cretaceous: Aptian–Albian) of Oklahoma. *Bulletin, Oklahoma Geological Survey* 146: 1–26.

Britt, B. B. 1991. Theropods of Dry Mesa Quarry (Morrison Formation, Late Jurassic), with emphasis on the osteology of *Torvosaurus tanneri. Brigham Young University Geology Studies* 37: 1–72.

Broderip, W. J. 1828. Observations on the jaw of a fossil mammiferous animal found in the Stonesfield Slate. *Zoological Journal of London* 3: 408–412.

Broom, R. 1910. On *Tritylodon,* and on the relationships of the Multituberculata. *Proceedings of the Zoological Society of London* 1910: 760–768.

Broom, R. 1914. On the structure and affinities of the Multituberculata. *Bulletin of the American Museum of Natural History* 33: 115–134.

Brunet, M., Jacobs, L. L., Congleton, Y., Dejax, J., Flynn, L. J., Hell, J., Jehenne, Y., Mouchelin, G., and Pilbeam, D. 1988. Première découverte d'un fragment de mandibule de Mammifère dans le Crétacé inférieur d'Afrique (Cameroun, Bassin de Koum). *Comptes Rendus de l'Académie des Sciences, Paris, sér. II,* 307: 1675–1680.

Brunet, M., Coppens, Y., Dejax, J., Flynn, L. J., Heintz, E., Hell, J., Jacobs, L. L., Jehenne, Y., Mouchelin, G., Pilbeam, D., and Sudre, J. 1990. Nouveaux mammifères du Crétacé inférieur du Cameroun, Afrique de l'Oueste. *Comptes Rendus de l'Académie des Sciences* 310: 1139–1146.

Bryant, H. N. 1994. Comments on the phylogenetic definition of taxon names and conventions regarding the naming of crown clades. *Systematic Biology* 43: 124–130.

Buckland, W. 1824. Notice on *Megalosaurus. Transactions of the Geological Society of London* 1: 390–396.

Buffetaut, E., and Wouters, G. 1986. Amphibian and reptile remains from the Upper Triassic of Saint-Nicolas-de-Port (eastern France) and their biostratigraphic significance. *Modern Geology* 10: 133–145.

Bugge, J. 1974. The cephalic arterial system in insectivores, primates, rodents, and lagomorphs, with special reference to the systematic classification. *Acta Anatomica* 87 (supplement 2): 1–160.

Bugge, J. 1979. Cephalic arterial pattern in New World edentates and Old World pangolins with special reference to their phylogenetic relationships and taxonomy. *Acta Anatomica* 105: 37–46.

Bull, J. J., Huelsenbeck, J. P., Cunningham, C. W., Swofford, D. L., and Waddell, P. J. 1993. Partitioning and combining data in phylogenetic analysis. *Systematic Biology* 42: 284–397.

Butler, A. B. 1994. The evolution of the dorsal pallium in telencephalon of amniotes: cladistic analysis and a new hypothesis. *Brain Research Reviews* 19: 66–101.

Butler, P. M. 1939. The teeth of the Jurassic mammals. *Proceedings of the Zoological Society of London* 109: 329–356.

Butler, P. M. 1952. Molarization of the premolars in Perissodactyla. *Proceedings of the Zoological Society of London* 121: 810–843.

Butler, P. M. 1956. The skull of *Ictops* and the classification of the Insectivora. *Proceedings of the Zoological Society of London* 126: 453–481.

Butler, P. M. 1961. Relationships between upper and lower molar patterns. *In* G. Vandebroek (ed.), *International Colloquium on the Evolution of Lower and Non Specialized Mammals,* 1: 117–126. Koninklijke Vlaamse Academie voor

Wetenschappen, Letteren en Schone Kunsten van België, Brussels.

Butler, P. M. 1972a. The problem of insectivore classification. *In* K. A. Joysey and T. S. Kemp (eds.), *Studies in Vertebrate Evolution,* 253–265. Oliver and Boyd, Edinburgh.

Butler, P. M. 1972b. Some functional aspects of molar evolution. *Evolution* 26: 474–483.

Butler, P. M. 1977. Evolutionary radiation of the cheek teeth of Cretaceous placentals. *Acta Palaeontologica Polonica* 22: 241–271.

Butler, P. M. 1978a. A new interpretation of the mammalian teeth of tribosphenic pattern from the Albian of Texas. *Breviora* 446: 1–27.

Butler, P. M. 1978b. Molar cusp nomenclature and homology. *In* P. M. Butler and K. A. Joysey (eds.), *Studies in the Development, Function and Evolution of Teeth,* 441–453. Academic Press, London.

Butler, P. M. 1988a. Docodont molars as tribosphenic analogues (Mammalia, Jurassic). *In* D. E. Russell, J.-P. Santoro, and D. Sigogneau-Russell (eds.), *Teeth Revisited: Proceedings of the VIIth International Symposium on Dental Morphology,* 329–340. Mémoirs du Muséum national d'Histoire naturelle C 53, Paris.

Butler, P. M. 1988b. Phylogeny of the insectivores. *In* M. J. Benton (ed.), *The Phylogeny and Classification of the Tetrapods, Vol. 2: Mammals,* 117–141. Clarendon Press, Oxford.

Butler, P. M. 1990a. Early trends in the evolution of tribosphenic molars. *Biological Reviews* 65: 529–552.

Butler, P. M. 1990b. Tribosphenic molars in the Cretaceous. *In* P. Smith and E. Tchernov (eds.), *Structure, Function and Evolution of Teeth,* 125–138. Freund Publishing House, Tel Aviv.

Butler, P. M. 1995. Another look at premolars. *In* R. J. Radlanski and H. Renz (eds.), *Proceedings of the 10th International Symposium on Dental Morphology,* 1–6. C. & M. Brunne, Berlin.

Butler, P. M. 1997. An alternative hypothesis on the origin of docodont molar teeth. *Journal of Vertebrate Paleontology* 17: 435–439.

Butler, P. M. 2000. Review of the early allotherian mammals. *Acta Palaeontologica Polonica* 45: 317–342.

Butler, P. M., and Clemens, W. A. 2001. Dental morphology of the Jurassic holotherian mammal *Amphitherium,* with a discussion of the evolution of mammalian post-canine dental formulae. *Palaeontology* 44: 1–20.

Butler, P. M., and Ford, R. 1977. Discovery of Cretaceous mammals on the Isle of Wight. *Proceedings of the Isle of Wight Natural History and Archaeological Society* for 1975, 6: 662–663.

Butler, P. M., and Kielan-Jaworowska, Z. 1973. Is *Deltatheridium* a marsupial? *Nature* 245: 105–106.

Butler, P. M., and Krebs, B. 1973. A pantotherian milk dentition. *Paläontologische Zeitschrift* 47: 256–258.

Butler, P. M., and MacIntyre, G. T. 1994. Review of the British Haramiyidae (?Mammalia, Allotheria), their molar occlusion and relationships. *Philosophical Transactions of the Royal Society of London* 345: 433–458.

Butler, R. F., and Lindsay, E. H. 1985. Mineralogy of magnetic minerals and revised magnetic polarity stratigraphy of continental sediments, San Juan Basin, New Mexico. *Journal of Geology* 93: 535–554.

Butler, R. F., Lindsay, E. H., Jacobs, L. L., and Johnson, N. M. 1977. Magnetostratigraphy of the Cretaceous–Tertiary boundary in the San Juan Basin, New Mexico. *Nature* 267: 318–323.

Callison, G. 1987. Fruita: a place for wee fossils. *In* W. R. Averett (ed.), *Paleontology and Geology of the Dinosaur Triangle,* 91–96. Museum of Western Colorado, Grand Junction.

Canudo, J. I., and Cuenca-Bescós, G. 1996. Two new mammalian teeth (Multituberculata and Peramura) from the Lower Cretaceous of Spain. *Cretaceous Research* 17: 215–228.

Canudo, J. I., Cuenca-Bescós, G., Ruiz-Omeñaca, J. I., and Soria, A. R. 1996a. Régistro fósil de vertebrados en el tránsito Jurásico–Cretácico de Galve (Teruel, España). *Revista de la Academia de Ciencias Zaragoza* 51: 221–236.

Canudo, J. I., Cuenca-Bescós, G., Ruiz-Omeñaca, J. I., and Soria, A. R. 1996b. Estratigrafía y paleoecología de los vertebrados del Barremiense superior (Cretácico inferior) de Vallipón (Castellote, Teruel). *Grupo de Estudios Masinos, Boletín* 15: 9–34.

Canudo, J. I., Amo, O., Cuenca-Bescós, G., Meléndez, A., Ruiz-Omeñaca, J. I., and Soria, A. R. 1997. Los vertebrados del Tithónico–Barremiense de Galve (Teruel, España). *Cuadernos de Geología Ibérica* 23: 209–241.

Carlson, S. J., and Krause, D. W. 1985. Enamel ultrastructure of multituberculate mammals: an investigation of variability. *Contributions from the Museum of Paleontology, University of Michigan* 27: 1–50.

Carpenter, K. 1979. Vertebrate fauna of the Laramie Formation (Maastrichtian), Weld County, Colorado. *Contributions to Geology, University of Wyoming* 17: 37–49.

Carpenter, K. 1998. Redescription of the multituberculate, *Zofiabaatar* and the paurodont, *Foxraptor,* from Pine Tree Ridge, Wyoming. *Modern Geology* 23: 393–405.

Carpenter, K., Chure, D. J., and Kirkland, J. I. (eds.). 1998. *The Upper Jurassic Morrison Formation: An Interdisciplinary Study. Part 1.* 533 pp. Modern Geology 22, Special Issue, Amsterdam.

Carroll, R. L. 1988. *Vertebrate Paleontology and Evolution.* 698 pp. Freeman and Company, New York.

Carvalho, I. I. 1994. *Candidodon:* um crocodilo com heterodontia (Notosuchia, Cretáceo Inferior, Brasil). *Anais do Academia Brasiliera de Ciências* 66: 331–346.

Carvalho, I. S., and Campos, D. A. 1988. Um mamífero Triconodonte do Cretáceo inferior do Maranhão, Brasil. *Anais da Academia Brasiliera de Ciências* 60: 437–446.

Casamiquela, R. M. 1964. *Estudios Icnológicos.* 229 pp. Imprenta del Colegio Industrial, Pío X, Buenos Aires.

Casey, R. 1963. The dawn of the Cretaceous Period in Britain. *Bulletin of the S-E Union of Scientific Societies* 117: 1–15.

Cassiliano, M. L., and Clemens, W. A. 1979. Symmetrodonta. *In* J. A. Lillegraven, Z. Kielan-Jaworowska, and W. A. Clemens (eds.), *Mesozoic Mammals: The First Two-thirds of Mammalian History,* 150–161. University of California Press, Berkeley.

Cave, A. J. E., and Haines, R. W. 1940. The paranasal sinuses of the anthropoid apes. *Journal of Anatomy, London* 74: 493–523.

Cheng, C. C. 1955. The development of the shoulder region in the opossum, *Didelphis virginiana,* with special reference to musculature. *Journal of Morphology* 97: 415–471.

Chow, M., and Qi, T. 1978. Paleocene mammalian fossils from the Nomogen Formation of Inner Mongolia. *Vertebrata PalAsiatica* 16: 77–85.

Chow, M., and Rich, T. H. 1982. *Shuotherium dongi,* n. gen. and sp., a therian with pseudo-tribosphenic molars from the Jurassic of Sichuan, China. *Australian Mammalogy* 5: 127–142.

Chow, M., and Rich, T. H. 1984a. A new triconodontan (Mammalia) from the Jurassic of China. *Journal of Vertebrate Paleontology* 4: 226–231.

Chow, M., and Rich, T. H. 1984b. New Mesozoic mammal sites from China. *Alcheringa* 8: 304.

Chure, D. J., Carpenter, K., Litwin, R., Hasiotis, S., and Evanoff, E. 1998. The flora and fauna of the Morrison Formation. *Modern Geology* 23: 507–537.

Cifelli, R. L. 1982. The petrosal structure of *Hyopsodus* with respect to that of some other ungulates, and its phylogenetic implications. *Journal of Paleontology* 56: 795–805.

Cifelli, R. L. 1983. The origin and affinities of the South American Condylarthra and early Tertiary Litopterna (Mammalia). *American Museum Novitates* 2772: 1–49.

Cifelli, R. L. 1990a. Cretaceous mammals of southern Utah. I. Marsupial mammals from the Kaiparowits Formation (Judithian). *Journal of Vertebrate Paleontology* 10: 295–319.

Cifelli, R. L. 1990b. Cretaceous mammals of southern Utah. II. Marsupials and marsupial-like mammals from the Wahweap Formation (early Campanian). *Journal of Vertebrate Paleontology* 10: 320–331.

Cifelli, R. L. 1990c. Cretaceous mammals of southern Utah. III. Therian mammals from the Turonian (early Late Cretaceous). *Journal of Vertebrate Paleontology* 10: 332–345.

Cifelli, R. L. 1990d. A primitive higher mammal from the Late Cretaceous of southern Utah. *Journal of Mammalogy* 71: 343–350.

Cifelli, R. L. 1990e. Cretaceous mammals of southern Utah. IV. Eutherian mammals from the Wahweap (Aquilan) and Kaiparowits (Judithian) formations. *Journal of Vertebrate Paleontology* 10: 346–360.

Cifelli, R. L. 1993a. Early Cretaceous mammal from North America and the evolution of marsupial dental characters. *Proceedings of the National Academy of Sciences USA* 90: 9413–9416.

Cifelli, R. L. 1993b. Theria of metatherian-eutherian grade and the origin of marsupials. *In* F. S. Szalay, M. J. Novacek, and M. C. McKenna (eds.), *Mammal Phylogeny: Mesozoic Differentiation, Multituberculates, Monotremes, Early Therians, and Marsupials,* 205–215. Springer-Verlag, New York.

Cifelli, R. L. 1993c. The phylogeny of the native South American ungulates. *In* F. S. Szalay, M. J. Novacek, and M. C. McKenna (eds.), *Mammal Phylogeny: Placentals,* 195–216. Springer-Verlag, New York.

Cifelli, R. L. 1994. Therian mammals of the Terlingua Local Fauna (Judithian), Aguja Formation, Big Bend of the Río Grande, Texas. *Contributions to Geology, University of Wyoming* 30: 117–136.

Cifelli, R. L. 1997. First notice on Mesozoic mammals from Oklahoma. *Oklahoma Geology Notes, Oklahoma Geological Survey* 57: 4–17.

Cifelli, R. L. 1999a. Therian teeth of unusual design from the medial Cretaceous (Albian–Cenomanian) Cedar Mountain Formation, Utah. *Journal of Mammalian Evolution* 6: 247–270.

Cifelli, R. L. 1999b. Tribosphenic mammal from the North American Early Cretaceous. *Nature* 401: 363–366.

Cifelli, R. L. 2000a. Counting premolars in early eutherian mammals. *Acta Palaeontologica Polonica* 45: 195–198.

Cifelli, R. L. 2000b. Tribosphenic mammal from the North American Early Cretaceous: correction. *Nature* 404: 904.

Cifelli, R. L. 2000c. Cretaceous mammals of Asia and North America. *Journal of the Paleontological Society of Korea, Special Publication* 4: 49–85.

Cifelli, R. L. 2001. Early mammalian radiations. *Journal of Paleontology* 75: 1214–1226.

Cifelli, R. L. 2004. Marsupial mammals from the Albian–Cenomanian (Early–Late Cretaceous) boundary, Utah. *In* G. C. Gould and S. K. Bell (eds.), *Tributes to Malcolm C. McKenna: his students, his legacy. Bulletin of the American Museum of Natural History* 341: 62–79.

Cifelli R. L., and Dykes T. D. 2001. *Phascolotheridium,* a new name for *Phascolodon* Simpson, 1925 (Mammalia) preoccupied by *Phascolodon* Stein, 1859 (Ciliphora, Phyllopharyngea). *Acta Palaeontologica Polonica* 46 (3): 392.

Cifelli, R. L., and Eaton, J. G. 1987. Marsupial from the earliest Late Cretaceous of western US. *Nature* 325: 520–522.

Cifelli, R. L., and Gordon, C. L. 1999. Symmetrodonts from the Late Cretaceous of southern Utah and distribution of archaic mammals in the Cretaceous of North America. *Geology Studies, Brigham Young University* 44: 1–16.

Cifelli, R. L., and Johanson, Z. 1994. New marsupial from the Upper Cretaceous of Utah. *Journal of Vertebrate Paleontology* 14: 292–295.

Cifelli, R. L., and Madsen, S. K. 1986. An Upper Cretaceous symmetrodont (Mammalia) from southern Utah. *Journal of Vertebrate Paleontology* 6: 258–263.

Cifelli, R. L., and Madsen, S. K. 1998. Triconodont mammals from the medial Cretaceous of Utah. *Journal of Vertebrate Paleontology* 18: 403–411.

Cifelli, R. L., and Madsen, S. K. 1999. Spalacotheriid symmetrodonts (Mammalia) from the medial Cretaceous (upper Albian or lower Cenomanian) Mussentuchit local fauna, Cedar Mountain Formation, Utah, USA. *Geodiversitas* 21: 167–214.

Cifelli, R. L., and Muizon, C. de. 1997. Dentition and jaw of *Kokopellia juddi,* a primitive marsupial or near marsupial from the medial Cretaceous of Utah. *Journal of Mammalian Evolution* 4: 241–258.

Cifelli, R. L., and Muizon, C. de. 1998a. Marsupial mammal from the Upper Cretaceous North Horn Formation, central Utah. *Journal of Paleontology* 72: 532–537.

Cifelli, R. L., and Muizon, C. de. 1998b. Tooth eruption and replacement pattern in early marsupials. *Comptes Rendus de l'Académie des Sciences, Paris* 326: 215–220.

Cifelli, R. L., Larson, E. M., and Sherburn, E. W. 1988. Dental wear and habitat preference in primitive tribosphenid mammals. *Journal of Vertebrate Paleontology* 8, supplement to no. 3: 12A.

Cifelli, R. L., Schaff, C. R., and McKenna, M. C. 1989. The relationships of the Arctostylopidae: new data and interpretation. *Bulletin of the Museum of Comparative Zoology* 152: 1–44.

Cifelli, R. L., Rowe, T. B., Luckett, W. P., Banta, J., Reyes, R., and Howes, R. I. 1996. Origin of marsupial pattern of tooth replacement: fossil evidence revealed by high resolution X-ray CT. *Nature* 379: 715–718.

Cifelli, R. L., Gardner, J. D., Nydam, R. L., and Brinkman, D. L. 1997. Additions to the vertebrate fauna of the Antlers Formation (Lower Cretaceous), southeastern Oklahoma. *Oklahoma Geology Notes, Oklahoma Geological Survey* 57: 124–131.

Cifelli, R. L., Kirkland, J. I., Weil, A., Deino, A. L., and Kowallis, B. J. 1997. High-precision $^{40}Ar/^{39}Ar$ geochronology and the advent of North America's Late Cretaceous terrestrial fauna. *Proceedings of the National Academy of Sciences USA* 94: 11163–11167.

Cifelli, R. L., Wible, J. R., and Jenkins, F. A., Jr. 1998. Triconodont mammals from the Cloverly Formation (Lower Cretaceous), Montana and Wyoming. *Journal of Vertebrate Paleontology* 18: 237–241.

Cifelli, R. L., Lipka, T. R., Schaff, C. R., and Rowe, T. B. 1999. First Early Cretaceous mammal from the eastern seaboard of the United States. *Journal of Vertebrate Paleontology* 19: 199–203.

Cifelli, R. L., Nydam, R. L., Eaton, J. G., Gardner, J. D., and Kirkland, J. I. 1999. Vertebrate faunas of the North Horn Formation (Upper Cretaceous–lower Paleocene), Emery and Sanpete counties. *In* D. D. Gillette (ed.), *Vertebrate Paleontology in Utah,* 377–388. Utah Geological Survey, Miscellaneous Publication 99-1, Salt Lake City.

Cifelli, R. L., Nydam, R. L., Weil, A., Gardner, J. D., Kirkland, J. I., Eaton, J. G., and Madsen, S. K. 1999. Vertebrate fauna of the upper Cedar Mountain Formation (Cretaceous; Albian–Cenomanian), Emery County: the Mussentuchit local fauna. *In* D. D. Gillette (ed.), *Vertebrate Paleontology in Utah,* 219–242. Utah Geological Survey, Miscellaneous Publication 99-1, Salt Lake City.

Cifelli, R. L., Madsen, S. K., Sigogneau-Russell, D., Kielan-Jaworowska, Z., and Maxwell, W. D. 2000. Early Cretaceous "symmetrodont" from Montana. *Journal of Vertebrate Paleontology* 20, supplement to no. 3: 35A– 36A.

Cifelli, R. L., Eberle, J. J., Lofgren, D. L., Lillegraven, J. A., and Clemens, W. A. 2004. Mammalian biochronology of the latest Cretaceous in North America. *In* M. O. Woodburne (ed.), *Late Cretaceous and Cenozoic Mammals of North America: Geochronology and Biostratigraphy,* 21–42. Columbia University Press, New York.

Clark, J. M., and Hopson, J. A. 1985. Distinctive mammal-like reptile from Mexico and its bearing on the phylogeny of the Tritylodontidae. *Nature* 315: 398–400.

Clark, J. M., Montellano, M., Hopson, J. A., Hernandez, R., and Fastovsky, D. E. 1994. An Early or Middle Jurassic tetrapod assemblage from the La Boca Formation, northeastern Mexico. *In* N. C. Fraser and H.-D. Sues (eds.), *In the Shadow of the Dinosaurs—Early Mesozoic Tetrapods,* 295–302. Cambridge University Press, Cambridge.

Clemens, W. A. 1961. A Late Cretaceous mammal from Dragon Canyon, Utah. *Journal of Paleontology* 35: 578–579.

Clemens, W. A. 1963a. Wealden mammalian fossils. *Palaeontology* 6: 55–69.

Clemens, W. A. 1963b. Fossil mammals of the type Lance Formation, Wyoming. Part I. Introduction and Multituberculata. *University of California Publications in Geological Sciences* 48: 1–105.

Clemens, W. A. 1966. Fossil mammals from the type Lance Formation, Wyoming. Part II. Marsupialia. *University of California Publications in Geological Sciences* 62: 1–122.

Clemens, W. A. 1968a. A mandible of *Didelphodon vorax* (Marsupialia, Mammalia). *Los Angeles County Museum, Contributions in Science* 133: 1–11.

Clemens, W. A. 1968b. Origin and evolution of marsupials. *Evolution* 22: 1–18.

Clemens, W. A. 1973a. Fossil mammals of the type Lance Formation, Wyoming. Part III. Eutheria and summary. *University of California Publications in Geological Sciences* 94: 1–102.

Clemens, W. A. 1973b. The roles of fossil vertebrates in interpretation of Late Cretaceous stratigraphy of the San Juan Basin, New Mexico. *Four Corners Geological Society Memoir* 1973: 154–167.

Clemens, W. A. 1979a. Marsupialia. *In* J. A. Lillegraven, Z. Kielan-Jaworowska, and W. A. Clemens (eds.), *Mesozoic Mammals: The First Two-thirds of Mammalian History,* 192–220. University of California Press, Berkeley.

Clemens, W. A. 1979b. A problem in morganucodontid taxonomy. *Zoological Journal of the Linnean Society* 66: 1–14.

Clemens, W. A. 1979c. Notes on the Monotremata. *In* J. A. Lillegraven, Z. Kielan-Jaworowska, and W. A. Clemens (eds.), *Mesozoic Mammals: The First Two-Thirds of Mammalian History,* 309–311. University of California Press, Berkeley.

Clemens, W. A. 1980a. Rhaeto-Liassic mammals from Switzerland and West Germany. *Zitteliana, Abhandlungen der Bayerischen Staatssammlung für Paläontologie und Historische Geologie* 5: 51–92.

Clemens, W. A. 1980b. *Gallolestes pachymandibularis* (Theria, incertae sedis) from Late Cretaceous deposits in Baja California del Norte, Mexico. *PaleoBios* 33: 1–10.

Clemens, W. A. 1982. Review of Crochet, J.-Y., Les Marsupiaux du Tertiaire d'Europe. *Journal of Mammalogy* 63: 724–725.

Clemens, W. A. 1986. On Triassic and Jurassic mammals. *In* K. Padian (ed.), *The Beginning of the Age of Dinosaurs,* 237–246. Cambridge University Press, Cambridge.

Clemens, W. A. 1995. A latest Cretaceous, high paleolatitude mammalian fauna from the north slope of Alaska. *In* Z. Kielan-Jaworowska, N. Heintz, and H. A. Nakrem (eds.), *Fifth Symposium on Mesozoic Terrestrial Ecosystems and Biota,* 15–16. Contributions from the Paleontological Museum, no. 364, University of Oslo, Oslo.

Clemens, W. A. 1997. Characterization of enamel microstructure and application of the origins of prismatic structures in systematic analyses. *In* W. von Koenigswald and P. M. Sanders (eds.), *Tooth Enamel Microstructure,* 85–112. A. A. Balkema, Rotterdam.

Clemens, W. A. 2002. Evolution of the mammalian fauna across the Cretaceous–Tertiary boundary in northeastern Montana and other areas of the Western Interior. *In* J. H. Hartman, K. R. Johnson, and D. J. Nichols (eds.), *The Hell Creek Formation and the Cretaceous–Tertiary Boundary in the Northern Great Plains: An Integrated Continental Record of the End of the Cretaceous,* 217–245. Geological Society of America, Special Paper, Denver.

Clemens, W. A., and Kielan-Jaworowska, Z. 1979. Multituberculata. *In* J. A. Lillegraven, Z. Kielan-Jaworowska, and W. A. Clemens (eds.), *Mesozoic Mammals: The First Two-thirds of Mammalian History,* 99–149. University of California Press, Berkeley.

Clemens, W. A., and Lees, P. M. 1971. A review of English Early Cretaceous mammals. *In* D. M. Kermack and K. A. Kermack (eds.), *Early Mammals,* 117–130. *Zoological Journal of the Linnean Society* 50, supplement 1, London.

Clemens, W. A., and Lillegraven, J. A. 1986. New Late Cretaceous, North American advanced therian mammals that fit neither the marsupial nor eutherian molds. *Contributions to Geology, University of Wyoming, Special Paper* 3: 55–85.

Clemens, W. A., and Mills, J. R. E. 1971. Review of *Peramus tenuirostris. Bulletin of the British Museum (Natural History), Geology* 20: 89–113.

Clemens, W. A., Lillegraven, J. A., Lindsay, E. H., and Simpson, G. G. 1979. Where, when, and what—a survey of known Mesozoic mammal distribution. *In* J. A. Lillegraven, Z. Kielan-Jaworowska, and W. A. Clemens (eds.), *Mesozoic Mammals: The First Two-thirds of Mammalian History,* 7–58. University of California Press, Berkeley.

Clemens, W. A., Wilson, G. P., and Molnar, R. 2003. An emigmatic (synapsid?) tooth from the Early Cretaceous of New South Wales, Australia. *Journal of Vertebrate Paleontology* 23: 232–237.

Clements, R. G. 1993. Type section of the Purbeck Limestone Group, Swanage, Dorset. *Proceedings of the Dorset Natural History and Archaeological Society* 114: 181–226.

Cluver, M. A. 1971. The cranial morphology of the dicynodont genus *Lystrosaurus. Annals of the South African Museum* 56: 155–274.

Cope, E. D. 1881a. A new type of Perissodactyla. *American Naturalist* 15: 117–118.

Cope, E. D. 1881b. Eocene Plagiaulacidae. *American Naturalist* 15: 921–922.

Cope, E. D. 1882a. A second genus of Eocene Plagiaulacidae. *American Naturalist* 16: 416–417.

Cope, E. D. 1882b. New genus of Taeniodonta. *American Naturalist* 16: 604–605.

Cope, E. D. 1882c. Mammalia in the Laramie Formation. *American Naturalist* 16: 830–831.

Cope, E. D. 1882d. The Periptychidae. *American Naturalist* 16: 832–833.

Cope, E. D. 1884. The Tertiary Marsupialia. *American Naturalist* 18: 686–697.

Cope, E. D. 1887. The marsupial genus *Chirox. American Naturalist* 22: 12–13.

Cope, E. D. 1892. On a new genus of Mammalia from the Laramie Formation. *American Naturalist* 26: 758–762.

Costa, R. L., Jr., and Greaves, W. S. 1981. Experimentally produced tooth wear facets and the direction of jaw movements. *Journal of Paleontology* 55: 635–638.

Cox, C. B. 1962. A natural cast of the inner ear of a dicynodont. *American Museum Novitates* 2116: 1–6.

Crochet, J.-Y. 1979a. Diversité systematique des Didelphidae (Marsupialia) européens tertiaires. *Géobios* 11: 365–378.

Crochet, J.-Y. 1979b. Données nouvelles sur l'histoire paléogeographique des Didelphidae (Marsupialia). *Comptes Rendus de l'Académie des Sciences* 288: 457–459.

Crochet, J.-Y. 1980. *Les Marsupiaux du Tertiaire d'Europe.* 279 pp. Editions de la Fondation Singer-Polignac, Paris.

Crochet, J.-Y., and Sigé, B. 1993. Les mammifères de Chulpas (Formation Umayo, transition Crétacé–Tertiare, Pérou). Données préliminaires. *Documents du Laboratoire de Géologie de Lyon* 125: 97–107.

Crochet, J.-Y., and Sigé, B. 1996. Un marsupial ancien (transition Crétacé–Tertiare) à denture évoluée en Amérique du Sud (Chulpas, formation Umayo, Pérou). *Neues Jahrbuch für Geologie und Paläontologie* (1996): 622–634.

Crompton, A. W. 1963. Tooth replacement in the cynodont *Thrinaxodon liorhinus* Seeley. *Annals of the South African Museum* 46: 479–521.

Crompton, A. W. 1964a. On the skull of *Oligokyphus. Bulletin of the British Museum (Natural History), Geology* 9: 70–82.

Crompton, A. W. 1964b. A preliminary description of a new mammal from the Upper Triassic of South Africa. *Proceedings of the Zoological Society of London* 142: 441–452.

Crompton, A. W. 1968. The enigma of the evolution of mammals. *Optima,* September 1968: 137–151.

Crompton, A. W. 1971. The origin of the tribosphenic molar. *In* D. M. Kermack and K. A. Kermack (eds.), *Early Mammals,* 65–87. *Zoological Journal of the Linnean Society* 50, supplement 1, London.

Crompton, A. W. 1972a. Evolution of the jaw articulation in cynodonts. *In* K. A. Joysey and T. S. Kemp (eds.), *Studies in Vertebrate Evolution,* 231–253. Oliver & Boyd, Edinburgh.

Crompton, A. W. 1972b. Postcanine occlusion in cynodonts and tritylodonts. *Bulletin of the British Museum (Natural History), Geology* 21: 30–71.

Crompton, A. W. 1974. The dentition and relationships of the southern African Triassic mammals, *Erythrotherium parringtoni* and *Megazostrodon rudnerae. Bulletin of the British Museum (Natural History), Geology* 24: 397–437.

Crompton, A. W. 1995. Masticatory function in nonmammalian cynodonts and early mammals. *In* J. J. Thomason (ed.), *Functional Morphology in Vertebrate Paleontology,* 55–75. Cambridge University Press, Cambridge.

Crompton, A. W., and Hiiemae, K. M. 1969. Functional occlusion in tribosphenic molars. *Nature* 222: 678–679.

Crompton, A. W., and Hiiemae, K. M. 1970. Molar occlusion and mandibular movements during occlusion in the American opossum, *Didelphis marsupialis* L. *Zoological Journal of the Linnean Society* 49: 21–47.

Crompton, A. W., and Hylander, W. L. 1986. Changes in mandibular function following the acquisition of a dentary-squamosal joint. *In* N. Hotton, III, P. D. MacLean, J. J. Roth, and E. C. Roth (eds.), *The Ecology and Biology of Mammal-like Reptiles*, 263–282. Smithsonian Institution Press, Washington, D. C.

Crompton, A. W., and Jenkins, F. A., Jr. 1967. American Jurassic symmetrodonts and Rhaetic "pantotheres." *Science* 155: 1006–1009.

Crompton, A. W., and Jenkins, F. A., Jr. 1968. Molar occlusion in Late Triassic mammals. *Biological Reviews* 43: 427–458.

Crompton, A. W., and Jenkins, F. A., Jr. 1979. Origin of mammals. *In* J. A. Lillegraven, Z. Kielan-Jaworowska, and W. A. Clemens (eds.), *Mesozoic Mammals: The First Two-thirds of Mammalian History*, 59–73. University of California Press, Berkeley.

Crompton, A. W., and Kielan-Jaworowska, Z. 1978. Molar structure and occlusion in Cretaceous therian mammals. *In* P. M. Butler and K. A. Joysey (eds.), *Studies in the Development, Function and Evolution of Teeth*, 249–287. Academic Press, London.

Crompton, A. W., and Luo, Z.-X. 1993. Relationships of the Liassic mammals *Sinoconodon, Morganucodon,* and *Dinnetherium.* *In* F. S. Szalay, M. J. Novacek, and M. C. McKenna (eds.), *Mammal Phylogeny: Mesozoic Differentiation, Multituberculates, Monotremes, Early Therians, and Marsupials*, 30–44. Springer-Verlag, New York.

Crompton, A. W., and Parker, P. 1978. Evolution of the mammalian masticatory apparatus. *American Scientist* 66: 192–201.

Crompton, A. W., and Sita-Lumsden, A. 1970. Functional significance of the therian molar pattern. *Nature* 227: 197–199.

Crompton, A. W., and Sun, A.-L. 1985. Cranial structure and relationships of the Liassic mammal *Sinoconodon. Zoological Journal of the Linnean Society* 85: 99–119.

Crompton, A. W., Taylor, C. R., and Jagger, J. A. 1978. Evolution of homeothermy in mammals. *Nature* 272: 333–336.

Crompton, A. W., Wood, C. B., and Stern, D. N. 1994. Differential wear of enamel: a mechanism for maintaining sharp cutting edges. *Advances in Comparative and Environmental Physiology* 18: 321–346.

Crusafont, M., and Gibert, J. 1976. Los primeros multituberculados de España: nota preliminar. *Acta Geológica Hispanica* 11: 57–64.

Crusafont-Pairó, M., and Adrover, R. 1966. El primer mamífero del Mesozoico Español. *Publicaciones de la Catedra de Paleontología Universidad de Barcelona* 13: 28–33.

Csiki, Z., and Grigorescu, D. 1998. Small theropods from the Late Cretaceous of the Haţeg Basin (western Romania)—an unexpected diversity at the top of the food chain. *Oryctos* 1: 87–104.

Csiki, Z., and Grigorescu, D. 2000. Teeth of multituberculate mammals from the Late Cretaceous of Romania. *Acta Palaeontologica Polonica* 45: 85–90.

Cuenca-Bescós, G., and Canudo, J. I. 1999. A Lower Cretaceous traveller: *Gobiconodon* ("triconodont," Mammalia) from Vallipón (upper Barremian, Teruel, Spain). *Servicio Publicaciones Universidad de Zaragoza, IV European Workshop on Vertebrate Palaeontology, Albarracin (Teruel, Spain), Programme and Abstracts Field Guide:* 41.

Cuenca-Bescós, G., Canudo, J. I., and Ruiz-Omeñaca, J. I. 1995. Los mamíferos del Barremiense superior (Cretácico inferior) de Vallipón (Castellote, Teruel). *Beca del Museo de las Matas. Año* 1995: 105–137.

Cui, G.-H., and Sun, A.-L. 1987. Postcanine root system of tritylodonts. *Vertebrata PalAsiatica* 25: 245–259.

Cuny, G., Godefroit, P., and Martin, M. 1995. Micro-restes de vertébrés dans le Trias Supéieur du Rinckebierg (Medernach, G-D Luxembourg). *Neues Jahrbuch für Geologie und Paläontologie, Abhandlungen* 196: 45–67.

Czelusniak, J., Goodman, M., Koop, B. F., Tagle, D. O., Shoshani, J., Braunitzer, G., Kleinschmidt, T. K., de Jong, W. W., and Matsuda, G. 1990. Perspectives from amino acid and nucleotide sequences on cladistic relationships among higher taxa of Eutheria. *In* H. H. Genoways (ed.), *Current Mammalogy*, 545–572. Plenum Press, New York.

Dashzeveg, D. 1975. New primitive therian from the Early Cretaceous of Mongolia. *Nature* 256: 402–403.

Dashzeveg, D. 1979. *Arguimus khosbajari* gen. n., sp. n. (Peramuridae, Eupantotheria) from the Lower Cretaceous of Mongolia. *Acta Palaeontologica Polonica* 24: 199–204.

Dashzeveg, D. 1994. Two previously unknown eupantotheres (Mammalia, Eupantotheria). *American Museum Novitates* 3107: 1–11.

Dashzeveg, D., and Kielan-Jaworowska, Z. 1984. The lower jaw of an aegialodontid mammal from the Early Cretaceous of Mongolia. *Zoological Journal of the Linnean Society* 82: 217–227.

Dashzeveg, D., Novacek, M. J., Norell, M. A., Clark, J. M., Chiappe, L. M., Davidson, A., McKenna, M. C., Dingus, L., Swisher, C. C., and Altangerel, P. 1995. Extraordinary preservation in a new vertebrate assemblage from the Late Cretaceous of Mongolia. *Nature* 374: 446–449.

Datta, P. M. 1981. The first Jurassic mammal from India. *Zoological Journal of the Linnean Society* 73: 307–312.

Datta, P. M., and Das, D. P. 1996. Discovery of the oldest fossil mammal from India. *India Minerals* 50: 217–222.

Datta, P. M., and Das, D. P. 2001. *Indozostrodon simpsoni*, gen. et sp. nov., an Early Jurassic megazostrodontid mammal from India. *Journal of Vertebrate Paleontology* 21: 528–534.

Datta, P. M., Yadagiri, P., and Jagannatha Rao, B. R. 1978. Discovery of Early Jurassic micromammals from Upper Gondwana sequence of Pranhita Godavari Valley, India. *Journal of the Geological Society of India* 19: 64–68.

Del Tredici, P. 1989. Ginkos and multituberculates: evolutionary interactions in the Tertiary. *BioSystems* 22: 327–339.

Demko, T. M., and Parrish, J. T. 1998. Paleoclimatic setting of the Upper Jurassic Morrison Formation. *Modern Geology* 22: 283–296.

Desmond, A. 1984. Interpreting the origin of mammals: new approaches to the history of palaeontology. *Zoological Journal of the Linnean Society* 82: 7–16.

Diem, S. 1999. *Vertebrate faunal analysis of the Upper Cretaceous Williams Fork Formation, Rio Blanco County, Colorado.* 187 pp. M.S. thesis, San Diego State University, San Diego.

Diem, S. D., and Archibald, J. D. 2000. Vertebrate fauna from the Upper Cretaceous Williams Fork Formation, northwestern Colorado. *Journal of Vertebrate Paleontology* 20, supplement to no. 3: 38A–39A.

Dietrich, W. O. 1927. *Brancatherulum* n. g., ein Proplacentalier aus dem obersten Jura des Tendaguru in Deutsch-Ostafrika. *Centralblatt für Mineralogie, Geologie und Paläontologie* 1927: 423–426.

Dollo, L. 1899. Les ancêtres des Marsupiaux était-ils arboricoles? *Traveaux de la Station Zoologique de Wimereux* 7: 188–203.

Dong, Z.-M. 1993. The field activities of the Sino-Canadian Dinosaur Project in China, 1987–1990. *Canadian Journal of Earth Sciences* 30: 1997–2001.

Doran, A. H. G. 1878. Morphology of the mammalian ossicula auditus. *Transactions of the Linnean Society of London* 1: 271–498.

Douglass, E. S. 1908. Vertebrate fossils from the Fort Union beds. *Annals of the Carnegie Museum* 5: 11–26.

Doyle, J. A. 1992. Revised palynological correlations of the Lower Potomac Group (USA) and the Cocobeach sequence of Gabon (Barremian–Aptian). *Cretaceous Research* 13: 337–349.

Dun, W. S. 1895. Notes on the occurrence of monotreme remains in the Pliocene of New South Wales. *Records of the Geological Survey of New South Wales* 4: 118–126.

Dyke, G. J. 2002. Should paleontologists use "phylogenetic" nomenclature? *Journal of Paleontology* 76: 793–796.

Eaton, J. G. 1987. *Stratigraphy, depositional environments, and age of Cretaceous mammal-bearing rocks in Utah, and systematics of the Multituberculata.* 294 pp. Ph.D. dissertation, University of Colorado, Boulder.

Eaton, J. G. 1990. Stratigraphic revision of Campanian (Upper Cretaceous) rocks in the Henry Basin, Utah. *Mountain Geologist* 27: 27–38.

Eaton, J. G. 1991. Biostratigraphic framework for the Upper Cretaceous rocks of the Kaiparowits Plateau, southern Utah. *Geological Society of America, Special Paper* 260: 47–63.

Eaton, J. G. 1993a. Mammalian paleontology and correlation of the uppermost Cretaceous rocks of the Paunsaugunt Plateau, Utah. *Museum of Northern Arizona Bulletin* 59: 163–180.

Eaton, J. G. 1993b. Therian mammals from the Cenomanian (Upper Cretaceous) Dakota Formation, southwestern Utah. *Journal of Vertebrate Paleontology* 13: 105–124.

Eaton, J. G. 1995. Cenomanian and Turonian (early Late Cretaceous) multituberculate mammals from southwestern Utah. *Journal of Vertebrate Paleontology* 15: 761–784.

Eaton, J. G. 1999a. Vertebrate paleontology of the Iron Springs Formation, Upper Cretaceous, southwestern Utah. *In* D. D. Gillette (ed.), *Vertebrate Paleontology in Utah,* 339–343. Utah Geological Survey, Miscellaneous Publication 99-1, Salt Lake City.

Eaton, J. G. 1999b. Vertebrate paleontology of the Paunsaugunt Plateau, Upper Cretaceous, southwestern Utah. *In* D. D. Gillette (ed.), *Vertebrate Paleontology in Utah,* 335–338. Utah Geological Survey, Miscellaneous Publication 99-1, Salt Lake City.

Eaton, J. G., and Cifelli, R. L. 1988. Preliminary report on Late Cretaceous mammals of the Kaiparowits Plateau, southern Utah. *Contributions to Geology, University of Wyoming* 26: 45–56.

Eaton, J. G., and Cifelli, R. L. 1998. Cretaceous vertebrates of the Grand Staircase–Escalante National Monument. *In* L. M. Hill (ed.), *Learning from the Land: Proceedings of the Grand Staircase–Escalante National Monument Science Symposium,* 365–372. Bureau of Land Management, Salt Lake City.

Eaton, J. G., and Cifelli, R. L. 2001. Multituberculate mammals from near the Early–Late Cretaceous boundary, Cedar Mountain Formation, Utah. *Acta Palaeontologica Polonica* 46: 453–518.

Eaton, J. G., and Nelson, M. E. 1991. Multituberculate mammals from the Lower Cretaceous Cedar Mountain Formation, San Rafael Swell, Utah. *Contributions to Geology, University of Wyoming* 29: 1–12.

Eaton, J. G., Kirkland, J. I., Hutchison, J. H., Denton, R., O'Neill, R. C., and Parrish, J. M. 1997. Nonmarine extinction across the Cenomanian–Turonian (C–T) boundary, southwestern Utah, with a comparison to the Cretaceous–Tertiary (K–T) extinction event. *Geological Society of America Bulletin* 109: 560–567.

Eaton, J. G., Munk, H., and Hardman, M. A. 1998. A new vertebrate fossil locality within the Wahweap Formation (Upper Cretaceous) of Bryce Canyon National Park and its bearing on the presence of the Kaiparowits Formation on the Paunsaugunt Plateau. *Technical Reports, National Park Service* 98/1: 36–40.

Eaton, J. G., Cifelli, R. L., Hutchison, J. H., Kirkland, J. I., and Parrish, J. M. 1999. Cretaceous vertebrate faunas from the Kaiparowits Plateau. *In* D. D. Gillette (ed.), *Vertebrate Fossils of Utah,* 345–354. Utah Geological Survey, Miscellaneous Publication 99-1, Salt Lake City.

Eaton, J. G., Diem, S., Archibald, J. D., Schierup, C., and Munk, H. 1999. Vertebrate paleontology of the Upper Cretaceous rocks of the Markagunt Plateau, southwestern Utah. *In* D. D. Gillette (ed.), *Vertebrate Paleontology in Utah,* 323–333. Utah Geological Survey, Miscellaneous Publication 99-1, Salt Lake City.

Eberle, J. J. 1999. Bridging the transition between didelphodonts and taeniodonts. *Journal of Paleontology* 73: 936–944.

Eberle, J. J., and Lillegraven, J. A. 1998a. A new important record of earliest Cenozoic mammalian history: geologic setting, Multituberculata, and Peradectia. *Rocky Mountain Geology* 33: 3–47.

Eberle, J. J., and Lillegraven, J. A. 1998b. A new important record of earliest Cenozoic mammalian history: Eutheria and paleogeographic/biostratigraphic summaries. *Rocky Mountain Geology* 33: 49–117.

Eberth, D. A., and Hamblin, A. P. 1993. Tectonic, stratigraphic, and sedimentologic significance of a regional discontinuity in the upper Judith River Group (Belly River wedge) of southern Alberta, Saskatchewan, and northern Montana. *Canadian Journal of Earth Sciences* 30: 174–200.

Eberth, D. A., Braman, D. R., and Tokaryk, T. T. 1990. Stratigraphy, sedimentology and vertebrate paleontology of the Judith River Formation (Campanian) near Muddy Lake, west-central Saskatchewan. *Bulletin of Canadian Petroleum Geology* 38: 387–406.

Edinger, T. 1955. The size of parietal foramen and organ in reptiles. *Bulletin of the Museum of Comparative Zoology* 114: 3–34.

Edinger, T. 1964. Midbrain exposure and overlap in mammals. *American Zoologist* 4: 5–19.

Edmund, A. G. 1960. Tooth replacement phenomena in the lower vertebrates. *Contributions of the Royal Ontario Museum, Life Science Division* 52: 1–190.

Eisenberg, J. F. 1981. *The Mammalian Radiations: An Analysis of Trends in Evolution, Adaptation, and Behavior.* 610 pp. University of Chicago Press, Chicago.

Eisenberg, J. F., and Wilson, D. E. 1981. Relative brain size and demographic strategies in *Didelphis marsupialis. American Naturalist* 118: 1–15.

Eizirik, E., Murphy, W. J., and O'Brien, S. J. 2001. Molecular dating and biogeography of early placental mammal radiation. *The Journal of Heredity* 92: 212–219.

Ellenberger, P. 1972. Contribution à la classification des pistes de Vertébrés du Trias: les types du Stormberg d'Afrique du Sud (I). *Palaeovertebrata, Mémoire Extraordinaire:* 1–117.

Ellenberger, P. 1974. Contribution à la classification des pistes des Vértebrés du Trias: les types du Stormberg d'Afrique du Sud (IIème partie: le Stormberg Supérieur—1. Le biome de la zone B/1 ou niveaux de Moyeni: ses biocénoses). *Palaeovertebrata, Mémoire Extraordinaire:* 1–147.

Ellenberger, P. 1975. L'explosion démographique de petits quadrupèdes à allure de mammifères dans le Stormberg Supérieur (Trias) d'Afrique du Sud: aperçus sur leur origine au Permien (France et Karroo). *Colloque International du Centre de la Recherche Scientifique* 218: 409–432.

Ellenberger, P. 1976. Une piste avec traces de soies épaisses dans le Trias Inférieur à Moyen de Lodève (Hérault, France): *Cynodontipus polythrix* nov. gen. nov. sp. Les cynodontes en France. *Géobios* 9: 769–787.

Else, P. L., and Hulbert, A. L. 1985. An allometric comparison of the mitochondria of mammalian and reptilian tissues: the implications for the evolution of endothermy. *Journal of Comparative Physiology* 248: R415–R421.

Emmons, E. 1857. *American Geology. Part VI.* 152 pp. Sprangue and Co., Albany, N.Y.

Emry, R. J., Archibald, J. D., and Smith, C. C. 1981. A mammalian molar from the Late Cretaceous of northern Mississippi. *Journal of Paleontology* 55: 953–956.

Engelmann, G. F., and Callison, G. 1998. Mammalian faunas of the Morrison Formation. *Modern Geology* 23: 343–379.

Engelmann, G. F., and Callison, G. 1999. *Glirodon grandis,* a new multituberculate mammal from the Upper Jurassic Morrison Formation. *In* D. D. Gillette (ed.), *Vertebrate Paleontology in Utah,* 161–177. Utah Geological Survey, Miscellaneous Publication 99-1, Salt Lake City.

Engelmann, G. F., Greenwald, N. S., Callison, G., and Chure, D. J. 1990. Cranial and dental morphology of a Late Jurassic multituberculate mammal from the Morrison Formation. *Journal of Vertebrate Paleontology* 10, supplement to no. 3: 22A.

Ensom, P. C. 2000. Mammals recorded for the first time from the Intermarine Limestone Member (Durlston Formation), Purbeck Limestone Group (Lower Cretaceous) of Dorset. *Proceedings of the Dorset Natural History and Archaeological Society* 121: 166–167.

Ensom, P. C., and Sigogneau-Russell, D. 1998. New dryolestoid mammals from the basal Cretaceous Purbeck Limestone group of southern England. *Palaeontology* 41: 35–55.

Ensom, P. C., and Sigogneau-Russell, D. 2000. New symmetrodonts (Mammalia, Theria) from the Purbeck Limestone Group, Early Cretaceous of southern England. *Cretaceous Research* 21: 767–779.

Ensom, P. C., Evans, S. E., Francis, J. E., Kielan-Jaworowska, Z., and Milner, A. M. 1994. The flora and fauna of the Sunnydown Farm footprint site, and associated sites: Purbeck Limestone Formation, Dorset. *Proceedings of the Dorset Natural History and Archaeological Society* 115: 181–182.

Erickson, G. M. 1996. Incremental lines of von Ebner in dinosaurs and the assessment of tooth replacement rates using growth line counts. *Proceedings of the National Academy of Sciences USA* 93: 14623–14627.

Erickson, G. M., and Tumanova, T. A. 2000. Growth curve and life history attributes of *Psittacosaurus mongoliensis* (Ceratopsia: Psittacosauridae) inferred from long bone histology. *Zoological Journal of the Linnean Society* 130: 551–566.

Erve, A., van, and Mohr, B. 1988. Palynological investigations of the Late Jurassic microflora from the vertebrate locality Guimarota coal mine (Leiria, central Portugal). *Neues Jahrbuch für Geologie und Paläontologie* 1988: 246–262.

Esteal, S. 1999. Molecular evidence for the early divergence of placentals. *BioEssays* 21: 1052–1058.

Estes, R. 1961. Cranial anatomy of the cynodont reptile *Thrinaxodon liorhinus. Bulletin of the Museum of Comparative Zoology* 125: 165–180.

Evans, A. R. 1995. *Miller's Anatomy of the Dog, 3rd Edition.* 1113 pp. W. B. Saunders Company, Philadelphia.

Evans, A. R., and Sanson, G. D. 1998. The effect of tooth shape on the breakdown of insects. *Journal of Zoology, London* 246: 391–400.

Evans, S. E. 1981. The postcranial skeleton of the Lower Jurassic eosuchian *Gephyrosaurus bridensis. Zoological Journal of the Linnean Society* 73: 81–116.

Evans, S. E., and Kermack, K. A. 1994. Assemblages of small tetrapods from the Early Jurassic of Britain. *In* N. C. Fraser and H.-D. Sues (eds.), *In the Shadow of the Dinosaurs—Early Mesozoic Tetrapods,* 271–283. Cambridge University Press, Cambridge.

Evans, S. E., and Milner, A. R. 1994. Middle Jurassic microvertebrate assemblages from the British Isles. *In* N. C. Fraser and H.-

D. Sues (eds.), *In the Shadow of the Dinosaurs—Early Mesozoic Tetrapods,* 303–321. Cambridge University Press, Cambridge.

Falconer, H. 1857. Description of two species of fossil mammalian genus *Plagiaulax* from Purbeck. *Quarterly Journal of the Geological Society of London* 13: 261–282.

Fassett, J. E., and Obradovich, J. D. 1996. Subsidence/sedimentation rates as Campanian Pictured Cliffs shoreline regressed across San Juan Basin, New Mexico and Colorado based on five new $^{40}Ar/^{39}Ar$ ages spanning 2.5 MY. *Geological Society of America, Abstracts with Programs* 28: 66.

Fassett, J. E., and Steiner, M. B. 1997. Precise age of C33n–C32r magnetic polarity reversal, San Juan Basin, New Mexico and Colorado. *New Mexico Geological Society Guidebook, 48th Field Conference, Mesozoic Geology and Paleontology of the Four Corners Region* 1997: 239–247.

Fastovsky, D. E., Clark, J. M., and Hopson, J. A. 1987. Preliminary report of a vertebrate fauna from an unusual paleoenvironmental setting, Huizchal Group, Early or Mid-Jurassic, Tamaulipas, Mexico. *In* P. J. Currie and E. H. Koster (eds.), *Fourth Symposium on Mesozoic Terrestrial Ecosystems,* 82–87. Tyrrell Museum of Palaeontology, Occasional Papers 3, Drumheller.

Fastovsky, D. E., Bowring, S. A., and Hermes, O. D. 1998. Radiometric age dates for the La Boca vertebrate assemblage (late Early Jurassic), Huizachal Canyon, Tamaulipas, Mexico. *Universidad Autónoma del Estado de Hidalgo, Avances en Investigación, Paleontología de Vertebrados, Publicación Especial* 1: 15–17.

Filan, S. L. 1991. Development of middle ear region in *Monodelphis domestica* (Marsupialia, Didelphidae): marsupial solution to early birth. *Journal of Zoology, London* 225: 577–588.

Fiorillo, A. R. 1989. The vertebrate fauna from the Judith River Formation (Late Cretaceous) of Wheatland and Golden Valley counties, Montana. *The Mosasaur* 4: 127–142.

Fiorillo, A. R., and Currie, P. J. 1994. Theropod teeth from the Judith River Formation (Upper Cretaceous) of south-central Montana. *Journal of Vertebrate Paleontology* 14: 74–80.

Fischer, M., and Tassy, P. 1993. The interrelation between Proboscidea, Sirenia, Hyracoidea, and Mesaxonia: the morphological evidence. *In* F. S. Szalay, M. J. Novacek, and M. C. McKenna (eds.), *Mammal Phylogeny: Placentals,* 217–234. Springer-Verlag, New York.

Flannery, T. F., Archer, M., Rich, T. H., and Jones, R. 1995. A new family of monotremes from the Cretaceous of Australia. *Nature* 377: 418–420.

Fleischer, G. 1973. Studien am Skelett des Gehörorgans der Säugetiere, einschliesslich des Menschen. *Säugetierkundliche Mitteilungen* 21: 131–239.

Flynn, J. J., Parrish, J. M., Rakotosamimanana, B., Simpson, W. F., and Wyss, A. E. 1999. A Middle Jurassic mammal from Madagascar. *Nature* 401: 57–60.

Flynn, L. J. 1986. Late Cretaceous mammal horizons from the San Juan Basin, New Mexico. *American Museum Novitates* 2845: 1–30.

Foote, M., Hunter, J. P., Janis, C. M., and Sepkoski, J. J. 1999. Evolutionary and preservational constraints on origins of biologic groups: divergence times of eutherian mammals. *Science* 283: 1310–1314.

Fosse, G. 2003. Calculation of numerical density of enamel prisms in multituberculate enamels: A review. *Acta Palaeontologica Polonica* 48 (4): 657–660.

Fosse, G., Risnes, S., and Holmbakken, N. 1973. Prisms and tubules in multituberculate enamel. *Calcified Tissues Research* 11: 133–150.

Fosse, G., Eskildsen, O., Risnes, S., and Sloan, R. E. 1978. Prism size in tooth enamel in some Late Cretaceous mammals and its value in multituberculate taxonomy. *Zoologica Scripta* 7: 7–61.

Fosse, G., Kielan-Jaworowska, Z., and Skaale, S. G. 1985. The microstructure of tooth enamel in multituberculate mammals. *Palaeontology* 28: 438–449.

Fosse, G., Kielan-Jaworowska, Z., and Ensom, P. C. 1991. Enamel ultrastructure of Late Jurassic multituberculate mammals. *In* Z. Kielan-Jaworowska, N. Heintz, and H. A. Nakerem (eds.), *Fifth Symposium on Mesozoic Terrestrial Ecosystems and Biota,* 25–26. Contributions from the Museum of Paleontology, University of Oslo, Oslo.

Fosse, G., Rădulescu, C., and Samson, P.-M. 2001. Enamel microstructure of the Late Cretaceous multituberculate *Kogiaonon. Acta Palaeontologica Polonica* 46: 437–440.

Fostowicz-Frelik, L., and Kielan-Jaworowska, Z. 2002. Lower incisor in zalambdalestid mammals (Eutheria) and its phylogenetic implications. *Acta Palaeontologica Polonica* 47: 177–180.

Fourie, S. 1974. The cranial morphology of *Thrinaxodon liorhinus* Seeley. *Annals of the South African Museum* 65: 337–400.

Fox, R. C. 1968. Early Campanian (Late Cretaceous) mammals from Alberta, Canada. *Nature* 220: 1046–1047.

Fox, R. C. 1969. Studies of Late Cretaceous vertebrates. III. A triconodont mammal from Alberta. *Canadian Journal of Zoology* 47: 1253–1256.

Fox, R. C. 1970. Eutherian mammal from the early Campanian (Late Cretaceous) of Alberta, Canada. *Nature* 227: 630–621.

Fox, R. C. 1971a. Early Campanian multituberculates (Mammalia: Allotheria) from the upper Milk River Formation, Alberta. *Canadian Journal of Earth Sciences* 8: 916–938.

Fox, R. C. 1971b. Marsupial mammals from the early Campanian Milk River Formation, Alberta, Canada. *In* D. M. Kermack and K. A. Kermack (eds.), *Early Mammals,* 145–164. Zoological Journal of the Linnean Society 50, supplement 1, London.

Fox, R. C. 1972a. An Upper Cretaceous symmetrodont from Alberta, Canada. *Nature* 239: 170–171.

Fox, R. C. 1972b. A primitive therian mammal from the Upper Cretaceous of Alberta. *Canadian Journal of Earth Sciences* 9: 1479–1494.

Fox, R. C. 1974a. *Deltatheroides*-like mammals from the Upper Cretaceous of North America. *Nature* 249: 392.

Fox, R. C. 1974b. A middle Campanian, nonmarine occurrence of the Cretaceous toothed bird *Hesperornis* Marsh. *Canadian Journal of Earth Sciences* 11: 1335–1338.

Fox, R. C. 1975. Molar structure and function in the Early Cretaceous mammal *Pappotherium:* evolutionary implications for

Mesozoic Theria. *Canadian Journal of Earth Sciences* 12: 412–442.

Fox, R. C. 1976a. Additions to the mammalian local fauna from the upper Milk River Formation (Upper Cretaceous), Alberta. *Canadian Journal of Earth Sciences* 13: 1105–1118.

Fox, R. C. 1976b. Cretaceous mammals (*Meniscoessus intermedius,* new species, and *Alphadon* sp.) from the lowermost Oldman Formation, Alberta. *Canadian Journal of Earth Sciences* 13: 1216–1222.

Fox, R. C. 1977. Notes on the dentition and relationships of the Late Cretaceous insectivore *Gypsonictops. Canadian Journal of Earth Sciences* 14: 1823–1831.

Fox, R. C. 1978. Upper Cretaceous terrestrial vertebrate stratigraphy of the Gobi Desert (Mongolian People's Republic) and western North America. *In* C. R. Stelck and D. E. Chatterton (eds.), *Western and Arctic Canadian Biostratigraphy,* 577–594. Geological Association of Canada, Special Paper 18.

Fox, R. C. 1979a. Ancestry of the "dog-like" marsupials. *Journal of Paleontology* 53: 733–735.

Fox, R. C. 1979b. Mammals from the Upper Cretaceous Oldman Formation, Alberta. I. *Alphadon* Simpson (Marsupialia). *Canadian Journal of Earth Sciences* 16: 91–102.

Fox, R. C. 1979c. Mammals from the Upper Cretaceous Oldman Formation, Alberta. II. *Pediomys* Marsh (Marsupialia). *Canadian Journal of Earth Sciences* 16: 103–113.

Fox, R. C. 1979d. Mammals from the Upper Cretaceous Oldman Formation, Alberta. III. Eutheria. *Canadian Journal of Earth Sciences* 16: 114–125.

Fox, R. C. 1980a. Mammals from the Upper Cretaceous Oldman Formation, Alberta. IV. *Meniscoessus* Cope. *Canadian Journal of Earth Sciences* 17: 1480–1488.

Fox, R. C. 1980b. *Picopsis pattersoni,* n. gen. and sp., an unusual therian from the Upper Cretaceous of Alberta, and the classification of primitive tribosphenic mammals. *Canadian Journal of Earth Sciences* 17: 1489–1498.

Fox, R. C. 1981. Mammals from the Upper Cretaceous Oldman Formation, Alberta. V. *Eodelphis* Matthew, and the evolution of the Stagodontidae. *Canadian Journal of Earth Sciences* 18: 350–365.

Fox, R. C. 1982. Evidence of new lineage of tribosphenic therians (Mammalia) from the Upper Cretaceous of Alberta, Canada. *Géobios, Mémoire Spécial* 6: 169–175.

Fox, R. C. 1983. Notes on the North American Tertiary marsupials *Herpetotherium* and *Peradectes. Canadian Journal of Earth Sciences* 20: 1568–1575.

Fox, R. C. 1984a. A primitive, "obtuse-angled" symmetrodont (Mammalia) from the Upper Cretaceous of Alberta, Canada. *Canadian Journal of Earth Sciences* 21: 1204–1207.

Fox, R. C. 1984b. *Paranyctoides maleficus* (new species), an early eutherian mammal from the Cretaceous of Alberta. *Special Publication, Carnegie Museum of Natural History* 9: 9–20.

Fox, R. C. 1985. Upper molar structure in the Late Cretaceous symmetrodont *Symmetrodontoides* Fox, and a classification of the Symmetrodonta. *Journal of Paleontology* 59: 21–26.

Fox, R. C. 1987a. An ancestral marsupial and its implications for early marsupial evolution. *In* P. J. Currie and E. H. Koster (eds.), *Fourth Symposium on Mesozoic Terrestrial Ecosystems,* 101–105. Tyrrell Museum, Drumheller.

Fox, R. C. 1987b. Palaeontology and the early evolution of marsupials. *In* M. Archer (ed.), *Possums and Opossums: Studies in Evolution,* 161–169. Surrey Beatty & Sons and the Royal Zoological Society of New South Wales, Sydney.

Fox, R. C. 1989. The Wounded Knee local fauna and mammalian evolution near the Cretaceous–Tertiary boundary, Saskatchewan, Canada. *Palaeontographica, Abteilung A* 208: 11–59.

Fox, R. C. 1990. The succession of Paleocene mammals in western Canada. *Geological Society of America, Special Paper* 343: 51–70.

Fox, R. C. 1997a. Late Cretaceous and Paleocene mammals, Cypress Hills region, Saskatchewan, and mammalian evolution across the Cretaceous–Tertiary boundary. *In* L. M. McAnally (ed.), *Upper Cretaceous and Tertiary Stratigraphy and Paleontology of Southern Saskatchewan,* 70–85. Geological Association of Canada, St. John's.

Fox, R. C. 1997b. Evolution of Cretaceous and Palaeocene mammals in western Canada. *Alberta* 1: 184–201.

Fox, R. C. 1999. The monophyly of the Taeniolabidoidea (Mammalia: Multituberculata). *In* H. A. Leanza (ed.), *VII International Symposium on Mesozoic Terrestrial Ecosystems, Abstracts,* 26. Buenos Aires.

Fox, R. C., and Meng, J. 1997. An X-radiographic and SEM study of the osseous inner ear of multituberculates and monotremes (Mammalia): implications for mammalian phylogeny and evolution of hearing. *Zoological Journal of the Linnean Society* 121: 249–291.

Fox, R. C., and Naylor, B. G. 1986. A new species of *Didelphodon* Marsh (Marsupialia) from the Upper Cretaceous of Alberta, Canada: paleobiology and phylogeny. *Neues Jahrbuch für Geologie und Paläontologie, Abhandlungen* 172: 357–380.

Fox, R. C., and Naylor, B. G. 1995. The relationships of the Stagodontidae, primitive North American Late Cretaceous mammals. *In* A.-L. Sun and Y. Wang (eds.), *Sixth Symposium on Mesozoic Terrestrial Ecosystems and Biota, Short Papers,* 247–250. China Ocean Press, Beijing.

Fox, R. C., Youzwyshyn, G. P., and Krause, D. W. 1992. Post-Jurassic mammal-like reptile from the Palaeocene. *Nature* 358: 233–235.

Frank, R. M., Sigogneau-Russell, D., and Voegel, J. C. 1984. Tooth ultrastructure of Late Triassic Haramiyidae. *Journal of Dental Research* 63: 661–664.

Fraser, N. C. 1989. A new pterosaur record from a Lower Jurassic fissure deposit in south Wales. *Neues Jahrbuch für Geologie und Paläontologie, Monatschefte* 1989: 129–135.

Fraser, N. C., Walkden, G. M., and Stewart, V. 1985. The first pre-Rhaetic therian mammal. *Nature* 314: 161–162.

Freeman, E. F. 1976a. A mammalian fossil from the Forest Marble (Middle Jurassic) of Dorset. *Proceedings of the Geologists' Association, London* 87: 231–236.

Freeman, E. F. 1976b. Mammal teeth from the Forest Marble (Middle Jurassic) of Oxfordshire, England. *Science* 194: 1053–1055.

Freeman, E. F. 1979. A Middle Jurassic mammal bed from Oxfordshire. *Palaeontology:* 135–166.

Freeman, E. F. 1982. Fossil bone recovery from sediment residues by the "interfacial method." *Palaeontology* 25: 738–743.

Freeman, P. W. 1979. Specialized insectivory: beetle-eating and moth-eating molossid bats. *Journal of Mammalogy* 60: 467–479.

Freeman, P. W. 1981. Correspondence of food habits and morphology in insectivorous bats. *Journal of Mammalogy* 62: 166–173.

Freeman, P. W. 1984. Functional cranial analysis of large animalivorous bats (Microchiroptera). *Biological Journal of the Linnean Society* 21: 387–408.

Friis, E. M., and Crepet, W. L. 1987. Time of appearance of floral features. *In* E. M. Friis, W. G. Chaloner, and P. R. Crane (eds.), *The Origins of Angiosperms and Their Biological Consequences,* 145–179. Cambridge University Press, Cambridge.

Friis, E. M., Chaloner, W. G., and Crane, P. R. (eds.). 1987. *The Origins of Angiosperms and Their Biological Consequences.* 358 pp. Cambridge University Press, Cambridge.

Gambaryan, P. P. 1960. *The Adaptive Features of the Locomotory Organs in Fossorial Mammals.* 195 pp. Izdatelstvo Akademii Nauk Armyanskoi SSR, Erevan (in Russian).

Gambaryan, P. P. 1967. The origin of the variety of gaits in mammals. *Zhurnal Obshchei Biologii* 28: 289–305 (in Russian).

Gambaryan, P. P. 1974. *How Mammals Run.* 367 pp. Wiley and Sons, New York.

Gambaryan, P. P. 1989. *The Evolution of Mammalian Facial Musculature.* 151 pp. Nauka, Leningrad (in Russian).

Gambaryan, P. P. 1998. Evolution of the scapulocoracoid in synapsid tetrapods. *Zhurnal Obshchei Biologii* 59: 623–635 (in Russian).

Gambaryan, P. P., and Averianov, A. O. 2001. Femur of a morganucodontid mammal from the Middle Jurassic of central Russia. *Acta Palaeontologica Polonica* 46: 99–112.

Gambaryan, P. P., and Kielan-Jaworowska, Z. 1995. Masticatory musculature of Asian taeniolabidoid multituberculate mammals. *Acta Palaeontologica Polonica* 40: 45–108.

Gambaryan, P. P., and Kielan-Jaworowska, Z. 1997. Sprawling versus parasagittal stance in multituberculate mammals. *Acta Palaeontologica Polonica* 42: 13–44.

Gardner, A. L. 1982. Virginia opossum. *In* J. A. Chapman and G. A. Feldhamer (eds.), *Wild Mammals of North America,* 3–36. Johns Hopkins University Press, Baltimore.

Gatesy, J., and O'Leary, M. A. 2001. Deciphering whale origins with molecules and fossils. *Trends in Ecology and Evolution* 16: 562–570.

Gaudin, T. J., Wible, J. R., Hopson, J. A., and Turnbull, W. D. 1996. Reexamination of the morphological evidence for the cohort Epitheria (Mammalia, Eutheria). *Journal of Mammalian Evolution* 3: 1–79.

Gaupp, E. 1902. Über die Ala temporalis des Saügetierschädels und die Regio orbitalis einiger anderer Wirbeltierschädels. *Anatomische Hefte* 19: 155–230.

Gaupp, E. 1913. Die Reichertsche Theorie (Hammer-, Amboss- und Kieferfrage). *Archiv für Anatomie und Entwicklungsgeschichte* 1912: 1–426.

Gauthier, J. A., Kluge, A., and Rowe, T. B. 1988. Amniote phylogeny and the importance of fossils. *Cladistics* 4: 105–209.

Gayet, M., Marshall, L. G., and Sempere, T. 1991. The Mesozoic and Paleocene vertebrates of Bolivia and their stratigraphic context: a review. *In* R. Suárez Soruco (ed.), *Fósiles y Facies de Bolivia. Volumen I. Vertebrados,* 393–434. Revista Técnica de Yacimientos Petrolíferos Fiscales Bolivianos, Santa Cruz.

Gayet, M., Marshall, L. G., Sempere, T., Meunier, F. J., Cappetta, H., and Rage, J.-C. 2001. Middle Maastrichtian vertebrates (fishes, amphibians, dinosaurs and other reptiles, mammals) from Pajcha Pata (Bolivia). Biostratigraphic, palaeoecologic and palaeobiogeographic implications. *Palaeogeography, Palaeoclimatology, Palaeoecology* 169: 39–68.

Gazin, C. L. 1941a. Paleocene mammals from the Denver Basin, Colorado. *Washington Academy of Sciences Journal* 31: 289–295.

Gazin, C. L. 1941b. The mammalian faunas of the Paleocene of central Utah, with notes on the geology. *Proceedings of the United States National Museum* 91: 1–53.

Gee, H. 1999. *In Search of Deep Time: Beyond the Fossil Record to a New History of Life.* 267 pp. Simon and Schuster, New York.

Geisler, J. H., and Luo, Z.-X. 1998. Cranial vasculature evolution in mesonychids and cetaceans and the phylogenetic position of Cete. *In* J. G. M. Thewissen (ed.), *Evolutionary Emergence of Whales,* 161–212. Plenum Press, New York.

Gelfo, J. N., and Pascual, R. 2001. *Peligrotherium tropicalis* (Mammalia, Dryolestida) from the early Paleocene of Patagonia, a survival from a Mesozoic Gondwanan radiation. *Geodiversitas* 23: 369–379.

Getty, R. 1975. *Sisson and Grossman's Anatomy of the Domestic Animals.* 2095 pp. W. B. Saunders, Philadelphia.

Gheerbrant, E., and Astibia, H. 1994. Un nouveau mammifère du Maastrichtien de Laño (Pays Basque espagnol). *Comptes Rendus de l'Académie des Sciences, Paris, sér. II,* 318: 1125–1131.

Gheerbrant, E., and Astibia, H. 1999. The Upper Cretaceous mammals from Laño (Spanish Basque country). *Estudios del Museo de Ciencias Naturales de Alava* 14: 295–323.

Gheerbrant, E., and Russell, D. E. 1991. *Bustylus cernaysi* nov. gen. nov. sp., nouvel adapisoriculidé (Mammalia, Eutheria), Paléocène d'Europe. *Géobios* 24: 467–481.

Gheerbrant, E., Abrial, C., and Cappetta, H. 1997. Nouveaux sites à microvertébrés continentaux du Crétacé terminal des Petites Pyrénées (Haute-Garonne et Ariège), France. *Géobios* 20: 257–269.

Gidley, J. W. 1909. Notes on the fossil mammalian genus *Ptilodus,* with description of a new species. *Proceedings of the United States National Museum* 36: 611–626.

Gidley, J. W. 1919. Significance of divergence of the first digit in the primitive mammalian foot. *Journal of the Washington Academy of Sciences* 9: 273–280.

Giebel, C. G. 1855. *Die Säugethiere in zoologischer, anatomischer und palaeontologischer Beziehung um fassend dargestellt.* 1108 pp. Ambrosius Abel, Leipzig.

Gilbert, N., and Labuda, D. 2000. Evolutionary inventions and continuity of CORE-SINES in mammals. *Journal of Molecular Biology* 298: 365–377.

Gill, P. G. 1974. Resorption of premolars in the early mammal *Kuehneotherium praecursoris. Archives of Oral Biology* 19: 327–328.

Gill, T. N. 1872. Arrangement of the families of mammals. With analytical tables. *Smithsonian Miscellaneous Collections* 11 (230): 1–98.

Gilmore, C. W. 1921. The fauna of the Arundel Formation of Maryland. *Proceedings of the United States National Museum* 59: 581–594.

Gilmore, C. W. 1942. Osteology of *Polyglyphanodon,* an Upper Cretaceous lizard from Utah. *Proceedings of the United States National Museum* 92: 229–265.

Gilmore, C. W. 1946. Reptilian fauna of the North Horn Formation of central Utah. *United States Geological Survey Professional Paper* 210-C: 21–53.

Gingerich, P. D. 1973. Molar occlusion and function in the Jurassic mammal *Docodon. Journal of Mammalogy* 254: 1008–1013.

Gingerich, P. D. 1977. Patterns of evolution in the mammalian fossil record. *In* A. Hallam (ed.), *Patterns of Evolution,* 469–500. Elsevier Science Publishers, Amsterdam.

Gingerich, P. D. 1990. Prediction of body size in mammalian species from long bone lengths and diameters. *Contributions from the Museum of Paleontology, University of Michigan* 28: 79–92.

Gingerich, P. D., ul Haq, M., Zalmout, I. S., Khan, I. H., and Malkani, M. S. 2001. Origin of whales from early artiodactyls: hands and feet of Eocene Protocetidae from Pakistan. *Science* 293: 2239–2242.

Godefroit, P. 1997. Reptilian, therapsid and mammalian teeth from the Upper Triassic of Varangéville (northeastern France). *Bulletin de l'Institut Royal des Sciences Naturelles de Belgique, Sciences de la Terre* 67: 83–102.

Godefroit, P., and Battail, B. 1997. Late Triassic cynodonts from Saint-Nicolas-de-Port (northeastern France). *Geodiversitas* 19: 567–631.

Godefroit, P., and Guo, D.-Y. 1999. A new amphilestid mammal from the Early Cretaceous of China. *Bulletin de l'Institut royal des Sciences naturelles de Belgique* 69, supplement B: 7–16.

Godefroit, P., and Sigogneau-Russell, D. 1995. Cynodontes et mammifères primitifs du Trias Supérieur, en region Lorraine et Luxembourgeoise. *Bulletin de la Société belge de Géologie* 104: 9–21.

Godefroit, P., and Sigogneau-Russell, D. 1999. Kuehneotheriids from Saint-Nicholas-de-Port (Late Triassic of France). *Geologica Belgica* 2: 181–196.

Godefroit, P., Cuny, G., Delsate, D., and Roche, M. 1998. Late Triassic vertebrates from Syren (Luxembourg). *Neues Jahrbuch für Geologie und Paläontologie, Abhandlungen* 210: 305–343.

Godinot, M., and Prasad, G. V. R. 1994. Discovery of Cretaceous arboreal eutherians. *Naturwissenschaften* 81: 79–81.

Goin, F. J., Carlini, A. A., and Pascual, R. 1986. Un probable marsupial del Cretácico tardio del norte de Patagonia, Argentina. *IV Congreso de Paleontología y Bioestratigrafía* 2: 43–47.

Goldberg, P. 2000. *Faunal composition, non-marine vertebrates, of the upper Cedar Mountain Formation (Cretaceous: Albian–Cenomanian), central Utah.* 243 pp. Ph.D. dissertation, University of Oklahoma, Norman.

Goodrich, E. S. 1930. *Studies on the Structure and Development of Vertebrates.* 837 pp. Macmillan and Company, London.

Goodwin, M. B., and Deino, A. L. 1989. The first radiometric ages from the Judith River Formation (Upper Cretaceous), Hill County, Montana. *Canadian Journal of Earth Sciences* 26: 1384–1391.

Gow, C. E. 1980. The dentitions of the Trithelodontidae (Therapsida: Cynodontia). *Proceedings of the Royal Society, London* 208: 461–481.

Gow, C. E. 1985. Apomorphies of the Mammalia. *South African Journal of Science* 81: 558–560.

Gow, C. E. 1986a. A new skull of *Megazostrodon* (Mammalia: Triconodonta) from the Elliot Formation (Lower Jurassic) of southern Africa. *Palaeontologia Africana* 26: 13–23.

Gow, C. E. 1986b. The sidewall of the braincase in cynodont therapsids and a note on homology of the mammalian promontorium. *South African Journal of Zoology* 21: 136–148.

Gow, C. E. 2001. A partial skeleton of the trithelodontid *Pachygenelus* (Therapsida, Cynodontia). *Palaeontologica Africana* 37: 93–97.

Gradstein, F. M., Agterberg, F. P., Ogg, J. G., Hardenbol, J., Van Veen, P., Thierry, J., and Huang, Z. 1995. A Triassic, Jurassic and Cretaceous time scale. *In* W. A. Berggren, D. V. Kent, M.-P. Aubrey, and J. Hardenbol (eds.), *Geochronology, Time Scales and Global Stratigraphic Correlation,* 95–126. SEPM (Society for Sedimentary Geology) Special Publication, Tulsa.

Gradziński, R. 1970. Sedimentation of dinosaur-bearing Upper Cretaceous deposits of the Nemegt Basin, Gobi Desert. *Palaeontologia Polonica* 21: 147–229.

Gradziński, R., and Jerzykiewicz, T. 1974. Sedimentation of the Barun Goyot Formation. *Palaeontologia Polonica* 30: 111–146.

Gradziński, R., Kaźmierczak, J., and Lefeld, J. 1968. Geographical and geological data from the Polish-Mongolian Palaeontological Expeditions. *Palaeontologia Polonica* 19: 33–82.

Gradziński, R., Kielan-Jaworowska, Z., and Maryańska, T. 1977. Upper Cretaceous Djadokhta, Barun Goyot and Nemegt formations of Mongolia, including remarks on previous subdivisions. *Acta Geologica Polonica* 27: 281–318.

Grambast, L., Martinez, M., Mattauer, M., and Thaler, L. 1967. *Perutherium altiplanense,* nov. gen., nov. sp., premier Mammifère mésozoïque d'Amérique du Sud. *Comptes Rendus de l'Académie des Sciences, Paris, sér. D,* 264: 707–710.

Grandstaff, B. S., Parris, D. C., and Denton, R. K. 1992. *Alphadon* (Marsupialia) and Multituberculata (Allotheria) in the Cretaceous of eastern North America. *Journal of Vertebrate Paleontology* 12: 217–222.

Grandstaff, B. S., Gallagher, W. B., Shannon, K., and Parris, D. C. 2000. New discoveries of Late Cretaceous mammals in eastern North America. *Journal of Vertebrate Paleontology* 20, supplement to no. 3: 46A.

Granger, W. 1915. New evidence of the affinities of the Multituberculata. *Bulletin of the Geological Society of America (Abstracts)* 26: 152.

Granger, W., and Simpson, G. G. 1928. Multituberculates in the Wasatch Formation. *American Museum Novitates* 112: 1–4.

Granger, W., and Simpson, G. G. 1929. A revision of the Tertiary Multituberculata. *Bulletin of the American Museum of Natural History* 56: 601–676.

Gray, J. E. 1821. On the natural arrangement of vertebrose animals. *London Medical Repository* 15: 296–310.

Graybeal, A., Rosowski, J. R., Ketten, D. L., and Crompton, A. W. 1989. Inner-ear structure in *Morganucodon,* an Early Jurassic mammal. *Zoological Journal of the Linnean Society* 96: 107–117.

Greaves, W. 1973. The inference of jaw motion from the tooth wear facets. *Journal of Paleontology* 47: 1000–1001.

Green, H. L. H. H. 1937. The development and morphology of the teeth of *Ornithorhynchus. Philosophical Transactions of the Royal Society of London* 288: 367–420.

Greenwald, N. S. 1988. Patterns of tooth eruption and replacement in multituberculate mammals. *Journal of Vertebrate Paleontology* 8: 265–277.

Gregory, H. E. 1951. Geology and geography of the Paunsaugunt region. *United States Geological Survey, Professional Paper* 226: 1–116.

Gregory, H. E., and Moore, R. C. 1931. The Kaiparowits region: A geographic and geologic reconnaissance of part of Arizona and Utah. *United States Geological Survey Professional Paper* 164: 1–161.

Gregory, W. K. 1910. The orders of mammals. *Bulletin of the American Museum of Natural History* 27: 1–524.

Gregory, W. K. 1947. The monotremes and the palimpsest theory. *Bulletin of the American Museum of Natural History* 88: 1–52.

Gregory, W. K., and Simpson, G. G. 1926. Cretaceous mammal skulls from Mongolia. *American Museum Novitates* 225: 1–20.

Griffiths, M. 1968. *Echidnas.* 282 pp. Pergamon Press, London.

Griffiths, M. 1978. *The Biology of the Monotremes.* 367 pp. Academic Press, New York.

Griffiths, M., Wells, T. T., and Barrie, D. J. 1991. Observations on the skulls of fossil and extant echidnas (Monotremata: Tachyglossidae). *Australian Mammalogy* 14: 87–101.

Grigorescu, D. 1984. New tetrapod group in the Maastrichtian of the Haţeg Basin: coelurosaurians and multituberculates. *In* W. Rief and F. Westphal (eds.), *Third Symposium on Mesozoic Terrestrial Ecosystems,* 99–104. Attempto Verlag Tübingen GmbH, Tübingen.

Grigorescu, D., and Hahn, G. 1987. The first multituberculate teeth from the Upper Cretaceous of Europe (Romania). *Geologica et Palaeontologica* 21: 237–243.

Grigorescu, D., Hartenberger, J.-L., Rădulescu, C., Samson, P., and Sudre, J. 1985. Découverte de Mammifères et Dinosaures dans le Crétacé supérieur de Pui (Roumanie). *Comptes Rendus de l'Académie des Sciences, Paris, sér. II,* 301: 1365–1368.

Grigorescu, D., Venczel, M., Csiki, Z., and Limberia, R. 1999. New latest Cretaceous microvertebrate fossil assemblages from the Haţeg Basin (Romania). *Geologie en Mijnbouw* 78: 301–314.

Grine, F. E., and Vrba, E. S. 1980. Prismatic enamel: a preadaptation for mammalian diphyodonty? *South African Journal of Science* 76: 139–141.

Gubin, Y. M., and Sinitza, S. M. 1996. Shar Teg: a unique Mesozoic locality of Asia. *In* M. Morales (ed.), *The Continental Jurassic,* 311–318. *Museum of Northern Arizona Bulletin* 60, Flagstaff.

Haeckel, E. 1866. *Generelle Morphologie der Organismen,* Vol. 2. 462 pp. Georg Reimer, Berlin.

Hahn, G. 1969. Beiträge zur Fauna der Grube Guimarota nr. 3. Die Multituberculata. *Palaeontographica, Abteilung A* 133: 1–100.

Hahn, G. 1971. The dentition of the Paulchoffatiidae (Multituberculata, Upper Jurassic). *Memória dos Serviços Geológicos de Portugal* 17: 7–39.

Hahn, G. 1973. Neue Zähne von Haramiyiden aus der Deutschen Ober-Trias und ihre Beziehungen zu den Multituberculaten. *Palaeontographica, Abteilung A* 142: 1–15.

Hahn, G. 1977a. Neue Schädel-Reste von Multituberculaten (Mamm.) aus dem Malm Portugals. *Geologica et Paleontologica* 11: 161–186.

Hahn, G. 1977b. Das Coronoid der Paulchoffatiidae (Multituberculata; Ober-Jura). *Paläontologische Zeitschrift* 51: 246–253.

Hahn, G. 1978a. Neue Unterkiefer von Multituberculaten aus dem Malm Portugals. *Geologica et Paleontologica* 12: 177–212.

Hahn, G. 1978b. Milch-Bezahnungen von Paulchoffatiidae (Multituberculata; Ober-Jura). *Neues Jahrbuch für Geologie und Paläontologie* 1978: 25–34.

Hahn, G. 1978c. Die Multituberculata, eine fosille Säugetier-Ordnung. *Sonderband der naturwissenschaftliche Verein, Hamburg* 3: 61–95.

Hahn, G. 1981. Zum Bau der Schädel-Basis der Paulchoffatiidae (Multituberculata, Ober-Jura). *Senckenbergiana Lethaea* 61: 227–245.

Hahn, G. 1983. Überblick ueber die Erforschungs-Geschichte der Multituberculaten (Mammalia). *Schriftenreihe für geologische Wissenschaften* 19/20: 217–246.

Hahn, G. 1985. Zum Bau des Infraorbital-Foramens bei den Paulchoffatiidae (Multituberculata, Ober-Jura). *Berliner geowissenschaftliche, Abhandlungen A* 60: 5–27.

Hahn, G. 1987a. Ein Multituberculaten–Zahn aus der Ober-Trias von Gaume (S-Belgien). *Bulletin de la Société belge de Géologie* 96: 39–47.

Hahn, G. 1987b. Neue Beobachtungen zum Schädel- und Gebiss-Bau der Paulchoffatiidae (Multituberculata, Ober-Jura). *Palaeovertebrata* 17: 155–196.

Hahn, G. 1988. Die Ohr-Region der Paulchoffatiidae (Multituberculata, Ober-Jura). *Palaeovertebrata* 18: 155–185.

Hahn, G. 1993. The systematic arrangement of the Paulchoffatiidae (Multituberculata) revisited. *Geologica et Paleontologica* 27: 201–214.

Hahn, G. 2001. Neue Beobachtungen an Schädel-Resten von Paulchoffatiidae (Multituberculata; Ober-Jura). *Geologica et Paleontologica* 35: 121–143.

Hahn, G., and Hahn, R. 1983. Multituberculata. *In* F. Westphal (ed.), *Fossilium Catalogus, I: Animalia, Pars 127,* 1–409. Kugler Publications, Amsterdam.

Hahn, G., and Hahn, R. 1992. Neue Multituberculaten-Zähne aus der Unter-Kreide (Barremium) von Spanien (Galve und Uña). *Geologica et Paleontologica* 26: 143–162.

Hahn, G., and Hahn, R. 1994. Nachweis des Septomaxillare bei *Pseudobolodon krebsi* n. sp. (Multituberculata) aus dem Malm

Portugals. *Berliner geowissenschaftliche, Abhandlungen E* 13: 9–29.

Hahn, G., and Hahn, R. 1998a. Neue Beobachtungen an Plagiaulacoidea (Multituberculata) des Ober-Juras 1. Zum Zahn-Wechsel bei *Kielanodon. Berliner geowissenschaftliche, Abhandlungen E* 28: 1–7.

Hahn, G., and Hahn, R. 1998b. Neue Beobachtungen an Plagiaulacoidea (Multituberculata) des Ober-Juras 2. Zum Bau des Unterkiefers und des Gebisses bei *Meketibolodon* und bei *Guimarotodon. Berliner geowissenschaftliche, Abhandlungen E* 28: 9–37.

Hahn, G., and Hahn, R. 1998c. Neue Beobachtungen an Plagiaulacoidea (Multituberculata) des Ober-Juras 3. Der Bau der Molaren bei den Paulchoffatiidae. *Berliner geowissenschaftliche, Abhandlungen E* 28: 39–84.

Hahn, G., and Hahn, R. 1998d. Neue Beobachtungen an Plagiaulacoidea (Multituberculate) des Ober-Juras 4. Ein Verteter der Albionbaataridae im Lusitanien Portugals. *Berliner geowissenschaftliche, Abhandlungen E* 28: 85–89.

Hahn, G., and Hahn, R. 1999a. Pinheirodontidae n. fam. (Multituberculata) (Mammalia) aus der tiefen Unter-Kreide Portugals. *Palaeontographica, Abteilung A* 253: 7–222.

Hahn, G., and Hahn, R. 1999b. Nomenklatorische Notiz: Namens-Änderung bei Multituberculata (Mammalia). *Geologica et Palaeontologica* 33: 156.

Hahn, G., and Hahn, R. 2001a. Multituberculaten-Zähne aus dem Ober-Jura von Porto das Barcas (Portugal). *Paläontologische Zeitschrift* 74: 583–586.

Hahn, G., and Hahn, R. 2001b. Multituberculaten-Zähne aus der Unter-Kreide (Barremium) von Pié Pajarón (Prov. Cuenca, Spanien). *Paläontologische Zeitschrift* 74: 587–589.

Hahn, G., and Hahn, R. 2002. Der Bau der Canini bei den Paulchoffatiidae (Multituberculata; Ober-Jura). *Paläontologische Zeitschrift* 76: 235–249.

Hahn, G., Lepage, J.-C., and Wouters, G. 1984. Cynodontier-Zähne aus der Ober-Trias von Medernach, Grossherzogtum Luxemburg. *Bulletin de la Société belge de Géologie* 93: 57–373.

Hahn, G., Lepage, J. C., and Wouters, G. 1987. Ein Multituberculaten-Zahn aus der Ober-Trias von Gaume (S-Belgien). *Bulletin de la Société belge de Géologie* 96: 39–47.

Hahn, G., Sigogneau-Russell, D., and Wouters, G. 1989. New data on Theroteinidae—their relations with Paulchoffatiidae and Haramiyidae. *Geologica et Paleontologica* 23: 205–215.

Hahn, G., Sigogneau-Russell, D., and Godefroit, P. 1991. New data on *Brachyzostrodon* (Mammalia; Upper Triassic). *Geologica et Paleontologica* 25: 237–249.

Hahn, G., Hahn, R., and Godefroit, P. 1994. Zur stellung der Dromatheriidae (Ober-Trias) zwischen den Cynodontia und den Mammalia. *Geologica et Paleontologica* 28: 141–159.

Haines, R. W. 1958. Arboreal or terrestrial ancestry of placental mammals. *Quarterly Review of Biology* 33: 1–23.

Harland, W. B., Armstrong, R. L., Cox, A. V., Craig, L. E., Smith, A. G., and Smith, D. G. 1989. *A Geologic Time Scale.* 263 pp. Cambridge University Press, Cambridge.

Hatcher, J. B. 1896. Some localities for Laramie mammals and horned dinosaurs. *American Naturalist* 30: 112–120.

Hayden, F. V. 1871. Geology of the Missouri Valley. *Fourth Annual Preliminary Report, U.S. Geological Survey of the Territories:* 1–511.

Hedges, S. B., Parker, P., Sibley, G., and Kumar, S. 1996. Continental breakup and ordinal diversification of birds and mammals. *Nature* 381: 226–228.

Heinrich, R. E., Hunter, J. P., and Weishampel, D. B. 1998. Mammals from the St. Mary River Formation (Upper Cretaceous), Montana: new insights on an "Edmontonian" land mammal age. *Journal of Vertebrate Paleontology* 18: 50A.

Heinrich, W.-D. 1991. Über *Brancatherulum tendagurense* Dietrich 1927 (Mammalia: Eupantotheria) aus dem Oberjura von Tendaguru, Tansania. *Mitteilungen aus dem Zoologischen Museum Berlin* 67(1): 97–104.

Heinrich, W.-D. 1998. Late Jurassic mammals from Tendaguru, Tanzania, east Africa. *Journal of Mammalian Evolution* 5: 269–290.

Heinrich, W.-D. 1999. First haramiyid (Mammalia, Allotheria) from the Mesozoic of Gondwana. *Mitteilungen des Museum für Naturkunde, Berlin, Geowissenschaften Reihe* 2: 159–170.

Heinrich, W.-D. 2001. New records of *Staffia aenigmatica* (Mammalia, Allotheria, Haramiyida) from the Upper Jurassic of Tendaguru in southeastern Tanzania, East Africa. *Mitteilungen aus dem Museum für Naturkunde Berlin, Geowissenschaftliche Reihe* 4: 239–255.

Heinrich, W.-D., Bussert, R., Aberhan, M., Hampe, O., Kapilima, S., Schrank, E., Schultka, S., Maier, G., Msaky, E., Sames, B., and Chami, R. 2001. The German-Tanzanian Tendaguru Expedition 2000. *Mitteilungen aus dem Museum für Naturkunde Berlin, Geowissenschaftliche Reihe* 4: 223–237.

Henkel, S., and Krebs, B. 1969. Zwei Säugetier-Unterkiefer aus der Unteren Kreide von Uńa (Prov. Cuenca, Spanien). *Neues Jahrbuch für Geologie und Paläontologie* 1969: 449–463.

Henkel, S., and Krebs, B. 1977. Der erste Fund eines Säugetier-Skelettes aus der Jura-Zeit. *Umschau in Wissenschaft und Technik* 77: 217–218.

Hennig, E. 1922. Die Säugerzähne des wüttembergischen Rhät-Lias-Bonebeds. *Neues Jahrbuch für Mineralogie, Geologie und Paläontologie* 46: 181–267.

Hennig, W. 1966. *Phylogenetic Systematics.* 263 pp. University of Illinois Press, Urbana.

Herring, S. W. 1993a. Epigenetic and functional influences of skull growth. *In* B. K. Hall (ed.), *The Skull. Vol. 1: Development,* 153–206. University of Chicago Press, Chicago.

Herring, S. W. 1993b. Formation of the vertebrate face: epigenetic and functional influences. *American Zoologist* 33: 472–483.

Hershkovitz, P. 1971. Basic crown patterns and cusp homologies of mammalian teeth. *In* A. A. Dahlberg (ed.), *Dental Morphology and Evolution,* 95–150. University of Chicago Press, Chicago.

Hershkovitz, P. 1972. The Recent mammals of the Neotropiocal region: a zoogeographic and ecological review. *In* A. Keast, F. C. Erk, and B. Glass (eds.), *Evolution, Mammals, and Southern Continents,* 311–431. State Univesity of New York Press, Albany.

Hershkovitz, P. 1982. The staggered marsupial lower third incisor (I_3). *Géobios, Mémoire Spécial* 6: 191–200.

Hershkovitz, P. 1995. The staggered marsupial third lower incisor: hallmark of cohort Didelphimorphia, and description of a new genus and species with staggered i3 from the Albian (Lower Cretaceous) of Texas. *Bonner Zoologische Beiträge* 45: 153–169.

Hicks, J. F., Brinkman, D. L., Nichols, D. J., and Watabe, M. 1999. Paleomagnetic and palynologic analyses of Albian to Santonian strata at Bayn Shireh, Burkhant, and Khuren Dukh, eastern Gobi Desert, Mongolia. *Cretaceous Research* 20(7): 829–850.

Hill, J. E. 1935. The cranial foramina in rodents. *Journal of Mammalogy* 16: 121–129.

Hill, J. E. 1937. Morphology of the pocket gopher mammalian genus *Thomomys. University of California Publications in Zoology* 42: 81–172.

Hillenius, W. J. 1992. The evolution of nasal turbinals and mammalian endothermy. *Paleobiology* 18: 17–29.

Hillenius, W. J. 1994. Turbinates in therapsids: evidence for endothermy in mammal-like reptiles. *Evolution* 48: 207–229.

Hillenius, W. J. 2000. The septomaxilla of non-mammalian synapsids: soft-tissue correlates and a new functional interpretation. *Journal of Morphology* 245: 29–50.

Hillis, D. M. 1995. Approaches for assessing phylogenetic accuracy. *Systematic Biology* 46: 3–16.

Hoffstetter, R. 1972. Donées et hypothèses concernant l'origine et l'histoire biogéographique des Marsupiaux. *Comptes Rendus de l'Académie des Sciences* 274: 2635–2638.

Hoffstetter, R. 1981. Historia biogeográfica de los mamíferos terrestres sudamericanos: problemas y enseñazas. *Acta Geologica Hispanica* 16: 71–88.

Hoganson, J. W., Campbell, J. M., and Murphy, E. C. 1994. Stratigraphy and paleontology of the Cretaceous Hell Creek Formation, Stumpf site, Morton County, North Dakota. *Proceedings of the North Dakota Academy of Science* 48: 95.

Holst, M.-C. 1986. *The olivocerebellar projection in a marsupial and a monotreme.* 167 pp. Ph.D. dissertation, University of New South Wales, Sydney.

Holtzman, R. C., and Wolberg, D. L. 1977. The Microcosmodontinae and *Microcosmodon woodi,* new multituberculate taxa (Mammalia) from the late Paleocene of North Dakota. *Scientific Publications of the Science Museum of Minnesota* 4: 1–13.

Hopson, J. A. 1964. The braincase of the advanced mammal-like reptile *Bienotherium. Postilla* 87: 1–30.

Hopson, J. A. 1965. *Tritylodontid therapsids from Yunnan and the cranial morphology of* Bienotherium. 295 pp. Ph.D. dissertation, University of Chicago, Chicago.

Hopson, J. A. 1966. The origin of the mammalian middle ear. *American Zoologist* 6: 437–450.

Hopson, J. A. 1967. Comments on the competitive inferiority of multituberculates. *Systematic Zoology* 16: 352–355.

Hopson, J. A. 1969. The origin and adaptive radiation of mammal-like reptiles and nontherian mammals. *Annals of the New York Academy of Sciences* 167: 199–216.

Hopson, J. A. 1970. The classification of non-therian mammals. *Journal of Mammalogy* 51: 1–9.

Hopson, J. A. 1971. Postcanine replacement in the gomphodont cynodont *Diademodon. In* D. M. Kermack and K. A. Kermack (eds.), *Early Mammals,* 199–216. *Zoological Journal of the Linnean Society* 50 supplement 1, London.

Hopson, J. A. 1973. Endothermy, small size, and the origin of mammalian reproduction. *The American Naturalist* 107: 446–462.

Hopson, J. A. 1977. Relative brain size and behavior in archosaurian reptiles. *Annual Review of Ecology and Systematics* 8: 429–448.

Hopson, J. A. 1979. Paleoneurology. *In* C. A. Gans, R. G. Northcutt, and P. Ulinsky (eds.), *Biology of the Reptilia, Neurology A, 10,* 39–145. Academic Press, London.

Hopson, J. A. 1991. Systematics of nonmammalian Synapsida and implications for patterns of evolution in synapsids. *In* H.-P. Schultze and L. Trueb (eds.), *Origins of the Higher Groups of Tetrapods—Controversy and Consensus,* 635–693. Cornell University Press, Ithaca.

Hopson, J. A. 1994. Synapsid evolution and the radiation of non-eutherian mammals. *In* R. S. Spencer (ed.), *Major Features of Vertebrate Evolution,* 190–219. The Paleontological Society, Knoxville.

Hopson, J. A. 1995. The Jurassic mammal *Shuotherium dongi:* "pseudo-tribosphenic therian," docodontid, or neither? *Journal of Vertebrate Paleontology* 15, supplement to no. 3: 36A.

Hopson, J. A. 1997. Is cusp C of the upper molars of *Kuehneotherium* homologous with the metacone of *Peramus* and tribosphenic mammals? *Journal of Vertebrate Paleontology* 17: supplement to no. 3: 53A.

Hopson, J. A., and Barghusen, H. 1986. An analysis of therapsid relationships. *In* N. Hotton, III, P. D. MacLean, J. J. Roth, and E. C. Roth (eds.), *The Ecology and Biology of Mammal-like Reptiles,* 83–106. Smithsonian Institution Press, Washington, D.C.

Hopson, J. A., and Crompton, A. W. 1969. Origin of mammals. *In* T. Dobzhansky, M. K. Hecht, and W. C. Steere (eds.), *Evolutionary Biology, Vol. 3,* 15–72. Appleton-Century-Crofts, New York.

Hopson, J. A., and Kitching, J. W. 2001. A probainognathian cynodont from South Africa and the phylogeny of nonmammalian cynodonts, 5–35. *In* F. A. Jenkins, Jr., T. Owerkowicz, and M. D. Shapiro (eds.), *Studies in Organismic and Evolutionary Biology in Honor of A. W. Crompton. Bulletin of the Museum of Comparative Zoology* 156, Cambridge.

Hopson, J. A., and Reif, W.-E. 1981. The status of *Archaeodon reuningi* von Huene, a supposed Late Triassic mammal from southern Africa. *Neues Jahrbuch für Geologie und Paläontologie* (1981): 307–310.

Hopson, J. A., and Rougier, G. W. 1993. Braincase structure in the oldest known skull of a therian mammal: implications for mammalian systematics and cranial evolution. *American Journal of Science* 293: 268–299.

Hopson, J. A., Kielan-Jaworowska, Z., and Allin, E. F. 1989. The cryptic jugal of multituberculates. *Journal of Vertebrate Paleontology* 9: 201–209.

Horner, J. R. 1982. Evidence for colonial nesting and "site fidelity" among ornithischian dinosaurs. *Nature* 297: 675–676.

Horner, J. R., and Makela, R. 1979. Nest of juveniles provides evidence of family structure among dinosaurs. *Nature* 282: 296–298.

Horovitz, I. 2000. The tarsus of *Ukhaatherium nessovi* (Eutheria, Mammalia) from the Late Cretaceous of Mongolia: an appraisal of the evolution of the ankle in basal therians. *Journal of Vertebrate Paleontology* 20: 547–560.

Howell, A. B. 1926. Anatomy of the wood rat. *American Society of Mammalogists, Monograph* 1: 1–225.

Howell, A. B. 1944. *Speed in Animals: Their Specialization for Running and Leaping.* 270 pp. University of Chicago Press, Chicago.

Hu, Y.-M., and Wang, X.-L. 2002a. *Sinobaatar* gen. nov.: first multituberculate from Jehol Biota of Liaoning, Northern China. *Chinese Science Bulletin* 46: 372–386 (in Chinese).

Hu, Y.-M., and Wang, Y. 2002b. *Sinobaatar* gen. nov.: first multituberculate from Jehol Biota of Liaoning, Northern China. *Chinese Science Bulletin* 47: 933–938.

Hu, Y.-M., Wang, Y.-Q., Luo, Z.-X., and Li, C.-K. 1997. A new symmetrodont mammal from China and its implications for mammalian evolution. *Nature* 390: 137–142.

Hu, Y.-M., Wang, Y.-Q., Li, C.-K., and Luo, Z.-X. 1998. Morphology of dentition and forelimb of *Zhangheotherium. Vertebrata PalAsiatica* 36: 102–125.

Huelsenbeck, J. P., Bull, J. J., and Cunningham, C. W. 1996. Combining data in phylogenetic analysis. *Trends in Ecology and Evolution* 11: 152–158.

Huene, F. v. 1933. Zur Kenntnis des württembergischen Rhätbonebeds mit Zahnfunden neuer Säuger und Säugerähnlicher Reptilien. *Jahreschefte der Vereins Vaterländische Naturkunde Württemberg* 89: 65–128.

Hunter, J. P., and Archibald, J. D. 2002. Mammals from the end of the age of dinosaurs in North Dakota and southeastern Montana. *In* J. H. Hartman, K. R. Johnson, and D. J. Nichols (eds.), *The Hell Creek Formation and the Cretaceous–Tertiary Boundary in the Northern Great Plains: An Integrated Continental Record of the End of the Cretaceous,* 191–216. *Geological Society of America Special Paper* 361, Denver.

Hunter, J. P., and Pearson, D. A. 1996. First record of Lancian (Late Cretaceous) mammals from the Hell Creek Formation of southwestern North Dakota. *Cretaceous Research* 17: 633–643.

Hunter, J. P., Hartman, J. H., and Krause, D. W. 1997. Mammals and molluscs across the Cretaceous–Tertiary boundary from Makoshika State Park and vicinity (Williston Basin), Montana. *Contributions to Geology, University of Wyoming* 32: 61–114.

Hurum, J. 1992. Earliest occurrence of sinus frontalis in Mammalia. *20 Nordiska Geologiska Vintermotet, Reykjvik:* 77.

Hurum, J. H. 1994. The snout and orbit of Mongolian multituberculates studied by serial sections. *Acta Palaeontologica Polonica* 39: 181–221.

Hurum, J. H. 1997. *Cranial structure and relationships of Mongolian Late Cretaceous multituberculate mammals.* Ph.D. dissertation, Paleontological Museum, University of Oslo, Oslo.

Hurum, J. H. 1998a. The braincase of two Late Cretaceous Asian multituberculates studied by serial sections. *Acta Palaeontologica Polonica* 43: 21–52.

Hurum, J. H. 1998b. The inner ear of two Late Cretaceous multituberculate mammals, and its implications for multituberculate hearing. *Journal of Mammalian Evolution* 5: 65–94.

Hurum, J. H., Presley, R., and Kielan-Jaworowska, Z. 1995. Multituberculate ear ossicles. *In* A.-L. Sun and Y. Wang (eds.), *Sixth Symposium on Mesozoic Terrestrial Ecosystems and Biota, Short Papers,* 243–246. China Ocean Press, Beijing.

Hurum, J. H., Presley, R., and Kielan-Jaworowska, Z. 1996. The middle ear in multituberculate mammals. *Acta Palaeontologica Polonica* 41: 253–275.

Huxley, T. H. 1880. On the application of the laws of evolution to the arrangement of the Vertebrata and more particularly of the Mammalia. *Proceedings of the Zoological Society of London* 43: 649–662.

ICZN 1999. *International Code of Zoological Nomenclature, 4th Edition.* 306 pp. International Trust for Zoological Nomenclature, c/o The Natural History Museum, London.

Illiger, C. 1811. *Prodromus systematis mammalium et avium additis terminis zoographicis utriusque classis.* C. Salfeld, Berlin.

Ivakhnenko, M. F., and Kurzanov, S. M. 1988. Geological structure and age of Udan-Sayr and Shara-Tsav localities. *Trudy Sovmestnoi Sovetsko-Mongol'skoi Paleontologicheskoi Ekspeditsii* 34: 100–105 (in Russian).

Jacobs, L. L., and Winkler, D. A. 1998. Mammals, archosaurs, and the Early to Late Cretaceous transition in north-central Texas. *In* Y. Tomida, L. J. Flynn, and L. L. Jacobs (eds.), *Advances in Vertebrate Paleontology and Geochronology,* 253–280. National Science Museum Monographs 14, Tokyo.

Jacobs, L. L., Congleton, J. D., Brunet, M., Djax, J., Flynn, L., Hell, J. V., and Mouchelin, G. 1988. Mammal teeth from the Cretaceous of Africa. *Nature* 336: 158–160.

Jacobs, L. L., Winkler, D. A., and Murry, P. A. 1989. Modern mammal origins: evolutionary grades in the Early Cretaceous of North America. *Proceedings of the National Academy of Sciences USA* 86: 4992–4995.

Jacobs, L. L., Winkler, D. A., and Murry, P. A. 1991. On the age and correlation of the Trinity mammals, Early Cretaceous of Texas, USA. *Newsletters on Stratigraphy* 24: 35–43.

Jaillard, E., Cappetta, H., Ellenberger, P., Feist, M., Grambast-Fessard, N., Lefranc, J. P., and Sigé, B. 1993. Sedimentology, palaeontology, biostratigraphy and correlation of the Late Cretaceous Vilquechico Group of southern Peru. *Cretaceous Research* 14: 623–661.

Jain, S. L. 1980. The continental Lower Jurassic fauna from the Kota Formation, India. *In* L. L. Jacobs (ed.), *Aspects of Vertebrate History,* 99–123. Museum of Northern Arizona Press, Flagstaff.

Jain, S. L., and Roy-Chowdhury, T. 1987. Fossil vertebrates from the Pranhita-Godvari Valley (India) and their stratigraphic correlation. *In* G. D. McKenzie (ed.), *Gondwana Six: Stratigraphy, Sedimentology, and Paleontology,* 219–228. American Geophysical Union, Geophysical Monograph 41, Washington, D.C.

Janensch, W. 1914. Bericht über den Verlauf der Tendaguru-Expedition. *Archiv für Biontologie* 3: 15–58.

Janke, A., Gemmell, N. J., Feldmaier-Fuchs, G., von Haesler, A., and Paabo, S. 1996. The mitochondrial gemone of a monotreme—the platypus (*Ornithorhynchus anatinus*). *Journal of Molecular Evolution* 42: 153–159.

Janke, A., Xu, X., and Arnason, U. 1997. The complete mitochondrial genome of the wallaroo (*Macropus robustus*) and the phylogenetic relationship among Monotremata, Marsupialia and Eutheria. *Proceedings of the National Academy of Sciences USA* 94: 1276–1281.

Janke, A., Magnell, O., Wieczorek, G., Westerman, M., and Arnason, U. 2002. Phylogenetic analysis of 18S RNA and the mitochondiral genomes of the wombat, *Vombatus ursinus*, and the spiny anteater, *Tachyglossus aculeatus:* increased support for the Marsupionta hypothesis. *Journal of Molecular Evolution* 54: 71–80.

Jenkins, F. A., Jr. 1969a. The evolution and development of the dens of the mammalian axis. *Anatomical Record* 164: 173–184.

Jenkins, F. A., Jr. 1969b. Occlusion in *Docodon* (Mammalia, Docodonta). *Postilla* 139: 1–24.

Jenkins, F. A., Jr. 1970a. The Chañares (Argentina) Triassic reptile fauna. VII. The postcranial skeleton of the traversodontid *Massetognathus pascuali* (Therapsida, Cynodontia). *Breviora* 352: 1–8.

Jenkins, F. A., Jr. 1970b. Cynodont postcranial anatomy and the "prototherian" level of mammalian organization. *Evolution* 24: 230–252.

Jenkins, F. A., Jr. 1970c. Limb movements in a monotreme (*Tachyglossus aculeatus*): a cineradiographic analysis. *Science* 168: 1473–1475.

Jenkins, F. A., Jr. 1971a. Limb posture and locomotion in the Virginia opossum (*Didelphis marsupialis*) and in other non-cursorial mammals. *Journal of Zoology, London* 165: 303–315.

Jenkins, F. A., Jr. 1971b. The postcranial skeleton of African cynodonts. *Peabody Museum of Natural History Bulletin* 36: 1–216.

Jenkins, F. A., Jr. 1973. The functional anatomy and evolution of the mammalian humero-ulnar joint. *American Journal of Anatomy* 137: 281–296.

Jenkins, F. A., Jr. 1974a. Tree shrew locomotion and arborealism. *In* F. A. Jenkins, Jr. (ed.), *Primate Locomotion,* 85–115. Academic Press, New York.

Jenkins, F. A., Jr. 1974b. The movement of the shoulder in claviculate and aclaviculate mammals. *Journal of Morphology* 144: 71–84.

Jenkins, F. A., Jr. 1984. A survey of mammalian origins. *In* P. D. Gingerich and C. E. Badgley (eds.), *Mammals: Notes for a Short Course,* 32–47. University of Tennessee, Department of Geological Sciences, Studies in Geology 8, Knoxville.

Jenkins, F. A., Jr. 1990. Monotremes and the biology of Mesozoic mammals. *Netherlands Journal of Zoology* 40: 5–31.

Jenkins, F. A., Jr., and Camazine, S. C. 1977. Hip structure and locomotion in ambulatory and cursorial carnivores. *Journal of Zoology, London* 181: 351–370.

Jenkins, F. A., Jr., and Crompton, A. W. 1979. Triconodonta. *In* J. A. Lillegraven, Z. Kielan-Jaworowska, and W. A. Clemens (eds.), *Mesozoic Mammals: The First Two-thirds of Mammalian History,* 74–90. University of California Press, Berkeley.

Jenkins, F. A., Jr., and McLearn, D. 1984. Mechanisms of hind foot reversal in climbing mammals. *Journal of Morphology* 182: 197–219.

Jenkins, F. A., Jr., and Parrington, F. R. 1976. The postcranial skeletons of the Triassic mammals *Eozostrodon, Megazostrodon* and *Erythrotherium. Philosophical Transactions of the Royal Society of London* 273: 387–431.

Jenkins, F. A., Jr., and Schaff, C. R. 1988. The Early Cretaceous mammal *Gobiconodon* (Mammalia, Triconodonta) from the Cloverly Formation in Montana. *Journal of Vertebrate Paleontology* 8: 1–24.

Jenkins, F. A., Jr., and Weijs, W. A. 1979. The functional anatomy of the shoulder in the Virginia opossum (*Didelphis virginiana*). *Journal of Zoology, London* 188: 379–410.

Jenkins, F. A., Jr., Crompton, A. W., and Downs, W. R. 1983. Mesozoic mammals from Arizona: new evidence on mammalian evolution. *Science* 222: 1233–1235.

Jenkins, F. A., Jr., Shubin, N. H., Amaral, W. W., Gatesy, S. M., Schaff, C. R., Clemmensen, L. B., Downs, W. R., Davidson, A. R., Bonde, N., and Osbaeck, F. 1994. Late Triassic continental vertebrates and depositional environments of the Fleming Fjord Formation, Jameson Land, East Greenland. *Meddelelser om Grønland* 32: 3–25.

Jenkins, F. A., Jr., Gatesy, S. M., Shubin, N. H., and Amaral, W. W. 1997. Haramiyids and Triassic mammalian evolution. *Nature* 385: 715–718.

Jensen, J. A. 1970. Fossil eggs in the Lower Cretaceous of Utah. *Brigham Young University, Geology Studies* 17: 51–66.

Jepsen, G. L. 1930a. New vertebrate fossils from the lower Eocene of the Bighorn Basin, Wyoming. *Proceedings of the American Philosophical Society* 69: 117–131.

Jepsen, G. L. 1930b. Stratigraphy and paleontology of the Paleocene of northeastern Park County, Wyoming. *Proceedings of the American Philosophical Society* 69: 463–528.

Jepsen, G. L. 1940. Paleocene faunas of the Polecat Bench Formation, Park County, Montana. *Proceedings of the American Philosophical Society* 83: 217–341.

Jerison, H. 1973. *Evolution of the Brain and Intelligence.* 482 pp. Academic Press, New York.

Jerison, H. J. 1977. Theory of encephalization. *Annals of the New York Academy of Sciences* 299: 146–160.

Jerison, H. J. 1990. Fossil evidence on the evolution of the neocortex. *In* E. G. Jones and A. Peters (eds.), *Cerebral Cortex, Vol. 8A,* 285–309. Plenum Press, New York.

Jerison, H. J. 1991. *Brain Size and the Evolution of Mind—59th James Arthur Lecture on the Evolution of the Human Brain (1989).* 99 pp. American Museum of Natural History, New York.

Jerzykiewicz, T. 1995. Cretaceous vertebrate-bearing strata of the Gobi and Ordos basins—a demise of the central Asian lacustrine dinosaur habitat. *Proceedings of the 15th International Symposium of Kyungpook National University* 1995: 233–256.

Jerzykiewicz, T., and Russell, D. A. 1991. Late Mesozoic stratigraphy and vertebrates of the Gobi Basin. *Cretaceous Research* 12: 345–377.

Jerzykiewicz, T., Currie, P. P. J., Johnston, P. A., Koster, E. H., and Gradziński, R. 1989. Upper Cretaceous dinosaur-bearing eolianites in the Mongolian Basin. *Abstracts, 28th International Geological Congress* 2: 122–123.

Jerzykiewicz, T., Currie, P. J., Eberth, D. A., Johnston, P. A., Koster, E. H., and Zheng, J.-J. 1993. Djadokhta Formation correlative strata in Chinese Inner Mongolia: an overview of stratigraphy, sedimentary geology, and paleontology and comparison with type locality in pre-Altaian Gobi. *Canadian Journal of Earth Sciences* 30: 2180–2195.

Ji, Q., Luo, Z.-X., and Ji, S. 1999. A Chinese triconodont mammal and mosaic evolution of the mammalian skeleton. *Nature* 398: 326–330.

Ji, Q., Luo, Z.-X., Yuan, C.-X., Wible, J. R., Zhang, J.-P., and Georgi, J. A. 2002. The earliest known eutherian mammal. *Nature* 416: 816–822.

Johanson, Z. 1993. A revision of the Late Cretaceous (Campanian) marsupial *Iqualadelphis lactea* Fox, 1987. *Journal of Vertebrate Paleontology* 13: 373–377.

Johanson, Z. 1994. New information concerning the Late Cretaceous marsupial *Albertatherium* Fox. *Journal of Vertebrate Paleontology* 14: 595–602.

Johanson, Z. 1996a. New marsupial from the Fort Union Formation, Swain Quarry, Wyoming. *Journal of Paleontology* 70: 1023–1031.

Johanson, Z. 1996b. Revision of the Late Cretaceous North American marsupial genus *Alphadon. Palaeontographica, Abteilung A* 242: 127–184.

Johnston, P. A. 1980. First record of Mesozoic mammals from Saskatchewan. *Canadian Journal of Earth Sciences* 17: 512–519.

Johnston, P. A., and Fox, R. C. 1984. Paleocene and Late Cretaceous mammals from Saskatchewan, Canada. *Palaeontographica, Abteilung A* 186: 163–222.

Jollie, M. 1962. *Chordate Morphology.* 478 pp. Chapman and Hall, Ltd., London.

Jouffroy, K. F. 1971. Musculature des membres. *In* P. P. Grassé (ed.), *Traité de Zoologie, 16, Fascicule* 2, 476pp. Masson et Cie, Paris.

Karczewska, J., and Ziembińska-Tworzydło, M. 1970. Upper Cretaceous Charophyta from the Nemegt Basin, Gobi Desert. *Palaeontologia Polonica* 21: 121–146.

Karczewska, J., and Ziembińska-Tworzydło, M. 1981. New Upper Cretaceous Charophyta from the Nemegt Basin, Gobi Desert. *Palaeontologica Polonica* 42: 97–146.

Kay, E. 1986. The phenotypic interdependence of the musculoskeletal characters of the mandibular arch in mice. *Journal of Embryology and Experimental Morphology* 98: 123–136.

Kay, R. F. 1975. The functional adaptations of primate molar teeth. *American Journal of Physical Anthropology* 43: 195–216.

Kay, R. F., and Cartmill, M. 1977. Cranial morphology and adaptations of *Palaechthon nacimienti* and other Paromomyidae (Plesiadapoidea, ?Primates), with a description of a new genus and species. *Journal of Human Evolution* 6: 19–53.

Kay, R. F., and Hiiemae, K. M. 1974. Jaw movement and tooth use in recent and fossil primates. *American Journal of Physical Anthropology* 40: 227–256.

Kay, R. F., and Simons, E. L. 1986. Dental formulae and dental eruption patterns in Parapithecidae (Primates, Anthropoidea). *American Journal of Physical Anthropology* 62: 363–375.

Kearney, M. 2001. Fragmentary taxa, missing data, and ambiguity: mistaken assumptions and conclusions. *Systematic Biology* 51: 369–381.

Kemp, T. S. 1972. Whaitsiid Therocephalia and the origin of cynodonts. *Philosophical Transactions of the Royal Society of London* 246: 83–103.

Kemp, T. S. 1980. Primitive cynodont *Procynosuchus:* structure, function and evolution of the postcranial skeleton. *Philosophical Transactions of the Royal Society of London* B288: 217–258.

Kemp, T. S. 1982. *Mammal-like Reptiles and the Origin of Mammals.* 363 pp. Academic Press, London.

Kemp, T. S. 1983. The relationships of mammals. *Zoological Journal of the Linnean Society* 77: 353–384.

Kemp, T. S. 1988. Interrelationships of the Synapsida. *In* M. J. Benton (ed.), *The Phylogeny and Classification of the Tetrapods, Vol. 2, Mammals,* 1–22. The Systematics Association Special Volume 35B, Clarendon Press, Oxford.

Kermack, D. M., and Kermack, K. A. 1984. *The Evolution of Mammalian Characters.* 149 pp. Croom Helm, London.

Kermack, D. M., Kermack, K. A., and Mussett, F. 1956. New Mesozoic mammals from south Wales. *Proceedings of the Geological Society of London* 1533: 31–32.

Kermack, D. M., Kermack, K. A., and Mussett, F. 1968. The Welsh pantothere *Kuehneotherium praecursoris. Journal of the Linnean Society (Zoology)* 47: 407–423.

Kermack, K. A. 1963. The cranial structure of the triconodonts. *Philosophical Transactions of the Royal Society of London* 246: 83–103.

Kermack, K. A. 1965. The origin of mammals. *Science Journal* 1: 66–72.

Kermack, K. A. 1967a. The interrelations of early mammals. *Journal of the Linnaean Society (Zoology)* 47: 241–249.

Kermack, K. A. 1967b. Molar evolution in Mesozoic mammals. *Journal of Dental Research* 46: 792–795.

Kermack, K. A. 1988. British Mesozoic mammal sites. *Special Papers in Palaeontology* 40: 85–93.

Kermack, K. A., and Kielan-Jaworowska, Z. 1971. Therian and non-therian mammals. *In* D. M. Kermack and K. A. Kermack (eds.), *Early Mammals,* 103–116. *Zoological Journal of the Linnean Society* 50, supplement 1, London.

Kermack, K. A., and Mussett, F. 1958a. The jaw articulation of the Docodonta and the classification of Mesozoic mammals. *Proceedings of the Royal Society, London* 149: 204–215.

Kermack, K. A., and Mussett, F. 1958b. The first mammals. *Discovery, London* 20: 144–151.

Kermack, K. A., and Mussett, F. 1983. The ear in mammal-like reptiles and early mammals. *Acta Palaeontologica Polonica* 28: 147–158.

Kermack, K. A., Lees, P. M., and Mussett, F. 1965. *Aegialodon dawsoni,* a new trituberculosectorial tooth from the lower Wealden. *Proceedings of the Royal Society of London, B* 162: 535–554.

Kermack, K. A., Mussett, F., and Rigney, H. W. 1973. The lower jaw of *Morganucodon. Zoological Journal of the Linnean Society* 53: 87–175.

Kermack, K. A., Mussett, F., and Rigney, H. W. 1981. The skull of *Morganucodon. Zoological Journal of the Linnean Society* 71: 1–158.

Kermack, K. A., Lee, A. J., Lees, P. M., and Mussett, F. 1987. A new docodont from the Forest Marble. *Zoological Journal of the Linnean Society* 89: 1–39.

Kermack, K. A., Kermack, D. M., Lees, P. M., and Mills, J. R. E. 1998. New multituberculate-like teeth from the Middle Jurassic of England. *Acta Palaeontologica Polonica* 43: 581–606.

Kerourio, P., and Sigé, B. 1984. L'apport des coquilles d'oeufs de dinosaures de Laguna Umayo à l'âge de la Formation Vilquechico (Pérou) et à la compréhension de *Perutherium altiplanense. Newsletters on Stratigraphy* 13: 133–142.

Khajuria, C. K., and Prasad, G. V. R. 1998. Taphonomy of a Late Cretaceous mammal-bearing microvertebrate assemblage from the Deccan inter-trappean beds of Naskal, peninsular India. *Palaeogeography, Palaeoclimatology, Palaeoecology* 137: 153–172.

Kielan-Jaworowska, Z. 1969a. Discovery of a multituberculate marsupial bone. *Nature* 222: 1091–1092.

Kielan-Jaworowska, Z. 1969b. Preliminary data on the Upper Cretaceous eutherian mammals from Bayn Dzak, Gobi Desert. *Palaeontologia Polonica* 19: 171–191.

Kielan-Jaworowska, Z. 1970a. New Upper Cretaceous multituberculate genera from Bayn Dzak, Gobi Desert. *Palaeontologia Polonica* 21: 35–49.

Kielan-Jaworowska, Z. 1970b. Unknown structures in multituberculate skull. *Nature* 226: 974–976.

Kielan-Jaworowska, Z. 1971. Skull structure and affinities of the Multituberculata. *Palaeontologia Polonica* 25: 5–41.

Kielan-Jaworowska, Z. 1974a. Multituberculate succession in the Late Cretaceous of the Gobi Desert (Mongolia). *Palaeontologica Polonica* 30: 23–44.

Kielan-Jaworowska, Z. 1974b. Migrations of the Multituberculata and the Late Cretaceous connections between Asia and North America. *Annals of the South African Museum* 64: 231–243.

Kielan-Jaworowska, Z. 1975a. Possible occurrence of marsupial bones in Cretaceous eutherian mammals. *Nature* 255: 698–699.

Kielan-Jaworowska, Z. 1975b. Preliminary description of two new eutherian genera from the Late Cretaceous of Mongolia. *Palaeontologia Polonica* 33: 5–16.

Kielan-Jaworowska, Z. 1975c. Evolution of the therian mammals in the Late Cretaceous of Asia. Part I. Deltatheridiidae. *Palaeontologia Polonica* 33: 103–132.

Kielan-Jaworowska, Z. 1977. Evolution of the therian mammals in the Late Cretaceous of Asia. Part II. Postcranial skeleton in *Kennalestes* and *Asioryctes. Palaeontologia Polonica* 37: 65–83.

Kielan-Jaworowska, Z. 1978. Evolution of the therian mammals in the Late Cretaceous of Asia. Part III. Postcranial skeleton in Zalambdalestidae. *Palaeontologia Polonica* 38: 5–41.

Kielan-Jaworowska, Z. 1979. Pelvic structure and nature of reproduction in Multituberculata. *Nature* 277: 402–403.

Kielan-Jaworowska, Z. 1980. Absence of ptilodontoidean multituberculates from Asia and its palaeogeographic implications. *Lethaia* 13: 169–173.

Kielan-Jaworowska, Z. 1981. Evolution of the therian mammals in the Late Cretaceous of Asia. Part IV. Skull structure in *Kennalestes* and *Asioryctes. Palaeontologia Polonica* 42: 25–78.

Kielan-Jaworowska, Z. 1982. Marsupial-placental dichotomy and paleogeography of Cretaceous Theria. *In* E. M. Gallitelli (ed.), *Palaeontology, Essential of Historical Geology,* 367–383. S.T.E.M. Mucci, Modena.

Kielan-Jaworowska, Z. 1983. Multituberculate endocranial casts. *Palaeovertebrata* 13: 1–12.

Kielan-Jaworowska, Z. 1984a. Evolution of the therian mammals in the Late Cretaceous of Asia. Part VII. Synopsis. *Palaeontologia Polonica* 46: 173–183.

Kielan-Jaworowska, Z. 1984b. Evolution of the therian mammals in the Late Cretaceous of Asia. Part V. Skull structure in Zalambdalestidae. *Palaeontologia Polonica* 46: 107–117.

Kielan-Jaworowska, Z. 1984c. Evolution of the therian mammals in the Late Cretaceous of Asia. Part VI. Endocranial casts of eutherian mammals. *Palaeontologia Polonica* 46: 157–171.

Kielan-Jaworowska, Z. 1986. Brain evolution in Mesozoic mammals. *Contributions to Geology, University of Wyoming, Special Paper* 3: 21–34.

Kielan-Jaworowska, Z. 1989. Postcranial skeleton of a Cretaceous multituberculate mammal. *Acta Palaeontologica Polonica* 34: 15–18.

Kielan-Jaworowska, Z. 1992. Interrelationships of Mesozoic mammals. *Historical Biology* 6: 185–202.

Kielan-Jaworowska, Z. 1994. A new generic name for the multituberculate mammal "*Djadochtatherium*" *catopsaloides. Acta Palaeontologica Polonica* 39: 134–136.

Kielan-Jaworowska, Z. 1997. Characters of multituberculates neglected in phylogenetic analyses of early mammals. *Lethaia* 29: 249–266.

Kielan-Jaworowska, Z. 1998. Humeral torsion in multituberculate mammals. *Acta Palaeontologica Polonica* 43: 131–134.

Kielan-Jaworowska, Z., and Bonaparte, J. F. 1996. Partial dentary of a multituberculate mammal from the Late Cretaceous of Argentina and its taxonomic implications. *Revista del Museo Argentino de Ciencias Naturales "Bernardino Rivadavia"* 145: 1–9.

Kielan-Jaworowska, Z., and Cifelli, R. L. 2001. Primitive boreosphenidan mammal (?Deltatheroida) from the Early Cretaceous of Oklahoma. *Acta Palaeontologica Polonica* 46: 377–391.

Kielan-Jaworowska, Z., and Dashzeveg, D. 1978. New Late Cretaceous locality in Mongolia and a description of a new multituberculate. *Acta Palaeontologica Polonica* 23: 115–130.

Kielan-Jaworowska, Z., and Dashzeveg, D. 1989. Eutherian mammals from the Early Cretaceous of Mongolia. *Zoologica Scripta* 18: 347–355.

Kielan-Jaworowska, Z., and Dashzeveg, D. 1998. Early Cretaceous amphilestid ("triconodont") mammals from Mongolia. *Acta Palaeontologica Polonica* 43: 413–438.

Kielan-Jaworowska, Z., and Ensom, P. C. 1991. Suprageneric taxa of Late Jurassic and Early Cretaceous multituberculate mammals. *In* Z. Kielan-Jaworowska, N. Heintz, and H.-A. Nakrem (eds.), *Fifth Symposium on Mesozoic Terrestrial Ecosystems and Biota,* 35–36. Contributions from the Paleontological Museum, University of Oslo, Oslo.

Kielan-Jaworowska, Z., and Ensom, P. C. 1992. Multituberculate mammals from the Upper Jurassic Purbeck Limestone Formation of southern England. *Palaeontology* 35: 95–126.

Kielan-Jaworowska, Z., and Ensom, P. C. 1994. Tiny plagiaulacoid mammals from the Purbeck Limestone Formation of Dorset, England. *Palaeontology* 37: 17–31.

Kielan-Jaworowska, Z., and Gambaryan, P. P. 1994. Postcranial anatomy and habits of Asian multituberculate mammals. *Fossils and Strata* 36: 1–92.

Kielan-Jaworowska, Z., and Hurum, J. H. 1997. Djadochtatheria—a new suborder of multituberculate mammals. *Acta Palaeontologica Polonica* 42: 201–242.

Kielan-Jaworowska, Z., and Hurum, J. H. 2001. Phylogeny and systematics of multituberculate mammals. *Palaeontology* 44: 389–429.

Kielan-Jaworowska, Z., and Lancaster, T. 2004. A new interpretation of multituberculate endocranial casts and encephalization quotient of *Kryptobaatar. Acta Palaeontologica Polonica* 49: 177–188.

Kielan-Jaworowska, Z., and Nessov, L. A. 1990. On the metatherian nature of the Deltatheroida, a sister group of the Marsupialia. *Lethaia* 23: 1–10.

Kielan-Jaworowska, Z., and Nessov, L. A. 1992. Multituberculate mammals from the Cretaceous of Uzbekistan. *Acta Palaeontologica Polonica* 37: 1–17.

Kielan-Jaworowska, Z., and Qi, T. 1990. Fossorial adaptations of a taeniolabidoid multituberculate mammal from the Eocene of China. *Vertebrata PalAsiatica* 28: 81–94.

Kielan-Jaworowska, Z., and Sloan, R. E. 1979. *Catopsalis* (Multituberculata) from Asia and North America and the problem of taeniolabidoid dispersal in the Late Cretaceous. *Acta Palaeontologica Polonica* 24: 187–197.

Kielan-Jaworowska, Z., and Sochava, A. V. 1969. The first multituberculate from the uppermost Cretaceous of the Gobi Desert (Mongolia). *Acta Palaeontologica Polonica* 14: 355–371.

Kielan-Jaworowska, Z., and Trofimov, B. A. 1980. Cranial morphology of the Cretaceous eutherian *Barunlestes. Acta Palaeontologica Polonica* 25: 167–185.

Kielan-Jaworowska, Z., and Trofimov, B. A. 1981. A new occurrence of Late Cretaceous eutherian mammal *Zalambdalestes. Acta Palaeontologica Polonica* 26: 3–7.

Kielan-Jaworowska, Z., and Trofimov, B. A. 1986. Endocranial cast of the Cretaceous eutherian mammal *Barunlestes. Acta Palaeontologica Polonica* 31: 137–144.

Kielan-Jaworowska, Z., Bown, T. M., and Lillegraven, J. A. 1979. Eutheria. *In* J. A. Lillegraven, Z. Kielan-Jaworowska, and W. A. Clemens (eds.), *Mesozoic Mammals: The First Two-thirds of Mammalian History,* 221–258. University of California Press, Berkeley.

Kielan-Jaworowska, Z., Eaton, J. G., and Bown, T. M. 1979. Theria of metatherian-eutherian grade. *In* J. A. Lillegraven, Z. Kielan-Jaworowska, and W. A. Clemens (eds.), *Mesozoic Mammals: The First Two-thirds of Mammalian History,* 182–191. University of California Press, Berkeley.

Kielan-Jaworowska, Z., Presley, R., and Poplin, C. 1986. The cranial vascular system in taeniolabidoid multituberculate mammals. *Philosophical Transactions of the Royal Society of London* 313: 525–602.

Kielan-Jaworowska, Z., Crompton, A. W., and Jenkins, F. A., Jr. 1987. The origin of egg-laying mammals. *Nature* 326: 871–873.

Kielan-Jaworowska, Z., Dashzeveg, D., and Trofimov, B. A. 1987. Early Cretaceous multituberculates from Mongolia and a comparison with Late Jurassic forms. *Acta Palaeontologica Polonica* 32: 3–47.

Kielan-Jaworowska, Z., Cifelli, R. L., and Luo, Z.-X. 1998. Alleged Cretaceous placental from down under. *Lethaia* 31: 267–268.

Kielan-Jaworowska, Z., Novacek, M. J., Trofimov, B. A., and Dashzeveg, D. 2000. Mammals from the Mesozoic of Mongolia. *In* M. J. Benton, M. A. Shishkin, E. N. Kurochkin, and D. M. Unwin (eds.), *The Age of Dinosaurs in Russia and Mongolia,* 573–652. Cambridge University Press, Cambridge.

Kielan-Jaworowska, Z., Cifelli, R. L., and Luo, Z.-X. 2002. Dentition and relationships of the Jurassic mammal *Shuotherium. Acta Palaeontologica Polonica* 47: 479–486.

Kielan-Jaworowska, Z., Hurum, J. H., Currie, P. J., and Barsbold, R. 2002. New data on anatomy of the Late Cretaceous multituberculate mammal *Catopsbaatar. Acta Palaeontologica Polonica* 47: 557–560.

Kielan-Jaworowska, Z., Hurum, J. H., and Badamgarav, D. 2003. Multituberculate mammal *Kryptobaatar* and the distribution of mammals in the Upper Cretaceous rocks of the Gobi Desert. *Acta Palaeontologica Polonica* 48: 161–166.

Killian, J. K., Burd, J. C., Jirtle, J. V., Mundy, B. L., Stoskopf, M. K., MacDonald, R. G., and Jirtle, R. L. 2000. M6P/IGF2R imprinting evolution in mammals. *Molecular Cell* 5: 707–716.

Killian, J. K., Buckley, T. R., Stewart, N., Munday, B. L., and Jirtle, R. L. 2001. Marsupials and eutherians reunited: genetic evidence for the Theria hypothesis of mammalian evolution. *Mammalian Genome* 12: 513–517.

Kindahl, M. E. 1963. On the embryonic development of the teeth in the golden mole, *Eremitalpa* (*Chryschloris*) *granti* Broom in South Africa. *Archiva Zoologia* 16: 97–115.

Kindahl, M. E. 1967. Some comparative aspects of reduction of the premolars in the Insectivora. *Journal of Dental Research* 46: 805–808.

Kirkland, J. I., Britt, B., Burge, D. L., Carpenter, K. C., Cifelli, R. L., DeCourten, F., Hasiotis, S., and Lawton, T. 1997. Lower to middle Cretaceous dinosaur faunas of the central Colorado Plateau: a key to understanding 35 million years of tectonics, sedimentology, evolution and biogeography. *Brigham Young University Geology Studies* 42: 69–103.

Kirsch, J. A. W., Lapointe, F.-J., and Springer, M. S. 1997. DNA-hybridisation studies of marsupials and their implications for metatherian classification. *Australian Journal of Zoology* 45: 211–280.

Klima, M. 1973. Die Frühentwicklung des Schültergürtels und des Brustbeins bei den Monotremen (Mammalia: Prototheria). *Advances in Anatomy, Embryology and Cell Biology* 47: 1–80.

Klima, M. 1987. Early development of the shoulder girdle and sternum in marsupials (Mammalia: Metatheria). *Advances in Anatomy, Embryology and Cell Biology* 109: 1–91.

Kluge, A. 1989. A concern for evidence and a phylogenetic hypothesis of relationships among *Epicrates* (Boidae, Serpentes). *Systematic Zoology* 38: 7–25.

Kobayashi, Y., Winkler, D. A., and Jacobs, L. L. 2002. Origin of the tooth-replacement pattern in therian mammals: evidence from a 110 Myr old fossil. *Proceedings of the Royal Society, London* 269: 369–373.

Koenigswald, W. von. 1980. Schmelzmuster und Morphologie in den Molaren des Arvicolidae (Rodentia). *Abhandlungen der Senckenbergischen naturforschenden Gesellschaft* 539: 1–129.

Koenigswald, W. von. 1997a. Brief survey of the enamel diversity at the Schmelzmuster level in Cenozoic placental mammals. *In* W. von Koenigswald and P. M. Sander (eds.), *Tooth Enamel Microstructure,* 137–161. A. A. Balkema, Rotterdam.

Koenigswald, W. von. 1997b. The variability of the enamel at the dentition level. *In* W. von Koenigswald and P. M. Sander (eds.), *Tooth Enamel Microstructure,* 193–201. A. A. Balkema, Rotterdam.

Koenigswald, W. von, and Clemens, W. A. 1992. Levels of complexity in the microstructure of mammalian enamel and their application in studies of systematics. *Scanning Microscopy* 6: 195–218.

Koenigswald, W. von, and Sander, P. M. (eds.). 1997a. *Tooth Enamel Microstructure.* 280 pp. A. A. Balkema, Rotterdam.

Koenigswald, W. von, and Sander, P. M. 1997b. Glossary of terms used for enamel microstructure. *In* W. v. Koenigswald and P. M. Sander (eds.), *Tooth Enamel Microstructure,* 267–280. A. A. Balkema, Rotterdam.

Koenigswald, W. von, Goin, F. J., and Pascual, R. 1999. Hypsodonty and enamel microstructure in the Paleocene gondwanatherian mammal *Sudamerica ameghinoi. Acta Palaeontologica Polonica* 44: 263–300.

Kowallis, B. J., Christiansen, E. H., Deino, A. L., Peterson, F., Turner, C. E., Kunk, M. J., and Obradovich, J. D. 1998. The age of the Morrison Formation. *Modern Geology* 23: 235–260.

Kranz, P. M. 1996. Notes on the sedimentary iron ores of Maryland and their dinosaurian fauna. *Maryland Geological Survey, Special Publication* 3: 87–115.

Kranz, P. M. 1998. Mostly dinosaurs: a review of the vertebrates of the Potomac Group (Aptian Arundel Formation), USA. *In* S. G. Lucas, J. I. Kirkland, and J. W. Estep (eds.), *Lower and Middle Cretaceous Terrestrial Ecosystems,* 235–238. New Mexico Museum of Natural History and Science, Albuquerque.

Kraus, M. J. 1979. Eupantotheria. *In* J. A. Lillegraven, Z. Kielan-Jaworowska, and W. A. Clemens (eds.), *Mesozoic Mammals: The First Two-thirds of Mammalian History,* 162–171. University of California Press, Berkeley.

Krause, D. W. 1977. Paleocene multituberculates (Mammalia) of the Roche Percée local fauna, Ravenscrag Formation, Saskatchewan, Canada. *Palaeontographica A* 159: 1–36.

Krause, D. W. 1980. Multituberculates from the Clarkforkian land-mammal age, late Paleocene–early Eocene. *Journal of Paleontology* 54: 1163–1183.

Krause, D. W. 1982a. Jaw movement, dental function, and diet in the Paleocene multituberculate *Ptilodus. Paleobiology* 8: 265–281.

Krause, D. W. 1982b. Multituberculates from the Wasatchian land-mammal age, early Eocene, of Western North America. *Journal of Paleontology* 56: 271–294.

Krause, D. W. 1984. Mammalian evolution in the Paleocene: beginning of an Era. *In* P. D. Gingerich and C. E. Badgley (eds.), *Mammals: Notes for a Short Course,* 87–109. University of Tennessee, Department of Geological Sciences, Studies in Geology, Knoxville.

Krause, D. W. 1986. Competitive exclusion and taxonomic displacement in the fossil record: the case of rodents and multituberculates in North America. *Contributions to Geology, University of Wyoming, Special Paper* 3: 95–117.

Krause, D. W. 1987a. Systematic revision of *Prochetodon* (Multituberculata, Mammalia) from the late Paleocene and early Eocene of western North America. *Contributions from the Museum of Paleontology, University of Michigan* 27: 221–236.

Krause, D. W. 1987b. *Baiotomeus,* a new ptilodontid multituberculate (Mammalia) from the middle Paleocene of western North America. *Journal of Paleontology* 61: 595–603.

Krause, D. W. 1992. *Clemensodon megaloba,* a new genus and species of Multituberculata (Mammalia) from the Upper Cretaceous type Lance Formation, Powder River Basin, Wyoming. *PaleoBios* 14: 1–8.

Krause, D. W. 1993. *Vucetichia* (Gondwanatheria) a junior synonym of *Ferugliotherium* (Multituberculata). *Journal of Paleontology* 67: 321–324.

Krause, D. W. 2000. New mammalian specimens from the Late Cretaceous of Madagascar. *Journal of Vertebrate Paleontology* 20, supplement to no. 3: 52A.

Krause, D. W. 2001. Fossil molar from a Madagascan marsupial. *Nature* 412: 497–498.

Krause, D. W., and Baird, D. 1979. Late Cretaceous mammals east of the North American Western Interior Seaway. *Journal of Paleontology* 53: 562–565.

Krause, D. W., and Bonaparte, J. F. 1990. The Gondwanatheria, a new suborder of Multituberculata from South America. *Journal of Vertebrate Paleontology* 9, supplement to no. 3: 48A.

Krause, D. W., and Bonaparte, J. F. 1993. Superfamily Gondwanatherioidea: a previously unrecognized radiation of multituberculate mammals in South America. *Proceedings of the National Academy of Sciences USA* 90: 9379–9383.

Krause, D. W., and Carlson, S. J. 1986. The enamel ultrastructure of multituberculate mammals: a review. *Scanning Electron Microscopy* 1986: 1591–1607.

Krause, D. W., and Carlson, S. J. 1987. Prismatic enamel in multituberculate mammals: tests of homology and polarity. *Journal of Mammalogy* 68: 755–765.

Krause, D. W., and Grine, F. E. 1996. The first multituberculates from Madagascar: implications for Cretaceous biogeography.

Journal of Vertebrate Paleontology 16, supplement to no. 3: 46A.

Krause, D. W., and Hahn, G. 1990. Systematic position of the Paulchoffatiinae (Multituberculata, Mammalia). *Journal of Paleontology* 64: 1051–1054.

Krause, D. W., and Jenkins, F. A. 1983. The postcranial skeleton of North American multituberculates. *Bulletin of the Museum of Comparative Zoology* 150: 199–246.

Krause, D. W., and Kielan-Jaworowska, Z. 1993. The endocranial cast and encephalization quotient of *Ptilodus* (Multituberculata, Mammalia). *Palaeovertebrata* 22: 99–112.

Krause, D. W., Kielan-Jaworowska, Z., and Turnbull, W. D. 1990. Early Cretaceous Multituberculata (Mammalia) from the Antlers Formation, Trinity Group, of southcentral Texas. *Journal of Vertebrate Paleontology* 10, supplement to no. 3: 31A.

Krause, D. W., Kielan-Jaworowska, Z., and Bonaparte, J. F. 1992. *Ferugliotherium* Bonaparte, the first known multituberculate from South America. *Journal of Vertebrate Paleontology* 12: 351–376.

Krause, D. W., Wunderlich, R., Hartman, J. H., and Kroeger, T. J. 1992. Latest Cretaceous and early Paleocene mammals from Makoshika State Park, Williston Basin, eastern Montana. *Journal of Vertebrate Paleontology* 12: 38A.

Krause, D. W., Hartman, J. H., Wells, N. A., Buckley, G. A., Lockwood, C. A., Wall, C. E., Wunderlich, R. E., Rabarison, J. A., and Randriamiararamanana, L. L. 1994. Late Cretaceous mammals. *Nature* 368: 298.

Krause, D. W., Hartman, J. H., and Wells, N. H. 1997. Late Cretaceous vertebrates from Madagascar. *In* S. M. Goodman and B. D. Patterson (eds.), *Natural Change and Human Impact in Madagascar,* 3–43. Smithsonian Institution Press, Washington, D.C.

Krause, D. W., Prasad, G. V. R., Koenigswald, W. von, Sahni, A., and Grine, F. E. 1997. Cosmopolitanism among Gondwanan Late Cretaceous mammals. *Nature* 390: 504–507.

Krause, D. W., Rogers, R. R., Forster, C. A., Hartman, J. H., Buckley, G. A., and Sampson, S. D. 1999. The Late Cretaceous vertebrate fauna of Madagascar: implications for Gondwanan paleobiogeography. *GSA Today* 9: 1–7.

Krebs, B. 1969. Nachweis eines rudimentären Coronoids im Unterkiefer der Pantotheria (Mammalia). *Paläontologische Zeitschrift* 43: 57–63.

Krebs, B. 1971. Evolution of the mandible and lower dentition in dryolestoids (Pantotheria, Mammalia). *In* D. M. Kermack and K. A. Kermack (eds.), *Early Mammals,* 89–102. *Zoological Journal of the Linnean Society* 50, supplement 1, London.

Krebs, B. 1980. The search for Mesozoic mammals in Spain and Portugal. *Mesozoic Vertebrate Life* 1: 23–25.

Krebs, B. 1985. Theria (Mammalia) aus der Unterkreide von Galve (Provinz Teruel, Spanien). *Berliner geowissenschaftliche, Abhandlungen A* 60: 29–48.

Krebs, B. 1987. The skeleton of a Jurassic eupantothere and the arboreal origin of modern mammals. *In* P. J. Currie and E. H. Koster (eds.), *Fourth Symposium on Mesozoic Terrestrial Ecosystems,* 132–137. Tyrrell Museum, Drumheller.

Krebs, B. 1991. Das Skelett von *Henkelotherium guimarotae* gen. et sp. nov. (Eupantotheria, Mammalia) aus dem Oberen Jura von Portugal. *Berliner geowissenschaftliche, Abhandlungen A* 133: 1–110.

Krebs, B. 1993. Das Gebiß von *Crusafontia* (Eupantotheria, Mammalia)—Funde aus der Unter-Kreide von Galve un Uña. *Berliner geowissenschaftliche, Abhandlungen E* 9: 233–252.

Krebs, B. 1995. The Barremian vertebrate locality Uña (Province of Cuenca): materials for a comparison with Las Hoyas. *II International Symposium on Lithographic Limestones,* 95–97. Universidad Autónoma de Madrid, Madrid.

Krebs, B. 1998. *Drescheratherium acutum* gen. et sp. nov., ein neuer Eupantotherier (Mammalia) aus dem Oberen Jura von Portugal. *Berliner geowissenschaftliche, Abhandlungen E* 28: 91–111.

Kretzoi, M. 1946. On Docodonta, a new order of Jurassic mammals. *Annales Historico-Naturales Musei Nationalis Hungarici* 39: 108–111.

Kretzoi, M. 1960. Zur Benennung des ältesten Symmetrodonten. *Vertebrata Hungarici* 2: 307–309.

Kretzoi, M., and Kretzoi, M. 2000. *Fossilium Catalogus 1: Animalia. Pars 137—Index Generum et Subgenerum Mammalium.* 726 pp. Backhuys Publishers, Leiden.

Krishtalka, L. 1976. Early Tertiary Adapiscoricidae and Erinaceidae (Mammalia, Insectivora) of North America. *Bulletin of the Carnegie Museum of Natural History* 1: 1–40.

Krishtalka, L., and Stucky, R. K. 1983a. Paleocene and Eocene marsupials of North America. *Annals of the Carnegie Museum* 52: 229–263.

Krishtalka, L., and Stucky, R. K. 1983b. Revision of the Wind River faunas, early Eocene of central Wyoming. Part 3. Marsupialia. *Annals of the Carnegie Museum* 52: 502–527.

Krishtalka, L., Emry, R. J., Storer, J. E., and Sutton, J. F. 1982. Oligocene multituberculates (Mammalia: Allotheria), youngest known record. *Journal of Paleontology* 56: 791–794.

Kron, D. G. 1979. Docodonta. *In* J. A. Lillegraven, Z. Kielan-Jaworowska, and W. A. Clemens (eds.), *Mesozoic Mammals: The First Two-thirds of Mammalian History,* 91–98. University of California Press, Berkeley.

Krusat, G. 1969. Ein Pantotherier-Molar mit dreispitzigen Talonid aus dem Kimmeridge von Portugal. *Paläontologische Zeitschrift* 43: 52–56.

Krusat, G. 1973. Beiträge zur Kimmeridge-Fauna der Kohlengrube Guimarota (Leiria, Portugal): *Haldanodon exspectatus* Kühne and Krusat 1972 (Mammalia, Docodonta). Inaugural-Dissertation zur Erlangung der Doktorwürde des Fachbereiches 24 (Geowissenschaften) der Freie Universität Berlin: 158 pp.

Krusat, G. 1980. Contribução para o conhecimento da fauna do Kimeridgiano da mina de lignito Guimarota (Leiria, Portugal). IV Parte. *Haldanodon exspectatus* Kühne & Krusat 1972 (Mammalia, Docodonta). *Memórias dos Serviços Geológicos de Portugal* 27: 1–79.

Krusat, G. 1989. Isolated molars of a triconodont and a symmetrodont (Mammalia) from the uppermost Jurassic of Portugal. *Berliner geowissenschaftliche, Abhandlungen A* 106: 277–289.

Krusat, G. 1991. Functional morphology of *Haldanodon exspectatus* (Mammalia, Docodonta) from the Upper Jurassic of Portugal. *In* Z. Kielan-Jaworowska, N. Heintz, and H. A. Nakrem (eds.), *Fifth Symposium on Mesozoic Terrestrial Ecosystems and Biota,* 37–38. Contributions from the Paleontological Museum, University of Oslo, 363, Oslo.

Kuhn, H.-J. 1971. Die Entwicklung und Morphologie des Schädels von *Tachyglossus aculeatus. Abhandlungen der Senckenbergischen Naturforschenden Gesellschaft* 528: 1–192.

Kuhn, H.-J., and Zeller, U. (eds.). 1987a. *Morphogenesis of the Mammalian Skull. Mammalia Depicta, Heft 13, Beihefte zur Zeitschrift für Säugetierekunde.* 144 pp. Paul Parey, Hamburg/Berlin.

Kuhn, H.-J., and Zeller, U. 1987b. The cavum epiptericum in monotremes and therian mammals. *In* H.-J. Kuhn and U. Zeller (eds.), *Mammalia Depicta, Heft 13, Beihefte zur Zeitschrift für Säugetierekunde,* 51–70. Paul Parey, Hamburg/Berlin.

Kuhn-Schnyder, E., and Rieber, H. 1986. *Handbook of Paleozoology.* 394 pp. Johns Hopkins University Press, Baltimore.

Kühne, W. G. 1946. The geology of the fissure-filling "Holwell 2"; the age determination of the mammalian teeth therein; and a report on the technique employed when collecting the teeth of *Eozostrodon* and Microcleptidae. *Proceedings of the Zoological Society of London* 116: 729–733.

Kühne, W. G. 1949. On a triconodont tooth of a new pattern from a fissure-filling in South Glamorgan. *Proceedings of the Zoological Society of London* 119: 345–350.

Kühne, W. G. 1950. A symmetrodont tooth from the Rhaeto–Liassic. *Nature* 166: 696–697.

Kühne, W. G. 1956. *The Liassic therapsid* Oligokyphus. 149 pp. British Museum (Natural History), London.

Kühne, W. G. 1958. Rhaetische Triconodonten aus Glamorgan ihre Stellung zwischen den Klassen Reptilia und Mammalia und ihre Bedeutung für die Reichert'sche Theorie. *Paläontologische Zeitschrift* 32: 197–235.

Kühne, W. G. 1961a. Eine Mammaliafauna aus dem Kimeridge Portugals. *Neues Jahrbuch für Geologie und Paläontologie, Monatshefte* 7: 374–381.

Kühne, W. G. 1961b. A mammalian fauna from the Kimmeridgian of Portugal. *Nature* 192: 274–275.

Kühne, W. G. 1968. Kimmeridge mammals and their bearing on the phylogeny of the Mammalia. *In* E. T. Drake (ed.), *Evolution and Environment,* 109–123. Yale University Press, New Haven.

Kühne, W. G. 1971. Collecting vertebrate fossils by the Henkel process. *Curator* 14: 175–179.

Kühne, W. G., and Krusat, G. 1972. Legaliesierung des taxon *Haldanodon* (Mammalia, Docodonta). *Neues Jahrbuch für Geologie, Paläontologie und Mineralogie, Monatshefte* 5: 300–302.

Kullander, K., Carlson, B., and Hallbrook, F. 1997. Molecular phylogeny and evolution of the neurotrophins from monotremes and marsupials. *Journal of Molecular Evolution* 45: 311–321.

Kumar, S., and Hedges, S. B. 1998. A molecular timescale for vertebrate evolution. *Nature* 392: 917–920.

Kurochkin, E. N., and Barsbold, R. 2000. The Russian-Mongolian expeditions and research in vertebrate palaeontology. *In* M. J. Benton, M. A. Shishkin, E. N. Kurochkin, and D. M. Unwin (eds.), *The Age of Dinosaurs in Russia and Mongolia,* 235–255. Cambridge University Press, Cambridge.

Lambe, L. 1902. On Vertebrata of the mid-Cretaceous of the North-West Territories. 2. New genera and species from the Belly River Series (mid-Cretaceous). *Contributions to Canadian Palaeontology* 3: 25–81.

Landry, S. O., Jr. 1963. The status of the theory of the replacement of the Multituberculata by the Rodentia. *Journal of Mammalogy* 46: 280–286.

Lawson, D. A. 1972. *Paleoecology of the Tornillo Formation, Big Bend National Park, Brewster County, Texas.* 182 pp. M.S. thesis, University of Texas, Austin.

Leahy, G. D., and Lerbekmo, J. F. 1995. Macrofossil magnetostratigraphy for the upper Santonian–lower Campanian interval in the Western Interior of North America: comparisons with European stage boundaries and planktonic foraminiferal zonal boundaries. *Canadian Journal of Earth Sciences* 32: 247–260.

Ledoux, J.-L., Hartenberger, J.-L., Michaux, J., Sudre, J., and Thaler, L. 1966. Découverte d'un Mammifère dans le Crétacé supérieur à Dinosaures de Champ-Garimond près de Fons (Gard). *Comptes Rendus de l'Académie des Sciences, Paris, sér. D,* 262: 1925–1928.

Lee, M.-H., Schroff, R., Cooper, S. J. B., and Hope, R. 1999. Evolution and molecular characterization of a B-globin gene from the Australian echidna *Tachyglossus aculeatus* (Monotremata). *Molecular Phylogenetics and Evolution* 12: 205–214.

Lefeld, J. 1971. Geology of the Djadokchta Formation at Bayn Dzak, Mongolia. *Palaeontologia Polonica* 25: 101–127.

Lehman, T. M. 1981. The Alamo Wash local fauna: a new look at the old Ojo Alamo fauna. *In* S. G. Lucas, J. K. Rigby, Jr., and D. Wolberg (eds.), *Advances in San Juan Basin Paleontology,* 189–221. University of New Mexico Press, Albuquerque.

Lehman, T. M. 1984. The multituberculate *Essonodon browni* from the Upper Cretaceous Naashoibito Member of the Kirtland Shale, San Juan Basin, New Mexico. *Journal of Vertebrate Paleontology* 4: 602–603.

Lehman, T. M. 1987. Late Maastrichtian paleoenvironments and dinosaur biogeography in the Western Interior of North America. *Palaeogeography, Palaeoclimatology, Palaeoecology* 60: 189–217.

Lehman, T. M. 1997. Late Campanian dinosaur biogeography in the Western Interior of North America. *Dinofest International Proceedings* 1997: 223–240.

Leidy, J. 1856. Notices of the remains of extinct reptiles and fishes discovered by Dr. F. V. Hayden in the badlands of the Judith River, Nebraska Territory. *Proceedings of the Academy of Natural Sciences of Philadelphia* 8: 72–73.

Leidy, J. 1860. Extinct Vertebrata from the Judith River and great lignite formations of Nebraska. *Transactions of the American Philosophical Society* 11: 139–154.

LeLoeff, J. 1991. The Campano–Maastrichtian vertebrate faunas from southern Europe and their relationships with other faunas in the world; palaeobiogeographical considerations. *Cretaceous Research* 12: 93–114.

Lemoine, V. 1880. Communication sur les ossements fossiles des terrains tertiaires inférieurs des environs de Reims. *Association Française pour l'Advancement des Sciences, Reims* 8: 585–594.

Lemoine, V. 1882. Sur deux *Plagiaulax* tertiares, recueillis aux environs de Reims. *Comptes Rendus de l'Académie des Sciences, Paris* 95: 1009–1011.

Leonardi, G. 1994. *Annotated Atlas of South American Tetrapod Footprints (Devonian to Holocene) with an Appendix on Mexico and Central America.* 246 pp. Republica Federativa do Brasil, Brasília.

Lerbekmo, J. F. 1985. Magnetostratigraphic and biostratigraphic correlations of Maastrichtian to early Paleocene strata between south-central Alberta and southwestern Saskatchewan. *Bulletin of Canadian Petroleum Geology* 33: 213–226.

Lerbekmo, J. F., and Coulter, K. C. 1985. Late Cretaceous to early Tertiary magnetostratigraphy of a continental sequence: Red Deer Valley, Alberta, Canada. *Canadian Journal of Earth Sciences* 22: 567–583.

Leshchinskiy, S. V., Voronkevich, A. V., and Fayngertz, A. V. 2001. Early Cretaceous vertebrate locality Shestakovo, western Siberia, Russia: a refugium for Jurassic relicts? *Journal of Vertebrate Paleontology* 21, supplement to no. 3: 73A.

Lessertisseur, J., and Saban, R. 1967. Squelette appendiculaire. *In* P. P. Grassé (ed.), *Traité de Zoologie, Tome 16, Fascicule 1. Mammifères, Téguments et Squelette,* 709–1078. Maison et Cie, Paris.

Lessertisseur, J., and Sigogneau, D. 1965. Sur l'acquisition des principales caracteristiques du squelette des mammifères. *Mammalia* 29: 95–168.

Lester, K. S., and Koenigswald, W. von. 1989. Crystallite orientation discontinuities and the evolution of mammalian enamel—or when is a prism? *Scanning Microscopy* 3: 645–663.

Lester, K. S., Boyde, A., Gilkeson, C., and Archer, M. 1987. Marsupial and monotreme enamel. *Scanning Microscopy* 1: 401–420.

Lewis, O. J. 1980. The joints of the evolving foot. Part I. The ankle joint. *Journal of Anatomy, London* 130: 527–543.

Li, C.-K., and Ting, S.-Y. 1985. Possible phylogenetic relationships of Asiatic eurymylids and rodents, with comments on mimotonids. *In* W. P. Luckett and J. J. Hartenberger (eds.), *Evolutionary Relationships Among Rodents,* 35–59. Plenum Press, New York.

Li, J.-L., Wang, Y., Wang, Y.-Q., and Li, C.-K. 2000. A new family of primitive mammals from the Mesozoic of western Liaoning, China (in Chinese). *Chinese Science Bulletin* 45: 2545–2549.

Li, J.-L., Wang, Y., Wang, Y.-Q., and Li, C.-K. 2001. A new family of primitive mammals from the Mesozoic of western Liaoning, China. *Chinese Science Bulletin* 46: 782–785.

Lillegraven, J. A. 1969. Latest Cretaceous mammals of upper part of Edmonton Formation of Alberta, Canada, and review of marsupial-placental dichotomy in mammalian evolution. *University of Kansas Paleontological Contributions* 50: 1–122.

Lillegraven, J. A. 1972. Preliminary report on Late Cretaceous mammals from the El Gallo Formation, Baja California del Norte, Mexico. *Natural History Museum of Los Angeles County, Contributions in Science* 232: 1–11.

Lillegraven, J. A. 1974. Biogeographical considerations of the marsupial-placental dichotomy. *Annual Review of Ecology and Systematics* 5: 263–283.

Lillegraven, J. A. 1975. Biological considerations of the marsupial-placental dichotomy. *Evolution* 29: 707–722.

Lillegraven, J. A. 1976. A new genus of therian mammal from the Late Cretaceous "El Gallo Formation," Baja California, Mexico. *Journal of Paleontology* 50: 437–443.

Lillegraven, J. A. 1979. Reproduction in Mesozoic mammals. *In* J. A. Lillegraven, Z. Kielan-Jaworowska, and W. A. Clemens (eds.), *Mesozoic Mammals: The First Two-thirds of Mammalian History,* 259–276. University of California Press, Berkeley.

Lillegraven, J. A. 1984. Why *was* there a marsupial-placental dichotomy? *University of Tennessee Studies in Geology* 8: 1–19.

Lillegraven, J. A. 1987. Stratigraphic and evolutionary implications of a new species of *Meniscoessus* (Multituberculata, Mammalia) from the Upper Cretaceous Williams Fork Formation, Moffat County, Colorado. *In* J. E. Martin and G. E. Ostrander (eds.), *Papers in Vertebrate Paleontology in Honor of Morton Green,* 46–56. South Dakota School of Mines and Technology, Dakoterra, Special Paper 3, Rapid City.

Lillegraven, J. A. 1991. Stratigraphic placement of the Santonian–Campanian boundary (Upper Cretaceous) in the North American Gulf Coastal Plain and Western Interior, with implications to global geochronology. *Cretaceous Research* 12: 115–136.

Lillegraven, J. A., and Eberle, J. J. 1999. Vertebrate faunal changes through Lancian and Puercan time in southern Wyoming. *Journal of Paleontology* 73: 691–710.

Lillegraven, J. A., and Hahn, G. 1993. Evolutionary analysis of the middle and inner ear of Late Jurassic multituberculates. *Journal of Mammalian Evolution* 1: 47–74.

Lillegraven, J. A., and Krusat, G. 1991. Cranio-mandibular anatomy of *Haldanodon exspectatus* (Docodonta; Mammalia) from the Late Jurassic of Portugal and its implications to the evolution of mammalian characters. *Contributions to Geology, University of Wyoming* 28: 39–138.

Lillegraven, J. A., and McKenna, M. C. 1986. Fossil mammals from the "Mesaverde" Formation (Late Cretaceous, Judithian) of the Bighorn and Wind River basins, Wyoming, with definitions of Late Cretaceous North American land-mammal "ages." *American Museum Novitates* 2840: 1–68.

Lillegraven, J. A., and Ostresh, L. M., Jr. 1990. Late Cretaceous (earliest Campanian/Maastrichtian) evolution of western shorelines of the North American western interior seaway in relation to known mammalian faunas. *Geological Society of America, Special Paper* 243: 1–30.

Lillegraven, J. A., Kielan-Jaworowska, Z., and Clemens, W. A. (eds.). 1979. *Mesozoic Mammals: The First Two-thirds of Mammalian History.* 311 pp. University of California Press, Berkeley.

Lillegraven, J. A., McKenna, M. C., and Krishtalka, L. 1981. Evolutionary relationships of middle Eocene and younger species of *Centetodon* (Mammalia, Insectivora, Geolabididae) with a description of the dentition of *Ankylodon* (Adapisoricidae). *University of Wyoming Publications* 45: 1–115.

Lillegraven, J. A., Thompson, S. D., McNab, B. K., and Patton, J. L. 1987. The origin of eutherian mammals. *Biological Journal of the Linnean Society* 32: 281–336.

Linnaeus, C. 1758. *Systema naturae per regna tria naturae, secundum classes, ordines, genera, species, cum characteribus, differentiis, synonymis, locis. Vol. 1: Regnum animale. Editio decima, reformata.* 824 pp. Laurentii Salvii, Stockholm.

Linnaeus, C. 1766. *Systema naturae per regna tria naturae, secundum classes, ordines, genera, species, cum characteribus, differentiis, synonymis, locis. Vol. 1. Regnum animale. Editio duodecima, reformata.* 532 pp. Laurentii Salvii, Stockholm.

Litwin, R. J., Turner, C. E., and Peterson, F. 1998. Palynological evidence on the age of the Morrison Formation, western interior US. *Modern Geology* 22: 297–319.

Liu, F.-G. R., and Miyamoto, M. M. 1999. Phylogenetic assessment of molecular and morphological data for eutherian mammals. *Systematic Biology* 48: 54–64.

Liu, F.-G. R., Miyamoto, M. M., Freire, N. P., Ong, P. Q., Tennant, M. R., Young, T. S., and Gugel, K. F. 2001. Molecular and morphological supertrees for eutherian (placental) mammals. *Science* 291: 1786–1789.

Lofgren, D. L. 1995. The Bug Creek problem and the Cretaceous–Tertiary transition at McGuire Creek, Montana. *University of California Publications in Geological Sciences* 140: 1–185.

Lofgren, D. L., Hotton, C. L., and Runkel, A. C. 1990. Reworking of Cretaceous dinosaurs into Paleocene channel deposits, upper Hell Creek Formation, Montana. *Geology* 18: 874–877.

Loope, D. B., Dingus, L., Swisher, C. C., and Minjin, C. 1998. Life and death in a Late Cretaceous dune field. *Geology* 26: 27–30.

Lucas, P. W. 1979. The dental-dietary adaptations of mammals. *Neues Jahrbuch für Geologie und Paläontologie* 8: 486–512.

Lucas, S. G. 1992. Extinction and the definition of the class Mammalia. *Systematic Biology* 41: 370–371.

Lucas, S. G. 1993. Vertebrate biochronology of the Jurassic–Cretaceous boundary, North American western interior. *Modern Geology* 18: 371–390.

Lucas, S. G. 1996. Vertebrate biochronology of the Jurassic of China. *In* M. Morales (ed.), *The Continental Jurassic,* 23–33. *Museum of Northern Arizona Bulletin* 60, Flagstaff.

Lucas, S. G., and Estep, J. W. 1998. Vertebrate biostratigraphy and biochronology of the Cretaceous of China. *New Mexico Museum of Natural History and Science Bulletin* 14: 1–20.

Lucas, S. G., and Hunt, A. P. 1989. *Alamosaurus* and the sauropod hiatus in the Cretaceous of the North American Western Interior. *Geological Society of America, Special Paper* 238: 75–85.

Lucas, S. G., and Hunt, A. P. 1990. The oldest mammal. *New Mexico Journal of Science* 30: 41–49.

Lucas, S. G., and Luo, Z.-X. 1993. *Adelobasileus* from the Upper Triassic of western Texas: the oldest mammal. *Journal of Vertebrate Paleontology* 13: 309–334.

Lucas, S. G., and Oakes, W. 1988. A Late Triassic cynodont from the American southwest. *Palaeontology* 31: 445–449.

Lucas, S. G., Heckert, A. B., Harris, J. D., Seegis, D., and Wild, R. 2001. Mammal-like tooth from the Upper Triassic of Germany. *Journal of Vertebrate Paleontology* 21: 397–399.

Luckett, W. P. 1977. Ontogeny of amniote fetal membranes and their application to phylogeny. *In* M. K. Hecht, P. C. Goody, and B. M. Hecht (eds.), *Major Patterns in Vertebrate Evolution,* 439–515. Plenum Press, New York.

Luckett, W. P. 1993. An ontogenetic assessment of dental homologies in therian mammals. *In* F. S. Szalay, M. J. Novacek, and M. C. McKenna (eds.), *Mammal Phylogeny: Mesozoic Differentiation, Multituberculates, Monotremes, Early Therians, and Marsupials,* 182–204. Springer-Verlag, New York.

Luckett, W. P., and Hong, N. 2000. Ontogenetic evidence for dental homologies and premolar replacement in fossil and extant caenolestids (Marsupialia). *Journal of Mammalian Evolution* 7: 109–127.

Luckett, W. P., and Woolly, P. A. 1996. Ontogeny and homology of the dentition in dasyurid marsupials: development in *Smithopsis virgininae. Journal of Mammalian Evolution* 3: 327–364.

Luckett, W. P., and Zeller, U. 1989. Developmental evidence for dental homologies in the monotreme *Ornithorhynchus* and its systematic implications. *Zeitschrift für Säugetierkunde* 54: 193–204.

Luo, Z.-X. 1989. *Structure of the petrosal of Multituberculata (Mammalia) and the morphology of the molars of early arctocyonids.* Ph.D. dissertation, University of California, Berkeley.

Luo, Z.-X. 1991. Variability of dental morphology and the relationships of the earliest arctocyonid species. *Journal of Vertebrate Paleontology* 11: 452–471.

Luo, Z.-X. 1994. Sister-group relationships of mammals and transformations of diagnostic mammalian characters. *In* N. C. Fraser and H.-D. Sues (eds.), *In the Shadow of the Dinosaurs—Early Mesozoic Tetrapods,* 98–128. Cambridge University Press, Cambridge.

Luo, Z.-X. 2001. Inner ear and its bony housing in tritylodonts and implications for evolution of mammalian ear. *In* F. A. Jenkins, Jr., T. Owerkowicz, and M. D. Shapiro (eds.), *Symposium in Honor of Professor Alfred W. Crompton,* 621–637. *Bulletin of the Museum of Comparative Zoology* 156, Cambridge.

Luo, Z.-X., and Crompton, A. W. 1994. Transformation of the quadrate (incus) through the transition from non-mammalian cynodonts to mammals. *Journal of Vertebrate Paleontology* 14: 341–374.

Luo, Z.-X., and Gingerich, P. D. 1999. Transition from terrestrial ungulates to aquatic whales: transformation of the basicranium and evolution of hearing. *Museum of Paleontology, University of Michigan, Papers on Paleontology* 31: 1–98.

Luo, Z.-X., and Ketten, D. R. 1991. CT scanning and computerized reconstructions of the inner ear of multituberculate mammals. *Journal of Vertebrate Paleontology* 11: 220–228.

Luo, Z.-X., and Wu, X.-C. 1994. The small tetrapods of the lower Lufeng Formation, Yunnan, China. *In* N. C. Fraser and H.-D. Sues (eds.), *In the Shadow of the Dinosaurs—Early Mesozoic Tetrapods,* 251–270. Cambridge University Press, Cambridge.

Luo, Z.-X., and Wu, X.-C. 1995. Correlation of vertebrate assemblage of the lower Lufeng Formation, Yunnan, China. *In* A.-L. Sun and Y.-Q. Wang (eds.), *Sixth Symposium on Mesozoic Terrestrial Ecosystems,* 83–85. China Ocean Press, Beijing.

Luo, Z.-X., Crompton, A. W., and Lucas, S. G. 1995. Evolutionary origins of the mammalian promontorium and cochlea. *Journal of Vertebrate Paleontology* 15: 113–121.

Luo, Z.-X., Cifelli, R. L., and Kielan-Jaworowska, Z. 2001. Dual origin of tribosphenic mammals. *Nature* 409: 53–57.

Luo, Z.-X., Crompton, A. W., and Sun, A.-L. 2001. A new mammal from the Early Jurassic and evolution of mammalian characteristics. *Science* 292: 1535–1540.

Luo, Z.-X., Ji, Q., and Ji, S. A. 2001. New evidence on dental replacement in symmetrodonts and its implications for mammalian evolution. *Journal of Morphology* 248: 256.

Luo, Z.-X., Kielan-Jaworowska, Z., and Cifelli, R. L. 2002. In quest for a phylogeny of Mesozoic mammals. *Acta Palaeontologica Polonica* 47: 1–78.

Luo, Z.-X., Kielan-Jaworowska, Z., and Cifelli, R. L. in press. Evolution of dental replacement in mammals. *In* M. R. Dawson and J. A. Lillegraven (eds.), *Fanfare for an Uncommon Paleontologist—Festschrift in Honor of Dr. Malcolm C. McKenna.* Bulletin of the Carnegie Museum of Natural History, New York.

Lydekker, R. 1887. *Catalogue of the Fossil Mammalia in the British Museum. Part V. Containing the Group Tillodontia, the Orders Sirenia, Cetacea, Edentata, Marsupialia, Monotremata, and Supplement.* 345 pp. British Museum (Natural History), London.

Lydekker, R. 1893. On the mammalian incisor from the Wealden of Hastings. *Quarterly Journal of the Geological Society of London* 49: 281–283.

MacDonald, D. (ed.) 2002. *The Encyclopedia of Mammals.* 930 pp. Barnes and Noble, New York.

MacIntyre, G. T. 1967. Foramen pseudovale and quasi-mammals. *Evolution* 21: 834–841.

MacIntyre, G. T. 1972. The trisulcate petrosal pattern of mammals. *In* T. Dobzhansky, M. K. Hecht, and W. C. Steere (eds.), *Evolutionary Biology, Vol. 6,* 275–303. Appleton-Century-Crofts, New York.

MacLean, P. D. 1986. Neurobehavioral significance of the mammal-like reptiles (therapsids). *In* N. Hotton, J. J. Roth, and E. C. Roth (eds.), *The Ecology and Biology of Mammal-like Reptiles,* 1–21. Smithsonian Institution Press, Washington, D. C.

MacLeod, N., and Rose, K. D. 1993. Inferring locomotor behavior in Paleogene mammals via eigenshape analysis. *American Journal of Science* 293: 300–355.

MacPhee, R. D. E. 1981. Auditory regions of primates and eutherian insectivores: morphology, ontogeny and character analysis. *Contributions to Primatology* 18: 1–282.

MacPhee, R. D. E., and Cartmill, M. 1986. Basicranial structures and primate systematics. *In* D. W. Swindler and J. Erwin (eds.), *Comparative Primate Biology, Vol. 1: Systematics, Evolution, and Anatomy,* 219–275. Alan R. Liss, Inc., New York.

MacPhee, R. D. E., and Novacek, M. J. 1993. Definition and relationships of Lipotyphla. *In* F. S. Szalay, M. J. Novacek, and M. C. McKenna (eds.), *Mammal Phylogeny: Placentals,* 13–31. Springer-Verlag, New York.

Madsen, O., Scally, M., Douady, C. J., Kao, D. J., DeBry, R. W., Adkins, R., Amrine, H. M., Stanhope, M. J., de Jong, W. W., and Springer, M. S. 2001. Parallel adaptive radiations in two major clades of placental mammals. *Nature* 409: 610–614.

Mahoney, J. A., and Ride, W. D. L. 1975. Index to the genera and species of fossil Mammalia described from Australia and New Guinea between 1818 and 1969. *Western Australian Museum Special Publication* 6: 1–247.

Maier, W. 1987. The ontogenetic development of the orbitotemporal region in the skull of *Monodelphis domestica* (Didelphidae, Marsupialia) and the problem of the mammalian alisphenoid. *In* H.-J. Kuhn and U. Zeller (eds.), *Mammalia Depicta, Heft 13, Beihefte zur Zeitschrift für Säugetierekunde,* 71–90. Paul Parey, Hamburg/Berlin.

Maier, W. 1989. Ala temporalis and alisphenoid in therian mammals. *In* H. Splechtna and H. Hilgers (eds.), *Trends in Vertebrate Morphology,* 396–400. Fischer, Stuttgart.

Maier, W. 1990. Phylogeny and ontogeny of mammalian middle ear structures. *Netherlands Journal of Zoology* 40: 55–75.

Maier, W. 1993. Cranial morphology of the therian common ancestor, as suggested by the adaptations of neonate marsupials. *In* F. S. Szalay, M. J. Novacek, and M. C. McKenna (eds.), *Mammal Phylogeny: Mesozoic Differentiation, Multituberculates, Monotremes, Early Therians, and Marsupials,* 165–181. Springer-Verlag, New York.

Maier, W. 1999. On the evolutionary biology of early mammals—with methodological remarks on the interaction between ontogenetic adaptation and phylogenetic transformation. *Zoologischer Anzeiger* 238: 55–74.

Maier, W., Den Heever, J., van, and Durand, J. 1996. New therapsid specimens and the origin of the secondary hard and soft palate of mammals. *Journal of Zoological, Systematic and Evolutionary Research* 34: 9–19.

Marsh, O. C. 1878. Fossil mammal from the Jurassic of the Rocky Mountains. *American Journal of Science* 15: 459.

Marsh, O. C. 1879a. Additional remains of Jurassic mammals. *American Journal of Science* 18: 215–216.

Marsh, O. C. 1879b. Notice of new Jurassic mammals. *American Journal of Science* 20: 396–398.

Marsh, O. C. 1879c. Notice of a new Jurassic mammal. *American Journal of Science* 18: 1–2.

Marsh, O. C. 1880. Notice on Jurassic mammals representing two new orders. *American Journal of Science* 20: 235–239.

Marsh, O. C. 1881. Notice of new Jurassic mammals. *American Journal of Science* 21: 511–513.

Marsh, O. C. 1887. American Jurassic mammals. *American Journal of Science* 33: 326–348.

Marsh, O. C. 1889a. Discovery of Cretaceous Mammalia. *American Journal of Science* 38: 81–92.

Marsh, O. C. 1889b. Discovery of Cretaceous Mammalia. Part II. *American Journal of Science* 38: 177–180.

Marsh, O. C. 1890. *Additional Genera Established by Professor O. C. Marsh, 1880–89.* 15 pp. Privately printed and distributed, New Haven.

Marsh, O. C. 1892. Discovery of Cretaceous Mammalia. Part III. *American Journal of Science* 43: 249–262.

Marshall, L. G. 1977. Cladistic analysis of borhyaenoid, dasyuroid, didelphoid, and thylacinid (Marsupialia: Mammalia) affinities. *Systematic Zoology* 26: 410–425.

Marshall, L. G. 1980. Marsupial paleobiogeography. *In* L. L. Jacobs (ed.), *Aspects of Vertebrate History: Essays in Honor of Edwin Harris Colbert,* 345–386. Museum of Northern Arizona Press, Flagstaff.

Marshall, L. G. 1982. Systematics of the South American marsupial family Microbiotheriidae. *Fieldiana: Geology* 10: 1–75.

Marshall, L. G. 1987. Systematics of Itaboraian (middle Paleocene) age "opossum-like" marsupials from the limestone quarry at Sao Jose de Itaboraí, Brazil. *In* M. Archer (ed.), *Possums and Opossums: Studies in Evolution,* 91–160. Surrey Beatty & Sons and the Royal Zoological Society of New South Wales, Sydney.

Marshall, L. G., and Kielan-Jaworowska, Z. 1992. Relationships of the dog-like marsupials, deltatheroidans and early tribosphenic mammals. *Lethaia* 25: 361–374.

Marshall, L. G., and Muizon, C. de. 1988. The dawn of the age of mammals in South America. *National Geographic Research* 4: 23–55.

Marshall, L. G., and Muizon, C. de. 1995. Part II: The skull. *In* C. de Muizon (ed.), Pucadelphys andinus *(Marsupialia, Mammalia) from the Early Paleocene of Bolivia,* 21–90. *Mémoires du Muséum national d'Histoire naturelle* 165, Paris.

Marshall, L. G., and Sigogneau-Russell, D. 1995. Part III: Postcranial skeleton. *In* C. de Muizon (ed.), *Pucadelphys andinus (Marsupialia, Mammalia) from the Early Paleocene of Bolivia,* 91–164. Mémoires du Muséum national d'Histoire naturelle 165, Paris.

Marshall, L. G., Muizon, C. de, and Sigé, B. 1983. *Perutherium altiplanense,* un notoungulé du Crétacé supérieur du Pérou. *Palaeovertebrata, Montpellier* 13: 145–155.

Marshall, L. G., Case, J. A., and Woodburne, M. O. 1990. Phylogenetic relationships of the families of marsupials. *Current Mammalogy* 2: 433–502.

Martin, A. J. 2001. Probable mammal burrows in the Two Medicine Formation (Late Cretaceous) of Choteau, Montana. *Journal of Vertebrate Paleontology* 21, supplement to no. 3: 78A.

Martin, J. E., and Foster, J. R. 1998. First Jurassic mammals from the Black Hills, northeastern Wyoming. *Modern Geology* 23: 381–392.

Martin, T. 1995. Dryolestidae from the Kimmeridge of the Guimarota coal mine (Portugal) and their implications for dryolestid systematics and phylogeny. *In* A.-L. Sun and Y. Wang (eds.), *Sixth Symposium on Mesozoic Terrestrial Ecosystems and Biota,* 229–231. Ocean Press, Beijing.

Martin, T. 1997. Tooth replacement in Late Jurassic Dryolestidae (Eupantotheria, Mammalia). *Journal of Mammalian Evolution* 4: 1–18.

Martin, T. 1998. The premolars of *Crusafontia cuencana* (Dryolestidae, Mammalia) from the Early Cretaceous (Barremian) of Spain. *Berliner geowissenschaftliche, Abhandlungen E* 28: 119–126.

Martin, T. 1999a. Dryolestidae (Dryolestoidea, Mammalia) aus dem Oberen Jura von Portugal. *Abhandlungen der Senckenbergischen Naturforschenden Gesellschaft* 550: 1–119.

Martin, T. 1999b. The mammal fauna of the Late Jurassic Guimarota ecosystem, Portugal. *In* H. A. Leanza (ed.), *VII International Symposium on Mesozoic Terrestrial Ecosystems,* 43. Asociación Paleontológica Argentina, Buenos Aires.

Martin, T. 2002. New stem-line representatives of Zatheria (Mammalia) from the Late Jurassic of Portugal. *Journal of Vertebrate Paleontology* 22: 332–348.

Martin, T., and Averianov, A. O. 2001. Phylogenetic integrity of Asiatic docodonts. *Journal of Vertebrate Paleontology* 21, supplement to no. 3: 78A.

Martin, T., and Krebs, B. (eds.). 2000. *Guimarota: A Jurassic Ecosystem.* 155 pp. Verlag Dr. Friedrich Pfeil, Munich.

Martin, T., and Nowotny, M. 2000. The docodont *Haldanodon* from the Guimarota Mine. *In* T. Martin and B. Krebs (eds.), *Guimarota: A Jurassic Ecosystem,* 91–96. Verlag Dr. Friedrich Pfeil, Munich.

Martinson, G. G. 1982. The Upper Cretaceous molluscs of Mongolia. *Trudy Sovmestnoi Sovetsko-Mongol'skoi Paleontologicheskoi Ekspeditsii* 17: 1–84 (in Russian).

Maschenko, E. N., and Lopatin, A. V. 1998. First record of an Early Cretaceous triconodont mammal in Siberia. *Bulletin de l'Institut royal des Sciences naturelles de Belgique, Sciences de la Terre* 68: 233–236.

Matthew, W. D. 1904. The arboreal ancestry of the Mammalia. *American Naturalist* 38: 811–818.

Matthew, W. D. 1909. The Carnivora and Insectivora of the Bridger Basin, middle Eocene. *Memoirs of the American Museum of Natural History* 9: 289–567.

Matthew, W. D. 1916. A marsupial of the Belly River Cretaceous. *Bulletin of the American Museum of Natural History* 35: 477–500.

Matthew, W. D. 1928. The evolution of mammals in the Eocene. *Proceedings of the Zoological Society of London* 1928: 947–985.

Matthew, W. D. 1937. Paleocene fauna of the San Juan Basin, New Mexico. *Transactions of the American Philosophical Society, New Series* 30: 1–510.

Matthew, W. D., and Granger, W. 1921. New genera of Paleocene mammals. *American Museum Novitates* 13: 1–7.

Matthew, W. D., and Granger, W. 1925. Fauna and correlation of the Gashato Formation of Mongolia. *American Museum Novitates* 189: 1–12.

Matthew, W. D., Granger, W., and Simpson, G. G. 1928. Paleocene multituberculates from Mongolia. *American Museum Novitates* 331: 1–4.

Matthew, W. D., Granger, W., and Simpson, G. G. 1929. Additions to the fauna of the Gashato Formation of Mongolia. *American Museum Novitates* 376: 1–12.

Mayer, H. von. 1832. *Palaeologica, zur Geschichte der Erde und ihrer Geschöpfe.* 560 pp. Schmerber, Frankfurt a/M.

Maynard Smith, J., and Savage, R. J. G. 1959. The mechanics of mammalian jaws. *School Science Review* 141: 289–301.

Mayr, E., and Ashlock, P. D. 1991. *Principles of Systematic Zoology, 2nd Edition,* xx + 475 pp. McGraw-Hill Book Co., New York.

McDowell, S. B. 1958. The Greater Antillean insectivores. *Bulletin of the American Museum of Natural History* 115: 113–214.

McKenna, M. C. 1961. On the shoulder girdle of the mammalian subclass Allotheria. *American Museum Novitates* 2066: 1–27.

McKenna, M. C. 1966. Paleontology and the origin of the Primates. *Folia Primatologica* 4: 1–25.

McKenna, M. C. 1969. The origin and early differentiation of therian mammals. *Annals of the New York Academy of Sciences* 167: 217–240.

McKenna, M. C. 1971. Multituberculata. *McGraw-Hill Encyclopedia of Science and Technology,* 736–737. McGraw-Hill Book Co., New York.

McKenna, M. C. 1975. Toward a phylogenetic classification of the Mammalia. *In* W. P. Luckett and F. S. Szalay (eds.), *Phylogeny of the Primates,* 21–46. Plenum Press, New York.

McKenna, M. C. 1987. Molecular and morphological analysis of high-level mammalian interrelationships. *In* C. Patterson (ed.), *Molecules and Morphology in Evolution: Conflict or Compromise?,* 55–95. Cambridge University Press, Cambridge.

McKenna, M. C. 1996. The multituberculate alloclavicle is not homologous with the interclavicle of monotremes. *Journal of Vertebrate Paleontology* 16, supplement to no. 3: 52A.

McKenna, M. C., and Bell, S. K. 1997. *Classification of Mammals Above the Species Level.* 631 pp. Columbia University Press, New York.

McKenna, M. C., Mellett, J. S., and Szalay, F. S. 1971. Relationships of the Cretaceous mammal *Deltatheridium. Journal of Paleontology* 45: 441–442.

McKenna, M. C., Kielan-Jaworowska, Z., and Meng, J. 2000. Earliest eutherian mammal skull from the Late Cretaceous (Coniacian) of Uzbekistan. *Acta Palaeontologica Polonica* 45: 1–54.

McLean, J. 1971. Stratigraphy of the Upper Cretaceous Judith River Formation in the Canadian Great Plains. *Saskatchewan Research Council, Geological Division Reports* 11: 1–96.

McNulty, C. J., and Slaughter, B. 1968. Stratigraphy of the Woodbine Formation, Tarrant County, Texas (locality 3). *Field Trip Guidebook, Second Annual Meeting, South-Central Section of the Geological Society of America,* 68–72.

Mehling, C. M. 2001. Additions to the Upper Cretaceous vertebrate fauna of Ramanessin Brook, New Jersey. *Journal of Vertebrate Paleontology* 21, supplement to no. 3: 80A.

Mellett, J. S. 1974. Scatological origin of microvertebrate fossil accumulations. *Science* 185: 349–350.

Meng, J. 1992. The stapes of *Lambdopsalis bulla* (Multituberculata) and transformational analyses on some stapedial features in Mammaliaformes. *Journal of Vertebrate Paleontology* 12: 459–471.

Meng, J., and Fox, R. C. 1995a. Evolution of the inner ear from non-therians to therians during the Mesozoic: implications for mammalian phylogeny and hearing. *In* A.-L. Sun and Y. Wang (eds.), *Sixth Symposium on Mesozoic Terrestrial Ecosystems and Biota,* 235–242. China Ocean Press, Beijing.

Meng, J., and Fox, R. C. 1995b. Osseous inner ear structures and hearing in early marsupials and placentals. *Zoological Journal of the Linnean Society* 115: 47–71.

Meng, J., and Fox, R. C. 1995c. Therian petrosals from the Oldman and Milk River formations (Late Cretaceous), Alberta, Canada. *Journal of Vertebrate Paleontology* 15: 122–130.

Meng, J., and Miao, D. 1992. The breast-shoulder apparatus of *Lambdopsalis bulla* (Multituberculata) and its systematic and functional implications. *Journal of Vertebrate Paleontology* 12 supplement to no. 3: 43A.

Meng, J., and Wyss, A. R. 1995. Monotreme affinities and low-frequency hearing suggested by multituberculate ear. *Nature* 377: 141–144.

Meng, J., and Wyss, A. R. 1997. Multituberculate and other mammal hair recovered from Palaeogene excreta. *Nature* 385: 712–714.

Meng, J., and Wyss, A. R. 2001. The morphology of *Tribosphenomys* (Rodentiaformes, Mammalia): phylogenetic implications for basal Glires. *Journal of Mammalian Evolution* 8: 1–72.

Messer, M., Weiss, A. S., Shaw, D. C., and Westerman, M. 1998. Evolution of the monotremes: phylogenetic relationship to marsupials and eutherians, and estimation of divergence dates based on α-lactalbumin amino acid sequences. *Journal of Mammalian Evolution* 5: 95–105.

Miao, D. 1986. Dental anatomy and ontogeny of *Lambdopsalis bulla* (Mammalia, Multituberculata). *Contributions to Geology, University of Wyoming* 24: 65–76.

Miao, D. 1988. Skull morphology of *Lambdopsalis bulla* (Mammalia, Multituberculata). *Contributions to Geology, University of Wyoming, Special Paper* 4: 1–104.

Miao, D. 1991. On the origin of mammals. *In* H.-P. Schultze and L. Trueb (eds.), *Origins of Major Groups of Tetrapods: Controversies and Consensus,* 579–597. Cornell University Press, Ithaca.

Miao, D. 1993. Cranial morphology and multituberculate relationships. *In* F. S. Szalay, M. J. Novacek, and M. C. McKenna (eds.), *Mammal Phylogeny: Mesozoic Differentiation, Multituberculates, Monotremes, Early Therians, and Marsupials,* 63–74. Springer-Verlag, New York.

Miao, D., and Lillegraven, J. A. 1986. Discovery of three ear ossicles in a multituberculate mammal. *National Geographic Research* 2: 500–507.

Middleton, M. D. 1982. A new species and additional material of *Catopsalis* (Mammalia, Multituberculata) from the Western Interior of North America. *Journal of Paleontology* 56: 1197–1206.

Middleton, M. D. 1983. *Early Paleocene vertebrates of the Denver Basin, Colorado.* 404 pp. Ph.D. dissertation, University of Colorado, Boulder.

Mills, J. R. E. 1964. The dentitions of *Peramus* and *Amphitherium. Proceedings of the Linnean Society of London* 175: 117–133.

Mills, J. R. E. 1966. The functional occlusion of the teeth of Insectivora. *Journal of the Linnean Society (Zoology)* 47: 1–25.

Mills, J. R. E. 1971. The dentition of *Morganucodon*. *In* D. M. Kermack and K. A. Kermack (eds.), *Early Mammals*, 29–63. *Zoological Journal of the Linnean Society* 50, supplement 1, London.

Mills, J. R. E. 1984. The molar dentition of a Welsh pantothere. *Zoological Journal of the Linnean Society* 82: 189–205.

Minjin, B. 2001. Evidence from the hand of *Kryptobaatar dashzevegi* for the phylogenetic position of Multituberculata. *Journal of Vertebrate Paleontology* 21(3 Supplement): 80A.

Miyamoto, M., and Fitch, W. M. 1995. Testing species phylogenies and phylogenetic methods with congruence. *Systematic Biology* 44: 64–76.

Mohr, B. 1989. New palynological information on the age and environment of Late Jurassic and Early Cretaceous vertebrate localities of the Iberian Peninsula (eastern Spain and Portugal). *Berliner geowissenschaftliche, Abhandlungen A* 106: 291–301.

Mones, A. 1987. Gondwanatheria, un nuevo orden de mamíferos sudamericanos (Mammalia: Edentata: ?Xenarthra). *Comunicaciones Paleontológicas del Museo de Historia Natural de Montevideo* 1: 237–240.

Montellano, M. 1988. *Alphadon halleyi* (Didelphidae, Marsupialia) from the Two Medicine Formation (Late Cretaceous, Judithian) of Montana. *Journal of Vertebrate Paleontology* 8: 378–382.

Montellano, M. 1992. Mammalian fauna of the Judith River Formation (Late Cretaceous, Judithian), northcentral Montana. *University of California Publications in Geological Sciences* 136: 1–115.

Montellano, M., Hopson, J. A., Clark, J. M., Fastovsky, D. E., and Hernández, R. 1998. Panorama de los mamíferos Mesozóicos en México. *Avances en Investigación, Paleontología de Vertebrados, Universidad Autónoma del Estado de Hidalgo, Publicación Especial* 1: 12–17.

Montellano, M., Weil, A., and Clemens, W. A. 2000. An exceptional specimen of *Cimexomys judithae* (Mammalia: Multituberculata) from the Campanian Two Medicine Formation of Montana, and the phylogenetic status of *Cimexomys*. *Journal of Vertebrate Paleontology* 20: 333–340.

Moore, C. 1867. On the abnormal condition of Secondary deposits. *Quarterly Journal of the Geological Society of London* 23: 449–568.

Moore, W. J. 1981. *The Mammalian Skull*. 369 pp. Cambridge University Press, Cambridge.

Morris, W. J. 1981. A new species of hadrosaurian dinosaur from the Upper Cretaceous of Baja California—?*Lambeosaurus laticaudus*. *Journal of Paleontology* 55: 453–462.

Moss, M. L. 1969. Evolution of mammalian dental enamel. *American Museum Novitates* 2360: 1–39.

Mourier, T., Jaillard, E., Laubacher, G., Noblet, C., Pardo, A., Sigé, B., and Taquet, P. 1986. Découverte de restes dinosauriens et mammalien d'âge Crétacé supérieur à la base des couches rouges du synclinal de Bagua (Andes nord-péruviennes): aspects stratigraphiques, sédimentologiques et paléogeographiques concernant la régression fini-crétacée. *Bulletin de la Société géologique de France* 2: 171–175.

Mourier, T., Bengston, P., Bonhomme, M., Buge, E., Cappetta, H., Crochet, J.-Y., Feist, M., Hirsch, K. F., Jaillard, E., Laubacher, G., LeFranc, J. P., Moullade, M., Noblet, C., Pons, D., Rey, J., Sigé, B., Tambareau, Y., and Taquet, P. 1988. The Upper Cretaceous–Lower Tertiary marine to continental transition in the Bagua Basin, northern Peru. *Newsletters on Stratigraphy* 19: 143–177.

Muizon, C. de. 1991. La fauna de mamíferos de Tiupampa (Paleoceno inferior de la Formación Santa Lucía), Bolivia. *In* R. Suarez-Soruco (ed.), *Fósiles y Fácies de Bolivia*, 575–624. Yacimiento Petrolíferos Fiscales Bolivianos, Santa Cruz.

Muizon, C. de. 1994. A new carnivorous marsupial from the Palaeocene of Bolivia and the problem of marsupial monophyly. *Nature* 370: 208–211.

Muizon, C. de. 1998. *Mayulestes ferox*, a borhyaenoid (Metatheria, Mammalia) from the early Palaeocene of Bolivia. Phylogenetic and palaeobiologic implications. *Geodiversitas* 20: 19–142.

Muizon, C. de, and Argot, C. 2003. Comparative anatomy of the didelphimorph marsupials from the early Palaeocene of Bolivia (*Pucadelphys, Andinodelphys*, and *Mayulestes*). *In* M. Jones, C. Dickman, and M. Archer (eds.), *Predators with Pouches: The Biology of the Carnivorous Marsupials*, 42–62. Surrey Beatty & Sons, Sydney.

Muizon, C. de, and Cifelli, R. L. 2000. The "condylarths" (archaic Ungulata, Mammalia) from the early Paleocene of Tiupampa (Bolivia): implications on the origin of the South American ungulates. *Geodiversitas* 22: 47–150.

Muizon, C. de, and Cifelli, R. L. 2001. A new basal "didelphoid" (Marsupialia, Mammalia) from the early Paleocene of Tiupampa (Bolivia). *Journal of Vertebrate Paleontology* 21: 87–97.

Muizon, C. de, and Lange-Badré, B. 1997. Carnivorous dental adaptations in tribosphenic mammals and phylogenetic reconstruction. *Lethaia* 30: 353–366.

Muizon, C. de, and Marshall, L. G. 1992. *Alcidedorbignya inopinata* (Mammalia: Pantodonta) from the early Paleocene of Bolivia: phylogenetic and paleobiogeographic implications. *Journal of Paleontology* 66: 499–520.

Muizon, C. de, Cifelli, R. L., and Céspedes, R. 1997. The origin of dog-like marsupials and the early evolution of Gondwanian marsupials. *Nature* 389: 486–489.

Murphy, E. C., Nichols, D. J., Hoganson, J. W., and Forsman, N. F. 1995. The Cretaceous/Tertiary boundary in south-central North Dakota. *North Dakota Geological Survey, Report of Investigation* 98: 1–74.

Murphy, W. J., Eizirik, E., Johnson, W. E., Zhang, Y. P., Ryder, O. A., and O'Brien, S. J. 2001. Molecular phylogenetics and the origins of placental mammals. *Nature* 409: 614–618.

Murphy, W. J., Eizirik, E., O'Brien, S. J., Madsen, O., Scally, M., Douday, C. J., Teeling, E., Ryder, O. A., Stanhope, M. J., de Jong, W. W., and Springer, M. S. 2001. Resolution of the early placental mammal radiation using Bayesian phylogenetics. *Science* 294: 2348–2351.

Murray, P. 1978a. A Pleistocene spiny anteater from Tasmania (Monotremata: Tachyglossidae, *Zaglossus*). *Papers and Proceedings of the Royal Society of Tasmania* 112: 39–68.

Murray, P. 1978b. Late Cenozoic monotreme anteaters. *Australian Zoologist* 20: 29–55.

Musser, A. M., and Archer, M. 1998. New information about the skull and dentary of the Miocene platypus *Obdurodon dicksoni*, and a discussion of ornithorhynchid relationships.

Philosophical Transactions of the Royal Society of London, Biological Sciences, Series B, 353: 1063–1079.

Napier, J. R. 1961. Prehensibility and opposability in the hands of primates. *Symposia of the Zoological Society of London* 5: 115–132.

Negus, V. 1958. *The Comparative Anatomy and Physiology of the Nose and Paranasal Sinuses.* 402 pp. E. and S. Livingstone Ltd., Edinburgh.

Nelson, G. J., and Platnick, N. 1981. *Systematics and biogeography: cladistics and vicariance.* 567 pp. Columbia University Press, New York.

Nelson, M. E., and Crooks, D. M. 1987. Stratigraphy and paleontology of the Cedar Mountain Formation (Lower Cretaceous), eastern Emery County, Utah. *In* W. R. Averett (ed.), *Paleontology and Geology of the Dinosaur Triangle,* 55–63. Museum of Western Colorado, Grand Junction.

Nessov, L. A. 1981. Cretaceous salamanders and frogs of Kyzylkum Desert. *Trudy Zoologicheskogo Instituta Akademii Nauk SSSR* 101: 57–88 (in Russian).

Nessov, L. A. 1982. Ancient mammals of the USSR. *Ezhegodnik Vsesoyuznogo Paleontologicheskogo Obshchestva* 35: 228–243 (in Russian).

Nessov, L. A. 1985a. New mammals from the Cretaceous of Kyzylkum. *Vestnik Leningradskogo Universiteta* 17: 8–18 (in Russian).

Nessov, L. A. 1985b. Rare osteichthyans, terrestrial lizards and mammals from the estuaries and coastal plain zones of the Cretaceous of the Kyzylkum Desert. *Ezhegodnik Vsesoyuznogo Paleontologicheskogo Obshchestva* 28: 199–219 (in Russian).

Nessov, L. A. 1987. Results of search and study of Cretaceous and early Paleogene mammals on the territory of SSSR. *Ezhegodnik Vsesoyuznogo Paleontologicheskogo Obshchestva* 30: 119–218 (in Russian).

Nessov, L. A. 1989. Mammals of the first half of the Late Cretaceous of Asia. *Operativno-informacionnye materialy k I Vsesoyuznomu soveshchaniyu po paleoteriologii:* 45–47 (in Russian).

Nessov, L. A. 1992. Mesozoic and Paleogene birds of the USSR and their paleoenvironments. *In* K. E. Campbell (ed.), *Papers in Avian Paleontology Honoring Pierce Brodkorb,* 465–478. Los Angeles County Museum, Los Angeles.

Nessov, L. A. 1993. New Mesozoic mammals of middle Asia and Kazakhstan, and comments about evolution of theriofaunas of Cretaceous coastal plains of ancient Asia. *Trudy Zoologicheskogo Instituta RAN* 249: 105–133 (in Russian).

Nessov, L. A. 1997. *Cretaceous Non-marine Vertebrates of Northern Eurasia.* 218 pp. (Posthumous paper, edited by L. B. Golovneva and A. O. Averianov.) University of Saint Petersburg, Institute of Earth Crust, Saint Petersburg (in Russian).

Nessov, L. A., and Khisarova, G. D. 1988. New data about Upper Cretaceous vertebrates from Shakh-Shakh and Baibolat localities (northeastern Aral region). *Materialy po istorii fauny i flory Kazakhstana* 10: 5–14 (in Russian).

Nessov, L. A., and Kielan-Jaworowska, Z. 1991. Evolution of the Cretaceous Asian therian mammals. *In* Z. Kielan-Jaworowska, N. Heintz, and H.-A. Nakrem (eds.), *Fifth Symposium on Mesozoic Terrestrial Ecosystems and Biota: Extended Abstracts,* 51–52. *Contributions from the Paleontological Museum, University of Oslo* 364, Oslo.

Nessov, L. A., and Trofimov, B. A. 1979. Ancient insectivores from the Cretaceous of Uzbek SSSR. *Doklady Akademii Nauk SSSR* 247: 952–955 (in Russian).

Nessov, L. A., Kaznyshkina, L. F., and Cherepanov, G. O. 1989. Ceratopsian dinosaurs and crocodiles of the Mesozoic of middle Asia. *In* T. N. Bogdanova and L. I. Khosatsky (eds.), *Teoreticheskie i prikladnye aspekty sovremennoi paleontologii,* 144–154. Nauka, Leningrad (in Russian).

Nessov, L. A., Kielan-Jaworowska, Z., Hurum, J. H., Averianov, A. O., Federov, P. V., Potapov, D. O., and Frøyland, M. 1994. First Jurassic mammals from Kyrgystan. *Acta Palaeontologica Polonica* 39: 315–326.

Nessov, L. A., Sigogneau-Russell, D., and Russell, D. E. 1994. A survey of Cretaceous tribosphenic mammals from middle Asia (Uzbekistan, Kazakhstan and Tajikistan), of their geological setting, age and faunal environment. *Palaeovertebrata* 23: 51–92.

Nessov, L. A., Archibald, J. D., and Kielan-Jaworowska, Z. 1998. Ungulate-like mammals from the Cretaceous of Uzbekistan and a phylogenetic analysis of Ungulatomorpha. *Bulletin of the Carnegie Museum of Natural History* 34: 40–88.

Nessov, L. A., Zhegallo, V. I., and Averianov, A. O. 1998. A new locality of Late Cretaceous snakes, mammals and other vertebrates in Africa (western Libya). *Annales de Paléontologie* 84: 265–274.

Netter, Frank H. 1983. *The CIBA Collection of Medical Illustrations, Vol. 1, Nervous System, Part I, Anatomy and Physiology.* xvi + 239 pp. CIBA Pharmaceutical Company, Summit, N.J.

Noblet, C., Leonardi, G., Taquet, P., Marocco, R., and Cordova, E. 1995. Nouvelle découverte d'empreintes laissées par des Dinosaures dans la Formation des Couches Rouges (bassin de Cuxco-Sicuani, Sud du Pérou): conséquences stratigraphiques et tectoniques. *Comptes Rendus de l'Académie des Sciences, Paris, sér. IIa,* 320: 785–791.

Norell, M. A. 1993. Tree-based approaches to understanding history: comments on ranks, rules and the quality of the fossil record. *American Journal of Science* 293-A: 407–417.

Novacek, M. J. 1977. A review of Paleocene and Eocene Leptictidae (Eutheria: Mammalia) from North America. *Paleobios* 24: 1–42.

Novacek, M. J. 1986a. The primitive eutherian dental formula. *Journal of Vertebrate Paleontology* 6: 191–196.

Novacek, M. J. 1986b. The skull of leptictid insectivorans and the higher-level classification of eutherian mammals. *Bulletin of the American Museum of Natural History* 183: 1–112.

Novacek, M. J. 1992a. Mammalian phylogeny: shaking the tree. *Nature* 356: 121–125.

Novacek, M. J. 1992b. Fossils, topologies, missing data, and the higher level phylogeny of eutherian mammals. *Systematic Biology* 41: 58–73.

Novacek, M. J. 1999. 100 million years of land vertebrate evolution: the Cretaceous–Early Tertiary transition. *Annals of Missouri Botanical Garden* 86: 230–258.

Novacek, M. J., and Clemens, W. A. 1977. Aspects of intrageneric variation and evolution of *Mesodma* (Multituberculata, Mammalia). *Journal of Paleontology* 51: 701–717.

Novacek, M. J., and Wyss, A. R. 1986. Higher-level relationships of the Recent eutherian orders: morphological evidence. *Cladistics* 2: 257–287.

Novacek, M. J., Bown, T. M., and Schankler, D. 1985. On the classification of the early Tertiary Erinaceomorpha (Insectivora, Mammalia). *American Museum Novitates* 2813: 1–22.

Novacek, M. J., Wyss, A. R., and McKenna, M. C. 1988. The major groups of eutherian mammals. *In* M. J. Benton (ed.), *The Phylogeny and Classification of the Tetrapods. Vol. 2: Mammals,* 31–72. The Systematics Association, Special Volume No. 35B, Oxford University Press, Oxford.

Novacek, M. J., Norell, M. J., McKenna, M. C., and Clark, J. M. 1994. Fossils of the Flaming Cliffs. *Scientific American* December: 60–69.

Novacek, M. J., Rougier, G. W., Wible, J. R., McKenna, M. C., Dashzeveg, D., and Horovitz, I. 1997. Epipubic bones in eutherian mammals from the Late Cretaceous of Mongolia. *Nature* 389: 483–486.

Novacek, M. J., Rougier, G. W., Dashzeveg, D., and McKenna, M. C. 2000. New eutherian mammal from the Late Cretaceous of Mongolia and its bearing on the origins of the modern placental radiation. *Journal of Vertebrate Paleontology* 20, supplement to no. 3: 61A.

Novikov, I. V., Lebedev, O. A., and Alifanov, V. R. 1998. New Mesozoic vertebrate fossil sites of Russia. *In* J. W. M. Jagt, P. H. Lambers, E. W. A. Mulder, and A. S. Schulp (eds.), *Third European Workshop on Vertebrate Palaeontology,* 58. Natuurhistorisch Museum Maastricht, Maastricht.

Nowak, R. M. 1991. *Walker's Mammals of the World, 2 vols.* 1629 pp. Johns Hopkins University Press, Baltimore.

Nowotny, M., Martin, T., and Fischer, M. 2001. Dental anatomy and tooth replacement of *Haldanodon exspectatus* (Docodonta, Mammalia) from the Upper Jurassic of Portugal. *Journal of Morphology* 248: 268.

Nummela, S. 1995. Scaling of the mammalian middle ear. *Hearing Research* 85: 18–30.

Nummela, S. 1997. Scaling and modeling of the mammalian middle ear. *Theoretical Biology* 4: 387–412.

O'Leary, M. A., and Geisler, J. H. 1999. The position of Cetacea within Mammalia: analysis of morphological data from extinct and extant taxa. *Systematic Biology* 48: 455–490.

Olsen, P. E. 1980. Fossil great lakes of the Newark Supergroup in New Jersey. *In* W. Manspeizer (ed.), *Field Studies of New Jersey Geology and Guide to Field Trips,* 352–398. New York State Geological Association, New York.

Olsen, P. E., and Galton, P. M. 1984. A review of the reptile and amphibian assemblages from the Stormberg of southern Africa with special emphasis on the footprints and the age of the Stormberg. *Palaeontologia Africana* 25: 87–110.

Olson, E. C. 1944. Origin of mammals based upon cranial morphology of the therapsid suborders. *Geological Society of America,* Special Paper 55: 1–136.

Olson, E. C. 1959. The evolution of mammalian characters. *Evolution* 13: 344–353.

Oron, U., and Crompton, A. W. 1985. A cineradiographic and electromyographic study of mastication in *Tenrec ecaudatus.* *Journal of Morphology* 185: 155–182.

Osborn, H. F. 1886a. A new mammal from the American Triassic. *Science* 8: 540.

Osborn, H. F. 1886b. Observations on the Upper Triassic mammals, *Dromatherium* and *Microconodon.* *Proceedings of the Academy of Natural Sciences, Philadelphia* 37: 359–363.

Osborn, H. F. 1887a. On the structure and classification of the Mesozoic Mammalia. *Proceedings of the Academy of Natural Sciences, Philadelphia* 38: 282–292.

Osborn, H. F. 1887b. The Triassic mammals *Dromatherium* and *Microconodon.* *Proceedings of the American Philosophical Society* 24: 109–111.

Osborn, H. F. 1888a. Additional observations upon the structure and classification of the Mesozoic Mammalia. *Proceedings of the Academy of Natural Sciences, Philadelphia* 1888: 292–301.

Osborn, H. F. 1888b. On the structure and classification of the Mesozoic Mammalia. *Journal of the Academy of Natural Sciences, Philadelphia* 9: 186–265.

Osborn, H. F. 1893. Fossil mammals from the Upper Cretaceous beds. *Bulletin of the American Museum of Natural History* 5: 311–330.

Osborn, H. F. 1898. Evolution of the Amblypoda. *Bulletin of the American Museum of Natural History* 10: 169–218.

Osborn, H. F. 1907. *Evolution of Mammalian Molar Teeth.* 250 pp. Macmillan and Company, New York.

Osborn, H. F. 1923. Two Lower Cretaceous dinosaurs from Mongolia. *American Museum Novitates* 95: 1–10.

Osborn, J. W. 1971. The ontogeny of tooth succession in *Lacerta vivipara* Jacquine (1787). *Proceedings of the Royal Society, London* B179: 261–289.

Osborn, J. W. 1973. The evolution of dentitions. *American Scientist* 61: 548–559.

Osborn, J. W. 1974a. On the control of tooth replacement in reptiles and its relationship to growth. *Journal of Theoretical Biology* 46: 509–527.

Osborn, J. W. 1974b. On the tooth succession in *Diademodon.* *Evolution* 28: 141–157.

Osborn, J. W., and Crompton, A. W. 1971. The evolution of mammalian from reptilian dentitions. *Breviora* 399: 1–18.

Osborn, J. W., and Sita-Lumsden, A. 1978. An alternative to "thegosis" and a re-examination of the ways in which mammalian molars work. *Neues Jahrbuch für Geologie und Paläontologie* 156: 371–392.

Osmólska, H. 1980. The Late Cretaceous vertebrate assemblages of the Gobi Desert, Mongolia. *Mémoires de la Société géologique de France* 59: 145–150.

Ostrom, J. H. 1970. Stratigraphy and paleontology of the Cloverly Formation (Lower Cretaceous) of the Bighorn Basin area, Montana and Wyoming. *Peabody Museum of Natural History Bulletin* 35: 1–234.

Owen, R. 1838. On the jaws of the *Thylacotherium prevostii* (Valenciennes) from Stonesfield. *Proceedings of the Geological Society of London* 3: 5–9.

Owen, R. 1842. Observations on the fossils representing the *Thylacotherium prevostii* (Valenciennes), with reference to the doubts of its mammalian and marsupial nature recently promulgated; and on the *Phascolotherium bucklandi. Transactions of the Geological Society of London* 2(6): 47–65.

Owen, R. 1845. *Odontography; or a Treatise on the Comparative Anatomy of the Teeth; Their Physiological Relations, Mode of Development, and Microscopic Structure, in the Vertebrate Animals. Part 3,* 289–655. Hippolyte Ballière, London.

Owen, R. 1846. *History of British Fossil Mammals and Birds.* 560 pp. John Van Voorst, London.

Owen, R. 1854. On some fossil reptilian and mammalian remains from the Purbecks. *Quarterly Journal of the Geological Society of London* 10: 420–433.

Owen, R. 1859. Palaeontology. *Encyclopaedia Britannica, 8th Edition,* Vol. 17, 91–176. Adam and Black, Edinburgh.

Owen, R. 1866. Description of part of the lower jaw and teeth of a small Oolithic mammal (*Stylodon pusillus* Ow.). *Geological Magazine* 3: 199–201.

Owen, R. 1868. *On the Anatomy of Vertebrates. Vol. III. Mammals.* Longman, Greens and Co., London.

Owen, R. 1871. Monograph of the fossil Mammalia of the Mesozoic formations. *Monograph of the Palaeontological Society* 33 (for 1870): 1–115.

Owen, R. 1884. On the skull and dentition of a Triassic mammal (*Tritylodon longaevus*) from South Africa. *Quarterly Journal of the Geological Society of London* 40: 146–152.

Padian, K. 1989. Presence of the dinosaur *Scelidosaurus* indicates Jurassic age for the Kayenta Formation (Glen Canyon Group, northern Arizona). *Geology* 17: 438–441.

Palmer, A. R., and Geissman, J. 1999. *Geologic Time Scale.* Geological Society of America.

Parker, T. J., and Haswell, W. A. 1897. *A Text-book of Zoology, Vol. 2.* 683 pp. Macmillan and Company, London.

Parrington, F. R. 1936. On the tooth replacement in theriodont reptiles. *Philosophical Transactions of the Royal Society of London* B226: 121–142.

Parrington, F. R. 1941. On two mammalian teeth from the lower Rhaetic of Somerset. *Annals and Magazine of Natural History, Series 11,* 8: 140–144.

Parrington, F. R. 1947. On the collection of Rhaetic mammalian teeth. *Proceedings of the Zoological Society of London (for 1946)* 116: 707–728.

Parrington, F. R. 1967. The origins of mammals. *Advancements in Science, London* 24: 165–173.

Parrington, F. R. 1971. On the Upper Triassic mammals. *Philosophical Transactions of the Royal Society B* 261: 231–272.

Parrington, F. R. 1973. The dentitions of the earliest mammals. *Zoological Journal of the Linnean Society* 52: 85–95.

Parrington, F. R. 1978. A further account of the Triassic mammals. *Philosophical Transactions of the Royal Society of London B* 282: 177–204.

Pascual, R., and Goin, F. J. 2001. Non-tribosphenic Gondwanan mammals, and the alternative attainment of molars with a reversed triangle cusp pattern. *In* H. A. Leanza (ed.), *VII International Symposium on Mesozoic Terrestrial Ecosystems,* 157–162. Asociación Paleontológica Argentina, Publicación Especial 7, Buenos Aires.

Pascual, R., Archer, M., Ortiz-Jaureguizar, E. O., Prado, J. L., Godthelp, H., and Hand, S. J. 1992a. First discovery of monotremes in South America. *Nature* 356: 704–705.

Pascual, R., Archer, M., Ortiz-Jaureguizar, E. O., Prado, J. L., Godthelp, H., and Hand, S. J. 1992b. The first non-Australian monotreme: an early Paleocene South American platypus (Monotremata, Ornithorhynchidae). *In* M. L. Augee (ed.), *Platypus and Echidnas,* 1–14. The Royal Zoological Society of New South Wales, Sydney.

Pascual, R., Goin, F. J., Krause, D. W., Ortiz-Jaureguizar, E., and Carlini, A. A. 1999. The first gnathic remains of *Sudamerica:* implications for gondwanathere relationships. *Journal of Vertebrate Paleontology* 19: 373–382.

Pascual, R., Goin, F. J., González, P., Ardolino, A., and Puerta, P. 2000. A highly derived docodont from the Patagonian Late Cretaceous: evolutionary implications for Gondwanan mammals. *Geodiversitas* 22: 395–414.

Pascual, R., Goin, F. J., Balarino, L., and Udrizar, D. E. 2002. New data on the Paleocene monotreme *Monotrematum sudamericanum,* and the convergent evolution of triangulate molars. *Acta Palaeontologica Polonica* 47: 487–492.

Patnaik, R., Sahni, A., and Prasad, G. V. R. 2001. Tooth enamel microstructure of a Late Cretaceous gondwanatherian mammal from India. *Journal of the Palaeontological Society of India* 46: 15–23.

Patterson, B. 1951. Early Cretaceous mammals from northern Texas. *American Journal of Science* 249: 31–46.

Patterson, B. 1955. A symmetrodont mammal from the Early Cretaceous of northern Texas. *Fieldiana: Zoology* 37: 689–693.

Patterson, B. 1956. Early Cretaceous mammals and the evolution of mammalian molar teeth. *Fieldiana: Geology* 13: 1–105.

Patterson, B., and Olson, E. C. 1961. A triconodontid mammal from the Triassic of Yunnan. *In* G. Vandebroek (ed.), *International Colloquium on the Evolution of Lower and Non-specialized Mammals,* 129–191. Koninklijke Vlaamse Academie voor Wetenschappen, Letteren en Schone Kunsten van Belgie, Brussels.

Paulli, S. 1900. Über die Pneumaticität des Schädel bei den Säugetieren. Eine Morphologische Studie. I. Über die Morphologie des Siebbeins und die Pneumaticität bei den Monotremen und den Marsupialiern. II. Über die Morphologie des Siebbeins und die Pneumaticität bei den Ungulaten und Proboscidien. III. Über die Morphologie des Siebbeins und die Pneumaticität bei den Insectivoren, Hyracoideen, Chiropteren, Carnivoren, Pinnipedien, Edentaten, Rodentiern, Prosimien und Primaten, nebst einer zusammenfassenden übersicht über die Morphologie des Siebbeins und die Pneumaticität des Schädels bei den Säugetieren. *Gegenbaurs Morphologisches Jahrbuch* 28: 147–178, 179–251, 483–564.

Peláez-Campomanes, P., López-Martínez, N., Álvarez-Sierra, M. A., and Daams, R. 2000. The earliest mammal of the European Paleocene: The multituberculate *Hainina*. *Journal of Paleontology* 74: 701–711.

Peng, J. 1997. *Palaeoecology of vertebrate assemblages from the Upper Cretaceous Judith River Group (Campanian) of southeastern Alberta, Canada.* 312 pp. Ph.D. dissertation, University of Calgary, Calgary.

Peng, J., Russell, A. P., and Brinkman, D. B. 2001. Vertebrate microsite assemblages (exclusive of mammals) from the Foremost and Oldman formations of the Judith River Group (Campanian) of southeastern Alberta: an illustrated guide. *Provincial Museum of Alberta, Natural History Occasional Paper* 25: 1–54.

Penny, D., and Hasegawa, M. 1997. The platypus put in its place. *Nature* 387: 549–550.

Penny, D., Hasegawa, M., Waddell, P. J., and Hendy, M. D. 1999. Mammalian evolution: timing and implications from using log-determinant transform for proteins of differing amino acid composition. *Systematic Biology* 48: 76–93.

Peyer, B. 1956. Über Zähne von Haramiyiden, von Triconodontiden und von wahrscheinlich synapsiden Reptilien aus dem Rhät von Hallau. *Schweizerischen Paläontologischen, Abhandlungen* 72: 1–72.

Platel, R. 1979. Brain weight-body weight relationships. *In* C. Gans, R. G. Northcutt, and P. Ulinski (eds.), *Biology of the Reptilia, Vol. 9*, 147–171. Academic Press, London.

Pleininger, W. 1847. *Microlestes antiquus* und *Sargodon tomicus* in der Grenzbreccie von Degerloch. *Jahreshefte der Vereins Vaterländische Naturkunde Württemburg* 3: 164–167.

Poche, F. 1908. Einige notwendige Änderungen in der mammalogischen Nomenklatur. *Zoologischen Annalen* 2: 269–272.

Pol, C., Buscalioni, A. D., Carballiera, J., Francés, V., Martinez, N. L., Marandat, B., Moratalla, J. J., Sanz, J. L., Sigé, B., and Villatte, J. 1992. Reptiles and mammals from the Late Cretaceous new locality Quintanilla del Coco (Burgos Province, Spain). *Neues Jahrbuch für Geologie und Paläontologie, Abhandlungen* 184: 279–314.

Pond, C. M. 1977. The significance of lactation in the evolution of mammals. *Evolution* 31: 177–199.

Prasad, G. V. R., and Godinot, M. 1994. Eutherian tarsals from the Late Cretaceous of India. *Journal of Paleontology* 68: 892–902.

Prasad, G. V. R., and Manhas, B. K. 1997. A new symmetrodont mammal from the Lower Jurassic Kota Formation, Pranhita-Godavari Valley, India. *Géobios* 30: 563–572.

Prasad, G. V. R., and Manhas, B. K. 1999. Discovery of a pseudotribosphenic-like mammal from the Early Cretaceous Kota Formation, peninsular India. *In* H. A. Leanza (ed.), *Abstracts, VII Symposium on Mesozoic Terrestrial Ecosystems* 55:3. Asociación Paleontológica Argentina, Publicación Especial 7, Buenos Aires.

Prasad, G. V. R., and Manhas, B. K. 2002. Triconodont mammals from the Jurassic Kota Formation of India. *Geodiversitas* 24: 445–464.

Prasad, G. V. R., and Sahni, A. 1988. First Cretaceous mammal from India. *Nature* 332: 638–640.

Prasad, G. V. R., Jaeger, J.-J., Sahni, A., Gheerbrant, E., and Khajuria, C. K. 1994. Eutherian mammals from the Upper Cretaceous (Maastrichtian) Intertrappean Beds of Naskal, Andhra Pradesh, India. *Journal of Vertebrate Paleontology* 14: 260–277.

Presley, R. 1979. The primitive course of the internal carotid artery in mammals. *Acta Anatomica* 103: 238–244.

Presley, R. 1980. The braincase in Recent and Mesozoic therapsids. *Mémoires de la Société géologique de France* 139: 159–162.

Presley, R. 1981. Alisphenoid equivalents in placentals, marsupials, monotremes and fossils. *Nature* 294: 668–670.

Presley, R. 1995. Some neglected relatives. *Nature* 377: 104–105.

Presley, R., and Steel, F. L. D. 1976. On the homology of the alisphenoid. *Journal of Anatomy, London* 121: 441–459.

Pridmore, P. A. 1985. Terrestrial locomotion in monotremes (Mammalia: Monotremata). *Journal of Zoology, London* 205: 53–73.

Prothero, D. R. 1981. New Jurassic mammals from Como Bluff, Wyoming, and the interrelationships of non-tribosphenic Theria. *Bulletin of the American Museum of Natural History* 167: 277–326.

Prothero, D. R. 1983. The oldest mammalian petrosals from North America. *Journal of Paleontology* 57: 1040–1046.

Prothero, D. R. 1993. Ungulate phylogeny: molecular vs. morphological evidence. *In* F. S. Szalay, M. J. Novacek, and M. C. McKenna (eds.), *Mammal Phylogeny: Placentals*, 173–181. Springer-Verlag, New York.

Prothero, D. R., and Jensen, J. A. 1983. A mammalian humerus from the Upper Jurassic of Colorado. *Great Basin Naturalist* 43: 551–553.

Prothero, D. R., Manning, E. M., and Fischer, M. 1988. The phylogeny of the ungulates. *In* M. J. Benton (ed.), *The Phylogeny and Classification of the Tetrapods. Vol. 2: Mammals*, 201–234. Clarendon Press, Oxford.

Quay, W. B. 1979. The parietal eye-pineal complex. *In* C. Gans, R. G. Northcutt, and P. Ulinski (eds.), *Biology of the Reptilia, Vol. 9*, 245–406. Academic Press, London.

Queiroz, A. de. 1993. For consensus (sometimes). *Systematic Biology* 42: 368–372.

Queiroz, A. de., Donoghue, M. J., and Kim, J. 1995. Separate versus combined analysis of phylogenetic evidence. *Annual Reviews of Ecology and Systematics* 26: 657–681.

Quiroga, J. C. 1979a. The brain of two mammal-like reptiles (Cynodontia, Therapsida). *Journal für Hirnforschung* 20: 341–350.

Quiroga, J. C. 1979b. Sobre un molde endocraneano del cinodonte *Probainognathus jenseni* Romer, 1970 (Reptilia, Therapsida) de la Formación Ischichuca (Trasico Medio), La Rioja, Argentina. *Ameghiniana* 17: 181–190.

Quiroga, J. C. 1979c. The inner ear of two cynodonts (Reptilia–Therapsida) and some comments on the evolution of the inner ear from pelycosaurs to mammals. *Gegenbaurs Morphologisches Jahrbuch*, Leipzig 125: 178–190.

Quiroga, J. C. 1980. The brain of the mammal-like reptile *Probainognathus jenseni* (Therapsida, Cynodontia). A correlative paleo-neurological approach to the neocortex at the reptile-mammal transition. *Journal für Hirnforschung* 21: 299–336.

Quiroga, J. C. 1984. The endocranial cast of the advanced mammal-like reptile *Therioherpeton cagnini* (Therapsida–Cynodontia) from the Middle Triassic of Brazil. *Journal für Hirnforschung* 25: 285–290.

Radinsky, L. 1976. Oldest horse brains: more advanced than previously thought. *Science* 194: 626–627.

Radinsky, L. 1977a. Brain of early carnivores. *Paleobiology* 3: 333–349.

Radinsky, L. 1977b. Early primate brains: facts and fiction. *Journal of Human Evolution* 6: 79–86.

Radinsky, L. 1981. Brain evolution in extinct South American ungulates. *Brain, Behavior and Evolution* 18: 169–187.

Radinsky, L. 1982. Some cautionary notes on making inferences about relative brain size. *In* E. Armstrong and D. Falk (eds.), *Primate Brain Evolution: Methods and Concepts,* 29–37. Plenum Press, New York.

Rădulescu, C., and Samson, P.-M. 1986. Précisions sur les affinités des Multituberculés (Mammalia) du Crétacé Supérieur de Roumanie. *Comptes Rendus de l'Académie des Sciences, Paris* 303: 1825–1830.

Rădulescu, C., and Samson, P.-M. 1996. The first multituberculate skull from the Late Cretaceous (Maastrichtian) of Europe (Haţeg Basin, Romania). *Anuarul Institutului Geologic al României* 69: 177.

Rădulescu, C., and Samson, P.-M. 1997. Late Cretaceous Multituberculata from the Haţeg Basin (Romania). *Sargetia* 17: 247–255.

Rainforth, E. C., and Lockley, M. G. 1996. Tracks of diminutive dinosaurs and hopping mammals from the Jurassic of North and South America. *Bulletin of the Museum of Northern Arizona* 60: 265–273.

Rasmussen, T. E., and Callison, G. 1981. A new species of triconodont mammal from the Upper Jurassic of Colorado. *Journal of Paleontology* 55: 628–634.

Rasnitsyn, A. P., Jarzembowski, E. A., and Ross, A. J. 1998. Wasps (Insecta: Vespida = Hymenoptera) from the Purbeck Wealden (Lower Cretaceous) of southern England and their biostratigraphical and palaeoenvironmental significance. *Cretaceous Research* 19: 329–391.

Rauhut, O. W. M., Martin, T., and Ortiz-Jaureguizar, E. O. 2002. The first Jurassic mammal from South America. *Nature* 416: 165–168.

Reichert, C. 1837. Über die Visceralbogen der Wirbeltiere in Allgemeinen und deren Metamorphosen bei den Vogeln und Saugetiereen. *Archiv für Anatomie, Physiologie und Wissentschaftlische Medicin* 1837: 120–220.

Reig, O. A. 1958. Notas para una actualización del conocimiento de la fauna de la Formación Chapadmalal. I. Lista faunistica preliminar. *Acta Geológica Lilloana* 2: 241–253.

Reig, O. A., Kirsch, J. A. W., and Marshall, L. G. 1987. Systematic relationships of the living and Neocenozoic American "opossum-like" marsupials (Suborder Didelphimorphia), with comments on the classification of these and of the Cretaceous and Paleogene New World and European metatherians. *In* M. Archer (ed.), *Possums and Opossums: Studies in Evolution,* 1–89. Surrey Beatty & Sons and the Royal Zoological Society of New South Wales, Sydney.

Renfree, M. B. 1981. Marsupials: alternative mammals. *Nature* 293: 100–101.

Renfree, M. B. 1983. Marsupial reproduction: the choice between placentation and lactation. *In* C. A. Finn (ed.), *Oxford Reviews of Reproductive Biology 5,* Oxford University Press, Oxford.

Renfree, M. B. 1993. Ontogeny, genetic control, and phylogeny of female reproduction in monotreme and therian mammals. *In* F. S. Szalay, M. J. Novacek, and M. C. McKenna (eds.), *Mammal Phylogeny: Mesozoic Differentiation, Multituberculates, Monotremes, Early Therians, and Marsupials,* 4–20. Springer-Verlag, New York.

Renne, P. R., Fulford, M. M., and Busby-Spera, C. 1991. High resolution $^{40}Ar/^{39}Ar$ chronostratigraphy of the Late Cretaceous El Gallo Formation, Baja California del Norte, Mexico. *Geophysical Research Letters* 18: 459–462.

Rensberger, J. M. 1973. An occlusion model for mastication and dental wear in herbivorous mammals. *Journal of Paleontology* 47: 515–528.

Reshetov, V. Y., and Trofimov, B. A. 1980. The main stages in the development of Asian mammals. *In* B. S. Sokolov (ed.), *Paleontology. Stratigraphy. International Geological Congress, 26th Session,* 103–114. Nauka, Moscow (in Russian).

Retief, J. D., Winkfein, R. J., and Dixon, G. H. 1993. Evolution of the monotremes: the sequences of the protamine P1 genes of platypus and echidna. *European Journal of Biochemistry* 218: 457–461.

Rich, T. H. 1991. The history of mammals in *Terra Australis. In* P. Vickers-Rich, J. M. Monaghan, R. F. Baird, and T. H. Rich (eds.), *Vertebrate Palaeontology of Australasia,* 893–1070. Pioneer Design Studio, Melbourne.

Rich, T. H., and Vickers-Rich, P. 1999. Palaeobiogeographic implications of Early Cretaceous Australian placental mammals. *Indian National Academy of Sciences Proceedings, Part A* 65: 315–327.

Rich, T. H., and Vickers-Rich, P. 2000. *Dinosaurs of Darkness.* 240 pp. Indiana University Press, Bloomington.

Rich, T. H., Flannery, T. F., and Archer, M. 1989. A second Cretaceous mammalian specimen from Lightning Ridge, N.S.W., Australia. *Alcheringa* 13: 85–88.

Rich, T. H., Vickers-Rich, P., Constantine, A., Flannery, T. F., Kool, L., and van Klaveren, N. 1997. A tribosphenic mammal from the Mesozoic of Australia. *Science* 278: 1438–1442.

Rich, T. H., Flannery, T. F., and Vickers-Rich, P. 1998. Alleged Cretaceous placental from down under: reply. *Lethaia* 31: 346–348.

Rich, T. H., Vickers-Rich, P., Constantine, A., Flannery, T. F., Kool, L., and van Klaveren, N. 1999. Early Cretaceous mammals from Flat Rocks, Victoria, Australia. *Records of the Queen Victoria Museum* 106: 1–35.

Rich, T. H., Flannery, T. F., Trusler, P., Constantine, A., Kool, L., van Klaveren, N., and Vickers-Rich, P. 2001. An advanced ausktribosphenid from the Early Cretaceous of Australia. *The Records of the Queen Victoria Museum* 110: 1–9.

Rich, T. H., Flannery, T. F., Trusler, P., and Vickers-Rich, P. 2001. Corroboration of the Garden of Eden hypothesis. *In* I. Metcalfe, J. M. B. Smith, M. Morwood, and I. Davidson (eds.), *Faunal and Floral Migrations and Evolution in SE Asia-Australia,* 323–332. A. A. Balkema, Rotterdam.

Rich, T. H., Vickers-Rich, P., Trusler, P., Flannery, T. F., Cifelli, R. L., Constantine, A., Kool, L., and van Klaveren, N. 2001. Monotreme nature of the Australian Early Cretaceous mammal *Teinolophos trusleri. Acta Palaeontologica Polonica* 46: 113–118.

Rich, T. H., Flannery, T. F., Trusler, P., Kool, L., van Klaveren, N., and Vickers-Rich, P. 2002. Evidence that monotremes and ausktribosphenids are not sister groups. *Journal of Vertebrate Paleontology* 22: 466–469.

Ricqlès, A., de. 1974. Evolution of endothermy: historical evidence. *Evolutionary Theory* 1: 51–80.

Ricqlès, A., de. 1976. On bone histology of fossil and living reptiles, with comments on its functional evolutionary significance. *In* A. de Bellairs and C. B. Cox (eds.), *Morphology and Biology of Reptiles,* 123–150. Academic Press, London.

Ride, W. D. L. 1957. The affinities of *Plagiaulax* (Multituberculata). *Proceedings of the Zoological Society of London* 128: 397–402.

Ride, W. D. L. 1959. Mastication and taxonomy in the macropodine skull. *In* A. L. Cain (ed.), *Function and Taxonomic Importance,* 33–59. The Systematics Association Publication No. 3, London.

Ride, W. D. L. 1964. A review of Australian fossil marsupials. *Journal of the Royal Society of Western Australia* 47: 97–131.

Rigby, J. K., Jr., and Wolberg, D. L. 1987. The therian mammalian fauna (Campanian) of Quarry 1, Fossil Forest study area, San Juan Basin, New Mexico. *Geological Society of America, Special Paper* 209: 51–79.

Rigney, H. W. 1963. A specimen of *Morganucodon* from Yunnan. *Nature* 197: 1122–1123.

Robinson, P. L. 1957. The Mesozoic fissures of the Bristol Channel area and their vertebrate faunas. *Journal of the Linnean Society (Zoology)* 43: 260–282.

Romer, A. S. 1966. *Vertebrate Paleontology, 3rd Edition.* 468 pp. University of Chicago Press, Chicago.

Romer, A. S. 1967. The Chañares (Argentina) Triassic reptile fauna. III. Two new gomphodonts, *Massetognathus pascuali* and *M. teruggii. Breviora* 264: 1–25.

Romer, A. S. 1968. *Notes and Comments on Vertebrate Paleontology.* 304 pp. University of Chicago Press, Chicago.

Romer, A. S. 1969a. The Chañares (Argentina) Triassic reptile fauna. V. A new chiniquodont, *Probeleson lewsi*—cynodont ancestry. *Breviora* 333: 1–24.

Romer, A. S. 1969b. The Brazilian Triassic cynodont reptiles *Belesodon* and *Chiniquodon. Breviora* 332: 1–16.

Romer, A. S. 1970a. The Chañares (Argentina) Triassic reptile fauna. VI. A chiniquodontid cynodont with an incipient squamosal-dentary articulation. *Breviora* 344: 1–18.

Romer, A. S. 1970b. *The Vertebrate Body, 4th Edition.* 601 pp. W. B. Saunders Company, Philadelphia.

Romer, A. S., and Lewis, A. D. 1973. The Chañares (Argentina) Triassic reptile fauna. XIX. Postcranial material of the cynodonts *Probelesodon* and *Probainognathus. Breviora* 407: 1–26.

Romer, A. S., and Parsons, T. S. 1986. *The Vertebrate Body, 6th Edition.* viii + 679 pp. Sounders College Publishing, Philadelphia, New York, Chicago.

Rose, K. D., and Emry, R. J. 1993. Relationships of Xenarthra, Pholidota, and fossil "edentates": the morphological evidence. *In* F. S. Szalay, M. J. Novacek, and M. C. McKenna (eds.), *Mammal Phylogeny: Placentals,* 81–102. Springer-Verlag, New York.

Rose, K. D., and Gingerich, P. D. 1987. A new insectivore from the Clarkforkian (earliest Eocene) of Wyoming. *Journal of Mammalogy* 68: 17–27.

Rose, K. D., Cifelli, R. L., and Lipka, T. R. 2001. Second triconodont dentary from the Early Cretaceous of Maryland. *Journal of Vertebrate Paleontology* 21: 628–632.

Rosowski, J. A. 1992. Hearing in transitional mammals: predictions from the middle-ear anatomy and hearing capabilities of extant mammals. *In* D. B. Webster, R. R. Fay, and A. N. Popper (eds.), *The Evolutionary Biology of Hearing,* 615–631. Springer-Verlag, New York.

Rosowski, J. A., and Graybeal, A. 1991. What did *Morganucodon* hear? *Zoological Journal of the Linnean Society* 101: 131–168.

Roth, J. J., Roth, E. C., and Hotton, N. 1986. The parietal foramen and eye: their function and fate in therapsids. *In* N. Hotton, P. D. MacLean, J. J. Roth, and E. C. Roth (eds.), *The Ecology and Biology of Mammal-like Reptiles,* 173–184. Smithsonian Institution Press, Washington, D.C.

Roth, S. 1903. Los ungulados sudamericanos. *Anales del Museo de La Plata* 5: 1–36.

Rougier, G. W. 1993. Vincelestes neuquenianus *Bonaparte (Mammalia, Theria) un primitivo mamífero del Cretácico Inferior de la Cuenca Neuquina.* 720 pp. Ph.D. dissertation, Universidad Nacional de Buenos Aires, Buenos Aires.

Rougier, G. W., and Bonaparte, J. F. 1988. La pared lateral del cráneo de *Vincelestes neuquenianus* (Mammalia, Eupantotheria) y su importancia en el estudio de los mamíferos mesozóicos. *Resumenes V Jornadas Argentina de Paleontológia de Vertebrados:* 14–15.

Rougier, G. W., and Novacek, M. J. 1998. Early mammals: teeth, jaws and finally . . . a skeleton! *Current Biology* 1998: R284–R287.

Rougier, G. W., Wible, J. R., and Hopson, J. A. 1992. Reconstruction of the cranial vessels in the Early Cretaceous mammal *Vincelestes neuquenianus:* implications for the evolution of the mammalian cranial system. *Journal of Vertebrate Paleontology* 12: 188–216.

Rougier, G. W., Wible, J. R., and Hopson, J. A. 1996. Basicranial anatomy of *Priacodon fruitaensis* (Triconodontidae, Mammalia) from the Late Jurassic of Colorado, and a reappraisal of mammaliaform interrelationships. *American Museum Novitates* 3183: 1–38.

Rougier, G. W., Wible, J. R., and Novacek, M. J. 1996a. Middle-ear ossicles of *Kryptobaatar dashzevegi* (Mammalia, Multituberculata): implications for mammaliamorph relationships and evolution of the auditory apparatus. *American Museum Novitates* 3187: 1–43.

Rougier, G. W., Wible, J. R., and Novacek, M. J. 1996b. Multituberculate phylogeny. *Nature* 379: 406.

Rougier, G. W., Novacek, M. J., and Dashzeveg, D. 1997. A new multituberculate from the Late Cretaceous locality Ukhaa Tolgod, Mongolia. Considerations on multituberculate relationships. *American Museum Novitates* 3193: 1–26.

Rougier, G. W., Wible, J. R., and Novacek, M. J. 1998. Implications of *Deltatheridium* specimens for early marsupial history. *Nature* 396: 459–463.

Rougier, G. W., Isaji, S., and Manabe, M. 1999. An Early Cretaceous Japanese triconodont and a revision of triconodont phylogeny. *Journal of Vertebrate Paleontology* 19, supplement to no. 3: 72A.

Rougier, G. W., Novacek, M. J., Pascual, R., and Gelfo, J. N. 2000. New Late Cretaceous mammals from Argentina and the survival of Mesozoic lineages into the Patagonian early Tertiary. *Journal of Vertebrate Paleontology* 20, supplement to no. 3: 65A.

Rougier, G. W., Novacek, M. J., McKenna, M. C., and Wible, J. R. 2001. Gobiconodonts from the Early Cretaceous of Oshih (Ashile), Mongolia. *American Museum Novitates* 3348: 1–30.

Rowe, T. B. 1986. *Osteological diagnosis of Mammalia, L. 1758, and its relationship to extinct Synapsida.* 446 pp. Ph.D. dissertation, University of California, Berkeley.

Rowe, T. B. 1988. Definition, diagnosis, and origin of Mammalia. *Journal of Vertebrate Paleontology* 8: 241–264.

Rowe, T. B. 1989. A new species of the theropod dinosaur *Syntarsus* from the Early Jurassic Kayenta Formation of Arizona. *Journal of Vertebrate Paleontology* 9: 125–136.

Rowe, T. B. 1993. Phylogenetic systematics and the early history of mammals. *In* F. S. Szalay, M. J. Novacek, and M. C. McKenna (eds.), *Mammal Phylogeny: Mesozoic Differentiation, Multituberculates, Monotremes, Early Therians, and Marsupials,* 129–145. Springer-Verlag, New York.

Rowe, T. B. 1996a. Brain heterochrony and origin of the mammalian middle ear. *Memoirs of the California Academy of Sciences* 20: 71–95.

Rowe, T. B. 1996b. Coevolution of the mammalian middle ear and neocortex. *Science* 273: 651–654.

Rowe, T. B. 1999. At the roots of the mammalian family tree. *Nature* 398: 283–284.

Rowe, T. B., and Gauthier, J. A. 1992. Ancestry, paleontology, and definition of the name Mammalia. *Systematic Biology* 41: 372–378.

Rowe, T. B., and Greenwald, N. S. 1987. The phylogenetic position and origin of Multituberculata. *Journal of Vertebrate Paleontology* 7, supplement to no. 3: 24A–25A.

Rowe, T. B., Cifelli, R. L., Lehman, T. M., and Weil, A. 1992. The Campanian age Terlingua local fauna, with a summary of other vertebrates from the Aguja Formation, Trans-Pecos Texas. *Journal of Vertebrate Paleontology* 12: 472–493.

Rowe, T. B., Carlson, W. D., and Bottorff, W. W. 1994. Thrinaxodon: *Digital Atlas of the Skull (CD-ROM).* University of Texas Press, Austin.

Rowe, T., Colbert, M. W., Ketchum, R. A., Maissano, J. A., and Denison, C. 1997. High-resolution X-ray computed tomography: a breakthrough technology for earth scientists. *Geotimes* 42: 23–27.

Russell, D. A. 1993. The role of central Asia in dinosaurian biogeography. *Canadian Journal of Earth Sciences* 30: 2002–2012.

Russell, D. A., Béland, P., and McIntosh, J. S. 1980. Paleoecology of the dinosaurs of Tendaguru (Tanzania). *Mémoirs de la Société Géologique de France* 139: 169–175.

Russell, D. E. 1964. Les mammifères Paléocènes d'Europe. *Mémoires du Muséum national d'Histoire naturelle, sér. C,* 13: 1–324.

Russell, D. E., and Sigogneau, D. 1965. Étude de moulages endocraniens de Mammifères Paléocènes. *Mémoires du Muséum national d'Histoire naturelle, Paris, sér. C,* 16: 1–36.

Russell, D. E., Russell, D., and Wouters, G. 1976. Une dent d'aspect mammalien en provenance du Rhetien francais. *Géobios* 9: 377–392.

Russell, L. S. 1926. A new species of the genus *Catopsalis* Cope from the Paskapoo Formation of Alberta. *American Journal of Science* 12: 230–234.

Russell, L. S. 1937. New interesting mammalian fossils from western Canada. *Proceedings and Transactions of the Royal Society of Canada (for 1936), sér. 3,* 30: 75–80.

Russell, L. S. 1952. Cretaceous mammals of Alberta. *Bulletin of the National Museum of Canada* 126: 110–119.

Russell, L. S. 1964. Cretaceous non-marine faunas of western North America. *Royal Ontario Museum, University of Toronto, Life Sciences Contribution* 61: 1–24.

Russell, L. S. 1975. Mammalian faunal succession in the Cretaceous System of western North America. *In* W. G. E. Caldwell (ed.), *The Cretaceous System in the Western Interior of North America,* 137–161. Geological Association of Canada Special Paper 13.

Ryszkiewicz, M. 1989. *How to Become a Man—an Evolutionary Recipe (in Polish).* 285 pp. Iskry, Warsaw.

Sahni, A. 1972. The vertebrate fauna of the Judith River Formation, Montana. *Bulletin of the American Museum of Natural History* 147: 321–412.

Sahni, A. 1979. Enamel ultrastructure of certain North American Cretaceous mammals. *Palaeontographica, Abteilung A* 166: 37–49.

Sahni, A. 1985. Enamel structure in early mammals and its role in evaluating relationships among rodents. *In* W. P. Luckett and J.-L. Hartenberger (eds.), *Evolutionary Relationships Among Rodents: A Multidisciplinary Analysis,* 133–150. Plenum Press, New York.

Sahni, A., and Lester, K. S. 1988. The nature and significance of enamel tubules in therapsids and mammals. *Mémoires du Muséum national d'Histoire naturelle* 53: 83–97.

Sánchez-Villagra, M. R., and Maier, W. 2001. Ontogenetic data challenges the current conception of the origin of the therian scapula. *Journal of Vertebrate Paleontology* 21, supplement to no. 3: 96A.

Sánchez-Villagra, M., and Maier, W. 2002. Ontogenetic data and the evolutionary origin of the mammalian scapula. *Naturwissenschaften* 89: 459–461.

Sánchez-Villagra, M. R., and Smith, K. K. 1997. Diversity and evolution of the marsupial mandibular angular process. *Journal of Mammalian Evolution* 4: 119–144.

Sánchez-Villagra, M. R., and Wible, J. R. 2000. Patterns of evolutionary transformation in the petrosal bone and some basicranial features in marsupial mammals, with special reference to didelphids. *Journal of Zoological, Systematic and Evolutionary Research* 40: 26–45.

Sánchez-Villagra, M. R., Gemballa, S., Nummela, S., Smith, K. K., and Maier, W. 2002. Ontogenetic and phylogenetic transformations of the ear ossicles in marsupial mammals. *Journal of Morphology* 251: 219–238.

Sander, P. M. 1997. Non-mammalian synapsid enamel and the origin of mammalian enamel prisms: the bottom-up perspective. *In* W. von Koenigswald and P. M. Sander (eds.), *Tooth Enamel Microstructure,* 41–62. A. A. Balkema, Rotterdam.

Sanderson, M., and Donoghue, M. J. 1989. Patterns of variation in levels of homoplasy. *Evolution* 43: 1781–1795.

Sankey, J. 1998. *Vertebrate paleontology and magnetostratigraphy of the upper Aguja Formation (late Campanian), Talley Mountain area, Big Bend National Park, Texas.* 251 pp. Ph.D. dissertation, Louisiana State University, Baton Rouge.

Sargis, E. J. 2000. *The functional morphology of the postcranium of* Ptilocercus *and tupaiines (Scandentia, Tupaiidae): implications for the relationships of primates and other archontan mammals.* 317 pp. Ph.D. dissertation, City University of New York, New York.

Sargis, E. J. 2002. Functional morphology of the forelimb of tupaiids (Mammalia, Scandentia) and its phylogenetic significance. *Journal of Morphology* 253: 10–42.

Sarjeant, W. A. 1975. A vertebrate footprint from the Stonesfield Slate (Middle Jurassic) of Oxfordshire. *Mercian Geologist* 5: 273–277.

Sarjeant, W. A. 2000. The Mesozoic mammal footprint record reconsidered, with an account of new discoveries in the Cretaceous of northwestern Alberta, Canada. *Bulletin of the Paleontological Society of Korea, Special Publication* 4: 153–168.

Sarjeant, W. A., and Thulborn, R. A. 1986. Probable marsupial footprints from the Cretaceous sediments of British Columbia. *Canadian Journal of Earth Sciences* 23: 1223–1227.

Savage, R. J. G. 1984. Mid-Jurassic mammals from Scotland. *In* W.-E. Reif and F. Westphal (eds.), *Third Symposium on Mesozoic Terrestrial Ecosystems,* 211–213. Attempto, Tübingen.

Schaeffer, B. 1941. The morphological and functional evolution of the tarsus in amphibians and reptiles. *Bulletin of the American Museum of Natural History* 78: 395–472.

Schaller, O. (ed.) 1992. *Illustrated Veterinary Anatomy Nomenclature.* 614 pp. Ferdinand Enke Verlag, Stuttgart.

Schmid, E. 1972. *Atlas of Animal Bones—for Prehistorians, Archaeologists and Quaternary Geologists.* 159 pp. Elsevier Science Publishers, Amsterdam.

Schudack, M. E. 1993. Charophyten aus dem Kimmerdgium der Kohlengrube Guimarota (Portugal) mit einer eingehenden Diskussion zur Datierung der Fundstelle. *Berliner geowissenschaftliche, Abhandlungen E* 9: 211–231.

Schudack, M. E. 2000. Geological setting and dating of the Guiamarota beds. In T. Martin and B. Krebs (eds.), *Guiamarota: a Jurassic Ecosystem,* 21–26. Verlag Dr. Friedrich Pfeil, München.

Scillato-Yané, G. R., and Pascual, R. 1984. Un peculiar Paratheria, Edentata (Mammalia) del Paleoceno de Patagonia. *Primeras Jornadas Argentinas de Paleontología Vertebrados, Resumenes:* 15. La Plata.

Scillato-Yané, G. R., and Pascual, R. 1985. Un peculiar Xenarthra del Paleoceno medio de Patagonia (Argentina). Su importancia en la sistematica de los Paratheria. *Ameghiniana* 21: 173–176.

Sereno, P. C., and McKenna, M. C. 1995. Cretaceous multituberculate skeleton and the early evolution of the mammalian shoulder girdle. *Nature* 377: 144–147.

Sereno, P. C., and McKenna, M. C. 1996. Reply. *Nature* 379: 406.

Setoguchi, T., Matsuoka, H., and Matsuda, M. 1999. New discovery of an Early Cretaceous tritylodontid (Reptilia, Therapsida) from Japan and the phylogenetic reconstruction of Tritylodontidae based on the dental characters. *In* Y.-Q. Wang and T. Deng (eds.), *Proceedings of the Seventh Annual Meeting of the Chinese Society of Vertebrate Paleontology,* 117–124. China Ocean Press, Beijing.

Setoguchi, T., Tsubamoto, T., Hanamura, H., and Hachiya, K. 1999. An early Late Cretaceous mammal from Japan, with reconsideration of the evolution of tribosphenic molars. *Paleontological Research* 3: 18–28.

Shapiro, M. D., and Jenkins, F. A., Jr. 2001. A cynodont from the Upper Triassic of east Greenland: tooth replacement and double-rootedness. *Bulletin of the Museum of Comparative Zoology* 156: 49–58.

Shikama, T. 1947. *Teilhardosaurus* and *Endotherium,* new Jurassic Reptilia and Mammalia from the Husin coal-field, south Manchuria. *Proceedings of the Japanese Academy* 23: 76–84.

Shoshani, J., and McKenna, M. C. 1998. Higher taxonomic relationships among extant mammals based on morphology, with selected comparisons of results from molecular data. *Molecular Phylogenetics and Evolution* 9: 572–584.

Shubin, N. H., Crompton, A. W., Sues, H.-D., and Olsen, P. E. 1991. New fossil evidence on the sister-group of mammals and early Mesozoic faunal distribution. *Science* 251: 1063–1065.

Shuvalov, V. F. 2000. The Cretaceous stratigraphy and palaeobiography in Mongolia. *In* M. J. Benton, M. A. Shishkin, E. N. Kurochkin, and D. M. Unwin (eds.), *The Age of Dinosaurs in Russia and Mongolia,* 256–278. Cambridge University Press, Cambridge.

Sidor, C. A. 2001. Simplification as a trend in synapsid cranial evolution. *Evolution* 55: 1419–1442.

Sidor, C. A., and Hopson, J. A. 1998. Ghost lineages and "mammalness": assessing the temporal pattern of character acquisition in the Synapsida. *Paleobiology* 24: 254–273.

Sigé, B. 1971. Les Didelphoidea de Laguna Umayo (Formation Vilquechico, Crétacé Supérieur, Pérou), et le peuplement marsupial d'Amérique du Sud. *Comptes Rendus de l'Académie des Sciences, Paris* 273: 2479–2481.

Sigé, B. 1972. La faunule des mammifères du Crétacé supérieur de Laguna Umayo (Andes péruviennes). *Bulletin du Muséum national d'Histoire naturelle, sér. 3, Sciences de la Terre* 99: 375–405.

Sigé, B., Archer, M., H., G., Hand, S., and Crochet, J.-Y. 1995. Peruvian-Australian Paleogene mammal connection. *Fifth Conference on Australian Vertebrate Evolution, Palaeontology and Systematics* 1: 2.

Sigé, B., Buscalioni, A. D., Duffaud, S., Gayet, M., Orth, B., Rage, J.-C., and Sanz, J. L. 1997. Etat des données sur le gisement Crétacé supérieur continental de Champ-Garimond (Gard, Sud de la France). *Münchner geowissenschaftliche, Abhandlungen A* 34: 111–130.

Sigogneau, D. 1974. The inner ear of *Gorgonops* (Reptilia, Therapsida, Gorgonopsia). *Annals of the South African Museum* 64: 53–69.

Sigogneau-Russell, D. 1983a. A new therian mammal from the Rhaetic locality of Saint-Nicolas-de-Port (France). *Zoological Journal of the Linnean Society* 78: 175–186.

Sigogneau-Russell, D. 1983b. Nouveaux taxons de Mammifères rhétiens. *Acta Palaeontologica Polonica* 28: 233–249.

Sigogneau-Russell, D. 1983c. Caractéristiques de la fauna mammalienne du Rhétien de Saint-Nicolas-de-Port (Meurthe-et-Moselle). *Bulletin d'Information des Geologues du Bassin de Paris* 20: 51–53.

Sigogneau-Russell, D. 1989a. Découverte du premier Symmétrodonte (Mammalia) du continent africain. *Comptes Rendus de l'Académie des Sciences Paris* 309: 921–926.

Sigogneau-Russell, D. 1989b. Haramiyidae (Mammalia, Allotheria) en provenance du Trias supérieur de Lorraine (France). *Palaeontographica, Abteilung A* 206: 137–198.

Sigogneau-Russell, D. 1990. Reconnaissance formelle d'une nouvelle espèce d'*Haramiya* dans l'hypodigme français des Haramiyidae (Mammalia, Allotheria). *Bulletin du Muséum national d'Histoire naturelle, Paris* 12: 85–88.

Sigogneau-Russell, D. 1991a. Découverte du premier mammifère tribosphénique du Mésozoïque africain. *Comptes Rendus de l'Académie des Sciences, Paris* 313: 1635–1640.

Sigogneau-Russell, D. 1991b. Nouveaux Mammifères theriens du Crétacé inférieur du Maroc. *Comptes Rendus de l'Académie des Sciences* 313: 279–285.

Sigogneau-Russell, D. 1991c. First evidence of Multituberculata (Mammalia) in the Mesozoic of Africa. *Neues Jahrbuch für Geologie und Paläontologie, Monatshefte* 1991: 119–125.

Sigogneau-Russell, D. 1991d. Early Cretaceous Moroccan Theria (Mammalia)—considerations on therian dental evolution. *In* Z. Kielan-Jaworowska, N. Heintz, and H. A. Nakrem (eds.), *Fifth Symposium on Mesozoic Terrestrial Ecosystems and Biota,* 61–62. Paleontological Museum, University of Oslo, Oslo.

Sigogneau-Russell, D. 1992. *Hypomylos phelizoni* nov. gen. nov. sp., une étape précoce de l'évolution de la molaire tribosphénique. *Géobios* 25: 389–393.

Sigogneau-Russell, D. 1994a. Further data and reflexions on the tribosphenid mammals (Tribotheria) from the Early Cretaceous of Morocco. *Bulletin du Muséum national d'Histoire naturelle* 16: 291–312.

Sigogneau-Russell, D. 1994b. Les Mammmifères dans la nuit des temps Mésozoïque. *La Vie des Sciences, Comptes Rendus, série générale* 11: 243–270.

Sigogneau-Russell, D. 1995a. Further data and reflexions on the tribosphenic mammals (Tribotheria) from the Early Cretaceous of Morocco. *Bulletin du Muséum national d'Histoire naturelle, Paris, sér. 4,* 16 (section C): 291–312.

Sigogneau-Russell, D. 1995b. Two possibly aquatic triconodont mammals from the Early Cretaceous of Morocco. *Acta Palaeontologica Polonica* 40: 149–162.

Sigogneau-Russell, D. 1998. Discovery of a Late Jurassic Chinese mammal in the upper Bathonian of England. *Comptes Rendus de l'Académie des Sciences, Paris* 327: 571–576.

Sigogneau-Russell, D. 1999. Réévaluation des Peramura (Mammalia, Theria) sur la base de nouveaux spécimens du Crétacé inférieur d'Angleterre et du Maroc. *Geodiversitas* 21: 93–127.

Sigogneau-Russell, D. 2001. Docodont nature of *Cyrtlatherium,* an upper Bathonian mammal from England. *Acta Palaeontologica Polonica* 46: 427–430.

Sigogneau-Russell, D., and Ensom, P. C. 1994. Découverte, dans le Groupe de Purbeck (Berriasian, Angleterre), de plus ancien témoinage de l'existence de mammifères tribosphéniques. *Comptes Rendus de l'Académie des Sciences, Paris* 319: 833–838.

Sigogneau-Russell, D., and Ensom, P. C. 1998. *Thereuodon* (Theria, Symmetrodonta) from the Lower Cretaceous of North Africa and Europe, and a brief review of symmetrodonts. *Cretaceous Research* 19: 1–26.

Sigogneau-Russell, D., and Godefroit, P. 1997. A primitive docodont (Mammalia) from the Upper Triassic of France and the possible therian affinities of the order. *Comptes Rendus de l'Académie des Sciences* 324: 135–140.

Sigogneau-Russell, D., and Hahn, G. 1994. Late Triassic microvertebrates from central Europe. *In* N. C. Fraser and H.-D. Sues (eds.), *In the Shadow of the Dinosaurs—Early Mesozoic Tetrapods,* 197–213. Cambridge University Press, Cambridge.

Sigogneau-Russell, D., and Hahn, R. 1995. Reassessment of the Late Triassic symmetrodont mammal *Woutersia. Acta Palaeontologica Polonica* 40: 245–260.

Sigogneau-Russell, D., and Kielan-Jaworowska, Z. 2002. Mammals from the Purbeck Limestone Group of Dorset, southern England. *In* A. R. Milner and D. J. Batten (eds.), *Life and Environments in Purbeck Times,* 241–255. Special Papers in Palaeontology, 68.

Sigogneau-Russell, D., and Sun, A.-L. 1981. A brief review of Chinese synapsids. *Géobios* 14: 275–279.

Sigogneau-Russell, D., Frank, D. R., and Hemmerlé, J. 1984. Enamel and dentine ultrastructure in the Early Jurassic therian *Kuehneotherium. Zoological Journal of the Linnean Society* 82: 207–215.

Sigogneau-Russell, D., Frank, R. M., and Hemmerlé, J. 1986. A new family of mammals from the lower part of the French Rhaetic. *In* K. Padian (ed.), *The Beginning of the Age of Dinosaurs,* 99–108. Cambridge University Press, Cambridge.

Sigogneau-Russell, D., Monbaron, M., and Russell, D. E. 1988. Découverte de Mammifères dans le Mésozoïque moyen d'Afrique. *Comptes Rendus de l'Académie des Sciences, Paris* 307: 1045–1050.

Sigogneau-Russell, D., Monbaron, M., and de Kaenel, E. 1990. Nouvelles données sur le gisement à mammifères Mésozoïques du Haut-Atlas Marocain. *Géobios* 23: 461–483.

Sigogneau-Russell, D., Bonaparte, J. F., Frank, R. M., and Escribano, V. 1991. Ultrastructure of dental hard tissues of *Gondwanatherium* and *Sudamerica* (Mammalia, Gondwanatheria). *Lethaia* 24: 27–38.

Sigogneau-Russell, D., Dashzeveg, D., and Russell, D. E. 1992. Further data on *Prokennalestes* (Mammalia, Eutheria inc. sed.) from the Early Cretaceous of Mongolia. *Zoologica Scripta* 21: 205–209.

Sigogneau-Russell, D., Evans, S. E., Levine, J. F., and Russell, D. A. 1998. The Early Cretaceous microvertebrate locality of Anoual, Morocco: a glimpse at the small vertebrate assemblages of Africa. *In* S. G. Lucas, J. I. Kirkland, and J. W. Estep (eds.), *Lower and Middle Cretaceous Terrestrial Ecosystems*, 177–182. *New Mexico Museum of Natural History and Science Bulletin* 14, Albuquerque.

Sigogneau-Russell, D., Hooker, J. J., and Ensom, P. C. 2001. The oldest tribosphenic mammal from Laurasia (Purbeck Limestone Group, Berriasian, Cretaceous, UK) and its bearing on the "dual origin" of Tribosphenida. *Comptes Rendus de l'Académie des Sciences* 333: 141–147.

Silva, M., and Downing, J. A. 1995. *CRC Handbook of Mammalian Body Masses.* 359 pp. CRC Press, Boca Raton.

Simmons, N. B. 1993. Phylogeny of Multituberculata. *In* F. S. Szalay, M. J. Novacek, and M. C. McKenna (eds.), *Mammal Phylogeny: Mesozoic Differentiation, Multituberculates, Monotremes, Early Therians, and Marsupials*, 146–164. Springer-Verlag, New York.

Simmons, N. B., and Miao, D. 1986. Paraphyly of *Catopsalis* (Mammalia: Multituberculata) and its biogeographic implications. *Contributions to Geology, University of Wyoming, Special Paper* 3: 87–94.

Simmons Greenwald, N. 1988. Patterns of tooth eruption and replacement in multituberculate mammals. *Journal of Vertebrate Paleontology* 8: 265–277.

Simpson, G. G. 1925a. A Mesozoic mammal skull from Mongolia. *American Museum Novitates* 201: 1–11.

Simpson, G. G. 1925b. Mesozoic Mammalia. I. American triconodonts, part 1. *American Journal of Science* 10: 145–165.

Simpson, G. G. 1925c. Mesozoic Mammalia. I. American triconodonts: part 2. *American Journal of Science* 10: 334–358.

Simpson, G. G. 1925d. Mesozoic Mammalia. II. *Tinodon* and its allies. *American Journal of Science* 10: 451–470.

Simpson, G. G. 1925e. Mesozoic Mammalia. III. Preliminary comparison of Jurassic mammals except multituberculates. *American Journal of Science* 10: 559–569.

Simpson, G. G. 1926a. Are *Dromatherium* and *Microconodon* mammals? *Science* 63: 548–549.

Simpson, G. G. 1926b. Mesozoic Mammalia. IV. The multituberculates as living animals. *American Journal of Science* 11: 228–250.

Simpson, G. G. 1926c. The fauna of Quarry Nine. *American Journal of Science* 12: 1–11.

Simpson, G. G. 1926d. Mesozoic Mammalia. V. *Dromatherium* and *Microconodon*. *American Journal of Science* 12: 87–108.

Simpson, G. G. 1927a. Mesozoic Mammalia. IX. The brain of Jurassic mammals. *American Journal of Science* 214: 259–268.

Simpson, G. G. 1927b. Mammalian fauna of the Hell Creek Formation of Montana. *American Museum Novitates* 267: 1–7.

Simpson, G. G. 1927c. Mesozoic Mammalia. VI. Genera of Morrison pantotheres. *American Journal of Science* 13: 409–416.

Simpson, G. G. 1927d. Mesozoic Mammalia. VIII. Genera of Lance mammals other than multituberculates. *American Journal of Science* 14: 121–130.

Simpson, G. G. 1928a. *A Catalogue of the Mesozoic Mammalia in the Geological Department of the British Museum.* 215 pp. Trustees of the British Museum, London.

Simpson, G. G. 1928b. Mesozoic Mammalia. XII. The internal mandibular groove of Jurassic mammals. *American Journal of Science* 15: 461–470.

Simpson, G. G. 1928c. Affinities of the Mongolian Cretaceous insectivores. *American Museum Novitates* 330: 1–11.

Simpson, G. G. 1928d. Further notes on Mongolian Cretaceous mammals. *American Museum Novitates* 329: 1–14.

Simpson, G. G. 1928e. Mesozoic Mammalia. XI. *Brancatherulum tendagurense* Dietrich. *American Journal of Science* 15: 303–308.

Simpson, G. G. 1928f. A new mammalian fauna from the Fort Union of southern Montana. *American Museum Novitates* 297: 1–15.

Simpson, G. G. 1929a. American Mesozoic Mammalia. *Memoirs of the Peabody Museum of Yale University* 3: 1–235.

Simpson, G. G. 1929b. The dentition of *Ornithorhynchus* as evidence of its affinities. *American Museum Novitates* 390: 1–15.

Simpson, G. G. 1931. A new classification of mammals. *Bulletin of the American Museum of Natural History* 59: 259–293.

Simpson, G. G. 1933a. Paleobiology of Jurassic mammals. *Palaeobiologica* 5: 127–158.

Simpson, G. G. 1933b. The "plagiaulacoid" type of mammalian dentition. *Journal of Mammalogy* 14: 97–107.

Simpson, G. G. 1936a. Studies of the earliest mammalian dentition. *The Dental Cosmos* 78 (8): 791–800.

Simpson, G. G. 1936b. Additions to the Puerco fauna, lower Paleocene. *American Museum Novitates* 849: 1–11.

Simpson, G. G. 1937a. Skull structure of the Multituberculata. *Bulletin of the American Museum of Natural History* 73: 727–763.

Simpson, G. G. 1937b. A new Jurassic mammal. *American Museum Novitates* 943: 1–6.

Simpson, G. G. 1945. The principles of classification and a classification of mammals. *Bulletin of the American Museum of Natural History* 85: 1–350.

Simpson, G. G. 1947. *Haramiya*, new name, replacing *Microcleptes* Simpson, 1928. *Journal of Paleontology* 21: 497.

Simpson, G. G. 1951. American Cretaceous insectivores. *American Museum Novitates* 1541: 1–19.

Simpson, G. G. 1953. *Evolution and Geography: An Essay on Historical Biogeography With Special Reference to Mammals.* 64 pp. Oregon State System of Higher Education, Eugene.

Simpson, G. G. 1959. Mesozoic mammals and the polyphyletic origin of mammals. *Evolution* 13: 405–414.

Simpson, G. G. 1961. Evolution of Mesozoic mammals. *In* G. Vandebroek (ed.), *International Colloquium on the Evolution of Lower and Non-specialized Mammals,* 57–95. Koninklijke Vlaamse Academie voor Wetenschappen, Letteren en Schone Kunsten van België, part 1, Leuven.

Simpson, G. G. 1970. The Argyrolagidae, extinct South American marsupials. *Bulletin of the Museum of Comparative Zoology* 139: 1–86.

Simpson, G. G. 1971. Concluding remarks: Mesozoic mammals revisited. *In* D. M. Kermack and K. A. Kermack (eds.), *Early Mammals,* 181–198. *Zoological Journal of the Linnean Society* 50, supplement 1, London.

Simpson, G. G. 1981. History of vertebrate paleontology in the San Juan Basin. *In* S. G. Lucas, J. K. Rigby, Jr., and B. S. Kues (eds.), *Advances in San Juan Basin Paleontology,* 3–25. University of New Mexico Press, Albuquerque.

Simpson, G. G., and Elftman, H. O. 1928. Hind limb musculature and habits of a Paleocene multituberculate. *American Museum Novitates* 333: 1–19.

Sinclair, W. J. 1906. Marsupialia of the Santa Cruz beds. *Reports of the Princeton University Expeditions to Patagonia* 4(3): 333–460.

Slaughter, B. H. 1965. A therian from the Lower Cretaceous (Albian) of Texas. *Postilla* 93: 1–18.

Slaughter, B. H. 1968a. Earliest known marsupials. *Science* 162: 254–255.

Slaughter, B. H. 1968b. Earliest known eutherian mammals and the evolution of premolar occlusion. *The Texas Journal of Science* 20: 3–12.

Slaughter, B. H. 1968c. *Holoclemensia* instead of *Clemensia. Science* 162: 1306.

Slaughter, B. H. 1969. *Astroconodon,* the Cretaceous triconodont. *Journal of Mammalogy* 50: 102–107.

Slaughter, B. H. 1971. Mid-Cretaceous (Albian) therians of the Butler Farm local fauna, Texas. *In* D. M. Kermack and K. A. Kermack (eds.), *Early Mammals,* 131–143. *Zoological Journal of the Linnean Society* 50, supplement 1, London.

Slaughter, B. H. 1981. The Trinity therians (Albian, mid-Cretaceous) as marsupials and placentals. *Journal of Paleontology* 55: 682–683.

Slijper, E. J. 1946. Comparative biologic-anatomical investigations on the vertebral column and spinal musculature of mammals. *Verhandlungen der Koninklijke Nederlandse Akademie van Wetenschappen, Afdeling Natuurkunde, Tweede Reels* 42: 1–128.

Sloan, R. E. 1966. Paleontology and geology of the Badwater Creek area, central Wyoming, Part 2. The Badwater Multituberculata. *Annals of the Carnegie Museum* 38: 309–315.

Sloan, R. E. 1979. Multituberculata. *In* R. W. Fairbridge and D. Jablonski (eds.), *The Encyclopedia of Paleontology,* 492–498. Gowden, Huthison & Ross, Inc., Stroudsburg.

Sloan, R. E. 1981. Systematics of Paleocene multituberculates from the San Juan Basin, New Mexico. *In* S. G. Lucas, J. K. J. Rigby, and B. S. Kues (eds.), *Advances in San Juan Basin Paleontology,* 127–160. University of New Mexico Press, Albuquerque.

Sloan, R. E. 1987. Paleocene and latest Cretaceous mammal ages, biozones, magnetozones, rates of sedimentation, and evolution. *Geological Society of America, Special Paper* 209: 165–200.

Sloan, R. E., and Russell, L. S. 1974. Mammals from the St. Mary River Formation (Cretaceous) of southwestern Alberta. *Royal Ontario Museum, Contributions in Life Sciences* 95: 1–21.

Sloan, R. E., and Van Valen, L. 1965. Cretaceous mammals from Montana. *Science* 148: 220–227.

Smith, B. H. 1994. Sequence and emergence of the permanent teeth in *Macaca, Pan, Homo,* and *Australopithecus:* its evolutionary significance. *American Journal of Human Biology* 6: 61–76.

Smith, B. H. 2000. "Schulz's rule" and the evolution of tooth emergence and replacement patterns in primates and ungulates. *In* M. F. Teaford, M. M. Smith, and M. W. J. Ferguson (eds.), *Development, Function and Evolution of Teeth,* 218–228. Cambridge University Press, Cambridge.

Small, B. J., and Carpenter, K. 1997. Upper Jurassic microvertebrates from the "Small Quarry," Garden Park, Colorado. *Journal of Vertebrate Paleontology* 17, supplement to no. 3: 76–77A.

Smith, K. K. 1992. On the evolution of the mammalian pharynx. *Zoological Journal of the Linnean Society* 104: 313–349.

Smith, K. K. 1996. Integration of craniofacial structures during development in mammals. *American Zoologist* 36: 70–79.

Smith, K. K. 1997. Comparative patterns of craniofacial development in eutherian and metatherian mammals. *Evolution* 51: 1663–1668.

Smith, T., Guo, D.-Y., and Sun, Y. 2001. A new species of *Kryptobaatar* (Multituberculata): the first Late Cretaceous mammal from Inner Mongolia (P. R. China). *Bulletin de l'Institut royal des Sciences naturelles de Belgique, Sciences de la Terre* 71, supplement: 29–50.

Springer, M. S. 1997. Molecular clocks and the timing of the placental and marsupial radiations in relation to the Cretaceous–Tertiary boundary. *Journal of Mammalian Evolution* 4: 285–302.

Springer, M. S., and Jong, W. W. de. 2001. Which mammalian supertree to bark up? *Science* 291: 1709–1711.

Springer, M. S., Amrine, H. M., Burk, A., and Stanhope, M. J. 1999. Additional support for Afrotheria and Paenungulata, the performance of mitochondrial versus nuclear genes, and the impact of data partitions with heterogeneous base composition. *Systematic Biology* 48: 65–75.

Springer, M. S., DeBry, R. W., Douady, C., Amrine, H. M., Madsen, O., de Jong, W. W., and Stanhope, M. J. 2001. Mitochondrial versus nuclear gene sequences in deep-level mammalian phylogeny reconstruction. *Molecular Biology and Evolution* 18: 132–143.

Standhardt, B. R. 1986. *Vertebrate paleontology of the Cretaceous/Tertiary transition of Big Bend National Park, Texas.* 299 pp. Ph.D. dissertation, Louisiana State University, Baton Rouge.

Starck, D. 1962. Die Evolution des Säugetier-Gehirns. *Sitzungsberichte der wissenschaftlichen Gesellschaft an der Wolfgang Goethe-Universität, Frankfurt am Main* 1: 1–60.

Starck, D. 1964. "Freiliegendes Tectum mesencephali" ein Kennzeichen des primitiven Säugetiergehirn? *Zoologischer Anzeiger* 171: 350–359.

Starck, D. 1967. Le crâne des Mammifères. *In* P. Grassé (ed.), *Traité de Zoölogie, Tome 16, Fascicule 1: Mammifères: Téguments et Squelette,* 405–549. Masson et Cie, Paris.

Starck, D. 1982. *Vergleichende Anatomie der Wirbeltiere auf evolutionsbiologischer Grundlage. Band 3.* 1110 pp. Springer-Verlag, Berlin.

Stein, F. 1859a. Characteristik der neuer Infusorien-Gattungen. *Lotos* 9: 2.

Steiner, M. B. 1998. Age, correlation, and tectonic implications of Morrison Formation paleomagnetic data, including rotation of the Colorado Plateau. *Modern Geology* 22: 261–281.

Storer, J. E. 1991. The mammals of the Gryde local fauna, Frenchman Formation (Maastrichtian: Lancian), Saskatchewan. *Journal of Vertebrate Paleontology* 11: 350–369.

Storer, J. E. 1993. Additions to the mammalian paleofauna of Saskatchewan. *Modern Geology* 18: 475–487.

Strait, S. G. 1993a. Molar microwear in extant small-bodied faunivorous mammals: an analysis of feature density and pit frequency. *American Journal of Physical Anthropology* 92: 63–79.

Strait, S. G. 1993b. Molar morphology and food texture among small-bodied insectivorous mammals. *Journal of Mammalogy* 74: 391–402.

Stucky, R. K., and McKenna, M. C. 1993. Mammalia. *In* M. J. Benton (ed.), *The Fossil Record 2,* 739–771. Chapman & Hall, London.

Sues, H.-D. 1983. *Advanced mammal-like reptiles from the Early Jurassic of Arizona.* 247 pp. Ph.D. dissertation, Harvard University, Cambridge.

Sues, H.-D. 1985a. First record of the tritylodontid *Oligokyphus* (Synapsida) from the Early Jurassic Kayenta Formation of Arizona. *Journal of Vertebrate Paleontology* 5: 328–339.

Sues, H.-D. 1985b. The relationships of the Tritylodontidae (Synapsida). *Zoological Journal of the Linnean Society* 85: 205–217.

Sues, H.-D. 1986. The skull and dentition of two tritylodontid synapsids from the Lower Jurassic of western North America. *Bulletin of the Museum of Comparative Zoology* 151: 217–268.

Sues, H.-D. 1992. No Palaeocene "mammal-like reptile." *Nature* 359: 278.

Sues, H.-D. 2001. On *Microconodon,* a Late Triassic cynodont from the Newark Supergroup of eastern North America. *In* F. A. Jenkins, Jr., T. Owerkowicz, and M. D. Shapiro (eds.), *Symposium in Honor of Professor Alfred W. Crompton. Bulletin of the Museum of Comparative Zoology* 156: 37–48.

Sukhanov, V. B. 1974. *Symmetrical Locomotion in Terrestrial Vertebrates and Some Features of Movement of Lower Tetrapods.* 274 pp. Amering Publishing Co., New Delhi.

Sullivan, R. M. 1999. *Nodocephalosaurus kirtlandensis,* gen. et sp. nov., a new ankylosaurid dinosaur (Ornithischia: Ankylosauria) from the Upper Cretaceous Kirtland Formation (upper Campanian), San Juan Basin, New Mexico. *Journal of Vertebrate Paleontology* 19: 126–139.

Sun, A.-L. 1984. Skull morphology of the tritylodont genus *Bienotheroides* of Sichuan. *Scientia Sinica, Series B* 27: 270–284.

Sun, A.-L., and Cui, G.-H. 1986. A brief introduction to the lower Lufeng saurischian fauna (Lower Jurassic: Lufeng, Yunnan, People's Republic of China). *In* K. Padian (ed.), *The Beginning of the Age of Dinosaurs—Faunal Change across the Triassic–Jurassic Boundary,* 275–278. Cambridge University Press, London/New York.

Sun, A.-L., and Cui, G.-H. 1987. Otic region in tritylodont *Yunnanodon. Vertebrata PalAsiatica* 25: 1–7.

Sun, A.-L., and Li, Y.-H. 1985. The postcranial skeleton of Jurassic tritylodonts from Sichuan Province. *Vertebrata PalAsiatica* 23: 135–151.

Sun, A.-L., Cui, G., Li, Y., and Wu, X. 1985. A verified list of the Lufeng saurischian fauna. *Vertebrate PalAsiatica* 22: 1–12 (in Chinese).

Swisher, C. C., III, and Prothero, D. R. 1990. Single-crystal $^{40}Ar/^{39}Ar$ dating of the Eocene-Oligocene transition in North America. *Science* 249: 760–762.

Swisher, C. C., III, Wang, Y.-Q., Wang, X.-L., Xu, X., and Wang, Y. 1999. Cretaceous age for the feathered dinosaurs of Liaoning, China. *Nature* 398: 58–61.

Swofford, D. L. 2000. *PAUP*. Phylogenetic Analysis Using Parsimony (*and Other Methods). Version 4.0.* Sinauer Associates, Sunderland.

Szalay, F. S. 1965. First evidence for tooth replacement in the subclass Allotheria (Mammalia). *American Museum Novitates* 2226: 1–12.

Szalay, F. S. 1966. The tarsus of the Paleocene leptictid *Prodiacodon* (Insectivora, Mammalia). *American Museum Novitates* 2267: 1–13.

Szalay, F. S. 1968. The beginnings of primates. *Evolution* 22: 19–36.

Szalay, F. S. 1975. Phylogeny of primate higher taxa: the basicranial evidence. *In* W. P. Luckett and F. S. Szalay (eds.), *Phylogeny of the Primates: A Multidisciplinary Approach,* 91–125. Plenum Press, New York.

Szalay, F. S. 1977. Phylogenetic relationships and a classification of eutherian Mammalia. *In* M. K. Hecht, O. C. Goody, and B. M. Hecht (eds.), *Major Patterns in Vertebrate Evolution,* 315–374. Plenum Press, New York.

Szalay, F. S. 1982. A new appraisal of marsupial phylogeny and classification. *In* M. Archer (ed.), *Carnivorous Marsupials,* 621–640. Royal Zoological Society of New South Wales, Sydney.

Szalay, F. S. 1984. Arboreality: is it homologous in metatherian and eutherian mammals? *In* M. K. Hecht (ed.), *Evolutionary Biology, Vol. 6,* 215–258. Plenum Press, New York.

Szalay, F. S. 1993a. Metatherian taxon phylogeny: evidence and interpretation from the cranioskeletal system. *In* F. S. Szalay, M. J. Novacek, and M. C. McKenna (eds.), *Mammal Phylogeny: Mesozoic Differentiation, Multituberculates, Monotremes, Early Therians, and Marsupials,* 216–242. Springer-Verlag, New York.

Szalay, F. S. 1993b. Pedal evolution of mammals in the Mesozoic: tests for taxic relationships. *In* F. S. Szalay, M. J. Novacek, and M. C. McKenna (eds.), *Mammal Phylogeny: Mesozoic Differentiation, Multituberculates, Monotremes, Early Therians, and Marsupials,* 108–128. Springer-Verlag, New York.

Szalay, F. S. 1994. *Evolutionary History of the Marsupials and an Analysis of Osteological Characters.* 481 pp. Cambridge University Press, Cambridge.

Szalay, F. S. 1999. M. C. McKenna and S. K. Bell: classification of mammals above the species level. *Journal of Vertebrate Paleontology* 19: 191–195.

Szalay, F. S., and Decker, R. L. 1974. Origin, evolution, and function of the tarsus in Late Cretaceous Eutheria and Paleocene Primates. *In* F. A. Jenkins, Jr. (ed.), *Primate Locomotion,* 223–259. Academic Press, London.

Szalay, F. S., and Lucas, S. G. 1996. The postcranial morphology of Paleocene *Chriacus* and *Mixodectes* and the phylogenetic relationships of archontan mammals. *New Mexico Museum of Natural History and Science, Bulletin* 7: 1–47.

Szalay, F. S., and McKenna, M. C. 1971. Beginning of the age of mammals in Asia: the late Paleocene Gashato fauna, Mongolia. *Bulletin of the American Museum of Natural History* 144: 269–318.

Szalay, F. S., and Sargis, E. J. 2001. Model-based analysis of postcranial osteology of marsupials from the Palaeocene of Itaboraí (Brazil) and the phylogenetics and biogeography of Metatheria. *Geodiversitas* 23: 139–302.

Szalay, F. S., and Trofimov, B. A. 1996. The Mongolian Late Cretaceous *Asiatherium,* and the early phylogeny and paleobiogeography of Metatheria. *Journal of Vertebrate Paleontology* 16: 474–509.

Szczechura, J. 1978. Fresh-water ostracodes from the Nemegt Formation (Upper Cretaceous) of Mongolia. *Palaeontologia Polonica* 38: 65–121.

Szczechura, J., and Błaszyk, J. 1970. Fresh-water Ostracoda from the Upper Cretaceous of the Nemegt Basin, Gobi Desert. *Palaeontologia Polonica* 21: 107–120.

Takada, T., Matsuoka, H., and Setoguchi, T. 2001. The first multituberculate from Japan. *In* T. Deng and Y. Wang (eds.), *Proceedings of the Eighth Annual Meeting of the Chinese Society of Vertebrate Paleontology,* 55–58. China Ocean Press, Beijing.

Tang, F., Luo, Z.-X., Zhou, Z.-H., You, H. L., Georgi, J. A., Tang, Z.-L., and Wang, X.-Z. 2001. Biostratigraphy and palaeoenvironment of the dinosaur-bearing sediments in Lower Cretaceous of Mazongshan area, Gansu Province, China. *Cretaceous Research* 22: 115–129.

Tatarinov, L. P. 1985. The present state of the problem of the ancestry of mammals. *Acta Zoologica Fennica* 170: 145–147.

Tatarinov, L. P. 1994. On an unusual mammalian tooth from the Mongolian Jurassic. *Paleontologicheskij Zhurnal* 1994: 97–105.

Tatarinov, L. P., and Maschenko, E. N. 1999. A find of an aberrant tritylodont (Reptilia, Cynodontia) in the Lower Cretaceous of the Kemerovo region. *Paleontologicheskii Zhurnal* 33: 422–428.

Tedford, R. H. 1974. Marsupials and the new paleogeography. *Society of Economic Paleontologists and Mineralogists, Special Publication* 21: 109–126.

Thewissen, J. G. M., and Gingerich, P. D. 1989. Skull and endocranial cast of *Eoryctes melanus,* a new palaeoryctid (Mammalia, Insectivora) from the early Eocene of western North America. *Journal of Vertebrate Paleontology* 9: 459–470.

Thewissen, J. G. M., Williams, E. M., Roe, L. J., and Hussain, S. T. 2001. Skeletons of terrestrial cetaceans and the relationship of whales to artiodactyls. *Nature* 413: 277–281.

Tokaryk, T. T. 1987. An eutherian from the Late Cretaceous (Maestrichtian, Kemp Clay Formation) of Texas. *Occasional Short Notes, Saskatchewan Museum of Natural History* 2: 1–6.

Tomes, J. 1849. On the structure of the dental tissues of marsupial animals and more especially of the enamel. *Philosophical Transactions of the Royal Society of London* 139: 402–412.

Trofimov, B. A. 1975. New data on *Buginbaatar* Kielan-Jaworowska and Sochava, 1969 (Mammalia, Multituberculata) from Mongolia. *Trudy Sovmestnoi Sovetsko-Mongol'skoi Paleontologicheskoi Ekspeditsii* 2: 7–13 (in Russian).

Trofimov, B. A. 1978. The first triconodonts (Mammalia, Triconodonta) from Mongolia. *Doklady Akademii Nauk SSSR* 243: 213–216 (in Russian).

Trofimov, B. A. 1980. Multituberculata and Symmetrodonta from the Lower Cretaceous of Mongolia. *Doklady Akademii Nauk SSSR* 251: 209–212 (in Russian).

Trofimov, B. A. 1997. A new generic name *Gobiotheriodon* for a symmetrodont mammal *Gobiodon* Trofimov. *Acta Palaeontologica Polonica* 42: 496.

Trofimov, B. A., and Szalay, F. S. 1994. New Cretaceous marsupial from Mongolia and the early radiation of the Metatheria. *Proceedings of the National Academy of Sciences USA* 91: 12569–12573.

Trouessart, E.-L. 1898. *Catalogus mammalium tam vivientium quam fossilium, tomus 2.* 665–1264 pp. R. Friedländer und Sohn, Berlin.

Trujillo, K. C. 1999. *Vertebrate paleontology, stratigraphy, and sedimentology of new microvertebrate localities in the Morrison (Upper Jurassic) and Cloverly (Lower Cretaceous) formations, Ninemile Hill, Carbon County, Wyoming.* 90 pp. M.S. thesis, University of Wyoming, Laramie.

Tumanova, T. A. 1987. The armored dinosaurs of Mongolia. *Trudy Sovmestnoi Sovetsko-Mongol'skoi Paleontologicheskoi Ekspeditsii* 32: 1–80 (in Russian).

Turnbull, W. D. 1970. Mammalian masticatory apparatus. *Fieldiana: Geology* 18: 147–356.

Turnbull, W. D. 1971. The Trinity therians: their bearing on evolution in marsupials and other therians. *In* A. A. Dahlberg (ed.), *Dental Morphology and Evolution,* 151–179. University of Chicago Press, Chicago.

Turnbull, W. D. 1995. Trinity mammal jaws from the late Early Cretaceous of Texas. *In* R. J. Radlonski and H. Renz (eds.), *Proceedings of the Tenth International Symposium on Dental Morphology,* 261–265. Christine und Michael Brünne, Berlin.

Turnbull, W. D., and Cifelli, R. L. 1999. Triconodont mammals of the Aptian–Albian Trinity Group, Texas and Oklahoma. *In* J. T. Mayhall and T. Heikkinen (eds.), *Dental Morphology 98,* 252–272. University of Oulu Press, Oulu.

Tyndale-Biscoe, C. H. 1973. *Life of Marsupials.* 254 pp. Edward Arnold Limited, London.

Tyndale-Biscoe, C. H., and Renfree, M. B. 1987. *Reproductive Physiology of Marsupials.* 476 pp. Cambridge University Press, Cambridge.

Ulinski, P. 1986. Neurobiology of the therapsid-mammal transition. *In* N. Hotton, P. D. MacLean, J. J. Roth, and E. C. Roth (eds.), *The Ecology and Biology of Mammal-like Reptiles,* 149–171. Smithsonian Institution Press, Washington, D.C.

Van Valen, L. 1966. Deltatheridia, a new order of mammals. *Bulletin of the American Museum of Natural History* 132: 1–126.

Van Valen, L. 1967a. The first discovery of a Cretaceous mammal. *American Museum Novitates* 2285: 1–4.

Van Valen, L. 1967b. New Paleocene insectivores and insectivore classification. *Bulletin of the American Museum of Natural History* 135: 217–284.

Van Valen, L. 1976. Notes on the origin of multituberculates (Mammalia). *Journal of Paleontology* 50: 198–199.

Van Valen, L. 1978. The beginning of the age of mammals. *Evolutionary Theory* 4: 45–80.

Van Valen, L. 1988. Paleocene dinosaurs or Cretaceous ungulates in South America? *Evolutionary Monographs* 10: 1–79.

Van Valen, L., and Sloan, R. E. 1965. The earliest primates. *Science* 150: 743–745.

Van Valen, L., and Sloan, R. E. 1966. The extinction of the multituberculates. *Systematic Zoology* 15: 261–278.

Vandebroek, G. 1961. The comparative anatomy of the teeth of lower and non-specialized mammals. *In* G. Vandebroek (ed.), *International Colloquium on the Evolution of Lower and Non-Specialized Mammals,* 1: 215–330; 2: 1–181. Koninklijke Vlaamse Academie voor Wetenschappen, Letteren en Schone Kunsten van België, Brussels.

Vázquez-Molinero, R., Martin, T., Fischer, M. S., and Frey, R. 2001. Comparative anatomical investigations of the postcranial skeleton of *Henkelotherium guimarotae* Krebs, 1991 (Eupantotheria, Mammalia) and their implications on its locomotion. *Mitteilungen aus dem Museum für Naturkunde in Berlin, Zoologische Reihe* 77: 207–216.

Vianey-Liaud, M. 1979. Les Mammifères montiens de Hainin (Paléocène moyen de Belgique). Part I. Multituberculés. *Palaeovertebrata* 9: 117–131.

Vianey-Liaud, M. 1986. Les Multituberculés Thanetiens de France, et leur rapports avec les Multituberculés Nord-Americains. *Palaeontographica, Abteilung A* 191: 85–171.

Vianey-Liaud, M., Hirsch, K., Sahni, A., and Sigé, B. 1995. Late Cretaceous Peruvian eggshells and their relationships with Laurasian and eastern Gondwanian material. *Géobios* 30: 75–90.

Waagé, K. M. 1968. The type Fox Hills Formation, Cretaceous (Maestrichtian), South Dakota. *Bulletin of the Peabody Museum of Natural History* 27: 1–175.

Waddell, P. J., Cao, Y., Hasegawa, M., and Mindell, D. P. 1999. Assessing the Cretaceous superordinal divergence times within birds and placental mammals by using whole mitochondrial protein sequences and an extended statistical framework. *Systematic Biology* 48: 119–137.

Waldman, M., and Savage, R. J. G. 1972. The first Jurassic mammal from Scotland. *Journal of the Geological Society* 128: 119–125.

Wall, C. E., and Krause, D. W. 1992. A biomechanical analysis of the masticatory apparatus of *Ptilodus* (Multituberculata). *Journal of Vertebrate Paleontology* 12: 172–187.

Wang, Y.-Q., Hu, Y.-M., Chow, M.-C., and Li, C.-K. 1995. Mesozoic mammal localities in western Liaoning, northeast China. *In* A.-L. Sun and Y.-Q. Wang (eds.), *Sixth Symposium on Mesozoic Terrestrial Ecosystems,* 221–228. China Ocean Press, Beijing.

Wang, Y.-Q., Clemens, W. A., Hu, Y.-M., and Li, C.-K. 1998. A probable pseudo-tribosphenic upper molar from the Late Jurassic of China and the early radiation of the Holotheria. *Journal of Vertebrate Paleontology* 18: 777–787.

Wang, X.-L., Wang, Y.-Q., Wang, Y., Xu, X., Tang, Z.-L., Zhang, F.-C., Hu, Y.-M., Gu, G., and Hao, Z.-L. 1998. Stratigraphic sequence and vertebrate-bearing beds of the lower part of the Yixian Formation in Sihetun and neighboring area, western Liaoning, China. *Vertebrata PalAsiatica* 36: 81–101.

Wang, Y.-Q., Hu, Y.-M., Li, C.-K., and Chang, Z.-L. 2001. Recent advances on the study of Mesozoic mammals in China. *In* H. A. Leanza (ed.), *VII International Symposium on Mesozoic Terrestrial Ecosystems,* 179–184. Asociación Paleontológica Argentina, Buenos Aires.

Wang, Y.-Q., Hu, Y.-M., Meng, J., and Li, C.-K. 2001. An ossified Meckel's cartilage in two Cretaceous mammals and origin of the mammalian middle ear. *Science* 294: 357–361.

Ward, D. J. 1984. Collecting isolated microvertebrate fossils. *Zoological Journal of the Linnean Society* 82: 245–259.

Watson, D. M. S. 1916. The monotreme skull: a contribution to mammalian morphogenesis. *Philosophical Transactions of the Royal Society of London* 207: 311–374.

Watson, D. M. S. 1920. On the Cynodontia. *Annals and Magazine of Natural History* 6: 506–524.

Watson, D. M. S. 1942. On Permian and Triassic tetrapods. *Geological Magazine* 79: 81–116.

Webb, M. W. 1998. A revised summary of Lancian (latest Cretaceous) mammal localities with introduction to a new Lancian locality (Lance Formation) in the southwestern Bighorn Basin. *In* W. R. Keefer and J. E. Goolsby (eds.), *Cretaceous and Lower Tertiary Rocks of the Bighorn Basin, Wyoming and Montana,* 131–136. Wyoming Geological Association, 49th Guidebook, Casper.

Webb, M. W. 2001. *Fluvial architecture and Late Cretaceous mammals of the Lance Formation, southwestern Bighorn Basin, Wyoming.* 172 pp. Ph.D. dissertation, University of Wyoming, Laramie.

Weber, M. 1927. *Die Säugetiere: Einführung in die Anatomie und Systemaik der Recenten und Fossilen Mammalia. Band I. Anatomischer Teil.* 444 pp. Gustav Fischer, Jena.

Weber, M. 1928. *Die Säugetiere: Einführung in die Anatomie und Systemaik der Recenten und Fossilen Mammalia. Band II. Systematischer Teil.* 898 pp. Gustav Fischer, Jena.

Weil, A. 1992. *The Terlingua local fauna: stratigraphy, paleontology, and multituberculate systematics.* 118 pp. M.S. thesis, University of Texas, Austin.

Weil, A. 1998. A new species of *Microcosmodon* (Mammalia: Multituberculata) from the Paleocene Tullock Formation of Montana, and an argument for the Microcosmodontinae. *PaleoBios* 18: 1–15.

Weil, A. 1999. *Multituberculate phylogeny and mammalian biogeography in the Late Cretaceous and earliest Paleocene Western Interior of North America.* Ph.D. dissertation, University of California, Berkeley.

Weil, A., and Clemens, W. A. 1998. Aliens in Montana: phylogenetically and biogeographically diverse lineages contributed to an earliest Cenozoic community. *Geological Society of America, Abstracts with Programs* 30(5): 69–70.

Weil, A., and Tomida, Y. 2001. First description of the skull of *Meniscoessus robustus* expands known morphological diversity of Multituberculata and deepens phylogenetic mystery. *Journal of Vertebrate Paleontology* 21, supplement to no. 3: 112A.

Weil, A., and Williamson, T. E. 2000. Diverse Maastrichtian terrestrial vertebrate fauna of the Naashoibito Member, Kirtland Formation (San Juan Basin, New Mexico) confirms "Lancian" faunal heterogeneity in western North America. *Geological Society of America, Abstracts with Programs* 32: 498.

Weishampel, D. B. 1990. Dinosaurian distribution. *In* D. B. Weishampel, P. Dodson, and H. Osmólska (eds.), *The Dinosauria*, 63–140. University of California Press, Berkeley.

Weishampel, D. B., Norman, D. B., and Grigorescu, D. 1991. The dinosaurs of Transylvania: island biogeography in the Late Cretaceous. *National Geographic Research and Exploration* 7: 68–87.

Werdelin, L. 1986. Comparison of skull shape in marsupial and placental carnivores. *Australian Journal of Zoology* 34: 109–117.

Werdelin, L. 1989. Constraint and adaptation in the bone-cracking canid *Osteoborus* (Mammalia: Canidae). *Paleobiology* 15: 387–401.

Westergaard, B., and Ferguson, M. W. J. 1986. Development of the dentition in *Alligator mississippiensis.* Early embryonic development in the lower jaw. *Journal of Zoology, London* A 210: 575–597.

Westergaard, B., and Ferguson, M. W. J. 1987. Development of the dentition in *Alligator mississippiensis.* Later development in the lower jaw of embryos, hatchlings and young juveniles. *Journal of Zoology, London* 212: 191–222.

Westerman, M., and Edwards, D. 1992. DNA hybridization and the phylogeny of monotremes. *In* M. L. Augee (ed.), *Platypus and Echidnas,* 28–34. The Royal Zoological Society of New South Wales, Sydney.

Wheeler, W. 1992. Extinction, sampling, and molecular phylogenetics. *In* M. J. Novacek and Q. Wheeler (eds.), *Extinction and Phylogeny,* 205–215. Columbia University Press, New York.

Whitmore, J. L. 1985. Fossil mammals from two sites in the Late Cretaceous Lance Formation in northern Niobrara County, Wyoming. *Dakoterra* 2: 353–367.

Whitmore, J. L., and Martin, J. E. 1986. Vertebrate fossils from the Greasewood Creek locality in the Late Cretaceous Lance Formation of Niobrara County, Wyoming. *Proceedings of the South Dakota Academy of Sciences* 65: 33–50.

Wible, J. R. 1983. The internal carotid artery in early eutherians. *Acta Palaeontologica Polonica* 28: 174–180.

Wible, J. R. 1984. *The ontogeny and phylogeny of the mammalian cranial arterial pattern.* 705 pp. Ph.D. dissertation, Duke University, Durham.

Wible, J. R. 1986. Transformation in the extracranial course of the internal carotid artery in mammalian phylogeny. *Journal of Vertebrate Paleontology* 6: 313–325.

Wible, J. R. 1987. The eutherian stapedial artery: character analysis and implications for superordinal relationships. *Zoological Journal of the Linnean Society* 91: 107–135.

Wible, J. R. 1990. Petrosals of Late Cretaceous marsupials from North America and a cladistic analysis of the petrosal in therian mammals. *Journal of Vertebrate Paleontology* 10: 183–205.

Wible, J. R. 1991. Origin of Mammalia: the craniodental evidence reexamined. *Journal of Vertebrate Paleontology* 11: 1–28.

Wible, J. R., and Hopson, J. A. 1993. Basicranial evidence for early mammal phylogeny. *In* F. S. Szalay, M. J. Novacek, and M. C. McKenna (eds.), *Mammal Phylogeny: Mesozoic Differentiation, Multituberculates, Monotremes, Early Therians, and Marsupials,* 45–62. Springer-Verlag, New York.

Wible, J. R., and Hopson, J. A. 1995. Homologies of the prootic canal in mammals and non-mammalian cynodonts. *Journal of Vertebrate Paleontology* 15: 331–336.

Wible, J. R., and Rougier, G. W. 2000. The cranial anatomy of *Kryptobaatar dashzevegi* (Mammalia, Multituberculata), and its bearing on the evolution of mammalian characters. *Bulletin of the American Museum of Natural History* 247: 1–124.

Wible, J. R., Miao, D., and Hopson, J. A. 1990. The septomaxilla of fossil and recent synapsids and the problem of the septomaxilla of monotremes and armadillos. *Zoological Journal of the Linnean Society* 98: 203–228.

Wible, J. R., Rougier, G. W., Novacek, M. J., McKenna, M. C., and Dashzeveg, D. 1995. A mammalian petrosal from the Early Cretaceous of Mongolia: implications for the evolution of the ear region and mammaliamorph relationships. *American Museum Novitates* 3149: 1–19.

Wible, J. R., Rougier, G. W., Novacek, M. J., and McKenna, M. C. 2001. Earliest eutherian ear region: a petrosal referred to *Prokennalestes* from the Early Cretaceous of Mongolia. *American Museum Novitates* 3322: 1–44.

Wiens, J. J. 1998a. Combining data sets with different phylogenetic histories. *Systematic Biology* 47: 568–581.

Wiens, J. J. 1998b. Does adding characters with missing data increase or decrease phylogenetic accuracy? *Systematic Biology* 47: 25–40.

Wiens, J. J. in press. Incomplete taxa, incomplete characters, and phylogenetic accuracy: is there a missing data problem? *Journal of Vertebrate Paleontology.*

Wiens, J. J., and Reeder, T. W. 1995. Combining data sets with different numbers of taxa for phylogenetic analysis. *Systematic Biology* 44: 548–558.

Wiley, E. O. 1979. An annotated Linnaean hierarchy, with comments on natural taxa and competing systems. *Systematic Biology* 28: 308–337.

Wiley, E. O. 1981. *Phylogenetics: The Theory and Practice of Phylogenetic Systematics.* 439 pp. John Wiley and Sons, New York.

Wiley, E. O., Siegel-Causey, D., Brooks, D. R., and Funk, V. A. 1991. The Compleat Cladist: A Primer of Phylogenetic Procedures. *The University of Kansas Museum of Natural History, Special Publication* 19: 1–158.

Wilkinson, M. 1995. Coping with abundant missing entries in phylogenetic inference using parsimony. *Systematic Biology* 44: 501–514.

Willett, E. W., and Willett, H. 1881. Notes on a mammalian jaw from the Purbeck beds at Swanage, Dorset. *Quarterly Journal of the Geological Society of London* 37: 376–380.

Williams, P. L., Williams, R., Dyson, M., and Bannister, L. H. 1989. *Gray's Anatomy, 37th Edition.* 1598 pp. Churchill Livingstone, New York.

Williamson, T. E., and Sullivan, R. M. 1998. A new local fauna, the Willow Wash local fauna, from the Upper Cretaceous (Campanian), Kirtland Formation, New Mexico. *Journal of Vertebrate Paleontology* 18, supplement to no. 3: 86A.

Wilson, D. E., and Reeder, D. M. 1993. *Mammal Species of the World. A Taxonomic and Geographic Reference. 2nd Edition.* 1206 pp. Smithsonian Institution Press, Washington, D. C.

Wilson, G. P., and Rana, R. S. 2001. Biogeographic implications of a new primitive tribosphenic mammal from the Intertrappean Beds of Rangapur, Andra Pradesh, India. *Journal of Vertebrate Paleontology* 21, supplement to no. 3: 114A.

Wilson, R. W. 1965. Type localities of Cope's Cretaceous mammals. *Proceedings of the South Dakota Academy of Sciences, University of South Dakota Bulletin* 44: 88–90.

Wilson, R. W. 1983. Late Cretaceous mammals of western South Dakota, 1974. *National Geographic Society Research Reports* 15: 749–752.

Wilson, R. W. 1987. Late Cretaceous (Fox Hills) multituberculates from the Red Owl local fauna of western South Dakota. *Dakoterra* 3: 118–122.

Wing, S. L., and Tiffney, B. H. 1987. The reciprocal interaction of angiosperm evolution and tetrapod herbivory. *Review of Palaeobotany and Palynology* 50: 179–210.

Winkler, D. A., Murry, P. A., and Jacobs, L. L. 1989. Vertebrate paleontology of the Trinity Group, Lower Cretaceous of central Texas. *In* D. A. Winkler, P. A. Murry, and L. L. Jacobs (eds.), *Field Guide for the 49th Annual Meeting of the Society of Vertebrate Paleontology,* 1–22. Institute for the Study of Earth and Man, Dallas.

Winkler, D. A., Murry, P. A., and Jacobs, L. L. 1990. Early Cretaceous (Comanchean) vertebrates of central Texas. *Journal of Vertebrate Paleontology* 10: 95–116.

Witschel, M. 1999. *Anatomy of the facial skull and of the nasal cavity of a tritylodontid synapsid from the Lower Jurassic of South Africa, with remarks on bone histology.* 64 pp. Diplomarbeit am Geologische-Paläontologischen, Institut der Eberhard-Karls-Universität, Tübingen.

Wood, C. B. 1992. *Comparative studies of enamel and functional morphology in selected mammals with tribosphenic molar teeth: phylogenetic applications.* 237 pp. Ph.D. dissertation, Harvard University, Cambridge.

Wood, C. B. 2000. Tooth enamel microstructure in *Deltatheridium* (Metatheria, Late Cretaceous of Mongolia), with comparison to some other Mesozoic mammals. *Bulletin of the Paleontological Society of Korea, Special Publication* 4: 127–152.

Wood, C. B., and Clemens, W. A. 2001. A new specimen and a functional reassociation of the molar dentition of *Batodon tenuis* (Placentalia, incertae sedis), latest Cretaceous (Lancian), North America. *Bulletin of the Museum of Comparative Zoology* 156: 99–118.

Wood, C. B., and Stern, D. N. 1997. The earliest prisms in mammalian and reptilian enamel. *In* W. von Koenigswald and P. M. Sander (eds.), *Tooth Enamel Microstructure,* 63–83. A. A. Balkema, Rotterdam.

Wood, C. B., Dumont, E. R., and Crompton, A. W. 1999. New studies of enamel microstructure in Mesozoic mammals: a review of enamel prisms as a mammalian synapomorphy. *Journal of Mammalian Evolution* 6: 177–213.

Wood, C. B., Zan, S., Jin, L., Lee, Y.-N., and Schaff, C. R. 2001. A new "middle" Cretaceous microvertebrate locality from Jilin Province, northeastern China. *Journal of Vertebrate Paleontology* 21, supplement to no. 3: 116A.

Woodburne, M. O., and Tedford, R. H. 1975. The first Tertiary monotreme from Australia. *American Museum Novitates* 2588: 1–11.

Woodburne, M. O., Rich, T. H., and Springer, M. S. 2003. The evolution of tribospheny and the antiquity of mammalian clades. *Molecular Phylogeny and Evolution* 28: 360–385.

Woodward, A. S. 1911. On some mammalian teeth of the Wealden of Hastings. *Quarterly Journal of the Geological Society of London* 67: 278–281.

Woodward, A. S. 1916. On a mammalian mandible from an Upper Cretaceous formation in Alberta, Canada. *Proceedings of the Zoological Society of London* 158: 525–528.

Wouters, G., Sigogneau-Russell, D., and LePage, J.-C. 1984. Découverte d'une dent d'Haramiyide (Mammalia) dans des niveaux rhétiens de la Gaume (en Lorraine belge). *Bulletin de la Société belge de Géologie* 93: 351–355.

Wu, X.-C., Sues, H.-D., and Brinkman, D. B. 1996. An atoposaurid neosuchian (Archosauria: Crocodyliformes) from the Lower Cretaceous of Inner Mongolia (People's Republic of China). *Canadian Journal of Earth Sciences* 33: 599–605.

Wyss, A. R., and Flynn, J. J. 1993. A phylogenetic analysis and definition of the Carnivora. *In* F. S. Szalay, M. J. Novacek, and M. C. McKenna (eds.), *Mammal Phylogeny: Placentals,* 32–52. Springer-Verlag, New York.

Yabe, H., and Shikama, T. 1938. A new Jurassic mammal from south Manchuria. *Proceedings of the Imperial Academy of Tokyo* 14: 353–359.

Yadagiri, P. 1984. New symmetrodonts from the Kota Formation. *Journal of the Geological Society of India* 25: 512–521.

Yadagiri, P. 1985. An amphidontid symmetrodont from the Early Jurassic Kota Formation, India. *Zoological Journal of the Linnean Society* 85: 411–417.

Yalden, D. W. 1972. The form and function of the carpal bones in some arboreally adapted mammals. *Acta Anatomica* 83: 383–406.

Young, C. C. 1940. Preliminary note on the Mesozoic mammals of Lufeng, Yunnan, China. *Bulletin, Geological Society of China* 20: 93–111.

Young, C. C. 1947. Mammal-like reptiles from Lufeng, Yunnan, China. *Proceedings of the Zoological Society of London* 117: 537–597.

Young, C. C. 1978. New materials of *Eozostrodon. Vertebrata PalAsiatica* 16: 1–3.

Young, C. C. 1982a. Two primitive mammals from Lufeng, Yunnan. *Selected Works of Yang Zhungjian,* 21–24. Science Press, Beijing.

Young, C. C. 1982b. *Selected Works of Yang Zhongjian (Young Chung-Chien).* 219 pp. (Edited by Editorial Committee for Selected Works of Yang Zhongjian, Chaired by M. C. Chow [M.-C. Zhou]). Science Press, Beijing.

Zeller, U. 1987. Morphogenesis of the mammalian skull with special reference to *Tupaia. In* H.-J. Kuhn and U. Zeller (eds.), *Morphogenesis of the Mammalian Skull,* 17–50. Verlag Paul Parey, Hamburg.

Zeller, U. 1988. The lamina cribrosa of *Ornithorhynchus* (Monotremata, Mammalia). *Anatomy and Embryology* 178: 513–519.

Zeller, U. 1989a. The braincase of *Ornithorhynchus. Fortschritte der Zoologie* 35: 386–391.

Zeller, U. 1989b. Die Enwicklung und Morphologie des Schädels von *Ornithorhynchus anatinus* (Mammalia: Prototheria: Monotremata). *Abhandlungen der Senckenbergischen Naturforschenden Gesellschaft* 545: 1–188.

Zeller, U. 1993. Ontogenetic evidence for cranial homologies in monotremes and therians, with special reference to *Ornithorhynchus. In* F. S. Szalay, M. J. Novacek, and M. C. McKenna (eds.), *Mammal Phylogeny, Mesozoic Differentiation, Multituberculates, Monotremes, Early Therians, and Marsupials,* 95–107. Springer-Verlag, New York.

Zeller, U. 1999a. Mammalian reproduction: origin and evolutionary transformations. *Zoologischer Anzeiger* 238: 117–130.

Zeller, U. 1999b. Phylogeny and systematic relations of the Monotremata: why we need an integrative approach. *Courier Forschungsinstitut Senckenberg* 215: 227–232.

Zeller, U., and Freyer, C. 2001. Early ontogeny and placentation of the grey short-tailed opossum, *Monodelphis domestica* (Didelphidae: Marsupialia): contribution to the reconstruction of the marsupial morphotype. *Journal of Zoological Systematics and Evolutionary Research* 39: 137–158.

Zhang, F. 1984. Fossil record of Mesozoic mammals of China. *Vertebrata PalAsiatica* 22: 29–38.

Zhang, F., and Cui, G. 1983. New material and new understanding of *Sinoconodon. Vertebrata PalAsiatica* 21: 32–41.

Zhang, F.-K., Crompton, A. W., Luo, Z.-X., and Schaff, C. R. 1998. Pattern of dental replacement of *Sinoconodon* and its implications for evolution of mammals. *Vertebrata PalAsiatica* 36: 197–217.

Zhou, M.-Z., Cheng, Z.-W., and Wang, Y.-Q. 1991. A mammalian lower jaw from the Jurassic of Lingyuan, Liaoning. *Vertebrata PalAsiatica* 29: 165–175.

Ziegler, A. C. 1971. A theory of the evolution of therian dental formulas and replacement pattern. *Quarterly Review of Biology* 46: 226–249.

Zittel, K. A. von. 1893. *Handbuch der Paläontologie. Abteilung 1. Paläozoologie. Band IV, Vertebrata (Mammalia).* 799 pp. R. Oldenbourg, Munich.

ADDITIONAL REFERENCES

This list includes papers published or accepted after March 31, 2002 (abstracts excluded) and not included in the list of references.

Archibald, J. D. 2003. Timing and biogeography of the eutherian radiation: Fossils and molecules compared. *Molecular Phylogeny and Evolution* 28: 350–359.

Archibald, J. D., and Averianov, A. O. 2003. The Late Cretaceous placental mammal *Kulbeckia*. *Journal of Vertebrate Paleontology* 23: 404–419.

Averianov, A. O., and Archibald, J. D. 2003. Mammals from the Upper Cretaceous Aitym Formation, Kyzylkum Desert, Uzbekistan. *Cretaceous Research* 24: 171–191.

Bonaparte, J. F., Martinelli, A. G., Schultz, C. L., and Rubert, R. 2003. The sister group of mammals: Small cynodonts from the Late Triassic of southern Brazil. *Revista Basileira de Paleontologia* 5: 5–27.

Cifelli, R. L., and Davis, B. M. 2003. Marsupial origins. *Science* 302: 1899–1900.

Cuenca-Bescós, G., and Canudo, J. I. 2003. New gobiconodontid mammal from the Early Cretaceous of Spain and its palaeogeographic implications. *Acta Palaeontologica Polonica* 48: 575–582.

Eaton, J. G. 2002. Multituberculate mammals from the Wahweap (Campanian, Aquilan) and Kaiparowits (Campanian, Judithian) Formations, within and near Grand Staircase-Escalante National Monument, Southern Utah. *Miscellaneous Publication, Utah Geological Survey* 02-4: 1–66.

Ekdale, E.G., Archibald, J. D., and Averianov, A. O. 2004. Late Cretaceous petrosal bones of zhelestid and zalambdalestid eutherian mammals from Uzbekistan. *Acta Palaeontologica Polonica* 49: 161–176.

Engelmann, G. F. in press. The anterior dentition of the Late Jurassic multituberculate *Ctenacodon*. *In* G. C. Gould and S. K. Bell (eds.), *Tributes to Malcolm C. McKenna: His Students, His Legacy.* Bulletin of the American Museum of Natural History 341.

Hahn, G., and Hahn, R. 2002. Neue Multituberculaten-Zähne aus dem Barremian (Unter-Kreide) von Galve (Spanien). *Paläontologische Zeitschrift* 76: 257–239.

Hahn, G., and Hahn, R. 2003. New multituberculate teeth from the Early Cretaceous of Morocco. *Acta Palaeontologica Polonica* 48: 349–356.

Hahn, G., and Hahn, R. *in press*. The dentition of the Plagiaulacida (Multituberculata, Late Jurassic to Early Cretaceous). *Geologica et Palaeontologica* 38.

Horovitz, I. 2003. Postcranial skeletal morphology of *Ukhaatherium nessovi* (Eutheria, Mammalia) from the Late Cretaceous of Mongolia. *Journal of Vertebrate Paleontology* 23: 857–868.

Krause, W. D., Gottfried, M. D., O'Connor, P. M., and Roberts, E. M. 2003. A Cretaceous mammal from Tanzania. *Acta Palaeontologica Polonica* 48: 321–330.

Luo, Z.-X., Ji, Q. Wible, J. R., and Yuan, Ch.-X. 2003. An Early Cretaceous tribosphenic mammal and metatherian evolution. *Science* 302: 1934–1940.

Martin, T., and Averianov, A. O. 2004. A new docodont (Mammalia) from the Middle Jurassic of Kyrgyzstan, Central Asia. *Journal of Vertebrate Paleontology* 24: 195–201.

Maschenko E. N., Lopatin A.V., and Voronkevich A.V. 2002. New Early Cretaceous mammals from Western Siberia. *Doklady Akademii Nauk* 386: 715–717 (in Russian).

Maschenko E. N., Lopatin A.V., and Voronkevich A.V. 2002. A new genus of the tegotheriid docodonts (Docodonta, Tegotheriidae) from the Early Cretaceous of West Siberia. *Russian Journal of Theriology.* 1: 75–81.

Meng, J., Hu, Y., Wang, Y., and Li, Ch. 2003. The ossified Meckel's cartilage and internal groove in Mesozoic mammaliaforms: Implications to origin of the definitive mammalian middle ear. *Zoological Journal of the Linnean Society* 138: 431–448.

Muser, A. M. 2003. Review of the monotreme fossil record and comparison of paleontological and molecular data. *Comparative Biochemistry and Physiology* Part A, 136: 927–942.

Pearson, D. A., Schaefer, T., Johnson, K. R., Nichols, D. J., and Hunter, J. P. 2002. Vertebrate biostratigraphy of the Hell Creek Formation in southwestern North Dakota and northwestern South Dakota. *Geological Society of America, Special Paper* 361:145–167.

Rana, R. S., and Wilson, G. P. 2003. New Late Cretaceous mammals from the Intertrappean beds of Rangapur, India. *Acta Palaeontologica Polonica* 48: 331–348.

Rich, T. R., and Vickers-Rich, P. 2004. Diversity of Early Cretaceous mammals from Victoria, Australia. *In* G. C. Gould and S. K. Bell (eds.), *Tributes to Malcolm C. McKenna: His Students, His Legacy.* Bulletin of the American Museum of Natural History 341.

Robertson, D. S., McKenna, M. C., Toon, O. B., Hope, S., and Lillegraven, J. A. 2004. Survival in the first hours of the Cenozoic. *Bulletin of the Geological Society of America* 116: 760–768.

Rougier, G. W., Ji, Q., and Novacek, M. 2003. A new symmetrodont mammal with fur impressions from the Mesozoic of China. *Acta Geologica Sinica* 77: 7–14.

Rougier, G. W., Spurlin, B. K., and Kik, P. 2003. A new specimen of *Eurylambda aequicrurius* and considerations on "symmetrodont" dentition and relationships. *American Museum Novitates* 3398: 1–15.

Rougier, G. W., and J. R. Wible. in press. Major changes in the mammalian ear region and basicranium. *In* M. T. Carrano, T. J. Gaudin, R. W. Blob, and J. R. Wible (eds.), *Amniote Paleobiology: Perspectives on the Evolution of Mammals, Birds, and Reptiles.* University of Chicago Press, Chicago.

Rougier, G. W., J. R. Wible, and M. J. Novacek. in press. New specimen of *Deltatheroides cretacicus* (Metatheria, Deltatheriididae) from the Late Cretaceous of Mongolia. *In* M. R. Dawson and J. A. Lillegraven (eds.), *Fanfare for an Uncommon Paleontologist: Papers in Honor of Malcolm C. McKenna.* Bulletin of Carnegie Museum of Natural History.

Sereno, P. 2004. Shoulder girdle and forelimb in multituberculates: Evolution of parasagittal forelimb posture in mammals. *In* M. T. Carrano, R. W., Blob, T. J., Gaudin, and J. R. Wible (eds.), *Amniote Paleobiology: Perspectives on the Evolution of Mammals, Birds, and Reptiles.* University of Chicago Press, Chicago.

Sigogneau-Russell, D. 2003. Docodonts from the British Mesozoic. *Acta Palaeontologica Polonica* 48: 357–374.

Sigogneau-Russell, D. 2003. Holotherian mammals from the Forest Marble (Middle Jurassic) of England. *Geodiversitas* 25: 501–537.

Sigogneau-Russell, D. 2003. Diversity of triconodont mammals from the Early Cretaceous of North Africa—affinities of the amphilestids. *Palaeovertebrata* 32: 27–55.

Tabuce, R., Vianey-Liaud, M., and Garcia, G. 2004. A eutherian mammal in the latest Cretaceous of Vitrolles, southern France. *Acta Palaeontologica Polonica* 49: 347–355.

Tsubamato, T., Rougier, G. W., Isaji, S., Manabe, M., and Forasiepi, A. M. 2004. New Early Cretaceous spalacotheriid "symmetrodont" mammal from Japan. *Acta Palaeontologica Polonica* 49: 329–346.

Underwood, C. J., and Rees, J. 2002. Selachian faunas from the lowermost Cretaceous Purbeck Group of Dorset, southern England. *In* A. R. Milner and D. J. Batten (eds.), *Life and Environments in Purbeck Times,* 83–101. Special Papers in Palaeontology 68.

Wible, J. R. 2003. On the cranial osteology of the short-tailed opossum *Monodelphis brevicaudata* (Didelphidae, Marsupialia). *Annals of Carnegie Museum* 72: 1–66.

Wible, J. R., Novacek, M. J., and Rougier, G. W. 2004. New data on the skull and dentition of the Mongolian Cretaceous eutherian mammal *Zalambdalestes. Bulletin of the American Museum of Natural History* 281: 1–144.

Wible, J. R., Rougier, G. W., and Novacek, M. J. in press. Overview of evidence for superordinal/ordinal eutherian taxa and timing of their diversification. *In* K. D. Rose and J. D. Archibald (eds.), *Placental Mammals: Origin, Timing, and Relationships of the Major Extant Clades.* Johns Hopkins University Press, Baltimore.

Wilson, G. P., Hilton, R. P., and Gohre, E. S. 2003. The first Mesozoic mammal from California. *PaleoBios* 23: 20–23.

Woodburne, M. O. 2003. Monotremes as pretribosphenic mammals. *Journal of Mammalian Evolution* 10: 195–248.

ILLUSTRATION CREDITS

Many of the figures included in this book were adapted (in part or whole) from illustrations previously published elsewhere. All sources for such material are documented in the figure legends; full literature citations appear at the end of the book. It is a pleasure to acknowledge, with heartfelt thanks, the many publishers and professional organizations that own current rights to these publications and that graciously permitted us to reproduce them herein. The following credits are listed alphabetically by copyright owner.

Academia Sinica Institute of Vertebrate Paleontology and Paleoanthropology: **Figures 3.22A,B, 7.11D, 9.4.** All reproduced by permission.

Académie des Sciences, Paris: **Figures 5.9A, 6.6C,D, 8.37B.** Reproduced by permission.

American Association for the Advancement of Science: **Figures 3.6A,B, 3.7, 3.15, 3.17C,E, 3.22C, 4.1, 8.41D,G, 8.43D, 11.5E.** All © American Association for the Advancement of Science and reproduced by permission.

American Journal of Science: **Figure 10.14A$_{2}$.** Reproduced with permission from the American Journal of Science.

American Museum of Natural History: **Figures 3.6D, 3.8E, 7.4A$_{1}$,C, 8.38C, 8.43E, 9.8B, 10.7C$_{1}$, 10.9A, B$_{1}$,E, 10.13A$_{1,2}$,B$_{3,4}$, 13.6B.** All reproduced with permission from the American Museum of Natural History.

Asociación Paleontológica Argentina: **Figure 7.10C,D.** Reproduced with permission from the Asociación Paleontológica Argentina.

Australian Mammal Society: **Figure 6.6A$_{1-3}$.** Reproduced with permission from the Australian Mammal Society.

Bayerischen Staatssammlung für Paläontologie und historische Geologie: **Figures 4.7A, 4.9E.** Both reproduced with permission.

Blackwell Publishing: **Figures 3.5E, 3.17D, 4.4C,D, 4.6A,B, 5.8C$_{1-3,5}$, 7.5A, 8.5B, 9.3A, 9.7A,B, 11.3, 13.19A.** All reproduced with permission from Blackwell Publishing.

California Academy of Sciences: **Figure 3.14.** Reproduced with permission from the California Academy of Sciences.

Cambridge University Press: **Figures 9.1A–E, 12.6B,C.** Both reproduced with permission from Cambridge University Press.

Carnegie Museum of Natural History: **Figures 3.25, 13.26A$_{2}$.** Both courtesy and copyright Carnegie Museum of Natural History.

Department of Geology and Geophysics, University of Wyoming: **Figures 3.9D,E,H, 5.1A–D, 5.2A,B, 7.3C, 8.4D, 11.6C.** All reproduced with permission of the Department of Geology and Geophysics, University of Wyoming.

Elsevier: **Figures 8.33F, 9.5A,B,C 9.7D, 11.4A$_{2-4}$, 11.5C, 13.11A,C.** All reproduced with permission of Elsevier.

The Geological Society of London: **Figure 5.7B.** Reproduced with permission of The Geological Society Publishing House.

Geological Survey of India: **Figure 4.7D.** Reproduced with permission of the Geological Survey of India.

The Institut für Geologie und Paläontologie im Fachbereich Geowissenschaften der Phillips-Universität Marburg: **Figures 4.9A,B, 8.8A, 8.30A–C,E, 8.31D.** All reproduced with permission of the Institut für Geologie und Paläontologie im Fachbereich Geowissenschaften der Phillips-Universität Marburg.

Institutul Geologic al României: **Figure 8.42A$_{3}$.** Reproduced with permission of the Institutul Geologic al României.

Instituto de Ciencias de la Terre "Jaume Almera" y la Facultad de Geologia de la Universidad de Barcelona: **Figure 8.34D.** Reproduced by permission.

Instituto Geologico e Mineiro: **Figures 5.3B$_{1-3}$, 5.7C.** Reproduced by permission of the Instituto Geologico e Mineiro.

Instytut Paleobiologii, Polska Akademia Nauk: **Figures 3.4E, 3.8D, 3.24E, 5.4A–C, 5.8B,C$_{4}$, 6.5B,C, 7.10A,B, 8.1, 8.3, 8.4E, 8.6A,B, 8.7A–D, 8.8C, 8.15, 8.20, 8.22, 8.34B,C,H,I–L, 8.35C$_{1}$,D$_{1,2}$, E,F$_{2}$,G$_{1-2}$, 8.38B,D–I, 8.39A–I, 8.40B–K, 9.9A,B, 12.4A,D, 12.7D,E, 13.1D,E, 13.2A,B, 13.3, 13.4, 13.5, 13.6A, 13.7, 13.8B,C,D, 13.12A–D, 13.14–16, 13.21A, 13.22, 13.23A, 15.1, 15.2.** All reproduced with permission of the Instytut Paleobiologii, Polska Akademia Nauk.

International Journal of Organic Evolution: **Figure 3.2B$_{2}$.** Reproduced with permission.

The Japan Academy: **Figure 9.6B.** Reproduced with permission of The Japan Academy.

Kluwer Academic/Plenum: **Figures 7.12E, 10.16C$_{1-2}$, 12.5A.** All reproduced with permission of Kluwer Academic/Plenum.

Macmillan Magazines: **Figures 6.3, 7.1A, 7.8A–D, 7.9A–F, 7.13, 8.2B, 8.5C, 8.12, 9.2A–C, 9.10D, 11.2A–C, 12.4B, 12.5C, 13.8A, 13.9, 13.10A,B, 13.17A, 13.18A, 13.19G, 13.23B.** All © Macmillan Magazines, Ltd., reproduced by permission.

Museum of Comparative Zoology, Harvard University: **Figures 3.1D, 3.2, 3.10E, 3.16B1, 3.17A, 3.18A–D, 3.19E,F.**

Museum National d'Histoire Naturelle, Paris: **Figures 4.7C, 5.9B$_{1-3}$, 9.1D, 9.3D, 9.10B$_{1-3}$,C$_{1-3}$, 10.13A$_{3,4}$,B$_{1,2}$, 10.16A$_{1,2}$,D$_{1,2}$, 10.17A,**

B_{1-4},C,D, 11.5A,B, 12.1A, 12.2, 12.11E.** All © Publications Scientifiques du Museum National d'Histoire Naturelle, reproduced by permission.

Museum für Naturkunde Berlin: **Figure 10.9D.** Reproduced by permission of the Museum für Naturkunde Berlin.

National Academy of Sciences USA: **Figures 12.8$A_{1,2}$, 14.1A,B_1,C.** Both © National Academy of Sciences USA, reproduced by permission.

National Research Council Press: **Figure 10.12$A_{2,3}$.** Reproduced by permission.

The Natural History Museum: **Figures 4.1A,B, 4.4B, 4.6C, 4.8A, 7.5B,C, 7.6D, 7.7, 7.12A,B, 8.35C_2, 9.10A_3, 10.5B, 10.8A,B, 10.12A_1, 10.16$B_{1,2}$.** Reproduced with permission of the Picture Library, The Natural History Museum, London.

The Palaeontological Association: **Figures 8.9, 8.10A,B, 8.33G, 8.34E,G, 8.35B, 8.36A–I, 8.45, 10.9$B_{2,3}$,C,F, 10.12 A_4,B.** Reproduced with permission.

Palaeovertebrata: **Figures 8.26B, 8.42B.** Reproduced by permission.

Paleontological Museum, University of Oslo: **Figure 5.5.** Reproduced by permission.

The Paleontological Society: **Figures 1.2, 7.2A,B,D,E, 8.8D, 8.21, 9.10$F_{1,2}$, 13.11B,D.** All reproduced with permission.

Queen Victoria Museum: **Figures 6.1B, 6.2A,B.** Reproduced by permission.

The Regents of the University of California: **Figures 8.41A–C, 8.43B,C, 12.11B,D_{1-5}, 12.12C_{1-4}, 12.13A–F.** All © The Regents of the University of California and reproduced by permission.

The Royal Society: **Figures 8.2A, 8.4A–C, 8.5A, 8.19, 11.4B.** All reproduced with permission of The Royal Society.

E. Schweizerbart, Borntraeger, Cramer, Science Publishers: **Figures 3.24B, 4.5, 8.8B, 8.26A, 8.27A,B,E, 8.29A,C, 8.30F, 8.31F, 8.32, 8.33A–E, 9.7C, 10.1A–C, 10.2B,D,E, 10.5A, 10.7A–C, 10.8C, 10.11E, 12.13G.** All reproduced by permission of E. Schweizerbart, Borntraeger, Cramer, Science Publishers.

Selbstverlag Fachbereich Geowissenschaften: **Figures 8.27C,D, 8.28, 8.29B,D, 8.30D, 8.31A,B, 8.34F, 10.4A,B, 10.5C_1, 10.9G, 10.10B.** All reproduced by permission of Selbstverlag Fachbereich Geowissenschaften.

Société belge de Géologie: **Figure 8.31E.** Reproduced by permission of the Société Belge de Géologie.

Society of Vertebrate Paleontology: **Figures 3.5A–D, 3.6$A_{1,2}$,B_2,C, 3.11A,B_2–F_2, 3.12, 3.16A_2,B_2, 3.23A,B, 4.9D, 6.6B,E, 7.1B, 7.5D, 7.11C, 8.18A,B, 8.37A, 8.43A, 10.7E, 10.13C_{1-5}, 10.15A,B, 12.3, 12.6A, 13.13, 14.2A.** All reproduced by permission of the Society of Vertebrate Paleontology.

South African Museum, Iziko Museums of Cape Town: **Figures 3.2$A_{1,2}$, 3.16A_1, 3.19B.** Reproduced by permission of the South African Museum, Iziko Museums of Cape Town.

Springer-Verlag: **Figures 3.1B,C,E,F, 3.9F,I, 3.10A_1,B,C, 3.13C, 3.21, 4.2, 4.5B,C_2, 4.8C,E.** All © Springer-Verlag, reproduced by permission.

Surrey Beatty and Sons: **Figures 12.10A, 12.12B.** Reproduced wih permission.

Taylor and Francis AS: **Figures 8.11, 8.14A–E, 8.16, 8.17, 8.24, 12.7A,B.** All reproduced by permission of Taylor and Francis AS, www.tandf.no/fossils.

Urban and Fisher Verlag: **Figure 3.1A.** Reproduced by permission of Urban and Fisher Verlag.

Verlag Dr. Friedrich Pfeil: **Figure 5.6.** Reproduced by permission.

Wetenschappen, Letteren en Schone Kunsten van België: **Figure 10.2A.** Reproduced with permission.

Wiley-VCH Verlag: **Figures 3.19C, 8.2C.** Both reproduced by permission of Wiley-VCH Verlag.

Yale Peabody Museum: **Figures 5.3A, 5.7A, 7.3B, 7.6A,B, 7.12C,D, 7.15A–E, 8.8E–H, 9.6A, 10.7D, 10.10C,D, 11.5D.** All reproduced with permission.

Zhurnal Obshchej Biologii: **Figure 6.4B,C.** Reproduced by permission of the chief editor of the journal.

INDEX

The suffix *f* on a page number indicates a figure; *n* indicates a footnote; *t* indicates a table.

GPSR Authorized Representative: Easy Access System Europe, Mustamäe tee
50, 10621 Tallinn, Estonia, gpsr.requests@easproject.com

www.ingramcontent.com/pod-product-compliance
Ingram Content Group UK Ltd.
Pitfield, Milton Keynes, MK11 3LW, UK
UKHW052140310526
471666UK00005B/14
9780231119184